SCIENCE FRONTIERS:

SOME ANOMALIES AND CURIOSITIES OF NATURE

COMPILED BY:

WILLIAM R. CORLISS

Published and Distributed by:

The Sourcebook Project P.O. Box 107 Glen Arm, MD 21057 USA

Library of Congress Catalog Number: 93-92800
ISBN 0-915554-28-3

First printing: January 1994

Printed in the United States of America

TABLE OF CONTENTS

DEDICATION

To my wife, Ginny, for her love and support these many years.

PREFACE

The primary intent of this book is entertainment. Do not look for profundities! All I claim here is an edited collection of naturally occurring anomalies and curiosities that I have winnowed mainly from scientific journals and magazines published between 1976 and 1993. With this eclectic sampling I hope to demonstrate that nature is amusing, beguiling, sometimes bizarre, and, most important, liberating. "Liberating?" Yes! If there is anything profound between these covers, it is the influence of anomalies on the stability of stifling scientific paradigms.

First, though, some statistics about my overall endeavor. The present collection consists of about 1500 items of science news and research originally published in the first 86 issues of Science Frontiers, my bimonthly newsletter. I have organized these items by scientific discipline (archeology, astronomy, etc.), updated them where required, and hopefully woven them into a coherent whole. Some bumpiness and gaps are to be expected because I selected only those tidbits that appealed to me. Complete coverage of all sciences was not a goal. Even so, I believe that most readers will be impressed by the vast panorama of nature laid out here before them. From 40,000-year-old archeological digs in the New World (definitely verboten), to the pseudofish displayed by some mussels, to the geological havoc wreaked by asteroid-raised tsunamis, the variety and richness of natural phenomena are to be seen on every page---and so are the scientific puzzles they pose.

I confess that my newsletter, Science Frontiers, is only a teaser to tempt its readers to partake in a much larger, more comprehensive banquet: the Catalog of Anomalies. This work, now comprising 13 volumes of a projected 30, represents my entire file of some 40,000 items gleaned from a survey of about 14,000 volumes of science journals and magazines from 1820 to date. This massive hoard of scientific engimas, paradoxes, and esoterica was assembled bit by bit from 363 volumes of Nature, 260 volumes of Science, 100 volumes of the Journal of Geophysical Research, and so on with other journals. I believe my collection is unique. It transcends modern computerized data bases in its very wide time frame and its focus on the anomalous and curious. The present book is a recent, limited sampling of the kind of material that goes into the Catalog of Anomalies.

The Catalog of Anomalies represents my personal attempt to assemble the riddles of science and, given a large array of them, to discern some meaning implicit in the melange. On the practical level, which as a self-employed researcher I cannot avoid, my priorities have had to be as follows: Goal #1 has been the satisfaction of my own curiosity; Goal #2 has been the marketing of enough books to support my research, for no government offices or private foundations seem at all interested in supporting this new discipline of "anomalistics"; Goal #3 has been more altruistic: the anticipation that there may be something scientifically useful in all this. But, even if there is not, the quest has been fulfilling in itself.

Some mainstream scientists may recoil at the thought of 40,000 anomalies and curiosities. Surely nature cannot be that enigmatic and cryptic! Actually, I must stress that my search is still far from complete. Anomalies---those observations that do not yield to mainstream explanations---are ubiquitous and proliferate. I have trawled through only a small fraction of the English-language scientific journals; thousands of volumes of specialized, less-known, publications gather dust untouched. Among them are: unexamined books, monographs, informal papers, and popular publications. Foreign-language sources have only been sampled, and I can attest that the fishing there is good, too. And in today's electronic milieu, anomalies travel from computer screen to computer screen, by E-mail, and by fax. What an immense untapped resource for the for the Catalog of Anomalies! From all this, I am certain that nature is even more anomalous than the following pages intimate.

I admit freely that this book and the Catalog of Anomalies harbor a scattering of fraudulent and questionable data. I try to weed these out; but no data base can be completely clean. On the other hand, I do not apologize for retaining phenomena upon which mainstream science has "closed the book." Didn't science do this to the idea of continental drift until the 1960s, only to canonize the the concept in the 1970s? Now, one believer has recommended that data contradicting plate tectonics no longer be published! (Be assured that the pages of Science Frontiers will always welcome such waifs. Scientific political correctness is anathema here.)

Mainstream science's response to my collections has been remarkably favorable despite my obvious iconoclastic tendencies. For example, Nature has reviewed five of my books without recommending their immediate incineration; other science journals have been likewise generous. The most annoying comment in the scores of reviews in my file has been that science should not waste time with esoterica! This reviewer apparently forgot about that tiny, esoteric advance of Mercury's perihelion that resisted explanation until Einstein came along. Also troubling have been warnings by more than one reviewer that undergraduates should not be exposed to my books lest the image of science be tarnished.

I began this Preface by warning against expecting anything profound to emerge from the simple process of collecting anomalies and curiosities. Data collection is, after all, only one part of the scientific process. I have avoided as far as possible the "fun" part of science: theorization. My purpose has been to keep the data base as value-free as possible. It is this value-free aspect of the Catalog of Anomalies plus the eclectic nature of my search that makes my endeavor not only entertaining but liberating. I will now explain what I mean by "liberating," and why this feature of anomalistics might be scientifically useful.

Unless you have been comatose the past several years, you must know that the entire outlook of science is in flux. The words "chaos" and "complexity" are the current buzz words. They betoken, finally, the formal recognition by science that nature is frequently:

- Unpredictable (as in weather forecasting beyond a few days)
- Complex (as in any life form)
- Nonlinear (as in just about all real natural phenomena)
- Discontinuous (as in saltations in the fossil record)
- Out-of-equilibrium (as in real economics and even the natural world)

Eroding fast are the philosophical foundation stones of the clockwork universe: the idea that nature is in balance, that geological processes are uniformitarian, that life evolved in small, random steps, and that the cosmos is deterministic.

My view is that anomaly research, while not science per se, has the potential to destabilize paradigms and accelerate scientific change. Anomalies reveal nature as it really is: complex, chaotic, possibly even unplumbable. Anomalies also encourage the framing of rogue paradigms, such as morphic resonance and the steady-state universe. Anomaly research often transcends current scientific currency by celebrating bizarre and incongruous facets of nature, such as coincidence and seriality. However iconoclastic the pages of this book, the history of science tells us that future students of nature will laugh at our conservatism and lack of vision.

Such heavy philosophical fare, however, is not the main diet of the anomalist. The search itself is everything. My greatest thrill, prolonged as it was, was in my forays through the long files of Nature, Science. the English Mechanic, the Monthly Weather Review, the Geological Magazine, and like journals. There, anomalies and curiosities lurked in many an issue, hidden under layers of library dust. These tedious searches were hard on the eyes, but they opened them to a universe not taught by my college professors!

December 1993
Glen Arm, Maryland

William R. Corliss

LIST OF PROJECT PUBLICATIONS

CATALOGS:

Lightning, Auroras, Nocturnal Lights (category GL)
Tornados, Dark Days, Anomalous Precipitation (category GW)
Earthquakes, Tides, Unidentified Sounds (categories GH, GQ, GS)
Rare Halos, Mirages, Anomalous Rainbows (category GE)

The Moon and the Planets (categories AE, AH, AJ, AL, AM, AN, AP, AR, AU, AV)
The Sun and Solar System Debris (categories AA, AB, AC, AE, AS, AX, AY, AZ)
Stars, Galaxies, Cosmos (categories AO, AQ, AT, AW)

Carolina Bays. Mima Mounds, Submarine Canyons (category ET)
Anomalies in Geology (category ES, in part)
Neglected Geological Anomalies (category ES, in part)
Inner Earth: A Search for Anomalies (categories EC, EQ, ES in part, EZ)

Biological Anomalies: Humans I (category BH in part)
Biological Anomalies: Humans II (category BH in part)

HANDBOOKS:

Handbook of Unusual Natural Phenomena
Ancient Man: A Handbook of Puzzling Artifacts
Mysterious Universe: A Handbook of Astronomical Anomalies
Unknown Earth: A Handbook of Geological Enigmas
Incredible Life: A Handbook of Biological Mysteries
The Unfathomed Mind: A Handbook of Unusual Mental Phenomena

SOURCEBOOKS:

Strange Phenomena, vols. G1 and G2
Strange Artifacts, vols. M1 and M2
Strange Universe, vols. A1 and A2
Strange Planet, vols. E1 and E2
Strange Life, vol. B1
Strange Minds, vol. P1

NEWSLETTER: Science Frontiers (current anomaly reports)

For information on the availability, prices, and ordering procedures write:

SOURCEBOOK PROJECT
P.O. Box 107
Glen Arm, MD 21057

Chapter 1

ARCHEOLOGY

ANCIENT ENGINEERING WORKS

SMALL ARTIFACTS

EPIGRAPHY AND ART

BONES AND FOOTPRINTS

DIFFUSION AND CULTURE

ANCIENT ENGINEERING WORKS

NORTH AMERICA

All of the sections in this chapter on archeology are divided first by geographical region and then by specific phenomena. Here, under the heading of Engineering Works, we begin our standardized trek in North America, move south through Mesoamerica, and thence to the rest of the world. Usually, the various ancient structures described in this section are considered mildly anomalous because they seem to indicate scientific and engineering sophistication and ambition beyond what we naively expect in ancient societies. However, the New England stone structures described below constitute an exception, for rather than unexpected sophistication they suggest extensive precolumbian contacts by the Old World; they are therefore highly anomalous, if they are what they are claimed to be.

For North America, we have broken down the subject matter collected in past issues of Science Frontiers as follows:

- Ancient astronomers. Alignments of stone structures, earthworks, etc.
- New England stone structures. Chambers, cairns, and other structures reminiscent of Old World constructions.
- Midwestern prehistoric works. Stone forts and pyramids, garden beds.
- Works of the Hohokam and Anasazi. Roads, canals, Chaco Canyon, transportation feats, etc.
- California's walls and cairns.
- Miscellaneous structures. Florida canals, rock rings, mud structures.

ANCIENT ASTRONOMERS

STONEHENGE IN QUEBEC?

"Are there carefully crafted stone structures in Quebec similar to that most mysterious of man-made structures, Stonehenge? The answer is yes, according to biology professor Gerard Leduc, who says he has found evidence of sundials in four different locations in the Laurentians and Eastern Townships.

.....

"The stone complexes, comprising a centre stone and others radiating toward the east and west, may have been used as calendars whereby farmers could, for example, have known when to plant and harvest crops." Leduc also claims to have discovered:

(1) Unexplained stone walls two to three feet high that begin and end with no apparent purpose, and which are not associated with the fields of farmers;

(2) Grass circles showing up as yellowish rings in green grassy fields, caused by a different type of vegetation. These grass circles are perfect in shape and associated with stone structures;

(3) Trilithons, located at the sundial sites, consisting of three closely grouped rocks. (Morrissy,John; "Stonehenge in Quebec?" Stonehenge Viewpoint, no. 79, p.3, Winter 1988.)

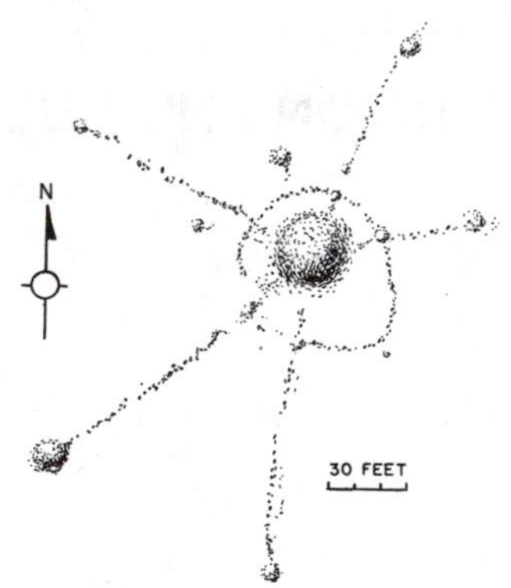

The Moose Mountain "medicine wheel," in Alberta, represents another type of North American astronomically aligned structure.

ANCIENT WISCONSIN ASTRONOMERS

Prof. James Scherz claims to have discovered an ancient Indian calendar site in a marshy region near Wisconsin Rapids, Wisconsin. Scherz was led to the site by aerial photographs taken during a wetlands mapping program. Strange "islands" of higher land detected within the bog were found, upon terrestrial inspection, to be unusually steep, possibly artificial. Some were round, some four-sided; others were shaped like a fish, a rabbit, and a snake. (Wisconsin has many similar effigy mounds in other areas.) Causeways connect some of the so-called islands. The most interesting features of the islands, however, are prominent rocks and rock cairns.

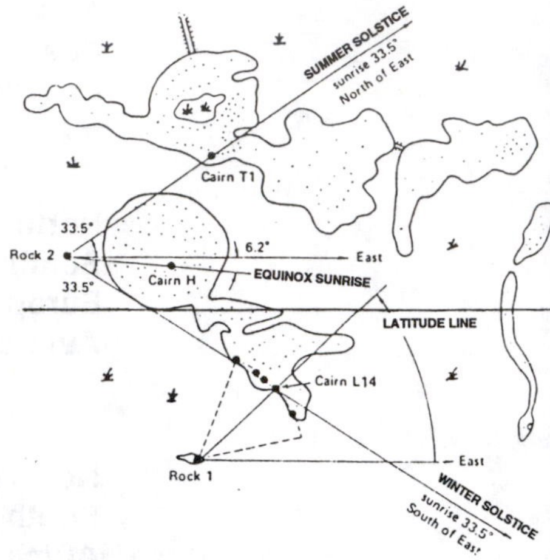

Map of the Wisconsin "calendar site," showing some of the rock-cairn alignments. The stippled islands are surrounded by marshes.

Braving hordes of mosquitoes and ticks, Scherz and an assistant mapped the islands, cairns, and rocks to determine if any astronomical alignments existed. Sure enough, the solstices and equinoxes were predictable from some of the alignments. Another alignment provided the site's latitude. The exploration of this site is incomplete, and further information is expected. Quite possibly, the site is associated with the famous prehistoric copper-mining activities around Lake Superior. (Murn, Thomas J.; "Portage County Cairns: Wisconsin's Rockhenge," NEARA Journal, 18:50, 1984. Originally published in Wisconsin Natural Resources, vol. 7, no. 2. NEARA = New England Antiquities Research Association,)

A POSSIBLE ASTRONOMICAL OBSERVATORY IN NEW ENGLAND

A recent Public TV documentary provided viewers with excellent footage of some of the domens, menhirs, and passage graves sprinkled throughout New England. Long scoffed at as merely the work of Colonial farmers, some of these structures yield radiocarbon dates 3000-4000 B.P. This sort of proscribed data is to be found only in the journals of "shadow archeology."

But now European archeologists are beginning to look at these structures. In 1990, the Belgian journal Kadath devoted an entire issue to the subject. This issue comprises 54 pages filled with many photos and drawings of walls, cairns, dolmens, and possible astronomical sites. We reproduce one of these drawings here. (Ferryn, Patrick; "Etranges Vestiges Megalithiques en Amerique du Nord," Kadath, no. 72, Spring 1990. Kadath is a Belgian archeological journal.)

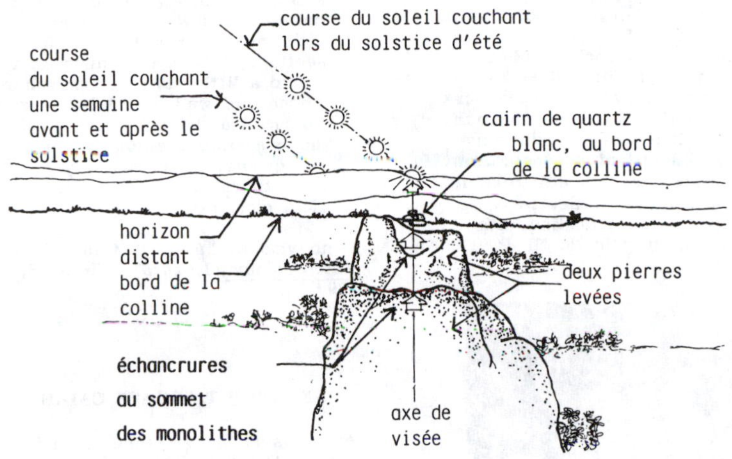

Sketch of the "archeoastronomical site" at Burnt Hill, in the Berkshires, in western Massachusetts. As the sun sets at the summer solstice, it is aligned with the notches in the two standing stones in the foreground, the cairn of white-quartz rocks in the background, and the notch in the skyline. (Kadath)

A SUN-AND-SPIRAL CLOCK

The astronomical sophistication of ancient man becomes more obvious each year. A novel method of keeping track of the seasons has been discovered on an isolated butte in New Mexico. Here, the Anasazi, who occupied Chaco Canyon between 400 and 1300 A.D., carved spiral petroglyphs into the face of a cliff. Then, they arranged stone slabs so that sharp slivers of sunlight fell on the spirals. The precise position of the sliver of light depends of course upon the location of the sun. The solstices and equinoxes are registered by unique configurations of light slivers and spirals. In contrast to other calendar sites, which rely upon the rising and setting points of the sun on the horizon, the New Mexico clock depends upon the altitude of the sun at midday. Slivers of moonlight on the spirals also seem to have astronomical significance. (Sofaer, Anna, et al; "A Unique Solar Marking Construct," Science, 206:283, 1979.)

Comment. Some archeologists have ventured that the sun-and-spiral clock is natural and purely fortuitous in origin.

ASTRONOMY AT POVERTY POINT

The astounding complex of six octagonal ridges, 4000 feet across, at Poverty Point, Louisiana, was not recognized until 1953, when aerial photos were analyzed. Roughly 3000 years old, the ridges are intersected by avenues that align with the summer and winter solstice points as well as some more obscure astronomical azimuths. These alignments represent remarkable astronomical sophistication for the New World in 1000 BC. (Anonymous; "Louisiana's 4,000-Foot Calendar," Science Digest, 90:22, July 1982.)

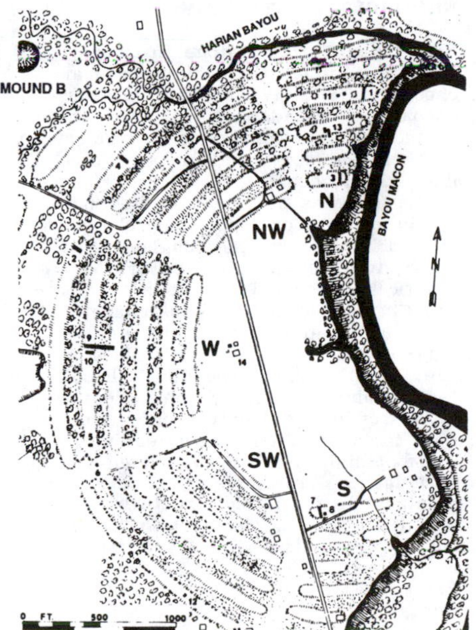

This sketch of the Poverty Point six earthen ridges reveals their octagonal geometry.

Comment. An incredible amount of labor was expended in constructing the six huge, concentric ridges. Actually, sighting lines could have been built with just a few mounds or simple markers. The Indians, if that is what they were, must have had something additional in mind to move all that dirt! Let's not be condescending and say they were for "ritual purposes," when we have no idea of their purpose. Note, too, that the better known hill-top earthen forts in Europe possess similar openings in their walls, undermining any theories that they were purely defensive works.

AN EARTHEN STONEHENGE

The six concentric octagonal (or circular) ridges at Poverty Point, Louisiana, are interrupted by four avenues, as shown in the illustration. Other avenues may have existed to the east, assuming the ridges actually continued to compplete the figures. Brecher and Haag have contended in earlier papers that two of the existing four avenues were solstice markers. Purrington, in the first of a pair of papers in American Antiquity, maintains that the Poverty Point ridges have been so badly eroded over the last 3000 years that sight lines cannot be determined with any accuracy.

In fact, the precise center of the octagonal figure is a matter of judgment. Purrington's reconstruction of the sight lines along the avenues, using his assumed center, does not support the idea that the avenues were solstice markers. Brecher and Haag respond, in the second paper, that their viewing center is 100 meters distant from Purrington's. With this change they claim good fits for two of the avenues as solstice markers. One of the remaining two avenues turns out to mark the setting of Canopus, the second brightest star in the sky. Even the last unassigned avenue has astronomical significance; it marks the setting of Gamma Draconis, a second magnitude star, which the ancients employed as a nocturnal hour hand as it swung around the pole star. (Purrington, Robert D.; "Supposed Solar Alignments at Poverty Point," American Antiquity, 48:157, 1983. Brecher, Kenneth, and Haag, William G.; "Astronomical Alignments at Poverty Point," American Antiquity, 48:161, 1983.)

Comment. It seems that just a difference of 100 meters in the viewing center makes the difference between a people who merely liked to build huge geometrically shaped ridges (Purrington's thought) and a race trying--Stonehenge fashion---to reflect the motions of the sun and stars in a colossal earthen "computer" over a mile in diameter. The real issue, of course, is not 100 meters but one's conception of ancient man!

NEW ENGLAND STONE STRUCTURES

THE STONE ENIGMAS OF NEW ENGLAND

Background. New England is dotted with an amazing array of stone structures that closely resemble the well-known, bona-fide megalithic ruins of Europe. Establishment archeologists deny that the New England standing stones, cairns, and stone chambers are anything more than Colonial root cellars and recent accumulations of stones removed from fields by farmers. There also exists, however, an enthusiastic group of researchers, mostly amateurs, who maintain that the Old World made contact with the New centuries before the Vikings and Columbus.

Science Frontiers coverage of New England's mysterious stone structures commenced with the maze of walls, courtyards, and chambers that characterize New Hampshire's Mystery Hill site, the various types of anomalous stone structures of the Northeast are reviewed. Three major classes are recognized:

(1) Covered passageways, up to 25 feet in length and analogous to the Cornish fogous

(2) Beehive chambers, such as the 10-foot-high chamber at Upton, MA, and

(3) Dolmen-like constructions, as exemplified by the 60-ton "balancing boulder" at North Salem.

Termed a "glacial erratic" by most, the North Salem stone seems distinctly unlike most erratics and more like some European dolmens. Noting that radiocarbon dates from Mystery Hill go back to more than 1000 BC, Doran and Kunnecke feel that these stone enigmas should receive professional attention in the context of world distributions of blood groups and other evidence of early, frequent transoceanic communication. (Doran, Michael F., and Kunnecke, Bernd H.; "The Stone Enigmas of New England," Anthropological Journal of Canada, 15:17, no. 2, 1977.)

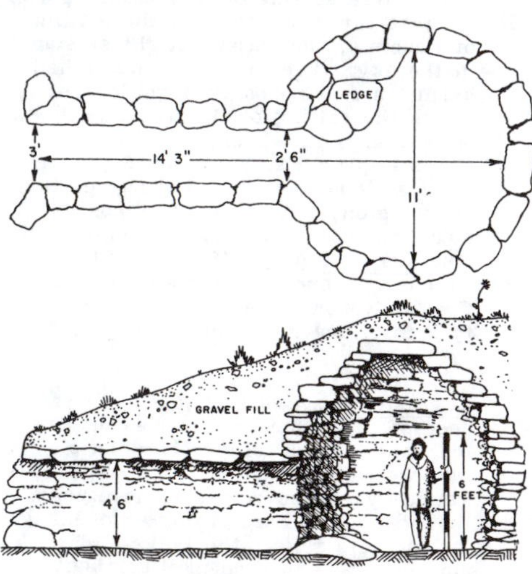

The dry masonry underground chamber at Upton, Massachusetts. (Adapted from ESRS Bulletin, 1:12, 1973.)

HOW ANCIENT IS VERMONT?

The present article from the British journal Antiquity is obviously an Establishment reaction to such books as Ancient Vermont and The Search for Lost America. Two aspects of the "problem" are discussed: (1) the supposed ogam writing discovered in New England; and (2) the many peculiar stone structures in the same region.

A. Ross and P. Reynolds have examined the claimed ogam inscriptions and are emphatic that they are really not of Celtic origin, although they are probably deliberately inscribed in many instances. On the other hand, the strange stone structures in New England, particularly Vermont, do bear some resemblance to megalithic remains in Europe. The authors are not as anxious to pass these off as Colonial root cellars as are their American allies within the Establishment. Ross and Reynolds suggest that much more work needs to be done here before the purpose of these chambers and standing stones can be identified. (Ross, Ann, and Reynolds, Peter;"Ancient Vermont," Antiquity, 52:100, 1978) See p. 31. for several other items dealing with possible ogam writing in North America.

Photograph of the interior of a stone chamber located in South Woodstock, Vermont. Note the rectangular geometry. (P. Ferryn, Kadath)

WHO BUILT THESE CHAMBERS?

New England's many stone chambers have long piqued the curiosities of archeologists and laymen alike. The archeologists are adamant that all of these structures were constructed by Colonial farmers, but some certainly seem unlikely potato cellars!

J. Egan has provided architectual details on 14 impressive stone chambers located in southern New England. Of these, two seem hardly the work of practical farmers. The first is the Pearson Chamber at Upton, Massachusetts. It is 10 feet high and 11 feet wide inside---pretty large for vegetable storage. The second is the Hunt's Brook "souterrain", Montville, Connecticut. It is 38 feet long and only about 3 feet high for most of its length, ending in a 5-foot-high chamber. We cannot visualize farmers crawling this distance for potatoes! In fact, this structure resembles the megalithic "souterrains" of Europe. (Egan, Jim; NEARA Journal, 22: 6, Summer/Fall, 1987. NEARA = New England Antiquities Research Association.)

A CONNECTICUT SOUTERRAIN?

Souterrains, such as that figured, are megalithic constructions usually considered to be exclusively of European origin. This look-alike, at Montville, CT, could indicate pre-Columbian contacts. Here are some details:

> This underground site is built into a rocky hillside in an isolated region. A 37½ ft. passage of straight-sided drywall stonework is interrupted after 8½ ft. by a 3½ ft. collapsed section. It then continues on for 20 ft. to a little corbelled chamber whose end wall is cut into a roughly quarried and levelled ledge. On the slope around the souterrain are about 100 cairns, some carefully constructed; others appear to be the result of field clearing.

(Anonymous; "An Arm-Chair Field Trip," NEARA Journal, 26:87, Winter/Spring 1992.)

CONNECTICUT "BOAT" CAIRN

An unusual, large stone cairn is located atop Rattlesnake Hill in Connecticut's Natchaug State Forest. At an elevation of 640 feet, it commands an almost 360° view. Its long axis is aligned with the Pole Star. The cairn seems to have been constructed according to some plan rather than just being a deposit of cleared stones. One's first impression is that it resembles a boat. Could it be a Norse "ship burial" such as found in Europe? It is impossible to prove such a conjecture without tearing the cairn apart. (Whittall, James P., II; "The 'Boat' Cairn, Chaplin, Connecticut," ESRS Bulletin, 12:39, December 1986. ESRS = Early Sites Research Society.)

A side view of the Connecticut "boat" cairn.

STANDING STONE CLUSTER IN EASTERN MASSACHUSETTS

"Driving up the roadway into LeBlanc Park in February 1984, I saw a sight I had not seen since my travels in the British Isles. Situated on a mound was a cluster of weathered megalithic stones. I was filled with disbelief---it just couldn't be---someone was having fun with my senses; Western Europe, yes,

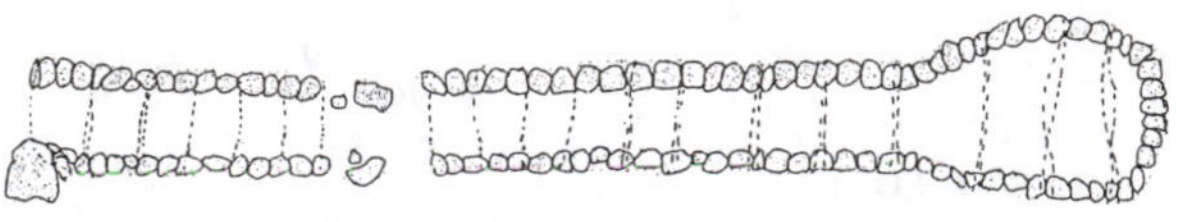

Plan view of the Hunt's Brook "souterrain." The dotted lines represent the capstones. This structure is almost 38 feet long.

but here, in Massachusetts, no. The reality of the scene before me was very difficult to focus on, the parallel with sites I had seen in Scotland and Ireland was astonishing."

Thus wrote James P. Whittall, II, when describing a group of standing stones on a mound called Druid Hill, at Lowell, Massachusetts. The mound itself is 112 feet long by 56 feet wide. The stones are separated into two groups, as shown. Since the site is near a highly populated area, it has seen some disturbance, and some stones have been moved. There is no historical record of the site's construction; the stones may have been there for centuries. Neither has there been any archeological investigation or site dating. Obviously, much more research must be done before we can get a clear idea as to who the builders were. Despite its close resemblance to European standing stone complexes, the Lowell cluster could be a recent construction---an intentional replica of European sites. Note that it is called Druid Hill! Other possible builders might have been American Indians (who are known to have built some stone structures), Iron Age Scandinavians, or Bronze Age wanderers from Europe. (Whittall, James P., II; "A Cluster of Standing Stones on Druid Hill, Lowell, Massachusetts," ESRS Bulletin, 11:19, no. 1, 1984.)

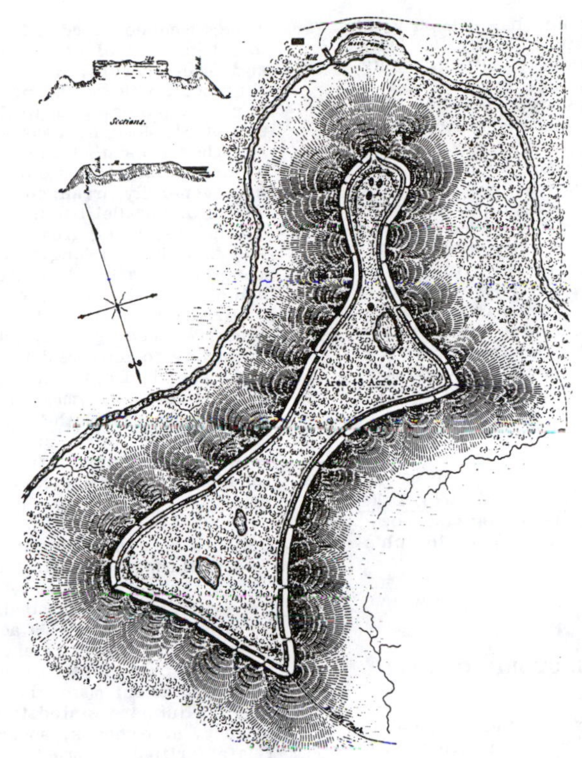

Fort Hill, Highland County, Ohio, as drawn by E.G. Squier and E.H. Davis in 1846. Scale: 1 inch = 1,000 feet. Area enclosed by the fort walls = 48 acres. (From: Ancient Monuments of the Mississippi Valley)

Disposition of the standing stones at Druid Hill, Lowell, Massachusetts.

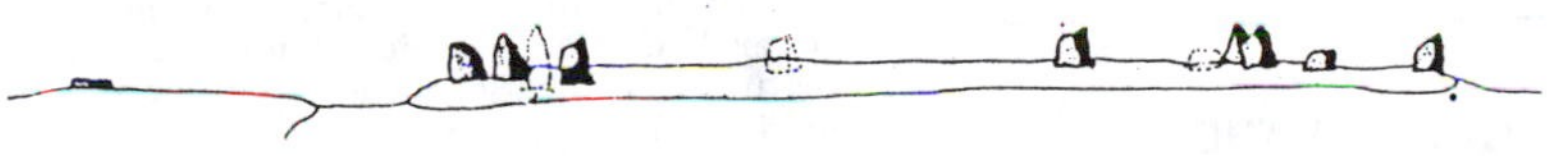

MIDWESTERN PREHISTORIC WORKS

ENIGMATIC STONE FORTS OF THE MIDWEST

From diverse sources, J. Singer has drawn together a fascinating compendium of large stone forts and walled structures west of the Alleghenies, From Ohio, Indiana, Wisconsin, and all across the midwest come descriptions of stone structures of almost heroic proportions: a wall 5-12 feet high and almost a mile long at Fort Hill, Ohio; two big walls at Fourteen Mile Creek, Indiana; two massive stone walls or pyramids under Rock Lake, Wisconsin; and many more.

Ignorance surrounds these structures. Who built them and when? Some are likely forts, for they are paralleled by ditches. Others seem to have no defensive value. The Moundbuilders may have had a hand in some. Madoc and his wandering Welshmen are blamed for at least one wall. (Singer, Jon Douglas; "Stone Forts of the Midwest," NEARA Journal, 13:63, and 13:91, 1979.)

Comment. Concerted field work and searches of local archives would doubtless multiply the number of unexplained stone structures in the midwest, as has happened in the northeast.

THE ROCK LAKE PYRAMIDS

Professor Sherz, of the University of Wisconsin, is trying to get to the bottom of the reputed stone pyramids submerged in Rock Lake, near Lake Mills, Wisconsin. Fishermen have hit the pyramids with their oars when the lake was very low; and others have spotted structures (perhaps as many as four) from the air. In 1937, a diver reported seeing a 29-foot-high pyramid in the murky waters. Recent divers have found boulder alignments; but the visibility is so poor that organized structures cannot be identified. Truncated earthen pyramids exist at nearby Pre-Columbian Aztalan; and strange rock piles exist elsewhere in Wisconsin. Consequently, one cannot brush off the idea that some edifices were built at Rock Lake when the water was much lower. (Smith, Susan Lampert; "Lake Mills' Lost Pyramids," Wisconsin State Journal, June 26, 1983. Cr. R. Heiden and L. Farish)

Comment. The 1937 dives were described in Ancient Man. It is relevant that sonar-imaging equipent has discovered obviously human-built rock structures on the bottom of Loch Ness, Scotland.

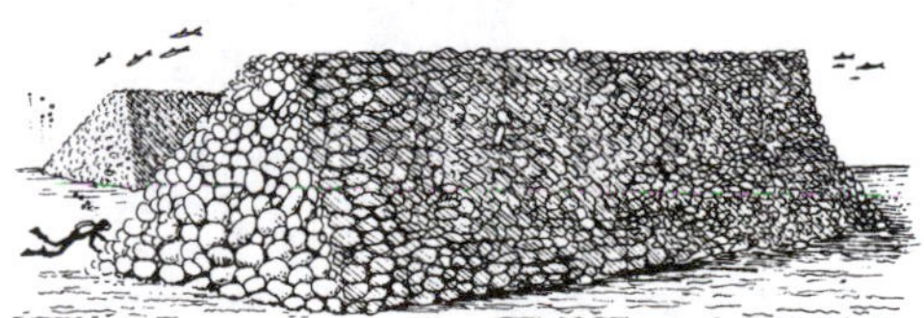

Artist's concept of the Rock Lake pyramids based on the early reconnaissance reported in 1970 in Skin Diver.

FOUND: THE LOST PYRAMIDS OF ROCK LAKE

The purported pyramids found on the muddy floor of Wisconsin's Rock Lake are so fascinating that we must include the following additional report, even though it comes from outside the group of publications we usually rely on. In fact, we have never seen anything on these pyramids in the scientific press. So, caveat emptor!

The author of this article, F. Joseph, states that beneath the surface of Rock Lake lie at least ten structures. Two of these have been mapped and photographed by skin divers and sonar. Structure #1, which has been dubbed the Limnatis Pyramid, has a base width of 60 feet, a length of about 100 feet, and a height of 18 feet, although only about 10 feet protrude from the silt and mud. It is a truncated pyramid, built largely out of round, black stones. On the truncated top, the stones are squarish. The remains of a plaster coating can be discerned.

The Rock Lake structures are made more believable by the presence, 3 miles away, of the Indian site of Aztalan. Here, there are two truncated, earthen pyramids, partially surrounded by a tall stockade, which was originally plastered. Aztalan seems to have been occupied as late as the Fourteenth Century. (Joseph, Francis; "Found: The Lost Pyramids of Rock Lake," Fate, 42: 88, October 1989.)

Comment. F. Joseph has also written a book on the Rock Lake structures entitled The Lost Pyramids of Rock Lake, 1992. Professional archeologists, however, remain unconvinced by Joseph's claims.

MICHIGAN'S PREHISTORIC GARDEN BEDS

The prehistoric ridged fields, canals, aqueducts, and other agricultural engineering feats found in South and Central America and even our own Southwest continue to amaze us. Almost totally forgotten, however, are the equally impressive "garden beds" of southern Michigan. Happily, the INFO Journal has just reprinted B. Hubbard's 1877 paper describing these works that stretched for miles along the Grand and St. Joseph Rivers. Of course, modern activities have obliterated them completely; and even in Hubbard's day they were mostly gone.

First, Hubbard's general description of the "garden beds":

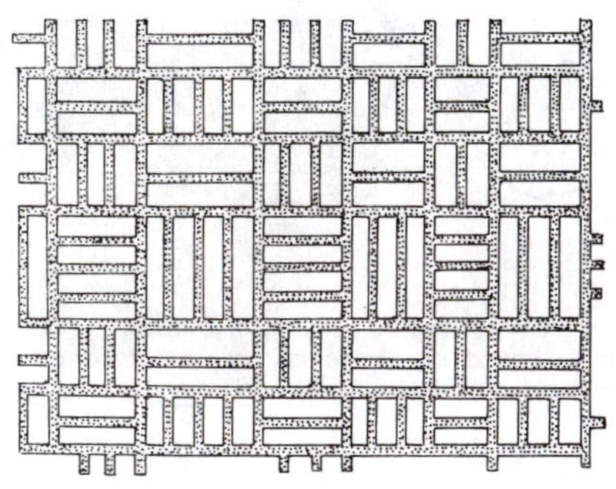

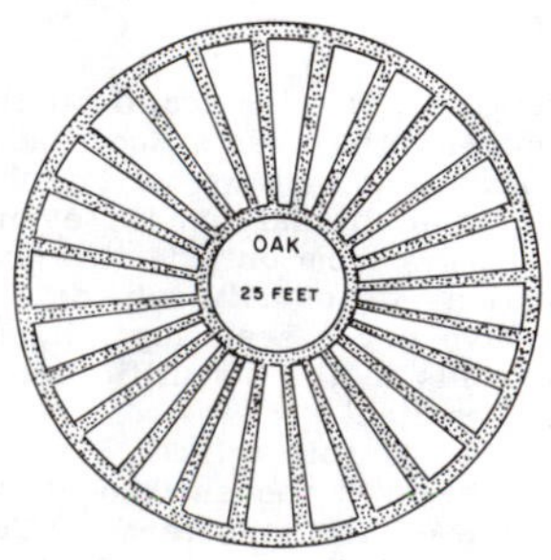

Two types of precolumbian garden beds located near Kalamazoo, Michigan.

"The so-called 'Garden Beds' were found in the valleys of the St. Joseph and Grand Rivers, where they occupied the most fertile of the prairie land and burr-oak plains, principally in the counties of St. Joseph, Cass and Kalamazoo.

"They consist of raised patches of ground separated by sunken paths, and were generally arranged in plats or blocks of parallel beds. These varied in dimensions, being from five to sixteen feet in width, in length from twelve to more than one hundred feet, and in height from six to eighteen inches.

"The tough sod of the prairie had preserved very sharply all the outlines. According to universal testimony, these beds were laid out and fashioned with skill, order and symmetry which distinguished them from the ordinary operations of agriculture, and were combined with some peculiar features that belong to no recognized system of horticultural art."

Hubbard recognized eight types of beds. Two of these are illustrated and described above. He gave no figure for the total extent of the beds. Individual plats ran from 20 to 300 acres. Considering that they stretched for miles through three counties, we are certainly talking about many thousands of acres. Hubbard stated that the usual pottery, arrowheads, spear points, and related artifacts seemed to be absent from the areas of the beds. (Hubbard, Bela; "Ancient Garden Beds of Michigan," reprinted in INFO Journal, 12:6, no. 58, 1989. INFO = International Fortean Organization, P.O. Box 367, Arlington, VA, 22210. Also reprinted in Ancient Man.

WORKS OF THE HOHOKAM AND ANASAZI

A PREHISTORIC TVA?

The Hohokam Indians lived along the Salt River Valley, in Arizona, about 300 BC to 1450 AD. They constructed a network of irrigation canals that certainly is one of the wonders of the Ancient New World. Early investigators recorded more than 500 kilometers of major canals and 1600 kilometers of smaller ones. (Less than 10 kilometers of these remain intact today.) One of the main canals was 3 meters deep, 11 meters wide at ground level, and 14 kilometers long. Old and recent aerial photos show traces of an incredibly complex irrigation network. Masse, the author of this article, is very impressed by the earth-moving task but even more so by the degree of social coordination and control that must have been exerted all along the Salt River Valley in building and regulating the use of this remarkable canal system. (Masse, W. Bruce; "Prehistoric Irrigation Systems in the Salt River Valley, Arizona," Science, 214:408, 1981.)

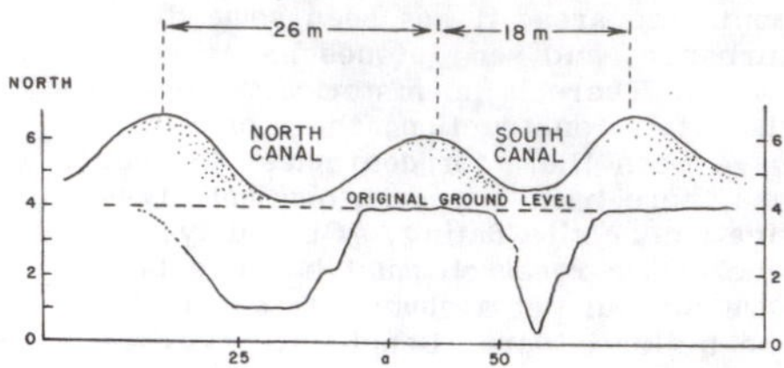

Section through two Hohokam canals, showing the original canal profile (bottom) and the final profile after long use (top). Sedimentation eventually raised the canal bottoms above the original ground level.

PHOENIX VS. THE HOHOKAM

Rapidly expanding Phoenix is gobbling up the vestiges of the Hohokam. Archeologists are striving to save as much information as possible before the bulldozers destroy the best artifacts left by this remarkable Indian civilization. This beautifully illustrated article touches on several of the precocious and puzzling features of the Hohokam Period, circa 0-1400 AD.

(1) The Hohokam apparently employed acid-etching to produce designs on shells. Acetic acid from fermented cactus juice was used to eat away portions of the shells not protected by tar.

(2) Four-story Casa Grande, which seems to have been an astronomical observatory, required at least 600 big wooden beams, all of which had to be transported over 50 miles from sources in the mountains.

(3) The Hohokam built an elaborate, well-engineered system of irrigation canals.

(4) Unexplained are many flat-bottomed oval pits up to 182 feet long, 55 feet wide, and 13-18 feet deep. Some surmise they were ball courts.

(5) Also puzzling are rectangular earthen mounds, 75 x 95 feet at the base, and 12 feet high, with flat adobe-

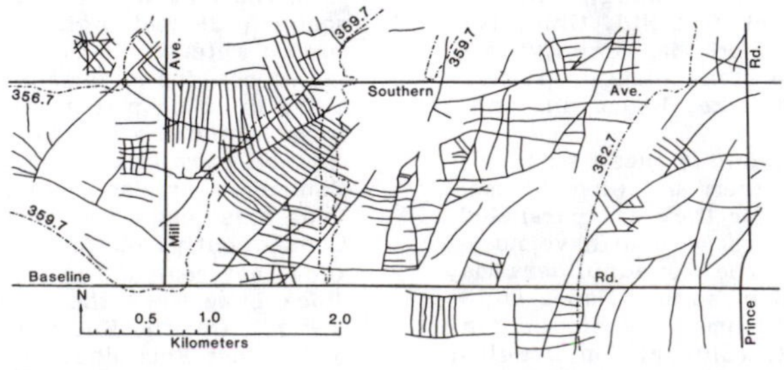

A composite map of Hohokam canals near Tempe, Arizona, as drawn from aerial photographs.

covered tops.

(Adams, Daniel B.; "Last Ditch Archeology," Science 83, 4:28, December 1983.)

THE CHACO CANYON ROAD SYSTEM

The Anasazi Indian culture of the American Southwest may not have employed quipus (p. 27), but they did build impressive works of civil engineering. The huge Great House built in Chaco Canyon has elicited coniderable admiration; but recent aerial photography and remote sensing have revealed an amazing pattern of straight roads radiating from Chaco Canyon. The purposes of these roads is still obscure. What is obvious is that we have much more to learn about these remarkable peoples.

(Anonymous; "The Chaco Canyon Road System," Archaeoastronomy, 4:50, October/December 1981.)

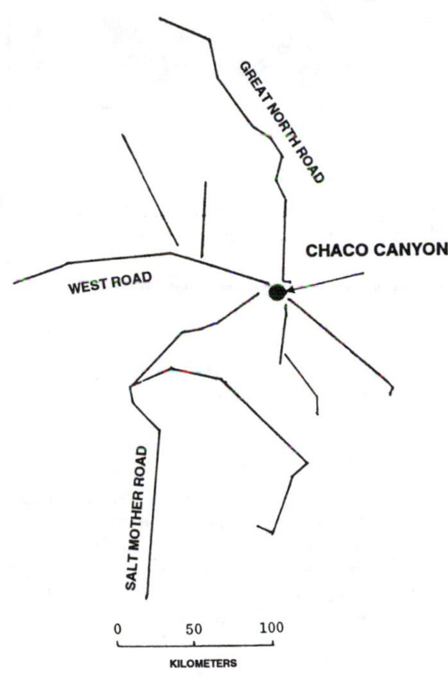

Some of the major Anasazi roads leading to Chaco Canyon. Nine Great Houses are located in the Canyon proper, more are scattered along the roads.

ALL ROADS LEAD TO CHACO CANYON

In 1987, L. L'Amour came out with his novel The Haunted Mesa. It's all about the Anasazi, a remarkable people of the old Southwest, circa 900-1200 AD, who, as far as we can tell, disappeared rather suddenly. L'Amour has the Anasazi returning to a parallel world through a space warp in a kiva window. Archeologists have not yet found this remarkable kiva, so we must be content with the things they left behind; but they are impressive enough.

A long article in Scientific American introduces us to the accomplishments of the Anasazi. We will concentrate here on their road system, but we cannot let a few statistics go by unnoticed.

Of the nine Great Houses of the Anasazi in Chaco Canyon, in northwestern New Mexico, Pueblo Bonito is the best studied. It covers three acres and once rose to at least five stories, with some 650 rooms. Constructed of tightly fitting sandstone blocks, each Great House required tens of millions of cut sandstone slabs. For floors, the Anasazi carried logs from forests 80 kilometers away. The Chaco Canyon Great House required about 215,000 trees---quite a problem in transportation! Strangely enough, the Great Houses seem to have been used only occasionally. In fact, Chaco Canyon was too poor agriculturally to support a large, permanent community. If this is so, what was the purpose of the Great Houses with their many kivas (large circular pits)? Obviously, they were for "ceremonial purposes"; the standard explanation of unexplained buildings and artifacts!

The Anasazi also built a marvelous system of roads leading to Chaco Canyon. The accompanying map reveals hundreds of miles of roads converging from all directions on Chaco Canyon. For long distances, these roads measure a uniform 9 meters wide. They are flanked by linear mounds of earth and are impressively straight. The Great North Road, for example, runs true north for almost 50 kilometers. What was the purpose of these roads? One theory is that they helped channel people to Chaco Canyon for the supposed ceremonies. But why does one need a 9-meter-wide road for a sparse population? And why did the Anasazi leave all this? (Lekson, Stephen H., et al; "The Chaco Canyon Community," Scientific American, 259:100, July 1988.)

TREE TOTING EXTRAORDINAIRE

Abstract. "Identification of spruce (Picea) and fir (Abies) construction timbers at Chetro Ketl in Chaco Canyon, New Mexico, implies that between A.D. 1031 and 1120 the Anasazi transported thousands of logs more than 75 km. These timbers came from high elevations, probably in mountains to the south (Mt. Taylor) and west (Chuska Mountains) where Chacoan interaction was well established. Survey in these mountains might disclose material evidence of these prehistoric logging activities."

The rest of the article contains even more amazing statistics. The ten major pueblos in Chaco Canyon consumed an estimated 200,000 trees. The average primary beam was 22 cm in diameter, 5 m in length, and weighed about 275 kg (600 pounds). Since these logs show no transportation scars, they were probably carried rather than dragged or rolled. Such labor required a large, complex sociocultural system. (Betancourt, Julio L., et al; "Prehistoric Long-Distance Transport of Construction Beams, Chaco Canyon, New Mexico," American Antiquity, 51:370, 1986.)

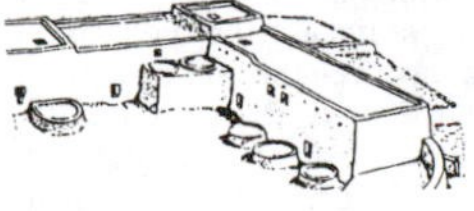

CALIFORNIA'S WALLS AND CAIRNS

WHO BUILT THE EAST BAY WALLS?

Ranging along the hills east of San Francisco Bay are long stretches of walls constructed from closely fitted basalt boulders. Some of these boulders weigh more than a ton. In some places, the walls reach five feet in height and three feet in width. They extend for miles along the hill crests from Berkeley to Milpitas and beyond. Russell Swanson, one of the few persons willing to pursue the walls in the field, estimates that all the walls strung together would run for at least 20 miles.

Naturally, time and civilization have destroyed some of the walls, but what remains is most impressive. Searches of property records going back to the Gold Rush and the studies of Spanish mission records give no hints of who built the walls or why. Evidently they are centuries old, possibly prehistoric. Why would anyone build miles of walls from ponderous boulders along miles of ridge crests? They would seem to serve no practical purpose.

Scientists seem to show no interest in the walls. One even stated, "I don't know of anyone who's come up with a credible explanation. I think what you're getting is an indication that there isn't any academic work on it." (Burress, Charles; "Unraveling the Old Mystery of East Bay Walls," San Francisco Chronicle, December 31, 1984. Cr. R. Swanson)

A section of one of the East Bay walls. The location is Mission Peak, near Fremont. N. Fink is shown measuring the wall, which is about 5 feet high here. It is said that these walls were in place when the Spanish arrived. (R. Swanson)

Comment. In a recent private communication, Russell Swanson has revealed that the walls now extend as far as San Jose, 50 miles from Berkeley, with more mysterious walls now reported in Marin County across the Bay. It would seem that academics would find plenty of raw material in all this!

ANCIENT ROCK CAIRNS IN THE CALIFORNIA DESERT

This paper's anomaly resides in the caliche ($CaCO_3$) adhering to the bones and rock undersides in a California burial cairn. The caliche has been dated at more than 21,000 BP by radiocarbon methods and 19,000 BP by thorium methods---ages much too large for conventional archeological histories of the Southwest. Wilke objects that only the age of the soil caliche has been measured and that the bones and their burial may be much younger. This cairn site, called the Yuha burial, possesses no other datable characteristics. On the other hand, many other California burial cairns (the California desert is amply provided with them) have been dated from 5000 BP to historic times. In this context, Wilke maintains that it is easier to believe that the Yuha bones and burial are merely 5000 years old and that something is wrong with the caliche dating. (Wilke, Philip J.; "Cairn Burials of the California Deserts," American Antiquity, 43:444, 1978.) See pp. 20 and 29 for additional evidence of New World contacts before 12,000 BP (pre-Clovis times).

MISCELLANEOUS STRUCTURES

MYSTERIOUS STONE RINGS

In Green Ridge State Forest, in western Maryland, one can find perhaps 150-200 annular piles of sandstone rocks. All lie on the western slope of Polish Mountain. No one seems to have a good explanation of their origin. Archeological digs have not unearthed any human artifacts. From a photograph of one ring, we estimate an outer diameter of 15 feet, and an inner hole 5 feet in diameter. The height of the rock ring is perhaps 2 feet. The sandstone rocks are generally slab-like. A popular theory states that the rocks were piled up to protect apple trees. (Anonymous; "Rings of Stone Pose Mystery in Md.," Washington Post, June 26, 1988. Cr. J. Judge)

Comment. This item might also be classified under geology because these rings could be periglacial phenomena; that is, akin to patterned ground. Periglacial structures are occasionally found in the Appalachians.

ANCIENT MUD STRUCTURES IN COLORADO

"The remains of 16 hardened mud structures recently discovered in the mountains northwest of Denver may prove to be the oldest-known buildings in North America. Found during construction of a pipeline, the remains are 4000 to 7000 years old, compared with 4700 years for the oldest Egyptian pyramid." Mud structures such as these usually disintegrate in a few hundred years, but these were fire-hardened. Because they appear to be permanent buildings, the current belief that the American Indians of this period were simple, nomadic hunter-gatherers may have to be reexamined. (Anonymous; Science Digest, 90:22, August 1982. Attributed to the Christian Science Monitor.)

Comment. The Rocky Mountain area also boasts enigmatic stone structures and graphic material attributed by some to ancient European, Asian, and African voyagers. See Ancient Man.

FLORIDA'S CIRCULAR CANALS

Circular canals up to 1,450 feet in diameter and 6 feet deep have been discovered in south central Florida. Dug in the savannas and flood plains around Lake Okeechobee, the man-made circles include gaps where drainage canals extend outwards. Forty of these circular earthworks have been located by R.S. Carr. Some are as old as 450 BC; others as recent as the 16th century. Mounds and large plazas are also part of this impressive example of Pre-Columbian engineering.

Carr supposes that the circular canals were fish traps, but no fish bones or other supporting evidence for this theory have appeared. Another thought is that the earthworks drained agricultural land, but no maize pollen has been found. Could they have been ceremonial sites? No one really knows. (Bower, B.; "Florida 'Circles' May Be Ancient Fisheries," Science News, 138: 6, 1990.)

Comment. Still other ancient canals are to be found in Florida, as sketched in the map below.

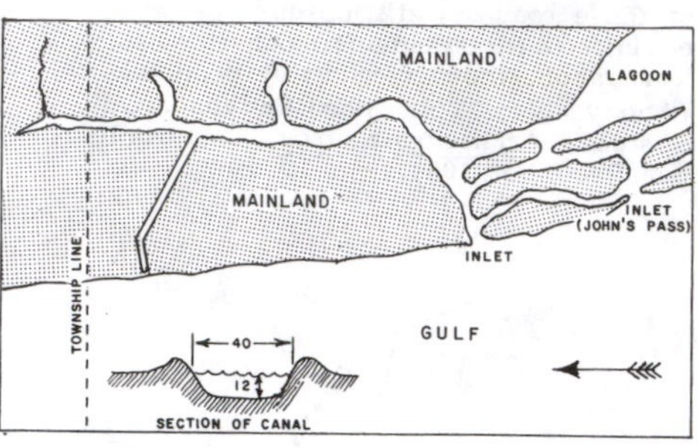

An ancient canal on the Florida Gulf Coast. See Ancient Man for details.

MESOAMERICA

From Tikal to Uxmal, Mesoamerica is liberally sprinkled with marvelous pyramids, ball courts, and other constructions. However, ever searching for the anomalous, only two Mesoamerican engineering works have been featured in Science Frontiers: the stone spheres of Costa Rica, which may be arranged in significant patterns in some cases, and the supposed underwater "roads" and "walls" submerged in shallow waters off Bimini, in the Bahamas. Curiously, the Bimini structures seem to be echoed far to the east in the Canary Islands!

COSTA RICA'S NEGLECTED STONE SPHERES

Books and articles on Stonehenge, the Easter Island statues, and the Egyptian pyramids are legion. Granted that these structures are important and intriguing, we still ask why Costa Rica's meticulously wrought stone spheres are languishing in the wings of science. They epitomize exquisite workmanship. Such geometric perfection rendered in granite is remarkable---for any ancient culture. Lastly, the stones' purpose completely escapes us. Why strew such masterpieces of stoneworking around the dark jungle floors?

M.T. Shoemaker also wonders about these things in a nice recapitulation of the stone-sphere mystery. His compilation of facts and figures only impels us to learn more about the spheres and what their shapers had in mind.

(1) The spheres are found on the

One of Costa Rica's precisely crafted stone spheres.

Diquis River delta, near the Pacific coast of southern Costa Rica.

(2) Sphere sizes range from an inch to 8 feet in diameter.

(3) At least 186 spheres have been recorded in the literature. Surely many more were destroyed and others remain undiscovered.

(4) No local source exists for the granite; and no stone-working tools have been found near the spheres.

(5) "The best spheres are perhaps the finest examples of precision stone-carving in the ancient world." The maximum circumference error in a 6 foot, 7 inch diameter sphere is only 0.5 inch, or 0.2%.

(6) The spheres are often grouped, but no general system or alignment mode seem to exist.

(7) "One very disturbing mystery emerges in examining the Diquis culture. The superb stone-carving skill necessary for the creation of the spheres was not applied to any other objects." Why?

Such are the salient facts. To our regret, they tell us little about How, Why, and perhaps, Who. (Shoemaker, Michael T.; "Strange Stone Spheres," Pursuit, 19:145, 1986.)

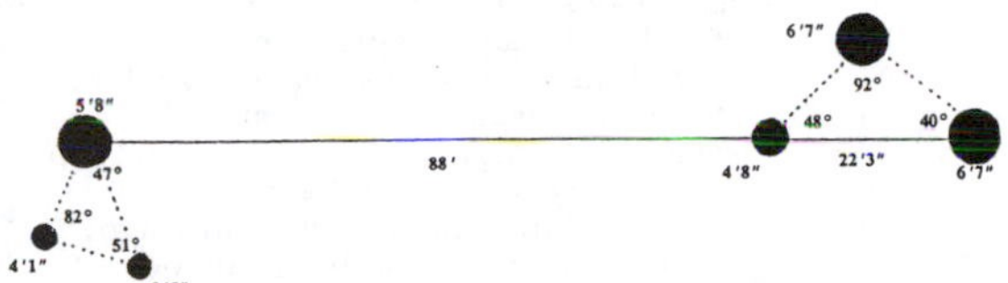

Disposition of six stone spheres on Farm 4-35C, in Costa Rica. The significance of this grouping, if any, is unclear.

GOODBYE TO THE BIMINI WALL AND ROAD?

A perennial fixture of sensational archeology has been the frequent reports of submerged walls and "roads" off North Bimini, in the Bahamas. That there are closely fitted, rectilinear stones under about 15 feet of water is not in question. The 1-to-10 ton blocks surely look manmade; but are they really? Shinn describes several "beach rock" formations in the area, some exposed and some submerged under a few feet of water. This beach rock, as his photos demonstrate, has a natural tendency to fracture in rectangular blocks, creating strips of pavement-like blocks essentially identical to the famous Bimini walls. Proponents of Atlantis and other radical archeological theories do not deny the similarity of the formations or even that the natural and supposedly manmade blocks are of the same composition. The Atlanteans obviously made use of readily available materials and beach rock was their choice.

Shinn goes on to prove to his satisfaction that the Bimini block formations are still in place right where geological forces placed them about 2200 years ago. Further, there are absolutely no traces of human workmanship and no human artifacts in the area. One mystery is admitted in this debunking article; and that is the question of how the Bimini rocks came to be submerged in 15 feet of water when considerable evidence indicates no such sea level changes occurred during the last 2200 years. (Shinn, E.A.; "Atlantis: Bimini Hoax," Sea Frontiers, 24:130, 1978.)

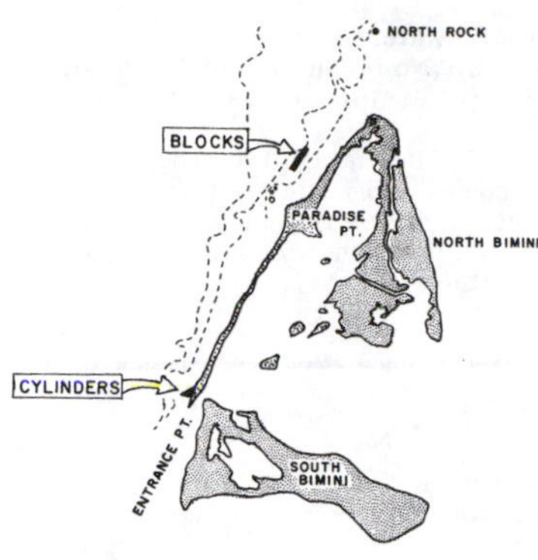

Locations of blocks and cylinders in shallow waters off Bimini.

Comment. Perhaps Shinn was too negative about Bimini. A decade later, D.G. Richards was able to identify many suspicious "regularities," as now related.

OTHER BIMINI ARCHEOLOGICAL ANOMALIES

The famous Bimini "wall" or "road", in the Bahamas, has engendered many a sensational article in the popular press. Atlanteans and even extraterrestrials have been credited with building the "road" and other constructions reported around this Caribbean island. The fact is that the "road" does exist, but the weight of opinion among those who have investigated it is that it is a natural formation of beach rock fractured in a disturbingly regular manner. But this assessment does not mean that all anomalies in the shallows around Bimini have been exorcised.

D.G. Richards, in a splendid article in the Journal of Scientific Exploration, gives us a blow-by-blow account of the investigations (both amateur and professional) of the Bimini waters. It is a curious panorama of wild claims by adherents of the Cayce-inspired Atlantis searchers and the knee-jerk academic scoffers---both of which go overboard! However, our purpose here is the recording of some of the features near Bimini that Richards thinks are still anomalous.

Three of these are located at A, B, and D in the accompanying drawing, which is based on an aerial photo taken at 6000 feet. A is a 90° bend in the reknowned "road". This bend is decidedly anomalous for a beachrock formation. B consists of a parallel row of stones. D is made up of regularly spaced piles of stones and extends over 1½ miles, cutting diagonally across ancient beach lines.

Richards also employed a satellite image of the area to locate other "regular" features, such as a triangle, a pentagon, and a sharp right-angle corner with mile-long sides. Inspecting these regularities from a small boat, Richards found no obvious structures of any kind. Rather the patterns were caused by sea grass and white sand. Even so, these superficial patterns may reflect the presence of artificial structures under the sediments. Certainly, if these regularities were observed in a photo taken over land, archeologists would rush to dig away the overburden. But this was Bimini, and everyone knows that no "high cultures" ever lived there! (Richards, Douglas G.; "Archaeological Anomalies in the Bahamas," Journal of Scientific Exploration, 2:181, 1988.)

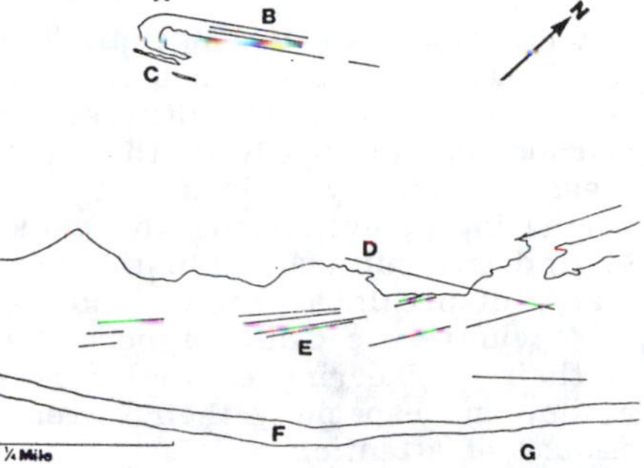

Map of some Bimini offshore features. C marks regular blocks; E indicates ancient beach lines; F is the current shore. See text for A, B, and D.

LANZAROTE: UN NOUVEAU BIMINI?

We are drawing on the French-language journal Kadath, an archeological publication from Belgium, for this item. The reason, of course, is that the mainstream English-language archeological journals are notoriously conservative and, well, mainstreamish!

The catchword in the title is "Bimini", a word which loses nothing in the translation, for it is well-known in the States as one of the Bahamian resort islands. It was in the waters off Bimini that divers found the famous Bimini "road" or "wall", which some maintain is constructed of human-sculpted stone blocks.

Lanzarote, on the other hand, is one of the Canary Islands. Here, too, one finds a submerged, Bimini-like row

(Right) Alignment of blocks in 22 meters of water off Lanzarote. (Left) Front view of the blocks, showing their stepped arrangement.

of apparently man-made blocks of stone. Some 22 meters down, the blocks are arranged in a sort of staircase, as shown in the figure. The steps, however, are 40 cm high, too big a step for humans. Is this structure a submerged pier, an altar, or something else? No one knows. Possibly relevant is a statuette, stylistically Olmec, which was also found in Lanzarote waters. (Bajocco, Alf; "Lanzarote: un Nouveau Bimini?" Kadath, no. 66, p. 6, Winter 1987.)

Comment. The name, Kadath, incidentally, comes from the writings of the American science-fiction writer, H.P. Lovecraft. Kadath was Lovecraft's great city of the ancients.

SOUTH AMERICA

When one contemplates ancient South America, the great civilization of the Incas invariably comes to mind. Yet, the Incas were preceded by still other remarkable cultures, which also built impressive edifices and learned how to cope with the rigors of the high Andes, Chief among these was the mysterious culture centered at Tiahuanaco. And, most surprisingly, even along the banks of the Amazon, where no one expected to find any cultural sophistication, archeologists are now excavating traces of pre-Inca peoples who were far from primitive.

Obviously we cannot ignore the Incas themselves, for we still marvel at their engineering expertise as seen in their stonework, their long roads, and especially their "Great Wall," which does not receive its fair amount of attention.

"HIGH" TECH FARMING AT TIAHUANACO

One of ancient Tiahuanaco's (now spelled "Tiwanaku") many puzzles has been how food for such a large city was grown at an altitude of circa 3,850 meters (12,600 feet) in the frosty, windswept Bolivian Andes. This problem along with the fabulous stonework and extensive ruins have precipitated theories involving extraterrestrial visitors and an age in the hundreds of thousands of years.

At least the food supply puzzle seems to be in hand. Stereoscopic aerial photographs show in startling detail "immense, curvilinear platforms of earth ....these fields form elevated planting surfaces ranging from five to 15 meters wide and up to 200 meters long....Extensive and nearly continuous tracts of these fields---all of which have been abandoned for centuries---run from the edge of Lake Titicaca to about 15 kilometers inland, and form virtually the only topographic relief in the broad, gradually sloping plain."

Some of the raised fields are remarkably sophisticated in design. At the base is a layer of cobblestones for stability. These are covered by a 10-centimeter layer of clay. On top of the clay are three distinct layers of sorted gravel; all capped by rich organic topsoil. These fields were simultaneously an aquifer for the fresh water percolating down from the surrounding hills and a barrier to the brackish water from Lake Titicaca. Even at Tiahuanaco's altitude these fields could have grown potatoes, oca or ulluco and the chenopod grains, as well as quinowa and caniwa. Tiahuanaco and its satellite cities could have been fed with enough left over for export. Not bad for farmers 2,000 years ago! (Kolata, Alan L.; "Tiwanaku and Its Hinterland," Archaeology, 40:36, January/February 1987.)

NEW WORLD CULTURE OLD

Archeologists working in Peru have unearthed stunning evidence that monumental architecture, complex societies and planned developments first appeared and flowered in the New World between 5,000 and 3,500 years ago---roughly the same period when the great pyramids were built in Egypt and the Sumerian city-states reached their zenith in Mesopotamia.

Among these edifices are great stepped pyramids, U-shaped temples over ten stories high, and broad plazas with adjacent residential areas. Scores of such sites built by an ancient Peruvian civilization are nested deep in narrow valleys leading from the Andes down to the Pacific. Archeologists date this civilization as thousands of years older than those that arose in Central America. The age and size of the Peruvian remains impelled Yale archeologist R. Burger to remark,"This idea of the Old World being ahead of the New World has to be put on hold." Of course, this Andean culture is not as old as that which developed in the Old World's Fertile Crescent; and these ancient Peruvians did not use the wheel and lacked writing. (Stevens, William K.; "Andean Culture Found To Be as Old as the Great Pyramids," New York Times, October 3, 1989. Cr. J. Covey)

Ruins of a stepped pyramid in the Chicana Valley, Peru.

IMMENSE COMPLEX OF STRUCTURES FOUND IN PERU

The well-known explorer Gene Savoy has discovered a "lost city" some 120 square miles in area in the jungle-covered mountains of Peru. This citadel, called Gran Vilaya, is located about 400 miles northeast of Lima on a 9000-foot mountain ridge.

Savoy "said the city's buildings ran along the ridge for at least 25 miles. He said the expedition calculated there were 10,350 stone structures in the defensive network along the ridge and 13,600 other stone buildings in three major city layouts. The stone structures, some measuring 140 feet in length, were built atop terraces that go up the mountain slopes like stairs, he said. He described them as 'complex units of circular buildings with doorways, windows and niched walls.' The walls, he said, 'soar up as high as a 15-story building'"

The city was built by the Chachapoyas Indians about 1000 years ago. The Chachapoyas empire is dated 800-1480 AD. The Incas, who finally conquered them, told Spanish explorers that the Chachapoyas were tall, fair-skinned people! (Anonymous; "Ruined City Found in Jungle in Peru," New York Times, July 7, 1985. Cr. M. Hall and L. Farish)

GREAT ANCIENT CIVILIZATIONS IN AMAZONIA? RIDICULOUS!

Even though the first Europeans who sailed up the Amazon circa 1540 reported large, well-populated cities on the Amazon flood plains, modern archeologists have generally traveled to the high Andes for "high" ancient civilizations in South America. In retrospect, this is not surprising. By 1700, the cities of the 1540s had been swallowed up by the jungle. Besides, most thought the conditions in the lowland tropics were too harsh to nourish advanced societies. Fortunately, a few archeologists have recently invaded Amazonia with aerial sensors, magnetometers, and the old-fashioned shovel. And, indeed, there once was a high civilization along the great river; and, some say, it may have spread from the lowlands to the Andes far to the west. What a turnabout in archeological thinking---if sustainable by facts.

One intriguing site in Amazonia is the island of Marajo, 15,000 square miles in area, located at the mouth of the Amazon. Here are found some 400 huge dirt mounds, including one with a surface area of 50 acres and a volume of a million cubic yards. Radiocarbon dates suggest that Marajo had been occupied for over a thousand years.

Nearby, on the Tapajos River in Brazil, A. Roosevelt found elaborate pottery, finely carved jade, and a culture going back perhaps 7000 years.

In other parts of Amazonia, surveys uncovered tens of thousands of acres of raised fields connected by causeways. There remains little doubt that an advanced, complex civilization dwelt in Amazonia for millennia. Archeologists are now asking where these people came from and how they were related to the Incas to the west and civilizations to the north in Central America. (Gibbons, Ann; "New View of Early Amazonia," Science, 248:1488, 1990.)

HOW THE INCAS MAY HAVE WORKED STONE

Inca stonework is famous for its large stones (some over 100 tons), which are fitted so precisely that "a knife blade cannot be inserted into the joints". An aura of mystery has always hung about the great "walls" at Saqsaywaman and Ollantaytambo (spellings vary). How could the Incas have quarried, dressed, transported, and fitted such huge stones? As usual with such remarkable ancient structures, the overzealous have proposed stone-softening agents, antigravity devices, and similar wild notions. In truth, as J. Protzen relates in the article under review, stonemasonry was surprisingly unsophisticated and efficient, although some mysteries do remain.

Protzen has spent many months in Inca country experimenting with different methods of shaping and fitting the same kinds of stones used by the Incas. He found that quarrying and dressing the stones were not problems at all using the stone hammers found in abundance in the area. Even the precision fitting of stones was a relatively simple matter. The concave depressions into which new stones were fit were pounded out by trial and error until a snug fit was achieved. Protzen's first-hand experience is impressive and convincing. Certainly no radical solutions were required.

The problems that Protzen was not able to solve to his satisfaction involved the transportation and handling of the large stones. The fitting process necessitated the repeated lowering and raising of the stone being fitted, with trial-and-error pounding in between. He does not know just how 100-ton stones were manipulated during this stage. To transport the stones from the quarries, some as far as 35 kilometers distant,

Much Inca stonework is characterized by close-fitting, odd-shaped blocks.

the Incas built special access roads and ramps. Many of the stones were dragged over gravel-covered roads, as evidenced by their polished surfaces. The largest stone at Ollantaytambo weighs about 140,000 kilograms. It could have been pulled up a ramp with a force of about 120,000 kilograms. Such a feat would have required some 2400 men. Getting the men was no problem, but where did they all stand? The ramps were only 8 meters wide at most. A minor problem perhaps, but still unsolved. Further, the stones used at Saqsaywaman were fine-dressed at the Rumiqolqa quarry and show no signs of dragging. Protzen does not know how they were transported 35 kilometers.

An intriguing observation by Protzen is that the cutting marks on some of the stone blocks are very similar to those found on the pyramidion of the unfinished obelisk from Aswan in Egypt. Is this a case of anomalous diffusion of Old World technology or simply independent invention? (Protzen, Jean-Pierre; "Inca Stonemasonry," Scientific American, 254:94, February 1986.)

THE GREAT WALL OF THE INCAS

The Inca's ability to build with stone is well-known. But one of their most ambitious projects is rarely mentioned in the literature and is poorly investigated in the field. This is the so-called Great Wall of the Incas. Probably no more than 150 miles in length, it cannot compare with China's Great Wall. Still, it is built at altitudes of 8,000-12,000 feet in extremely rugged terrain. It runs along high ridges and is studded with stone forts at strategic intervals. Even though the Inca Wall is only a few feet high, it would certainly slow down a force charging up precipitous terrain at two miles altitude. The true extent and condition of the Great Wall is not accurately known. Only a few accessible sections have been checked out. The theory is that the Incas built it to discourage invasions by lowland Indians. Like all Great Walls, it seems to have met with small success. (Paddock, Franklin K.; "The Great Wall of the Inca," Archaeology, 37:62, July/August 1984.)

Comment. The Great Wall of the Incas is located in Bolivia. There is another Great Wall in Peru that seems to have been built by the Chimu people, perhaps to defend themselves against the Incas! The Chimu Great Wall averages 7 feet in height and reaches 20-30 feet where it crosses gullies. Also incompletely explored, it may run some 60 miles.

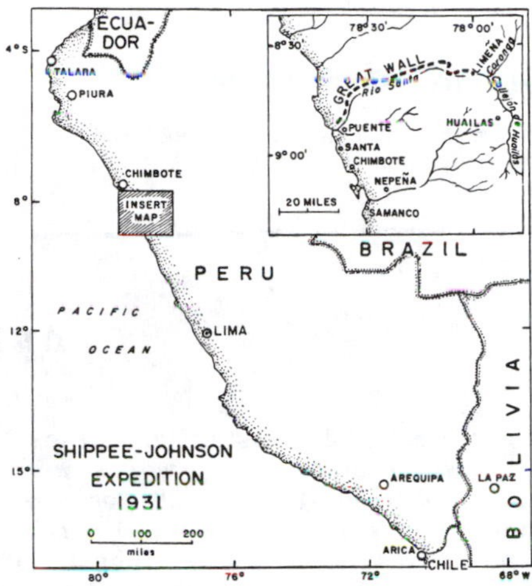

No discussion of precolumbian South American walls would be complete without a mention of this Great Wall of Peru, thought to have been built by the Chimu people. Stretching for many miles, the Great Wall was interrupted by forts, some with walls 15 feet high and 5 feet thick.

INCA WALLS AND ROCKWALL, TEXAS

First, we have what "seems" to be a new Inca wall of impressive proportions. This story began when R. Chohfi, a UCLA graduate student, was examining aerial photos of the Machu Picchu region in Peru. He noticed a straight line where no archeological ruins had been recorded. Friends put up money for Chohfi to journey to Peru and investigate. His hunch was that since straight lines are rare in the jungle, something manmade must there. He was right. He found a wall more than 7 feet thick, at least that high, and more than 1000 feet long. Other structures were also found in the area, suggesting the existence of a major new archeological site. (Dye, Lee; "Incas; UCLA Student May Have Opened a New Door," Los Angeles Times, October 4, 1986. Cr. E. Krupp)

Next, let us consider Rockwall, Texas, a small town named for a strange rock wall, mostly buried, that exists in the area. We have had inquiries about this structure but have little in the way of substantial data. Just arrived is a facetious newspaper item that relates how, some 50 years ago, R.F. Canup excavated part of this wall. He dug 8 feet down and eventually unearthed about 100 feet of the wall. That was enough to convince him that it was the masonry wall of an ancient city. Geologists, on the other hand, ridicule his idea, saying it is only a natural rock formation. (Streater, Don; "Geologists Burst Rockwall's Bubble," Beaumont Enterprise, September 8, 1986. Cr. S. Parker via L. Farish)

Comment. What we really need in both instances above are authoritative geological and archeological reports. Have any professionals visited the sites? In particular, it seems incredible that Canup could have mistaken a natural rock wall for an artificial one!

THE INKA ROAD SYSTEM

An important new archeological book bears the above title (and alternate spelling of Inca). As one reviewer puts it: "The Imperial Inka road system must rank alongside the Great Wall of China and the Egyptian Pyramids as one of the greatest achievements of an ancient civilization. Yet despite this, relatively little is known about the nature, extent and functioning of this vast communications network." Some impressive statistics: the Inka road system runs for more than 23,000 kilometers through Peru, Bolivia, Ecuador, Chile, and Argentina. Generally, the roads were 11-25 meters wide. They were greatly superior to anything built in Europe at that time. One reviewer notes that many of the so-called Inka highways had a non-Inkan origin---and then leaves us hanging. What pre-Inkan civilization built such roads? (Saunders, Nick; "Monumental Roads," New Scientist, p. 31, June 8, 1985. Also: Lyon, Patricia J.; "Imperial Connection?" Science, 228: 1420, 1985.)

AFRICA

Many books of substantial masses have been published dealing with the impressive engineering works we have inherited from the ancient Egyptians. During its first 20 years of publication, Science Frontiers has netted several curious tidbits relating to the Great Pyramid and the Sphinx, as noted below. These, of course, are just the tips of the icebergs that have been sighted in the literature surveyed during this short period. Further, Africa is also home to many megalithic structures, whose purposes are not yet clearly understood. They, too, are included in this section.

•The Great Pyramid. Was it astronomically aligned? Could its millions of stone blocks been poured like concrete rather than hewn from quarries?

•The Sphinx. Its possible great antiquity. The question of whose visage is really represented.

•African stone alignments. Were they for astronomical, ritual, or some other purpose?

THE GREAT PYRAMID

GREAT PYRAMID ENTRANCE TUNNEL NOT ASTRONOMICALLY ALIGNED

Early in the Nineteenth Century, astronomer John Herschel speculated that the Ancient Egyptians had constructed the Great Pyramid so that the downwardly slanting entrance would be aligned precisely with the pole star, Thuban (Alpha Draconis), when the star was at its lowest culmination. Over 70 years ago, Percival Lowell ran through the calculations and found that Thuban was not near the tunnel's line of sight when the pyramid was constructed (about 2800 BC). No one seems to have listened to Lowell, even though he was quite correct. Most books on the Great Pyramid still insist on the fancied pole star alignment.

If the entrance tunnel wasn't pointing at the pole star, what other esoteric reason did the pyramid builders have for the 26°.523 angle. (It seems that everyone expects all dimensions of the Great Pyramid to have special significance!) R.L. Walker, of the Naval Observatory, has come to the rescue. He observes that the tangent of 26°.523 is almost exactly ½ (actually 0.4991). Although there may be some occult significance to ½, this fraction signals us that 26°.523 is also the angle created when two cubical blocks are laid horizontally for every one installed vertically, as in the sketch. It seems that 26°.523 is simply the natural consequence of the internal pyramid construction process. (Anonymous; "End of a Pyramid Myth," Sky and Telescope, 69:496, 1985.)

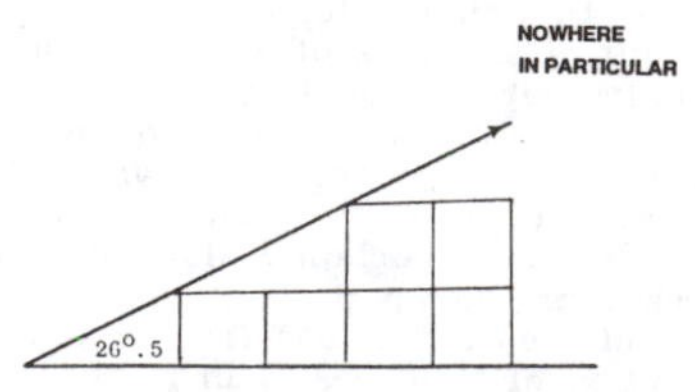

Comment. Lowell also showed that Thuban did cross the tunnel entrance centuries before and after the accepted date of construction. Could the date of the pyramid be in error, or were the builders planning for the future? Anything is possible in Pyramidology!

CURIOUS SAND-FILLED CAVITIES IN THE GREAT PYRAMID

"Japanese and French experts are investigating a new mystery at the 4,000-year-old pyramids---why the pharoahs built geometrical cavities inside the Great Pyramid of Cheops and filled them with mineral-enriched sifted sand.

.....

"From the outside, the pyramid appears to be built of solid blocks of limestone. But two French architects, Gilles Dormion and Jean Patrice Goidin, discovered cavities which could total 15 to 20 per cent of the structure.

.....

"The French team used an instrument which measures differences in gravity to find the internal spaces. Then they drilled small holes through the 1.8-metre blocks and found sand---but not ordinary sand from the nearby desert.

"Laboratory tests showed it came from another part of Egypt and was sifted and enriched with minerals before being placed inside the pyramid by the ancient architects." (Fouad, Ashraf; Vancouver Sun, March 7, 1987. Cr. G. Conway via L. Farish) The French results were subsequently confirmed by a Japanese team. (Buccianti, Alexandre; "Des Scientifiques Japonais Confirment les Travaux de la Mission Francaise," Le Monde, p. 18, February 4, 1987. Cr. C. Maugé)

POURING A PYRAMID

The ancient Egyptians may have been more clever than we thought. Instead of chipping away laboriously in limestone quarries to precisely shape the stones constituting the pyramids, they may have cast the stones from a slurry of crushed limestone and a special mineral binder. Polymer chemist Joseph Davidovits has examined the limestone casing stones that were used to face some of the pyramids.(Most of the original casing stones were removed long ago for use in construction projects.) Davidovits claims that the casing stones contain minerals not found in the quarries, and that they contain as much as 13% binder material. In addition the casing stones have a millimeter-thick coating of this binder. This theory might help explain the precise fitting of the stones. Others have analyzed the stones, too, and oppose Davidovits' claims. (Peterson, I.; "Ancient Technology: Pouring a Pyramid," Science News, 125:327, 1984.)

EGYPTIAN PYRAMIDS ACTUALLY MADE OF SYNTHETIC STONE?

Three years after his first announcement about the artificial nature of the pyramid stones, Davidovits provided the following additional information in a newspaper article:

"An authority on ancient building materials reported evidence Friday that the Egyptian pyramids were built of a remarkable synthetic stone that was cast on the site like concrete.

"The new theory challenges the widely accepted belief that the pyramids were built from natural stone obtained from quarries and laboriously moved to the site on wooden rollers.

"Dr. Joseph Davidovits said that the synthetic pyramid stone was made with cement far stronger than modern portland cement, which binds together the rock and sand in concrete. Portland cement has an average span of about 150 years, he said, but cements like those used in the pyramids last thousands of years.

.....

"Davidovits said that a new deciphering of an ancient hieroglyphic text now provides some direct information about pyramid construction and supports his theory that synthetic stone was the construction material.

.....

"Davidovits said the cement used in pyramid stone binds the aggregate and other ingredients together chemically in a process similar to that involved in the formation of natural stone.

"Portland cement, in contrast, involves mechanical rather than molecular bonding of the ingredients. Thus, pyramid stone is extremely difficult to distinquish from natural stone.

"He cites a number of other pieces of evidence to support his theory. Chemical analyses of stone from pyramids, for example, show it contains minerals not found in Egyptian quarry stone.

"Laboratory analyses also have revealed indications of organic fibers---possibly human or animal hair---inside the stone used to build the pyramids. Davidovits said he believes the materials accidentally fell into the forms when ancient Egyptians were casting the stone."

(Anonymous; "Pyramids Made of Synthetic Stone, Researcher Reports," Orange County Register, April 11, 1987. Cr. S. Yaple via L. Farish)

DID THE PHARAOHS CHEAT WITH CONCRETE?

"The Great Pyramid of Egypt, one of the seven wonders of the ancient world, may have been at least partially constructed with man-made concrete.

"Edward Zeller, cirector of the radiation physics laboratory at Univ. of Kansas Space Technology Center, Lawrence, recently examined a stone sample, taken from a pyramid passageway, under a binocular microscope and discovered that it was filled with oval air bubbles, 'like you'd expect to see in plaster,' he says.

"The finding supports a theory proposed by French geochemist Joseph Davidovits, who says Egyptians built the pyramids by pouring concrete into forms. Such technology was not thought to have existed during the construction of the pyramids around 2690 BC.

"Zeller still has his doubts.

"'The sample is clearly made of manufactured stone and it's part of the pyramid,' he says. 'But a single specimen does not a pyramid make.'" (Anonymous; "Did the Pharaohs Cheat with Concrete?" R&D Magazine, p. 5, December 1990. Cr. J. Wenskus)

Comment. Davidovits' contentions have generally been ignored by the archeological establishment, as is readily apparent from the lack of references above from the professional literature. For elaboration of Davidovits' heresy see his and M. Morris' 1988 book The Pyramids: An Enigma Solved.

THE SPHINX

A TERRESTRIAL RIDDLE

The ancient Egyptians apparently built the enigmatic Sphinx by first excavating a limestone formation and then clearing away the debris to expose a huge stone block over 240 feet long and 66 feet high. From this, they carved a lion with a human head out of the soft natural rock.

A drawing from 1825 showing Napoleon's soldiers examining the Sphinx.

Once the soft limestone was exposed, the rain and atmosphere began to erode it. R.M. Schoch, a Boston University geologist, studying the weathering patterns on the Sphinx, found signs of water action up to 8 feet deep in the front and sides of the colossal statue. Other structures in the vicinity, made from the same limestone, supposedly at the same time (about 2500 BC), do not display such deep erosion. Based upon the depth of the weathering, Schoch dates the Sphinx at 5000-7000 BC---much older than the mainstream date of 2500 BC. In fact, Schoch opines that work on the Sphinx could have begun as early as 10,000 BC. Egyptologists, of course, will have none of this. C. Redmount, a Univerisity of California archeologist specializing in Egyptian artifacts, said, "There's just no way that could be true."

Some non-establishment archeologists, such as J.A. West, have long maintained that the Sphinx is much older than 2500 BC. Supporting the claims of much earlier dates is the massive stone wall and tower of Jericho, whose construction is now placed in the ninth millennium BC. Who knows, the Neolithic peoples of 10,000 BC might have been more precocious than we give them credit for. (Wilford, John Noble; "A Very Old Sphinx May Be Older Yet," New York Times, October 25, 1991. Cr. J. Covey. Anonymous; Experts at War over Age of Sphinx," Los Angeles Times News Service, October 24, 1991. C. F. Hurlburt)

FLAT FACED CHEPHREN NOT SPHINX!

J.A. West, mentioned above, holds that the Sphinx was not built by the pharaoh Chephren circa 2500 BC. Rather, the revisionists say, it was constructed some 5000 years earlier by predecessors of the "classical" Ancient Egyptians.

Adding fuel to the controversy is a reconstruction of the Sphinx's visage by F. Domingo, top forensic artist of the New York Police Department. Domingo traveled to Egypt to measure first-hand the exact dimensions of the statue's facial features. When Domingo compared the Sphinx's profile as it probably appeared originally with the profile of Chephren taken from a statue in the

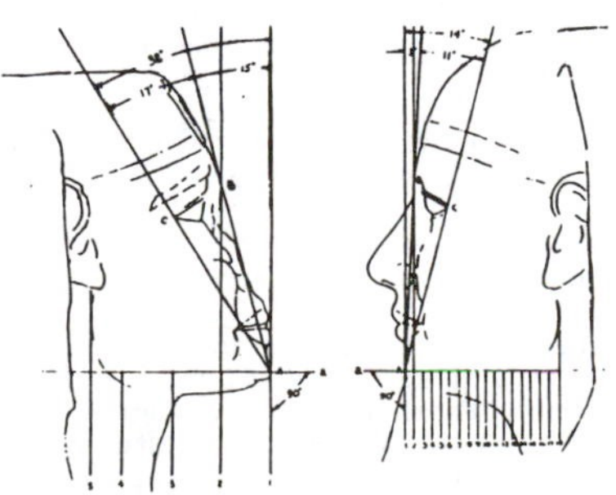

Profile of a reconstructed Sphinx (left) compared with pharaoh Chephren's profile (right). (Drawing by F. Domingo)

Cairo Museum, there was scant resemblance. (West, John Anthony; "The Case of the Missing Pharaoh," New York Times, June 27, 1992. Cr. J. Covey)

But the story doesn't end there. Orthodontist S. Peck responded to the Times article suggesting an even more radical notion:

> The analytical techniques he [West] and Detective Frank Domingo used on facial photographs are not unlike methods orthodontists and surgeons use to study facial disfigurements. From the right lateral tracing of the statue's worn profile a pattern of bimaxilliary prognathism is clearly detectable. This is an anatomical condition of forward development in both jaws, more frequently found in people of African ancestry than in those from Asian or Indo-European stock. The carving of Chephren in the Cairo Museum has the facial proportions expected of a proto-European.

(Peck, Sheldon; "Sphinx May Really Be a Black African," New York Times, July 18, 1992. Cr. J. Covey)

A MARTIAN RIDDLE

Background. One of the most radical notions introduced in the 1980s was the so-called Cydonian Hypothesis that holds that giant, intelligently designed structures can be discerned in photos of the Martian landscape taken from the Viking spacecraft. While most of the Science Frontiers material pertinent to this rash claim can be found in the Astronomy chapter (p. 64) , it is pertinent to mention here the curious resemblance thought to exist between the Martian "face" and the Sphinx.

Let us now look at cold, desert-like Mars, which still clings to a thin, oxygen-less atmosphere and where, some say, the artifacts of a long-dead intelligent race may be seen.

A livable Mars in past eons is not a physical impossibility. Some scientists argue that Martian geological and geochemical data:

> ...are consistent with past conditions on Mars that were favorable to earth-like life forms: Abundant liquid water and an atmosphere that was dense and warm, and possibly rich in oxygen.

That life---intelligent life---once thrived on Mars is suggested by photos taken of the Martian surface by Viking spacecraft:

> Images of the surface of Mars showing, at several sites what appear to be three carved humanoid faces, of kilometer scale, and having similar anatomical and ornamental details between all three. Appearing with these objects are numerous other objects and surface features that resemble Earth-like archaeological ruins, of a Bronze Age culture, with no evidence of advanced technology or civilization.

The Martian faces, pyramids, and cities are the foundation of the Cydonian Hypothesis:

Adaptation of J. Channon's sketch of the "Face on Mars" emphasizing its similarities to the Sphinx. (From: Pozos, Randolfo Raphael; The Face on Mars, Chicago, 1986, p. 50)

> That Mars once lived as the Earth now lives, and that it was once the home of an indigenous humanoid intelligence.

(Brandenburg, John E., et al; "The Cydonian Hypothesis," Journal of Scientific Exploration, 5:1, 1991. This journal is published by the Society for Scientific Exploration, which is composed primarily of established scientists.)

AFRICAN MEGALITHS

STRANGE MEGALITHIC MONUMENTS IN THE CENTRAL SAHARA

Archeologists have recently made some spectacular discoveries of megalithic structures in the central Sahara. This region boasts many V-shaped prehistoric monuments as well as spiral and meander-type carvings. There are even a few cup-and-ring markings, like those so prevalent in northern Europe. Most curious are the so-called "axle-type" monuments, which consist of a central hub with two straight projecting arms. Early in 1981, the monument shown in the sketch was found in the Immidir district. It is basically V-shaped, with sets of "auxiliary" arms, one V-type and one axle-type, both of which are detached from the hub like "spare parts." The tips of the north-south arms in the sketch are about 75 meters apart. Clearly, orientation was important to the builders of the monuments, but the asymmetry and auxiliary arms are puzzling. The stonework in these central Saharan monuments is good. Dates are elusive, but all indications are that the sites are ancient. (Milburn, Mark; "Multi-Arm Stone Tombs of Central Sahara," Antiquity, 55:210, 1981.)

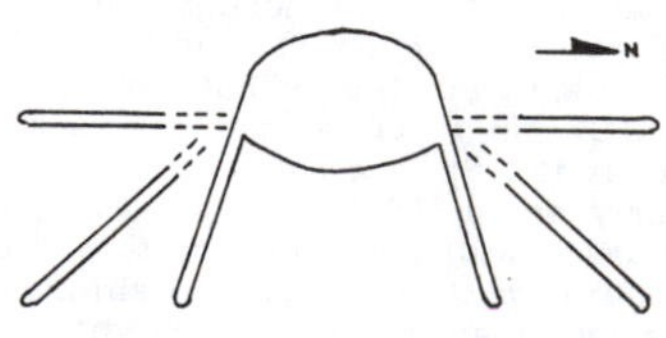

A peculiar megalithic monument found in the central Sahara.

THE SENEGAMBIAN MEGALITHIC MONUMENT COMPLEX

When one thinks of megaliths, one's thoughts usually turn to Britain and Brittany, forgetting that North Africa is covered with them. Hill sketches out in this paper the full extent of the great tract of megalithic remains on the Atlantic coast of Africa near Cape Verde, which he calls the Senegambian Monument Complex since it sits astride both Senegal and Gambia. An archeological inventory of the region discloses 212 pillar-circle sites and 251 "tombelles," which are stone cairns or heaps often surrounded by a ring-like stone wall. Hundreds of sites with tumuli also exist in the area. One of the pillar-circle sites boasts all of 50 individual pillar circles. Some of the pillars are topped with cupules, raised discs, or balls. The fanciest pillars are V-or Y-shaped with crossbars. Archeological exploration of these impressive sites is incomplete. Preliminary dating makes the Senegambian Complex over 1,000 years old. The functions of this vast array of megalithic sites is unknown, although it is not obviously astronomical. (Hill, Matthew H.; "The Senegambian Monument Complex; Current Status and Prospects for Research,: in Megaliths to Medicine Wheels: Boulder Structures in Archaeology, Michael Wilson, et al, eds., p. 419.)

Sketch of a postage stamp depicting lyre-shaped monoliths at Kaffrine, in Senegal.

STONE ALIGNMENTS IN SUBSAHARAN AFRICA

Megalithic sites are found everywhere; many were apparently used for calendar reckoning. Although numerous megalithic circles and other arrangements are known in Africa, particularly Ethiopia, astronomy does not seem to have been a primary objective of African sites. Now, however, a stone alignment in northwestern Kenya called Namoratunga has been found with unmistakable astronomical overtones. At Namoratunga, 19 large basalt pillars are arranged in rows forming a suggestive pattern. Since the site is dated at approximately 300 BC, archeologists have taken sightings on seven prominent stars as they would have appeared during this period. (The azimuths of some of these stars had changed as much as 12° in 2,200 years.) The stars chosen are those employed by Eastern Cushites, the present inhabitants of the region, in calculating their rather sophisticated calendar. Pairs and frequently triads of these pillars line up very accurately (to less than 1°) with the seven key stars. The people occupying this part of Kenya around 300 BC, therefore, probably possessed detailed astronomical information. (Lynch, B.M., and Robbins, L. H.; "Namoratunga: The First Archaeoastronomical Evidence in Sub-Saharan Africa," Science, 200:766, 1978.)

Comment. This astronomical sophistication is consistent with the claimed, but contested, remarkable celestial knowledge of the Dogon Tribe. See Robert K.G. Temple's 1976 book The Sirius Mystery.

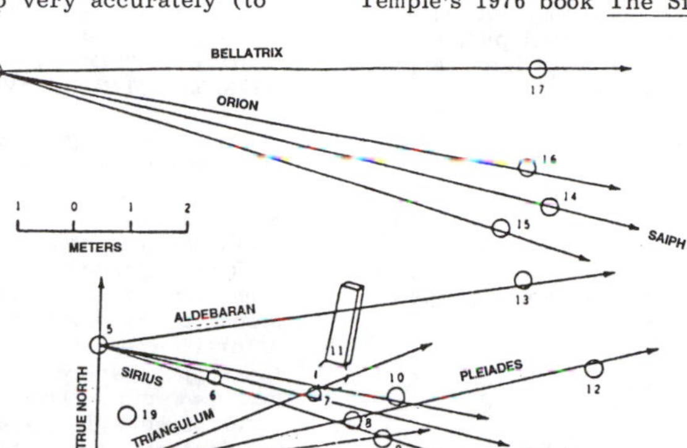

Basalt-pillar alignments at the Namoratunga-II site in Kenya.

EUROPE

Europe's inventory of megalithic sites includes hundreds of dolmens, stone circles, stone alignments, and the like. Most of these, while impressive, are not particularly anomalous and have been described almost too often. Science Frontiers, in its quest for the unusual and anomalous, selected only a few sites for their size and workmanship (the Hambledon Hill fortress and the Sardinian nuraghi). Several entries were chosen because of their use of wood rather than stone and earth (the Sweet Track and the well of Kuckhoven). Finally, there are some potentially anomalous characteristics of some European archeological sites that warranted inclusion, even though they are usually scorned by the archeological establishment: (1) The supposed astronomical features of Stonehenge and other stone circles that may have permitted the prediction of eclipses by these "primitive" peoples; and (2) The claimed "magnetic" and "psychic" properties of some sites that are supposed to be manifestations of "earth energy." The implication of the second supposed characteristic is that ancient humans could somehow sense these "energies." Of course, modern science ridicules such notions, but many cultures had and still have psychic feelings about certain locations, such as the so-called "sacred" sites.

SARDINIA'S PREHISTORIC TOWERS

Sardinia is home to an immense population of mysterious prehistoric stone towers called "nuraghi." Over 7000 of these remarkable dry-stone edifices exist---a concentration of monumental stone architecture unparalleled in Europe.

"'Nuraghe' derives from the prehistoric Sardinian root 'nur' which means both 'hollow' and 'heap.' But the nuraghi are neither hollow nor are they haphazard heaps of stone. The nuraghe interior often presents a complex plan of chambers, winding staircases, dead-end corridors, concealed rooms with trap doors, and a variety of niches and compartments. Standing up to three stories high with magnificently corbelled domes one on top of the other, some structures have as many as 18 subsidiary towers attached to the main keep. Large complexes were sometimes completely enclosed by enormous stonewalls punctuated with still more towers."

Over 3000 years old, the nuraghi have withstood the depredations of weather and later humans by virtue of their excellent design and construction. As with many other such ancient structures, one is impressed with the size of the stones used. How were they moved? How were the stones (usually hard basalt) cut and dressed by artesans with no metal tools harder than copper and bronze? And what was the purpose of the nuraghi? A quick answer to the last question is that they were fortresses; but they might also have been dwellings or storehouses. (Gallin, Lenore; "The Prehistoric Towers of Sardinia," Archaeology, 40:26, September/October 1987.

THE HAMBLEDON HILL NEOLITHIC FORTRESS

Hambledon Hill sits astride the Stour River in the chalklands of southwestern England. Almost 6000 years ago, Neolithic people began erecting a great funeral center and fortress here. When the ramparts were complete, they were visible for miles. The southern and western sides were rimmed by a timber-framed rampart 2500 meters long. The northern flank was protected by a 1200-meter multi-ditch outwork. "A Neolithic herdsman who looked up to the hilltop in about 3400 B.C. would have seen an impressive sight. Crowning Hambledon Hill was a huge defensive enclosure with three concentric ramparts. The inner rampart, the most formidable of the three, was supported by 10,000 oak beams as thick as telephone poles. In the ditch around the ramparts human skulls placed at intervals added an eerie note to the appearance of the fortifications."

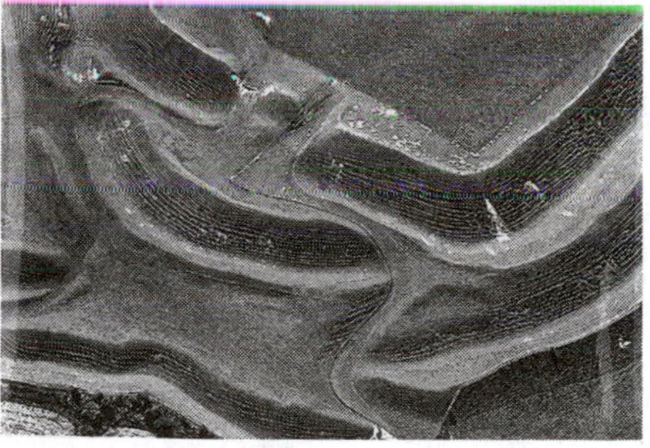

Aerial view of the entrance to Maiden Castle, a neolithic hill fort, showing the scale of the earthworks and their strange geometry. (J.D.H. Radford/ Janet & Colin Bord)

Such a construction feat must have taken considerable organization and community energy, much like the pyramids then under construction in Egypt. In the absence of stone quarries and with plenty of forests, Hambledon Hill's fortress was simple wood and dirt, but nonetheless very impressive. Even its great size, however, did not save it from conquest and burning. (Mercer, R.J.; "A Neolithic Fortress and Funeral Center," Scientific American, 252:94, March 1985.)

THE SWEET TRACK

About 6000 years old, the Sweet Track is one of the oldest manmade structures in Europe. The Sweet Track is a single line of wooden planks that runs for about 1800 meters (well over a mile) across a swamp in Somerset, England. Evidence is that this ancient road was overwhelmed by the swamp after only 10 years of use. A jadite axehead found along the Sweet Track probably came from the Swiss Alps, indicating cross-Channel commerce even 6000 years ago. (Lloyd, Philippa; "A Long and Ancient Road," Nature, 345:577, 1990.) See also Antiquity, 64:210, 1990.

THE GREAT WOODEN WELL OF KUCKHOVEN

While on the subject of ancient hydrological engineering, it is appropriate to mention a remarkable wooden well found in northwest Germany. Over 200 oaken planks have been discovered so far. These are up to 15 centimeters thick and 50 wide. Fairly large oaks had been cut and split with stone axes and then worked into planks. Mortises were cut in some way so that the planks could be joined. It is quite clear that the Neolithic peoples of the region were skilled carpenters. The size of the well, too, is impressive: it was more than 15 meters deep. The tree rings on the planks permitted very accurate dating: 5303 BC---well over 7000 years ago. (Bahn, Paul G.; "The Great Wooden Well of Kuckhoven," Nature, 354:269, 1991.)

6,000 YEAR-OLD WOODEN STRUCTURE IN SCOTLAND

Timber fragments from a building 78 feet long, 39 feet wide, and 30 feet high, have been radiocarbon-dated at 4000 BC. The size and method of construction of this ancient building on the edge of the Scottish river Dee indicate a high level of civilization 1000 years before Stonehenge. At the same time civilization was supposed to be getting its start in the Middle East, the precocious Scots were evidently constructing large wooden structures, cultivating barley, and probably tending domesticated farm animals. (Anonymous. "An Epic Find," Time, 64, June 26, 1978.)

WERE THE BRITISH MEGALITHS BUILT AS SCIENTIFIC INSTRUMENTS?

Alexander Thom and his son have meticulously surveyed nearly 100 megalithic sites in Britain and nearby Europe. Archeologists generally applaud the Thoms' careful work but vehemently attack their conclusions. The Thoms see in their surveys evidence that the early Britons built megalithic astronomical instruments with scientific capabilities far beyond their needs for calendar-keeping. Actually, they suggest that these "primitive" men built a society so strong and productive that it could devote time and labor to a program of astronomical research generations in extent. In short, they were curious, precociously bright, and socially strong---they could indulge their scientific desires.

The Thoms' prehistoric scenario departs radically from that of the current archeological establishment, which has searched for flaws in the Thoms' work. Naturally, some defects have emerged. Clive Ruggles, the author of this article, is one of the skeptics. He feels that the megalithic sites are impressive and intriguing but not the work of mental giants. After all, Ruggles says, 72 points of the compass have some lunar significance. Almost any circle of stones built for simple ritual purposes would have some significant lunar alignment! (Ruggles, Clive; "Prehistoric Astronomy: How Far Did It Go?" New Scientist, 90:750, 1981.)

Comment. This kind of statistical argument reminds one of those monkeys who will eventually type out the works of Shakespeare. Presumably, the same monkeys could construct Stonehenge, given enough time.

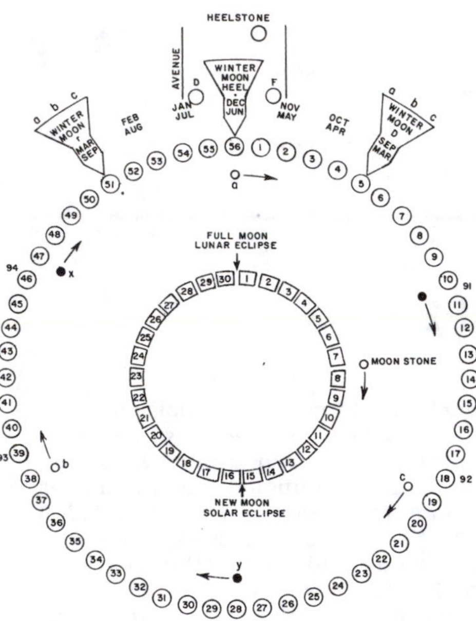

In 1964, G.S. Hawkins suggested that Stonehenge was actually a neolithic eclipse predictor, with many meaningful alignments. For details on its operation see his long article in Nature, 202:1258, 1964, or it's condensation in Ancient Man. Some authorities reject Hawkins' interpretation.

HIDDEN STONEHENGE

The configuration of standing stones at Stonehenge has been etched in everyone's mind by thousands of photographs and drawings down the centuries. There is also a hidden Stonehenge: traces of earlier configurations, hints of trials and errors, and just plain enigmas. The hidden or cryptic Stonehenge is uncovered mostly by accident when some chance digging reveals a previously unrecorded hole, ditch, or buried artifact. To illustrate, in May and June 1979, a 24-meter trench was excavated along the path of a proposed telephone cable. The diggers found a completely unexpected, backfilled pit showing the impression of the standing stone that once occupied it. Now called Stone 97, this feature was "erased" thousands of years ago for some unknown reason. Several such fortuitous finds hint that Stonehenge is like one of those paintings painted over an earlier painting. Apparently we are just beginning to comprehend the history of this remarkable site. (Pitts, M.W.; "Stones, Pits and Stonehenge," Nature, 290:16, 1981.)

MANIFESTATIONS OF EARTH ENERGY AT MEGALITHIC SITES?

Stories have long circulated that strange phenomena cluster around megalithic sites such as Stonehenge. Those who claim psychic powers state that earth energies (whatever they are) seem to focus at these ancient constructions. The story goes that the builders of the stone circles could also detect these natural forces and intentionally chose these spots where the energies were most powerful. "Proper" siting and orientation were doubtless important to the builders of the megalithic structures, but can modern, no-nonsense science even begin to explore these mystical, psychical claims?

Given today's scientific impatience with all psychic subjects, one would not expect a scientific journal, even a popular one, to touch "earth energies." Yet, here is an article describing the use of ultrasound detectors and Geiger counters in surveying megalithic monuments for foci of earth energies. Sure enough, curious enhancements of ultrasound intensity were discovered at the Rollright Stones. At another site, the natural radiation background level was anomalously depressed. It is all very mystifying. (Robins, Don; "The Dragon Project and the Talking Stones," New Scientist, 96:166, 1982.)

Comment. In truth, of course, there conceivably could be something to the "earth energy" concept; and there is nothing wrong with exploring it scientifically. It is just such a surprise to see the subject discussed in a mainstream scientific publication.

With respect to ultrasound enhancement, see p. 43, where echos from rock-art sites are discussed.

ARCHEOLOGY IN BRITAIN: STRAYING FROM THE PARTY LINE

In the final 1982 issue of New Scientist (see also the above item), Paul Devereux and Robert Forrest relate the history of "leys" in England. "Leys" are supposedly intentional clusterings of megalithic sites along straight lines extending several kilometers. Some alignments occur through chance, but several leys seem statistically significant. The puzzle is why ancient peoples aligned their edifices, assuming they did. Most professional archeologists believe leys to be figments of the imagination of amateurs. This being so, the claims made two weeks later in New Scientist, must have incensed them. Stimulated by the report on leys, a retired engineer presented his measurements of the magnetic fields around the Rollright Stones. He maintained that he was able to detect magnetically several converging leys and, in addition, a spiral pattern inside the stones. A psychic accompanying him independently perceived the leys and spiral. (Devereux, Paul and Forrest, Robert; "Straight Lines on an Ancient Landscape, "New Scientist, 96:822, 1982. Brooker, Charles; "Magnetism and the Standing Stones, "New Scientist, 97:105, 1983.)

Comment. Psychics, especially dowsers, have long maintained that megalithic sites are the foci of mysterious forces, notably spirals. This is pretty wild stuff for a respected science journal to print. The editors would be well-advised to send someone with a magnetometer down to the Rollright Stones to straighten out this matter.

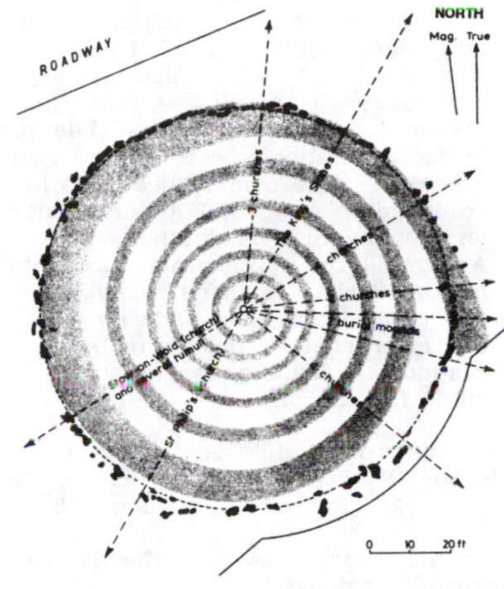

The spiral magnetic pattern detected around the Rollright Stones.

MEGALITHIC RECYCLING

During restoration work on a capstone at the tomb of Gavrinis, located on an island just off the Brittany coast, C. -T. Le Roux discovered carvings that had long been concealed. The carvings and the rock itself fit perfectly with the capstone of the Table des Marchands, another famous megalithic monument some 4 kilometers away. Stimulated by this discovery, a third capstone on another monument nearby was found to fit like a jigsaw puzzle piece. The result is a huge stela 14 meters high, 3.7 meters wide, and 0.8 meters thick. The total stela would weigh about 100 tons. It is decorated on one side with animals (bovids) and other devices. Apparently, this stela once stood near the even larger stela called Grand Menhir Brise or 'er-Grah.' The Grand Menhir Brise is also broken into pieces. Evidently, the period of megalithic tomb building, which probably began about 5200 BP, was preceded by a period when giant, decorated stelae were erected. These stelae were later pulled down and broken up for use in constructing tombs.

The civilization that raised the stela is not well-understood; and one wonders why such impressive monuments were torn down and their engravings concealed. Incidentally, the stela chunk found at Gavrinis weighs about 20 tons and was transported over 4 kilometers across several watercourses. (Bahn, Paul G.; "Megalithic Recycling in Brittany," Nature, 314:671, 1985.)

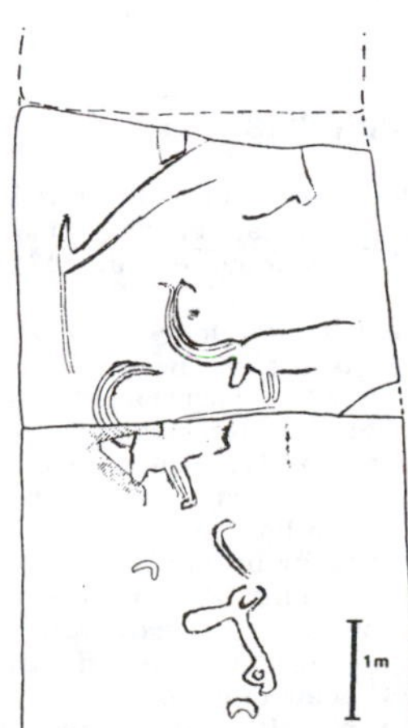

Reconstruction of a decorated stela found on capstones taken from megalithic tombs in Brittany.

ATLANTIS FOUND---AGAIN

When Soviet oceanographers examined their underwater photos taken of the Ampere Seamount, they discovered what seem to be walls, stairways, and other artificial stonework. The Ampere Seamount is 450 miles west of Gibraltar, just the area where Plato placed Atlantis. (Anonymous; "Undersea Discovery May Be Atlantis," Baltimore Sun, p. A1, April 5, 1981.

ANCIENT GREEK PYRAMIDS?

Yes, the ancient Greeks had their pyramids, too, only they had a very practical purpose: They were water-catchers. They had learned that piles of porous rocks could, in desert climes, capture and condense surprisingly large quantities of water. Take, for example, the 13 pyramids of loose limestone rocks that the Greeks constructed some 2500 years ago at Theodosia in the Crimea:

> The pyramids averaged nearly 40 feet high and were placed on hills around the city. As wind moved air through the heaps of stone, the day's cycle of rising and falling temperatures caused moisture to condense, run down, and feed a network of clay pipes.
>
> One archaeologist calculated a water flow of 14,400 gallons per pyramid per day, based on the size of the clay pipes leading from each device.

Weren't the ancient Greeks clever? But perhaps they had observed how some mice in the Sahara pile small heaps of rocks in front of their burrows and lick the condensed moisture off in the morning. Possibly we should have classified this item under "Biology"! (Dietrich, Bill; "Water from Stones: Greeks Found a Way," Arizona Republic, p. AA1, December 22, 1991. Cr. T.W. Colvin.)

Comment. In a curious parallel evolution of behavior, Australian native mice also construct piles of pebbles to wring moisture from the air. See p. 137.

ASIA

STONE CIRCLE IN SAUDI ARABIA

Enigmatic circular stone formations reminiscent of those in Europe are found on remote hilltops and valleys throughout Saudi Arabia. The rings are 3 to 100 meters in diameter and are surrounded by stone walls a foot or two tall. Some of the rings have "tails" that stretch out for hundreds of meters. From the air, the patterns have a striking resemblance to designs etched in Peru's Nazca plateau. Little is known about the circles and virtually nothing about their purpose. (Anonymous; "Saudis Seek Experts to Solve a Desert Mystery," Kayhan International, January 1, 1978, p. 7.)

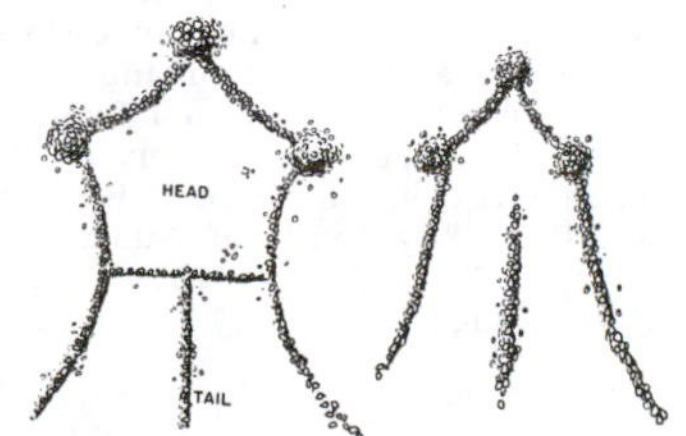

Comment. Some scientists theorize that these stone structures were employed by herdsmen as corrals and as aids in shepherding their flocks.

UNBELIEVABLE BAALBEK

The city of Baalbek, called Heliopolis by the ancient Greeks, lies some 50 miles northeast of Beirut. Here are the ruins of the greatest temple the Romans ever tried to construct. However, we must focus not on mundane Roman temples but upon a great assemblage of precisely cut and fitted stones, called the Temple today, which the Romans found ready-made for them when they arrived at Baalbek. It was upon this Temple, or stone foundation, that the Romans reared their Temple of Jupiter. No one knows the purpose of the much older Temple underneath the Roman work.

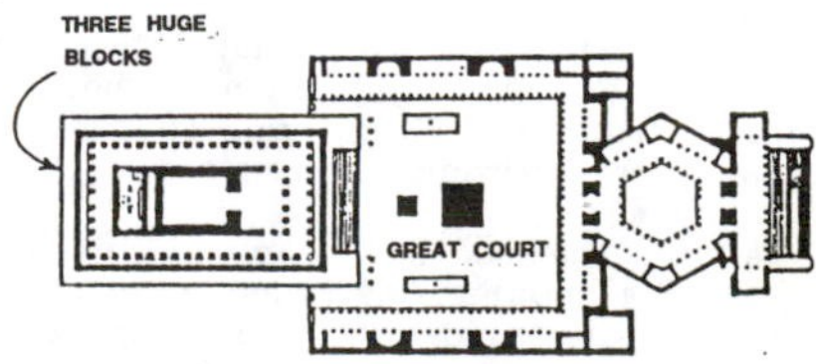

The Romans built their Temple of Jupiter upon a preexisting foundation which included several gigantic worked stones.

J. Theisen has reviewed the basic facts known about the Temple's construction---and they are impressive, perhaps even anomalous. Being 2,500 feet long on each side, the Temple is one of the largest stone structures in the world. Some 26 feet above the structure's base are found three of the largest stones ever employed by man. Each of these stones measures 10 feet thick, 13 feet high, and over 60 feet long. Knowing the density of limestone permits weight estimates over 1.2 million pounds. Some people with impressive engineering skills cut, dressed and moved these immense stone blocks from a quarry 3/4 of a mile away.

Some of the immense foundation stones of the Baalbek Temple can be seen in this photograph.

A walk to this quarry introduces the observer to the Monolith, an even larger block of limestone, nicely dressed, and with the dimensions: 13 feet, 5 inches; 15 feet, 6 inches; and 69 feet, 11 inches. The monolith weighs in at over 2,000,000 pounds. In comparison, the largest stones used in the Great Pyramid tip the scales at only 400,000 pounds. Not until NASA moved the giant Saturn V rocket to its launch pad on a gigantic tracked vehicle has man transported such a large object.

Today, one sees no evidence of a road connecting the quarry and Temple. Even if a road existed, logs employed as rollers would have been crushed to a pulp. But obviously, someone way back then knew how to transport million-pound stones. Just how, we do not know. (Theisen, Jim; "Unbelievable Baalbek," INFO Journal, no. 55, p. 5, 1988.)

ANCIENT ENGINEERING FEAT

Current excavations at Shringaverapura, in Uttar Pradesh, India, have brought to light a huge, well-preserved water tank about 800 feet long. At its widest point, the tank is about 60 feet wide and 12 feet deep. The entire complex is brick-lined and includes a large settling tank for water diverted from the River Ganga. During the dry season, additional water was supplied from a series of dug wells. This immense system of channels, walls, access stairways, etc, is about 2,000 years old. (Lal, B.B.; "A 2.000-Year-Old Feat of Hydraulic Engineering in India," Archaeology, 38:48, January/February 1985.)

AUSTRALIA

Astounding archeological finds have been claimed in Australia. In many such instances, further investigation usually forces them to evaporate into nonanomalousness. Queensland's Gympie Pyramid and Mystery Craters seem to fall into this category.

THE AUSTRALIAN PYRAMIDS

> Standing in the bushland some distance from the town of Gympie in southern Queensland, is a crudely-built, 40-metre-tall terraced stone pyramidal structure which, I believe, will one day help to alter the history of Australia---to prove that 3000 years ago, joint Egyptian and Phoenician mineral-seeking expeditions established mining colonies here.

Thus runs the lead paragraph of this article in a popular Australian publication. This pyramid boasts 18 recognizable terraces. The bottom 14 terraces are built from rather small stones; but the top four consist of slabs weighing up to 2 tons. Trees as old as 600 years poke up through the stones, attesting to a pre-European origin. Another much larger pyramid inhabits dense scrubland near Sydney. The claim that these admittedly crude structures are Egyptian is based upon the discovery of artifacts in the area with Egyptian and Phoenician characteristics; i.e., a stone idol resembling a squatting ape, an onyx scarab beetle, and cave paintings with Egyptian symbols. Aboriginal legends also tell of 'culture heros' arriving at Gympie in large ships shaped like birds. (Gilroy, Rex; "Pyramids of Australia," Australasian Post, August 30, 1984, p. 9, Cr. A. Jones)

CHECKING OUT THOSE AUSTRALIAN PYRAMIDS

Several Australian readers have been kind enough to send this item about the Australian pyramids reported above, especially at Gympie, Queensland. The article's author, T. Wheeler, personally checked out the Gympie pyramid, the other supposedly ancient artifacts found in the area, and the testimonies of authorities and old-timers in the Gympie area. We quote his conclusions below:

> What are we left with: The facts are (probably) that the Gympie Golden pyramid is actually an ordinary hill terraced by early Italian immigrants for viniculture that has been disfigured by erosion and the removal of stone from the retaining walls for use elsewhere; the stone wall around Gympie's Surface Hill Uniting Church is exactly what the Rev Mr Geddes says it is---a wall made from irregular, freshly quarried stone. The 'Gympie Ape/Iron Man' statue was carved by a Chinese gold prospector and later abandoned. The sun symbol and snakes were carved quite recently. The prickly pear was introduced to Australia by early settlers journeying via South America. As for all the supporting statements by the various authorities, all but a few unimportant ones fade away as one after another proves to be a misquote, a falsification or an outright fabrication.

(Wheeler, Tony; "In Quest of Australia's Lost Pyramids," Omega Science Digest, p. 22, November/December 1985.)

Comment. Today, few take the Gympie "pyramid" seriously.

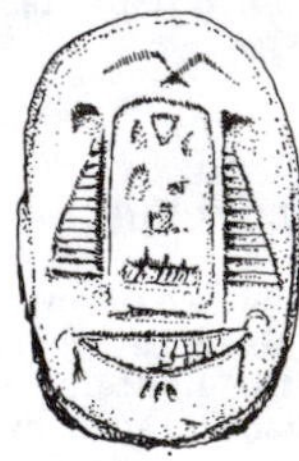

Underside of a stone scarab reportedly dug up in an Australian cane field.

STRANGE CRATERS

We have just found the following item from the Australasian Post. Can any of our Australian readers elaborate?

> For more than 10 years, scientists and geologists have been baffled by the discovery of some 30 strange craters between Bundaberg and Gin Gin (Queensland). Made of sandstone, siltstone, and red ochre and at least 25 million years old, the craters contain unidentifiable markings which could be manmade!
>
> The holes were discovered by a farmer clearing his land and they have now been opened to the public who offer a multitude of theories about their origin, ranging from natural formations---although the craters are not of volcanic origin---to the work of visitors from outer space.

(Anonymous; "Strange Craters," Australasian Post, July 31, 1986. Cr. R. Collyns via L. Farish)

DEMYSTIFYING THOSE AUSTRALIAN CRATERS

Above, we reported on some mysterious craters in Queensland, Australia. Were they excavated by ancient man? Do they display strange inscriptions? Australian readers were quick to supply additional information. It turns out that several years ago geologists did inspect the so-called "Mystery Craters." This appellation was applied by the owners of the land, who have made the craters into a tourist attraction. (This fact alone is enough to raise suspicion!) The geologists' report completely dispelled any aura of mystery. Here follows their Summary:

> A geological investigation of the 'Mystery Craters' adjacent to Lines Road, South Kolan, indicates that these structures are sinkholes in a laterite profile. The sinkholes have been caused by the collapse of overlying strata into underground voids produced by tunnel erosion.

(Robertson, A.D.; "Origin of the 'Mystery Craters' of South Kolan, Bundaberg Area," Queensland Government Mining Journal, p.448, September 1979. Cr. R. Molnar) (No mention was made of any ancient inscriptions in this report.)

A portion of the "temple" at Nan Madol. It was constructed from naturally occurring basalt prisms. (Bishop Museum)

OCEANIA

THE LATTE STONES

When one thinks of megaliths, Stonehenge, Carnac, and the other classic sites of Europe come to mind. In actuality, megalithic ruins of sorts are to be found almost everywhere. In Oceania, for example, we have the Latte Stones found decorating many of the Mariana Islands in Micronesia:

> Looking a bit like immense birdbaths, the Latte are shaped like truncated pyramids (halenges) capped with enormous hemispherical stones (tasas). They range in height from 5 to 20 feet with circumferences of up to 18 feet and weights that have been estimated at up to 30 tons. The columns are found throughout the Marianas, usually in double rows of up to six stones.

One of the smaller Latte stones found on Guam.

Are the Latte Stones burial monuments or, as some are convinced, merely supports for the dwellings important personages in ancient Micronesia? But why would anyone build a house 20 feet above the ground? Even the Micronesians scoff at the idea. Instead they attribute the Latte Stones to the taotaomona, the spirits of the "before-time" people!

The Latte Stones are so old that even when Magellan discovered the Marianas in 1521, no one remembered what they had been used for. (Davis, Esther Payne; "The Strange Latte Stones of Guam," World Explorer, 1:23, Spring/Summer 1992.)

THE LOST CITY OF NAN MADOL

Nan Madol has never been lost as the title of this item implies, but this fantastic complex of 92 artificial islets sees few tourists. Located on Pohnpei (formerly spelled Ponape) in Micronesia, Nan Madol lacks the well-publicized glamour of the Pyramids and Chichen Itza. If Nan Madol is not glamourous, it is certainly incongruous. Who would expect such huge stone structures to rise in the middle of nowhere?

William Ayres, a University of Oregon anthropologist sponsored by the National Geographic Society, has been a recent researcher at Nan Madol. In an interview, Ayres described Nan Madol in these terms:

> To withstand time and the sea, the artificial platforms were built in a staggeringly laborious process. Multi-ton basalt columns, naturally formed by volcanic activity, were stacked horizontally, log cabin-style, to form outer walls. The inside then was filled with coral rubble to form a dry surface several feet above high-tide level. Radiocarbon testing finds signs of human habitation at Nan Madol as early as A.D. 500, and the megalithic construction was completed by about 1500.

Besides incongruity and a certain bizarreness, Nan Madol does pose several problems: How were the huge, very heavy prismatic columns of basalt quarried and transported? Why was Nan Madol built at all? Why about 1400 AD did the inhabitants stop building their massive ocean canoes and begin a decline? (Hanley, Charles J.; "Oregon Anthropologist Unravels Story of Lost City of Pacific," The Oregonian, February 3, 1986. Cr. D.A. Dispenza)

Comment. An associated question asks why Nan Madol, the Mayas, the Hohokams, the Moundbuilders, and other cultures all apparently declined precipitously at about the same time.

THE MYSTERIOUS TUMULI OF NEW CALEDONIA

The Isle of Pines, New Caledonia, is spangled with about 400 large tumuli or mounds, ranging from 30 to 165 feet in diameter. Their heights are 2 to about 15 feet. All of the material making up the mounds seems to come from the immediate surroundings: coral debris, earth, and grains of iron oxide. The larger tumuli enclose a block of tuff, about 5 feet high and 6 feet in diameter, comprised of tumulus material held together by a calcareous cement or mortar. Some who have investigated these mounds believe that the presence of cement, presumably manmade, is proof-positive that the tumuli are the product of human activity. Other archeologists doubt this, because the early settlers of New Caledonia did not use cement. Besides, there seem to be no other signs of human involvement. This has led to the hypothesis that the mounds were built by huge, now-extinct, flightless birds for the purpose of incubating their eggs. Some birds do incubate their eggs in this manner today; and some 5000 years ago New Caledonia did boast a giant bird (Sylviornis neocaledoniae), which was 5-6 feet tall! The authors feel that the giant bird hypothesis is just as reasonable as the theory that the mounds were built by humans. (Mourer-Chauvire, Cecile, and Poplin, Francois; "Le Mystere des Tumulus de Nouvelle-Caledonie," La Recherche, 16:76, September 1985. Cr. C.

Maugé)

Comment 1. We find in Ancient Man an article by A. Rothovius entitled "The Mysterious Cement Cylinders of New Caledonia". This 1967 article covers much the same ground as that in La Recherche, but sans the giant bird theory. Rothovius states that the cylinders inside the tumuli "are of a very hard, homogeneous lime-mortar, containing bits of shells which yield a radiocarbon dating of from 5,120 to 10,950 B.C.---even the lowest date being some 3,000 years earlier than man is believed to have reached the Southwest Pacific from the area of Indonesia."

Comment 2. Several species of birds in the South Pacific incubate their eggs in piles of rotting vegetation. While often several feet in diameter, these piles of vegetable debris bear little resemblance to the tumuli of coral fragments and earth described above.

SMALL ARTIFACTS

NORTH AMERICA

The most common vestiges of ancient humanity are not pyramids and other engineering works, but rather arrowheads, pottery fragments, and the like. These "small" artifacts may suggest at least two sorts of anomalies: (1) unexpectedly "high" technology for their time frame; and (2) evidence for cultural diffusion that challenges prevailing paradigms concerning the peopling of the planet.

Beginning with North America, we find that the back files of Science Frontiers contain many entries in the second category listed above; i.e., anomalous diffusion. North American diffusion anomalies can be divided further into two groups:

- **Precolumbian contacts. Small artifacts, such as Old World pottery, that seem to betoken precolumbian visits to North America.**

- **Pre-Clovis artifacts. Evidence of human habitation in North America prior to the 12,000 BP (Before Present) Clovis culture. The date of 12,000 BP is sacrosanct in mainstream American archeology. The reigning paradigm does not allow earlier contacts from other continents.**

PRECOLUMBIAN CONTACTS

THE ANCIENT WORLD WAS SMALLER THAN YOU THINK

Background. If the artifacts described below are authentic, the oceans of the world were crisscrossed frequently by various peoples well before Columbus and the Vikings set sail. Such ancient, long-distance sea voyages are definitely frowned upon by the great majority of archeologists; which is, of course, why they are included in this book.

ANCIENT IBERIAN JARS RECOVERED OFF MAINE COAST

While skindiving in the Bay of Castine in 1971, Norman Bakeman discovered two peculiar ceramic storage jars in 12 meters of water. These jars were recovered and have since been compared to Portugese "anforetas" used during the Roman period for the storage of wine, oil, honey, etc. A similar anforeta (or "amphora" in English) has also been recovered in Jonesboro, Maine. The clay paste and grit of the Maine jars closely resemble those used in Iberia almost 2,000 years ago. The possibility that these containers might be Spanish olive jars circa 1800 is also discussed. (Whittall, James P., II; "Anforetas Recovered in Maine," Early Sites Research Society, Bulletin, 5:1, 1977.)

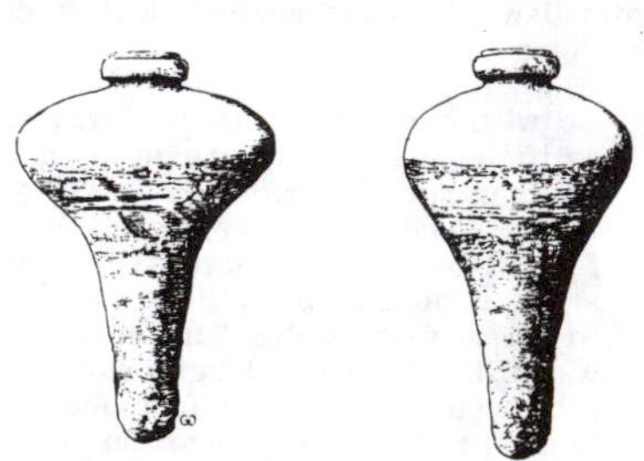

Ancient Iberian jars recovered from the floor of Maine's Bay of Castine.

ANCIENT OLD-WORLD LAMPS TURN UP IN NEW ENGLAND

In 1980, at a Boston antique fair, a Greek lamp, probably dating from the Sixth Century B.C. turned up. The antique dealer stated that the lamp had been dug up at an Indian site in Manchester, NH. (Totten, Norman; "Late Archaic Greek Lamp Excavated at Amoskeag Falls," Early Sites Research Bulletin, 10:25, no. 2, 1983.)

In 1952, a Byzantine oil lamp was found at the Clinton, Connecticut harbor shell-midden after plowing. The finder described it as an Indian pipe, but it is typical of the Mediterranean area circa 750-800 A.D. (Whittall, James P., II; "Byzantine Oil Lamp from Connecticut," Early Sites Research Society, Bulletin, 10:26, no 2, 1983.)

THE ENIGMATIC "MOORING STONES"

Archeologists love to puzzle over pyramids, stone axes, and such straightforward productions of ancient man. In contrast, simple holes in boulders are hardly the things important scientific papers are written about. Yet, scattered about the Great Plains are some 300 boulders of very hard rock, each possessing the same rounded triangular holes. Surely such a phenomenon would pique some archeologist's curiosity!

The holes are made with high precision to the dimensions shown in the figure. They are 6 inches deep, plus or minus an inch. Holes with a rounded triangular shape represent a sophisticated drilling technology. Steel tools and high craftmanship are indicated.

Even though the holes have been known for over a century, only amateurs have shown much interest. A few such enthusiasts have tracked down hundreds in Minnesota, North Dakota, South Dakota, Illinois, and the eastern seaboard. All of them seem to be located on present-day lakes and rivers and now-dry waterways. This marine affinity has let to the theory that they are "mooring stones", especially Viking mooring stones! In truth their real purpose is unknown.

How old are the holes? Weathering of those in granite suggests ages of at least several hundred years---well before the westward push of American settlers. The peculiar shape of the holes seems to rule out production by modern drills (usually round) for purposes of blasting or installation of surveyors' markers.

Another puzzle, probably related to

the purpose of the holes, is the presence of large, smooth grooves on some of the boulders bearing triangular holes.

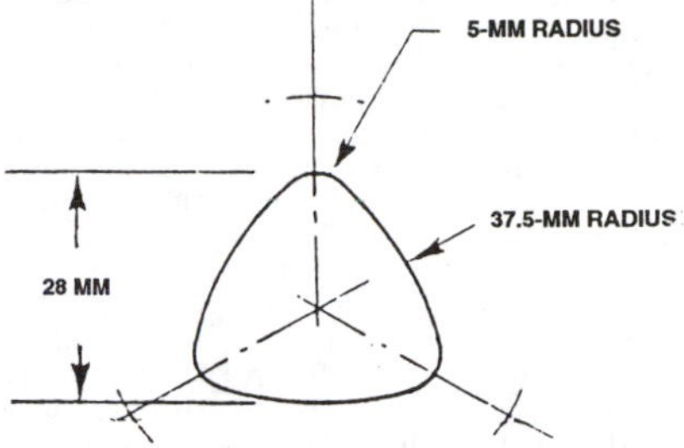

Cross section of a possible Norse mooring stone hole. Many of these strange holes have been found in the northern tier of states.

The technique of "cam-wedging" may lead us to the purpose of the holes. If one inserts a triangular shaft into such a hole and rotates it part of a turn, the shaft becomes firmly wedged in place. A rotation in the opposite direction quickly frees the shaft. It does seem that the mooring-stone hypothesis is consistent with cam-wedging. (Olson, John J.; "'Mooring Stones': An Enigma Deserving More Attention," Epigraphic Society, Occasional Papers, 18:253, 1989.)

Comment. Before climbing on the Viking bandwagon, it is reasonable to ask whether the Vikings drilled the same holes in Europe and whether voyagers even earlier than the Vikings might have drilled the mooring holes.

THE CHINA SYNDROME IN ARCHEOLOGY

Bit by bit evidence accumulates showing that Chinese and Japanese ships visited the American Pacific coast long before Europeans. Indian traditions tell of many "houses" seen on the Pacific waters. Chinese history, too, tells a charming account of voyages to the land of Fusang. Even old Spanish documents describe oriental ships off the Mexican coast in 1576. Japanese explorers and traders evidently left steel blades in Alaska and their distinctive pottery in Ecuador. Recent underwater explorations off the California coast have yielded stone artifacts that seem to be anchors and line weights (messenger stones?). One line weight found at 2,000 fathoms is covered with enough manganese to suggest great antiquity. The style and type of stone point to Chinese origins for all these artifacts.

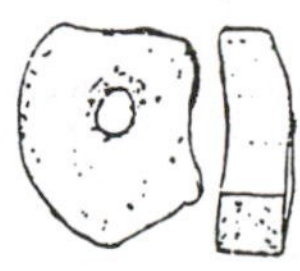

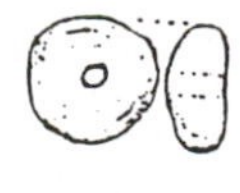

Possible Asian anchor and messenger stones found off the west coast of North America.

Apparently, vessels from the Orient were riding the Japanese Current to North American shores long before the Vikings and Columbus. (Pierson, Larry J., and Moriarty, James R.; "Stone Anchors: Asiatic Shipwrecks off the California Coast," Anthropological Journal of Canada, 18:17, 1980.)

MORE ON THOSE CHINESE ANCHORS IN CALIFORNIA WATERS

In the above item, the discovery of Chinese-type anchors off the California coast was described, suggesting a Chinese presence in America centuries ago. This is not really a respectable notion among some archeologists, for we now have a strong rebuttal that begins by raising the "horrible" spectres of Hyerdahl and von Dainiken. (Should both of these names be used to scare archeologists?) First of all, F.J. Frost sinks the Land of Fusang legend by relating how Gustaaf Schlegel showed in 1892 that the ancient Chinese mapmakers knew perfectly well that Fusang was an island just off the northeast Asian coast. Next, Frost tells how a recent attempt to duplicate the voyage from China to America in a Chinese junk riding the Kuroshio Current was a dismal failure. Then, how about those stone anchors found in shallow waters off Palos Verdes, California? They are legitimate Chinese anchors all right, but they are modern, having been lost by local California fishermen of Chinese extraction. History tells how Chinese immigrants quickly applied the techniques of their native land to the California coast. After all this debunking, Frost identifies some "genuine" unsolved mysteries off Palos Verdes. It seems that some of the stones found underwater are most curious indeed. Near where the stone anchors were found are two grooved columnar stones over a meter long with drilled holes. There is also a ton-sized stone sphere with a groove around its circumference. (Frost, Frank J.; "The Palos Verdes Chinese Anchor Mystery," Archaeology, 35:23, January/February 1982.)

AMERICA B.C. AND EVEN EARLIER

The thought that the Atlantic might have been a thoroughfare long before Columbus and the Vikings has been ridiculed by most archeologists for decades. New England megaliths and B. Fell's translations of purported Celtic ogham inscriptions have met only with derision in the professional literature. But times are changing---at least we hope so.

The Red Paint People. Public TV recently aired a program on North America's Red Paint People, so-called because they added brilliant red iron oxide to their graves. It also seems they knew how to sail the deep ocean, as G.F. Carter now relates.

> Decades ago, Gutorn Gjessing pointed out that the identical [Red Paint] culture was found in Norway. No one paid much attention to that, but more recent carbon-14 dating has shown that the identical cultures had identical dates, and people began to pay more attention. It is now admitted that this is a high latitude culture that obviously sailed the stormy north Atlantic and stretched from northwest Europe over to America. It seemingly extends from along the Atlantic coast of Europe to America and in America from the high latitudes of Labrador down into New York state.
>
> The dates are mind-boggling: 7,000 years ago both in Europe and America. That is 2,000 years earlier than the Great Pyramids of Egypt. It is at least 4,000 years earlier than the Mound Builders of the Ohio Valley. The evidence is cummulative, varied in nature, and most probably highly reliable. (Carter, George F.; "Before Columbus," Ellsworth American, November 23, 1990. Cr. R. Strong.)

PRE-CLOVIS ARTIFACTS

WHEN WERE THE AMERICAS PEOPLED?

For more than 50 years, archeological dogma has had the first Americans trooping across the Bering Land Bridge about 12,000 years ago as the Ice Ages were waning. Despite very tight discipline among most professional archeologists (jobs and grants go only to approved individuals), a few cracks are beginning to appear. A Brazilian site has now been reliably dated at 32,000 years. If these early Brazilians came over the Bering Land Bridge, they must have left even earlier traces in North America. In fact, there are two hotly debated North American sites that seem to be very much older than that in Brazil: the Calico Site in California; and a spot along the Old Crow River in the Yukon.

Thousands of stone artifacts, apparently showing signs of being shaped by humans, have been recovered at Calico over the past two decades. The Calico artifacts are usually contemptuously dismissed as naturally fractured chert flakes. But at the other end of the belief spectrum (Don't laugh, much of science is just as much a belief system as religion!) are those who see a long human history at Calico.

> "Two periods of human occupation have been dated at Calico. From about 15,000 to 20,000 years ago the area was inhabited by what [R.D.] Simpson suggests was a hunting-gathering people with more sophisticated tools, including stones flaked on both sides. In deeper layers estimated to be at least 200,000 years old are the simpler flakes of people, she says, who probably gathered plants and other foods."

Much farther north, along the Yukon's Old Crow River, nearly 10,000 horse and mammoth bone artifacts have been picked up and dug out of the river banks. W.N. Irving, from the University of Toronto, claims that the last five seasons of archeological research have uncovered a 'bone industry' of extremely great age---100,000 years or more. (Bower, Bruce; "Flakes, Breaks, and the First Americans," Science News, 131:172, 1987.)

Comment. It seems significant that French archeologists explored the Brazilian site, and Canadians the Old Crow site. American archeologists, with few exceptions here and there, scoff at the whole business.

For more on the Calico site, see entries beginning on p. 24.

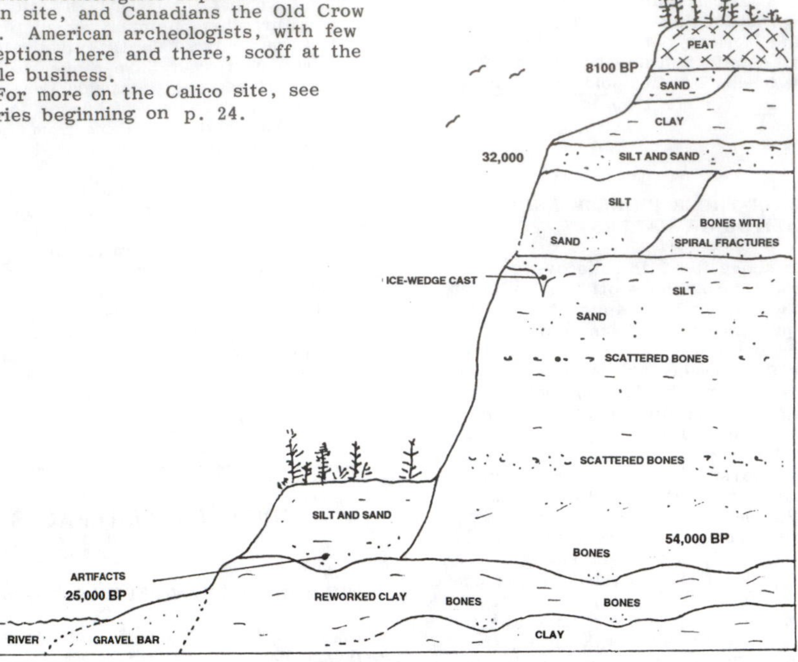

Composite section of the Old Crow site. Notice the complexity of the stratigraphy. (Adapted from R.E. Morlan's paper in Early Man in America, A.L. Bryan, ed., p. 81.)

A RELUCTANT, LONG-OVERDUE PARADIGM SHIFT

In late March, 1990, at a conference in Boulder, Colorado, D. Stanford led off with: "It's time to acknowledge that we do have a pre-Clovis culture in the New World." (Pre-Clovis = before 12,000 BP.)

It's all true! A long-standing consensus has collapsed---the 12,000-year barrier, like the Berlin Wall, has disintegrated. The two most important demolition charges were the now-accepted dates of 16,000 BP. from the Meadowcroft Rockshelter, Pennsylvania, and 13,000 BP. from the Monte Verde site in Chile. The Monte Verde date probably represents an American entry date of at least 20,000 years, if one accepts that the first Americans trekked all the way down from the Bering Strait to Chile.

Will there be a "domino effect" in American archeology? Radiocarbon dates of 33,000 BP. have already been accepted by some non-American archeologists for the Monte Verde site. Add 7,000 years for the trip south from Alaska and the entry date is pushed back to 40,000 BP. There are even older dates ---over 100,000 years---suggested by mavericks such as G.F. Carter. It appears that the American past is going to be exciting in the future. (Morell, Virginia; "Confusion in Earliest America," Science, 248:439, 1990.)

Coming back to reality, the archeological picture is really not changing as rapidly as we have suggested above. Paradigms don't collapse overnight. Rather, as the great physicist Max Planck ventured, their proponents die off. In the latest issue of American Antiquity, T.F. Lynch critically reviews the same data available to D. Stanford in the forgoing item. Lynch is unconvinced by the South American findings. As for the "best" South American data, Lynch says:

> I am not ready to reject Monte Verde as a pre-Clovis archaeological site, but I have strong doubts. One, or possibly two, of the 26 'modified stones' from the oldest deposits would be acceptable to me as an artifact, had these stones come from a more clearly cultural context---and I have handled only part of the collection. The pictures of the 'hearth-like basins' do not convince me, however. Most of all I find it improbable that 13,000- and 33,000-year-old sites would be found, one nearly on top of the other. (Lynch, Thomas F.; "Glacial-Age Man in South America? A Critical Review," American Antiquity, 55:12, 1990.)

Comments. The Monte Verde site is located in Chile. See p. 26.

The archeological turmoil over the claimed pre-Clovis evidence in obvious in the following item.

THE CLOVIS POLICE

A new group of law-enforcers has been formed. Although the Clovis Police do not carry guns, they will make sure that all who stray from the archeological mainstream will be held up for censure. (Does this mean denial of funds and access to some journals?) The "law" that the Clovis Police will enforce says that humans did not enter the New World before 12,000 BP---the oldest date of the artifacts attributed to the Clovis people.

Perhaps we have dwelt on this subject too long, but the whole idea of the Clovis Police is counter to the spirit of science. The members of the Clovis squad and their objectives can be found in a recent issue of Science. (Marshall, Eliot; "Clovis Counterrevolution," Science, 249:738, 1990.)

Somehow, two important articles escaped the Clovis Police.

Meadowcroft Rockshelter, Pennsylvania. Responding to mainstream criticism of Meadowcroft radiocarbon dates (Some people just refuse to believe them!), J.M. Adovasio et al report that they now have 50 internally consistent dates, some made using accelerator mass spectrometry, that place humans at Meadowcroft at least 14,000-14,500 years ago. (Adovasio, J.M., et al; "The Meadowcroft Rockshelter Radiocarbon Chronology 1975-1990," American Antiquity, 55:348, 1990.)

Monte Verde, Chile. Another recent issue of Science reviews the first of two volumes on the Monte Verde site. This volume deals with the site itself. The artifacts themselves are reserved for Vol. 2. The reviewer states: "Even without a detailed consideration of artifacts and cultural features, it presents convincing evidence of 12,000-to-13,000-year-old human occupation in southern Chile." If these ancient Chileans came across the Bering land bridge no earlier than 12,000 BP, they made excellent time down to Monte Verde! The Monte Verde site has also produced some apparent tools radiocarbon-dated at 33,000 BP. The book's title is: Monte Verde. A Late Pleistocene Settlement in Chile. Tom D. Dillehay. Smithsonian Institution Press, Washington, 1989, 306 pp., $49.95. (Morlan, Richard E.; "Pleistocene South Americans," Science, 249: 937, 1990.)

T. Lynch, one of the Clovis Police, responds to such research: "'no indisputable or completely convincing cases' have come to light in America." (From E. Marshall's article.)

See p. 26 for more on the Monte Verde archeological site.

UPDATING MAN-IN-THE-AMERICAS

Archeologist P.S. Martin stated in 1987 that, "If humans lived in the New World more than 12,000 years ago, there'd be no secret about it." This quotation came from an article in which archeologists (some of them at least) were trying to roll forward the date humans entered the Americas from 12,000 to 11,500 years ago. In just a few months, we have collected three new items from the general scientific literature (not the U.S. archeological literature) that seem to be swimming against the current.

1. The Meadowcroft Rockshelter. J.M. Adovasio and R.C. Carlisle, in a letter to Science, argue that better dating techniques have consistently pushed back the date at which humans arrived in the New World. Using their own work at the Meadowcroft Rockshelter, in Pennsylvania, they cite many radiometric

dates earlier than 12,000 years. At the Pennsylvania site, the six oldest dates definitely associated with cultural material indicate that humans arrived here 13,955 to 14,555 years ago. (Adovasio, J.M., and Carlisle, Ronald C.; "The Meadowcroft Rockshelter, "Science, 239: 713, 1988.)

2. Kansas River skeletal remains. Using electron spin resonance to date a piece of a human skull, W. Dort and L.D. Martin affirm a date of 15,400 years before the present. (Bower, B.; "Skeletal Aging of New World Settlers," Science News, 133:215, 1988.)

3. A review of the paleoindian debate. W. Bray recounts in Nature what happened at a meeting at the Smithsonian last September. Various controversial sites were discussed, such as Calico Hills (200,000 years claimed) and Toca de Esperanca, Brazil (3,000,000 years claimed). But oddly enough, the Meadowcraft Rockshelter and Kansas River data were not mentioned. Chiefly, though, Bray was concerned with what did and did not constitute generally accepted proof in archeological dating. That this matter goes beyond idealized science is evident in Bray's quote of anthropologist E. Leach:

> Justification in terms of scientific methodology is in part self-deception, for when the figures turn out wrong the true believer will always shuffle the figures; when contrary evidence turns up he throws doubt upon the credentials of the investigator.

(Bray, Warwick; "The Palaeoindian Debate," Nature, 332:107, 1988.)

THE OROGRANDE, NM, SITE

The discovery in a New Mexico cave of numerous stone artifacts, hearths, butchered animal bones and a clay fragment dating back at least 35,000 years could provide proof that the Americas were inhabited long before the generally accepted date of 12,000 years ago, believes Richard MacNeish, research director of the Andover, Mass., Foundation for Archaeological Research. An article in the Baltimore Sun stated:

> The most solid proof of human presence earlier than 12,000 years ago may be a piece of a clay pot that appears to have a human fingerprint. The shard was found in a layer of sediment that has been dated as being 35,000 years old. If confirmed as human, it could be the key to the findings, some archaeologists say.

(Chandler, David L.; "Dig Finds Signs of Humans in N.M. 35,000 Years Ago," Baltimore Sun, p. 3A, May 6, 1991.)

Comment. It is certain that these discoveries will be disputed---and rightfully so. Even if they stand, it takes a generation to erase a false paradigm from the roster of science.

WINNING BY A HAIR

Archeologists have been very skeptical about the purported human artifacts and handprint found in the Orogrande Cave, New Mexico. The chief archeologist working at the site, R. MacNeish, has now found several hairs embedded in the cement-hard layers of the cave's floor. One of these hairs, less than an inch long, has definitely been labeled as human by Canadian forensic experts. Carbon-14 dating of a nearby piece of charcoal from the same layer has yielded a date of 19,180 BP---considerably more ancient than the passionately defended 12,000-BP date for first arrivals in the New World. MacNeish is confident that his claims will now be accepted, joking, "It looks like I'm going to win this one by a hair." Other archeologists, however, are not laughing. Handprints and hairs are insufficient; they want human bones. (Chandler, David L.; "Strand of Hair May Be Proof of Much Earlier Americans," Boston Globe, June 28, 1992. (Cr. R. Coltman)

LONG BEFORE THE VIKINGS AND POLYNESIANS

Scene: Orogrande Cave, New Mexico. R. MacNeish, a respected archeologist from the Andover Foundation for Archeological Research, has charted a 30,000-40,000-year-old paleonotological record of ancient camels, horses, tapirs, and other fauna found while excavating this cave. Intermixed with the animal bones are layers of charcoal (easily carbon-dated) and hints of human occupation. But don't we all know that humans did not arrive in the New World until 12,000 years ago? Nevertheless, there they are: (1) rude human tools; and (2) a possible human palm print. Mainstream archeologists are stonewalling again; there must be some mistake! (Appenzeller, Tim; " A High Five from the First New World Settlers?" Science, 255:920, 1992.)

Scene: In the Bluefish Caves in the northern Yukon.

> Arctic caves in the northern Yukon have yielded apparent bone tools carved 24,000 years ago, more than 13,000 years earlier than the earliest confirmed human habitation of the Americas, a Canadian archeologist [R.E. Morian] reported yesterday.

(Petit, Charles; "24,000-Year-Old Tools Found in Yukon," San Francisco Chronicle, February 10, 1992. Cr. D.H. Palmer.)

Comment. These bone tools "appear" to be worked by humans, but it is always possible that they were naturally fractured.

BERING STRAIT THEORY AGAIN IN TROUBLE

If humans first populated North America via the Bering land bridge 12,000 years ago, how did human bones and artifacts get buried under a 50,000-year-old alluvial fan in California? Dogma demands that such finds be discredited. Thus, "Pleistocene man at San Diego," the Calaveras Skull, and dozens of other archeological anomalies have been dismissed as hoaxes and misidentifications of nonprofessionals. The latest hint of truly ancient man in America came after heavy rains in 1976 cut through 21 m of deposits at Yuha Pinto Wash, just north of the Mexican border in Imperial Valley. The artifacts, still firmly in place, and associated bones are undeniably human. The overlying sediments are dated at more than 50,000 years old. (Childers, W. Morlin, and Minshall, Herbert L; "Evidence of Early Man Exposed at Yuha Pinto Wash," American Antiquity, 45:297, 1980.)

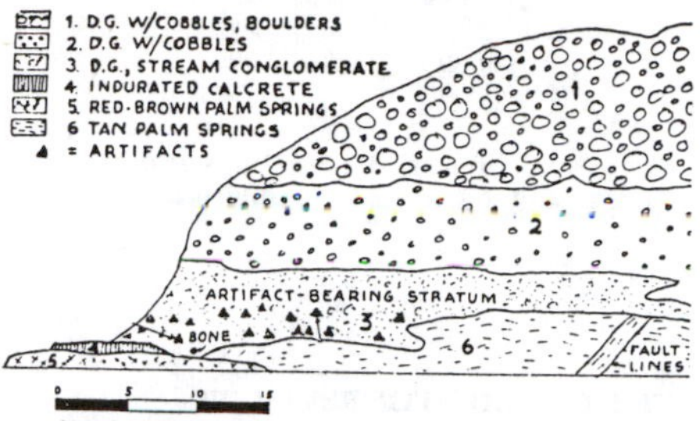

Cross section of deposits at Yuha Pinto Wash in the Imperial Valley.

ANCIENT CAMP FOUND 40 FEET BELOW COLORADO SURFACE

Outside Fort Morgan, Colorado, workmen digging pits for a landfill uncovered a prehistoric campsite 40 feet under the sandy bed of an ancient stream. The diggers found bones, worked flints, and burnt stones arranged in a ring. Excavations were stopped when the importance of the site became obvious. Estimates of the campsite age were as old as 30,000 years---a figure that would have been heresy a decade or two ago. (Anonymous; "Ancient Camp Unearthed at Colo. Landfill," Baltimore Sun, p. A3, December 14, 1980.)

Comment. Actually, the 30,000-year figure remains heresy.

THE HUEYATLACO DILEMMA

Beds containing human artifacts at Valsequillo, Mexico, have been dated at approximately 250,000 years before the present by fission-track dating of volcanic material and uranium dating of a camel pelvis. The dilemma posed by such dates is clearly stated in this quotation from the Conclusions of this article:

"The evidence outlined here consistently indicates that the Hueyatlaco site is about 250,000 yr old. We who have worked on geological aspects of the Hueyatlaco area are painfully aware that so great an age poses an archeological dilemma. If the geologic dating is correct, sophisticated stone tools were used at Valsequillo long before analogous tools are thought to have been developed in Europe and Asia. Thus, our colleague, Cynthia Irwin-Williams, has criticized the dating methods we have used, and she wishes us to emphasize that an age of 250,000 yr is essentially impossible."

(Steen-McIntyre, Virginia, et al; "Geologic Evidence for Age of Deposits at Hueyatlaco Archeological Site, Valsequillo, Mexico," Quaternary Research, 16:1, 1981.)

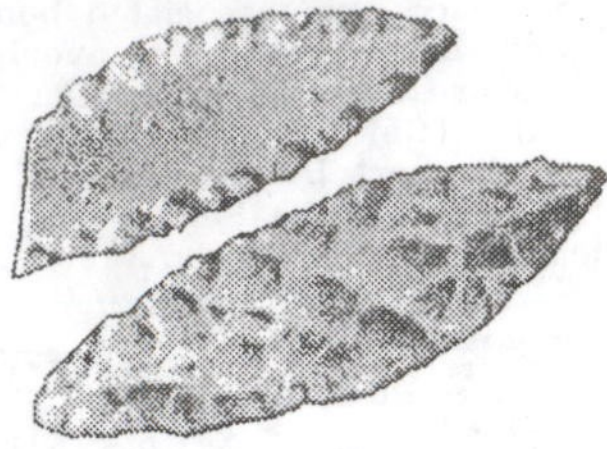

Stone tools found at the Hueyatlaco site. These may be 250,000 years old!

THE CALICO SITE REVISITED

Background. The so-called "Calico" site, in California, has been the focus of considerable debate over whether the suggestively shaped stones found there are the products of human labor or merely the work of natural forces.

Abstract. "Continuation of the Calico investigation, both in field and laboratory, has conclusively established the presence of Early Man, through the demonstration of numerous tools in several categories, as proven by a number of significant traits or attributes familiar to archaeologists. Microscopic examination reveals use-wear patterns. Uranium thorium tests yield a date of 200,000 ± 20,000 years for the artifacts." (Evidence or no evidence, the archeological community is not yet ready to believe 200,000 years. Ed.) (Simpson, Ruth D.;"Updating the Early Man Calico Site, California," Anthropological Journal of Canada, 20:8, no. 2, 1982.)

THE CALICO DEBATE, PLUS A LITTLE EDITORIALIZING

Passions run higher in archeology than in most fields of scientific endeavor. Favored hypotheses mesmerize some, despite contradictory data and cogent arguments. In this respect, much science verges on religion.

The foregoing observation from the editor was occasioned by letters written to Science News in response to B. Bower's article on the probability of human artifacts---as old as 100,000 BP---having been found at the Calico site, in California.

First, J.G. Duvall III attacked Bower's article, asserting that the human origin of the Calico artifacts had long ago been shown to be untenable. For a reference, he cited an article by himself and W.T. Venner in the Journal of Field Archeology. Duvall's major point was that the Calico tools do not resemble proven Paleoindian tools.

Responding to Duvall, G.F. Carter first pointed out that the Duvall-Venner paper was "almost instantly shown to be erroneous" by L.W. Patterson in the pages of the very same journal. As for the differences in artifacts, Carter asked why one should expect 12.000-year-old Paleoindian artifacts to look like 200,000-year-old artifacts from an entirely different culture. (Duvall, James G., III; "Calico Revisited," Science News, 131:227, 1987. Carter, George F.; "Calico Defended," Science News, 131:339, 1987.)

Comment. We don't really know whether or not the Calico artifacts were really made by humans 200,000 years ago. No one really does! One may opine or theorize, and that's it. The really annoying aspect of the Calico business is the tendency of scientists to make absolute statements in the face of contradicting evidence. This desire for certainty extends to all of science. Reviewers of our books of anomalies often remark that such-and-such an anomaly was explained long ago, despite the many references to contradictory facts and well-founded minority opinions. So anxious are some scientists to stifle dissent that one geologist recently asserted that continental drift was now so well-proven that no more contradictory data should be published!

ARTIFACTS OF THE AURIFEROUS GRAVELS

Now dismissed entirely and even ridiculed by establishment archeologists is the evidence of ancient human activity found in California's auriferous gravels. R.E. Gentet sets the geological stage in the following paragraph.

> The 1849 gold rush to the state of California was the beginning of some of the most unusual reported finds of early man in North America. The gold-bearing gravels of California are recognized as being Tertiary in age, ranging from oldest to youngest Tertiary, depending upon the exact geological setting. At the time these gravels were deposited, volcanic eruptions also laid down lava beds often tens or scores of feet thick. This occurred a number of times, and together with much erosion since then, have now resulted in table mountains, that is, lava-capped hills where the harder lava has better withstood erosion stresses while surrounding softer material has been swept away. It is under the hard lava beds, in the gold-bearing (auriferous) gravels, where the reported human bones and artifacts were found, not just once or twice, but hundreds of times by miners during the span of time from the 1850s through the 1890s while engaging in mining operations. Findings were spread over a wide geographical area.

During the late 1800s, several books and many papers recorded the discoveries. Some of the finds were made by respected scientists of the day. Human skulls were found embedded over 130 feet below the surface underneath thick lava beds. Also retrieved were many mortars and pestles, stone sinkers, strange double-headed stones, and the doughnut-like object pictured here. (Gentet, Robert E.; "Geological Evidence of Early Man," Creation Research Society Quarterly, 27:122, 1991.)

Comment. One can understand why creationists would evidence interest in Tertiary man, because they reject con-

Doughnut-shaped stone artifact recovered from California's auriferous gravels.

ventional geological dating and human evolutionary timetables. But why aren't today's archeologists interested? Because their reputations would be in jeopardy. Everyone knows the first humans didn't reach California until 12,000 years ago; and the Tertiary Period ended 1.6 million years ago! All those bones and artifacts must have been planted by mischievous miners or somehow deposited by flood waters. See also the Yuha Pinto Wash item on p. 23.

A 400-MILLION-YEAR-OLD HAMMER?

Background. Every once in a while, someone will claim to have found highly technical artifacts far more ancient than allowable given the accepted dates for the origin of hominids (10-15 million years BP). Scientific creationists have made much out of a supposed 400-million-year-old hammer, as described in a 1984 article in Ex Nihilo, an Australian creationist magazine.

This article begins with a startling photograph of an obvious hammer partly embedded in rock. Data: "1. The hammer was discovered on the Llano uplift, south west of the Paluxy River, Texas, U.S.A. The Llano uplift is a granite intrusion covered by Ordovician sandstone. 2. The hammer was discovered within a concretion of shell-bearing sandstone. (Initial reports incorrectly labelled it as limestone.) 3. The hammer handle is probably of spruce wood. 4. The interior of the handle is partly coalified. 5. The handle contains pockets of fluid. 6. The wood in the handle was hard and fibrously intact when discovered. 7. When the stone surface was first removed, the iron (alloy?) head was shiny and began to corrode only several months later. 8. The concretion contained fossil shells which can just be seen at the top left of the picture. (Nucula Pelecypods) 9. When the concretion was first broken open, there was a significant space around the hammer." (Anonymous; "Ordovician Hammer Report," Ex Nihilo, 6:16, 1984.)

Comments. If the hammer was really deposited with the sandstone, it would be about 400 million years old, according to present geological dating methods. This item was taken from a creationist publication, which has an obvious stake in undermining the prevailing scheme of geological dating. Nowhere does the report say the concretion was found in situ in the Ordovician sandstone. It may have been loose on the surface. Furthermore, concretions often contain peculiar things, as described in Unknown Earth. And finally, the discovery was made near the Paluxy River, where one also finds intermingled dinosaur and human-like tracks! The whole business is at once fascinating and suspicious. See p. 44.

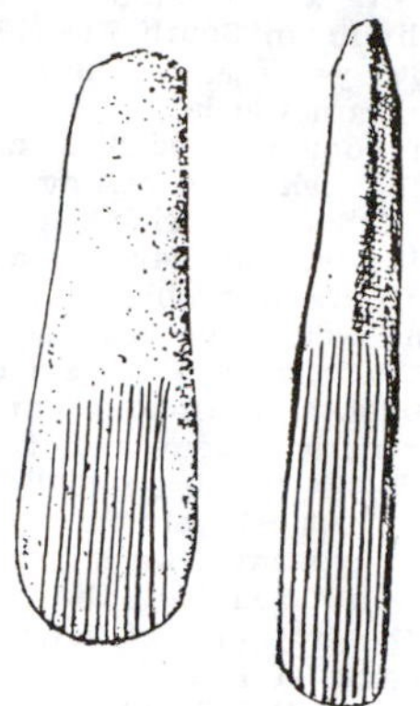

Stone beaters used in making bark paper from Mesoamerica (left) and southeast Asia (right).

SOUTH AND MESOAMERICA

As in North America, a few small artifacts found in South and Mesoamerica hint at Precolumbian contacts; that is, relatively recent transoceanic diffusion; and, in addition and even more anomalous, the prence of humans long before 12,000 BP---the so-called pre-Clovis limit established by North American archeologists. In actuality, the evidence for pre-Clovis humans in South America is much more common and persuasive than it is in North America. The implication of this startling observation is that South America may have been peopled before North America. This is, of course, contrary to the widely accepted ideas that the New World was first settled by migration across the Bering Land Bridge and thence south to Mesoamerica and into South America.

This section also introduces a third class of anomalies presented by small artifacts; namely, "precocious" technology or technical sophistication well beyond that normally expected from specific cultures.

Here, then, are the types of anomalies covered in this section:

- Precolumbian contacts. Old World pottery (both European and Asian) and Asian paper-making techniques.

- Pre-Clovis artifacts. Evidence of human habitation in South America prior to the 12,000 BP time boundary---long before in some instances.

- Precocious technology. Calculating devices (quipus) and metallurgy (use of bismuth).

PRECOLUMBIAN CONTACTS

ROMANS IN RIO?

In 1976, diver Jose Roberto Texeira salvaged two intact amphoras from the bottom of Guanabara Bay, 15 km from Rio de Janeiro. Six years later, archeologist Robert Marx found thousands of pottery fragments in the same locality, including 200 necks from amphoras. Amphoras are tall storage vessels that were used widely throughout ancient Europe. These particular amphoras are of Roman manufacture, circa the second century B.C. Much controversy erupted around the finds because Spain and Portugal both claim to have discovered Brazil around 1500 A.D. Roman artifacts were distinctly unwelcome. More objectively, the thought of an ancient Roman vessel crossing the Atlantic is not really so far-fetched. Roman wrecks have been discovered in the Azores; and the shortest way across the Atlantic is from Africa to Brazil---only 18 days using modern sailing vessels. (Sheckley, Robert, "Romans in Rio?" Omni, 5:43, June 1983.)

Comment. Photographs of these amphoras can be found in Marx's book: In Quest of the Great White Gods, New York, 1992.

A PAPER TRAIL FROM ASIA TO THE AMERICAS

The Mayan codices were made from bark paper as opposed to ordinary paper. To make bark paper, one first takes the inner layer of bark, or bast, from a tree. This material is then thinned, widened, and made flexible by soaking it in water and beating it. The final product retains much of the bark's structure with its interconnecting fibers. Ordinary paper today is also made of wood fibers, but the original fiber interconnections are destroyed in the pulping process.

The manufacture of bark paper requires characteristic grooved beaters, specimens of which have been found in both Mesoamerica and Southeast Asia. Were bark paper and the tools required to make it invented independently on both sides of the Pacific, or were they transported across the Pacific by early navigators? If the latter, the flow was probably from Asia to America because the paper-making tools first appeared in Southeast Asia 4-5000 years ago and in Mesoamerica only 2500 years ago. Even so, trans-Pacific voyages 2500 years ago are definitely not part of acceptable archeology.

Anthropologist P. Tolstoy, swimming against the mainstream, has surveyed the manufacturing technology of both bark paper and ordinary paper on a worldwide basis. He identified some 300 variable features in the process, 140 uses of the final products, and 100 specific details of bark beaters. Tolstoy concluded:

> All this points to the direct transfer of technology from Southeast Asia to Mesoamerica, apparently by a sea voyage that took place about 2500 years ago.

Tolstoy rejects the tapa (bark cloth) of Polynesia as a credible link between Southeast Asian and Mesoamerican bark-paper making. The technology transfer was not island-to-island but direct! Invoking Kon Tiki and the prevailing currents and winds, he postulates a 2500-year-old voyage swinging north of Hawaii along an islandless route to Mesoamerica. (Tolstoy, Paul; "Paper Route," Natural History, 100:6, June 1991.)

JAPANESE-STYLE POTTERY FOUND IN PERU

5000-year-old pottery found in coastal Peru bears an uncanny resemblance to pottery made in Japan during the same period. How could the Japanese have reached Peru circa 3000 BC. Easy! Storms could have blown fishermen into the trans-Pacific current. (Wilford, John Noble; "Case for Other Pre-Columbian Voyagers," New York Times, July 7, 1992.)

Update. Fascinating evidence also exists indicating an infusion of Japanese blood and culture into the Zunis of New Mexico around the 12th. century. See: Davis, Nancy Yaw; "The Zuni Enigma," NEARA Journal, 27:39, Summer/Fall 1993.

PRE-CLOVIS ARTIFACTS

HUMANS IN THE AMERICAS 32,000 YEARS AGO

Abstract. "The view that man did not arrive on the American continent before the last glaciation has been supported by the fact that until now the known and dated archaeological sites have not been of very great antiquity. But now we report radiocarbon dates from a

Brazilian site which establish that early man was living in South America at least 32,000 years ago. These new findings come from the large painted rock-shelter of Boqueirao do Sitio da Pedra Furada, the walls and ceiling of which are decorated with a rich set of prehistoric paintings. We have excavated a sequence containing abundant lithic industry and well-structured hearths at all levels. Carbon-14 dates from charcoal establish a continuous chronology indicating human occupation from 6,160 ± 130 to 32,160 ± 100 years BP. A date of 17,000 ± 400 BP, obtained from charcoal found in a level with fragments of a pictograph fallen from the walls, testifies to the antiquity of rupestral art in this region of Brazil." (Guidon, N., and Delibrias, G.; "Carbon-14 Dates Point to Man in the Americas 32,000 Years Ago," Nature, 321:769, 1986.)

Comment. The Nature article just abstracted and other reports of the Brazilian discovery do not mention or discount the considerable evidence for ancient man in North America. The dogma that man entered North America about 12,000 years ago across the Bering Strait has dominated archeology so forcibly that this evidence has been largely suppressed.

BREAKING THE 12,000-BP BARRIER

We have reported above evidence for humans occupying the Americas prior to 12,000 BP, in some cases long prior. The American archeological establishment has been very skeptical about such claims, but now the 12,000-BP barrier seems to be collapsing.

The turning point may have occurred at a recent meeting at the University of Maine's Center for the Study of the First Americans. The skeptics were bombarded by radiocarbon dates, tools, hearths, and bones from the Monte Verde site in Chile. Previously unbelieving archeologists are now ready to admit dates around 13,000 BP for Monte Verde---at least some are. (Lewin, Roger; "Skepticism Fades over Pre-Clovis Man," Science, 244:1140, 1989.) It was in 1987 that R. Lewin wrote an article in Science entitled, "The First Americans Are Getting Younger." Quite a turnaround!!

Nevertheless, the Science article above did not even mention some other presentations at the Maine conference. But the New York Times did:

> At a conference here this week at the University of Maine, Niede Guidon, an archeologist at the Institute of Advanced Social Science Studies in Paris, startled scientists by reporting new results that she said showed the Brazilian rock shelters were occupied by humans at least as long ago as 45,000 years. The 'quantity, diversity and preservation' of materials at the sites, she said, should lead to 'profound changes in the knowledge of prehistorical America.'

The Brazilian rock shelters boast elaborate paintings, fireplaces, tools, and butchered-animal bones, as described earlier.

Although some American archeologists have edged back to 13,000 BP, others are stonewalling at 12,000 or less. Actually, the debate has become unscientific on occasion, as revealed by R. Bonnichsen, of the University of Maine.

> 'Numerous meritorious grant proposals have been rejected because their goals and objectives were incompatable with entrenched academic opinion,'he said. 'At least five South American archeologists admitted that they are suppressing pre-12,000-year-old data out of fear that their funds would be cut off by American colleagues who endorse the short-chronology school of thought.'

(Wilford, John Noble; "Findings Plunge Archeology of the Americas into Turmoil," New York Times, May 30, 1989. Cr. J. Covey.)

Update. In a more recent article in Nature, F. Parenti and N. Guidon assert human habitation at Pedra Furada at 50,000 BP or earlier. (Bahn, Paul G.; "50,000-Year-Old Americans of Pedra Furada," Nature, 362:114, 1993.)

THE MONTE VERDE SITE, CHILE

We need quote here only the last two sentences from the Abstract of a paper in Nature by T.D. Dillehay and M.B. Collins:

> We report here two carbon-14 dates from charcoal taken from cultural features associated with the older materials of 33,000 yr BP. These findings provide additional evidence that people colonized the Americas much earlier than previously thought.

(Dillehay, Tom D., and Collins, Michael B.; "Early Cultural Evidence from Monte Verde, Chile," Nature, 332:150, 1988,) See pp. 22 and 26 for more details on the Monte Verde site.

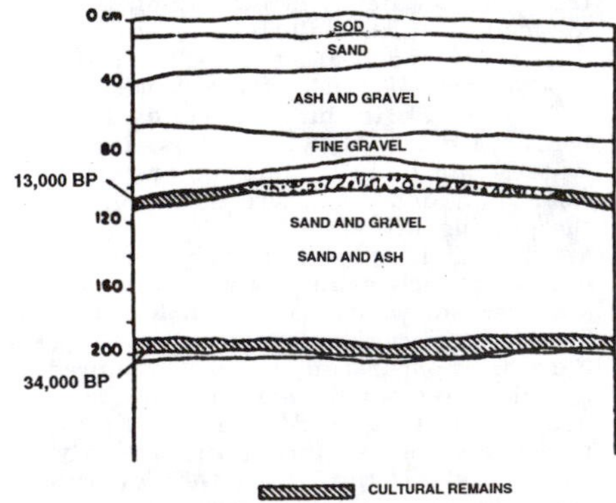

Stratigraphy at Monte Verde, Chile. Human artifacts may have been found in the 34,000-BP layer.

300,000-YEAR-OLD SITE IN BRAZIL?

"Central, Brazil---Archeologists excavating a cave in Brazil's remote northeastern backlands say that they have found evidence that man has lived in the New World for at least 300,000 years.

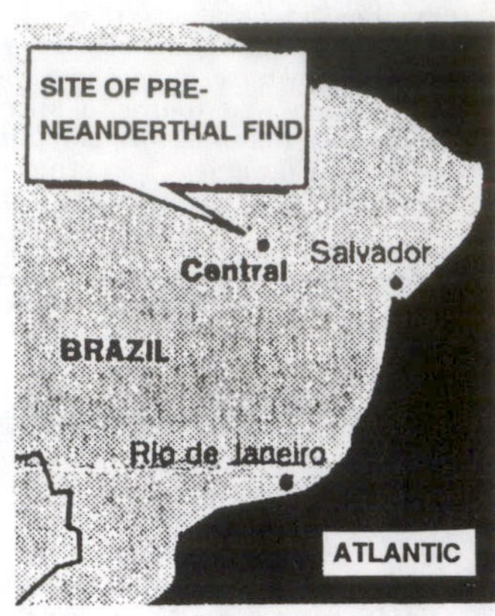

Location of the (claimed) 300,000-year-old site in Brazil.

If confirmed, it would be the first proof of pre-Neanderthal man in the Americas and a severe blow to current theories that the first humans came here from Asia during the last Ice Age, only about 35,000 years ago.

The scientists also report they have discovered what may be the world's oldest astronomical observatory.

.....

The signs of man were found in a cave called Toca de Esperanca (Grotto of Hope), deep in the black limestone cliffs of the Serra Negra mountains, 1,100 miles northeast of Rio de Janeiro.

The site caught the interest of the scientific community after archaeologist Maria Beltrao reported finding a stone implement and the cut bones of an extinct species of horse in a dig last year.

The bones were so old that they could not be dated by carbon 14, which can measure about 40,000 years. The Weak Radiation Laboratory in France tested them by the more sensitive uranium-thorium method, and came back with a staggering date of 300,000 years.

.....

A cave called Grotto of the Cosmos at nearby Xique-Xique contained paintings of suns, stars and comets and this is what archaeologists believe is the oldest astronomical observatory in the Americas.

'There probably were at least two cultures here,' said (J.) Labeyrie. 'One, about 10,000 years ago, made the paintings. Another much older was responsible for the bones and artifacts.'

In the grotto's dim light, a red comet 4.5 feet long stretches across the low ceiling, against a painted backdrop of stars. Red suns rise and set amid figures of lizards, a creature traditionally associated with the sun.

.....

Near the entrance of the cave is a notch where every year, precisely on the winter solstice (June 21 in the Southern Hemisphere), the sunlight enters and illuminates a red sun painted on the slanted ceiling." (Muello, Peter; "Find Puts Man in America at Least 300,000 Years Ago," Dallas Times Herald, p. A-1, June 16, 1987.)

Comment. No announcement or confirmation of this remarkable discovery has been found in the literature we regularly monitor!

PRECOCIOUS TECHNOLOGY

THE INCA'S USE OF BISMUTH

From Machu Picchu, Peru, comes a unique artifact: a llama-head knife made from two types of bronze. The knife blade and stem are made from low-tin bronze (not at all unusual), but the llama-head is bronze with 18% bismuth. Bismuth occurs as a native metal in Peru and it is not surprising that the Incas knew of it. This is the first artifact, however, containing any appreciable proportion of bismuth in bronze. The authors of this paper believe that the use of bismuth was intentional for at least two reasons. First, it gave the llama head an attractive lighter tone than the rest of the knife. Second, the handle was cast directly on the stem, and the use of the bismuth in the bronze would prevent the bronze from expanding too much during solidification. The handle would therefore be more securely attached to the stem. The Incas seem to have been better metallurgists than we have supposed. (Gordon, Robert B., and Rutledge, John W.; "Bismuth Bronze from Machu Picchu, Peru," Science, 223:585, 1984.)

CODE OF THE QUIPU

In a 1982 issue of Science Gary Urton reviewed a new book with the above title. The authors are Marcia and Robert Ascher, who have studied roughly 200 Inca quipus, demonstrating in the process that the Incas did indeed have a "written" language as well as a surprisingly sophisticated system of mathematical notation. A quipu appears to the uninitiated as a meaningless jumble of strings. To an Inca quipu reader, though, the positioning and colors of the secondary and tertiary strings appended to the primary cord all have meaning. The knots along each string also convey messages. Quipus incorporated, in a sense, three-dimensional notation, as opposed to the two-dimensional text on this page. Inca mathematical developments are inherent in quipu notation, which clearly reveals base-of-10 positional notation and the use of the zero. Instead of a tangle of colored strings, the quipu actually displays sophisticated concepts of number, geometrical configuration, and logic. (Urton, Gary; "Inca Encodements," Science, 216: 869, 1982.)

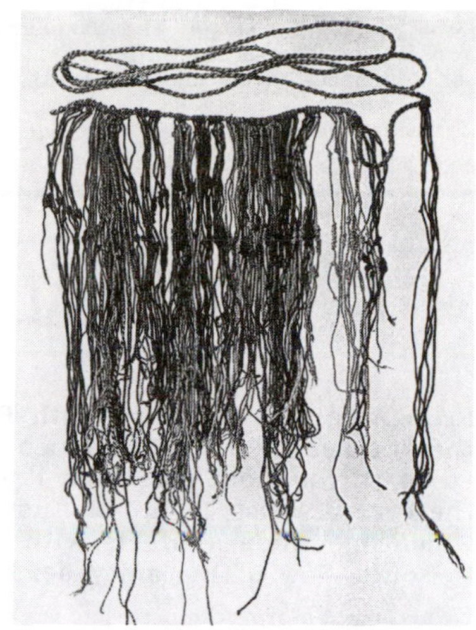

A Peruvian quipu. Messages are incorporated in the knotting, colors, and arrangemnt of the strings. (Smithsonian Institution)

EUROPE

In Europe, there exists a situation analogous to the 12,000-year-BP barrier found in the New World. The European barrier, however, is erected at about 1 million years, and is not as sacrosanct. As in the Americas, this barrier is breached by artifacts that imply a much earlier spread of humanity into and across this continent.

Most of the anomalous European artifacts described below fall into the category of "precocious technology"; that is, they exhibit technology of unexpected sophistication for their age.

THE ANCIENT-HUMANS-IN-EUROPE CONTROVERSY

Many times in this book appears the question: "When did humans first arrive in the Americas?" It is amusing to find that European archeologists have an analogous problem, only there the accepted date is about 1 million years ago, compared to 12,000 years for the Americas. Most European archeologists believe that primitive humans migrated northward from Africa into Europe, but the timing has always been a little fuzzy. The Europeans are willing to consider minor adjustments in the million-year figure. However, there are now several sites seeming to boast human artifacts that are about 2.5 million years old. This is just too old, and a debate has commenced.

The most controversial site is Saint-Eble, just below Mont Coupet, in south-central France. Here one finds quartz fragments that look manmade to some archeologists, but seem products of natural fracturing to others. These crude objects are what some American archeologists call "Carterfacts", after G.F. Carter, who has found similar rock fragments in the Americas and dates them much, much before 12,000 years. In Europe, there is little argument about the 2.5-million-year date for the stratum in which the controversial rocks are found. The debate is over whether they are natural or products of human manufacture. The French champion of human manufacture is E. Bonifay, an archeologist at the National Center for Scientific Research, in Marseilles.

At stake here is the mainstream view that modern man is the last in a succession of three species. The first was Homo habilis, which arose in Africa about 2 million years ago. (See pp. 47 and 126 where a southeast Asian origin is championed.) The second species in the series is Homo erectus, which appeared about 1.6 million years ago, also in Africa, and migrated into Europe about 1 million years ago. Homo sapiens, "our" species, appeared about 500,000 years ago in "archaic" form, to be succeeded by "modern" Homo sapiens about 200,000 years ago. Obviously a 2.5-million-year date in Europe undermines this scenario. There seems to be no fossil evidence at all that Homo habilis ever made it to Europe. (Acherman, Sandra; "European History Gets Even Older," Science, 246:28, 1989.)

WHO WAS MANUFACTURING WHAT?

Between 240,000 and 750,000 years ago, someone in the northern Jordan Valley made a flat, polished plank, 25 centimeters long, from a willow tree. The area where the plank was found is Middle Pleistocene in age and rich in stone tools as well as fragments of wood. (Anonymous; "Mollusc Confirms Dating of Oldest Known Plank," New Scientist, p. 14, July 20, 1991.)

EARLY BOOMERANG

"Scientists who found a curved piece of mammoth tusk in a cave in southern Poland have dubbed it the world's oldest known boomerang, dating to about 23,000 years ago.

"The claim is based on the artifact's shape, curvature and flattening at both ends, report Pael Valde-Nowak and his colleagues of the Polish Academy of Sciences in Krakow. It spans about 27 inches and is up to 2.3 inches wide and 0.6 inches thick. One side preserves the external, rounded surface of the tusk, while the other has been polished almost flat." (Bower, B.; "Prehistoric Tusk: Early Boomerang?" Science News 132:215, 1987.)

THE MALLIA TABLE

The Mallia Table was discovered in the Central Court of the Minoan Palace of Mallia in Crete. It is a large limestone disk 90 cm in diameter and 36 cm thick. Around its circumference are 33 cups of equal size. A 34th cup is larger and is located in a sort of ear that extends beyond the normal circumference of the disk. The larger cup is oriented due south. The disk is set in the stone pavement of a small terrace that is slightly elevated above the level of the

Central Court. This strange monolith, which dates circa 1900-1750 B.C., has been a puzzle to scholars since its discovery in 1926 by French excavators.

Herberger's thesis is that the disk is a lunisolar clock. The 33 small cups provide a convenient and symmetrical division of the 99 lunations of the 8-year cycle. By moving markers from cup to cup with each lunation, one could have a fairly accurate lunisolar clock. The 34th cup by virtue of its larger size would announce the need for an intercalated month. This sort of clock, even though arrived at empirically, represents a remarkable innovation for a period almost 4,000 years ago. (Herberger, Charles F.; "The Mallia Table: Kernos or Clock?" Archaeoastronomy, 6:114, 1983.)

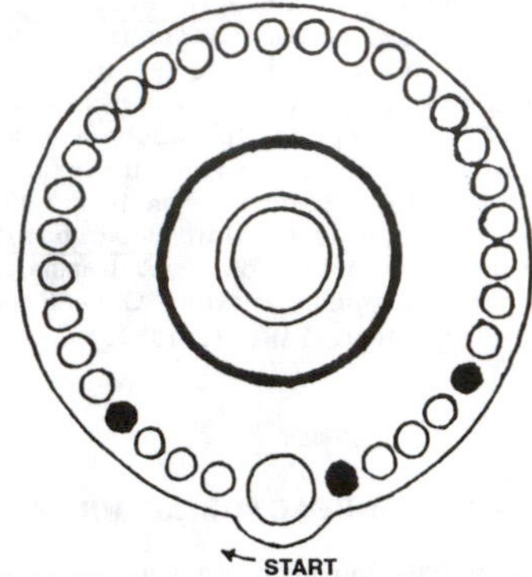

Diagram of the Mallia Table, with intercalcation cups in black. Overall diameter is about 1 meter.

LENSES IN ANTIQUITY

The ancient Greeks seem to have thought of just about everything. True, they didn't conceive of silicon chips or H-bombs, but they did know rudimentary optics. Excavations down the years have yielded hundreds of lenses ground from quartz crystals. (Later, the Romans used glass.) Many of these early lenses were articles of high craftsmanship, being accurately spherical and well-polished. Lathes were evidently available for grinding the rock crystal into appropriate shapes. Some ancient lenses had holes drilled through their centers, possibly so that they could be carried around the neck on cords. These seem to have been used for kindling fires. Most lenses, though, were probably magnifiers for authenticating seals and for carving gems. (Sines, George, and Sakellarakis, Yannis A.; "Lenses in Antiquity," American Journal of Archaeology, 91:191, 1987)

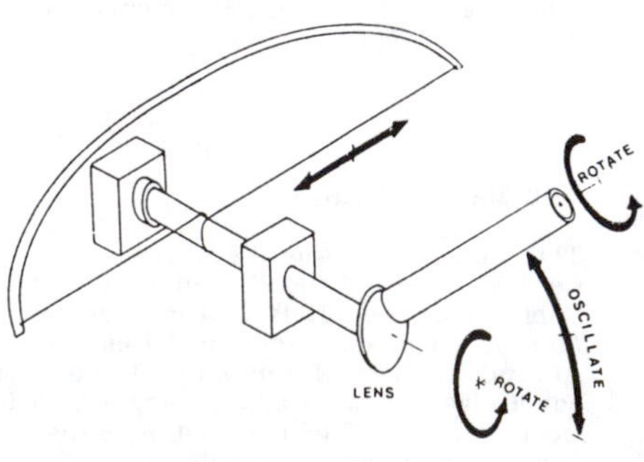

The ancients probably ground their lenses with the aid of bow-driven spindles.

Comment. We wonder if any ancient Greek ever put two of these lenses together to make a telescope. Such a tandem arrangement of lenses seems such a natural human experiment; i.e., if one is good, two will be better!

AUSTRALIA AND OCEANIA

From Australia and the South Pacific come artifacts that seem to betray the anomalous diffusion of peoples across this great ocean and, in particular, throughout Australia itself. In a strange parallel to the Americas, there also seems to be an Australian "artifact gap," on either side of which one finds artifacts with dates that challenge the anticipated north-to-south flow of humanity across this island continent.

SOUTH AMERICAN-POLYNESIAN CONTACTS AT EASTER ISLAND

In a letter to Archaeoastronomy, Jim Wheeler lists three bits of evidence suggesting that there were ancient contacts between South America and the Easter Islanders. (1) The Rapa Nui legends mention the arrival of strange men (about 25) from the east. (2) Excavation of the ancient Easter Island tombs in 1981 revealed that some of the skeletons belonged to American Indians. (3) The wall of carved stone at Vinapu on Easter Island is almost identical with the South American stone structures at Pisac and Machu Picchu. (Wheeler, Jim; "Comment on Ben Finney's Review, Archaeoastronomy, 5:8, July-September, 1983.)

EARLY CHINESE CONTACTS WITH AUSTRALIA?

In Chapter 1, there are three separate articles relating to the Australian 'pyramids'. In the final analysis, these pyramids do not seem to be pyramids at all, at least in the archeological sense, nor are they truly ancient. All of this pyramid excitement was precipitated by Rex Gilroy, an amateur Australian archeologist. Well, Gilroy is at it again. This time he claims evidence of ancient Chinese visits to Australia---long before the Dutch explorers and Captain Cook. Although Australian contacts have warned us about Gilroy and although his pyramid evidence has been debunked, his latest data should at least be made open for inspection, with caveats attached of course.

Since China is much closer to Australia than Egypt and the way is paved with handy islands, early Chinese contacts would not be as anomalous as Egyptian-built pyramids. Gilroy's major claims are:

- A carved stone head unearthed near Milton, New South Wales, seems to represent an ancient Chinese goddess.
- An old Chinese record, Atlas of Foreign Countries, describes the north coast of a great land to the south inhabited by pygmies, evidence of which has been found in Queensland.
- A 6th. century copper Chinese scroll includes a crude map of Australia. A 2000-year-old vase also seems to show a crude map of Australia.
- In 1948, fragments of Ming porcelain were dug up on Winchelsea Island. 35 years ago a jade Buddha was unearthed near Cooktown, Queensland.

Are these data accurate and as convincing as Gilroy claims? We will await further communications from Down Under. (Gilroy, Rex: "Were the Chinese First to Discover Australia?" Australiasian Post, p.8, May 1,1986, Cr. A.L. Jones) For more a possible Egyptian connection, see p. 18.

EARLY MAN IN AUSTRALIA EVEN EARLIER

In an alluvial deposit in Western Australia, Mance Lofgren and John Clarke have discovered more than 30 stone tools possibly dating back more than 100,000 years. Previously, the first men were thought to have invaded Australia only 40,000 years ago. (Anonymous; "Man's Arrival in Australia Put Back 60,000 Years," New Scientist, 78: 734, 1978.)

Comment. This 100,000-year date obviously is in conflict with the Australian date mentioned above. An analogous situation prevails in North America where 100,000-year-old tools have been claimed, even though the archeological establishment has set a 12,000 year limit.

EXPLAINING THE "ARTIFACT GAPS"

Earthquake researchers often speak of "seismic gaps", where earthquakes "should" occur but don't. Well, artifact gaps exist, too, namely in North America and northern Australia. Both Australia and the Americas seem to have been peopled late; and both regions seem to have been invaded from the north---according to conventional thinking. Serious anomalies arise because the dates of human occupation from South America and southern Australia are considerably older than those from North America and northern Australia---that is, if we dismiss those North American dates considered "unreliable" by the archeological establishment. In the Americas, all "archeologically acceptable" dates older than 12,000 years come from South America; in Australia, all dates exceeding 24,000 years are found along the southern coast. The "artifact gaps" are clearly established! Humans were supposed to have migrated from north to south, but the artifacts say otherwise.

R.G. Bednarik, in a recent paper in Antiquity, offers a related observation:

> It is generally agreed that both regions must have been settled from the north, yet no trace has been found of the first settlers in either North America or northern Australia. In northern Australia one finds ground-edge axes at up to 23,000 years b.p., which have no counterparts in the probable catchment area of the first colonizers, Indonesia; while in North America the earliest human settlers used elaborate projectile points, which have no counterparts in the final Pleistocene of eastern Siberia. When and where did such innovations then evolve?

Bednarik's solution to this double dilemma assumes that the first colonizers in both cases did indeed come from the north---some 40,000 years ago in both Australia and North America---but that their invasion paths were confined to the low coastal areas. Then, as the Pleistocene ended and the ice sheets melted, the sea level rose to obliterate the signs of earliest occupation. Therefore, if archeological digs could be conducted under a few hundred feet of water along the continental shelves, the artifact gap could be filled! (Bednarik, Robert G.; "On the Pleistocene Settlement of South America," Antiquity, 63: 101, 1989.)

Comment. Dredges have brought up mammoth's teeth from the continental shelves, but we are not aware of any human artifacts. Also the continental shelf along North America's west coast is hardly a thoroughfare, since it is very narrow.

EPIGRAPHY AND ART

NORTH AMERICA

The search for and the study of anomalous inscriptions is a popular avocation in the United States, where this activity is termed "epigraphy." Of course, professional archeologists indulge in epigraphy, too, but they generally look askance at those inscriptions and symbols favored by the amateurs. In mainstream archeology, those tablets and cave walls marked with characters and motifs suggesting the Precolumbian presence of Romans, Hebrews, Libyans, Celts, Iberians, Chinese, etc., are usually deemed to be fraudulent or misinterpretations.

The first 86 issues of Science Frontiers have amassed a fascinating collection of anomalous epigraphy and artwork. However, because of mainstream skepticism toward the genre, many (but not all) of the entries come from the publications of amateur groups. This situation does not mean that the inscriptions and artwork are nonexistent; rather, it is the interpretations that are at issue.

The materials presented below, which represent only a very small fraction of those available, are divided into three groups:

- Controverted inscriptions. Tablets, graven stones, and cave writing that betray the Precolumbian presence of Vikings, Romans, and sundry other peoples. Some of these inscriptions have received wide pubilicity.

- Ogam writing. Ogam is a simple form of writing characteristic of some ancient Old World inscriptions. If ogam is widespread in North America, as sometimes asserted, Frequent Precolumbian contacts are implied.

- Curious symbols and art. Various motifs, art styles, depictions of non-native animals, etc., that suggest Precolumbian diffusion to North America from other continents.

CONTROVERTED INSCRIPTIONS

A FAR-WANDERING LOST TRIBE?

Imagine hiking near Los Lunas, New Mexico, and coming upon a huge basalt boulder inscribed as follows:

This is obviously not an Indian petroglyph. Rather, it is the Ten Commandments set down in an old Hebrew script. The script and its translation seem unmysterious. What everyone wants to know is who chiseled it and when. It was apparently discovered in the 1880s. Harvard anthropologist Frank C. Hibben visited the site in 1930 and pronounced the inscription to be at least 100 years old. Who in New Mexico in 1830 knew ancient Hebrew? The inscription may be much older, for the whole boulder, weighing 60-80 tons, is tipped 20-30°, so that the lines of script are tilted. Is it all a hoax? Some think so. It is easier to live with a hoax than the thought of a Hebrew outpost in New Mexico a couple thousand years ago. (Underwood, L. Lyle; "The Los Lunas Inscription," Epigraphic Society, Occasional Papers, 10:57, no. 237, 1982.)

Comment. It should be remarked that there are many purported Hebrew and Roman finds in the American Southwest; viz., the Tucson lead crosses with Roman inscriptions.

HOW OLD IS THE LOS LUNAS INSCRIPTION?

Volume 13 of the Epigraphic Society's Occasional Papers (one of two large volumes for 1985) contains several articles of great interest to anomalists with an archeological bent.

In the first of these, Barry Fell deals with the criticism that the now-famous Los Lunas (New Mexico) inscription cannot be the work of ancient Hebrew-writing visitors to the New World because it employs modern punctuation marks. Fell counters this by reproducing several ancient texts that use similar punctuation conventions, thus blunting this particular attack on the antiquity of the Los Lunas inscription.

In a second paper, geologist G.E. Morehouse comes to grip with a second criticism leveled at the inscription; namely, that the engraving looks fresh and lacks the patination characteristic of great age. Morehouse concludes that the freshness actually derives from the frequent recent scrubbing of the inscription (with wire brushes on some occasions!) to improve its visibility. Taking this into account, Morehouse estimates the age of the Los Lunas inscription by comparing its weathering with a nearby 1930 incription. Conclusion: the Los Lunas inscription is much older. Any time from 500-2000 years or more would be "quite reasonable". (Fell, Barry; "Ancient Punctuation and the Los Lunas Text," Epigraphic Society, Occasional Papers, 13:35, 1985; and Morehouse, George E.; "The Los Lunas Inscriptions, A Geological Study," Epigraphic Society, Occasional Papers, 13:44, 1985.)

Comment. We are therefore still left with the possibility that Old World travelers with a knowledge of ancient Hebrew visited what is now New Mexico, perhaps around the time of Christ.

A NEW LOOK AT THE BAT CREEK INSCRIPTION

The January 1989 issue of the Tennessee Anthropologist contains a long article on the Bat Creek Stone by J.H. Mc Cullock, of Ohio State University. We rely here upon a summary written by R. Strong.

"The Bat Creek Stone has generated so much controversy, yet it was excavated in an undisturbed burial mound in 1889 under the direction of Cyrus Thomas, Project Director of the Bureau of American Ethnology's Mound Survey, a part of the Smithsonian Institution. There could be no question of forgery because it was found under the head of one of the nine skeletons that were excavated. Pieces of wood presumed to be the remains of wooden earspools were preserved in the Smithsonian's collections as were a pair of brass C-shaped bracelets. Thomas immediately declared the nine characters on the small stone to be Cherokee and the burial assumed to be post-contact---what else could the bracelets be but trade items or native copper?"

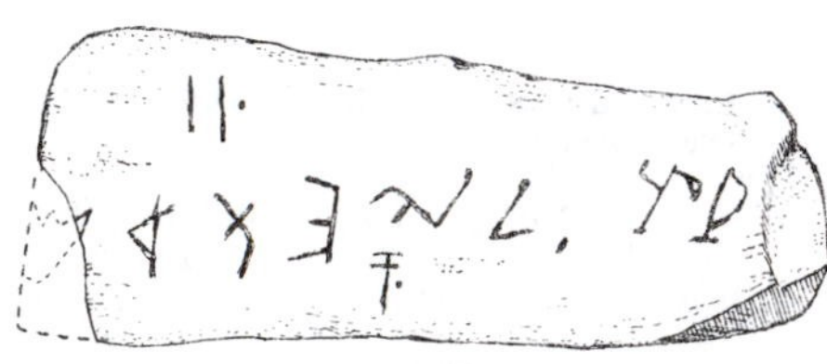

The Bat Creek stone with its supposed Hebrew letters. Which side is up has been a problem!

That would seem to be the end of the story, but some language students failed to see any resemblance between the Bat Creek inscription and the written Cherokee language. Further, C. Gordon, admittedly a proponent of early Phoenician contacts with the New World, declared that the Bat Creek characters were Paleo-Hebrew, a family of languages that includes Phoenician. Then in 1987, the wood accompanying the Bat Creek Stone was radiocarbon dated in the range 32-769 AD---definitely not "post-contact". Modern analysis was also applied to the bracelets, leading to the discovery that they had the same proportions of lead and zinc as the brass made by the Romans between 45 BC and 200 AD. In sum, the Bat Creek Stone now seems more likely to be something inscribed by early Phoenician visitors to North America. (Strong, Roslyn; "A New Look at the Bat Creek Inscription," NEARA Journal, 23:26, Summer/Fall 1988.)

THE "AMERICA BEFORE COLUMBUS" CONFERENCE

In the summer of 1992, the New England Antiquities Research Association (NEARA) sponsored the above-named conference at Brown University. One of the papers discussed in the New York Times report on the meeting concerned the so-called Bat Creek stone. Here is what appeared in Science Frontiers:

Jewish refugees from the Roman Empire may have somehow reached eastern Tennessee, if the famous Bat Creek Stone really bears an ancient Hebrew inscription. The grave in which the stone was found has been carbon-dated between 32 and 769 AD.

(Wilford, John Noble; "Case for Other Pre-Columbian Voyagers," New York Times, July 7, 1992.)

THE KENSINGTON STONE: A MYSTERY NOT SOLVED

It is safe to say that the great majority of professional archeologists consider the Kensington Stone, with its runic inscription, to be a hoax. E. Wahlgren's 1958 book bears the title: The Kensington Stone: A Mystery Solved. This title reflects the professional attitude, although not the disdain, even contempt, with which academics now view this controversial artifact.

Two major challenges to the authenticity of the Kensington Stone are: (1) Its use of Arabic number placement (that is, decimal placement); and (2) Its use of the symbol Φ for 10. Both usages are illustrated by the following numbers taken from the Stone.

for 2

for 8

for 10

for 14

for 22

for 1362

Professional archeologists and epigraphers maintain that genuine runic inscriptions did not use these Arabic innovations.

In a more recent article, R. Nielsen, University of Denmark, demonstrates with actual, well-established runic inscriptions that the above criticisms are without foundation. Such notation and convention were employed. In fact, the use of Arabic innovations actually supports the authenticity of the Kensington Stone. (Nielsen, Richard; "The Arabic Numbering System on the Kensington Rune Stone," Epigraphic Society, Occasional Papers, 15:47, 1986.)

MYSTERIOUS COPY OF THE GRAVE CREEK STONE

The so-called Morristown Tablet was apparently discovered near Morristown, Tennessee. A symbol-by-symbol comparison with the famous Grave Creek Stone, reveals that both are inscribed with the same message, probably in a Semitic language. B. Fell renders the message thus: "Tumulus in honor of Tadach. His wife caused this engraved tile to be inscribed." Why would anyone wish to copy such a message? The Grave Creek Stone was associated with a burial in West Virginia; could there have been two Tadachs? Are either or both hoaxes? (Buchanan, Donal; "Report on the Morristown Tablet," Early Sites Research Society, Bulletin, 10:22, no. 1, 1982.)

Comment. The real anomaly, assuming authenticity, is the presence of Semitic inscriptions in ancient American graves.

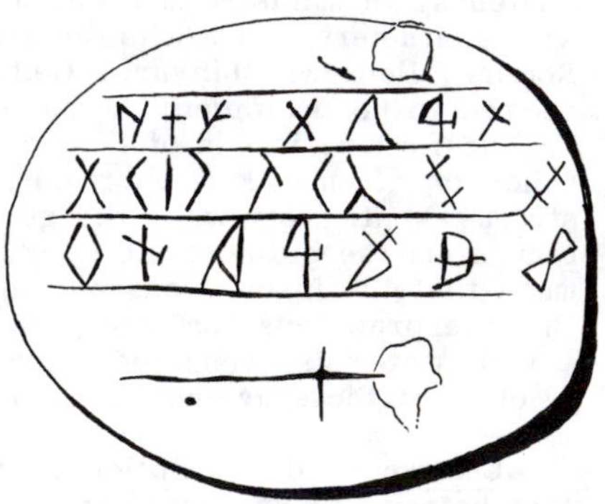

The Grave Creek stone.

LIBYAN SIGNS FROM SOUTHEASTERN KENTUCKY

Curious images and writings turn up with surprising frequency in North America. The sculpture illustrated was among several found in a cave on the Virginia-Kentucky border. Made life-sized from hard sandstone, it bears curious symbols on its back. Quite obviously, it is the head of an Indian chief, but the symbols might be Libyan, according to Barry Fell. He translates them as: "luminous, radiant, sun-like." Interestingly enough, the ancient leaders of the Natchez Tribe were called "Suns" and wore feathered headdresses like those of the later Plains Indians. Further, the Natchez Suns apparently maintained that their ancestors came to North America from North Africa---thus the Libyan symbols. (Calhoun, Vernon J.; "Libyan Evidence in Southeast Kentucky." Epigraphic Society, Occasional Papers, 7:127, 1978.)

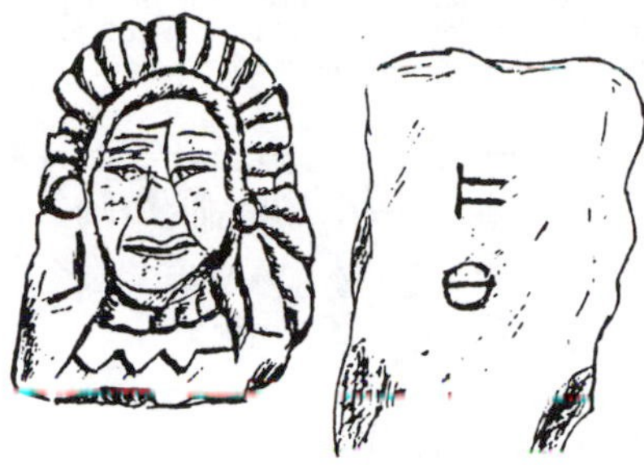

A sculpture from Kentucky bearing symbols, claimed to be Libyan, on its back.

WHERE DID CHIEF JOSEPH GET A CUNEIFORM TABLET?

Among the effects of Chief Joseph, the famed leader of the Nez Perce Indians, was a clay tablet bearing a cuneiform inscription. The tablet transmits no startling message, being merely a receipt for one lamb changing hands. But where did a Northwest Indian chief get a 3000-year-old clay tablet? The tablet first came to light around 1878, long before cuneiform tablets became common on the artifacts market. Still, it could have been a gift from some missionary or tourist---or even planted as a hoax. (Park, Edwards; "Where Did Chief Joseph Get a Cuneiform Tablet?" Smithsonian Magazine, 9:36, February 1979.)

BASQUES IN SUSQUEHANNA VALLEY 2500 YEARS AGO?

Back in the 1940s, W.W. Strong collected about 400 inscribed stones from Pennsylvania's Susquehanna Valley. Called the Mechanicsburg Stones, they seemed to bear Phoenician characters---at least Strong interpreted them as such. Naturally, Strong was ridiculed, for the Columbus dogma was dominant then. More recently, however, Barry Fell claimed that the Mechanicsburg Stones were the work of Basque settlers circa 600 BC. The Basque theory has fared no better than the Phoenician. Now a noted authority on the Basque language, Imanol Agire, has strongly supported Fell's conclusion that ancient Basques carved the stones. (Anonymous; "Noted Basque Scholar Supports Claim That Mechanicsburg Stones Were Cut by Ancient Basques," NEARA Journal, 15: 68, 1981.)

An enigmatic inscription from one of the Mechanicsburg stones.

OGAM WRITING

OGAM, OGAM, EVERYWHERE

Ogam is a curious sort of writing characterized by a long reference line decorated with shorter perpendicular lines. The lengths and arrangements of the shorter lines and their crossing and non-crossing of the reference line constitute alphabetical symbols. The ancient Irish are credited with inventing Ogam; and Ogam inscriptions are common in that part of the world.

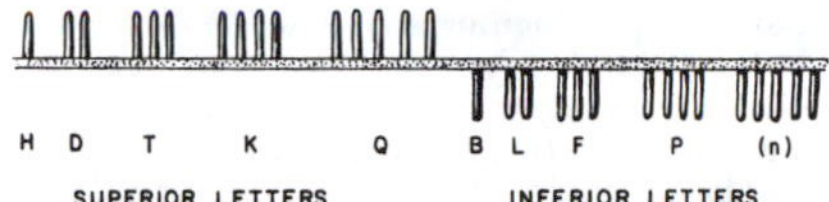

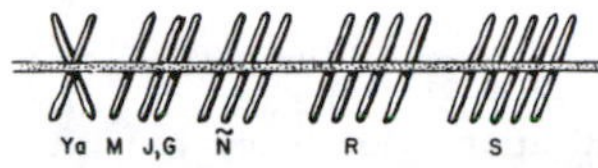

One version of the ogam alphabet.

During the last two decades, B. Fell and members of the Epigraphic Society have discovered many possible Ogam inscriptions in the continental United States. A few of these anomalous finds are described below.

OGAM INSCRIPTIONS IN WEST VIRGINIA?

Several petroglyphs in Wyoming and Boone counties, West Virginia, long-identified as random Indian doodling with little message content, may actually be Celtic Ogam writing. Translations of the petroglyphs reveal several Christian messages, as in the segment illustrated below. Based upon the style of the Ogam, these petroglyphs may have been chiseled some time between the early Sixth and late Eighth centuries. The Ogam writers may have been Irish monks who, after the fashion of St. Brendan, sailed west from Europe during this period. (Pyle, Robert L., "A Message from the Past," p. 3. Gallagher, Ida Jane; "Light Dawns on West Virginia History," p. 7, Fell, Barry; "Christian Messages in Old Irish Script Deciphered from Rock Carvings in W. Va.," p. 12. All three articles in Wonderful West Virginia, 47:3, March 1983.)

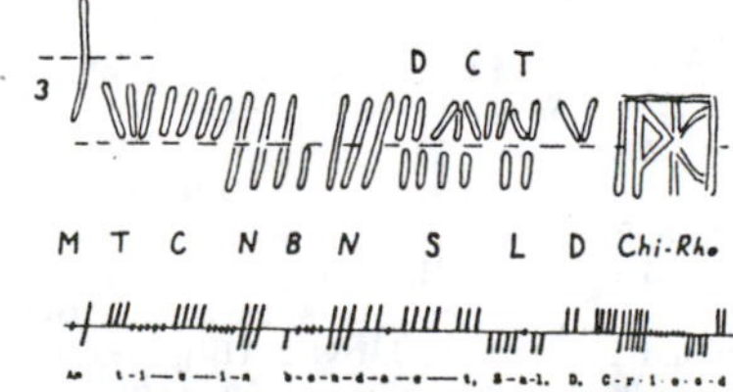

One translation of the West Virginia ogam inscription: "The season of the blessed advent of the Saviour, Lord Christ (Salvatoris Domini Christi).

Comment. Although Wonderful West Virginia is not a scientific journal, Pyle is identified as an archeologist. The articles contain many excellent color photographs of the inscriptions, so their reality can hardly be doubted.

TWO REMARKABLE INSCRIBED STONES

The first stone is located in western Colorado on a remote canyon ledge, overlooking a broad valley with a stream.

"The dolmen is four feet across the top and has three placed stones holding it above the ledge in a level position approximately six feet from the cliff face. The Ogam on top of the capstone is intermixed with cupule-like depressions ranging in size from $7\frac{1}{2}$" - $9\frac{1}{2}$" long, 3" - $3\frac{1}{2}$" wide and $1\frac{1}{4}$" - $1\frac{1}{2}$" deep in the center. The cupule-like depressions are very striking because of their uniformity, smoothness, and peculiar shape. The Ogam on the side of the capstone is abundant and occasionally connecting with lines on the top. The surface of the dolmen was obviously smoothed and prepared for the inscriptions. The actual age is unknown but the desert varnish on the Ogam, the depressions, and the smoothed surface is substantial."

The Colorado inscribed dolmen in situ. The top of the upper stone is also inscribed. (Adapted from the referenced Epigraphic Society paper)

B. Fell has translated the markings, which in his view are in Arabic Ogam, as:

Top: God is strong. Strong to help his right hand.
Front: The Koran is the unique achievement of the prophet pious and tender.

(Morehouse, Judy; "A Colorado Dolmen Inscribed with Ogam," Epigraphic Society, Occasional Papers, 11:209, no. 269, 1983.)

Comment. A photograph accompanying the article shows a striking artifact with strongly engraved markings.

The second stone was discovered at a petroglyph site in south central Alabama. This stone has an apparent jumble of scratches or grooves, on two faces, which one might easily ascribe to plows. Barry Fell, however, considers them obvious Iberic letters. His translation of the Arabic:

Top: A vegetable garden. Cession of land, a conveyance of property.
Side: Cession of land, a conveyance of property.

(Henson, B. Bart, and Fell, Barry; "Inscribed Rocks in South Central Alabama," Epigraphic Society, Occasional Papers, 11:235, 1983.)

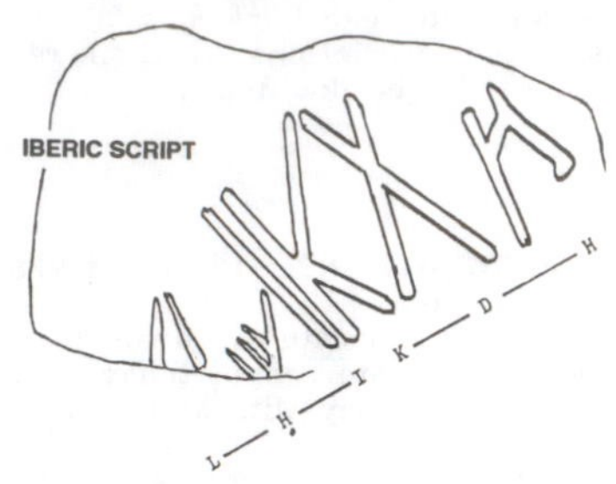

The upper, or principal, face of the inscribed stone from Alabama.

Comment. The large number of North American sites with enigmatic marks documented by the Epigraphic Society elicits several questions:

(1) Are the sites and artifacts genuine? In view of the large number discovered at various times, in various places, by different people; some would certainly seem to be.

(2) Are the markings really ancient Ogam, Libyan, and similar brands of old writing? Admittedly, some grooves and scratches on small stones may have been created by random processes; but others, like the extensive series in West Virginia reported above, must be manmade.

(3) Were there really ancient Celts, Romans, Arabs, Egyptians, and other peoples in North America well before Columbus? The large number of old inscriptions would argue for a "yes"; but one must also wonder what these old explorers or colonists did except carve symbols on rocks. Where are the expected other artifacts, such as pottery, campsites, etc.? Certainly the conventional archeological literature, which we survey, is devoid of any references to such artifacts. The data reported above are definitely not being incorporated into mainstream science.

A DEMURRER FROM THE EPIGRAPHIC SOCIETY

The following letter was received in response to the forgoing item. Since it contains much interesting information we decided to reprint it, with the writer's permission.

"I am writing to mildly protest the substance and style of your article 'Two Remarkable Inscribed Stones.' Of course substance and style are interrelated, but I'll try to separate them as best I can. First, let me address style. The condescending 'fragile hypothesis treatment' ('...which in his view..,"...Fell, however, considers them...,"...Comparisons..suggest...' is totally inappropriate when one considers the mathematical probabilities of thousands of petroglyphs possessing markings coincidentally identical to those of ancient languages of the Old World. And one has to marvel at the cutting power and linguisitic talent of certain plows.

"In the substance area vis-a-vis the alleged absence of artifacts: While it wouldn't be fair to expect the author to have been familiar with Professor Fell's three books on the subject, and the previous volumes of ESOP, with their many references to artifacts (loomweights, amphorettas, Roman lamps, countless Roman and other coins, and various other Old World items) it would not seem unreasonable to expect the author to have been familiar with the articles about Roman artifacts in the same volume he extracted portions from. Concerning the glib question about '...what these old explorers or colonists did except carve symbols on rocks..', it's instructive to remember that while many Old World artifacts have been found, many artifacts were of a bio-degradable nature--wood, leather, fabric, etc. But perhaps the most cogent remnants of the Old World visitors can be found in the numerous Indian languages which still use Old World alphabets and vocabularies." (Radloff, David M.; private communication, April 16, 1984.)

Comment. Radloff is an Associate Editor of the Epigraphic Society. In a separate communication, Fell concurred with Radloff's comments, adding that while the American archeological establishment ignores Old World artifacts in North America, European and North African journals do report them.

CELTIC FRONTIER SITE IN COLORADO?

In 1980, P.M. Leonard and J.L. Glenn, from the Hogle Zoological Gardens, Salt Lake City, Utah, visited a rock outcropping in Colorado that was reputed to be inscribed with 'peculiar markings.' The markings were peculiar all right, for Leonard and Glenn believe they are excellent examples of Consainne ogam writing, a type ascribed to ancient Celts. Translation by Barry Fell suggests that the Colorado site was a shelter for Celtic travellers long before Columbus! One of the many inscriptions was translated as: "Route Guide: To the west is the frontier town with standing stones as boundary markers." (Leonard, Phillip M., and Glenn, James L.; "A Celtic Frontier Site in Colorado," Epigraphic Society, Occasional Papers, 9:175, 1081.)

Comment. Although the Colorado ogam cannot be written off as plow scratches, as it is in the East, one should beware of the highly controversial nature of ogam inscriptions found in North America.

MORE FELL FALLOUT

J.H. Bradner and H. Laudin present a readable synopsis of Fell's ideas about precolumbian expeditions to North America. In this article, as in Fell's books, the data form the core of the controversy. Reviewed are the Blanchard Stone (Celtic writing in Vermont); a ceramic tablet inscribed in ancient Libyan (Big Bend National Park, Texas); the Massacre Lake petroplyphs (apparent Carthaginian writing in Nevada); and two Roman coins from a group picked up along a Massachusetts beach. Traditionalists denounce these finds and Fell's interpretations with a fervor once reserved for von Daniken. (Bradner, John H., and Laudin, Harvey; "America's Prehistoric Pilgrims," Science Digest, 89:90, May 1981.)

Comment. The fact is that if any one of Fell's many, many identifications and translations of North American inscriptions is correct, our whole view of ancient seafaring will have to change.

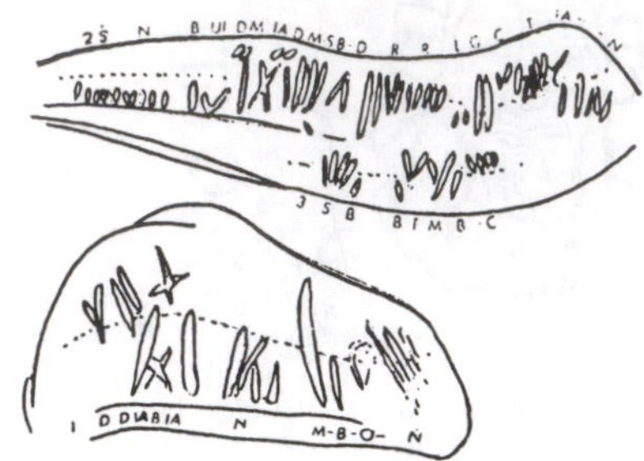

The Blanchard stone, discovered in Vermont, is thought to be a prayer for rain inscribed in a form of Gaelic used by Iberian Celts, according to B. Fell.

OGAM IN CONNECTICUT

A stone inscribed in ogam was recently reported in Connecticut. (See the Bulletin of the Early Sites Research Society, vol. 12, 1985. B. Fell translates the ogam as follows: "In this small stone lies the power of averting sickness/The ogam protects from debilitation of the Evil Eye." (p. 18)

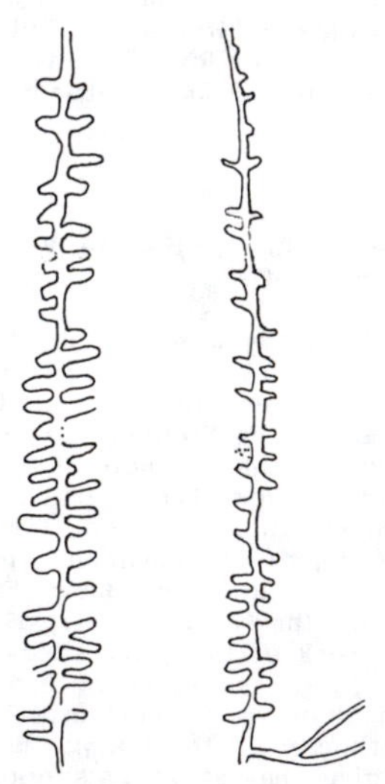

An inscribed stone from Connecticut. Interpreting the marks as ogam writing: Face 1 (left) reads up; Face 2 (right) reads down. (Adapted from the Bulletin of the Early Sites Research Society and the Occasional Papers of the Epigraphic Society.

In Wyoming an extraordinary rebus/ogam panel was discovered in 1986. An excellent photograph of the inscription may be found among the Occasional Papers of the Epigraphic Society, (16: 304) in an article by R.E. Walker.

SOME NEWLY DISCOVERED ARCHEOLOGICAL ANOMALIES FROM NORTH AMERICA

The 1980 volume of the Epigraphic Society's Occasional Publications contains some fascinating, but not yet thoroughly verified, tidbits:

(1) A letter from a Cherokee Indian describes the Cherokee tradition of "Little People" or pygmies, who once lived in the southern Appalachians. Interestingly enough, the Cherokee language has a word for pygmies that resembles words in several European languages that mean dwarfs or pygmies.

(2) Another letter describes ogam writing on a large rock panel at Cedar Canyon, near Rock Springs, Wyoming. (Ogam writing implies very early European contact with the New World.)

(3) An inscribed lead disk has been found in a small cache of Indian artifacts from Adams County, Ohio. The inscription indicates that it was an Iberian traveler's amulet. (Epigraphic Society, Occasional Papers, 15:33 and 15:77, 1986.)

DID THE ANCIENT EGYPTIANS EXPLORE OKLAHOMA?

The drawing reproduced below was made by C. Keeler when he visited the famous inscription-filled Anubis Cave. Some of the figures have Egyptian overtones, as remarked by D. and A. Buchanan:

> This "cave" (really a cave-like rock shelter located in Northwestern Oklahoma close to the Colorado line), as well as others like it in the vicinity, was first recorded by Gloria Farley after her visit to the site in June 1978. She especially remarked upon the Anubis figure you see here as well as the figure with the rayed head surmounting the "cube-in-perspective" or "3-D Cube" (as some have called it). Besides the Egyptian motifs, she also noted the ogam-like strokes and a number of other apparent Celtic connections.

A portion of the inscriptions from the so-called Anubis cave.

Translation of the ogam by B. Fell indicated that the site was used for Celtic rites. (Buchanan, Donal, and Buchanan, Ann; "The Anubis Cave in Old World Iconography," ESRS Bulletin, 18:27, October 1991.)

Comment. Anubis was the Egyptian jackal god. It is the stylized figure in the top center of the drawing. Such interpretations are ignored by the archeological establishment, and almost all research on such sites is carried out by amateurs.

AN EVALUATION OF NORTH AMERICAN OGAM WRITING

The editor of the Review of Archaeology describes D.H. Kelley as "an epigrapher of considerable reputation." And what is the subject of this respected professional journal and reputable epigrapher? It is B. Fell's work on North American ogam inscriptions!

Kelley is concerned by the strange lack of supporting archeological evidence at the inscription sites, but as the following quotation demonstrates, he dares to admit an ancient Celtic presence in North America.

> I have no personal doubts that some of the inscriptions which have been reported are genuine Celtic ogham. Despite my occasional harsh criticism of Fell's treatment of individual inscriptions, it should be recognized that without Fell's work there would be no ogham problem to perplex us. We need to ask not only what Fell has done wrong in his epigraphy, but also where we have gone wrong as archaeologists in not recognizing such an extensive European presence in the New World." (Kelley, David H.; "Proto-Tifinagh and Proto-Ogham in the Americas," Review of Archaeology, 11:1, 1990.)

CURIOUS SYMBOLS AND ART

INSCRIBED STONE FROM TENNESSEE NECROPOLIS

The pictured stone was found in the early 1890s in an extensive cemetery near Nashville. The stone's back was hollowed out like a cupstone, while the front was inscribed with symbols. Barry Fell considers the symbols Libyan, pre-100 AD style. He translates the stone thus: "The colonists pledge to redeem." (Whittall, James, P., II; "An Inscribed Libyan Token from a Necropolis in Tennessee," Early Sites Research Society, Bulletin, 6:37, 1978.)

A STONE FACE FROM UNGAVA

September 1976, Lac Guerard, Ungava, Canada. A stone face was found on the lake shore by caribou hunters. The back of the sculpture was covered with moss and stained underneath with age; the front was well-weathered. It was a crude sandstone carving---almost a doodle in stone---but the facial features were unmistakably Norse. Stylistically the face resembles nothing carved by Eskimos or the local Indians. The apparent antiquity of the stone and the strongly Nordic features suggest past Norse exploration of this desolate tundra near Hudson Bay. (Lee, Thomas E.; "Who Is This Man?" Anthropolgical Journal of Canada, 17:45, 1979.)

Comment. Once into Hudson Bay, why not on to Minnesota (and the Kensington Stone), then down the Mississippi to Oklahoma, where Viking signs are claimed to exist by some.

ARCHAEOLOGICAL RIDDLE

W. Elliott, of Grand Lake Stream, Maine, reports the discovery of some truly remarkable stone artifacts in eastern Washington County, Maine. These artifacts were extracted from a small cist, or tomb, constructed of slabs of slate laid over a hollow space between a ledge and row of stones. As the following description attests, these artifacts are anomalous to Maine (and perhaps anywhere in the States).

"Elliott has three stone artifacts that he says came to light when he lifted two or three of the slabs covering one end of the tomb, which lies east-to-west. The grave items consist of a smooth rectan-

Maine amulet with unusual symbols. On the other side is an "eye of God," an Old World motif.

gular green stone resembling a whetstone but bearing four letters or symbols; a four-inch pendant that is a flat stone oval bearing on one side an eye and on the other side a face of the sun with four rays, a crescent above, and six or seven letters in an undetermined script below; and a 15-inch ceremonial slate spear point showing on one side a bearded, trousered man in a hat or helmet with one arm severed and one foot missing, and on the other side a bear-like animal with two spears sticking out of him. In front of the bear are marks resembling the Roman numerals for eight, with the V tipped to one side."

Members of NEARA (New England Antiquities Research Association) visited the site; and professional archeologists have been invited to inspect the finds. (Wiggins, John R.; "Archaeological Riddle," Ellsworth American, August 3, 1989. Cr. J. Covey.)

Comment. Obviously, we have here either a hoax or an important anomaly.

THE GRAND LAKE STREAM ENIGMA

A little over a year ago, we reported on W. Elliott's discovery of a small, stone-covered cist in eastern Washington County, Maine. This cist contained several stone artifacts bearing remarkable symbols, writing, and portrayals of a man and a slain animal. Naturally, mainstream archeologists look askance when amateur archeologists come across such anomalous materials. Happily in this instance, a professional archeologist, J.B. Petersen, Director of the University of Maine's Archaeological Laboratory, took an interest in the site near Grand Lake Stream. After careful study of the site and its artifacts, he has prepared a preliminary report.

Petersen's report is accompanied by many photos and sketches made during his excavations. On p. 33, we reproduce a sketch of the amulet with its strange epigraphy. Now, we add a sketch of the "elongated hafted ground biface, with human figure." Over 13 inches long, this artifact depicts a trousered, bearded man of European countenance, who is missing one arm and a foot. Petersen asserts that the artifacts have no affinities with American Indian artifacts: rather they have a European flavor.

An "elongated, hafted, ground biface with human figure" from Grand Lake Stream, Maine.

What can one make out of all this? Petersen is only able to state:

> Although the site is undoubtedly human-made, its function, antiquity and cultural attribution cannot be precisely specified on the basis of the unique characteristics of both the artifacts and the cist. Tentative interpretations allow suggestion that it is attributable to some portion of the historical period, a European cultural tradition, and probably is contemporaneous with or postdates local stone working at the site.

In other words, we could have anything from a pre-Columbian European contact to rock doodling by Colonial stoneworkers. (Petersen, James B.; "Grand Lake Stream, The Elliott II Site: An Archaeologist's Preliminary Report," NEARA Journal, 25:3, Summer/Fall, 1990.)

WHO LEFT THESE ARTIFACTS IN BURROWS CAVE?

Below are pictured two artifacts extracted from a controversial site in Illinois, called Burrows Cave. A large quantity of epigraphic material has been found there, and it certainly doesn't seem to be fare digestible by mainstream archeologists. The artifacts are somewhat reminiscent of the Davenport, Iowa, and Wilmington, Ohio, tablets. (See Ancient Man for these just-mentioned sites.) The illustrations below were taken from the Mid-Atlantic Chapter of the Epigraphic Society, Occasional News Letter, no. 2, May 1, 1989.

Some of the controversial artifacts from Burrows Cave.

Comment. The Burrows Cave artifacts are so fantastic that their validity is questioned by most amateurs and all professional archeologists.

THE CUP-AND-RING MOTIF IN AMERICA

The 1988 volume of the Occasional Papers of the Epigraphic Society is, as usual, chock full of ancient symbols, motifs, and writing, many of which come from anomalous times and/or places. R. W.B. Morris, an authority on prehistoric rock art, has contributed an article comparing the cup-and-ring motif as found in Great Britain with that found in North America. Since this stereotype motif decorates the rocks on all continents, save Antarctica, and since the hey-day of cup-and-ring engraving was 3-5 millennia ago, this unique design suggests the worldwide diffusion of culture thousands of years ago.

A cup-and-ring engraving consists of a hollow or cup anywhere from 4 to 30 inches in diameter, surrounded by 1 to 9 rings. The rings may be gapped, with a groove running through the gaps from the outside. See the illustration. Cups-and-rings have been found at over 700 sites in Great Britain. Most date between 2200 and 1600 BC.

The cup-and-ring motif is much rarer in the States. A few are known from Alabama, California, Georgia, Hawaii, Texas, and doubtless other states. In contrast to the British cups-and-rings, the American ones are ungapped though otherwise indistinguishable.

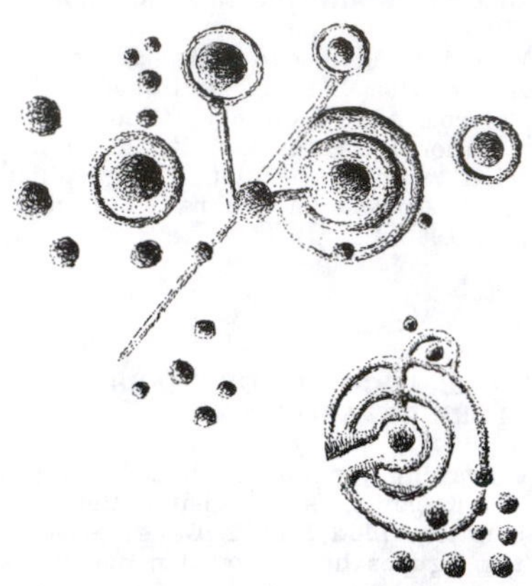

Typical cups-and-rings from a site in Ireland. These motifs are found worldwide.

What is the significance of the motif? Of course no one can say for sure. In Britain, many are near copper and gold workings. Others are associated with burials and astronomical alignments. Some European archeologists think they represent the sun or sun god. For the anomalist, the cups-and-rings hint at an ancient worldwide culture that left its signature on rocks just about everywhere. (Morris, Ronald W.B.; "The Cup-and-Ring Motif in the Rock Art of the British Isles and in America." Epigraphic Society, Occasional Papers, 17:19, 1988.

STELE WITH UNKNOWN GLYPHS FOUND NEAR VERA CRUZ

A basalt stele found submerged in the Acula River, 40 miles southeast of Vera Cruz, Mexico, is "the most important stele found in America to date," says F.W. Capitaine, Director of the Jalapa Museum of Anthropology. The stele is 7.8 feet high, weighs 4 tons, and is adorned with 16 columns of glyphs.

> The Veracruz stele has the same enumeration symbols used by the Mayas ---small circles and bars---which enabled Mr. Winfield to identify two dates among the hieroglyphics: May 22, 143, and July 13, 156.
>
> The remaining glyphs probably

record events between those dates. Although there are 20 glyph types similar to the ones used by the Mayas, 100 more are new. The stone carries a total of 600 glyphs.

Winfield hopes that the newly found stele will help explain what happened during the transition between Olmec and Maya cultures. He thinks it possible that the stele is the product of a previously unrecognized civilization. (Anonymous; "Inscribed Stone May Hold Secrets of Mexican Culture," Baltimore Sun, p. 9A, June 8, 1988.)

THE PECKED CROSS SYMBOL IN ANCIENT AMERICA

Twenty-nine instances of the so-called pecked cross have been collected by the authors of this article. Usually consisting of two concentric circles centered on orthogonal axes, this cross design is found carved on rocks and in the floors of ceremonial buildings throughout Mesoamerica. Such a motif would ordinarily not evoke much comment, but here the figure is formed from many small, evenly spaced depressions so arranged as to hint at larger meanings. For example, many pecked crosses have 260 depressions, suggesting a calendric interpretation (i.e., the 260-day Mesoamerican cycle). On some occasions, the cross arms are astronomically oriented. In addition, the holes may have been used to hold pieces of ritual games similar to patolli.

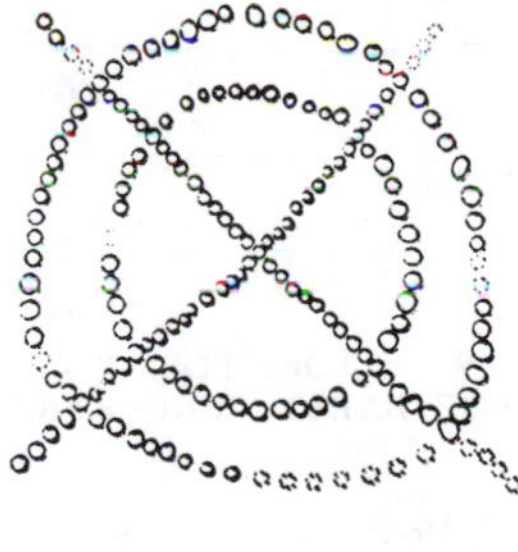

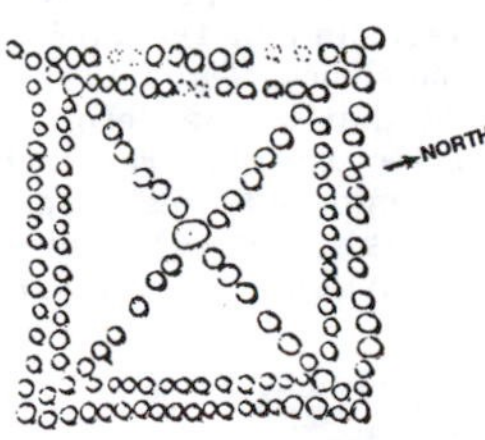

Representative pecked crosses.

The pecked crosses are wide-spread and were apparently quite significant to the ancient Mesoamericans. Perhaps, the authors suggest, the figures had a composite astronomical, calendric, and ritual purpose. This would be consistent with the Mesoamerican cosmological belief that everything was interlinked and that the works of man must be too. (Aveni, Anthony F., et al; "The Pecked Cross Symbol in Ancient Mesoamerica," Science, 202:267, 1978.)

THE RABBIT IN THE MOON: MORE EVIDENCE OF DIFFUSION?

Diffusionists seize upon all manner of artifacts to prove that peoples of the various continents made frequent contacts long before the European exploration of the planet. In the latest issue of Archaeoastronomy (dated 1984 but published in 1986), C.R. Wicke analyzes the rabbit-in-the-moon motif:

> Representations of a hare or rabbit on the moon are found in the art of ancient China and in Pre-Columbian Mexico. Mythologies of both areas place a rabbit in the moon. Although such linkage might appear to be arbitrary, a comparison of the visible surface of the full moon with the silhouette of a rabbit reveals a degree of congruence. Not only the distinctive ears of the rabbit but other features as well appear to be delineated on the moon's surface.

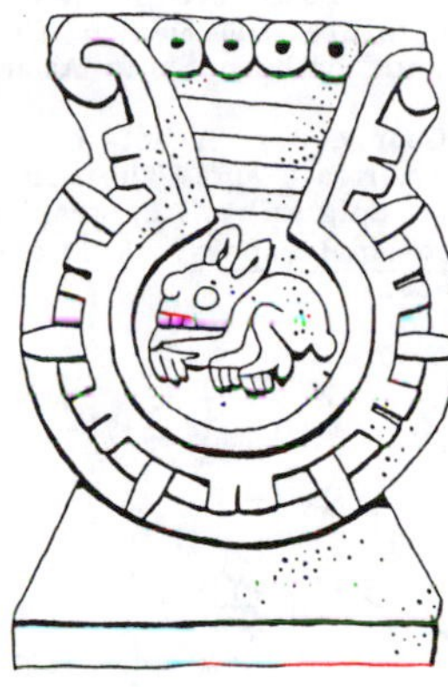

A Mixtec stela from Tiaxiaco, Mexico, showing the rabbit-in-the-moon motif.

Could the parallelisms in art and myth in China and ancient Mexico not be a simple coincidence helped along by the rabbit-like visage of the full moon? Wicke's article deals with this possibility in depth, but he discounts it in the following paragraph:

> Moreover, if one delves into the complexities of the association of hare and moon as manifest in mythology as well as in graphic imagery, correspondence between those of China and Mexico seem both too complex and too arbitrary to have been arrived at independently. Indeed, the mythology and imagery of the hare on the moon in Mesoamerica would seem to derive from Transpacific sources.

(Wicke, Charles R.; "The Mesoamerican Rabbit in the Moon; An Influence from Han China?" Archaeoastronomy, 7:46, 1984.)

A RECENTLY DISCOVERED "BOOK" OF PETROGLYPHS

Unfortunately, we cannot reproduce this huge assemblage of marvelously intricate petroglyphs here; but be assured that they are not haphazard doodlings of unaccomplished primitives. We quote from the author's abstract:

"A recently discovered sheltered rock scar with red pictographs, in Jalisco, west Mexico, is a major addition to the rather meager data on pictographs in Mesoamerica. It appears to contain a complex set of data pertaining to the cosmology of the relatively unknown Indians who inhabited the Jalisco coast during the last Pre-Hispanic period. Analysis of the scar has incorporated both the artistic symbolism of the nearby Huichol Indians, and concepts developed through archaeoastronomy. This analysis suggests that the ceiling pictographs record the use of sky transits of the sun, Venus, or the constellation Orion as wet season/dry season calendrical markers. Wall pictographs show the sun on the mountainous horizon, below which is the earth filled with symbols of plants and animals, among these stand shamans calling down the life-giving rain from the god(s) of the sky. I also explore the possibility that one of the ceiling pictographs is a record of the appearance of the Crab supernova in the sky in AD 1054." (Mountjoy, Joseph B.; "An Interpretation of the Pictographs at La Pena Pintada, Jalisco, Mexico, American Antiquity, 47:110, 1982.)

A JAPANESE PRESENCE IN ANCIENT MEXICO?

A. von Wuthenau, a specialist in pre-Columbian art at the University of the Americas in Mexico City, has long been a champion of ancient contacts between the New World and Africa, the Orient, and the Mediterranean region. For example, his book Unexpected Faces in Ancient America, contains hundreds of photographs of pre-Columbian figurines and other artwork showing facial features typical of the Old World and Asia. His latest find consists of a terra cotta model of an ancient sailing ship manned by figurines of ten oarsmen, all with striking Japanese features. The model boat is one foot long; the oarsmen, two inches high. It was discovered at a burial site in the Guerrero region of Mexico. Von Wuthenau has tentatively dated the boat as 2500 years old.

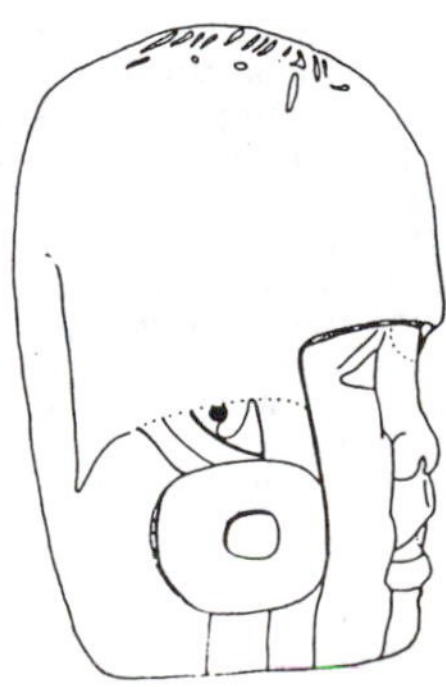

A sketch of one of the giant Olmec stone heads from von Wuthenau's book. He believes that this particular head, La Venta III, displays Asiatic features. Other such heads seem African.

(Anonymous; "Sailors in a Model of an Ancient Ship Found in Mexico Have Asian Features," Boston Sunday Globe, November 10, 1985. Cr. J. Whittall)

THE UBIQUITY OF AMERICAN ARCHEOLOGICAL ANOMALIES

One would think that all North American archeological anomalies worth mentioning would already be firmly ensconced in the professional literature. This does not seem to be true, unless one is very conservative about defining "worth mentioning". Ancient coins, anomalous inscriptions, and other intriguing tidbits are being found all the time.

The conventional journals, such as American Antiquity and American Anthropologist disdain such discoveries. One place to find them is in the Occasional Papers of the Epigraphic Society. The 1987 collection of papers is at hand, and it is chock full of fascinating things. The following is taken from volume 16, 1987.

Ancient coins. A bronze coin of the ancient Greek city of Amisos was found about six years ago by Doyle Ellis, who was searching for gold with a metal detector in the channel of the Snake River in Idaho. It was deeply imbedded in the gravel.

In a small Indian mound at Deer Creek, near Chilicothe, Ohio, a Numidian bronze coin was recently uncovered. It has a BC date.

> Oddly, those same coins, regarded in the Old World as artifacts of the highest importance, are never regarded at all by archeologists in America. who blithely declaim the "absence" of Old World artifacts in America. (p. 14)

NUMISMATIC UFOs

Jeremiah Epstein has assembled an absolutely fascinating analysis of some 40 "discoveries" of pre-Columbian coins in the United States. The lengthy table detailing the finds and the long list of references are alone enough to make this article a classic. Epstein carefully scrutinizes each find with admirable dispassion. His conclusion: frauds, counterfeits, and recent losses of imported ancient coins suffice to explain all the data. Supporting and disagreeing comments from researchers active in the field follow. The advocates of pre-Columbian diffusion naturally take issue with Epstein, claiming that there is a residue of cases not adequately explained. (Epstein, Jeremiah F., et al; "Pre-Columbian Old World Coins in America; An Examination of the Evidence," Current Anthropology, 21:1, 1980.)

RECENT SURVIVAL OF THE ELEPHANT IN THE AMERICAS

Elephants were supposed to have disappeared from the Americas about 10,000 years ago, as the Ice Age waned. This date is another of those "consensus" scientific facts that no one dares challenge any longer if he or she wishes to get published or win research grants. Although this subject remains "closed off" in normal scientific intercourse, there remain tantalizing hints that elephants roamed the Americas until very recently---perhaps even a few hundred years ago!

The following snippets are culled from two articles written by G. Carter, Texas A&M, now emeritus, but always heretical:

1. Numerous folk memories of the elephant were retained by the American Indians.
2. A mastadon was killed, cooked, and eaten by humans in Ecuador circa 1500 BC.
3. Indians told Thomas Jefferson that elephants could still be seen in the region of the Great Lakes.
4. In Florida, a cache of extinct animals, including elephants, was carbon-dated at 2000 BP.
5. Elephant heads are prominent in art and sculpture from Mexico, Central America, and northern South America.

(Carter, George F.; "A Note on the Elephant in America," and "The Mammoth in American Epigraphy," Epigraphic Society, Occasional Papers, 18:90 and 18:213, 1989.)

Mayan "elephant" motif. Some authorities do not see the elephant here, but perhaps a macaw instead!

A MAMMOTH FRAUD IN SCIENCE

The Holly Oak pendant, shown in the accompanying sketch, depicts a mammoth incised on a piece of seashell. Said to have been discovered in 1864 at a Delaware archeological site, it has been employed to "prove" two different theories: (1) that humans were in North America as the Ice Ages waned and mammoths still roamed the continent; and (2) that the mammoth survived in North America well into the Christian era.

In an article in American Antiquity, J.B. Griffin et al marshal considerable evidence implying that the Holly Oak is a fraud. Much of this evidence seems weak: (1) the discoverer of the pendant, H.Y. Cresson, was not highly regarded in American archeological circles of the time; (2) the pendant was not taken seriously by other archeologists; (3) the drawing of the mammoth looks like it was copied from a European engraved tusk; (4) the shell from which the Holly Oak pendant was made looks like shells found in other archeological sites of more recent dates; and so on. The only "hard" evidence that the pendant is a fake comes from radio-carbon dating, which suggests the shell is only 1530 ± 110 years old. The authors state that since mammoths positively did not survive this recently, the pendant must be a fraud.

Griffin et al thus dump the Holly Oak pendant into the archeological wastebasket of "proven" frauds. This rather large wastebasket, they say, also contains the Calaveras skull, the Davenport elephant pipes, the Lenape stone, and the Nampa Image! (Griffin, James B., et al; "A Mammoth Fraud in Science," American Antiquity, 53:578, 1988. Also: Lewin, Roger; "Mammoth Fraud Exposed," Science, 242:1246, 1988.)

The Holly Oak pendant.

THE MEXICAN SELLOS: POSSIBLE EVIDENCE FOR EARLY EUROPEAN CONTACTS

Thousands of Mexican sellos (seals), both of the cylinder and flat types, have been collected in the world's archeological museums. Many bear striking designs, like that of the heron shown below. For too many years, archeologists have considered these designs to be only designs. But with more discerning examination, bolstered by

A Mexican sello featuring a stylized heron along with apparent Libyan characters, such as those listed at the right.

knowledge of ancient Old World alphabets, it now appears that these seals incorporate many letters from Libyan, Iberian, Punic, and other alphabets.

Since many of the Mexican sellos date back to 1200 BC, the implication is that Old World-New World contacts occurred long before the Christian era.

But, asks G.F. Carter in his long article on the sellos, if the ancient peoples of Mexico and Central America did absorb Old World symbols, "Why was the alphabet not developed instead of the clumsy hieroglyphics?" (Carter, George F.; "Mexican Sellos: Writing in America, or the Growth of an Idea," Epigraphic Society, Occasional Papers, 19:159, 1990.)

Comment. Hieroglyphics do appear clumsy to us, but perhaps it is because we do not fully appreciate them. There may be nuances of shape and rendition that convey more than the bare translation of the glyph, just as a sonnet is more than the sum of its words.

CURIOUS SILVER CROSSES FROM A GEORGIA MOUND

In November of 1832, two silver crosses were extracted from an Indian mound in Murray County, Georgia, along with more usual Indian relics. The crosses are exquisitely wrought and were most likely brought to the Americas by the expedition of Hernando de Soto. Some of de Soto's men, under Adelantado, ventured into what is now Georgia trying, among other things, to Christianize the Indians.

The puzzle of the silver crosses is not in their source but in the crude figures and inscription added to one of them. The cross shown in the figure depicts a horse on one side and an owl on the other. The inscription (too small to be read on the figure) is within the central ring and states: IYNKICIDU, which makes no sense in any known language.

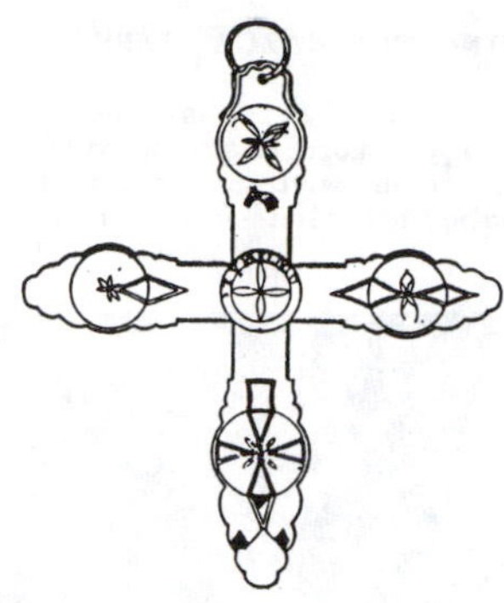

One of the silver crosses found in a Georgia burial mound.

This minor mystery was first revealed in the 1881 Annual Report of the Smithsonian Institution. Charles Fort took note of it in his Book of the Damned, where he pointed out that the letters C, D, and K are turned the wrong way in the inscription and, further, that the crosses, having equal arms, are not conventional crucifixes. (Pontolillo, James; "The Silver Indian Crosses of Murray County, Georgia," INFO Journal, no. 63, p. 26, June 1991.)

SOUTH AMERICA

Once again, as in North America, we find in South America evidence of Precolumbian contacts. The evidence is not nearly as compelling as it is in North America, but Science Frontiers has duly recorded claims for Phoenician and even Viking signs below the New World Equator.

Still mysterious as to their purpose, even after decades of archeological research, are the famed Nazca Lines of Peru. To these must now be added the so-called "geoglyphs" located some 1000 kilometers north of Nazca. The geoglyphs resemble the Nazca Lines but are said to be "less abstract" though just as impressive.

At the end of this section on South America, we append a general discussion of human diffusion in the Americas and Precolumbian contacts. This item applies to many of the examples of American engineering structures, small artifacts, epigraphy, and arts.

VIKINGS IN SOUTH AMERICA?

The American archeological establishment admits that the Vikings made it as far as Greenland and probably had a settlement in northeastern Canada at L'Anse aux Meadows; but the Kensington Stone, the Newport Tower, Oklahoma runes, etc., and other evidence of further penetration into the New World are viewed with approbation, even contempt. Nevertheless, the latest number of the Belgian journal Kadath is devoted entirely to Viking (hyperboréenne) contacts in South America! Now that's a far piece from Greenland.

This long article (40 pages) is replete with photographs, interpretations, and translations of runic inscriptions found in Argentina, Brazil, and Paraguay. It is impossible to do justice to this mass of inscriptions here, but we will reproduce one of the figures below. (de Mahieu, Jacques; "Corpus des Inscriptions Runiques d'Amerique du Sud," Kadath, no. 68, p. 11, 1988.)

Comment. To American anomalists, the frustrating part of this whole business is the need to go to foreign-language journals to escape the prison of archeological orthodoxy. South American runes are rarely mentioned in American archeological journals, and we doubt that the treasure house of material just presented in Kadath will make much of an impression on this side of the Atlantic. Why risk one's career for a few scratches on South American rocks?

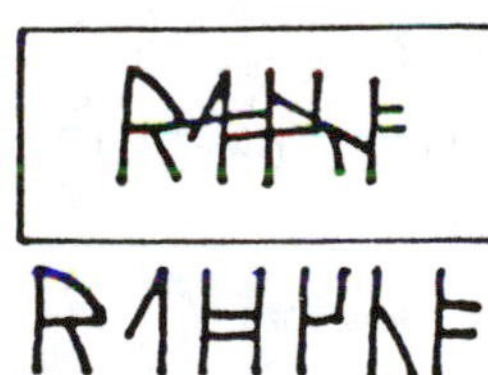

(Top) Runes on the "coiffure" of a statue from San Agustin, Columbia. (Bottom) Runes found on a Nazca urn in Peru, followed by their "normalization."

FANTASTIC CLAIM BY EXPLORER

> An American explorer said yesterday that he had found three ancient stone tablets in Peru's highland jungle that may prove that the area was King Solomon's legendary gold mine.

After being startled by this introductory paragraph, it is anticlimatic to discover that the explorer is G. Savoy. He states that he has found three tablets, each weighing several tons, measuring about 5 x 10 feet. The site is in a cave near Gran Vilaya, in the Peruvian Andes. Engraved on these hefty tablets are inscriptions that appear to be Phoenician or Semitic hieroglyphics. (Anonymous; "Mysterious Tablets Found in the Andes," San Francisco Chronicle, December 7, 1989. Cr. J. Covey)

Comment. Combining this item with the preceding one, there is an implication that advanced Peruvian civilizations may have benefited from contacts with early voyagers from the Old World!

ON CUNA WRITING AND ITS AFFINITIES

When Europeans "officially" reached the New World some 500 years ago, the "official" account states that they found that only the Mayans and Aztecs possessed writing. However, all anomalists recognize that "official" stories often sweep untidy facts under what has come to be an immense rug. One seldom mentioned and rather awkward bump beneath this rug is Cuna writing. The Cuna Indians occupied Panama and some nearby Caribbean islands at the Time of Contact. That the Cuna carved symbols of sorts on wooden boards and scribbled with natural pigments on bark cloth and paper is generally admitted, but this is not considered in the same league as Mayan writing.

Cuna writing is ideographic. Today's average Cuna Indian can usually identify each ideogram as a bird, plant, or some other object. However, to those skilled in Cuna writing, each ideogram actually represents a phrase of about 8-10 words. The symbols thus have mnemonic value. Each wooden tablet is actually read from the lower right corner to the left. The next line up reads left to right, in so-called "boustrophedon" style. The tablets are usually songs for healing, histories, etc.

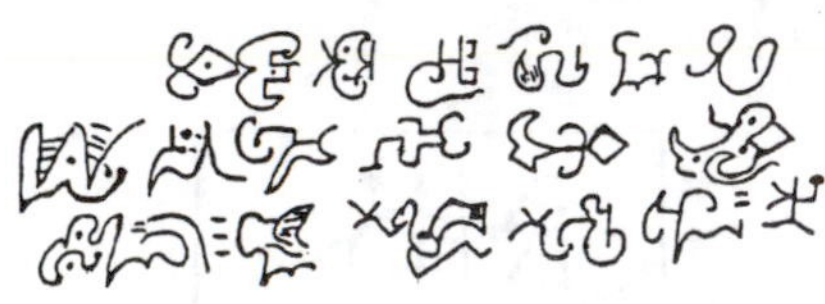

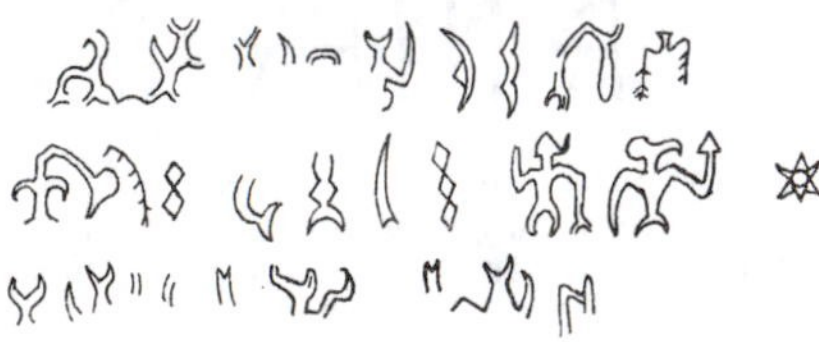

(Top) A sample of Cuna "writing." (Bottom) "Writing" on an Easter Island "talking board."

In these features and general appearance, Cuna writing resembles the "writing" found on the "talking boards" of Easter Island, which in turn seem to have affinities with the ancient script of the Indus Valley in India. To a diffusionist, these affinities or similarities can only mean that pre-Columbian contacts may have occurred between ancient India, Easter Island, and Panama! Such precocious voyages are not considered possible by mainstream archeologists. (Carter, George F., and Case, James; "On Cuna Writing," Epigraphic Society, Occasional Papers, 20:232, 1991.)

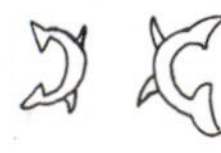

EXPLAINING THE NAZCA LINES

The Nazca lines of Peru have been one of archeology's more enduring anomalies. Books have been written about them; theories abound; what more can be said?

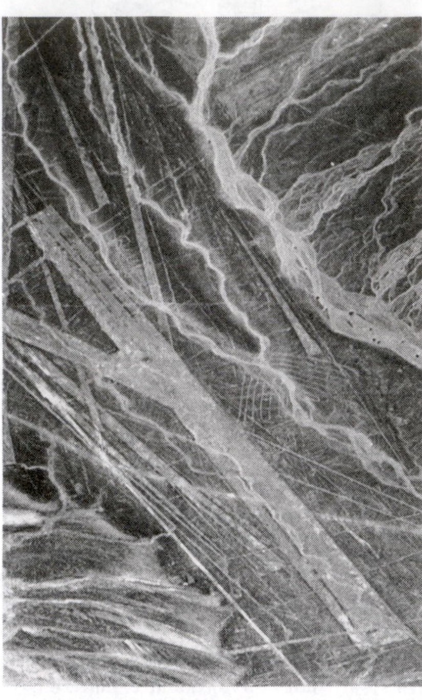

Aerial photograph of some of the Nazca Lines.

A.F. Aveni's recent article in Archaeology sets the Nazca lines in perspective and adds some new observations. First, Aveni deflates their mystery a bit. You do not need to be in an airplane to appreciate the lines; most can be viewed from ground level, even better, from nearby foothills. Although there are some 1300 kilometers of lines and about 300 geometric figures, their construction did not really involve much labor or special engineering skills. Even so, the Nazca lines are remarkable, and we really do not know for certain why they were etched in the Peruvian pampa.

In his research on the Nazca lines, Aveni noted early their strong similarity to the ceque system of 41 imaginary straight lines radiating outwards from the Inca's Temple of the Sun, at Cuzco ---the "navel" of the Inca universe, ".. the ceque system was a highly ordered hierarchical cosmographical map, a mnemonic scheme that incorporated virtually all important matters connected with the Inca world view." Could the Nazca lines have been a forerunner of the ceque system? Aveni also noticed that the Nazca lines and geometrical figures were closely related to watercourses. Also, many of the lines definitely functioned as footpaths. It was also apparent that the animal figures, which were laid down much earlier than the line systems, were not related conceptually to the line scheme.

Aveni concludes, "...whatever the final answer may be to the mystery of the Nazca lines, this much is certain; the pampa is not a confused and meaningless maze of lines, and it was no more intended to be viewed from the air than an Iowa wheatfield. The lines and line centers give evidence of a great deal of order, and the well-entrenched concept of radiality offers affinities between the ceque system of Cuzco and the lines on the pampa. All the clues point to a ritual scheme involving water, irrigation and planting; but as we might expect of these ancient cultures, elements of astronomy and calendar were also evident." (Aveni, Anthony F.; "The Nazca Lines: Patterns in the Desert," Archaeology, 39:33, August 1986.)

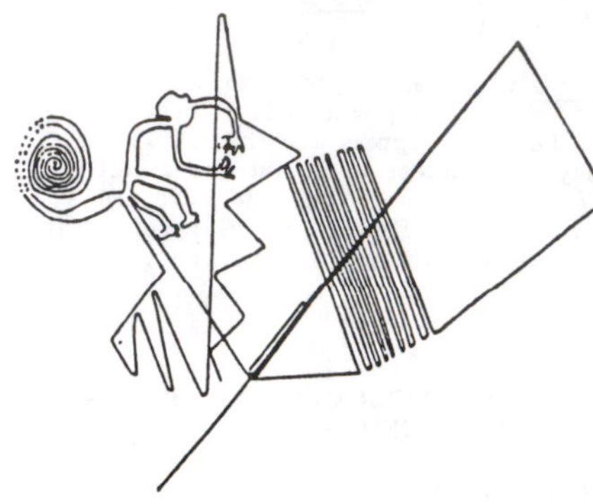

Monkey effigy and abstract patterns in the Nazca lines.

NAZCA FIGURES DUPLICATED

Sensational literature makes much of the huge geometrical figures and outlines of animals etched on high Peruvian tablelands. Most of these figures and lines are so large that one can fully appreciate them only from the air. This fact has led some writers to invoke ancient astronauts and even ancient Peruvian balloon technology. The fact of the matter is that archeological evidence is pretty convincing that the Nazca Indians constructed the lines about 500 AD. But did they have extraterrestrial help or use some advanced technology? The author of this article, with the help of a few friends answered these questions quite convincingly---in the negative---in the summer of 1982. Using sticks and string, they scaled up a small copy of the famous Nazca "condor" into a 440-foot-long replica of the real thing on a Kentucky landfill. The scaling up involved no high technology; in fact, even angular measurements were avoided. Instead of removing stones to make the lines, as the Nazcans did, lime was applied. The whole figure was laid out

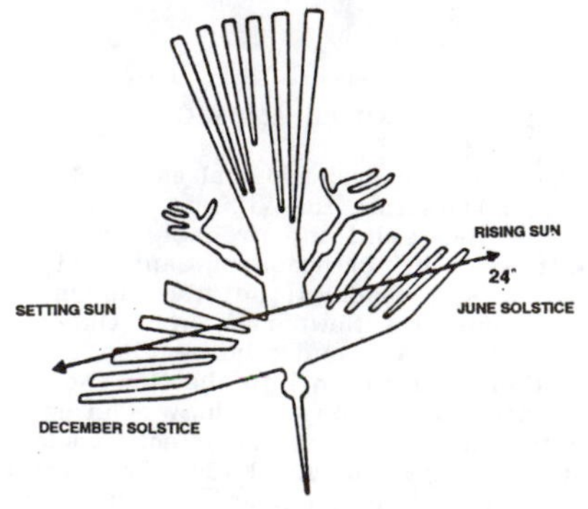

A huge Nazca bird figure.

in just a few manhours. Then, from an aircraft at 1000 feet, photos were taken. Lo and behold, a very convincing replica of the Nazca condor appeared on the prints. Still unanswered, though, is why anyone would want to make such huge drawings. (Nickell, Joe; "The Nazca Drawings Revisited: Creation of a Full-Sized Duplicate," Skeptical Inquirer, 7:36, Spring 1983.)

Comment. While some of the animal figures and arrays of lines may have had ritual or archaeoastronomical purposes, others seem too complex for such simple explanations.

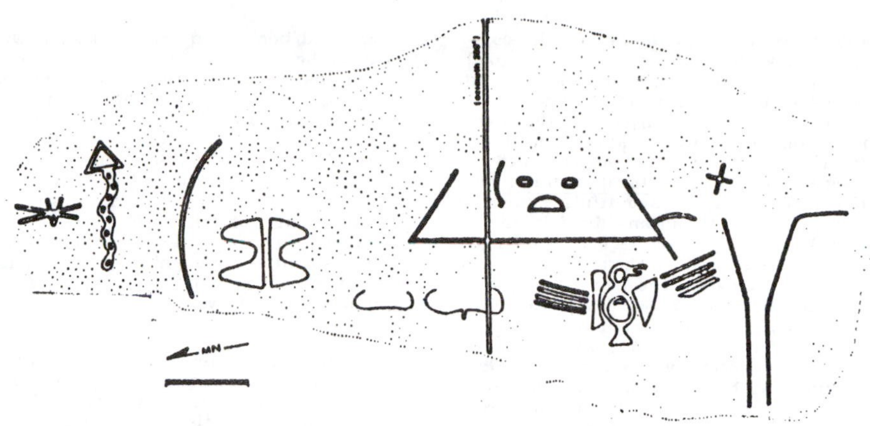

Plan view of some Santa Valley geoglyphs from Peru. The scale at the lower left represents 10 meters. The vertical double line is not a geoglyph but the azimuth heading of 280°. It is difficult to imagine how such huge figures could be used for ritual purposes.

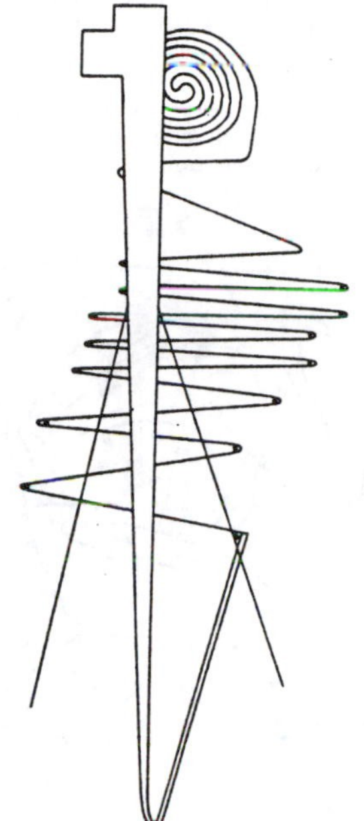

One of the more puzzling abstract Nazca figures.

PERUVIAN GEOGLYPHS

Just about everyone has seen the very impressive aerial photographs of the famous Nazca lines in Peru. Not as well known are the many other "geoglyphs" found elsewhere in South America.

> For example, several areas that contain crisscrossing lines and figures similar to those of Nazca recently have been studied on the central coast between the Fortaleza-Pativilca and Rimac Valleys. Additional lines have been reported for Viru Valley, on the north coast, and for Zana Valley, over 1000 km to the north of Nazca. Interestingly, most coastal ground drawings that can be dated tentatively, either by associated ceramic remains and sites or by their similarity to diagnostic pottery motifs, fall in the earlier part of the Early Intermediate period (ca. 350 BC to AD 650)---i.e., to a time following the establishment of irrigation agriculture as the primary subsistence focus, but prior to the rise of state societies.
>
> .....
>
> As is well known several studies have been conducted that involved mapping and computer analysis of the Nazca lines to examine the hypothesis that they were related to astronomical phenomena. This theory is now discounted, at least as it applies to the great majority of the lines which do not appear to have been oriented toward the sky. More recent studies dealing with the lines have provided convincing arguments, primarily through comparison with lines currently in use on the Bolivian altiplano, that they were constructed as part of elaborate ritual ceremonies related to agricultural fertility.
>
> (Wilson, David J.; "Desert Ground Drawings in the Lower Santa Valley, North Coast of Peru," American Antiquity, 53:794, 1988.)

Comment. Not surprisingly, the author of this article, D.J. Wilson, has neglected the popular theory that the Nazca lines were etched to attract and guide extraterrestrial visitors. Mainstream archeology always opts for explanations involving rituals and society-unifying activities. But why make geoglyphs that cover hundreds of square miles which can hardly be appreciated at all from the ground?

Getting back to the article, Wilson continues in some detail with the Santa Valley geoglyphs. These differ from the classical Nazca lines in that they display fewer abstract designs and more representations of llamas, condors, and other animals. They are, however, of impressive size and well-executed. A small sample is illustrated here.

MARCAHUASI: A MYSTERY IN STONE

In the Louisiana Mounds Society Newsletter of October 1, 1991, J. Hunt publishes a letter from B. Cote that tells briefly of an eerie Peruvian site:

> In June of 1989, a group of us traveled to Peru and visited a 12,500-foot plateau called Marcahuasi. We spent only one night there, but what we saw was so exciting that we decided to go back and make a film of it. The entire plateau seems to be populated with hundreds of figures carved out ot stone, some of them 90 feet tall. Yet this unique spot is relatively unknown to the outside world.
>
> What little is written about Marcahuasi indicates a certain reluctance on the part of archaeologists to say that the figures are man-made. Indeed, many of them are subtle and not always obvious to the viewer. But that is precisely what contributes to the mystery. There are so many recognizable forms there, that one is tempted to say they must be man-made, or else nature is having a great joke on us.
>
> Daniel Ruzo, a 90-year-old archaeologist who lives near Mexico City, aided us. The figures we saw and filmed in 1989 were both strange and fascinating. We were first greeted by a 60-foot rock called by Ruzo The Monument to Humanity because several different races are recognizable on it. They overlap each other in a unique way, but one can clearly discern a Caucasian youth, a Semitic man, a skull-like face that could be Negroid, and several others.
>
> There are many other faces on the plateau, as well as animals. Some of the animals depicted never existed on the continent, such as the rhino, lion, camel, and a turtle-like creature.

(Cote, Bill; "Marcahuasi--A Mystery in Stone," Louisiana Mounds Society Newsletter, no. 42, p. 1, October 1, 1991.)

Comment. The photographs we have seen of the Marcahuasi site are not especially persuasive.

HUMAN DIFFUSION AND PRECOLUMBIAN CONTACTS IN THE AMERICAS: THE NATURE OF THE DEBATE

Above we have presented considerable evidence for the existence of man in the Americas well before 12,000 years ago--the "acceptable" limit. For example, we mentioned the 300,000 BP site in Brazil. There are many more. Of course, controversy hangs over all these sites and the dates assigned to them. The controversies about the specifics are good; but now the archeological establishment seems to be trying to enforce the 12,000-year dogma through authoritarian pronouncements in key publications. By way of illustration, we have P.S. Martin's article in Natural History, entitled "Clovisia the Beautiful!", bearing the subtitle, "If humans lived in the New World more than 12,000 years ago, there'd be no secret about it." Now some archeologists are even trying to roll forward the 12,000-year date. See, for example, R. Lewin's review in Science (referenced below), which is subtitled: "In recent years anthropological opinion has been shifting in favor of a relatively recent date (not much more than 11,500 years ago) for the first human colonization of the Americas." In all of these articles, anomalous data are simply labelled "erroneous". (Martin, Paul S.; "Clovisia the Beautiful!" Natural History, 96:10, October 1987; and Lewin, Roger; "The First Americans Are Getting Younger," Science, 238: 1230, 1987.)

The practical effect of this whole business is that a discipline of "shadow archeology" is forming outside the establishment. In this relatively undisciplined and unrefereed environment, we find books and reports loaded with anomalies, dealing with not only early humans in the Americas, but pre-Viking European contacts, expeditions to the Americas from the Orient, ancient pyramids in Australia, etc. As a matter of fact, all scientific disciplines are paralleled by "shadow disciplines", which are "staffed" by amateurs and mavericks. But enough of this musing. The consequences are now being recognized by a few scientists and philosophers. A long article in a recent issue of Nature provides some pithy, pertinent comments:

> The current predicament of British science is but one consequence of a deep and widespread malaise. In response, scientists must reassert the pre-eminence of the concepts of objectivity and truth.
>
> .....
>
> By denying truth and reality science is reduced to a pointless, if entertaining, game; a meaningless, if exacting, exercise; and a destinationless, if enjoyable, journey.

(Theocharis, T., and Psimopoulos, M.; "Where Science Has Gone Wrong," Nature, 329:595, 1987.)

AFRICA

AFRICAN OGAM!

G. Carter describes a tablet in the possession of a South African Zulu, which pictures a giraffe and a zebra along with inscriptions in Egyptian, Arabic, and Ogam! Carter continues:

> I will put below a picture of a giraffe with Ogam alongside. The Ogam letters are RZRF. Add vowels and this becomes: Rai Za Ra Fa; old Arabic for 'behold the giraffe.' Alongside a zebra figure one finds Ogam letters ZBDB, which in Arabic reads 'painted ass.' These animal figures are usually considered to date to the Upper Paleolithic. Apparently Arabic speakers added the inscriptions much later. But when, pray, did the Arabs write in Ogam?

(Carter, George F.; "Before Columbus," NEARA Journal, 22:61, 1988.)

A Zulu tablet with apparent ogam writing.

BIRDS OF BURDEN

Anthropologists long ago decided that the ostrich was domesticated only in historical times. They pooh-poohed a prehistory sketch showing an ostrich carrying a human rider and pictographs of ostriches apparently fitted with pack saddles. The latest discovery may change their minds. It is a Neolithic figure (5000-7000 years old), deeply engraved on rocks along the River Blaka, in Niger, Africa. Here, the ostrich definitely appears to be loaded with cargo that is strapped on. The bird's legs are folded in a resting position.

The Egyptians occasionally captured young ostriches and broke them to harness, but this engraving seems to prove that this practice had been going on long before. (Bahn, Paul; "A Head in the Sands of Time," Nature, 346:794, 1990.)

Comment. One wonders what Neolithic goods the ostrich caravans carried and where they were bound.

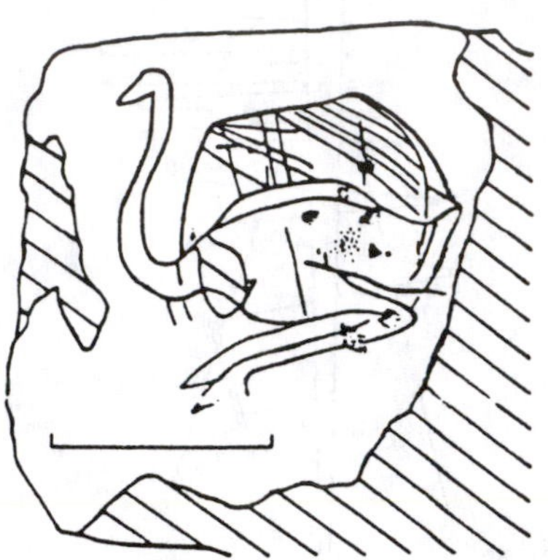

A Neolithic rock figure from Nigeria depicting what seems to be an ostrich laden with cargo.

EUROPE

Turning now to European graphic artifacts, Science Frontiers has garnered three classes of material; some of which is anomalous, some merely curious:

- **Three-dimensional symbols. Clay artifacts, painted pebbles, sculptured stone spheres.**
- **Calendars and maps. Suprisingly sophisticated astronomical devices and old maps suggesting, but hardly proving, ancient Antarctic exploration!**
- **Art anomalies. Very ancient sculptured stones and possible Precolumbian drawings of New World turkeys in Europe (reverse diffusion).**

THREE-DIMENSIONAL SYMBOLS

FIRST WRITING MAY HAVE BEEN THREE-DIMENSIONAL

Archeologists have long been puzzled by large numbers of small, fired-clay objects found in the Middle East. Denise Schmandt-Besserat, University of Texas at Austin, believes that these small geometrical shapes (cones, spheres, disks, etc.) were actually symbols used in commerce to indicate numbers and types of commodities (sheep, oil, and so on). Generally less than an inch in size, the clay objects were apparently sealed in hollow clay spheres to make bills of lading as early as 8500 BC. This is 5000 years before two-dimensional clay tablets were introduced. (Anonymous; "From Reckoning to Writing," Scientific American. 237:58, August 1977.)

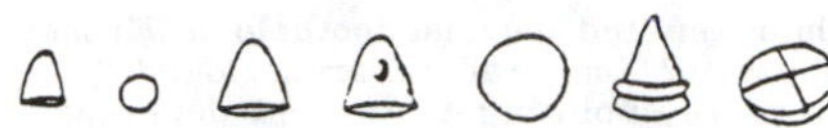

Typical clay symbols used in ancient business transactions.

THE AZILIAN PEBBLES

Although P.G. Bahn's article is primarily about how to spot fake Azilian pebbles, he also provides some fascinating facts about the real ones. Genuine Azilian pebbles seem to have been carefully selected for size and shape. Most are oval or oblong and may be up to 8 cm long, 3 cm wide, and 1 cm thick. Over a thousand have been found around Europe. They have been considered primitive art, markers for prehistoric games, and possibly early forms of notation. Claude Couraude has analyzed the markings and concluded that particular combinations of signs recur frequently; the same for certain numbers of dots. There are 16 simple types of signs (dots and lines) but only 41 of the many possible combinations have been found. Evidence supports the existence of some sort of 'syntax'. Couraud speculates that some sort of cyclic notation is involved, perhaps lunar in nature, like some of the markings found on bones from the same general period. (Bahn, Paul G.; "How To Spot a Fake Azilian Pebble," Nature, 308:229, 1984.)

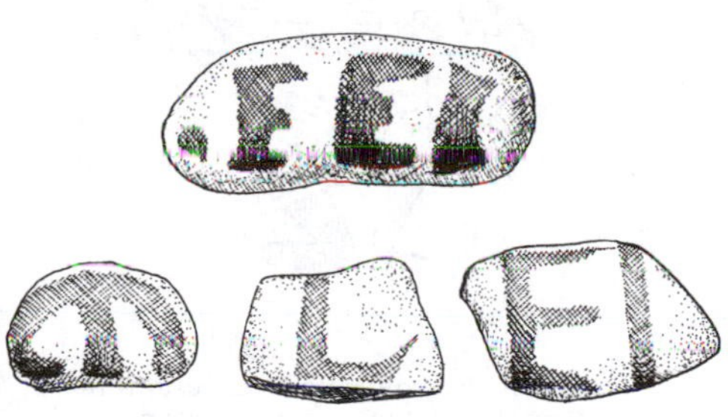

Typical Azilian painted pebbles.

HARDBALL FOR KEEPS!

"Archeologists call them 'balls' for want of a better word; but after several centuries of intensive collection, scrutiny, and study, nobody really knows what they are.

"Imagine, if you will, a spherical piece of carved rock a little smaller than a baseball. The shape bespeaks artifice. Something---somebody---made it.

"More than 500 of these objects have been found in Great Britain and Ireland most of them in Scotland, near prehistoric dwelling places, passage graves and the mysterious rings of standing stones whose specific purpose also eludes the experts."

Archeologists believe the balls are more than 4000 years old. All are different; all are symmetrical with projecting knobs, six in most cases.

So much for the basic data. Now let us progress (?) to theory. D.B. Wilson suggests that the balls were really hand-thrown missiles used in bloody games played at standing-stone sites during astronomically decreed rites. (Remember the Mayans had their grisly ballgames, too!) The stone balls are indeed perfectly weighted, shaped, and textured for throwing at the heads of opposing players. Perhaps, says Wilson, the games had rules such that you were safe when touching a standing stone, but to score you had to run to another stone while fair game for the first IPMs (Interpersonal Missiles). And so on and so on. You now get the gist of this clever little piece. (Wilson, David B.; "Hardball for Keeps," Boston Globe, October 12, 1986.)

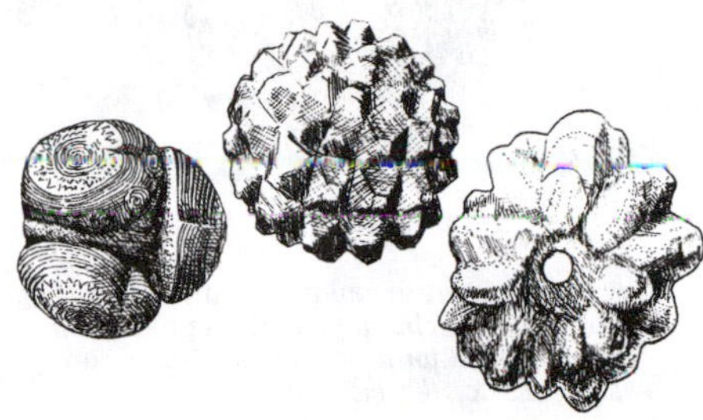

Three of the enigmatic carved stone balls found at some megalithic sites.

CALENDARS AND MAPS

A GOLDEN CALENDAR FOR USE AT STONEHENGE?

"The lozenge, of 0.5-mm beaten gold, was excavated in 1808 AD from the Bush Barrow, 1 km from Stonehenge.

"Until now, it has been assumed that the plaque was only decorative. After examination and measurement, the patterns of its carefully inscribed markings are believed to be identifiable as a calendar fashioned for use at Stonehenge. Found over the breast of a skeleton of a tall man, its symmetrical shape and correct corner angles make it appear probable that the plaque had something to do with the four cardinal points and solstitial sunrises and sunsets.

"By fixing the flat lozenge on a table at eye level and orientating it with its shorter diagonal on the meridian, an observer could use an alidade while watching sunrise or sunset throughout the year. Were the bronze rivets, found nearby, the remains of the alidade? Markings exist on the plaque which indicate that the 16-month calendar was in use. Guide lines exist for inserting the intercalary leap day. Eight additional lines can be identified as indicating moonrise and moonset at the equinoxes' standstills. Using actual horizon altitudes at Stonehenge and azimuths shown by the lozenge, calculation shows that the average discrepancy of the solar lines is 0.36 days and that it was made about 1600 BC.

"Was this ceremonially buried gold artifact a copy of a more robust working calendar, or was it the original master copy?

"The lozenge was a means whereby observed angular measurements could be recorded and subsequently retrieved years later without recourse to writing. It was essentially a textbook for making the calendar, a reference encyclopedia." (Thom, A.S.; "The Bush Barrow Gold Lozenge: Is It a Solar and Lunar Calendar for Stonehenge?" Louisiana Mounds Society Newsletter, no. 37, February 14, 1991.)

THOSE OLD MAPS OF ANTARCTICA

Was Antarctica nearly ice-free within the last few thousand years? Did the old navigators sail into these now-frigid waters and map this great southern continent? One way to answer such questions is by turning to old maps of this part of the world and asking the geophysicists if most of the continent's ice cover could have disappeared fairly recently.

C.P. Hapgood, well-known for his book, Maps of the Ancient Sea Kings, believed that ancient mariners did indeed map Antarctica when those climes were warmer. More recently, J.G. Weihaupt came to similar conclusions by a different route, which included geophysical considerations. (See pp. 203 and 220.) Now, D.C. Jolly has put the whole problem in perspective in an excellent review.

Jolly has studied the data in depth, as indicated by his 51 sources. In his

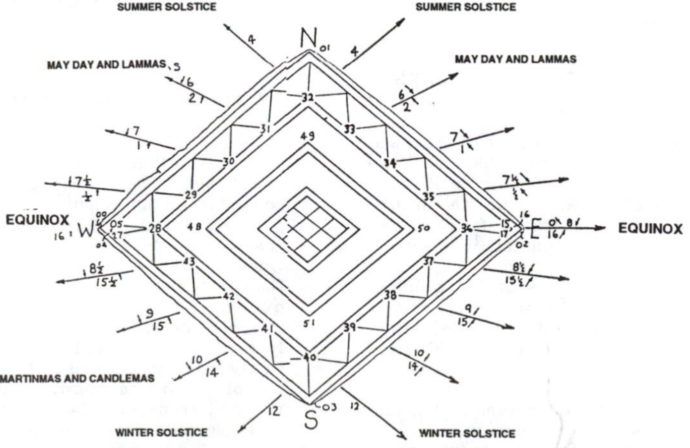

Markings on a gold lozenge excavated near Stonehenge. Some interpret the lines as indicators of solar and lunar positions on astronomically significant days. If so, this lozenge represents surprising astronomical sophistication 3,000 years ago.

view, it boils down to the fact that the old maps, which everyone uses, are often incomplete and ambiguous. One can read a lot into them. To claim an ice-free Antarctica, one has only to make a few assumptions. For example, one reduces the size of a feature here, and rotates another there. Those old map makers didn't get things quite right! Jolly is fair about the whole business and even admits admiration for Hapgood. His conclusion (in part):

"Our knowledge of early cartography is limited, since much of the material from the sixteenth century is now lost. While this affords ample opportunity for speculation, there have been many scholarly studies of this period. These studies were not done by dunces, but by individuals who spent years acquiring the skills and perspective necessary to interpret the evidence. Professor Hapgood to his credit, spent almost ten years studying the evidence and consulting experts in the field. His ideas were rejected in scholarly circles not because of animus but because had had not proved his case. Too many leaps of faith were needed to establish his thesis. I fear it is impossible to be equally charitable toward some later advocates of the Hapgood thesis, whose methods do little credit to his memory." (Jolly, David C.; "Was Antarctica Mapped by the Ancients?" Skeptical Inquirer, 11:32, 1986.)

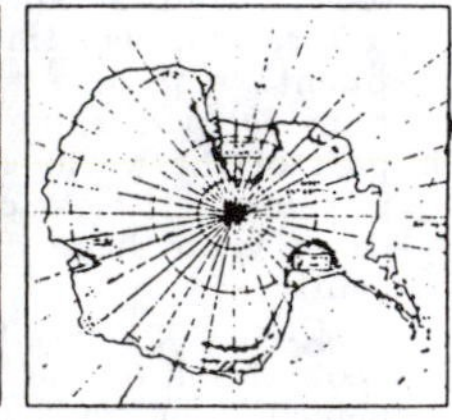

"Antarctica is in the eye of the beholder. Do these two maps show the same thing? Note that the Finaeus map (1531, left) would be much larger than the modern map (right) if the scales were equal." (D.C. Jolly)

ART ANOMALIES

PLIOCENE SCULPTURES OR FREAKS OF NATURE?

We have at hand Number 20 of Archaeologische Berichten, 1990, 108 pp. This thick booklet bears the subtitle Picture Book of the Stone Age. And a picture book it is, with hundreds of drawings interspersed with 26 pasted-in color photos. The text is English.

What do these photos and drawings show? Basically, they portray stones and pebbles picked out of gravel pits and similar accumulations of rocky debris that look like human heads, ape heads, primitive tools, etc. Some of the pebbles do indeed resemble human-made artifacts. (See accompanying sketch.) Most, though, require some imagination. The import of these artifacts, if that is what they really are, comes home when one learns that they come from deposits that are millions of years old! If any of these pebbles are really human-made, anthropology will

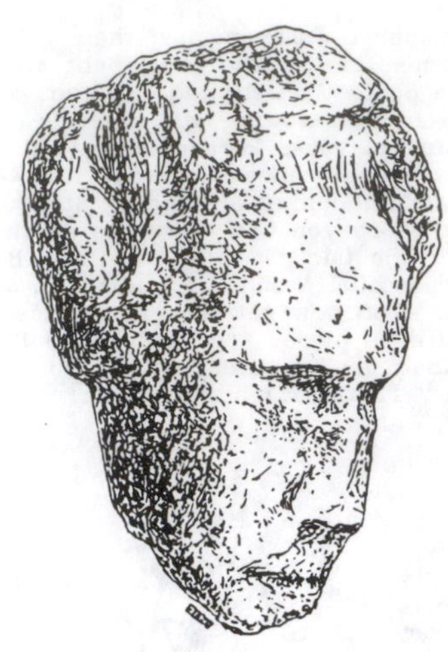

Remarkable human-like features on a stone found in a gravel deposit supposedly millions of years old. Head size: 12 x 9½ cm.

be stood on its head.

Since the present report is the 20th in a series, one can assume that the contributors to Archaeologische Berichten have amassed incredibly large collections of ancient stones and pebbles that "look like" artifacts.

Comment. Similar purported artifacts have been found at the controversial Calico Hills site, California, in Pleistocene deposits that may be 100,000 years old. And don't forget that "Face on Mars"!

TALKING TURKEYS?

Did the Vikings ship American turkeys back to Europe circa 1010 AD from their reputed colonial foothold in Massachusetts? Some radical archeologists think so, pointing to two old depictions of turkey-like birds from Precolumbian Europe. The upper figure was painted on a wall in Schleswig Cathedral about 1280. The lower sketch is from the Bayeux Tapestry, which dates back to 1066-1077. (Anonymous; "Talking Turkey," Fortean Times, no. 61, p. 27, February-March 1992.)

Comment. The Bayeux Tapestry turkey, in particular, is questionable.

Supposed Precolumbian representations of New World turkeys in Europe.

AUSTRALIA AND OCEANIA

AN ANCIENT EGYPTIAN SHIP IN AUSTRALIA?

The illustration below was taken from a newspaper, The Australian. It is a computer-enhancement of a badly faded painting found on Booby Island off the coast of Australia. R. Coleman, the Queensland Museum's curator of Maritime History and Archaeology, was quoted in the paper as saying: "Using this technique we are able to selectively neutralize confusing background virtually making the original image pop out of the background...this system... will add tremendously to our knowledge of those cultures prior to European settlement." (Anonymous; "An Ancient Egyptian Ship in Australia?" Epigraphic Society, Occasional Papers, 19:211, 1990.)

Comment. The vessel in the sketch does seem to have Egyptian lines. However, as our friends in Australia often remind us, we must be wary of what we read in Australian papers. See p. 50 for a drawing of an supposed Egyptian scarab dug up in Australia!

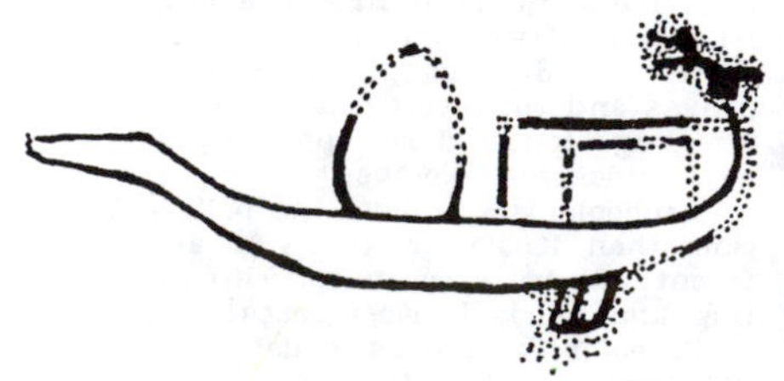

A computer-enhanced rock painting from Australia claimed to be that of an ancient Egyptian vessel.

ANCIENT EGYPTIANS IN HAWAII

Most people tend to think of the ancient Egyptians as stay-at-homes who were too busy building pyramids to explore far lands. But many artifacts from the South Pacific and even Hawaii hint that they were otherwise. Some Hawaiian rock carvings include well-known Egyptian motifs and even a few hieroglyphics. The three main sites are: (1) the great boulders at Luahiwa, Lanai; (2) the old landing at Anaehoomalu; and (3) at Kii, Kauai.

The evidence for an Egyptian presence is even stronger in New Guinea, where the Egyptians may have had a gold-mining colony. Other ancient cultures also frequented New Guinea, where Sumerian beads and bronze weapons have been found by Australian archeologists. Further, there seems to have been a thriving market in the ancient Middle East for bird-of-paradise skins, which could only have come from New Guinea. (Knudsen, Ruth; "Egyptian Signs in the Hawaiian Islands," Epigraphic Society, Occasional Papers, 12:190, 1984.)

Bighorn sheep portrayed on the walls of Nine-Mile Canyon, Utah. Do echoes from this rock wall sound like the clatter of hooves?

WORLDWIDE

THE DIFFUSION OF SCIENCE IN PREHISTORIC TIMES

In contrast to many archeologists who tend to play down the intelligence of prehistoric man, B.A Frolov insists that these "primitive hunters" constructed surprisingly sophisticated models of the natural world, especially the motions of celestial bodies. Many of these models seem to have been nonutilitarian; that is, built to satisfy intellectual curiosity. Furthermore, some scientific notions were widespread geographically, indicating perhaps long lines of communication. To illustrate, Frolov cites the similar astronomical sophistication revealed by the Lake Onega petroglyphs in Russia and Stonehenge. He also points out that the aborigines of North America, Australia, and Siberia all called the Pleiades the "Seven Sisters." Coincidence is very unlikely here, he says; this and other notions must have existed before Australia and North America were peopled.

The absence of writing, as we know it, would not have deterred ancient man from developing and communicating mathematical and scientific skills and accumulating knowledge, possibly in the form of myth. (Frolov, B.A.; "On Astronomy in the Stone Age," Current Anthropology, 22:585, 1981.)

Comment. A passing thought; may not writing and today's omnipresent computers be crutches that permit our memories and mental skills to deteriorate? There are many recorded cases of remarkable memory and information-processing ability. Such skills could be common but suppressed by technology.

THE ACOUSTICS OF ROCK ART

S. Waller has visited rock art sites in Europe, North America, and Australia. Standing well back from the painted walls, he claps or creates percussion sounds, and records the echos bouncing back. A casual observer might be tempted to call 911. It turns out, though, that rock art seems to be placed intentionally where echos are not only unusually loud but are also related to the pictured subject matter. Where hooved animals are depicted, one easily evokes echos of a running herd. If a person is drawn, the echos of voices seem to emanate from the picture itself!

> At open air sites with paintings, Waller found that echos reverberate on average at a level 8 decibels above the level of the background. At sites without art the average was 3 decibels. In deep caves such as Lascaux and Font-de-Gaume in France, echos in painted chambers produce sound levels of between 23 and 31 decibels. Deep cave walls painted with cats produce sounds from about 1 to 7 decibels. In contrast, surfaces without paint are "totally flat".

What did the ancient artists have against cats? (Dayton, Leigh; "Rock Art Evokes Beastly Echos of the Past," New Scientist, p. 14, November 28, 1992.)

BONES AND FOOTPRINTS

NORTH AMERICA

One particular human skeleton (from Guadeloupe) and one specific group of human-like footprints (from the Paluxy River, Texas) have fuelled considerable debate among some scientists. Both the bones and footprints are said, by some, to be millions of years old. Not even the most ardent of the pre-Clovis archeologists speak in terms of millions of years for the age of modern humans in the New World. Actually, some scientific creationists have seized on the Guadeloupe skeleton and Paluxy footprints as evidence that the whole geological dating scheme is incorrect, and that the human race is actually only a few thousand years old and even coexisted with the dinosaurs. As we shall see, mainstream scientists have some rather convincing responses to these claims.

HUSHING UP THE GUADELOUPE SKELETONS

Background. American archeologists vigorously insist that the first humans to reach the New World arrived via the Bering Land Bridge some 12,000 years ago. Therefore, the discovery of human skeletons in rock that might be 25 million years old is cause for alarm to mainstream scientists.

Just offshore of Guadeloupe, in the West Indies, lies a kilometer-long formation of extremely hard limestone dated as Miocene, or about 25-million years old. Nothing surprising so far! However, history records that in the late 1700s, many human skeletons---all indistinguishable from

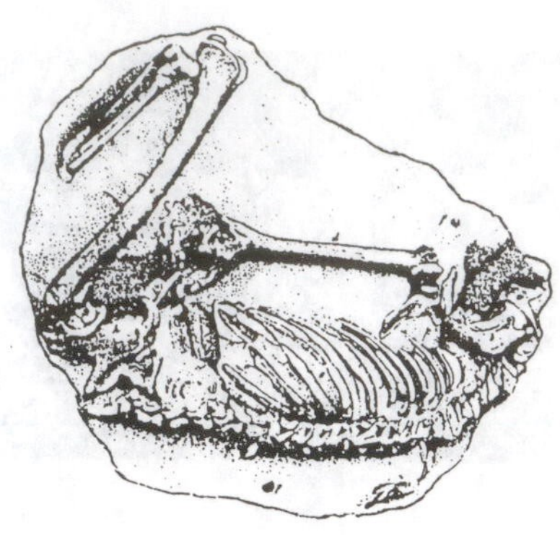

Sketch of a fossilized human skeleton from Guadeloupe.

modern man---were excavated from this limestone. One of the quarried specimens, ensconced in a 2-ton slab, was shipped to the British Museum. It arrived in 1812 and was placed on public display. With the ascendance of Darwinism, the fossil skeleton was quietly spirited away to the basement. The discovery of these human remains has been well-documented in the scientific literature.

Here is another pertinent geological fact: the limestone formation in question is situated 2-3 meters below a 1-million-year-old coral reef. If the limestone is truly 25 million years old, the human evolutionary timetable is grossly in error. Even if this is not the case, and the bones are merely 1 million years old or so, as required by the coral reef, fully developed men lived in the New World long before the Bering Land Bridge went into service. The only way a serious geological or archeological anomaly can be avoided is to predicate that the limestone formation was laid down in the last 10-20,000 years---something that doesn't seem too likely. (Cooper, Bill; "Human Fossils from Noah's Flood," Ex Nihilo, 7:6, no. 3, 1985.)

Comment. This sort of dating puzzle is manna to the scientific creationists. It is thus not surprising to discover that Ex Nihilo is published by the Creation Science Foundation of Australia. Nevertheless, the Guadeloupe skeletons truly exist---it's just that the creationists seem to be the only ones talking about them.

THE GUADELOUPE SKELETONS REVISITED

The forgoing article describes the discovery of the famous Guadeloupe skeleton in limestone that seemed very ancient indeed from all indications. The article also states that the British Museum suppressed discussion of this paradigm-shifting discovery by hiding it away somewhere.

It seems that the skeleton was actually never hidden and, in fact, was on public display between 1882 and 1967. The claimed Miocene dating of the skeleton has also been challenged, although no one seems to agree on just how old the bones may really be. The geological facts mentioned above are not discussed at all in the present article. A post-Columbian date was suggested on the basis that implements and a dog skeleton were also found with the Guadeloupe skeleton. The whole business has split the ranks of British scientific creationists. (Howgate, Michael, and Lewis, Alan; "The Case of Miocene Man," New Scientist, p. 44, March 29, 1984.)

Comment. The facts presented in the New Scientist and Ex Nihilo are so discordant that we await further developments with great interest and some amusement. Beach rock forms quite rapidly in some areas; and the skeleton could be very recent. However, the 'facts' presented in Ex Nihilo speak for great antiquity!

BACK TO GUADELOUPE AGAIN

Just how old are those modern-looking human skeletons in those chunks of Guadeloupe limestone? Opposing views are discussed above. The basic problem is the dating of the limestone in which the skeletons are embedded. If the limestone is truly of Miocene age (about 25 million years old), the presence of human skeletons represents a major scientific anomaly, since modern man apparently arrived on the scene only about 5 million years ago. Most scientists say the limestone is only recently formed beach rock a few hundred years old, and that radioactive dating proves this. Doubters have pointed to 3-million-year-old coral reefs stratigraphically above the limestone. In a recent issue of Ex Nihilo, a few more cans of gasoline have been thrown on the fire: (1) The radioactive date usually served up actually came from another island in the area; (2) Beach rock is not now forming at the skeleton site, rather the skeleton limestone is being eroded; (3) The skeleton limestone is harder than marble and not loosely consolidated beach rock; (4) True Miocene limestone does exist in the area; and (5) Geologists have carefully described and mapped the rest of Guadeloupe but have omitted the skeleton site---presumably because of the anomalies involved. (Tyler, David J., et al; Ex Nihilo, 7:41, no. 3, 1985.)

FOSSIL FOOTPRINTS OF MAN OR DINOSAUR?

Background. The prevailing theory of evolution requires that the dinosaurs met their demise about 60 million years before any creatures even resembling modern humans evolved. It is not surprising then to find that the seeming appearance of human and dinosaurs tracks in the same geological sediments is hotly challenged.

BACKTRACKING ALONG THE PALUXY: OR IS THERE A DEEPER MYSTERY?

Ostensibly, the facts are as follows. Several series of tracks in the sedimentary rocks along the Paluxy River, in Texas, which many creationists have considered to be of human origin, have recently changed appearance, apparently due to erosion.

"Due to an unknown cause, certain of the prints once labeled human are taking on a completely different character. The prints in the trail which I have called the 'Taylor Trail,' consisting of numerous readily visible impressions in a left-right sequence, have changed into what appear to be tridactyl (three-toed) prints, evidently of some unidentified dinosaur. The changes in the impressions themselves are mostly confined to lengthening in the downriver direction. The most significant change, however, is that surrounding the toe area. In almost each of the prints in the trail, three large 'toes' have appeared, similar to nearby dinosaur tracks. These toes, typically, are coloration phenomena only with no impressions, in most cases. Frequently the 'mud pushup' surrounding the original elongated track is crossed by this red coloration. The shape of the entire track, including both impression and coloration, is unlike any known dinosaur print."

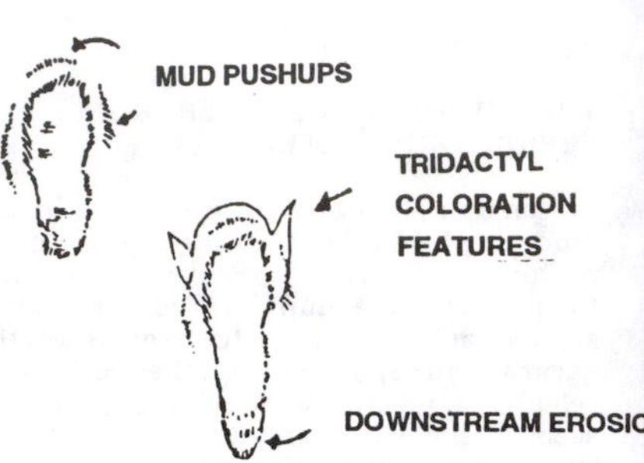

(Left) Original Paluxy footprint, (Right) Present appearance. The coloration continues right across the mud pushup.

J. Morris, the author of this article and a creationist, suggests that creationists no longer use the Paluxy tracks as evidence that humans and dinosaurs once coexisted. But he adds that several mysteries still exist, two of which are: (1) Why is erosion improving the quality of the tracks instead of the reverse? and (2) Why aren't those tracks exposed earlier also acquiring stain marks? Morris also points out that the reddish stains could have easily been added artificially! (Morris, John D.; "The Paluxy River Mystery," ICR Impact Series, no. 151, January 1986.)

Comment. Fraud is certainly implied by the suggestion of artificiality. Would the enemies of creationism go out at night with a brush to destroy a key datum supporting creationism? Of course, we don't know. But the whole business bears a fascinating resemblance to the case of Paul Kammerer, who in the 1920s reported lab data supporting the inheritance of acquired characteristics. Evolutionists were aghast and attacked Kammerer vigorously. Then one scientist discovered that some of Kammerer's specimens had been deliberately tampered with. Kammerer eventually committed suicide. However, A. Koestler, in his book The Case of the Midwife Toad, reviewed the affair and suggested that Kammerer may have been framed by someone who surreptitiously injected India ink into the feet of some of Kammerer's specimens. A curious parallelism, is it not? Moral: Theories are more important than facts!?

MORE PALUXY IMPRESSIONS

The response of readers to the preceding item on the Paluxy mingling of dinosaur and purported human footprints was immediate, copious, and sometimes emotional. Even though we regularly survey 100-or-so scientific journals, it seems that the considerable Paluxy field work has never attained those hallowed pages---and probably it never will! Even though the above report was rather negative on the issue of the validity of the claims of the creationists, it evidently wasn't negative enough. We now have some documentation with which to clarify some points.

G.J. Kuban has been in the forefront of Paluxy research for several years. He has submitted a long letter plus the Spring/Summer issue of a publication entitled Origins Research, published by Students for Origins Research. This particular issue contains a lengthy article by Kuban plus a shorter contribution by J. Morris (author of the ICR article digested above).

First, we quote from Kuban's personal communication:

"As is explained in the enclosed Origins Research issue, the tracks never did merit a human interpretation, and presently are not as 'mysterious' as ICR and some other creationist groups would have us believe. Indeed, whereas the geo-chemistry of the colorations is still being studied, the color distinctions are definitely part of the rock material, and show no evidence of a 'painting' hoax. Further, even without the colorations, the other features of the tracks thoroughly refute the human interpretation. The drawings in the ICR Impact article (and reproduced in the Science Frontiers newsletter) are not accurate. These drawings are discussed in depth in the enclosed article; however, I would like to emphasize a few points here. First, although in many cases the color distinctions are not associated with significant depressions in the rock surface (apparently due to an infilling of secondary material), where there are depressions or elevations in the rock surface, the colorations generally conform to the borders of the contours quite naturally, rather than in the unnatural manner shown in the ICR article drawing. Second, most of the tracks do not have rounded mud pushups at the anterior (as is suggested by the ICR drawing), but most show evidence of a more pointed middle digit, as well as lateral dinosaurian digits---with the entire anterior typically showing a widely-splayed 'V' or tridactyl shape. Indeed, the anterior of many of the 'man tracks' always did show shallow impressions in a tridactyl pattern and other dinosaurian features were noticed by researchers as early as 1970, and have been well-documented by myself and others in recent years." (Kuban, Glen J.; personal communication, May 30, 1986; buttressed by "The Taylor Site 'Man Tracks'," Origins Research, 9:1, Spring/Summer 1986.)

As to the implication of fraud mentioned in the first item, J. Morris retracts the charge: "Many have suggested that someone may have stained the surface surrounding some of the human-like tracks to give them a reptilian appearance. However, no evidence of fraud has been found, and some hints of these dinosaur toe stains have now possibly been discerned on photos taken when the prints in question were originally discovered. Furthermore, on the same site, over 100 clear dinosaur tracks have now appeared with similar surface stains, which had never before been discovered. Evolutionists have long falsely accused creationists of fraudulently carving their tracks (with no evidence), but we must not resort to their tactics by similarly charging them with fraud without clear evidence." (Morris, John; "Follow up on the Paluxy Mystery," Origins Research, 9:14, Spring/Summer 1986.)

Comment. Many, but not all, creationists have now ceased using the Paluxy tracks as evidence that humans and dinosaurs were contemporaries. Some creationists in fact maintain that new infrared evidence proves that the tridactyl patterns are only stains and not clawmarks at all. In their view, the tracks are still human. More later when this infrared evidence is published!!

CALIFORNIA SKELETONS NOT SO OLD AFTER ALL

About a decade ago, considerable controversy erupted when some obviously old human skeletons were dated at 37,000, 44,000, and even 70,000 BP by a new dating technique called Aspartic Acid Racemization (AAR). Conventional radioactive dating had put all of these skeletons at well under 10,000 BP and, consequently, well within bounds of the Bering Strait migration hypothesis. The early AAR dates were thus at odds with current archeological thinking and, in addition, very encouraging to those who believed that humans occupied North America long before 10,000 BP. Jeffrey L. Bada, a proponent of AAR dating, now states that the controversial dates were based on calibration skeletons which had been erroneously dated by radioactive methods. It seems that AAR dating requires an accurately dated reference skeleton. With the reference skeleton dates now known to be incorrect, Bada had to recalibrate his AAR dating scheme. The reference skeletons were redated using a more accurate radioactive technique. All of the incredibly ancient skeletons have now been redated by AAR methods using the revised reference skeletons. The 37,000 BP date now becomes a reasonable 5100 ± 2000 BP figure. The AAR dating crisis seems to be over. All anomalies have been expunged. (Bada, Jeffrey L.; "Aspartic Acid Racemization Ages of California Paleoindian Skeletons," American Antiquity, 50:645, 1985.)

AMERICAN PYGMIES

Today's anthropological texts say little about pygmies populating ancient North America, but a century ago, when tiny graves replete with tiny skeletons were discovered in Tennessee, controversy erupted. Were they the bones of pygmies or children of normal-sized tribes? The latter choice was made, and we hear no more on the subject---at least on the standard academic circuits.

But a few reverberations are still detectable elsewhere. V.R. Pilapil, for example, asserts that the disputed Tennessee graves really did contain pygmy remains. Not only that, but he hypothesizes that the pygmies arrived in ancient times from southeast Asia, probably the Philippines, where today's diminutive Aetas live.

To support his case, Pilapil recalls B. Fell's examination of the Tennessee skeletal material. Fell noted that: (1) The skull brain capacity was equivalent to only about 950 cubic centimeters, about the volume of a non-pygmy 7-year-old; (2) The teeth were completely developed and showed severe wear characteristic of mature individuals; and (3) The skulls were brachycephalic with projecting jaws. Fell had, in fact, described skulls very much like those of today's adult Philippine Aetas.

Another line of evidence adduced by Pilapil involved the traditions of British Columbia tribes, which recognized a tribe of very small people called the Et-nane. More significant is the oral history of the Cherokees, which mentions the existence of "little people" in eastern North America. (Pilapil, Virgilio R.; "Was There a Prehistoric Migration of the Philippine Aetas to America?" Epigraphic Society, Occasional Papers, 20:150, 1991.)

REST OF WORLD

SCIENTIFICALLY ACCEPTABLE FOSSIL FOOTPRINTS

Mary Leakey has announced the discovery of fossil footprints made by prehuman ancestors more than 3.5 million years ago at Laetolil, 25 miles southwest of Olduvai Gorge in Tanzania. The hominid apparently walked across slightly moist sand and the prints were soon filled with volcanic ash. The prints were made by feet shorter and wider than those of modern humans; but the big toe definitely points forward and is not splayed as in apes. (Anonymous; "Footprints in the Sands of Time," New Scientist, 77:483, 1978.)

A RUSSIAN PALUXY

Meanwhile, a similar mixture of supposedly human tracks and those of dinosaurs has been discovered in the Soviet Union. The sometimes sensationalistic Moscow News had the following to say.

"This spring, an expedition from the Institute of Geology of the Turkmen SSR Academy of Sciences found over 1,500 tracks left by dinosaurs in the mountains in the south-east of the Republic. Impressions resembling in shape a human footprint were discovered next to the tracks of the prehistoric animals." Professor Kurban Amanniyazov, leader of the expedition elaborated, "We've discovered imprints resembling human footprints, but to date have failed to determine, with any scientific veracity, whom they belong to, after all. Of course, if we could prove that they do belong to a humanoid, then it would create a revolution in the science of man. Humanity would 'grow older' thirty-fold and its history would be at least 150 million years long."(Anonymous; "Tracking Dinosaurs," Moscow News, no 24, p. 10, 1983. Cr. V. Rubtsov)

ANOTHER REMARKABLE SPECIMEN OF ANCIENT MAN

In August 1984, Alan Walker from Johns Hopkins discovered near Lake Turkana, in Kenya, most of the skeleton of a 12-year-old-boy, which is estimated to be 1.6 million years old. Classified as a specimen of Homo erectus, the boy was 5 feet 5 inches tall and would probably have grown to 6 feet at maturity. Until this find, our supposed ancestors were thought to be generally small and puny, but here is a strapping fellow, looking much like a modern human, although his skull and jawbone resemble those of a Neanderthal. (Joyce, Christopher, "Now Pekin Man Turns Up in Kenya," New Scientist, p. 8, October 25, 1984.)

Comment. Many giant and pygmy skeletons have been found in North America. See the handbook Ancient Man.

INTERPROXIMAL GROOVING OF TEETH

The following illustration appeared recently in a respected scientific journal. No, it was not a dentistry journal, nor was it an ad for a new toothpaste. It was Current Anthropology.

It seems that some human and near-human skeletons possess teeth with the peculiar grooves shown above. The age range is huge: 1.84 million years to comparatively modern skeletons 10,000 years old. These teeth are found on several continents. Some archeologists say simply that ancient humans just picked their teeth a lot to remove trapped food particles. But the grooves do not seem to be correlated with dental decay problems. This fact has led to the so-called "cultural" theory, which holds that the picking of teeth was just another bad human habit, probably a sort of stereotyped behavior. (Formicola, Vincenzo; "Interproximal Grooving of Teeth: Additional Evidence and Interpretation," Current Anthropology, 29:663, 1988. Also "Ancient Tooth Grooves: Take Your Pick," Science News, 134:237, 1988.)

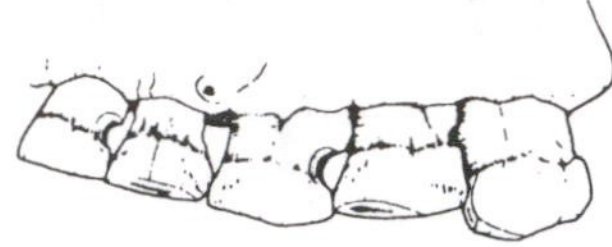

Grooved teeth on ancient skeletons.

Comment. Could toothpicking have been a religious rite? Are there cave drawings showing humans picking their teeth? Well, you can see from the Science News title above that this publication also had fun titling this item. Yet, the geographical and temporal reach of this phenomenon indicates that ancient man was doing something rather strange for reasons we can only guess at.

ANOTHER ANOMALY BITES THE DUST

Some mysterious cultural practice of ancient peoples all over the world resulted in curious grooves on tooth surfaces. The grooves occur near the cemento-enamel junction, mostly on molars and premolars, and usually on males. The diameter of the channel between adjacent teeth varies from 1-4 millimeters.

Proposed solutions to this riddle range from bacterial attack to gritty saliva propelled through the teeth, to the overenthusiastic use of bone toothpicks. But Australian aborigines have provided a more convincing explanation. When the aborigines want thin, strong cords for fashioning spears and spearthrowers, they take a pliable, thinned, kangaroo sinew, pull it down between their molars like dental floss and begin "stripping" it, by pulling it back and forth. They get their thin cords but also grooved teeth. (Eckhardt, Robert B.; "The Solution for Teething Problems," Nature, 345:578, 1990.)

Comment. Unless someone comes up with a fatal objection to this theory, we must de-anomalize the grooved-teeth phenomenon.

EARLY CHINESE VOYAGES TO AUSTRALIA

Dr. Alan Thorne, at the Australian National University, after studying early human fossils from both Australia and China, concludes that there was a significant movement of people from the Chinese coast to north Australia at least 10,000 years ago. He hypothesizes that the Chinese built sea-going rafts of bamboo and explored Indonesia as well as the Australian Coast. (Anonymous; "Chinese 'First to Australia'," Melbourne Sun, August 14, 1982. Cr. G.D. Thompson)

Comment. The China-Australia trip can be simplified by island-hopping, but the existence of a Chinese sea-faring capability may have significance for the Americas.

DIFFUSION AND CULTURE

HUMAN DIFFUSION

Here, we return to the subject of anomalous human diffusion, but in the contexts of languages, physical traits, and plant-and-animal diffusion.

- **African-origin debate. Physical traits and fossils tend to disprove humanity's exclusive African origin---the prevailing paradigm.**
- **Origin-of-Americans debate. Native American languages and blood types indicate multiple origins, contrary to the popular theory.**
- **Origin-of-the-Ainus debate. The testimony of physical traits.**
- **Biological evidence for anomalous diffusion. Plants, animals, etc.**

AFRICAN-ORIGIN DEBATE

AFRICA NOT HUMANITY'S ORIGIN

Background. Current theory has it that modern humans evolved in Africa and spread to the other continents from there. A few lonely dissenters have pointed to other possible Gardens of Eden, especially Southeast Asia. Such suggestions have been emphatically rejected.

Abstract. "It has been suggested that the evolution of man took place in Africa. This suggestion results from the unusual abundance of fossil material in Africa that is quite ancient in comparison with what is known elsewhere. The theory of an African origin has influenced the interpretation of the age of some non-African archaeological sites. A case in point is the 'Ubeidiya locality in Israel, which is generally considered to be about 700,000 yr old because it has been assumed by a few that the associated Early Acheulian tool industry, and the persons who used it, would have taken considerable time to disperse from Olduval Gorge to this non-African site in Israel. Here we evaluate fossil mammals from 'Ubeidiya, which are stratigraphically and directly associated with Early Acheulian artifacts, and find no substantial reason for considering the locality younger than 2Myr, and possibly as much as 500,000 yr older than any record of Early Acheulian artifacts or Homo erectus in Africa." (Repenning, Charles A., and Fejfar, Oldrich; "Evidence for Earlier Date of 'Ubeidiya, Israel, Hominid Site," Nature, 299:344, 1982.)

Comment. In sum, the Israeli site is much older than previously believed possible and "could" indicated a non-African origin of humans.

DO YOU THINK HUMANS ORIGINATED IN AFRICA?

We have heard for so long now that modern humans got their start in Africa. (This assertion is as hackneyed as "Life began in the sea"!) There is, of course, some evidence for this claim: Studies of genetic material and the fossil record are suggestive, although the latter includes the Middle East as a possible birthplace. Other data, however, put the "founding group" of modern humans in Southeast Asia.

The iconoclast here is C.G. Turner, II, an anthropologist at Arizona State University. He has analyzed 28 secondary dental traits (number of roots, bumps, etc.) of 12,000 individuals from around the world---both ancient and modern. Turner believes that the "great web of humanity" originated in Southeast Asia. Since then, two large populations, each recognizable by their dental features, have evolved: (1) northeast Asians and the ancient residents of the Americas; and (2) southeast Asians, Europeans, ancient Australians, and Africans. Also of note is the close resemblance between native Australians and Africans. (Bower, B.; "Asian Human-Origin Theory Gets New Teeth," Science News, 136:100, 1989.)

NOW IT'S GREECE!

The general consensus is that modern humans first emerged in Africa. This assertion was challenged above, where an origin in Southeast Asia was championed. Now it's Greece.

"The immediate ancestors to the human family---the hominids---might have been living in Greece, rather than Africa, some 10 million years ago in the late Miocene, according to the French palaeontologist Luis de Bonis.

"In September 1989, de Bonis and George Koufos of the University of Thessaloniki discovered the fossilized face of an ape-like creature, Ouranopithecus, at a site in the Valley of Rain, 40 kilometres northwest of Thessaloniki. Although the fossil has not yet reached the scientific press, de Bonis has publicly described it as a possible precursor of the earliest known hominid species, Australopithecs afarensis, from Africa 3.5 million years ago." (Lewin, Roger; "Humans May Have Come from Greece, Not Africa," New Scientist, p. 35, January 27, 1990.)

ORIGIN-OF-AMERICANS DEBATE

HOW AND WHEN THE AMERICAS WERE PEOPLED

Background. Once again in this book the Bering Strait dogma is challenged; that is, the theory that the New World was peopled by migrations across the Bering Strait beginning no earlier than 12,000 years ago. Here, the analysis of New World languages disputes this fanatically defended dogma.

Abstract. "A study of aboriginal language distributions supports Knut Fladmark's hypothesis that the initial route of entry of peoples into the New World was along the Pacific coast rather than through the interior ice-free corridor. The greatest diversification of aboriginal languages, as indicated by number of language isolates and major subdivisions of language phyla, is observed on the Pacific Northwest Coast, in California, on the northern Gulf of Mexico Coast, in Middle America, and in South America. Following a conventional principle of historical linguistics, it is assumed that the development of language diversification is proportional to time depth of human occupation of an area. A review of the archeological evidence from the areas of greatest language diversification indicates a time depth of at least 35,000 years for human occupation of most of the Americas." (Gruhn, Ruth; "Linguistic Evidence in Support of the Coastal Route of Earliest Entry into the New World," Man, 23:77,1988. Cr. E. Fegert)

Comment. Did the last sentence of that Abstract say "35,000 years"? Surely this cannot be an American archeological publication. It isn't! Man is produced by the Royal Anthropological Institute, in London. In the States, 12,000 years remains the maximum age of entry of humans into the New World. While the above article focuses on the analysis of languages, many radiocarbon dates greater than 12,000 years are quoted from North and South America.

HOW MANY MIGRATIONS WERE THERE?

One way of determining the directions and strengths of human migrations is through language analysis. People carry words along with them and, even after centuries of modification, traces of their original languages survive. In 1492, an estimated 30-40 million Native Americans spoke more than 1,000 different languages. Can anyone discern patterns in such a hodge-podge? Careful study reveals many similarities. For example, all New World languages can be classified into three groups:

1. The Eskimo-Aleut or Eurasiatic group, which is related to Indo-European, Japanese, Ainu, Korean, and other languages.

2. The Na-Dene family, related to a different set of Old World languages, such as Sino-Tibetan, Basque, (North) Caucasian, and others.

3. The Amerind family. "The origins of the Amerind family are the most baffling, but there are a number of apparent cognates with language families of Africa, Europe, Asia, Australia, and Oceania. For example, the root 'tik', meaning 'finger, one, to point,' is found in Africa, Europe, and Asia, as well as in the Americas. The Amerind words for dog bear a striking resemblance to the Proto-Indo-European word..."

Can the language analysts answer the question in our title above? Based upon the above grouping, they say, "No more than three." (Ruhlin, Merritt; "Voices from the Past," Natural History, 96:6, March 1987.)

Comment. While the people carrying the roots of the Eskimo-Aleut and Na-Dene language groups may well have come across the Bering Land Bridge, those bringing the Amerind languages could have come from just about anywhere!

LONG BEFORE THE VIKINGS AND POLYNESIANS

Scene: Inside Amerind cells. DNA analyses of the mitochondria present in the cells of North American Indian populations indicate that the Eskimo-Aleut and Nadene populations arrived about 7,500 years ago. The more geographically widespread Amerind population, however, seems to be descended from two separate influxes; the first about 30,000 years ago, the second about 10,000 years ago. D. Wallace, from Emory University, surmises that the sharply defined rise of the Clovis culture, conventionally dated from 12,000 years ago, may have resulted from the second Amerind immigration. (Lewin, Roger; "Mitochondria Tell the Tale of Migrations to America," New Scientist, p. 16, February 22, 1992.)

Comment. The 30,000-year date, however, is consistent with MacNeish's discoveries at the Orogrande site. (See p. 23.) Hang in there archeology anomalists, the 12,000-year paradigm is melting in the warm spring sun!

ORIGIN-OF-THE-AINUS DEBATE

THE SAMURAI AND THE AINU

Findings by American anthropologist C. Loring Brace, University of Michigan, will surely be controversial in race-conscious Japan. The eye of the predicted storm will be the Ainu, a "racially different" group of some 18,000 people now living on the northern island of Hokkaido. Pure-blooded Ainu are easy to spot: they have lighter skin, more body hair, and higher-bridged noses than most Japanese. Most Japanese tend to look down on the Ainu.

Brace has studied the skeletons of some 1100 Japanese, Ainu, and other Asian ethnic groups and has concluded that the revered samurai of Japan are actually descendants of the Ainu, not of the Yayoi from whom most modern Japanese are descended. In fact, Brace threw more fuel on the fire with:

> Dr. Brace said this interpretation also explains why the facial features of the Japanese ruling class are so often unlike those of typical modern Japanese. The Ainu-related samurai achieved such power and prestige in medieval Japan that they intermarried with royalty and nobility, passing on Jomon-Ainu blood in the upper classes, while other Japanese were primarily descended from the Yayoi.

The reactions of Japanese scientists have been muted so far. One Japanese anthropologist did say to Brace, "I hope you are wrong."

The Ainu and their origin have always been rather mysterious, with some claiming that the Ainu are really Caucasian or proto-Caucasian--in other words, "white". At present, Brace's study denies this possiblity. (Wilford, John Noble; "Exalted Warriors, Humble Roots," New York Times, June 6, 1989. Cr.J. Covey)

Comment. Fringe anthropology notes many "white" races in strange places; viz., the white Indians of Panama and the Mandans of the American West.

BIOLOGICAL EVIDENCE FOR ANOMALOUS DIFFUSION

MAIZE IN ANCIENT INDIA

Conventional wisdom is clear on two accounts: (1) Maize originated in the New World; and (2) There were no cultural maize-bearing contacts between the New and Old Worlds in the lengthy period between the (hypothetical) dash across the Bering Land Bridge circa the waning of the Ice Ages and the (hypothetical) Viking incursions into North American waters.

But C.L. Johannessen is certain that the ancient Indians (i.e., those in India) were enjoying corn-on-the-cob at least as early as the Twelfth Century BC.

"Goddesses and gods in sculptured soapstone friezes in Hoysala temples of the twelfth and thirteenth centuries BC near Mysore, India, hold in their hands representations of maize ears. There are more than 63 of these large ears at Somnanthpur, and maize is represented at three other temples I have visited.

"In the Hoysala tradition, worshippers must have used maize as a golden coloured and a many-seeded fertility symbol in their religious rites. That the ears are modelled on maize is shown by the ear length-to-diameter ratio, the ear sizes in relation to parts of human figures, and the wide variation of anatomical detail in the carvings that all belong to maize: the ears have either parallel, highly tapered or bulging sides, their tips are pointed, and their axes may be straight or warped, depending on the moisture at the time of picking and the way maize dries....No other plant or object has the extensive intricacy and variation of highly segregated maize that could serve as a model for the sculptures. No other fruits have the same number and shape of the closely packed kernels that are arranged in parallel rows in the sculptures." (Johannessen, Carl L.; "Indian Maize in the Twelfth Century BC," Nature, 332:587, 1988. Cr. R. Noyes)

THE ANCIENT DISPERSAL OF USEFUL PLANTS

George F. Carter, a noted geographer, has summarized the botanical evidence for early transoceanic voyages. Domestic cotton: Thousands of years old in the Americas; believed to be a hybrid between New World wild cotton and species from southwest Africa. The Bottle Gourd: Of African origin but known in Peru about 11,000 years ago; dispersable by ocean currents but appeared in Peru only after man learned to navigate on the sea. The Sweet Potato: A New World plant that has been known in Polynesia for at least 500 years; but the South American name for the sweet potato (kumara) turns out to be a Sanskrit word from India, which is most perplexing. Coconuts: Arrived in the Americas from the Indian Ocean region via Polynesia; can be dispersed by ocean currents. Peanuts: Well-established on the Peruvian coast thousands of years ago, but the same variety known in pre-Shang China before 1500 BC. (Carter, George F.; "Kilmer's Law: Plant Evidence of Early Voyages," Oceans, 12:8, 1979.)

RAMSES HOOKED ON TOBACCO

French scientists examining the stomach of Egyptian Pharaoh Ramses II found fragments of tobacco leaves. Further analysis of the 3,200-year-old mummy indicated the presence of nicotine in the body. Conventional wisdom has it that tobacco was unknown in the Old World until the Spanish brought it back from the Americas in the 16th. Century. (L., T.E.; "Tobacco in Egypt, "Anthropological Journal of Canada, 16:10, 1978.)

AN ARCHEOLOGICAL HOT POTATO

Mangaia is a small volcanic isle in the Cook Island group. During the excavation of a rock shelter here, large fragments of sweet potato were discovered. These were subsequently carbon-dated at about 1000 AD.

> The prehistoric transferral of this South American domesticate into Polynesia obviously raises issues of cultural contact between the coast of South America and the Polynesian Islands. In our view, the most likely transferrors would have been the seafaring Polynesians, on a voyage of exploration to South America and return.

(Hather, Jon, and Kirch, P.V.; "Prehistoric Sweet Potato (Ipomoea batatas) from Mangaia Island, Central Polynesia," Antiquity, 65:887, 1991.)

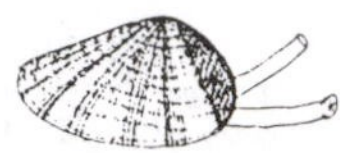

CLAMS BEFORE COLUMBUS

> Zoologists and geologists agree that specimens of Mya arenaria (the American soft-shell clam) from the Holocene found in Europe first appeared in the sixteenth century, after the voyage of Columbus. But we have dated a sample from the Kattegat region on the east coast of the Skaw in northern Jutland, Denmark, that predates Columbus's voyage. This result implies that contact between America and Europe existed before the sixteenth century.

In fact, one sample of the American soft-shell clam was carbon-dated by the authors of the above paragraph at 1245-1295 ± 1 standard deviation. This particular shell was discovered in the sand barrier farthest from the Danish coast. Since it seems highly unlikely that the larvae of the American clam could have swam across the Atlantic, the authors wonder if the Vikings might have carried them home. (Petersen, K.S., et al; "Clams before Columbus," Nature, 359: 679, 1992.)

Comment. Certainly the Vikings might have consumed these clams on their voyages to the New World, but why would they or any other explorer carry live clams or their larvae back to the Old World? Or was the transfer accidental in ships' ballast?

CULTURAL ENIGMAS

In this last section of our chapter on archeology, we offer a pot pourri of cultural observations that, in the main, stray far from long-held mainstream positions:

•Hunting, agriculture, animal domestication. Noble hunter or merely a scavenger? The drawbacks of agriculture.

•Language development and its diffusion. The invention of phonetic languages may have initiated civilization! The unknown roots of the Basque language. Nile-language affinities in North America!

•Precocious science. Mayan math and astronomy.

•Role of natural phenomena in human culture. Some effects of tsunamis, volcanos, and wind.

HUNTING, AGRICULTURE, ANIMAL DOMESTICATION

MAN THE SCAVENGER

Some 2 million years ago, man's supposed ancestors were meat-eaters. But were they noble hunters with dominion over other life forms? Probably not! The analysis of tool marks on ancient animal bones tells us that human tool marks predominate in regions of the bones where there is little meat, as if ancient humans were dismembering the animals for skins and other products. On the meat-bearing portions of the bones, the tooth marks of non-human carnivores predominate. Where the tool marks overlap the tooth marks of other carnivores, the tool marks are mostly on top of the tooth marks. The gist of the tool-mark analysis is that humans got to the animals second---after the non-human carnivores. In other words, ancient humans were probably meat scavengers---opportunists rather than the noble hunters often portrayed. As a matter of fact, one characteristic of a scavenger species is its ability to cover wide areas with little expenditure of energy, like the vultures. Now human bipedalism is pitifully poor for running down game but great for searching far and wide with minimum physical effort. Tooth-wear studies of ancient human skulls indicate that humans were vegetarians first and meat-eaters second. This situation was suddenly reversed when Homo erectus came along. Again according to tooth-wear patterns, there was a shift to a mainly meat diet. This was also the time when the human territory expanded greatly geographically. The reason for this change is unknown. (Lewin, Roger; "Man the Scavenger," Science, 224:861, 1984.)

REMARKABLE EARLY DATES FOR AGRICULTURE

B.K. Maloney, of the British Museum, describes his pollen analysis of sediments that have accumulated in the Toba Highlands of North Sumatra, Indonesia. The base of a 9.7 m core from the Pea Sim-sim Swamp has yielded a radiocarbon date of 18,496 years. Pollen studies of the core indicate a brief decline of forest pollen around 17,800 BP along with increased sedimentation characteristic of cleared land. Taken by themselves, these data would probably be interpreted in terms of natural climatic changes. But extremely early dates for human activity exist nearby: 14,000 BP for agriculture in Thailand and 11,000 BP for forest clearance on Taiwan. It is possible, therefore, that men were clearing land for planting in North Sumatra almost 18,000 years ago. (Maloney, B.K.; "Pollen Analytical Evidence for Early Forest Clearance in North Sumatra," Nature, 287:324, 1980.)

Comment. Some archeologists hotly dispute these early dates.

WHERE DID AGRICULTURE REALLY BEGIN?

The archeological party line points to the banks of the Tigris and Euphrates rivers. There, the authorities say that, about 10,000 years ago, humans suddenly learned how to sow and harvest such crops as wheat and barley. There, civilization really began. Or was it there?

Wadi Kubbaniya, Egypt. At this site, G. Hillman, of the Institute of Archeology, London, has found grinding stones and tubers. This site is dated at 17,000-18,000 years old.

New Guinea highlands. J. Golson, formerly of the Australian National University, has found ditches and crude fields in this area. The implication is that humans were tending plants here between 7,000 and 10,000 years ago.

Buka Island, Solomons. While excavating Kilu cave, M. Spriggs and S. Wickler unearthed small flake tools with surfaces displaying starch grains and other plant residues. Evidently, these tools were used for processing taro. Further, the starch grains resembled those of cultivated rather than wild taro. Date: about 28,000 years ago.

(Dayton, Leigh; "Pacific Islanders Were World's First Farmers," New Scientist, p. 14, December 12, 1992.)

INVENTION OF AGRICULTURE MAY HAVE BEEN A STEP BACKWARD

Anthropological texts have always ballyhooed the development of agriculture as one of man's greatest achievements. Not so, says Mark Cohen, from SUNY Plattsburgh. The switch from hunting and gathering to sedentary agriculture, it seems, occurred rather suddenly and was attended by a sharp drop in life expectancy. Ancient human bones reveal much more disease, fewer old people, and more violent deaths for centuries after the adoption of agriculture. Why did humanity give up the surprising security, freedom, and leisure of hunting and gathering? Cohen claims that population pressure was the cause. Unable to do anything about the population explosion, ancient man was forced to adopt a life of toil, disease, and stress. (Lewin, Roger; "Disease Clue to Dawn of Agriculture, "Science, 211:41, 1981.)

Comment. Note the parallel to the Garden of Eden story.

THE ANCIENT HORSEMEN

Over 70 years ago paleontologist Henri Martin found horse teeth estimated to be 30,000 years old that showed clear signs of "crib biting." Crib biting occurs when captive horses, perhaps out of boredom, bit ropes, enclosure structures, and even rocks---wild horses don't do it. The implication was that man had domesticated the horse long before archeologists believed possible. The theory languished until recently, when Paul Bahn brought it out of limbo. He has found additional teeth showing evidences of crib biting. Bahn maintains that man may have been riding the horse for 100,000 years! (Perrin, Timothy; "Prehistoric Horsemen," Omni, 5:37, August 1983.)

Comment. The ancient Africans also seem to have domesticated the ostrich. See p. 40.

LANGUAGE DEVELOPMENT AND DIFFUSION

SINISTER DEVELOPMENT IN ANCIENT GREECE

The unprecedented genius of Ancient Greece remains unexplained. Why the sudden surge of "civilized" activities: drama, poetry, philosophy, mathematics, and even science? J.R. Skoyles has an interesting answer. It was all because the Ancient Greeks developed an alphabet that included vowels in addition to consonants. The Greek language became a fully phonetic representation of language. The left side of our brain, it seems, is much more capable than the right in matters phonetic. In contrast, other forms of writing in the ancient world, such as hieroglyphics and vowelless alphabets, are better handled by the right side of the brain. (As an aside, it is interesting that in the modern world Japanese and Chinese are better processed by the right side of the brain, while the phonetic representations of language, such as English, are handles better sinistrally.) Back in Ancient Greece, the new alphabet shifted language activities from the right to the left side of the brain. According to Skoyles, this "unlocked" left-brain competences that had previously been dominated by analogous right-brain competences. The newly liberated competences involved rational, analytical, and logical faculties. Therefore, from the addition of a few vowels sprang Ancient Greece and in time modern civilization. (Another aside: each side of the brain seems to have the potential for performing all necessary functions, but the left side is better at performing some duties than the right and vice versa. Sometimes the dominant half produces better answers, sometimes not.) Skoyles' point is that the Greek invention of a phonetic language unlocked or made dominant the left side with its superior civilizing capabilities! (Skoyles, John R.; "Alphabet and the Western Mind," Nature, 309:409, 1984.)

DID THE ANCIENT EGYPTIANS SAIL UP THE MISSISSIPPI?

No one has found chariot wheels or pyramids attributable to the ancient Egyptians in the Lower Mississippi Valley, nor are their hieroglyphics carved on the rocks in that area. However, there are striking correspondences between the languages of ancient Egypt and those of the Indians that inhabited the areas around Louisiana about the time of Christ!

B. Fell, the main pillar of the Epigraphic Society, has stated that the language of the Atakapas, and to a lesser extent those of the Tunica and Chitimacha tribes, are unique in the sense that they seem to be related to no known languages. But there are affinities with Nile Valley languages. In fact, the similarities involve just those words one would associate with Egyptian trading communities of 2,000 years ago.

As would be expected, most archeologists will have none of this. "Where are the coins, the buildings, the piers?" they ask. Countering such criticism, W. Rudersdorf notes that no artifacts have ever been found from Coronado's expedition, only 450 years ago, when thousands of Spanish soldiers marched across the South. (Anonymous; "Professor Believes Egyptians Sailed Mississippi, Left Culture," Northwest Florida Daily News, December 27, 1991. Cr. R. Reid via L. Farish. Also see: Fell, Barry; Epigraphic Society Occasional Papers, 19:35, 1990.)

EUROPE'S MYSTERY PEOPLE

The researches of R. Frank, a scholar at the University of Iowa, suggest that the Basques were far-advanced in navigational skills and other aspects of technology long before the rise of the Roman Empire. The Basques, she believes, are the last remnants of the megalith builders, who left behind dolmens, standing stones, and other rock structures all across Europe and perhaps even in eastern North America.

Two facts set the Basque peoples apart from the other Europeans who have dominated the continent the past 3,000 years: (1) The Basque language is distinctly different; and (2) The Basques have the highest recorded level of Rh-negative blood (roughly twice that of most Europeans), as well as substantially lower levels of Type B blood and a higher incidence of Type O blood.

Some probable technological feats of the Basques or their ancestors are:

- Stonehenge and similar megalithic structures
- A unique system of measurement based on the number 7 instead of 10, 12, or 60
- Regular visits to North America long before Columbus to fish and to trade for beaver skins. Recently unearthed British customs records show large Basque imports of beaver pelts from 1380-1433.
- The invention of a sophisticated navigational device called an "abacus." (No relation to the common abacus.)

(Haddingham, Evan; "Europe's Mystery People," World Monitor, p. 34, September 1992. Cr. A. Rothovius.)

PRECOCIOUS SCIENCE

A DIFFERENT WAY OF LOOKING AT THE UNIVERSE

A. Aveni has provided us with an excellent survey of American archeoastronomy from Canadian medicine wheels to Mesoamerican aligned structures. To the anomalist, the most interesting part of Aveni's review paper is found in his comments about the world view of the pre-Columbian Americans, particularly the Mayas.

That the Mayas were acute astronomers is beyond question. They had even developed a correction scheme to keep Venus' 584-day canonic cycle on track with its true synodic period of 583.92 days. Their predictions of the positions of Venus were accurate to 2 hours in 481 years! Not bad for a civilization that did not seem to have any conception that the planets revolved around the sun.

The Mayas, in fact, did not seem to care what made heavenly bodies move; they only wished to predict their appearances accurately. Instead of developing celestial mechanics based on gravitation and laws of motion, as we have done, they were content with numerical algorithms; that is, ways of computing a desired result. Their astonishingly accurate predictions of Venus, solar eclipses, and other astronomical phenomena evolved from cycles of numbers advancing each day like interlocking gear teeth. These algorithms gave them no insight into cause and effect, but they got the right answers. They needed no physical laws, just patterns of numbers. It was a different way of comprehending and dealing with the universe. (Aveni, Anthony F.; "Native American Astronomy," Physics Today, 37:24, June 1984.)

ROLE OF NATURAL PHENOMENA IN HUMAN CULTURE

A TSUNAMI AND A PERUVIAN CULTURAL GLITCH

Abstract: "While investigating the archaeological background of early maize on the coast of Peru, I realized that several factors affect interpretation. The estimated date for the start of common use of maize there is close to the apparent dates of a large tsunami, the abandonment of many coastal sites, and the start of occupation at Chavin de Huantar in the highlands. While investigating the possible relations between the principal pre-tsunami coastal culture and Chavin, I discovered that depictions of a monstrous head link the two cultures."

The "monstrous head" is thought by the author to be a stylized representation of a tsunami wave. The physical evidences of tsunami damage along the Peruvian coast are not mentioned at all in the Abstract above. The article, however, portrays the possible effects on the man-made structures in the region as well as the widespread deposits of sand, cobblestones, and other sediments. (Bird, Robert McK.; "A Postulated Tsunami and Its Effects on Cultural Development in the Peruvian Early Horizon," American Antiquity, 52:285, 1987.)

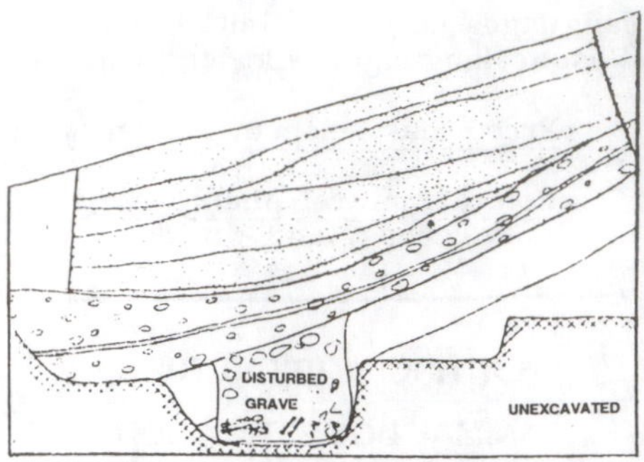

Profile of a north-south trench at a site on the Peruvian coast, showing thick strata covering a human grave. These strata may be tsunami deposits.

A DISASTER-DRIVEN EARLY CIVILIZATION

Archeologists had long recognized the existence of a highly sophisticated early civilization in the Cauca Valley region stretching 550 miles from northern Ecuador into Columbia. This civilization produced a distinctive pottery and spectacular gold artifacts. It was obviously a highly advanced culture technologically and socially. But it was dated at 400-800 AD; and for this period in South American history these accomplishments did not seem out of line. Recently, though, additional evidence of this civilization was discovered beneath a datable volcanic ash. The new dates for the civilization are 600-1500 BC, putting it about 1000 years ahead of Mayan and Incan achievements. The "digs" show further that this culture was frequently beset by devastating outbursts of volcanic activity, which often rendered large areas of land uninhabitable. Rather than suppressing this remarkable culture, Donald Lathrap, a University of Illinois archeologist, says, "Those disasters pushed people from the region and led to upward leaps in social evolution..." (Anonymous;"Key to a Vanished Empire," San Francisco Chronicle, June 14, 1984. Cr. J. Covey.)

DID THE ERUPTION OF THERA DO THE MINOANS IN?

According to popular archeological doctrine, the eruption of a volcano on the island of Thera destroyed the great Minoan civilization of Crete. Tidal waves, a thick ash blanket, and fires set when quakes overturned oil lamps did the job. This vivid, riveting scenario has been repeated again and again in the media until it seems to be a fact instead of a theory.

"Unfortunately, it seems to have been pure myth. Over the past decade or so evidence against Marinatos's theory has been piling up. Much of it has come from unlikely sources---the Greenland ice sheet, for instance, and trees in California and Ireland. But nearly all of it points to the same conclusion: whatever precipitated the Minoan collapse, it was probably not Thera. The volcano seems to have erupted more than a century before Minoan civilization died."

Briefly, although the explosion of Thera certainly occurred, it was too early. Further, its tidal waves were greatly exaggerated (really about 300 feet high instead of 600 feet), and most of Thera's ash fell east of Thera, with less than a half inch on Crete itself. The shock of Thera's eruption, 70 miles away, would have been slight---hardly enough to knock many lamps over, although fire does seem to have been a factor in the demise of Cretean civilization.

In sum, the cause for the collapse of the Minoan culture still eludes us. (Chen, Allan; "The Thera Theory," Discover, 10:77, February 1989.)

Comment. Perhaps pertinent are the signs of conflagration present in the ruins of Troy and other ancient cities.

COMPUTER CONFIRMS CROSSING!

Exodus states quite clearly that the Red Sea parted allowing Moses and the Israelites to escape the pursuing soldiers of the Pharaoh. This may not have been as miraculous as we learned in Sunday school.

> Because of the peculiar geography of the northern end of the Red Sea, researchers report in the Bulletin of the American Meteorological Society, dated last Sunday, that a moderate wind blowing constantly for about 10 hours could have caused the sea to recede about a mile and the water level to drop 10 feet, leaving dry [?] land in the area where many Biblical scholars believe the crossing occurred.

Some artists and movie makers have portrayed huge walls of water being held apart by a supernatural force, but it seems that the Biblical account may have been more accurate after all:

> The Lord caused the sea to go back by a strong east wind all the night, and made the sea dry land, and the waters were divided. And the children of Israel went into the midst of the sea on the dry ground. (Exodus)

If the wind has shifted suddenly, the displaced waters might well have returned quickly to inundate the Egyptian army. (Maugh, Thomas H., II; "Red Sea May Have Parted, Scientists Say," Denver Post, p. A1, March 14, 1992.)

Chapter 2

ASTRONOMY

PLANETS AND MOONS

SOLAR SYSTEM DEBRIS

STARS

GALAXIES AND QUASARS

COSMOLOGY

PLANETS AND MOONS

MERCURY

Mercury is an unusual planet in many ways. Its small size and heavily cratered surface give it a distinctly moon-like appearance. Yet, its high density (5.44) is radically different from that of our moon (3.34), rather it is more like that of earth (5.53). Mercury's orbit is scarcely better behaved. Only Pluto's is more eccentric and more highly inclined. These facts suggest that Mercury may not have been formed in the orbit it now occupies. The curious 2:3 resonance between Mercury's revolution around the sun and its axial spin only strengthens our suspicion concerning the pedigree of this planet. One of the greatest surprises about Mercury has been its apparent intrinsic magnetic field. It is the only planet in the inner solar system besides the earth to have what seems to be a self-generated magnetic field.

The pages of Science Frontiers have touched several times upon the anomalies of Mercury's orbit, it deviant density, and its unexpected magnetic field. In addition, the final item presented below wonders whether the advance of Mercury's perihelion, long taken as one proof of the General Theory of Relativity, may not be explained just as well in another way.

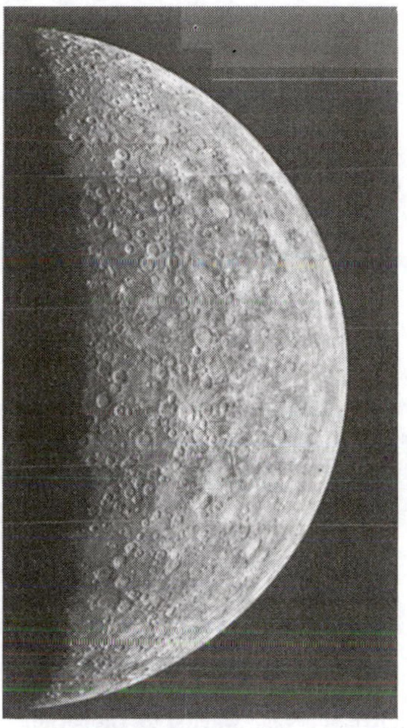

The 1974 flyby of Mercury by Mariner 10 revealed a heavily cratered planet. (NASA)

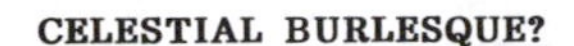

CELESTIAL BURLESQUE?

Astronomers have long wondered about Mercury. Its density (5.44) is unusually high for such a small planet, and its orbital inclination (7°) and eccentricity (0.206) are anomalously high. In one blow, W. Benz, A.G.W. Cameron, and W. Slattery may have solved all three problems. In their scenario, Mercury's original, lighter, silicate outer layers were stripped off during a collision with one of the small protoplanets that are thought to have swirled around the inner solar system shortly after its formation. Computations on a supercomputer revealed to these three researchers that, if the protoplanet hit Mercury at between 20 and 30 kilometers/second, the dense iron core would survive pretty much intact. A lower velocity would not strip off the lighter outer layers; anything higher would blast the whole planet into smithereens.

Calculations of this type also suggest that if a protoplanet the size of Mars hit protoearth, it would likewise rip off its light silicate mantle. After this removed material gravitationally sphericized itself in orbit around the earth, it became---you guessed it---our moon. (Stewart, Glen R.; "A Violent Birth for Mercury," Nature, 335:496, 1988. Also: Anonymous; "Mercury Stripped by Blow from Meteorite," New Scientist, p. ~~00,~~ November 5, 1988.)

Comment. It seems that there were Velikovskian overtones in the early solar system, with many celestial missiles flying about! (Note: I. Velikovsky hypothesized in his controversial 1950 book Worlds in Collision that the inner solar system was in disarray within the time of human myths and records!) See p. 60 for the possible catastrophic origin of our moon, and go to p. 73 for material on the long-term stability of the solar system.

Four frames from a computer simulation of proto-Mercury being stripped of its lighter outer crust by a collision with another protoplanet. Frame times are: -1, +2.3, +7.7, and +41.5 minutes after impact. In Frame #4, the molten sheet of mostly iron became Mercury, while the light silicates formed our moon.

MERCURY: THE IMPOSSIBLE PLANET

Mercury, largely hidden in the sun's glare, also conceals beneath its baked, cratered surface: (1) far more iron than solar-system theory allows; and (perhaps) (2) a dynamo that should not exist.

Let us take the excess-iron problem first. Mercury's density is 5.44 (compared to earth's 5.52), so that it very likely contains much iron. Our moon, which resembles Mercury in size and external appearance, only has a density of 3.34, implying an altogether different origin. In the currently accepted theory of solar-system formation, all of the planets and their satellites condensed from a primordial disk of dust surrounding the just-formed sun. The planets closer to the solar inferno lost more of their easily vaporized constituents due to the sun's heat. The cooler, outer planets were able to retain large amounts of ices. In this scenario, we would expect Mercury to be rich in iron and rocks. This seems to be the case, but it has too much iron to fit the theory. Astronomers have tried to save the theory by supposing that a large asteroid sideswiped Mercury tearing off part of its outer layer of lighter rocks, leaving the heavier iron core untouched. As mentioned above, this debris may have coalesced into our moon.

As for Mercury's magnetic field, it is small, only 1% that of the earth. But where does it originate? All of the other planets with magnetic fields (earth, Jupiter, Saturn, Neptune, and Uranus) rotate rapidly and are believed to have molten interiors, allowing fluid dynamos to form. Mercury, in contrast, spins very slowly and seems solid throughout. Therefore, the magnetic dynamos that supposedly create the fields of other planets cannot exist inside Mercury. (Crosswell, Ken; "Mercury---the Impossible Planet," New Scientist, p. 26, ~~June~~ 1, 1991.

Comment. Could Mercury be a permanent magnet? M. Stock suggested this in a letter in the July 13, 1991, issue of the New Scientist.

TWO HOT SPOTS ON MERCURY

Radio telescopes can give planetary astronomers a rough idea of the temperature existing several feet below the surface of a distant planet. Scrutinizing Mercury with their big electronic "ears", they have found two spots on the planet where the temperatures are several hundred degrees higher than in the sur-

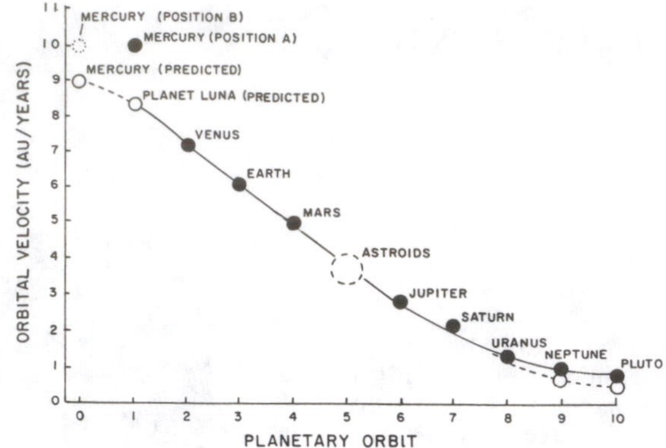

This graph of planetary orbital velocities illustrates how far present-day Mercury (solid circle) is from the general trend. The fit is better if Mercury was formed along with the planet Luna at the "predicted" positions. In this scheme, the earth captured Luna in subsequent solar-system upheavals and Mercury moved to its present position. (See AHB2 in The Moon and the Planets)

rounding areas. Actually, these hot spots are easy to understand; because, to a Mercurian, the sun seems to stop in the sky over one of these points and then move backwards to the other point 180° away. As the sun tarries over these two spots, it heats them preferentially. The strange apparent motion of the sun is due to the 3:2 ratio between Mercury's period of revolution around the sun (88 days) and its axial spin period (59.6 days).

What is surprising is that the energy detected radiating from the two hot spots is all reradiated solar energy; that is, there seems to be no contribution at all from Mercury's core! If no heat is leaking out of Mercury's core, the core itself is very likely solid. If it is solid, it cannot establish convection cells and thus generate a magnetic field through dynamo action. But back in 1975, the Mariner 10 spacecraft radioed back that Mercury actually did possess a magnetic field and a surprisingly large one at that. (Wilford, John Noble; "Theory of Mercury's Hot Poles Is Shown to Be Fact," New York Times, June 13, 1990. Cr. J. Covey)

Comment. It also seems that Mercury, despite its proximity to the sun, possesses some very cold spots, as described below.

MERCURY'S POLAR CAPS AND ICY MINICOMETS

Mercury, closest planet to the sun, should be baked bone dry, seeing that equatorial temperatures reach 800°F. When Mariner 10 flew past Mercury in 1974, its camera eye reinforced the baked-cinder model. To everyone's surprise, recent radar images obtained with powerful earth-based antennas, revealed a highly reflective patch at Mercury's north pole. Could it be ice, for ice reflects radar waves well? Quite possibly, for when Mercury's polar temperatures are calculated, away from the sun's direct glare, they plunge to -235°F. This means that some of the water vapor in the planet's thin atmosphere might freeze out in the poles, creating ice or frost caps.

Wouldn't Mariner 10 have seen such a remarkable deposit? Not necessarily, for the spacecraft viewed only half the planet and, if the 640 x 300-kilometer ice patch were covered with dust, it could have been invisible to the camera. But it would still be a bright patch on terrestrial radar scopes, because radars see through thin dust layers.

So, polar ice is not physically impossible on Mercury, although it is definitely surprising so close to the sun. All that is needed is a little water in the planet's atmosphere. Mainstream thinking is that "passing comets and asteroids" might bequeath Mercury some of their H_2O cargos. (Cowen, R.; "Icy Clues from Mercury's Other Half," Science News, 140:295, 1991. Wilford, John Noble; "Photographs by Radar Hint of Ice on Poles of Mercury," New York Times, p. A14, November 7, 1991. Cr. J. Covey)

Comment. What the above references do not mention is the possibility that the requisite water vapor for the formation of Mercury's polar caps might come from a steady rain of icy minicomets. L. Frank has suggested that 100-meter icy minicomets continuously pepper solar-system planets. They might even have contributed to the formation of the earth's oceans. Icy comets are anathema here on earth and are equally detestable at Mercury's orbit. See p. 275 for additional discussion of L. Frank's icy comets and the data that support their existence.

MERCURY'S ORBIT EXPLAINED WITHOUT RELATIVITY

A most satisfying element of support for Einstein's General Theory of Relativity (GR) has been its accounting for the residual precession of Mercury's orbit. In recent years, however, a rival explanation has been found in the nonsymmetric gravitational field of the sun. Surface oscillations of the sun betray hidden internal rotation which produces asymmetry in its gravitational field. By applying this distorted field in predicting the orbit of Mercury and the minor planet Icarus, astronomers are more successful than when they use the GR. The authors of this paper claim that GR averages some 2¼ standard deviations off the mark, while results using the nonsymmetrical gravitational field of the sun are right on the money! (Campbell, L., et al; "The Sun's Quadrupole Moment and Perihelion Precession of Mercury," Nature, 305:508, 1983.)

VENUS

Venus, its crescent hanging bright in the evening or morning sky, is like no other solar system planet. Its thick, hot atmosphere cloaks a parched surface seen only by a few spacecraft that have survived the descent through kilometers of searing gases and by radars on the earth and orbiting satellites. In spite of its atmospheric shield, Venus presents many enigmas to the terrestrial telescope user. Many of these mysteries seem to involve optical and subjective factors: the eerie ashen light, the blunted cusps or horns, the fleeting radial spoke system, the famous phase anomaly, and several other unexpected apparitions. These phenomena are treated in The Moon and the Planets. Here, except for the curious Maedler phenomenon, our entries are based mainly upon data acquired recently from spacecraft sent to Venus or from highly precise terrestrial radar measurements.

Of particular interest to the anomalist are the several hints that Venus may have been a recent acquisition of the inner solar system and, in addition, that Venus may once have had a closer relationship with earth. This latter claim is based upon the fascinating discovery that, when the earth and Venus are closest (inferior conjunction), Venus always points the same face toward the earth. Why?

As if Venus-seen-through-the telescope and terrestrial radars were not puzzling enough, several U.S. and Russian spacecraft have orbited Venus, sampled its atmosphere, and landed upon its surface. The anomalies described below include layered rocks, an immense gash in the planet's surface, and a strange mix of primordial gases. All in all, Venus seen close-up appears to be radically different from earth, Mercury, and Mars---its neighbors in the inner solar system. Perhaps Venus had a different origin and evolutionary history!

COMETARY APPEARANCE OF VENUS

The famed astronomer Johann Maedler was the first to record strange brushes of light emanating from the bright limb of Venus. Two luminous fans opening sunward gave the planet the appearance of a multi-tailed comet. Maedler made this curious observation on April 7, 1833; and it has come to be known as The Maedler Phenomenon. If it were not for Maedler's fame as an honest, meticulous observer, the event would have been forgotten long ago. A similar observation was made by Gadbury in 1686. The only explanation offered so far involves some obscure type of halo or sun pillar phenomenon. (Baum, Richard; "The Maedler Phenomenon," Strolling Astronomer, 27:118, 1978.)

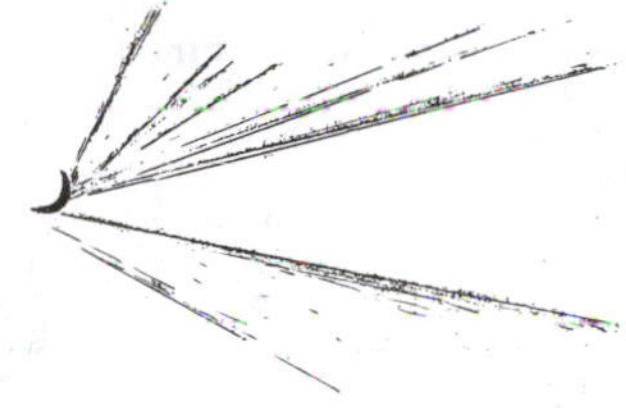

The Maedler phenomenon is characterized by faint brushes of light fanning out from the cusps of Venus' bright crescent.

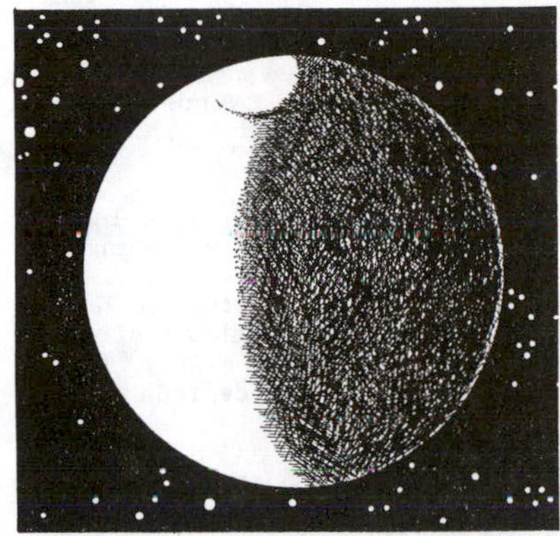

Venus as seen through the telescope, illustrating the blunted-cusp phenomenon. (See AOV1 in The Moon and the Planets for further discussion.)

VENUS HAS UNCERTAIN PEDIGREE

The five instrumented Pioneer probes that plunged into the thick Venusian atmosphere in late 1978 discovered unexpectedly large quantities of the isotope argon-36. The significance of argon-36 is that it is supposed to be primordial argon; that is, an argon isotope formed when the solar system was created. Since argon-36 is radioactive, most of the originally created supply should have disintegrated and disappeared over the four-billion-year history of the solar system. Indeed, the atmospheres of earth and Mars have much, much smaller quantities of argon-36 than Venus. Venus, therefore, may have an origin different from those of earth and Mars---either a much more recent birth (so that the argon-36 has not disintegrated), or an altogether different kind of origin in which more argon-36 was created than for earth and Mars. (Anonymous; "Venus Probes Solar System Birth," New Scientist, 80:916, 1978.)

VENUS: HIGHLY RADIOACTIVE OR JUST COOLING DOWN?

The surface temperature of Venus is about 480°C, higher than any other solar system planet. While Venus does trap solar radiation in its atmosphere greenhouse fashion, data from the Pioneer Venus Orbiter show that the planet radiates 15% more energy than it receives from the sun. In other words, Venus' surface is hotter than it would be if only the greenhouse effect were operating. Where could this extra energy come from? If it arises from the decay of naturally occurring radioactivity, as some of the earth's heat does, Venus would have to have 10,000 times as much radioactivity as the earth. If this is the case, Venus must have had an origin radically different from earth's. (Anonymous; "The Mystery of Venus's Internal Heat," New Scientist, 88:437, 1980.)

SEDIMENTARY ROCKS ON VENUS?

The Soviet spacecraft Veneras 13 and 14 took photos of the hot surface of Venus from their landing spots beneath the planet's thick cloud cover. The rocky debris surrounding the spacecraft shows, to nearly everyone's surprise, strong evidence of sedimentary, layered structure. The rock formations display ripple marks, thin layering, differential erosion, and even hints of cross-bedding. The vision of ancient seas on Venus leaps to the mind; but according to Russian scientists it is far more likely that the sediments were created by winds, episodic volcanism, or repeated meteor strikes. (Florensky, C.P., et al; "Venera 13 and Venera 14: Sedimentary Rocks on Venus?" Science, 221:57, 1983.)

SALT STRUCTURES ON VENUS?

Chemical analyses of the Venusian atmosphere have caused C.A. Wood and D. Amsbury to speculate as follows:

"The discovery of a surprisingly high deuterium/hydrogen ratio on Venus immediately led to the speculation that Venus may have once had a volume of surface water comparable to that of the terrestrial oceans. We propose that the evaporation of this putative ocean may have yielded residual salt deposits that formed various terrain features depicted in Venera 15 and 16 radar images.

"By analogy with models for the total evaporation of the terrestrial oceans, evaporite deposits on Venus should be at least tens to hundreds of meters thick. From photogeologic evidence and insitu chemical analyses, it appears that the salt plains were later buried by lava flows. On Earth, salt diapirism leads to the formation of salt domes, anticlines, and elongated salt intrusions ---features have dimensions of roughly 1 to 100 km. Due to the rapid erosion of salt by water, surface evaporite landforms are only common in dry regions such as the Zagros Mountains of Iran, where salt plugs and glaciers exist. Venus possesses a variety of circular landforms, tens to hundreds of kilometers wide, which could be either megasalt domes or salt intrusions colonizing impact craters. Additionally, arcuate bands seen in the Maxwell area of Venus could be salt intrusions formed in a region of tectonic stress. These large structures may not be salt features; nonetheless, salt features should exist on Venus." (Wood, C.A., and Amsbury, D., American Association of Petroleum Geologists, Bulletin, 70:664, 1986.)

Comment. Perhaps Venus had its "salt

Two pictures of the Venusian terrain taken by the Russian spacecraft Venera 14. Note the layered character of the rocks in the area of the lander.

ages" just as the earth had its "ice ages"! In fact, radar mapping of the Venusian surface from the orbiting spacecraft Magellan revealed a very long, river-like channel. Instead of water, which could not survive in liquid form on Venus, it may be that Venus possessed rivers and oceans of some other chemical, such as molten sulphur.

WHAT FLUID CUT THE STYX?

> One of the most bizarre features yet identified on Venus is a remarkably long and narrow channel that [the spacecraft] Magellan scientists have nicknamed the River Styx. Although it is only half a mile wide, Styx is 4,800 miles long. What could have carved such a channel in unclear. Water, of course, is out of the question. Flowing lava is a possibility, but it would have to have been extremely hot, thin, and fluid.

Another suggested fluid is sulphur, but there is still room for speculating about exotic fluids, given Venus's high surface temperatures. Another point of interest: the River Styx does not run steadily downhill. It takes an up-and-down course. Either the Venusian topography has shifted since the Styx was cut, or the channel is not a river at all but rather some bizarre geological feature. (Chaikin, Andrew; "Magellan Pierces the Venusian Veil," Discover, 13:22, January 1992.)

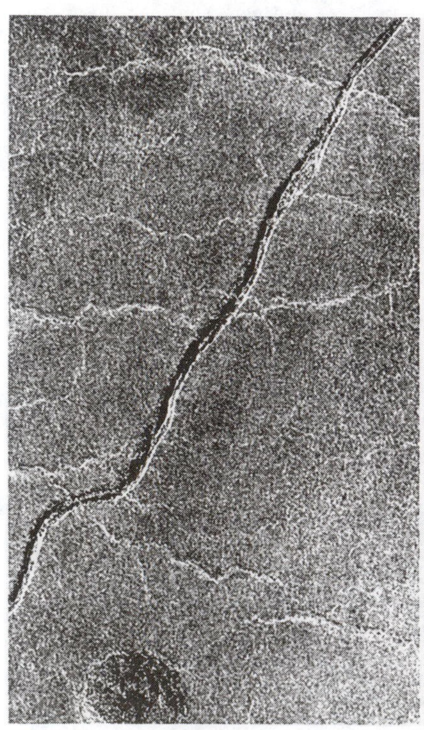

The 4,800-mile-groove in the surface of Venus named the River Styx, although one cannot be certain that it was actually cut by fluids. (NASA)

VENUS TOO PRISTINE

In the November 16 issue of Science, R.A. Kerr remarks:

> The planetary geologists who are studying the radar images streaming back from Magellan (the space probe) find that they have an enigma on their hands. When they read the geological clock that tells them how old the Venusian surface is, they find a planet on the brink of adolescence. But when they look at the surface itself, they see a newborn babe. (Kerr, Richard A.; "Venus is Looking Too Pristine," Science, 250: 913, 1990.)

Of the 75 craters mapped so far by Magellan, only one shows any signs of aging; i.e., tectonic movements, lava-filling, etc. The surface of Venus should be hundreds of millions of years old, yet it looks freshly minted. The anticipated spectrum of degradation has not yet been seen.

One theory is that recent lava flooding erased the old craters, and we now see only recent impact scars. But why would a planet's volcanism turn off so completely and so abruptly? Our earth, Venus's sister planet in many ways, still perks away, leaving craters of various ages. Why is Venus different?

Comment. One idea not advanced by Kerr in Science is that Venus might be a recently acquired member of the solar system.

VENUS AND EARTH: ENGAGED OR DIVORCED?

Proponents of astronomical catastrophism have made much of the supposed resonance between the earth and Venus, in which Venus rotates exactly four times between the times of closest approach (inferior conjunction). Astronomers have maintained that the gravitational forces are too slight to force a resonance lock. This has led to speculation that the two planets were once much closer. The most recent measurements of the rotational period of Venus show it to be 243.01 days rather than the 243.16 days for precise resonance. I.I. Shapiro (MIT) and his colleagues, who made the measurements, wonder whether Venus may be just approaching or just escaping a resonance lock. Whatever the case, the near-resonance is not likely to be a coincidence. (Anonymous, "Venus and Earth: Engaged or Divorced?" Astronomy, 7:58, October 1979.)

Comment. If Venus is just escaping resonance lock, how long ago was the lock exact---a few thousand years, or a few hundred million years?

THE LONG ARMS OF VENUS AND JUPITER

Many times in the two or three "scientific" centuries now behind us, investigators have discovered, almost against their wills, that the moon and planets affect the earth. The moon's influence is understandable, but the planets are too far away for their gravitational fields to influence one terrestrial dust mote. Well, here is one more study showing that the planets (Venus and Jupiter, in this case) do affect the peak electron density in the earth's ionosphere. The effect is most noticeable when these planets are close to earth and dwindles as they swing around the other side of the sun.

The authors are at a loss to explain this effect in terms of gravitation, suggesting that perhaps Venus and Jupiter may affect solar activity, which in turn modifies the terrestrial ionosphere. (Harnischmacher, E., and Rawer, K.; "Lunar and Planetary Influences Upon the Peak Electron Density of the Ionosphere," Journal of Atmospheric and Terrestrial Electricity, 43:643, 1981.)

Comment. Actually, no one has shown how the planets can possibly influence the sun with known action-at-a-distance fields. Electrical fields are taboo. There are no other "recognized forces." This controversial subject is treated in depth under ASO9 in The Sun and Solar System Debris.

THE EARTH

The earth itself is an astronomical object only for astronauts and extraterrestrial beings. Even so, we do make many observations of the earth and her environs that have astronomical import. Besides our moon (discussed in the next section), we detect considerable natural space debris circling the earth. In fact the earth may have now or have had in the past a retinue of moonlets and even a thin, Saturn-like ring of cosmic flotsam and jetsam. Other observables of astronomical interest are the large inclination of earth's magnetic field with respect to its spin axis, and also a hint of a recent shift of the spin axis itself.

THE EARTH'S OTHER MOONS

Over the past two centuries, night-sky observers have recorded a number of objects that moved too fast to be asteroids and too slowly to be meteors. John P. Bagby has studied this problem for over 20 years, publishing several hotly debated papers during this period.

His latest contribution summarizes evidence supporting his contention that the earth has captured chunks of space debris, some of which have disintegrated, some of which are still in orbit amidst tons of artificial satellite debris. The supporting observations have come from optical surveillance programs, space tracking networks, radio propagation anomalies, and (most interesting to the anomaly collector) old reports of bright objects near the sun (especially the August 1921 object) and the curious group of retrograde objects that passed over Germany in 1880. (Bagby, J.P.; "Natural Earth Satellites," British Interplanetary Society Journal, 34:289, 1981.)

Comment. Many mysterious bright objects have been seen near the sun, some of which were interpreted as an intramercurial planet, named and placed in the textbooks of the time as the planet Vulcan! See AEO1 in The Sun and Solar System Debris.

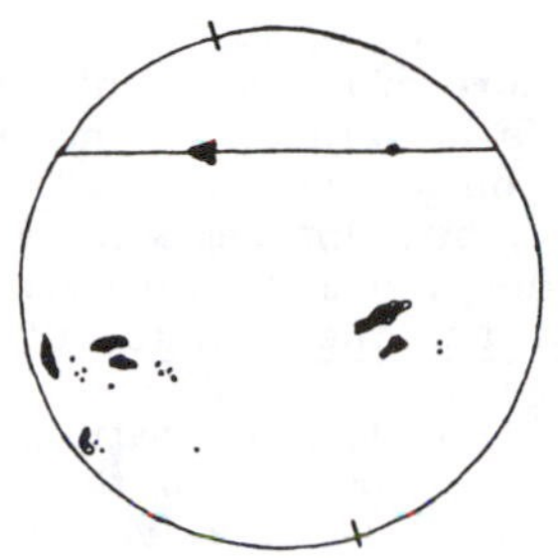

On January 14, 1983, a black spot drifted across the sun's face along the arrow. Was it an unrecognized moonlet of earth?

THE EARTH'S RING

The most profound climatic event of the Tertiary Period was the terminal Eocene event 34 million years ago. The sudden change in the abundance of forest plants suggests that the winters became much more severe while the summers remained about the same. At about the same time, the radiolaria were devastated by some sort of disaster. This was also the time when the North American tektite strewn field was deposited--a field that stretches half way around the world. John O'Keefe hypothesizes that some of the tektites and microtektites that rained down during this period missed the earth and went into orbit around it, forming an opaque Saturn-like ring. This ring might have lasted a million years or more; its shadow could have caused the extra-severe winters postulated from botanical data. (O'Keefe, John A.; "The Terminal Eocene Event: Formation of a Ring System around the Earth," Nature, 285:309, 1980.)

Comment. Many nonscientists who have previously speculated about terrestrial ring systems, such as I.N. Vail, were called pseudoscientists!

WILL EARTH'S RINGS RETURN?

In the past, the Earth had a ring system just like Saturn, Uranus and Neptune, according to a Danish astronomer. He has gone so far as to say that our planet boasted rings on 16 separate occasions in the past 2800 years.

Kaare Rasmussen of the National Museum in Copenhagen carried out a survey of all reports of meteorite falls, fireballs and showers of shooting stars from 800 BC to 1750 AD. He then carried out a statistical analysis of the data. He discovered that there were many distinct periods of intense activity.

Rasmussen found that the active periods, during which he believes a ring existed about the earth, began with a peak of high meteoritic activity as the ring formed, as a consequence of the earth's capture of a comet or asteroid. This was typically followed by a decrease of activity indicating ring stability. Finally, the ring broke up as its particles were decelerated by the earth's upper atmosphere, leading to another peak of activity. Thus, the plot of meteorite activity is U-shaped, as in the figure, with the bottom of the U a bit higher than the normal background. (Gribbin, John; "Will the Earth's Rings Return?" New Scientist, p. 19, April 6, 1991.)

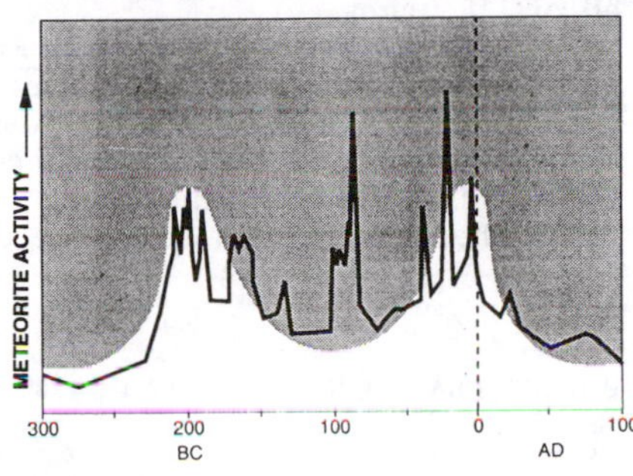

Meteorite activity seems to peak when a ring forms around the earth and again when it breaks up.

WHAT HAPPENED TO THE EARTH'S AXIS IN 2345 BC?

Australian astronomer G.F. Dodwell has analyzed the observational records compiled for the many gnomons erected all over the civilized world during the past 4,000 years. Gnomons are vertical markers that cast shadows from which the local latitudes can be computed. (All one needs are the measurements of shadow lengths on the longest and shortest days of the year.) The earth's tilt or obliquity of the ecliptic may also be calculated from gnomon data---and therein lies the anomaly.

The tilt of the earth's axis is supposed to vary cyclically between 22 and 24.5° over a period of some 40,000 years due to the pulls of the moon, sun and planets on the earth's equatorial bulge, as indicated in the diagram. Tilt angles computed from ancient gnomon observations deviated markedly from the theoretical curve. The alignment of the Egyptian temple at Karnak and other oriented sites extend the deviation toward the date 2345 BC. Either the ancient observations were systematically in error all over the world or the earth's tilt angle changed in historical times. (Bowden, M.; "The Recent Change in the Tilt of the Earth's Axis," Pamphlet No. 236, July 1983. Creation Science Movement.)

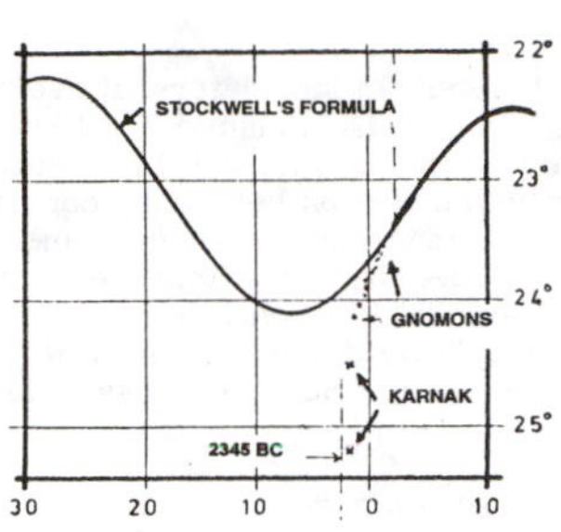

Earth's tilt angle versus millennia as plotted from theory and ancient megalithic alignments. An axial shift is implied.

Comment. One would think that such startling data, compiled by a recognized astronomer, would be the subject of intense study in archeoastronomical circles; instead, it is an English creationist tract that discussed the subject. No amplification of Dodwell's work has been uncovered as yet.

TILTED PLANETARY MAGNETIC FIELDS

The axis of the earth's magnetic field is tilted about 11° away from our planet's axis of rotation. This tilt is embarrassing to the reigning "dynamo theory", which is supposed to explain the origin of the earth's magnetic field. In this theory, both the axes of the magnetic field and of rotation are presumed to be aligned. The situation is even worse with some of the other planets; viz., Uranus with 60° and Neptune with 47°. Jupiter's tilt is 10°, while Saturn's is only 1°. Clearly something is awry.

Two possibilities are: (1) The dynamo theory is incorrect; or (2) The magnetic fields measured by spacecraft are actually combinations of central, dynamo-created dipole fields, which are aligned with the axes of rotation, and "other" dipole fields located in the outermost portions of the planetary cores. If the latter solution to the dilemma is correct, we must account for the origin and disposition of these "other" dipoles. (Eberhart, J.; "Straightening the Magnetic Tilts of Planets," Science News, 137:294, 1990.)

Cross reference. The anomaly of offset magnetic axes of solar system planets can be pursued further on pp. 70 and 235.

NATURAL LASERS IN THE TERRESTRIAL AND MARTIAN SKIES?

In a current laser patent dispute, one side claims that a certain laser patent is invalid because natural phenomena cannot be patented under U.S. law. It seems that last year Michael Mumma and colleagues at Goddard Space Flight Center and the University of Maryland discovered a 10 micrometer (infrared) laser in the Martian atmosphere. This laser is located about 75 kilometers above the surface, is optically pumped by the sun, and radiates an astonishing 10^{12} watts. The terrestrial atmosphere may contain a natural 4.3 micrometer laser---auroras are accompanied by very intense molecular emissions at this wavelength. (Raloff, J.; "Gould Laser Patent Ruled Invalid---So Far," Science News, 121: 199, 1982.)

Comment. An elaboration of the foregoing brief note appeared in Nature a few months later. See below.

NATURAL LASER BEACONS

"It has long been realised that the Earth's upper atmosphere cannot be in thermodynamic equilibrium, and during the last decade astronomers have made telescopic observations of nonequilibrium processes taking place in the upper atmospheres of our Earth-like neighbors, Mars and Venus. A preliminary analysis by Michael Mumma of Goddard Space Flight Center in Maryland, and his colleagues, indicated that processes analogous to optical pumpimg in laboratory lasers were taking place and led to the coining of the term 'natural lasers'. Now, Deming and colleagues from the Goddard team have taken new observations of emission from Mars and Venus at wavelengths near 10 micrometers, and modelled them to show that stimulated emissions---the effect which makes lasers so powerful---accounts for up to seven per cent of the total emission. This is not a large amplification factor by laboratory standards, particularly for a CO_2 laser. But, the authors speculate that the sheer size of the natural lasers could make them useful tools in the future for communicating with distant civilizations beyond our own planetary system."

The atmospheres of Mars and Venus are almost pure CO_2. The CO_2 molecules are excited by the absorption of energetic solar photons; then, thermally emitted photons at about 10 micrometers from lower reaches of the atmosphere collide with the excited molecules, stimulating them to emit another 10 micrometer photon, thus doubling the number of photons. This is typical laser action. Deming and Mumma speculate that the natural laser action existing in the Martian atmosphere could be intensified and focussed into an intense beam of infrared radiation of enormous power by placing two large mirrors in orbit, creating a space-borne analog of a laboratory laser. With this huge laser, interstellar communication might be possible. (Taylor, F.W.; "Natural Lasers on Venus and Mars," Nature, 306:640, 1983.)

PLANETS AS SUN-TRIGGERED LASERS

Apparently the earth, Mars and Jupiter emit radio energy when triggered by bursts of radio waves arriving from the sun. The outer atmospheres of these planets act like radio lasers, which store radio energy and then release it suddenly when stimulated by much weaker solar signals. The earth's laser operates between frequencies of 50 and 600 kilokertz. Its emissions are known as the auroral kilometric radiation (AKR). While some of these terrestrial emissions are spontaneous, others are stimulated by Type III solar radio bursts. The newly discovered Jovian laser operates at hectometric wavelengths and is also triggered by the solar radio bursts. (Calvert, W.; "Triggered Jovian Radio Emissions," Geophysical Research Letters, 12:179, 1985.)

Comment. Earth and Jupiter thus act like radio transponders, releasing large bursts in response to small solar stimuli. The role of electricity in the history of the solar system is only beginning to be appreciated. Of course, the lasers mentioned above are not very powerful, but what might have occurred during the formative stages of the solar system? Could electromagnetic forces have been more important than they are now? In this regard, note that electrical forces may be strongly involved in the dynamics of Saturn's rings. And Saturn's rings may themselves resemble a miniature solar system in the accretion phase.

THE MOON

Down the centuries, astronomers have lavished more effort in scrutinizing the moon than any other astronomical object. This is natural for the moon is close-by, intriguing, and a spectacular object on a clear night. We know a lot about the moon from all this attention, but this same scrutiny has also uncovered a wide array of lunar anomalies and curiosities. In fact, the chapter on lunar anomalies in The Moon and the Planets takes all of 110 pages.

In the past pages of Science Frontiers, we have collected a sampling of these unresolved lunar enigmas: the still unsolved problem of the moon's origin, the non-random distribution of its craters, it swirl patterns and magcons (magnetic concentrations), and last, but not least, a flash of light seen on the moon, one of the hundreds of TLPs (transient lunar phenomena) seen down the centuries.

BRIGHT FLASH ON THE MOON IN 1985

Over the years, moon-watchers have recorded hundreds of flashes, glows, and color changes occurring on the moon. One of these transient lunar phenomena was seen in 1985, as G. Kolovos et al recount:

"We present photographic evidence of a very short duration, strong flash from the surface of the Moon (near an irregularly shaped crater in Palus Somni). The flash covered a region roughly 22 by 18 km wide with a total energy of the order of 10^{17} erg. The event is established to be slightly above the surface of the Moon. An explanation is proposed involving outgassing and a subsequent electrical discharge caused by a piezoelectric effect." (Kolovos, G., et all: "Photographic Evidence of a Short Duration, Strong Flash from the Surface of the Moon," Icarus, 76:525, 1988.)

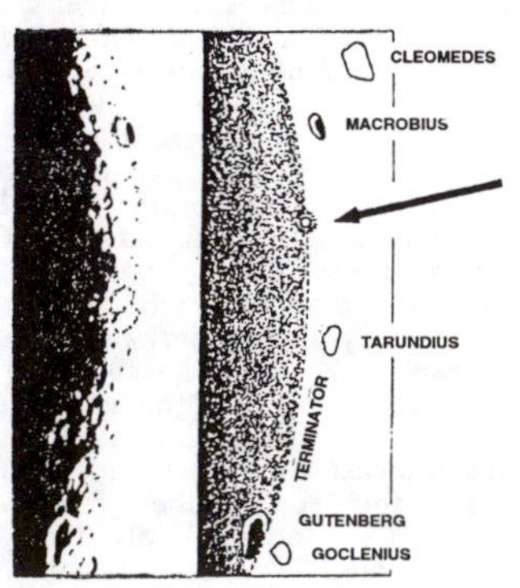

Left. Photo of the moon's limb. Right. A sketch based on the photo. The arrow marks the position of the flash. Some large craters are named for reference, but they do not show up well on the photo.

Comment. Of special interest above is the suggestion that the flash was generated by the electrical ignition of expelled gases. It has been proposed that terrestrial earthquake lights are kindled the same way. Further, the presence of methane on the moon is compatible with T. Gold's theory that the earth retains huge amounts of methane beneath the crust. See p. 206.

EXPLAINING LUNAR FLASHES WITH LIFE-SAVERS

Two recent items in the literature suggest ways in which flashes of light can be generated on the face of the moon. The first enlists the "Life-Saver effect" (or "sugar-cube effect"). When Life-Savers, sugar cubes, and rocks are fractured, light may flash from the broken surfaces. R.R. Zito, from Lockheed, thinks these flashes occur when energetic electrons are emitted from freshly fractured surfaces. Lunar rocks cracking under stress might very well produce flashes visible from earth. Zito also states that newly fractured rocks emit a curious burst of radio waves in the frequency range of 900-5000 hertz. (Eberhart, J.; "Does the Moon Spark like a Life-Saver?" Science News, 136: 375, 1989.)

The second explanation of lunar flashes blames them on earth satellites passing in front of the moon. Satellite surfaces can flash like a car's windshield in sunlight, thus simulating a lunar flash. It was just this mechanism that was used to explain the mysterious "flasher" in Perseus. The May 23, 1985 lunar flash mentioned above may have been a reflection from a large military weather satellite that was transiting the moon at the time of the flash, say R.H. Rast and P. Maley. However, the timing and position are not as precise as some would like. Besides, the outline of the flash on the photograph is "confined by features of the lunar terrain." (Anonymous; Lunar Flash Mystery: Solved or Deepened?" Sky and Telescope, 78:461, 1989.)

Comment. Lunar flashes were noted long before the satellite era. See p. 86 for the Perseus Flasher.

ANOMALOUS DISTRIBUTION OF LARGE, FRESH LUNAR CRATERS

The overwhelming majority of astronomers favors a meteor impact origin for the giant, fresh lunar craters. (Here, "fresh" means post-mare formation.) Such an origin would seem to favor a random distribution of these craters.

> However, it appears that the distribution of these large, fresh craters is far from random, contrary to what would be expected if their mechanism of formation was by impact. Even the most casual observer of the Moon cannot help but note that the maria contain very few large craters. The more experienced observer will take note of several apparent anomalies. Six magnificent post-mare craters are almost fortuitously located immediately adjacent to mare regions, these being Langrenus, Theophilus, Cavalerius, Aristoteles, Aristarchus, and Copernicus.

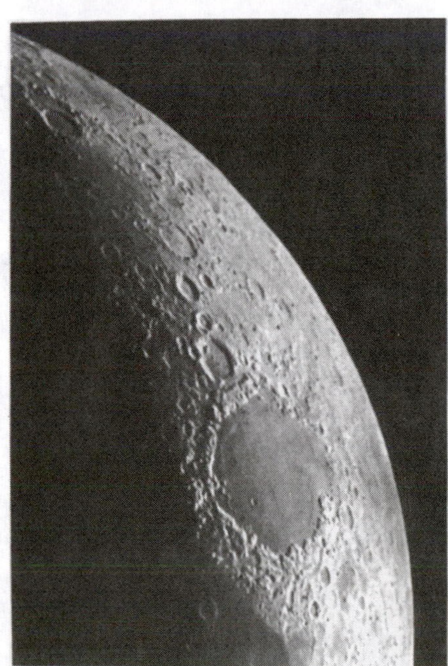

By comparison with the surrounding terrain, dark-floored Mare Crisium is hardly pock-marked at all by large meteor craters.

M.T. Kitt then buttresses these superficial observations with a statistical analysis, which indicates a strong nonrandom distribution of these fresh craters. Apparently, the volcano-meteorite controversy is not completely settled after all these years. (Kitt, Michael T.; "Anomalous Distribution of Large, Fresh Lunar Craters," Strolling Astronomer, 31:22, 1985.)

Comment. Some of the fresh craters on the mare borders, such as Aristarchus and Copernicus, are well-known sites of lunar transient phenomena. Could they be analogous to the terrestrial volcanos constituting the "ring of fire" around the Pacific Basin?

THE MOON'S MOONLETS

The great lunar basins are not arranged randomly. They occur in bands---not one band but several. How can this geometry be explained? One hypothetical scenario has the primitive moon surrounded by many moonlets, 60 miles and larger in diameter, plying equatorial orbits that are unstable. As the moonlets' orbits decayed, some crashed into the moon's equatorial regions, blasting out a band of huge craters. The force of the impacts also caused the lunar crust to slide over the still-liquid core by as much as 90°. When the next group of moonlets crashed, they gouged out a new belt of craters and shifted the crust still more. Magnetic measurements of lunar rocks tend to confirm that the lunar crust did indeed shift by large angles---several times. (Anonymous; "Did the Moon Have Moonlets?" Science Digest, 92:20, January 1984.)

Comment. Such events could have also happened on earth, which would account for tropical-zone fossils being found at the present poles.

METEOROID IMPACTS: THE OTHER SIDE OF THE STORY

Astronomers have long puzzled over the origin of localized magnetic anomalies on the moon. These magnetic concentrations (called "magcons") are located precisely on the opposite side of the moon from the larger lunar basins. How could an impact on the moon magnetize the antipodal region?

The impact of a large silicate meteoroid at speeds of 10 kilometers/second would not only blast out a big crater but it would also create a huge cloud of hot, partially ionized gas. This hot gas or plasma will conduct electricity and interact with lunar magnetic fields. As the plasma cloud spreads away from the impact site, it acts like a bulldozer, compressing the lunar magnetic fields ahead of it, as it envelopes the whole moon and rushes towards the antipodal point. It drives the compressed magnetic field into the surface, permanently magnetizing the rocks at the antipodal point. Voila! Magcons. (Hood, L.L., and Huang, Z.; "Formation of Magnetic Anomalies Antipodal to Lunar Impact Basins: Two-Dimensional Model Calculations," Journal of Geophysical Research, 96:9837, 1991.)

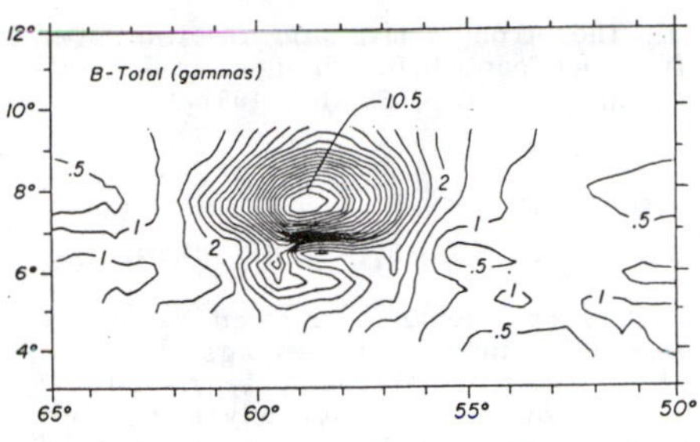

The Reiner gamma magnetic anomaly is defined by these contours of constant magnetic field strength. Nearby there is one of the curious swirl markings.

Comment. The earth also sports scars from the impacts of large meteoroids. Are there magnetic anomalies opposite these craters? Even more interesting to check out would be the holes blasted in the earth's biosphere by the converging masses of hot gases at the antipodal points. Wouldn't there be extinctions seen in the fossil record at these antipodal points? For example, the astronomical projectile that presumably caused the dinosaurs' extinction at the end of the Cretaceous Period (65 million years ago) might well have disturbed the fossil record on the opposite side of the earth from its impact point.

MYSTERIOUS SWIRL PATTERNS ON THE MOON

In at least three lunar locations, enigmatic bright and dark swirl patterns drape craters and mare terrains. Ranging from 10 km across to less than 50 m, they may be ribbon-like, open-looped, or closed-looped. The swirls are sharply defined but do not appear to scour or otherwise disturb the terrain where they occur. Similar swirl patterns have been recognized on Mercury. Two intriguing characteristics of the lunar swirls are: (1) they coincide with strong magnetic anomalies; and (2) they appear to be very young, being superimposed on top of essentially all lunar features of all ages. Schultz and Srnka suggest that recent cometary impacts created the patterns. (Schultz, Peter H., and Srnka, Leonard J.; "Cometary Collisions on the Moon and Mercury," Nature, 284:22, 1980.)

THE MOON'S MAGNETIC SWIRLS

The impressively strong magnetic anomalies discovered on the lunar surface remain enigmas. They appear to be superficial patches of highly magnetic material rather than deep-seated manifestations of basic lunar structure. Instead of being associated with gravity anomalies, the magnetic patches seem coincident with strange swirl-like markings on the moon's surface. The logical inference is that the swirls are surface patterns of a highly magnetic substance ---but why peculiar patterns and where did the strongly magnetic material come from? (Hood, L.L.; "Enigma of Lunar Magnetism," Eos, 62:161, 1981.)

COMETARY SCARS ON THE MOON?

More information has surfaced on the enigmatic lunar swirl markings. These whitish blotches are not only visually incongruous, being obviously different from the debris splashes around craters, but they also exhibit curious magnetic properties. J.F. Bell and B.R. Hawke, of the University of Hawaii, have acquired near-infrared spectra of the swirl designated Reiner Gamma. They report that the composition of the swirl material does not match crater ejecta; and, also, that a previously undetected reddish halo surrounds the swirl. Best guess at present: the swirls are the scars of comets---probably less than 100 million years old. (Anonymous; "Cometary Scars on the Moon? Sky and Telescope, 75:11, 1988.)

Comment. Does the earth also bear cometary scars? Some think that the 1908 Tunguska event or Siberian Meteor was a comet-caused phenomenon.

THE PUZZLE OF THE MOON'S ORIGIN

The moon is the closest and best-studied astronomical object. Yet, there is no agreement as to its mode of origin. One might say that planetary scientists have just about thrown in the towel on the three major theories of lunar origin. Two recent articles attest to this discouraging situation.

A Sky and Telescope article provides an excellent review of all three theories, indicating the reasons why each fails to convince a majority of scientists. The theories and the primary reasons for their rejection are:

1. Fission from earth (lack of sufficient angular momentum in the earth-moon system and the fact that the moon does not orbit in the plane of the earth's equator).
2. Gravitational capture (the capture of such a large object in a nearly circular orbit is considered too improbable).
3. Earth-moon accretion as a double planet (the compositions of the earth and moon are too different).

This article concludes that the resolution of the problem of lunar origin must await our return to the moon for more scientific exploration. (Rubin, Alan E.; "Whence Came the Moon?" Sky and Telescope, 68:389, 1984.)

An article in Science also discusses the classical theories of lunar origin and quickly disposes of them for the same reasons. However, a fourth theory makes an appearance, which we might call the Big Splash Theory. The idea is that a Mars-sized object (1/10 the earth's mass) made a grazing collision with the earth, when the solar system was more heavily populated with debris. The collision vaporized the projectile's rocky mantle and a similar quantity from the earth. This material was flung into orbit by the force of the impact and the expanding hot gases. Beyond the Roche Limit, the expelled debris coalesced into the moon. This fourth theory is now seriously considered by planetary scientists who---so far---have found no serious defects. (Kerr, Richard A.; "Making the Moon from a Big Splash," Science, 226:1060, 1984.)

ORIGIN OF THE MOON DEBATED

In October of 1984, the Conference on the Origin of the Moon convened in Kona, HI. The clear favorite among the four contending hypotheses was the earth impact scenario, which may be stated as follows:

> Near the end of the Earth's accretion, after its core had formed and while the growing planet was still molten, an object at least the size of Mars smashed into it at an oblique angle. The cataclysm put large quantities of vaporized or partly vaporized impactor and Earth into orbit. The primitive Moon formed from that material.

Conferees turned thumbs down on the theory that the moon was captured by the earth. Still not ruled out are the double-planet hypothesis (earth and moon accreted in their present configuration) and the fission-from-earth theory. (Taylor, G. Jeffrey; "Lunar Origin Meeting Favors Impact Theory." Geotimes, 30:16, April 1985.)

EARTH-MOON FISSION: A SLIGHT HINT

Background. One of the favorite theories for the origin of the moon states that it split off or fissioned from the earth. Here is some evidence for this catastrophic event.

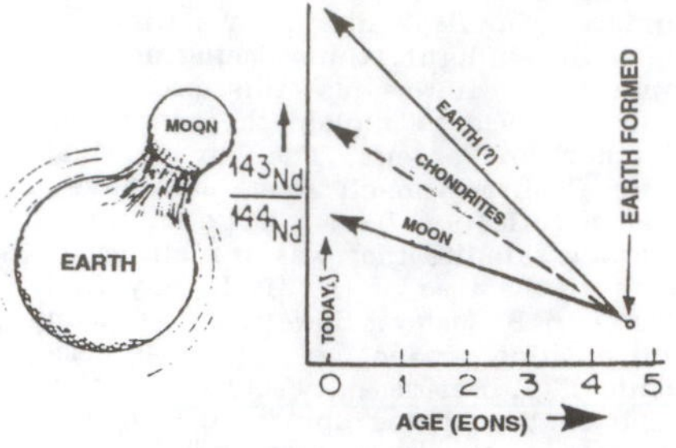

The hypothetical fission of the earth-moon could have shifted the Nd isotope ratio relative to that of chondrites.

Although the Sm/Nd isotope ratios of the earth and the chondrites (a type of meteorite) seem pretty close in value, the possibility of a slight but significant discrepancy remains. If this small difference is confirmed, it would strongly imply that the earth's Sm/Nd ration was shifted after its formation. The most reasonable event that could shift the ratio is the long-debated earth-moon fission, wherein the earth's lighter surface material was somehow torn off to form the moon. (DePaolo, Donald J.; "Nd Isotopic Studies: Some New Perspectives on Earth Structure and Evolution," Eos, 62:139, 1981.)

Comment. The moon's density is markedly less than the earth's, so the idea is not as wild as it seems.

MARS

Of all the other planets that circle the sun, we know Mars the best. Although not as conspicuous in the night sky as Venus, Mars possesses no opaque atmosphere to hide its surface details from terrestrial telescopes and the eyes of spacecraft cameras. Thanks to the Mariner and Viking missions to Mars, we now have high-quality, close-up photos of the Martian surface. Under such scrutiny, one would suppose that all of the mysteries of Mars would have melted away long ago. But newer mysteries have replaced some of the old ones, and many of the old enigmas persist down to the present day.

The Martian anomalies and curiosities corralled by Science Frontiers divide easily into the following five categories:

- Canals and craters. The continuing observation and photographing of linear features through terrestrial telescopes that cannot be identified on close-up spacecraft photographs.
- Rivers and lakes. Dry channels, empty lakes, and layered strata hint that Mars was much wetter in the past and perhaps more conducive to life.
- Curious geological formations. The "white rock," the "face," and puzzling ridges.
- The possibility of life. Did the Viking experiments really rule out life on Mars? Rethinking the question.
- The puzzles of Phobos. The parallel grooves and a "mystery object."

CANALS AND CRATERS

EPHERMERAL LINES ON MARS

Background. For more than a century, a few astronomers have been seeing strange lines on the Martian surface. P. Lowell, an American astronomer, thought they were water-carrying canals. Some people still see the lines, but others do not. Certainly, spacecraft sent to Mars don't record them. What could these lines be?

At first, the close-up Viking and Mariner photos of the Martian surface seemed to dispose of the famous "canals." Few permanent linear features were discovered---certainly nothing like the complex grid of straight lines sketched and photographed by Lowell. Lowell may be vindicated yet, for at least one sharp, dark line has been photographed by a Viking Orbiter during three Martian springs just north of the great volcano Arsia Mons. Called a "weather wave," this line appears only in the spring when Lowell's canals darkened. This year a second long line, slightly curved, joined up with the first line at a triangular junction looking suspiciously like one of Lowell's "oases." (Anonymous; "Rare Martian Weather Wave---with a Kink," Science News, 118:7, 1980.)

Comment. Could it be that the notorious canals are atmospheric features that come and go? This is actually unlikely because many observers down the years have drawn the same canals in the same locations.

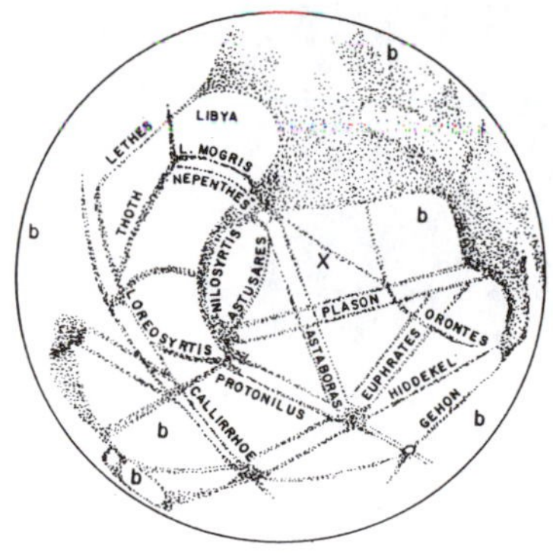

A drawing of Mars made by the astronomer G.V. Schiaparelli in June 1888, using an 18-inch refracting telescope. The "canals" were rendered clearly.

HYPNOTIC MARS

Scientists (and most people, in fact) have a strong innate urge to "close the book on problems"; that is, come up with final, absolute solutions. Apparently nature---Mars, at least---is not cooperating. When the Mariner and Viking spacecraft found no traces of Martian canals, most astronomers "closed the book" on the century-old Martian canals. Percival Lowell and all the other astronomers, who also saw canal networks, were obviously deluded.

Wouldn't you know it, those canals still haven't gone away! Consider this testimony of I. Dryer: "As staff photographer and observer at Lowell Observatory during the 1960-61 apparition of Mars, I spent several nights scrutinizing the planet's surface through the 24-inch Clark refractor. At instants of steady seeing, I saw and attempted to photograph an apparent network of fine lines. Unfortunately, I was unable to duplicate clearly what I saw. Still, several of the more visually distinct 'canals' can be traced on my original prints. Each is a composite of the finest four to eight images out of 49. Such prints suppress grain, remove artifacts, and enhance detail." The canals thus photographed match some of Lowell's well, although detail is lacking. (Dyer, Ivan; "Martian Canals," Sky and Telescope, 73:605, 1987.)

Comment. Let's open that "book" on the canals. Even better, let's never close books on any phenomena.

MARTIAN CANALS: IS LOWELL VINDICATED?

Whenever we get the opportunity, we try to clear Percival Lowell's name. Although he may have gone too far in claiming that the canals of Mars were the labors of intelligent beings, he definitely saw something. Earthbound observers still see and photograph Martian canals, despite the acknowledged fact that Martian orbiters and landers saw nothing resembling canals.

R. Gordon now relates how on June 6, 1967, he and a friend, W.H. McHugh, were viewing Mars through an 8-inch f/9 reflecting telescope. The thick haze reduced atmospheric transparency, but the seeing was excellent. The infamous canals were there! "Two canals stretched clearly from Sabaeus Sinus and Meridiani Sinus to the northern deserts, where they faded. A most interesting canal was Deuteronilus-Protonilus---originating in Niliacus Lacus which ran both east and west until I lost sight of it near the limb---we counted at least six oases on this one, strung out like beads on a string!" (Gordon, Rodger; "Martian Canals: Is Lowell Vindicated?" Sky and Telescope, 75:348, 1988.)

Comment. Yes, the canals Lowell and others drew are still there---not physically perhaps---but possibly as an anomaly of perception and camera/telescope aberrations.

DID BARNARD & MELLISH REALLY SEE CRATERS ON MARS?

The answers are "No" and "Probably," respectively. Well, so what? Everyone knows from spacecraft photos that Mars is definitely peppered with craters; and who are Barnard and Mellish anyway?

E.E. Barnard was one of the great American telescopic observers. J.E. Mellish was an amateur astronomer and a protege of Barnard. Both men may have seen Martian craters; Barnard at Lick Observatory in the early 1890s, and Mellish at Yerkes in 1915. These early dates are what make this story interesting, because prior to the Mariner-4 flyby of Mars in 1965, anyone claiming to have seen craters on Mars would have been labeled a crackpot. Just a mere three decades ago, planetary catastrophism was a ridiculous notion.

Barnard never dared publish his drawings of Martian craters for fear of ruining his reputation. Mellish was not so reticent. He wrote and lectured widely on his anomalous observations. No one believed him because his observa- conflicted with reigning paradigms.

Once the paradigm shifted and craters on other planets were legitimized, astronomers looked back and wondered if Barnard and Mellish really did see craters. After all, nobody else had, although several reknowned astronomers had drawn networks of canals they had definitely seen. Some of Barnard's early sketches of Mars surfaced in 1987. They show known volcanos and the huge canyon complex called Valles Marineris, but the spots (thought to be craters) do not coincide with any known craters. Unfortunately, Mellish's drawings of his craters were destroyed by fire a year before the Mariner-4 flyby. However, Mellish's verbal descriptions of the craters are very convincing; and his honesty and accuracy are well-known. So, if anyone really did see pre-Mariner Martian craters, it was probably Mellish. (Sheehan, William; "Did Barnard & Mellish Really See Craters on Mars?" Sky and Telescope, 84:23, 1992.)

WHY AREN'T THE MARTIAN CRATERS WORN DOWN?

Although several of the articles that follow will recall Lowell's vision of a watery Mars, these inferred conditions are long gone. The Mariner and Viking spacecraft radioed back pictures of a dry planet. Wind rather than water is now the major erosional force. The sand dunes and drifts, the wind-shadows be-

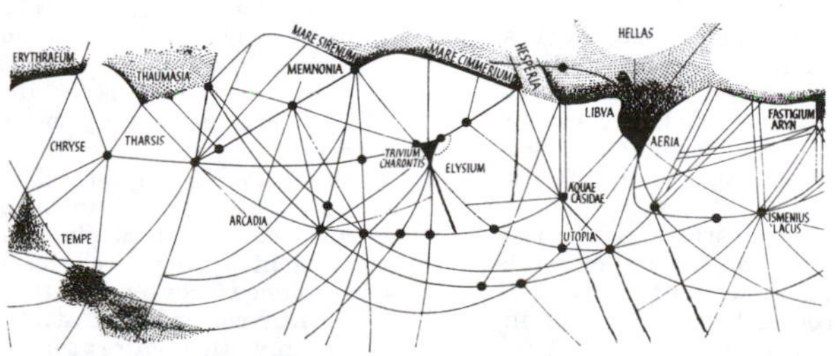

A Mercator Projection of Mars and its major canals drawn by Lowell. The black dots are "oases."

hind rocks, and the sand-blasted surfaces all attest to the desertification of Mars.

Ronald Greeley, of Arizona State University, and his colleagues have simulated Martian winds in a special wind tunnel at NASA's Ames Research Center. Using spacecraft-measured wind velocities and patterns, they tried to duplicate the Martian erosional environment. The results were a surprise. They implied that the Martian surface should be worn down by wind-driven sand and dust at rates up to 2 centimeters per century. But at this rate, the Martian craters, which are estimated to be hundreds of millions of years old, would have been worn level long ago. The researchers are now wondering what is wrong with their simulation. They venture that the Martian sand may not be "normal," or perhaps the eroding particles do not travel as fast as they figured. (Anonymous; "The Windblown Planet Mars," Sky and Telescope, 68: 507, 1984.)

Comment. Perhaps Mars has not been a desert for as long as thought!

RIVERS AND LAKES

AN OASIS ON MARS---NO PALM TREES BUT.....

Most of the data returned from the Viking landers and orbiters confirm a highly dessicated surface for Mars. Life-as-we-know-it would seem impossible in such an ultradry environment. The Solis Lacus region is an exception. On occasion, orbiter photos of this region have revealed heavy frosts and fogs. Further, the clouds here have more moisture in them than elsewhere on Mars. Conclusion: considerably more water exists near the surface of the Solis Lacus region. Since this area was the source of the great 1971 Martian dust storm, one wonders whether the unusual concentration of water has been revealed only because winds have stripped off the normal dry surface layer. (Huguenin, R.L., et al; "Mars An Oasis in Solis Lacus," Eos, Transactions of the American Geophysical Union, 60:306, 1979.)

Comment. Close-up photos of Mars show many signs of fluid erosion. Considerable fluid may still remain well below the surface.

WHEN MARS HAD LAKES

Rhythmic layered deposits can be seen in the Valles Marineris, a large Martian valley. The strata are erosional remnants up to 5 kilometers high, with individual layers 170-220 meters thick. They can be traced on spacecraft photographs for some 50 kilometers. The material making up the strata is clearly different from that of the valley walls. After the layers were deposited, they were deeply eroded by some event in Martian history that seems related to the formation of the great outflow channels associated with Valles Marineris. The author of this American Geophysical Union paper concludes, "The morphology and history of the sediments are consistent with deposition in standing bodies of water early in Martian history." (Nedell, Susan S., and Squyres, Steven W.; "Geology of the Layered Deposits in the Valles Marineris, Mars," Eos, 65:979, 1984.)

Viking photo of probable sedimentary layers along the side of a Martian ridge. (NASA)

Comment. The "event" that deeply eroded the Martian deposits may have been similar to the catastrophic emptying of Lake Missoula, which carved out the channelled scablands of eastern Washington state as the Ice Ages waned. For details, see ETV5 in Carolina Bays, Mima Mounds, Submarine Canyons.

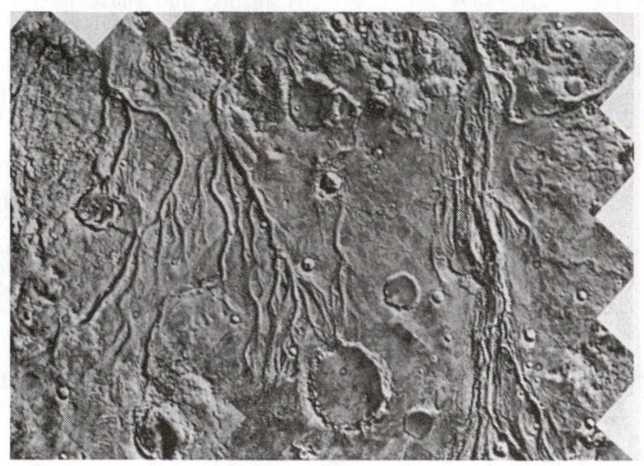

These Martian channels photographed by Viking lie just west of Chryse Planitia. (NASA)

RECENT MARTIAN RIVERS ERODE ALBA PATERA

Most of the Martian surface is thought to be more than 3.8 billion years old. This portion is densely cratered from a period of heavy meteorite bombardment. It is also carved by many channels that are thought to have been cut in ancient times by flowing water, water which quickly escaped into space or combined chemically with Martian minerals. The present atmosphere of Mars, in consequence, contains little water vapor.

But some of the Martian landscape, notably Alba Patera, raises questions about the above scenario. The anomalous characteristic of Alba Patera is its relative smoothness and scarcity of impact craters. This Martian real estate is believed to be 2 billion years younger than the rest of the planet. Even so, it too is marked by "fluvial" features that resemble stream beds.

Question #1. How did Alba Patera get smoothed out or "reworked"? In other words, what happened to the ancient craters that must have pocked its surface, as they do everywhere else?

Question #2. Where did the water come from to cut Alba Patera's stream beds if all of the Martian water disappeared 2 billion years earlier?

One line of thought maintains that fluvial does not mean pluvial, and that Martian water has come from below rather than as rain from the atmosphere. Both fluvial episodes, in this view, occurred when something caused the Martian crust to release huge quantities of stored water. Hydrothermal activity is mentioned as a possibility. (Eberhart, J.; "The Martian Atmosphere: Old Versus New," Science News, 135:21, 1989.)

Comment. Another speculation is that immense quantities of Martian water are tied up in methane hydrate and is catastrophically released when the ambient temperature is somehow increased. Immense quantities of methane hydrate also exist in terrestrial marine sediments.

PETROL CHANNELS ON MARS?

The many channels on Mars closely resemble terrestrial river beds. But Martian models that assume water to be the eroding agent encounter difficulties because the Martian gravity is too weak to hold hydrogen when water is dissociated by solar radiation. A better bet, say Y.L. Yung and J.P. Pinto is liquid hydrocarbons; i.e., petrol. Starting with a methane atmosphere, at one-tenth of an earth atmosphere pressure, the natural loss of hydrogen would lead to the polymerization of hydrocarbons and eventual condensation. "Petrofalls" from this atmosphere could cover the Martian surface to a depth of one meter and lead to heavy erosion. (Anonymous; "Martian Surface in Good Spirits," New Scientist, 79:19, 1978.)

CATASTROPHIC FLOODING ON MARS?

Could that parched red planet seen in the Viking pictures have been the site of a colossal flood---a wall of water greater than anything ever seen on earth?

Terrestrial geologists point to the Channeled Scablands in eastern Washington State as evidence of what the sudden release of a huge lake's water can do to the landscape. Everywhere in this part of Washington are deeply incised grooves and dry cataracts separated by water-streamlined bars. Exactly this sort of harsh, scoured topography can be found at Kasei Valles, Mars.

> The upper part of the channel system is typically less than 1 km deep and descends from Echus Chasma about 1 km over a distance of 1000 km; it then splits into north and south channels. On the basis of a stereomodel of Viking images, we have measured the geometry of a steep, constricted reach of the north channel that drops 900 m in only 100 km. A late-stage flood is hypothesized to have scoured the channel. If we assume that channel striations indicate water levels, then the flood had a minimum cross-sectional area of 3.12×10^7 m^2 (the putative flood had a width of 83 km, an average depth of 373 m, and a maximum depth of 1280 m). These channel measurements suggest that flood velocities ranged from 32 to 75 $m \cdot s^{-1}$ and that discharge was greater than 1 $km^3 \cdot s^{-1}$, values larger than those calculated for any other flood event on Mars or Earth.

That maximum flood velocity is equal to 170 miles/hour! The erosion features included a ¼-mile-deep pothole. Some temporary lake was likely the source of the flood, although no one sees any surface water on Mars today. (Robinson, Mark S., and Tanaka, Kenneth L.; "Magnitude of a Catastrophic Flood Event at Kasei Valles, Mars," Geology, 18:902, 1990.)

THE MARTIAN GREAT LAKES

Percival Lowell hasn't been vindicated by the discovery of sedimentary formations on Mars, but his spirit must be pleased. Lowell's geometrical network of artificial canals has been replaced by great arroyos, flood-created deposits, and now evidence that Mars was once host to ice-covered lakes up to 3 miles deep and as large as Lake Superior. Photos from the Viking spacecraft reveal sedimentary layers up to 250 feet thick that seem to have been laid down by liquid water. The source of the sediments and mode of deposition are unknown. (Anonymous; "Great Lakes on Mars," Science 86, 7:13, April 1986.)

A CHILLY MARTIAN NIGHT

Viking Lander 2 photographed frost on Mars in September 1977 during the Martian winter. A planet-wide dust storm had just subsided, and the theory evolved that both water and carbon-dioxide ice had frozen on dust particles in the atmosphere. Such particles were heavy enough to fall and give the scene around Viking a snow-like coating. However, frost was again photographed in 1979 (one Martian winter later) without the benefit of a dust storm. So Mars theorists are in a quandry---no dust, no frost theory. (Anonymous; "Viking, Three Years Later,: Eos, American Geophysical Union, Transactions, 60:635, 1979.)

Some Martian channels are associated with collapse features. (NASA)

Comment. Evidently, frost cannot form directly on the Martian surface as it does on earth due to the very low vapor pressure of water on the planet.

A MARTIAN ICE AGE?

Long, sinuous depressions called "outflow channels" are concentrated in the equatorial regions of Mars. They clearly resemble terrestrial stream beds and have been attributed to water action. The ancient Spokane Flood that carved out Washington State's scablands was seen as an apt terrestrial analogy. The Martian channels, however, are much larger than the water-eroded terrestrial analogs. The authors of this paper suggest that ice rather than water created the Martian channels, pointing out that terrestrial ice-stream and glacier landforms are much closer in scale to the Martian features. (Lucchitta, Baerbel K., et al; "Did Ice Streams Carve Martian Outflow Channels?" Nature, 290: 759, 1981.)

Comment. This suggests the possibility that earth and Mars may have had synchronous Ice Ages due to solar perturbations or an encounter between the solar system and a cloud of absorbing matter.

GAIA ON MARS?

Background. The Gaia hypothesis states that the whole earth is really a living entity that adjusts geophysical parameters to preserve itself and even expands the domain of life. Perhaps Gaia also exists on Mars.

H.L. Heifer, University of Rochester, noting the absence of extensive cratering on the northern plains of Mars, suggests that some 2-3.5 billion years ago these plains were covered with oceans. These ancient seas, perhaps as much as 700 meters deep, protected the plains from direct impacts. Further, crater density counts for Chryse and the Martian highlands imply that Mars possessed a fairly dense atmosphere until about 1.5 billion years ago. In this Abstract Heifer speculates as follows:

> With both early Earth and early Mars having similar atmospheric compositions and not too dissimilar atmospheric structures, it is reasonable to suppose that the warm Martian oceans, like the ancient oceans of Earth, would develop anaerobic and aerobic photosynthesizing prokaryotes and structures like stromatolites. Their development might have changed the Martian atmosphere. Their fossils might be found along the fringes of the old oceans, the northern lowland plains. (Heifer, H.L.; "Of Martian Atmospheres, Oceans, and Fossils," Icarus, 87: 228, 1990.)

Comment. The Gaia influence is seen in the molding of the Martian atmosphere into something more conductive to the development of life. One can also speculate that if life did develop on Mars, it could have seeded the earth via bits of debris blasted off by meteorite impacts. Several meteorites picked up in Antarctica are thought to have come from Mars originally. See pp. 76 and 220.

CURIOUS GEOLOGICAL FORMATIONS

STRANGE HILLOCKS AND RIDGES ON MARS

On the northwestern flanks of the huge Martian volcano Arsia Mons are countless hillocks of undetermined origin. Mostly 100 to 500 meters in diameter, the hillocks cover the edge of the volcano flank. In addition, the outer edge of the flank is surrounded by dozens of parallel ridges that stretch lengthwise for hundreds of kilometers. A peculiar feature of the ridges is that they have not been disturbed by craters or flow features; there are not even any variations in surface brightness. One explanation suggests that both hillocks and ridges were created by a huge landslide. (Anonymous; Science News, 113;43, 1978.)

Comment. The hillocks resemble the much smaller terrestrial Mima Mounds. See p. 201.

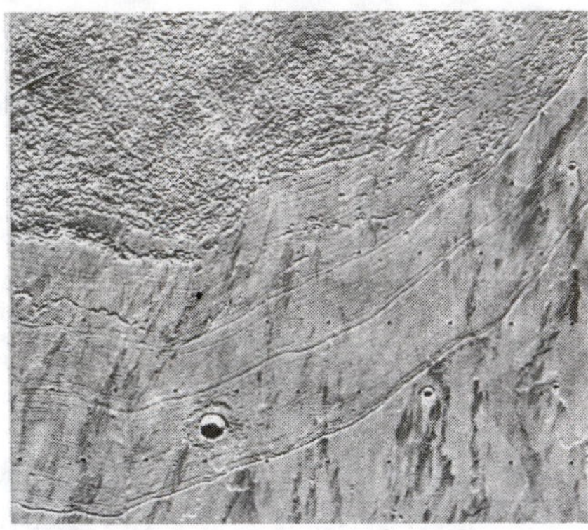

Puzzling hillocks and parallel ridges near the Martian volcano Arsia Mons. (NASA)

WHITE AREA IN BOTTOM OF MARTIAN CRATER

Near the Martian equator, in the bowl of a 58-mile crater, Viking Orbiter snapped a peculiar white region. Called the White Rock, the formation is 8.5 x 11 miles in size and possesses an unusual grooved surface. White Rock is too close to the equator to be ice or snow. It is a unique and unexplained feature. (Anonymous; "A Martian Mystery," Astronomy, 7:64, January 1979.)

Comment. White Rock looks a bit like an eroded salt plug!

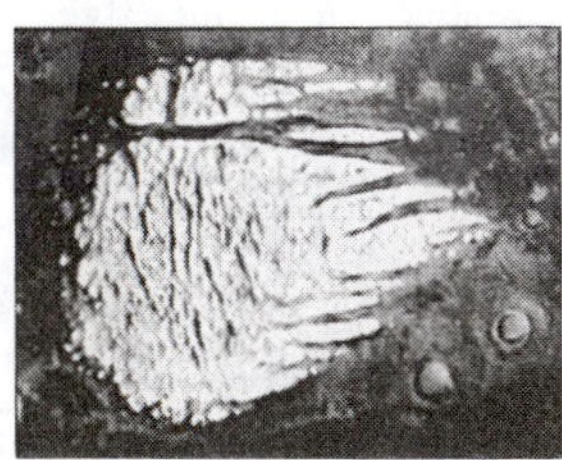

The "white rock" photographed by Viking. (JPL)

DARK PATCHES ON MARS

Inside the vast Valles Marineris Canyon complex, Viking Orbiter photos have picked out wind-blown patches of dark material. These patches are strung out along faults for some 200 kilometers. Astronomers believe they are volcanic vents, which are only a few million years old. (Anonymous; "Recent Volcanism on Mars?" Sky and Telescope, 73:602, 1987.)

A NEW FACE ON MARS?

Background. Some photos taken by the Viking Spacecraft seem to show a huge human face sculptured on the Martian surface. This is popularly known as "the face on Mars." Scientists scoff at the whole idea, saying the face is simply natural erosion, and its resemblance to a face is purely coincidental.

That face on Mars does not seem to go away! Could it, after all, really be artificial? M. Carlotto, writing in Applied Optics, presents the results of his analysis of the photos of the enigmatic "face":

> The image enhancement results indicate that a second eye socket may be present on the right, shadowed side of the face; fine structure in the mouth suggests teeth are apparent.

"Teeth?" This Martian face is becoming too human-like to be an accident of nature! Carlotto also summarized his impression of the face's topography:

> ....results to date suggest that they may not be natural. (Hecht, Jeff; "Computer Does a Double-Take on the Face of Mars," New Scientist, p. 39, July 7, 1988. Also see Applied Optics, 27:1926, 1988.)

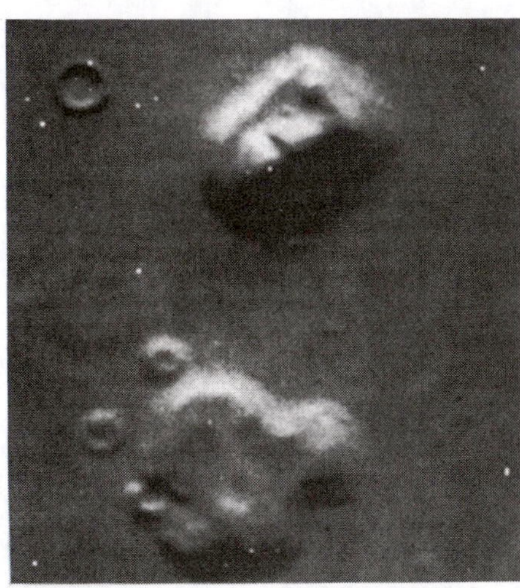

The Martian "face" as seen in NASA frame 70A13, as photographed from the Viking orbiter. (NASA)

In a more light-hearted manner, P. Jones asserts that a second Martian face exists, and that somehow scientists have managed to keep it under wraps. A photo of the second face appears in the August 25, 1988, issue of New Scientist. Jones remarks:

> If faces are what you are looking for, it's reasonably convincing, perhaps more convincing than its better-known sibling. It has symmetry, two eyes, a nose, a mouth, a chin; it even has a comical shadow to provide a beard, if you happen to like beards. (Jones, Pat; "Mars Reveals Its Second Face," New Scientist, p. 62, August 25, 1988.)

Comment. What can one make of all this? Two faces might just mean that nature on Mars is as prolific a face-maker as she is here on earth. But as always, we could be treating the "face" too lightly. On p. 14, this strange Martian geological feature appears in a different context.

POSSIBILITY OF LIFE

UNEARTHLY LIFE ON MARS

From the media standpoint---and therefore that of most people---the Viking Martian biological experiments were uncompromisingly negative. Lewis points out that this is simply not so. The labelled-release experiments on both landers produced positive results every time a nutrient was added to fresh Martian soil. (The nutrient was tagged with carbon-14 and radioactive carbon dioxide always evolved, suggesting biological metabolism.) Further, the soil samples, when sterilized by heat, gave uniformly negative results.

On earth, such repeatable experiments would be considered strong evidence that life existed in the samples. The reason the Viking experiments were described as negative is that the other two "life-detection" experiments produced negative or equivocal results. The gas chromatograph, for example, detected no organic molecules in the Martian soil: and it is difficult to conceive of life without organic molecules.

At first, most scientists preferred to explain the ambiguous life-detection experiments in terms of strange extraterrestrial chemistry. Nevertheless, strange extraterrestrial life would explain the data equally well. Everyone should be aware that the Viking biology team still considers life on Mars as a real possibility. (Lewis, Richard; "Yes There Is Life on Mars," New Scientist, 80:106, 1978.)

IS THERE LIFE ON MARS AFTER ALL?

G. Levin and P. Straat, who designed one of the three life-detection experiments on the Mars Viking landers, have always maintained that the positive results obtained with their experiments were unreasonably overruled by the negative data from the other two experiments. At a recent scientific meeting in Washington, they stated, "It is more likely than not that our experiment detected life on Mars." Their research in the decade following the Mars landings has only strengthened their belief. Further, they have demonstrated that one of the other life-detection experiments producing negative results was not sensitive enough to detect low population levels of microorganisms. Realizing that the no-life-on-Mars dogma is well-entrenched, they looked for other kinds of evidence for life.

"In support of their claims, the two researchers presented two photographs of a Martian rock taken years apart by a camera on one of the landers. The photographs show greenish patches which had changed over time. Spectral analysis of the photographs compared favorably with the spectra given out by lichen-bearing rocks on Earth, as seen through a replica of the lander's camera." (Anonymous; "Is There Life on Mars After All?" New Scientist, p. 19, July 31, 1986.)

WARM, WET, FERTILE MARS

The geological history of Mars is looking more and more as if it could have supported or perhaps still does support life, although these Martian life forms might be very simple, such as bacteria.

> A large number of anomalous landforms on Mars can be attributed to glaciation, including the action of ice and meltwater. Glacial landscapes are concentrated south of lat -33° and in the Northern Plains suggesting vast Austral and Boreal ice sheets. Crater densities on the glaciated terrains indicate that the final glacial epoch occurred late in Martian history. Thus, Mars may have had a relatively warm, moist climate and dense atmosphere much later than previously believed.

(Kargel, Jeffrey S., and Strom, Robert G.; "Ancient Glaciation on Mars," Geology, 20:3, 1992.)

If Mars was warm and wet not too long ago, as implied above, perhaps life did gain a foothold there through either independent invention or, perhaps, through seeding by template-carrying comets or meteorites. P.J. Boston et al have investigated one possible Martian ecosystem:

> We have reexamined the question of extant microbial life on Mars in light of the most recent information about the planet and recently discovered nonphotosynthetic ecosystems on Earth---deep sea hydrothermal vent communities and deep subsurface aquifer communities. On Mars, protected subsurface niches associated with hydrothermal activity could have continued to support life even after surface conditions became inhospitable. Geochemical evidence from the SNC meteorites and geomorphological evidence for recent volcanism suggest that such habitats could persist to the present time...We suggest a possible deep subsurface microbial ecology similar to those discovered to depths of several kilometers below the surface of the Earth.

(Boston, Penelope J., et al; "On the Possibility of Chemosynthetic Ecosystems in Subsurface Habitats on Mars," Icarus, 95:300, 1992.)

Comment. Although Boston et al speak in terms of microscopic Martian life, there is no reason why chemosynthetic life forms could not be large---perhaps even large enough to leave traces on the Martian surface!

THE MARS-ANTARCTICA CONNECTION

"A study of ice-covered lakes in Antarctica has provided scientists with clues as to what conditions were like on Mars billions of years ago. Sufficient heat and gas would have been trapped beneath the Martian surface to have generated living organisms such as algae, bacteria, protozoa, and fungi. But life would have died out as the planet cooled and much of its atmosphere was dissipated. 'It is highly unlikely life could exist on Mars today,' (C.) McKay said.

.....

"However, some scientists have not dismissed the possibility that primitive life may still exist on Mars. 'The chances are remote but life may be located in slushy brines well below the surface, or even inside Martian rocks,' said Howard Klein, who headed the biological experiments on board Viking. 'Living microorganisms have been found just below the surface rocks in Antarctica,' Klein said." (Anonymous; "Antarctica Hints at Why There May be Fossils on Mars," New Scientist, p. 20, September 4, 1986.)

Comment. It is curious that some of the meteorites picked up in Antarctica are thought to have originated on Mars and been blasted off by meteoric impacts. This observation leads to the speculation that terrestrial life might have been seeded from Mars! Meteoric panspermia! Are we all Martians?

If not, could we become Martians? We might if we did some planetary engineering, as described in the next item.

TERRAFORMING MARS

The concept of terraforming a planet is an old standby of science fiction; it is the process by which a technologically advanced race manipulates the surface and atmosphere of an uninhabitable planet so that it becomes inhabitable. We humans know to our dismay that we have the capacity to modify the earth's environment, but could we perhaps exercise better judgment and terraform Mars? C.P. McKay et al have looked into this possibility:

> From our analysis, one could propose the following sequence of events: production of CFCs (or other greenhouse gases) starts on Mars and the surface temperature warms up by about 20°K. The regolith and polar caps release their CO_2 and the pressure rises to 100 mbar. One of two things could then happen. If there were large regolith and polar CO_2 reservoirs, the pressure would continue to rise on its own. If these were absent, the CO_2 pressure would stabilize, and additional CO_2 would have to be released from carbonate minerals. At this point (perhaps between 100 and 10^5 years) Mars may be suitable for plants. If there was a mechanism for sequestering the reduced carbon, these plants could slowly transform the CO_2 to produce an O_2-rich atmosphere in perhaps 100,000 years. If sufficient N_2 could also be released from putative soil deposits, and the CO_2 level kept low enough, then a human-breathable atmosphere could be produced. (McKay, Christopher P., et al; "Making Mars Habitable," Nature, 352:489, 1991.)

Comment. There is more to terraforming, but you get the idea from the above quote.

Now. J. Lovelock and others have speculated that our earth is much like a living organism that, with the help of the biosphere, maintains conditions suitable for life; that is, the Gaia concept. Terraformers of Mars would somehow have to induce Gaia to emigrate to Mars to regulate things there.

Or, since we are on a science-fiction kick, could it be that the earth itself was intentionally "terraformed" in the distant past and Gaia installed for our benefit---or even someone else's benefit?

NEW LIFE FOR MARTIAN LIFE

After the negative (some say "ambiguous") results from the Viking spacecraft life-detection experiments in 1976, astronomers and biologists have proclaimed that Mars is sterile. This pronouncement may have been premature.

A meteorite discovered in Antarctica in 1979 may change a few minds on this matter. This particular meteorite is one of the handful thought to have been blasted off into space by an oblique impact of an asteroid on the surface of Mars. Somehow they found their way to the Antarctic snows. But what is really exciting is the recent discovery that chemical analysis of one of these purported Martian meteorites revealed a high concentration of organic material deep within. The implication is that Martian life existed, perhaps still does exist, beneath the Martian surface, where the Viking's scoop could not get at it. (Anonymous; "Life under Mars?" Sky and Telescope, 78:461, 1989.)

Comment. In other words, Mars, like the earth, may harbor an unappreciated fauna in crevicular structure beneath the environmentally rigorous surface.

PUZZLES OF PHOBOS

WHAT CAUSED THE GROOVES ON PHOBOS?

Photographs from the Viking Orbiters show that the Martian satellite Phobos displays a heavily grooved surface. Enough high-resolution photos have been taken to prove that these grooves emanate from the large crater Stickney and run around the satellite to the opposite side where they die out. This suggests that the origin of the crater and the grooves are related. Further, the widest and deepest grooves (700 meters wide and 90 meters deep) are located close to Stickney. On the other side of Phobos, grooves are consistently less than 100 meters wide.

Despite these hints of impact origin, the grooves are not quite what one would expect from simple fracturing by collision. Some show beaded or pitted structures. Other grooves are composed of irregularly bounded segments. Final-

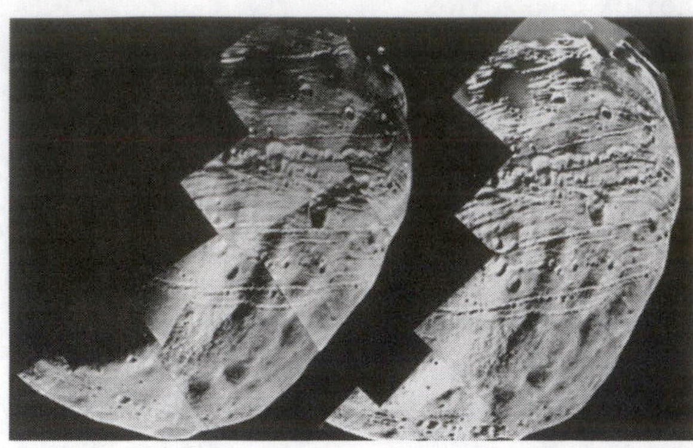

Photos of Phobos taken by Viking in 1977. The right-hand photo has been computer-enhanced. (NASA)

ly some of the straight-walled sections seem to have slightly raised rims. Evidently some internal forces, perhaps stimulated by the formation of Stickney, also played a part. (Thomas, P., et al; "Origin of the Grooves on Phobos," Nature, 273:282, 1978.)

GROOVES OF PHOBOS STILL UNEXPLAINED

The Martian satellite Phobos is etched by curious grooves. Initially, the grooves were thought to be fracture lines formed by the impact that blasted out the crater Stickney. Studies of the grooves reveal at least three families of grooves of different ages, with the members of each family located on parallel planes cutting right through the body of the satellite.

Two different papers have proposed radically different explanations. A. Horvath and E. Illes wonder whether Phobos might not be a layered structure, having once been part of a larger stratified body. J.B. Murray thinks the families of grooves might have been scraped out by disciplined formations of meteorites launched into orbit by Martian volcanos. (Horvath, A., and Illes, E.; "On the Possibility of the Layered Structure of Phobos," Eos, 62:203, 1981, and Murray, J.B.; "Grooved Terrains on Planetary Satellites," Eos, 62:202, 1981.)

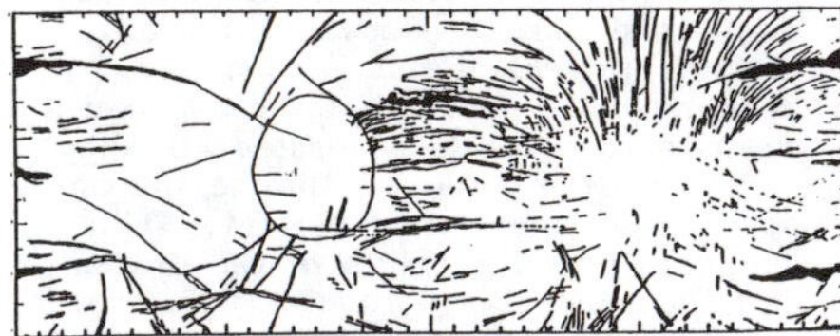

Map of Phobos' grooves. The crater Stickney is just left of center.

THE PHOBOS MYSTERY OBJECT

On March 28, 1989, Russian ground controllers suddenly and unexpectedly lost contact with their spacecraft that was shadowing the Martian moon Phobos. The last close-up photo of Phobos snapped by the spacecraft contained "an object which shouldn't have been there."

The "mystery object" photographed near Phobos.

Naturally, this Phobos Mystery Object (PMO) was quickly dubbed a UFO by some. It was even speculated that the Russian mission had been deliberately terminated by aliens! Such a scenario dovetailed neatly with the old speculations that Phobos is actually an artificial satellite of Mars, which is being used as a base of operations by someone or something.

The final photo of Phobos, taken in infrared light just three days before the communication failure, reveals the outlines of both Phobos and the PMO. All surface detail is washed out, as is common in infrared photographs. If the PMO was at the same distance as Phobos itself, it would be about 2 kilometers wide and 20 long. Its surface brightness is the same as that of Phobos. The sides of the PMO are perfectly parallel; it is rounded at both ends; the end towards Phobos narrows slightly; the other end seems to have a slight protrusion.

Since the PMO does not appear to have a metallic surface and displays no antennas or other indicators of artificiality, it is reasonable to ask whether it might be some natural phenomenon. One possibility is that the PMO image is only a "trailed moonlet;" that is, the smeared image of a small piece of debris also in orbit about Mars but moving at a slightly different velocity from that of the spacecraft and Phobos. Since the exposure time of the last photo was 8 seconds, a smeared image due to the PMO's relative motion is reasonable. (Anonymous; "Mystery Object Encountered by Russian Phobos Spacecraft," Meta Research Bulletin, 1:1, March 15, 1992.)

JUPITER

The giant planet of our solar system is Jupiter. The naked eye sees Jupiter as a bright jewel in the night sky; and some observers claim they can even see Jupiter's four, large "Galilean" satellites without a telescope. With a little optical help, however, these four moons (Io, Europa, Ganymede, and Callisto) become bright planets circling a miniature sun---a solar system within the solar system. It is, in fact, the Galilean satellites that generate all of the Jovian anomalies garnered by Science Frontiers since 1976.

SEEING DOUBLE AND EVEN TRIPLE ON JUPITER

Sometimes the shadows cast by Jupiter's satellites on the face of the planet during transit are doubled or, more rarely, tripled. This article draws attention to several recent observations of this most perplexing phenomenon. Fox points out that these multiple shadows are really not all that uncommon. He does not try to explain them. (Fox, W.E.; "Jupiter; Double and Triple Satellite Phenomena," British Astronomical Association, Journal, 88:360, 1978.)

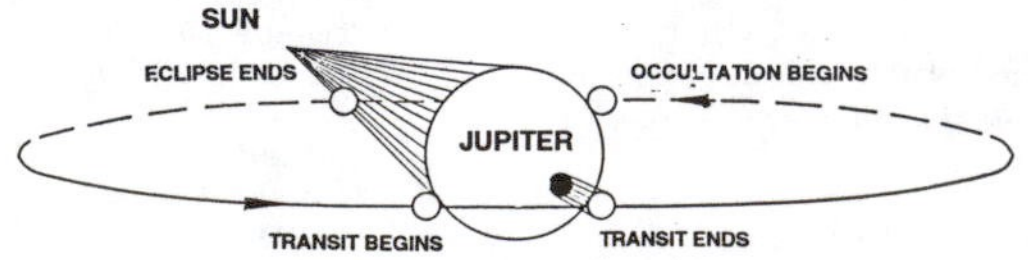

The Galilean satellites of Jupiter sometimes display eclipse and transit anomalies. For example, Io on rare occasions casts double shadows on Jupiter.

POST-ECLIPSE BRIGHTENING OF IO CONFIRMED

For about 15 minutes after Jupiter's satellite Io emerges from the planet's shadow after an eclipse, it unaccountably brightens far beyond its normal level. Observing Io with a spectrophotometer in 1978, Witteborn et al measured a brightness increase between 4.7 and 5.4 micrometers that was three to five times the brightness at other phase angles. Long a controversial phenomenon, this confirmation of Io's post-eclipse brightening has led to a search for possible explanations. Witteborn et al suggest that the transient flare up is a complex thermoluminescent effect excited by interaction with Jupiter's magnetosphere followed by solar heating as Io emerges from the shadow. (Witteborn, F.C., et al; "Io; An Intense Brightening near 5 Micrometers," Science, 203:643, 1979.)

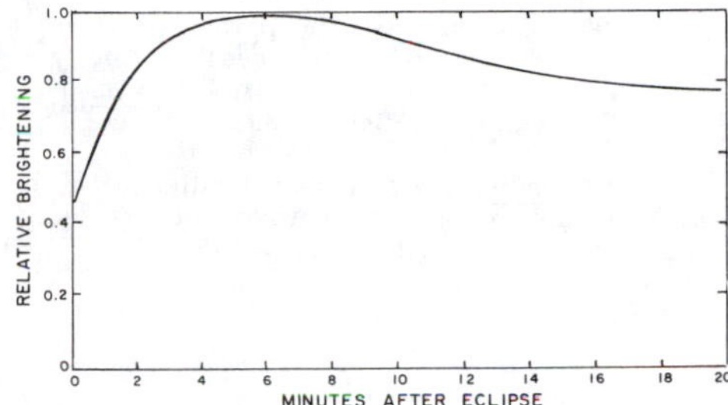

Typical brightening curve of Io after emerging from Jupiter's shadow.

DARK SECRET BEHIND JUPITER

When Jupiter's satellite Io ducks into Jupiter's shadow, something mysterious happens. Some of the time, not always, Io emerges from the shadow about 10% brighter than when it entered. In 10-20 minutes its brightness decays to normal levels. One suspicion is that SO_2 in Io's atmosphere condenses on the planet's surface in the cold shadow, coating some dark areas with a bright sulfurous 'frost.' A recent measure of Io's post-eclipse brightness detected no brightness change whatever. Obviously we have a real but rather unreliable phenomenon. (Morrison, Nancy D., and Morrison, David; "Io: Post-Eclipse Brightening Still Mysterious," Mercury, 11:27, 1982.)

IO'S ELECTRICAL VOLCANOS

Gold, of Cornell, long known for his provocative theories, has not disappointed us in this paper. Jupiter's moon, Io, exhibits an anomaly that seems to call for a radical explanation. Io's volcanos erupt with such violence that molten material is flung to heights of 250 kilometers. These outbursts proceed from caldera, and one is led to assume that normal volcanic action is to blame. Unfortunately for this simplistic solution, Io does not seem to possess low-molecular-weight substances, such as water, that could serve as a good propellant at reasonable temperatures. Sulphur is common, but its atomic weight is so high that temperatures exceeding 6000°K would be required to shoot matter out to 250 kilometers.

Gold suggests that Io's volcanos get their fire power from electrical sources. He points out that Io short-circuits Jupiter's ring current periodically. Gold estimates that 5 million amperes flow through Io when it passes through the ring current. The energetic eruptions and caldera might thus be electric-arc phenomena. The electrical energies available are sufficient to account for the observed outbursts. (Gold, Thomas; "Electrical Origin of the Outbursts on Io," Science, 206:1071, 1979.)

Comment. Several scientists and non-scientists have proposed in the past that the sunspots and even some planetary craters result from electric-arc action. They have always been ridiculed for their heresy.

RADAR GLORIES ON JUPITER'S MOONS

"Three ice-covered moons of Jupiter, in comparison with rocky planets and the earth's moon, produce radar echoes of astounding strengths and bizarre polarizations. Scattering from buried craters can explain these and other anomalous properties of the echoes. The role of such craters is analogous to that of the water droplets that create the apparition known as 'the glory,' the optically bright region surrounding an observer's shadow on a cloud." (Eshleman, Von R.; "Radar Glory from Buried Craters on Icy Moons," Science, 234:587, 1986.)

EUROPA'S "SUBTERRANEAN" OCEANS

The surface of Europa, one of Jupiter's large Galilean satellites, seems to be covered with a relatively smooth veneer of ice. Beneath this frigid skin, according to one theory, lie about 100 kilometers of liquid water. Why hasn't this water frozen completely, given the scant sunlight at Jupiter's distance from the sun? Tidal stresses provide some heat but not enough; unless, of course, Europa's orbit were much more eccentric in recent times. (Anonymous; "Oceans under the Crust of Europa," Sky and Telescope, 73:602, 1987.)

Comment. An alternate possibility is that Europa's ice and water inventories were recently acquired, say, via an influx of icy comets. See p. 275.

THE FIRST FOOD: THOLIN

You've just had a hard day evolving into the first life-form on your primitive planet, and you're ready to chow down. Problem: What can you eat? A quick survey of the food chain isn't promising: you're it. Do you simply starve to death, ending your world's brief experiment with life? Not if a rust-colored substance tholin is within reach. Tholin may have served as breakfast, lunch, and dinner for the first life on earth.

The tholins are hard, red-brownish substances made of complex organic compounds. They do not exist naturally on earth because our present oxidizing atmosphere blocks their synthesis. However, tholins can be made in the lab by subjecting mixtures of methane, ammonia, and water vapor to simulated lightning discharges. Conditions like this probably exist many places in the universe. In fact, the icy moons of the outer-solar-system planets appear ideal places for tholin manufacturing.

What would eat such stuff? Lab tests show that many kinds of bacteria love it and thrive on it. (Chaikin, Andrew; "First Foods," Discover, p. 18, February 1991.) See also p. 70.

Comment. A purposeless universe that just happens to create sustenance for primitive life? Strange that things should be this way.

SATURN

In a good telescope, Saturn is a spectacular sight. The rings, of course, make all the difference. Although other planets possess ring systems, Saturn's is bright and vivid, even for amateur astronomers. The three major rings are thousands of kilometers in width, but their thicknesses seem to be measured only in meters. Such obvious features are duly registered in all the astronomy books. But the idiosyncracies of the rings---those recorded before the encounters of the Voyager spacecraft with Saturn---are rarely mentioned. Saturn has a long telescopic history of bright spots appearing on its rings and changes in ring size and brightness. The Voyager missions lengthened the list of ring anomalies considerably. These space probes sent back photos showing that the rings are actually composed of thousands of tiny ringlets, some of which are eccentric, others twisted, and many spiral. The material of the rings seems to be predominantly water ice, but where all this ice came from and when are not known. Several ring phenomena hint that Saturn's rings may be young relative to the age of the solar system.

Saturn's strongly banded sphere displays few anomalies of consequence. Temporary white spots do appear occasionally. The planet does emit considerable infrared energy, indicating that like Jupiter this planet is generates energy over and above that received from the sun. How, no one knows. Of all Saturn's electromagnetic radiations, the intense bursts of radio energy called SEDs (Saturn Electrostatic Discharges) evoke the most scientific excitement. The SED bursts peak about every 10 hours. Astronomers are not sure whether the SEDs emanate from Saturn's atmosphere or rings.

Saturn also possesses a large retinue of moons and moonlets. Here, we remark only upon Hyperion's chaotic rotation and Phoebe's unusual "unlocked" spin.

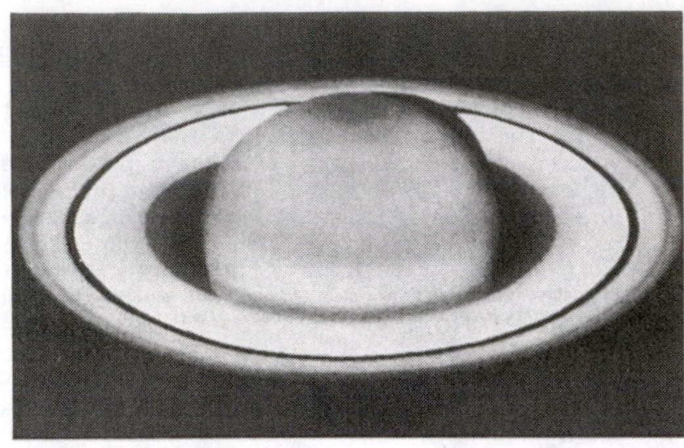

Saturn's highly inclined rings as seen through terrestrial telescopes.

SATURN IS STILL COOKING AWAY

Observations from the earth and the Pioneer spacecraft have long puzzled astronomers because they indicate that Saturn is radiating much more internal heat than it should. This anomaly leads to the outrageous possibility that Saturn is actually very young and has not cooled off yet.

The infrared detector on the recent Voyager 1 flyby seems to show that the atmosphere of Saturn possesses only about half as much helium as theory would have. The surmise is that the missing helium is still residing in the planet; this might account for some of the abnormal heat generation. (Anonymous; "Puzzling over Saturn's Internal Heat," Eos, 62:538, 1981.)

Comment. Thus, the excess heat anomaly may be replaced by the missing helium anomaly. Curiously, some ancient documents refer to Saturn as the "sun of night."

SATURN'S LATEST BURP

In late September, a large white spot appeared on Saturn. Soon this blemish spread into an oval 21,000 kilometers in length. By early November, it had developed into a wide planet-encircling band. Apparently Saturn had "burped," expelling hot gases from its interior. (Saturn emits 50% more heat than it absorbs from the sun.) So far, this is not too beguiling to the anomalist. But now it seems that other white spots, not as large, have been recorded in 1876, 1903, 1933, and 1960. Could the white-spot phenomenon be periodic---like a percolator? More food for thought is Saturn's orbital period around the sun: 29.4 years. ("New White Spot on Saturn Grows, Changes," Science News, 138: 325, 1990. Brown, William; "Giant Bubble of Gas Rises through Saturn's Atmosphere," New Scientist, p. 22, October 20, 1990.)

A HEX ON SATURN

By piecing together pictures of Saturn taken by the Voyager 2 spacecraft, David A. Godfrey (National Optical Astronomy Observatories) has discovered an unusual feature in the planet's atmosphere. His mosaics reveal a hexagon centered on the north pole. Since neither probe flew directly over the pole, the complete scene has to be created by 'stretching' a series of photographs taken from the side as Saturn rotated, then fitting them to a polar projection.

The straight sides of the hexagon are each about 13,800 kilometers long. The entire structure rotates with a period of 10h 39m 24s, the same period as that of the planet's radio emissions, which is assumed to be equal to the period of rotation of Saturn's interior. The hexagonal feature does not shift in longitude like the other clouds in the visible atmosphere.

The pattern's origin is a matter of much speculation. Most astronomers seem to favor some sort of standing wave pattern in the atmosphere; but the hexagon might be a novel sort of aurora. More extreme speculation has Saturn's radio emissions emanating from the hexagon (something we can see and which has the right rotation period) rather than from the planet's interior (something we cannot see). (Anonymous; "A Hex on Saturn," Sky and Telescope, 77:246, 1989.)

Comment. Note the similarities with the radial spoke structure sometimes seen in the atmosphere of Venus.

MYSTERIOUS "THING" IN ORBIT AROUND SATURN

When the two Voyager spacecraft flew past Saturn, both detected strong bursts of radio emissions recurring every 10 hours and 10 minutes. Now termed SEDs (Saturn Electrostatic Discharges), the period of these bursts would be matched by the period of an object rotating around Saturn at a distance of about 109,000 kilometers. Is there anything visible at this distance? Sure enough, Voyager optical instrumentation detected a thinning, possibly an actual gap, about 150 meters wide in the B-ring at this radius. The big puzzle is why a thinness or gap is maintained over a long period of time and how it is

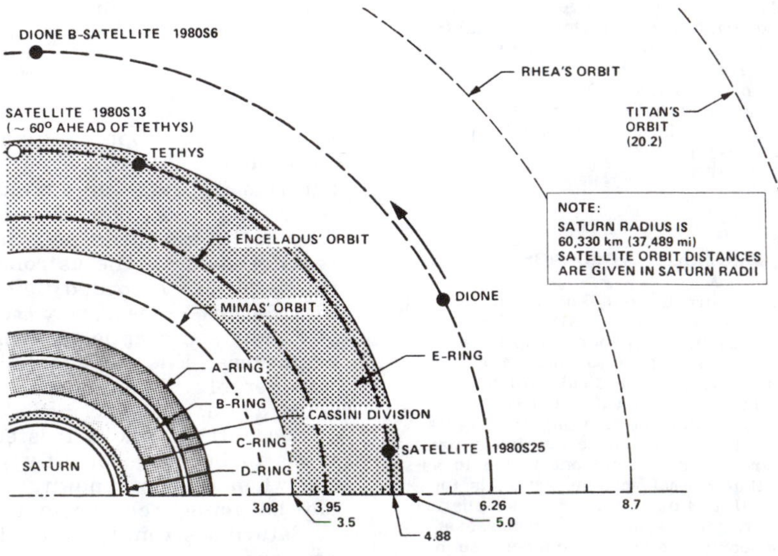

The ring system of Saturn is complex and encompasses the orbits of several satellites. (D. Morrison)

associated with the SEDs. (Evans,D.R., et al; "The Source of Saturn Electrostatic Discharges," Nature, 299:236, 1982.)

Comment. Could Saturn and its rings, which may be electrically charged, be some sort of electromagnetic machine, with arcing occurring at the gap?

RADIAL SPOKES IN SATURN'S RINGS

Among all the recently discovered complexities of Saturn's rings, the dark spokes are perhaps the most challenging to astronomers. These dark areas seem to rotate with the rings and are likely regions nearly devoid of the particles that constitute the rings. The "normal" annular gaps between the rings can be explained in part as due to the gravitational influences of Saturn's moons. The dark spokes, however, do not succumb so easily. There are no obvious gravitational nuances that can sweep particles selectively from radially aligned areas. (Anonymous; "Voyager Discovers Spokes in Saturn's Rings," New Scientist, 88:276, 1980.)

Comment. Any theory accounting for radial gaps may also explain the "knots" of brightness occasionally seen through the telescope down the years. Incidentally, a few observers in the past have also claimed to have seen radial gaps in the rings; so the dark spokes are not exactly new. See ARL5 in The Moon and the Planets.

Voyager photograph of dark spokes in the rings of Saturn.

MAGNETIC TUNE PLAYED ON SATURN'S RINGS

The strange dark radial spokes seen in Voyager's photos of Saturn's rings wax and wane with a period of about 621 minutes. This is very close to the rotation period of Saturn's magnetic field. Somehow, the rotating magnetic field interacts with the particles making up the rings, forcing density or reflectivity changes that we see as transitory spokes. It is not yet clear how this magnetic tune is played. (Proco, C.C., and Danielson, G.E.; "The Periodic Variation of Spokes in Saturn's Rings," Eos, 63:156, 1982.)

Comment. Brightness changes have been noted in the zodiacal light, and one wonders whether solar-system magnetic fields may also be at the bottom of these mysterious variations.

SATURN'S RINGS MAY BE YOUNG

When the Voyager spacecraft swept past Saturn, they radioed back photos of a complex, very dynamic system of rings ---thousands of rings. Studies of these rings have led some astronomers to wonder if they are really as old as Saturn itself. Two lines of thinking suggest a recent origin:

(1) The rings are composed of both light material (very likely water ice) and dark material (probably rocks and dust). The rocky fragments, according to the prevailing nebular theory, should have condensed early in solar system history, and then been swept gravitationally into the planet as they were slowed by friction with the uncondensed nebular material. Yet, dark material is present in the rings.

(2) The incessant bombardment of the rings by meteorites should have pulverized the rings, sending fragments and vaporized material in all directions. In just 10 million years, the rings should have been largely erased. (Cuzzi, Jeffrey N.; "Ringed Planets: Still Mysterious---II," Sky and Telescope, 69:19, 1985.)

WHY DIDN'T GALILEO RESOLVE SATURN'S RINGS?

Several times we have intimated that Saturn's rings may be of recent vintage or have changed in historical times. In this vein, K. Fabian writes about an interesting inconsistency:

> In the early 17th Century, Galileo discovered that the planet Mars goes through a minor gibbous phase. Even in its maximum gibbous phase, Mars is 88% illuminated. Quoting James Muirden in the Amateur Astronomer's Handbook,'It is remarkable that Galileo was able to make out the phase with his tiny telescope.'
>
> Even more amazing, in my opinion, is that Galileo, while he was able to resolve the slight phase of Mars, was unable to resolve the major ring around Saturn. Mars is a difficult object in a small telescope, while Saturn is easily resolved as a ringed planet in even a 40 mm spotting scope at 30X. Why did the rings of Saturn elude Galileo, while the more difficult Martian phases did not? Perhaps at the time of Galileo the rings of Saturn were much more difficult to observe than they are today. (Fabian, Karl; personal communication, September 9, 1988.)

WAITING FOR SATURN'S RINGS TO COLLAPSE

The more we learn about Saturn's rings, the stranger they seem. One of the latest theoretical models of the rings has them composed of balls of hard ice, which interact through mutual collision and are herded gravitationally by small moons. The success of this model has been tempered by its implication that Saturn's rings are very young:

"Theorists would have no problem with a broad, featureless ring surviving the 4.5 billion years since the early days of the solar system; but features such as spiral density waves are clear evidence that satellites, including the profusion of small ones found near the rings, are draining angular momentum from the rings. The satellites should be spiraling outward into ever larger orbits as they gain angular momentum, and the A-ring should collapse inward into the B-ring in just 100 million years as its particles lose angular momentum." (Kerr, Richard A.; "Making Better Planetary Rings," Science, 229:1376, 1985.)

MICROMETEORITES ERODE SATURN'S RINGS

Micrometeorites constantly chip away at Saturn's C-ring. Using current micrometeorite flux estimates, the age of the C-ring is between 4.4 and 67 million years. Compared to the purported age of the solar system, 4.5 billion years, Saturn's C-ring (and perhaps the other rings, too) is a brandnew feature. Where did it come from? Is it related to the icy comets that seem to be raining steadily down on the earth's atmosphere? (Northrop, T.G., and Connerney, J.E.P.; "A Micrometeorite Erosion Model and the Age of Saturn's Rings," Icarus, 70:124, 1987.)

SOMETHING HOT BENEATH SMALL SATURN SATELLITE SURFACES

Crater density studies of the small icy Saturn satellites Rhea, Dione, Mimas, and Tethys reveal important nonuniformities in crater distribution and age. The anomalies are so large that astronomers have concluded that these objects must have undergone considerable evolution after they were formed by accretion (the currently accepted mode of formation). Unfortunately, these four satellites are so small that they could not accommodate any reasonable energy source capable of causing the observed crustal evolution. The authors suggest strong local concentrations of radioactive heat generators rather than uniformly distributed radiogenic substances, such as those that helped mold the earth's surface. (Plescia, J.B., and Boyce, J.M.; "Crater Densities and Geological Histories of Rhea, Dione, Mimas and Tethys," Nature, 295:285,1982)

Comment. Interestingly enough, local concentrations of radioactivity have been discovered on the moon. (See ALE13 in The Moon and the Planets.)

PHOEBE NOT LOCKED TO SATURN

When Voyager 2 passed through the Saturn system a few months ago, it snapped pictures of Phoebe, Saturn's outermost moon. Phoebe is nicely rounded, 200 kilometers in diameter, and swings around Saturn in a retrograde orbit 550 days long---nothing anomalous so far. Phoebe, however, turns out to be the only known solar system satellite whose axial rotation is not about equal to its period of rotation about its parent planet. All other moons, including our own, are gravitationally "locked" so that they always point the same hemisphere at the parent planet. (Anonymous; "Voyager's Fleeting Glimpse of Phoebe," New Scientist, 91: 799, 1981.)

Comment. One inference here is that Phoebe is a relatively recent addition to the Saturn system and has not yet become gravitationally locked.

DIRECT OBSERVATIONS OF HYPERION'S CHAOTIC MOTION

Hyperion is a 150-kilometer-diameter satellite of Saturn. Hyperion's irregular shape and the gravitational pull of Titan, a larger satellite of Saturn, make it a prime candidate for chaotic motion. After accumulating 52 Hyperion-days of observation, J. Klavetter has confirmed this theoretical suspicion. The satellite's brightness varies wildly from day to day, as it spins unpredictably. The laws of motion and the largest computers are helpless here; although computer simulation can identify situations where chaos can develop.

More alarmingly, some "subtle" chaos appears in computer simulations of Pluto's motion and "perhaps other planets". (Kerr, Richard A.; "First Direct View of Solar System Chaos," Science, 246: 998, 1989.)

Comment. Just contemplate what might happen to the earth of any of the major planets did move chaotically! Solar system stability is discussed further on p. 73.

PREBIOLOGICAL CHEMISTRY IN TITAN'S ATMOSPHERE

Background. The experimental work abstracted below relates to the formation of prebiotic compounds that might lead either directly to new life forms or to sustenance for new life forms.

"An organic heteropolymer (Titan tholin) was produced by continuous dc discharge through a 0.9 N_2/0.1 CH_4 gas mixture at 0.2 mbar pressure, roughly simulating the cloudtop atmosphere of Titan (one of Saturn's moons). Treatment of this tholin with 6N HCl yielded 16 amino acids by gas chromatography after derivatization to N-trifluroacetyl isopropyl esters on two different capillary columns....The presence of 'nonbiological' amino acids, the absence of serine, and the fact that the amino acids are racemic within experimental error together indicate that these molecules are not due to microbiol or other contamination, but are derived from the tholin....These results suggest that episodes of liquid water in the past or future of Titan might lead to major further steps in the prebiological organic chemistry on that body." (Khare, Bishun N., et al; "Amino Acids Derived from Titan Tholins," Icarus, 66:176, 1986.)

Comment. Thiolin may also be a component of other solar system satellites. See pp. 67 and 70.

NEPTUNE

Neptune is only one-third the diameter of Jupiter, and it is much more dense. Like Jupiter and Saturn, Neptune radiates more energy than it receives from the sun, suggesting that it possesses an internal heat source of some undetermined type. To add to this planet's dossier of mysteries, its magnetic field is tilted some 50° from its axis of rotation. This tilt casts doubt upon the dynamo theory customarily applied to explain the generation of planetary magnetic fields. Even Neptune's orbit is anomalous, for its seems to be perturbed by some undiscovered solar-system body---perhaps the notorious Planet X. Most bizarre of all are the broken arcs of matter circling Neptune. Where did they come from?

PROBLEMS AT THE RIM OF THE SOLAR SYSTEM

Neptune is an undisciplined member of the solar system. No one has been able to predict its future course accurately. Already this maverick planet is drifting off the orbit predicted just 10 years ago using the best data and solar system models. All of the outer planets, in fact, confound predictions to some degree. In addition, some long-period comets have anomalous orbits. Astronomers have been aware that something was wrong for decades and anticipated finding a trans-Neptunian planet large enough to perturb the outer solar system. The discovery of Pluto did not help matters; it is much too small.

The most popular explanation of the orbital anomalies relies on a large, still undetected planet, possibly 3-5 times the mass of the earth, swinging around the sun at some 80-100 Astronomical Units. Although many have searched, no one has found anything. Planet X, as it is often called, is just another bit of "missing mass." Thomas C.Van Flandern and Robert Harrington propose that all the obvious damage in the outer solar system was the result of a single encounter between Neptune and a body, call it Planet X if you wish, that was passing through the outer reaches of the solar system. (Frazier, Kendrick; "A Planet beyond Pluto," Mosaic, 12:27, September/October 1981.)

NEPTUNE SPINS TOO FAST AND ITS MAGNETIC FIELD IS AWRY

Some pre-Voyager theories about Neptune have been severely tried by the data trickling back to earth across the great gulf separating us from what is now the most distant planet.

Before Voyager, Neptune's spin period was believed to be about 17 hours. This was just the spin rate needed by theorists to explain why Neptune radiates much more heat than Uranus. It seems that spin rate is related to the mixing of a planet's molten innards, which in turn affects the rate at which heat reaches the surface where it is radiated away. With Neptune's period now put at 16 hours by Voyager measurements, the mixing-cooling theory is in trouble.

The magnetic-field situation is even worse. When planetary scientists found that Uranus' magnetic field was tilted 60° to the axis of rotation, they worried a bit but didn't think that one exception would overthrow the favored dynamo theory. After all, the fields of Jupiter, Saturn, and earth are reasonably well-behaved. But Neptune's field is misaligned by 50°! The confidence of the planetologists has been shaken. What, if anything, is different about Neptune and Uranus? It just may be that we don't know how the magnetic field of any planet is generated. (Kerr, Richard A.; "The Neptune System in Voyager's Afterglow," Science, 245: 1450, 1989.)

NEPTUNE'S PARTIAL RINGS

Neptune's rings cannot be seen directly. Instead, earth-based astronomers watch for occultations or dimmings of stars as they pass behind the rings. This seems straightforward enough in theory, but the occultations have been perplexing in practice. First, one member of a closely spaced double star will be occulted normally by the rings but its companion won't. Second, some terrestrial observatories will record an occultation but another a few thousand miles away will not. Such experiences have led to the hypothesis that the rings are discontinuous; that is, arcs rather than complete rings.

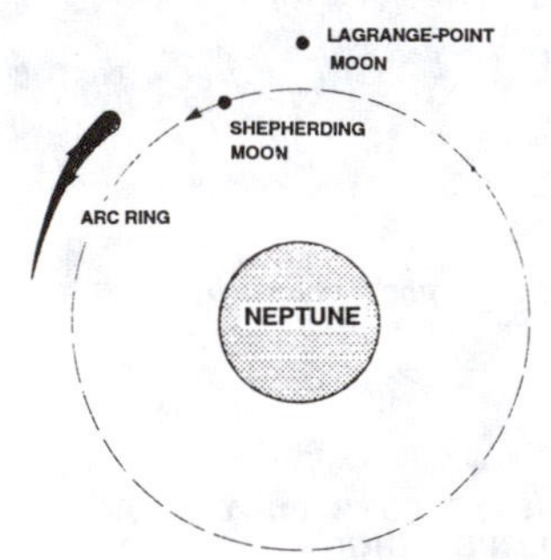

Neptune's broken satellite arcs may be shaped by two moons, a so-called "shepherd" moon and another at a point of stability termed a "Lagrangian point."

Why should Neptune's rings be different from those of the other major planets? One speculation maintains that the arcs are the consequence of one or more recently disrupted satellites. Another hypothesis, by J.J. Lissauer, has the arcs shaped and maintained by two moons, one of the shepherd type, the other at a Lagrangian point in the arc's orbit. (Kerr, Richard A.; "Neptune's Ring Arcs Confirmed," Science, 230: 1150, 1985; and Lissauer, Jack J.;

"Shepherding Model for Neptune's Arc Ring," Nature, 318:544, 1985.)

NEPTUNE'S INCOMPLETE RING

When the star SAO 186001 had a "close" encounter with Neptune on July 22, 1984, a number of astronomers were watching it carefully to see if its light was diminished by an encircling, Saturn-like ring of particles. Uranus' ring system was discovered by studies of stellar occultations. Sure enough, astronomers at the European Southern Observatory, in Chile, and the Cerro Tololo Observatory, also in Chile 90 kilometers away, detected a 1 second, 35% reduction in the star's light at the same instant. These data indicate the presence of an object 10-20 kilometers wide ---hardly an undiscovered satellite, but possibly a ring. But given the geometry shown, there should have been two occultations, but only the one on the right was registered. Speculation is now rife that Neptune has a partial ring or a grotesquely twisted one. (Eberhart, J.; "Signs of a Puzzling Ring around Neptune," Science News, 127:37, 1985.)

Comment. Of course the geometry of the ring could have been such that the path of the star was tangent at one point. It should also be noted that modern astronomers have always laughed off the 1846-1847 observations of a Neptunian ring by W. Lassell and J. Challis.

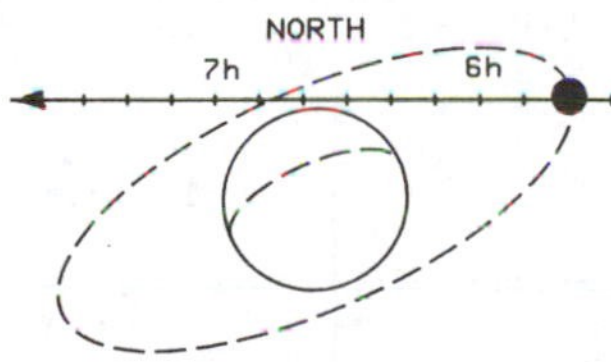

The arrow shows the apparent path of star SAO 186001 behind Neptune. This star's light was dimmed at the solid black circle, apparently by ring matter.

NEPTUNE'S ARCS: EMBRYONIC MOONS?

The publicity given to the 1984 observations of possible discontinuous rings around Neptune have brought to light two other enigmatic observations. The 1981 sighting of a "third satellite" of Neptune have now been interpreted as another discontinuous ring at a different radius. A third discontinuous ring seems to be indicated by the reanalysis of some 1968 occultation data. Astronomer Bill Hibbard, at the University of Arizona, speculates that the three separate arcs of material are "trying to decide whether to become a satellite." (Hecht, Jeff, and Henbest, Nigel; "Neptune's Arcs---A Satellite in Formation?" New Scientist, 19, April 25, 1985.)

Comment. If the debris around Neptune is just now accreting into satellites and Saturn's rings really do have youthful features, one has to consider some disquieting possibilities: (1) Saturn and Neptune have been recently "disturbed," or (2) the entire solar system is not as old as the conventional scenario demands.

NEPTUNE'S STRANGE NECKLACE

The puzzling occultations of stars by Neptune have led scientists to postulate that discontinuous rings of debris rotate around the planet. But given the number of recent failures to detect the rings at all, astronomers have been reduced to thinking about even weirder configurations of matter. The most recent model, by P. Goldreich et al, envisions a necklace of arcs in orbit, as illustrated. They calculate that the resonant effects of a yet undiscovered satellite in an inclined orbit could produce this strange pattern. (Murray, Carl D.; "Arcs around Neptune," Nature, 324:209, 1986.)

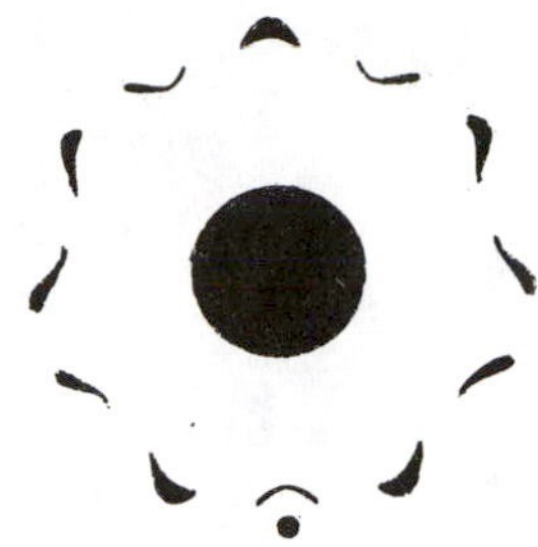

A possible configuration of ring arcs and a shepherding satellite orbiting Neptune, according to Goldreich et al. (Would any astronomer, even 10 years ago, have countenanced such a spectacle in the solar system?)

NEREID: GROTESQUE SHAPE OR TWO-FACED?

Nereid, a satellite of Neptune, is peculiar in several ways:

(1) Its orbit is retrograde and highly elliptical (1.4 x 9.7 million kilometers);
(2) Its brightness changes by a factor of four as it rotates; and
(3) Its diameter, according to M.W. and B.E. Schaefer (Nature, 333:435, 1988), is thought to be at least 660 kilometers.

None of these facts taken alone is anomalous, but (2) and (3) taken together seem incompatible. If the large brightness changes are due to a highly irregular shape, Nereid's 660-kilometer size is too large, because astronomers agree that gravitational forces will sphericize all objects larger than 400 kilometers. On the other hand, if Nereid is two-faced, like Saturn's Iapetus (carbon black on one side, light-colored on the other), astronomers are again faced with trying to explain how such a large solar-system object can acquire so much carbonaceous material on one side only. Also Nereid's eccentric, retrograde orbit surely hints at a history of capture or orbit disruption. (Weisburd, S.; "Neptune's Nereid: Another Mysterious Moon," Science News, 133:374, 1988. Also: Veverka, J.; "Taking a Dim View of Nereid," Nature, 333:391, 1988.)

URANUS

As in the case of Neptune, Uranus is smaller and more dense than the giant planets Jupiter and Saturn. Actually, Uranus is so far from earth that we really know very little about it. Below, we have collected only three items from the recent literature that seem to record anomalies. Two of these concern the ring system of the planet, and possibly other unrecognized debris (moons and moonlets?) in orbit; the third anomaly involves the "enigma of the Uranian satellites' orbital eccentricities."

RINGS OF URANUS: INVISIBLE AND IMPOSSIBLE?

Now that they have discovered nine rings around Uranus, astronomers are having trouble explaining them. First, if they are made up of small chunks of matter, the laws of celestial mechanics dictate that they should quickly spread out radially into much wider rings in just a decade or two. In other words, if the rings are ancient they should not have maintained their present form. Second, the rings are invisible when one would expect them to be bright like Saturn's. Yet, they reflect less light than the blackest coal dust.

The authors propose that each ring is actually a single satellite, so small that we cannot see it, that sheds gases as it orbits. The small solid body would make the celestial mechanics happy, and the gases would be invisible to the eye but still absorb light, making the ring detectable when Uranus occults a star. (Van Flandern, Thomas C.; "Rings of Uranus: Invisible and Impossible?" Science, 204:1076, 1979.)

Comment. An alternative explanation is that the rings are recently acquired and will soon disappear. An 1787 observation of rings around Uranus exists, but this has always been considered illusory, even though it was made by W. Herschel. (For more on this, see AUL1 in The Moon and the Planets.)

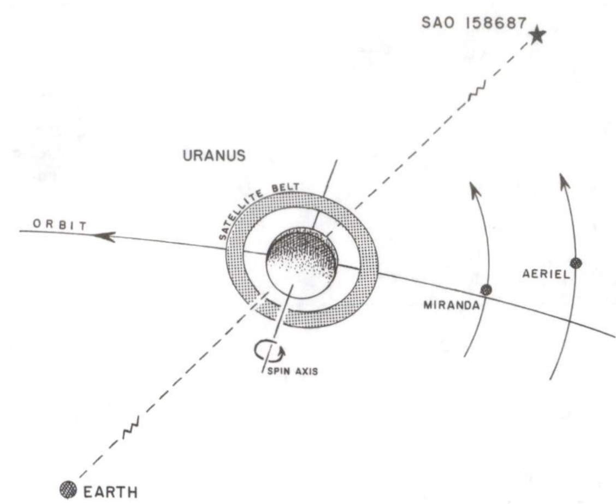

A ring on Uranus was detected on March 10, 1977, when the star SAO 158687 was occulted. Note that the spin axis of Uranus lies nearly in the plane of its orbit rather than perpendicular to it like other planets.

MYSTERIES AROUND URANUS

August 15, 1980. European Southern Observatory, La Silla, Chile. Using the 3.6-meter reflector and photoelectric detector, astronomers recorded the occultation of a star by Uranus. The currently recognized rings of Uranus were duly noted as they dimmed the star's light, but so were seven other "objects." Observers at Las Campanas and Cerro Tololo, who were also monitoring the occultations, did not see the seven extra occultations of the star. Clouds and faulty equipment have been ruled out. No one knows what caused the anomalies. (M., R.A.; "More Mysteries of Uranus' Rings," Solar System Today, 3:56, 1981.)

NEXT, LET US CONSIDER URANUS

Uranus is so distant that its satellites are difficult to observe. What astronomers do see is unsettling. The orbital eccentricities of the three inner satellites, using reasonable assumptions about tidal interactions, should decay to zero (perfect circles) in 10^7-10^8 years. If the observational data are correct, one inference is that the Uranian satellite system should be evolving rapidly from a state of higher eccentricity. (Squyres, Steven W., et al; "The Enigma of the Uranian Satellites' Orbital Eccentricities," Icarus, 61:218, 1985.)

Comment. Here we have one more sign of recent disturbance or solar-system youth. Time spans of 10^7-10^8 years are very small compared to the estimated solar-system age of 5 X 10^9 years.

WERE TITIUS AND BODE RIGHT?

For a couple of centuries, astronomers have been trying to explain in physical terms why the empirical and very simple formula of Titius and Bode works so well. It is only an unimposing geometric progression, which if one inserts the earth's average distance from the sun yields the distances of the other planets with enough accuracy to perturb astronomers. They must insist that the Titius-Bode Law has physical underpinnings.

S. Weidenschilling and D. Davis now propose that the planets owe their present positions to the combination of two effects:

(1) The frictional drag of the gas in the solar nebula, which favors the presence of small planets in the inner solar system; and

(2) Gravitational perturbations, which create favored places for the coalescing of planets.

C. Patterson has expanded on this suggestion and finds that a model based on these effects works quite well for Jupiter and beyond, which are "bound together" in an interlocking system of orbital-period resonances. (See diagram.)

Several important anomalies persist, however. That vacant niche between Uranus and Neptune is presently inexplicable. In the inner solar system, the presence of Mercury is embarrasing. (Anonymous; "Were Titius and Bode Right?" Sky & Telescope, 73:371, 1987.)

Comment. The problem of Mercury is treated on p. 53 and in The Moon and the Planets.

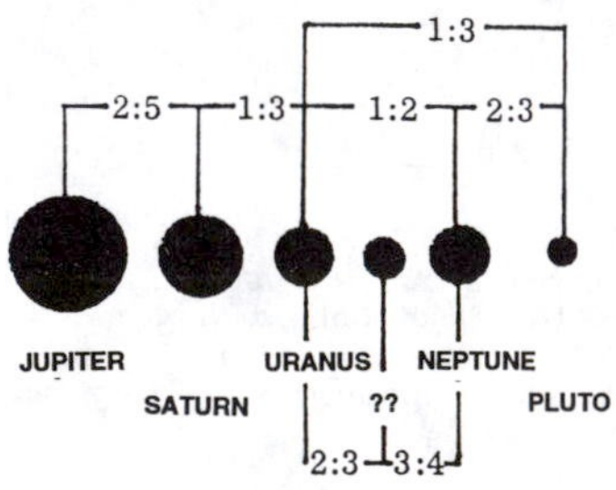

In C. Patterson's model, the outer planets accreted at those spots where their orbital periods formed simple ratios. Planet ?? predicted by this theory has never been found.

SOLAR SYSTEM DYNAMICS

The solar system as-a-whole possesses many interesting and curious properties that provide clues about its origin and evolution. Some of these properties are anomalous because they imply a history that is different from the accepted scenario of condensation and accretion followed by long-term stability.

Solar system dynamic properties engender several questions: Why do the planets possess so much more angular momentum than the sun? Even more disturbing is the possibility that the postulated Oort Cloud of comets has much more angular momentum than all the planets taken together. Also, why are there so many resonances among the members of the solar system? Is the solar system really stable?

A second class of anomalies includes the Titius-Bode Law and similar relationships among solar system parameters that may or may not be real physical "laws." All we know is that some of these formulas work quite well, but we don't know why.

Finally, a most profound question asks if the rise of intelligent life on earth (or any planet in a system like ours) requires a defensive barrier of giant planets like Jupiter and Saturn.

A DIFFERENT WAY OF LOOKING AT THE SOLAR SYSTEM

The current scientific consensus has the sun and its planets forming during the same process of accretion/condensation. In this view, the inner terrestrial planets differ from the outer giant planets only because their volatile elements were driven off by heat. This scenario has many problems.

G.H.A. Cole thinks that astronomers might have more success in explaining the origin of the solar system if they considered it a system of five large bodies (the sun, Jupiter, Saturn, Uranus, and Neptune---all consisting of light elements), each surrounded by its own retinue of high density satellites (the sun's four satellites would be the inner planets). In effect we would have a quintuple star system in which only one member (the sun) collected enough star stuff to make it to incandescence.

The advantages of this change of perspective are threefold:

(1) All five central bodies are now compositionally similar as a class;
(2) In each of the five systems, the angular momentum of the central body is greater than that of its satellites, whereas in the unitary solar system the angular momentum of the nine planets is much greater than that of the sun---an embarrassing anomaly;
(3) A final "bonus" appears when the distances of the satellites in the five systems are plotted, as indicated, and compared. The arrangement of the four terrestrial planets (solar satellites) closely resembles the distribution of Jupiter's Galilean satellites, etc.

There are loose ends, to be sure, like Saturn's rings and Pluto; but the idea seems worth studying further. (Cole, G.H.A.; "Dynamical Form of the Solar System," Observatory, 105:96, 1985.)

Comment. The arrangements of the satellites in the figure may have no physical significance, but if you like Bode's Law you will appreciate the situation.

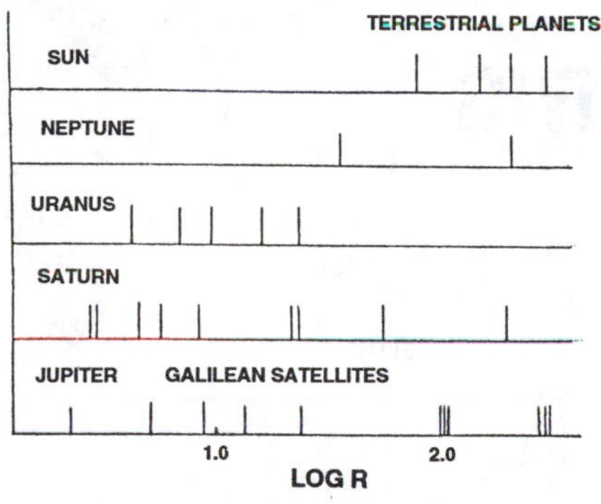

Distributions of orbital radii (R) of central-body satellites, where R is measured in units of central-body radii. Cole terms the similarities "remarkable."

WHY INTELLIGENT LIFE NEEDS GIANT PLANETS

The two giant planets, Jupiter and Saturn, are 318 and 95 times more massive than the earth, respectively. Being so weighty they strongly perturb the orbits of comets, deflecting many away from the inner solar system, where we reside. Calculations by G. Wetherill, at the Carnegie Institution, reveal that if Jupiter and Saturn were only 15 times the mass of the earth, the earth would have been devastated every 100,000 years by giant comets, instead of about every 100,000,000 years, as indicated by the geological record. Under such intense bombardment, it would probably have been difficult for advanced life forms to develop. (Croswell, Ken; "Why Intelligent Life Needs Giant Planets," New Scientist, p. 18, October 24, 1992.)

Comment. Reasonable as the foregoing assertion sounds, we do not really know what stimulates the development of new life forms. Actually, the fossil record reveals that some biological "radiations" occurred soon after great geological upheavals. That the Jupiter-Saturn "shield" was and is not completely effective is indicated by the heavy debris traffic mentioned on p. 74.

BAD SPIN SPLIT

Astronomers have long realized that the angular momentum of the sun is only 1/180th that of the solar system as-a-whole. The overwhelming majority of the angular momentum is tied up in planetary motion. To make matters even more puzzling, the angular momentum vectors of the sun and the planetary system are 7° apart. The implication is that the sun and planets could not have been formed by the rapid condensation of a molecular cloud---the present theory. Rapid condensation requires that the sun get a much bigger share of the angular momentum.

These anomalies have led Thomas Gold to propose a slow condensation model, in which several hundred million years are required rather than the tens of thousands of years in the current scenario. Another unexpected feature of Gold's model makes the sun a degenerate object, perhaps a neutron star, for part of its history. As the author of this article states, "Gold has stood the conventional view of the origin of the solar system on its head." (Maddox, John; "Origin of Solar System Redefined," Nature, 308:223, 1984.)

THE PLANETS AS FRAGMENTS OF AN ANCIENT COMPANION OF THE SUN

J. Webb has called attention to a fascinating feature of the solar system:

> If one calculates the total energy and the total orbital angular momentum of the planets, the numbers turn out to be very nearly the same as those of a single planet having a mass essentially the same as the total mass of all the planets, and orbiting the sun in an orbit which is near to the present day center of mass of all the planets. The possibility that the solar system was once a binary star (or is in the process of becoming one) needs to be examined more closely.

(Webb, Jerry; "The Solar System and a Binary Star: Is There a Connection?" American Journal of Physics, 53:938, 1985.)

"TAIL WAGS DOG" IN SOLAR SYSTEM

The Oort Cloud of comets hovering at the far frontiers of the solar system is not without its anomalies. Here, let us assume that it really does exist, even though we cannot see it.

From a new book by I. Asimov (title below), we learn that this remote haze if icy fluff, the Oort Cloud, may really have about 90% of the angular momentum of the entire solar system. It was already sufficiently anomalous to discover that the planets possess fifty times the angular momentum of the much more massive sun. Astronomers have been attempting for years to explain this 50: 1 split. Now, with the Oort Cloud apparently having ten times the angular momentum of the planets, the situation is much worse. According to Asimov, the solar-system angular momentum is now:

Oort Cloud	90%
All of the planets	9.8%
The sun	0.2%

The total mass of the Oort Cloud is estimated as roughly that of Saturn.

The recent flyby of Halley's Comet created this dilemma. It was discovered that Halley was a chunk containing 140 cubic miles of ice---much larger than anticipated for this "typical" comet. If the estimated 2 trillion comets are, on the average, Halley's size, the Oort Cloud is a thousand times more massive than previously thought. This combined with estimates of Oort Cloud distance and angular velocity leads to the almost ridiculous distribution of solar-system angular momentum. This will keep the theorists busy a while. (Asimov, Isaac; Frontiers; New Discoveries about Man and His Planet, Outer Space and the Universe, New York, 1989, p.270, Cr. C. Ginenthal.)

Comment. Perhaps the Oort Cloud isn't there after all. Or, if it is, it isn't rotating as rapidly as thought. In any case, the accepted theory of solar system formation is in trouble.

SUN-EARTH-MOON SYSTEM MAY NOT BE STABLE

The application of zero-velocity surfaces (a mathematical technique) to the sun-earth-moon three-body system indicates that the eccentricity of the earth's orbit renders the system unstable. The conclusion is that the moon may one day escape the earth and become a planet and vice versa, that the origin of the moon by capture is a strong possibility. (Szebehely, V., and McKenzie, R.; "Stability of the Sun-Earth-Moon System," Astronomical Journal, 82:303, 1977.)

Comment. This paper is typical of several recent papers in celestial mechanics that throw doubt on long-held dogmas about the long-term stability of the solar system.

CHAOTIC DYNAMICS IN THE SOLAR SYSTEM

"Newton's equations have chaotic solutions as well as regular solutions. The solar system is generally perceived as evolving with clockwork regularity, yet there are several physical situations in the solar system where chaotic solutions of Newton's equations play an important role. There are physical examples of both chaotic rotation and chaotic orbital evolution.

"Saturn's satellite Hyperion is currently tumbling chaotically, its rotation and spin axis orientation undergo significant irregular variations on the time scale of only a couple of orbit periods. Many other satellites in the solar system have had chaotic rotations in the past. It is not possible to tidally evolve into synchronous rotation without passing through a chaotic zone. For irregularly shaped satellites this chaotic zone is attitude-unstable and chaotic tumbling ensues. This episode of chaotic tumbbling probably lasts on the order of the tidal despinning timescale. For example, the Martian satellites Phobos and Deimos tumbled before they were captured into synchronous rotation for a time interval on the order of 10 million years and 100 million years, respectively. This episode of chaotic tumbling could have had a significant effect on the orbital histories of these satellites."

This Abstract continues, naming as other candidates for chaotic histories: some of the asteroids, Miranda (a satellite of Uranus), and Pluto. (Wisdom, J.; "Chaotic Dynamics in the Solar System," Eos, 69:300, 1988.) See also: Kerr, Richard A.; "Pluto's Orbital Motion Looks Chaotic," Science, 240: 986, 1988.

Comment. We have been assured often, particularly during the days of Velikovsky, that the solar system has been stable for billions of years! Yet, Wisdom states very clearly above that synchrony cannot be evolved without passing through a chaotic zone. The solar system abounds with resonances, not the least interesting of which is the earth-Venus resonance. See our Catalog volume The Sun and Solar System Debris for more. More recently, the advent of chaos theory assures us that physical systems as complex as the solar system cannot be stable over millions of years.

THE PLANETS ARE UNPREDICTABLE

It was only about 40 years ago when astronomers, aghast at Velikovsky's vision of worlds in collision, stated very firmly that the solar system was presently stable and had been so for eons. Now it seems that they may have been a bit hasty and all-encompassing. Not that the Velikovsky scenario is correct or that Mars might at any moment depart its present orbit, but rather that astronomers must now admit an inability to predict planetary motion over billions of years---even tens of millions of years! For the solar system, if those ubiquitous computers are correct, is not a well-behaved family of planets. J. Laskar concludes from extensive numerical experiments:

> The motion of the Solar System is thus shown to be chaotic, not quasiperiodic. In particular, predictability of the orbits of the inner planets, including the Earth, is lost within a few tens of millions of years.

(Laskar, J.; "A Numerical Experiment on the Chaotic Behavior of the Solar System," Nature, 338:237, 1989.)

Comment. Laskar's remarks are directed toward the future, but the same conclusions should apply if we ran the solar system backwards in time. From this very narrow perspective of celestial mechanics, one cannot say positively that the planets, the inner ones especially, could not have radically altered their orbits within the past few million years.

SOLAR SYSTEM DEBRIS

METEORS AND METEORITES

Meteors and meteorites constitute our prime sources of extraterrestrial matter. Where do they come from and what can they tell us about the origin and history of the solar system and, perhaps, even life on earth?

The meteor and meteorite anomalies appearing in past issues of Science Frontiers can be classified into four groups: (1) The anomalies of meteor celestial mechanics, particularly the possible periodicity of very large meteors; (2) The anomalies of meteor flight through our atmosphere; (3) The chemical and physical properties of recovered meteorites, especially the carbonaceous chondtites; and (4) The peculiar distribution of recovered meteorites, with particular attention to those picked up in Antarctica. Entries in each of these classifications will tell us something unexpected about these messengers from outer space.

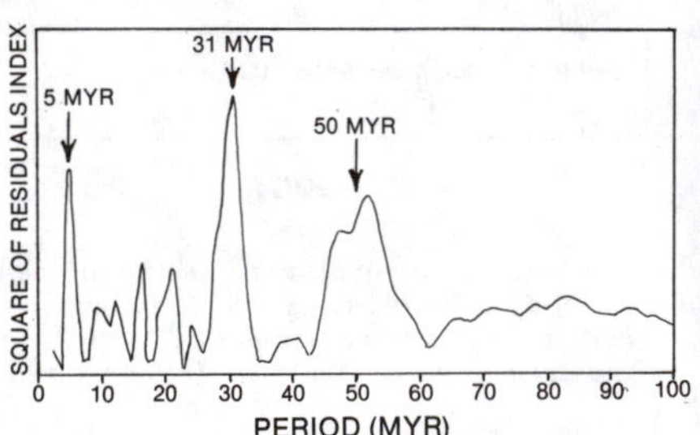

Periodicity of terrestrial impact crater ages. The 5- and 50-million-year peaks are analytical artefacts. (After M.R. Rampino and R.B. Stothers)

DO LARGE METEORS/COMETS COME IN CYCLES?

Only a few years ago, geologists refused to recognize any terrestrial meteor craters larger than Arizona's Meteor Crater, which is merely one mile across. Now we have a long list of craters or astroblemes (star wounds), some of which measure hundreds of miles in diameter. In fact, there are enough dated craters so that some scientists have taken up a time-honored human pastime: Looking for cycles or periodicities in the data. (Humans can find cyclicities in almost any collection of data!) To be specific, some have claimed that meteor craters come in clusters 28-31 million years apart. These catastrophic events have been correlated with biological extinctions, magnetic field reversals, and basalt flooding from volcanoes. The astronomical causes of the periodicity range from the solar system's crossing of the galactic plane, the perturbations of an unseen solar companion, to regular perturbations of the Oort Cloud of comets that hovers at the fringe of the solar system. In short, a large, interlocking edifice of speculation has been erected upon a foundation of terrestrial crater dates.

But how well do we really know the ages of these craters? How complete is the cratering record? The answer to the first question is: "Not well at all." Further, we can be sure that many craters still lie undiscovered beneath sediments; and most meteors/comets splashed into the sea, leaving no record at all. An updating of the most recent crater data available, such as they are, greatly weakens the case for the popular 28-31 million year period and strengthens support for a 19-22 million year period. But neither cycle is in synchronism with the famous K-T (Cretaceous-Tertiary) boundary, with its perplexing iridium layer and massive biological extinctions. In fact, say V.L. Sharpton et al, the entire known cratering record could well be the consequence of chance encounters between the earth and stray meteors and comets. Thus pass the periodicities of this world! (Sharpton, V.L., et al; "Periodicity in the Earth's Cratering Record?" Eos, 68:344, 1987.)

DANCING TO THE COMETS' TUNE

When planetary scientists examine one kind of meteorite rich in iron, the H-chondrites, they find that the meteorites' ages do not spread evenly through time. Instead, the ages appear to cluster at 7 million and 30 million years.

Astronomers have hithertofore been content to attribute these clumped ages to collisions among the meteorites' parent bodies---the asteroids---which ply periodic orbits. However, S. Perlmutter and R.A. Muller, at Berkeley, point to the apparent 26-30-million-year periodicities of three terrestrial phenomena: (1) biological extinctions in the fossil record; (2) terrestrial crater ages; and (3) magnetic field reversals. Could there be a connection between the clumped meteorite ages and these terrestrial phenomena?

Perlmutter and Muller propose that all four phenomena are the consequence of periodic storms of comets that invade the inner solar system from the direction of the Oort Cloud of comets that purportedly hovers at the fringe of the solar system. These comets not only devastate the earth but also collide with the asteroids, knocking off those bits and pieces we call meteorites. (Anonymous; "Do Meteorite Ages Tell of Comet Storms?" Astronomy, 17:12, January 1989.)

Comment. Unanswered above is the question of why comet storms should be periodic. One hypothesis is that Nemesis, the so-called Death Star, lurks out there, periodically nudging the Oort Cloud of comets.

A MYSTERIOUS OBJECT

January 14, 1983. Winnipeg, Manitoba. "I observed a perfectly round black orb crossing the sun. It started at $17^h54^m23^s$ Universal Time and ended at $17^h54^m26^s$ Universal Time and lasted three seconds. On a projected solar disk with a diameter of 18 centimetres, the object had a diameter of one-half centimetre." (Lohvinenko, Todd; "A Mysterious Object," Royal Astronomical Society of Canada, National Newsletter, 77:L19, 1983.)

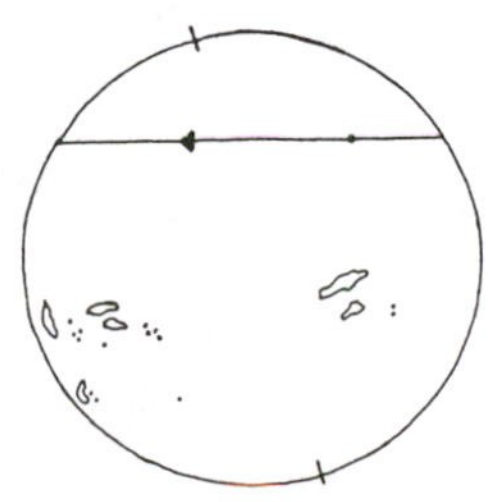

The arrow indicates that path taken by a round spot crossing the sun on January 14, 1983.

Comment. The object travelled too fast to be an intramercurial planet, too slowly for a meteor.

AN UNEXPLAINED EVENT

September 6, 1990, 2029 UT. Hornchurch, England. "The sky was 'crystal clear' and there was brilliant moonlight, the Moon being 1 day past Full. An object approximately 0.5-0.75 the size of the Full Moon, but a dull, mottled red in colour, was observed to cross the sky from west to east in approximately 3 seconds. As it approached the area of the Moon it faded away, giving the impression of being drowned out by the moonlight. Mr Scarlioli observed the object to be of an irregular shape and says that he could see it 'turning' as it moved along. No trail was left behind the object and there were no 'sparks' normally associated with the fragmentation of a fireball during the ablation process. No sound was heard from this object. There is currently no explanation for this event." (Anonymous; "Unexplained Event Sep 6, 2029 UT," Meteoros, 20:44, Autumn 1990.)

METEORIC "DUST BUNNIES"

All around the world, watching through the long nights, is a band of dedicated meteor observers---mainly amateur astronomers. Collectively, they record many meteor showers and fireballs; all quite respectable astronomical objects. But sometimes fuzzy, rapidly moving luminosities appear, as in the following paragraph written by J.S. Gallagher:

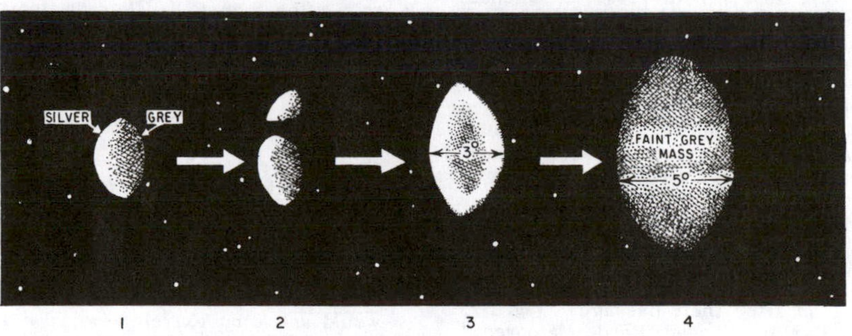

Evolution of the July 29, 1970, nebulous meteor over Dover---a typical meteoric "dust bunny." (R.A. Mackenzie)

Diffuse luminous objects moving at angular velocities similar to those of meteors were observed during over 200 hours of meteor watching in 1991. They fell in three broad categories: arcs, patches, and "meteors" similar in appearance to comet comas. Though I at first dismissed the possibility of their being related to meteors, I reconsidered this relation after eliminating other possible causes such as reflections from aircraft lights and tricks of vision. Their meteor-like behavior suggested that perhaps these events might be caused by clouds of exceedingly small meteoroids, visible only because of their numbers and compact grouping. Because such a formation would be unlikely to be maintained long in space, it appeared necessary that the particles involved must have maintained some weak physical contact until just prior to becoming visible. Perhaps some type of "cosmic dust bunny," disrupted by air resistance, might be the cause of these events.

These moving patches of light also resemble the "auroral meteors" cataloged under GLA3 in Lightning, Auroras, Nocturnal Lights. Physically, they might be related to the small, icy comets postulated by L.A. Frank to account for the transient "holes" seen in satellite ultraviolet images of the earth. (See p. 275 for more.) (Gallagher, John S.; "Diffuse Luminous Objects Having Angular Velocities Similar to Meteors," Strolling Astronomer, 36:115, 1992. Cr. P. Huyghe)

A TRIO OF STRANGE METEORS

January 25, 1990. Western North America. "While residents from Anoka to Mankato were calling radio stations Thursday to report the sighting a bright, slow-moving light in the predawn sky, people across the western half of the nation were doing the same thing."

.....

"Most eyewitnesses in the spectacle which was reported over a 12-hour period from a number of locations, said it was greenish, although some said it was turquoise, or white, or had an orange tail. Others said it didn't have a tail." (McAuliffe, Bill; "Was It Junk? Maybe So, But It Sure Lit Up the Sky," Minneapolis Star-Tribune, January 26, 1990. Cr. R. PanLener via L. Farish)

Comment. The color changes are not uncommon for a meteor sighting, but this phenomenon was very slow-moving for a meteor.

January 27, 1990. U.S. Midatlantic States. "Thousands of people in the Eastern United States reported seeing a strange bluish-green light in the sky Saturday night which some experts said could have been an unusually large meteorite."

.....

"In North Carolina, Jim Iodice, who was flying a Cessna 172 over Pilot Mountain Saturday night, said that he saw a 'glowing, yellowish-blue light' between 7 and 7:30 p.m. that appeared to be near the plane. The object was descending in a northeast direction toward Martinsville, Va., but it leveled off at some 3,000 feet, flew at the same altitude for several hundred yards, then changed to a southward direction, Iodice said." (Anonymous; "In the Dark," Winston-Salem Journal, January 29, 1990. Cr. G. Fawcett via L. Farish.)

Comment. Here, it is the change of direction that is peculiar.

February 18, 1990. Northeastern U.S. "Reports of a fireball that blazed through the skies over the Northeast on Sunday, changing colors and even executing a fiery loop before vanishing, have been filtering into local agencies, a Museum of Science official said yesterday.

"Observers from Nova Scotia to New Jersey reported the spectacular fireball, which they said was visible for more than 10 seconds at 7:50 p.m. Sunday in the southeastern sky." (Saltus, Richard; "Looping Fireball Dazzled Observers in Northeast," Boston Globe, February 23, 1990. Cr. B. Greenwood)

Comment. Fireballs or meteors do not execute loops.

MEMORIES OF 1913

March 31, 1991, about 7:10 PM, Quebec. G. Morisette and his wife were driving along the road to Sept-Iles, when his wife asked him to stop the car to watch a strange luminous phenomenon. Thinking that it was only Venus or an aircraft, Morisette pulled off the road and got out. To his surprise, it was a formation of five or six meteors cruising leisurely toward the north on parallel paths. This fascinating spectacle lasted about 15-20 seconds---long as meteor events go. The fireballs disappeared simultaneously. No sounds were heard during or after their passage. (Morisette, Gartan; "Escadrille de Meteores," Astronomie-Quebec, July-August 1991. Cr. F. St. Laurent)

Sketch of the 1991 meteor procession

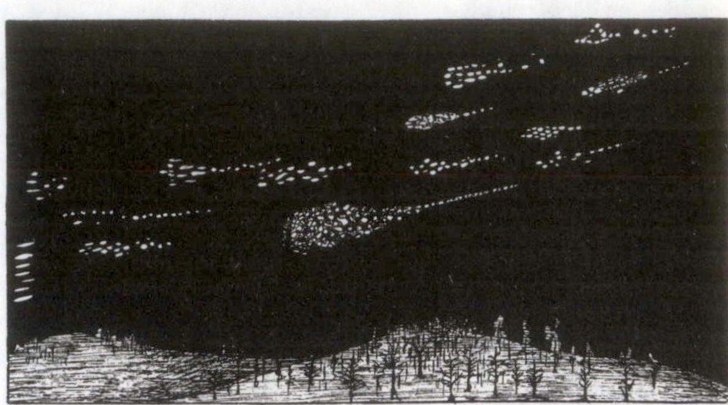

One Canadian's sketch of the February 9, 1913, meteor procession.

Comment. The slow progress and disciplined motion of the Quebec meteors remind one of the famous meteor procession of February 9, 1913, which was also a predominantly Canadian event. However, the 1913 procession headed southeast over the northeastern states and out into the Atlantic. See AY07 in Sun and Solar System Debris.

ANTARCTIC METEORITES ARE DIFFERENT

The thousands of meteorites rescued from the Antarctic ice are markedly different from those collected elsewhere on our planet. First, the Antarctic collections contain rare types of that seem to have come from the moon and perhaps Mars. Second, the trace elements in the Antarctic specimens differ substantially from those found elsewhere. Age is another distinguishing parameter. The Antarctic specimens seem to have been laying about for some 300,000 years. Almost all other meteorites are less than 200 years old, having been picked up soon after they fell. (Note: meteorites have two kinds of ages: the length of time they have lain on the earth's surface and the length of time since their formation in space. Here, we are talking about the former.) The implication is that those extraterrestrial projectiles that have accumulated in Antarctica had a different source. (Dennison, Jane E., et al; "Antarctic and Non-Antarctic Meteorites Form Different Populations," Nature, 319:391, 1986.)

Comment. A dedicated catastrophist would ask what extraterrestrial event occurred 300,000 years ago? Did it involve the moon? Was terrestrial life, including man, affected? See p. 220.

WHY ARE ANTARCTIC METEORITES DIFFERENT?

Here's the problem: "Differences exist between Antarctic and non-Antarctic meteorites, and the significance of this is only now beginning to be recognized. Dennison et al point out that relative to non-Antarctic falls, the Antarctic population is underabundant in iron and stony-iron meteorites, among others." Trace element studies "demonstrate a statistical unlikelihood that both sample populations drive from the same parent population." One reason for the differences is that the Antarctic ice has been accumulating meteorites for many thousands of years longer than modern man has been picking up non-Antarctic meteorites. (Lipschutz, Michael E., and Cassidy, William A.; "Antarctic Meteorites: A Progress Report," Eos, 67:1339, 1986.)

Comment. If Antarctic meteorites differ because they impacted the earth over a longer span of time, it must be that the meteor population in the vicinity of the earth has been changing. Why? In this connection, read next about an immense accumulation of unusual meteorites strewn across South Australia.

THE NULLARBOR LODE

For the last few hundred years people have been picking up sparsely strewn meteorites all over the planet. But Antarctic explorers, within the last few decades, found that thousands of meteorites have been concentrated in the ice of the southernmost continent. Even more recently, the desolate, desert-like Nullarbor ("no-trees") Plain, in Southern Australia, has been discovered to be another concentrated source of of meteorites. There may be millions there. The problem is that only 2.9% of them are iron meteorites, whereas those picked up in recent years around the planet-at-large are 4.8% irons. The meteorites from the Antarctic lode, on the other hand, weigh in with only 2.2% irons. Why the marked differences? Could it be age? The Antarctic meteorites seem to be up to a million years old; those of Nullarbor, perhaps 16,000-18,000 years. (Anonymous; "A Meteorite Bounty from Down Under," Sky and Telescope, 82: 461, 1991.)

Comment. Perhaps pertinent is the observation that fossil meteorites are essentially nonexistent in geological formations older than a million years. This is an anomaly of itself!

SHERGOTTITES AND NAKHLITES: YOUNG AND MYSTERIOUS

The shergottites and nakhlites are two types of meteorites that have scientists scratching their heads. Both types have been dated by various radioactive clocks as having been created 1300 million years ago or less. At this age they are far younger than all other meteorites. Where could such young meteorites have originated? The asteroids and moon's surface are far too old. A current guess is the surface of Mars. There, an impacting meteor could have blasted pieces of young lava sheets into space and thence to earth. The shergottites have a shocked structure and could well have originated in such catastrophism, but the nakhlites show no signs of violence and seem to require a separate explanation. (Anonymous; Mystery Meteorites May Come from Mars," New Scientist, 91:219, 1981.)

FOSSIL FROM MARS?

Who would have thought that the dreary Antarctic wastes would harbor pieces of Mars, much less samples of Martian prebiotic material or, thinking more radically, Martian fossils? I.P. White et al, in Nature, do profess such ideas:

> The meteorite EETA 79001, which many believe to have originated on Mars, contains carbonate minerals thought to be martian weathering or alteration products. Accompanying the carbonates are unexpectedly high concentrations of organic materials (defined here as carbonaceous matter that has a low stability towards oxidation, and so combusts at < 600°C; the term 'organic' does not necessarily imply an origin by biogenic processes.) Although the carbon isotope composition of these materials is indistinguishable from terrestrial biogenic components, and so cannot be used to assess the source, we argue that their occurrence in an interior sample of a clean Antarctic meteorite militates against a wholly terrestrial origin. A sample of martian organic materials may thus be available for further study in the laboratory. (Wright, I.P. et al: "Organic Materials in a Martian Meteorite," Nature, 340:220, 1989.)

But there are many "buts":

1. Meteorite EETA 79001 may not have come from Mars after all, even though many scientists think it did.
2. The organic material in EETA 79001 may have come instead from the comet that supposedly blasted the meteorite into space from the Martian surface, even though the carbon ratios do not favor a cometary origin.
3. The organic material may only be terrestrial contamination, despite careful handling of the meteorite.

Nevertheless, EETA 79001 has revived speculation about life on Mars. Could not the calcium carbonate, for example, have come from the shell of some Martian water creature? I.P. Wright does not avoid this possibility:

> There is a remote chance that we're looking at some (extraterrestrial) fossil life form." (Amato, I.; "Meteorite May Carry Organic Martian Cargo," Science News, 136:53, 1989.)

METEORITES ALSO TRANSPORT ORGANIC PAYLOADS!

Excerpts follow from the Abstract of an article that appeared in Nature in 1987:

"Much effort has been directed to analyses of organic compounds in carbonaceous chondrites because of their implications for organic chemical evolution and the origin of life. We have determined the isotopic composition of hydrogen, nitrogen and carbon in amino acid and monocarboxylic extracts from the Murchison meteorite...These results confirm the extraterrestrial origin of both classes of compound, and provide the first evidence suggesting a direct relationship between the massive organo-synthesis occurring in interstellar clouds and the presence of pre-biotic compounds in primitive planetary bodies." (Epstein, S., et al; "Unusual Stable Isotope Ratios in Amino Acid and Carboxylic Acid Extracts from the Murchison Meteorite," Nature, 326:477, 1987.)

A HINT OF EXTRATERRESTRIAL OCEANS

The Allende carbonaceous chondrite (a well-known meteorite) contains a layered mineral related to serpentine, which seems to have been formed under aqueous conditions before it was incorporated into the meteoric mass. In the sometimes obscure language of science, the authors say that the unusual characteristics of this mineral may "reflect undetermined extraterrestrial conditions experienced by some chondrules and aggregates." (Tomeoka, Kazushige, and Buseck, Peter R.; "An Unusual Layered Mineral in Chrondrules and Aggregates of the Allende Carbonaceous Chondrite," Nature, 299:327, 1982.)

Comment. The "undetermined extraterrestrial conditions" might have involved other water-covered bodies in the solar system.

INTERPLANETARY DUST

OF DUST CLOUDS AND ICE AGES

Ice cores from the polar ice at Camp Century, in Greenland, have been analyzed for the presence of cosmic dust:

"It is concluded that on five occasions during the interval 20,000-14,000 years BP, cosmic dust mass concentrations in the solar system rose by one and two orders of magnitude above present day levels. Moreover if the particles found in these ice core samples are indicative of the particle size distribution which prevailed in the interplanetary medium at that time, then it may be concluded that the space number density of submicron sized particles must have increased by a factor of 10^5 or more. During these times the light transmission properties of the solar system would have been significantly altered resulting in major adverse effects to the earth's climate. Thus it is quite possible that these dust congestion episodes were responsible for the abrupt climatic variations which occurred toward the end of the Last Ice Age."

Whence these interplanetary dust clouds? The author ruled out terrestrial volcanism (an insufficient source of iridium) and encounters with asteroids and cometary tails (too infrequent to account for the long periods of high dust levels). Rather, the dust source may have been the same event that created the recently discovered dust ring between Mars and Jupiter, which is believed to be only a few tens of thousands of years old. The nature of the "event" is not specified. (LaViolette, Paul A.; "Evidence of High Cosmic Dust Concentrations in Late Pleistocene Polar Ice (20,000-14,000 Years BP)," Meteoritics, 20:545, 1985.)

YOUNG INTERPLANETARY DUST

"Nuclear tracks have been identified in interplanetary dust particles (IDP's) collected from the stratosphere. The presence of tracks unambiguously confirms the extraterrestrial nature of IDP's, and the high track densities (10^{10} to 10^{11} per square centimeter) suggest an exposure age of approximately 10^4 years within the inner solar system." (Bradley, J.P., et al; "Discovery of Nuclear Tracks in Interplanetary Dust," Science, 226:1432, 1984.)

Comment. Where does the young dust come from? The Poynting-Robertson drag is supposed to sweep the inner solar system clean of dust fairly quickly. If comets supply a steady stream of dust, the particles should display a wide range of exposure ages.

ASTEROIDS

Well over 2,000 asteroids have been officially recognized. Their sizes normally range from a few kilometers in diameter to 960 kilometers for Ceres, the largest of the "minor planets." Most asteroids orbit the sun between Mars and Jupiter in the so-called asteroid belt, although it is becoming more apparent that many smaller bodies ply paths that pervade the entire solar system, including near-earth space. For many years asteroids and meteors were thought to be closely related, the latter being the debris from asteroid collisions. Lately, however, this connection has become less certain. Some asteroids, in fact, seem to have comet-like characteristics and may in fact be burnt-out comets.

DISPARITY BETWEEN ASTEROIDS AND METEORITES

Since the meteorites we pick up on earth (Antarctica and elsewhere) are thought to have come from pulverized asteroids, it is something of a shock to find profound dissimilarities between asteroids and meteorites:

"The problem is that while reflectance spectra of some meteorites measured in the laboratory appear to correspond to spectra of various asteroids, the S class, which makes up about half of all asteroids in the inner belt, does not appear to match any common meteorite class. Conversely, common meteorite classes (e.g., ordinary chondrites) appear to match only a few asteroids." (Harris, Alan W.; "Asteroid 29 Amphitrite Is a Topic of Interest," Geotimes, 30:25, June 1985.)

Comment. Note that the visual meteors or shooting stars that burn up high in the atmosphere are believed to be cometary debris and mostly ice and dust. Those meteors large and substantial enough to make it through the atmosphere and arrive at the surface as meteorites, must have a different source---something more palpable, such as the asteroids, although the spectral disparities reported above may force a reevaluation of this theory.

COSMIC SNOWBALLS AND MAGNETIC ASTEROIDS

The great diversity of the debris swirling around the solar system is making life difficult for scientists trying to reconstruct solar-system history. At the high end of the density spectrum, we now have an asteroid that seems to be mostly metal (probably iron). This is the asteroid Gaspra, some 13 kilometers across, that the Galileo spacecraft encountered in August 1992 on its way to Jupiter. Scientists had not expected Galileo's magnetometer to flicker as it passed Gaspra at a distance of 1600 kilometers---but it did. In fact, considering the inverse square law and Gaspra's small size, it was a magnetic wallop. Thus, Gaspra is the first known magnetic asteroid; and it is probably mostly metal. (Kerr, Richard A.; "Magnetic Ripple Hints Gaspra Is Metallic," Science, 259:176, 1993.)

At the low end of the density spectrum, we now find that Pluto's moon, Charon, and some of Saturn's moons have very low densities (1.2-1.4), meaning they are probably mostly water ice. Such density figures come from direct observation of these objects' volumes combined with mass estimates from their orbital dynamics. (Crosswell, Ken; "Pluto's Moon Is a Giant Snowball," New Scientist, p. 16, November 21, 1992.)

CHIRON: BLACK SHEEP OF THE SOLAR SYSTEM

Charles Kowal discovered the smooth, very dark sphere called Chiron in the 1970s. Only a little more is known about it today. Chiron is 200-400 km in diameter---asteroid-sized. But its orbit (aphelion, 18.9 A.U.: perihelion, 8.5 A.U.) is definitely anomalous for asteroids. One would expect to find only comets in this region of the solar system. To compound the mystery, Chiron's orbit is unstable. The planetoid was originally somewhere else (no one says where) and was nudged into its present orbit by a major planet. One group of researchers calculates that Saturn could have been the nudger, and that the event might have happened as recently as 1664!! (Lipscomb, R.; "Chiron," Astronomy, 11:62, March 1983.)

Note: A.U. = Astronomical Unit, which is the average distance of the earth from the sun.

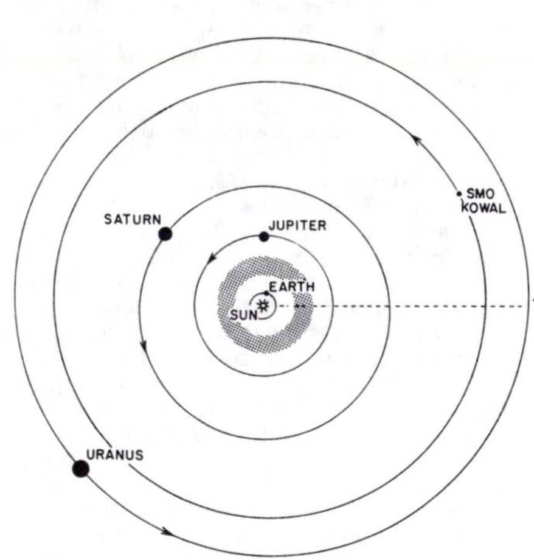

The asteroid Chiron, originally called SMO Kowal, orbits well beyond the classical asteroid belt indicated by the shading.

ASTEROIDS THAT TURN INTO COMETS

Even though Chiron is large (about 200 kilometers in diameter) and a bit dark for a rocky asteroid, astronomers have been quite comfortable with its classification. True, its orbit between Saturn and Uranus is unusual, but what else could it be but an asteroid? A comet, that's what! Recently, the brightness of Chiron has doubled, suggesting that it has expelled considerable gas and dust---a characteristic of comets, according to mainstream thinking. Another peculiar aspect of the phenomenon is that Chiron is now located 12 A.U. from the sun (12 earth orbits out). Conventional wisdom has it that solar heating is too weak at that distance to vaporize cometary ices. However, other comets, such as Schwassmann-Wachmann 1, have displayed comas even farther away from the sun. Thus, we have two possible anomalies here:

(1) the existence of a huge comet-like asteroid in a peculiar orbit; and
(2) a mechanism that expels gas and dust from comets at great distances from the sun.

The blurring of the distinctions between asteroids and comets is aggravated by the recognition that some other asteroids produce streams of particles that create meteor showers; that is, some asteroids are not simply associated with meteor streams, they actually create them, just as comets expel ice and rocky debris. Some bold astronomers now ask whether asteroids are not all burnt-out comets. (Kerr, Richard A.; "Another Asteroid Turned Comet," Science, 241:1161, 1988.)

Comment. Taking the matter a step further, could not the four inner, terrestrial planets also be burnt-out comets?

ASTEROIDS WITH MOONS?

Several recent observations made of asteroids as they occult stars suggest that some asteroids are circled by moonlets. The observational technique resembles that used in the discovery of the now-famous rings of Uranus. Briefly, the star being observed blinks out not once in a clean-cut fashion but in a complex scenario that may indicate the presence of a second body. To illustrate, during the 1975 occultation of a star by the asteroid Eros "all sorts of people saw things," meaning secondary events or extra dimmings. Another kind of supporting evidence comes from the light curve of 44 Nysa, which closely resembles that of an eclipsing binary star. (Anonymous; "Asteroids with Moons?" Science News, 114:36, 1978.)

Comment. In 1993 radar images of a passing asteroid showed it to consist of two separate bodies in near contact.

HEAVY TRAFFIC IN NEAR-EARTH SPACE

That region of outer space near the earth carries a heavier load of flotsam and jetsam than scientists expected. The devastation of the 1908 Tunguska impact in Siberia warns us that this space debris---meteors, comets, asteroids---is an active threat. A recent spate of articles paints an ominous future. Although this item from Science Frontiers straddles three categories of solar system debris, its message is more forceful if we keep it intact.

The earth's retinue of mini-asteroids.

> Asteroids as big as houses pass near the Earth 100 times more often than anyone suspected. On an average day, about 50 asteroids measuring at least 10 metres across come closer to the Earth than the Moon, and each

year about five such objects may hit the planet. (Second reference below.)

These startling data come from D. Rabinowitz and coworkers at the University of Arizona, who have been scanning nearby space with a telescope fitted with supersensitive charge-coupled devices (CCDs). They have picked up astronomical objects that have escaped conventional instruments. Several sources have been suggested for this unexpected, threateningly large population of small asteroids: (1) debris hurled earthward from collisions within the asteroid belt located between Mars and Jupiter; (2) the breakup of a large object formerly in orbit about the earth; and (3) fragments blasted off the moon by impacts of large asteroids there. (Kerr, Richard A.; "Earth Gains a Retinue of Mini-Asteroids," Science, 258: 403, 1992. Also: Mitton, Simon; "House-Sized Asteroids Home In on the Earth," New Scientist, p. 16, October 31, 1992.)

Comet Smith-Tuttle. At a conference in Sydney last October, astronomer D. Steel announced that comet Smith-Tuttle is heading towards a possible impact with earth on August 14, 2116. This 3.1-mile-diameter chunk of ice would have the destructive power of 20 million megatons (1.6 million Hiroshima bombs). (Anonymous; "Astronomer Predicts Comet Collision," Baltimore Sun, October 26, 1992.)

Some recent meteorite impacts. Turning from the dire consequences discussed above, just what sort of astronomical debris actually does hit the earth on a day-to-day basis? Fist-sized meteorites strike our planet about every two hours. These are the ones we read about in the newspapers; and they have left a surprisingly large legacy of damage to human structures. C. Spratt and S. Stephens, in a survey published in Mercury in 1992, listed 61 verified meteorite strikes since 1790 in which buildings and other human works were damaged. (Of course most fell harmlessly in the sea and unpopulated areas.) Spratt and Stephens also provide a table of 26 near-misses of humans plus one confirmed human impact. At least one horse and a dog have been killed by meteorites. These lists make engrossing reading, but we cannot take the space to reproduce them here. (Spratt, Christopher, and Stephens, Sally; "Against All Odds," Mercury, 21:50, March/April 1992.)

COMETS

With their long, spectacular tails, great comets, such as Halley's, attract worldwide attention as they approach the sun and flare up to naked-eye visibility. The anomalies of comets, however, have little to do with their brilliancy or tail lengths; rather, the scientific problems are found mainly in their orbital parameters, erratic flare-ups, and unusual compositions.

Until recently, comet theory rested upon three hypothesis: (1) Comets are composed of a nucleus of ices and rocky material, popularly called a "dirty snowball"; (2) Solar heating and the solar wind stimulate sun-approaching comets to flare up; and (3) Most comets originate in the Oort Cloud of pristine comets located some 50,000-100,000 AU from the sun, where the perturbations of passing stars occasionally nudge some of them toward the sun.

The past issues of Science Frontiers have focussed upon four categories of cometary anomalies:

- Halley's anomalies. The unexpected flare-ups of this comet while far from the sun and its apparent cargo of organic materials.
- Kinked tails and smoke rings. Some comets do not obey the rules!
- Comet clouds. Doubts about the existence of the Oort Cloud of comets and the possible existence of other cometary clouds.
- Comets and life. Do comets carry the templates of life or perhaps even primitive life forms?

HALLEY'S ANOMALIES

HALLEY'S COMET IS WINKING AT US (1984)

Halley's Comet, still a billion kilometers away, is just beginning to emit gases at the urging of the sun's rays. It should, therefore, be getting brighter---and it is---but its brightness pulsates. A French team of scientists, led by Jean Lecacheux, has determined that Halley's Comet flares up at regular intervals just over 24 hours apart.

We usually do not study comets carefully until they are very close to the sun, so we don't know if this blinking behavior is typical or not. The most reasonable explanation is that Halley's Comet rotates about every 24 hours and that its surface is not uniform. One portion of its surface may be brighter or emit more luminous gases than the rest. In any event, we have a new astronomical curiosity. (Lloyd, Andrew; "Halley's Comet Is Blinking," New Scientist, p. 20, May 24, 1984.)

HALLEY'S CONFOUNDING FIRWORKS (1986)

"Serendipitous discovery in science so often comes as a surprise, even when it might reasonably be expected. Such was the case with Comet Halley in the wake of an observation program that dwarfed all preceding efforts. The surprises included sudden outbursts in the presumably steady vaporization of its icy nucleus and a periodic, complex pulsation of the comet's brightness. Whether this pulsation reflects the rotation of the nucleus, wobbling of the nucleus, or some still unimagined phenomenon became the controversial focus of the recent meeting on the Exploration of Halley's Comet in Heidelberg."

Some data and attendees at this conference supported a rotational period of 2.2 days; others favored 7.4 days; a few liked a 2.2-day period with a 7.4-day wobble superimposed. (Kerr, Richard A,; "Halley's Confounding Fireworks," Science, 234:1196, 1986. Also: Campbell, Philip; "How Fast Does Halley Spin?" Nature, 324:213, 1986.)

Halley's fireworks (visible outbursts) have also stimulated comet model-making. The latest is an icy-glue construction of sorts. See the accompanying figure and the accurate but not-particularly-enlightening caption. (Gombosi, Tamas I., and Houpis, Harry, L.F.; "An Icy-Glue Model of Cometary Nuclei," Nature, 324:43, 1986.)

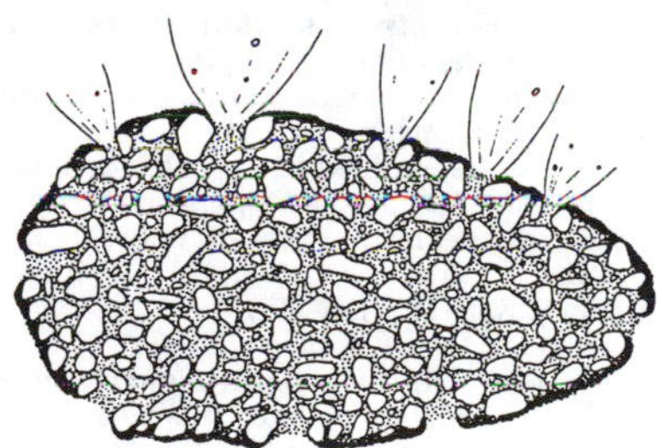

A section through the icy-glue model of a comet's nucleus, showing active and inactive micropores, small boulders, and the outer dust covering.

HALLEY: A YOUNG, COMBUSTING, ALIEN INTERLOPER! (1989)

Can this be the comet Halley of the textbooks? Comets are supposed to be as old as the solar system itself, 4.6 billion years, born of solar system stuff when a gaseous cloud condensed. Above all, comets do not "burn" or combust!

The vision of a "burning" comet was advanced by recent observations that the velocity and temperature of the gases escaping from Halley are higher than one would expect from the sublimation of ices under solar radiation. Also, the concentration of expelled material in large, hypersonic jets carrying large quantities of fine dust further undermine the sublimation model. E.M. Drobyshevski has concluded:

"The new observations, together with some earlier data still poorly understood (e.g. the appearance in the coma of large amounts of C_3) can be accounted for by assuming the cometary ices to contain, apart from hydrocarbons, nitrogen-containing compounds, etc. also of free oxygen ($<$ 15 wt. %). Under these conditions, burning should occur in the products of sublimation under deficiency of oxidizer accompanied by the production of 'soot,' 'smoke,' etc. The burning should propagate under the surface crust and localize at a few sites.

"The presence of oxygen in cometary ices follows from a new eruption theory assuming the minor bodies of the Solar System to have formed in explosions of the massive ice envelopes saturated by electrolysis products on distant moonlike bodies of the type of Ganymede and Callisto." (Drobyshevski, E.M.; "Combustion as the Cause of Comet P/Halley's Activity," Earth, Moon, and Planets, 43:87, 1988. Cr. L. Ellenberger)

Drobyshevski's combustion theory assumes a "local" origin (within the solar system) for Halley. But measurements of the ratio of carbon-12 to carbon-13, made during the 1986 flyby, produced a ratio of 65:1. This compares to 89:1 for solar system material. The 65:1 ratio, it turns out, is more typical of interstellar material. This datum seems to place Halley's birthplace somewhere outside the solar system. (Weiss, R.; "Carbon Ratio Shows Halley May be Alien," Science News, 135: 214, 1989.)

As if all this were not bad enough, calculations of the amount of matter expelled from Halley and incorporated in known meteor streams allow an estimate of Halley's residence time in the inner solar system. (One has, of course, Halley's present mass, but its original mass must be estimated!) The conclusion is that Halley has spent only 23,000 years in its present orbit! (Maddox, John; "Halley's Comet Is Quite Young," Nature, 339:95, 1989.)

Comment. This is a pretty shaky calculation, admittedly, but it accords with the calculation that Halley had a close encounter with Jupiter about 20,000 years ago. The manifest contradictions in the inferences made above from the recent observations of Halley mean that we still have a lot to learn about comets, Halley in particular. One should also recall that the solar system has other features that may be youthful, such as Saturn's rings.

HALLEY REAPPEARS! (1991)

No, comet Halley has not reversed direction for an anomalous encore. We'll have to wait another 70-or-so years for that. However, Halley did make a surprise reappearance on February 12, 1991. Astronomers were startled by a sudden flare-up. It was not a trivial brightening; the width of the flare was a remarkable 300,000 kilometers.

Comets often flare up as they swing close to the sun and absorb its heat and radiation. But Halley is now billions of kilometers away in the frigid reaches of the outer solar system. No one knows what happened. (Pease, Roland;

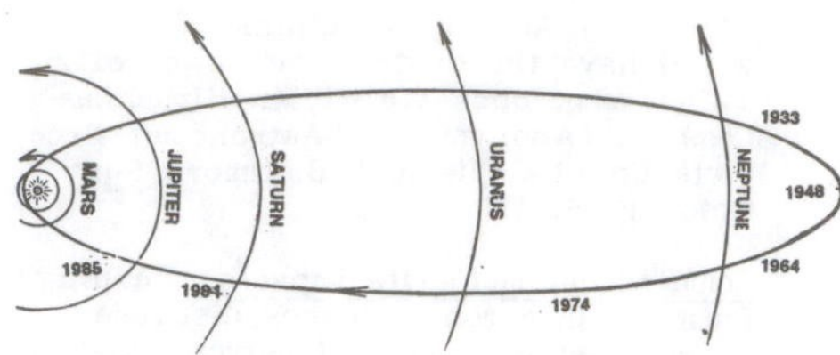

The orbit of Halley's comet.

"Halley at Large," Nature, 349:732, 1991.)

Comment. Other comets have mysteriously flared up far from the sun. See ACO2 in The Sun and Solar System Debris. Apparently, comets harbor considerable pent-up energy. If proximity to the sun is not required to stimulate gas releases or chemical reactions, comets may have their own energy agenda. Comets seem to be little more than chunks of dirty ice. Where could the flare energy come from? If only cold fusion were a viable source!

KINKED TAILS AND SMOKE RINGS

COMET PUFFS A SMOKE RING

When the periodic comet Kopff was photographed on August 13, 1983, using the 4-meter Mayall reflector at Kitt Peak, the black-and-white photo showed nothing out-of-the-ordinary. But digitization and computer enhancement revealed an unnoticed cloud of matter ½ million miles long trailing the nucleus. Curiously, the cloud resembled a huge smoke ring---a distinct departure from the long flowing tails shown in the textbooks. (Anonymous; "The Unusual Dust Cloud of Comet Kopff," Sky and Telescope, 67:226, 1984.)

"?" !?

Photographs of Comet Bradfield taken on October 10, 1987, show an odd kink in its tail that does not appear on photos taken the nights before and after. This kink was shaped like a backwards "?". Kinks in the tails of comets are well-known phenomena. A comet's tail is electrically charged, and it flaps in the solar wind like a flag on a gusty day. So why run this observation up the anomaly flagpole? First of all, the Bradfield kink is 10 million kilometers long; second, it appeared and disappeared in a matter of hours. Both size and speed of formation are difficult-to-explain in terms of existing solar wind velocity and the shifting interplanetary magnetic field. (Anonymous; "Did Anyone Photograph This Comet?" Astronomy, 16:16, July 1988.)

COMET CLOUDS

MYSTERY OF THE MISSING COMETS

A key feature of our solar system is the Oort Cloud of comets that surrounds the sun and its family of planets. No one has yet seen the Oort Cloud directly, but the textbooks say that it must be there. In fact, all stars like our sun with planetary systems should have their own private Oort Clouds of comets, if the prevailing theory of planetary system formation is correct.

When we see a comet looping around the sun, it is because it has been jostled loose from the Oort Cloud by a passing star or molecular cloud. Further, some of these jostled comets should be kicked outwards and thus escape the solar system. Continuing with this reasoning, we on earth should sometimes see interstellar comets that have been shaken loose from other stellar systems. But we don't! T.A. McGlynn and R.D. Chapman worry about this:

> This lack of detections of extrasolar comets is becoming an embarrassment to the theories of solar system and cometary formation.

McGlynn and Chapman calculate that we should have seen six interstellar comets in the past 150 years, but the actual number is zero. Such interstellar comets would be easy to spot because they would be moving much faster than our own comets. Two possible explanations for the missing interstellar comets are: (1) the Oort Cloud theory is wrong; and (2) solar systems like ours are rarer than supposed. (Anonymous; "Mystery of the Missing Comets," Sky and Telescope, 79:254, 1990.)

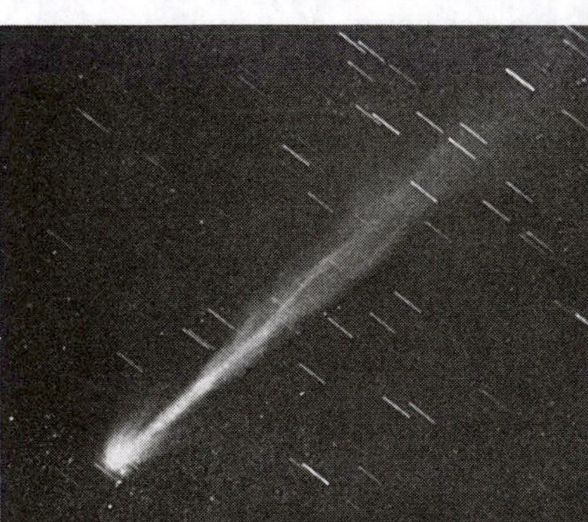

Morehouse's Comet, September 29, 1908. The star trails imply that the comet's direction of travel is nearly at right angles to its tail, which is directed away from the sun. (Royal Observatory)

TOO MANY SHORT-PERIOD COMETS

Some comets, such as Halley's, have periods less than 200 years. Scientists have postulated that these comets orbiting relatively close to the sun originally came from the far distant Oort Cloud on parabolic (non-returning) orbits around the sun. Perturbations by the planets, notably Jupiter, deflected them into the tighter orbits we see today.

The problem is that the number of parabolic comets entering the inner solar system from the Oort Cloud of comets, located at the outermost fringes of the solar system, is 100 times too small to account for the existing population of short-period comets. M.E. Bailey believes this discrepancy can be removed if the Oort Cloud possesses a massive inner core of comets. (Bailey, M.E.; "The Near-Parabolic Flux and the Origin of Short-Period Comets," Nature, 324:350, 1986.)

Comment. The Oort comet cloud is an entrenched part of astronomical dogma. See ACB15 in The Sun and Solar System Debris for other suggestions that it might not in fact exist.

A NEARBY RING OF COMETS?

Some 589 long-period comets are known. They ply orbits around the sun that take millions of years to complete. Astronomers are generally agreed that these bodies originate in a very distant (100,000 AU) halo of cometary material surrounding the entire solar system. J. Oort proposed this cloud, and it is named after him. Of course, we anomalists become wary when scientists "generally agree" on a hypothetical entity which no one can see. The Oort Cloud of comets, like the unseeable black holes, are given substance only by the effects they have on other solar system denizens and seeable cosmic objects.

But there may be another cloud of comets that we can view directly. It is called the Kuiper Cloud (after G. Kuiper), and it is concentrated in the plane of the ecliptic just beyond the orbit of Neptune. Like the Oort Cloud, the Kuiper Cloud has not been seen yet, but we might just be able to! Its existence is hypothesized from the parameters of a different group of comets ---the short-period comets, as exemplified by the 76-year Halley's Comet. About 120 short-period comets have been discerned; and our computers now tell us that they cannot have originated in the distant Oort Cloud. Something closer and on the ecliptic is required. Thus the Kuiper Belt was born. It is thought to be debris left over after the formation of the solar system. (Kerr, Richard A.; "Comet Source: Close to Neptune," Science, 239:1372, 1988.)

Comment. Just before the above Science article appeared, a piece on comet origins was printed in the New Scientist. Curiously, the Kuiper Belt was not even mentioned. Instead, we find: "Astronomers have discovered about 200 of these 'short-period' comets, with Halley's Comet the best known, but their orbits are of little use to astronomers studying the origin of comets." (Theokas, Andrew; "The Origin of Comets," New Scientist, p. 42, February 11, 1988.)

COMETS AND LIFE

STAR SLUDGE

All of a sudden it seems that astronomers are finding dark---even "charcoal-black"---materials in unlikely places in the solar system. Three "sludgy" sites have been highlighted in the recent literature:

1. Comets. "Black" comets certainly defy our expectations. Are all those pictures of white, flaming apparitions wrong? Not really. The comets approaching the sun are made visible by sunlight reflected from the gases and dust in the coma and tail, plus some direct emission. However, the heart of a comet, its nucleus, has long been considered a "dirty snowball," i.e., a mixture of dirt and ices. Now it appears that comets are more like "icy dirtballs"! New measurements of the bare nuclei of comets, using a visual-infrared technique, find that the nuclei reflect as little as 2% of the incident sunlight. They are indeed charcoal black. Comet nuclei, according to W. Hartman and his colleagues, are colored by a brownish black primordial organic sludge, and have the appearance of "a very dark Hershey bar." The use of the adjective "organic" may be premature, but maybe not in the light of the next item.

2. Carbonaceous chondrites. This well-known class of meteorites sometimes appears tarry and is characterized by carbon contents of up to 2% and more. The Japanese have just reported the analysis of Yamoto 791198, a carbonaceous chondrite picked up in the Antarctic. This meteorite is loaded with amino acids, 20 kinds of them. These extraterrestrial amino acids are not considered biogenic because they are half left-handed and half right-handed; whereas all amino acids synthesized by terrestrial animals, plants, etc., are left-handed. (This is obviously presumptuous because we have no notion how extraterrestrial life works, assuming it exists at all! WRC)

3. The rings of Uranus. Voyager photos reveal that the rings are composed of unknown dark material quite unlike the high-albedo (bright) rings of Saturn.

References. (Kerr, Richard A.; "A Comet's Heart May Be Big But Black," Science, 229:372, 1985.) (Emsley, John; "Amino Acids from Outer Space," New Scientist, p.30, December 19/26, 1985.) (Anonymous; "Fine Particles Viewed in Uranus' Rings Leave Scientists 'Happily Bewildered'," Baltimore Sun, p. 5A, January 28, 1986.)

Comment. How and where is star sludge manufactured? Again we have to venture that the venerable "primordial soup," in which life mysteriouly assembled itself, is located in outer space rather than in warm, sunlit earthly ponds. It also seems strange that all these amino acids just seem to "fall together," with little urging, in seemingly hostile environments. More on this in following items.

HALLEY'S COMET INFECTED BY BACTERIA?

"Halley's comet is coated with organic molecules. Two astronomers working with the Anglo-Australian Telescope, David Allen and Dayal Wickramasinghe, have found strong evidence in an infrared spectrum of the comet, taken two weeks ago. The spectrum, spanning the wavelengths 2 to 4 micrometres (μm), shows a prominent feature centred at 3.4 μm which the two astronomers attribute to emission by carbon-hydrogen bonds in a solid."

Chandra Wickramisinghe (Dayal's brother) states that the emissivity of Halley's Comet matches exactly the emissivity of bacteria as observed in the laboratory. This observation supports the Hoyle-Wickramasinghe suggestion that comets transport life forms around the universe. Of course, more conservative scientists contend that rather complex organic molecules can by synthesized in space abiogenically. (Chown, Marcus; "Organics or Organisms in Halley's Nucleus," New Scientist, p. 23, April 17, 1986.)

Comment. This article seems to be accompanied by a bit of scientific revisionism. In response to this discovery, one English scientist remarked that these results "only confirm what everyone has always suspected." Now it is true that comets have long been termed "dirty snowballs," but until very recently virtually no one maintained that comets were covered with dark, organic sludge!

CARBON IN A NEW COMET

"Astronomers in Australia have confirmed that organic matter exists in Comet Wilson, which is on what is believed to be its first and only visit to the Solar System.

"It is the first time that organic matter has been found on a comet new to the Solar System. The astronomers, who observed the comet from the Anglo-Australian Telescope at Siding Spring Observatory in New South Wales, say that this finding lends weight to the theory that comets brought to Earth the carbon-based chemicals from which life evolved." (Anonymous; "Astronomers Spot Carbon in New Comet," New Scientist, p. 23, May 28, 1987.)

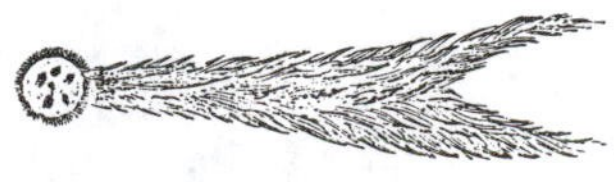

STARS

THE SUN

As our closest star, the sun is our key to understanding the other stellar denizens of the cosmos. Our increasingly rich store of knowledge about the sun has led to some of the most interesting anomalies in all of astronomy:

- Eclipse anomalies. While the oft-observed shadow bands are more curious that anomalous, the effect of a solar eclipse on the motion of a pendulum is definitely scientifically disturbing.
- The sunspot cycle. The origin and vagaries of the sunspot cycle are not well understood.
- Solar neutrinos. Terrestrial instruments still measure fewer of these than stellar theory predicts.
- Solar oscillations. Some of these may have origins external to the solar system.
- A massive solar companion. Is our sun a binary star?

ECLIPSE ANOMALIES

ECLIPSE SHADOW-BAND ANOMALIES

J.L. Codona, in a long article in Sky and Telescope, described eclipse shadow bands in these words:

> ....mysterious gray ripples are sometimes seen flitting over the ground within a minute or two of totality. The bands are initially faint and jumbled; but as totality approaches, they become more organized, their spacing decreases to a few centimeters, and their visibility improves. After totality ends the bands can reappear and become progressively fainter and more disorganized until they disappear.
>
> Shadow bands seem to move perpendicularly to their length, but this is only an illusion. It stems from a lack of features that allow the eye to track motion along the length of the bands.

Codona explains the shadow bands as basically a twinkling effect involving the thin solar crescent just before and after totality. The twinkling is created by turbulence only tens or hundreds of feet above the ground.

The eclipse shadow bands, like so many other "well-explained" phenomena, display idiosyncracies that do not dovetail well with theory. Codona mentions two of these: (1) Bands of different colors, travelling at different speeds, are sometimes seen superimposed on each other; and (2) Bands of giant size have been observed. (Codona, Johana L.; "The Enigma of Shadow Bands," Sky and Telescope, 81:482, 1991.)

Comment. Still other shadow-band anomalies are cataloged in GES1, in Rare Halos.

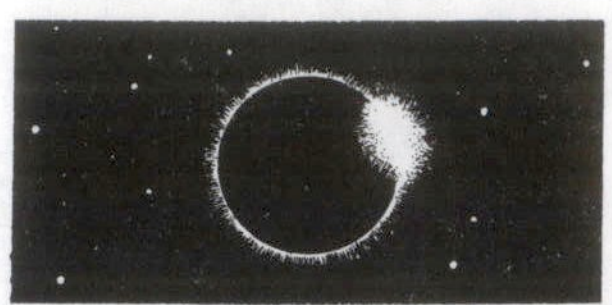

SOLAR ECLIPSE AFFECTS A PENDULUM ---AGAIN!

The period of a Foucault pendulum located at Jassy University, Romania, was carefully monitored during the solar eclipse of February 15, 1981. The pendulum's length was 25.008 meters; its spherical bob weighed 5.5 kilograms. The eclipse commenced at $8^h49^m3^s$ and terminated at $11^h16^m50^s$. Observations are recorded below:

Time	Period (sec)
8:49	10.028±0.004
9:13	10.028±0.004
9:43	10.024±0.004
10:00	10.019±0.004
10:12	10.020±0.004
10:24	10.024±0.004
10:58	10.028±0.004

If the above effect of the eclipse on the pendulum period is not strange enough, consider what happened at 10:08:

"At that moment a surprising fact occurred, the pendulum produced a perturbation by describing an ellipse whose major axis deviated in relation to the initial plane by approximately 15°. The eccentricity of the ellipse was 0.18. At the end of the eclipse the pendulum continued to maintain the elliptical oscillation, but the major axis approached increasingly to its initial plane." (Jeverdan, G.T., et al; "Experiments Using the Foucault Pendulum during the Solar Eclipse of 15 February, 1981," Biblical Astronomer, 1:18, Winter 1991. This journal was formerly the Bulletin of the Tyconian Society.

Comment. Such observations are naturally hard to accept, but similar effects have been reported before by M.F.C. Allais (see figure) and E.J. Saxl, at Harvard. See ASX6 in The Sun and Solar System Debris.

Shadow bands moving across the front of a house during the solar eclipse of December 22, 1870.

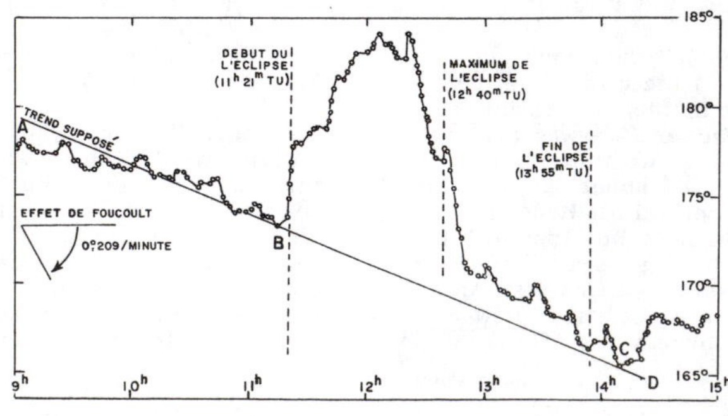

Effect of a solar eclipse upon a paraconical pendulum. (After M.F.C. Allais)

WHERE DID THE 1780 ECLIPSE GO?

October 27, 1780. The Maine Coast. An expedition from Harvard and the American Academy of Arts and Sciences, led by Samuel Williams, had set up equipment to observe the predicted total eclipse of the sun. A solar eclipse occurred all right, but the expedition was shocked to find itself outside the path of totality. They saw a thin arc of the sun instead of the expected complete obscuration by the moon.

Modern analyses of this embarrassing incident for embryonic American science blame Williams for miscalculating the path of totality. Actually, recent computations compound the mystery. The expedition measured the time of the eclipse as 40 seconds later than it should have been for their (erroneously selected) site. Modern analysts insist that the expedition should have seen an arc 10 arcsec wide subtending an angle of 89°; instead Williams and his colleagues measured an arc of less than 24° and so thin it could not be measured. Finally, the measured duration of the eclipse was far different than it should have been.

The members of the expedition were skilled and their instruments excellent. What happened? Was it human error? Are modern eclipse calculating methods in error? Did something astronomical happen, such as a temporary change in the earth's period of rotation? The mystery persists. (Rochschild, Robert F.; "Where Did the 1780 Eclipse Go?" Sky and Telescope, 63:558, 1982.)

SUNSPOT CYCLE

WHAT CAUSES THE SUNSPOT CYCLE?

Ever since the sunspot cycle was discovered, people have been trying to prove that it is caused by the influence of the planets, particularly Jupiter with its 11.86-year period. A century of various correlations has convinced almost no one. John P. Bagby has introduced a new piece to the puzzle of solar system cyclic behavior. While searching for possible perturbations of the planets due to a tenth major planet or a dark massive solar companion (MSC), he discovered that the perihelia of the outer planets (orbital points closest to the sun) were being disturbed with an average period of 11.2 years. This is almost exactly the sunspot period.

This serendipitous finding caused Bagby to wonder whether some common influence was causing not only the sunspot sycle and the perturbations in outer-planet perihelia but also cyclic volcanic and seismic activity on earth. Some correlations indeed do indicate a sun-earth link of some kind. Bagby suggests two possiblilities: (1) Mutual resonance effects between the planets; and (2) The effects of a massive solar companion. (Bagby, John P.; "New Support for the Planetary Theory of Sunspots," privately circulated paper, 1983.)

Comment. Even "farther out" is the thought that gravitational waves or some unrecognized influence from the galaxy or beyond causes the whole solar system to pulse at the sunspot frequency. In other words, the solar cycle might be externally forced.

THE MISSING SUNSPOT PEAK

"Abstract. An analysis of mean annual sunspot numbers is made with particular emphasis on cycles having periodicities near 21 years. The results are compared not only with the original sunspot data but also with long-term geomagnetic and economic data. It is concluded that the '11-year' solar cycle periodicity increased during the 19th century, during which time there were only 8 peaks when 9 peaks might have been expected. Doubt is cast on the reality of a 22-year sunspot cycle during the past three centuries, and the likelihood is shown that the reliable 21.2-year sunspot cycle is also the Hale magnetic cycle and that several of its harmonics are present in the economic data." (Robbins, Roger W.; "The Case of the Missing Sunspot Peak," Cycles, 36:53, 1985.)

Comment. While sunspot numbers do wax and wane in a cycle that is approximately 11 years long, the solar magnetic field actually changes polarity on a 22-year cycle called the Hale cycle, during which there are two sunspot maxima and minima.

CHANGES IN SOLAR ROTATION

John A. Eddy has continued his historical studies of sunspots from the earliest records to date. Analysis of spot drawings suggest that between 1625 and 1645 the equatorial velocity of the sun was significantly higher than it was earlier and is now. Eddy believes that this acceleration presaged the onset of the peculiar Maunder Minimum, 1645-1715, when the sun was virtually clear of spots. (Eddy, John A., et al; "Anomalous Solar Rotation in the Early 17th Century," Science, 198:824, 1977.)

ARE THE SUN'S FIRES GOING OUT?

After reviewing solar measurements from the last 200 years, John A. Eddy, of the Smithsonian Observatory, has concluded that the sun's diameter shrinks by about 10 km per year. This observation fits nicely with the ten-year-old enigma of missing solar neutrinos. A shrinking diameter implies a cooling sun, while the paucity of neutrinos suggests that the sun's nuclear fires are at a low ebb. Something similar may have happened in the 17th. Century during the Maunder Minimum, when the so-called Little Ice Age prevailed. (Anonymous; "Scientists Shrink from Smaller Sun," New Scientist, 82:982, 1979.)

A LARGER SUN DURING THE MAUNDER MINIMUM

Europe's so-called "Little Ice Age" (1645-1715) coincided with the Maunder Minimum---a period during which sunspots were exceedingly rare. How was the sun different during the Maunder Minimum? This subject of solar variability (in both diameter and period of rotation) has been long debated. Some early measurements of solar diameter, begun at Greenwich in 1830, seemed to some to show a slowly shrinking sun, but others found cyclical patterns.

E. Ribes, et al, have just presented some data on solar diameter actually taken during the Maunder Minimum. "By analysing a unique 53-year record of regular observations of the solar diameter and sunspot positions during the seventeenth century, we have shown for the first time that the angular diameter was larger and rotation slower during the Maunder Minimum." A larger sun might be cooler, providing less heat thus accounting for climatic changes. (Ribes, E., et al; "Evidence for a Larger Sun with a Slower Rotation during the Seventeenth Century," Nature, 326:52, 1987.)

Comment. Just why the sun contracts and expands over a period measured in hundreds of years is a major astrophysical conundrum.

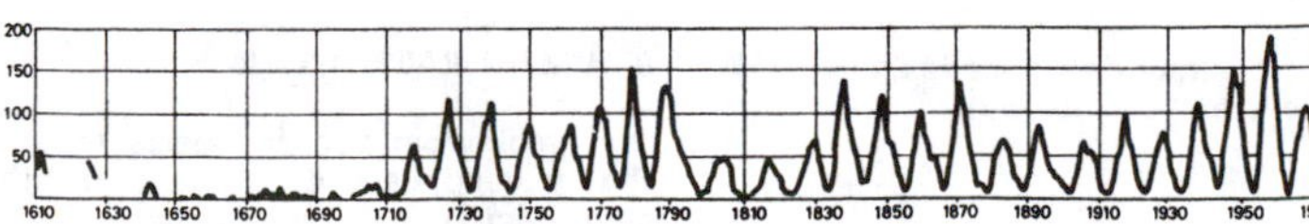

The average number of sunspots recorded between 1610 and 1975.

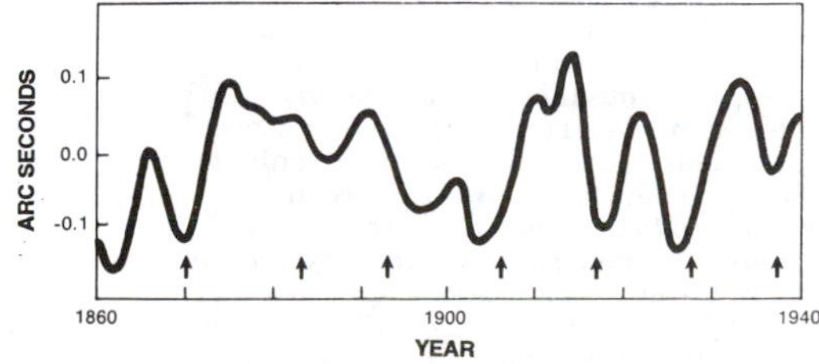

Variation in solar diameter, 1860-1940. Arrows indicate sunspot maxima. (From ASO8 in The Sun and Solar System Debris.)

SOLAR NEUTRINOS

WIMPS IN THE SUN?

For over two decades now, physicists have been measuring the neutrino flux emitted by the sun---and despite all attempts this flux is much too low. It just doesn't jibe with what theorists say should be happening in the thermonuclear powerhouse in the sun's interior.

John Faulkner and Ron Gilliland have conceived a solution to this dilemma. They postulate a large population of WIMPS (Weakly Interacting Massive Particles) orbiting the sun's core, but still well beneath the sun's visible surface. The WIMPS help convey heat out of the core, thereby cooling it to temperatures significantly less than those predicted by the astrophysicists. A cooler core emits fewer neutrinos, bringing theory into line with reality.

And just what are these WIMPS? One suggestion is that they are photinos, a particle suggested (but not proved) by recent experiments at CERN. (Thomsen, D.E.; "Weak Sun Blamed on WIMPS," Science News, 128:23, 1985.)

Comment. WIMPS represent just the kind of wild theorizing that Dewey Larson rails against in his book, The Universe of Motion. He maintains that astronomers have to engage in such ridiculous theoretical gymnastics and invention only because they have picked the wrong energy-generating mechanism for stars and refuse to give it up! Larson's theory (his Reciprocal Theory) solves this and many other astronomical problems, but only at the initial cost of a radical change in one's conception of the universe. See p. 328.

SOLAR NEUTRINO UPDATE

Our terrestrial neutrino detectors catch only about 1/3 as many solar neutrinos as stellar theory requires. We highlight this anomaly because at risk here is our basic theory of how stars work. Is our knowledge of stellar furnaces fundamentally in error or are some of the solar neutrinos somehow removed from the stream of neutrinos bound for earth?

Recent calculations by Hans Bethe have brought sighs of relief to astrophysicists. Without going into the details, Bethe finds that the interactions of the electron-neutrinos emitted by solar thermonuclear reactions with the atoms constituting the solar mass change a substantial fraction of them into muon-neutrinos. Since our terrestrial neutrino detectors detect only the electron-neutrinos, we may really be seeing only a fraction of the actual number of neutrinos emitted by the sun.

If Bethe's calculations are correct, he may have eliminated a Class 1 anomaly. But at a price! It seems that his calculations also predict a mass of only 0.008 ev for the muon-neutrino. This is much too small for neutrinos to account for the "missing mass" of the universe ---something cosmologists had devoutly hoped for. (Maddox, John; "Hans Bethe on Solar Neutrinos," Nature, 320:677, 1986.)

SOLAR OSCILLATIONS

FLATTENED SUN MEANS TROUBLE FOR EINSTEIN

Recent measurements of the sun's diameter by P. Goode and H. Hill show:

(1) That the sun is "ringing like a bell" due to constant sunquakes; and
(2) That it is flattened enough at the poles to affect significantly the precession of Mercury's orbit.

The gravitational influence of the sun's bulge is so large that it, in effect, competes with relativistic effects as an explanation of Mercury's precession. Goode, in fact, has suggested that the Theory of Relativity must be in error. (Anonymous; "Reputed Mistake Found in Einstein's Theory of Relativity," Baltimore Sun, p. A9, April 6, 1982.)

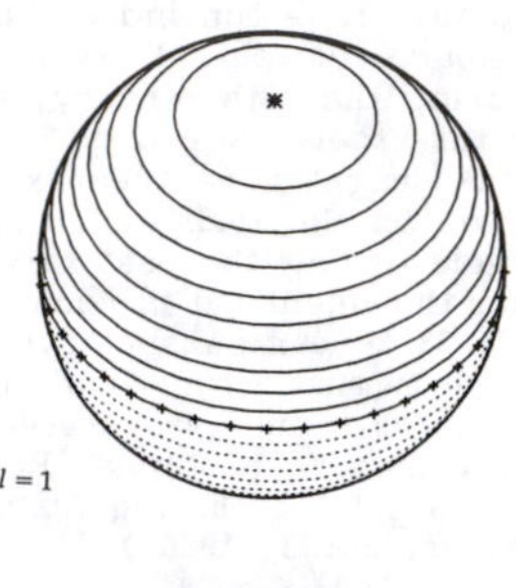

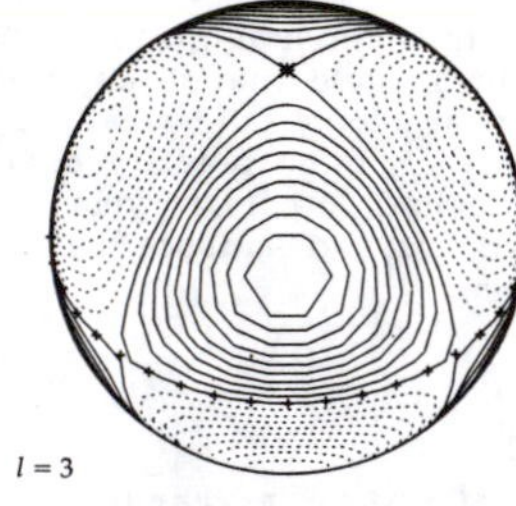

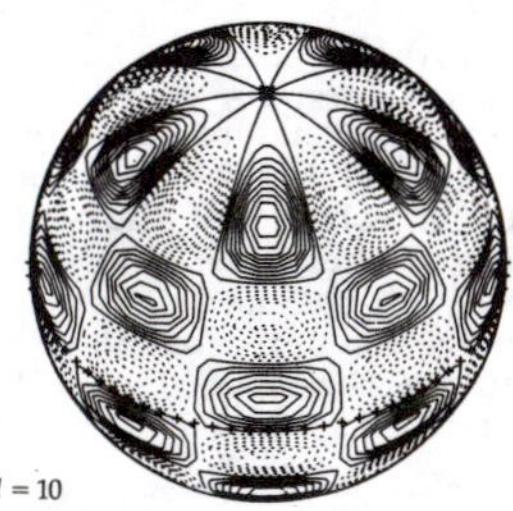

Three of the thousands of modes of oscillation possible on the sun.

THE SUN AS A SCIENTIFIC INSTRUMENT

In connection with the preceding item on solar oscillations and asymmetry, a few brave astrophysicists are now proposing that one mode of solar oscillation (the 160-minute period) is really a manifestation of the sun "ringing" in response to gravity waves sweeping through it! A nearby binary star, Geminga, has a period of this length. It seems that the 160-minute oscillation of the sun is far too long to be a solar pressure wave, and external forces could conceivably be involved.

This article also mentions "the throbbing earth," an effect which may result from gravitational waves emanating from the center of our galaxy. (Walgate, Robert; "Gravitational Waves on the Sun?" Nature, 305:665, 1983.)

MASSIVE SOLAR COMPANION

THE MASSIVE SOLAR COMPANION

Something big out there beyond Neptune perturbs the orbits of the sun's outer fringe of planets. In addition, there are unexplained perturbations in the orbits of earth satellites, peculiar periodicities in the sunspot cycle, and equally puzzling regularities in earthquake frequency. Infrared detectors have also picked up unidentified objects in the sky. These anomalies might be explained by the existence of a large dark planet with several moons---or, if the mystery object turns out to be far away and very large, a dark stellar companion of our sun, with its own system of planets.

Several astronomers have been trying to pin down the properties of this Planet X or Massive Solar Companion (MSC). John P. Bagby has recently published a novel solution to this nagging puzzle in celestial mechanics. He suggests that the Massive Solar Companion is actually a distributed system; that is, part of its mass is distributed between the stable Lagrangian points. The total MSC mass might be as much as half the sun's mass, but spread out at perhaps 100 Astronomical Units (100 times the earth's distance from the sun).

If the MSC and its attendants are this massive, astronomers will have to revise the mass and density of the sun downward by a good bit. (What they have done is estimate the mass of the solar system as a whole and assumed it mostly resides in the sun.) This would require a large change in our model of the sun and its system of planets. (Bagby, John P.; "Evidence for a Tenth Planet or Massive Stellar Companion beyond Uranus," Paper given at the Tomorrow Starts Here Conference, September 1982.)

A REAL DEATH STAR

The geological record seems to show that widespread biological extinctions have occurred about every 26 million years. Coupled with this is Walter Alvarez's recent observation that terrestrial impact craters 10-km-diameter and up have been blasted out episodically--every 28.4 million years on the average. This figure is close enough to 26 million years to impel some astronomers to search for a periodic source of cosmic projectiles.

R.A. Muller and M. Davis, at Berkeley, think they have found one. They postulate that the solar system is really a double star system. Our sun's companion star has only about a 0.1 solar mass and is so faintly luminous that we have not visually found it. It does, however, now cruise along its orbit some 2.4 light years away. But it will be back! In fact, it returns every 26 million years to jostle the Oort cloud of comets that hovers on the fringe of the solar system. This nudge periodically sends a large shower of comets careening around the inner solar system. The earth intercepts one or more of these projectiles and---bang---we have a new crater and another biological catastrophe. (Anonymous; "A Star Named George," Scientific American, 250:66, April 1984.) See pp. 74 and 84 for possible periodicities in the terrestrial cratering record.

SUPERNOVAS

Novas occur when stars suddenly and temporarily flare into great brilliance. In keeping with its name, a supernova is a star that explodes into even greater brilliance than a nova---to absolute magnitudes of -16 or more. A supernova may even be brighter than the galaxy in which it resides. We can observe novas in our own galaxy, but supernovas are usually seen only in other galaxies. Type-I supernovas form a distinctive class with well-defined characteristics; all nonconforming supernovas are thrown into Type II. Type-II supernovas display diverse light curves and spectra. Supernova taxonomy is a subjective business at best. As for the cause(s) of supernovas, they may be the result of binary stars, perhaps pairs of white dwarves, falling into one another and detonating. The following items from Science Frontiers focus on supernova 1987A, but they demonstrate that our understanding of supernovas is shaky.

SUPERNOVA PROBLEMS

Everyone has been talking and writing about the new supernova, 1987A, so we might as well, too. In fact, we really must, because 1987A is more than usually anomalous. The newspapers have oohed and aahed about this rare opportunity scientists have to study a nearby supernova. However, instead of getting closer to a final understanding of supernovas, 1987A seems to be confounding the theorists:

(1) No one can determine which star, if any, blew up. The 12th. magnitude star Sanduleak -69 202 was first fingered, for it is located in the proper spot. But it is still there, apparently unchanged, as is a still fainter companion. The problem is that if 1987A really originated with an even fainter star, such a star would not have enough mass to go supernova, according to theory.

(2) Part of 1987A's spectrum, its luminosity evolution, and precursory neutrino burst all indicate a Type-II supernova. Unfortunately, its ultraviolet spectrum is that of a Type-I supernova. Also anomalous are its high rate of evolution and low luminosity (only magnitude of 4.5 instead of the predicted 2-plus).

(3) Japanese apparatus located in a deep mine near Kamioka detected a burst of neutrinos about 22 hours before 1987A flared into the visible spectrum. This fits Type-II supernova theory, but the Mont Blanc equipment detected a neutrino burst just 4.6 hours before the flare-up. Which detector is correct; and why didn't both detectors detect both bursts, if such there were? At the moment, some scientists think the Mont Blanc event was spurious, in view of previous unexplained bursts recorded there.

(Waldrop, M. Mitchell; "The Supernova 1987A Shows a Mind of Its Own---and a Burst of Neutrinos," Science, 235:1322, 1987.)

SUPERNOVA CONFUSION AND MYSTERIES

In the above item, we reported some of the anomalies surrounding supernova 1987A, the first nearby supernova since 1604 AD. 1987A's nearness has given astronomers the opportunity to identify just which star exploded. (Other supernovas have been much too far away.) The prime suspect was the star Sanduleak -69 202. But then it was claimed that old -69 202 was still alive and well. This presented a quandry because no other star in the area was large enough to go supernova. But now, it seems there was a mistake, and it was Sanduleak -69 202 all along that detonated. (Waldrop, M. Mitchell; "Supernova 1987A: Notes from All Over," Science, 236:522, 1987.)

Comment. Naturally all of astronomy heaved a sigh of relief over this. Unfortunately, the relief was short-lived and a second dose of antacid seems needed!

Just a few days after the above, researchers reported in the New York Times discomforting news:

"According to Dr. Robert W. Noyes of Harvard-Smithsonian (Center for Astrophysics), the observations gathered by an extraordinarily sensitive camera show that the bright exploding star, or supernova, is actually two points of light, very close together, one about 10 times brighter than its companion. Since neither was present before the explosion, astronomers assume they both arose from the same blast, but how this could happen is a mystery.

......

"The supernova and its apparent companion lie in the Greater Magellanic Cloud, a satellite galaxy of our Milky Way, at a distance of about 150,000 light years from earth. The instrument used by the Harvard-Smithsonian group measured the distance between the two supernova elements as about one-twentieth of an arc second. This is about the separation a human eye would see between the headlights of a car some 5,000 miles away.

"Dr. Noyes said this distance was equivalent to an actual distance between the two bright objects of only about 3,000 Astronomical Units, where one Astronomical Unit is the distance from the sun to the earth, about 93,000,000 miles. This is a tiny distance in astronomical terms, but if both objects stemmed from the same explosion, Dr. Noyes said, they must have been moving apart at more than half the speed of light---an immense and surprising speed." (Browne, Malcolm W.; "Stellar Explosion Reported to Spawn Mysterious Twin," New York Times, May 23, 1987. Cr. J. Covey)

Comment. In modern supernova theory there is no place for stellar fission with fragments flying off at more than half the speed of light!

INDIGESTIBLE SUPERNOVA LEFTOVERS

There seems to be a mysterious "central compact object" lurking amid the debris of Supernova SN1987A. Prevailing supernova paradigms cannot account for this high density remnant. While some aspects of standard supernova theory were supported by observations made during and since the 1987 explosion, astrophysicists are left with several puzzles in addition to the mystery object itself:

> Other puzzles include the large-scale asymmetries observed in the heavy element ejecta (Fe-group line emission), the supernova envelope (optical polarization), and the circumstellar medium ([O III] ring), which are in addition to the complex structures resulting from hydrodynamic instabilities.

(Chevalier, Roger A.; "Supernova 1987A at Five Years of Age," Nature, 355:691, 1992.)

SUPERNOVA THEORY EXPLODED

What happens when two white dwarf stars in close orbit finally fall into one another? Theory says you get a colossal explosion called a Type-I supernova. But this hypothesis is in trouble because a recent survey of white dwarfs revealed absolutely no double white dwarfs in a sample of 25 from the Milky Way. Even if a few pairs are eventually found, they do not appear to be numerous enough to account for the rate at which supernovas are observed. (Crosswell, Ken; "Supernova Theory

Exploded by Solitary White Dwarfs," New Scientist, p. 23, March 23, 1991.)

OPTICAL BURSTERS

Against the background of the night sky, the careful observer can on rare occasions discern brief flashes of light lasting a few seconds or less. These phenomena are termed "optical bursters." Too brief to be novas, optical bursters could signal some new astronomical phenomenon.

MYSTERIOUS SPATE OF SKY FLASHES

Background. The first hint of something unusual came in 1985 in a brief note in Astronomy.

Bill Katz and a small group of Canadian amateur astronomers have accumulated a total of 14 bright flashes in Aries in just a year or so. "Point" meteors (meteors seen head-on) usually appear as flashes like this, but to see 14 in the same region of the sky in such a short span of time is truly remarkable. (Katz, Bill; "Chasing the Ogre," Astronomy, 13:24, April 1985.)

OPTICAL BURSTERS

For years astronomers have been puzzling over the significance of "bursters"; i.e., short bursts of radiation from various spots in the heavens. With sophisticated terrestrial and satellite-borne instruments, they have detected gamma-ray, X-ray, and infrared bursters. The visible portion of the spectrum has been neglected because of the slow development of sensitive, high-time-resolution detectors capable of monitoring large areas of the sky. Of course, the human eye is an excellent intrument for searching for optical bursters, but professional naked-eye astronomers are few and far between nowadays. It has fallen to amateur astronomers to pioneer this field, as mentioned above. At last, the professional astronomers are taking more interest in these bright, unexplained flashes in the night sky. Those amateur astronomers, with their "primitive" instrumentation have actually had a paper published in the highly technical Astrophysical Journal:

"Abstract. Between 1984 July and 1985 July, 24 bright flashes were detected visually near the Aries-Perseus border by eight different observers at a total of 12 sites across Canada. One flash was photographed, and another was seen, by two observers at different locations. Their duration was usually less than 1 s. The estimated positions of 20 of the events and another seen in 1983 were close enough in the sky to suggest a common celestial origin. The brightest of the flashes was of magnitude -1 and lasted about 0.25 second." (Katz, Bill, et al; "Optical Flashes in Perseus," Astrophysical Journal, 307: L33, 1986.)

Comment. Hurray for Katz and the cooperating amateurs in the U.S. and Canada. One can wade through a 10-foot pile of the Astrophysical Journal and not find another paper based on naked-eye astronomy. Does this mean that science is at last going to take an interest in other transient luminous phenomena on earth? Unfortunately, this does not seem likely.

THE PERSEUS FLASHER: MYSTERY SOLVED!

So goes the title of an item in Sky and Telescope. Instead of ending in an exclamation point, a question mark would have been more appropriate. The Perseus Flasher (a recurring flash of light in Perseus), is a topic upon which astronomers also want to "close the book." So, we must ask ourselves how accurate the above title is.

(1) Photographic records of the Perseus Flasher site show no flashes at all in 3,288 hours of monitoring. In fact, a Harvard plate, which was being exposed at the same time that a 0-magnitude flash was reported by naked-eye observers, revealed nothing!

(2) The passages of some artificial satellites through the Perseus site have been correlated with naked-eye-observed flashes, suggesting the flashes are only the sun glinting off the metallic surfaces of spacecraft in earth orbit.

(3) One meteor observer, N. McLeod, claims that there is a background level of flashes from other parts of the sky which can also be attributed to satellite glints.

The Sky and Telescope item concludes "So the mystery is solved." (Anonymous: "The Perseus Flasher: Mystery Solved," Sky and Telescope, 73:604, 1987.)

Comment. So, science in its relentless, inerrant progress has positively solved still another mystery. (Triumphal background music here!) In case you haven't noticed, the three "exhibits" above do not hang together. First, it is implied that the Perseus flashes do not exist at all, since they have not been detected through photographic monitoring. Then, the flashes are said to be only sun glints from satellites, which is an admission that the flashes are real after all. In all probability, the photographic plates may not be capable of recording such brief flashes, but nothing was said on this matter. Further, many Perseus flashes are apparently not correlated with satellites. And we have no indication that the satellite in question had a reflecting surface properly oriented at just the proper moment.

MISSING STARS

Stars do disappear! Novas fade from the optical grasp of our telescopes; other stars are obscured by dust clouds. Here, though, we deal with stars that "should" theoretically be there but which are not.

WHERE ARE THE PRIMORDIAL STARS?

Down the years, astronomers have been able to divide almost all stars into two groups: Population-I, made up of young stars enriched by the products of their ancestors; and Population-II, the relatively older ancestors containing more hydrogen and fewer heavier elements. Population-I, according to present thinking, was formed out of the "ashes" of Population-II stars.

What is missing in this picture are Population-III stars---stars almost devoid of elements heavier than hydrogen and helium, and formed while the Big Bang was still echoing throughout the cosmos. Current astrophysical theory requires "ashes" from Population-III to create Population-II.

Do the astronomers find any primordial Population-III stars still kicking around? Hardly a handful; not nearly enough to satisfy the prevailing model of stellar evolution. One explanation is that Population-III stars have been around long enough to collect a camouflaging veneer of metallic debris. Some surmise that Population-III stars are truly extinct for some unknown reason. (Anonymous; "Where Is Population III?" Sky and Telescope, 64:19, 1982.)

THE COSMOLOGICAL ATLANTIS

Current cosmological theory states that immediately after the Big Bang, the only elements existing in any signigicant quantities were hydrogen and helium.

Yet, we observe stars today with various amounts of the heavier elements. How did the present stars, which are divided into Population-I and II, ever acquire their heavier elements? By thermonuclear synthesis, of course.

The primordial hydrogen and helium condensed into primitive stars, now labelled Population-III, where the first heavier elements were synthesized. The "ashes" of the Population-III stars provided the makings of the later stars with heavier elements. However, no matter how hard astronomers have looked, very few Population-III stars seem to be left anywhere---not even way out in the universe, which we see in terms of light billions of years old. A vital "transitional form" is missing in astronomy's fossil record! (Maran, Stephen P.; "Stellar Old-Timers." Natural History, 96:80, February 1987.)

GLOBULAR CLUSTERS

Globular clusters are nicely spherical aggregations of stars containing from about 20,000 to upwards of 1 million stars. Cluster diameters range from 5 to 25 parsecs. Within our own galaxy, astronomers can count about 200 globular clusters. Other galaxies seem to possess comparable populations.

Globular clusters present the astronomer with many anomalies. Past issues of Science Frontiers have touched only upon two of these: (1) the apparent "reincarnation" of some globular clusters; and (2) the spherical distribution of globular clusters in galaxies that are otherwise flat spirals. For several other globular cluster anomalies consult the index in Stars, Galaxies, Cosmos.

The globular cluster Messier 13 in Hercules.

BEAUTIFUL OBJECTS, BEAUTIFUL THEORIES

Imagine a million brilliant stars densely packed in a tight sphere by gravity. In the telescope these globular clusters are spectacular objects; a million points of light in disciplined motion around a center so closely packed with stars that they cannot all be resolved. Surely such an orderly assemblage of matter should be easy to model given the laws of celestial mechanics and high-speed computers. Not so! Both theory and computer models predict that a few stars will escape a globular cluster during its lifetime of several billion years, but that most most will be drawn inevitably inward as the cluster collapses. However, observation, the final arbiter, reveals that globular clusters do not follow this scenario. Indeed, some clusters seem to have collapsed already and are again evolving in a sort of "reincarnated" state that our best theories refuse to predict. (Lightman, Alan; "Misty Patches in the Sky," Science 83, 4:24, June 1983.)

GLOBULAR CLUSTERS UPSET THEORY OF GALAXY FORMATION

Globular clusters through the telescope are beautiful spherical aggregations of bright stars that seem to get ever denser toward the cluster's center. Globular clusters harbor many anomalies; here we mention just one.

Many spiral galaxies, like our own Milky Way, spin ponderously in the center of a spherical cloud of scores, even hundreds, of globular clusters (see sketch). Not only do the globular clusters surrounding us display a different spatial distribution (spherical vs. spiral), but their individual ages undercut galaxy theory. All of the Milky Way's globular clusters were supposed to have been formed when our galaxy was created. Yet, the ages of our clusters vary by as much as 5 billion years, according to the following reference. (Dayton, Leigh; " Globular Clusters Upset Theory of Galaxy," New Scientist, p. 34, May 13, 1989.)

Comment. It should also be noted that the globular clusters do not participate in the galaxy's general rotation. Where did these oddballs come from?

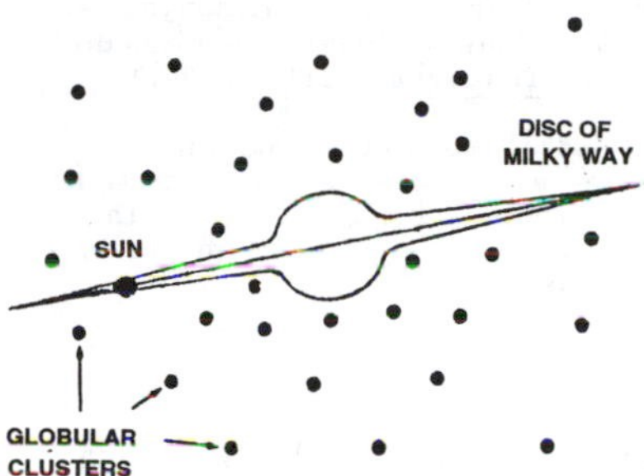

A spherical cloud of globular clusters surrounds our Milky Way. Each cluster is itself a spherical aggregation of stars.

DIFFUSE STRUCTURES

Most stars cannot be optically resolved to reveal their structures; we see them only as points of light. A few objects in our galaxy, however, are extensive enough to show up as patches, rings, jets, and other structures. The constituents of these extended objects usually turn out to be dust and gas. The anomalies in this category are generally concerned with the shapes and origins of these diffuse objects.

ENORMOUS SHELL OF MATTER RAISES THEORETICAL QUESTIONS

"Stars tend to lose material to the space that surrounds them. Some of this loss is gradual and continuous---the so-called stellar winds. Some is abrupt--the sudden blowing off of a surface layer that then forms a shell around the star. A group of astronomers now reports in the Nov. 15 Astrophysical Journal the discovery of an especially large, cool shell around the star R Coronae Borealis. How this shell was formed and what makes it glow are both mysteries for which current theory does not seem to have answers."

The newly discovered shell is 26 light-years across, roughly 20 times larger than previously discovered shells. If our sun were in the center, the shell would encompass the nearest 50 stars! The usual stellar shells glow as they absorb and reemit radiation from their parent stars but the R Coronae Borealis shell, given its distance, cannot be explained this way. (Thomsen, D.E.; "Enormous Stellar Shell Raises Theoretical Questions," Science News, 130:333, 1986.)

THE MESSAGE OF ALUMINUM-26

"Our solar system may be inside the cloud of debris from a star that exploded 10,000 to 1,000,000 years ago. This startling conclusion was reached by Donald Clayton of Rice University after studying observations of the amount of aluminum-26 (^{26}Al) in the interstellar medium."

Instruments on satellites (gamma-ray spectrometers) have detected so much aluminum-26 that radical hypotheses seem required. The problem is that aluminum-26 is radioactive with a half life of only about 1 million years---a very short time astronomically speaking. The aluminum-26 cannot be primordial solar system stuff; it cannot even be 10 million years old. It had to be created somewhere nearby recently. The best aluminum-26 factory conceived so far is a nova in our vicinity. (Anonymous; "Are We Inside a Supernova Remnant?" Sky and Telescope, 69:13, 1985.)

Comment. A nova close enough to engulf the earth with its debris must have had a profound effect on the earth and its cargo of life---perhaps on Saturn's rings, too.

THERE ARE COLD ANOMALIES "OUT THERE"

As data from the IRAS (Infrared Astronomy Satellite) pile up (at 700 million bits per day), astronomers are seeing a new universe---one consisting of cold gas, dust, and debris that emit little or no visible light. Here are just four of the new enigmas revealed:

-Infrared "cirrus clouds." A network of faint wisps of cold matter cover the whole sky.

-Galactic matter of an unknown nature. This material has been observed only on the 100-micrometer IRAS scans.

-A ring of solid particles around the star Vega.

-"Blank fields." IRAS scans have found infrared sources where no visible objects exist.

(Waldrop, M. Mitchell, and Kerr, Richard A.; "IRAS Science Briefing," Science, 222:916, 1983.)

"COMPACT STRUCTURES": WHAT NEXT?

We know that immense molecular clouds drift through interstellar space; but a new denizen has now made itself known through its ability to diffract quasar radio signals. Although constituted only of ionized gases, these new objects are called "compact structures."

"'Compact' means that these objects are about as big as the earth's orbit around the sun, and therefore larger than all but the biggest stars. They are, however, much smaller than the clouds that previous observations have detected in interstellar space. They reveal their presence by diffracting the radio waves coming from distant quasars.

......

The objects move too fast to be near the quasar---to be that far away, they would have to go at 500 times the speed of light---so the observers conclude they are in our own galaxy. Previous observers didn't see them, (R.L.) Fiedler says, because they didn't observe the same quasar at close enough intervals."

If these ionized clouds are spherical, they have masses comparable to the asteroids; but, if they are elongated, their mass is anyone's guess. No one knows how they are formed, how long they last, or where the energy comes from to maintain them in an ionized state. Extrapolating from the five instances recorded so far, the observers speculate that these compact structures may be 500-1000 times more numerous than stars! (Thorrsen, D.E.; "Oodles of 'Noodles' Found in Galaxy," Science News, 131:247, 1987.)

CARTWHEELS IN SPACE

An intriguing and totally unexpected wheel-shaped structure has been discovered by K. Taylor (Royal Greenwich Observatory) and D. Axon (Sussex University). Plates made with a 1-meter telescope show this curious dark pattern silhouetted against the Great Nebula in Orion. The circularity and neat set of six spokes make it seem like a stellar UFO. No one knows its distance, age, or constitution. Speculation is that some interstellar winds may have created this artificial-looking object. (Anonymous; "Cartwheels in the Sky," New Scientist, 82:804, 1979.)

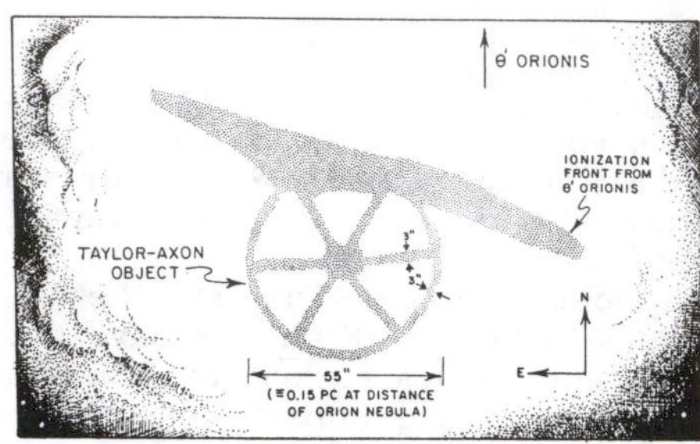

Curious cartwheel structure silhouetted against a bright nebula.

STELLAR MISCELLANY

THE 11-MINUTE BINARY

4U1820-30 is an X-ray binary star located in the heart of the globular cluster NGC 6624. Its claim to fame is a very short period of rotation---only 685 seconds. Thus, according to current theory, in just a shade over 11 minutes, a neutron star orbits a white dwarf; and therein may lie an anomaly. The orbit diameter is only 1/7 the radius of our sun, which implies the stars themselves are also small. 4U1820-30 is the shortest-period binary ever found--so short that astronomers are looking for other explanations. Another curious fact mentioned in this item is that X-ray binaries are much more common in globular clusters than elsewhere in the universe. What is so different about globular clusters? (King, A.R., and Watson, M.G.; "The Shortest Period Binary Star," Nature, 323:105, 1986.)

A RECENT TRANSFORMATION OF SIRIUS?

Many Greek, Roman, and Babylonian sources definitely label Sirius as a red star. Some dispute these old accounts because today Sirius is white, with a bluish tinge, and is classified as a white dwarf. W. Schlosser and W. Bergmann have now found a "new," apparently independent reference to Sirius' red color. It is in a manuscript of Lombardic origin, which contains the otherwise lost "De Cursu Stellarum" by Gregory of Tours (about AD 538-593). This new source reiterates that Sirius was once a red star:

> Thus, Sirius B might well have changed from a red giant to the white dwarf as it appears today. However, the rapidity and smoothness of this transformation are quite unexpected, and its timescale is surprisingly short. Furthermore, no traces of catastrophic effects connected with such an event have ever been found. The only indication that something has happened is the somewhat higher metallicity of Sirius A, believed to have resulted from contamination by the giant's blown-off shell.

(Schlosser, Wolfhard, and Bergmann, Werner; "An Early-Medieval Account on the Red Color of Sirius and Its Astrophysical Implications," Nature, 318:45, 1985.)

Comment. See BHT5 in Biological Anomalies: Humans I for speculations about the color vision of ancient peoples.

RECENT EXPLOSION ON SIRIUS?

The apparent recent color change of Sirius is now receiving theoretical attention. Sirius seems to have been a bright red star in medieval times, whereas it now shines with bluish-white light. The latest theory is that Sirius B recently experienced a thermonuclear runaway event. (Bruhweiler, Frederick C., et al; "The Historical Record for Sirius: Evidence for a White-Dwarf Thermonuclear Runaway?" Nature, 324: 235, 1986.)

Comment. Such an event should have been visible from earth and recorded!

SOCKET STARS

"A picture book hardly seems a likely source of an astronomical discovery, especially in a world where mysteries of the universe usually tumble from sophisticated electronic instruments attached to huge telescopes. Nevertheless, while recently paging through Exploring the Southern Sky, by C. Madsen and R. West, Walter A. Feibelman (NASA Goddard Space Flight Center) recognized something that had caught his attention decades before. High-resolution photographss of nebulae show large numbers of faint stars surrounded by circular or oval 'empty' regions, giving the impression that the stars are sitting in 'sockets' a few arc seconds across, swept free of nebulosity."

Feibelman rules out photographic effects, such as halation, but has no ready explanation for socket stars.

W.W. Castelaz (Allegheny Observatory) has added to the mystery by pointing out that many socket stars show excess infrared emission. He considers this strong evidence that socket stars are surrounded by shells of dust and may represent an unrecognized stage of stellar evolution. (Anonymous; "Socket Stars," Sky and Telescope, 79: 476, 1990.)

MONSTER STAR LURKS NEARBY!

Astrophysicists have long believed that the upper limit for stellar masses was 100 times that of the sun. This rule seems to be violated right on our doorstep, in the Large Cloud of Magellan. The nonconforming object is designated R136; and it resides in the southern constellation Doradus. The central part of R136 radiates about a million times more visible light than our sun, and 50 million times more if the ultraviolet wavelengths are included. If R136 is a single object, its mass may be 1,000 times that of the sun. (Mathis, John S., et al; "A Superluminous Object in the Large Cloud of Magellan," Scientific American, 251:52, August 1984.)

BLUE STRAGGLER STARS

The stars comprising a star cluster are usually assigned the same age since it is thought that they were all created at the same time the cluster was formed. Lurking in many star clusters, however, are brighter, bluer, nonconformists called "blue straggler stars." These stars seem to have about twice the mass of the "normal" cluster members; and they appear to be only about one-fifth as old as their compatriots. The motions of the blue stragglers are consistent with those of bona fide cluster members, implying that they are not interlopers or foreground objects.

Several explanations have been suggested to explain the presence of blue stragglers. One thought is that they harbor asteroid-sized blackholes at their cores. So far, all of the explanations have serious flaws. (Fogg, Martyn J.; "Blue Straggler Stars: A Cosmic Anomaly," The Explorer, 6:4, Spring 1990.)

DOUBLE-STAR SYSTEM DEFIES RELATIVITY

We all know that chapter from the Bible of Science that tells how Einstein's General Theory of Relativity triumphed over Newtonian celestial mechanics by accounting for the residual advance of Mercury's perihelion. The General Theory should also explain the precessions of double-star orbits, but a serious anomaly has been found. The double star DI Herculis has a Newtonian precession rate of 1.93 degrees per century, with another 2.34 added by relativistic effects. With more than 3000 well-observed orbits of this star system on the books, astronomers come up with only 0.64 degrees per century, instead of the 4.27 predicted by theory. Something is obviously awry. All searches for errors and other influences have been negative. (Anonymous; "Double-Star System Defies Relativity," New Scientist, p. 23, August 29, 1985.)

Comment. As a matter of record, Newtonian mechanics can account for Mercury's perihelion advance if the sun is actually an oblate spheroid instead of the mathematically perfect sphere usually assumed. Also, the gravitational theory of J. Moffat seems to explain the motions of both Mercury and DI Herculis.

GALAXIES AND QUASARS

GALAXIES

Galaxies are thought by many astronomers to be the "natural unit" in astronomy, something akin to the "species" of biology. It is true that most galaxies seem to be simple, isolated islands of stars in space, independent aggregations of anywhere from a few million to trillions of stars. In fact, however, galaxies are morphologically complex, displaying a wide variety of spiral, barred, and elliptical configurations. Further, they unite to form large-scale structures and participate in coordinated movement. Here, we examine only the following abbreviated list of galactic anomalies:

- The winding dilemma. Spiral galaxies should not retain their spiral geometries for as long as they do!

- Galactic shell game. Whence the interleaved partial shells of stars around the small ends of elliptical galaxies?

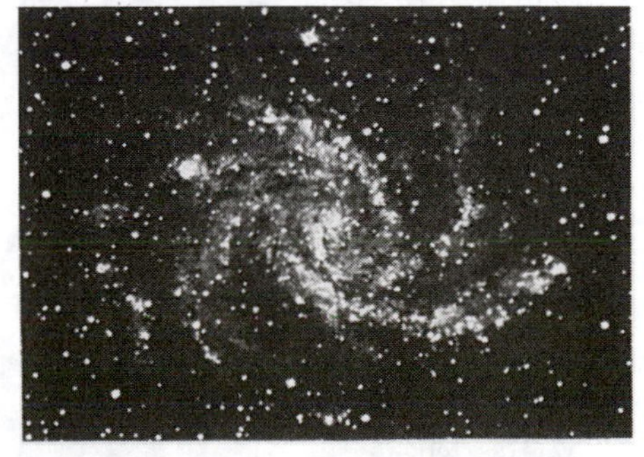

Spiral galaxy NGC 6946. (Mount Wilson)

•Flip-flop galaxies. Possible alternating jets emanating from galaxies.

•Maverick galaxies. Gyroscopic galaxies, wrong-way galaxies, crouching giants, etc.

Galaxies also enter into later discussions of the redshift controversy, where galaxies that seem physically connected have disparate redshifts, and in general cosmology, since many galaxies seem to be arranged in walls, shells, and other geometries.

Finally, we separate galaxies and quasars in this book, even though it is recognized that they may be manifestations of the same astronomical phenomenon.

THE WINDING DILEMMA

THE RIMS OF GALAXIES SPIN TOO FAST

According to the physical laws governing orbital motion in a central force field, like that of our sun, distant objects should rotate more slowly than those closer to the central mass. Thus, the outer planets of the solar system swing around the sun more slowly than the earth and the other inner planets.

Astronomers always expected that the stars rotating around the massive hubs of the galaxies would obey these laws in a like manner. No such luck! Doppler measurements of galaxy rotations made by Vera C. Rubin and her colleagues, at the Carnegie Institution, indicate that most stars out near the galactic rims rotate just as fast or faster than those closer to the hub. Astronomers suppose that this anomalous rotation may be due to halos of undetected burntout stars and/or gas fringing the galaxies. Such halos would distort the inverse-square-law field, permitting the observed anomalous rotation. (Anonymous; "Fast-Spinning Galaxies," Science Digest, 89:18, November 1981.)

Comment. No one has yet discovered these hypothetical halos of mass. The universe seems to be full of "missing mass." An unpopular alternative would be the admission that Newton's Law of Gravitation does not hold on the galactic scale.

WHY DO SPIRAL GALAXIES STAY THAT WAY? OR DO THEY?

Sometimes the simplest of observations produces the stickiest of dilemmas. Take, for instance, a well-formed spiral galaxy, of which there are a great many. When astronomers measure the circumferential velocities of the stars, as they circle around the galaxy's hub, they find that all the stars orbit at about the same velocity, regardless of how far out from the hub they are. Their speeds do not drop off with increasing distance, as the velocities of the planets do in the solar system. This observation is anomalous itself, because it seems that the laws of orbital motion have been violated. The anomaly we are after now is termed "The Winding Dilemma." N. Comins and L. Marschall elaborate as follows:

"Stars closer to the center of a spiral galaxy don't have as far to go to complete an orbit as stars located farther from the center. Thus, inner stars should orbit more frequently than outer stars, resulting in a spiral that gradually winds up as the galaxy ages. But observations of spiral galaxies at various distances---and thus at different stages in their evolution---have shown that this is not the case. Astronomers believe density waves, stochastic star formation, or perhaps a combination of both processes may sustain or regenerate the spiral pattern."

Density waves have recently been applied to explain the spiral rings of Saturn, and now to the arms of spiral galaxies. The density waves are thought to stimulate the condensation of bright new stars as they move through space. A good analogy is the bioluminescent wake of ship in tropical waters. The density waves in a galaxy maintain the spiral pattern with new stars, while the old stars die out (in much less time than it takes for them to orbit the hub) as they orbit out of the spiral pattern.

Postulating density waves just raises more questions, as is often the case in science. What causes the density waves? Theory says that the density waves should damp out in under a billion years, yet we see spiral galaxies over a wide range of ages. (Comins, Neil, and Marschall, Laurence; "How Do Spiral Galaxies Spiral?" Astronomy, 15:7, December 1987.)

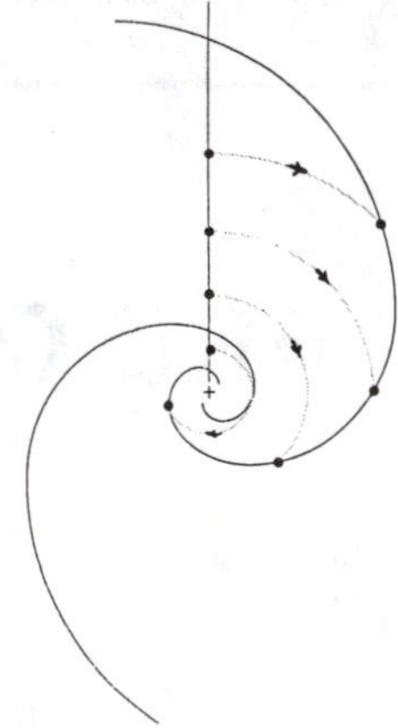

The "winding dilemma." If stars start in a line (top), they should form tighter spirals as time passes.

Comment. The scientifically outrageous resolution of the winding dilemma is to assert that the universe is so young that the spiral patterns have not yet been dispersed. Interestingly enough, Saturn's rings may turn out to be very young, too! See pp. 68-69.

GALACTIC SHELL GAME

LITTLE BIG BANGS!

The photographic enhancement of plates taken by the UK Schmidt and Anglo-Australian telescopes has revealed that several normal elliptical galaxies are surrounded by shell-like structures. D.F. Malin and D. Carter report that these envelopes are vast---up to 180 kiloparsecs in diameter. Furthermore, some galaxies are wrapped in a series of thin shells. Malin and Carter believe that the colossal shells are really thin layers of stars either created by a powerful shock wave during galaxy formation or comprised of a debris layer of old stars blown out of the galaxy during some cataclysmic event. (Malin, David F., and Carter, David; "Giant Shells around Normal Elliptical Galaxies," Nature, 285:643, 1980.)

Comment. This article typifies the emergence of "catastrophic astronomy" which contrasts sharply with the older vision of a leisurely evolution of stars and galaxies.

GALACTIC SHELL GAME

Elliptical galaxies are immense assemblages of stars. There may be a trillion stars like our sun in one of these monster galaxies. But it is not the mind-boggling number of stars that is anomalous (astronomy dotes on big numbers); rather the anomaly at hand concerns the 11% of the elliptical galaxies that are partly girdled by strange low-luminosity shells.

First reported in 1980, these sharply defined shells seem to be composed of still more stars---vast ellipsoidal sheets of stars emplaced along the long axis of

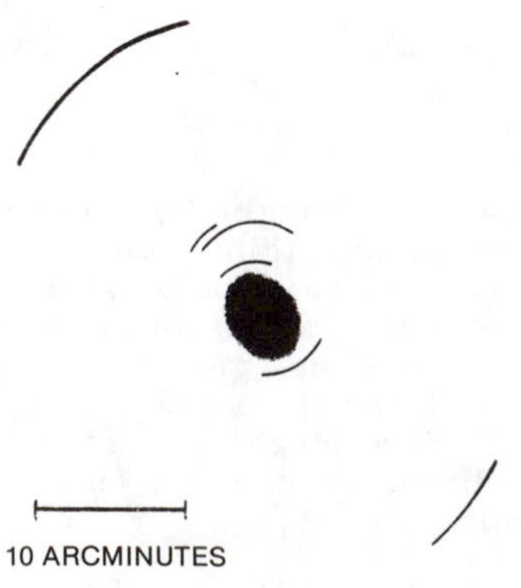

Partial shells of stars around the ends of elliptical galaxy NGC 3923.

the elliptical galaxy. Some elliptical galaxies have up to twenty partial shells divided between the two ends of the ellipsoid. What is most intriguing is the fact that the shells are systematically arranged. The closest partial shell will be at one end of the ellipsoid, while the second closest will be at the opposite end. The third closest will be just beyond the first closest, and so on. The shells "interleave" or alternate ends as their distances increase.

If the alternating partial shells of stars belong to the elliptical galaxy (they seem to, agewise), did the elliptical galaxy shoot the first wave out one end and then expel the second wave out the opposite end? Or did the alternating shells form in situ from the primordial gas and dust that made the galaxy? Another possibility is that a small galaxy collided with the monster elliptical galaxy, and its constituent stars were scattered in regular waves. There is some physical and mathematical support for such a regular scattering of a burst of stars by a dense elliptical mass of target stars. (Edmunds, M.G.; "Galaxies in Collision," Nature, 311:10, 1984.)

Comment. This star-scattering process reminds one of how electrons are scattered by crystals and other subatomic scattering situations. We may have here another place where quantum mechanics applies to macroscopic nature. D.M. Greenberger speculates further on astronomical quantization on p. 105.

FLIP-FLOP GALAXIES

FLIP-FLOP RADIO JETS?

Many radio galaxies and quasars are found to have a double-lobed structure, with one lobe on one side of the nucleus and another diametrically opposite. When examined in detail, these lobes turn out to be quite different in size, shape, and intensity. In particular, very bright regions on one lobe often correspond to gaps or regions of low brightness on the other. So striking are these asymmetries that astronomers think that these huge, tremendously energetic systems are ejecting material first from one side then the other.

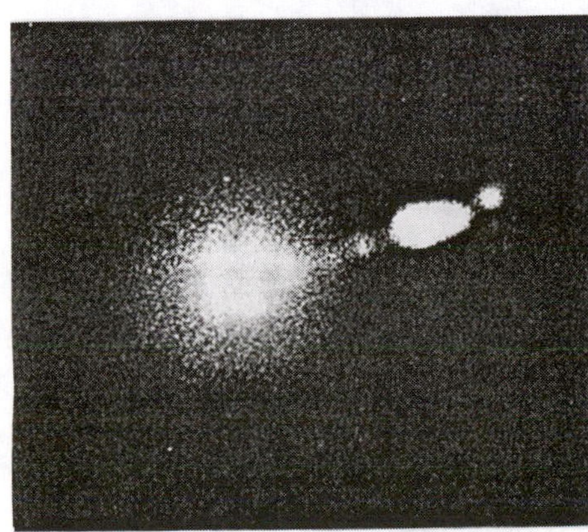

One-sided jet apparently shooting out from the galaxy M87.

Somehow, one side of the galaxy or quasar communicates with the other, which may be many light years away, and coordinates a flip-flop action. How and why radio galaxies and quasars should flip-flop is a major mystery. (Anonymous; "Flip-Flop Radio Jets?" Sky and Telescope, 68:506, 1984.)

Comment. This flip-flop action immediately recalls the great elliptical galaxies which seem to be shooting out shells of stars first from one end, then the other. See p. 90.

MAVERICK GALAXIES

GYROSCOPIC GALAXIES

The popular conception of a galaxy draws it in pancake shape, with a spiral structure consisting of many millions of stars. Oddities and deviants exist, but hardly anything as bizarre as a handful of recently discovered ringed spirals. Although the spiral sections of the ringed spirals seem normal enough, the rings are perpendicular to the plane of the spiral---an inclination hardly countenanced by theories of galactic evolution. The ringed spirals look superficially like toy gyroscopes. One suggestion is that two galaxies collided at an angle, but there is no evidence of such a cataclysm. Ringed galaxies are eminently anomalous. (Anonymous; "Ringed Galaxy Clue to Cosmic Riddle," Science Digest, 91:22, February 1983.)

GALAXY SPINS WRONG WAY

We have all seen impressive photographs of spiral galaxies with their whirlpool patterns. One can easily imagine them spinning ponderously, with streams of stars spiraling out behind the rotating core. This picture is intuitively satisfying from a mechanical point of view. But:

> A galaxy with a spiral arm going the wrong way has been found by astronomers in the U.S. Spiral arms usually trail behind as a galaxy turns, but in NGC 4622 an inner arm is wrapped round the galaxy in the direction it is rotating.

(Mitton, Simon; "Puzzle of Galaxy That Points Two ways," New Scientist, p. 22, February 29, 1992.)

GALACTIC RADIATION BELT?

Both Jupiter and the earth boast radiation belts consisting of electrically charged particles tethered by the planet's magnetic fields. Recent radio astronomical studies of the Milky Way reveal long filaments of ionized gas about 150 light years long curving up out of the galactic disk, at a point about 30,000 light years from earth. These filaments emit radio energy just like the planetary radiation belts and are presumably held in the grip of a galactic magnetic field. There have been previous hints of a weak and disorganized galactic magnetic field, but this is the first evidence for a strong polar field in our own Milky Way or any other galaxy.

The unexpected filaments were discovered by chance in a study of star formation in the core of the Milky Way. The radio energy emitted by the belts was originally thought to come from the galactic machinery that makes new stars; but now it looks like that machinery is not grinding out nearly as many new stars as once thought. (Thompsen, D.E.; "Galactic Dynamoism: A Radiation Belt?" Science News, 126:20, 1984.)

SUPERMASSES THAT COME AND GO

Quasars, black holes, violently active Seyfert galaxies, jets of matter expelled from galaxies, and many similar puzzles of modern astronomy fall into place (and reason) if one unthinkable assumption is made: the cyclic appearance and disappearance of supermasses inside galaxies. Normal galaxies seem to have masses of about 10^{11} times that of the sun. The unthinkable assumption suggests that every 10^{8} years or so, these ordinary, unassuming galaxies become supermassive (about 10^{13} solar masses) for several million years.

When the core of a galaxy becomes supermassive, its stars are tugged into tight new orbits. The subsequent switching off of the supermass allows the galaxy to expand outwards again. V. Clube claims to have found just such expansion effects among the globular clusters in our own galaxy. His data are striking and quite convincing. The notorious "missing mass" problem of cosmology disappears with the cyclic supermass assumption because the time-averaged mass of each galaxy will be much higher than that observed in its normal enervated state.

Doesn't this sudden temporary appearance of mass violate the laws of physics? No, says the author, physicists habitually assume a superfluid, superconducting vacuum state, which is the ultimate source of all mass-energy, when they develop their theories of fundamental particles. If particle physicists can (and must) evoke such magic, so can astronomers. (Clube, Victor; "Do We Need a Revolution in Astronomy?" New Scientist, 80:284, 1978.)

Comment. The magic of supermassive injections is, of course, no more magical than the existence of gravitational force or God.

ICEBERGS AND CROUCHING GIANTS

"Astronomers have discovered what they believe is the largest, darkest, most gas-rich spiral galaxy known in a 'void' beyond the Virgo Cluster of Galaxies. Discovered by accident on long-exposure photographic plates, the object, called Malin 1, appears to be still forming.

......

'We didn't expect to find anything like this,' (C.) Impey says, referring to Malin 1. 'This galaxy is far removed from our ideas of what a normal galaxy should look like. There could be more of these things that haven't been discovered yet.'" (Anonymous; "Massive, Dark Galaxy Found in Void," Astronomy, 15:75, September 1987.)

The same discovery is dealt with, in a more technical way, in Nature. A revealing paragraph from the Nature article follows:

"Although Malin 1 could be a unique case there is in fact a significant body of circumstantial evidence to suggest otherwise. Our present knowledge of the galaxy population is so biased by a single insidious selection effect that it is entirely possible that Malin 1 is just the first example of a class of such low-surface-brightness giant galaxies that forms a significant constituent of the universe."

The "selection effect" mentioned above is a consequence of the very bright sky that confronts astronomers. Astronomical interest and instrumentation have focussed on those astronomical objects that are bright. In truth, the universe may be full of "icebergs and crouching giants." (Disney, Michael, and Phillipps, Steven; "Icebergs and Crouching Giants," Nature, 329:203, 1987.)

Comment. Since astronomers have built their models of galaxies, galactic distribution, and the evolution of the universe on an unrepresentative portion of the actual cosmos, we may be in for some major changes in our overall view of the universe and its beginning---if it actually had one.

QUASARS

The salient observational features of quasars (also called quasi-stellar objects and QSOs) are: (1) high redshifts; (2) starlike appearance; (3) occasional association with radio sources; and (4) brightness variations on a scale of months. The current interpretation of these observations places quasars at great distances from the earth, gives them very small dimensions (a few light months compared to the thousands of light years assigned to ordinary galaxies), and confers incredible power densities. Quasars may be a species of galaxy, but no one is sure. Two anomalous aspects of quasars are treated below:

- Anomalous distribution. Distance, inhomogeneity.

- Superluminal velocities. Some parts of quasars seem to attain speeds faster than that of light.

Like the galaxies, quasars figure prominently in the later discussions of redshift problems and general cosmology.

ANOMALOUS DISTRIBUTION

A NEW QUASAR DISTANCE RECORD: A NEW EMBARRASSMENT

A quasar has been detected with a redshift of 4.73. If this redshift is interpreted as a measure of the quasar's distance (Who would risk his reputation by suggesting otherwise?), it is 14 or so billion light years away. If the Big Bang is assumed (Who would risk..etc?), this quasar is only about a billion light years from the edge of the universe. Its age, then, is only a billion years. But this stripling of a quasar appears perfectly "normal" with no signs of youth! Its spectrum indicates that even at this young age, the elements were present in the same abundances found in older quasars. And, of course, at this quasar's core there must be a billion-solar-mass black hole (Who would risk..etc.?). Current theory is hardpressed to explain this very rapid evolution of a "normal" quasar with its immense black hole. (Peterson, I.; "Quasar Illuminates the Most Distant Past," Science News, 136:340, 1989.)

Comment. Could it be that our fanatically held ideas about redshifts, black holes, and Big Bangs are wrong? You bet it could!

A GATHERING OF QUASARS

The universe is supposed to be approximately uniform in all directions---the evenly distributed smoke from the Big Bang. Halton Arp, an energetic opponent of the standard cosmological view, points out that quasars are socializing in disgracefully large numbers in one region of the sky. In the direction of the so-called Local Cluster of galaxies, between redshifts 1.2-2.5, there are roughly four times as many quasars per unit volume as in the other parts of the sky. This unexpected clumping of quasars affects a region 1,300 million light years in diameter and 4,875 million light years deep, a rather substantial chunk of the cosmos. Arp's discovery places astronomy in a no-win situation. Either the distribution of quasars is too clumpy for current theory or the redshift/distance law is wrong. Neither situation makes astronomers very happy. (Anonymous; "Quasars and Quasi Quasars," New Scientist, p. 20, May 17, 1984.)

A LARGE QUASAR INHOMOGENEITY IN THE SKY

"Abstract. In an area roughly 20° x 70° on the sky, there exists an excess of bright, high-redshift quasars. Quasars with this distribution of apparent magnitude and redshift have a negligible chance of being drawn from the population of quasars present in other areas of the sky. At a mean redshift distance corresponding to their average z = 2, these quasars would represent an unprecedented inhomogeneity over enormous volumes of space in the universe."

It is difficult for astronomers to accept such a large "bubble" in the cosmos, because the Big Bang Theory basically produces a "smooth" universe. The author of this paper, H. Arp, comments that the size of the inhomogeneity could be shrunk considerably if redshifts were not taken as measures of distance. (Arp, Halton; "A Large Quasar Inhomogeneity in the Sky," Astrophysical Journal, 277:L27, 1984.)

SUPERLUMINAL VELOCITIES

FOUR EXTRAGALACTIC SOURCES EXPAND FASTER THAN LIGHT

Three quasars and one galaxy possess structures that apparently expand faster than light. The sizes of the three quasars were measured over periods of time by Very Long Baseline Interferometers (VLBIs). In the case of quasar 3C279, the apparent velocity of expansion was ten times that of light. The quasars all have rather large redshifts, indicating great distances from earth, but the lone galaxy displaying "superluminal" expansion has a redshift of only 0.032. This fact suggests that superluminal velocities cannot be employed as arguments against redshifts being cosmological; that is, measures of distances from earth. Therefore, if the redshift is

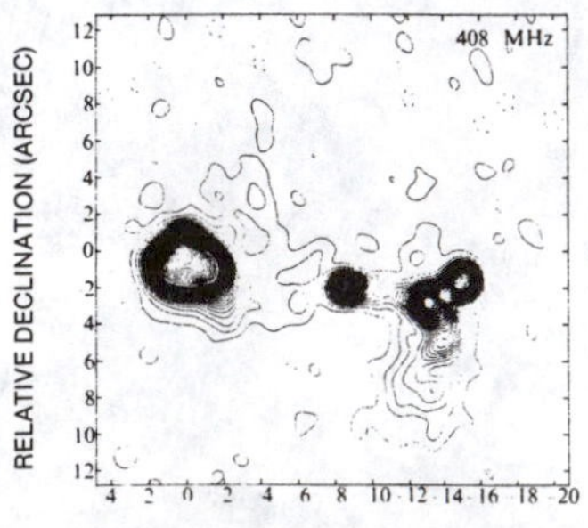

Radio map of quasar 3C179, showing its various components, some of which seem to move at speeds greater than that of light.

truly a measure of distance (as it seems to be), some astronomical structures (perhaps not matter itself) seem to grow faster than the velocity of light. (Cohen, M.H., et al; "Radio Sources with Superluminal Velocities," Nature, 268:405, 1977.)

ANOTHER QUASAR ENIGMA

Astronomer A.P. Fairall has reported an "impossible" physical situation---a strong scientific anomaly. It seems that a Seyfert galaxy (a very low-luminosity quasar) emits forbidden spectral lines that vary in intensity about every 30 minutes. It is not the "forbidden" character of the spectral lines that disturbs the astronomers. ("Forbidden" simply means "improbable" in spectroscopy!) Rather, it is the half-hour variations that pose the dilemma. This is because most astronomers are convinced that the said forbidden lines are emitted by regions of the quasar that are 100-10,000 light years across.

A disturbance or physical change leading to a variation in intensity, even travelling at the speed of light, could not begin to move across this huge region in half an hour. Yet, the changes in intensity seem to be there, inferring a physical change that travels perhaps a million times faster than physics permits. Variations in intensity decades long would be acceptable, but half an hour is out-of-the-question! The author of this referenced comment in Nature believes that the observational procedures employed must be at fault. (Gaskell, C. Martin; "Spectra That Defy Explanation," Nature, 304:212, 1983.)

Comment. This possible anomaly is closely related to the so-called superluminal velocities also observed in quasars, in which physical effects seem to travel faster than light. Apparently something is very wrong in our model of a quasar, or our distance scale, or even our basic physics.

AN ORPHAN SUPERLUMINAL GLOB?

Radioastronomical observations of quasar 3C345 have picked out a glob of material that seems to be travelling at between 13 and 17 times the velocity of light. (Such "superluminal" speeds may be apparent and not physically real.) This speedy mass of material is now moving radially away from 3C345 and even seems to be accelerating! Although this glob may have been ejected from 3C345 and followed a curved path, its present path may imply a different origin. (Moore, R.L., et al; "Superluminal Acceleration in 3C345," Nature, 306:44, 1983.)

Comment. We classify superluminal velocities as anomalous until their real nature is established. But here we have the added anomalies of acceleration and a possible extra-quasar origin.

A QUICK QUASAR

Quasar 4C29.45 threatens to make quasars even harder to understand. Quasars have always appeared to emit too much energy for their size; that is, our present knowledge of physics does not provide us with a mechanism for generating such huge energy densities. Now, quasar 4C29.45 comes along and pulsates on time scales of 30 minutes and less. These pulsations are sharp and not spread out timewise, implying that quasar 4C29.45 must be smaller than 30 light-minutes in size---otherwise the disturbance causing the pulsation would have to travel faster than light.

On the night of April 10, 1981, the situation (already bad) worsened, when brightness jumps of 0.2 magnitude occurred nearly instantaneously. Conclusion: quasar 4C29.45 may be only light-seconds in diameter, which should really by physically impossible. The anonymous author of this item ventures that "... since the real nature of quasars is unknown, it is uncertain how they can or cannot behave." (Anonymous; "A Quick Quasar," Sky and Telescope, 68:115, 1984.)

Comment. Perhaps we have been naive in thinking that the laws of physics determined how things can and cannot behave. Evidently these laws are not as secure as we have been led to believe! Note in passing: the quasar impasse would be easier to bridge if quasars were very close instead of as distant as their redshifts demand. Of course, we wouldn't dare to scuttle the redshift/distance law and the expanding universe!

COSMOLOGY

REDSHIFT CONTROVERSY

The nature of the redshift controversy is adequately---perhaps "over-adequately"---presented in the several items from Science Frontiers reproduced below. The subject divides rather naturally into three parts:

- Discordant redshifts. Some apparently physically related astronomical objects have wildly different redshifts.
- Quantized redshifts. Certain redshift values are "favored" by galaxies and quasars!
- Hubble troubles. Hubble's law relates an object's redshift to its velocity and, in turn, to its distance from earth.

DISCORDANT REDSHIFTS

A NEW COSMIC HERESY

Often the simplest of observations will have the most profound consequences. It has long been a cornerstone of modern science, to say nothing of man's cosmic outlook, that the earth attends a modest star that shines in an undistinguished part of a run-of-the-mill galaxy. Life arose spontaneously and man evolved on this miscellaneous clump of matter and now directs his own destiny without outside help. This cosmic model is supported by the Big-Bang and Expanding Universe concepts, which in turn are buttressed by the simple observation that astronomers see redshifts wherever they look.

These redshifts are due, of course, to matter flying away from us under the impetus of the Big Bang. But redshifts can also arise from the gravitational attraction of mass. If the earth were at the center of the universe, the attraction of the surrounding mass of stars would also produce redshifts wherever we looked! The argument advanced by George Ellis in the article referenced below is considerably more complex than this, but Ellis' basic thrust is to put man back into a favored position in the cosmos. His new theory seems quite consistent with our astronomical observations, even though it clashes with the thought that we are godless and making it on our own. (Davies, P.C.W.; "Cosmic Heresy?" Nature, 273:336, 1978.)

A REDSHIFT UNDERMINES THE DOGMA OF AN EXPANDING UNIVERSE

Halton Arp has closely studied the galaxy NGC-1199, which is the brightest member of a small cluster of galaxies. One of its companions is a galaxy so dense that it appears to be a star. This compact object sports a circular shadow and seems to be silhouetted against the central galaxy NGC-1199.

Arp's analysis of the absorption ring seems to prove that the compact galaxy is in front of the central galaxy. This would normally be permissible, but here the central galaxy has a redshift of 2,600 km/sec compared to 13,300 km/sec for the galaxy in front of it. This is astounding because the farther away an object is, the greater its redshift is supposed to be. (Arp, Halton C.; "NGC-1199," Astronomy, 6:15, September 1978.)

Comment. Other examples of such anomalous redshifts are known. Three possible conclusions are: (1) the redshift-distance law is wrong, upsetting the Big-Bang Theory; (2) some galaxies and other objects have acquired anomalous velocities through some unknown mechanism; and (3) these unusual redshifts do not indicate velocities at all.

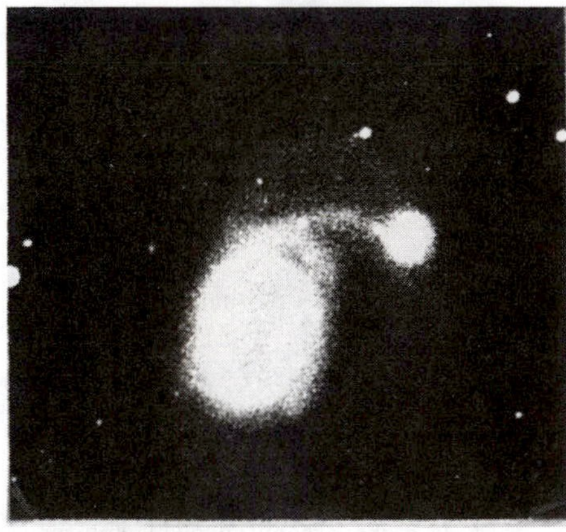

The larger galaxy shown (NGC 7603) is apparently connected to a companion galaxy by a luminous bridge. The redshifts are 8,700 and 16,900 kilometers per second respectively.

MORE ANOMALOUS REDSHIFTS

Halton Arp, of the Mount Wilson and Las Campanas Observatories, has discovered three more pairs of galaxies that seem to threaten that cornerstone of astronomy, the redshift distance scale. The new pairs are all in the Southern Hemisphere and, like others on Arp's list, seem to be interacting physically. For example, the filaments of one pair member seem to reach out and connect with the companion. Surely, these dynamically connected galaxies should be equidistant from earth. Such distances are measured by the object's redshift, which is supposedly proportional to its recessional velocity. Thus, each member of a pair should have the same redshift. This does not occur with these three pairs. In one pair, the recessional velocity appears to be 4,600 km/sec for one galaxy and 37,300 km/sec for the other. Arp's conclusion is that at least some of the redshift must be intrinsic; that is, not due to recessional velocity alone. If this is true, the basic cosmological distance scale is suspect. (Anonymous; "X-Ray Quasars Fit Theories...But Some Galaxies Refuse to Play Ball," New Scientist, 88:22, 1980.)

ANOMALOUS REDSHIFTS (AGAIN)

For years Halton Arp has been searching the skies for anomalous astronomical objects. Among the many maverick items in his catalog are objects that seem physically related but have radically different redshifts. Examine for instance the illustration, which shows the large galaxy NGC 7603 (possessing a supposed recessional velocity of 8,700 km/sec) and its smaller companion at the end of the filament, which recedes at the much higher velocity of 16,900 km/sec. According to the Theory of the Expanding Universe, the greater the recessional velocity (as measured by the red or Doppler shift), the farther away the object is. But Arp's many examples of physically connected galaxies with wildly different redshifts suggest that some of the disparities in redshifts may actually be due to the related objects flying apart from each other. In the illustration, the small companion might have been "shot out" of its parent galaxy at high velocity by some unappreciated galactic gun. (Arp, Halton C.; "Related Galaxies with Different Redshifts?" Sky and Telescope, 65:307, 1983.)

Comment. Left unsaid in Arp's article is the possibility that redshifts are not good measures of distance. If they are not, doubt is cast on the Theory of the Expanding Universe and the reality of the Big Bang itself. Some astronomers, according to news items in scientific publications, have heard enough about discordant redshifts and would rather see scarce telescope time used for other types of work!

QUASAR, QUASAR, BURNING BRIGHT; WHAT SHIFTS YOUR SPECTRAL LIGHT?

T. Heckman, of the University of Maryland, has ordained that the apostles of noncosmological redshifts must now recant. He and his colleagues believe that they now have the most convincing demonstration to date that quasar redshifts are of cosmological origin; that is, the larger the redshift, the faster the quasar is receding and the farther away it is.

"Availing themselves of the extraordinary new imaging and spectroscopic capabilities of charge-coupled-device (CCD) detectors, they have measured the redshifts of 19 nebulous objects that appear to be companion galaxies of 15 relatively low-redshift quasars. Observing at the Kitt Peak 4-meter telescope, they have determined that, in 18 of these 19 cases, the apparent companion has a redshift very close to that of the quasar. While Burbidge, Arp and their partisans may argue that quasars are so peculiar that they can generate redshifts of unknown origin, this position becomes difficult to maintain for companion galaxies that otherwise look perfectly ordinary."

H. Arp and G. Burbidge, chief among those ordered to recant, are not convinced. Arp points to many cases where bridges of luminous material connect high-redshift quasars with low-redshift galaxies. Also, the clustering of quasars around nearby galaxies supports the nearness of quasars. As Burbidge has observed, if "just one large redshift is not due to the universal expansion, Pandora's box is open. Much of our currently claimed knowledge of the extragalactic universe would be at risk, as would a number of scientific reputations."

Indeed, reputations are on the line as are scientific ethics. Telescope time has been cut off from those supporting noncosmological redshifts. The papers written by scientists who haven't recanted are held up, refereed forever, and rejected. Come to think of it, the theological overtones of the call for "recantation" fit the redshift situation very well. (Anonymous; "Companion Galaxies Match Quasar Redshifts: The Debate Goes On," Physics Today, 37:17, December 1984.)

HOW TO BE UNFAMOUS IN ASTRONOMY

When Sky and Telescope devotes almost five full pages to a new book, you may be sure that something important has happened. The book is H. Arp's Quasars, Redshifts, and Controversies.

We know that we have perhaps overplayed the shakiness of the redshift-distance hypothesis and the fizzling of the Big Bang, but our whole cosmological outlook is at stake. Now, rather than review again the scientific pros and cons (you can read Arp's book for that), we will be content here with a few comments about how science has failed to work well in Arp's case.

G. Burbidge, who reviews the book, recalls how the politics of science works in the following quotation:

> ...the important factors for a successful career are your sponsors (where and with whom did you get your Ph.D); field of research (popular or unpopular); and diplomatic skills (always speak quietly with great conviction, and, when in doubt, agree with the wisest person present, who by definition must come from one of the the very few [recognized] institutions). Look upon new ideas with great disapproval and never discover a phenomenon for which no explanation exists, and certainly not one for which an explanation within the framework of known physics does not appear to be possible.

Arp played this game for 29 years at the Mount Wilson and Palomar Observatories. He compiled the marvelous Atlas of Peculiar Galaxies, and was once rated among the top 20 astronomers. But he kept finding anomalies---apparently-associated celestial objects with different redshifts. More and more he began to believe and (perhaps recklessly) assert that some redshifts are not cosmological; that is, a measure of recessional velocity and distance. Soon, his rating dropped from the "upper 20" to "under 200". The final (and disgraceful) blow came about four years ago, when he received an unsigned letter stating that his work was without value and that he could have no more telescope time! Arp now lives in West Germany. (Burbidge, Geoffrey; "Quasars, Reshifts, and Controversies," Sky and Telescope, 75:38, 1988.)

Comment. More political details may be found in Arp's book. Is Arp a martyr-in-the-making? You bet he is! Burbidge, an admitted Arp sympathizer, suggests that the "Arp Effect" is only the tip of the iceberg. In closing his review, he invokes the ghost of Alfred Wegener, who had the temerity to suggest that continents could drift.

SWEEPING ANOMALIES UNDER THE RUG

A series of articles in the science magazine Mercury so slavishly followed the scientific party line on the meaning of the redshift that G. Burbidge was prompted to pen a rejoinder.

Burbidge reviewed the considerable observational evidence supporting a non-cosmological interpretation of some redshifts. (Such data are also reported in our book Stars, Galaxies, Cosmos.) A typical observation is the apparent physical connection (streams of connecting matter) between quasars and galaxies with radically different redshifts. Burbidge remarks:

> Evidence of this kind exists. If it is accepted it means:
>
> 1. That at least some quasars don't lie at so-called cosmological distances.
> 2. That at least some parts of the redshifts of quasars are due to some effect other than the expansion of the universe.
> 3. That quasars are physically related to bright, comparatively nearby galaxies.

Burbidge is not concerned by the fact that some astronomers find the data unconvincing, rather he objects to the so-obvious attempts to brush such anomalous data under the rug. His concluding remarks are pertinent to all of science:

> I cannot end this part of the discussion without making two points which are rarely made, but which are important:
>
> 1. Evidence of the kind just mentioned which is favorable to the cosmological interpretations of the redshifts does not negate the other evidence. It simply means that the world is a complicated place.
>
> 2. Only in articles of this kind is one expected to describe such results. In articles such as that by Weedman, it is somehow considered all right to totally ignore the non-cosmological hypothesis.
>
> The fairest way to deal with the problem is not to fall back on authority (what eminent authorities believe or don't believe) but to examine the evidence for oneself. The most extensive collection of this evidence is in the book by Halton C. Arp. (Quasars, Redshifts, and Controversies) ...If, after examining the statistics yourself and understanding the evidence, you are unconvinced, so be it. Remember, if the conventional view is correct, all of these apparent juxtapositions must be accidental. Above all, do not be swayed by the views of the authorities, be they Dan Weedman, Allan Sandage, Maarton Schmidt, Chip Arp, or myself. We are fallible, too, and some of us (ask the others!) have axes to grind."

(Burbidge, Geoffrey; "Quasars in the Balance," Mercury, 17:136, 1988.)

Comment. Burbidge, a professor of physics at the University of California at San Diego, has had a long and distinguished career in science. He can write and get articles like the one above published. How many young, aspiring astronomers would dare?

QUANTIZED REDSHIFTS

GALAXY REDSHIFTS COME IN CLUMPS

"Red shifts in the light from distant galaxies seem to favor certain values, at intervals corresponding to a spacing of 72 kilometres per second, if the red shifts are interpreted as indicating the recession velocities of the galaxies. According to the latest evidence, this provides a yardstick against which we can measure the absolute motion of the Sun through space. This rather startling indication that the red shifts of galaxies are quantised, rather like the atomic spectral lines by which the red shifts themselves are measured, has a pedigree that goes back more than 10 years. Since 1972, W.G. Tifft, of the University of Arizona, has been producing evidence for the red shift quantisation from analyses of various catalogues of galaxy red shifts and, as his collection of data has mounted, the idea, unpalatable though it seemed at first, has become steadily more respectable. It has not taken the astronomical world by storm, and even Tifft has no definite idea why the red shifts should be grouped like this. But it is no longer possible to dismiss the evidence out of hand."

Actually, Tifft has speculated that galaxy redshifts might represent an intrinsic property of galaxies, which takes on quantized values, like the energy states of an atom! (Gribbin, John; "Galaxy Red Shifts Come in Clumps," New Scientist, p. 20, June 22, 1985.)

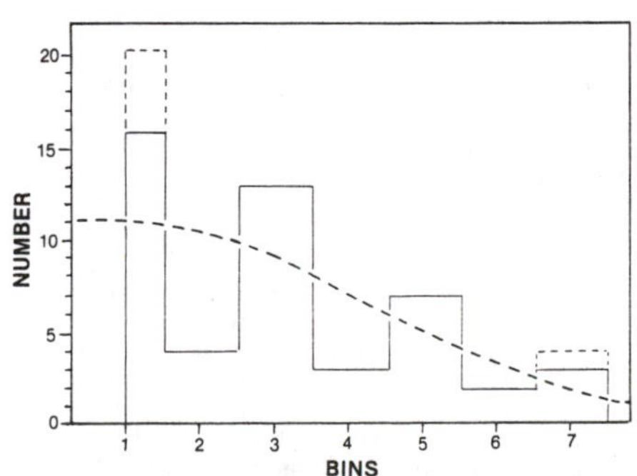

Tifft compared the redshifts of physically associated galaxies and found the differences to be integral multiples of 72 kilometers per second.

QUANTIZED GALAXY REDSHIFTS

> The history of science relates many examples where the conventional view ultimately was proven wrong.

Tifft and Cocke begin their article with this sentence. Wisely, they followed with the tale of how vehemently the quantization of the atom was resisted earlier in this century. They were wise because without such a reminder to be open-minded, many astronomers would automatically toss their article in the wastebasket! In fact, when Tifft's first paper on redshift quantization appeared in the Astrophysical Journal, the Editor felt constrained to add a note to the effect that the referees "neither could find obvious errors with the analysis nor felt that they could enthusiastically endorse publication."

Even today, after much more evidence for redshift quantization has accumulated, scientific resistance to the idea is extreme. We shall now see what all this fuss is about.

Tifft first became suspicious that the redshifts of galaxies might be quantized; that is, take on discrete values; when he found that galaxies in the same clusters possessed redshifts that were related to the shapes of the galaxies. The obvious inference was that the redshifts were at least partly dependent upon the galaxy itself rather than entirely upon the galaxy's speed of recession (or distance) from the earth. Then, he found more suggestions of quantization. The redshifts of pairs of galaxies differed by quantized amounts (see figure). More evidence exists for galactic quantization, but this should give the reader a feeling for the conceptual disaster waiting on the wings of astronomy.

Can galaxies, like atoms and molecules, possess quantized states? And do the findings of Tifft and Cocke undermine the redshift-distance relationship? The answer might be YES; and then all of astronomy and our entire view of the universe and its history would have to be reformulated. (Tifft, William G., and Cocke, W. John; "Quantized Galaxy Redshifts," Sky and Telescope, 73:19, 1987.)

MORE EVIDENCE FOR GALACTIC "SHELLS" OR "SOMETHING ELSE"

Measurements of periodic red-shift bunching appeared in the literature at least as far back as 1977 in the work of W.G. Tifft. (See AWF8 in Stars, Galaxies, Cosmos.) The implications of this phenomenon are apparently too terrible to contemplate, for astrophysicists have not taken up the challenge. They may be forced to take the phenomenon more seriously, because two new reports of red-shift bunching have surfaced.

First, B. Guthrie and W, Napier, at Edinburgh's Royal Observatory, have checked Tifft's "bunching" claim using accurately known red shifts of some nearby galaxies. They found a periodicity of 37.5 kilometers/second---no matter in which direction the galaxies lay. (Gribbin, John; "'Bunched' Red Shifts Question Cosmology," New Scientist, p. 10, December 21/28, 1991.) The work of Guthrie and Napier is elaborated upon in the next item.

Second, B. Koo and R. Krone, at the University of Chicago, using optical red-shift measurements, discovered that, in one direction at least, "the clusters of galaxies, each containing hundreds of millions of stars, seemed to be concentrated in evenly spaced layers." (Browne, Malcolm W.; "In Chile, Galaxy-Watching Robot Seeks Measure of Universe," New York Times, December 17, 1991. Cr. P. Gunkel.)

Comment. Explanations for the unexpected bunching vary and are highly controversial:

- There are systematic defects in the radiotelescopes and/or the observational techniques. But, as just reported by Koo and Krone, the phenomenon is also seen with optical instruments.
- The red shifts are not entirely due to the Doppler Effect and the recessional velocities of galaxies. If this is so, the dimensions and age of the universe would have to be revised.
- The red-shift bunching occurs because some galaxies are arranged in shells surrounding the earth. To some, this would be philosophically disastrous, because it would place humanity in a favored spot in the cosmos.

GALACTIC SHELL GAME

W.G. Tifft, an astronomer at the University of Arizona, has maintained for some two decades that the redshifts of the galaxies do not fall on a smooth curve as one would expect. Instead, Tifft asserts, redshifts are bunched at intervals of 72 kilometers/second and at one-half and one-third that value. Mainstream astronomers insist that redshifts be interpreted as Doppler shifts due to the expanding universe. Quantized redshifts just don't fit into this view of the cosmos, for they imply concentric shells of galaxies expanding away from a central point---earth!

Even though more recent redshift data have supported the notion of quantized redshifts, cosmologists find them undigestible, even pathogenic. But replication and non-replication are the essence of science, so B. Guthrie and W.M. Napier, at the Royal Observatory at Edinburgh, undertook another study. They selected 89 nearby spiral galaxies that had not been incorporated in any of the previous surveys. These galaxies had very accurately measured redshifts and were distributed all over the celestial sphere.

As expected, the galaxies' redshifts showed a smooth distribution. Clearly, no quantization was being introduced by the radio telescopes or the data reduction process. But after Guthrie and Napier corrected each redshift to account for the Earth's motion around the center of the Milky Way ---a different correction for each location in the sky---out popped a periodicity of 37 km/sec, close to one of Tifft's values. It was so strong that the chance of it being a statistical fluke was less than 1 in 3,000.

Tifft's work therefore seems to have been verified again. But Tifft is now waxing even more iconoclastic, claiming that galactic redshifts have actually changed slightly in just a few years! (Anonymous; "Quantized Redshifts: What's Going on Here?" Sky and Telescope, 84:128, 1992.)

Comment. A strange geometrical concordance exists between quantized redshifts and the shells of stars surrounding some elliptical galaxies. See p. 90.

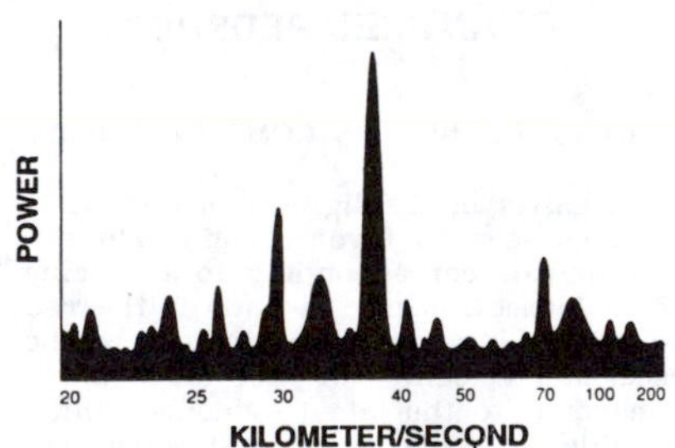

The quantized power spectrum computed by Guthrie and Napier.

QUASAR REDSHIFT CLUSTERS AND (EVEN WORSE) MULTIPLE REDSHIFTS

At the XIIIth Krakow Summer School of Cosmology, September 7-12, 1992, many of the world's top cosmologists experienced the disorientation that accompanies both earthquakes and shifting paradigms. Two of many cosmoseisms felt during the meeting in Poland are recorded below:

> Halton Arp, Max-Planck Institute for Astrophysics, spoke about his "Variable Mass" cosmology. He pointed out the need for cosmologies to explain why quasar redshifts cluster near 0.3, 0.6..., with the grouping just below z = 1.2 dominating all others; and why certain classes of stars have significant excess redshifts. He also pointed out the inconsistency that local galaxy groups seem to have velocity dispersions of less than 100 km/s, while distant groups seem to have members with dispersions up to 1000 km/s.
>
> Jack Sulentic spoke about multiple redshifts seen in some quasars and AGNs. Line profiles come in all types; symmetric, double-peaked, and asymmetric. Relative shifts are both toward the red and the blue. Arguments against an accretion disk/black-hole model were reviewed. Apparently a non-Doppler redshift-blueshift mechanism is needed. For example, one broad line (in 1404 + 28) shifts back and forth by 1000 km/s relative to another narrow H-line, with an average offset of 2000 km/s. These shifts correlate perfectly with intensity.

Less technically speaking, the long-held belief that redshifts are solely due to the Doppler effect is receding along with the expanding universe!

(Van Flandern, T.; "Recent Meeting: XIIIth Krakow Summer School of Cosmology," Meta Research Bulletin, 1:25, September 15, 1992.)

HUBBLE TROUBLES

DOUBLE HUBBLE: AGE IN TROUBLE

A key concept in modern astronomy---the distance scale---has been challenged by a new measurement technique. In recent years the so-called Hubble Constant has been used to determine the distances of the farthest observable galaxies and, by assuming they are at the periphery of the expanding universe, obtaining the age of the cosmos by dividing distance by the speed of light. Until this current challenge, the age of the universe was generally taken as about 20 billion years.

The "old" Hubble Constant, however, was determined from measurements of the distances to rather close galaxies and then assuming that the Constant held for the entire universe. The new yardstick reaches farther out into space. It is based on the observation that the broadening of a galaxy's 21-centimeter line depends upon its rate of rotation, plus the belief that the rate of rotation is proportional to its brightness! The "new" Hubble Constant is 95 kilometers/second/megaparsec, which translates into an age of only 10 billion years for the universe. Both the 10-billion-year and 20-billion-year camps claim strong supporting evidence for themselves and point to serious difficulties in the opposing method. The stage is set for a delightful controversy. (Hartline, Beverly Karplus; "Double Hubble: Age in Trouble," Science, 207:167, 1979.)

1981: "TIRED LIGHT" THEORY REVIVED

The Expanding Universe Theory depends to a large degree upon the correctness of Hubble's Law; viz., the redshifts of distant objects are directly proportional to their distances from earth. Unfortunately for the Expanding Universe, some redshift measurements indicate a quadratic rather than linear relationship between redshift and distance. I.E. Segal's chronometric theory of the cosmos, however, does predict a quadratic relationship. In Segal's theory redshifts are due to the gravitational slowing of light rather than any gereral expansion of the universe.

Even if most astrophysicists are finally persuaded that the quadratic relationship is real, they will be loath to abandon the philosophically appealing Expanding Universe. Not only is the Expanding Universe consistent with Relativity but it states unequivocally that the earth (and man) does not occupy a preferred place in the universe. (Hanes, David A.; "Is the Universe Expanding?" Nature, 289:745, 1981.)

Comment. A geocentric theory would intimate a supernatural force favoring humanity.

1986: "TIRED LIGHT" REVIVED AGAIN

Back in 1929, F. Zwicky proposed that the redshifts astronomers observed in the spectra of celestial objects might not be due to universal expansion but rather to "tired light." In other words, the wavelengths of the photons entering our telescopes are redshifted because they have lost energy through interactions with matter en route to earth. The "tired light" theory was eclipsed by the esthetically appealing concepts of the Big Bang and Expanding Universe.

But not everyone has forgotten Zwicky's tired light. P. LaViolette has "...compared the tired light cosmology to the standard model of an expanding universe on four different observational tests and has found that on each one the tired-light hypothesis was superior. The differences between the rival cosmologies are most apparent at large redshifts, however, and it is in this region that observations are most difficult to make." (Anonymous; "New Study Questions Expanding Universe," Astronomy, 14:64, August 1986.)

MISSING MASS QUESTION

Over 50 years ago, F. Zwicky noticed that the galaxies in some clusters had such high velocities that they should easily escape the gravitational bonds of the cluster, but they have not. The same effect was duly noted for stars in galaxies. Some unseen force was apparently constraining them. This unseen force is generally conceded to be the gravitational force due to dark or "missing" matter. This is, of course, an assumption. Some other, more bizarre phenomenon may be the cause.

NINE-TENTHS OF THE UNIVERSE IS UNSEEN

In a rather lengthy article on the disturbing discoveries of modern astronomy, Ivan King is quoted as follows:

> The most serious problem in extragalactic astronomy today is the notorious 'missing mass.' There are rich clusters of galaxies where it is quite clear that the total gravitating mass is about ten times what we can account for in the conventional masses of the individual galaxies...
>
> The other missing mass problem shows up in the outer parts of spiral galaxies, where the rotational curves have clearly never heard of Kepler... The rotation curves say there is a large amount of mass out there, [but] it emits no light by which we can study its nature.
>
> The missing mass problem is extremely disquieting...We are talking about 90 percent of the mass of the universe, present but not speaking. Can we really claim to know anything about the nature of the universe if we don't know the properties, or even the nature, of 90 percent of its material?

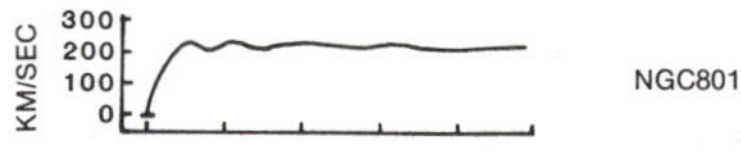

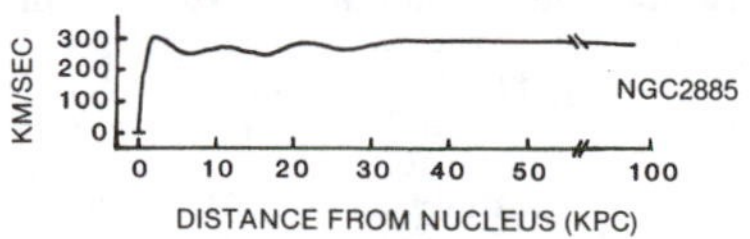

Velocity profiles of matter in two galaxies. According to Newtonian dynamics, velocities should decrease as distance from the center increases, just as planetary velocities decrease with distance from the sun in the solar system. Because they do not decrease in galaxies and clusters of galaxies, astronomers postulate the existence of hidden mass. See p. 90 for the "winding dilemma."

(Anonymous; "The Extragalactic Ferment," Mosaic, 9:18, May/June 1978.)

HAS THE UNIVERSE'S MISSING MASS BEEN FOUND?

In the above item, an article from Mosaic was quoted to the effect that 90% of the universe is "unseen." In pursuit of this missing mass, a U.S. team of astronomers has now detected previously unseen halos around several spiral galaxies. The halo luminosities are comparable to the brighter imbedded disks when integrated over a large area surrounding the galaxies. The halo masses, however, as inferred from the galaxies' rotation curves far exceed the masses of the bright spiral cores.

The big question is "What are the dim but massive halos composed of?" They might consist of small, faint stars or nonluminous matter of some sort. The researchers had to conclude, though, that the halos are galactic components of "totally unknown nature." (Anonymous; "Has the Universe's Missing Mass Been Found?" New Scientist, 80:174, 1978.)

EXORCISING THE HIDDEN MASS

These days the astronomical publications are full of discussions of the "missing mass" problem. It seems that for galaxies to move the way they do, there has to be some "dark matter" out there, assuming Newton's Laws of Gravitation and Motion are valid. Something unseen is tugging on galaxies and the stars that comprise them.

This is a sad situation, according to Moto Milgrom, an Israeli astrophysicist. Maybe there is nothing hidden and Newton's Law of Gravitation is wrong. After all, it was derived solely on the basis of solar-system observations. On a larger scale, it might be incorrect. Milgrom offers a startling alternative: for accelerations greater than a^{o}, let Newton's Law be; below that value, let the square of the acceleration be proportional to the mass of the attracting body and the inverse square of the distance.

This done and presto chango the need for missing mass disappears. Even more remarkable is the fact that a particle with the acceleration a^{o} just reaches the speed of light over the age of the universe. (Milgrom, Moto; "Newtonian Gravity Falls Down," New Scientist, p. 45, March 7, 1985.)

Comment. It would be more than passing strange for cosmic laws to suddenly shift gears so radically at a specific value of acceleration.

HOLD EVERYTHING: IT MAY BE A NONPROBLEM

Newtonian gravitation may not have to be tampered with, as described above. Galaxies may not need "missing mass" to stabilize them after all. B. Byrd and M. Valtonen estimate that many clusters of galaxies are really flying apart.

> Clusters of galaxies can eject members by a gravitational slingshot process, with one galaxy after another being accelerated through the dense centre of the cluster and fired out into the Universe at large. If this happens, the ejected galaxies are moving at more than the escape velocity from the system, so estimates of the total mass in the system based on the assumption that all the galaxies are in bound orbits will be incorrect.

(Anonymous; "Expanding Clusters Confuse Astronomers," New Scientist, p. 13, March 21, 1985.)

Comment. Previously (p. 91), we have seen jets of stars being squirted into space, immense shells of stars being ejected by elliptical galaxies, and other cosmic sowings of astronomical systems. Now, entire galactic clusters are being thrown around the universe. This hardly seems a universe that is "running down," as the Laws of Thermodynamics would have us believe. Somebody or something is stirring the pot---a pot in which biological systems and perhaps super-biological systems are ingredients in the stew.

BLACK HOLES

General Relativity predicts that when stars of greater than three solar masses burn up all their nuclear fuel, their matter collapses into an ultradense state. Such aggregations of dense matter are called "black holes." Space-time is so distorted in the vicinities of black holes that no matter or light can escape them, making them appear "black" or in a sense invisible. Black holes can, however, attract matter to them, in effect "swallowing" it up. Black holes can in principle be detected by the effects they have on their surroundings. Whether black holes have indeed been detected in this way is debatable.

WHAT IS IT? A BLACK HOLE, OF COURSE!

Radio-telescope measurements of a compact radio source churning away in the center of our galaxy reveal that it is only 20 AU in diameter at radio wavelengths of 1.35 centimeters. This is roughly the size of the solar system inside Saturn's orbit. This tiny radio source is so energetic that there seems no escaping the conclusion that it is a black hole. No other astronomical object is capable of generating so much energy in so small a volume. Since other galaxies also seem to harbor small, but very powerful radio sources in their centers, astronomers wouldn't be too surprised if all galaxies had black-hole cores. Quasars, in fact, might be galaxies with spectacularly active centers.

Would these unseeable black holes be the notorious "missing mass" in the universe? Not likely. The mass of the purported black hole in our Galaxy is only about several million solar masses--not even close to what is needed. (Maddox, John; "Black Hole at the Galactic Centre," Nature, 315:93, 1985.)

Comment. Actually, it would be rather amusing if the problem of the missing mass, which we cannot see, were solved by black holes, which we cannot see either!

DO BLACK HOLES EXIST?

Can we believe our eyes? Dare anyone suggest that black holes do not lurk out in the cosmos sucking in stars and unwary spaceships? It's all true; an article bearing the above title appeared in the January 1988 number of Sky and Telescope. Doubts do surface once in a while, despite all the TV documentaries, all the textbooks, and all the newspaper jottings, where black holes are described in the hushed tones used only with profound truths of nature.

To set the stage, we quote a paragraph from said article:

> There is, however, a serious problem with black holes, one that leaves some scientists skeptical about their existence. The overarching mystery lies hidden at a hole's center. Einstein's general theory of relativity predicts that we will find there an object more massive than a million Earths and yet smaller than an atom ---so small, in fact, that its density approches infinity. The idea of any physical quantity becoming infinite flies in the face of everything we know about how nature behaves. So there is good reason to be skeptical that such a nasty thing could happen anywhere at all.

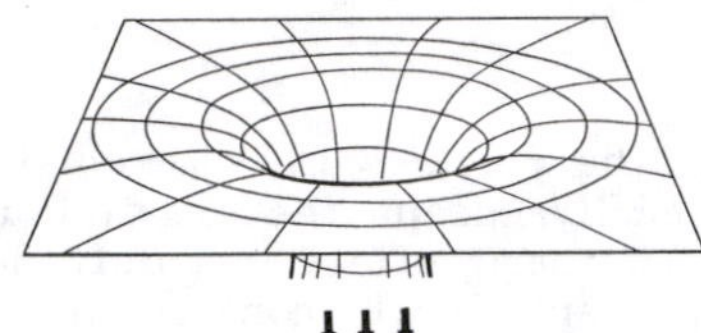

In the vicinity of a black hole space-time is distorted such that matter and light can flow in but cannot escape!

Among the observations that hint at the reality of black holes are the X-ray binaries. In a typical X-ray binary, prodigious, flickering fluxes of X-rays reveal the presence of an ultradense star and an orbiting companion. The rapid orbital motion of the companion star tells us that the central X-ray star has a mass of more than three suns. General Relativity assures us that such a star can only collapse further to form a black hole. Therefore, black holes must exist.

J. McClintock, the author of this article, does not buy this reasoning. General Relativity, he says, has been shown to be valid so far only in weak gravitational fields, not in the very powerful gravitational fields of an X-ray star.

> ...we presume that Einstein's theory correctly describes strong gravity when we argue that certain X-ray stars are black holes, yet, at the same time, these alleged black holes are the acid test of Einstein's theory of strong gravity.

(McClintock, Jeffrey; "Do Black Holes Exist?" Sky and Telescope, 75:30, 1988.)

Comment. The long history of science teaches us that all theories are eventually displaced by more accurate, more all-inclusive formulations. Unfortunately, this makes textbook writing difficult.

WANTED: A BONA FIDE BLACK HOLE

Don't you get tired of all those science books, newspaper articles, TV documentaries, and commentators gushing at length about black holes as if they were well-verified denizens of the universe? Black holes are popularly presented as "fact"; no doubts permitted; here the Book of Science is closed! It was like a breath of fresh air to read this sentence in Sky and Telescope:

> Scientists are still unable to confirm the existence of even a single black hole, despite a widespread belief that such things should, and indeed must, exist.

This single sentence won't change anything, because everyone is comfortable with black holes. They are part of the (often false) reality that the media smothers us with.

Actually, there are two places where black holes "might" dwell, based upon the anomalous behavior of matter around these regions: (1) at the centers of some galaxies, including our own Milky Way; and (2) as unseen components of some close double stars, where the mass of the unseen companion is too great for it to be an ordinary neutron star. W. Kundt and D. Fischer, at Bonn University, have recently concluded that the second possibility is better explained without resorting to black holes. For example, a neutron star with a massive accretion disk might suffice. As for black holes at the centers of galaxies, with masses of several million suns, gravitationally sucking in surrounding matter and careless spaceships---well, they are possible. Unfortunately, galactic centers are too far away and obscured by dust for us to be certain what lies at their cores. Black holes are really only surmise; although they make good copy! (Anonymous; "No Black Holes?" Sky and Telescope, 78: 572, 1989.)

SPACE RADIATION

The electromagnetic and particulate radiation we receive from the universe (with the emanations of known sources edited out) has embedded in it some engaging anomalies. This section is conerned with several components of this space radiation, in particular the cosmic rays, including gamma rays and X-rays. Of special concern is the steady gamma-ray background upon which are sometimes imposed strong bursts from unidentifiable sources. The puzzle here involves the remarkably uniform distribution of these gamma-ray "bursters." These were expected to be confined to the disk of our galaxy. Another enigma for the astronomers is the origin of the diffuse background of X-rays pervading outer space.

There have been other recently identified space radiation anomalies that were not mentioned in Science Frontiers. It was, however, impossible to miss the many measurements of the cosmic background microwave radiation. These are treated separately in the discussions of the validity of the Big Bang hypothesis beginning on p. 102.

COSMIC RAYS NOT RANDOM

Conventional wisdom maintains that the cosmic rays intercepted by the earth are randomly distributed in space and time because of the smoothing action of the galactic magnetic field. But a cosmic-ray telescope buried beneath 600 meters of rock has recently detected bursts of cosmic rays emanating from specific directions. The two major sources are in the the direction of the galactic north pole and the constellation Cygnus. Since the galactic magnetic field seems sufficient to randomize all charged particles during their long flights through space, pristine cosmic rays may not be charged particles at all. (Hecht, Jeff, and Torrey, Lee; "Scientists Find Sources of Cosmic Rays," New Scientist, 99:764, 1983.)

MESSENGERS OF A "NEW PHYSICS"

In a "garage" off the road tunnel running deep under Mont Blanc sits a huge particle detector called Nusex. A second, complementary experiment resides 600 meters below the surface in a Minnesota mine. Both experiments are tuned to measure charged particles of very high energy, especially muons, which penetrate their high rocky ceilings with ease. These two arrays of buried detectors have both picked up fluxes of muons coming from the direction of Cygnus X-3

Now Cygnus X-3 is already classed as a remarkable object because it spews out pulses of X-rays and gamma-rays. It turns out that the muon fluxes arrive in phase with the pulses of gamma-rays and X-rays, and are thus definitely linked to Cygnus X-3. The problem here is that muons are electrically charged particles that would assuredly be thrown far off course by intergalactic magnetic fields if they originated at Cygnus X-3.

The muons, therefore, must be created by electrically neutral particles arriving at the earth's atmosphere from Cygnus X-3. Neutrons can be ruled out because they would decay in transit. X-ray photons and neutrinos have also been ruled out. The only alternative left seems to be some unknown neutral particle generated at Cygnus X-3. Cygnus X-3 may be a huge particle accelerator which "may operate in a realm of physics inaccessible on Earth, and the high-energy muons may be the first messengers of that new physics." (Sutton, Christine; "Subatomic Particles from Space, "New Scientist, p. 18, May 23, 1985.)

NEW KINDS OF MATTER TURNS UP IN COSMIC RAYS

Japanese physicists claim to have found evidence of 'strange matter' in cosmic rays. Their detectors have recorded two separate events, each of which can be explained by the arrival of a particle with a charge 14 times as great as the charge on a proton, and a mass 170 times the proton's mass. No atomic nucleus---made of protons and neutrons---exists that matches this description, but these properties are precisely in the range predicted for so-called quark nuggets, which physicists believe may be made of a type of material dubbed strange matter. (Gribbin, John; "New Kind of Matter Turns Up in Cosmic Rays," New Scientist, p. 22, November 10, 1990.)

The original report appeared in Physical Review Letters, 65:2094, 1990. In it, the Japanese scientists describe their balloon-borne equipment, proving that one does not need fancy spacecraft to make important discoveries.

The key feature of the quark nugget is its very high mass-to-charge ratio. Where do quark nuggets come from? The theoreticians surmise that they may be created when neutron stars collide or, perhaps, they are left over from the hypothetical Big Bang.

TIDAL WAVE OF GAMMAS SWEEPS SOLAR SYSTEM

On March 5, 1979, a colossal burst of gamma rays swept through the solar system, triggering radiation detectors on nine different spacecraft. By comparing the times of arrival of the burst, the direction of the source was narrowed down to a "box" a couple of arc minutes across. Gamma-ray bursts have never before been correlated with visible sources, but this time the box contained the remnants of a supernova in the Large Magellanic Cloud, a satellite galaxy of our own Milky Way.

The anomaly that arises involves the immense distance of the supposed source and the strength of the burst when it reached the solar system. The power level of the supernova remnant gamma flash would have had to be about 10^{37} watts---a stupendous figure. If the supernova remnant is a neutron star, as current theories suggest, the neutron star would have to be 10 to 100 times the size of the usual neutron stars. (Anonymous; "Gamma-Ray Burst Comes from Outside the Galaxy," New Scientist, 87:776, 1980.)

WHEN ISOTROPY CONFOUNDS

Take a look at the distribution of 153 gamma-ray bursts registered by the Gamma-Ray Observatory (a satellite). There is no pattern, gamma-ray bursters seem to be evenly distributed in all directions. This is not what the astronomers expected, and the implications of this isotropy are staggering.

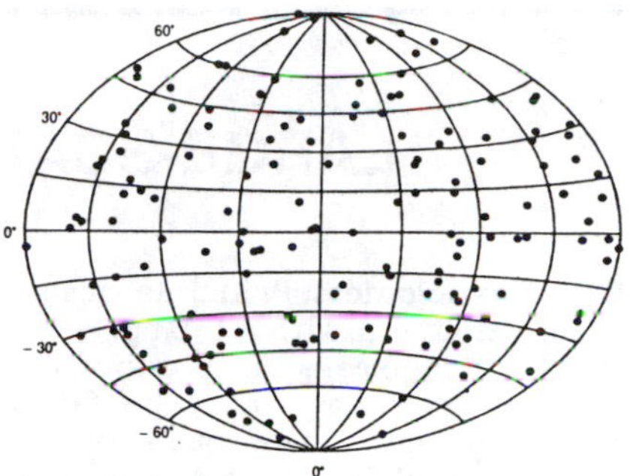

Angular distribution of 153 gamma-ray bursts detected by the GRO satellite in 1991.

Gamma-ray bursts emanate from highly localized unseen sources. They may last for a few milliseconds or stretch out for several minutes. The energy in the bursts ranges over 26 orders of magnitude. The rise-times of the bursts are so short that the sources can only be a few hundred kilometers across.

Before the accompanying map appeared, most scientists thought that the bursters were nearby, probably in the disk of the galaxy, and were due to asteroids being digested by neutron stars or possibly neutron-star quakes. If such were so, the bursters would be concentrated in the plane of the galaxy (the Milky Way), which clearly they are not. Another theory places the bursters in a distant spherical halo about our galaxy. But, in this case, the bursters would have to be much more energetic than astronomers care to contemplate. In fact, if they exist in a galactic halo, we should also be able to detect the bursters in our neighboring galaxies ---but we do not!

A more exciting suggestion is that gamma-ray bursters are really very close! This would be consistent with the failure to find cosmological redshifts in the burster spectra. Could they be really close, just a few hundred light hours away? Perhaps arranged in a spherical halo about our solar system in the vicinity of the postulated Oort Cloud of comets? If this were so, they would not have to be nearly as powerful as they would in the neutron-star model. If the gamma-ray bursters really do lurk just at the fringes of the solar system, they must, given their power and small size, be objects completely new to astronomy. (Schwarzschild; Bertram; "Compton Observatory Data Deepen the Gamma Ray Burster Mystery," Physics Today, 45:21, February 1992.)

Comment. Historically speaking, the gamma-ray bursters were discovered accidentally by satellites launched to detect surreptitious tests of nuclear weapons. Wouldn't it be ironical if our satellites are really monitoring artificial phenomena, generated by battles we can only write science fiction stories about?

WHENCE THE X-RAY BACKGROUND?

A persistent problem needling astronomers has been the diffuse X-ray background radiation; that is, that flux of X-rays that pervades the universe but which seems to come from no place in particular. Distant quasars are thought to contribute some of this diffuse X-ray flux but, even with recent quasar discoveries, there are just not enough of them to account for the X-rays observed. To make matters worse, quasar X-ray spectra do not match that of the X-ray background either, particularly at very short wavelengths.

Superimposed on the general X-ray background are discrete X-ray sources separated by extended blobs of X-ray emitting material. If these blobs are really clumps of clumps of quasars too close to be separated by our instruments, the Big Bang model is at risk, for it cannot account for large, organized assemblages of quasars. (Powell, Corey S.; "X-Ray Riddle," Scientific American, 264:26, March 1991.)

explained by supposing this mass to be large enough to pull the earth (and our Galaxy) toward it. This mass would have to be about 10 billion light years away and weigh as much as 100 million Galaxies.

Such a gigantic blob or inhomogeneity in the universe would be very difficult to explain. As it is, the aggregation of stars into galaxies after the Big Bang remains poorly understood. The bigger the inhomogeneity, the harder it is to account for. The Big Bang should have spread matter out pretty evenly. (Anonymous; "Large Mass May Pull Earth Through Space," New Scientist, 83:21, 1979.)

LARGE-SCALE STRUCTURES

The Cosmological Principle asserts that the universe is everywhere the same; that is, without large-scale inhomogeneities. Guided by this outlook, astronomers fully expected that the stars and galaxies, as products of the Big Bang, would be fairly uniformly distributed---a sort of cosmological puree. To be sure, stars are concentrated in galaxies (a fact not easy to explain), but the galaxies themselves should be evenly distributed. Observations have denied this expectation; galaxies display considerable large-scale ordering. There are clusters of galaxies, and even superclusters. Beyond this scale, galaxies are arranged in "walls" and skeins. The universe also incorporates large voids. Coordinated motions of large assemblages of galaxies are also remarked. Such large-scale order is inimical to the Big Bang hypothesis. The larger the structures and the younger the structures, the more they challenge the Big Bang and the Cosmological Principle.

LUMPS, CLUMPS, AND JUMPS

> Astronomers have already discovered lumps, motion, and structure never suspected in a universe once considered smooth and expanding uniformly in all directions. Two researchers now say the universe is even lumpier, has faster relative motion and shows larger structures than previously believed.

N. Bahcall and R. Soneira have been studying the structures and motions of superclusters of galaxies. Each supercluster consists of clusters of clusters of galaxies and contains upwards of hundreds of billions of stars. (Obviously, these are not inconsequential entities!) By analyzing the redshifts of galaxies, Bahcall and Soneira have found that the universe is much more dynamic and inhomogeneous than expected. (1) The clusters of galaxies are larger and more extensive. Superclusters can be 500 million light years across---about 1% of the known universe; (2) Relative motions within the clusters are as high as 2,000 kilometers per second more than one can account for using gravitational attraction alone. (Kleist; T.; "Lumps, Clumps and Jumps in the Universe," Science News, 130:7, 1986.)

LARGE, UNSEEN MASS IS PULLING EARTH TOWARD IT

Recent measurements of the cosmic microwave background indicate that the earth moves relative to it. New cosmic X-ray data from the satellite Ariel 5 suggests that a large, hitherto unsuspected mass is located in the same direction that the earth is moving. Thus, both X-ray and microwave data could be

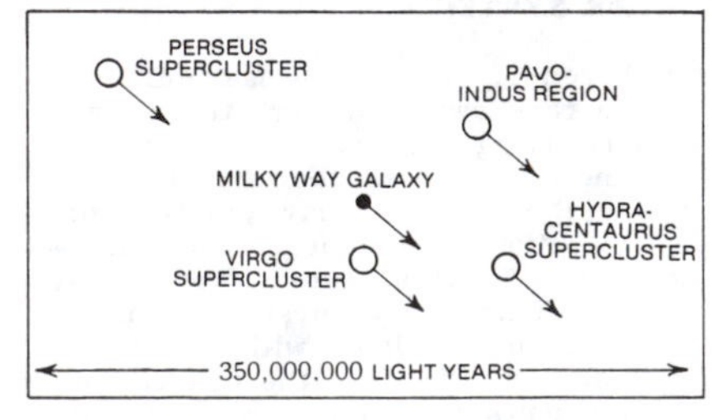

The galaxies in the Local Group, including our Milky Way, are streaming in about the same direction at roughly the same speed relative to the cosmic background radiation.

COSMIC CURRENTS

A survey of 390 elliptical galaxies have identified an unexpected streaming effect superimposed on the (postulated) general outward expansion of the universe. A team of seven astronomers first measured the velocities of the elliptical galaxies relative to the earth. Next they subtracted out the velocities of universal expansion and, lastly, the velocity of the earth relative to the 3°K cosmic background radiation. These subtractions enable us to determine how the 390 elliptical galaxies move relative to the cosmic background radiation---which is about the best fixed reference frame we can come up with.

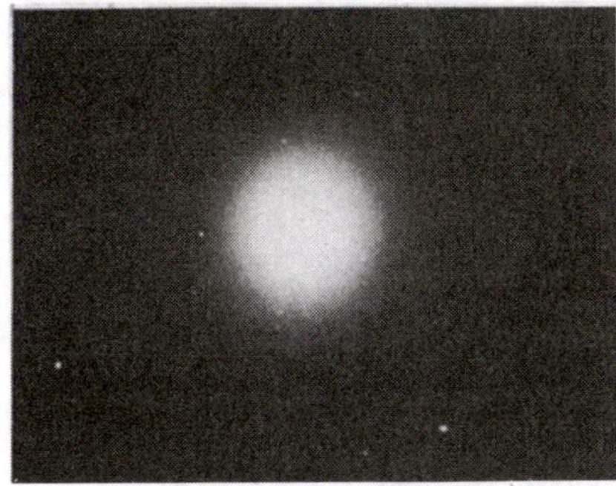

A giant elliptical galaxy containing many billions of stars. Hundreds of these immense assemblages are moving collectively through the cosmos. What propels them?

For roughly 50 million parsecs in all directions from earth, the clusters and superclusters of galaxies are streaming through the cosmos in a group. What's more, they lie in and move parallel to a fairly well-defined plane. Just what this coordinated motion means in terms of the origin and evolution of the universe is anyone's guess. There doesn't seem to be anything in the Big Bang theory that would explain these cosmic currents. Something set all these galaxies in motion---but what? (Waldrop, M. Mitchell; "The Currents of Space," Science, 232:26, 1986.)

A BIG VOID IN SPACE OR A DEFECTIVE YARDSTICK?

R. Kirshner and colleagues have discovered an immense void almost completely devoid of galaxies. Smaller voids have been found in other surveys of the heavens, but this one is too big to explain away in terms of random variations in galaxy distribution. Kirshner et al carefully measured galactic redshifts in three widely separated regions of the sky and found almost no galaxies in the redshift velocity interval 12,000 to 18,000 km/sec in all three areas. One interpretation of this huge gap is that the initial post-Big-Bang distribution of matter in the universe was unexpectedly lumpy. A further problem arising is that such a large void should show up as a blip in the 3°K cosmic background radiation---but it doesn't. (Anonymous; "Deep Redshift Survey of Galaxies Suggests Million MPC^3 Void," Physics Today, 35:17, January 1982.)

Comment. A less popular possibility is that galaxy redshifts do not measure distance at all and that no void exists. The above phenomenon is obviously similar to the quantized redshifts mentioned on p. 95.

THE FOAMY COSMOS

According to recent observations, there seem to be a lot of spherical voids in the cosmos. These vast nothingnesses are closed in by nearly spherical shells of galaxies. In other words, the cosmos consists of foam; homogeneous in the large, but bubbly in the small---if spherical shells of galaxies can be considered "small." The only way, say the theorists, that such a foamy structure can arise is for the universe to be closed, not flat as counts of galaxies seem to indicate. Thus, the foamy cosmos infers that we do not detect a lot of the mass existing in the universe. (Anonymous; "Honeycomb Universe Must Be Closed," New Scientist, 98: 540, 1983.)

SPACE SPUME

The celestial pot seems to have boiled over "in the beginning." New surveys of the galaxies suggest that they are mostly located on the surfaces of bubbles, not as we thought for so long distributed uniformly throughout the cosmos, the expanding debris of the Big Bang. If further surveys confirm a bubbly universe, the "conventional explanations for the evolution of large-scale structure in which gravity played a dominant role may have to be modified or abandoned."

To explain the bubbles, a new scenario has Big Bang #1 creating a population of uniformly distributed, extremely massive stars, which eventually burned out and exploded in a crescendo of supernovas. One stellar detonation stimulating adjacent giant stars to explode in a chain reaction. The bubble-like shock waves expanding outward from these explosions stimulated the condensations of the stars we now see in the heavens. Naturally, these stars and galaxies are concentrated on the surfaces of the shock wave bubbles. (Anonymous; "New 3-D Map Shows the Cosmos with a 'Bubble Bath' Appearance," Baltimore Sun, p. 3A, January 5, 1986.)

Comment. The space bubbles are mapped using redshifts as measurements of distance. As all-too-frequently asserted in this book, some redshifts may not be distance yardsticks, in which case these theoretical bubbles would burst. As the structures of the cosmos and the subatomic worlds become more and more foreign to everyday experience, we have to ask whether such bizarre constructions may not be the consequence of incorrect physical theories, such as Relativity, the Big Bang Hypothesis, and so on.

CLUMP OF ANTIMATTER

The clumpiness of the universe described above assumes that all observed matter is "ordinary" matter. Perhaps there are inhomogeneities on a different, more basic level---matter vs. antimatter.

According to one popular theory, the universe began with equal amounts of matter and antimatter. If so, where did all the antimatter go? We assume we observe a universe that is virtually 100% matter. Of course, we cannot really tell for certain because an antimatter galaxy would appear to us just like a galaxy composed of ordinary matter. The only clues revealing substantial pockets of antimatter would be the annihilation radiation produced where matter and antimatter regions rubbed against one another. The two types of matter always annihilate one another in bursts of very distinctive radiation.

Well, there seems to be at least one region of antimatter near the center of our galaxy. The HEOS3 satellite and balloon-borne instruments have pinpointed a source of 511-kev gamma rays that can come only from a spot where electrons and positrons are mutually annihilating each other. (The positrons are antimatter analogs of electrons.) This region of mutual destruction is about 10^{13} kilometers across. Is it a pocket of antimatter left over after the Big Bang that a sea of surrounding matter is finally wiping out, or is it newly created antimatter in the vicinity of a black hole? No one knows. The mystery has deepened with the discovery that the intensity of the annihilation radiation varies with time. Something strange is going on out there. (Anonymous; "Galactic Positronium Mystery Deepens," Science News, 130:40, 1986.)

AN ASTRONOMICAL PARADOX

Just a few years ago, most astonomers would have predicted that, as they examined larger and larger volumes of the universe, they would find more and more homogeneity. The Big Bang Theory predicts this; and it is seconded by the isotropy of the microwave background radiation. The mapping of the universe, however, has actually turned up all manner of galactic clusters, superclusters, and great skeins of superclusters strung across the heavens. Instead of a puree of matter, there is more and more structure the farther we peer into space.

R.B. Tully, at the University of Hawaii, now charts a billion-light-year structure that he calls the Pisces-Cetus complex. This aggregation of galaxies includes us (the Milky Way), our Local Supercluster, and many neighboring superclusters. In actuality, the Pisces-Cetus complex is not a continuous structure. Rather, it is defined by a plane---one containing a host of superclusters as well as voids. The problem posed for theorists is that they can suggest no way in which such a far flung manifestation of order could have evolved in the time available since the Big Bang. (Waldrop, M. Mitchell; "The Large-Scale Structure of the Universe Gets Larger--Maybe," Science, 238:804, 1987.)

ASTRONOMERS UP AGAINST THE "GREAT WALL"

> For more than a decade now, astronomers have been haunted by a sense that the universe is controlled by forces they don't understand. And now comes a striking confirmation: 'The Great Wall.'

The Great Wall is the largest known structure in the universe at present, having superceded sundry superclusters and clusters of superclusters. The Wall is a "thin" (15 million-light-year) sheet of galaxies 500 million light years long by 200 wide; and it may extend even farther. It is emplaced some 200-300 million light years from earth. It helps outline contiguous parts of vast "bubbles" of nearly empty space. Both the Wall and the adjacent voids are just too large for current theories to deal with. All popular theories have great difficulties in accounting for such large inhomogeneities. To illustrate, an important observable---the 2.7°K cosmic background radiation---which is usually described as the afterglow of the Big Bang, argues for a very smooth, uniform distribution of galaxies. Great Walls are definitely anomalous in this context.

M.J. Geller, codiscoverer of the Great Wall with J.P. Huchra, remarked:

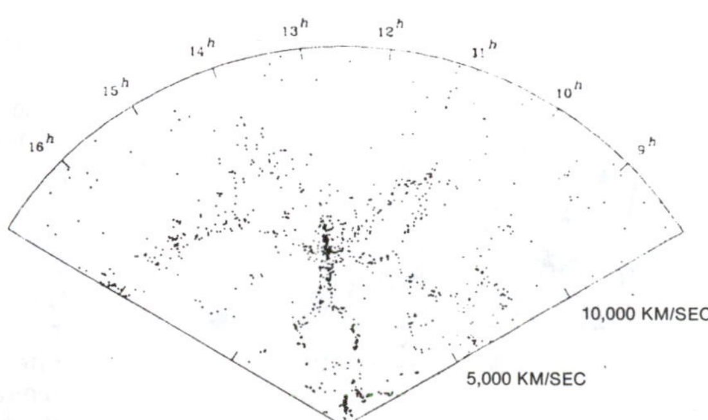

Cone diagram for 2,500 galaxies between declinations 26.5° and 44.5°. The Great Wall runs across the map. (M.J. Geller and J.P. Huchra)

> My view is that there is something fundamentally wrong in our approach to understanding such large-scale structure---some key piece of the puzzle that we're missing.

(Waldrop, M. Mitchell; "Astronomers Go Up against the Great Wall," Science, 246:885, 1989.) Also: Geller, Margaret J., and Huchra, John P.; "Mapping the Universe," Science, 246:897, 1989. And: McKenzie, A.; "Cosmic Cartographers Find 'Great Wall,'" Science News, 136:340, 1989.)

The discovery of the Great Wall of galaxies and the regular clumping of galactic matter has greatly surprised astronomers, who have been emphasizing how uniformly distributed galactic matter should be---according to theory, at least. Now, D.C. Koo, at the University of California at Santa Cruz, says, "The regularity is just mind-boggling." M. Davis, an astrophysicist at Berkeley, admits that if the distribution of galaxies is truly so regular, "...it is safe to say we understand less than zero about the early universe." (Wilford, John Noble; "Unexpected Order in Universe Confuses Scientists," Pittsburgh Post Gazette, May 28, 1990. Cr. E. Fegert)

MEGAWALLS ACROSS THE COSMOS

The Great Wall described above turned out to be only one of many similar cosmic structures, as remarked by N. Henbest:

> The universe is crossed by at least 13 vast 'walls' of galaxies, separated by about 420 million light years, according to a team of British and American researchers. The walls seem to be spaced in a very regular way that current theories of the origin of the universe cannot explain.

The astronomers have collected observations of galaxy redshifts along a linear "borehole" through the universe 7 billion light years long centered on the earth. If the redshifts are assumed to be measures of distance (as mainstream thinking demands), one gets the clumping effect seen in the accompanying illustration. (Henbest, Nigel; "Galaxies Form 'Megawalls' across Space," New Scientist, p. 37, March 17, 1990.)

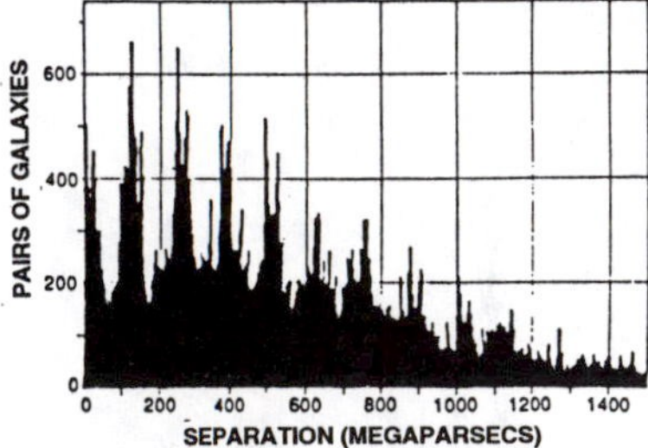

"Walls" of galaxies emerge when galaxy separation distance is plotted against the number of galaxies possessing specific separation distances.

Comment. Not mentioned in the above article are the papers by W.G. Tifft on quantized redshifts. (See page 95.)

It will be interesting to learn if "boreholes" pointed in other directions will encounter the same megawalls. If they do, the earth will be enclosed by shells of galaxies, much as some elliptical galaxies are surrounded by shells of stars. Wouldn't it be hilarious if the earth were at the center of these concentric shells? Some measurements of the universe's rotation also seem to imply geocentrism! (p. 105) It is more likely, however, that redshifts are just poor cosmological yardsticks. See our Catalog volume Stars, Galaxies, Cosmos for more on these subjects.

MYSTERIOUS BRIGHT ARCS MAY BE THE LARGEST OBJECTS IN THE UNIVERSE

Several brilliant bluish arcs, some 300,000 light years long, were unexpectedly discovered during a survey of galactic clusters. R. Lynds, of Kitt Peak National Observatory, estimates that the arcs are as luminous as 100 billion suns. The nice circularity of the arcs is perplexing; and it is stated that nothing like them has been reported before. The arcs might be incandescent gas, but many astronomers opt instead for swaths of bright young stars. Spectroscopic tests will decide this point.

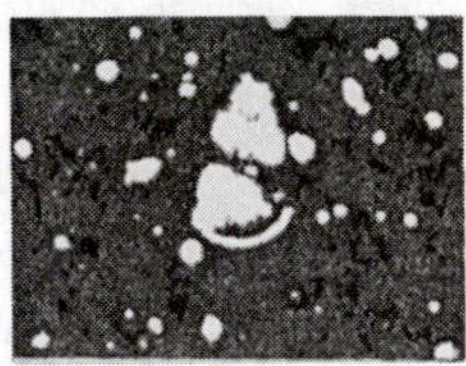

A giant blue arc located near the galaxy cluster 2242 - 02. (NOAO)

It has been difficult to conceive of an origin for the arcs. Are they blast waves or the results of tidal action between galaxies? No one knows, for all suggestions seem flawed. Something out there not only manipulates stupendous amounts of mass and energy but also does it with a draftsman's compass. (Anderson, Ian; "Astronomers Spot the Biggest Objects in the Universe," New Scientist, p. 23, January 15, 1987.)

Comment. In the interest of accuracy, it should be noted that some superclusters of galaxies are larger than the arcs. Also, some similar phenomena are described in our Catalog volume Stars, Galaxies, Cosmos, viz., the stacked, interleaved arcs of stars around elliptical galaxies (AWO5) and ring galaxies without significant nuclei (AWO6).

CELESTIAL MIRAGES?

Those mysterious giant blue arcs mentioned above have at least two explanations. The less-favored states that the arcs are formed when a spherical light pulse from a briefly flaring quasar encounters a plane of gas and dust. This matter scatters the light, making it visible to us as a huge arc. The preferred model employs gravitational lensing. Here, the arcs are simply distorted images---mirages, if you want---of distant galaxies. (Two of the many reports on this subject are: Waldrop, M. Mitchell; "The Giant Arcs Are Gravitational Mirages" Science, 238:1351, 1987; and Anonymous; "Giant Arcs: Light Echoes or Lensed Galaxies?" Sky and Telescope, 75:7, 1988.)

Comment. A bit of background: according to Einstein, the presence of matter can bend light rays, just as our atmosphere does when mirages are created by refraction. Gravitational mirages of celestial objects are thus predicted by Relativity.

BIG-BANG CHALLENGES

Anyone remotely cognizant of today's hierarchy of scientific hypotheses knows that the Big Bang hypothesis reigns supreme in astronomy. It is taught to students in schools and is treated as undeniable fact almost everywhere. Nevertheless, a few objectors do exist. Down the years, Science Frontiers has recorded many objections to the Big Bang and one of its supporting ideas, the Expanding Universe. These demurrers are presented below in chronological order.

Before launching into doubts concerning the Big Bang, it seems pertinent to mention how important the Big Bang idea is in modern scientific and general philosophical thought. It is almost universally believed that the universe (and life, too) had to have a beginning, and that that beginning was simple, a product of chance, and located at a point in space. The Big Bang may have begun as a "vacuum fluctuation"; life as a chance creation of prebiotic chemicals. Both matter and life then evolved, not purposefully but by random processes. The concepts of creation and subsequent evolution are ingrained in our thinking. For these reasons, it will be difficult to overturn the Big Bang and Expanding Universe paradigms, if indeed they need to be.

WAS THERE REALLY A BIG BANG?

Narlikar says, "Maybe not," and proceeds to tick off observational evidence against it. He begins, however, by pointing out the philosophical impasse encountered as Big Bang proponents look backward to time = 0 and earlier. Where did the matter/energy of the Big Bang come from? Was the venerable Law of Conservation of Mass/Energy violated? Big Bangers loftily dismiss such questions as "nonsense." Narlikar follows with some observational problems of the Big Bang:

1. There seem to be objects in the universe that are older than the Big Bang age of the universe (9-13 billion years);
2. Quasar redshifts used to support the Big Bang theory may not arise from the general expansion of the universe;
3. The microwave background radiation of 3°K, which was gleefully embraced by Big Bangers as an echo of their version of creation, is actually of the same energy density as starlight, cosmic rays, etc., and need not have anything to do with the Big Bang; and
4. The Big Bang Theory and General Relativity assume a constant G (the gravitational constant), but some recent lunar orbit measurements suggest that G is slowly decreasing!

(Narlikar, Jayant; "Was There a Big Bang?" New Scientist, 91:19, 1981.)

Comment. Perhaps the most disturbing aspect of the whole Big Bang business is the contempt with which theory supporters dismiss all objections. Is the Big Bang a scientific theory or part of a belief system?

DISTANT GALAXIES LOOK LIKE THOSE CLOSE-BY

Apropos the preceding item that the Expanding Universe Theory may be flawed, astronomers have discovered that supposedly distant galaxies look pretty much like those in our immediate neighborhood. Specifically, galaxies 10 billion light years away differ little spectrally speaking from those only a billion light years away. The point is that the distant galaxies should appear 9 billion years younger because their light took that long getting here. They look the same, and that fact could imply:

1. Galaxies mature rapidly and do not change much after a billion years;
2. Our cosmic time scale is all wrong;
3. There was no Big Bang and galaxies may have widely varying ages; or
4. None of the above.

(Anonymous; "Most Distant Galaxies: Surprisingly Mature," Science News, 119:148, 1981.)

A BUMP IN THE COSMIC BACKGROUND

The accepted explanation of the microwave cosmic background is that it is the "echo" of the Big Bang that created the cosmos as we now know it. Ideally, this background radiation should be uniform in all directions and follow the intensity curve of a black body radiating at 2.7°K. Spatial anomalies have already been reported, and now an embarrassing bump has been found on the intensity curve at 0.5-1.0 millimeters wavelength. No explanation for this departure from the black body curve has been provided except to say that the Big Bang deviated from the perfect uniformity. (Anonymous; "Cosmic Background Not So Perfect," New Scientist, 92:23, 1981.)

Update. In 1992, the cosmic microwave background was found to contain "fluctuations" that, according to some astronomers, proved once and forever the validity of the Big Bang. See p. 104. for an item on this development.

In 1993, very precise measurements of the microwave background showed it to have a spectrum almost exactly that of a black body; that is, no "bumps" such as mentioned above.

THE BETTER, BIGGER BIG BANG

Astronomers are ever more discomfitted by the Big Bang hypothesis for the creation of the universe. The reasons are several:

1. The observed universe is extremely homogeneous, even though theory says that distant parts of the universe could never have been causally connected;
2. No satisfactory explanation exists for the density fluctuations that had to occur for galaxies to be formed; and
3. The universe seems to be flat, not curved, and the Big Bang does not explain why.

Paul Steinhardt and Andreas Albrecht, at the University of Pennsylvania, have developed a radically different Big Bang---a two-stage one, with hot and super-cooled states. The three objections listed above are neatly disposed of in the new version, but at the cost of a radically new view of the cosmos. The "new" universe is about 10^{100} times as big as the 12 billion light years assigned to the cozy universe we used to know---and it is presumably correspondingly older.

This means that the portion of the cosmos we see is only a negligible fraction of the whole---a fraction that just happens to be homogeneous. Somewhere, way out beyond the farthest quasar, things could be---well---different! (Anonymous; "A Bigger, Better Big Bang," Astronomy, 11:62, February 1983.)

THE BIG BANG AS AN ILLUSION

That the universe began with the Big Bang is now so ingrained in our thinking that we almost never search for plausible alternatives. Perhaps the Big Bang is just a facade that diverts us from theories that better explain the observed characteristics of the universe. A trio of American and British astronomers (B.J. Carr, J.R. Bond, W.D. Arnett) are exploring the possibility that the cosmos began with a generation of very massive stars rather than the debris of the Big Bang. These huge stars would have had masses 100 or so times that of the sun. By virtue of the much higher pressures and temperatures at their cores, they would have burnt up their fuel inventories much faster than sun-sized stars. Thus, they would have burned themselves out long ago, probably surviving as black holes. Such an ancient generation of massive stars can explain four puzzling features of the universe:

1. The amount and character of the microwave background radiation.
2. The identify of the "missing mass" needed to hold the universe together.
3. The primordial abundance of helium.
4. The near-absence of heavy elements in the universe.

Although the success of this hypothesis is far from total, it may help wean us away from the Big Bang! (Maddox, John; "Alternatives to the Big Bang," Nature, 308:491, 1984.)

Comment. Note that, like the Big Bang itself, the generation of massive stars hypothesized above came from nowhere, like something that was pulled out of a magician's hat!

DOWN WITH THE BIG BANG

We might have concocted the above title, but we didn't! Rather, J. Maddox, the Editor of Nature, raised that red flag. To make things even worse, he headed his editorial with:

> Apart from being philosophically unacceptable, the Big Bang is an over-simple view of how the Universe began, and it is unlikely to survive the decade ahead.

His philosophical objcections to the Big Bang are powerful:

> For one thing, the implication is that there was an instant at which time literally began and, so, by extension, an instant before which there was no time. That in turn implies that even if the origin of the Universe may be successfully supposed to lie in the Big Bang, the origin of the Big Bang itself is not susceptible to discussion.

The Big Bang, Maddox says, is no more scientific than Biblical creation!

The scientific objections involve space, time, and the curvature of space. The Big Bang further fails at explaining quasars and the hidden mass of the Universe. Maddox doubts that the Big Bang will survive the new data to be provided by the Hubble telescope. (Maddox, John; "Down with the Big Bang," Nature, 340: 425, 1989.)

NATURE, HYPOTHESIS, AND THE BIG BANG

As noted above, J. Maddox, Editor of the preeminent journal Nature, seems intent on muffling the Big Bang. Now we see a newly added section in Nature bearing the heading Hypothesis. Hypothesis "is intended as an occasional vehicle for scientific papers that fail to win the full-throated approval of the referees to whom they have been sent, but which are nevertheless judged to be of sufficient importance to command the attention of readers..." Certainly, this is a commendable development. But not surprisingly, the first paper is an attack on the Big Bang.

Most of the authors of this first article are familiar to readers of Science Frontiers: H. Arp (Not all redshifts are measures of receding velocity.); G. Burbidge (Quasars are not as far away as they seem.); and F. Hoyle (The multidisciplinary iconoclast who helped develop the Steady State theory of the universe.) None of these scientists has recanted, even in face of not-so-subtle pressures to conform.

The first paper in Hypothesis. Arp et al summarize in two sentences:

> We discuss evidence to show that the generally accepted view of the Big Bang model for the origin of the Universe is unsatisfactory. We suggest an alternative model that satisfies the constraints better.

Most of the paper sets out observational evidence for the authors' main themes, as stated parenthetically above following their names. Space is also devoted to the contention that the vaunted "proofs" of the Big Bang are really not. Since these themes have appeared repeatedly in Science Frontiers, we will bypass details here. The paper concludes with suggestions for an alternative to the Big Bang, which is based upon multiple creation events---thousands of them, each on the scale of superclusters of galaxies! (Arp, H.C., et al; "The Extragalactic Universe: an Alternative View," Nature, 346:807, 1990.)

Comment. There are two ironies:

Irony #1. J. Maddox, Nature's editor, while trying to encourage alternatives to the Big Bang on one hand, has been most fierce in suppressing Benveniste's infinite-dilution research and cold fusion, although perhaps with some justification in these instances.

Irony #2. In their conclusion, Arp et al remark: "Geology progressed favourably from the time Hutton's principle of uniformity was adopted, according to which everything in geology is to be explained by observable ongoing processes." They then suggest that cosmology and cosmogony might well adopt such an outlook! Hasn't geology actually been imprisoned by Hutton's uniformitarianism? (See p. 217.) Certainly the large meteor/asteroid/comet impacts now strongly favored as a major mechanism of biological extinction events in the fossil record are a far cry from uniformitarianism.

OF IRON WHISKERS AND PARTICLES THAT INCREASE MASS WITH AGE!

Two pillars of the Big Bang hypothesis are: (1) Redshifts of galaxies support the notion of an expanding universe; and (2) The background microwave radiation can be interpreted as the dying embers of the Big Bang itself. Proponents of the Big Bang feel secure atop these pillars. But should they?

A few Big-Bang skeptics, who have survived considerable establishment pressure, see growing cracks in those pillars. J. Narlikar identified two such cracks and, best of all, offered exciting remedies: (1) The redshift relationship, which works well with galaxies, falls apart when applied to quasars (see graphs); and (2) The background microwave radiation is much too smooth to come from the lumpy universe we observe. Narlikar opines as follows:

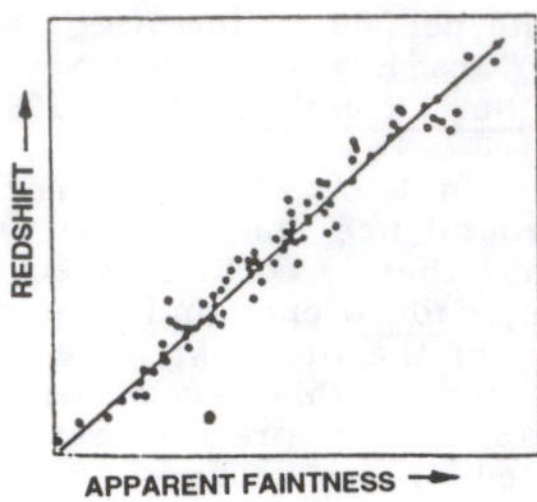

Plot of red shift vs galaxy faintness supports the proposition that red shift is proportional to distance.

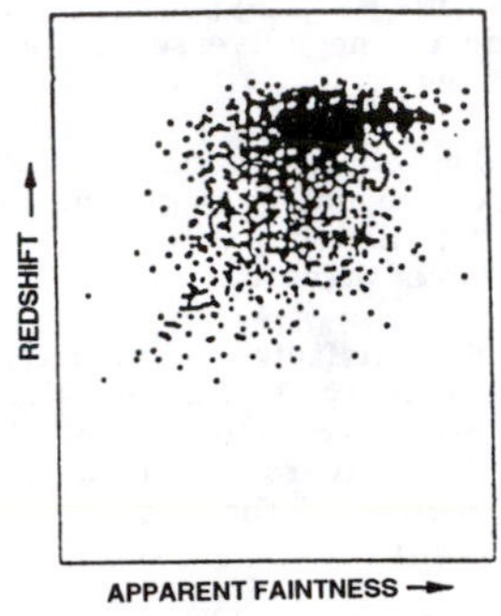

The same plot for quasars produces a scatter of points, suggesting thet here red shifts have nothing to do with distance.

> Given these problems, it is not a sound strategy to put all of our cosmic eggs in one big-bang basket. Rather, we should explore other possibilities. Thirty years ago, there was a more open debate on alternative theories, which made valuable contributions to our understanding of cosmology. For a healthy growth of the subject, the big bang hypothesis needs competition from other ideas.

Some of Narlikar's suggestions are heresies of the first order:

1. Particles are born with zero mass and acquire mass only as they interact with other matter in the universe (consistent with Mach's principle). The radiation from the lighter particles making up young quasars would thus be red-shifted when compared to radiation from old galaxies.
2. The background microwave radiation from a lumpy universe could easily be smoothed out by tiny (1 mm) iron whiskers condensing in the expanding envelopes of supernovae. (Narlikar, Jayant; "What If the Big Bang Didn't Happen?" New Scientist, p.48, March 2, 1991.)

Comment. These ideas are almost as audacious as the suggestion that continents can actually drift apart! But are they enough to lift the intellectual pall created by a hypothesis-enshrined-as-fact? Probably not!

BIG-BANG BROUHAHA

Unless you have been in a coma the past couple months, you have heard that the Big Bang has now been elevated from a theory to a fact. The reason for the media hullabaloo was the announcement that minute fluctuations had been detected in the cosmic microwave background. The media hype was notably chauvinistic. Some Big-Bang proponents declared that discovery was the greatest scientific advance of the century, completely ignoring the genetic code, continental drift, nuclear fission, and so on and so on.

More sober scientists rejected such extravagant claims. They pointed out that independent confirmation of the fluctuations was yet to come and that, after all, the fluctuations were very small (only some 30 millionths of a °K). And which of the many variations of Big Bang was going to be enthroned? Even Nature advised extreme caution, quoting H. Bondi in this regard:

> ...the data in cosmology are so likely to be wrong that I propose to ignore them.

(Anonymous; "Big Bang Brouhaha," Nature, 356:731, 1992.)

COSMIC SPECULATIONS

Speculations about the real nature of the universe undoubtedly began early in hominid history. Today, astronomers are still speculating, even though they know considerably more about the cosmos. Instead of attributing the sun's motion across the sky to Apollo and his chariot, we now wonder if the universe might not be held together with "cosmic string." Obviously, our speculations have become a bit more abstruse! Scientists also ask if quantum theory also applies to stars and galaxies, or whether chaos actually rules the macroscopic universe. Supposing that the uni-

verse really is expanding, has it always been so, or have there been cycles of expansion and contraction? These and topics like them are covered in this section. Let us begin with the age-old speculation that humans actually occupy a favored place in the cosmos. Of course, modern cosmologists struggle to avoid such anthropomorphic interpretations, although there are some intriguing observations in this vein!

THE COSMIC WHIRL

From the analysis of position angles and polarizations of some radio stars, it seems that the entire cosmos is rotating with an angular velocity of about 10^{-13} radians per year. The Big Bang scenario allows for uniform, universal expansion but certainly no general rotation. P. Birch, the author of this article, puts his finger on the problem in his abstract:

> This would have drastic cosmological consequences, since it would violate Mach's principle and the widely held assumption of large-scale isotropy."

(Birch, P.; "Is the Universe Rotating?" Nature, 298:451, 1982.)

Comment. Since Birch's indicators of rotation are positive in one half of the sky and negative in the other, are we really seeing an entire universe rotating about the earth (and humankind) as a center? What an anti-Copernican thought; we are the focus of everything after all!

THE SPIN WE'RE IN

From planets to stars to entire galaxies, the rate of spin is related to mass. If one defines Q (see graph) as angular momentum divided by the object's mass and a density factor, it is roughly equal to the object's $(\text{mass})^{0.7}$. Such empirical equations are generally accurate only over a limited region, but this newly discovered relationship holds for almost 50 orders of magnitude. For some unfathomed reason, all astronomical entities spin at rates determined by their masses. Even clusters of galaxies fall into step, slowly pirouetting in intergalactic space to some unknown tune. (Anonymous; "How Things Spin," Sky and Telescope, 64:228, 1982.)

Comment. Above, under the title "The Cosmic Whirl," a team of investigators reported that the entire cosmos may spin. Does the whole cosmos, too, obey this new momentum-mass formula?

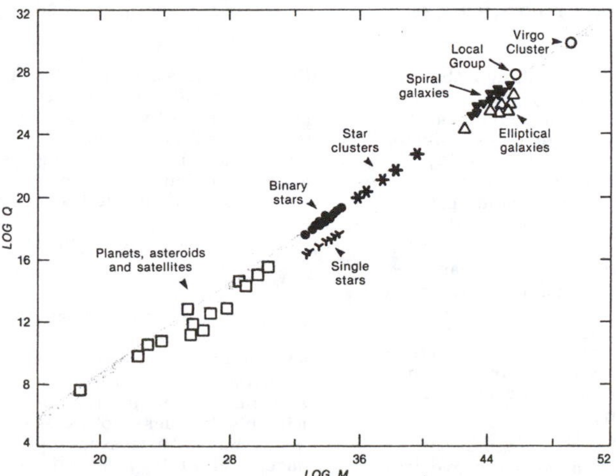

A possible universal relationship between the angular momentum and mass of astronomical objects.

DOES STRING HOLD THE UNIVERSE TOGETHER?

Cosmological speculation is getting more and more bizarre. Astronomers are now postulating a kind of cosmic 'string' that is very, very thin (10^{-30} cm), enormously massive (10^{22} grams per centimeter), and very taut (10^{42} dynes tension). This string exists only in closed loops of infinite strands. Such string in loop form could have seeded galaxies and even black holes of solar mass. But these are not the major reasons why astronomers like the string hypothesis. It turns out that this bizarre string can tie the universe together gravitationally; that is, provide the long-sought 'missing mass.'

The so-called 'missing-mass problem' is two-fold:

1. Astronomers cannot see, with eye and instrument, enough mass to keep the universe from expanding indefinitely. If the kinetic energy of cosmic expansion is to be balanced by gravitational potential energy (an apparent philosophical imperative), we have so far identified only 15% of the required mass.

2. On a smaller scale, galaxies in large galactic clusters are moving too fast. They should have flown apart long ago, but some unseen 'stuff' holds them together. Is it cosmic string? (Waldrop, M. Mitchell; "New Light on Dark Matter? Science, 224:971, 1984.)

Comment. Since cosmic string weighs about 2 x 10^{15} tons per inch, the whole business is beginning to sound a bit silly. Actually, all action-at-a-distance forces, which we readily accept as real, are only artificial constructs of the human mind. Gluons, colored 'particles,' top quarks, cosmic string; where will it all end?

THOU CANST NOT STIR A FLOWER, WITHOUT TROUBLING OF A STAR

This poetic title from Francis Thompson tries to express the unity of nature from the smallest to the largest realms. One characteristic of the realms even smaller than that of the flower is the quantization typical of the subatomic world---that is, microscopic nature. At the human locus in the dimensional scheme of things, quantization is difficult to detect outside the physics laboratory.

Daniel M. Greenberger, perhaps with the above title in mind, asked whether quantization might not also exist in astronomy and cosmology---that is, macroscopic nature. He has applied the principles of quantum mechanics to nature in-the-large where gravitational forces are dominant. (Gravitational forces are negligible in the subatomic world.) His math cannot be reproduced here. Suffice it to say that Greenberger has applied his findings to the absorption lines of quasars and the elliptical rings surrounding normal galaxies. Now, quasars and galaxies are far from atomic nuclei, being vast assemblages of diverse matter. Somewhat surprisingly, his equations are successful in predicting some features of these two macroscopic entities. (Greenberger, Daniel M.; "Quantization in the Large," Foundations of Physics, 13:903, 1983.)

Comment. At the very least it is mind-stretching to find that complex systems with millions of stars may exhibit quantum effects. With some relief, we note that like microscopic quantization effects, the consequences of macroscopic quantization will be hard to discern in our comfortable "smooth" world.

ORDER FROM DISORDER?

The apparent creation of order in a universe that is supposed to be running down is the subject of a thoughtful report by Wallace and Karen Tucker. They give additional examples of order arising from disorder in astronomy (spiral galaxies, superclusters), from chemistry (Belousav-Zhabotinsky reactions), and biology (embryonic development). Again and again, they ask why order persists on increasing when the Second Law of Thermodynamics seems to demand more disorder.

Throughout the article, the Tuckers employ the card-shuffling analogy. If nature is shuffling the cards of the universe, why are so many royal flushes being dealt? They introduce the works of Ilya Prigogine and others which focus on chemical situations (the Belousov-Zhabotinsky reactions), where reaction by-products actually make the reaction more likely and in which large, stable spiral and ring-shaped structures appear spontaneously. At the macroscopic level, shock waves from supernovas can (at least in computer models) stimulate the formation of spiral arms in galaxies.

The article concludes with a quote from astronomer David Layzer, "The universe is unfolding in time but not unraveling; on the contrary, it is becoming constantly more complex and richer in formation." (Tucker, Wallace, and Tucker, Karen; "Against All Odds: Matter and Evolution in the Universe," Astronomy, September 1984.)

Comment. Now Layzer's statement seems a clear denial of the Second Law. It says, rather, that there is something intrinsic in the universe that creates order (spiral galaxies, amino acids, humans). We don't explain this tendency (assuming it really exists) just by identifying Belousov-Zhabotinsky reactions and saying, "That's the way things are." We can go one step further and say that the fundamental particles of physics have just the right properties so that they fall together into atoms, molecules, galaxies, and life forms. Can science go any further?

ASTRONOMERS COPE WITH BOTH CHAOS AND TOO MUCH ORDER IN THE UNIVERSE

The solar system. The recent advent and fast rise in popularity of chaos theory is destroying some favorite, long-sworn-to notions of astronomers. One in particular is solar-system stability. Could any of the planets pop out of their orbits and embark upon wild and unpredictable trajectories? "We can't rule it out," stated J. Wisdom, an MIT planetary scientist. (Freedman, David H.; "Gravity's Revenge," Discover, 11:54, May 1990.)

Comment. Well, OK, there's a tiny theoretical chance such an event might occur in the future, but it certainly never happened in the past. To admit such a possibility would open that Pandora's box of vigorously suppressed catastrophic scenarios.

THAT'S THE WAY THE UNIVERSE BOUNCES

What follows is a chain of ideas (perhaps "speculations" is a better word) that was recently unleashed by L. Smolin in the journal Classical and Quantum Gravity (9:173). At stake here is the very nature of Nature herself.

We begin with the notion of anthropic cosmology, in which the physical constants of the universe are identified as having just the "right" values to allow the existence of stars, planets, carbon compounds, and the other ingredients of human life. (Just why this state of affairs prevails is a question rarely addressed!) Adherents of anthropic cosmology hold that our "human-friendly" universe is just one of many universes populating a larger metauniverse. These "other" universes are thought to have different values of the fundamental physical constants (viz., the mass of the proton) and, in consequence, wildly different forms of life. In nonhuman universes, there could even be entities for which our word "life" is inadequate.

The second idea is that of an oscillating universe. In this concept, universes expand just so far and then collapse back into the "singularities" (i.e., black holes) from which they arose. Then, Phoenix-like, they bounce back and reexpand into new universes---ones with slightly different physical constants. These rebounding universes are in a sense mutated universes, which have been slightly modified during the physical trauma of collapsing into singularities.

Now comes a stimulating thought. The most abundant sort of universe occupying the metauniverse will be that type that generates the most new black holes during its expansion and contraction phases, for each of its "progeny" can spawn a new universe of its own. As in biological Darwinism, these are the "selected" universes. Some universes may fail to reproduce at all. Thus, with the help of small mutations occurring during each bounce, the metauniverse and its constituent universes are evolving like biological life---but towards what? (Gribbin, John; "Evolution of the Universe by Natural Selection?" New Scientise, p. 22, February 1, 1992.)

Comments. There do seem to a few black holes in our own universe (the Milky Way), perhaps many of them. So, universes like ours could well be highly successful in the cauldron of cosmological evolution. Since life and humans are possible in our type of universe, humanity seems to be favored not only by the evolutionary forces existing on earth but also by those permeating the metauniverse. What we do not know is: (1) Whether other life forms (or entities) are even more favored in their universes; (2) Whether meta-metauniverses exist; and (3) Who or what is "Master of the Show."

Whoever or whatever evidently plies its trade at the molecular level, too. See "Mutant Molecules" on p. 321.

ALL THINGS APPEAR TO THOSE WHO ACCELERATE

We have said some nasty things about WIMPS (Weakly Interacting Massive Particles). But perhaps the existence of WIMPS is only a matter of one's private trajectory in the cosmos! Recently, theorists fiddling with the possible consequences of quantum mechanics came up with some shockers:

1. An accelerating observer can detect particles that a stationary observer will swear do not exist;
2. An accelerating reflecting surface (theoretically) creates a flux of particles that stream away from its surface; and
3. An accelerating mirror can carry away negative energy!

After cogitating on such discoveries, Paul Davies wondered, "Might it even be that the apparently solid matter of the Universe around us is only the consequence of our particular motion?" Particles of matter, in this theoretical scheme, would be only chimeras of our individual motions. (Davies, Paul; "Do Particles Really Exist?" New Scientist, p. 40, May 2, 1985.)

Comment. Omar Khayyam practically predicted this curious state of affairs when he wrote: "We are no other than a moving row/Of magic shadow-shapes that come and go."

THE DEFLATIONARY UNIVERSE

One of our major astronomical targets in Science Frontiers has been the cosmological redshift; that is, the assumption that an object's redshift is entirely a Doppler effect and, when coupled to the expanding universe concept, is proportional to distance. Well, we don't have any more contradicting data (of which there is plenty), but we do have:

1. A new theory which shows how noncosmological redshifts can occur; and
2. Laboratory demonstrations of "spectral noninvariance" that show how a non-Doppler component can be added to light's redshift.

The physicist behind this new research is E. Wolf, at the University of Rochester. His theoretical work was reported in the March 31, 1986, issue of Physical Review Letters. There he showed how quasars and so-called "superluminary" astronomical sources might emit light with a spectrum that evolves as it travels through space. Scientists have always assumed that once light left its source its spectrum remained unchanged. But Wolf shows how spectral changes are "sort of coded into the light due to correlations in the source." Meanwhile, two of Wolf's colleagues have backed up his theory in the lab.

The consequences of Wolf's work would in effect shrink the universe, because objects would not be as far away as we now calculate from their redshifts. The size of the universe might contract "by a factor of 100 or more," says Wolf. If this much deflation is accepted by other scientists (It could be quite a fight!), then the age of the universe will also shrink, since it is based in part on our observations of the outer fringe of the universe and the speed of light. (Amato, I.; "Spectral Variations on a Universal Theme,: Science News, 130:166, 1986.)

Comment. If we divide the currently accepted age of the universe, about 15 billion years, by 100, we are left with only 150 million years. But the radioactive clocks of the geologists register about 5 billion for the earth. There seems to be a problem somewhere!

AT LAST, A THEORY OF EVERYTHING!

"A Soviet astrophysicist has made the startling claim that the Earth and other astronomical bodies may be riddled with mini black holes---objects smaller than atoms but with masses which, in some cases, might be as great as a planet. Such objects, he claims, could account for volcanic hot spots, gravitational anomalies, concentrations of mass on the Moon (mascons), the existence of the rings of Saturn, and even the observations that gave rise to the notion of a 'fifth force.'"

J. Gribbin, whose article begins with the above paragraph, is quick to proclaim that this "theory of everything" is not just silly-season kite flying. Rather, it was proposed by A.P. Trofimenko in the well-respected Astrophysics and Space Science (168:277)

Restricting ourselves to speculations concerning the earth, Trofimenko sees our planet as a sphere of low-density material enclosing 126 mini black holes that account, first, for the many gravity anomalies we measure on the surface; and, second, the earth's high density. That's right, there's no iron core in this model! Some of the mini black holes near the surface create local hot spots (plumes, volcanos, etc.) through the emission of Hawking radiation. Trofimenko's scheme encompasses the planets, the stars, and, as advertised, "everything." (Gribbin, John; "Could Mini Black Holes Provide a 'Theory of Everything?'" New Scientist, p. 25, September 1, 1990.)

Chapter 3

BIOLOGY

HUMANS

OTHER MAMMALS

BIRDS

REPTILES AND AMPHIBIANS

FISH

ARTHROPODS

INVERTEBRATES

PLANTS AND FUNGI

MICROORGANISMS

GENETICS

ORIGIN OF LIFE

EVOLUTION

HUMANS

EXTERNAL APPEARANCE

The most obvious characteristics of living organisms are size, color, shape, limbs, etc. It is understandable, therefore, that most sections in this chapter commence with "external appearance phenomena" and their potential anomaly and curiosity value. Since the first 86 issues of Science Frontiers could not possibly encompass all appearance phenomena, the reader is urged to consult the Biological Anomalies catalog volumes for more thorough coverage.

The aspects of human appearance found in Science Frontiers and assembled below are confined to hair anomalies and curiosities, bipedalism, and luminous phenomena. (Compare this impoverished list to the 56 categories of appearance phenomena treated in Biological Anomalies: Humans I.) However, even this short list calls into question the still-promoted concept of recapitulation in the embryo, the Aquatic Ape Hypothesis, and so-called "initial bipedalism."

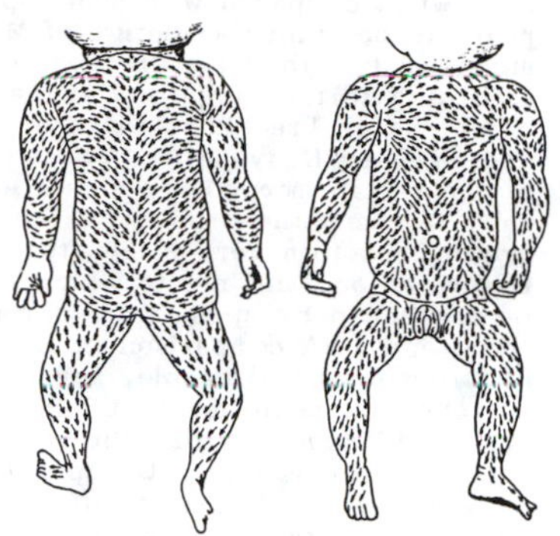

Hair tracts on the human fetus seem to indicate water-flow patterns. This phenomenon is often used to support the Aquatic Ape Hypothesis.

A HAIR-RAISING PHENOMENON

The Creation Research Society Quarterly often touches on subjects avoided by the establishment scientific journals. In the latest issue, a medical doctor reviews the hoary ontogeny-recapitulates-phylogeny hypothesis. As classically stated by evolutionists, the human embryo passes through stages in which it looks like creatures that preceded it in evolution. The doctor, G.R. Culp, remarks that although evolutionists maintain that reputable scientists no longer employ this argument as evidence for evolution, the "recapitulation" claim is still being made in some classrooms and even during some of the recent creationist-evolutionist debates.

Culp then shows why the recapitulation claim failed in five stages in the development of the human embryo. A rich lode of anomalies exists here: cells migrate purposefully, mysterious structures grow and then disappear; it is a kaleidoscope of changing structures and processes. Take, for example, the "hair stage":

> The embryo is covered with very fine hair at about the seventh month of development of the embryo. The evolutionist claims that this is evidence that men came from hairy mammals like the apes. However, these hairs are unlike the hair found on apes, as they are very small in diameter and always soft and unpigmented. This hair disappears from the body soon after birth. It is called lanugo and is quite unlike the permanent hair that grows on the human body and head...

To the anomalist, the battle between the evolutionists and creationists is secondary to the anomalies that keep cropping up without satisfactory explanations from either side. Here, we ask the purpose of the lanugo. Does it have to have a purpose? Just being there suggests purpose. Why does it disappear? (Culp, G. Richard; "Embryology---Overlooked Facts You Should Know," Creation Research Society Quarterly, 25:100, 1988.)

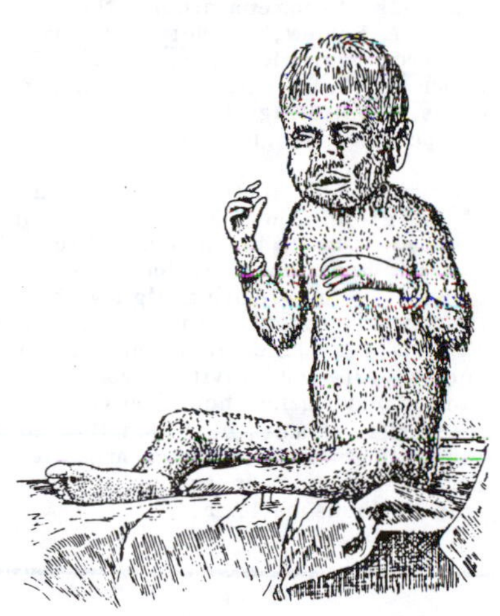

In some humans, such as the "hairy child" sketched above, the lanugo or natal hair persists beyond the womb.

Comment. Lanugo appears on other animals, too; sometimes only in the embryo, more rarely after birth.

OUR AQUATIC PHASE!

Elaine Morgan, author of The Aquatic Ape, reviews new evidence supporting the Aquatic Ape Hypothesis. Sir Alister Hardy suggested this hypothesis in 1960 in an attempt to account for several human characteristics that are unique among primates but common in aquatic mammals. Some of these are: position of fetal hair, loss of body hair, subcutaneous fat, face-to-tace copulation, weeping, etc. The combination of hairlessness and subcutaneous fat seems almost totally confined to aquatic mammals and humans.

This article also deals with certain skeletal features of early man that seem to indicate an aquatic stage. For example, the Lucy skeleton has a shoulder joint that indicates that Lucy spent a lot of time with her hands above her head, as she would have if aquatic or tree-swinging! (Morgan, Elaine; "The Aquatic Hypothesis," New Scientist, p. 11, April 12, 1984.) See p. 122 for additional observations from this article.

Comment. The Aquatic Ape Hypothesis states, in essence, that humans went through an aquatic phase in their development during which evolution favored us with some characteristics common in marine mammals. Needless to say, this hypothesis is not entrained in the mainstream of scientific belief!

DID WE LEARN TO SWIM BEFORE WE LEARNED TO WALK?

This item adds another facet to the Aquatic Ape Hypothesis:

> Humans may have first walked upright because they had to carry the baby---not because it was born less developed than other primates, but because its parents were ex-aquatic apes.

(Morgan, Elaine; "Lucy's Child," New Scientist, p. 13, December 25, 1986.)

INITIAL BIPEDALISM!

Our anomaly-collecting net has pulled in an interesting catch; namely, CERBI (Centre d'Etudes et de Recherches sur la Bipedie Initiale). This Center, operated by F. de Sarre, publishes a little journal called Bipedia. In the first issue of Bipedia, de Sarre set out, in English, his basic thesis:

> Abstract. The explanation of Man's special nature is to be sought in the original combination formed by a primordial brain, the globular form of the skull and initial bipedalism. The ape, when compared with Man, appears to be rather a vestige of Man's ancestral line than his predecessor, according to the views of Max Westenhofer, Serge Frechkop, Klaas de Snoo and Bernard Heuvelmans. The study of the human morphology allows logically to carry the problem of Man's origin back to a very early stage of the evolution, and not to which has been reached by apes. From chromosomal and DNA comparison in the cells of living apes and people, several researches argue to-day that humans are genetically more like the common ancestor than is either Chimpanzees or other apes. The array of facts and considerations should be sufficient for an unbiased mind to discount away any idea of simian antecedents in Man's ascent.

The body of the article supports de Sarre's thesis with observations from embryogenesis, comparative anatomy (skull, hand, foot), and phylogenesis. (de Sarre, Francois; "Initial Bipedalism: An Inquiry into Zoological Evidence," Bipedia, 1:3, September 1988.)

Comment. Obviously, de Sarre is taking an extreme position, and any observations supporting his position are anomalous by definition.

OBSERVATIONS OF LUMINOUS PHENOMENA AROUND THE HUMAN BODY

In front of us is a 13-page paper dealing at length with phenomena normally considered impossible. The paper's abstract is not as informative as the introductory paragraph, which we now quote:

> Mysterious and unexplained luminous phenomena have fascinated humankind since ancient times. The psychical research literature offers a variety of unexplained luminous phenomena such as reports of glows seen around magnets, crystals and minerals, lights reportedly seen with mediums, and luminous apparitions, among others.

The paper then goes directly into a review of specific observations. We shall have to be satisfied here with just one of the many examples:

> D.D. Home was reported to produce many striking luminous phenomena. In one seance Dunraven observed that one of Home's hands 'became quite luminous,' and that two persons 'saw tongues or jets of flame proceeding from Home's head.' On another occasion: 'He was elongated slightly.. and raised in the air, his head became quite luminous at the top, giving the appearance of having a halo round it. When he was raised in the air, he waved his arms about, and in each hand there came a little globe of fire (to my eyes blue)...

(Alvarado, Carlos S.; "Observations of Luminous Phenomena around the Human Body: A Review," Society for Psychical Research Journal, 54:38, 1987.) This paper concludes with 5 pages of references, illustrating the great extent of the parapsychological literature.

Comment. D.D. Home, to provide a bit of background, was a famous English medium. One of his favorite "stunts" was self-levitation. We have seen sketches of him floating high with his head nearly touching the ceiling. It is perhaps a bit snide to remark that if one can conquer gravity, producing luminous phenomena should be easy!

Finally, it must be added that some human luminous phenomena are scientifically credible, as in wounds infected with luminous bacteria.

BEHAVIOR

Human behavior is another key observable when searching out the anomalies and curiosities of the human race. We have here collected a fascinating group of phenomena:

- Twin phenomena. The remarkably similar behavior of identical twins reared apart.
- Handedness phenomena. The difficulties and life expectancy of lefties in a right-handed world.
- Behavior and eye color. Baseball skills and eye color---with a bit of tongue in cheek!
- Behavior and photoperiod. The length of the day seems to affect the levels of human hostility.
- Behavior and solar activity. The correlation of solar activity with human aggressiveness and cultural activity.
- Behavior and astronomy. The purported effect of the moon and stars upon human behavior, especially the so-called "Mars effect."
- Genes and culture.

TWIN PHENOMENA

UNITED BY AN INVISIBLE CORD

The largest study of separately raised identical twins has discovered incredible similarities among twins who never set eyes on each other before. There are differences, too, but not as many as expected by psychologists who hold that we are shaped primarily by our environments. Only a few of the astounding (and strange) similarities can be recounted here.

Two 39-year-old twins, meeting for the first time, were both wearing seven rings each, two bracelets on one wrist, a watch and one bracelet on the other wrist.

Two men both had dogs named Toy, had married and divorced women named Linda, remarried women named Betty, and named their sons James Allan and James Alan.

Two other males had similar medical histories: hemorrhoids, same pulse rates and blood pressures, same sleep patterns. Both had put on 10 pounds at the same times in their lives.

(Holden, Constance; "Identical Twins Reared Apart," Science, 207:323, 1980.)

THE MOST IDENTICAL OF IDENTICAL TWINS

In the foregoing item, some remarkable similarities between identical twins reared apart were recounted. A truly fantastic case has now come to light where identical twins (reared together in this case) behave synchronously. "They do everything together, scream or sulk if parted and, most uncannily, talk in unison when under stress, speaking the same words in identical voice patterns that create a weird echo effect." Doctors say that Greta and Freda Chaplin are so close that they seem linked by telepathy. Talking or working, they function in unison. Otherwise, they are of normal intelligence and suffer no mental illness. (Anonymous; "British Twins Too Close for Trucker's Comfort," Baltimore Sun, p. A3, December 8, 1980.)

Comment. Animals often move in remarkable synchrony; e.g., flocks of wheeling birds, schooling fish, tropical fireflies, and even human chorus lines. The "invisible cords" controlling these synchronies are thought to be the faculties of sight, sound, and, in the case of fish, electrical fields.

WHEN IDENTICAL TWINS ARE NOT IDENTICAL

Past studies of identical twins separated at birth have documented remarkable similarities between them, despite the fact that they were reared under radically different circumstances. Their physical appearances, habits, vocations, health histories, and other factors are often eerily the same. For example, as mentioned above, the two female identical twins, who had never seen each other before, each wore seven rings! The upshot of such investigations is that most of a person's characteristics are genetic in origin; that is, nature dominates nurture.

But what about identical twins who are remarkably different? They can, for instance, differ appreciably in size, intellect, and behavior. In such cases, does nurture dominate nature? No! Identical twins may diverge even in the womb, where one may receive more oxygen and nutrients than the other. One also may be assailed in utero by viruses, bacteria, or drugs, while the other escapes. Even more drastic is the possibility that one twin may pick up an extra chromosome soon after the original egg has split. Also, mutations may doom one twin to Down's syndrome or some other genetic affliction, while the other is unscathed. Identical twins may even be of different sex! Of course, such twins are genetically different, but they are still monozygotic (from the same egg). Blood tests will show them to be identical.

It used to be thought that the small differences that did exist between identical twins separated at birth were surely due to nurture, not nature. But, considering all the differences that can accrue in utero, it seems that the role of nurture in shaping individuals is much smaller than thought, possibly negligible. (Horgan, John; "Double Trouble," Scientific American, 263:25, December 1990.)

HANDEDNESS PHENOMENA

WHY MOST PEOPLE ARE RIGHT-HANDED

Depending upon which estimate you believe, 6 to 16% of us are left-handed. On the surface, left-handers seem little different from right-handers. So why the highly skewed population? Why not 50% of each? Dr. Peter Irwin, at Sandoz, Ltd., of Basle, Switzerland, has what is certain to be a controversial answer. He was led to his conclusion by a series of experiments with psychoactive drugs, performed in collaboration with Professor Max Fink, SUNY, Stony Brook. Left-handers, the study demonstrated, are much more sensitive to drugs that act upon the central nervous system. Irwin believes that this finding is consistent with the known association of left-handedness with epilepsy and learning disorders.

> But perhaps the most exciting aspect of Irwin's hypothesis is that it makes evolutionary sense. 'A greater resistance of right-handers to centrally active substances, when Man was a forager and before he learned to identify non-toxic edibles, would have favoured right-handed survival. This might account for the skew in the present handedness distribution that is unique to humans.'

And why should left-handers be more sensitive to psychoactive substances? Irwin thinks they must absorb or metabolize them differently, or perhaps there is a difference in the blood-brain barrier that affects the transport of substances into the brain. (Grist, Liz; "Why Most People Are Right-Handed," New Scientist, 22, August 16, 1984.)

DO RIGHT-HANDERS LIVE LONGER?

> Several findings suggest that left-handedness may be associated with reduced longevity. For instance, Porac and Coren reported that 13 per cent of 20-year-olds are left-handed but only 5 per cent of those in their fifties and virtually nobody of 80 or above. We believe that this absence of left-handers from the oldest age groups reflects higher biological and environmental risk.
>
> To investigate this asymmetry further, D.F. Halpern and S. Coren repaired to The Baseball Encyclopedia, where longevity and handedness are duly recorded for many players. Here again, they found that, although mortality is about the same up to age 33, thereafter about 2% more right-handers than left-handers survive at each age.
>
> Halpern and Coren suggest a few possible causes: (1) Prenatal and perinatal birth stressors are more probable in left-handers; (2) The immune systems of lefthanders may be reduced by genetic effects and intra-uterine hormones; and (3) Left-handers may suffer more accidents in a world designed for right-handers! (Halpern, Diane F., and Coren, Stanley; "Do Right-Handers Live Longer?" Nature, 333:213, 1988.)

GAME OF LIFE FAVORS RIGHT-HANDERS

With condolences to our southpaw readers, things do not look too good for them. Take accidents for example:

> Left-handed people are almost twice as likely to suffer a serious accident as right-handers, according to a recent study. Stanley Coren, an experimental psychologist at the University of British Columbia in Vancouver, Canada, claims that his finding helps to explain why less than 1 per cent of all 80-year-olds are 'southpaws,' whereas they comprise nearly 13 per cent of all people aged 20.

Coren surveyed students at his University for four years and found that the probability of a left-hander having a car accident was 85% higher; accidents with tools were 54% higher; home accidents were 49% higher; etc.

Coren blames these lopsided statistics on the fact that the world is ordered for right-handers, not that left-handers are innately more clumsy. (Dayton, Leigh; "The Perils of Living in a Right-Handed World," New Scientist, p. 32, October 28, 1989.)

But Coren's study, above, omits the "health" factor, which we now supply from a different source.

> Summary. ---Halpern and Coren recently described an association between left-handedness and a lower life expectancy. This finding is not unexpected because left-handedness has been linked to three leading causes of death in our society---alcoholism, smoking, and breast cancer---as well as to several neurological and immune disorders.

(London, Wayne P.; "Left-Handedness and Life Expectancy," Perceptual and Motor Skills, 68: 1040, 1989.)

Comment. Later studies have brought happier tidings to southpaws, calling into question the above conclusions.

BEHAVIOR AND EYE COLOR

SCIENCE AND BUBBLEGUM CARDS

> Summary.---139 professional baseball players who appeared on Topps bubble gum cards (copyright 1987) were subjects. The players, whose printed eye colors could be identified from their photographs, were sorted into three categories of 45 dark-eyed white players, 27 light-eyed players, and 67 black players. The statistics on the backs of the cards were dependent measures and included: Games, At Bat, Runs, Hits, Second Base, Third Base, Home Runs, Runs Batted In, Stolen Bases, SLG, Bunts, Strike Outs, and Batting Average.

The researchers then performed analyses of variance with these data. The most important findings were that black players scored more triples, stole more bases, and boasted better batting averages. Eye color did not seem to be an important factor!! (Beer, John, Beer, Joe; "Relationship of Eye Color to Professional Baseball Players' Batting Statistics Given on Bubblegum Cards," Perceptual and Motor skills, 69:632, 1989.)

Facetious Comment. Why must we spend billions on the Supercollider and Space Station when the equipment for important scientific research can be had for pennies at the corner store?

BEHAVIOR AND PHOTOPERIOD

MORE LIGHT, MORE FIGHT

Using 2,131 acts of hostility recorded over the last 3500 years, G. Schreiber et al have shown that these conflicts did not begin at random. Instead onsets of hostility are nicely correlated with the number of hours of sunlight in the day each war began.

> In the Northern Hemisphere, latitudes 30-60° N., the annual rhythm in the opening dates of wars shows a peak in August and a nadir in January (a in the figure). An inverse pattern in the annual rhythm of wars with a peak in December-February and a nadir in July was found in the Southern Hemisphere latitudes 30-60° S (c in the figure)....The results in the Northern Hemisphere suggest that there is a phase-shift of about one month between the two rhythms. We found a constant rate of acts of hostility throughout the year around the line of the Equator (b in the

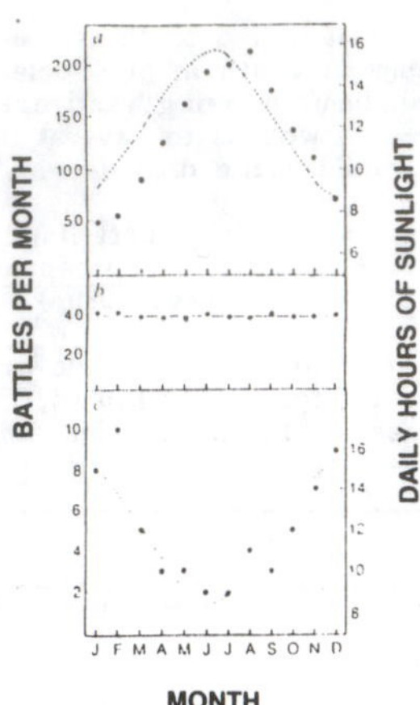

Battles-per-month and length-of-day plotted by month for: (a) the Northern Hemisphere; (b) Equatorial latitudes; and (c) the Southern Hemisphere.

figure).

(Schreiber, Gabriel, et al: "Rhythms of War," Nature, 352:574, 1991.)

Comment. From the curves, it appears that inhabitants of the Northern Hemisphere are about 20 times more bellicose than those below the Equator (a population effect?).

BEHAVIOR AND SOLAR ACTIVITY

PEACE AND SUNSPOTS

"Periods of international peace were found to occur in nearly regular cycles of 11 years by Edward Dewey in 1957 by analyzing the earlier data of Raymond Wheeler. In this paper the phase relationship between sunspot cycles and international battles was investigated. It was found that peaceful periods ended 7 out of 11 times within two years prior to sunspot peaks. The probability of this occurring by chance is less than .008.

"Geomagnetic storms are postulated as the triggering event since:

1. Geomagnetic storms are known to occur with greater frequency and intensity near sunspot peaks; and
2. Geomagnetic storms have been found by other researchers to be associated with increased frequency of accidents, illness, psychiatric hospital admissions, and crimes."

(Payne, Buryl; "Cycles of Peace, Sunspots, and Geomagnetic Activity," Cycles, 35:101, 1984.)

SOLAR ACTIVITY AND BURSTS OF HUMAN CREATIVITY

"In a previous paper, S. Ertel reported evidence that suggests a link between the historical oscillations of scientific creativity and solar cyclic variation. Eddy's discovery of abnormal secular periods of solar inactivity ('Maunder minimum' type) offered the opportunity to put the present hypothesis to a crucial test. Using time series of flourish years of creators in science, literature, and painting (AD600-AD1800), it was found that, as expected:

1. Cultural flourish curves show marked discontinuities (bursts) after the onset of secular solar excursions, synchronously in Europe and China;
2. During periods of extended solar excursions, bursts of creativity in painting, literature, and science succeeded one another with lags of about 10-15 years;
3. The reported regularities of cultural output are prominent throughout with eminent creators. They decrease with ordinary professionals.

"The hypothesized extraterrestrial connection of human cultural history has thus been considerably strengthened." (Ertel, Suitbert; "Synchronous Bursts of Creativity in Independent Cultures; Evidence for an Extraterrestrial Connection," The Explorer, 5:12, Fall 1989.)

Comment. With apologies to the author, a few minor changes in punctuation have been made above.

RHYTHMS IN RHYTHM

> We are currently living in the last quarter-century of the fifth 500-year cycle, which began with the Italian Renaissance in the fifteenth century. All the harmonies we hear today were developed in this last cycle.

So begins the final section of a recent reprinting of W.D. Allen's sweeping 1951 overview of the human fascination with music. But what's this about a 500-year cycle in music? It turns out that not only is there a 500-year pulse in musical creativity, but nested within the long swings are 100-year subcycles!

Allen's article, as it appeared originally in the Journal of Human Ecology (1:1, 1951), ran 41 pages. We can hit only a few high notes here. And, since we are concerned mainly with anomalies, we shall concentrate on this unexpected periodicity in musical creativity.

Allen describes how musical theorists have proposed both supernatural and evolutionary explanations for this periodicity, which commenced some 2,500 years ago with the Ancient Greeks. He is not convinced by either class of explanations. Instead, Allen has been beguiled by the long-period tones of environmental cycles:

> Now we have knowledge of a constantly operating cyclic factor in our cosmos, scientifically based on a mass of inductive evidence that goes beyond recorded history into the tree-ring records from centuries B.C. For the first time, we are provided with a powerful conditioning factor, if not a determinant, in the creation of music.

Here are two statements reflecting Allen's observations on the subject:

> After 1590, as a new warm period began in the 100-year cycle, a new Golden Age began in music, as in Science.

> In our own day, some composers have been extremely sensitive to cyclic changes. Stravinsky, notably in his return to neoclassicism after 1920, reflected the warm trend,

(Allen, Warren Dwight; "The 500-Year Cycle in Music: The Modern Period," Cycles, 42:100, 1991. A reprinting.)

Comment. Left unexplained in the "weather theory" of culture is just how warm trends inspire creativity. If warmth alone were the crucial factor, we would expect to see an inspiring outpouring of great music from today's Equatorial regions!

BEHAVIOR AND ASTRONOMY

DOES THE MOON REALLY FAZE PEOPLE?

Folklore strongly supports the power of the full moon to disturb people's minds, as underscored by the term "lunatic." The many scientific studies of this supposed lunar effect, however, have come to conflicting conclusions. Templer and Veleber have surveyed previous studies and believe that the discrepancies arise because of different methodologies. By combining new and older data and using a common approach, they confirm folklore by finding a disproportionate frequency of abnormal behavior occurring at the times of full moon, new moon, and the last half of the lunar phase. (Templer, Donald I., and Veleber, David M.; "The Moon and Madness: A Comprehensive Perspective," Journal of Clinical Psychology, 36:865, 1980.)

THE MOON, THE STARS, AND HUMAN BEHAVIOR

We humans have an inherited penchant for observing the heavens and wondering if the stars can affect our daily lives. Secular humanists hate astrology with a passion because, like Cassius in Julius Caesar, they believe we are masters of our own destinies. Nevertheless, astrology columns are still prominent in most newspapers. In the scientific press, however, we have to score a big plus for the anti-astrologers. First, Nature has just published a detailed analysis of the predictive power of astrology, and astrology has come up very short.

The Nature study is by S. Carlson, a physicist at the University of California at Berkeley. At the end of seven data-packed pages Carlson concludes:

"We are now in a position to argue a surprisingly strong case against natal astrology as practised by reputable astrologers. Great pains were taken to insure that the experiment was unbiased and to make sure that astrology was given every reasonable chance to succeed. It failed. Despite the fact that we worked with some of the best astrologers in the country, recommended by the advising astrologers for their expertise in astrology and their ability to use CPI (California Personality Inventory), despite the fact that every reasonable suggestion made by the advising astrologers was worked into the experiment, despite the fact that the astrologers approved the design and predicted 50 per cent as the 'minimum' effect they would expect to see, astrology failed to perform at a level better than chance. Tested using double-blind methods, the astrologers' predictions proved to be wrong. Their predicted connection between the positions of the planets and other astronomical objects at the time of birth and the personalities of test subjects did not exist. The experiment clearly refutes the astrological hypothesis." (Carlson, Shawn; "A Double-Blind Test of Astrology," Nature, 318:419, 1985.)

Next, if overkill is required, the Skeptical Inquirer, matches the Nature article with one on the effect of the moon on human behavior. The authors (two psychologists and an astronomer) conclude:

"This article outlines the results of a meta-analysis of 37 studies and several more recent studies that examined lunar variables and mental behavior. Our review supports the view that there is no causal relationship between lunar phenomena and human behavior. We also speculate on why belief in such relationships is prevalent in our society. A lack of understanding of physics, psychological biases, and slanted media reporting are suggested as some possible reasons." (Kelly, I.W., et al; "The Moon Was Full and Nothing Happened: A Review of Studies on the Moon and Human Behavior and Lunar Beliefs," Skeptical Inquirer, 10;129, 1985.)

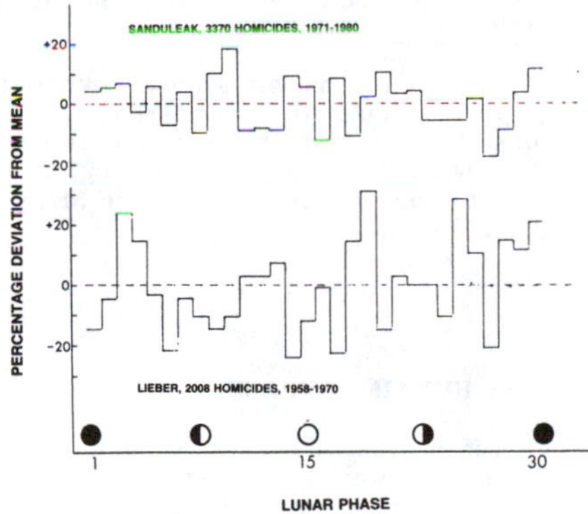

Comparison of two studies of Cuyahoga County homicide-assault-rate versus lunar phase. There is no obvious lunar effect on human behavior.

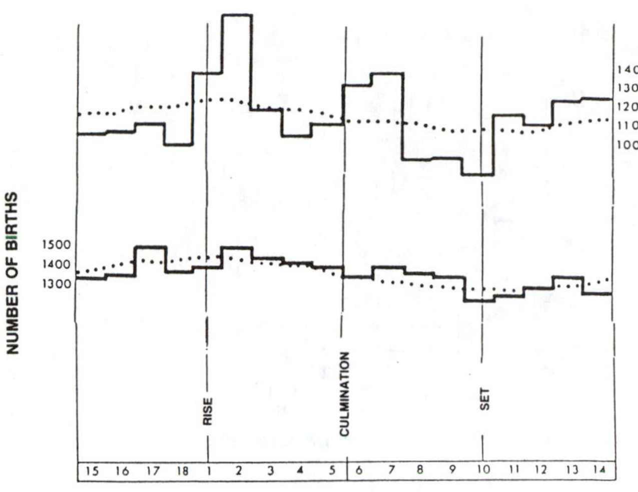

(Top) Distribution of births in an 18-sector Martian day for 2,088 sports champions. (Bottom) Ditto for 24,961 ordinary persons. (After M. Gauquelin)

CELESTIAL INFLUENCES

The "Mars Effect" refers to the "significant tendency for champion athletes to have been born at the time of either the rise or the upper culmination of the planet Mars." What has the position of Mars in the earth's sky to do with an athlete's prowess? No one knows! And certainly no scientist who wants to remain employed will try to find out.

We have M. Gauquelin to thank for discovering the Mars Effect. Gauquelin began his research with checking out the claims of conventional astrology. He "found no truth whatever behind certain major tenets of the horoscope, including the alleged influence of the signs of the zodiac, the reality of the astrological 'aspects,' the reported role of the 'houses,' or the prediction of future events." That's the good news, now here's the bad. During his work, he recorded the birth dates of thousands of French men and women who were especially successful in various endeavors, especially sports. These data, easily verified, do seem to demonstrate that the Mars effect truly exists!

G. Abell, who was a reknowned astronomer, did not categorically reject Gauquelin's claims. But he was not happy with them:

> To be honest, I am highly skeptical of Gauquelin's findings and his hypothesis. The main reason is that I cannot imagine a mechanism whereby the effect can be produced. However, I do not know that the effect is not there; my skepticism cannot be considered closed-mindedness, any more than a gullible acceptance of astrology should be regarded as open-mindedness. If the planetary effects suggested by Gauquelin are real, then his discovery is of profound importance. Consequently, I think the Gauquelin evidence, based on a great mass of data collected over many years, deserves to be checked out.

While Gauquelin has concentrated on sports champions, he has found similar effects for other vocations. Musicians, for example, have a negative Mars Effect. (Obviously, violinists should not compete in the broad jump!) Scientists are positive with Saturn; actors are positive with Jupiter. And so on. (Gauquelin, Michel; "Is There a Mars Effect?" Journal of Scientific Exploration, 2:29, 1988.)

Comment. In the indented quotation above, Abell could not imagine a "mechanism" for the Mars Effect. Yet, Mars and the other planets are drawn to the sun by gravitation, which is just as "mechanismless" as are the other action-at-a-distance forces. We can write an equation for gravitational force, and perhaps because of this and long familiarity, we feel comfortable with gravitation. But is it any less spooky than the Mars Effect?

THE SCIENTIFIC BASIS OF ASTROLOGY

At a recent meeting of the Society for Scientific Exploration, S. Ertel, a German scientist, reported on his inquiry into the so-called "Mars Effect," discovered by Michel Gauquelin. Here are two excerpts from his Summary:

"Since 1955 Gauquelin claims to have discovered planetary effects on human births: After rise of a planet and after its crossing of the meridian, birth frequencies of eminent men may either increase beyond or decrease below chance level.

.....

"In order to find out how clean Gauquelin's database is, the author travelled to Gauquelin's Paris laboratory and checked the files, including data which had been separated from publication, especially athletes' data. Using all obtainable data, Gauquelin's strongest hypothesis was tested, that planetary effects are more pronounced the greater the person's professional success. This claim was objectified with the help of citation frequencies, a sensitive procedure Gauquelin himself had not yet used. The total of 2089 athletes was subjected to this procedure. The results clearly supported Gauquelin's eminence claim." (Ertel, Suitbert; "An Assessment of the Mars Effect," The Explorer, 4:8, October 1987.)

Comment. Is all this simply astrology with scientific trappings? It certainly sounds like it is! Debunking groups, such as CSICOP (Committee for the Sci-

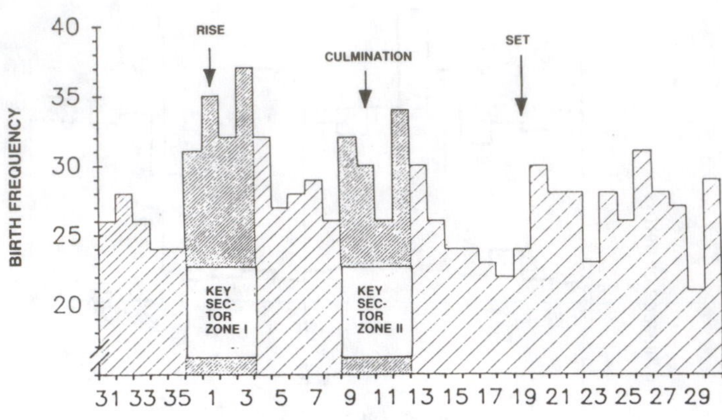

Birth frequencies of eminent athletes plotted for a 36-sector Martian day. (After S. Ertel)

entific Investigation of Claims of the Paranormal), have spent considerable effort trying to disprove the Mars Effect, without, according to Ertel, convincing results.

GENES AND CULTURE

HOORAY, ANOTHER "DANGEROUS" BOOK!

The May 22, 1981, issue of Science devotes three entire pages to a discussion of the issues raised in the book Genes, Minds, and Culture, written by Edward Wilson and Charles Lumsden. The subject of this book is "gene-culture coevolution," which infers that human culture is controlled not so much by "free will" as by rapidly changing human genes. The authors propose that as few as 1000 years are sufficient for important genetic shifts. Such shifts might, for example, impel humans to break out of the Middle Ages and bring on the Industrial Revolution.

The most controversial facets of the theory are: (1) The tight genetic control over human culture with little room for free will; and (2) The rapid blossoming of many cultures as genes shift about. As one scientist remarked, this book is "dangerous." Others describe it as marvelous. The Science article deals not so much with the book as with the reactions to it---and the reactions have been powerful, both pro and con. (Lewin, Roger; "Cultural Diversity Tied to Genetic Differences," Science, 212:908, 1981.)

Comment. The impression one gets from the synopsis of the book is that humankind is diversifying rapidly into new cultural configurations not through human volition but because of those imperious "selfish genes" we all carry.

ANOMALOUS TALENTS

Humans possess many remarkable talents verging on the anomalous, particularly in connection with seeing and hearing. In fact, our catalog Biological Anomalies: Humans I explores 25 categories of unusual human capabilities. In Science Frontiers, however, the fare has been limited to a few items associated with hearing and direction finding that we found in the recent current literature.

A NOTE ON PERFECT PITCH

An individual with perfect pitch can identify a tone without hearing a second tone for comparison. It is difficult to avoid the conclusion that such individuals possess a set of frequency standards somewhere in their permanent memories. The less fortunate of us do not have the genes that lead to the construction of these cerebral frequency standards. (Anonymous; "A Note on Perfect Pitch," Scientific American, 250:82, June 1984.)

Comment. Question 1. Does perfect pitch have any survival value that would stimulate its evolution? Question 2. How are these frequency standards coded on the genes?

DO YOU HEAR WHAT I HEAR?

No wonder music critics often disagree. They do not necessarily hear the same thing, even sitting side by side in the concert hall!

> The discovery, made by psychologist Diana Deutsch of the University of California at San Diego, concerns pairs of tones that are a half octave apart. When one tone of a pair, followed by a second, is played, some listeners hear the second tone as higher in pitch than the first. Other people, hearing the same tones, insist that the second tone appears to be lower in pitch.

Other differences in perception also exist. (Peterson, I., "Do You Hear What I Hear?" Science News, 130:391, 1986.)

MUSIC IN THE EAR

For three weeks a 70-year-old woman had been complaining about hearing music when there was no music within normal earshot. Since the woman wore a hearing aid in each ear, it was first thought that she might be picking up local radio stations; but a check showed that none was playing the repertoire she reported. Mostly she heard songs from the 1930s and 1940s. Finally, it was discovered that she was taking 12 aspirins a day. When this dosage was halved, the music stopped. Doctors have known that too much aspirin can cause ringing in the ears, but this is the first time that specific songs were induced. (Anonymous; "Stop the Music," Science News, 128:168, 1985.)

Comment. One is left wondering how the aspirin stimulated the apparent recall of old music from the woman's memory. Can some drugs improve memory? See p. 307 for this possible effects of hypnosis on memory.

THE HUMAN COMPASS

In recent years, scientists have found magnetic material (magnetite) in birds, snails, porpoises, bacteria, and other animals. The utility of these biologically manufactured compasses is obvious. Humans, too, seem to have a magnetic sense, although no one has yet dissected the human head to search for magnetite crystals. Rather, the proof of a magnetic sense comes from direction-finding experiments by Robin R. Baker, in England.

In a series of tests involving many subjects, blindfolded humans have been taken far afield and then asked, while still blindfolded, to point "home" and north. The results were surprising. Sense of direction was not lost despite long journeys. Furthermore, tests after removal of the blindfolds showed a marked deterioration of the direction-finding ability. The attachment of magnets and simulated magnets to the subjects proved that the magnets upset direction-finding capabilities. The controls with brass "magnets" retained their magnetic sense. (Baker, Robin R.; "A Sense of Magnetism," New Scientist, 87:844, 1980.)

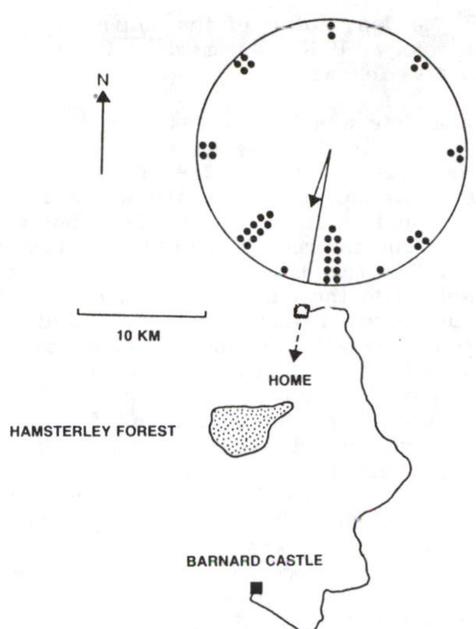

In the first set of Barnard Castle experiments conducted by R.R. Baker, subjects were taken---blindfolded and earmuffed---on the devious route mapped in the lower portion of the figure. At the top, each solid circle represents an individual's estimate of the direction of "home." The Australian experiments described herein were similar in concept.

HUMAN DIRECTION FINDING

"Summary. The fact that humans have an innate sense of direction is well established. Proof of this skill has usually been demonstrated in experiments in which subjects have been called upon to estimate the direction of the point of origin of a journey. This note extends such work by describing an experiment which showed that blindfolded humans, deprived of environmental cues, also have an ability to estimate accurately the direction of their place of residence within a town, even when driven around that town in such a way as to render them unable to identify where they are. The experiment throws into question the explanation usually offered for the existence of an innate sense of direction, namely, its value to the species, in an evolutionary sense, in facilitating a return to the starting point of exploratory journeys."

A fascinating facet of this experiment does not appear in the above Summary. All of the subjects in the study first assembled at the University of New England (in Armidale, NSW, Australia) and were there blindfolded and driven in a circuitous route 19.4 kilometers long to a spot 5.2 kilometers from the University. The sun had set and audible cues were suppressed. Very few of the 35 subjects could guess the direction of the University, the spot from which the journey began. Thus, this experiment does not really support that which is said to be "well established" in the Summary. On the other hand, the subjects were uncannily accurate in "guessing" the directions of their homes ---a fact duly reported in the Summary. What kind of sense could be involved in such a capability? (Walmsley, D.J., and Epps, W.R.; "Direction-Finding in Humans: Ability of Individuals to Orient towards Their Place of Residence," Perceptual and Motor Skills, 64:744, 1987.)

A COMPASS IN OUR SINUSES?

You may not feel any north-dircted nasal twinges, but the thin hard bones lining the human sinuses contain deposits of magnetic ferric iron. This discovery adds man to a long list of organisms from bacteria to birds known to possess localized accumulations of magnetic material. Experiments with these animals, including humans, seem to indicate a widespread ability to detect ambient magnetic fields. Some animals appear to use this sense for navigation. Whether humans do or do not is still a moot question. (Baker, Robin R., et al; "Magnetic Bones in Human Sinuses," Nature, 301:78, 1983.)

CHEMICAL REACTIONS

Of all the bizarre chemical reactions involving humans, SHC (Spontaneous Human Combustion) has to lead the list. Mainstream scientists who deign to examine the phenomenon maintain that SHC is: (1) never spontaneous; and (2) easy to explain in terms of the slow combustion of body fat--- the so-called "human candle" effect. Some cases of SHC, though, do not seem entirely amenable to this explanation. So, the controversy simmers away, not in Academia but rather among Forteans and anomalists.

This section concludes with items on blood type and chemotherapy.

SPONTANEOUS HUMAN COMBUSTION

Background. The data supporting the reality of SHC (Spontaneous Human Combustion) are largely anecdotal and, therefore, suspect to most scientists. The bizarre and improbable nature of SHC has also helped steer mainstream science away from the phenomenon. Besides, who in today's scientific milieu would ever supply funds for research into mysteriously burning human bodies? On top of these factors, mainstream science has come forth with an appealing explanation for SHC: obese humans, usually drunk or sedated, ensconced in overstuffed chairs or beds make in effect "human candles." If ignited by a dropped cigarette or spark from a fireplace, they burn and burn---slowly but almost completely, with rendered body fat feeding the wick-like upholstery.

An old sketch of the SHC incident related by Dickens in his novel Bleak House.

Although hundreds of cases of SHC have been recorded, we find very few in the modern scientific publications that we monitor in preparing Science Frontiers. Still, those that we have recorded display all of the characteristics typical of SHC down the centuries; and they are sufficient to raise doubts concerning the wisdom of science's condescending attitude toward SHC.

THE GWENT SHC CASE

John Heymer's job with the Gwent Police was attending the scenes of serious crimes and sudden deaths to gather forensic evidence. On January 6, 1980, he investigated a "rather unusual" death at a Gwent council house.

When he opened the door of the room, steamy, sauna-like heat still remained. The walls still radiated heat; condensation was running down the window; all surfaces were covered with a greasy black soot.

"On the floor, about one metre from the hearth, was a pile of ashes. On the perimeter of the ashes, furthest from the hearth, was a partially burnt armchair. Emerging from the ashes were a pair of human feet clothed in socks. The feet were attached to short lengths of lower leg, encased in trouser leg bottoms. The feet and socks were undamaged. Protruding from what was left of the trousers were calcined leg bones which merged into the ashes. The ashes were the incinerated remains of a man.

"Of the torso and arms nothing remained but ash. Opposite the feet was a blackened skull. Though the rug and carpet below the ashes were charred, the damage did not extend more than a few centimetres beyond the perimeter of the ashes. Less than a metre away, a settee, fitted with loose covers, was not even scorched. Plastic tiles which covered the floor beneath the carpet were undamaged."

We do not have space for additional details and must conclude with two questions posed by Heymer: (1) Human combustion requires a temperature of 1600°C (assuming no draft) applied for many hours; how was such heat achieved in a closed room without scorching nearby materials? (2) In the hottest of fires, the extremities are consumed but the torso remains; why did the reverse happen here and in other reported cases of human combustion? (Heymer, John; "A Case of Spontaneous Human Combustion? New Scientist, p. 70, May 15, 1986.)

Remains of a "combusted" human, as sketched in the above New Scientist article.

But every tale has two (or more) sides. The following letter appeared in response to the above article:

"John Heymer will no doubt assume that I am suffering from the 'Lavoisier Syndrome' if I disagree with the conclusion he had reached from his meticulous observations. His mistake is in trying to draw a parallel between the extensive burning to the body which he examined and the processes of cremation, when they can be distinguished by one critical factor. Cremation is intended to destroy a body in the shortest possible time and is therefore carried out under extreme conditions, but a relatively small fire can consume flesh and calcine bone if it is allowed to burn for a long time.

"This process, which I prefer to call prolonged human combustion, is usually fuelled by fat rendered from the body by the fire. It is no coincidence that in many of the cases this unit has encountered the victim was obese, and there was always a long delay before the fire was discovered. Examples of prolonged human combustion are, admittedly, rare but this should not be taken as evidence than an unusual source of ignition is involved. Indeed, all cases investigated by this unit have been resolved to the satisfaction of the courts without recourse to the excuse of 'spontaneous' human combustion." (Halliday, D.J.X.; "Human Combustion," New Scientist, p. 63, May 29, 1986.)

Comment. Halliday works for the Fire Investigation Unit of the London police.

JOHN HEYMER STILL DOESN'T BELIEVE THE STOCK SHC EXPLANATIONS!

Above, John Heymer described the death scene of Henry Thomas---a suspected case of SHC (Spontaneous Human Combustion). He now gives details for a remarkably similar case, that of an Annie Webb, of Newport, Gwent, U.K.

"The two deaths had amazing similarities, not the least of which was the fact that both people had reduced the intake of air into their rooms by draught-proofing them. Thomas had sealed both doors to his room with a standard draught excluder, while Webb had inserted strips of newspaper into every possible gap around both the door and window of her room.

"The torsos of both persons were completely destroyed. Not a single organ survived except a leather-like shrunken left lung in the case of Webb. All the bones were reduced to ash from the neck to the mid-thigh.

"In both cases the blackened skulls and untouched lower portions of legs remained. Webb's right arm was also intact. She had been incinerated on the floor with her arm outflung from her torso, hence its survival.

"As in the case of Thomas, furniture in Webb's room, which had commenced to burn, stopped burning due to the lack of oxygen. Yet again a complete human torso was reduced to ash in an atmosphere too devoid of oxygen to support the continued combustion of readily combustible materials."

Heymer considers this last fact very strange. Further, he claims that in these cases the usual pattern of human burning was reversed. Normally, the extremities are consumed, leaving a charred but recognizable torso. Here, only the extremities remain and the torso is gone! (Heymer, John; "A Burnt-Out Case?" New Scientist, p. 68, May 19, 1988.)

NOT THE NORMAL TYPE OF FIRE

December 5, 1987. Greensboro, Maryland. A man was attempting to light a propane stove, when his clothes caught fire. He died in a curious manner.

> The Dec. 5 fire was unusual because it burned half of the man's body and the floor directly beneath him but nothing else in the house.

Bob Thomas, the deputy state fire marshal, stated:

> This is not the normal type of fire we see when someone's clothes catch on fire.

Thomas thought that it was not spontaneous human combustion (SHC) because the entire body was not consumed. (Anonymous; "Spontaneous Combustion Debunked in Man's Death." Baltimore Sun, p. 2B. January 10, 1988,)

Comment. Actually, in some cases of supposed SHC, a portion of the body, a good perhaps, may survive. The very localized burning is also typical of "classical" SHC.

SPONTANEOUS HUMAN COMBUSTION AND BALL LIGHTNING?

Mainstream science scarcely acknowledges ball lightning; spontaneous human combustion it ridicules. Recently, G. Egely of the Central Institute of Physics, in Budapest, investigated a case where both phenomena may have been involved. G.T. Meaden, editor of the Journal of Meteorology, U.K., summarized Egely's report as follows:

> The date was 25 May 1989, and the place a field by the roadside near Kerecsend, a village 109 kilometers from Budapest. The victim was a 27-year old engineer within whose body, it is conjectured, ball lightning formed. The man had stopped his car and walked to the edge of a field about ten metres distant to urinate. Suddenly his wife who had remained behind in the car saw that the young man was surrounded by a blue light. He opened his arms wide and fell to the ground. His wife ran to him, noticing that one of his tennis shoes had been torn off. Although it looked hopeless she tried to help him, but soon after she was able to stop a passing bus. Amazingly, the bus was filled with medical doctors returning from a meeting; unhappily they immediately pronounced that the man was dead.
>
> At the autopsy a hole was found in the man's heel where the shoe had been. The lungs were torn and damaged, and the stomach and belly were carbonized! This is indicative of internal combustion, just as the blue light is proof of atmospheric electricity, while the damaged heel and shoe are indicative of electrical earthing.

(Anonymous; "Spontaneous Combustion," Journal of Meteorology, U.K., 15:320, 1990.)

Comment. Although the sky was cloudy, there were no thunderstorms in the immediate area. The body was not mostly consumed as in classical cases of human combustion, nor was a fireball observed. Still, this incident strengthens the suggestion that ball lightning may kindle spontaneous human combustion.

SHC OR H/T HOMICIDE?

The possible case of SHC (Spontaneous Human Combustion) just reported above may have a rational but bizarre explanation. S.L. Wernokoff commented as follows in the Journal of Meteorology:

> The very first thing that comes to mind is whether overloaded high-tension [H/T] or low-tension wires were anywhere nearby. Since this incident occurred at a roadside, nearby power lines may well have been present. If such lines are overloaded or badly insulated, fatal arcing can occur from the ground at a considerable distance from the power lines. This has happened often in our country [the U.S.], in rural areas where public utilities have quietly exceeded the capacity of their lines. The resulting discharges can easily electrocute livestock over ¼ mile from the "leaky" H/T lines. I would wager that the Hungarian utility agencies are guilty of the same practice. Personally, I suspect that this unfortunate young man may have been electrocuted through his own urine! The "blue light" witnessed by the victim's wife may have been St. Elmo's Fire---an ungrounded luminous corona visible around the victim in the humid, pre-thunder-

storm conditions. The hole in his heel and tennis shoe indicate where the current finally grounded itself.

Wernikoff goes on to tell of a case in Canada where a man washing up at an outdoor table, 100 yards from overhead power lines, was electrocuted when he emptied the basin onto the ground. He, too, had a hole burned through the heel of his boot! (Wernikoff, Sheldon L.; "The 'Hungarian Spontaneous Combustion' Case---Another Explanation," Journal of Meteorology, U.K., 17:22, 1992.)

ARE BLUEBLOODS MORE OFTEN TYPE A?

In the 1983 issue of Nature (303:522), J.A. Beardmore and F. Karimi-Booshehri reported that, based on a study of a specific British population, A-blood groups are significantly more common among the higher socio-economic groups. As one might predict whenever someone asserts that human success is genetically determined, an avalanche of mail descended on the Nature office. Two other studies that did not show the blueblood effect were offered, although somewhat different populations were involved. Many letters tried to find an explanation for this anomaly in the constitution of the sample. By the time one got to the response by the authors, the whole issue was clouded. (Mascle-Taylor, C.G. N., et al; "Blood Group and Socio-Economic Class," Nature, 309:395, 1984.)

CIRCADIAN RHYTHMS AND CHEMOTHERAPY

The toxicities of many commonly used anticancer drugs depend upon when they are administered during the day. This phenomenon occurs in humans and other animals. The effect is not trivial but "profound." (Hrushesky, W.J.M.; "Circadian Timing of Chemotherapy," Science, 228:73, 1985.)

Comment. This "profound" effect should, by extrapolation, also apply to drug potency, the workings of the immune system, and all biochemical reactions. The location of and reason for the circadian clock are matters of conjecture.

HEALTH

Since the pages of Science Frontiers aim at highlighting the major controversies and anomalies of science, it is not surprising to find many items in this "health" section devoted to cancer and AIDS. Both of these "diseases" are complex and poorly understood and deserve such special attention. Almost equal space is given to epidemic phenomena, especially the possible role of extraterrestrial influences, such as solar activity. A smattering of other health phenomena complete the outline of this section.

- Cancer. Is it the "price" of higher life? The efficacy of imaging.
- AIDS. Is HIV the sole cause of AIDS?
- Epidemic phenomena. The cyclic characteric of many diseases and the possible role of solar activity and other astronomical phenomena.
- Malaria, diabetes, plague. Sundry phenomena.
- Health and evolution. The curious survival of motion sickness. Natural selection in action today.
- Placebos and homeopathy. Placebo addiction. The efficacy of homeopathy demonstrated!

CANCER

HOW DO CANCERS ATTRACT A SUPPORTING CAST

The ability of a cancer cell to grow depends upon getting a good supply of blood from its host's capillary system. Cancer cells, always insidious, seem to be able to con its host's body into constructing a special system of capillaries just to support tumor growth. First, the cancer sends out a chemical signal that attracts the host's mast cells. As the mast cells work their ways to the cancer, they apparently leave a heparin-lined tunnel for the capillary cells to follow. Before long, the body has provided the blood supply the cancer needs to grow and possibly, eventually kill the host. (Gunby, Phil; "How Do Cancers Attract a Supporting Cast?" American Medical Association, Journal, 245:1994, 1981.)

Comment. In view of all of Nature's marvelous adaptations, why hasn't the body evolved a counter strategy to foil cancer sabotage? And how did cancer cells evolve in the first place, seeing that they usually cause the demise of their host, and are not subject to natural selection?

WHY DON'T WE ALL HAVE CANCER?

Biologists have just found that the difference between a normal human gene responsible for manufacturing a specific protein and a gene causing cancer is the replacement of a single nucleotide by another in a very long string of nucleotides. This is a very delicate situation. The difference between cancer and no-cancer is simply too tiny. Given the high frequency of random changes (mutations), we should all have cancer. One implication is that humans (and other animals, too) have come up with some method of preventing or correcting these minor mutations---otherwise we would have become extinct long ago. No one knows what this mechanism is or why it sometimes fails. (Anonymous; "More Speculation about Oncogenes," Nature, 300:213, 1982.)

CANCER: THE PRICE FOR HIGHER LIFE?

For unknown reasons, plants and the simpler animals, such as sponges and jellyfish, do not get cancer. But all laterally symmetric organisms are prone to cancer. According to James Graham, the acquisition of cancer-initiating oncogenes by organisms (also an unexplained event) has forced these afflicted organisms to develop all sorts of defenses against external forces which might, with the help of the oncogenes, trigger cancer.

Typical biological defenses include systems to insure accurate replication of cells, to destroy transformed cells, and to protect or immunize the organism against invading systems. Efficient immune systems in turn permitted life to invade mutagenic environments (such as sunlight) and to shed restrictive body coverings. In other words, cancer may have been a blessing in disguise---the price of higher life! (Anonymous; "Cancer: The Price for Higher Life?" New Scientist, 99:766, 1983.)

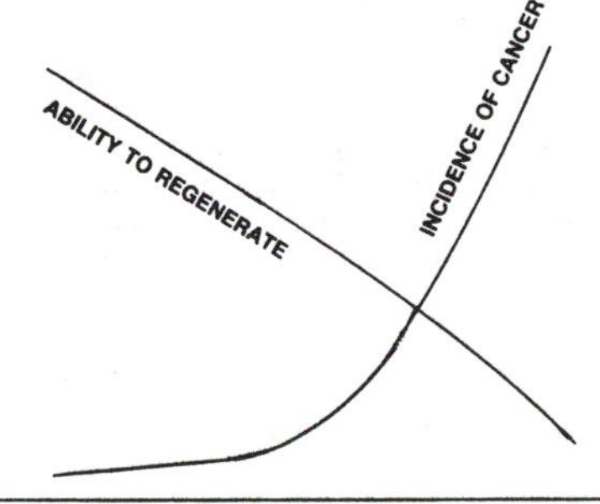

As the complexity of an organism increases, its ability to regenerate tissue recedes while its susceptibility to cancer increases.

Comment. It is so easy to say that evolution "developed" this or that characteristic in response to some applied force. Exactly how such responses are made is a major mystery. And why do oncogenes exist? Are they a product of chance? They hardly confer short-term survival capability.

IMAGING CANCER AWAY

Anna had been given three months to live. The malignant tumor, growing rapidly at the back of her neck, had virtually crippled her. Her upper body was hunched over, her head was forced painfully to one side, and her right arm was contracted and paralyzed. The best thing she could do, said her doctor, was to go home and make arrangements for the future of her young son and daughter.

Instead Anna learned how to "image." She conceived the tumor to be a dragon on her back and her white blood cells as knights attacking the dragon with swords. A year later, the tumor had shrunk. Later, it disappeared completely. Can "imaging" work? Obviously, this is a very controversial question.

Admittedly, little real scientific research has been done on imaging per se ---it is a bit too radical a concept. But a few scientists are beginning to chart the chemistry and information flow in the mind-body relationship. For example, the death of a spouse has long been associated with the increased mortality of the surviving spouse. Clinical studies of bereaved spouses reveal fewer circulating lymphocytes, which help the body fight disease, and significantly higher levels of cortisol, a substance that suppresses the immune system's response to disease.

Although it is very early in the game, there are verifiable correlations between state-of-mind and body chemistry. Further, other researchers have found that there are sympathetic nerve terminals in such organs as the spleen and lymph nodes, both of which play important parts in defending the body. Imaging just might send the right signals through these terminals, while depression might tend to shut the defense system down. (Hammer, Signe; "The Mind as Healer," Science Digest, 92:47, April 1984.)

Comment. Imaging is only the latest psychological device humans have tried in fighting disease and promoting health. History is full of such ploys. See also p. 298 on the use of "visualization" in activating the immune system.

TAKING THE RADON CURE

"An internationally respected radon researcher has uncovered some surprising and perplexing evidence. High levels of radon exposure are known to cause lung cancer. Studies of people exposed to such levels in mines demonstrate that conclusively. Then it follows, most people have thought, that the best level is zero, and any increase should produce an increasing rate of lung cancer. But controversial studies show that this may not be so.

"In one study Dr. Bernard Cohen of the University of Pittsburgh compared average lung-cancer rates in many counties with the average radon rates found in the respective counties. It's an ambitious study---39,000 measurements in 415 counties.

"The results: In counties where lung cancer in women would have been expected to be up 25 percent from the radon levels, the incidences of cancer were actually down 30 percent. There are others. Finland has average indoor-radon levels of 2.5 picocuries per liter ---2.5 times higher than the world average. Yet the female lung-cancer rate in Finland is only 70 percent that of other industrialized countries." (Gilmore, C.P.; "Radon: Cancer Killer?" Popular Science, p. 8, May 1989. Cr. R.W Schiller)

AIDS

AIDS: ANOTHER GREAT DECEIVER

In most diseases, we can count on the presence of antibodies as proof positive of infection. Thus, the usual test for AIDS registers the presence of antibodies and not the virus itself. But, researchers at the Johns Hopkins School of Public Health, in Baltimore, have discovered four AIDS victims in a group of 1000, who seem to have lost their AIDS antibodies but not the AIDS virus itself. Curiously, two of the four later lost the AIDS virus, too. It is possible that the AIDS virus is not really "lost" but merely hiding out somewhere, perhaps in the brain where tests of circulating blood cannot detect it. (Anonymous; "Antibodies Can Disappear from Infected People," New Scientist, p. 41, June 9, 1988.)

Comment. Another possibility, of course, is that of a spontaneous cure. Whatever the answer, AIDS is a tricky disease.

DOES THE AIDS VIRUS REALLY CAUSE AIDS?

All but a tiny minority of scientists accept as fact that an organism called the human immunodeficiency virus (HIV) is the cause of AIDS. This fact is hallowed and defended as vigorously as the facts of evolution, the Big Bang, and continental drift. Extremely nasty things are being said about a handful of heretics who attack this position.

> One leading dissident, UC Berkeley molecular biologist Peter H. Duesberg, believes that HIV is not the cause of AIDS---at least not the sole cause.
>
> He thinks the virus may be an opportunistic organism that found a willing host in the AIDS patient who became sick from something else. That is, he believes HIV is the result of the disease, not the cause. Duesberg thinks the cause of AIDS has more to do with the life style of most of the AIDS patients, but he admits that he doesn't know exactly what.
>
> Duesberg points out that three things must be true before a microorganism can be blamed for causing a disease. These are called Koch's Postulates, after R. Koch, who formulated them a century ago:
>
> 1. Every patient who has the disease must also harbor the suspected microorganism. Some AIDS sufferers do not have the AIDS virus, although it is debated whether as many as half don't or very few don't.
> 2. The microorganism must cause the disease when injected into research animals---primates for example. The AIDS virus does not; although some other diseases, such as small pox, do not affect other animals either.
> 3. The suspect microorganism must be isolated from the patient and grown in a culture.
>
> Duesberg claims that HIV definitely fails the first two Koch tests. (Shurkin, Joel N.; "The AIDS Debate: Another View," Los Angeles Times, January 18, 1988. Cr. J.M. Ward)

Three months after the above article was published, the journal Science jumped into the fray. Additional points of interest:

1. Duesberg considers the HIV to be such a "pussycat" that he would gladly be injected with the virus.
2. Duesberg has published his reservations in Cancer Research, but no formal response from the scientific community has resulted, although there has been plenty of unpublished name-calling.
3. The HIV behaves like no other known virus; viz., its long latency and its persistence despite the production of antibodies. (Actually, the herpes family of viruses is also known for its long latency.) (Booth, William; "A Rebel without a Cause of AIDS," Science, 239: 1485, 1988.)

Comment. There is much more to this controversy than we can cover here, including charges of financial improprieties and the existence of an AIDS Mafia.

DUESBERG REVISITED

P. Duesberg is a molecular biologist at the University of California, Berkeley. He contends that the human immunodeficiency virus (HIV) is not the cause of AIDS and is, instead, a harmless "passenger" in the bodies of AIDS victims. Naturally, this stance is controversial, and just as naturally we have had cause to mention Duesberg before.

Duesberg is back in the news again because his iconoclastic views were prominently featured in a TV documentary entitled "The AIDS Catch" seen in Britain in June. The scientific community was furious, claiming that the documentary was one-sided and selective. Further, it was maintained that Duesberg's arguments have been completely refuted.

Briefly, Duesberg believes that AIDS is not an infectious disease because:

1. Too few T-lymphocytes in the peripheral blood are infected to cause the disease;
2. HIV carriers without symptoms exist; and
3. HIV in pure form doesn't seem to induce AIDS in humans or animals.

Rather, says Duesberg, AIDS is a collection of symptoms arising from such factors as the repeated use of intravenous drugs and malnutrition. Mainstream researchers think that Duesberg is wrong on (1); that (2) is irrevelant, since asymptomatic carriers of typhoid and cholera exist; and that (3) may be incorrect, since SIV (Simian Immunodeficiency Virus) does induce simian AIDS in monkeys. (Weiss, Robin A., and Jaffe, Harold W.; "Duesberg, HIV and AIDS," Nature, 345:659, 1990.) Brown, Phyllida; "'Selective' TV Documentary Attacked by AIDS Researchers," New Scientist, p. 23, June 16, 1990.)

Comment. However self-assured the mainstreamers are, they must have flinched at a paper given by L. Montagnier, of the Pasteur Institute, at the recent AIDS conference in San Francisco:

> Montagnier says research conducted in his lab suggests HIV initially exists peacefully within the CD4 T-lymphocytes, white blood cells that assist in immune defenses. But co-infection with a mycoplasma, he contends, may transform the slowly replicating HIV into a killer.

(Fackelmann, K.A.; "Data and Dispute Mark AIDS Meeting," Science News, 137:404, 1990.)

RETHINKING AIDS

The columns of Science Frontiers have frequently publicized the heresy of P. Duesberg, who holds that the so-called AIDS virus, HIV, is not the sole cause of AIDS. Echoing many of Duesberg's assertions is biochemist and immunologist R.S. Root-Bernstein. He points out that:

> People all over the world are getting AIDS without being exposed to or infected with HIV.

Root-Bernstein continues with:

> The implications of this revelation are truly astounding. Essentially there are only three possibilities. The HIV may really be there, but everyone has missed it. This is unlikely, since many of the researchers reporting HIV-negative cases of AIDS are the top HIV experts in the world. Another possibility is that there is a new virus that everyone has missed. This is again unlikely given the huge amount of retroviral research that has been performed in the past decade on AIDS patients. Finally, these may be the cases that demonstrate that AIDS can be produced by the types of synergistic, multifactorial assaults on the immune system that Joseph Sonnabend and I have been proposing for years.

Although most AIDS researchers are still wedded to the theory that HIV is the sole and only cause of AIDS, cracks in the stonewalling are beginning to appear. In fact, C.A. Thomas, Jr., formerly a Professor of Biochemistry at Harvard, has organized the Group for the Scientific Reassessment of the HIV/AIDS Hypothesis. (Root-Bernstein, Robert S.; "Rethinking AIDS," Frontier Perspectives, 3:11, Fall 1992.

Comment. Overzealous defense of the HIV paradigm may have cost billions in misdirected research.

EPIDEMIC PHENOMENA

SUNSPOTS AND FLU

The last six sunspot peaks have coincided with flu pandemics. During the sunspot maxima of 1947, 1957, and 1968, the influenza-A virus underwent antigenic shifts that allowed the virus to bypass the immunity built up in the populace. In 1937, a pandemic occurred but no genetic change was detected, although one might have gone unnoticed. The deadly worldwide 1918-1919 epidemic transpired just after the 1917 sunspot peak and before the discovery of the flu virus. The sunspot maximum of 1928 may have signaled a major shift from the virus causing the 1918-1919 pandemic to the type now afflicting us. (Hope-Simpson, R.E.; "Sunspots and Flu: A Correlation," Nature, 275:86, 1978.)

PERIODICAL INVASIONS OF ALIENS

Forget those contemporary tales of UFO landings and human contacts with their alien navigators. Aliens have been landing here and mixing with the human populace for centuries. In fact, their traffic peaks about every 11 years, just when the solar cycle reaches its maximum. By now, you've probably guessed that F. Hoyle and N.C. Wickramasinghe are again talking about flu pandemics and sunspots. You must admit, however, that their correlation is becoming more and more convincing.

First, we have their graph covering the past 70 years which speaks for itself. You can add the 1990 flu outbreak to the curve yourself! To strengthen the correlation Hoyle and Wickramasinghe tabulate flu and sunspot data back to 1761. They find that flu pandemics and sunspot maxima have kept in step for the last 17 cycles.

Key to the Hoyle-Wickramasinghe argument is their contention that simple life forms (viruses, bacteria, etc.) not only exist in outer space but likely evolved there. If so, how do they ride in to afflict us on the peaks of the solar cycle? Here's how, in their words:

> In conclusion, we note that electrical fields associated with intense solar winds can rapidly drive charged particles of the size of viruses down through the exposed upper atmosphere into the shelter of the lower atmosphere, the charging of such particles being due to the photoelectric effect. This could define one possible causal link between influenza pandemics and solar activity.

(Hoyle, F., and Wickramasinghe, N. C.; "Sunspots and Influenza," Nature, 343:304, 1990.)

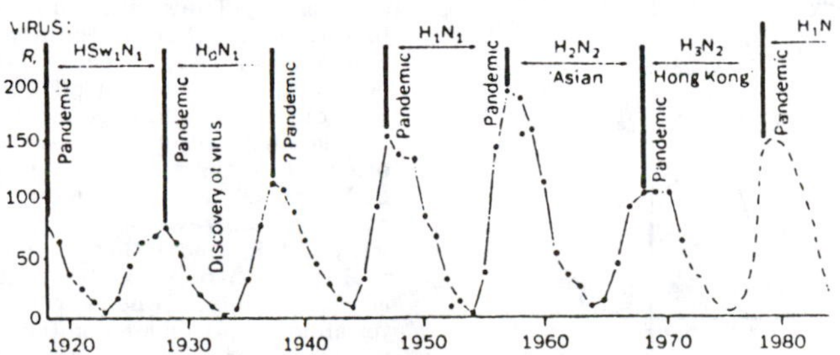

Yearly means of daily sunspot numbers correlated with dates of flu pandemics.

SUNSPOTS AND DISEASE

Six of the major influenza epidemics, at least as far back as 1917, were synchronized with the sunspot cycle. Furthermore, all but one of these epidemics involved an antigenic shift, wherein the flu virus developed a new coat of protein, which made it resistant to the immunities the population had built up over the years. There is no known mechanism by which solar activity can abet virus evolution, except penetrating radiation, which is inherently destructive.

Lowered human immunity may also be a consequence of solar activity, according to Solco W. Tromp, director of the Biometeorological Research Center in the Netherlands. Over 30 years of research, using blood data from 730,000 male donors, led Tromp to the conclusion that the blood sedimentation rate varies with the sunspot cycle. Since this rate parallels the amount of albumin and gamma globulin, resistance to infection may also follow the lead of the sun. (Freitas, Robert A., Jr.; "Sunspots and Disease," Omni, 6:40, May 1984.)

THE FAULT, DEAR READER, IS NOT IN OUR STARS BUT OUR PIGS!

Fred Hoyle, in his usual maverick style, has hypothesized that some human flu epidemics are caused by new viruses injected into the biosphere from outer space. (See his book Diseases from Space.) In yet another book, Evolution from Space, he goes further, stating that the evolution of terrestrial life can also be affected by the extraterrestrial inoculation of genetic material.

But, just maybe, influenza pandemics are due to pigs! Every 10-20 years, new flu viruses seem to crop up against which humans have little resistance. The latest theory is that there exists a human-duck-pig connection. It seems that human flu viruses can multiply in ducks, but are not transmitted among ducks. It is also likely that duck viruses multiply in humans, but are not transmitted from one person to another. But enter the pigs:

> There is firm evidence that pigs can become infected by and may transmit both human and avian influenza viruses not only amongst other pigs but also back to the original hosts. Therefore, pigs seem to be 'mixing vessels' where two separate reservoirs meet and where reassortment between avian and human influenza A viruses occurs, giving rise to the antigenic shift by creating new human pandemic influenza strains with new surface antigens.

The article stimulating this discussion worries about new aquaculture practices, especially in Asia (the so-called Blue Revolution), in which duck and pig manure is dumped into fish ponds as fertilizer. The dense concentration of humans, ducks, and pigs threatens to be a factor for constructing new strains of flu. (Scholtissek, Christoph, and Naylor, Ernest; "Fish Farming and Influenza Pandemics," Nature, 331:215, 1988.)

Comment. Noting that the AIDS virus may have originated and still be mutating in African monkeys, and coupling this with the above discussion of flu, we can speculate a la Evolution from Space that terrestrial life itself is its own evolutionary engine! Going still another step further, we can wonder if life-as-a-whole (the Gaia concept) is not trying to check the burgeoning human population by biological warfare---a check and balance arrangement. Isn't it amazing how much speculation a couple simple facts can engender?

THE DEADLY SUN

Sunspottery, or the linking of seemingly unrelated phenomena to solar activity, has been a popular pastime for as long as sunspot records have been kept. Usually pooh-poohed by scientists because the link between cause and effect seems absent, some impressive statistical evidence now associates heart attacks with geomagnetic and solar activity. Malin and Srivastava have shown that the number of cardiac emergencies in their area of India is very closely tied to geomagnetic activity, which in turn is controlled by the sun. Standard statistical tests confirm an especially strong correlation.

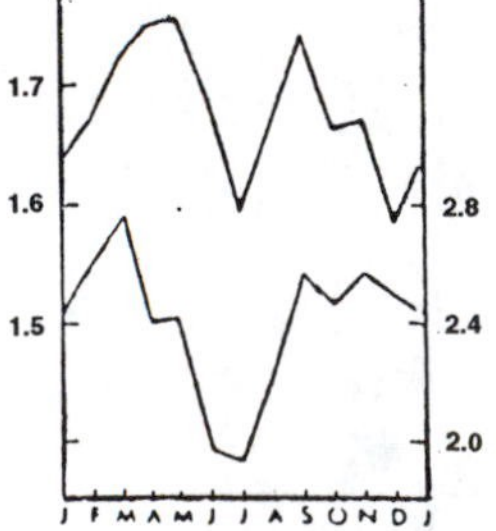

(Top) Magnetic activity index. (Bottom) Daily admissions of cardiac emergencies.

But why should the two observables be associated at all? The authors' concluding sentence reads: "The possibility that there is some other cause (or solar origin?) responsible for both the magnetic and medical phenomena should not be ignored." (Malin, S.R.C., and Srivastava, B.J.; "Correlation between Heart Attacks and Magnetic Activity," Nature, 277:646, 1979.)

INFECTIONS FROM COMETS

Astronomer Fred Hoyle notes that pandemics and plagues have generally appeared very suddenly and unexpectedly, sweeping the globe with hard-to-explain swiftness. Noting in passing that comets have always been considered bad omens, he postulates that cometary matter may actually contain bacteria and viruses that infect the earth's populace as the planet passes through cometary tails.

Recent spectroscopic studies of interstellar matter and comets themselves indicate a richness of life-associated compounds that infers that outer space might well be the breeding ground of simple life. The authors review some of the strange history of epidemic diseases and wonder if their theory of germ-bearing comets might not satisfy the data as well as the more common hypothesis of random mutation and genetic recombination. (Hoyle, Fred, and Wickramasinghe, Chandra; "Does Epidemic Disease Come from Space?" New Scientist, 76:402, 1977.)

MEASLES EPIDEMICS: NOISY OR CHAOTIC?

We should talk about chaos more. This subject threatens to undermine the popular notion that nature is fully deterministic. We like to think that if we are given enough data that scientific laws will allow us to predict the future accurately. But, unhappily, determinism stumbles when trying to cope with the weather, asteroid motion, the heart's electrical activity, and an increasing number of natural systems. Chaos lurks everywhere!

The growing split in scientific outlook is seen very clearly in the statistics of New York City measles epidemics before mass vaccinations. Take a look at the graph of recorded cases. The expected peaks occur each winter, but there is a strong tendency toward alternate mild and severe years.

Very nice mathematical models exist that purport to predict the progress of epidemics. They take into account such factors as the human contact rate, disease latency period, the existing immune population, etc. It is all very methodical, but it fails to account for the irregularities in actual data. Deterministic scientists claim that just by adding a little "noise" they could duplicate the observed curve. On the other hand, a very simple model that acknowledges the reality of chaos easily duplicates the measured data. Who is right? The determinists and chaosists (chaosians?) are now fighting it out. (Pool, Robert; "Is It Chaos, or Is It Just Noise?" Science, 243:25, 1989.)

Comment. Much more of nature may be

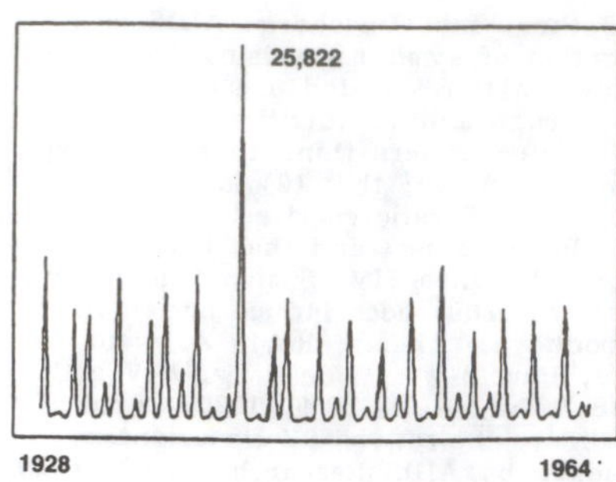

Incidence of measles in New York City (thousands of cases per month), 1928-1964. Notice the tendency to peak in alternate years.

chaotic. Even evolution itself may be so. Are we merely a blip on a biological diversity curve, with a future that is unpredictable, regardless of what actions we take? Fortunately, lack of space prohibits the mention of Free Will!

MALARIA, DIABETES, PLAGUE

EVOLUTION OF MAN AND MALARIA

Malarial parasites are customarily classified according to the species infected and then further subdivided by morphology and biological characteristics. The two assumptions implicit in this classification procedure, which is supposed to mirror actual historical evolution, are:

1. Malarial parasites evolved in parallel with their hosts; and
2. Morphology is a measure of evolutionary relatedness.

With modern biochemical techniques it is possible to test these assumptions by comparing the DNA structures of the different malarial parasites. P. falciparum, the parasite transmitting the most deadly human malaria, turns out to be more closely related to rodent and avian malaria than the other primate malarias. Therefore, assumption #1 above is incorrect in this view. Assumption #2 is also wrong because some species of malaria parasites which are very similar morphologically are quite different DNA-wise. (McCutchan, Thomas F., et al; "Evolutionary Relatedness of Plasmodium Species as Determined by the Structure of DNA," Science, 225:808, 1984.)

Comment. The article does not draw attention to still another assumption; namely, that similarities are measures of evolutionary relatedness. If this assumption isn't correct, evolutionary family trees based on bodily structure, which means most of the family trees in the textbooks, may not truly reflect what really happened in the development of life. Further, if malarial parasites did evolve along with their hosts, human evolution seems farther removed from the evolution of the other primates than usually supposed.

ANOMALOUS GEOGRAPHICAL DISTRIBUTION OF DIABETES MELLITUS

The incidence of diabetes mellitus among children varies dramatically with geography. For children under 15, it is only 1.7 per 100,000 in Japan but rises to 29.5 in Finland. Within the States, it is 9.4 per 100,000 in San Diego and peaks at 20.8 in Rochester, Minnesota. Children of European descent in New Zealand contract it three times as often as Maori children. U.S. whites get the disease more frequently than blacks and Hispanics. "Causes of these 'extraordinary' distribution differences remain unknown....Both genetic and environmental factors appear necessary for the disease." (Eron, C.; "Cold Facts on Diabetes," Science News, 134:117, 1988.)

BUBONIC PLAGUE AS AN INDICATOR OF DIFFUSION?

Every year a few people in the Arizona-New Mexico region contract bubonic plague. Where did this persistent pocket of infection come from? One school of thought has the germ arriving with the rats on ships docking in California during the Gold Rush of 1849. But how could the plague have crossed the mountains across several radically different ecosystems? One would anticipate finding records of the plague as it made its way into the Southwest. It is true that a less virulent disease, the sylvatic plague, transmitted by similar mechanisms, does exist in the Pacific coast area; but the bubonic plague does seem highly localized in Arizona and New Mexico.

Perhaps another explanation can be discovered in the history of the bubonic plague and the settlement of the Southwest. The plague seems to have commenced in Athens about 430 BC. More or less isolated epidemics followed, but from 1334 to 1351 the disease decimated most of the known world; Europe, Asia, North Africa. Of course, the American Southwest was not part of the "known world" of 1334-1351. But, coincidentally (?), this was just about the time that the Hohokam and Anasazi cultures began to decline rapidly in the Southwest. Link this observation to the purported Roman and Hebrew artifacts in the region and one sees the possibility that Old World travellers brought the bubonic plague to the New World well before Columbus! (Underwood, L. Lyle; "Bubonic Plague in the Southwest," Epigraphic Society, Occasional Papers, 14:207, 1985.)

HEALTH AND EVOLUTION

MOTION SICKNESS DIFFICULT TO EXPLAIN IN TERMS OF EVOLUTION

Motion sickness has been called an evolutionary anomaly because it seems highly disadvantageous to those who suffer from it. Yet, motion sickness occurs in many species. Why should it have evolved at all? Recognizing this problem, Michel Treisman seeks to explain the anomaly by noting that neurotoxins accidently ingested by animals cause essentially the same symptoms as motion sickness. To survive, animals must eliminate ingested neurotoxins by vomiting or defecation, both of which also accompany motion sickness. It is simply coincidental that modern vehicles duplicate these symptoms through their motions. The body interprets the signals created by motion as due to dangerous ingested material and acts accordingly. (Treisman, Michel; "Motion Sickness: An Evolutionary Hypothesis," Science, 197:493, 1977.)

EVEN TODAY NATURAL SELECTION IS MOLDING HUMAN POPULATIONS

Nauru is a remote Pacific atoll with a population of 5,000 Micronesians. Formerly, the Nauruans led energetic lives ---fishing, subsistence farming---and they were slim and healthy. Then came colonization and phosphate mining; with these came wealth, imported calorie-packed food, sedentary lives, obesity, and, unhappily for this tropical paradise, diabetes. The incidence of diabetes mellitus shot up to 60%, an astounding statistic by world standards. On one of the wealthiest of the Pacific islands, the inhabitants have the shortest life spans! The same scenario is being played out in other parts of the world where life styles have changed drastically; for example, some Polynesians, American Indians, and Australian aborigines are similarly afflicted. Furthermore, an epidemic of diabetes mellitus is anticipated as the "benefits" of civilization are brought to India and China.

Two questions must be answered:

1. Why is the incidence of diabetes mellitus only 8% among American junk-food-eating couch potatoes? Probable answer: natural selection has already modified the American genotype by eliminating those who are supersensitive to diabetes mellitus under conditions of rich diets and sedentary lives.
2. Why are modern populations still living under Spartan conditions so sensitive to diabetes in the first place?

Possible answer: The so-called "thrifty genotype" hypothesis. In this view, the genotype that is sensitive to diabetes also confers survival advantages in societes where food supplies are meager and unpredictable. This genotype provides for a hair-trigger release of insulin for the rapid conversion of rare food gluts into body fat deposits that will sustain the individual during the next famine. Unfortunately, when rich food is continuously available, people with this "hair-trigger" genotype succumb to diabetes. (Diamond, Jared M.; "Diabetes Running Wild," Nature, 357:362, 1992.)

PLACEBOS AND HOMEOPATHY

ADDICTION TO PLACEBOS

A 38-year-old married schizophrenic was in psychotherapy for severe depression and multiple suicide attempts. She was addicted to methylphenidate, taking 25 to 35 10-mg pills per day. She was incredibly adept at persuading pharmacists to refill old prescriptions. With the help of her husband and a drug company, placebos were gradually substituted for the real pills to the point where only two real pills and 25-30 placebos were taken each day. The patient never noticed, indicating that the placebos satisfied the patient's real need----something to fill an inner void. (Muntz, Ira; "A Note on the Addictive Personality: Addiction to Placebos." American Journal of Psychiatry, 134: 327, 1977.)

EFFICACY OF HOMEOPATHY

Over the last several years we have been following the feud between the homeopathists and mainstream medicine, particularly the saga of J. Benveniste. Despite what the media writers say, homeopathy continues to produce positive results, as confirmed in the following item from the British Medical Journal:

> Many doctors do not believe that homoeopathy [sic] is an efficacious treatment as it is highly implausible that infinitesimally diluted substances retain their biological effects. It is also often said that homoeopathy has not been evaluated with modern methods---that is, controlled trials. The first argument may be true, the second is not. [J.] Kleijnen et al searched the literature and found 96 reports containing 107 controlled trials of homoeopathy. Most trials turned out to be of very low quality, but there were many exceptions. The results show the same trend regardless of the quality of the trial or the variety of homoeopathy used. Overall. of the 105 trials with interpretable results, 81 showed positive results of homoeopathic treatment. A complicating factor in such reviews, especially of controversial subjects such as homoeopathy, is publication bias. If the results of Kleijnen et al do not reflect the true state of affairs, publication bias must be considered a great problem in evaluations of homoeopathy. In any event, there is a legitimate case for further evaluation of homoeopathy, but only by means of trials with sound methodology.

(Anonymous; "Clinical Trials of Homoeopathy," British Medical Journal, February 9, 1991. Cr. M. Truzzi.) The lengthy article of Kleijnen et al follows this item on p. 316 of the above reference.

Comment. Of course publication bias applies to most of the subjects presented in Science Frontiers since we intentionally concentrate on mainstream journals. Even so, anomalies are ubiquitous!

The research of J. Benveniste, mentioned above, also appeared to demonstrate the effectiveness of homeopathy. See p. 317 for more on this controversy.

BODILY FUNCTIONS

GROWTH SPURTS IN CHILDREN

Despite much anecdotal evidence and the convictions of many parents, biologists have not generally recognized the reality of short, sharp growth spurts on the order of 1 centimeter in a single 24-hour period. Rather, the consensus has been that child growth was divided into three stages (infancy, childhood, adolescence), each characterized by different, but steady rates of growth. This conclusion was based upon annual and quarterly length measurements. However, when children are measured more often (weekly or daily), the growth curve is seen to be step-like rather than smooth, as in the accompanying illustration. Indeed, the mean amplitude of the growth spurts was found to be about 1 centimeter; and the duration of the spurts, about one day. These spurts punctuated long intervals of no growth. In infants, for example, 90-95% of their development is growth-free! (Lampl, M., et al; "Saltation and Stasis: A Model of Human Growth," Science, 258:801, 1992.)

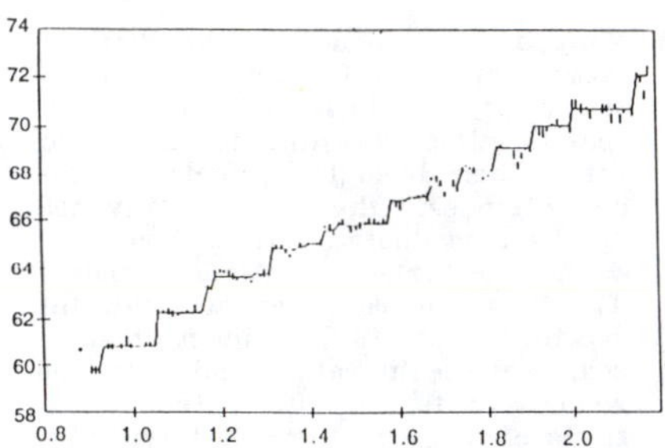

Daily length measurements (in centimeters) versus age (in years) for a male infant, showing growth spurts.

ONCE MORE SCIENCE FICTION PREDICTS THE FUTURE!

In the preceding item, the paper by M. Lampl et al demonstrated how children actually grow in sharp spurts, contrary to scientific expectations. It seems that H.G. Wells foresaw this discovery in his novel The Food of the Gods, published first in 1904. Wells had his Professor Redwood working with a measuring tape, too:

> Redwood, you know, had been measuring growing things of all sorts, kittens, puppies, sunflowers, mushrooms, bean plants and (until his wife put a stop to it) his baby, and he showed that growth went on, not at a regular pace, or, as he put it, so

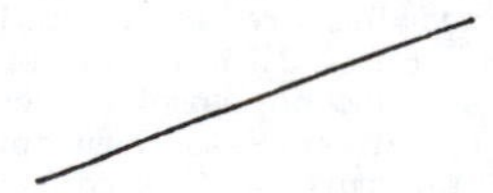

> but with bursts and intermissions of this sort,

> and that apparently nothing grew regularly and steadily, and so far as he could make out nothing could grow regularly and steadily: it was as if every living thing had first to accumulate force to grow, grew with vigour only for a time and then had to wait for a space before it could go on growing again.

The Wells omnibus Seven Famous Novels, Knopf, 1934, also contains this story. (Cr. A. Mebane)

BIOLOGICAL REGENERATION: TWO ANOMALIES

Anomaly 1: Contrary to the popular belief that mammals do not regenerate lost digits like the "lower" vertebrates, not only do mice regrow the tips of their foretoes, but young humans can regrow cosmetically perfect fingertips. However, the amputation cannot be too far back, and herein lies the second anomaly.

Anomaly 2: Foretoe regeneration in mice is astoundingly sensitive to the site of amputation. Move the site only 0.2-0.3 millimeters farther back and no regrowth will occur. No one understands why such a tiny change in distance completely changes the body's response. (Borgens, Richard B.; "Mice Regrow the Tips of Their Foretoes," Science, 217:747, 1982.)

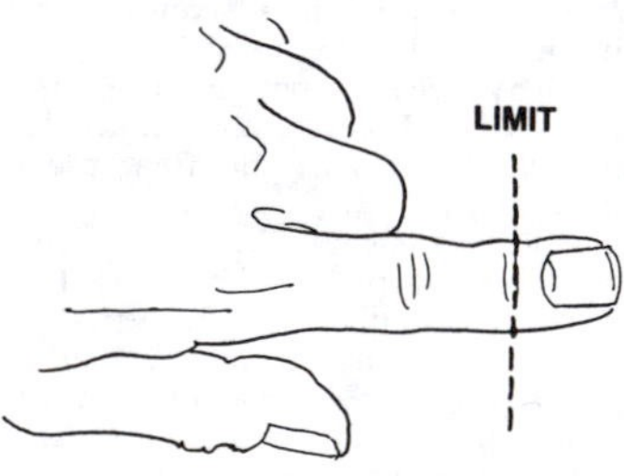

Limb regeneration in human children is limited to the fingertips.

NOSE NEWS

"Renewed discussion of a nasal breathing cycle, first discovered 5000 years ago, has recently been documented in the November 1986 issue of American Health by David Shannahoff-Khalsa of the Khalsa Foundation for Medical Science. Apparently the yogis of ancient India were the first to notice that breathing is dominated by either the right or left nostril for short cycle spans of one to three hours. (Cycles of this duration are known as ultradian rhythms, and are common to many biological functions.) By simply placing a mirror under your nostrils and watching for the larger amount of condensation, one can determine which nostril is in use.

"What are the ramifications of this seemingly insignificant phenomenon? The yogis reportedly have said that improved sleeping, more satisfying sex, enhanced digestion, and appropriate thought patterns were controlled by the use of a certain nostril." It is further maintained that one can force a change in nostril breathing through meditation. In this way, it is possible to enhance sleeping, sex, digestion, and mental acuity! (LeBow, Howard A.; "Have You Heard about This One?" Cycles, 37:191, 1986.)

OUR AQUATIC PHASE

Reviewing the evidence supporting the Aquatic Ape Hypothesis, which supposes that humans went through an aquatic phase during their evolution, E. Morgan explored the following bodily functions in depth:

1. The discovery that some prehistoric shell middens consist of deep-water shellfish, which must be the result of breath-held diving. This human skill, again unique among primates, is obviously quite ancient. Furthermore, recent experiments suggest that in humans, in addition to seals and ducks, vascular constriction is not limited to the arterioles but extends to the larger arteries, too. This indicates some degree of specialized adaptation to a diving life.

2. Most animals with a sodium deficiency display an active craving for salt which, when satisfied, disappears. In humans, salt intake has little or no relation to the body's needs. Some Inuit tribes avoid salt almost completely, while people in the Western world consume 15-20 times the amount needed for health. In other works, a single African species (assuming humans have an African origin) possesses a wildly different scheme of salt management. Humans are also the only primates to regulate body temperature by sweat-cooling, a system profligate in the use of sodium. Proponents of the Aquatic Ape Hypothesis believe that sweat-cooling could not have developed anywhere except near the sea where diets contain considerable salt, in fact much more salt than the body requires.

(Morgan, Elaine; "The Aquatic Hypothesis," New Scientist, p. 11, April 12, 1984.)

RHYTHMS IN 5,927,978 FRENCH BIRTHS

"Summary. Is there any relationship between the times when babies are born and the synodic lunar cycle? There are published works that show that there is such a relationship. We have looked at 5,927,978 French births occurring between the months of January 1968 and the 31st December 1974. Using Fourier's spectral analysis we have been able to show that there are two different rhythms is birth frequencies:

1. A weekly rhythm characterized by the lowest number of births on a Sunday and the largest number on a Tuesday;
2. An annual rhythm with the maximum number of births in May and the minimum in September-October.

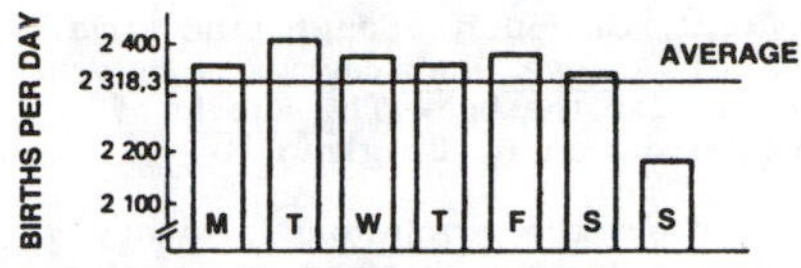

Average number of births per day, for each day of the week, found in a French study of 5,927,978 births.

"A statistical analysis of the distribution of births in the lunar month shows that more are born between the last quarter and the new moon, and fewer are born in the first quarter of the moon. The differences between the distribution observed during the lunar month and the theoretical distribution are statistically significant." (Guillion, P., et al; "Naissances, Fertilité, Rhthmes et Cycle Lunair," Journal de Gynècologie, Obstètrique et Biologic de la Reproduction, 15:265, 1986. Cr. C. Maugé)

DEATH AND SOCIAL CLASS

A 10-year study of 17,530 London civil servants showed a strong relation between mortality rate and employment grade---the higher the grade level, the lower the mortality rate. The mortality rate for unskilled laborers was three times that of high-level administrators.

Part of the disparity is doubtless due to differences in weight-to-height ratio, cigarette consumption, and amount of leisure-time exercise, which are also strongly correlated with mortality rate. But such personal habits tell only part of the story. Coronary heart disease, which accounted for 43% of all the deaths, was much more prevalent among the lower employment grades, even among nonsmokers. Childhood nutrition and other "early life factors" also play roles. Nevertheless, a factor-of-three is a whopping difference in mortality rate. (Anonymous; "Death, Be Not Proud," Scientific American, 253:68, July 1985.)

Comment. There are so many contributing factors here that we cannot be sure if a biological anomaly exists.

BLIND MAN RUNS ON LUNAR TIME

A psychologically normal blind man, living and working in normal society, was found to have circadian rhythms of body temperature, alertness, cortisol excretion, etc., that were out-of-step with society's normal 24-hour schedule. The periods of these biological cycles were about 24.84 hours and indistinguishable from the lunar day. (Miles, L.E.M., et al; "Blind Man Living in Normal Society Has Circadian Rhythms of 24.9 Hours," Science, 198:421, 1977.)

Comment. How are these apparently lunar influences communicated to the body; or are they innate?

ORGANS

On the subject of human organ anomalies, the focus of Science Frontiers has been the brain, as the number of "brain" entries below attests. Of course, we consider our brain to be evolution's crowning achievement, so why should it not get most of the attention? Having rationalized this prejudice shown in our newsletter, we hasten to assure readers that our Catalog volume Biological Anomalies: Humans II accords the vertebrate eye roughly equal billing with the human brain. Here, however, our lopsided coverage of human organs plays out as follows:

- The human brain. The effects of missing portions of the brain; the functions of the corpus callosum; the possibility of holographic memory; the brain's reprogramming capabilities.
- Heart. Aortic arch homologies in mammals.
- Ear. Sound emission by the human ear and acoustic holograms.

THE HUMAN BRAIN

IS YOUR BRAIN REALLY NECESSARY?

John Lorber, a British neurologist, has studied many cases of hydrocephalus (water on the brain) and concluded that the loss of nearly all of the cerebral cortex (the brain's convoluted outer layer) does not necessarily lead to mental impairment. He cites the case of a student at Sheffield University, who has an IQ of 126 and won first-class honors in mathematics. Yet, this boy has virtually no brain; his cortex measures only a millimeter or so thick compared to the normal 4.5 centimeters.

Although the deeper brain structures may carry on much of the body's work, the cortex is supposed to be a late evolutionary development that gave humans their vaunted mental powers and superiority over the other animals. If the cortex can be removed with little mental impairment, what is it for in the first place? (Lewin, Roger; "Is Your Brain Really Necessary?" Science, 210:1232, 1980.)

HALF A BRAIN SOMETIMES BETTER THAN A WHOLE ONE

The orthodox view of the human brain holds that the left or dominant half governs the right side of the body and is concerned with logical thought, verbal analysis, etc. The right side of the brain controls the left side of the body and is responsible for spatial and intuitive thinking. The right side supposedly cannot even participate in verbal expression. The two halves of the brain are connected by the corpus callosum. That this interconnection sometimes creates problems is evident from the fact that its severance often leads to dramatic improvement in some types of epilepsy. These split-brain individuals, however, must contend with such bizarre situations as not being able to verbally identify objects seen or felt by the left eye and hand, even though they know what the objects are.

Such situations merely confirm the orthodox view of the brain. But when half of the brain is completely removed, the conventional picture of the brain is upset. In one case, a woman with partial paralysis and frequent epileptic seizures had the left side of her brain removed. Her seizures and paralysis disappeared permanently; even more, her personality improved markedly. The half of the brain that remained assumed all brain functions and performed them better than the complete brain had. Conclusion: each half of the human brain has the intrinsic capability of operating as a whole brain despite the usual specialization of the halves. (Gooch, Stan; "Right Brain, Left Brain," New Scientist, 87:790, 1980.)

LEFT-HANDERS HAVE LARGER INTER-BRAIN CONNECTIONS

The two halves of the human brain are connected by a bundle of nerve fibers called the corpus callosum. The corpus callosum is thought to help integrate the activities of the right and left brains which, for reasons unknown, seem to specialize in different kinds of mental operations. Studies of the corpus callosum reveal that it is about 11% larger in left-handers than in right-handers. In terms of interconnecting nerve fibers this comes to 25,000,000 more for the left-handers. Just what sort of information flows along these myriad pathways is not known, although we do know that left-handers have a greater bihemi-

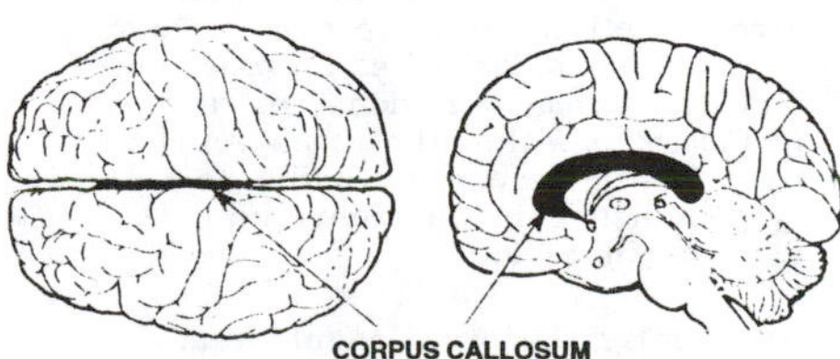

Top view and section through the human brain showing the location of the corpus callosum.

spheric representation of cognitive functions; i.e., the brain functions are not so specialized in each half of the brain.

But why should left-handers and right-handers be different at all? Are they born with unequal corpus callosa? Or are these nerve highways equal at birth and atrophy in right-handers? (Witelson, Sandra F.; "The Brain Connection: The Corpus Callosum Is Larger in Left-Handers," Science, 229:665, 1985.)

ANOMALOUS EEG DISCHARGES

Two researchers at the Mayo Clinic have discovered a new, distinctive type of brain-wave burst in the EEG recordings of 65 patients with a wide variety of neurological and brain complaints. These EEGs, which were otherwise normal, would commence quietly but suddenly erupt into strong, rapid, rhythmic pulses. Episodes would last 40-80 seconds, sometimes even longer. The anomalous, high-amplitude waves occasionally occurred while the subjects were asleep. The mystery surrounding this discovery is that the subjects did not notice the discharges and showed no outward signs of this violent brain activity. The 65 patients had little in common except being ill in diverse ways. (Anonymous; "Red Herring," Cycles, 32:200, 1981. As digested from: Electroencephalography and Clinical Neurophysiology, 51:186, 1981.)

BRAINS NOT HARDWIRED

The prevalent conception of the brain compares it to a hardwired computer in which all the wires and components are all permanently soldered together. An equivalent situation would prevail in the brain if all sensory pathways and cells had fixed duties and memories to handle. If the portion of the brain dedicated to speech were damaged, as in a stroke, it could never repair itself. This dogma is now being challenged.

A pertinent line of brain research is now underway at the Coleman Laboratory of the University of California in San Francisco, where Michael Merzenich and his associates are studying the brains of monkeys.

> Merzenich's findings challenge a prevailing notion that most sensory pathways in the nervous system are 'fixed' or 'hardwired' by the maturation of anatomic connections, either just before or soon after birth. They also address the puzzling question of what forces may be at work when stroke victims partly recover. Do 'redundant copies' of skills exist outside the damaged regions, or is physical damage within the brain repaired over time? Or can old skills be newly established in different, undamaged brain regions.

Apparently the brain should really be compared with a reprogrammable computer. Perhaps the brain even stores duplicates of critical "programs"; i.e., skills. Merzenich's findings go even farther. He finds that the parts of the brain associated with certain skills or data processing move and change shape spontaneously. The brain, it seems, continually reorganizes itself. Fading fast is the idea that each data point is recorded in a specific cell or neuron interconnection. (Fox, Jeffrey L.; "The Brain's Dynamic Way of Keeping in Touch,: Science, 225:820; 1984.)

MILLION-CELL MEMORIES?

Brain researchers have long believed that simple memories were stored as "traces" in chains of brain cells. However, E.R. John and his colleagues have confounded such thinking. They have prepared metabolic memory maps of cats' brains using carbon-tagging. In this way, they discern which parts of the brains have been activated during the recall of an element of memory. Such experimentation has demonstrated that huge numbers of brain cells actively participate in the recall of a simple thought. John stated, "I thought we'd find maybe 20,000 to 40,000 cells involved in the learned memory....The shock was that it was so easy to see wide-spread metabolic change....The number of brain cells [between 5 million and 100 million] involved in the memory for a simple learned discrimination made up about one-tenth of the whole brain."

The findings of John et al are hotly contested by some brain researchers. One obvious conflict is that if up to 100 million brain cells are involved in storing just one simple memory, the brain will quickly use up all available cells. It must be that individual brain cells can participate in the storage of many different memories. The conventional memory-trace theory would have to be replaced by a new type of memory architecture. (Bower, Bruce; "Million-Cell Memories," Science News, 130:313, 1986.)

Comment. Our thinking about biological memory may be controlled by our preoccupation with the two-dimensional circuits of computer memories. Biological memories might be three-dimensional, or of even higher order. Some scientists have ventured that memory might entail electrical charge distribution patterns in the brain; such need not be limited to two dimensions. The same thinking can be applied to the storage of genetic information. While DNA, RNA, etc., may be pieces in the puzzle, the complete solution may include the ways in which these molecules are bent, twisted, convoluted, arrayed, juxtaposed, and so on.

In a letter relating to the above article, P.J. Rosch points out that John's results are consistent with the holographic theory of brain function supported by Pribram and Bohm.

> In a hologram, every element of the subject is distributed throughout the photographic plate, making it possible to reconstruct the entire original image from any portion of the picture. In this paradigm, the brain stores memory and deals with interactions by interpreting and integrating frequencies, retaining the data not in a localized area but dispersed throughout its substance.

(Rosch, Paul J.; "The Brain as Hologram," Science News, 130:355, 1986.)

Comment. Curiously enough, the same issue of Science News carries an advertisement for the book The Fabric of Mind, in which R. Bergland:

> ...offers the revolutionary theory, already stirring controversy among fellow researchers, that the brain is actually a gland and depends on changes of hormones and molecules for its function....While he does not deny that electrical impulses occur in the brain, they are only superficial signals, he says, and not as important in conveying messages to the brain and within it, as hormones are.

BRAIN ARCHITECTURE: BEYOND GENES

> The human brain probably contains more than 10^{14} synapses, and there are simply not enough genes to account for this complexity. .... 'Neuroscientists at a recent meeting highlighted how extragenetic factors---including neuronal activity, contact with other cells, radiation, and chemical factors---influence brain circuitry, especially during development.

In computer terms, our brains are constantly reprogramming themselves in response to internal and external forces. (Barnes, Deborah M.; "Brain Architecture: Beyond Genes," Science, 233: 155, 1986.)

Comment. Musing in a Lamarckian way, can the brain, as reprogrammed by external influences, ever feed back information to the genes? Certainly, we do not know now of any way in which this could happen. However, since the brain is known to influence the immune system and other bodily functions, the idea cannot be dismissed arbitrarily. See p. 297.

COOLER HEADS, BIGGER BRAINS?

When anthropologist D. Falk discovered that an automobile's engine was limited in power by its radiator's capacity to cool it, he applied this thinking to the human brain.

The human brain, like the automobile engine, must be kept cool if it is to function well. It follows that if the brain of an animal is not functioning well, the body that brain controls will not perform well either. Overheated brains, then, are sure roads to extinction in the highly competitive natural world. A couple million years ago, two groups of human precursors were competing for dominance in Africa. The group that won and subsequently evolved into Homo sapiens had, according to Falk, a better brain-cooling system. The evolutionary development that probably led to this advantage was a more extensive network of emissary veins, which permitted better dissipation of heat. This, in turn, allowed the evolution of larger brains and dominance by Homo sapiens. Other anthropologists, however, doubt that such a minor change in the circulatory system could account for the emergence of modern man. (Shipman, Pat; "Hotheads," Discover, 12:18, April 1991.)

Comment. What an intriguing concept! Perhaps human male baldness also confers more cooling efficiency and is setting the stage for a new expansion of the human brain---at least the male brain. Sorry, girls! More seriously, did the better blood-cooling system develop in response to an enlarging brain, or vice versa? Even more seriously, it is simplistic to say that an organism just went ahead and evolved this way or that way. A bigger brain requires not only more cooling but a bigger skull, more neurons, more connections between them, and additional infrastructure. Saying simply that "a larger brain evolved" obscures the fact that many different, inheritable changes had to take place in synchronism, and that we really have no inkling as to how the size change was carried out on the microscopic, cause-and-effect level.

HEART

THE AORTIC ARCH AND EVOLUTION

Comparative anatomy is supposed to tell us which creatures are closely related so that we can draw those familiar evolutionary family trees. That anatomical similarities may be misleading is proved by the various configurations of the mammalian aortic arch---certainly one of the major body structures. Five principal configurations of mammalian aortic arches are sketched in the accompanying figure. The species possessing these various configurations make kindling of the usual evolutionary family trees.

1. Horses, pigs, deer;
2. Whales, shrews;
3. Marsupials, rats, dogs, apes, monkeys;
4. The platypus, sea cows, some bats, humans;
5. African elephants, walruses.

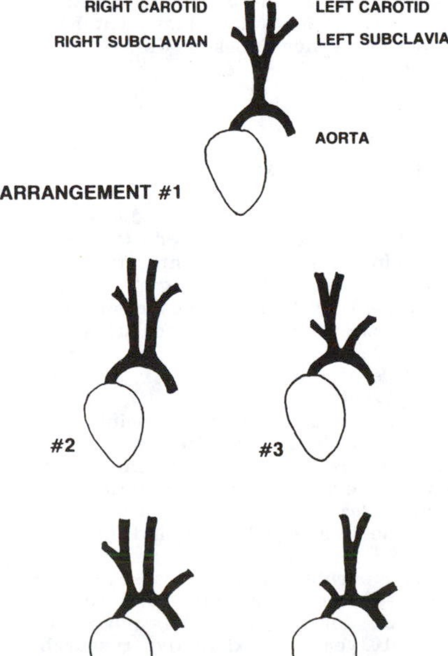

Various configurations of mammalian aortic arches. See text for discussion.

(Davidheiser, Bolton; "The Aortic Arch," Creation Research Society Quarterly, 20:15, 1983.)

Comment. On this basis alone, humans are more closely related to seacows than the apes! Why aren't such discrepancies highlighted in the mainstream scientific literature?

EAR

HEARING VIA ACOUSTIC HOLOGRAMS

Humans and the higher primates can locate the source of a sound without turning their ears or heads. Other animals are not so fortunate. Current theories of hearing, according to Hugo Zuccarelli, cannot explain this human capability, which we all take for granted. He has come up with a new theory that pictures our ears as truly remarkable organs. First, our ear itself is a sound emitter. It emits a reference sound that combines with incoming sound to form an interference pattern inside the ear. The nature of this pattern is sensitive to the direction of the incoming sound. Our ear's cochlea detects and analyzes this pattern as if it were an acoustic hologram. The brain then interprets this data and infers the direction of the sound. (Zuccarelli, Hugo; "Ears Hear by Making Sounds," New Scientist, 100:438, 1983.)

Comment. We have been able to appreciate this slick biological trick only after we "discovered" holograms. We should wonder if we are missing anything else!

Two letters quickly appeared casting doubt on not only Zuccarelli's Theory but his personal scientific capabilities. (Baxter, A.J., and Kemp, David T.; "Zuccarelli's Theory," New Scientist, 100:606, 1983.)

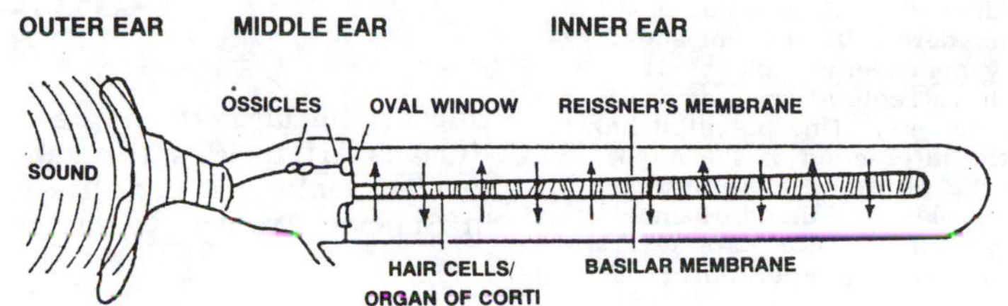

(Above) In the most popular theory of the ear's operation, the basilar membrane alone activates the hair cells and creates the sensation of sound. (Below) In Zuccarelli's theory, both Reissner's membrane and the basilar membrane work together to form an acoustic hologram with the aid of a reference sound signal created by the ear itself.

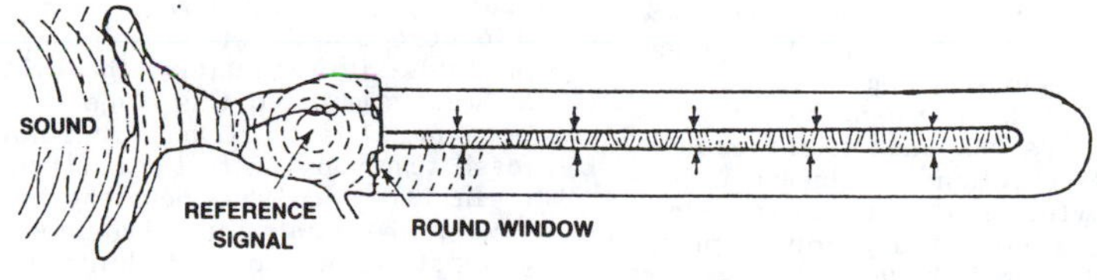

INTERNAL SYSTEMS

The term "internal systems," as used in the Catalog of Anomalies, includes the immune system, the central nervous system, the lymph system, etc. In the first 86 issues of Science Frontiers, only the immune system has received any attention, and then primarily for its possible association with psychological factors. See Biological Anomalies: Humans II for a more complete treatment.

MIND MARSHALS WHITE BLOOD CELLS

If a hypnotist suggests to a receptive subject that his white blood cells are attacking cancer cells in his body, the population of white blood cells ranging through the subject's body will increase. Such "visualization" research is being conducted by Howard R. Hall at Penn State. The results seem to demonstrate the direct influence of hypnotic suggestion on the body's immune system. (Anonymous; "Hypnotism May Help Antibody Production," Baltimore Sun, October 19, 1981. AP item)

Comment. Since hypnotic suggestion is known to affect warts, body temperature, etc., its enhancement of the white blood cell population is merely one more example of the power of the mind over the body. The real mystery is just how the brain's electrical signals are converted into specific biological activity.

THE IMMUNE SYSTEM AS A SENSORY ORGAN

John Maddox, the editor of Nature, has written a remarkable editorial on psychoimmunology; that is, the science of the brain's effects on the body's immune system. It is basically a running commentary on new discoveries that are helping us to understand this poorly appreciated relationship. Maddox begins by mentioning the 20-plus-year collection compiled by Professor G.W. Brown, University of London, of life-events that affect the health of outwardly normal people. Typical life-events are the death of a spouse, imprisonment, personal bankruptcy, etc. Everyone seems to recognize---if only through anecdotes ---that mental states affect health, but how this brain-body link is maintained is hard to pin down. D. Maclean and S. Reichlin (Psychoneuroimmunology, 12: 475, 1981.) have reviewd some of the possible connections. One potential link is through the interaction of the hypothalamus on the pituitary. The pituitary is a source of materials that influence the immune system. Maddox lists several specific candidates, and then observes:

> The more radical psychoimmunologists talk as if there is no state of mind which is not faithfully reflected by a state of the immune system.

So far, not too radical! But then Maddox comes to an article by J.E. Blalock, University of Texas (Journal of Immunology, 132:1067, 1984.) bearing the title, "The Immune System as a Sensory Organ." Blalock argues that the interaction between the central nervous system and immune system must be reciprocal. By this he means that the immune system's response to infection, through the secretions of disease-fighting lymphocytes, gets back to the central nervous system and produces physiological and even behavioral changes in the infected animal.

Applicable studies of animals have been reported recently. For example, rats under stress are found to have less easily stimulated immune systems. (Science, 221:568, 1983.) Also, men who have recently lost their wives to breast cancer have immune systems less responsive to mitogens. (Journal of the American Medical Association, 250:374, 1984.) (Maddox, John; "Psychoimmunology Before Its Time," Nature, 309:400, 1984.)

Comment. This is an appropriate time to suggest that "psychoevolution" may be physiologically possible. If the brain can fight disease and even control cell growth, why not a role for the mind in stimulating the development of new species, perhaps in response to extreme environmental pressures, and perhaps not on the conscious level? The body's sensory system would detect great external stresses, the brain would process the information, and direct some astute genetic shuffling. The genetic inheritance of an organism is not sacrosanct. Radiation, chemicals, and various others mutagens are recognized. There seems to be no a priori reason why the brain-body combination cannot generate mutagens---possibly not randomly but inteligently! (We ignore here selfish DNA and Sheldrake's morphogenic fields.) Does this mean that if we wish to mutate, we can? Well, it's probably not as simple as wishing warts away, but Maddox's editorial underscores the complexity and subtlety of the brain-body combination.

PERSONALITY AND IMMUNITY

Abstract. "Natural killer (NK) cells are important in immune function and appear, in part, to be regulated by the CNS (central nervous system). The authors compared NK cell activity and MMPI scores of 111 healthy college students and found weak but statistically significant correlations between NK values and psychopathology for 10 of 12 scales. Students with the highest NK values had a 'healthier' MMPI profile than those with the lowest. Students with high MMPI scores (T > 70) had NK values below the sample median. These findings support theories of interaction between mental state and immune status, but the mechanisms and direction of interaction remain largely unexplored." (Heisel, J. Stephen, et al; "Natural Killer Cell Activity and MMPI Scores of a Cohort of College Students," American Journal of Psychiatry, 143:1382, 1986. Also: Bower, B.; "Personality Linked to Immunity," Science News, 130:310, 1986.)

SKELETAL REMAINS

One of the largest chapters in the Catalog volume Biological Anomalies: Humans III tackles the subject of anomalous and controversial human fossils. Only two such items from this immense collection have appeared in Science Frontiers; and one of these is highly suspect.

FLAT-FACED HOMINID SKULLS FROM CHINA

The "African Eve" theory of human evolution was given much play in the media a few years back. According to the "African" view, modern humans arose exclusively in Africa and, about 100,000 years ago, expanded rapidly from there into Europe and Asia, displacing "lesser" hominids. Unfortunately, the DNA studies that stimulated this conjecture have been found to be flawed. And now new fossil testimony casts further doubt.

In 1989 and 1990, near the Han River, in China's Hube Province, anthropologists found hominid skulls with the characteristic flat faces of modern humans. These skulls seem to be about 350,000 years old. Although they apparently retain some primitive features, paleoanthropologist D. Erler, of the University of California, asserted, "This shows that modern features were emerging in different parts of the world." In other words, all of the evolutionary action was not confined to Africa. Proponents of the "African Eve" theory retort that the dating of the Chinese skulls is questionable and that flat faces alone are not enough to support the idea that modern humans arose separately in widely separated locales? (Gibbons, Ann; "An About-Face for Modern Human Origins," Science, 256:1521, 1992. Also: Bower, Bruce; "Erectus Unhinged," Science News, 141:408, 1992.)

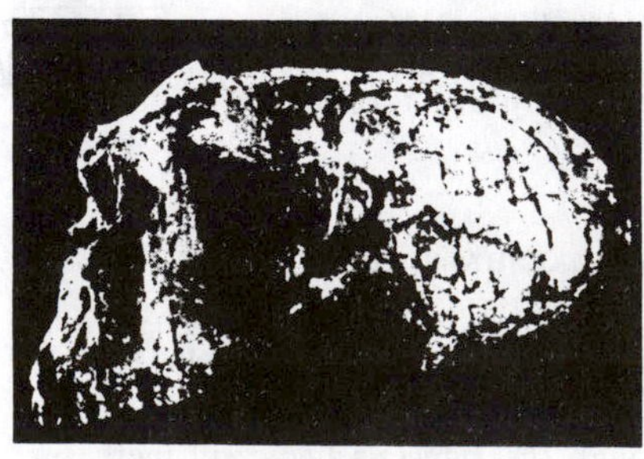

Flat-faced skull from China. (UC, Berkeley)

Comment. Could the African and Asian fossils imply that so-called "parallel" or "convergent" evolution has occurred in the human lineage, too, just as it has in so many other forms of life?

ANTHRACITE MAN?

It is not surprising that the discovery described below has not made its way into mainstream scientific literature. Most mainstream anthropologists would shy away from human bones reputed to come from anthracite coal measures hundreds of millions of years old! Our source is a small newspaper in eastern Pennsylvania.

Scientifically acceptable hominid fossils are no older than a few million years at most. So, when anyone cognizant of prevailing paradigms enters the Greater Hazelton Historical Society Museum, he is astonished to find an:

> ...elaborate display of rock-like objects found in the anthracite region by Ed Conrad who insists, based on his 10 years of exhaustive research and scientific testing, that he possesses undeniable evidence that they are petrified bones.

Society officials undoubtedly are impressed because a small sign displayed on a front window carries some very large words: "This is the only museum in the world where petrified bones, found between coal veins, are on display."

The photos accompanying the newspaper article certainly portray objects that "look like" human skulls. E. Conrad also asserts that he had also found hominid jawbones, teeth, a femur, and even a petrified brain! (Anonymous; "Bone Display Draws National Interest," Hazelton Standard-Speaker, December 8, 1990. Cr. L. Farish)

Comment. From the newspaper article it is impossible to learn what professional geologists and anthropologists think about these purported fossils. So, caveat emptor.

Actually, hominid fossils have been reported from coal veins before, notably the "Abominable Coalman" found in an Italian coal mine. See Ancient Man.

GENETICS

DESCENT OF MAN---OR ASCENT OF APE?

New Scientist has just published a controversial pair of articles by John Gribbin and Jeremy Cherfas. Summarizing mightily, it seems that:

1. There are no fossils that are unequivocally ancestral to chimpanzees and gorillas but not to man;
2. Therefore, the only good measure of the time when these three species split from one another is the comparison of genetic material;
3. Genetic dating and serological techniques are unanimous in dating the chimp-gorilla-man split at about 5 million years ago.

The conclusion that chimpanzees, gorillas, and humans diverged from a common ancestor only 5 million years ago is opposed to the widely accepted 20 million years. This conflict in dating is controversial enough, but Gribbin and Cherfas, after considerable fossil analysis, take one more giant step: they suggest that chimps, gorillas, and man descended from an ancestor that was more man-like than ape-like. Chimpanzees and gorillas in this view are descended from man rather than vice versa. (Cherfas, Jeremy, and Gribbin, John; "The Molecular Making of Mankind," and "Descent of Man---Or Ascent of Ape?" New Scientist, 91:518 and 91:592, 1981.)

Comment. This hypothesis is inflammatory enough without our adding more fuel, but the possible connection to the Sasquatch/Abominable Snowman problem should not be overlooked.

THE ALIEN PRESENCE

Hidden behind an obscure technical title is a most curious discovery. I.C. Eperon and his coworkers at the MRC Laboratory of Molecular Biology, England, have shown that "human mitochondria did not originate from recognizable relatives of present day organisms." The authors go even further, describing human mitochondria as a "radical departure." (Eperon, I.C., et al; "Distinctive Sequence of Human Mitochondrial Ribosomal RNA Genes," Nature, 286:460, 1980.)

Comment. The inferences above may be far-reaching. Mitochondria are vital components in the cells of the so-called higher organisms. Apparently possessing their own genetic material, they are suspected of being descendants of ancient bacteria that invaded and took up residence in cells. If human mitochondria are radically different, could changes in mitochondria be the source of the purported wide gap between humans and other animals? Did the mitochondria change ("evolve") in existing ancient mammals, converting them suddenly into humans? Or did a new "species" of mitochondria infect terrestrial cells, perhaps coming to earth on cosmic debris, as Fred Hoyle has suggested?

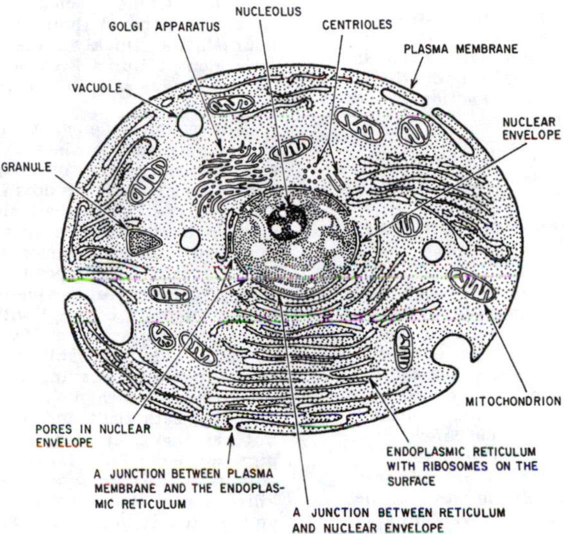

One mitochrondrion is indicated at the lower right, but inspection reveals several others with the cell.

THE CHROMOSOME GAP

"Compared with the chromosomes of humans and other great apes, the pygmy chimpanzee's chromosomes are...the most specialised...they have changed more over time than have the others. 'Surprisingly,' Stanyon and his colleagues conclude, 'the human karyotype (chromosome complement) is the most conservative...It has more unchanged ancestral chromosomes.'" (Anonymous; Chromosomes Show Apes 'More Evolved Than Man'," New Scientist, p. 24, July 17, 1986.)

UNRECOGNIZED HOMINIDS

NEANDERTHAL MAN MAY STILL SURVIVE IN ASIA

All continents have their tales of wild men, abominable snowmen, sasquatch, etc. Most anthropologists give little credence to these stories. Shackley, however, has assembled considerable evidence for the reality of the so-called Almas (Plural form: Almasti), primitive men that closely resemble Neanderthal Man, or at least what we think Neanderthal Man looked like. Abundant, internally consistent data come from an east-west band running from the Caucacus, across the Pamir Mountains, through the Altai Mountains, to Inner Mongolia. Even today, sightings of these creatures are rather common; and several scientists have seen them. One incident occurred in 1917, when the Reds were pursuing White Army forces through the Pamirs.

The troops of Major General Mikail Stephanovitch Topilsky shot an Almas as he was emerging from a cave.

> The eyes were dark and the teeth were large and even and shaped like human teeth. The forehead was slanting and the eyebrows were very powerful. The protruding jawbones made the face resemble the Mongol type of face. The nose was very flat ...the lower jaws were very massive.

In some instances the Almasti have even associated with modern man; and cases of successful interbreeding have been reported. After reviewing the mountains of evidence, Shackley feels that the Almasti are very likely surviving Neanderthals, because the physical characteristics of the Almasti and reconstructed Neanderthals are basically identical. This long review article also discusses the many Chuchunaa sightings from northern Russia---perhaps another relict population of Neanderthals. (Shackley, Myra; "The Case for Neanderthal Survival: Fact, Fiction or Faction?" Antiquity, 56:31, 1982.)

CHECKLIST OF APPARENTLY UNKNOWN ANIMALS

B. Heuvelmans, who operates the Center for Cryptozoology, in France, has compiled an annotated checklist of between 110 and 138 animals (some questions remain about how many are distinct species) which do not seem to be recognized by science. His list is based upon his collection of 20,000 references. We pass along this interesting one.

Hairy "wild men," known as satyrs in classical antiquity. These were probably Neanderthals that survived into historical times. The most recent sightings were in 1774, in the Pyrenees, and 1784, in the Carpathians. (Heuvelmans, Bernard; "Annotated Checklist of Apparently Unknown Animals with Which Cryptozoology Is Concerned," Cryptozoology, 5:1, 1986.)

YETI OR WILD MAN IN SIBERIA?

Reports from Russia tell of a creature known locally as the "Chuchunaa" which is over 2 m tall, clad in deerskin, and unable to talk, although it does utter a piercing whistle. A man-eater, the Chuchunaa often steals food from settlements. Observers say that the creature has a protruding brow, long matted hair, a full beard, and walks with its hands hanging below its knees. Soviet scientists speculate that the Chuchunaa represents the last surviving remnant of the Siberian paleoasiatic aborigines that retreated to the upper reaches of the Yana and Indigirka rivers. The last reliable sightings were in the 1950s, and this animal may now be extinct. (Anonymous; Nature, 271:603, 1978.)

CHINESE HUNT RED-HAIRED BIGFOOT

Spurred by reports of large (6½-feet-tall) animals with wavy red hair walking on two legs, Chinese scientists have been combing the thick forests of Shennongjia, in Hubei Province. Many footprints 12-16 inches long as well as samples of hair and feces have been found. So far, though, no photos or specimens. (Anonymous; "It's Tall, It Has Wavy Red Hair and Chinese Keep Hunting for It," New York Times, p. 5, January 5, 1980.)

THE CHINESE WILD MAN

The Chinese Wild Man seems to have much in common with the North American Sasquatch or Bigfoot, if we are to believe all the reports coming out of China these days. From western Yunnan and northwestern Hubei provinces come hundreds of recent sightings. Since 1976, four Chinese scientific expeditions have concentrated their attentions in the mountainous, thickly forested Shennongjia region of Hubei Province. So far, though, there are no specimens or even good photos.

The major evidence for the existence of the Wild Man consists of anecdotal reports, many casts of footprints (18 inches long), hair (reddish), and samples of feces. The same situation prevails in North America as far as Sasquatch evidence is concerned. Summarrizing recent sightings: the Wild Man is a bipedal creature, seven-feet-plus in height, usually covered with reddish hair, possessing human features, with no tail, having the ability to laugh and cry, capable of weaving bamboo sleeping couches, and with no fear of fire. The Wild Man eats fruit and small animals, but has also been known to steal small pigs and corn from farmers. An anecdote from the 1940s: a band of hunters killed a Wild Man with a machine gun and cooked it in a pot. The taste was so foul that no one would eat it! (Wren, Christopher S.; "On the Trail of the 'Wild Man' of China," New York Times, June 5, 1984, p. C1. Cr. P. Gunkel)

A reconstruction of a yeti based upon many sightings. (Adapted from Heuvelmans, Bernard; On the Track of Unknown Animals, New York, 1958)

FIRST YETI PHOTOS?

A.B. Wooldridge claims that he observed and photographed a yeti in the Himalayas in March 1986. Travelling alone toward Hemkund, at about 11,200 feet, in an area with steep wooded slopes, he encountered strange 10-inch tracks, which he duly photographed. Pushing on, he was crossing an exposed snow slope at 13,000 feet, when his run was halted by a wet snow avalanche. Moving closer to the avalanche to assess the snow's stability, he again saw the strange tracks heading across the slope to a small bush. "Behind the bush stood an erect entity over 6 feet tall. The figure, of general human proportions and stance, remained immobile, seemingly looking down the slope. 'The head was large and squarish, and the whole body appeared to be covered with dark hair.'"

Wooldridge quickly snapped several photographs. He then advanced to within 500 feet of the entity and took more pictures. After 45 minutes of observation, Wooldridge decided to continue his journey. When asked why he did not approach the figure to force it to move or react, he stated that he got as close as he felt it was safe, being concerned about snow stability, the creature itself, and his solitary situation. (Anonymous; "First Yeti Photos Spark Renewed Interest," ISC Newsletter, 5:1, Winter 1986. ISC = International Society for Cryptozoology.)

Comment. The photos and sketch drawn under Wooldridge's guidance certainly do show a human-like creature. The maddening aspect of this whole business is the near motionlessness of the entity. If only it had moved significantly during the picture-taking. Instead of a smoking gun, we have just smoke; that is, enticing but still unconvincing data.

YETI EVIDENCE TOO HARD!

Above, we printed a news item about photographs taken in the Himalayas by A.B. Wooldridge that were thought to show a Yeti. The latest issue of Cryptozoology contains a letter from Wooldridge in which he tells of his return, in 1987, to the site where the 1986 photos were taken. The bush in the 1986 photos was still there; the Yeti wasn't; the snow was somewhat deeper. Wooldridge and some companions took more pictures of the site:

> "Stereo pairs of photos taken in 1987 have been used to produce a three-dimensional map of the terrain near the bush. When this is used to derive an absolute scale for pairs of photos from 1986, it shows that, whatever I photographed in 1986, lies below the snow level in the 1987 photos. The object is leaning slightly uphill, and no movement can be detected when comparing photos taken at different times in 1986. The apparent change in position relative to the bush in some photos taken from different camera positions is caused by parallax. This evidence demonstrates beyond a reasonable doubt that, what I had believed to be a stationary, living creature was, in reality, a rock."

(Wooldridge, Anthony B.; "The Yeti: A Rock After All?" Cryptozoology, 6:135, 1987.)

OTHER MAMMALS

EXTERNAL APPEARANCE

Not surprisingly, Australian fauna account for almost half of the mammal anomalies and curiosities collected here. Perhaps some of the bizarre characteristics of echidnas, the platypus, and kangaroos do not seem so bizarre to Australians who see these creatures frequently; but to us they do. Familiarity aside, these mammals from Down Under do pose some interesting problems to biologists; for example, the apparent parallel evolution of form and function, as evidenced below by the platypus/beaver and echidna/anteater "convergences."

The remainder of the section is concerned with some remarkable features of placental mammals; i.e., the babirusa's bizarre tusks and the polar bear's "invention" of the thermal diode!

ECHIDNA ECCENTRICITIES

The echidna is one of the monotremes--an egg-laying mammal. Like its relative, the platypus, it is a strange mixture of mammalian, marsupial, and reptilian characteristics. For example, echidna eggs are soft and leathery, like those of reptiles, but they are brooded in a marsupial-like pouch. The emerging baby echidna has an egg tooth like the birds and reptiles, while the adult has no teeth at all. Rather, it has a narrow snout through which it ingests ants and termites caught on its sticky tongue. In this it resembles the mammalian anteaters, which are also toothless but an ocean away from Australia. In fact, the echidna is often called a "spiny anteater" for it has the sharp spines of a hedgehog or porcupine.

There are more anatomical peculiarities, but let us focus on the echidna's strange behavior during the mating season. At this time, 2 to 8 echidnas can be seen roaming the Australian bush in "trains" headed by a female with the smallest male acting as a caboose. When mating time arrives, the female anchors herself to a tree with her forelegs. Together the males dig a circular "mating rut" up to 10 inches deep around the tree. (Australians have puzzled over these circular trenches for years.) Eventually the strongest male evicts the other males from the trench, the purpose of which now becomes apparent. As the old saying goes, porcupines make love very carefully. Well, the echidna has an interesting technique; he simple lays on his side in the trench under the female! (Rismiller, Peggy D., and Seymour, Roger S.; "The Echidna," Scientific American, 264:96, February 1991.)

Two monotremes. (Above) One of the two species of echidnas. (Below) The platypus.

PLATYPUS PARADOXES

After elucidating echidna eccentricities in the preceding item, we now provide platypus paradoxes.

•Did you know that the platypus bill is a finely tuned instrument with approximately 850,000 electrical and tactile receptors? These are far more sophisticated than those found in fish. When the platypus goes foraging underwater, a furry groove closes, covering its eyes and ears, and the nostrils on the bill are sealed shut. It becomes a high-tech predator---despite all those snide remarks about its primitive nature.

•The poison spurs on the back legs of the male platypus are nothing to fool around with. They can cause humans severe pain and weeks of paralysis. And a dog can lose its life when a platypus clamps its legs around its muzzle and drives in its spurs. But, ask evolutionists, how did this poison apparatus get on the hind legs? The supposed ancestors of the platypus, the reptiles, modified their salivary glands for venom delivery. How did the platypusses break from this evolutionary mold and innovate?

•The fossil record reveals that a platypus-like creature lived long before the Age of Mammals. These early platypusses had teeth in the adult phase, whereas their modern relatives replace their baby teeth with horny plates---another innovation. Therefore, far from being a hodgepodge of parts left over from bird and reptile evolution, the platypus has actually pioneered several zoological features.

•Very curious is the fact that the platypus is in many ways like the beaver---a very, very distant relative both in distance and position on the Tree of Life. Both platypus and beaver are furry, aquatic creatures with webbed feet and a large, flat tail. We have saved the strangest part for the end! Platypusses, being Monotremes (one-enders) have a common vent for waste and reproduction. Beavers, it turns out, are among the very rare placental mammals that (like the birds) possess a cloaca---a common vent for urine and excrement. (Hoffman, Eric; "Paradoxes of the Platypus," Scientific American, 264:18, March 1991.)

Comment. The "beaver" part just mentioned was not in Hoffman's article; we added it because it underscores the apparent convergent evolution of platypus and beaver.

GETTING THE POUCH RIGHT

When we think of kangaroos hopping about Australia (which isn't very often), we know that all baby kangaroos are safely "buckled in" their mothers' pouches---nature's own vehicle restraint system. How fortunate it is that kangaroo pouches open at the top; otherwise, little kangaroos would be falling out all over the place. While some marsupials have pouches opening "up" or "forward" in quite a few others evolution got the directions for pouch manufacture reversed. The koala, the wombat, the thylacine, and the marsupial mole all have backward-opening pouches. Obviously, a forward-opening pouch on the mole would act like a dirt scoop, to the great inconvenience of any occupants. On the other quadrupeds, the backward-opening pouch may protect the young from branches and vegetation. (Marshall, Jeremy H.; "Directional Pouches," Nature, 309:300, 1984.)

Evolution gone wrong? Nature's cartoon of a kangaroo with a "wrong-way" pouch. Some female marsupials have no pouch at all; then again, one male marsupial (the yapok) sports a pouch!

Comment. This is an example of the so-called "problem of perfection," where life seems marvelously attuned to its environment; that is, "fittest." Somewhere among the millions of species alive today, there must be one out-and-out failure. Of course, if full-scale nuclear war breaks out, we will know that evolution did make at least one mistake!

THE BABIRUSA: A QUASI-RUMINANT PIG

The babirusa, an inhabitant of Indonesia, looks like a thin pig, but its stomach is like that of a sheep, which is a "simple" ruminant. The babirusa's stomach possesses an extra sac; and the animal often browses on leaves and shoots. It does not, however, chew a cud.

Taxonomists are a bit puzzled over the babirusa. They aren't sure whether its closest relatives are modern pigs, peccaries, or hippos.

The babirusa's "tusks" pose more questions about its evolution:

> The creature's oddest characteristic is the two impressive pairs of curving tusks grown by the males. One pair are simply extended lower canines, but the second are actually upper canines, the sockets of which have rotated, resulting in tusks that grow through the top of the muzzle and emerge from the middle of the animals's face. The effect is bizarre and startling. The males fight with their dagger-like lower canines and probably deflect opponents' blows with the upper set, thus protecting their eyes. Indonesians say the tusks are similar to deer antlers, giving the babirusa its name, which means 'pig deer.' (Rice, Ellen K.; "The Babirusa: A Most Unusual Southeast Asian Pig," Animal Kingdom, 91:46, March/April 1988.)

The Indonesian babirusa, a pig-like mammal with curious tusks and (for a pig) an unusual digestive system.

Comment. Turning a pair of teeth 180° in the upper jaw is a fascinating evolutionary accomplishment. It is difficult-to-explain on the basis of random mutations, especially in view of the fact that many pigs, with lower tusks only, get along quite well.

POLAR BEAR COATS ARE THERMAL DIODES

"A polar bear's hairs are completely transparent. The bear appears white because visible light reflects from the rough inner surface of each hollow hair. However, the hairs are designed to trap ultraviolet light. Like light within an optical fiber, the radiation is conducted along the hairs to the skin. This summertime energy supplement provides up to a quarter of the bear's needs. Thus, even while actively pursuing prey, the bear can still concentrate on building up its blubber layers in preparation for winter." In other words, the bear's fur lets heat in but not out---in effect a thermal diode. (Anonymous; "Solar Bear Technology," Science News, 129:153, 1986.)

Comment. How come polar bears are favored with this "marvelous adaptation" while the arctic foxes and other mammals shiver?

SQUIRRELS AS MEASURES OF GEOLOGICAL TIME

Over a century ago, when the truth of biological evolution via natural selection was hotly debated, the proponents of Darwinism were delighted when the geologists presented them with almost endless periods of time in which evolution could progress in small steps from species to species. Now, in a strange turnabout, a creationist writer is using evolutionary theory to infer a very short history for the formation where geologists want a good deal of time. We quote from the conclusion of J.R. Meyer:

"If any group of animals were ever going to undergo significant degrees of evolution from parent stock and obtain resultant speciation, surely the Kaibab squirrel would be one of the more likely candidates. Supposedly isolated from their neighbors for hundreds of thousands of generations over a period of at least several million years, and significantly violating virtually every restriction of the Hardy-Weinberg equilibrium for the non-evolving population, these organisms, even by creationist standards, should have undergone significant and detectable changes. In reality all they show are moderate changes, primarily in two coat color characteristics for part of their population. To make things even worse, this species is known to have a highly variable coat-color polymorphism throughout its range. Thus, even the differences displayed appear to be easily accounted for by several mutations and a slight change in gene frequency for one or two loci, all occurring in a limited period of time.

"If an organism such as the Kaibab squirrel is able to escape all but a few minor changes in coat color (and these of dubious survival value) given the supposed immense time of rather complete isolation and violation of the Hardy-Weinberg equilibrium, we must ask wherein lies the fault. Is it in the violation of the Hardy-Weinberg concept, or is it in the time alloted for gene pool changes to occur?"

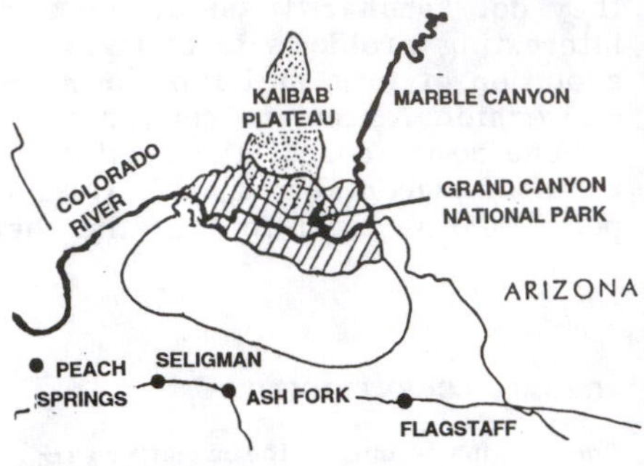

The Kaibab squirrel is isolated on the Kaibab Plateau (60 x 35 miles in extent) north of the Grand Canyon. South of the Grand Canyon and elsewhere, the similar Abert squirrel is found.

Meyer answers his questions by stating that the Hardy-Weinberg equilibrium concept seems reasonably well-founded, so that the problem must lie in the length of time estimated for the Kaibab squirrel's isolation from surrounding populations. In effect, Meyer concludes that the Grand Canyon is much younger than generally supposed. (Meyer, John R.; "Origin of the Kaibab Squirrel," Creation Research Society Quarterly, 22:68, 1985.)

BEHAVIOR

We have divided ten Science Frontiers items on mammal behavior into two subsections:

- **Curious behavior. Bovid stotting, hedgehog anointing, sheep circling, whale pitting, and elephant transferring.**

- **Intelligent behavior. Cat cleverness, ape insight, cattle credulity, and rat learning.**

Not all of these items are anomalous by our criteria. Some sorts of curious behavior, for example, have good explanations once one inquires into the phenomenon. Intelligent behavior, too, need not be anomalous, because biological laws do not rule out mammals acting wisely and rationally. Even so, anomalies do lurk below, for some aspects of mammal behavior can be employed to support two controversial theories: (1) the Gaia Hypothesis; and (2) R. Sheldrake's speculations concerning morphic resonance.

CURIOUS BEHAVIOR

HOW THE CHEETAH LOST ITS STOTTS

Faced with a predator, for example, a cheetah, many deer, antelope and other bovids turn tail and run. But they also go in for a very curious display, before and during the run. They bounce up in the air, keeping all four legs straight. Stotting, as the display is known, must make the animal visible, and presumably also vulnerable to the predator. It certainly attracts the human observer's attention, and there has been no shortage of 'explanations' for this strange behavior.

Actually, at least 11 hypotheses have been proposed. T. Caro has observed Thomson's gazelles stott on more than 200 occasions, usually in response to a cheetah or himself. Caro thinks that adult gazelles stott to proclaim to the cheetah that it has been detected and no longer has surprise in its favor. Cheetahs often do give up after stotting. Further, stotting gazelles have never been seen to be caught---so far. (Anonymous; "How the Cheetah Lost Its Stotts," New Scientist, p. 34, June 19, 1986.)

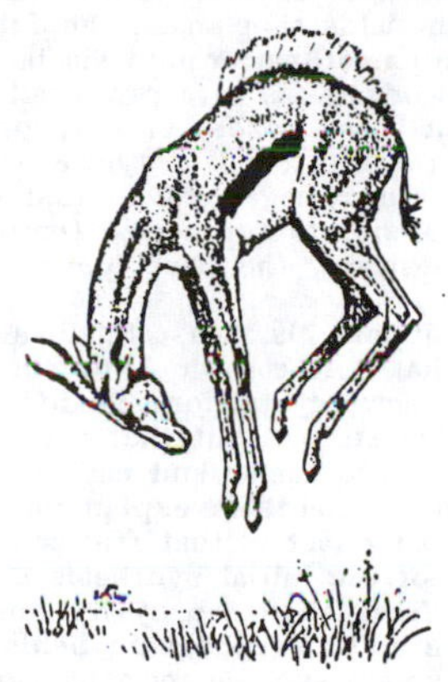

A springbok stotting or "pronking."

WHALES AND SEAFLOOR PITS

The focus of a 1987 paper in Scientific American, by C.H. Nelson and K.R. Johnson, is the northeastern Bering Sea, where sensitive side-scanning sonar has sketched large numbers of pits and furrows in the shallow sands. The pits range from 1-10 meters in length, 0.5-7 meters in width, and 0.1-0.4 meters in depth. No known geological processes seem responsible. Farther east, in Norton Sound, methane eruptions from buried organic matter do blow out circular craters; but the elongated pits investigated by Nelson and Johnson are gouged in sand considered too permeable for gas-crater formation.

Rather surprisingly, the gray whale has become suspect as a pit excavator. They feed in the area of the pits; and the pits, before enlargement by currents, are just the size of the whales' mouths. The whales apparently dredge up sediment and, with their baleen, strain out amphipods (shrimp-like crustaceans) from the sand. The coexisting

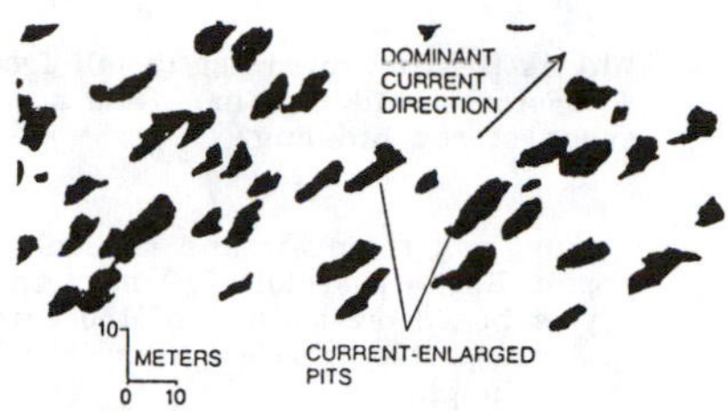

Typical sizes, shapes, and disposition of whale-excavated pits on the floor of the Bering Sea.

narrow furrows turn out to be the work of walruses digging for clams. (Nelson, C. Hans, and Johnson, Kirk R.;"Whales and Walruses as Tillers of the Sea Floor," Scientific American, 256:112, February 1987.)

Comment. The whale tale seems a reasonable explanation of the pits described by Nelson and Johnson, but how far in time and space can it be stretched? Whales do frequent the North Sea, but we do not know whether they or methane eruptions excavate the many craters observed there. As for the much larger Carolina Bays, which exist by the thousands in sandy, coastal terrain, who can say without further study. The Carolina Bays, like the whale-made pits of the Bering Sea, are oriented. One can imagine that, when the oceans stood higher, whale pits were subsequently enlarged by swift currents. See Carolina Bays, Mima Mounds, Submarine Canyons for background material on seafloor pits.

SHEEP CIRCLES!

Britain has been plagued lately with circular areas incised in fields of cereal crops. G.T. Meaden has collected scores of such events, some of which he has published in his Journal of Meteorology. On the theory that one kind of circle might somehow be related to another kind of circle (the same reasoning biologists employ to draw the Tree of Life), J.C. Belcher submitted a most interesting letter to Meaden.

"'By their very nature sheep tend to be stubborn self-willed animals exhibiting individual characteristics not suggestive of good group co-ordination. For example, when disturbed by a potential predator, a flock of sheep tends to mass protectively in a group of irregular outline, the group being formed of individual groups of small numbers of sheep. When grazing undisturbed, sheep tend to fan out from a given point, sometimes following a dominant group leader. Progress is usually unco-ordinated and ragged. In general patterns presented by sheep en masse are seen to be haphazard, indeed generally random in nature. If follows that any suggestion of flocks of sheep forming geometric patterns would appear to be highly improbable, since this would call for group co-ordination only to be found in such as wolves and wild-dogs. In view of this it would appear that certain exceptional observations made on Sunday 21 August 1988 would be worthy of recording.'

"Out on an afternoon drive M. Belcher parked his car near the trigonometric survey point on Baildon Moor, near Leeds, in Yorkshire, at approximately 1430 GMT facing north-east. His wife suddenly exclaimed: 'Look at that circle of sheep in that field!' That was Sheep Circle 1 on the plan where 'a hundred or so sheep were in a circular formation, each sheep being more or less equidistant from the next. At the north end of the field some 20 or 30 cows were standing, grazing and chewing cud in the usual haphazard manner. The circular formation of these sheep

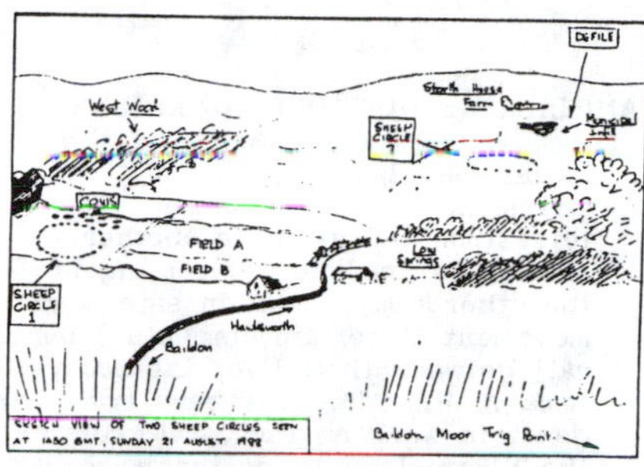

J.C. Belcher's sketch of sheep circles on Baildon Moor.

was so unusual that I thought I was looking at bales of hay set out in the field by the farmer. Indeed, a stone-age stone circle might have been appropriate on this occasion. I looked around from north-west to north-east, and then espied a similar sheep circle (2) on a plateau opposite...In the sector between north and north-east, flocks of sheep were in other fields but in no case exhibited the circular formation, being in typically haphazard groups.' In a second letter Mr Belcher emphasized that the sheep of the two circles were 'variously standing, laying down, or grazing. All quietly occupied, but nevertheless forming this very regular circular formation.'" (Meaden, G.T.; "Sheep in Circular Flocks: Is There a Meteorological, or Some Other, Connection?" Journal of Meteorology, 14:54, 1989.)

Comment. In his letter, Belcher wondered what mysterious force had arrayed the two flocks of sheep so neatly. Meaden, very timidly to be sure, ventured that the same sort of whirlwinds that he believes cuts the crop circles might cause sheep to drift outwards into a circle! In the interest of completeness, some grazing animals, such as the muskoxen, will form into a circle when threatened. But are there appropriate whirlwinds in the Arctic? So much for deprecating taxonomy-through-morphology! Our exposition on "crop circles" begins on p. 206.

HEDGEHOGS USE TOAD VENOM FOR DEFENSE

European hedgehogs chew toad skins to extract venom from the paratoid glands. They then lick their spines with the saliva-venom mixture. Experiments with human volunteers prove that the venom-anointed spines are much more painful and irritating than clean ones. Such hedgehog behavior is innate and fully developed before the juveniles leave the nest. Tenrecs, which are similar to hedgehogs but in an entirely different family, display a somewhat different self-anointing type of behavior that must have developed independently. Conclusion: self-anointing with toad venom is so useful that it developed twice under evolutionary pressures. (Brodie, Edmund D., Jr.; "Hedgehogs Use Toad Venom in Their Own Defense," Nature, 268:627, 1977.)

ANIMALS AS NUTRIENT CARRIERS

> It has long been recognized that the movement of grazing animals from one terrestrial ecosystem to another, feeding in one and defaecating in the other, may result in significant movement of certain elements [chemical] between them (i.e. the ecosystems). What has now been made evident, in work on the coral reefs of the Virgin Islands, is that a similar process takes place in aquatic ecosystems.
>
> Several examples, terrestrial (such as elephants) and aquatic, follow this introductory paragraph. (Moore, Peter D.; "Animals As Nutrient Carriers," Nature, 305:763, 1983.)

Comment. This process may emphasize the fine-tuning of the Gaia Hypothesis, in which life-as-a-whole operates in ways that make the planet-as-a-whole more productive of life.

INTELLIGENT BEHAVIOR

CAT**CATS

Clever cat

Tales about clever animals abound, but the following is too good to pass up.

It seems that C.G. Martin, residing in Stoke-on-Trent, had to trap a feral cat. He wrote as follows:

> Although I have made an adequate living as a mechanical design engineer, it took me a couple of minutes to work out how to position the various rods and links to set and bait the trap, which done, I observed from a concealed position. The cat duly arrived, studied the trap suspiciously from different angles, retired, sat and contemplated. Then, after less time than it had taken me to work it out, she entered the trap purposefully, placed her paws underneath the trip plate, took the food and backed out.

(Martin, C.G.; "Clever Cat," New Scientist, p. 53, August 29, 1992.)

An even cleverer cat

Yes, it's true that cats can circumvent our specially designed traps, but we did not realize that they also knew their aerodynamics.

> Why is it safer for a cat to fall from a 32-storey building than from a seven-storey building?
>
> .....
>
> Just ask scientific and medical reporter Karl Kruszelnicki, whose theory is based on a study of 150 cats that plummeted from windows at different heights.
>
> Falling from 32 storeys, a cat had more time to work out a plan of action, because once it reached terminal velocity and stopped accelerating, it started to relax, he said in Sydney yesterday.
>
> Once the moggie [?] reached top speed of 100 kmh and realised it was not speeding up any more, it spread-eagled its limbs in the perfect position for maximum wind resistance.
>
> "Once it reaches the ground, the cat just kisses the ground on all four paws simultaneously and the shock is absorbed," Dr. Kruszelnicki told his bemused audience at the University of New South Wales during a talk organized by the Alumni Association.
>
> Of the 150 cats that fell from high-rise buildings in New York over a five-month period, 10 per cent died, with the chances of survival rising with the distance of the fall.

It seems that at least one cat per day takes the plunge in New York City, but do they jump...or are they pushed? Dr. Kruszelnicki supposed that some may have leaped at passing birds! (Anonymous; "High-Flying Cats Have the Big Drop Licked," Wellington, New Zealand, The Dominion, September 17, 1992. Cr. P. Hassall)

Comment. Would "really clever" cats really jump at birds flying around skyscrapers?

PLANTS OF THE APES

Many biologists are convinced that apes, bears, cats, and dogs eat plants---many of them obviously distasteful---in order to medicate themselves for diseases and parasites. What also seems likely, according to K. Strier, of the University of Wisconsin, Madison, is that some monkeys regulate their fertility by the judicious consumption of certain plants. Going even farther, K. Glander, Duke University, suggests that howler monkeys control the sex of their offspring through their diets.

Glander divides howler monkey females into three groups. In the first are the high-ranking females that predominantly produce male offspring. This "male-offspring" strategy favors these females because the males they produce tend to become dominant adults that will pass on more of the females' genes than would female offspring, who are limited in the number of infants they can engender in comparison to the males. Similar optimization strategies, according to Glander, induce middle-ranking females to produce mainly female progeny, and low-ranking females to birth almost all males.

These howler monkeys seem to control the sex of their offspring pharmologically by selecting certain plants to eat. These plants, in turn, control the electrical conditions in the females' reproductive tracts to either attract or repel sperm carrying the male Y-chromosomes, which are thought to carry different electrical charges than the X-carrying sperm! (Lewin, Roger; "What Monkeys Chew to Choose Their Children's Sex," New Scientist, p. 15, February 22, 1992. Gibbons, Ann; "Plants of the Apes," Science, 255:921, 1992.)

Comments. It will take much more research to validate these startling assertions. We also have to ask how these instincts (or conscious, calculated strategies?) evolved. Since so many of the medicinal plants are distasteful, why would the monkeys eat them in the first place and thus learn, instinctively or consciously, their value in advancing the prospects for their genes?

WHAT WAS, IS, AND SHALL BE

Rupert Sheldrake, an English plant physiologist, has written a new book entitled A New Science of Life; The Hypothesis of Formative Causation. In it, he revives and expands the theory of morphogenic fields. Basically, this theory states that existing organized structures, such as crystals and organisms, establish fields that shape the future organization of matter into similar crystals and organisms in a probabalistic way. In other words, once a specific crystal (or life form) is synthesized, it sets up a morphogenic field that will make it easier to synthesize further the same, or nearly the same, crystal (or life form).

To support his ideas, Sheldrake claims that it is common knowledge that a brand-new crystal form is difficult to synthesize at first but that further syntheses become easier and easier. The prevailing "scientific" explanation of this amazing fact is that fragments (seeds) of the initial synthesis are carried from lab to lab by humans and even the air! Morphogenic fields, however, explain such phenomena very nicely without postulating tiny crystal seeds in scientists' beards.

Sheldrake then goes on to review McDougall's experiments in the 1920s in which trained rats from water mazes apparently passed their new knowledge on to their progeny. McDougall thought

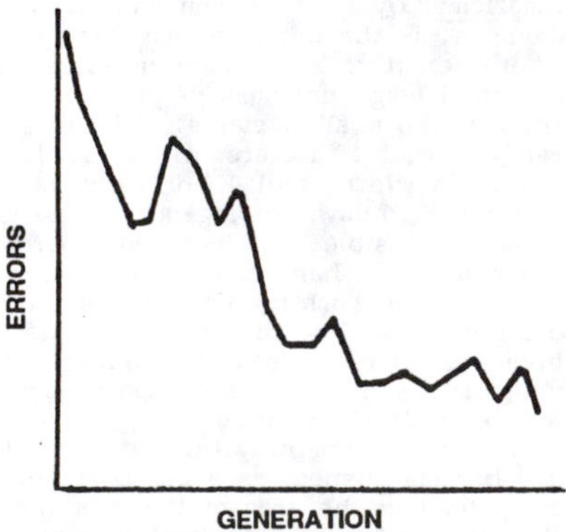

Decreasing number of maze errors made by rats selected for slowness in learning. One would expect such rats to do worse and worse, instead the reverse is true.

that he had proved the inheritance of acquired characteristics. Other biologists repeating his heretical experiments found that their first-generation rats solved the same water mazes much faster than had McDougall's rats. In addition, the progeny of untrained rats used as controls showed improved abilities in maze-solving with each generation, just as if their parents had been trained. Current theory has not explained these curious results, but they are consistent with Sheldrake's Theory of Formative Causation. (Sheldrake, Rupert; "A New Science of Life," New Scientist, 90:768, 1981.)

Comment. The Theory also seems to explain the many cases of simultaneous invention and even telepathy, assuming it exists. Apparently, we are all immersed in overlapping morphogenic fields created by all other humans.

LEARNING BY INJECTION

Abstract. "In an attempt to replicate previous findings that learned information could be transferred from trained donor animals to untrained recipient animals by means of brain extracts, two groups of rats were trained to approach a food cup in response to a discriminative stimulus (click or light). RNA extracted from the brains of these animals was injected intraperitoneally into untrained rats. The two untrained groups showed a significant tendency to respond specifically to the stimulus employed during the training. The results support the conclusion that acquired behaviors can be transferred between animals by transferring brain DNA, and further suggest that the transfer effect is dependent upon and specific to the learning of the donors." (Oden, Brett B., et al; "Interanimal Transfer of Learned Behavior through Injection of Brain RNA," Psychological Record, 32: 281, 1982.)

Comment. Of course, morphogenic fields as described in R. Sheldrake's A New Science of Life, could explain this effect.

YOU CAN FOOL SOME OF THE ANIMALS SOME OF THE TIME, BUT....

R. Sheldrake has found another possible example of "morphic resonance," as he relates in the New Scientist:

> Ranchers throughout the American West have found that they can save money on cattle grids by using fake grids instead, consisting of stripes painted across the road. Real cattle grids, known as cattle guards in the U.S., are usually made of a series of parallel steel tubes or rails with gaps in between, which make it physically impossible for cattle to walk across them. However, cattle do not usually try to cross them; they avoid them. The illusory grids work just like the real ones. When cattle approach them, they 'put on the brakes with all four feet,' as one rancher expressed it to me.
>
> .....
>
> According to my hypothesis of formative causation..., organisms inherit habits from previous members of their species. This collective memory, I suggest, is inherent in fields, called morphic fields, and is transmitted through both time and space by morphic resonance, a process which takes place on the basis of similarity. From this point of view, cattle confronted for the first time by grids, or by things that look like grids, would tend to avoid them because of morphic resonance from other cattle that had learnt by experience not to try to cross them.

Sheldrake follows these two paragraphs with an examination of more conventional explanations. He is wise to do this because morphic resonance does seem pretty farfetched. Do new naive cattle somehow acquire this aversion to cattle grids, real or fake, from an experienced member of the herd? Apparently not, because a herd of all naive cattle will avoid the painted grids. Can the spell of fake grids be broken? Yes, due to fear or the desire for food, some cattle will jump over the fake grids, but others will inspect them carefully, see what they really are, and proceed to walk across. Thereafter the fake grids are useless with an "educated" herd. (Sheldrake, Rupert; "Cattle Fooled by Phoney Grids," New Scientist, p. 65, February 11, 1988.)

Comment. Animal behavior provides much more ammunition for Sheldrake; viz., the spontaneous spread of the milk-bottle-opening talent among some British birds. His newest book, The Presence of the Past, is crammed with evidence. Nevertheless most scientists wince when Sheldrake and morphic resonance are injected into a conversation. We suspect that this wincing habit has been communicated to all scientists through morphic resonance!

DISTRIBUTION

The word "distribution" in anomalistics is employed in situations such as: (1) When animals are found where they are not expected to be, either because they are not presently indigenous to the area in question, or because their permanent presence is difficult to explain by prevailing evolutionary paradigms; (2) When population levels show unexplained cyclicities; (3) When large numbers of animals appear suddenly under remarkable circumstances, as in "bird falls"; and (4) When aggregations of animals are found in bizarre situations, such as "rat kings." Below, we present examples of all four of these "distribution" curiosities and anomalies. The reader, by consulting the other sections under the "distribution" heading, will discover additional phenomena of this type.

BAIKAL'S DEEP SECRETS

Lake Baikal, in Siberia, requires many superlatives in its description. It is the deepest lake, 1637 meters; the oldest lake, 20-25 million years; and home to the richest array of lake life, both in terms of biomass and recorded species. There are found here 1550 species and variants of animals plus 1085 plants. Over 1000 of these species of life are found nowhere else. The sediments deposited on the lake floor are of astounding thickness. Bedrock lies 7 kilometers below the lake surface in some spots. With a maximum depth of 1637 meters, we find by subtraction places where more than 5 kilometers of sediment have collected.

The diversity of Baikal's life is remarkable in itself, but there are two aspects of it that approach the anomalous: (1) Baikal's seals are 1000 kilometers of so from salt water. How did they get there and when? (2) Hydrothermal-vent communities have been discovered at a depth of about 400 meters in the northern part of the lake. These communities contain sponges, bacterial mats, snails, transparent shrimp, and fish; some of which are new to science. Baikal's thermal vents are the only ones known in freshwater lakes. Their relation to saltwater vent communities has not yet been explored. (Stewart, John Massey; "Baikal's Hidden Depths," New Scientist, p. 42, June 23, 1990. Monastersky, R.; "Life Blooms on Floor of Deep Siberian Lake," Science News, 138:103, 1990.)

Comment. Despite its inland position, the suspicion develops that Baikal was connected to the oceans in recent geological times.

GROUNDED BATS NICHELESS

New Zealand boasts two bat species (Mystacina) which can fly but really prefer to clamber around on the ground hunting for insects, pollen, and fruit. So far, these bats have defied classification. They are distinctly different from their flying cousins in the area. Blood protein analysis links them to tropical bats in South America. No other bats from New Zealand and Australia show such a relationship. Although these two oddball species may have flown in from South America some 80 million years ago, when the land masses were thought to be much closer, one then has to explain how these tropical bats survived the Ice Ages that afflicted New Zealand. (Anonymous; "Grounded Bats," New Scientist, p. 25, October 2, 1986.)

STATIC ON THE HARE-LYNX CYCLE SIGNAL

Almost all ecology textbooks present the 10-11 year hare-lynx cycle as a classic case of prey-predator oscillations. The major data source for such population studies is the record of pelt sales rather than actual field observations. Looking beyond such superficial information, researchers have discovered that the quantity of pelts offered for sale by the Indians depends upon the amount of time they can divert to hunting pelts. This, in turn, is affected by the abundance of food animals, such as moose and hares. It is food first and pelts second. Furthermore, when the plants consumed by hares are overbrowsed during periods of dense hare population, they defend themselves by generating resins and other compounds toxic or repellent to hares. Thus, the hare abundance cycle is affected by: (1) plant defenses; (2) Indian hunting strategies; and (3) the lynx. (May, Robert M.; "Cree-Ojibwa Hunting and the Hare-Lynx Cycle,: Nature, 286:108, 1980.)

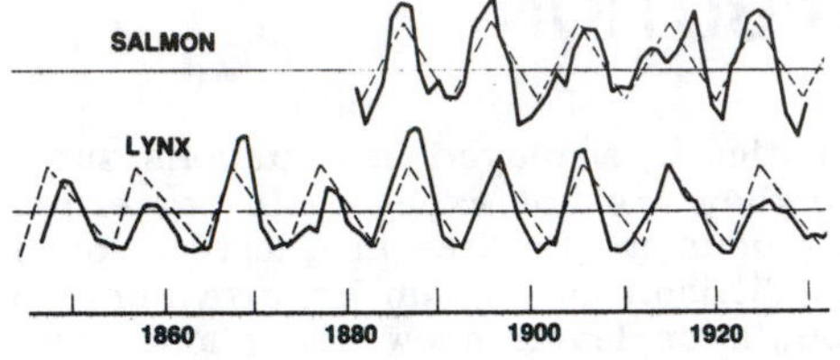

Curiously, the 9.6-year cycle in the Canadian lynx population is in phase with the New Brunswick salmon cycle. (After: E.R. Dewey; Cycles, New York, 1971.)

A BAT FALL

September 6, 1989. Fort Worth, Texas.

> Pedestrians dodged hundreds of bats that fell onto downtown sidewalks yesterday afternoon. The winged mammals were sick and dying, and no one knows why.
>
> 'I have never seen bats on the sidewalk at 4 o'clock in the afternoon before,' said restauranteur Chris Farkas after encountering the bats in the 600 block of Main Street. 'About half of them were crawling on the ground. There were about 50 in the air flying around.'

Many of the bats subsequently died. Two possible causes advanced were heat-stroke and building fumigation. Neither could be shown correct. (Gilberto, Julie; "Scores of Bats Rain on Downtown," Fort Worth Telegram, September 7, 1989. Cr. R.L Anderson.)

Comment. Bat falls and bird falls are rare in the Fortean literature. Storms, intensely cold weather, and sheer exhaustion are the most common causes.

TANGLED-TAILS TALES

"Rat kings" have always been a favorite Fortean phenomenon. They are clusters of rats whose tails have somehow become knotted or glued together. Naturalists also find "squirrel kings" in the wild. However improbable these "kings" may seem, new cases keep coming to the fore. Here follows the first of two, as recounted in the Fortean Times;

> The first incident occurred in Easton, Pennsylvania, in 1989. As 16-year-old Crystal Cresseveur set off for church around midday on Sunday 24 September, she noticed a commotion in the hedge outside her house, it was a writhing furry bundle of six young squirrels all squeaking at once. At first she thought they were playing but she soon realized they were in a panic, and as they pulled in all directions at once they had become firmly stuck among the trunk of the bushes. She called her father, Paul, and their neighbour, Charles Kootares, and with help from the growing crowd of onlookers, managed to extract the frantic cluster from the hedge.
>
> In this case, the squirrels' tails could not be disentangled, and the poor animals were put to sleep. The second incident occurred in Baltimore on September 18, 1991. Here, the squirrels' tails were tangled and stuck together by tree sap, hair, and nesting debris.

(Anonymous; "Tangled Tales," Fortean Times, no. 63, p. 13, 1992.)

Comment. Squirrel kings have even received a modicum of attention in the scientific literature; viz., Animal Kingdom, 55:46, 1952. (See Incredible Life for quotations.) Some involve several adult squirrels, and it is hard to imagine how such active animals could become mutually tied and/or stuck together.

TALENTS AND FACULTIES

Down through the course of evolution, animals have acquired a wide spectrum of signal sensors and transmitters that improve their abilities to cope in a competitive and unpredictable environment. Mammals have not inherited all of the talents and faculties of the more "primitive" animals, such as the pit viper's heat sensors or the box jellyfish's deadly venom; nevertheless, their capabilities are remarkable. A few of these have been recorded in Science Frontiers:

- **Flight. Bats may have invented this capability twice!**
- **Echolocation. This faculty is highly developed in bats and the cetacea and is feeble in a few shrews.**
- **Seismic sensitivity. Dogs and many other animals seem to be able to perceive the weak precursory activity of earthquakes.**
- **Navigational senses. Many mammals, perhaps even humans (p. 114). can sense subtle environmental forces and cues.**
- **Hunting talents. Acoustical stunning, electrical foraging.**
- **Singing and acoustical communication. Whale songs and dialects.**
- **Engineering capabilities. Mouse pyramids, orang tool-making.**

FLIGHT

BATS MAY HAVE INVENTED FLIGHT TWICE (AT LEAST!)

Bats are divided into megas and micros. The "megabats," represented by the fruit bats, possess an "advanced" connection between their eyes and midbrains. No other mammals except the primates possess this type of advanced visual organization. In contrast, the "microbats," the common echo-locating insect-eaters, have a "primitive" eye-brain connection. This deep division in the bat family---mega/micro, vegetarian/carnivorous, sight-dependent/echolocating---suggests that mammal flight has developed at least twice. (Pettigrew, John D.; "Flying Primates? Megabats Have the Advanced Pathway from Eye to Midbrain," Science, 231:1304, 1986.)

ECHOLOCATION

BAT ECHO-LOCATION

Bats navigate by somehow constructing an image of the external world from the echoes of their squeaks. Since bats have but two ears, one wonders how they can develop a three-dimensional image from a two-dimensional sensor; i.e., two ears give right-and-left information only. The moustache bat makes up for this deficiency by generating echo-locating pulses at three distinct harmonics: 30, 60, and 90 kilohertz. Its external ears are so shaped that each of these three frequencies has a different acoustic axis, giving the bat in effect three separate sets of ears pointing in three different directions. Inside the bat's head, in the inferior colliculus of the brain, are three separate sets of neurons sensitive to the three different frequencies. No one knows how the bat processes such information into a "display" it can use in swooping after insects at night. (Anonymous; "The Ins and Outs of a Bat's Ears," New Scientist, 20, August 30, 1984.)

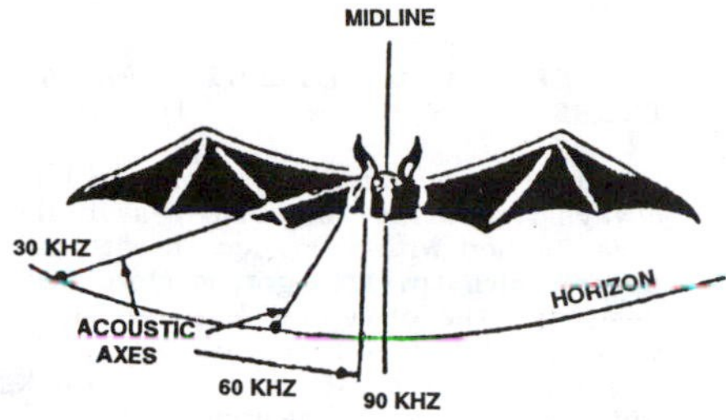

The three acoustic axes of the moustache bat.

ARE FRUIT BATS PRIMATES?

The profound differences between the fruit bats (megabats) and echo-locating, insect-eating bats (microbats) were mentioned above. The primate-like eye-brain system of the fruit bats suggests two possibilities:

1. Primitive bats first developed flight and then the fruit bats developed their primate-like eye-brain systems through "convergent evolution"; or
2. The fruit bats inherited their eye-brain system from closely related primates and then developed flight through "convergent evolution."

R.D. Martin, a physical anthropologist and author of this article, has reviewed the morphological characteristics of the fruit bats and primates. On this basis, he doubts that the fruit bats are primates. Furthermore, molecular studies are also negative. Therefore, possibility #1 above is the more likely one. (Martin, R.D.; "Are Fruit Bats Primates?" Nature, 320:482, 1986.)

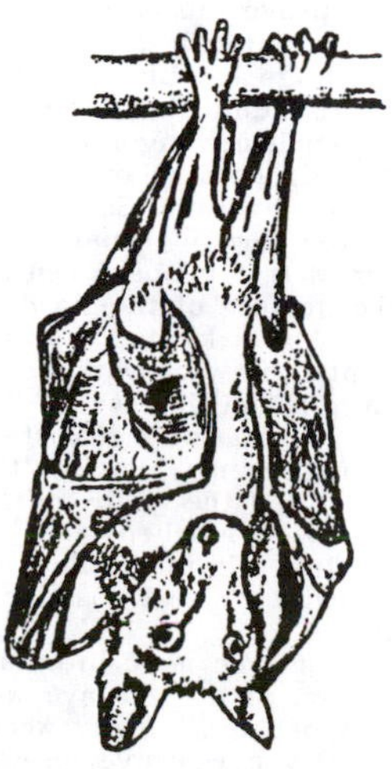

An epaulleted fruit bat, one of the megabats.

Comment. In either case, #1 or #2, we must acknowledge convergent evolution and the likelihood that some subroutine in the genetic code repeats itself in divergent species. Speculative as always, we must ask if the genetic instructions for human or even superhuman intelligence do not reside dormant-for-now in other species.

SEISMIC SENSITIVITY

HARK, HARK, THE DOGS DO BARK

American geophysicists have been slow to take unusual animal behavior prior to earthquakes seriously. Spurred by Chinese work, a network of observers has been set up in California in an earthquake-prone area. Most reports of strange animal behavior have been after-the-fact. Furthermore, the "strange behaviors" frequently turn out to be common during quake-free periods, but simply not remarked upon. Nevertheless, geophysicists did observe some clear-cut instances of animals' supersensitivity to quake phenomena.

Studying aftershocks in the Mohave Desert in 1979, Donald Stierman and his colleagues often heard earthquake booms 4-10 sec after feeling the shock and seeing their portable seismometers record the tremor. Two dogs nearby inevitably responded with a chorus of barking. Sometimes though, the human observers heard and felt nothing when the seismometers and dogs announced another aftershock. (Kerr, Richard A.; "Quake Prediction by Animals Gaining Respect," Science, 208:695, 1980.)

ANIMAL BEHAVIOR PRIOR TO THE HAICHENG EARTHQUAKE

The catastrophic Chinese Haicheng earthquake of 1975 was preceded by many reports of unusual animal behavior. Beginning in December 1974, lay observers noted dazed rats and snakes that appeared to be "frozen" to the roads. In February, reports of this type increased markedly, including observations of general restlessness and agitation of the larger animals, such as cows and horses. Rats now appeared as if drunk. Chickens refused to enter their coops and geese frequently took to flight.

Chinese scientists seem convinced that such animal behavior might help predict some of the larger earthquakes. Further research is being undertaken at the Institute of Biophysics in Peking and at Peking University. (Molnar, Peter, et al; "Prediction of the Haicheng Earthquake," Eos, 58:254, 1977.

NAVIGATION SENSES

MAGNETIC "DEAD" RECKONING

Try on this theory for size. Whales subconsciously strand and kill themselves in order to maintain their populations at optimum levels! Well, the dozen or so other theories that have been advanced to account for whale strandings haven't been much better.

M. Klinowska thinks that she has some clues indicating a better theory. First, all whale strandings (in Britain, at least) occur where magnetic field contours are perpendicular to the shoreline. Second, strandings are also correlated with irregular changes in the magnetic field. You will see the significance of these facts after you hear her theory:

"Cetaceans use the total geomagnetic field of the Earth as a map. A timer, also based on this field, allows them to monitor their position and progress on the map. They are not using the directional information of the Earth's field, as we do with our compasses, but small relative differences in the total local field. I arrived at this explanation after a detailed analysis of the records of strandings in Britain, but it has so far been confirmed by two groups working in the U.S. Similar work is in progress in other parts of the world.

"The total magnetic field of the Earth is not uniform. It is distorted by the underlying geology, forming a topography of magnetic 'hills and valleys.' My analysis shows that the animals move along the contours of these magnetic slopes, and that in certain circumstances this can lead them to strand themselves. In the oceans, sea-floor spreading has produced a set of almost parallel hills and valleys. Whales could use these as undersea motorways, but might swim into problems when they came near the shore, because the magnetic contours do not stop at the beach. They continue onto the land, and sometimes so do the whales."

In addition to stranding because of land-intersecting contours, unpredictable changes in the earth's magnetic field can upset the whales' timing mechanism, causing them to lose their true position on their magnetic dead-reckoning maps. Magnetically speaking, they become lost. (Klinowska, Margaret; "No Through Road for the Misguided Whale," New Scientist, p. 46, February 12, 1987. Also Ellis, Richard; "Why Do Whales Strand?" Oceans, 20:24, June 1987.)

Comment. Klinowska's theory is attractive in the sense that one can test it by checking strandings against magnetic contours and magnetic variations. However, the theory requires whales to sense changes in the earth's field of only 1 nanotesla (that is, one part in 50,000). No one has any idea how this can be accomplished biologically. Furthermore, how are the world-wide magnetic reference maps constructed and stored in the whales' brains?

WHALES AND DOLPHINS TRAPPED MAGNETICALLY

Joseph L. Kirschvink, of the California Institute of Technology, has plotted the hundreds of beachings of whales and dolphins along the U.S. east coast. He finds that these cetaceans tend to run aground at spots where the earth's magnetic field is diminished by the local magnetic fields of rocks. These coastal magnetic lows are at the ends of long, continuous channels of magnetic minima that run for great distances along the

A mass stranding of short-finned pilot whales on the Florida Gulf Coast in 1971. (William K. Fehring)

ocean floors. Kirschvink believes that the stranded whales and dolphins were using these magnetic troughs for navigation and failed to see the stop sign at the beaches and ran aground. The magnetic troughs in this view are superhighways for animals equipped with a magnetic sense.

If Kirschvink's theory is correct, the magnetic sensors of the whales and dolphins are extremely sensitive, because the deepest magnetic troughs are only about 4% weaker than the background magnetic field. Magnetite crystals have been found in birds, fish, and insects, where they are thought to contribute to a magnetic sense of some sort. So far, no magnetite has shown up in whales and dolphins. (Weisburd, S.; "Whales and Dolphins Use Magnetic 'Roads,' Science News, 126:389, 1984.)

A MAGNETIC SENSE IN MICE

"Abstract. We displaced white-footed mice (Peromyscus leucopus) 40 m away from their home areas and released them in a circular arena. Mice concentrated their exploratory and escape activity in the portion of the arena corresponding to home direction. In another group of

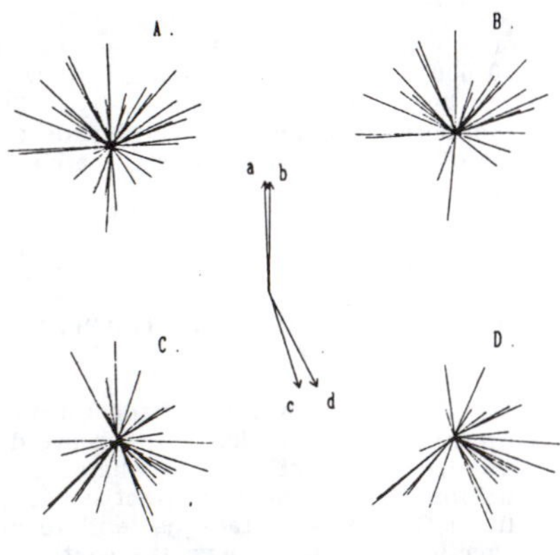

Orientation responses of mice displaced in a normal geomagnetic field (A and B) and a reversed field (C and D). Mean directions are indicated in the center (a through d).

mice, we reversed the horizontal component of the geomagnetic field surrounding them during displacement. These individuals concentrated their activity in areas of the circular arena opposite home direction. Mice were not likely using visual, chemical, or kinesthetic cues to establish home direction. Tissues of P. leucopus exhibit strong isothermal remanant magnetization and may contain biogenic ferromagnetic material. Our results suggest that white-footed mice have a magnetic sense and use the geomagnetic field as a compass cue." (August, Peter V., et al; "Magnetic Orientation in a Small Mammal, Peromyscus Leucopus," Journal of Mammalogy, 70:1, 1989.)

HUNTING TALENTS

PORPOISE STUN GUN

Just about everyone knows that some whales and porpoises have oil/wax-filled sound lenses in their foreheads. These biological lenses focus clicks and other sounds sonar-fashion ahead of the swimming animal, which then listens for echoes from prey and other targets. But what if these bursts of sound could be made very powerful---could they be employed to stun and disorient prey?

Bits of evidence are accumulating to support the theory that some whales and porpoises actually have acoustic stun guns in their foreheads. First, there are visual observations of fish being hunted by whales and porpoises suddenly giving up flight, becoming passive, and almost asking to be snapped up by their pursuers. Second, the stomachs of whales often contain much faster and more mobile prey---often without any teeth marks. Finally, bottlenose dolphins are known to have the capability of producing bursts of sound five orders of magnitude more intense than their usual navigating clicks. This is more than enough to kill small fish. (Norris, Kenneth S., and Mohl, Bertel; "Can Odontocetes Debilitate Prey with Sound?" American Naturalist, 122:85, 1983.)

Comment. Here is another instance of the "problem of perfection." An existing organ of great complexity seems utterly useless if only fractionally developed. One would think that the complicated sound lenses, the muscular sound-generating tissues, and their containing structures would have to have developed in a single step in order to have any survival value.

PLATYPUS BILL AN ELECTRICAL PROBE

The bill of the duck-billed platypus has always looked kind of dumb---as if Nature flushed with her success with polar bears' hairs (p. 130) got careless when designing the platypus! How ignorant we were of Nature's genius. The platypus didn't borrow its snout from ducks but rather from the electric fishes.

> That evolutionary enigma, the duck-billed platypus, has more than its egglaying to distinguish it from other mammals. It now appears that in common with some species of fish and amphibians, it can detect weak electric fields (of a few hundred microvolts or less). Not only that, but it uses its electric sense to locate its prey, picking up the tiny electrical signals passing between nerves and muscles in the tail of a shrimp.

(Anonymous; "The Battery-Operated Duck Billed Platypus," New Scientist, p. 25, February 13, 1986.)

THE AYE-AYE, A PERCUSSIVE FORAGER

> The aye-aye, one of the strangest and rarest species of primates in the world, has an equally unusual method of finding food. Zoologists have discovered that it taps wood to locate cavities under the surface. Its skills are so well developed that it can tell holes containing grubs from those that are empty. It is the only mammal known to use such a technique.

To improve the efficiency of its "percussive foraging," the aye-aye has evolved huge bat-like ears and a highly elongated middle finger on each hand. This specialized finger does the tapping and the big ears relay the nuances of sound to the brain. So sensitive is this specialized form of sonar that the aye-aye can detect grubs 2 centimeters below the surface of the wood. Once a grub has been located, the aye-aye tears into the wood with its forward-curving, chisel-like teeth. The incisors are remarkable for a primate, for they keep on growing, just like those of rodents. When the grub-containing chamber has been reached, the long, narrow middle finger is inserted and the grub is retrieved. A neat combination of attributes.

What is even more interesting is a comparison of the aye-aye with many of the woodpeckers. Many woodpeckers also employ percussive foraging, have special bills for chiselling, and possess very long, spiny tongues for extracting grubs. In other words, the aye-aye is a primate that occupies the niche of a woodpecker. As luck (?) would have it, the aye-aye lives on Madagascar where there are no woodpeckers! (Mason, Georgia; "Grubs on Tap for the Aye-aye," New Scientist, p. 23, June 22, 1991.)

A hand-held aye-aye. Note the big ears and long fingers.

SINGING AND ACOUSTICAL COMMUNICATION

POETS AT SEA: OR WHY DO WHALES RHYME?

When scientists talk about whales singing songs, they're not talking about mere noise. They're talking about intricate, stylized compositions---some longer than symphonic movements---performed in medleys that can last up to 22 hours. The songs of humpback whales can change dramatically from year to year, yet each whale in an oceanwide population always sings the same song as the others. How, with the form changing so fast, does everyone keep the verses straight? Biologists Linda Guinee and Katharine Payne have been looking into the matter, and they have come up with an intriguing possibility. It seems that humpbacks, like humans, use rhyme.

Guinee and Payne suspect that whales rhyme because they have detected particular subphrases turning up in the same position in adjacent themes. (Cowley, Geoffrey; "Rap Songs from the Deep," Newsweek, p. 63, March 20, 1989. Cr. J. Covey)

Comment. This is all wonderfully fascinating, but why do whales rhyme at all, or sing such long complex songs? Biologists fall back on that hackneyed old theory that it has something to do with mating and/or dominance displays. Next, we'll hear that human poets write poems only to improve their chances of breeding and passing their genes on to their progeny!

KILLER WHALE DIALECTS

J. Ford is the curator of marine mammals at Vancouver's Public Aquarium. For years, he has been listening to killer whales converse as they hunt along the coast of the Pacific Northwest. About 350 whales in the area are divided into two communities, each of which is subdivided into several pods. Each pod has its own dialect of sounds used in communication. Some of the dialects are regional, like Bostonian or Texan; others are more divergent, like English and Japanese. This discovery promotes killer whales to the level of some primates and harbor seals. Usually, Ford says, the sounds made by animals are determined genetically. (Dayton, Leigh; "Killer Whales Communicate in Distinct 'Dialects,'" New Scientist, p. 35, March 10, 1990.)

ENGINEERING CAPABILITIES

MORE MOUSE ENGINEERING

Shortly after writing (on p. 117) about the "Ancient Greek Pyramids" and the Saharan mice that construct small pyramids of pebbles to extract moisture from the air, we serendipitously ran across the following:

> Australian Native Mice. The species P. chapmani builds low mounds of pebbles over its burrow systems, and P. hermannsburgensis may use these mounds after they are constructed. The pebbles are of a uniform size and cover a large area, often a meter in diameter. The pebbles are probably collected both by excavation and from the surface. Some local mammalogists believe these are used as dew traps. Since the air around the pebbles warms more rapidly as the sun rises than do the pebbles themselves, dew forms on the pebbles by condensation. As the areas in which these mounds are found are quite dry, except after a heavy rain, these dew traps solve the problem of water shortage. Local farmers use the many pebble mounds for mixing concrete. It is believed that the ancient people of the Mediterranean region used a dew trap method comparable to that of P. chapmani.

(Nowak, Ronald M.; "Australian Native Mice," Walker's Mammals of the World, Baltimore, 1991, p. 820.)

Comment. Now we must decide between at least three possibilities. Since the Australian native mice and Saharan mice are many thousands of miles apart, we have: (1) independent mouse inventions; (2) mouse telepathy; or, worst of all, (3) an example of Sheldrake's morphic resonance!

NEW DEFINITION FOR HUMANS NEEDED

One scientist has defined humans as "tool makers" as distinguished from "tool users." This distinction is necessary because several animals employ tools for simple tasks, such as fishing termites out of holes. However, Kitahara-Frisch points out in this paper that experiments by Wright with a young orangutan proved that at least one animal can actually make tools; that is, use one tool to make another. More specifically, Wright taught the orangutan to strike sharp flint flakes from a core and then use them to cut a cord and gain access to its favorite food. (Kitahara-Frisch, J.; "Apes and the Making of Stone Tools," Current Anthropology, 21:359, 1980.)

Comment. Apparently, with orangutans, at least, no manipulative or cognitive barriers exist to prevent them from entering their own Stone Age.

HEALTH

FOR SOME, SEX = DEATH

It has long been known that the males of some species of marsupial mice mate in their first year and then die off completely, leaving the perpetuation of their species to their male progeny. Females of these species usually survive to breed a second and even third year. The poor males, however, succumb due to "elevated levels of free corticosteroids in the blood and associated disease such as hemorrhagic ulceration of the gastric mucosa, anemia, and parasite infestation." In short, they seem programmed to die after mating, like the male octopus. And one wonders why evolution has wrought this mass die-off.

In their studies of marsupial mice, C.R. Dickman and R.W. Braithwaite have extended the phenomenon to two new genera: Dasyurus and Parantechinus. They have also found that the phenomenon is a bit more complex. First, in P. apicalis, the male die-off occurs in some populations and not others. In D. hallucatus, the die-off may occur in the same population in some years and not others. Furthermore, the females of this species may on occasion all die off, too---but after giving birth, of course. (Dickman, C.R., and Braithwaite, R.W.; "Postmating Mortality of Males in the Dasyurid Marsupials, Dasyurus and Parantechinus," Journal of Mammalogy, 73:143, 1992.)

SCRAPIE TRANSMITTED BY PRIONS

Scrapie is an infectious neurological disorder in sheep. The infectious agent has been isolated, but it consists only of geneless prions. Somehow, these prions, which are merely protein filaments, get into a sheep's brain and replicate themselves to cause scrapie. With no genetic material of their own, how do the prions multiply? Recent laboratory work suggests that the prions subvert a gene that normally dwells in the brain. With the help of this gene, an endless stream of prions emerges, and the animal is sick. In hamsters, which are employed in laboratory research on scrapie, a gene demarcated PrP has been implicated in scrapie. PrP is present in both healthy and infected hamster brains, but no one knows what its normal function is, if indeed it has one. (Anonymous; "Prion Gene," Scientific American, 253:60, July 1985.)

Comment. One can make an immediate connection between the traitorous PrP genes in the hamster brains and the excess genetic material in humans and all life forms. Biologists commonly call excess genetic material "nonsense DNA" which only means that they haven't devined its purpose. But, as already suggested, these unused blueprints may have had some past purpose or will be called into action in the future. The purpose may be insidious, as in the case of scrapie, or vital to the organism's survival in some unrecognized biological Armageddons.

GENETICS

MICE TRANSMIT HUMAN GENE SEQUENCES TO THEIR PROGENY

Viruses habitually subvert the manufacturing facilities of host cells so that they turn out viruses instead of material useful to the host. Stewart and his colleagues injected fertilized mouse eggs with human β-globin gene sequences. One of the resulting adult mice carried the human gene sequence intact; one of the others carried at least part of the sequence. More significantly, the latter mouse transmitted the human gene sequence to its progeny in a Mendelian ratio. (Stewart, Timothy A., et al; "Human β-Globin Gene Sequences Injected into Mouse Eggs, Retained in Adults, and Transmitted to Progeny," Science, 217:1046, 1982.)

Comment. Animal cells are therefore not too fastidious about what they manufacture and what is transmitted to progeny. The unanswered questions are: How far can this proxy replication and transmission of genes go; and, most important, can it occur in nature to a degree sufficient to contribute to the evolution of new species?

SKELETAL EVIDENCE

Recent paleontological discoveries have brought into question the standard evolutionary scenarios for primates, horses, and whales. We have also recorded the finding of the remains of sea mammals under mysterious circumstances in South Pacific sea caves.

WRONG-WAY PRIMATE MIGRATION

Scientists have long assumed that the primates originated in Africa, spread to Europe, and then jumped to North America. This hypothesis may be overturned by the discovery of fossils of early primates in Wyoming. The revised route would be Africa-Asia-North America, the reverse of the prevailing theory. This item also remarks that "Remains of early 'true' primates have not been uncovered in Africa." The reason is that the deposits that should contain them have all been eroded away. (Bower, B.; "Wyoming Fossils Shake Up Views of Early Primate Migration," Science News, 129:71, 1986.)

Comment. One wonders about the "missing" African strata and fossils. Perhaps they never existed. Here is an instance where theory requires certain data, as with evolution's missing links. Sometimes "missing" data is really missing!

A GLITCH IN THE EVOLUTION OF WHALES

S.A. McLeod, a paleontologist at the Los Angeles County Museum of Natural History, has been studying the bones of a 10-million-year-old whale found in a backyard in Southern California. It is claimed that these bones "will fill an important gap in (the) knowledge of the evolution of whales." Actually, the opposite seems to be the case.

> One of the interesting things about the discovery is that it appears 'this guy didn't follow the same evolutionary path as living whales,' McLeod says. Sperm whales today have well-developed teeth only in the lower jaw, whereas the fossil whale shows evidence of very large, very well-developed teeth in both upper and lower jaws.

(Tyndall, Katie; "A Whale's Legacy," Insight, 49, June 15, 1987. Cr. C. Stiles.)

Comment. One would think that a full mouth of teeth would serve sperm whales better, especially in their battles with the giant squid they prey upon. Is evolution reversing for whales?

HORSING AROUND WITH EVOLUTION

In the Borrego Badlands of California, Barbara Quinan has stumbled upon the fossilized skull of a modern horse, E. equus. The skull was found in situ, partly mineralized, a process usually requiring hundreds of thousands of years. Mammoth bones punctuate the strata immediately above and below those containing the horse fossil. The paleontological anomaly is that modern horses were supposed to have evolved in Asia and not brought to the New World until the Spanish explorers landed. The only way to evade rewriting horse history is to: (1) Cast doubt on the dating of the strata, or (2) Insist that the fossil is not really a horse at all but a similar animal, such as the long-headed zebra. (Smith, Gordon; "E. Equus: Immigrant or Emigrant?" Science 84, 5:76, April 1984.)

MONSTER SKELETONS FOUND IN UNDERWATER FIJI CAVE

The scene is a remote underwater cave 51.5 kilometers from one of Fiji's resort islands. K. Deacon, a professional diver from Sydney is the source of the following report.

> We have found what appear to be two adults, one adolescent and one juvenile,' he said. 'They bear no resemblance to any marine creature I know.'
>
> The adult skulls were about 1 m long with a total body length of 8 m to 10 m. They looked prehistoric and perhaps were land animals or amphibious species.
>
> The cave, now a part of a reef about 50 m under water, may once have been above sea level.
>
> .....
>
> Mr Deacon believed the creatures were either prehistoric or contempory animals unknown to science---or, if they were some known kind of animal, then how they had found their way into the cave was a mystery.

(Spencer, Geoff; "Monsters Found in the Fiji Deep," New Zealand Herald, May 31, 1990, Cr. R. Collyns via L. Farish.)

NO UNKNOWN MONSTERS IN THOSE FIJI UNDERWATER CAVES: NEVERTHELESS, THE MYSTERY DEEPENS

A videotape of those unusual skeletons in the Fiji underwater caves mentioned above has been studied by scientists at the Queensland Museum. The are not the remains of unknown monsters---they are only dolphin bones! But one mystery has been replaced by several. It seems that there were three different species of dolphins, and they were found at the closed ends of underwater passages that were just a bit larger than the living animals themselves. Why did three different species go into the caves at all? Why did they go all the way to the ends of the closed passages, given their excellent echo-location systems?

Pertinent here is the discovery of skeletons and recent carcasses of green turtles in similar situations in underwater caves in Indonesia. Turtles lack the dolphins' echo-location equipment, but they are still excellent navigators. (Molnar, R.E.; personal communication, July 2, 1991. Molnar is a scientist at the Queensland Museum.)

Comment. Another question comes to mind: Could the demise of the dolphins in the Fiji caves be related to the occasional strandings of whales and other cetacea on beaches all over the world? Is there a common failure in perception and/or navigation?

UNRECOGNIZED SPECIES

THE RETURN OF THE TASMANIAN TIGER

In March 1982, a park ranger in northwestern Tasmania awoke in the dead of night. From force of habit, he scanned the woods, his spotlight punching through black walls of rain. And there in the beam was one of the strangest creatures he had ever seen. About the size and shape of a dog, it was covered with stripes that ran from its shoulders across its back to its thick, rigid, tail.

The animal stood still as the startled ranger counted the stripes, then it nonchalantly gave an enormous jaw-stretching yawn. But when the ranger reached for his camera, the creature faded into the undergrowth, leaving nothing but a rank smell. It also left a trail of excitement, for the bizarre beast looked exactly like a Tasmanian tiger---also called a thylacine or Tasmanian wolf---an animal thought to have been extinct nearly 50 years ago.

Hundreds of people claim they have spotted the Tasmanian tiger since the last captive died in 1936, but we have no good photos or other "proofs." Mediaman Ted Turner has offered a prize of $100,000 for "verifiable evidence" that the Tasmanian tiger still lives. Consequently, the Tasmanian wilds are being combed diligently and automatic cameras, triggered by infrared beams, are being set up in likely spots. (Bunk, Steve; "Just How Extinct Is Tasmania's Tiger?" International Wildlife, 15:37, July-August 1985. Cr. M.J. Shields)

TIGERS IN WESTERN AUSTRALIA?

The title of course refers to the Tasmanian tiger or wolf or thylacine. We reported above on the possibility of a small relict population of Tasmanian tigers in Tasmania, where the supposedly last specimen expired in a Hobart zoo in 1936. There is now good evidence that the thylacine also roams Western Australia, where it has been believed extinct for thousands of years! At hand are photographs, casts of footprints, a carcass that may be very recent, and many eye-witness reports.

Much of the recent evidence has been gathered by Kevin Cameron, a first-rate bushman with two superbly trained dogs. A.M. Douglas, the author of this article and formerly Senior Experimental Officer at the Wetern Australian Museum in Perth was skeptical about living thylacines at first but is now a firm believer. He states, "I think Kevin Cameron has made the single most important wildlife discovery of this century." (Douglas, Athol M.; "Tigers in Western Australia?" New Scientist, p. 44, April 24, 1986.)

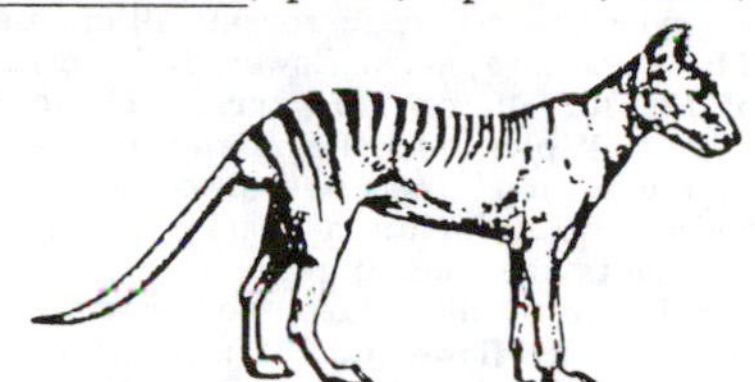

The thylacine or Tasmanian tiger.

EXTINCTION DISCOUNTED

A computer analysis has left little doubt that the supposedly extinct Tasmanian tiger or wolf still exists in remote areas of Australia's island state.

The last captive specimen of the marsupial tiger or thylacine died in a Tasmanian zoo in 1986. No living specimen has been verified since, but sporadic reports persist in Tasmania, Victoria, and Western Australia.

H. Nix, of the Australian National University's Centre for Resource and Environmental Studies, has a computer program based upon detailed descriptions of climatic, topographic, and environmental factors that identifies areas where a particular animal or plant could flourish. Nix gathered the environmental requirements of the thylacine from records of where they had been shot and trapped in the past. This plus the computer program allowed Nix to identify prime thylacine territory. Comparing this information with the best sightings over the past 60 years, Nix found perfect agreement. In other words, post-extinction reports of thylacines come from just those areas where one would expect them to! (Anonymous; "Computers Help to Hunt the Tasmanian Tiger," New Scientist, p. 24, March 10, 1990.)

Comment. This all sounds a bit tautological; that is, "circular reasoning"!

DOES RI = MERMAID?

The astounding item that follows is taken from the new journal Cryptozoology, which is the official journal of the International Society of Cryptozoology. The Society was founded by a group of scientists interested in unrecognized species of animals.

Abstract. An aquatic creature roughly resembling the traditional 'mermaid,' and sometimes identified with it, is reportedly known through a variety of encounters with natives of Central New Ireland. The ri, as they are called, are frequently sighted by fishermen, occasionally netted or found dead on beaches, and sometimes eaten. Males, females and juveniles are reported, subsisting on fish in the shallow seas around the Bismarck and Solomon archipelagos. It is unlikely that the animals are dugongs or porpoises, both of which are known to, and readily identified by, the natives.

New Ireland is northeast of Papua-New Guinea. The article proper goes on to describe the ri as an air-breathing mammal with human-like head, arms, genitalia, and upper trunk. The lower trunk is legless and terminates in a pair of lateral fins. (Wagner, Roy; "The Ri---Unidentified Aquatic Animals of New Ireland," Cryptozoology, 1:33, 1982.)

RI SEEN

In June 1983, a three-man team travelled to New Ireland, off the coast of New Guinea, to track down the ri, an unrecognized aquatic mammal with some mermaid overtones. The natives of New Ireland kill and eat the ri, which they insist is different from the dugong. The team was fortunate to observe a ri from as close as 50 feet as it hunted fish in Elizabeth Bay. The animal was 5-7 feet long, skinny and fast. No dorsal fin was seen, and the tail flukes were mammilian (i.e., horizontal). The creature surfaced about every 10 minutes. Such behavior is quite unlike that of known cetaceans and sirenians. (Anonymous; "New Guinea Expedition Observes Ri," ISC Newsletter, 2:1, Summer 1983.)

DOUBTS ABOUT THE RI SURFACE

On the other side of the world, the ri question has surfaced again. Above, the observations of a group from the ISC (International Society of Cryptzoology) to Papua-New Guinea were described. This expedition was searching for a mermaid-like mammal called the ri. The group returned with one ambiguous sighting and considerable anecdotal evidence. Now, Jon Beckjord relates how he spent two weeks in the same area of New Guinea with negative results. He believes that dugongs can account for all observations and anecdotes. (Sheaffer, Robert; "Psychic Vibrations," Skeptical Inquirer, 8:221, 1984.)

Comment. In fairness, it must be noted that Beckjord belongs to the National Cryptozoological Society and not the ISC! The whole field of anomaly research, from UFOs to archeology, suffers from infighting among various groups. In contrast, the Skeptical Inquirer provides a central focus for responses to claims of paranormal phenomena.

all in one article! This is rich grist for the anomaly mill. But can we believe any of it?

"HOPEFUL MONSTERS" IN ICELAND?

A UPI item from Reykjavik is appropriate here after the heavy dose of speculation above. Two Icelandic bird hunters say they saw a pair of unidentifiable creatures playing on a beach. The creatures, said to be bigger than horses, emerged from Lake Kleiffarvvatn, 20 miles south of Reykjavik. The mysterious animals swam like seals but ran about the beach like dogs. Their footprints were larger than horse hoofs but split into three cloves. An Icelandic scientist commented (quite safely) that, "there is more in nature than we know." (Anonymous; "Icelandic Hunters Claim Sighting of Two Unidentifiable Creatures," Houston Chronicle, November 16, 1984. Cr. J.B. Burns)

RI = DUGONG; DOGGONE!

The ri has been reputed to be a mermaid-like mammal frequenting the shores of New Ireland, Papua New Guinea. Some cryptozoologists have insisted that the ri is not a new species but probably just a commonplace dugong. The skeptics are right. A February 1985 expedition took underwater photos of the animal that some natives call the ri, and it was definitely a dugong. (Anonymous; "New Expedition Identified Ri as Dugong," ISC Newsletter, 4:1, Spring 1985.)

A MAMMOTH TALE!

In the discussion of the genuineness of the Holly Oak Pendant on p. 36, the possibility was raised that mammoths might have survived in North America until just a few centuries ago. Such survival is contrary to all mainstream thinking. Thus, when a datum comes along, even though it appears rather far-fetched, that testifies for the recent survival of mammoths, we must at least examine it.

The datum in question (and it really is questionable) comes from the The National Tombstone Epitaph, hardly part of the scientific literature! The article develops the theme that Chinese explorers landed in North America several millennia ago. The basis for such speculation is an ancient Chinese work called the Shun-Hai Ching, which is reputed to be about 3500 years old. In it, the Chinese explorers mention encounters with several strange animals. One is easily recognized as the collared peccary, known only in the New World, thus establishing the reality of a trans-Pacific contact. Now, here is the piece de resistance:

> Here we met a creature as tall as three men and so great that the earth trembled as he walked. He had a voice as loud as thunder. He was red like fire. From his mouth he spat spears of pearl, and he had but one long arm. He was wont to take up men in his hand and dash their brains out against rocks.

Could this creature be anything but a mammoth? Incidentally, the frozen Siberian mammoths are reported to be covered with reddish hair. (Eckhardt, C.F.; "Prehistoric Explorers of the West?" National Tombstone Epitaph, p. 17, October 1988. Cr. H.J. Hanson)

Comment. Ancient Chinese in America and the late survival of the mammoth---

INTERSPECIES PHENOMENA

KILLER BAMBOOS

There are more than 500 species of bamboo. Together, they have conspired ---in a vegetative way---to exterminate the pandas. Why pick on such a cute, lovable animal? Pandas, you see, eat nothing but bamboos; and the bamboos have had enough! The bamboos' strategy is to flower only once in a lifetime. When the appointed time arrives for each species, all plants of the species all over the world flower simultaneously. The various species flower at intervals of 15, 30, 60, or 120 years. (These 15-year multiples and the unknown clocks that determine them are anomalies in themselves.) After a species flowers, all plants die, leaving the fate of the species to a thick carpet of seeds. Until the next flowering, it will extend its domain via vegetative reproduction only. Ten years will pass before the bamboos have grown enough to be a viable panda food source. The pandas' only hope is to find a species of bamboo that did not flower.

It is hard to think of a plant as malevolent, but here is how P. Shipman describes the situation:

> Green and slender, deceptively innocent-looking, it spreads out slowly. year by year, until it has its victims surrounded. Meanwhile the pandas, poor patsies, are eating out of the bamboo's hand. Only when the pandas are well and truly dependent on it does the bamboo deal its coup de grace. It flowers and seeds, thus ensuring its own survival as a species. And then, in an act of sweet self-sacrifice, it dies, taking its archenemy with it.

If the pandas manage to survive after the 15-year bamboos have had a try at death-by-death, a time will come when 15- and 30-year bamboos will flower simultaneously. In the mid-1970s, several species, with different periodicities, all flowered simultaneously in China. The panda population was decimated. (Shipman, Pat; "Killer Bamboo," Discover, 11:22, February 1990. Cr. R. Dorion.)

The pandas may be saved by the human-conceived strategem of feeding plant hormones to bamboo shoots. Researchers in the lab have been able to break the lockstep bamboo cycle in this manner. (Johnson, Julie; "Hormonal Clue to Bamboo's Elusive Blossomings," New Scientist, p. 31, March 31, 1990.)

Comment. The panda, it seems is not without its own evolutionary strategy. By evolving into a cute, cuddly, teddy-bear, it has so enthralled a third species, that this more advanced (?) life form has found a way to save the panda from its cyclic dilemma via science!

DOLPHINS TO THE RESCUE---AGAIN!

September 1983. Tokerau Beach, Northland, New Zealand. A pod of 80 pilot whales ran aground and were stranded by the ebbing tide. Local townspeople followed a new technique developed for aiding stranded cetaceans. They waded out, talking soothingly to the whales, and keeping their skins wet. When the tide came back in and refloated the whales, the New Zealanders turned them around and tried to guide them to deeper waters.

Sometimes refloated cetaceans just turn around and reground themselves again, but this time the pilot whales were fortunate. A school of dolphins fishing offshore somehow apprehended the situation and swam into the shallows around the pilot whales. The dolphins then guided them out to sea. 76 of the pilot whales were thus saved. In a similar incident 5 years earlier at Whangarei harbor, a helicopter followed the dolphins and whales several miles out to sea, confirming interspecies aid. Such stories are reminiscent of those where drowning humans are helped by dolphins. (Anonymous; "Dolphin Pilots," Oceans, 17:50, 1984.)

TWO-FACED INDIANS TRICK TIGERS

A significant hazard for fishermen and forest workers in western Bengal is a tiger attack. As these Indians go about their fishing, wood chopping, and honey gathering, tigers are wont to sneak up from behind, spring, and carry off a good-sized meal. But in recent experiments, some 900 volunteers have been wearing human masks on the backs of their heads. This strategem has cut tiger attacks drastically. The idea is that tigers, trailing a potential supper, see that human face and figure that the person is alert and watchful. In fact, tigers have been known to track mask-wearers for hours without attacking. Pretty clever! How long before the tigers catch on? (Anonymous; "Protective Mimicry in Humans," BioScience, 39: 750, 1989.)

BIRDS

EXTERNAL APPEARANCE

WHY BIRDS ARE PRETTY

Darwin believed that many male birds are brightly colored because females prefer flashy finery and thus put evolutionary pressure on the development of these characteristics. A large-scale study by Baker and Parker indicates that Darwin erred and that the evolutionary pressure comes instead from predators avoiding brightly colored targets. Instinct tells the predators---incorrectly in many cases---that colorful prey taste bad or are noxious.

The remarkable (possibly anomalous) aspect of bird coloration is the incredible external similarity of unrelated birds occupying similar habitats. For example, in the accompanying figure, the American eastern meadowlark (left) closely resembles the African yellow-throated long claw (right). (Krebs, John R.; "Bird Colours," Nature, 282: 14, 1979.)

Comment. Two questions cannot be repressed: How do the genes orchestrate this amazing convergence in response to environmental factors? Why was evolution not equally clever in equipping predators with countermeasures to see through these ruses?

UPSIDE-DOWN ANIMALS

Stephen Jay Gould's recent essay, "The Flamingo's Smile," like all his writing, is thought-provoking. The essay goes far beyond the happy flamingo. It is about unusual adaptations in nature, as illustrated by three inverted or partially inverted creatures.

1. The flamingo is a filter-feeder that strains food out of the water with its bill while its head is upside-down, The flamingo's bill and tongue are (and must be) radically different from those of other birds to succeed in this strange behavior.

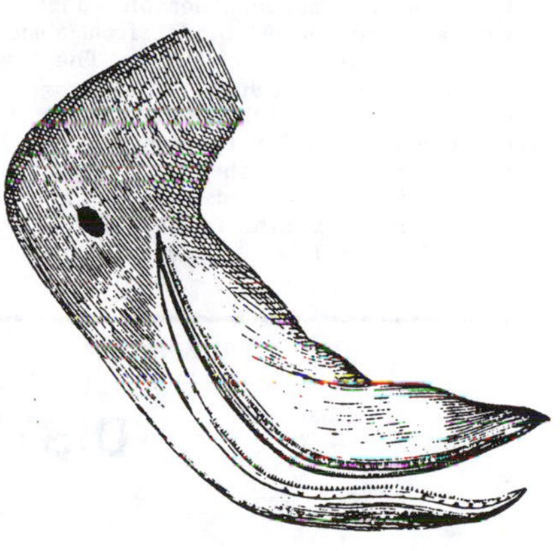

A flamingo's head in its feeding position.

2. One type of jellyfish, rather than swimming around with its pulsating bell on top, plunks itself upside-down on the bottom and uses its bell as a suction cup to anchor itself. It then shoots poisonous darts attached to strings of mucous at passing targets and reels them in.

3. Some African catfish graze on algae on the undersides of water plants. They swim upside down all the time and display a reversed color scheme, being black on the bottom and light on top.

Gould employs these three examples to argue that changes in animal behavior must have preceded the many changes in form, function, color, etc. that make upside down living profitable. In other words, the proto-flamingos tried feeding with their heads upside down; and it didn't work too well. But "nature" responded with a series of random biological changes, some of which were just what was needed for efficient upside down feeding. In this way, we end up with admirably adapted, inverted flamingos, jellyfish, and catfish. (Gould, Stephen Jay; "The Flamingo's Smile," Natural History, 94:7, March 1985.)

Comment. If we were a bit flip above, it is because Gould and most biologists believe that such examples of "perfection" are nicely explained by evolutionary theory. We cannot be so sanguine, for we are still left with too many unanswered questions. Why would animals indulge in such bizarre behavior in the first place? A proto-flamingo experimenting with inverted feeding sans the proper equipment---huge tongue, fantastic bill, straining devices, etc.---would not be very efficient and would probably get a snootful of water in the bargain! A half-flamingoized bill wouldn't be much better; and a perfect flamingo bill is pretty useless without the pumping and raking action of the special tongue. Can all these changes be orchestrated by random mutations? Finally, are there any fossils of transition-stage flamingos? Antievolutionists have been using such arguments for years. The creationist literature is full of them. In reviewing the "answers" proclaimed by both evolutionists and creationists, it seems as if we really deserve some real explanations rather than superficial, philosophically shaped dogmas.

GENES AREN'T EVERYTHING

Many species of birds show considerable variation in size, shape, and coloration in various parts of their ranges. Hithertofore, such geographical variations had been considered to be genetically controlled; that is, the genes also varied with the geography. But experiments in which the eggs of redwinged blackbirds were interchanged between nests in northern and southern Florida and also between Colorado and Minnesota seem to show that the environment is more important than genetic endowment.

In the redwinged blackbird, the bill size and shape and body weight of the transplanted birds (after hatching, of course) were characteristic of the locale in which they were raised rather than that where the eggs were laid. (James, Frances C.; "Environmental Component of Morphological Differentiation in Birds," Science, 221:184, 1983.)

BEHAVIOR

MEMORY IN FOOD-HOARDING BIRDS

Birds are not popularly thought to possess superlative memories, but the behavior of food-hoarding birds proves this is untrue. Several species of birds gather huge quantities of food when it is abundant and cache it for later use. On the surface this trait does not seem remarkable, but a look at the numbers of caches involved belies this superficial evaluation, especially to a species who forgot where he put the car keys a few hours ago.

Take Clark's nutcracker as an example. A bird of the U.S. Southwest, Clark's nutcracker harvests conifer seeds when in season and buries them for future use during the rest of the year. One bird may bury as many as 33,000 seeds in thousands of caches, 4-5 seeds per cache. Its memory guides it back to these caches during the next year. (Shettleworth, Sara J.; "Memory in Food-Hoarding Birds,; Scientific American, 248:102, March 1983.)

A Clark's nutcracker burying a seed.

Comment. Such capabilities should not be filed away under "Isn't nature wonderful?" or "Gee whiz!" The import of these special characteristics is suggested in a quote from the article: "... certain species have adaptive specializations that make them particularly good at learning and remembering things it is important for them to know." Then read about calculating prodigies in the Handbook Unfathomed Mind.

BIRD BRAIN

Alex can name 80 things, tell colors, and even seems to be able to handle a few abstract ideas. Alex is not a chimp or porpoise. Alex is an African grey parrot, who has been living in an avian Sesame Street for 13 years. Parrots are wonderful mimics and pretty bright as birds go. Nevertheless, skeptics scoff at Alex's accomplishments as only the product of long, intense training. Alex can't be too dumb. Once when he couldn't lift a cup covering a tasty nut, he turned to the nearest human assistant and demanded, "Go pick up cup." Who's training whom? (Stipp, David; "Einstein Bird Has Scientists Atwitter over Mental Feats," Wall Street Journal, May 9, 1990. Cr. J. Covey.)

MYSTERIOUS BIRD DEATHS

> An eight-year study by Indian zoologists has failed to establish why birds commit suicide year after year at the small village of Jatinga in the northeastern state of Assam.
>
> Attracted by the lights, birds converge on Jatinga at night and on landing become immobile, stop feeding and starve. They neither resist capture nor try to fly away.

The mysterious phenomenon dates back to 1905. It peaks in September and October, as the monsoon season wanes, with as many as 500 birds, from some 36 species, dying each night. The birds alight at the same spot each year---a one-kilometer stretch in the town. No one can account for the selection of this precise spot or for the dazed condition of the birds. (Jayaraman, K.S.; "Mystery of Birds Deaths in Assam," Nature, 331:556, 1988.)

WHY DO FLAMINGOS STAND ON ONE LEG?

This profound question elicited a wide range of replies from readers of New Scientist, some of which are worth recording here.

W. Smith supposed that because the flamingo has exceptionally long, thin legs that it was difficult for its heart to return blood from its feet. Therefore, by standing on one leg and occasionally switching, the flamingo prevents blood from collecting in its feet.

L.J. Los replied with reference to a phenomenon of which we were unaware:

> Farm animals are well known for letting sleep be linked to half of their brain at a time. In this way they can maintain a measure of alertness ---even while looking fast asleep.
>
> Flamingos roost upon one of their legs while the other half of their body is in the sleep stage. When the other half of their brain and body earns a rest, they change legs. A leg that is in the sleep stage would not support the bird as a whole.

But P. Hardy had the best answer:

> Why do flamingos stand on one leg? So ducks only bump into them half the time.

(Various authors; "Flamingo File," New Scientist, p. 52, August 17, 1991.)

DISTRIBUTION

THOUSANDS OF GREBES FALL FROM THE SKIES

December 10, 1991. Minersville, Utah. About 9:30 PM, the skies of Minersville were filled with the cries of birds. According to V. Hollingshead

> They were just falling out of the sky, hitting the church, cars, the ball parks. Hundreds of them fell all over the streets. You could hear them hitting each other in the air, and hitting the ground.

Minersville Elementary School Secretary

S. Taylor reported that the birds landed everywhere, including the roofs of houses; they even broke some automobile windshields. Hundreds were killed, but many survived their fall and were taken to bodies of water where they could rest and take off. (Grebes cannot take off from land.)

The birds were identified as eared grebes, which were migrating from Great Salt Lake to Baja California. It was theorized that a snowstorm and fog had exhausted and disoriented them. (Christensen, Kathleen; "Thousands of Grebes Fall from the Skies," Spectrum, December 12, 1991. Cr. D.H. Palmer.)

TALENTS AND FACULTIES

Birds manifest many remarkable talents and capabilities, a handful of which have been noticed in Science Frontiers. Included below are bowerbird "art," albatross long-distance flights, bird navigation conundrums, and egg mimicry in cuckoos. Ornithologists may complain that there are many more impressive avian capabilities, such as hummingbird flight and complex bird duets. The ornithologists are correct. We have only captured a few examples in the first 86 issues of Science Frontiers. Those readers who wish more should direct their attentions to the Series-B catalog volumes.

BOWERBIRD ART FOR ART'S SAKE

The bowerbirds of New Guinea and Australia build and decorate marvelously intricate and esthetic works of art. One must classify these impressive structures as works of art, even by human standards. For example, the bowers have geometrical organization and may be oriented to a specific compass direction, depending upon the species. Colorful berries, stones, and, when available, human artifacts are systematically arranged around the bower. Some bowerbirds even take a piece of bark in their bills and paint their bowers with colored berry juices.

Each species has a certain style, but the bowers vary from individual to individual and with the age of the bird. Manifestly, these birds use tools for artistic purposes. Or do they? Is it all instinct?

Some animal behaviorists believe the bower's purpose is to attract mates, but the males often chase females away, although mating does eventually occur within the bowers. A second explanation is that the bowers symbolize territorial rights. In this context, bowerbirds frequently raid and destroy neighboring bowers, stealing choice decorations---all very human-like behavior. (Diamond, Jared M.; "Evolution of Bowerbirds' Bowers: Animal Origins of the Esthetic Sense," Nature, 297:99, 1982.)

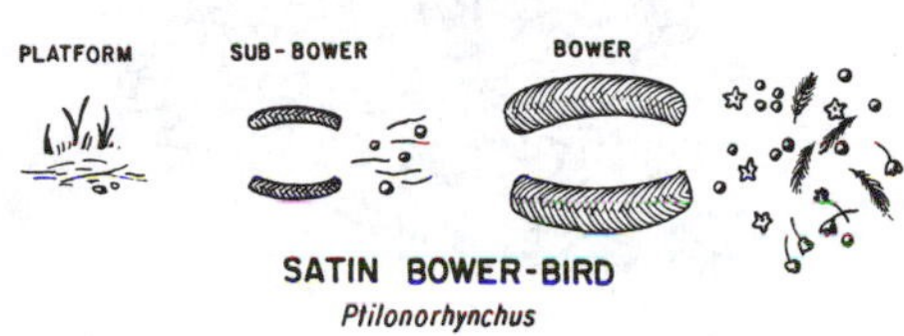

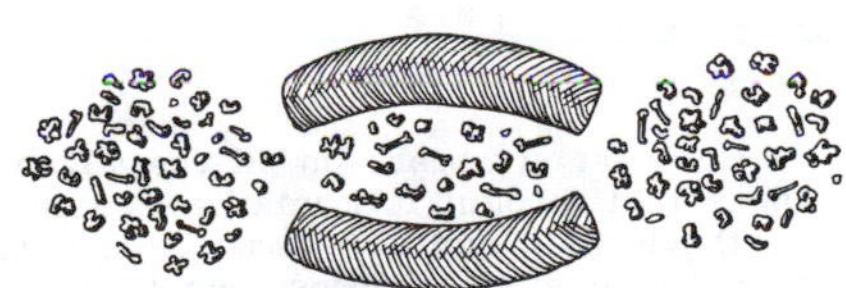

SPOTTED BOWER-BIRD
Chlamydera maculata

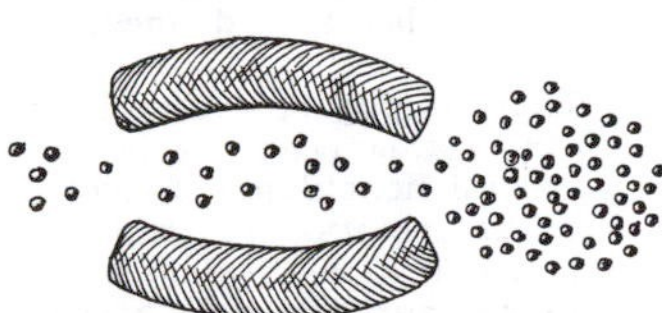

FAWN-BREASTED BOWER-BIRD
Chlamydera cerviniventris

Plan views of three bowerbird bowers.

Comment. Is human art any more profound than that of the bowerbirds? Human artists doubtless feel they are doing something more than attracting mates or proclaiming territory. Unfortunately, we cannot ask the bowerbirds what they are thinking as they carefully select colors and develop designs.

WANDERING ALBATROSSES REALLY WANDER

Six male wandering albatrosses nesting on Crozet Island, between South Africa and Antarctica, were fitted with tiny (180-gram) transmitters and tracked by satellite. Their flights were amazing:

> Tracks of wandering albatrosses in the southwestern Indian Ocean showed that they covered between 3,600 and 15,000 km in a single foraging trip during an incubation shift. They flew at speeds of up to 80 km per h and over distances of up to 900 km per day. They remained active at night, particularly on moonlit nights...

(Jouventin, Pierre, and Weimerskirch, Henri; "Satellite Tracking of Wandering Albatrosses,; Nature, 343:746, 1990.)

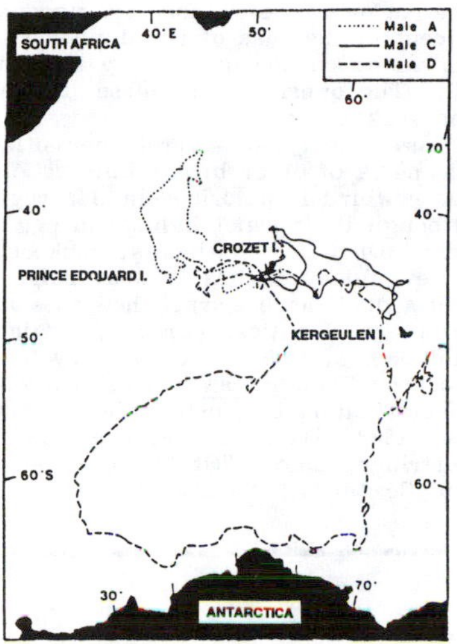

Paths flown by three wandering albatrosses that were tracked in the southern Indian Ocean.

HAVE MAGNETS, WILL TRAVEL

Homing pigeons seem to possess at least two direction sensors. Years of experiments with released birds have proved that they use sun compasses on sunny days but have magnetic backups for cloudy days. But how do they sense the earth's magnetic field? Paired-coil tests suggested that the pigeon compass resided in the neck or back of the head. Narrowing the search with sensitive magnetometers and two dozen dissected pigeons, the authors discovered tiny bits of tissue containing magnetite crystals. The same tissues contained yellow crystals likely made by the iron-storage protein ferritin, which was probably used in the biological synthesis of the magnetite. (Walcott, Charles, et al; "Pigeons Have Magnets," Science, 205:1027, 1979)

Comment. Many species of mud bacteria also synthesize magnetite for purposes of orientation, indicating that nature or some directive force used the same strategy in two widely separated species.

MAGNETIC BUMP

Do birds utilize the earth's magnetic field for navigation? Many have so surmised; and there exists anecdotal evidence for it. A Swedish ecologist, T. Almberstam, decided to attempt scientific observations. At Norberg, in central Sweden, a huge deposit of magnetite creates a powerful magnetic anomaly. The deposit is 12 kilometers long by several wide. At low altitudes, the total magnetic intensity of the earth's field is 60% higher than normal. What happens when migrating birds fly into this magnetic bump?

> Although Almerstam found that many migrating birds showed no signs of avoiding the Norberg anomaly, and often managed to keep on the right course as they passed through it, there were definite indications that the birds' orientation might be affected under special circumstances. Some migrants flying at low altitudes, where the magnetic intensity was greatest and the inclination and declination distorted greatly, became disoriented briefly. They nervously landed and then circled around before taking off again. Other birds changed their altitude abruptly, dropping 100 metres in two minutes and breaking up their flock formations.

Certainly something is happening, but no one knows what. (Anonymous; "Magnetic Anomaly Upsets Migrating Birds," New Scientist, p. 32, November 5, 1987.)

DO BIRDS USE GENETIC MAPS DURING MIGRATION?

Many young birds migrate successfully without help from older birds who have made the trip before. The implication is that migration instructions, perhaps even some sort of map of astronomical or geographical references, are somehow written upon the genes inherited from their parents. Just how maps can be coded into gene structure is anyone's guess. (In fact, since the DNA in the genes seems to code only for protein synthesis, the locations and characters of inheritable maps and other biological instructions are not immediately obvious.)

The problem has been exacerbated by recent experiments with German and Austrian blackcaps. These two common European warbler species take different

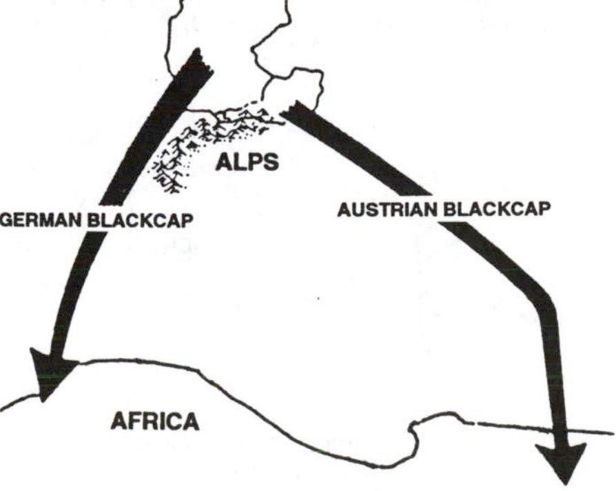

Routes taken by migrating German and Austrian blackcaps en route to Africa. Hybrids bisected these initial flight path directions and would have ended up in the Alps.

routes to Africa in the winter. The German blackcaps fly southwest and the Austrian southeast--routes 50° apart. A. Helbig has crossed the German and Austrian blackcaps to see what route(s) their hybrid offspring would take. Curiously, they favored a route intermediate between those of their parents. The hybrids' route--- bisecting those of the parents'---would take the hybrids right into the Alps, where survival would be unlikely. (Day, Stephen; "Migrating Birds Use Genetic Maps to Navigate," New Scientist, p. 21, April 21, 1991.)

Comment. The puzzle at hand concerns those purported genetic maps. Presumably, the hybridization of the two blackcaps involves the melding to two different, highly specific, maps and sets of migration instructions; possibly including compass directions (astronomically or magnetically determined), landmark locations, and even characteristic pathway odors. How can such instructions be combined (perhaps averaged) to draw up entirely new navigation instructions for a route the parents have never taken?

EGG MIMICRY IN CUCKOOS

In Britain, cuckoos mainly parasitize five species of smaller birds. They do this by laying their eggs in the hosts' nests. After hatching, the young cuckoos grow much faster than the young of the host species. Soon the cuckoo is able to eject the host's young from the nest and get all the food brought by the parents. Actually, the cuckoo is so aggressive in this business of parasitization that, when it finds a host nest with eggs so far along in incubation that parasitization is impractical, it destroys the whole nest. This usually forces the host birds to lay fresh eggs, giving the cuckoo a chance to parasitize the nest.

The most remarkable thing about cuckoo parasitism is the birds' ability to match the eggs of the host species in size, spottedness, background color, and darkness. The eggs of all five species commonly parasitized in Britain are much smaller than a bird the size of the cuckoo would normally lay; but the cuckoo still manages to lay eggs of just the right size. When the bona fide eggs of the five parasitized species are placed side-by-side with the mimics laid by the cuckoos, the matches are uncanny---except in the case of the dunnock, which the cuckoo doesn't try to mimic at all. The question, of course, is how the cuckoos do it.

American cuckoos rarely parasitize the nests of other birds; but the American cowbird is notorious in this regard, although it does not indulge in egg mimicry. On other continents, cuckoos, honeyguides, finches, a weaverbird, and a duck have learned how to slough off parental duties. (Brooke, M. de L., and Davies, N.B.; "Egg Mimicry by Cuckoos Cuculus canorus in Relation to Discrimination by Hosts," Nature, 335: 630, 1988. Also: Harvey, Paul H., and Partridge, Linda; "Of Cuckoo Clocks and Cowbirds," Nature, 335:586, 1988.)

THE BIRD THAT SMELLS LIKE COW MANURE

Pity the poor hoatzin. This "extremely primitive" bird is usually described as being just a step beyond the reptiles. The young hoatzins clamber about the jungle foliage using functional claws on their wings. This certainly sounds primitive; then, there's that awful smell!

But perhaps we have been wrong about the hoatzin. It's all just bad press. A. Grajal et al have just discovered that this South American bird utilizes foregut fermentation in digesting its diet of leaves. In fact, the hoatzin is the only bird that has evolved this useful capability. Cows, sloths, and a few other mammals and marsupials evolved foregut fermentation. It is hardly a primitive development! Aside from the smell of fermenting vegetable matter, the hoatzin is a rather remarkable animal---more advanced and well-adapted to its environment than previously thought.

A hoatzin feeds its chick a regurgitated mush of partially digested leaves.

BODILY FUNCTIONS

WHEN A BIRD IN THE HAND IS WORSE THAN TWO IN THE BUSH

When J. Dumbacher, an ornithologist working in Papua New Guinea, scratched his hand while freeing a hooded pitohui from a collecting net, his first instinct was to suck the wound. This was a bad move, for he immediately experienced a numbing and burning in his mouth. The reason for this, it turned out, was because the skin and feathers of pitohuis are loaded with homobatrachotoxin, a type of poison. This discovery makes the pitohuis the first known poisonous birds. Like many other poisonous animals, the pitohuis also emit a foul odor and advertise their unsavory nature with bright colors. (Dumbacher, John P., et al; "Homobatrachotoxin in the Genus Pitohui: Chemical Defense in Birds?" Science, 258:799, 1992. Also: Anonymous; "Bird with a Sting in Its Tail," New Scientist, p. 10, October 31, 1992.)

The chemical structure of homobatrachotoxin.

Comment. As we see from the diagram, homobatrachotoxin possesses a rather complex chemical structure. One wonders how the pitohuis acquired the ability to synthesize it through random mutations. The puzzle deepens when one discovers that homobatrachotoxin is also manufactured by the New World poison-dart frogs. Although far-separated taxonomically, both species traveled along the same path of random mutations to achieve this evolutionary convergence.

Grajal et al remark on all the advantages that foregut fermentation confer on the hoatzin and how remarkable it is that this digestive process can be accomplished in such a small volume (cows have huge stomachs). How did the hoatzin hit upon this mechanism before the mammals did? Why didn't other birds "adopt" it? Grajal et al speculate about this hoatzin advance:

> Their highly specialized digestive strategy may have arisen from an ancestral nonobligate folivore because of an evolutionary trade-off between detoxification of plant chemical defenses and enhanced use of cell wall as a nutritional resource.

(Grajal, Alejandro, et al; "Foregut Fermentation in the Hoatzin, a Neotropical Leaf-Eating Bird," Science, 245:1236, 1989.)

Comment. The rather murky quotation above is only speculative. Leaves are abundant in the tropics, and it is fair to ask why other birds did not develop foregut fermentation.

AVIAN PALEONTOLOGY

Dominating avian paleontology is the belief that birds evolved from the reptiles. Unfortunately, fossils that would conclusively demonstrate this contention are exceedingly rare or, in the view of some, nonexistent. The main fossil weapon of the evolutionists is Archaeopteryx, a fossilized, nicely feathered, bird-like creature that also possesses some reptilian features. The several Archaeopteryx fossils that have been found draw criticism like a lightning rod attracts thunderbolts. Some claim that Archaeopteryx is an outright forgery by overenthusiastic evolutionists; others insist that it not a transitional fossil at all, but rather a true bird. Muddying the waters further has been the discovery of Protoavis, which is older than Archaeopteryx and more bird-like to boot. Finally, there is the long-fought debate over just how feathers and flight arose in birds. Here, the central question is: What good is half a wing? This is the so-called Problem of Perfection.

A WEAK MISSING LINK

Evolutionists have always pointed to the Archaeopteryx as a most convincing missing link between the reptiles and birds. The modern study of some really excellent fossil specimens of Archaeopteryx have clouded this issue. The feathers of Archaeopteryx, as preserved in fine limestone, are found to be asymmetrical as required for efficient flight. (Flightless birds have symmetric feathers!) The skull is more birdlike than previously thought. In fact, some aspects of Archaeopteryx are like those in "advanced" birds; others are "primitive" There are now three strongly held views among scientists: Archaeopteryx is related to (1) crocodiles; (2) theropod dinosaurs; and (3) thecodontians (other reptiles). (Benton, Michael J.; "No Consensus on Archaeopteryx," Nature, 305:99, 1983.)

One guess as to how Archaeopteryx might have looked.

STATUS OF ARCHAEOPTERYX UP IN THE AIR!

That famous missing link, Archaeopteryx, the flying reptile, continues to make headlines. The major argument at the 1984 International Archaeopteryx Conference, in Eichstaett, was about whether Archaeopteryx could fly at all, despite its advanced, aerodynamically shaped feathers. It certainly could not have flown well since it lacks the supracoracoideus pulley-system that acts as a wing elevator in birds. Archaeopteryx could not have raised its wings above the horizontal, making it a poor flier at best. It also lacked the birds' keel bone to which the wing muscles are anchored. But those exquisitely designed feathers, so modern in appearance, tilted the scales. The consensus of the Conference was that Archaeopteryx could indeed fly. (Howgate, Michael E.; "Back to the Trees for Archaeopteryx in Bavaria," Nature, 313:435, 1985.)

The really interesting part of the continuing Archaeopteryx saga comes from the recent charge of Fred Hoyle and others that the Archaeopteryx fossil is an outright forgery. Hoyle et al insist that Archaeopteryx could not have flown at all, given its bones and musculature. Archaeopteryx looks like a reptile and was a reptile. As for the modern-looking feathers, they were probably added to the fossil fraudulently. And there do seem to be parts of the fossils on display in London and East Berlin that look highly suspicious. Conventional paleontologists are, of course, aghast that anyone would question the validity of these key transition fossils. (Vines, Gail; "Strange Case of Archaeopteryx 'Fraud'," New Scientist, 3, Mar.14,1985.)

Comment. A wonderful tempest seems to be brewing. Could Archaeopteryx be another Piltdown Man? To put the matter in proper context, we must remember that Archaeopteryx is in all the evolution books alongside the family tree of the horse. It is an emotional issue. On the other hand, Fred Hoyle seems equally convinced that evolution is statistically impossible, and an Archaeopteryx fraud would fit well with his predispositions.

FEATHERS FLY OVER FOSSIL 'FRAUD'

How many plays on words do you count in the following paragraph? "Britain's own scientific knight-errant, Sir Fred Hoyle, has fallen fowl of the palaeontologists. He has flocked together with those who think that the best-ever missing link, the reptile-bird Archaeopteryx, is a convenient fake. Creationists are understandably in high feather about it all. But now, it looks as though they may be left with egg on their faces."

We introduced the possibility of fraud regarding Archaeopteryx fossils above. In the present article by Ted Nield, the evolutionists seem to be responding to Hoyle's claim with ridicule and innuendo. Referring to the claim of fossil forgery, which Hoyle based on photos taken with a low-angle flash and EN100 film, Nield wonders why it was published in the British Journal of Photography instead of Nature or Science, implying that Hoyle's group didn't dare submit their report to high-class journals! As for the "discovery" of double-struck feathers in the Archaeopteryx fossil, which Hoyle thinks were the result of inexpert forgers, Nield remarks that these were noted by naked eye as long ago as 1954, and are due to two rows of slightly overlapping feathers with faint "through-printing." And while it is true that the two halves of the fossil studied by the Hoyle group are not perfect positive-negative pairs, this is but an artifact due to the complexity of the break.

Evidently the charge of forgery-to-save-Darwinism cut geologists to the quick, for Nield revives the old bones of contention between physicists and geologists, "...geologists are especially twitchy about physicists, who for years told them continental drift was impossible, but---after stumbling on the proof ---have strutted around ever since as though it had been their idea all along." (Nield, Ted; "Feathers Fly over Fossil 'Fraud'," New Scientist, p. 49, August 1, 1985.)

Comment. We must not forget that when geologists wanted hundreds of millions of years to account for the strata they saw in the field, Lord Kelvin told them they had to settle for 100,000 years because that was as long as the sun could run on gravitational energy. Nuclear energy came along later. Mercifully we omit more bird/feather jokes.

FEATHERED FLIGHTS OF FANCY

Some more salvos have been fired in an endlessly fascinating controversy (at least it is that to us). First, F. Hoyle and C. Wickramasinghe, hardly strangers to these columns, have published Archaeopteryx, The Primordial Bird: A Case of Fossil Forgery. The book elaborates their theory that the Archaeopteryx fossils, much ballyhooed as "proofs" of evolution, are outright forgeries. Second, T. Kemp, a zoologist on the staff of the University Museum, has returned the fire with a mean-minded review. He states that Hoyle and Wickramasinghe "exhibit a staggering ignorance about the nature of fossils and fossilization processes." Kemp concludes his review with an admission that the possibility of forgery should indeed be investigated. "But it should be done by those who actually understand fossils, fossilization and fossil preparation, not by a couple of people who exhibit nothing more than a gargantuan conceit that they are clever enough to solve other people's problems for them when they do not even begin to recognize the nature and complexity of the problems." (Kemp, Tom; "Feather Flights of Fancy, Nature, 324:185, 1986.

Finally, Hoyle and Wickramasinghe reply in a letter to Nature that L.M. Spetner and his colleagues in Israel have analyzed samples of the Archaeopteryx fossil with a scanning electron microscope and X-ray spectroscopy. Results: the rock matrix and the feathers thought to be spurious are radically different. "These striking differences in texture and composition between the suspect regions and the native matrix are, in our view, a strong indication that this dispute will eventually be resolved in our favour." (Wickramasinghe, N., and Hoyle, F.; "Archaeopteryx, the Primordial Bird?" Nature, 324:622, 1986.

Comment. If you think the controversy over, you do not understand the passions involved! Incidentally, would the paleontologists themselves ever thought about forgery and considered applying X-ray spectroscopy and electron microscopes to the problem? A judge should not sit on the bench during his own trial.

ARCHAEOPTERYX AND FORGERY: ANOTHER VIEWPOINT

We have here what must be considered the evolutionists' reply to the claim of Hoyle and Wickramasinghe that overzealous followers of Darwin deliberately tampered with scientific evidence.

Abstract. "Archaeopteryx lithographica might be regarded as the most important zoological species known, fossil or recent. Its importance lies not in that its transitional nature is unique---there are many such transitional forms at all taxonomic levels---but in the fact that it is an obvious and comprehensible example of organic evolution. There have been recent allegations that the feather impressions on Archaeopteryx are a forgery. In this report, proof of authenticity is provided by exactly matching hairline cracks and dendrites on the feathered areas of the opposing slabs, which show the absence of the artificial cement layer into which modern feathers could have been pressed by a forger." (Charig, Alan J., et al; "Archaeopteryx Is Not a Forgery," Science, 232:622, 1986.)

Comment. The article itself offers some new evidence, but seems to fall a bit short of the proof promised in the Abstract. Let us wait for a rebuttal by Hoyle & Co. The Abstract's claim that many transitional forms exist at all taxonomic levels certainly does not square with the fossil record described by the punctuated evolutionists! In any event, the fossil record gap between dinosaurs and sophisticatedly feathered Archaeopteryx is still a Marianas Trench.

THE NEW ARCHAEOPERYX FOSSIL

A new fossil of Archaeopteryx has been found in a private collection, where it was misclassified as a small dinosaur. The specimen was actually found many years ago by an amateur in the Upper Jurassic Solnhofen limestone in Bavaria, in about the same area as the Berlin and Eichstatt Archaeopteryx fossils.

Under low-angle illumination, the new specimen shows parallel impressions originating from the lower arm of the left "wing." These impressions are "interpreted" as imprints of feather shafts. Thus, the new fossil reinforces the mainstream position that Archaeopteryx really did have feathers and was a link between reptiles and birds. Evolutionists will rest easier now!

Two bothersome observations intrude, however. First, although the report on the new specimen states that the question of forgery does not arise here, even though the specimen's tail has been restored to the length deemed proper by the owner. In addition, the new Archaeopteryx is 10% larger than the London specimen, 30% larger than the Berlin specimen, and fully twice the size of the Eichstatt specimen. Is there more than one Archaeopteryx species? (Wellnhofer, Peter; "A New Specimen of Archaeopteryx," Science, 240:1790, 1988. Also: Wilford, John Noble; "Fossil May Help Tie Reptiles to Birds," New York Times, June 24, 1988. Cr. J. Covey)

Comment. We wonder if Hoyle and Wickramasinghe will be allowed to examine the new specimen! Of course, this new discovery does not disprove the forgery claim for the two specimens studied by Hoyle and Wickramasinghe; it merely weakens their case.

Update. In 1993, A. Feduccia asserted that the claws of Archaeopteryx were sharp and curved like those of modern aboreal birds and not at all like those of terrestrial birds or those theropod dinosaurs that many claim were the immediate ancestors of birds. (Science, 259:790, 1993)

ARCHAEOPTERYX A DEAD END?

"Just as the 150-million-year-old Archaeopteryx fossil is being reinstated as the earliest known bird after considerable controversy, along come two crow-size skeletons that are not only 75 million years older than Archaeopteryx but also more birdlike, according to the paleontologists who discovered them. The Washington, D.C.-based National Geographic Society, which funded the work, announced this week that Sankar Chatterjee and his colleagues at Texas Tech University in Lubbock found the 225-million-year-old fossils near Post, Tex." (Weisburd, S.; "Oldest Bird and Longest Dinosaur," Science News, 130:103, 1986.)

Chatterjee has named the new fossil Protoavis. "Protoavis seems certain to reopen the long-running controversy on the evolution of birds. In particular whether the common ancestor of birds and dinosaurs was itself a dinosaur. Protoavis, from the late Triassic, appears at the time of the earliest dinosaurs, and if the identification is upheld it seems likely that it will be used to argue against the view of John Ostrom of Yale University that birds are descended from the dinosaurs. It also tends to confirm what many paleontologists have long suspected, that Archaeopteryx is not on the direct line to modern birds. It is in some ways more reptilian than Protoavis, and the period between the late Jurassic Archaeopteryx and the world-wide radiation of birds in the Cretaceous has to some seemed suspiciously brief." (Anonymous; "Fossil Bird Shakes Evolutionary Hypotheses," Nature, 322: 677, 1986.)

One guess as to how Protoavis might have looked.

Comment. But what about all those textbooks that assure us positively that birds descended from dinosaurs and that Archaeopteryx is a classic missing link?

FOSSIL IDENTITY STILL UP IN THE AIR

In 1986, we reported the discovery of bird-like fossils in Texas by S. Chatterjee, a paleontologist at Texas Technical University. Chatterjee was so certain that the fossils (two specimens exist) were primitive birds that he named the species Protoavis texensis (first bird from Texas). During the past five years, the scientific community has chafed while Chatterjee studied his finds and wrote them up. It seems that many paleontologists do not think that Protoavis is really a bird at all, and Chatterjee has been slow in releasing details. But now his first paper has appeared in the Philosophical Transactions of the Royal Society. Result: Many doubts still remain about the status of Protoavis.

A. Feduccia: "Calling this the original bird is irresponsible." (1)

J.H. Ostrom: "Sad to say, for all its length, little support for the claim is to be found in this paper." (2)

J. Gauthier: While some of the bones appear bird-like, they also look dinosaurian and could represent a new type of theropod dinosaur. (3)

For his part, Chjatterjee asserts that Protoavis' skull has 23 features that are fundamentally bird-like, as are the forelimbs, the shoulders, and the hip girdle.

> His reconstruction also shows a flexible neck, large brain, binocular vision, and, crucially, portals running from the rear of the skull to the eye socket---a feature seen in modern birds but not dinosaurs. (1)

Just why is there so much fuss over a handful of poorly preserved bones? If Protoavis is really a bird, it places the origin of birds 75 million years earlier and dethrones Archaeopteryx as a transitional link between dinosaurs and birds. In fact, Protoavis essentially denies that birds evolved from the dinosaurs. In short, Protoavis could change a limb or two on that Tree of Life you see in all the textbooks.

References

1. Anderson, Alun; "Early Bird Threatens Archaeopteryx's Perch," Science, 253:35, 1991.
2. Ostrom, John H.; "The Bird in the Bush," Nature, 353:212, 1991.
3. Monastersky, Richard; "The Lonely Bird," Science News, 140:104, 1991.

EVOLVING ON HALF A WING (AND A PRAYER?)

Just about everyone agrees that half a wing is of little use to an animal "straining" to develop the capability of flight. So, how did the marvelously crafted wings of birds, insects, and mammals evolve in infinitesimal steps? Biologists, including Darwin himself, have long puzzled over this. Stephen Jay Gould in a recent article explores a currently favored way of circumventing the negligible additional survival value of half a wing, or even 90% of a wing. This solution (?) maintains that protowings were not "intended" for flight at all but were developed initially as aerodynamic stabilizers, thermoregulatory systems, sexual attractors or other functions requiring large areas.

Gould describes the experiments of Kingsolver and Koehl in which protowings were modelled and tested for their thermoregulatory and flight values. Surprisingly, there was a sharp transition, as the size of the protowing increased, from good thermoregulation but poor flight capability to the reverse---good flight capability and poor thermoregulation. In other words, a structure developed for one purpose, if enlarged, might be useful for something else! (Gould, Stephen Jay; "Not Necessarily a Wing," Natural History, 94:14, October 1985. See also: Lewin, Roger, "How Does Half a Bird Fly?" Science, 230:530, 1985.)

Comment. The work of Kingsolver and Koehl, though doubtless of high quality, does not come to grips with the fact that a wing for flight is a highly sophisticated combination of skeleton, feathers, membrane, muscles, nervous system, control system, aerodynamic design, etc. ---most of which have nothing to do with thermoregulation. Even some crustaceans have evolved feathery appendages and can almost fly! Where did their "feathers" come from? Somewhere we---all of us---are missing vital clues that tell us how organisms really develop.

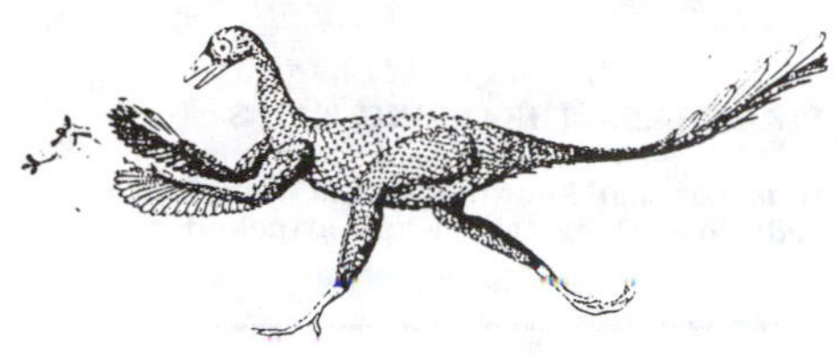

Evolutionists sometimes get carried away in inventing protobirds. The above conceptual protobird was "developing" feathers---supposedly via random mutations---on its forelimbs or proto-wings to help catch insects prior to "developing" flight! (From: Denton, Michael; Evolution: A Theory in Crisis)

REPTILES AND AMPHIBIANS

EXTERNAL APPEARANCE

EIGHT LEATHERBACK MYSTERIES

Our subject here is the leatherback turtle. Weighing up to 1600 pounds, it is the largest of the sea turtles. It is also the fastest turtle, hitting 9 miles per hour at times. But weight and speed are not necessarily mysterious; here are some characteristics that are:

•The leatherback is the only turtle without a rigid shell. Why? Perhaps it needs a flexible shell for its very deep dives. What looks like a shell is its thick, leathery carapace---a strange streamlined structure with five to seven odd "keels" running lengthwise.

•These turtles are warm-blooded, and able to maintain their temperatures as much as 10°F above the ambient water, just as the dinosaurs apparently could.

•The bones of the leatherback are more like those of the marine mammals (dolphins and whales) than the reptiles. "No one seems to understand the evolutionary implications of this."

•Leatherbacks dive as deep as 3000 feet which is strange because they seem to subside almost exclusively on jellyfish, most of which are surface feeders.

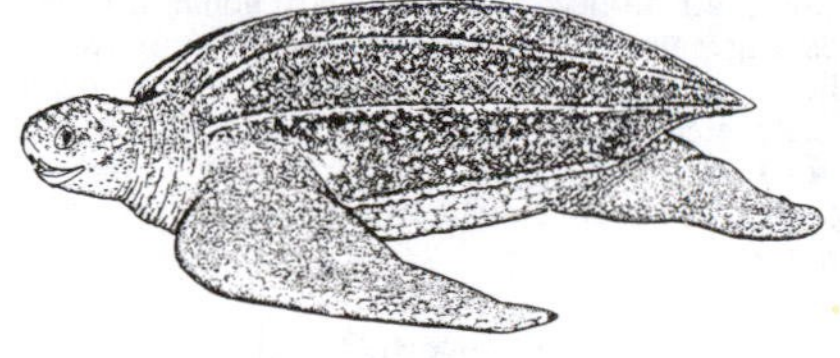

The leatherback turtle with its characteristic strongly keeled shell.

•Like all turtles, leatherbacks can stay submerged for up to 48 hours. Just how they do this is unexplained.

•Their brains are miniscule. A 60-pound turtle possessed a brain weighing only 4 grams---a rat's weighs 8!

•Leatherbacks' intestines contain waxy balls, recalling the ambergris found in the intestines of sperm whales.

•The stomachs of leatherbacks seem to contain nothing but jellyfish, which are 97% water. Biologists wonder how the huge, far-ranging leatherback can find enough jellyfish to sustain itself.

(McClintock, Jack; "Deep-Diving, Warm-Blooded Turtle," Sea Frontiers, 37:8, February 1991.)

THE MITE POCKETS OF LIZARDS

Many lizards are infested by chiggers, the larvae of trombiculid mites, which feed on tissue fluid and cell debris. Surprisingly, lizards seem to go out of their way to attract the chiggers---they have special mite pockets that provide a protected, warm and humid site. In many cases, the skin of the lizard also has smaller scales than normal and a good blood supply in the pocket, which enables the parasites to feed more readily.

There does not seem to be any advantage to the lizards providing plush accommodations for the chiggers. The chiggers can wreak havoc on their hosts in the form of skin lesions, allergic reactions, secondary infections, and the transmission of diseases. Nevertheless, some 150 species in 5 distinct lizard families possess mite pockets, which are often located in different places in different lizard species. Apparently, the mite pockets evolved separately several times. But why? (Benton, Michael J.; "The Mite Pockets of Lizards," Nature, 325:391, 1987.)

Comment. Why haven't the lizards evolved thicker skin or some sort of chemical defense instead of reducing their fitness with mite pockets? Or, are other factors operating?

LIZARDLESS THRASHING TAILS

It is common knowledge that many lizards lose their tails when attacked by a predator. In some lizard species, the released tail is a live thing, thrashing violently, and deluding the predator into thinking he has caught the real animal. Predators, even if not completely fooled by the struggling tail, are diverted into subduing it, giving the lizard time to escape. The detached tails contain their own autonomous nervous system and energy supply. (Dial, Benjamin E., and Fitzpatrick, Lloyd C.; "Lizard Tail Autonomy,..." Science, 219:391, 1983.)

Comment. Once again we have a biological system requiring several simultaneous evolutionary developments to be successful. Such complex biological evolution in response to predator-prey feedback is indeed marvelous.

BEHAVIOR

THE DECEITFUL SHE-MALES

Abstract. "In many diverse taxa, males of the same species often exhibit multiple mating strategies. One well-documented alternative male reproductive pattern is 'female mimicry,' whereby males assume a female-like morphology or mimic female behavior patterns. In some species males mimic both female morphology and behavior. We report here female mimicry in a reptile, the red-sided garter snake (Thamnophis sirtalis parietalis). This form of mimicry is unique in that it is expressed as a physiological feminization. Courting male red-sided garter snakes detect a female-specific pheromone and normally avoid courting other males. However, a small proportion of males release a pheromone that attracts other males, as though they were females. In the field, mating aggregations of 5-17 males were observed formed around these individual attractive males, which we have termed 'she-males.' In competitive mating trials, she-males mated with females significantly more often than did normal males, demonstrating not only reproductive competence but also a possible selective advantage to males with this female-like pheromone."

In the competitive mating trials, the she-males were successful in 29 out of 42 trials. The normal males won out in only 13! The authors ask the following question: Why aren't all males she-males given such an advantage? (Mason, Robert T., and Crews, David; "Female Mimicry in Garter Snakes," Nature, 316: 59, 1985.)

Comment. Among the fishes, bluegills and salmon (and probably many others) have female-appearing males competing with normal males.

FROG MOTHERS DO SO CARE!

We usually think of reptiles and amphibians as bad parents, leaving their eggs unguarded and their young to fend for themselves. The strawberry poison-dart frog of Panama and Columbia seems to be an exception. The parents stand guard over the eggs, moistening them until the tadpoles emerge. Then, the mother allows the tadpoles to wriggle onto her back and, one at a time, she carries them to separate little pools of water trapped in bromeliad fronds. She even goes one remarkable step further. Remembering the location of each tadpole, she makes the rounds, depositing infertile eggs for them to eat! (Anonymous; "Gallery," Discovery, 6:55, May 1988)

THE WOOD TURTLE STOMP

J.H. Kaufmann is a zoologist with strong proclivities for wood-turtle watching. Not a very strenuous vocation you say! Be that as it may, wood turtles make up for their lack of speed with some interesting talents. Besides being able to home accurately over unfamiliar terrain, they also know how to "grunt"---not vocal grunting, but a much more curious activity. Kaufmann relates one of his observations:

> I came upon an adult male. When I first saw him he was sitting quietly beside a creek, but he soon wandered into a damp thicket of alder, spicebush, and false hellebore. Before disappearing from sight, however, he began to rock back and forth. I followed, trying to stay just close enough to see what he was up to without disturbing him. Fortunately, he did not scare easily, which allowed me to approach within a few yards as he meandered, walking and rocking. First, I noticed that the rocking was caused by short bouts of stomping with the front feet, alternating between left and right. Then he suddenly jabbed his head at the ground and ate something. This behavior continued for a half hour, and several times I caught a glimpse of the prey---earthworms snatched from the surface. I suddenly realized the turtle was 'grunting' for worms!"

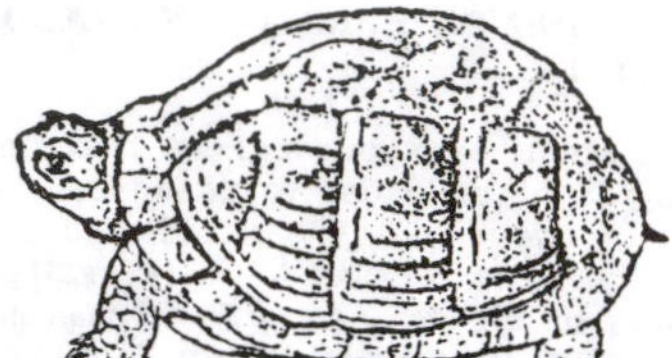

A wood turtle preparing to stomp or "grunt"!

Earthworms, for reasons we cannot fathom, pop out of the ground and flee in panic when vibrations flood their milieu. The wood turtle has apparently learned] how to take advantage of this weakness of worms. So have some gulls and plovers. In fact, some humans make their living grunting for earthworms. They simply drive a stake into the ground and draw a notched stick across it to make the worms surface. In Apalachicola National Forest, in Florida, about 700 permits ($30 each) are issued each year to earthworm grunters!

Back to wood turtles, how do they acquire this skill? Even naive, artificially raised wood turtles know the technique! And what great fear drives the earthworms out of the ground? Some say that the vibrations make the worms think moles are chasing them. (Kaufmann, John H.; "The Wood Turtle Stomp," Natural History, 98:8, August 1989.)

DISTRIBUTION

As customary in this book, the "distribution" heading brings together tales of unusual occurrences and concentrations of animals. In the case of reptiles and amphibians, we have such bizarre and typically Fortean phenomena as alligators in sewers and toads apparently falling from the skies. On a more serious note is the report of sharply declining amphibian populations all over the world---a phenomenon also seen in bird populations.

A SINUOUS LINE OF SEA SNAKES

May 4, 1932. In the Malacca Straits. a surface congregation of sea snakes 10 feet wide and 60 miles long was observed. Helicopter pilots off Viet Nam and Pakistan have reported similar but much smaller concentrations. Groups of several thousand have also been noted in Panama Bay. This tendency to gather in great numbers at the surface is an enigmatic aspect of sea snakes. One possible answer may lie in the surface feeding habits of some species, such as the yellow-bellied sea snake. These creatures seem to float passively on the sea surface, feeding and reproducing, letting the winds and currents accumulate them in long drift lines. (Minton, Sherman A., and Heatwole, Harold; "Snakes and the Sea," Oceans, 11:53, April 1978.)

GREAT BALLS OF SNAKES

Most garter snakes in the northern states spend the winter in communal dens below the frost line. Some dens host as many as 10,000 to 15,000 red-lined garter snakes, which emerge en masse in the spring. Although garter snakes cannot survive freezing temperatures, they apparently do not congregate in such enormous numbers to keep warm, for sexually immature garter snakes commonly hibernate alone.

Big concentrations of sexually mature garter snakes seem to be part of the reproduction strategy of the species. In the big aggregations, males usually outnumber females by 50-1. As each female emerges in the spring, she is immediately mobbed by dozens of males. So-called "mating balls" of up to 100 males and a single female are formed. Naturalists commonly explain the wintering concentrations and mating balls as clever schemes evolved to maximize reproduction with minimum expenditure of energy. This article accepts this theme uncritically. (Lynch, Wayne; "Great Balls of Snakes," Natural History, 92: 65, April 1983.)

Comment. Evolutionists tend to "explain" facts in a circular fashion; that is, only the most efficient reproducers (or "fittest") survive, therefore those that survive *must* be the best reproducers. While the garter snake strategy has some advantages in terms of getting male and female together, things may have gone too far. For example, one communal den was flooded, killing 10,000 snakes. Predators have a field day when emergence occurs. One would think that dispersed hibernating snakes, with 1:1 male-to-female ratio, might prove to be an even better strategy. The point here is that blind application of the theory of evolution may mask other natural strategies. Survival of the fittest (or the most efficient reproducers) may not be the Master Plan.

THE HAZARDS OF SEWER EXPLORATION

A modern bit of folklore tells of discarded pet baby alligators flushed down toilets into the sewers of New York. There they grew fat on rats and confronted startled sanitation workers. Is there factual basis for such wild tales? Coleman states that he has compiled a list of 77 encounters with erratic or out-of-place alligators for the period 1843-1973, including one 5.5-foot specimen found frozen to death in Wisconsin in 1892. Only one in the 77 is a sewer specimen, but it is from New York City. The New York Times of February 10, 1935, reported a 125-pound alligator, almost 8-feet long, pulled out of a snow-clogged sewer on East 123rd Street. Obviously half-frozen from the cold, the animal snapped weakly at its captors. "Let 'im have it!" the cry went up. The only known sewer alligator perished under flailing snow shovels. No one could explain how the alligator got into or survived in a New York sewer. (Coleman, Loren; "Alligators-in-the-Sewers: A Journalistic Origin," Journal of American Folklore, 92:335, 1979.)

TOADS FALL TO SQUASHY FATE

June 1949. On Route 66 near Gallup, New Mexico. Temperature 104°. Absolute blue sunny skies. No clouds anywhere to be seen, from one horizon to the other for 360°.

"Out of nowhere, without warning, it poured extremely hard rain, hail, and toads. The hail balls were maybe the size of grapes to the size of peas. The toads were a medium brown in color and approximately the size of an adult's thumbnail. This whole incident lasted for less than 5 minutes, if my memory is correct.

.....

"The highway and the desert sands seemed to be one and the same, and the whole area seemed to be alive and moving. By now, we were down to a very slow speed, and under closer observation we noticed that the area was littered with millions of hailstones and those toads hopping all over.

"The storm stopped as fast as it started, and the toads disappeared just as fast. I'll never forget how slippery the road was as we drove over those toads, and the popping of their bodies under the tires of my automobile." (Schuler, Richard A.; personal communication, July 23, 1987.)

EXTINCTION COUNTDOWN

Some plants may, as mentioned on p. 173, have environment-sensitive genes that help them adjust to external pressures. Amphibians and birds do not seem to be so pliable.

The worldwide precipitous decline of amphibian populations is alarming. Herpetologists are literally seeing species disappear before their eyes. Here is a typical anecdote:

> In 1974, Michael Tyler of the University of Adelaide, Australia, described a newly discovered frog species that broods its young in its stomach. The frog was once so common "an agile collector could have picked up 100 in a single night," Tyler says. By 1980 it had completely disappeared from its habitat (a 100-square-kilometer area in the Conondale Ranges, 100 miles north of Brisbane). It has not been seen since.

Similar stories emanate from Brazil, Japan, Mexico, Norway, and elsewhere. Many environmental causes have been proposed, but it is significant that the frogs are also disappearing from nature preserves where environmental pressures are small. D. Wake, a biologist at Berkeley, has remarked:

> [Amphibians] were here when the dinosaurs were here, and [they] survived the age of mammals. If they're checking out now, I think it is significant.

In this context, Wake believes that there is a single, global, still-unidentified cause operating. (Barinaga, Marcia; "Where Have All the Froggies Gone?" Science, 247:1033, 1990. Also: Cowen, Ron; "Tales from the Froglog and Others," Science News, 137:158, 1990.)

In the same issue of Science, S.A. Temple reviewed the book Where Have All the Birds Gone? The situation for North American forest-dwelling song birds is not as critical as that for frogs and toads, but it is still very serious. The populations of warblers, vireos, and thrushes are declining rapidly, even though the North American forests are now expanding. (Temple, Stanley A.; "Winter Absences," Science, 247: 1128, 1990.)

TALENTS AND FACULTIES

NEW VERTEBRATE DEPTH RECORD

S. Eckert, of the University of Georgia, has reported that a leatherback turtle fitted with a recording device dove to 1200 meters. This exceeds the previous record for air-breathing vertebrates (sperm whales). Leatherbacks also hold other records, being the largest of living turtles (over 600 kilograms) and the most widely distributed reptile in the world. They are also capable of maintaining their body temperatures substantially above the ambient water temperature, although no one has as yet claimed that they are warm-blooded. (Mrosovsky, N.; "Leatherback Turtle Off Scale," Nature, 327:286, 1987.)

LISTENING WITH THE FEET

Male white-lipped frogs exhibit conspicuous behavioral responses to calling conspecific males that are nearby but out of view. Since the calls often are accompanied by strong seismic signals (thumps), and since the male white-lipped frog exhibits the most acute sensitivity to seismic stimuli yet observed in any animal, these animals may use seismic signals as well as auditory signals for intraspecific communication.

(Lewis, Edwin R., and Narins, Peter M.; "Do Frogs Communicate with Seismic Signals?" Science, 227:187, 1985.)

Comment. Rabbits often thump the ground to communicate, but their thumps are thought to be sound generators rather than vibration generators.

FLYING, PARACHUTING, AND FALLING FROGS

Falling frogs have always been a Fortean favorite. These particular frogs plummet to earth in uncontrolled, unchecked free fall after (presumably) being lofted by whirlwinds. But there are frogs that are aerodynamically more sophisticated; these creatures glide and parachute through the dense tropical forests. S.B. Emerson and M.A.R Koehl have inquired into (even resorting to models in wind tunnels) the morphological and behavioral changes that have accompanied the repeated evolutions of these airworthy amphibians.

"This paper reports an examination of the shift from aboreal to 'flying' frogs where we evaluate the the aerodynamic performance consequences of both a behavioral and morphological change. 'Flying' frogs have evolved independently several times among the 3,400 species of anurans. Although the particular nonflying sister species to each flying form remains unknown, in all cases flyers are distinguished from related, nonaerial, aboreal frogs by a similar suite of morphological characters: enlarged hands and feet, full webbing on the fingers and toes, and accessory skin flaps on the lateral margins of the

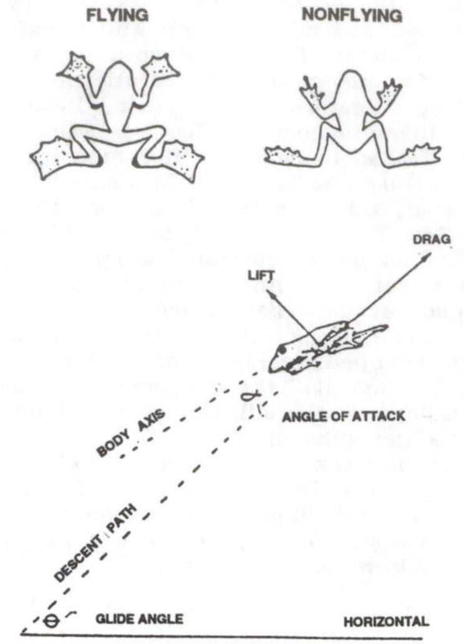

(Top) Shapes of flying and non-flying frogs. (Bottom) Aerodynamic diagram for a flying frog.

arms and legs. 'Flying' frogs are not capable of powered flight, but do travel considerable horizontal distances during vertical descent. They are technically classified as gliders because they can descend at an angle less than 45° to the horizontal. Interestingly, aboreal frog species lacking particular morphological specializations (= nonflying frogs) drop from vertical heights as well. These animals descend at glide angles greater than 45° and are, by definition, parachuting." (Emerson, Sharon B., and Koehl, M.A.R.; "The Interaction of Behavioral and Morphological Change in the Evolution of a Novel Locomotor Type: 'Flying' Frogs," Evolution, 44: 1931, 1990.)

DO WE REALLY UNDERSTAND THE DINOSAURS?

Until very recently, the standard dinosaur scene in the books and magazines showed huge, ungainly beasts shuffling around in lush swamps. Things are changing. Dinosaurs are now becoming more lively and talented; they may even have been warm-blooded!

A recent paleontological expedition to the Gobi Desert by some Canadians will change the dinosaur stereotype even more. The Gobi dinosaur-bone sites are incredibly rich---comparable to those in Alberta. What is most impressive, however, is the environment the Gobi dinosaurs lived in.

> The dinosaurs of China and Mongolia did not live in the same type of lush, well-watered environment that existed in North America during the Mesozoic era, when dinosaurs dominated the globe. The dinosaurs of Alberta flourished on a great swampy coastal plain on the edge of a vast inland sea. In ancient China, conditions were much harsher. A modern-day equivalent would be the Great Salt Lake Basin of Utah. Water did exist in vast shallow lakes, but it was often alkaline and high in soda. The vegetation was scrubland with coniferous forests on the higher ground.

(Anderson, Ian; "Chinese Unearth a Dinosaurs' Graveyard," New Scientist, p. 26, November 12, 1987.)

Comment. To these Gobi observations should be added those below from northern Alaska, all at 70° north latitude, which suggest that dinosaurs also survived in a land where darkness reigned almost six months of the year. It seems that these great beasts could live almost anywhere. Why, then, do most scientists maintain that climatic changes wiped them out?

PALEONTOLOGICAL PROBLEMS

THE NIGHT OF THE POLAR DINOSAUR

Somewhere west of Deadhorse, a small town on Prudhoe Bay in northern Alaska, paleontologists have found the bones of at least three species of dinosaurs. But wait, the latitude there is 70° north today and, according to magnetic measurements of the rocks, it was about the same when the dinosaurs met their demise. At these high latitudes the dinosaurs either had to contend with several months of darkness each year or they had to migrate many hundreds of miles over the rough Alaskan landscape. The visions of dinosaurs groping for tons of vegetable food in the polar night is about as incongruous as imagining them trekking down to the Lower 48!

Scientists are now maintaining that these dinosaurs did prosper on the shore of the Arctic Ocean, even in the dark, because the climate then was semitropical or temperate. This was because the earth's climate was more equable or uniform. They are, however, surprised by the lack of mineral deposition in the dinosaur bones, which look rather "modern." (Anderson, Ian; "Alaskan Dinosaurs Confound Catastrophe Theorists," New Scientist, p. 18, August 22, 1985.)

Comment. The apparent survival of dinosaurs during two months of darkness is being used as an argument against asteroidal catastrophism, which it is claimed wiped out the dinosaurs with a long-lived dust cloud that blocked the sun. More on this in the two following entries.

THE DINOSAURS OF WINTER AND THE POLAR FORESTS

It seems appropriate after suggesting above that the dinosaurs might have been frozen to death in a cosmic winter to remind the reader that some of the dinosaurs were pretty tough animals. Many dinosaur fossils have been dug up in Alaska, northern Canada, Siberia, New Zealand, and Antarctica. Not only were some dinosaurs cold-resistant but, seeing many were herbivorous, they were also able to migrate to more temperate climes as the long days of the polar summers waned. The point here is that the dinosaurs as a clan were very adaptable and should have survived severe environmental stress. (Vickers-Rich Patricia, and Rich, Thomas H.; "The Dinosaurs of Winter," Natural History, 100:33, April 1991.)

That the polar regions were once covered by lush forests has been underscored by recent discoveries in both polar regions. Stumps of huge trees 45 million years old dot the now-bleak landscape of Axel Heiberg Island far north of the Arctic Circle. In Antarctica, heaps of 3-million-year-old fossil leaves have been found within 400 kilometers of the South Pole. (Francis,

Jane E.; "Arctic Eden," Natural History, 100:57, January 1991, and Peterson, Christian; "Leafing through Antarctica's Balmy Past," New Scientist, p. 20, February 9, 1991.) Coal beds are also known from Spitzbergen and Antarctica.

Comment. The vision of dinosaurs roaming polar Edens evokes many questions. If the polar regions were indeed 10-20°C warmer than now, what could have survived at the Equator? The dates given above (3 and 45 million years ago) for polar heat waves are well after the 65-million-year demise of the dinosaurs. This suggests that the biosphere recovered very well from the apparent great catastrophe that ended the Cretaceous period. Why, then, did the dinosaurs succumb so completely?

BONES OF CONTENTION

The following item requires an introductory comment, because it demonstrates how a very tiny and obscure brick in the Temple of Science, long thought to be structurally sound, might lead to a widespread collapse. Are there other such sleepers?

A group of burrowing lizards (the amphisbaenians) possess a heavy bone in their heads that helps them ram their ways through the soil. In surface-living lizards this structure is merely a soft, flimsy cartilage. It was long assumed that the bone in the burrowing lizard developed from the cartilage of its surface-living kin. But a study of embryos now shows that the head bone of the burrowing lizard actually developed from a membrane instead of cartilage. The two similarly located structures are not homologous after all. They had different origins.

Superficially this doesn't seem very anomalous and especially not very exciting. But vertebrate evolution in particular has been charted on the basis of homologous structures. If these structures have different biological origins---even in just some cases---the evolutionary family trees may be drawn wrong. ("Lizard Bone Shakes World of Taxonomy," New Scientist, 98:221, 1983.)

Comment. No one yet knows how serious this problem really is. Basically it means that some animals that look alike (at least bonewise) need not be closely related. To use an analogy, if nature has the plans for a house stored in genetic material, it may be able to build that house out of wood, brick, or what ever material is available.

UNRECOGNIZED SPECIES

This section is where sea- and lake-serpents are accumulated. The obvious assumption in this assignment is that all of these "serpents" are really reptiles. Actually some, such as Caddy, off the British Columbia coast, may be mammals. Even the Loch Ness monster, assuming it exists, could be a mammal. We will not know until we have better data---something we may never have in some instances, for misidentifications are rife in this branch of anomalistics.

Our collection, which comprises only a few of the recent observations, commences on the shores of Loch Ness, moves to various land-locked bodies of water, and thence to the open seas.

BAD YEAR FOR WATER MONSTERS

R. Razdan and A. Kielar describe in a recent issue of the Skeptical Inquirer the results of their 1983 experiments at Loch Ness with a sonar tracking array. Here are their conclusions:

> We have shown that continuous sonar monitoring for seven weeks to a depth of 33 meters in an area where many previous sonar contacts had been reported showed no evidence of anything larger than a 1-meter fish. The circumstances under which previous expeditions had obtained sonar and photographic evidence in support of the existence of the Loch Ness monster could not withstand scrutiny. The evidence itself revealed discrepancies. This is especially true of the Academy's flipper photographs, the published versions of which differ from the original computer-enhanced photographs. Careless deployment of equipment and over-zealous interpretation of the data account for much of the so-called scientific evidence.
>
> While it is not possible to prove definitely that the monster does not exist, the evidence so far advanced strongly suggests that the Loch Ness monster is nothing more than a long-lived and extremely entertaining legend.

The "Academy" mentioned above is the Academy of Applied Science. This article reproduced both the original JPL computer-enhanced photo of the famed flipper and the photo that was widely published. The "retouching" seems extensive. (Razdan, Rikki, and Kielar, Alan; "Sonar and Photographic Searches for the Loch Ness Monster: A Reassessment," Skeptical Inquirer, 9:147, 1984.)

NESSIE PHOTOS NOT RETOUCHED

Above, we reported the Skeptical Inquirer charge, by Razdan and Kielar, that the famous "flipper" photos taken at Loch Ness were retouched fraudently. This charge has now been answered by R.H. Rines, et al, of the Academy of Applied Science. It is true that the very convincing "flipper" photo is not one of the computer-enhanced photos provided by JPL (Jet Propulsion Laboratory). Rather it is a composite of the original negative and several JPL computer-enhanced negatives.

Negative compositing is another kind of enhancing; and it is considered perfectly proper and ethical since new information is not added. It works because a detail that is faint in one negative may show up better in another. No fraud was involved and the innuendos in the Skeptical Inquirer (also Discover) were uncalled for. Discover has refused to publish any rebuttal. (Anonymous; "Retouching of Nessie Flipper Photo Claimed---Denied," ISC Newsletter, 3:1, no. 4, Winter 1984)

LATEST EPISODE: LOCH NESS

The continuing saga of the purportedly retouched photo of the flipper of the Loch Ness monster chalked up another episode in the Summer issue of the Skeptical Inquirer. First, there is a response by Robert Rines to the debunking article by Razdan and Kielar in the preceding Winter issue, in which the charge is made that retouching had taken place. This is followed by a reply by Razdan and Kielar. To top it all off, there is a nasty letter printed later on about the Academy of Applied Science, of which Rines is a member. With all the charges and countercharges, it is impossible to tell whether or not the flipper photograph was "subjectively" enhanced or not. (Rines, Robert; "Loch Ness Reanalysis: Rines Responds," Skeptical Inquirer, 9:382, 1985. Razdan, Rikki, and Kielar, Alan; "Loch Ness Reanalysis: Authors Reply," Skeptical Inquirer, 9:387, 1985.)

Comment. The whole business is now as murky as Loch Ness itself. The use of obfuscation and character assasination is common in the anomaly business.

ANOTHER TALE OF OGOPOGO

Another report of Lake Okanagan's monster, palindromically named Ogopogo, has surfaced. A Canadian woman, Mrs. B. Clark, actually bumped into Ogopogo while swimming in the British Columbia lake in July of 1974.

> Mrs. Clark's report states: 'I did not see it (the animal) first. I felt it. I was swimming towards a raft/diving platform located about a quarter of a mile offshore, when something big and heavy bumped my legs. At this point, I was about 3 feet from the raft, and I made a mad dash for it and got out of the water. It was then that I saw it.' The report goes on to describe the observation: 'When I first saw it, it was about 15-20 feet away. I could see a hump or coil which was 8 feet long and 4 feet above the water moving in a forward motion. It was traveling north, away from me. It did not seem to be in much of a rush, and it swam very slowly. The water was

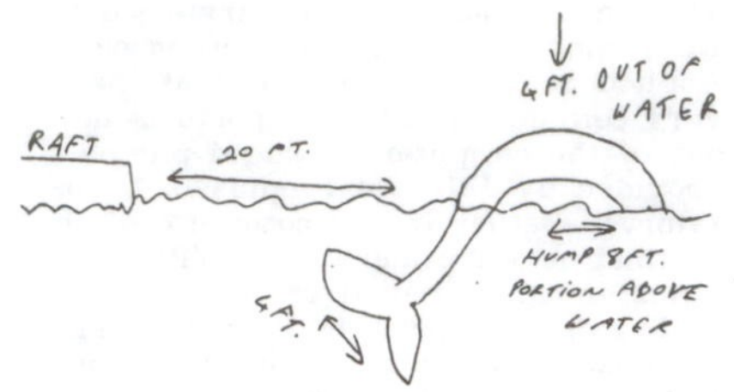

Drawing by Ogopogo observer and "bumpee," Mrs. B. Clark.

very clear, and 5 to 10 feet behind the hump, about 5 to 8 feet below the surface, I could see its tail. The tail was forked and horizontal like a whale's, and it was 4 to 6 feet wide. As the hump submerged, the tail came to the surface until its tip poked above the water about a foot. ...About 4 or 5 minutes passed from the time it bumped me until the time it swam from view.'

Ogopogo's estimated length was 25-30 feet; breadth, 3-4 feet. No fins or hair were seen. The animal was serpentine---seemingly without a neck. Its vertical undulations and horizontal tail proc-claimed it to be a mammal, possibly a primitive type of whale. (Anonymous; "Close Encounter in Lake Okanagan Revealed," ISC Newsletter, 6:1, Spring 1987.)

1989 SIGHTINGS OF OGOPOGO

Okanagan Lake, in south central British Columbia, is the home of Ogopogo. At least this is where a large, elusive lake monster has been reported for many years. During the summer of 1989, the British Columbia Cryptozoology Club (BCCC) made two expeditions to Okanagan to search out Ogopogo. Several sightings of the animal were made, as well as a video tape. The first sighting, on July 30, was quite detailed, and we quote here from the BCCC report.

"The focus of the investigation turned to Summerland, and a particularly good vantage point was located at Peach Orchard Beach, Lower Summerland, on July 30. All four members of the investigating team were stationed at various points on the beachfront when, at 3:55 p.m., a most extraordinary occurrence took place. A large patch of white water materialized close to a headland at the southern end of the beach, drawing the attention of the BCCC observers. It was about 1,000 feet distant at this point, and it was clear that a large animal was swimming in a northerly direction against the prevailing wind and slight swell. At a distance of about 600 feet, Kirk Sr. was able to see clearly through a Bushnell 40X telescope that this was the classic Ogopogo, with its humps well above the water level. Both Clarks were also able to see the object clearly through binoculars. The animal displayed, variously, five and sometimes six humps.

"Kirk's telescope allowed him to see that the animal's skin was whale-like, and that there were what appeared to be random calcium-like deposits under the skin which appeared to be similar to barnacles in shape. All the team members agree that the animal was between 30 and 35 feet in length, and was almost 3 feet above the surface at its highest point---that being the middle hump." (Kirk, John; "BCCC Report on Okanagan Lake, 1989," Cryptozoology, 8:75, 1989.)

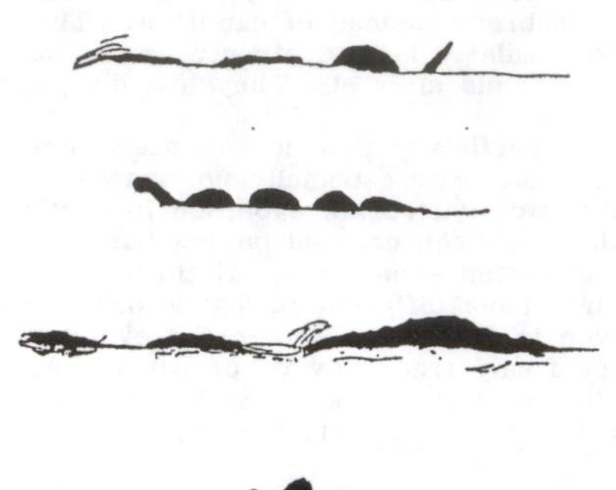

July 30, 1989. Four sketches of Ogopogo as seen from four different observation points.

SIBERIAN LAKE MONSTER

This report comes from a remote Siberian village via Tass. So, make of it what you will! It concerns a giant green snake with a sheeplike head seen patrolling a lake near Sharipovo. Tass said:

> Dozens of people have seen this green monster, which has the girth of a large tree trunk and is around 6 or 7 yards long. One of them even managed to take a photograph of it. It swims along with its head held high in the air.

The creature makes tracks in the grass along the shoreline resembling those from the runners of a large sleigh. (Anonymous; "Snake with Sheep Head Is Spotted in a Lake," Baltimore Sun, p. 5A, November 21, 1991.)

CHESSIE CAPTURED ON VIDEOTAPE

May 31, 1982. Kent Island, Maryland. Robert Frew was enjoying a holiday meal with family and friends at his home on Love Point, overlooking Chesapeake Bay, when a large object was seen moving against the tide about 100 feet offshore. Inspection through binoculars revealed a sinuous, humped creature, over 20 feet long, swimming lazily along. As others watched, Frew popped a videotape into his camera and shot a remarkable three-minute documentary, which was aired later on TV. Witnesses estimated the creature's length at about 30 feet; width, about 10 inches. Two other recent sightings of "Chessie" occurred in September 1980 at the other end of Kent Island, and in July 1978 near Heathsville, Virginia. (Robinson, Russ; "Chessie May Have Made Video Debut," Baltimore Sun, July 11, 1982. p. A1.)

CHAMP IN 1985

Below we quote two of the 14 digests of Champ sightings in 1985. Champ, as all cryptobiologists know, is the oft-reported monster of Lake Champlain.

> June 29, 1985: Peg McGeoch and Jane Temple; off Scotch Bonnet, south of Basin Harbor, Vermont; "length well over 30 feet"; head/neck similar to a brontosaurus, with head held "about 5 feet above surface"; body was snakelike.
>
> August 8, 1985: Jean and Becky Joppru; in Mullen Bay, New York; 4 or 5 black humps protruding 2 or 3 feet from water; total length, 30 feet.

(Zarzynski, Joseph W.; "LCPI Work at Lake Champlain, 1985," Cryptozoology, 4:69, 1985.)

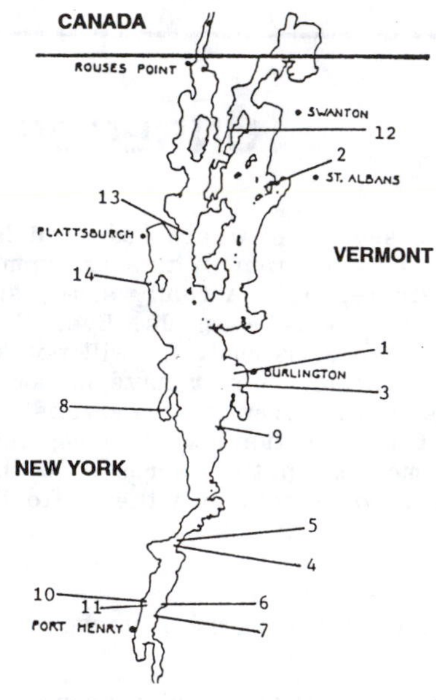

Locations of the 14 sightings of Champ during 1985.

CALIFORNIA SEA SERPENT FLAP

During October and November 1983, several sightings of a dark, eel-like creature came from the California coast. (Stinson Beach, north of San Francisco, and Costa Mesa). Three humps (just like in the classic sea serpents on old

What Chessie might look like!

maps) followed a small head, which rose above the surface to look around. Many individuals saw the serpent, some with binoculars. At Stinson Beach, the animal was followed by about 100 birds and two dozen sea lions. (Anonymous; "'Sea Serpents' Seen off California Coast," ISC Newsletter, 2:9, Winter 1983,)

Comment. Of the vertebrates, only mammals are built so that they can easily flex vertically.

BC SEA SERPENTS

Marine biologist E. Bousfield and oceanographer P. LeBlond have amassed impressive evidence for the reality of a large marine vertebrate presently unrecognized by science.

> They say that in the past 60 years at least six specimens of the sea creature have been discovered, including a live baby and a dead youngster found undigested in the stomach of a whale.
>
> .....
>
> Since World War II, two apparent skeletons and two carcasses have washed up on the shores of British Columbia and neighboring Washington State.

This British Columbia sea serpent, named "Caddy," is believed to measure 40-60 feet in length. It has been clocked at speeds up to 25 miles/hour---fast enough to leave killer whales in its wake. (Anonymous; "Sea Serpent Sightings Substantiated by British Columbia Scientists," Baltimore Sun, July 30, 1992.)

THE UBIQUITY OF SEA SERPENTS

Public interest is usually focussed (by the media) upon the supposed monsters in Loch Ness, Lake Champlain, the Chesapeake Bay, etc. Actually, an immense body of sea serpent reports also exists. B. Heuvelmans collected many of these in his 1965 classic In the Wake of the Sea-Serpents.

P.H. LeBlond, a professor at the University of British Columbia, is extending Heuvelman's work, concentrating on the thousand miles of Pacific Coast between Alaska and Oregon. Since 1812, there have been 53 sightings of sea serpents or other unidentified animals along this narrow strip of ocean. Some of these are very impressive. Take this one for example:

> In January 1984 a mechanical engineer named J.N. Thompson from Bellingham, Washington, was fishing for Chinook salmon from his kayak on the Spanish Banks about three-quarters of a mile off Vancouver, British Columbia, when an animal surfaced between 100 and 200 feet away. It appeared to be about 18-20 feet long and about two feet wide, with a "whitish-tan throat and lower front" body. It had stubby horns like those of a giraffe, large ("twelve to fifteen inches long") floppy ears, and a "somewhat pointed black snout." The creature appeared to Thompson to be "uniquely streamlined for aquatic life," and to swim "very efficiently and primarily by up and down rather than sideways wriggling motion..."

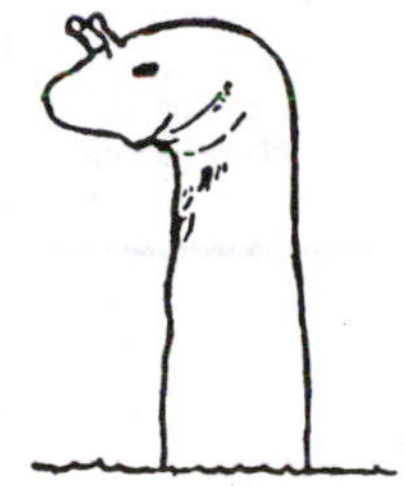

Heuvelmans' sketch of the long-necked sea serpent with its characteristic giraffe-like horns.

LeBlond and biologist J. Sibert have analyzed all of the 53 sightings in a 68-page report entitled "Observations of Large Unidentified Marine Mammals in British Columbia and Adjacent Waters," published by the University of British Columbia's Institute of Oceanography. Of the 53 sightings, 23 "could not definately or even speculatively be accounted for by animals known to science." The authors of the report emphasize that the reports are of high quality, made by people knowledgeable about the sea and its denizens. (Gordon, David G.; "What Is That?" Oceans, 20:44, August 1987.)

MOKELE-MBEMBE

Lake Tele is a shallow, oval lake about 4 by 5 kilometers in the People's Republic of the Congo. Swamp forest reaches to the very edge of the water. From this difficult-to-reach body of water, almost right on the equator, come reports of a large unidentified animal---Mokele-Mbembe. In 1983, Marcelin Agnagna, a biologist, led an expedition to Lake Tele in hopes of observing Mokele-Mbembe. The expedition was successful, but we don't know too much more than we did before. Agnagna and others observed the animal from a distance of about 240 meters. Unfortunately no photos were taken. Mokele-Mbembe appeared as shown in the accompanying sketch. What is it? (Agnagna, Marcelin; "Results of the First Congolese Mokele-Mbembe Expedition," Cryptozoology, 2:103, 1983.)

Comment. Vague rumors have long floated around that dinosaurs survive in deepest Africa! Maybe they are true.

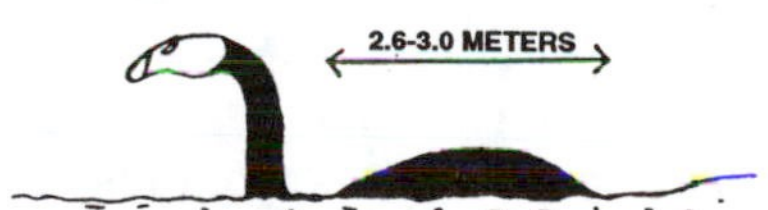

Sketch of a large animal seen at Lake Tele that might have been the famous African dinosaur Mokele-Mbembe.

FISH

EXTERNAL APPEARANCE

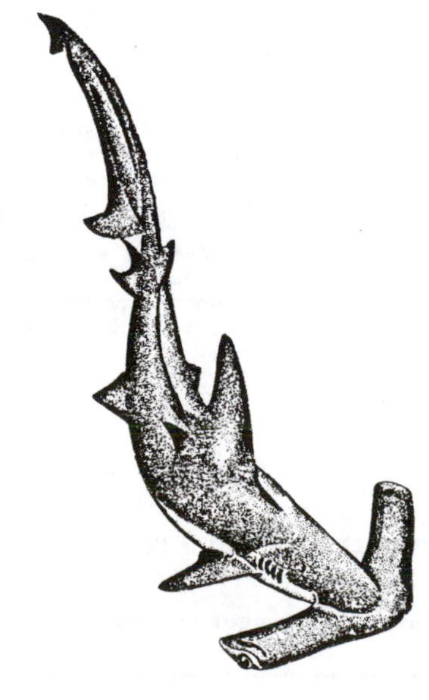

WHY THE HAMMER HEAD?

You probably thought, as we did, that the function of the hammerhead shark's weirdly shaped head was to separate the eyes and thus improve binocular vision. This is not the case. The visual fields of the hammerhead's eyes do not overlap at all. Each eye presents the brain with a separate, completely different image to integrate. What, then, could be the purpose of the hammer head? No one really knows, but three suggestions are as follows:

1. The head acts as a hydrofoil and

gives the heavier-than-water, swim-bladderless shark better swimming control.

2. Grooves on the hammer head channel water toward the nostrils, providing "stereoscopic sniffing."

3. The head is a platform for electromagnetic sensors that help locate prey. Stingrays are a favorite food of the hammerhead, and the shark may detect them electromagnetically, as surmised by the author of this article in the following encounters:

> I have observed great hammerheads swimming close to the bottom, swinging their heads in wide arcs (a motion common, in a lesser degree, to all large sharks) as if using the increased electroreceptive area of their hammer-like head like the sensor plate of a metal detector. Sometimes, these animals would double back to scoop up one of several stingrays hiding in the bottom silt. The minute electrical pulse that keeps the stingray's heart and spiracles operating betrays their presence to a hungry hammerhead.

(Martin, Richard; "Why the Hammer Head?" Sea Frontiers, 35:142, 1989.)

FISH CREATES FISH

A Philippine anglerfish, looking for all the world like a rock or shell, waves before its maw a piece of bait resembling small fish found in this region. The bait, which is part of the anglerfish's body, has fins, a tail, and black spots for eyes. The waving about of the bait attracts predatory fish close enough for the anglerfish to snap them up. The authors surmise that the anglerfish evolved this realistic bait (and rod and reel) in order to save energy in acquiring food. (Pietsch, Theodore W., and Grobecker, David B.; "The Complete Angler: Aggressive Mimicry in an Antennariid Anglerfish," Science, 201:369, 1978.)

Comment. One wonders how many unfishlike baits were evolved before just the right shape and coloration were achieved.

An angler fish waving its fish-like bait to entice potential victims.

DISTRIBUTION

EEL ODDITIES

Garter snakes are reknowned for their habit of congregating in large, writhing masses, but we never heard of "eel balls" until A. Gardiner mentioned them in a recent issue of the Fortean Times.

"These [eel balls] are recorded in Christopher Moriarty's excellent Eels: a Natural and Unnatural History (David and Charles, 1978). Moriarty cites Pliny as the earliest historical reference. According to him, Eel Balls occur in Lake Garda, Italy, when it has been Storm-tossed by the effects of the October 'Autumn star'. Smitt in his Scandanavian Fishes (1895) says that eels knot themselves together in bunches 'up to a fathom in circumference' and are seen rolling along the stream beds, or, strangely, resting in this position. On 17 August 1935, fishery scientist J.C. Medcof observed, in the outflow of Lake Ainslie in Nova Scotia, 'three splendid clumps of Eels, half a metre in diameter, 30 to a clump, knotted tightly and remaining motionless in the rushes.' Medcof mentions that Eel Balls are sometimes free floating on the surface, which suggests formation with an air pocket or some communal control of air bladders. He says that this behavior occurs before eels 'silver' prior to the spawning migration. The record of Eel Balls in Nova Scotia proves that this behaviour is not confined to the European Eel." (Gardiner, Alan; "Eel Oddities," Fortean Times, no. 56, p. 53, Winter 1990.)

BEHAVIOR

THE PROBLEM OF THE PRECOCIOUS PARR

A startling shift in the breeding patterns of the Atlantic salmon is taking place in the rivers of North America and Europe. Male parr (parr are young salmon that have not yet been to sea) are becoming sexually precocious in large numbers. The small males have up to 20% of their body weight allocated to testes and, in consequence, seem to be bulging from full meals. Actually, they are ready to dart in under mature spawning salmon and fertilize some of the newly laid eggs. Experiments show that the parr are rather successful in their furtive endeavors. The question at hand is why the parr are developing precociously when there are sufficient mature male salmon around to do the job. (Montgomery, W. Linn; "Parr Excellence," Natural History, 92:59, June 1983.)

Comment. See also the item on p. 148 on "she-males."

THE GULPER EEL AND ITS KNOTTY PROBLEM

Occasionally brought up from great depths in nets, the gulper eel consists mainly of a huge mouth followed by a large bag of skin and, finally, a very long thin tail. The eel, often 6 feet long, can swallow prey as large or larger than itself. Such features are not particularly rare in deep-sea creatures, but we do have to briefly describe this grotesque fish to get a delightful anomaly. It seems that in a few recovered specimens, the thin tail is tied in several overhand knots! Now moray eels can knot themselves, but the gulper eel is just a floating stomach with negligible musculature in its whip-like tail. So, just how did the knots get there? (de Sylva, Donald P.; "The Gulper Eel and Its Knotty Problem," Sea Frontiers, 32:104, March-April 1986.)

Comment. We cannot resist mentioning the occasional discovery of groups of rats all tied together by their tails. Called "rat kings," these hapless snarls of rodents are usually dismissed as pranks or outright prevarication. However, in recent years, respected naturalists have found "squirrel kings" in the wild. Gulper eels are not the only animals with knotty problems. See p. 134 for an item on "squirrel kings."

THOSE SLIPPERY (ADULT) EELS

Every year untold millions of adult eels swim down the rivers of the continents toward the sea, where they are literally swallowed up. They are never seen again! In the Atlantic, the oft-told scientific tale is that all the adult eels from Europe and eastern North America converge on the Sargasso Sea. Here, they mate and die. It is in this area of the Atlantic that one finds high concentrations of eel larva, called leptocephali; and this alone is why the eels are thought to spawn here.

In a long article in Science News, E. Pennisi is the latest to wonder where the adult eels are. She relates how, despite several ambitious expeditions well-armed with nets, traps, and sundry eel-catching devices, "...no one has ever spotted adult eels in the spawning grounds."

Actually, Pennisi's article focusses on the Pacific and a 1991 Japanese expedition that searched for the spawning grounds of Anguilla Japonica, the Japanese eel. Earlier searches had been inconclusive. The 1991 attempt, after arduous labors and 16,000 kilometers of cruising, found the highest concentrations of leptocephali east of the Philippines. But, as in the Atlantic, even though many larvae were captured, no adult eels turned up in the nets. (Pennisi, Elizabeth; "Gone Eeling," Science News, 140:297, 1991.)

Comment. It is our understanding that adult eels are never caught anywhere once they leave their home rivers. Can anyone refute this?

PUZZLING GROUP BEHAVIOR OF SHARKS

For some unknown reason, sharks often congregate in immense groups. Approximately 2,000 sharks took over 24 kilometers of the surf zone near Corpus Christi during June 1977. Some courageous divers decided to study large groups of the scalloped hammerhead that regularly gather in the Gulf of California. Happily, the hammerheads were not aggressive when so occupied and could be approached closely. They swam pointed in roughly the same direction, maintaining about the same spacing through the groups, which sometimes numbered 100 or more. They did not feed, mate, or do anything collectively; but once in a while an individual would suddenly engage briefly in acrobatic behavior---one common type was dubbed the "shimmy dance." The researchers concluded that these shark groups had no obvious purpose and that, for reasons beyond the ken of man, this behavior somehow contributed to their evolutionary success. (Klimley, A. Peter; "Grouping Behavior in the Scalloped Hammerhead," Oceanus, 24:65, Winter 1981/1982.)

Comment. The sharks might be much "farther along" without complex, time-wasting group behavior. What do sharks know about evolution anyway?

ORGANS

THE FOUR-EYED FISH SEES ALL

Anableps, the four-eyed fish, frequents the rivers and estuaries from southern Mexico to northern South America. Actually, this curious fish has only two eyes, but each is divided in half horizontally; that is, each eye had two separate optical systems, each with its own focal length. The top half is for seeing in the air; the bottom half is for underwater. Thus equipped, Anableps can see prey and predators above and below the surface at the same time and increase its opportunities to get meals as well as escape from its enemies. (Zahl, Paul Al; National Geographic Magazine, 153:390, 1978.)

Comment. Nature is full of such "marvelous" adaptations, but it is hard to see how fish bifocals could develop gradually, given the intolerance of optical systems to minute changes in dimensions, position, and refractive index.

EYES OF DEEP-SEA FISH HAVE SPARE PARTS

The sunlight that filters down into the depths of the sea is exceedingly weak. It is so dark down there that one would expect deep-sea fish to be blind like many cave-dwelling animals. They are not blind; rather many have eyes of fantastic size and novel construction. An unusual feature of some deep-sea eyes is a layered retina. In the conger eel, five layers of photoreceptors are plastered on top of one another. Yet, experiments with conger eel eyes reveal that only one layer of photoreceptors is active at any one time. R. Shapley and J. Gordon, who carried out these experiments at the Plymouth Lab., surmise that the extra retinal layers are being held in reserve, much like the rows of spare teeth found in sharks' mouths. If so, deep-sea fish are the only animals that have evolved spare stores of visual pigments. (Anonymous; "The Mystery of the Non-Functioning Receptors," New Scientist, 88:366, 1980)

Comment. Why haven't cave-dwelling fish taken the same evolutionary route?

FISH CHANGE GENDER WHEN NECESSARY

Here follows the Abstract from an article in Science by D.Y. Shapiro:

> The simultaneous removal of three to nine males from large social groups of Anthias squamipinnis led to close to a one-to-one replacement of the removed males by sex-reversing females. The females changed sex serially within each group with a mean interval between successive onset times of 1.9 days. The timing of sex change is thus not independent for each fish but is influenced by the events surrounding other sex reversals within the group.

(Shapiro, Douglas Y.; "Serial Female Sex Changes after Simultaneous Removal of Males..." Science, 209:1136, 1980.)

PISCATORIAL DATA PROCESSING

Mammals such as bats and porpoises have their acoustical navigational gear, while many fish have opted for electrical methods of scanning their surroundings. The short-range "radars" of these fish are marvelously sophisticated, considering the low limb fish occupy on the Tree of Life. In fact, the following introductory paragraph from an article in Nature sounds almost as if it came from a textbook on electronic signal processing.

"Behavioural experiments have demonstrated that certain species of fish can perform remarkable analyses of the temporal structure of electrical signals. These animals produce an electrical signal within a species-specific frequency range via an electric organ, and they detect these signals by electroreceptors located throughout the body surface. In the context of one electrosensory behaviour, the jamming avoidance response (JAR), the fish Eigenmannia determines whether a neighbour's electric organ discharge (EOD), which is jamming its own signal, is higher or lower in frequency than its own. The fish then decreases or increases its frequency, respectively. To determine the sign of the frequency difference, the fish must detect the modulations in the amplitude and in the differential timing, or temporal disparity, of signals received by different regions of its body surface. The fish is able to shift its discharge frequency in the appropriate direction in at least 90% of all trials for temporal disparities as small as 400 ns." (Rose, Gary, and Heiligenberg, Walter; "Temporal Hyperacuity in the Electric Sense of Fish," Nature, 318:178, 1985.)

Comment. ns = nanosecond = 10^{-9} second. There must be some chips in those fish! Sorry, couldn't resist that one.

REMARKABLE ENGINEERING DESIGN IN NATURE

An unusual example of inspired design in nature has been described recently: The swordfish possesses special tissues rich in mitochrondria and cytochrome-C that generate heat for the animal's eye and brain. Not only do these heating elements keep the swordfish eye and brain significantly warmer than the surrounding water but they also keep these organs warm and thus more effective during deep dives into the cold ocean depths. (Carey, Francis G.; "A Brain Heater in the Swordfish," Science, 216:1327, 1982.)

THE EELS STRIKE BACK

Credit cards and bank cards are commonly kept in holders made from eelskin. So what! Who likes eels anyway? Well, there may be more to eelskin than meets the eye. Thousands of bank cards, when taken out of their eelskin holders, have failed to work in bank machines. The electronic coding on the cards has somehow been erased or scrambled. Perhaps, says one theory, the eelskins have bits of magnetite in their skins for navigational purposes. (Some other animals have such magnetic particles in their bodies to help orient them.) But could these tiny particles be powerful enough to erase card information? Another theory is that magnetic clamps on purses and handbags are the culprits. (Anonymous; "Credit Cards Fall Prey to Primitive Fish," New Scientist, p. 30, March 3, 1988.)

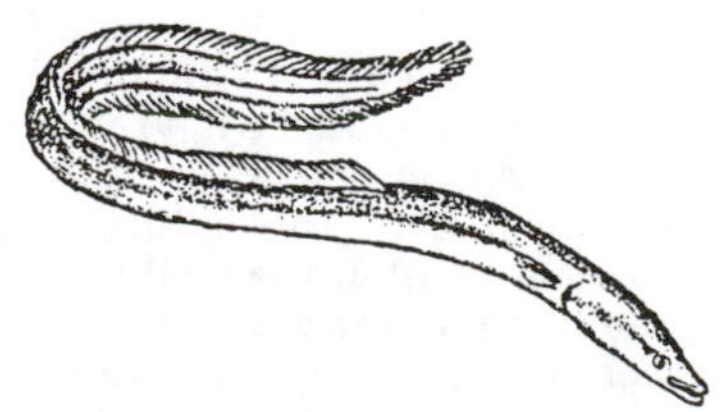

Banking-business bane!

GENETICS

HOW ANIMALS MIGHT GET INVERTED

The above title is just a literary ploy. We don't know how upside-down animals get that way; and, obviously, we don't think anyone else does either. Nevertheless, biologists are now discovering some radical things about life that may lead to better "answers."

First, we have a case of genetic material being transferred from a fish to a bacterium. The case at hand is the light-producing bacterium that provides the ponyfish with its luminous organ. In this symbiotic arrangement, the fish somehow passes genetic instructions to its retinue of bacteria. (Lewin, Roger; "Fish to Bacterium Gene Transfer," Science, 227:1020, 1985.)

Comment. Perhaps symbiotic relationships are fine-tuned by the mutual exchange of information!

UNRECOGNIZED SPECIES

SOMETHING BIG DOWN THERE!

Off Bermuda, while working large traps at depths between 1,000 and 2,000 fathoms, fishermen proclaim that some huge sea creature has been breaking heavy lines and towing fishing boats about. Some of these deep-sea traps measure 6 x 6 x 3 feet and are used to catch large shrimp (about 1 foot long) and crabs (2 feet, claw to claw). Something down there grabs these traps and refuses to let go. A giant octopus is believed to be the culprit. (Anonymous; "Giant Octopus Blamed for Deep Sea Fishing Disruptions," ISC Newsletter, 4:1, Autumn 1986.)

SEARCHING FOR MONSTER SHARKS

Tantalizing reports surface now and then lending credibility to the claim that there exists a very rare, deep-water shark that rivals the blue whale in size. We are talking 50-foot sharks and larger here; sizes that make the hero (or heroine) of the Jaws series seem minnow-like.

All of these hints come from the Pacific and focus on the possible survival of the shark Carcharodon megalodon, a monster relative of the great white shark. Megalodon is thought to have met its demise a million or so years ago. The word megalodon means "big tooth," and indeed the fossil teeth of this monster approach 6 inches in length. Sharks sporting teeth of this size could be as long as 50 feet. Measurements of the manganese dioxide layers accumulated on megalodon teeth dredged up from the seafloor suggest that it might actually have survived the Ice Ages and terrorized the Pacific as late as 10,000 years ago. Actually, some unfossilized teeth 5 inches long have been brought up by dredges, implying an even more recent existence.

Do scuba divers have anything to fear today? There are rare reports of huge versions of a shark resembling the great white but without the high dorsal fin. So, if the shark of Jaws scared you, think what a 50-foot version with 5-inch, serrated teeth could do to you and your boat. (Shuker, Karl P.N.; Fate, 44:41, March 1991.)

Comment. Admittedly, these recent data are soft, but there's no error about those teeth in the museums. New "living fossils" are being found all the time.

GIANT FISH REPORTED IN CHINA

In August. 1985, both AP and UPI wired dispatches from Beijing, People's Republic of China, reporting the discovery of enormous red fish in a remote lake in the Xinjiang Autonomous Region of northwestern China. The report, originating with the Xinhua News Agency, quoted Professor Xiang Ligai, of the biology department of Xinjiang University, as stating: 'The mystery of monsters in Lake Hanas has been solved.' The fish, he stated, were a giant species of salmon reaching a size of over 30 feet.

The ISC Newsletter originally declined to relay the AP/UPI stories without confirmation by more responsible Chinese sources. Then, in April of 1986, the official Chinese magazine China Reconstructs, presented a short article on the subject, which included additional information. For example, large nets have been destroyed and horses going down to the lake to drink have disappeared! (Anonymous; "Giant Fish Reported in China," ISC Newsletter, 5:7, Autumn 1986.)

ARTHROPODS

EXTERNAL APPEARANCE

It may seem strange to lump crabs and ants together, but they are both legitimate arthropods by virtue of having "jointed feet."

In external appearance, arthropods are incredibly diverse. Insect species number in the millions, perhaps tens of millions! Most of them are unnamed and undescribed. In fact, their morphologies are so wild in concept that anomalousness loses its meaning here. This section, in fact, focuses not on diversity but similarity: the phenomenon of mimicry, which encompasses amazing look-alikes and "clever" camouflage jobs. The phenomenon of mimicry is usually tossed off as the consequence of "convergent" or "parallel" evolution. These processes are always deemed the

A butterfly with a tail that resembles its head.

consequence of random mutation modulated by natural selection. Mimicries confer evolutionary advantages; and things are the way they are because that is the way they are!

HEADS OR TAILS?

Fake-head butterflies often escape predators that make passes at their cleverly designed tails, which as the figure shows, look awfully like butterfly heads, with eye spot, fake antennas, etc. Further, the converging stripes direct attention to the false head. The result is that this type of butterfly may lose its tail but save its real head. This effective stripe pattern has evolved (?) independently at least six times in the Neotropics. (Robbins, Robert K.; "The 'False Head' Hypothesis: Predation and Wing Pattern Variation of Lycaenid Butterflies," American Naturalist, 118:770, 1981.)

Comment. One wonders why all butterflies haven't evolved this neat ploy?

PREDACEOUS INSECT LARVAE DON "SHEEP'S CLOTHING"

The larva of the green lacewing lives in colonies of the wooly alder aphid upon which it feeds. The aphids, however, are protected by "shepherd" ants which normally remove any undisguised predatory larva. To foil the ants, the larva plucks some of the waxy wool from nearby aphids and sticks it on its own back. Thus disguised, the larva continues to consume aphids without the ants being the wiser. (The aphids know of course.) Artificially denuded larva are immediately spotted by ants and ejected from the colony. Cases are known where animals protect themselves from predators by covering themselves with vegetable matter and other debris, but the lacewing larva is unusual in that it mimics its prey by stealing the prey's own clothing. (Eisner, Thomas, et al; "Wolf-in Sheep's Clothing Strategy of a Predaceous Insect Larva," Science, 199:790, 1978.)

Comment. Does evolution satisfactorily explain the existence of such a trait as this?

CONVERGENT EVOLUTION OR CHANCE LOOK-ALIKES

Why should caddis fly larvae and a species of aquatic snail look alike? Mimicry is rather common in nature for it often confers some sort of advantage to one or both of the species in the turmoil of evolutionary pressures. Or so the theory goes. Most examples of convergent evolution involve closely related species. In the present case, though, the species are in different phyla. The caddis fly larva builds its snail-like shell by cementing grains of sand together with a silk-like secretion, while the snail's shell is a calcareous excretion. One would expect a strong advantage to be conferred on one or the other species, especially in the matter of predation. Using brook trout as predators, however, proved perplexing, for the trout would eat only the snails, avoiding the carbon-copy larvae. (Berger, Joel, and Kaster, Jerry; "Convergent Evolution between Phyla...." Evolution, 33:511, 1979.)

Comment. This is a remarkable case of mimicry. One wonders how the caddisfly larvae know exactly what snails look like, and how the unique shell constructing methods were coded into its genes by evolution. Or did the snail emulate the larvae?

CATERPILLARS THAT LOOK LIKE WHAT THEY EAT

While E. Greene was studying insect-eating birds, he was startled when an oak tree catkin started to crawl away from him. The crawling catkin turned out to be a cleverly camouflaged caterpillar (Nemoria arizonaria). When these caterpillars start eating oak catkins in the spring, they soon take on the golden color and fuzzy appearance of the catkins. However, the second brood, which matures after the catkins have disappeared, develops instead a twig-like appearance after consuming oak leaves. Thus, both broods acquire the proper protective camouflage for each season. Experiments show that plant chemicals control the appearance of the caterpillars. (Green, Erick; "A Diet-Induced Developmental Polymorphism in a Caterpillar," Science, 243:643, 1989. Also: Wickelgren, I.; "Caterpillar Disguise; You Are What You Eat," Science News, 135:70, 1989.

Comment. Is it naive to wonder why the oaks contribute to their own destruction by providing the caterpillars with chemicals that help conceal them from predators? Plants are usually very clever about producing insect-discouraging chemicals in their leaves. One would expect that "evolutionary forces" would have produced chemicals that would have made the caterpillars more obvious to their predators instead of vice versa.

A 'CLEVER' ADAPTATION

According to theory, many butterflies have wing patterns that evolved in response to predation. Some wing patterns blend into the background, making the butterfly hard to spot; other wings have prominent eyespots that are supposed to deter predators or trick them into striking at the wings instead of the soft, vulnerable body. But some tropical butterflies have a double problem; their predators change from dry to wet season. One butterfly, Orsotriena medus, masters this situation by changing wing patterns with the season. In the dry season, it is dark brown and inconspicuous; in the wet season, it switches to black wings with ostentatious eyespots and white bands. (Anonymous; "Cryptic Butterflies," New Scientist, p. 20, September 13, 1984.)

NATURE COMMUNICATES IN MYSTERIOUS WAYS

Most of us will recall that the wings of butterflies and moths sometimes display eyespots, which, according to current thinking, are designed to startle potential predators. Perhaps so, but butterfly and moth wings can convey a wide range of "signals." K.B. Sandved, a nature photographer, has also found remarkable renditions of all the letters in the English alphabet (one at a time, of course) on the wings of these insects. In fact, he has accomplished this several times over using different species. He has found all the Arabic numerals, too, as well as ampersands, question marks---you name it! Although Greek pi and capital omega have turned up, butterflies and moths are clearly trying to impress people who utilize the Roman alphabet. After all, it is difficult enough to evolve an ampersand; generating Chinese characters would strain credulity too much. (Amato, Ivan; "Insect Inscriptions," Science News, 137: 376, 1990.)

Comment. Incidentally, of what survival value are these wing symbols? Obviously, the butterflies and moths have not got their act completely together as yet. Words and phrases will come soon, we are certain. Look at the eggplants for example. They have specialized in Arabic. It has recently been reported in British newspapers and on BBC Radio 4 that when the Kassam family sliced up an eggplant, the patterns of seeds spelled out "Ya-Allah" (God is everywhere.). Donnelly, Steve; "Egregarious Eggplants," The Skeptic, 4:4, May/June 1990.)

Comment. Readers: please be alert when you carve up watermelons this year! Of course we shouldn't mix plants and insects as we have done here. However, the two items complement each other so well---and it's all in fun anyway. The Skeptic is a British publication resembling The Skeptical Inquirer.

THE MELANIC MOTH MYTH

The peppered moth remains one of the best examples of evolution in action. But as in so many other cases, the real story is turning out to be more complicated than the biologists first thought.

Several details don't match the moth propaganda. For example, all the photos show the moths resting out in the open on tree trunks, whereas they actually rest inconspicuously under branches and where branches join the tree trunk. (Cherfas, Jeremy; "Exploding the Myth of the Melanic Moth," New Scientist, p. 25, December 25, 1986.)

Comment. Since no new species of moth have arisen (merely population shifts between dark and light phases), why do evolutionists make so much out of this?

AGGRESSIVE MIMICRY

Field studies have revealed that bolas spiders can mimic the odor of female moths, thus attracting for consumption the male moths. More specifically, the hunting adult female spider, Mastophora cornigera, releases volatile substances containing three moth sex pheromone compounds. (Stowe, Mark K., et al; "Chemical Mimicry: Bolas Spiders Emit Components of Moth Prey Species Sex Pheromones," Science, 236:964, 1987.)

Comment. As in many other cases of mimicry, one wonders how the spider's capability developed by chance and in small steps.

BEHAVIOR

Except for the bee that stings you or the crab that nips your toe at the shore, the most compelling arthropod behavior phenomena are of the collective type. Insects, especially, tend to act en masse, as this section's organization testifies:

•Migration. Doubts about the migration of monarchs.

•Cooperative behavior. Synchronous fireflies, cicadas, and crickets. Shrimp and lobster trains.

•Colonial phenomena. Slave ants and hints of superorganisms.

•Miscellaneous. Ants like amps!

MIGRATION

MONARCH MIGRATION AN ILLUSION

The epic autumn migration of the eastern monarch butterfly to wintering grounds in Mexico, where millions cluster on trees in semi-dormancy to await spring, has become known as one of the standard 'wonders of nature' in the decade since the Mexican winter clusters were found.

There are, however, some flies in this ointment:

1. Monarchs tagged in the north have never been found in the Mexican clusters.
2. Fall-fattened monarchs can store only enough energy for a flight of about 200 miles---far too short, unless they refuel along the way (no one knows if they do or not); and
3. The monarchs seen in Mexico are almost always in pristine condition and show no wing wear or tattering.

A.M. Wenner, University of California at Santa Barbara, thinks that the "appearance" of mass migration reported frequently from many locales may just be due to a curious fall habit of the monarchs. It seems that widely scattered individuals begin to fly into the wind, and the wind concentrates and channels them to local roosts where they spend the winter. In other words, there is no long distance migration at all. (Rensberger, Boyce; Washington Post, September 15, 1986.

MONARCHS SLIGHTED---SORRY!

Contrary to the above report, at least one tagged monarch butterfly has been found among the wintering colonies in Mexico. In early 1976, a tagged individual from Chaska, Minnesota, was indeed found among a huge cluster in Mexico. Also, a few butterflies tagged in the northern states have turned up in Texas, well on their way to Mexico. (Urquhart, Fred A.; "Found at Last: The Monarch's Winter Home," National Geographic Magazine, 150:161, August 1976. Cr. B. Ickes.)

Comment. So, it seems that A.M. Wenner, the University of California researcher, will have to correct his records and perhaps modify his theory.

COOPERATIVE BEHAVIOR

SYNCHRONOUS RHYTHMIC FLASHING OF FIREFLIES

We humans are pretty smug about our ability to communicate complex messages via sound waves. Of course, we recognize that whales and other cetaceans also seem to "talk" to one another, and that other animals employ their sense of smell for relaying messages. But most of us do not realize that lowly fireflies congregate to communicate en masse, with untold thousands of individuals cooperating in huge synchronized light displays. In reading some of the descriptions of these great natural phenomena, one recalls the light displays used to communicate with the aliens in the movie Close Encounters of the Third Kind.

J. Buck has been studying flashing fireflies for over half a century. In fact, his first review paper was published in 1938. Buck has now brought that paper up to date in the current Quarterly Review of Biology with a 24-page contribution. It is difficult to do justice to this impressive work in a newsletter. Our readers will have to be satisfied with a mere two paragraphs, in which Buck summarizes some of the incredible synchronies.

"More than three centuries later Porter observed a very different behavior in far southwestern Indiana in which, from the ends of a long row of tall riverbank trees, synchronized flashes '...began moving toward each other, met at the middle, crossed and traveled to the ends, as when two pebbles are dropped simultaneously into the ends of a long narrow tank of water...'

"In 1961 Adamson described a still different type of display, the first from Africa: 'It is then too that one sees the great belt of light, some ten feet wide, formed by thousands upon thousands of fireflies whose green phosphorescence bridges the shoulder-high grass. The fluorescent band composed of these tiny organisms lights up and goes out with a precision that is perfectly synchronized, and one is left wondering what means of communication they possess which enables them to coordinate their shining as though controlled by a mechanical device.' A generation later, a flurry of full-dress bioluminescence expeditions had obtained photometric, cinematographic and electrophysiological measurements from congregational displays in Thailand, New Britain, New Guinea and Malaysia, confirming the reality of mass synchrony and uncovering a variety of types. Contemporaneously, Otto and Smiley photographed group wave synchrony of flying fireflies in central Texas, Ohba recorded two frequencies of synchrony in a Japanese species and Cicero described spectacular and enigmatic bouts of chain-flashing, tightening into synchronized strings, by fireflies on the ground in, of all places, the Arizona desert. Thus, work of the past 20 years has shown that 'synchrony' is a complex of behaviors." (Buck, John; "Synchronous Rhythmic Flashing of Fireflies. II," Quarterly Review of Biology, 63:265, 1988.)

Comment. Most theorists take the simplistic view that firefly displays are connected with mating and reproduction. Perhaps related are the complex, geometrical, luminous displays seen at sea, and attributed to bioluminescent organisms in the water. See GLW in the Catalog, Lightning, Auroras, Nocturnal Lights.

WHAT DRUMMER DO PERIODICAL CICADAS HEAR?

Periodical cicadas have the longest life cycles of the insects. Every 13 or 17 years they emerge in vast numbers. How did such long life cycles evolve? How is such precise periodicity maintained. Evolutionists answer the second question with ease. Periodical cicadas are successful in life because their overwhelming numbers, at such widely separated times, completely saturate the appetites of predators, whose populations are not synchronized with the cicada's. Any deviant cicadas emerging a year or so early or late are quickly snapped up, thus promoting synchronicity. So far, so good; but how did such a novel method of coping with predators evolve? There seems to be no way that the cicada's "adaptive peak" of evolutionary success could have been attained from an initial nonperiodic origin. In other words, the cicada cyclic prison is so strong that evolutionists cannot imagine how the prison was made in the first place. (May, Robert M.; "Periodical Cicadas," Nature, 277:347, 1979.)

Comment. Was it a giant, blind evolutionary step that just happened to succeed?

CRICKET COORDINATION

In the August 31, 1991, issue of Science News, there appeared an item on the famous synchronously flashing fireflies of Southeast Asia. W. Clements, writing in response to the firefly story, asserts that Indian crickets chirping in unison are much more impressive. He wrote:

> I once rode on the back of a truck at night along mountain roads in India. There the crickets sound out quite loudly. The sound swells and diminishes with a persistent beat. As we drove along mile after mile, there was not the tiniest perceptible change in the rhythm. In other words, the insects we listened to at any point were modulating their sound at exactly the same frequency, if not phase, maintained by their contemporaries many miles back. Considering the vast areas that must be represented wherever it occurs, the phenomenon must involve unimaginable millions of insects all acting in concert. This is vastly more impressive than the spectacle of fireflies performing together in a single tree.

Picture, if you will, millions, perhaps billions, of crickets all moving their limbs together in unison over many square miles! (Clements, Warner; "Flashy Displays," Science News, 140: 323, 1991.)

SHRIMP TRAINS ARE A'COMING

In March's "Gallery" pages of Discover, several incredibly colored and patterned shrimp stun the eyes of the reader. Some of these shrimp put the gaudiest butterflies and birds to shame. We won't stop here to dwell on why some shrimp are so colorful while others are so tasty. The anomaly at hand is buried in the caption describing the red-and-white striped peppermint shrimp, which decorates the Great Barrier Reef. It turns out that this shrimp, like the Atlantic spiny lobster, sometimes joins up with others of its species to form long moving trains or chains of animals. This behavior remains very puzzling to biologists. (Anonymous; "Shrimp You Won't Find in Your Cocktail," Discover, 6:55, March 1985.)

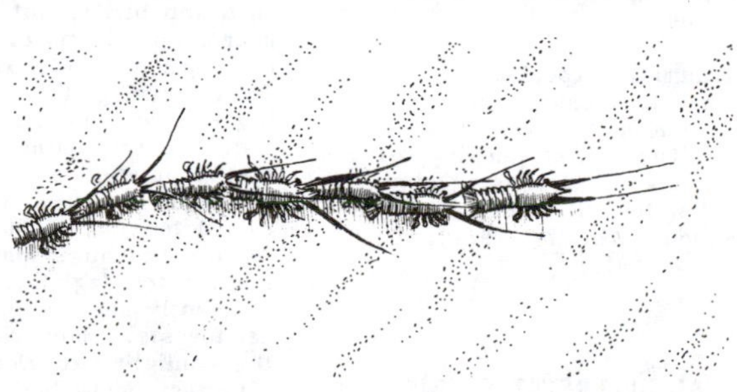

Spiny lobsters often form trains during their migrations.

COLONIAL PHENOMENA

ANTS AS "EXCITABLE SUBUNITS"

Abstract. Activity levels within ant colonies are monitored by using a solid-state automatically digitizing camera. The movement-activity levels of whole colonies and of isolated groups of workers are studied. Whole colonies of Leptothorax allardycei show rhythmic changes in movement-activity level. Fourier and autocorrelation analyses indicate that the activity levels of colonies are periodic, with an average period of 26 min. Single, isolated workers do not show the pattern of periodic changes in activity level. Single workers become active spontaneously, but at no particular interval. Pairs of workers, confined together, also do not show periodicity in activity level. One worker can stimulate another worker to become active, thus coupling their movement-activity patterns. As ants are placed in larger groups, the variation in the interval between activity peaks declines in a manner predicted by coupled-oscillator theory. It is argued that the colony can be regarded as a population of "excitable subunits."

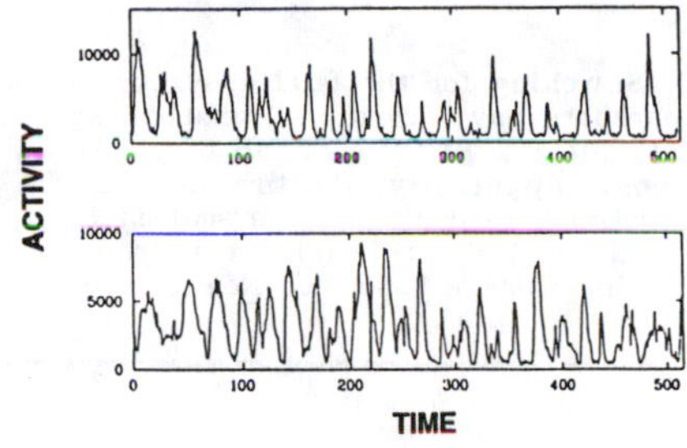

Activity records from two ant colonies. Time is measured in 30-second intervals.

(Cole, Blaine J.; "Short-Term Activity Cycles in Ants: Generation of Periodicity by Worker Interaction," American Naturalist, 137:244, 1991.)

Comment. The author also pointed out the "formal" or mathematical similarity of the ant movement-activity levels and the dynamics of epidemics! This makes us wonder whether wars, economic cycles, etc. might be explained by considering humans as "excitable subunits."

ARMY ANTS: A COLLECTIVE INTELLIGENCE?

Put a hundred army ants on a flat surface and they will walk around in never decreasing circles until they die from exhaustion. But a colony of a million army ants is a sophisticated "superorganism." The colony carries out its legendary raids and can even keep nest temperatures constant to within a degree. An army ant colony seems endowed with an intelligence far beyond that of any individual ant. N.R.Franks speculates thus:

> It seems that intelligence, natural or artificial, is an emergent property of collective communication. Human consciousness itself may be an epiphenomenon of extraordinary processing power. Although experts prefer to avoid simplistic definitions of intelligence, it seems clear that all intelligence involves the rational manipulation of symbolic information. This is exactly what happens when army ants pass information from individual to individual through the "writing" and "reading" of symbols, often in the form of chemical messengers or trail pheromones, which act as stimuli for changing behavior patterns.

In the body of his article, Franks describes two remarkable capabilities of an army ant colony: time-keeping and navigation. The outward manisfestation of time-keeping is in the precise timing of the colony's nomadic phase of 15 days (during which larvae are growing) and the 20-day stationary phase (during which pupae develop). The queen's egg-laying also conforms with this schedule. Raids into the rain forest occur in both phases.

Perhaps more remarkable is the systematic orientation of the raids in the stationary phase. These raids are separated by an average 123°, as diagrammed. This scattering allows time

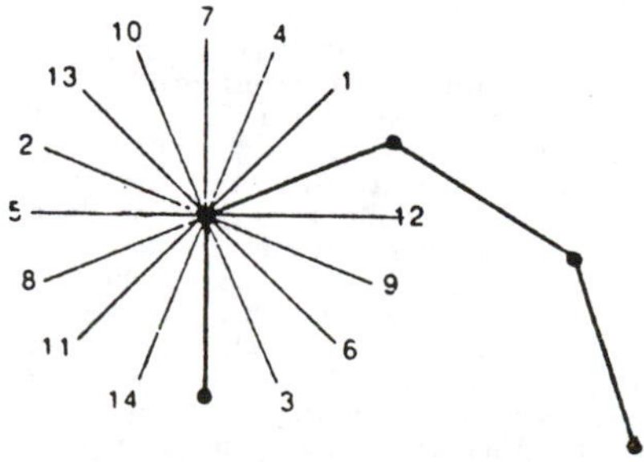

During its 20-day stationary phase, an army ant colony scatters about 14 foraging raids directed about 120° apart. The heavy line indicates the colony's path during its nomadic phase.

for new prey to enter the previously raided areas.

But how does the colony determine direction in the dense rain forest? Probably from polarized sunlight, thinks Franks. But, here we have a problem: each army ant, instead of having multi-faceted compound eyes like most insects, has just a single facet in each eye.

> The mystery is how the colony can navigate with each of its workers having such rudimentary eyesight. In my wildest dreams, I imagine that the whole swarm behaves like a huge compound eye, with each of the ants in the swarm front contributing two lenses to a 10- or 20-m wide "eye" with hundreds of thousands of facets.

(Franks, Nigel R.; "Army Ants: A Collective Intelligence," American Scientist, 77:139, 1989.)

Comment. By analogy, the human body is a colony of individual cells, most of which are specialized in some way. Individual human cells can be grown alone, but they are as directionless as the 100 ants on the flat surface.

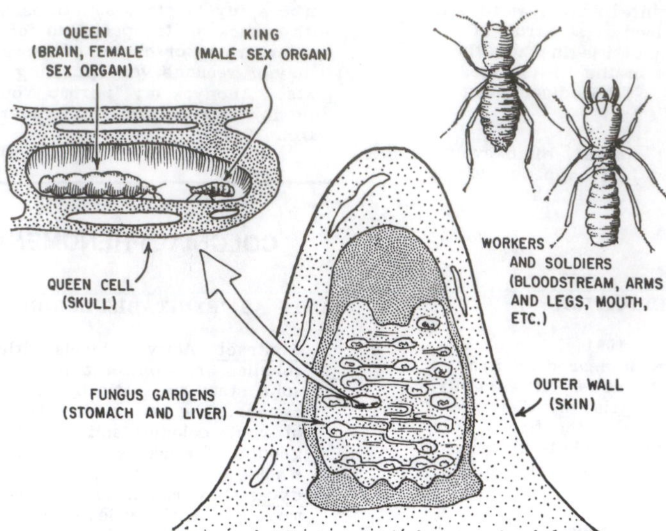

Other insects, particularly the termites, live in colonies which may be thought of as superorganisms.

WHY ARE THERE NO SLAVE ANT REBELLIONS?

Some ant species regularly raid the nests of other ants and carry some of them off into slavery. An oft-discussed question is: Why don't the slaves rebel or at least try to escape back to their home nest? There seems to be no evolutionary advantage in remaining in passive slavery promoting the fortunes of the slave-makers. The opportunities to run off and rebel are never taken. It would seem that slave-ant passivity is a marked disadvantage that has slipped through evolution's net. Regardless of the reason, an imbalance exists. (Gladstone, Douglas E.; "Why There Are No Ant Slave Rebellions," American Naturalist, 117:779, 1981.)

Comment. Perhaps evolution just hasn't had time to redress the situation and even now is preparing clever mutations to rescue the slave ants; i.e., improve their fitness.

MISCELLANEOUS

ANTS LIKE AMPS

Those pesky fire ants that plague man and beast alike in the southern states have a curious weakness: they dote on electrical equipment. They don't eat it; they just "like" it! Why, nobody knows. They invade a wide variety of outdoor electrical devices: airport runway lights, stoplight control boxes, household electrical meters, etc. In particular, they favor relays, where they congregate in masses, interfering with current flow and damaging circuitry.

The phenomenon is made stranger by the fire ants' complete abandonment of their usual search for food and water (and ankles). They starve in droves and clog up everything. It's a moth-and-candle story.

Searching for the fatal attractor, researchers have already eliminated magnetic fields, vibrations, and ozone emissions. Apparently, the fire ants "see" something we can't and are smitten by it. (Weiss, Rick; "Ants Get a Transforming Charge," Science News, 136: 412, 1989.)

Update. As reported in Science Frontiers #87, a wide range of mammals and arthropods seem attracted to electrical equipment.

DISTRIBUTION

THE LOUSE LINE

We all learned about the Tropics of Capricorn and Cancer in high school, but the voyagers of old also recognized a "louse line."

> They wrote about arriving at a longitudinal point in the tropics---the so-called "louse line"---where fleas and lice abandoned even healthy humans. Although a specific line is a myth, cultural entomologist Charles Hogue of the Natural History Museum of Los Angeles believes part of the story. "Voyagers in the tropics often experience a rise in body temperature of as much as 4 degrees F." That's enough, he says, to kill some species of fleas.

Also, the traumatic experience of being caught in a trap causes the body temperature of animals to rise. Thus, trappers often witness fleas jumping off a trapped animal by the dozens. (Johnson, Donna; "How to Tell Time by a Cat's Eye," National WIldlife, 29:12, October/November 1991.)

THE BIOLOGICAL DIVERSITY CRISIS

If life fills all available energy niches, life must be capable of transforming itself (or of being transformed) into a multitude of different energy transducers or energy utilizers. E.O. Wilson has outlined the diversity of terrestrial life in a recent issue of BioScience. The earth, it appears, is a veritable Genesis Machine; and it is only one planet among a possible infinitude.

So many terrestrial species have already been described that one could easily believe that biological collectors roaming the planet's wild places have just about completed their task. Some recent totals: 47,000 species of vertebrates, 440,000 plants, and 751,000 insects. But we may not even be close to grasping life's diversity on earth! We do well in counting the large mammals and birds, but most insects and microscopic forms of life have escaped description. To illustrate, in 1964, the British ecologist C.B. Williams, combining intensive local sampling and mathematical extrapolation, extimated the insect population as 3 million species. However, by 1985, this figure has been raised ten-fold to 30 million species.

Why the huge jump? For the first time, entomologists had found a way to efficiently sample the canopies of tropical forests. This rich stratum between the sunlight and gloomy forest floor 100+ feet below had been largely neglected before. The slick tree trunks and the attacking swarms of wasps and stinging ants deterred the insect counters. What the collectors did was to fire projectiles with ropes over the high branches and then haul up canisters of a knockdown gas. Insects rained down ---a cloudburst of new species---neatly collected on sheets spread out below. Such techniques led to the 30-million figure. As Wilson put it, "The pool of diversity is a challenge to basic science and a vast reservoir of genetic information." (Wilson, Edward O.; "The Biological Diversity Crisis," BioScience, 35:700, 1985.)

Comment. Are there other "hot spots of diversity" waiting to be discovered? Probably, but they will be under our feet, in the deepest waters---places we do not frequent or suspect. We do know of an ancient mudbank that gave birth to multitudes of new and fantastic creatures. It is now lithified as the Burgess shale, in Canada. See p. 163 for more on this formation.

MUST WE DIE? THE MEDFLY'S ANSWER

In the early 1800s, B. Gompertz, an actuary, crafted an empirical law stating that mortality rates increase exponentially with age. Later analyses of census records indicated that the situation was not quite as bad as Gompertz had supposed. Nevertheless, the death rate does increase with age; but we might be able to do something about it. Immortality might be achievable---if we take recent medfly studies seriously.

> Growing old does not increase your immediate risk of dying---at least if you are a fruit fly. The chances of a Mediterranean fruit fly (Ceratitis capitata) dying on a particular day reaches a peak and then declines, according to James Carey of the University of California at Davis and James Vaupel of Duke University, North Carolina, and Odense University in Denmark. Their results contradict the notion that the death rate rises with age in all species.

The upshot is that there may be no genetic limit to an individual medfly's lifetime. And, if these results can be extended to humans, "then medical advances might eventually allow the elderly to live indefinitely." (Bradley, David; "Who Wants to Live Forever?" New Scientist, p. 16, November 14, 1992.)

TALENTS AND CAPABILITIES

PRISONERS OF THE BOUNDARY LAYER

Wings were an inspired evolutionary development. They permit the geographical dispersal of many species, especially insects. But nature, ever-innovative, has other aeronautical techniques up her sleeve. Consider the tiniest insects that do not possess wings. It is difficult for large animals like ourselves to realize that these tiny creatures are actually prisoners of the so-called "boundary layer" of air hugging all surfaces. The thin boundary layer is stagnant very close to the surface. Any tiny insect wishing to take advantage of wind-dispersal to propagate the species farther afield must somehow breach this layer. Some of the scale insects have in their instar phases developed the trick of rearing up on their hind legs, penetrating the boundary layer, and presenting a high-drag surface to the wind. (Many climb along plant surfaces inside the boundary layer to exposed areas before exhibiting this behavior.) The wind plucks them off the plant and carries them off to new territories. The authors think this may be convergent evolutionary strategy for many minute insects. (Washburn, Jan O., and Washburn, Libe; "Active Aerial Dispersal of Minute Wingless Arthropods.....," Science, 223:1088, 1984.)

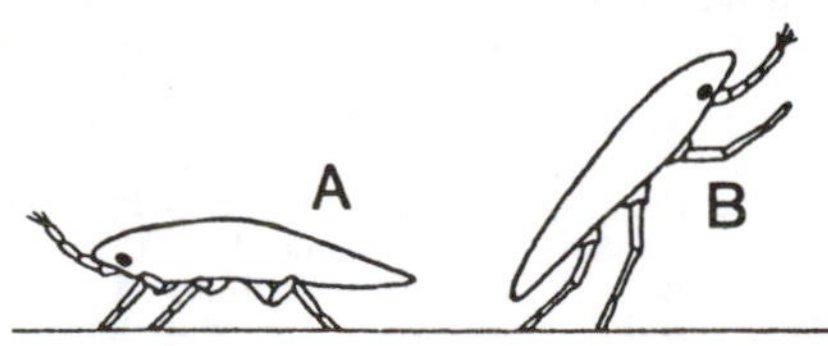

(A) Scale insect (first instar phase) in the slow-moving boundary layer. (B) Rearing through the boundary layer to present a high-drag cross section to the wind.

Comment. The fact that these insects are shaped like airfoils (i.e., aircraft wings) is also interesting.

SIGNALS IN THE NIGHT

Consider the figure below and the four sets of signals (plain blips) and responses (blips with circles over them). Are these from the radar screen of a fighter closing in on an enemy aircraft? Or perhaps the electrical signals generated by the fish mentioned on p. 154? Of course the answer is: None of the above. We have a different story to tell. These blips, representing queries and responses, are not generated by human-built radars or by electrical fish, but rather by animals much 'lower' on the evolutionary ladder---fireflies. This illustration is Fig. 3 in a lengthy review article and carries the following unilluminating caption: "Examples of Entrainment of femme fatale C's (see Table 3) Responses to Multiple Counterfeit Flashes." It seems that we have some sort of electronic warfare between the femme fatales (predatory female fireflies that lure other fireflies with false signals) and the preyed-upon species. The many pages describe all sorts of feints, verification signals, and other stratagems. (Carlson, Albert D., and Copeland, Jonathan; "Communication in Insects," Quarterly Review of Biology, 60:415, 1985.)

Comment. It is impossible to do justice to this paper in this short review, but two things should be mentioned: (1) Fireflies may be considered "low" on the evolutionary ladder, but their tiny brains certainly process a lot of data in complex ways; and (2) In southeast Asia, massed fireflies flash in synchronism along some riverbanks, creating one of the great spectacles of nature. See Incredible Life for details, also p. 158.

BEETLES MAKE SCENTS

Termite nests frequently host foreign species that seem to be accepted as fellow termites. Can't termites recognize the invaders? The authors believe that termites probably recognize one another by specific hydrocarbon labels synthesized on their cuticles. If the alien species were to be somehow marked with similar chemical identifiers, the blind termites might not know the difference. Howard et al think this may be the case with a species of beetle often found integrated into termite society. By chemically analyzing beetle and termite cuticles, they have found both wearing the same hydrocarbon labels. Furthermore, the beetles synthesize their own chemical masks. This is an astounding instance of parallel or convergent evolution between remotely related species. (Howard, R.W. et al, "Chemical Mimicry as an Integrating Mechanism....," Science, 210:431, 1980.)

Comment. Synthesizing exactly the right hydrocarbons was certainly a great stroke of good fortune for the beetles!

SPIDER SWORDPLAY

D. raptor, a Hawaiian spider, has lost its ability to spin webs and therewith capture prey. This unusual spider, however, has evolved:

> ...one of the most remarkable morphological features ever found in spiders (immense elongations of the tarsal claws).

These claws, just visible on the two lowermost pairs of legs in the sketch, are employed to skewer passing insects in flight:

> The spider is strictly nocturnal, spending most of the activity-period hanging upside down from silk threads. Small insects are snagged directly from the air using a single long claw. For larger insects the spider uses both long claws on legs I, or sometimes all the long claws.

(Gillespie, Rosemary G.; "Impaled Prey," Nature, 355:212, 1992.)

Comment. Nature has produced many remarkable creatures. They become anomalous only if they cannot be explained as the products of small, random, cumulative mutations.

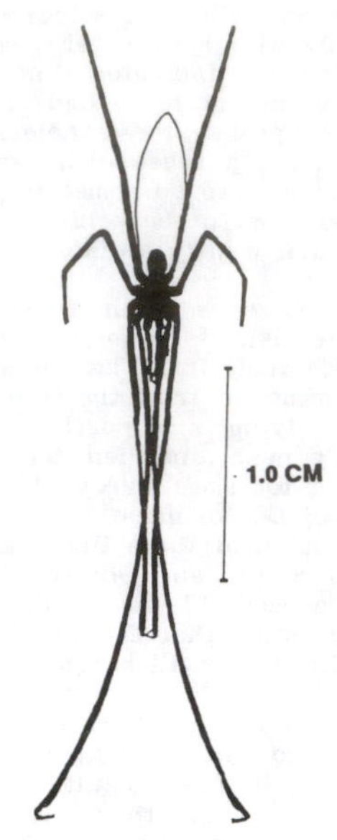

Ventral view of D. raptor. The claws are on the tips of the bottom two pairs of legs. Greatly enlarged photos in the article show them to be wicked-looking, fang-like structures well suited to snare prey.

ORGANS

THE BOMBARDIER BEETLE PULSE-JET

Creationists have long pointed to the bombardier beetle's jet-like defensive spray mechanism as a device that could not have evolved in many small steps. It must be complete and perfect to work at all. New high-speed photos and related research demonstrate that:

> The ejection system of the beetle shows basic similarity to the pulse jet propulsion mechanism of the German V-1 'buzz' bomb of World War II.

What the beetle has "evolved" is an intermittent explosive process that fires about 500 pulses per second. The explosive energy comes from the mixing of two separate fluids (hydroquinones and hydrogen peroxide with oxidative enzymes). (Dean, Jeffrey, et al; "Defensive Spray of the Bombardier Beetle: A Biological Pulse Jet," Science, 248: 1219, 1990.)

Comment. The fundamental question is, of course, how can many, small, random mutations contribute to the development of the mechanisms of the pulse jet, its two fuels, the pumps, the fuel reservoirs, the control system, etc., when only the complete, perfected system has survival value. Although creationists argue that the theories of evolution and natural section are unconvincing here; it is still possible that atheistic factors still beyond our ken are operating, and that what we really need is a better theory of evolution.

PRESCIENT EVOLUTION

Lately, a fossil moth egg was found in 75-million-year-old sediments in Massachusetts. The egg is positively assigned to the moth family Noctuidae and extends the fossil record of this family back into the Cretaceous. So what? Well, it turns out that Noctuidae family moths have special organs for detecting the ultrasonic cries of insect-hunting bats. The fossil record of the bats, however, only goes back to the early Eocene, perhaps 20 million years after the Noctuidae moths. Sinse no other insect predators like bats existed, it would seem that the moths developed these special organs in anticipation of the bats! (Gall, Lawrence F., and Tiffney, Bruce H.; "A Fossil Noctuid Moth Egg from the Late Cretaceous of Eastern North America," Science, 219:507, 1983.)

Comment. Do humans have talents that seem unimportant now but which may be useful some day? Calculating prodigies, eidetic imagers, etc.

FOSSIL PHENOMENA

Two important phenomena in the development of life are apparent in the fossil record of the arthropods: (1) Saltations or quantum jumps in evolution, as seen in the sudden appearance of terrestrial insects and the power of flight; and (2) Episodes of greatly accelerated biological innovation, as evident in the Burgess shale fossils and the curious Ediacaran fauna. (Note: the Ediacaran fauna does not seem to belong to any of the recognized phyla.)

LAND ANIMALS: EARLIER AND EARLIER

Two biologists looking for plant fossils in the Catskills found instead the remains of ancient centipedes, mites, and spider-like creatures---a classical case of serendipity. These animals were in a Devonian formation dated at 380 million years. It turned out that they were the oldest fossils ever found of purely land animals. (Some fossil animals of about the same age are known in European rocks, but in semiaquatic environments.) Two aspects of the fossils are of special interest:

1. The animals found were already well-adapted to terrestrial life, inferring that the (assumed) invasion of the land from the sea has to be pushed back much farther in time; and
2. Many of the fossil animals are essentially identical to modern forms, suggesting that little if any evolution has occurred in 380 million years.

(Anonymous; "Fossils Found in N.Y. Alter Scientists' View," Baltimore Sun, p. A3, May 29, 1983. Supplied by the New York Times News Service.)

Comment. Note the sudden jump from no land animals to well-developed, frozen-in-time land animals.

THE OBSCURE ORIGIN OF INSECTS AND THEIR WINGS

The earliest fossil insect is a wingless springtail found in Scotland's Devonian cherts, which conventional dating schemes tell us are about 350 million years old. Some biologists doubt that springtails should be classified as true insects. In any event, these ancient springtails are considered too specialized to be the ancestors of modern winged insects. The next insects in the fossil record appear suddenly in the Upper Carboniferous (300 million years ago) with fully developed wings. There exists an embarrassing 50-million-year gap between the fossil springtails and the more specialized insects.

Evolution requires that this gap be filled with many random experiments at insect construction, including the first attempts at fashioning wings. Whalley admits the gap and the total mystification of paleontologists about how insects and biological flight first developed. Perhaps, he surmises, wings may have been the natural extrapolation of flap-like outgrowths required for body cooling. Random mutations would have added the muscles needed to orient the flaps and move them to improve circulation! (Whalley, Paul; "Derbyshire's Darning Needle," New Scientist, 78:740, 1978.)

THE BURGESS SHALE PUZZLE

In British Columbia, a middle Cambrian (550 million years BP) formation called the Burgess Shale has miraculously preserved a vast assemblage of soft-bodied sea creatures, especially arthropods. Does this rich and unusual deposit help elucidate arthropod evolution? No, it has complicated the problem. Few of the fossil arthropods can be easily related to groups now living. The Burgess Shale arthropod population is primitive in some ways but remarkably specialized in others. Some of the fossils have body segments like those in one recognized arthropod group but display limbs resembling those of an entirely different group. (Fortey, R.A.; "The Burgess Shale: A Unique Cambrian Fauna," Nature, 293:189, 1981.)

Comment. It appears that Nature was shuffling the gene deck, or that there was rampant hybridization, or that confusing programs for evolutionary change were drifting in from the cosmos a la Hoyle and Wickramasinghe!

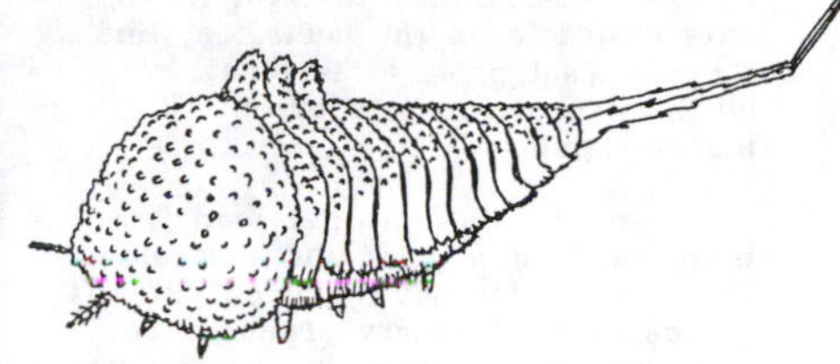

A bizarre arthropod from the Burgess shale.

TREASURES IN A TOXONOMIC WASTEBASKET

The Burgess Shale, in British Columbia, ia all that is left of a Middle Cambrian mudbank that adjoined a massive algal reef. Here, many "experimental" forms of life prospered and succumbed. The Burgess Shale is probably the world's greatest depository of fossils of soft-bodied creatures. Quoting Stephen Jay Gould, "The Burgess (fortunately for us) occupies a crucial time in life's history. It represents our only 'window' upon the first great radiation of complex life on earth. All but one or two modern phyla originated in a burst of evolutionary activity associated with the so-called Cambrian explosion some 570 million years ago. The Burgess provides our only peek at the soft-bodied forms of this first flowering. All other soft-bodied fossil assemblages are much younger; they represent faunas well past the initial burst and sorting out of Cambrian times."

The morphological diversity of the Burgess Shale, incorporating many bizarre forms of life, represents a true biological revolution. Here are found a dozen genera that do not fit into any modern phylum. Most of the novelties never survived into modern times. (Gould, Stephen Jay; "Treasures in a Taxonomic Wastebasket," Natural History, 94:24, December 1985.)

Comment. Somewhere on today's earth, there must be mudbanks washed by nutrient-rich waters and bathed in tropical sunlight. Is some ingredient missing, or perhaps present, in today's mudbands that suppresses the wild speciation seen in the Burgess Shale?

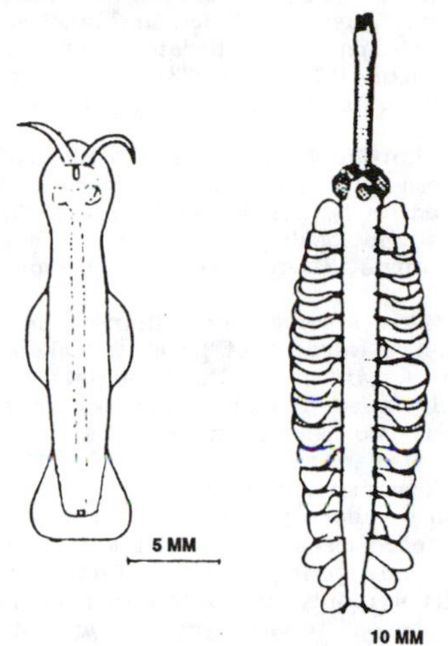

Two of the many mysterious fossils found in the Burgess shale. At the right is Opabina regalis, with five eyes at the base of a nose-like structure ending in teeth. On the left is Amiskwia sagittiformis. Although these creatures are named, no one really knows what they were like in life.

THE FAILURE OF TWO-DIMENSIONAL LIFE

The fossil record tells us that just prior to the Cambrian explosion of life, the earth was populated by a diverse assemblage of soft-bodied, shallow-water marine invertebrates, some with dimensions as large as 1 meter. This whole group of animals did not survive into Cambrian times, thus ending what has been termed The Ediacaran Experiment. Some paleontologists have tried to find similarities between the Ediacaran and Cambrian life forms to preserve the continuity of life. This has proved difficult, and some scientists now feel that Ediacaran life, in which "largeness" was achieved by increasing surface area failed. The Ediacarans were shaped like pancakes, tapes, fans, etc. This enabled them to present large areas to the environment for respiration, feeding, and other biological functions.

In contrast, many present life forms achieve "largeness" by increasing internal areas, as in the lungs, folded intestines, etc., along with the forced circulation of air, blood, and other substances. This latter approach survived, while the two-dimensional Ediacaran Experiment did not. The demise or extinction of the Ediacarans led Gould, the author of this far-ranging article, to the influence of extinctions on life in general---a hot topic these days. Gould stated that with natural selection operating, one would expect continual "improvement" in life forms, but that this had not happened.

> I regard the failure to find a clear "vector of progress" in life's history as the most puzzling fact of the fossil record. But I also believe that we are now on the verge of a solution, thanks to a better understanding of evolution in both normal and catastrophic times. We need a two-tiered explanation for patterns (or nonpatterns) in the history of life.

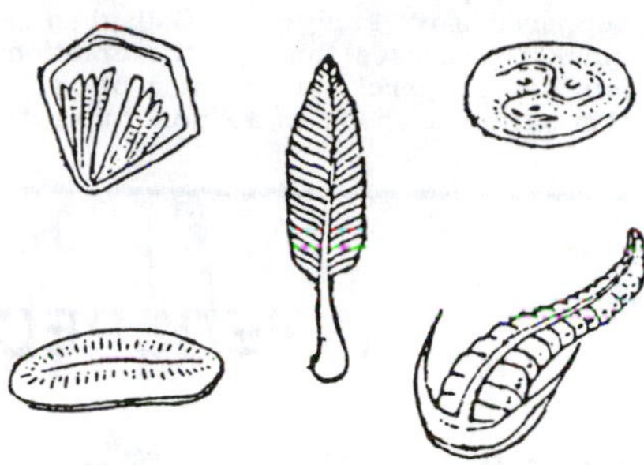

Ediacarian life forms were typically flattish to maximize their external areas. They failed anyway!

The first tier of explanation involves the theory of punctuated equilibrium, as championed by Gould and Eldredge. Gould says he once thought that punctuated evolution would be sufficient to explain all of life's development, but that now a second tier seemed required; namely, a general theory of mass extinction or what catastrophism does to life and its development. (Gould, Stephen Jay; "The Ediacaran Experiment," Natural History, 93:14, February 1984.)

GENETICS

SCANT ANT CHROMOSOMES

The Australian ant Myrmecia pilosula, called the "bulldog ant" because of its viciousness, carries all its genetic information in a single pair of chromosomes. (Males are haploid and have just one chromosome.) Although classified as a "primitive" ant, the bulldog ant exhibits complex social behavior and is obviously far from a simple biological entity. Biologists were therefore surprised to find all genetic instruction residing in a single chromosome pair. Social insects tend to have higher chromosome numbers. It is also interesting that Myrmecia pilosula, originally described as a single species, actually consists of several distinct sibling species with chromosome numbers (i.e., pairs) of 9, 10, 16, 24, 30, 31, and 32. Yet, they all look pretty much alike. (Crosland, Michael W.J., and

Crozier, Ross H.; Myrmecia pilosula, an Ant with Only One Pair of Chromosomes," Science, 231:1278, 1986.

Comment. Chromosome number or the sheer quantity of genetic material seems poorly correlated with biological complexity.

INSTANCES OF OBSERVED SPECIATION

Creationists have long maintained that no one has ever observed the creation of a new species in nature. C.A. Callaghan has sought to counter this attack on evolution with a paper bearing the above title. Her concluding paragraph is:

> I have cited several instances of observed speciation that can be used as illustrative examples in the classroom. They should also silence at least one common creationist argument against evolution.

The paper begins with the well-worn peppered-moth story; but Callaghan quickly dismisses this, as the creationists do, as merely an example of variation within a species. We now quote the lead sentences from her discussions of the next two candidates:

> An incipient neospecies of Drosophila may have developed in Theodosius Dobzhansky's laboratory sometime between 1958 and 1963 in a strain of D. paulistorum....

> A probable instance of a naturally emerging plant species was discovered on both sides of Highway 205 at a single locality 25.5 miles south of Burns, Harney County, Oregon....

We have inserted underlining beneath the two words that greatly weaken the paper. In short, the biologists are not really sure that speciation occurred in these two cases. The reasons for doubt are also presented.

The paper concludes with a discussion of allopolyploidy in plants, in which the chromosomes of a sterile hybrid are doubled, giving rise to a fertile variety. Allopolyploidy has been observed in primroses, tobacco, cotton, and other plants.

And that's it; two questionable examples and allopolyploidy. (Callaghan, Catherine A.; "Instances of Observed Speciation," American Biology Teacher, 49:34, January 1987.)

INTERSPECIES PHENOMENA

All of the earth's phyla interact, but the arthropods seem more prone to do this than most. Science Frontiers has been able to accumulate only a handful of arthropod examples of interspecies phenomena from a pool which must number many thousands. To illustrate, we have nothing on insects as vectors of human diseases, but there is an item on the unique susceptibility of humans to one particular spider's venom. The defoliating capabilities of gypsy moths are not mentioned, but we do relate how insects can "outwit plants! This limited coverage of interspecies phenomena will at least pique the reader's curiosity. The final entry brings to the fore once again, this time via the termite clan, the Gaia hypothesis and the apparent unity of all life.

GLITCH IN THE EVOLUTION OF FUNNELWEB SPIDER VENOM?

The Australian funnelweb spider has a venom that appears to be effective only against humans, monkeys, baby rats, and fruit flies. None of these animals is normally on the spider's menu; those prey that are seem unaffected by the venom. Did the evolution of the poison miss its intended targets or did the spider's usual prey evolve resistance? It is interesting that mature rats are immune to the venom, although neonatal rats are not. (Anonymous; "Did You Know?" Ex Nihilo, 7:16, no. 3, 1988. Facts taken from The Australian Doctor, January 20, 1984.)

THOSE AMAZING INSECTS

Scientific creationists often point out marvels of biological engineering as proofs that evolution by small chance-directed steps is, to say the least, improbable. A sharp arrow for their quiver has just appeared in Science.

Caterpillar-ant vibrational communication. Some species of caterpillars cannot survive predation unless they are protected by ants. To attract the ants, who just happen to dote on special secretions of the caterpillars, the caterpillars send out vibrational signals across leaves and twigs. In addition to their secretory structures, the caterpillars have also evolved novel vibrators to send out their calls for protection. A few butterfly species from all continents (except Antarctica) have evolved these devices. Looking at this geographical spread, P. J. DeVries thinks that the two sets of organs must have developed independently at least three times. (DeVries, P.J.; "Enhancement of Symbioses between Butterfly Caterpillars and Ants by Vibrational Communication," Science, 248:1104, 1990.)

SINGING CATERPILLARS

Actually, the singing caterpillars are not particularly tuneful. They really generate a vibration that is transmitted through the material they are resting on. You and I cannot hear caterpillar songs, but some ants can, and they are attracted to these insect sirens.

The singing caterpillars belong to the Lycaenidae, which include such butterflies as the hairstreaks and blues. It is not only the singing or vibrating of this group of caterpillars that makes them remarkable, it is the complexity of their symbiotic relationships with several species of ants and a plant. Since both the ants and the caterpillars favor the Croton plant, they could well meet by chance, but the caterpillars' singing serves to accelerate contact. Once met on the Croton plant, a fascinating triangle is completed:

Player 1. The Croton plant provides nourishment to the caterpillars through both its leaves and specially evolved nectaries (nectar-producing organs), but receives nothing in return. The ants also dote on the nectaries, but they at least protect the plant from all herbivorous insects except the singing caterpillars.

Player 2. The ants get food from both the Croton plant and the caterpillars. The latter have evolved extrudable glands called "nectary organs." For their part of the bargain, the ants protect the caterpillar from predatory wasps, just as they defend the Croton plant from its enemies.

Player 3. The caterpillars, though seemingly benign, are the heavies in this menage-a-trois! They get both leaves and nectar from the plant for nothing. They do supply the ants with nectar in exchange for protection, but subtle subversion prevails here! First they attract the ants with their songs; then, they seduce them with nectar that is much more nutritious and attractive than that produced by the Croton plant. Finally, they chemically force the ants into defensive postures against predatory wasps by spraying them with a mesmerizing substance from special "tentacle organs" near their heads.

Why is all this subversive on the part of the caterpillars? It appears that the caterpillars have invaded and undermined the normal ant-plant symbiosis---a very common, mutually beneficial arrangement. The ants have been seduced into letting the caterpillars feast on the Croton plant, although the ant-plant compact originally required that the ants repel all herbivorous insects.

What makes this tale of subterfuge so remarkable is that the caterpillars had to evolve three separate organs in order to accomplish it: (1) their vibratory papillae; (2) their nectary glands; and (3) their mesmerizing tentacle organs. (DeVries, Philip J.; "Singing Caterpillars, Ants and Symbiosis," Scientific American, 267:76, October 1992.)

THE INSECTS' REVENGE

Plants may fool some insects with their mimicry and deter others with toxic chemicals, but the insects have their tricks, too, as seen in the following item from Science:

> Many mandibulate insects that feed on milkweeds, or other latex-producing plants, cut leaf veins before feeding distal to the cuts. Vein cutting blocks latex flow to intended feeding sites and can be viewed as an insect counteradaptation to the plants' defensive secretion.

(Dussord, David E., and Eisner, Thomas; "Vein-Cutting Behavior: Insect Counterploy to the Latex Defense of Plants," Science, 237:898, 1987.)

Comment. Right now, even as we write this, the plants are evolving counterploys---high-voltage veins perhaps! But this may not work. See the final item in this section.

A TALE OF TWO ECOSYSTEMS---OR MAYBE MANY MORE

Comment. It was quite fortuitous that the following two pieces, found on the same day, fit together so nicely. (Nature is fortuitous!)

A group of scientists studying termites have isolated over 100 species of protozoa and bacteria living cooperatively inside the termites gut. Some of the bacteria even live inside the protozoa and other bacteria forming ecosystems of several symbiotic levels within each termite. Each termite itself is part of a complex social "superorganism," the termite colony. That termites had bugs inside them has long been known; but the new-found complexity and interdependency of life systems within life systems is remarkable. The researchers believe that the life forms inside the termite work together to create the uniform internal environment needed by all inhabitants, just as the termites themselves cooperate to maintain a favorable environment inside their hill. (Anonymous; "And Littler Bugs inside Them," Scientific American, 246:78, February 1982.)

The termites, though, are only part of a much larger ecosystem, the earth itself. J.E. Lovelock, in his Gaia, A New Look at Life On Earth, has observed that our planet's environment has actually changed little down the eons despite solar variations. Lovelock's hypothesis is that all terrestrial life---animals, plants, termites, etc.---work symbiotically to maintain planetary temperatures, atmospheric constituents, etc., conducive to life, just like the termite's internal residents on a much smaller scale. The Gaia concept was restated by R.A. de Bie in a letter to Nature about the possibility of life in outer space. He observes that terrestrial life, according to Gaia-thought, will naturally develop a species that can carry life off the planet to new and safer environs. Humankind, of course, is the first attempt we know of to create such an agent. Other planetary systems might have differently constituted agents. (de Bie, Roeland A.; "The Ultimate Question," Nature, 295:8, 1982.)

Comment. To extend the biological analogy, the appearance of humankind represents the "fruiting" or "spore" stage of composite earth life. Humankind or some equivalent life form would thus be an inevitable development of any life colony in the broad and seething universe of galaxies.

INVERTEBRATES

EXTERNAL APPEARANCE

STRANGE PATTERNS IN ANOTHER OCEANIC HABITAT

The sea-floor vents and their unique assemblages of animals are just beginning to be explored.

> Perhaps the most intriguing biological mystery in the vent area, however, was the finding of thousands of highly symmetric, Chinese-checkerboard-like patterns on the sea floor, which were first photographed several years ago. (P.) Rona thinks the patterns may be either an animal itself or the burrows made by an animal. He says the patterns are "dead ringers for a 70-million-year-old....trace fossil that is exposed in the Alps."

(Weisburd, Stefi; "Hydrothermal Discoveries from the Deep," Science News, 130:389, 1986.)

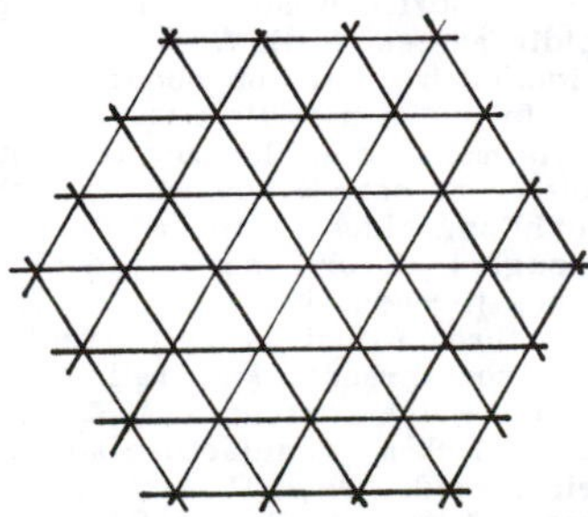

Thousands of these Chinese checkerboard patterns have been spotted on seafloors. Presumably, they are made by invertebrates.

RUBBERNECKIA

The long-standing belief that unlimited rotary motion is impossible in animals has been shattered. It was, after all, a very reasonable assumption, because necks and other appendages turn only so far before bones and muscles begin to snap. Well, it seems that inside termite guts there resides a single-celled animal with a head that rotates constantly 30 times a minute. Since none of its membranes shear during rotation, we must infer that membranes are basically fluid structures rather than solids as supposed.

The animal, called Rubberneckia, has a shaft running the full length of its body plus a motor of undetermined character. To make Rubberneckia even more bizarre, thousands of tiny, rod-like bacteria occupy long grooves on the cell's surface. Like galley slaves, the bacteria row with their flagella to keep Rubberneckia moving---a curious symbiotic relationship. (Cooke, Robert; "A Tale to Make Your Head Spin,: Boston Globe, March 20, 1984, p. 1. Cr. P. Gunkel)

THE LURES OF MUSSELS

Mussels entrust their larvae to the vagaries of the waters in which they live. How, then, are mussels ever able to colonize rivers, whose currents would always sweep their larvae downstream?

> The riverine pioneers ran this roadblock by custom designing their baby mussels to hitchhike on fish. Knee-high to a pinhead, the larval mussel, or glochidium, is nurtured by the thousands or millions in their mother's gills, and spewed in teeming puffs to the open waters. They cling as benign parasites to passing fish, and take a one- to three-week trip, drawing nutrients through their host's membranes and a free ride to new dwellings. They then drop to the bottom and begin their independent lives, some of which will span a half century or more.
>
> Glochidia that do not hook up with a host fish are doomed. To cover these stakes, the pocketbook mussel and its relatives have evolved a fleshy appendage that flaps in the currents and, to a smallmouth bass, looks like a breakfast minnow. Taking the bait. the duped fish gets doused with glochidia. Another resourceful mussel sends its glochidia out in

pulsating little packets resembling worms.

(Stolzenburg, William; "The Mussels' Message," Nature Conservancy, p. 17, November/December 1992.)

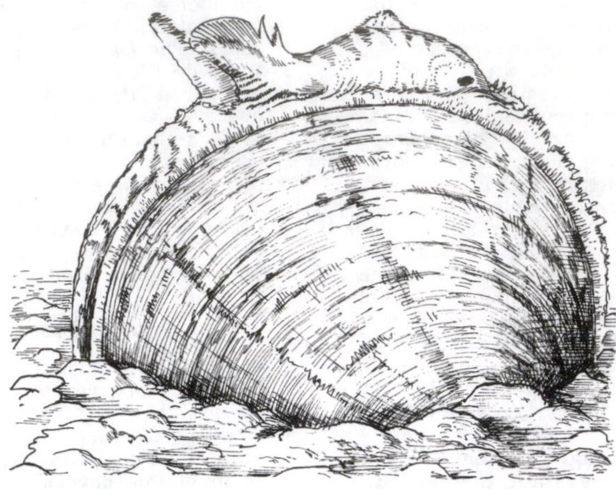

A pseudofish with tail, fins, and eye spot displayed by a mussel to lure fish closer.

DISTRIBUTION

REMARKABLE DISTRIBUTION OF HYDROTHERMAL VENT ANIMALS

Hydrothermal vents support a bizarre array of large clams, mussels, worms and other curious species. These biological communities are unique in that they are supported not by solar energy but rather the earth's thermal energy. What verges on the anomalous is the appearance in both Atlantic and Pacific Oceans of very similar vent communities, with similar or identical species. How did these continent-separated communities originate? In the words of the author of the present article:

> The cooccurence of a clam, a mussel, and a vestimentiferan worm at widely separated sites in the Pacific and Atlantic represents either an unusual distribution from a single lineage or, even more remarkably, cases of parallel evolution.

(Grassle, J. Frederick; "Hydrothermal Vent Animals: Distribution and Biology," Science, 229:713, 1985.)

BEHAVIOR

THE MOON AND LIFE

There are so many examples of lunar rhythms in terrestrial life that we tend to assume that these phenomena are understood. Obviously, evolution "created" these rhythms to further the cause of each moon-tuned organism. Palmer and Goodenough recount the classic example of the lunar synchronism of the palolo worm and add the even-more-amazing tale of P. megalops, another marine worm. Sure enough, the moon-modulated matings of these worms seem to improve reproductive efficiency.

Less well known are many other moon-synchronous biological rhythms; viz., the sizes of the pits dug by ant lions to trap ants and the angles flatworms assume in swimming away from light. Many such lunar rhythms apparently have no adaptive value whatsoever. So, why do they exist? Even more disconcerting is the fact that lunar rhythms persist in the lab where the moon is not visible. Are internal clocks responsible here? If so, how do they work and how are they set? These questions are hard to answer if the rhythms have no value to the organism's success. (Palmer, John D., and Goodenough, Judith E.; "Mysterious Monthly Rhythms," Natural History, 87:64, December 1978.)

Comment. It would, or course, be outright heresy to suggest that heavenly bodies may be the sources of unrecognized but biologically significant forces.

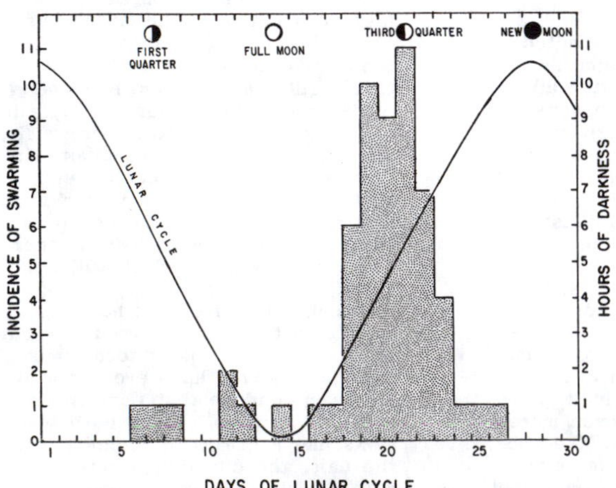

Swarming incidence of the Atlantic palolo worm (shaded histogram) compared with the lunar cycle. (Adapted from Science News Letter, 39:219, 1941.)

WANDERING MOLLUSCS

> How could a mollusc which lived all its adult life cemented fast to the seabed in shallow water occur thousands of kilometers apart in the Arabian Gulf and the Caribbean? Almost identical forms of the extinct bivalve Torreites sanchezi have been found in rocks from the Caribbean and the Gulf. Peter Skelton of the Open University and Paul Wright of the University of Bristol suggest that the larval stage of the bivalve must have "island hopped."

The two researchers rule out convergent evolution and note that the two seas were never any closer together. They suppose that in Cretaceous times there was an equatorial current that swept the larval forms long distances. (Anonymous; "Wandering Molluscs," New Scientist, p. 33, October 15, 1987.)

Comment. The same situation prevails for other species, such as some of the amphipods and the unique life forms dependent on seafloor vents.

The Asiatic freshwater clam has spread rapidly across North America since its accidental introduction about 50 years ago. In addition to its natural dispersal via its more mobile larvae, the young adult clams have a surprising method of hitchhiking rides on the water currents. Through their siphons they deploy long mucous threads. Water currents pull on these threads just as air currents catch the silken threads of migrating spiders. Given a water current of 10-20 cm/sec, the small clams manufacture and deploy their threads, and off they go downstream. (Prezant, Robert S., and Chalermwat, Kashane; "Flotation of the Bivalve Corbicula Fluminea as a Means of Dispersal," Science, 225:1491, 1984.)

HUMPS OF PARTICLES IN THE GULF STREAM

Peculiar humps or hills of particulate material, thousands of feet long and hundreds high, have been been observed in the Florida Current with ultrasensitive, 20-kHz sonar. The humps extend from the bottom of the oceanic mixed layer to near the surface. Sediment and plankton probably make up the humps, but the actual constitution and cause of the concentrations are unknown. Conceivably currents could concentrate the particulate matter; so could some sort of coordinated biological activity. (Proni, John R., et al; "Vertical Particulate Spires or Walls within the Florida Current and near the Antilles Current," Nature, 276:360, 1978.)

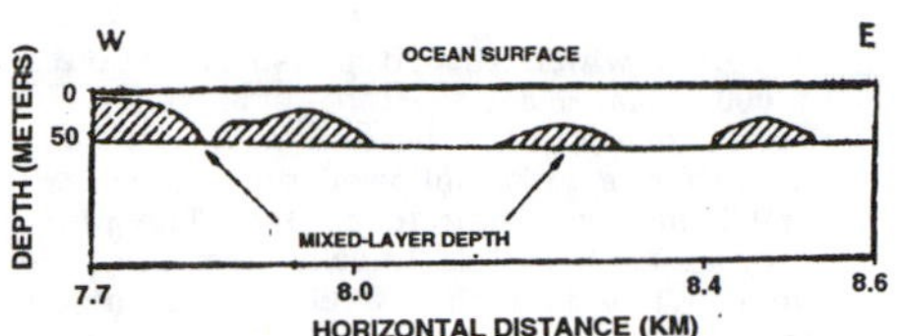

Humps of particulate matter (shaded regions) detected by sonar in the Gulf Stream.

CAPABILITIES AND FACULTIES

LIFE IN THE DARK

The scene here is deep down at 10,000 feet in the Gulf of Mexico. There, scientists in the submersible Alvin found a well-developed community of large clams, crabs, mussels, and tube worms, which closely resembles those around the Pacific hydrothermal vents. These life colonies do not use sunlight at all, nor do they depend on other life forms based on solar energy. They employ chemosynthesis, and the hydrogen sulfide and other substances in the vented waters replace sunlight. Although there are no obvious vents at the Gulf of Mexico site, the waters there contain plenty of hydrogen sulfide, indicating seepage from somewhere. The life forms are all new to science, although they resemble those in the Pacific. (Anonymous; "Worms without Vents, " Oceans 17:50, September/October 1984.)

Comment. Question: how do non-mobile life forms travel the great distances from one vent or seepage locale to another? It seems as if we are just beginning to appreciate life's colonizing capabilities. Who knows what life forms subsist in the hot geothermal fluids circulating deep in the earth's crust?

LIFE SEEKS OUT ENERGY SOURCES WHEREVER THEY MAY BE

Life is opportunistic; it siphons off energy wherever it can find it. That life utilizes solar energy we all know. And, of course, humans have tapped the atom for energy. In just the past few years, remarkable colonies of life forms have been discovered congregated around deep-sea hydrothermal vents where sunlight is essentially nonexistent. Still more recently, similar life forms have been found clustered around oil seeps in the Gulf of Mexico. As at the hydrothermal vents, the clams, worms, crabs, and other organisms depend mainly upon the ability of bacteria to chemosynthesize---the primary energy source being hydrogen sulfide in the vented water. (Paull, C.K., et al; "Stable Isotope Evidence for Chemosynthesis in an Abyssal Seep Community," Nature, 317:709, 1985; and, Weisburd, S.; "Clams and Worms Fueled by Gas?" Science News, 128:231, 1985.)

LUNAR MAGNETIC MOLLUSC

From the Abstract of an article published in Science:

> Behavioral experiments indicated that the marine opisthobranch mollusk Tritonia diomedea can derive directional cues from the magnetic field of the earth. The magnetic direction toward which nudibrachs spontaneously oriented in the geomagnetic field showed recurring patterns of variation correlated with lunar phase, suggesting that the behavioral response to magnetism is modulated by cira-lunar rhythm.

The magnetic and lunar-phase detectors of this mollusc are not known. In fact, the authors remark in their introductory paragraph that, even in organisms possessing ferromagnetic materials in their systems, there exists no "direct neurophysiological evidence implicating ferromagnetic particles in the detection of magnetic fields." (Lochman, Kenneth J., and Willows, A.O. Dennis; "Lunar-Modulated Geomagnetic Orientation by a Marine Mollusk," Science, 235:331, 1987.)

AN AMUSING ASSEMBLAGE OF ANOMALIES

We don't read much about "waterguns" in the modern scientific literature, but a century ago Nature published many ear-witness accounts of them. These muffled detonations heard near the coasts of almost all the continents are believed by some to be caused by eruptions of methane from the seafloor. The same eruptions probably also account for the myriads of "pockmarks" found in the sediments of shallow seas. Whether this outgassing of methane comes from shallow accumulations of organic matter or from deep within the crust is still debated.

Here, geophysics merges with biology. Recently, a group of researchers discovered a large (540 square meters) patch of chemosynthetic mussels in a brine-filled pockmark, at a depth of 650 meters, off the Louisiana coast. The mussels grew in a ring around the concentrated brine. The mussels harbor symbionts which consume the methane still seeping up through the brine from a salt diapir (a massive finger-like intrusion 500 meters below the brine pool. The origin of some diapirs is not well-understood.) The mussels get the oxygen they require from the ordinary seawater covering the dense brine. Like the biological communities surrounding the "black smokers" and other ocean-floor seeps, the brine-filled pockmark community includes several species of shrimp, crabs, and tube worms. We have here another example of the astounding ability of lifeforms to take advantage of unusual, even bizarre niches. (MacDonald, I. Rosman, et al; "Chemosynthetic Mussels at a Brine-Filled Pockmark in the Northern Gulf of Mexico," Science, 248:1096, 1990.)

Comment. Such examples of life's adaptability are so common one hesitates to label them as anomalous. Yet, one wonders how and why life acquired this property. Is the human urge to go to the planets a genetically derived extension of this urge to colonize new territories.

BODILY FUNCTIONS

WORMS WITH INSIDE-OUT STOMACHS

The recently discovered tube worms, living near the hot water vents on the ocean bottom off the Galapagos, have no mouths or guts. Their bodies are covered with thousands of feathery tentacles, each packed with blood vessels. Apparently, the tube worms extract nutrients directly from the sea water and expel wastes the same way---having in effect external stomachs. These worms, which may be many feet long, contain enzymes that permit them to extract carbon dioxide from the seawater and fix it much like plants do during photosynthesis. George Somero, at Scripps, estimates that the enzyme levels in the worms are similar to those in a spinach leaf. (Anonymous; "15-Foot Sea Worm Has Plant Qualities," San Diego Evening Tribune, May 22, 1980. UPI dispatch.)

Comment. This curious biological anomaly developed in an ecological niche where the primary energy source for sustaining life is geothermal rather than solar. How did this remarkable situation arise? Do the tube worms have relatives in the fossil record showing a step-by-step development of inside-out stomachs?

AN ANIMAL THAT PHOTOSYNTHESIZES

At a recent meeting of the American Society for Photobiology, chemist Pill-Soon Song, of Texas Tech University, reported the discovery of a blue-green, trumpet-shaped protozoan that employs photosynthesis to sustain itself. Called Stentor coeruleus, this protozoan is only 0.2 mm long and swims backward by rotating its cilia. According to the article, this is the first instance of a photosynthesizing animal. (Anonymous; "Animal That Lives on Light," San Francisco Chronicle, June 28, 1985, p. 2. Cr. J. Covey)

Comment. Nothing was said about whether the protozoan also ate food in the conventional manner. If verified, this is not a trivial discovery. Of course, some plants eat meat, but animals seem to have found sunlight too weak to utilize for mobility and other energy-rich processes and activities.

CAENORHABDITIS ELEGANS

The creature with this formidable name is only about a millimeter long and develops from egg to adult in about 3.5 days, at which time it possesses about 1,000 somatic cells. C. elegans is a roundworm, but a famous one. Its growth has been followed on a cell-by-cell basis from egg to adult. The history of each cell is known from birth to death. The fact that C.elegans is nicely transparent helps the cell-watcher.

Here are some of the interesting things to be seen as cells proliferate, live, and die. First, C. elegans is bilaterally symmetrical, but the pattern of cell generation on the right differs from that on the left. Nevertheless, the creature ends up symmetrical, making one wonder where the directions for symmetry come from. Some cells are transients, dying when their jobs are done. A few doomed cells are generated only because they produce sister cells that are needed in the final animal. Such a programmed loss of cells may be a method of modifying an organism during evolution. John Sulston, one of the researchers, says, "Within the lineage you can see the fossil of its past." (Marx, Jean L.; "Caenorhabditis Elegans: Getting to Know You," Science, 225:40, 1984.)

Comment. Sulston's statement reminds one of "Ontogeny recapitulates phylogeny," which we thought had been discredited long ago.

PALEONTOLOGICAL PROBLEMS

AFRICAN FOSSIL SEQUENCES SUPPORT PUNCTUATED EVOLUTION

East of Lake Turkana, in northern Kenya, the geologist finds exceptionally fine sequences of fossil molluscs in old lake deposits. Williamson has scrutinized the distribution of some 190 faunas with high stratigraphic resolution; that is, he believes he has been able to sketch for the first time evolutionary events on a fine time scale. Williamson underlines three important observations:

1. Species seemed to arise suddenly, as predicted by the "punctuated evolution" model;
2. The formation of new species was accompanied by marked developmental instability in the transitional forms; and
3. All lineages were morphologically stable for long periods---they did not change form!

The biological implications of this important study are summarized in the following item. (Williamson, P.G.; "Palaeontological Documentation in Cenozoic Molluscs from Turkana Basin," Nature, 293:437, 1981.)

Comment. Evolutionists have often bewailed the obvious lack of transitional forms (missing links) in the stratigraphic record. According to Williamson's results, transitional forms would be few in number and display considerable morphological instability. In essence, this means that missing links may not exist in a practical sense. If this is true, one wonders whether those famous evolutionary family trees in all the textbooks, such as that of the horse, are really misleading.

SPECIES STABILITY IS A REAL PROBLEM

The reader should refer to the above item for the basic paleontological facts discussed by Williamson below. The biological implications of the mollusc lineages that he examines are rather profound. In the present item, Williamson complains that scientists and critics have focussed primarily upon his claim that his mollusc lineages support the punctuated evolution model (which they do) but avoid his main point: namely, that the lineages are static over very long periods of time. They do not change slowly, bit by morphological bit, into new species as an evolutionist would expect. Instead, they remain unchanged until they become extinct. This striking aspect of the fossil record is not predicted by neo-Darwinism---and there is the rub! (Williamson, Peter G.; "Morphological Stasis and Developmental Constraint: Real Problems for Neo-Darwinism," Nature, 294:214, 1981.)

Comment. In neo-Darwinism, evolution unfolds by small accumulated changes, the causes of which may be chemicals in the environment, nuclear radiation, and other "stresses." Neo-Darwinism goes hand-in-hand with geological Uniformitarianism, both of which are favored philosophically by scientists because slow change is more amenable to scientific explanation. The large sidewise steps of punctuated evolution are difficult to explain in terms of known "forces." In this context, the radical concepts of directed panspermia and the impact of viruses on evolution may be important!

INTERSPECIES PHENOMENA

ARE PARASITES REALLY THE MASTERS?

All animals harbor parasites; and some parasites even have their own parasites. The usual effect of a parasite upon its host is debilitation, often to the point of death. But parasites have to reproduce, and some settle for the modification of their hosts in ways that improve their chances. Parasites can change the size, color, and even the behavior of their host. The object is usually to encourage a specific predator to eat the host so the parasite can continue its life cycle. A classic example is the lancet fluke which infests ants and then sheep. The problem is that sheep don't normally eat ants, giving the flukes a chance to switch vehicles. So, the innovative flukes somehow force the ants to crawl to the tops of plants and lock themselves there with their jaws. The next hungry sheep that comes along has his meal seasoned with ants.

The bulk of the present article deals with thorny-headed worms, which are not as endearing as the lancet flukes. These parasites are merely bags of reproductive organs attached to a thorny probiscus, by which they attach themselves to the intestinal walls of vertebrates. Living in a sea of processed nutrients, the worms don't even have a digestive tract. Part of the life cycle of this parasite is spent in arthropods (insects, crustaceans).

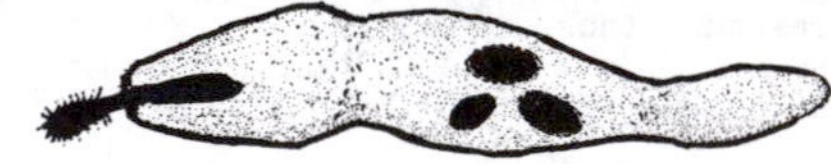

A thorny-headed worm that cycles between ducks and crustaceans. (Adapted from Scientific American)

As with the lancet fluke, the thorny-headed worm's big challenge is getting the arthropod eaten by a vertebrate. In most instances, it alters the behavior of the arthropod in a way that makes it more conspicuous to the predators. For example, infested pill bugs do not hide from birds, as they normally do, and are snapped up. Infested crustaceans move towards the light where ducks consume them. No one knows how a parasite floating in the body cavity of its host can control the host's behavior. (Moore, Janice; "Parasites That Change the Behavior of Their Host," Scientific American, 250:108, May 1984.)

PARASITES CONTROL SNAIL BEHAVIOR

A species of estuarine snail bearing the larvae of the trematode parasite Gynaecotyla adunca behaves differently from what it does when not infected. It lets itself become stranded high on beaches and sandbars, where it becomes easy prey to crustaceans living in this region. These crustaceans serve as the parasite's next host. Somehow, the parasite is able to modify the snail's behavior in a way that enhances its own chances for success. The question, as always in such cases, is how? And if it is a chemically induced change in behavior, how did it evolve? (Curtis, Lawrence A.; "Vertical Distribution of an Estuarine Snail Altered by a Parasite," Science, 235:1509, 1987.)

Comment. One cannot but wonder if human behavior is somehow controlled by parasites. Obviously we deny such dominance. Yet, some have speculated that our urge for space travel is only DNA's way of expanding its dominion.

PLANTS AND FUNGI

EXTERNAL APPEARANCE

Except for the final item on "humongous organisms," this section dwells on mimicry in plants and fungi. Not only do plants mimic one another, as with some Australian mistletoes, but they may try to look like insects or even produce fake insect eggs! One species of fungus even propagates itself by forcing blueberry plants to generate fake flowers! As we shall see, evolutionists have come up with explanations for many instances of mimicry; but they often stretch credulity. And, as stressed often in this book, we have selected only a few from a huge reservoir of examples.

PLANT MIMICRY AND EVOLUTION

In the September issue of Scientific American, S.C.H. Barrett presents an excellent review of mimicry in the plant world. All sorts of wondrous mimicry are described, involving form, color, odor, texture and even synchrony of life cycles. Plants mimic insects, stones, other plants, and substrates (backgrounds). Repeatedly, Barrett asserts that all of these remarkable developments are the consequence of small, randome mutations guided by the forces of natural selection. To Barrett, plant mimicry is proof positive that evolution is true. It should not surprise the readers of Science Frontiers that this very same article is a goldmine of biological anomalies, that is, data that seem to challenge ruling paradigms. (Barrett, Spencer C.H.; "Mimicry in Plants," Scientific American, 257:76, September 1987.)

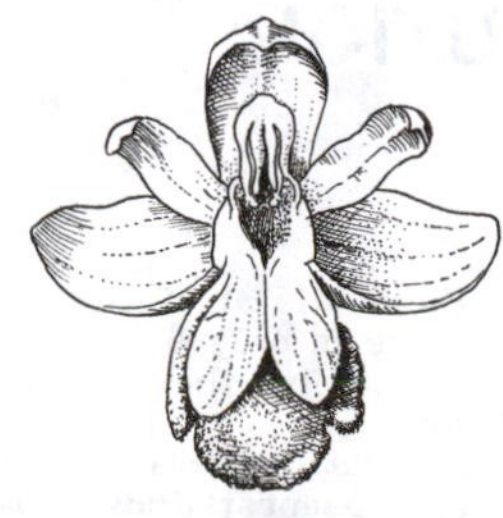

The orchid flower on the right mimics a a female wasp (left) and induces pseudocopulation by male wasps that subsequently transfer its pollen.

Comment. Evolution, like beauty, must be in the eye of the beholder! At this point, we could easily launch into a lengthy harangue about why it seems highly improbable that a plant, through chance mutations, could hit upon just the right combination of form, color, odor, and flowering time to dupe an insect pollinator---even with the aid of natural selection and a billion years. The point we wish to stress here is that the author of this paper sees the same facts and comes to diametrically opposite conclusions!

PLANTS MANUFACTURE FAKE INSECT EGGS

Plants are usually considered rather passive to environmental forces, but careful observation show that they fight back against predators in subtle ways. Williams and Gilbert, for example, have found that a number of Passiflora species, which are heavily defoliated by the larvae of Heliconius butterflies, have developed tiny structures that closely resemble in size, shape, and color the eggs of these butterflies. Heliconius butterflies, when searching for likely plants on which to lay eggs, tend to avoid plants that already have eggs on them. The plants' fake eggs, then, help protect the plant from predation. (Wiliams, Kathy S., and Gilbert, Lawrence E.; "Insects as Selective Agents on Plant Vegetative Morphology....."; Science, 212:467, 1981.)

Comment. We have heard over and over again about Nature's "marvelous adaptations," but it is still difficult to imagine chance-driven evolution of fake eggs of just the right size, shape, and color. How many shapes and colors were tried before the plants got it right?

FUNGUS MANUFACTURES PHONY BLUEBERRY FLOWERS

Mummy-berry disease is a fungus that preys on blueberries. It propagates itself by turning blueberry leaves into whitish, bell-like structures resembling true blueberry flowers. Bees deceived by this ruse land on the fake blossoms, pause for a moment to sip a sugary fluid (fortuitously) exuding from lesions on the leaves, accidentally pick up some fungus spores, and then fly off to true blueberry blossoms. The transferred spores infect other blueberry plants, causing them to produce white mummyberries rather than blueberries. When spring comes round, the fungus-filled mummy-berries release the fungus to the leaves, and the cycle continues. (Anonymous; "A Fungus That Courts with Phony Flowers," Science-85, 6:10, September 1985.)

Comment. The explanations usually served up for such remarkable adaptations are: (1) They are products of chance and natural selection; and (2) The Creator made things this way. Are there not other possibilities? Perhaps the fungus somehow stole the blueprints for the flower from the blueberry's genome; i.e., genetic endowment. After all, viruses are always subverting cell machinery.

AUSTRALIAN MISTLETOES MIMIC THEIR HOSTS

Many species of Australian mistletoes closely mimic their hosts in leaf form and general appearance, blending deceptively into the host's foliage. Plant mimicry for purposes of protection (for example, stone plants) and for propagation are well known and, in the logic of evolutionists, have evolved because of the advantages conferred on the species. Since the Australian mistletoes are evidently highly palatable to arboreal marsupials, the tenets of evolution hold that it is only natural that these mistletoes should develop so as to resemble their hosts for purposes of protection. This is called cryptic or camouflage mimicry. Australian mistletoes parasitize a wide variety of plants, and it is truly marvelous how they can detect and imitate the leaves and appearances of such a wide variety of hosts. (Barlow, Bryan, and Wiens, Deibert; "Host-Parasite Resemblance in Australian Misletoes," Evolution, 31:69, 1977.)

THE HUMONGOUS ORGANISM CONTEST!

When we read of that 10,000-kilogram fungus discovered in Michigan, it sounded like a good item for Science Frontiers. But then it was described as the "largest and oldest living organism," and we knew that if we waited a couple months this Guinness-like record would be eclipsed. (Superlatives are risky in this business!)

A killer fungus in Washington State. In the foothills of Mount Adams, a specimen of the fungus Armillaria ostoyae covers 1,500 acres and seems to be 400-1,000 years old, as compared to the 38-acre, 1,500-year-old Michigan fungus. Although younger than its Michigan counterpart, the Washington fungus is lethal and can wipe out whole populations of trees. (Anonymous; "The Great Fungus," Nature, 357:179, 1992.) (It has also been reported that a huge, spreading, pathological growth flourishes in Washington, DC!)

Some even more humongous plants.

> A grass clone, Holcus mollis, has been found with a diameter of 900 metres and an age of over 1000 years. A clone of box-huckleberry has been found with a diameter of 2000 metres and an age of 13 000 years. The big granddaddy is, however, an aspen (Populus fremaloides) covering 81 hectares and over 10 000 years old.

(Bullock, James; "Huge Organisms," New Scientist, p. 54, May 30, 1992.)

BEHAVIOR

The word "behavior" usually implies stimulus and response. Certainly plants and fungi are not famous for fast physical reactions, but on occasion they do respond quickly to stimuli; as in the case of some parasitic fungi. The phenomenon of circumnutation, on the other hand, is definitely in the slow-motion category! Slower yet are plant responses that depend upon feedback through the genome.

NATURE'S BALLISTIC MISSILE

Abstract:

> The parasitic fungus Haptoglossa mirabilis infects the rotifer host by means of a gun-shaped attack cell. The anterior end of the cell is elongated to form a barrel; the wall at the mouth is invaginated deep into the cell to form a bore. A walled chamber at the base of the bore houses a complex, missile-like attack apparatus. The projectile is fired from the gun cell at high speed to accomplish initial penetration of the host.

(Robb, E. Jane, and Barron, G.L.; "Nature's Ballistic Missile," Science, 218:1221, 1982.)

GRAVITY AND GOING AROUND IN ELLIPSES

We thought that our readers might like to know that the force of gravity apparently has no significant effect on circumnutation. Now circumnutation is the result of an "impressively ubiquitous mechanism" in all elongating plant organs. More simply, it is the elliptical weaving motion seen in the tips of growing leaves, shoots, flower stalks, branch roots, etc. In a 4- to 5-day-old sunflower seedling, the ellipse traced is 6-8 millimeters long and takes about 110 minutes. The ellipses result from differential growth in the elongating plants. No one knew whether the force of gravity played a role in circumnutation until some sunflower seedlings were flown on Spacelab 1. Zero-g did not affect circumnutation at all. (Brown, Allan H. and Chapman, David K.; "Circumnutation Observed without Significant Gravitational Force in Spaceflight," Science, 225:230, 1984.)

Comment. Nature seldom indulges in frivolous actions, but we just may have a phenomenon here that has absolutely no deeper significance.

THE GENOME'S RESPONSES TO CHALLENGES

The genome is an organism's genetic endowment. It contains instructions for the organism's growth and development, but it is not like a rigid, uncompromising computer program. Rather, the genome "...is a highly sensitive organ of the cell that monitors genomic activities and corrects common errors, senses unusual and unexpected events, and responds to them, often by restructuring the genome. We know about the components of genomes that could be made available for such restructuring. We know nothing, however, about how the cell senses danger and instigates responses to it that often are truly remarkable." Thus Barbara McClintock ends the paper she delivered in Stockholm when she received a Nobel Prize in 1983.

Most of McClintock's paper reviews her pioneering work with the corn genome, but she adds some examples of other genomic responses to external stresses. One such stress is applied to an oak tree when a wasp lays its egg in a leaf. The stress causes the oak genome to reprogram itself and construct a wholly new and unplanned plant structure to house and feed the developing insect. Some of these structures (galls) are very elaborate and are precisely tailored to each different wasp species. From such examples, it is apparent that the genome of an organism somehow perceives stresses and reacts to them---often in completely unanticipated ways. The stresses may be mechanical, thermal, chemical; in fact, almost anything. McClintock's conclusion is "that stress, and the genome's reaction to it may underlie many formations of new species." (McClintock, Barbara; "The Significance of Responses of the Genome to Challenge,: Science, 226:792, 1984.)

Comment. The implications here are broad and deep. Evolution can be driven by external stresses. The new species thus produced may differ substantially from the original organism, eliminating the need to look for "missing links" in the fossil record. What "hopeful monsters" are latent in our human genome, awaiting only the right stresses to manifest themselves? And is the genomes's malleability reversible; that is, can extinct species be recovered when the engendering stresses are removed?

DISTRIBUTION

LIFE IN THE DARK

A most interesting discovery has been made recently in deep ocean waters. Abundant plant life has been found at depths of up to 268 meters, well beyond the 200-meter limit biologists had set based on the availability of sunlight. It wasn't difficult to discount photosynthetic life at 268 meters, because light there is only 0.0005% that at the surface. But there it was; and it may be found even deeper now that we've taken off the blinders. (Littler, Mark M., et al; "Deepest Known Planet Life Discovered on an Uncharted Seamount," Science, 227:57, 1985.)

TALENTS AND FACULTIES

Plants and fungi are not supposed to be sentient organisms; yet, we are all familiar with flowers that open and close with the sun and even follow its path across the sky. Even so, it comes as something of a shock to find that plants can even distinguish colors in their environment! Even more startling is the notion that they can communicate with one another across significant distances via chemical signals.

PLANTS ARE NOT COLOR BLIND!

"Scientists have long known from laboratory experiments that various colors of light can affect stem size, root structure and other aspects of plant growth. But they have remained largely in the dark about the potential practical benefits of the phenomenon.

"Using colored mulch to bathe plants in reflected light of certain hues, the South Carolina group (Clemson University) has begun to explore what colors plants prefer in agricultural growing conditions. Last year, for example, the group found that tomatoes grown with red mulch---made with plastic sheets painted red---had 20% higher yields than those with black mulch. Preliminary results this year show that potatoes and bell peppers grow best with white mulch..." (Anonymous; "Plants' Colors," Wall Street Journal, September 16, 1987. Cr. J. Covey.)

Comment. Many questions arise here, but we'll take only three: (1) How do plants sense colors? (2) How do different colors mediate growth differently? (3) Is all this explicable in terms of evolution?

HOW TREES TALK TO ONE ANOTHER

Trees talk only in children's cartoons---that's currently accepted wisdom. But when trees attacked by caterpillars sound an alarm that other trees in the vicinity detect and heed, some sort of communication system seems required. The evidence is found in trees that react to caterpillar attack or leaf damage by making their leaves harder to digest. When one tree is attacked, not only does it start making less nutritious leaves but so do other trees as far as 200 feet away. No root connections have been found.

In tests with potted maples and poplars inside plexiglass enclosures, the attack warning got through to trees in the same chamber but not to control trees outside the plexiglass. Thus, the warning seems to be transmitted by air ---probably chemically. David Rhoades is conducting further research at the University of Washington. (Boling, Rick; "Tree ESP," Omni, p. 42, December 1982. Cr. P. Gunkel)

Comment. Question: How is the message received, decoded, and turned into biological action? If we could set up chemical "antennas" in the air around us, what other revealing messages would we "hear"?

TREES MAY NOT CONVERSE AFTER ALL!

We reported above how some evidence suggested that trees might communicate with one another in connection with insect attacks. S.V. Fowler and J.H. Lawton contest this conclusion, and they have experimental evidence to back them up. Working with birch trees, they defoliated saplings 5% and 25% and looked for signs of intertree communication. They found none. As for previous claims for this phenomenon, Fowler and Lawton believe that one study was statistically flawed, and the other due to an infectious disease transmitted between caterpillars rather than talking of trees. (Fowler, Simon V., and Lawton, John H.; "Rapidly Induced Defense and Talking Trees: The Devil's Advocate Position," American Naturalist, 126:181, 1985.)

TREES TALK IN W-WAVES

We quote below from as Associated Press dispatch:

> Grants Pass, Ore. (AP)---Physicist Ed Wagner says he has found evidence that trees talk to each other in a language he calls W-waves.
>
> "If you chop into a tree, you can see that adjacent trees put out an electrical pulse," said Wagner. "This indicates that they communicated directly."
>
> Explaining the phenomenon, Wagner pointed to a blip on a strip chart recording of the electrical pulse.
>
> "It put out a tremendous cry of alarm," he said. "The adjacent trees put out smaller ones."
>
> .....
>
> "People have known there was communication between trees for several years, but they've explained it by the chemicals trees produce," Wagner said.
>
> "But I think the real communication is much quicker and more dramatic than that," he said. "These trees know within a few seconds what is happening. This is an automatic response."
>
> Wagner has measured the speed of W-waves at about 3 feet per second through the air.
>
> "They travel much too slowly for electrical waves," he said. "They seem to be an altogether different entity. That's what makes them so intriguing. They don't seem to be electromagnetic waves at all."

(Anonymous; "Physicist Says Blip Proves Trees Talk," Seattle Sun Times, February 12, 1989. Cr. R.L. Simmons)

Comment. In addition to the above discovery, Wagner, who holds a PhD in physics from the University of Tennessee, has detected electrical standing waves in trees. The voltage measured by electrodes implanted in trees goes up and down as one goes higher and higher up the trees. Wagner's work has been published in Northwest Science, but we have not yet seen it. Incidentally, electricity does seem to affect plant growth, as described in Incredible Life.

KILLER TREES THAT TALK AMONG THEMSELVES

> Acacia trees pass on an "alarm signal" to other trees when antelope browse on their leaves, according to a zoologist from Pretoria University. Wouter Van Hoven says that acacias nibbled by antelope produce leaf tannin in quantities lethal to the browsers, and emit ethylene into the air which can travel up to 50 yards. The ethylene warns other trees of impending danger, which then step up their own production of leaf tannin within just five to ten minutes.

(Hughes, Sylvia; "Antelope Activate the Acacia's Alarm System," New Scientist, p. 19, September 29, 1990.)

Free-ranging antelope can always find unwarned acacias with low levels of tannin to browse on, but on some South African game ranges, they are forced to consume high tannin leaves. Too much tannin inactivates the antelopes' liver enzymes, and they die in about two weeks. Hundreds may perish in a very dry season. (See also: Vincent, Catherine; "Les Arbres Communiquent entre Eux," Le Monde, p. 11, September 14, 1990. Cr. C. Maugé)

BODILY FUNCTIONS

Expanding on the theme of sentient plants, we now present items showing how plants and fungi have somehow "learned" to use effectively light and electricity. Furthermore, plants no longer can be considered to be "cold-blooded," because they can actually generate remarkable quantities of heat as they metabolize their food. Thus, do we dispose of that old nostrum that plants and fungi and very low on the Tree of Life!

LIGHT PIPES IN PLANTS

Plants, it seems, developed light pipes long before humans. Certain plant tissues (etiolated or dark-grown) act as multiple bundles of optical fibers and coherently transfer light over distances of at least 2 cm. Optical tests show that these natural light pipes are much more effective transmitters of light than media that simply scatter light. This unsuspected sophistication of Nature's design may require significant revisions in photobiology, which did not allow for such ingenuity. (Smith, Harry; "Light-Piping by Plant Tissues," Nature, 298: 423, 1982.)

Comment. Since some plants are known to emit light (see our Incredible Life), we would not be surprised, the way things are going, to learn of natural plant lasers!

ELECTRIC-POWER PLANTS!

Referring to a previous Science News item (132:53, 1987) on the electrostatic dispersal of fungal spores, A.F. Kah describes another use of electricity by plants:

> In a similar manner, the same-charged fluffy fibers of milkweed (and presumably other fuzzy seeds) spring apart from electrostatic repulsion when the fibers have dried out. This explosive fiber spreading at the right moment is beautiful and fascinating to watch, and is certainly effective in getting them airborne!

(Kah, Ann F.; "Fluffy Explosion," Science News, 132:163, 1987.)

Comment. See below for an item on heat production in plants. It must have been a serendipitous series of tiny random mutations that led to this electrostatic phenomenon. Of course, we can say the same for electric catfish, too.

THE CURRENTS OF LIFE

Danny Brower and Richard McIntosh of the University of Colorado at Boulder have discovered that growing cells apparently generate electrical fields that control the shapes of living organisms. They have been experimenting with a disc-shaped alga with a lobed edge. Normally the alga reproduces by splitting in half, with each half regenerating the lost half. Nicely symmetric discs are manufactured. But if an external electrical field (about 14 volts/cm) is applied across the nutrient medium, the regeneration geometry is distorted. The experimenters surmise that the membrane chemistry is affected by the external field which augments or reduces cell-created electric fields. (Anonymous; "Electric Charges May Shape Living Tissue," New Scientist, 86:245, 1980.)

Comment. Natural external electric fields, such as the atmospheric potential gradient, may therefore have some biological effects, as some experiments with electricity and plant growth have proven.

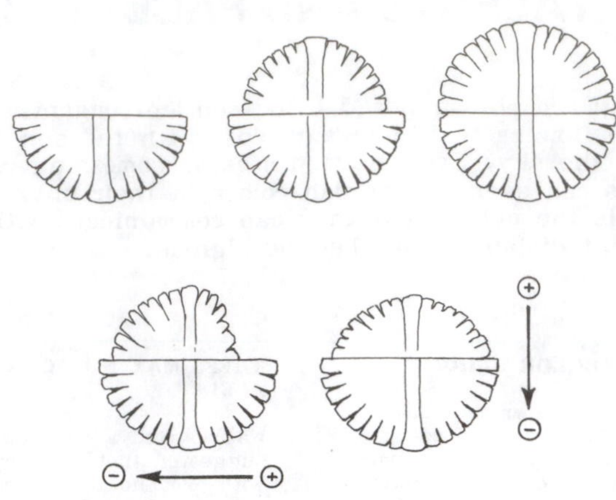

(Top) Disc-shaped alga normally split and regenerate symmetrical discs. Bottom) The presence of electrical fields distorts the process.

HOT PLANTS

You've heard of hot potatoes, but they aren't naturally hot. However, in the early spring skunk cabbages are and so are some philodendrons during their flowering periods. In fact, some philodendrons burn fat to generate their heat, just like animals. Metabolism based on fats allows some philodendrons to reach temperatures of 124°F. In terms of their rates of metabolism, they rival those of hummingbirds. Furthermore, philodendrons are able to regulate their chemical fires, whereas skunk cabbages, which burn only starch, consume all their stored energy like a rocket in one snow-melting crescendo.

Why do plants generate heat? Apparently to attract pollinating insects. The hot skunk cabbage poking through the snow is the only food in sight for early spring insects, while the philodendrons may attract pollinating insects who like to bask or mate in warm places. (Blakeslee, Sandra; New York Times, August 9, 1983, p. C4. Cr. P. Gunkel)

Comment. Are plants really "lower" forms of life, seeing how successful and well-tuned to their environments they are?

GENETICS

PROMISCUOUS DNA

The cells of plants (photosynthetic eukaryotes) are genetically the most complex that biologists have discovered. Each cell has three genetic systems: its own, that of the chloroplasts; and that of the mitochondria. It is supposed that the chloroplasts and mitochondria were once free-living cells that linked up with the embryonic plant cell to form a symbiotic partnership, with the host "plant" cell being the dominant member.

Up until now, the three genetic systems were thought to be discrete, each going down its own pathway. But chloroplast genes have now been found inside plant mitochondria, overturning conventional wisdom. To sum it all up, DNA seems promiscuous---no respecter of privacy and breaking down all isolating genetic barriers.

This discovery at once raises a dozen questions. For example, are mitochondria genes in chloroplast cells? How far does this promiscuity go? Can the same thing happen in higher organisms; say, with humans and symbiotic microorganisms or even not-so-symbiotic disease organisms? Is there no stopping this DNA? (Ellis, John; "Promiscuous DNA---Chloroplast Genes inside Plant Mitochondria," Nature, 299:678, 1982.)

PLANT PALEONTOLOGY

TERRESTRIAL LIFE OLDER THAN EXPECTED

The oldest fossils found so far are dated at 3.5 billion years ago. Discovered in Australia, these life forms seem to have been capable of synthesizing their own food, much like present-day plants. In other words, these life forms were not so primitive after all. If we are to assume that still simpler life and pre-life chemical evolution preceded these fossils, the advent of life on earth must be pushed back much farther to allow time for gestation. (G., L.: "Oldest Life-Forms May Have Made Own Food," Eos, 61:578, 1980.)

Comment. Another possibility is that life did not originate on earth and was instead brought here on comets, meteorites, etc.

PRETERNATURALLY RAPID DEVELOPMENT OF PHOTOSYNTHESIS?

Abstract:

> An increased ratio of ^{12}C to ^{13}C, an indicator of the principal carbon-fixing reaction of photosynthesis, is found in sedimentary organic matter dating back to almost four thousand million years ago---a sign of prolific microbial life not long after the Earth's formation. Partial biological control of the terrestrial carbon cycle must have been established very early and was in full operation when the oldest sediments were formed.

(Schidlowski, Manfred; "A 3,800-Million-Year Isotopic Record of Life from Carbon in Sedimentary Rocks," Nature, 333: 313, 1988.)

Comment. Photosynthesis is not a simple biological process. To discover that it and life forms using it developed so quickly on the primitive earth is surprising. Did this complexity and "biological control" arise so quickly: (1) by chance; (2) by inoculation from extraterrestrial sources; (3) by act of God; or (4) by Gaia in nascent form? (Note that Gaian overtones emanate above from the words "Biological control.")

THE ARCTIC WOMB

Abstract:

> Magnetostratigraphic correlation of Eureka Sound Formation in the Canadian High Arctic reveals profound difference between the time of appearance of fossil land plants and vertebrates in the Arctic and in mid-northern latitudes. Latest Cretaceous plant fossils in the Arctic predate mid-latitude occurrences by as much as 18 million years, while typical Eocene vertebrate fossils appear some 2 to 4 million years early.

(Hickey, Leo J., et al; "Arctic Terrestrial Biota: Paleomagnetic Evidence of Age Disparity with Mid-Northern Latitudes During the Late Cretaceous and Early Tertiary," Science, 221:1153, 1983.)

Comment. The anomaly here is in the vision of the high Arctic lands basking in the warm sun busily evolving new life forms well in advance of their appearance in lands closer to the Equator. What happened to the earth's axial tilt. These fecund polar territories should have been engulfed in darkness almost half of the year---hardly an environment for precocious plant evolution. Further, trees found buried in the Arctic muck could never have grown where found due to the long polar darkness.

INTERSPECIES PHENOMENA

DON'T PET YOUR HOUSE PLANTS!

Researchers at Stanford's Medical Center were spraying Arabidopsis plants (in the mustard family) with hormones to see if they could trigger any of the plants' genes. They could, and the treated plants grew up stunted. But, it was serendipitously discovered, the same genes could also be triggered by spraying with water, by gusts of wind, and even by the human touch. Evidently, some of the genes in these plants can be turned on by various environmental stimuli, and thus affect future plant development. This mechanism perhaps explains why trees along the seacoast and timberline are stunted. (Crawford, Mark H., ed.; "Nolo Me Tangere," Science, 247:1036, 1990.)

Comment. One is tempted to ask how widespread this phenomenon is in biology. Are humans, for example, born with a console of gene-buttons that the environment can push---as in cancer? Or, even in evolution itself?

UNDERGROUND ORCHIDS

Australia boasts many peculiar animals and plants. One of these is an orchid that grows underground. Obviously this orchid cannot employ photosynthesis. Rather, it grows in conjunction with a fungus that obtains nutrients from surrounding roots. The fungal threads penetrate both roots and orchids. The orchids make ends meet by systematically killing and digesting the nutrient-laden fungal threads. When the orchid flowers, it pushes toward the surface just enough to open some tiny cracks in the earth. In these cracks, still below the surface, appear the tiny burgundy red flowers. These flowers are pollinated by minute flies, but just how the orchid's seeds are dispersed is still a mystery. (Cooke, John; "Hidden Assets," Natural History, 93:75, October 1984.)

PARASITES MAY REPROGRAM HOST'S CELL

The long, segmented filament shown in the illustration consists of the cells of a parasite that preys on the cells of red algae. Two such cells abut the parasitic filament. The small black circles are parasitic cell nuclei which, when confronted with a red alga cell, become wrapped in small "conjuctor cells" which are then somehow transferred to the host cell on the right. The actual transfer involves the formation on the host cell wall of a sort of dimple called a "secondary pit connection." (Why this forms is not known.) Once inside the host cell, the parasite nucleus and/or the cytoplasm transferred with it dramatically reprograms host cell operations. The host cell shifts into high-gear food production, enlarging up to twenty times its normal size. The host cell wall thickens, its nuclei (large black circles) and chloroplasts multiply. Adjacent cells (left) remain unaffected. (Lewin, Roger; "New Regulatory Mechanism of Parasitism," Science, 226:427, 1984.)

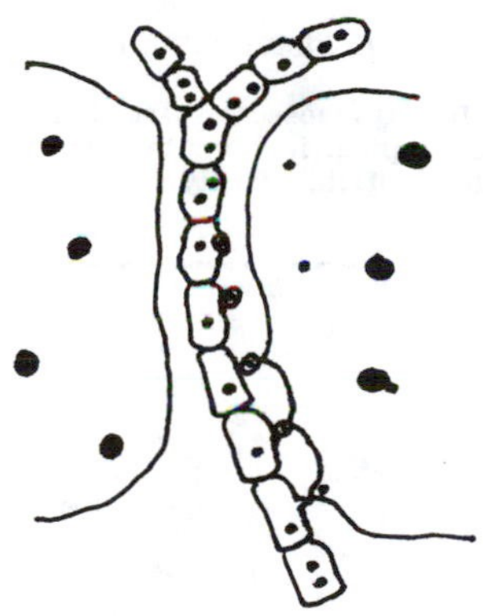

A parasitic filament growing between host cells, illustrating how parasite nuclei (black spots) are first enclosed in bud-like conjunctor cells and then inserted into the host cell.

Comment. It is one thing for external stresses to stimulate cell reprogramming, but quite another when another species inserts its programs into those of the host. Is this another way of producing "hopeful monsters?"

KILLER FUNGI CAST STICKY NETS

Your garden soil likely contains nematodes (popularly called eelworms) that will gnaw away at your crops. Nematodes are about a millimeter long and very active, thrashing through the soil like fish through water. Their numbers are kept in check by a surprisingly sophisticated fungus which thrives on them. If nematodes are around (not otherwise), the fungus sets out two kinds of traps. The first is the sticky net made of threads sent out by the fungus. Any nematode that brushes against these sticky strands is held while the fungus rams special feeding pegs into it. The second kind of trap is even more marvelous. It is an array of rings, each consisting of three unique cells that are sensitive to touch. Attracted by alluring chemicals secreted by the fungus, the nematodes probe around the rings. In a tenth of a second after they are touched, the fungus rings contract around the interloping nematodes. Again the nematode is doomed as the terrible feeding pegs penetrate its body. Another chemical is then released by the fungus to keep other fungi away from its kill. (Simons, Paul; "The World of the Killer Fungi," New Scientist, 20, March 1, 1984.)

Comment. Does anyone really believe that even the "simplest" form of life is really simple?

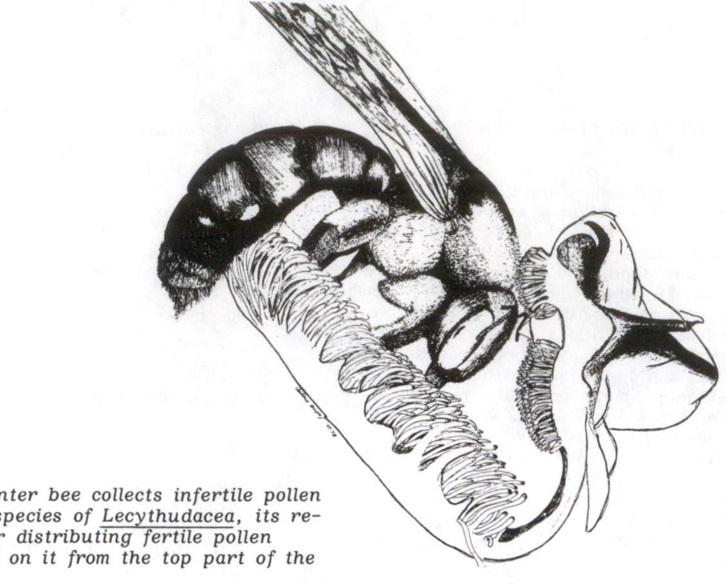

A carpenter bee collects infertile pollen from a species of Lecythudacea, its reward for distributing fertile pollen brushed on it from the top part of the flower.

OH, THOSE CLEVER PLANTS

The Lecythidaceae or Brazil nut family is pantropical in distribution. Some members of this family produce two different kinds of pollen: (1) normal pollen for fertilization; and (2) nongerminating pollen that is collected by insects for food. The latter variety of pollen is considered (anthropomorphically) as the plant's way of rewarding insects for carrying the fertile pollen to other plants. As in so many of Nature's remarkable adaptations, the two types of pollen are located in exactly the right portions of the flower to match the anatomy of the foraging insect. In the figure, a carpenter bee collects infertile pollen from the bottom of the flower while being dusted on the head and back by the regions of the fertile pollen. (Mori, Scott Alan, et al; "Intrafloral Pollen Differentiation in the New World.. ..." Science, 209:400, 1980.)

Comment. How can the flower, even over many generations, determine that only the pollen from the upper portion is being used for fertilization and that the lower area of pollen may safely "be allowed" to become infertile?

MICROORGANISMS

CELLS

This short section is devoted to the possible anomalous behavior of individual cells in or from multicellular organisms. Not only do some of these cells possess some autonomy of movement; but, as they acquire malignant tendencies during the onset of cancer, they are endowed with seemingly purposeful behavior inimical to their parent organism.

Bacteria, which are independent unicellular organisms, will be discussed in the following section.

BLEBS AND RUFFLES

Single cells taken from multicellular organisms tend to inch along like independent amoebas---almost as if they were looking for companionship or trying to fulfill some destiny. This surprising volition of isolated cells becomes an even more remarkable property when the individual cells are fragmented. Guenter Albrecht-Buehler, at Cold Spring Harbor Laboratory, has found that even tiny cell fragments, perhaps just a couple percent of the whole cell, will tend to move about. They develop blebs (bubbles) or ruffles and extend questing filopodia. They have all the migratory urges of the single cells but cannot pull it off. Cell fragments will bleb or ruffle, but not both. Why? Where are they trying to go? (Anonymous; "The Blebs and Ruffles of Cellular Fortune," New Scientist, 90:87, 1981.)

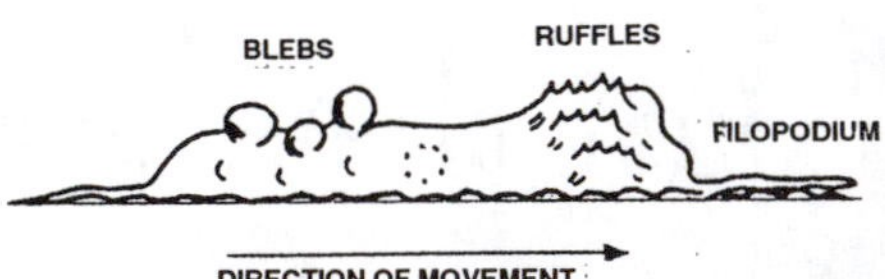

Blebs and ruffles on a cell in motion.

GUIDING CELL MIGRATION

"Cell movement has always been thought to be independent of the extracellular matrix which encourages and guides movement; the cells were thought to move along the pathways under their own power, rather as a train moves along a railway track. New research by Stuart Newman, Dorothy Frenz and James Thomsack, of the New York Medical College, now suggests that the extracellular matrix itself may help to propel cells along." (Experiment details omitted here.) "No one knows whether this matrix-driven movement actually occurs in living organisms. But potentially, it could achieve the high-speed movement of cells depending on the size and characteristics of the cells. Considering that many of the events of development occur remarkably rapidly, this is an attractive possibility for cell migration." (Anonymous; "More Clues to Cell Movement," New Scientist, p. 30, September 5, 1985.)

Comment. This phenomenon is not trivial, for it implies that non-living substances (the extracellular matrix) are not necessarily passive but might even be cooperative, if such an adjective can be applied to the non-living.

WHY CANCER?

> However, the most formidable obstacle to the successful treatment of disseminated cancer may well be the fact that the cells of a tumor are biologically heterogeneous. This phenotypic diversity, which allows selected variants to develop from the primary tumor, means not only that primary tumors and metastases can differ in their responses to treatment but also that individual metastases differ from one another. This diversity can be generated rapidly even when the tumors originate from a single transformed cell.

(Fidler, Isaiah J., and Hart, Ian R.; "Biological Diversity in Metastatic Neoplasma: Origins and Implications," Science, 217:998, 1982.)

Comment. The ability of single cancer cells to multiply into different kinds of cells, as well as propagate throughout an organism, seems to betoken an insidious biological entity, whose origin and purpose (?) we have hardly begun to comprehend. How could cancer have evolved if it leaves no progeny? How could natural selection leave us so susceptible to cancer?

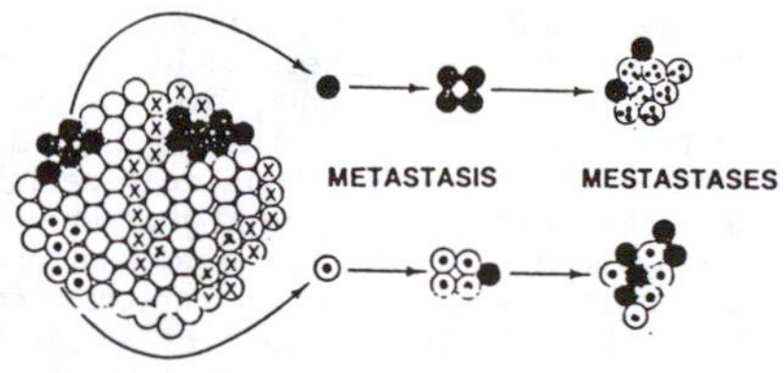

During cancer metastasis, some of the cells in the primary cancer break away. As indicated by the differently coded circles, these metastases may differ among themselves.

CANCER EVEN MORE INSIDIOUS

A recent advance in cancer research has been the discovery that the development of some cancers is initiated by oncogenes---genes which have been "switched on" by biological or environmental forces. That humans and other organisms harbor such Trojan Horses is unsettling enough, but it now seems that the development of cancer may require the stepwise cooperation of several different oncogenes. In other words, one oncogene controls the action of another and so on in cascade until the cancer is finally initiated. (Marx, Jean L.; "Cooperation between Oncogenes," Science, 222:602, 1983.)

Comment. How and why would such a complex mechanism, without obvious short-term survival value, ever have developed? See "Cancer: The Price for Higher Life," on p. 117.

HOW CANCERS FIGHT CHEMOTHERAPY

How do cancer cells develop resistance to lethal chemicals? The clues seem to reside in extrachromosomal DNA that carries drug-resistance conferring genes from once cancer cell to another. Cancer cells dying from chemotherapy may, for example, cast off extrachromosomal DNA with information on how to combat the chemicals. Other factors may also be at work, but basically we have only suspicions. (Silberner, Joanne; "Resisting Cancer Chemotherapy," Science News, 131:12, 1987.)

Comment. Insects and other organisms also acquire resistance to chemical poisons. Does extrachromosomal DNA play a role in these instances, too? Can information encoded in extrachromosomal DNA be passed from one species to another, say via insect bites?

KAMIKAZE SPERM

Sperm is popularly thought to have but a single purpose---fertilization of the egg. This is not so!

> Nonfertilizing sperm with special morphologies have long been known to exist in invertebrates. Until recently, abnormal sperm in mammals were considered errors in production. Now, however, Baker and Bellis have proposed that mammalian sperm, like some invertebrate sperm are polymorphic and adapted to a variety of nonfertilizing roles in sperm competition, including prevention of passage of sperm inseminated by another male. More specifically, their "kamikaze" hypothesis proposes that deformed mammalian sperm are adapted to facilitate the formation and functioning of copulatory plugs.

The author of the present paper, A. H. Harcourt, thinks that although some 20% of mammalian sperm, on the average, is abnormal (two heads, no heads, two tails, no tails, coiled tails, etc.) such sperm represents only errors on the assembly line. These abnormal sperm have no special purpose, at least in mammals. (Harcourt, A.H.; "Sperm Competition and the Evolution of Nonfertilizing Sperm in Mammals," Evolution, 45:314, 1991.)

Comment. Even if mammals haven't yet developed kamikaze sperm, some animals have; and one must wonder exactly how multipurpose sperm (and ova, too) evolved. For a copulatory plug to be effective, large numbers of mutant sperm with special plugging properties and knowing just where to go would have to be generated all at once. That would seem to require a lot of chance mutations all at once! The "perfection" problem once again.

THINGS THAT AIN'T SO

Back in 1953, Irving Langmuir, the famous American physicist, gave a talk at General Electric on the subject of "pathological science." He discussed several things that could not be so. Of course, UFOs were on the list, but so was mitogenetic radiation---radiation supposedly emitted by cells when they divide. In the context of Langmuir's talk, despite his emphatic interment of the subject, mitogentic radiation was resurrected recently in the pages of Physics Today. A letter from V.B. Shirley included several recent scientific references that suggest that Langmuir was premature.

> The gist of these articles is that many cell systems emit ultraviolet light during or immediately before cell division and that the total effect of this emission on neighboring cells is still not known.

(Shirley, Vestel B.; "Mitogenetic Radiation: Pathology or Biology?" Physics Today, 43:130, October 1990.)

Comment. This is still another example of the scientific tendency to "close the book" on topics that are too far off the mainstream. Langmuir did mention UFOs, too; will we ever see them resurrected?

BACTERIA

That bacteria are everywhere from boiling springs to human stomachs is widely recognized. It is not as well appreciated that bacteria are found at very great depths inside the earth and, apparently, in the deepest regions of space, too. In fact, the following list of subheadings not only reiterates the ubiquity of bacteria but suggests that they are unexpectedly influential in geological and biological processes.

•Behavior phenomena. Collective action, "thinking."

•Geological influence and presence at great depths.

•Cosmic presence. Survivability in space, biological evolution, panspermia.

•"Directed" evolution. Do bacteria influence their own evolution?

BEHAVIOR PHENOMENA

BACTERIA

BACTERIA: "widely distributed group of microscopic, one-celled vegetable organisms..." One dictionary's definition.

As a matter of fact, the nearly universal image of a bacterium is that of a simple, single-cell organism. But:

> That view is now being challenged. Investigators are finding that in many ways an individual bacterium is more analogous to a component cell of a multicellular organism than it is to a free-living, autonomous organism. Bacteria form complex communities, hunt prey in groups and secrete chemical trails for the directed movement of thousands of individuals.

J.A. Shapiro, author of the preceding quote, attributes the simplistic picture of bacteria to medical bacteriology, in which disease-causing bacteria are classically identified by isolating single cells, growing cultures from them, and then showing that they cause the disease in question. In the microscopic real world, bacteria virtually always live in colonies, which possess collective properties quite different and much more impressive than those of the single-cell-in-a-dish! That old human urge for reductionism has led us astray again.

Shapiro seeks to remove the blinders of reductionism in a wonderful article in the June, 1988, issue of Scientific American. We have room here to mention only the Myxobacteria, many of which never exist as single cells in nature. Even those that do are "social" in the sense that, when two cells meet, they align themselves side by side and go through ritual motions that seem foreign to such "simple" organisms! (Where is this "dance" encoded in the single cells? Do they have 'memories'? WRC)

Movements within colonies of Myxobacteria are highly coordinated. "Trails of extracellular slime are secreted and serve as highways for the directed movement of thousands of cells, rhythmic waves pulse through the entire population, streams of bacteria move to and from the center and edges of a spreading colony, and bacteria aggregate at specific places within the colony to construct cysts or, in some species, to form elaborate fruiting bodies." The Myxobacteria also collectively form bag-like traps to engulf and digest prey. It is apparent now that as simple as a single bacterium may seem, bacterial colonies are pretty complex. (Shapiro, James A.; "Bacteria as Multicellular Organisms," Scientific American, 258:82, June 1988.)

Comment. How are the collective actions of bacteria effected? Is there a central control center? Oh, oh! We are falling into the reductionist trap again! Life forms do not have to operate like computers do, nor do their functions have to be defined in those block diagrams we like so well. There may be holistic forces involved that escape reductionist thought. It is all very remarkable that bacteria which have been around for billions of years (See article below.) should have acquired such sophistication so quickly, waiting only for the earth to cool a bit.

IF BACTERIA DON'T THINK, NEITHER DO WE

As one goes step-by-step down the ladder of biological complexity, one discovers that flatworms, plants, and even bacteria display purposeful behavior. Bacteria, which are usually regarded as fairly inert when it comes to responding to environmental pressures, actually react in different ways to dozens of different stimuli. This ability at the very least requires sensory equipment, internal clocks, a memory, and a decision-making capability.

If bacterial activity is all preprogrammed (the reductionist view), are not humans also preprogrammed? Human programs are larger and more complex, of course, but still devoid of "thinking." Conversely, if humans really do think; that is, transcend preprogramming (free will, if you wish), then bacteria must also think. The third possibility is that at some step in the ladder of life, "higher" life forms begin to think. There is little evidence that life is split so profoundly between thinkers and non-thinkers. (Morowitz, Harold J.; "Do Bacteria Think?" Psychology Today, 15:10, February 1981.)

Comment. This ancient controversy about determinism has been revived as:

1. Simple life forms have been found to be not-so-simple;
2. All life seems unified by a single (or small number of) genetic codes; and
3. "Higher" life forms seem more and more to be just composites of simpler, cooperating biological entities.

GEOLOGICAL INFLUENCE AND PRESENCE AT GREAT DEPTHS

EARLY LIFE AND MAGNETISM

The tiny granules of magnetite found in magnetized sediments come in various crystalline forms. Inorganic magnetite precipitated from molten rock is octahedral, while the particles manufactured by bacteria are cubes, hexagonal prisms, or noncrystalline teardrops. The magnetite found in marine sediments appears to be organically formed---at least the shapes of the particles are characteristic of bacterial manufacture. Apparently these industrious bacteria have been busy producing magnetite ever since "lowly" life forms appeared in the Precambrian.

These facts pose at least four questions:

1. How much of the earth's iron ore has been concentrated biologically and is there a connection with the Gaia Hypothesis?
2. Is it possible that magnetic field reversals, now believed to be of purely geophysical origin, might be biological artifacts (that is, due to population and/or species changes of magnetic bacteria)?
3. If magnetic field reversals are of geophysical origin, how do the magnetic bacteria find their food sources during the long periods of near-zero field?
4. Lab experiments prove that magnetic bacteria require free oxygen to secrete magnetite, but the Precambrian atmosphere and oceans were supposedly devoid of oxygen until 2.3 billion years ago. How did the magnetic bacteria prosper before then?

(Simon, C.; "Tiniest Fossils May Record Magnetic Field," Science News, 124:308, 1983.)

Crystal engineering. Left to themselves, molecules of calcium carbonate tend to crystallize into neat rhombohedrons. But when a sea urchin gets hold of the same molecules, its biological machinery coaxes them into crystallizing into long spines, complete with pores and curved edges. X-ray diffraction patterns prove that the spines are all one crystal. In like fashion, bacteria mold miniature single-crystal bar magnets for navigational purposes. Many animals indulge in crystal engineering. Now if we can only train organisms to build crystalline electronic devices for us. Biogenic chips? (Mann, Stephen; "Crystal Engineering: The Natural Way," New Scientist, p. 42, March 10, 1990.)

MAGNETIC BACTERIA IN THE SOIL AND WHO KNOWS WHERE ELSE?

It is already well-established that salt- and fresh-water sediments harbor bacteria that synthesize grains of magnetite ---presumably for the purpose of sensing the ambient magnetic field and orienting themselves. Similar bacteria have recently been discovered living in ordinary soil in Bavaria. It is near-certain that they will now be found just about everywhere. J.W.E. Fassbinder et al, who reported the Bavarian bacteria, conclude their Abstract with: "We suggest that the magnetic bacteria and their magnetofossils can contribute to the magnetic properties of soils." (Fassbinder, Jorg W.E., et al; "Occurrence of Magnetic Bacteria in Soil." Nature, 343:161, 1990.)

Comment. It is easy to reach great heights of speculation given the facts that: (1) magnetic bacteria exist; (2) bacteria in general are exceedingly abundant; and (3) bacteria are found deep inside the earth's crust and, seemingly, just about anywhere one cares to look. Now, let's see how ridiculous one can get:

1. Magnetic bacteria and/or their fossils contribute heavily to the magnetic properties of sedimentary rocks and unlithified sediments, such as deep-sea sediments. In fact, magnetostratigraphy and paleomagnetism in general may be based upon bioartifacts and be suspect.
2. Magnetic bacteria and/or their fossils are present in such immense numbers deep in the crust that they contribute significantly to the earth's magnetic field. They "might" even be responsible for most of it, including its historical behavior.
3. Magnetic bacteria, as agents of Gaia, actually constructed the earth's magnetic field for the specific purpose of erecting a shield against space radiation, and thereby allowing the development of more complex life forms on the planet's surface.

Imagine the consequences if any one of the above speculations is even close to the mark!

LIFE BEYOND 100°C

Bacteria can survive and multiply in hot springs near and slightly above 100°C---the boiling point of water at atmospheric pressure. Few scientists have contemplated the possibility of life forms prospering at temperatures well beyond 100°. Recently, however, the discovery of many new and frequently bizarre organisms clustered around deep-sea vents has forced a reexamination of high-temperature life. It seems that bacteria actually flourish in the 350°C water streams from the deep-sea vents. In the lab, these same bacteria multiply rapidly in water at 250°C kept liquid by pressures of 265 atmospheres of pressure. What a surprise! Quoting a concluding sentence from this article: "This greatly increases the number of environments and conditions both on Earth and elsewhere in the Universe where life can exist." (Baross, John A., and Deming, Jody W.; "Growth of 'Black Smoker' Bacteria at Temperatures of at Least 250°C," Nature, 303:423, 1983.)

Comment. Ignoring for the moment the extraterrestrial possibilities, the earth is riddled like a Swiss cheese with hot, fluid environments, which we may now consider potential abodes of life. Subterranean life represents a new biological frontier. Who knows what kinds of organisms have developed to feed upon the planet's heat? Could they have contributed to our supplies of petroleum and natural gas?

EXPLORING THE SUBERRANEAN WORLD OF LIFE

Examining fluid inclusions in hydrothermal quartz crystals obtained from a drill hole in Yellowstone National Park, K.E. Barger and his colleagues from the U.S. Geological Survey noted many rodlike and threadlike particles that closely resembled bacteria. Although these particles move, as if alive, they are only in Brownian motion. But even in death, they tell us that life forms can prosper deep underground at very high pressures and temperatures. The crystals that ultimately grew around the fluid particles came from fractures in Pleistocene rhyolite hundreds of feet below the surface. The authors concluded:

> Thermophilic microorganisms may hold the key to an understanding of several biological and geochemical processes, including the origin of life. The discovery of possible microorganisms in these fluid inclusions from the Yellowstone volcanic area enlarges the range of potential environments over which subsequent investigations should be conducted.

(Barger, Keith E., et al; "Particles in Fluid Inclusions from Yellowstone National Park---Bacteria?" Geology, 13:483, 1985.)

Comment. It is appropriate to note that similar "organized elements" have been noticed in meteorites for over a century.

LIVING STALACTITES! SUBTERRANEAN LIFE!

Translation of the Introduction of an article from Science et Vie:

> One has always held that the calcareous concretions in caves are the work of water and the chemical constituents of the rock. Surprise! The true workers in the kingdom of darkness are living organisms.

It's all true. All the references we have state unequivocally that stalactites and stalagmites are created by dripping water that is charged with minerals, calcium carbonate in particular.

That stalactites contain crystals of calcite is not denied in the Science et Vie article. Indeed, an electron microscope photograph shows them clearly; but it also shows that a web of mineralized bacteria is also an intergral part of the stalactite's structure. Laboratory simulations have shown that microorganisms take an active role in the process of mineralization. (Dupont, George; "Et Si les Stalactites Etaient Vivantes?" Science et Vie, p. 86, August 1987. Cr. C. Maugé.)

Besides being a surprising adjustment of our ideas about stalactite growth, the recognition that microorganisms may play an active role in the subterranean world stimulates two new questions: (1) Can we believe any longer that stalactite size is a measure of age, as is often claimed? (2) Is the immense network of known caves (some as long as 500 kilometers) the consequence only of chemical actions?

It turns out that the earth beneath our feet is not so solid after all. Some 40,000 caves are known in the United States alone. There are thought to be ten times that number that have no surface openings and therefore escape spelunking census takers. And besides caves big enough for humans to crawl into, there exists an immensely greater continuum of cracks, crevices, channels, and pores which circulate air, water, and chemicals in solution. This "crevicular structure" may be continuous for thousands of miles, possibly around the world. Furthermore, it is filled with life forms of great variety, usually blind, and usually related to creatures of the light. A recent article in American Scientist focuses on the evolution of the larger forms of subterranean life, especially the amphipods. Interestingly enough, it doesn't even mention microorganisms. (Holsinger, John R.; "Troglogbites: The Evolution of Cave-Dwelling Organisms," (American Scientist, 76:147, 1988.)

Comment. We have juxtaposed these two articles because together they underscore the great extent of the crevicular domain, the "kingdom of the darkness" of the French article, and also the fact that this crevicular realm teems with life forms, some of which are involved in its construction.

MICROORGANISMS COMPLICATE THE K-T BOUNDARY

Ancient bacteria, it appears, have tampered with the K-T (Cretaceous-Tertiary) boundary of some 65 million years ago. A key marker of this boundary is a thin "spike" of iridium that is found worldwide, and which was supposedly deposited by the asteroid impact that helped finish off the dinosaurs. For many scientists, the asteroid-impact scenario has become a "non-negotiable" brick in the Temple of Science. The problem they have faced is that the iridium layer is variable in thickness and concentration from site to site. Sometimes iridium can be detected well above and below the K-T boundary. This variability has tended to undermine the asteroid-impact theory.

Recent experiments at Wheaton College by B.D. Dyer et al have demonstrated that bacteria in ground water can both concentrate and disperse iridium deposits. In other words, bacteria could smear out an iridium spike, perhaps partially erase it, or even move it to a deeper or shallower layer of sediment. (Monastersky, R.; "Microbes Complicate the K-T Mystery," Science News, 136:341, 1989.)

Comment. An obvious question now is how bacteria might have affected other chemicals, such as oxygen and carbon isotopes, widely used in stratigraphy.

SPORES STILL VIABLE AFTER 7,000 YEARS

Bacterial spores embedded in muds lining Minnesota's Elk Lake---the muds were carbon-dated at 7518 years (!)---grew vigorously when placed in a nutrient-rich solution. (Anonymous; "Spores Still Viable after 7,000 Years," Science News, 124: 280, 1983.)

Comment. Such long periods of suspended animation support the panspermia concept, which states that life can be transported through outer space on meteorites and other debris.

GAIA AT WORK UNDER THE HUDSON

Oxygen-starved bacteria working in sediment beneath the Hudson River are transforming toxic PCB compounds into less dangerous forms, raising hopes of a significant easing of a major PCB cleanup problem nationwide, researchers said yesterday.

The resulting types of PCBs do not accumulate in living tissue, a government scientist said.

Researchers at the University of Michigan stated that it appeared that these bacteria had evolved specifically to attack PCBs. (Anonymous; "Hudson Bacteria Transform PCBs into Safer Forms," Baltimore Sun, November 4, 1988.)

Comment. Thus does Gaia protect humanity from its own mistakes!

SUBTERRANEAN LIFE!

We almost forgot this recent tidbit from Science News that mentions microscopic life forms:

> In recent years, scientists have found bacteria, as far down as 1,150 feet, in wells that penetrate deeply buried aquifers---porous layers of rock that hold underground water. Such finds have forced hydrologists to question their traditional belief that deep aquifers were devoid of life. But it was not clear whether these bacteria were native residents of the aquifers or just contaminants from the world above, living solely within the wells. Moreover, no one had established how the bacteria were affecting the environment, if at all.

Experiments have now shown that these subterranean bacteria are indigenous and are important to groundwater chemistry. The bacteria feed on organic molecules and display a curious propensity for metabolizing the carbon-13 isotope rather than carbon-12. Thus, carbon dissolved in some deep aquifer water is enriched in carbon-13 compared to surface water. None of the bacteria found so far seems dangerous to humans. (Monastersky, R.; "Bacteria Alive and Thriving at Depth," Science News, 133: 149, 1988.)

Comment. Subterranean bacteria may be associated with the creation of oil and natural gas.

MICROORGANISMS AT GREAT DEPTHS

It was a surprise when diverse biological communities were discovered around deep-sea thermal vents, where sunlight is nonexistent and the energy for sustaining life must be extracted from the mineral-charged water gushing from the vents. An analogous situation occurs at great depths in the earth's crust itself, as proven by sampling at three deep boreholes in South Carolina.

The concentration and diversity of microorganisms (mostly bacteria) at depths as great as 520 meters (1610 feet) below the ground's surface are remarkably high. It makes one wonder what will be found even farther down. To illustrate, more than 3000 different microorganisms have been found in the boreholes. Many of the bacteria are new to science.

As the following two paragraphs demonstrate, subterranean life consists of many well-adapted microorganisms working together.

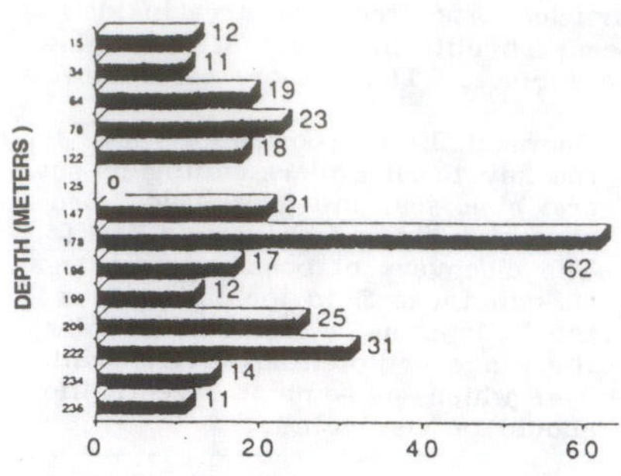

Number of microorganism colony types at various depths at Site P28.

> The traditional scientific concept of an abiological terrestrial subsurface is not valid. The reported investigation has demonstrated that the terrestrial deep subsurface is a habitat of great biological diversity and activity that does not decrease significantly with increasing depth.
>
> The enormous diversity of the microbiological communities in deep terrestrial sediments is most striking. The organisms vary widely in structure and function, and they are capable of transforming a variety of organic and inorganic compounds. Regardless of the depth sampled, the microorganisms were able to perform the cycling of carbon, nitrogen, sulfur, manganese, iron, and phosphorous. Although the organisms were not of the same physiological types, each niche contained a basic cast of microbiological players capable of these nutrient transformations. Such versatility is surprising, and contrary to traditional thinking in soil microbiology, because the deep subsurface is presumably a nutrient-limited environment where photosynthesis and photosynthates are not abundant.

The scientists writing this article speculate that similar microorganisms could well be prospering far below the Martian surface, where they are sheltered from radiation and the cruel surface environment. (Fliermans, Carl B., and Balkwill, David L.; "Microbial Life in Deep Terrestrial Subsurfaces," BioScience, 39:370, 1989.)

DEEPLY-BURIED LIFE

Deep in the Gulf of Mexico, along the edge of the great carbonate platform that breaks the surface as Florida and the Bahamas, thrives a diverse community of animals that does not depend upon the sun for energy. Instead, they feast on carbohydrates provided by symbiotic bacteria. Since there are no ocean-floor vents spewing nutrients and hot water in the area, scientists have wondered where these bacteria obtain the methane and sulfides that nourish them. C.S. Martens and C.K. Paull, of the University of North Carolina, propose that bacteria living miles down within the carbonate platform generate the methane and sulfides as they consume organic matter buried long ago in the limestone. These excreted, energy-rich gases and fluids seep upward and outward, sustaining biological communities along the edge of the platform. (Monastersky, R.; "Buried Rock, Bacteria Yield Deep-Sea Feast," Science News, 140:103, 1991.)

Comment. (1) Looking far back in time, the sun was, of course, the energy source, because it helped create the buried organic matter. (2) However, there is always the possibility that the methane seeping out of the earth is abiogenic. (See p. 206.) (3) How deeply into the crust has life penetrated? The Soviets reported bacteria 12 kilometers down in their drill hole on the Kola Peninsula.

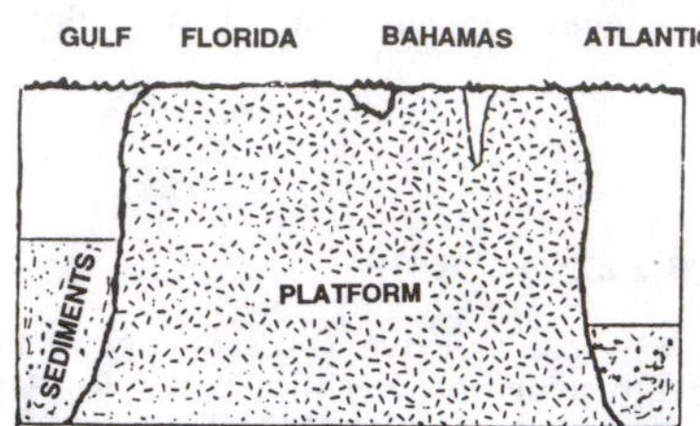

West-to-east profile of the Florida-Bahamas carbonate platform.

COSMIC PRESENCE

CAN SPORES SURVIVE IN INTERSTELLAR SPACE?

There is good evidence that life appeared on earth just 200-400 million years after the crust had cooled (assuming conventional methods of measuring age). Two hundred million years seems a bit on the short side for the spontaneous generation of life, although no one really knows just how long this process should take (forever?). The apparent rapidity of the onset of terrestrial life has led to a reexamination of the old panspermia hypothesis, in which spores, bacteria, or even nonliving "templates" of life descended on the lifeless but fertile earth from interstellar space.

P. Weber and J.M. Greenberg have now tested spores (actually Bacillus subtilis) under temperature and ultraviolet radiation levels expected in interstellar space. They found that 90% of the spores under test would be killed in times on the order of hundreds of years ---far too short for panspermia to work at interstellar distances. However, if the spores are transported in dark, molecular clouds, which are not uncommon between the stars, survival times of tens or hundreds of million years are indicated by the experiments. Under such conditions, the interstellar transportation of life is possible.

But perhaps the injection and capture phases of panspermia might be lethal to spores. Weber and Greenberg think not---under certain conditions. The collision of a large comet or meteorite could inject spores from a life-endowed planet into space safely, particularly if the impacting object glanced off into space pulling ejecta after it. The terminal phase, the capture of spores from a passing molecular cloud by the solar system and then the earth, would be nonlethal if the spores were somehow coated with a thin veneer of ultraviolet absorbing material. In sum, the experiments place limits on panspermia, but do not rule it out by any means. (Weber, Peter, and Greenberg, J. Mayo; "Can Spores Survive in Interstellar Space?" Nature, 316:403, 1985.)

Comment. Weber and Greenberg do not discuss the possible existence of dense, low-temperature regions in molecular clouds where conditions might be conducive to the development of large molecules. Does life have to have the proverbial warm, sunlit pond to develop?

ARTIFICIAL PANSPERMIA ON THE MOON

A colony of earth bacteria, Streptococcus mitis, apparently survived on the moon's surface between April 1967 and November 1969. The organisms were discovered in a piece of insulating foam in the TV camera retrieved from Surveyor 3 by Apollo astronauts. (Anonymous; Science Digest, 90:19, April 1982.)

ODD GROWTHS FOUND ON SATELLITE

We cautiously classify the following phenomenon as "biological," although it might well be inorganic in nature---perhaps something akin to "whisker growth" seen in metals under some conditions.

> Scientists at the National Aeronautics and Space Administration are scratching their heads over how a tiny patch of something managed to grow even though it was exposed to the harshness of outer space for nearly six years.
>
> The mystery growth has been found in a toothpick-sized region on what is known as the Long Duration Exposure Facility. The bus-sized LDEF was launched in 1984 and was retrieved by a space shuttle in January 1990, a few weeks before its decaying orbit would have sent it crashing back to earth.
>
> .....
>
> NASA scientists in Huntsville, Ala., discovered the growth while examining a brownish discoloration on a Teflon-covered section of the satellite.
>
> Using an electron scanning microscope, they saw tiny, stalactite-like structures on the Teflon. Tiny means the longest were about seven microns in size. That's about one-tenth the width of a human hair.
>
> At first NASA scientists thought the growth might be a fungus or a mold that had contaminated the LDEF upon its return. However, their tests came up negative,

(Anonymous; "Odd Space Growth on Satellite Baffles NASA," Arkansas Democrat-Gazette, September 9, 1992. Cr. L. Farish)

OR DID IT DRIFT IN FROM WITHOUT?

Hoyle and Wickramasinghe conceive the cosmos as a seething retort of energy, gases, dust, and, most significantly, organic molecules and microbes. The space between the stars is more important than the stars themselves, for this thin soup is, in their view, the real "swamp" where life originated! The main evidence supporting their radical hypothesis consists of spectrograms, particularly in the infrared, which are difficult to account for on an inorganic basis, but which are fitted nicely by some organic materials, especially microbes. Hoyle and Wickramasinghe devote most of the present article to making a pectroscopic case for their theory, but near the end they shake the Temple of Science a bit:

> Precious little in the way of biochemical evolution could have happened on the earth. It is easy to show that the two thousand or so enzymes that span the whole of life could not have evolved on the Earth. If one counts the number of trial assemblies of amino acids that are needed to give rise to the enzymes, the probability of their discovery by random shufflings turns out to be less than 1 in $10^{40\ 000}$.

They conclude that the genes that control the development of terrestrial life must have evolved on a cosmic scale, where there has been more time and much more room for shufflings. (Hoyle, Fred, and Wickramasinghe, Chandra; "Where Microbes Boldly Went," New Scientist, 91:412, 1981.)

Comment. Could not the "new" bacteria that appeared in the Mt. St. Helens area, as described on p. 185, have drifted down through the atmosphere into the lakes and ponds---a sort of modern, ever-continuing panspermia? It is interesting to note here that even Hoyle, who has espoused the Steady State theory of the cosmos, seems to require the creation of life followed by evolution. This need for an origin of life is a human philosophical weakness. In principle, matter and life, too, could have always existed.

AN EVEN LARGER OCEAN OF LIFE

Fred Hoyle has written another book, The Intelligent Universe; A New View of Creation and Evolution. The subject matter is irresistible and, to make things more interesting, New Scientist has published a scathing review of it, castigating Hoyle for his doubts about evolution and terming his approach "dispicable." (My, how conventional people hate unconventional people!)

In this new book, Hoyle goes far beyond his previous thesis, which in essence declared that from statistical considerations life could not have arisen and evolved on earth. Rather, life had to come from outer space, probably in the form of bacteria and viruses. Evolution was and is dependent upon new information arriving from outer space on tiny bits of life. Hoyle now greatly extends his theory:

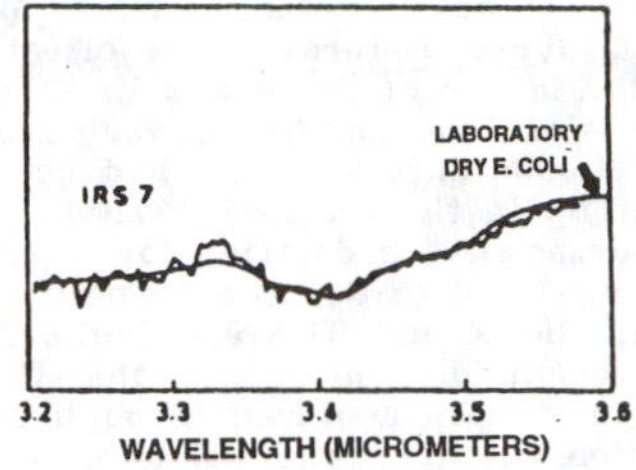

Hoyle and his colleague, N.C. Wickramasinghe, believe that this spectrum of the astronomical object GC-IRS7 closely follows the laboratory absorption spectrum of the bacterium E. coli and thus constitutes strong evidence for the presence of bacteria in outer space.

> But where did a knowledge of amino acid chains of enzymes come from? To use a geological analogy, the knowledge came from the cosmological equivalent of a previous era, from a previously existing creature if you like, a creature that was not carbon-based, one that was permitted by an environment that existed long ago. So information is handed on in a Universe where the lower symmetries of physics---and characteristics of particles and atoms---are slowly changing, forcing the manner of storage of the information to change also in such a way as to match the physics. It is this process that is responsible for our present existence, and it is the one which our descendants would be fated to continue.

To continue his search for the ultimate, Hoyle recognizes that, contrary to what transpires in the inorganic world, life as-a-whole is actually gaining order and information. He sees life leading the universe forward to a remarkable future:

> That biological systems are able in some way to utilise the opposite time-sense in which radiation propagates from future to past. Biology works backwards in time. Living matter responds to quantum signals from the future, instead of the Universe being committed to increasing disorder and decay, the opposite could be true. The ultimate cause being a source of information, an intelligence if you like, placed in the remote future.

(Halstead, Beverly; "Fred Hoyle's Gods," New Scientist, 100:940, 1983.)

Comment. In most religions, the great act of creation by a supreme intelligence occurred in the distant past. Hoyle sees this supreme intelligence residing in the future beckoning us on. No wonder the the book was treated harshly.

LIFE AS A COSMIC PHENOMENON

> The arguments in support of life as a cosmic phenomenon are not readily accepted by a culture in which a geocentric theory of biology is seen as the norm.

The quotation above heads a revealing discussion by F. Hoyle and N.C. Wickramasinghe as to why their conclusions about cosmic life have not been accepted by the scientific community. Stimulating this article was a statement by J. Maddox, Editor of Nature, to the effect that the labors of Hoyle and Wickramasinghe were not convincing very many scientists because they "...had become caught up in the eccentric doctrine of panspermiology, a doctrine for which there was said to be little or no evidence." Hoyle and Wickramasinghe deny their eccentricity and affirm that their ideas have been acquired through observation and the generally accepted methods of deductive science.

Basically, Hoyle and Wickramasinghe maintain that evidence supports the concept of life originating, evolving, and being transported from place to place in outer space. We have published many of their results in past issues of SF and see no need to cover the same ground again. Rather, we wish to dwell on the scientific reception of their work. We do this with two quotations from their Nature article.

These quotations are embedded in their review of the infrared evidence for biological material in outer space:

> Still persuing the infrared problem, we eventually found that among organic materials polysaccharides gave the best correspondence to the astronomical data, and it was exactly at this point in our work that we began to experience hostility from the referees of journals and from the assessors of grant applications at what was then the Science Research Council. We realize now that because polysaccharides on the Earth are a biological product we had unwittingly made a contact that is deeply forbidden in our scientific culture, a contact between biology and astronomy.

And now the second quote:

> We are aware that astronomers and chemists can be found who will claim that these results are not impressive, because equally good results could be obtained using plausible non-biological materials. Our answer is that equally good results have not been obtained using plausible non-biological materials. Such claims are advanced and listened to only because they are designed to be culturally acceptable, whereas our results, although based on careful observations, experiments and calculations are not culturally acceptable. In such a situation the critic is permitted to say anything at all without being weighed in the balance and found wanting."

(Hoyle, F., and Wickramasinghe, N.G.; "The Case for Life As a Cosmic Phenomenon," Nature, 322:509, 1986.)

Comment. Hoyle and Wickramasinghe may be correct in their comments on cultural acceptance; but they are fortunate to even get a chance to make their case in a preeminent scientific journal. Hoyle will not lose his job for promoting revolutionary ideas, but those with lesser reputations might. To be a scientific revolutionary you have to be an already famous scientist or outside the scientific community completely.

IS THE EARTH SEEDING THE REST OF THE SOLAR SYSTEM?

We begin with the lead paragraph from a recent letter to Nature from H.J. Melosh;

> Recent evidence that the SNC meteorites originated on Mars raises the question of whether large impacts on Earth may eject rocks that could fall on Mars (or other planets in the Solar System) and, if so whether they might contain spores or some sort of viable microorganisms that would have the opportunity to colonize Mars.

After some computations Melosh concludes:

> It seems likely that the impacts that produced craters on Earth that are greater than 100 km in diameter would each have ejected millions of tons of near-surface rocks carrying viable microorganisms into interplanetary space, much in the form of boulders large enough to shield those organisms from ultraviolet radiation, low-energy cosmic rays, and even galactic cosmic rays. Under such circumstances spores might remain viable for long periods of time.

(Melosh, H.J.; "The Rocky Road to Panspermia," Nature, 332:687, 1988.)

Comment. Next we need a reasonable mechanism that spreads life through interstellar space. Light pressure, that's it; and the idea is over a century old! Incidentally, SNC is short for Shergottites, Nakhalites, Chassignites; all rare classes of meteorites.

LIFE CURRENTS IN SPACE

A few of the hundreds of meteorites picked up in the Antarctic wastes have chemical properties consistent with a Martian origin. Calculations, too, support the notion that a large meteoric impact could propel bits of the Martian surface into space where, statistically speaking, a tiny fraction would be captured by the earth's gravitational field. Some of these would fall to earth; others would remain in orbit.

Now the reverse scenario has been investigated numerically. S.A. Phinney and colleagues at the University of Arizona have investigated what would happen to small chunks of the earth's crust if a large meteor impact excavated a 60-mile-wide crater. "Phinney's group used a computer to calculate where 1,000 particles would go if ejected from Earth in random directions, moving about 2.5 kilometers per second faster than the minimum speed necessary to escape. Of the 1,000 hypothetical particles, 291 hit Venus and 165 returned to earth; 20 went to Mercury, 17 to Mars, 14 to Jupiter and 1 to Saturn. Another 492 left the solar system completely, primarily due to gravitational close encounters with either Jupiter or Mercury that 'slingshot' them on their way." (Eberhart, Jonathan; "Have Earth Rocks Gone to Mars?" Science News, 135:191, 1989.)

Comment. One implication from the preceding analysis is that terrestrial bacteria and spores could well have infected every planet in the solar system and perhaps even planets in nearby star systems! Conceivably, if other star systems had histories like ours, biological traffic might be quite heavy in interstellar space. In fact, extraterrestrial life forms may be arriving continually; and we may be such ourselves!

"DIRECTED" EVOLUTION

PURPOSEFUL EVOLUTION?

One seemingly unassailable dogma of evolutionary biology insists that natural selection involves, first, the continuous, random, environment-independent generation of genetic mutations; and, second, the subsequent fixation of those mutations that are favored by prevailing conditions. In other words, the genetic mutations cannot be influenced by external events and conditions. But in recent experiments with bacteria (E. coli), J. Cairns et al, at the Harvard School of Public Health, find they actually do produce mutations in direct response to changes in their environment. The adjective "purposeful" has even been applied to the action of these bacteria! Can anything be more heretical?

> One of the experiments involves taking colonies of E. coli that are incapable of metabolizing lactose and exposing them to the sugar. If the lactose-utilizing mutants simply arise spontaneously in the population and are then favored by prevailing conditions, then this would lead to one pattern of new colony growth. A distinctly different pattern is produced if, under the new conditions, the rate of production of lactose-utilizing mutants is enhanced. The observation is something of a mixture of patterns, indicating that directed mutation appears to be occurring. "This experiment suggests that populations of bacteria...have some way of producing (or selectively retaining) only the most appropriate mutations," note Cairns and his colleagures. (Lewin, Roger; "A Heresy in Evolutionary Biology," Science, 241:1431, 1988.)

Research with E. coli at other labs is producing similar heresy. Cairns does not doubt that some mutations arise spontaneously and randomly, but some bacteria have found a way to do a little better. (Hendricks, M.; "Experiments Challenge Genetic Theory," Science News, 134:166, 1988. Cherfas, Jeremy; "Bacteria Take the Chance out of Evolution," New Scientist, p. 34, September 22, 1988.)

Comment. This discovery seems at least as "impossible" as the "infinite dilution" experiments discussed elsewhere. Will Nature now dispatch a "hit squad" to Harvard?

DIRECTED MUTATION

Dear reader, things have a way of working out serially. For several months, we have had in our possession a paper from Nature, by J. Cairns, of Harvard, plus some passionate correspondence stimulated by the paper. Now that the preceding item has provided the basic facts, we are in a good position to jump into the fray.

Basically, Cairns (in Nature) and B. H. Hall (in Genetics) say that organisms can respond to environmental stresses by somehow modifying their genomes in a purposeful way. Such "directed mutation" shifts the course of evolution in a non-random way.

Such a conclusion was like waving a red flag in front of the evolutionists. R. May, at the University of Oxford, complained, "The work is so flawed, I am reluctant to comment." On the other side, a University of Maryland geneticust, S. Benson, comments, "Many people have had such observations, but they have problems getting them published."

Our template in this discussion is an article by A.S. Moffat in American Scientist. She says, "The stakes in this dispute are high, indeed. If directed mutations are real, the explanations of evolutionary biology that depend on random events must be thrown out. This would have broad implications. For example, directed mutation would shatter the belief that organisms are related to some ancestor if they share traits. Instead, they may simply share exposure to the same environmental cues. Also, different organisms may have different mutation rates based on their ability to respond to the environment. And the discipline of molecular taxonomy, where an organism's position on the evolutionary tree is fixed by comparing its genome to those of others, would need extreme revision."

What sort of experiment did Cairns do to cause such a ruckus? In particular, he studied E. Coli bacteria. Normally, these bacteria cannot metabolize the sugar lactose. Cairns exposed the E. Coli to a sudden dose of lactose, demonstrating that if the bacteria must have lactose to survive, they quickly cast off the two genes that inhibit their metabolizing of lactose. Of course, the experiments were more complicated than this, but the fundamental finding was that the bacteria mutated so that they could use lactose much, much faster than chance mutation would permit, stastically speaking.

The battle lines are forming. A supporter of directed mutation, J. Shapiro, of the University of Chicago, is quoted as follows in Moffat's article:

> The genome is smart. It can respond to selective conditions. The significance of the Cairns paper is not in the presentation of new data but in the framing of the questions and in changing the psychology of the situation. He has taken the question "Are mutations directed?" which was taboo, and made it an issue that people will now do experiments on.

(Moffat, Anne Simon; "A Challenge to Evolutionary Biology," American Scientist, 77:224, 1989.)

HYPERMUTATION RATHER THAN DIRECTED MUTATION?

The year 1988 saw a flurry of excitement over the purported discovery of "directed mutation" in the laboratory; that is, mutations of organisms that seemed goal-directed rather than random and independent of environmental pressures. The concept of directed mutation is counter to prevailing biological dogma, and, naturally, this research bore severe scrutiny. Objections and arguments against directed mutation arose, but the original research was never shown to be faulty.

New experiments along this line, by B. Hall at the University of Rochester, have supported the original Harvard work and refuted some of the objections that had been leveled. Working with a special strain of Escherichia coli, Hall starved the bacteria of a certain amino acid, (tryptophan) that they usually got from the environment and could not synthesize. In the days that followed, many of the bacteria mutated so that they could synthesize their own tryptophan. Hall concluded: "Mutations that occur more when they're useful than when they're not: That I can document any day, every day, in the laboratory."

Rather than assign any "conscious" goal-seeking attributes to the bacteria, Hall prefers to think that the environmental stress induced a state of "hypermutation." In this state all sorts of mutations occurred in abundance. Only those that synthesized tryptophan survived the starvation program; even potentially favorable mutations died quickly. (Stolzenburg, W.; "Hypermutation: Evolutionary Fast Track," Science News, 137:391, 1990.)

Comment. Besides the implication that environmental stresses have initiated the mutations that gave us today's fauna and flora, we see that the idea of hypermutation is merely an acceleration of standard evolutionary processes. But how do environmental stresses turn on the biological switch that starts the mutation machine going? What sort of environmental stresses would cause humans to mutate? What would we turn into if, say, global temperatures rose 5°?

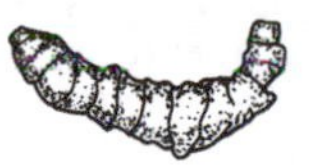

CAN ORGANISMS DIRECT THEIR EVOLUTION?

> In 1988. Harvard molecular biologist John Cairns committed an act of scientific heresy. He proposed in a Nature article that bacteria living in an unfavorable environment are able to choose which mutations to produce to adapt to the stressful situation. Cairns made his directed-mutation hypothesis in response to an unusual finding---data that strongly hinted bacterial mutations might occur more often when beneficial.

The above quotation is the lead paragraph in a long BioScience article that details the consternation Cairns' results have created in the biological community.

The problem that biology-as-a-discipline has is that it has deified a paradigm: neo-Darwinism. Now, neo-Darwinism is supported by many experiments showing that some mutations are indeed random. Consequently, as M. Gillis remarks in her BioScience article, the biological community "got locked into its belief that an organism cannot control its own mutation." Furthermore, Cairns' claims recall the long battle with Lamarckism, a subject that biology has closed-the-book-on. In a nutshell, Lamarckism has been interred since the 1950s, and "Nobody wants to give the appearance of straying from the neo-Darwinism fold."

Gillis goes on to review some recent experiments supporting those of Cairns. But, impressive though these may be, there have been neo-Darwinian explanations for some of the results. Even so, more and more biologists are now willing to accept at least the possibility of nonrandom mutation of bacteria.

But, in the end, all participants in the debate recognize a great void: There exists no acceptable mechanism by which a life form can steer its own evolutionary way; that is, shape its own genome. What besides natural selection can do this? (Gillis, Anna Maria; "Can Organisms Direct Their Own Evolution?" BioScience, 41:202, 1991.)

VIRUSES

SUBTLE IS THE VIRUS

> Without causing noticeable structural damage, a virus administered to laboratory mice has been found to disrupt hormone production in a particular type of pituitary cell. This novel observation---that viruses are able to injure their hosts in ways not previously suspected---may trigger a far-reaching search for viruses as causes of many unexplained human diseases.

Some of the other types of diseases mentioned as possible consequences of virus infection are those involving the faulty manufacture of insulin, neurotransmitters, hormones, and immune system regulators. (Miller, J.A.; "Subtle is the Virus: Cells Stay Intact," Science News, 125:70, 1984.)

Comment. This item dwells on the negative aspects of viral infections. Indeed, we automatically assume every infection by any virus or bacterium to be bad for the organism. This may not be so. Now that we have discovered that viruses can cause bodily changes without damaging the cells of the infected organism, we should ask whether favorable physical changes might not be caused by viruses, but not recognized as such. Going a few steps further: Is intelligence a disease? Could evolution be accelerated or directed through the mediation of viruses? See below for more on this.

AN OCEAN FULL OF VIRUSES

> A decade ago, veterinarian Alvin Smith, now at Oregon State University, found that a virus causing lesions

and spontaneous abortions in California sea lions was "indistinguishable" from one that ravaged pigs nationwide in 1952. New varieties of the culprit ---called a calicivirus---have since turned up in diverse hosts: whales, cats, snakes and even primates. To reach such a variety of hosts, they either jump from organism to organism, Smith proposes, or they escape from bubbles popping on the ocean surface, waft ashore and enter a food chain. If he is right, the seas may be a bottomless reservoir for viruses ---and our attempts to combat diseases on land may be nullified by legions of new strains waiting to come ashore. In fact, some flu viruses are said to be spread by wild ducks.

(Anonymous; "Are the World's Oceans a Viral Breeding Ground?" Science Digest, 92:20, February 1984.)

CARE FOR A CUP OF VIRUSES?

Abstract:

The concentration of bacteriophages in natural unpolluted waters is in general believed to be low, and they have therefore been considered ecologically unimportant. Using a new method for quantitative enumeration, we have found up to 2.5×10^8 virus particles per millilitre in natural waters. These concentrations indicate that virus infection may be an important factor in the ecological control of planktonic microorganisms, and that viruses might mediate genetic exchange among bacteria in natural aquatic environments.

(Bergh, Øivind, et al; "High Abundance of Viruses Found in Aquatic Environments," Nature, 340:467, 1989.)

A sip of water could therefore introduce a billion virus particles into your stomach! This level of virus density in natural water is about 10 million times that formerly estimated.

Besides reducing your thirst, what are the implications of this discovery? First, it suggests that bacteria in natural waters are probably kept in check by viruses as well as protozoans. So far, this sounds good. Second, since viruses can ferry genetic material between organisms via transduction (i.e., host DNA is carried to the next host). This means that genes for antibody resistance and increased bacterial virulence (as present in sewage) may be spread quickly and widely. Also, "engineered bacteria" proposed for use in agriculture, viz., the ice-minus bacterium created to protect strawberries, may die, but their new genes will soon be everywhere. (Weiss, R.; "Aquatic Viruses Unexpectedly Abundant," Science News, 136:100, 1989.)

Comment. Since bacteria are just about everywhere---deepsea vents, thousands of feet underground---we suppose that viruses are there, too. Life, as the Gaia hypothesis suggests, pervades the whole planet from deep within the solid earth to the fringes of the atmosphere and, perhaps, even into outer space itself.

GENETICS

The first 86 issues issues of Science Frontiers have delved into the broad subject of genetics eleven times. The major question addressed has been whether the genetic code completely determines heredity. For example, what is the purpose of nonsense DNA? Is the genetic code universal? Does DNA exist and function outside cells? Can genes be passed from one species to another? Such questions suggest that heredity may require more than just long strings of four-letter genetic code words.

SHORT-CIRCUITING HEREDITY

Genes, those carriers of heredity, have turned out to be great gadabouts. Not only do they jump about within a species, but also between species, especially the simple prokaryotic organisms, such as bacteria. It is now well accepted that genes flow between different species of bacteria in physical contact, thus stirring the evolutionary pot. Until recently, scientists believed that the higher organisms, the eukaryotic species, including you and me, did not indulge in such "horizontal" traffic between species. But a few cases have now been found, one involving humans and a microorganism associated with tumors. And the search is just beginning, as biologists look for something they never thought of looking for before. (Lewin, Roger; "Can Genes Jump between Eukaryotic Species?" Science, 217:42, 1982.)

Comment. This apparent short-circuiting of classical heredity channels supports the radical notion that evolutionary blueprints may be transmitted between divergent species. In the long view, maybe we should not malign viruses, germs, and biting insects!

DNA EVEN MORE PROMISCUOUS

It was a surprise when DNA sequences from mitochondria in yeast cells were discovered setting up shop in the nuclear genomes (i.e., the normal genetic endowment of the cell nucleus). Now biologists find that DNA sequences in many species regularly and frequently hop from one genome to another. Genetic material from cell chloroplasts mix with that of the mitochondria and that of the normal nucleus in what seems to be a free-for-all. This genome hopping has earned DNA the adjective "promiscuous."

The significance of DNA promiscuity is to be found in the general belief that the cell's mitochondria and chloroplasts were once independent biological entities that, in the course of life's development, invaded or were captured by cells and have led a symbiotic life ever since. The mitochondria and chloroplasts perform certain important functions in the cell but were thought, until now, to retain considerable genetic independence. (Lewin, Roger; "No Genome Barriers to Promiscuous DNA," Science, 224:970, 1984.)

Comment. The promiscuity of DNA raises speculation that other DNA-bearing entities that invade the body, especially the viruses, may transfer their DNA to the host, and conceivably vice versa. With DNA apparently much more promiscuous than believed earlier, the role of disease in the development of life takes on a new importance. In other words, all species can potentially exchange genetic information with all others. In fact, in a broad sense, sperm are infectious agents, and pregnancy a disease! DNA will stop at nothing to spread itself around.

DYNAMIC DNA

Smugness over our discovery of the genetic code and some simple features of biological synthesis has recently been undermined by the recognition that the so-called nonsense segments in genes may be important after all. Now comes the realization that the DNA molecule may not be a staid, static construction. Travelling kinks and other disturbances seem to play some unknown role in biological recognition. Some biochemists have even suggested that DNA winds and unwinds or "breathes" like a living thing as it helps to manufacture biological substances. (Spencer, Michael; "Bent DNA," Nature, 281:631, 1979.)

Comment. The addition of the time dimension to biological synthesis evokes thoughts of oscillating systems, frequency dependence, filters, etc. Perhaps Nature has invented something better than the silicon chip.

DNA ON CELL SURFACES

DNA attached to a cell's surface? Such a notion was shocking to scientific orthodoxy in the 1970s. At that time, observations of the phenomenon were rejected. Even worse, funding to continue the work was not forthcoming. Happily, other researchers have later stumbled onto cell-surface DNA; and this startling phenomenon has been rescued from conformity's wastebasket.

Now that cell-surface DNA can be talked about, we can wonder aloud where it comes from and what its significance is. First, this out-of-place DNA ---thought to amount to about 1% of a cell's total DNA---could come from either inside the cell itself or from blood-borne cellular debris. There is considerable argument on this point. Second, this cell-surface DNA does not appear to undergo replication nor does it perform any genetic coding function. Speculation is that it may somehow be involved in the immunological response of the body; for its position on the cell surface is ideal for such a role. Some researchers think that cell-surface DNA may aid in the drug treatment of T-cell lymphoma, a type of cancer. On the other side of the coin, it may mask those molecules on tumor cells that provoke immune responses. Such divergence of opinion indicates how much there is to learn here. (Wickelgren, Ingrid; "DNA's Extended Domain," Science News, 136:234, 1989.)

Comment. If the cell-surface DNA does not come from within the cell itself, is there a possibility it might be alien DNA that somehow got into the blood stream?

HIERARCHIES OF EVOLUTION

All organisms from man to mouse to amoeba are merely DNA's way of manufacturing still more DNA---so goes the modern ramification of molecular biology and the Genetic Code. In other words, DNA and genes are selfish, and ultimate parasites, directing the evolution of life only to maximize the production of DNA.

This theme is not the subject of this paper by Doolittle and Sapienza. Rather, they wonder about those nonsense DNA sequences that do not code for protein. The presence of these "useless" bits of genetic material is often explained in terms of gene "expression." Emphasis is always on maximizing the "fitness" of the organism (phenotype). Perhaps this seemingly excess genetic material actually maximizes the fitness (survivability) of the DNA itself. Evolution thus occurs at DNA and gene (genome) levels, despite what transpires at the organism (phenotype) level. (Doolittle, W. Ford, and Sapienza, Carmen; "Selfish Genes, the Phenotype Paradigm and Genome Evolution," Nature, 284:601, 1980.)

Comment. We know that mitochondria and chloroplasts have their own genetic material; evolution may be occurring at this level, too, independent of pressures for change on the organisms. Waxing speculative, may there not be other hierarchies where systems are trying to maximize their own survivability, even at molecular, atomic, and subatomic levels? Don't laugh! Is not all life implicitly encoded in the properties of the most fundamental particles? If not, reductionism is a lie.

WHY CONSERVE JUNK?

A. Jeffries, working at Leicester University with globin genes from man and related primates, has been studying how these genes direct the blood cells to make the alpha and beta chains of hemoglobin. Jeffries' analyses seem to indicate that the genes now coding for these hemoglobin chains are almost identical to those existing in human ancestors some 500 million years ago.

Two curious facts have cropped up, however. First, about 200 million years ago, these genes were modified very slightly and relocated to entirely different chromosomes. Second, 95% of the DNA associated with these genes is "junk"---with no known use. Why did nature conserve junk for 500 million years? Are vital genes in the habit of jumping from one chromosome to another? (Yanchinski, Stephanie; "DNA: Ignorant, Selfish and Junk," New Scientist, 91:154, 1981.)

THE IMPORTANCE OF NONSENSE

One of the greatest surprises of modern molecular biology has been the discovery of "split genes" in higher forms of life. In the chromosomes of lowly bacteria, genes march along one behind the other, but in more complex organisms the genes are separated by segments of genetic material that apparently have nothing to do with the manufacture of protein. Because there seems no need for these inserted jumbles of genetic information, they are characterized as "nonsense." But evolutionists insist that this nonsense must have some survival value or it wouldn't be there!

Present speculation is that the nonsense segments separate mini-genes that contain the blueprints for assembling well-defined parts of proteins that possess specific functions. To illustrate, the main part of the immunoglobulin molecule has four functional parts (one for interacting with cell membranes, another that functions as a hinge, and so on). Lo and behold, the immunoglobulin gene consists of four mini-genes separated by three segments of nonsense. The suspicion is that the evolution of higher life forms has been accelerated by keeping these prefabricated, functionally oriented mini-genes apart and shuffling them as integral units. The shuffling of entire functional elements rather than smaller bits and pieces of genetic information might speed up organic evolution. (Lewin, Roger; "Why Split Genes?" New Scientist, 82:452, 1979.)

GENETIC GARRULOUSNESS

It is tempting to predict that those cells with the most genetic material will belong to the most advanced organisms. One would, for example, expect to find more DNA or nucleotide pairs in human cells than the cells of bacteria or plants. In the case of the bacteria, this expectation is realized. Some plants, however, have one hundred times more DNA per cell than humans. Some fish and salamanders do, too.

One reason why there is no simple relationship between a cell's genetic complement and the organism's complexity is that a lot of genetic material is apparently useless, with no known functions. Human genes, by way of illustration, possess about 300,000 copies of a short sequence called Alu. The Alu sequences seem to be simply dead weight ---functionless---yet continuously reproduced along with useful sequences. One purposeless mouse gene sequence is repeated a million times in each cell. (Stebbins, G. Ledyard, and Ayala, Francisco J.; "The Evolution of Darwinism," Scientific American, 253:72, July 1985.)

Comment. Why so much redundance? Or is there some purpose for this excess genetic material that we haven't yet descried? The "useless" sequences may merely be left over from ancient gene shufflings; or they may be awaiting future calls to action. The above tidbits come from a long review article that is generally supportive of the modern theory of evolution. (Would it have been printed in Scientific American if it weren't?) The article also treats the Neutral Theory of Evolution, also mentioned on p. 190, noting that this theory still depends upon natural selection operating at the phenotype (organism) level.

GENETIC CODE NOT UNIVERSAL!

Most current textbooks pronounce that the genetic code is not only universal to life but that it must be. Alterations in the code, the reasoning goes, would garble genetic messages. Such dogma, however, is based on another dogma, which states that all life on earth derives from a single ancestral line in which the genetic code is fixed and has always been fixed---a "frozen" accident hit upon "by chance" on the primitive earth. Recently, though, small deviations from the code have turned up in ciliated protozoans and mycoplasmas, much to everyone's surprise. (Scott, Andrew; "Genetic Code Is Not So Universal," New Scientist, p. 21, April 11, 1985.)

Comment. No major deviations from the standard genetic code have been found; but then no one has really looked, because everyone knew the code was universal. It would be quite a shock if terrestrial life were found to have several genetic foundations and as many schemes of evolution.

THE LANGUAGE OF LIFE

Popular writers on biology are fond of saying that the genes and their DNA carry all information necessary for the development of an organism and the transfer of inherited characteristics. With the advent of the multibillion-dollar project to map the human genome (our genetic inventory), we have been seeing this extreme claim more often. The truth is that a map of the human genome will not tell us everything. By way of confirmation, we quote the lead paragraph from a recent article in New Scientist:

> In the early days of molecular biology, during the 1950s and 1960s, scientists as much as journalists fuelled the euphoria that surrounded the cracking of the genetic code. The secret of life was revealed, so many people thought. As our understanding has grown, however, so has our awareness of our ignorance. Research at the forefront of the molecular sciences has shown that we can no longer regard DNA---the stuff of genes---as a direct and complete set of instructions for the synthesis of proteins. The evidence begins to suggest that messages in the DNA are, in themselves, no more precise than the symbols and sounds with which we communicate. As in the languages with which we are familiar, the correct sense of a message written in DNA seems to depend on the rigorous checking and correction of errors, and on the context in which they are read.

The final sentence of the article is really paradigm-shaking:

> Thus genetic and evolutionary changes are no longer confined solely to the genome at the pinnacle of a hierarchy of information and control, but reside also in the interplay between DNA and the other components of cells.

(Tapper, Richard; "Changing Messages in the Genes," New Scientist, p. 53, March 25, 1989.)

Comment. If DNA can be read in more than one way, depending upon the context, our current concept of evolution may be in jeopardy. For example, how does an organism transmit "context" to its offspring; the same thought applies to error-correcting capabilities.

THE MUSIC OF THE GENES

> S. Ohno has cracked a new genetic code. The 58-year-old geneticist doesn't have the whole thing worked out yet, but when he sets the genes to music---or music to genes---some strange and wonderful things occur. To wit:
>
> The SARC oncogene, a malignant gene first discovered in chickens, causes cancer in humans as well. When Ohno translated the gene into music, it sounded very much like Chopin's Funeral March.
>
> An enzyme called phosphoglycerate kinase, which breaks down glucose, or sugar, in the body revealed itself to Ohno as a lullaby.

Seeing this item is from a newspaper, it was nearly consigned to the wastebasket. But wait a moment, Susumo Ohno is a Distinguished Scientist at the Beckman Research Institute of the City of Hope National Medical Center, Duarte, CA. Could there be something to it?

Reading further; we find that Ohno believes that the structure of music seems to parallel that of the genes. He translates genes into music by assigning notes according to molecular weights. His ultimate goal is the discovery of some basic pattern (melody?) that governs all life. (Anonymous: "Scientist Tunes in to Gene Compositions," San Jose Mercury News, p. E1, May 13, 1986. Cr. P. Bartindale.)

Comment. Not too long ago the motions of the planets were supposed to conform to an esthetically pleasing Music of the Spheres. Ohno, it seems, has found a way to express the Music of the Genes. Are simple organisms just short tunes and humans full-fledged operas? Are some refrains repeated in different organisms? All this is not entirely frivilous because a fundamental tenet of science expects nature to be describable in terms of a few laws that are not only simple but esthetically pleasing as well.

ORIGIN OF LIFE

TERRESTRIAL ORIGINS

You will find only doubts in the text below about that renowned "warm little pond" where earth's life supposedly got its start. Instead, some iconoclasts have looked much deeper in the earth's crust and along the ocean floor's hydrothermal vents for life's point of origin. Others have wondered whether life's chemicals and templates could have drifted to earth on cometary debris and about the potential biological creativity of the crucibles of asteroidal impacts on the terrestrial surface. An even more radical notion---that life was synthesized in outer space itself---is the subject of the section after this one.

EARTH'S WOMB

Three recent items indicate that scientists are now recognizing how the earth's crust is tailor-made for biochemical reactions of great variety and complexity.

First, E.G. Nisbet explains how subsurface hydrothermal systems are ideal places to make biochemical products, particularly in the light of the discovery that RNA molecules can extrude introns and then behave like enzymes.

> The most likely site for the inorganic construction of an RNA chain, which would have occurred in the Archaean, is in a hydrothermal system. Only in such a setting would the necessary basic components (CH_4, NH_3, and phosphates) be freely available. Suitable pH (fluctuating around 8) and temperatures around 40°C are characteristic of hydrothermal systems on land. Furthermore, altered lavas in the zeolite metamorphic facies, which are rich in zeolites, clays and heavy metal sulphides, would provide catalytic surfaces, pores and molecular sieves in which RNA molecules could be assembled and contained. If the RNA could then replicate with the aid of ribozymes and without proteins, the chance of creating life becomes not impossible but merely wildly unlikely.

The article concludes with a statement that self-replicating molecules synthesized in hydrothermal systems would be pre-adapted to "life" in the open ocean if they "learned" to surround themselves with bags of lipids. (Bag of lipids = a membrane.) (Nisbet, E.G.; "RNA and Hot-Water Springs," Nature, 322:206, 1986.)

It just so happens that D.W. Deamer, University of California, Davis, has now found that the 4.5-billion-year-old Murchison meteorite from Australia contains lipid-like organic chemicals that can self-assemble into membrane-like films. His paper was presented before the International Society for the Study of Origins of Life. (Raloff, J.; "Clues to Life's Cellular Origins," Science News, 130: 71, 1986.)

Comment. Strange that the earth should be "tailor-made" for biochemical operations and that outer space teems with meteorites transporting other ingredients of life-synthesis.

That the earth's crust and deep soil are conducive to life is apparent in recent work done sponsored by DuPont and the Department of Energy. This effort has found that life is abundant at least 850 feet below the surface---a realm hardly suspected to harbor life.

> 'There is life down there, and it is very diverse,' says Carl Fliermans of Dupont's Savannah River Laboratory in Aiken, S.C. The numbers are high enough to affect the chemis-

try of the environment: Some of the samples contained as many as 10 million organisms per gram of soil. But even more surprising than the high concentrations is the diversity of the microorganisms, according to David Balkwill of Florida State University in Tallahassee. Many varieties of bacteria and fungi have been seen, and there have been indications of amoeba. And the diversity---which doesn't appear to decrease with depth---may force a reappraisal of the environment that lies between soil and bedrock.

(Anonymous; "The Bugs beneath Us," Science News, 130:58, 1986.)

THE OKLO PHENOMENON AND EVOLUTION

A decade ago, French scientists discovered the remains of a natural nuclear reactor at Oklo, Gabon, in Africa. Somehow nature had concentrated enough uranium-235 in one place to start a chain reaction, with the attendant production of heat and radiation. Now U-235 is radioactive, and there is now much less around than in past geological eras. This has led some scientists to speculate that many more Okla phenomena may have flamed momentarily in earlier times, especially Precambrian days. The mutagenic radiation from such natural reactors could have been a major driving force in evolution. (Anonymous; "Natural Reactors Helped Evolution," New Scientist, 100:737, 1983.)

ARCHAEA: THE LIVING ANCESTORS OF ALL LIFE FORMS

Life's place of origin may soon shift from that long-favored "warm little pond" to undersea hydrothermal vents.

> Important new discoveries on the properties of the early earth and atmosphere, including the frequency and size of bolide impacts, have strongly implicated submarine hydrothermal vent systems as the likely habitat for the earliest organisms and ecosystems, while stimulating considerable discussion, hypotheses and experiments related to chemical and biochemical evolution. Some of the key questions regarding the origins of life at submarine hydrothermal vent environments are focussed on the effects of temperature on synthesis and stability of organic compounds and the characteristics of the earliest organisms on earth. There is strong molecular and physiological evidence from present-day mircoorganisms that the earliest organisms on earth were capable of growing at high temperatures (~ 90°C) and under conditions found in volcanic environments. These "Archaea", the living ancestors of all life forms, display a variety of strategies for growth and survival at high temperatures, including thermostable enzymes active at temperatures ~140°C. Further molecular and biochemical characterization of the presently cultured thermophiles, as well as future work with the many species, particularly from subsurface crustal environments, not yet isolated in culture, may help resolve some of the important questions regarding the nature of the first organisms that evolved on earth.

(Baross, J.A.; "Hyperthermophilic Archaea: Implications for the Origin and Early Evolution of Life at Submarine Hydrothermal Vents," Eos, 72:59, 1991.)

IT CAME FROM WITHIN

Life did, that is! Forget that warm little pond where life incubated according to all the textbooks. Instead, says T. Gold, an iconoclastic Cornell physicist, life began in rocky fissures deep down in the earth's crust. The idea is not as unlikely as it sounds. Look at the most primitive life forms we know, the archaebacteria. They like heat, need neither air nor sunlight, and prosper on sulfur compounds for sustenance. Such bacteria are today found in boreholes as deep as 500 meters, in thermal springs, and around deepsea vents. Gold surmises that these archaebacteria migrated to the surface long ago, where they evolved into higher forms of life.

> Gold argues, moreover, that the earth's interior would have provided a much more hospitable environment for proto-life four billion years ago than the surface would have, ravaged as it was by asteroids and cosmic radiation. And if life emerged within the earth, then why not within other planets? "Deep, chemically supplied life," Gold says, "may be very common in the universe.

(Horgan, John; "It Came from Within," Scientific American, 267:20, September 1992.)

LIFE'S ORIGIN WITHIN THE EARTH?

Biologists usually hark back to warm, sunlit swamps and tidal pools when contemplating the origin of life. Lately, Hoyle has proposed a cosmochemical origin (see p. 180). Few look within the earth. Yet, when Mt. St. Helens erupted it essentially sterilized all lakes and ponds in the immediate area as far as known life forms were concerned, and then introduced previously unknown chemosynthetic bacteria. At least, this is one interpretation. Scientists at Oregon State University found the waters around the volcano to be teeming with these bacteria, up to a billion per drop. The bacteria resemble nothing in the local soil but do seem related to bacteria existing around Precambrian volcanos. (Anonymous; "Secrets of Life in a Volcano?" Boston Globe, July 14, 1981.)

Comment. Were the new bacteria in the volcanic ejecta or had they just gone unnoticed in the soil? Could the hot rocks, geothermal brines, and restless magmas beneath our feet be the real cradle of terrestrial life, with photosynthesis-dependent surface species being relatively unimportant to the big picture?

COMETS AND LIFE

"New simulations suggest that large amounts of the organic molecules needed to form the first life on Earth could have been brought by comets that bombarded the planet early in its history. The models show that comets of moderate size would have slowed down enough during entry into Earth's atmosphere for their organic component to survive the impact intact."

.....

"The idea that comets supplied the Earth with the organic material needed to create life has been around for more than 20 years, but as often as some scientists have put forward the hypothesis, others have shot it down.

"About 20% of comet nuclei are composed of organic matter, the rest ice and dust. Most of the organic component appears to be in a complex, polymerized form similar to kerogen, which is found in sedimentary rocks on Earth. From 4.3 - 3.7 billion years ago, during the period of heaviest bombardment by meteorites and comets, an estimated 10^{23}g of organic material would have been added to Earth by comet impacts had they made up 10% of the extraterrestrial flux. That is 100 - 1000 times larger than most estimates of the photochemical production of the compounds over the same period." (Anonymous; "Comets and Life," Eos, 70:196, 1989.)

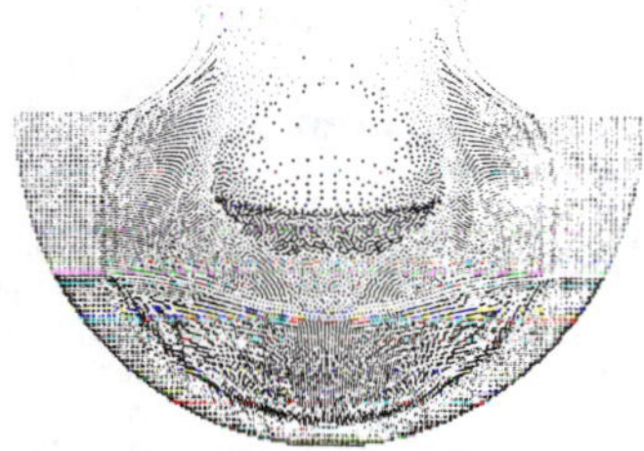

This comet has a radius of about 1 kilometer and an impact velocity of 15 kilometers/second. In this mathematical simulation, the initial spherical shape has become flattened a bit and pieces have begun to separate from the main body. These pieces and the back of the comet would, according to theory, remain cool enough to permit the survival of some biological cargo.

CELESTIAL CRUCIBLE

> Catastrophic extinctions caused by impacts would change the rules governing who is most fit, who becomes extinct, and who survives. 'If much of the patterning of life's history is not set by Darwin's slow biotic mechanisms, then I think Darwin is in trouble. Is catastrophic mass extinction a major agent of patterning?' If so, 'impacts are a quirky aspect' of the process.

Who is speaking within the single quotes above? S.J. Gould, a proponent of the punctuated equilibrium view of the evolutionary scenario. He added:

> 'The history of life is enormously more quirky than we imagined.'
>
> In fact, the geological record shows

so many quirk-inducing impacts that there is little room left for slow, plodding, uniformitarian evolution of the earth itself, life-in-general, and humanity. Mammals, for example, may not have survived the postulated (but now assumed factual) Cretaceous-Tertiary impact event simply because they were small in size---not smarter. (Kerr, Richard A.; "Huge Impact is Favored K-T Boundary Killer," Science, 242:865, 1988.)

Comment. It now seems that Cassius was wrong about the stars when he was lining up Brutus to help assassinate Julius Caesar. And the "celestial" situation gets even worse below.

LARGE MOON ESSENTIAL TO THE DEVELOPMENT OF LIFE?

Even before its well known effect on romance, our moon was the key to the development of life on earth, according to J. Pearson. He ventures that our ponderous moon (much larger than the moons of Mars) was the key to the melting of the earth's core via tidal friction. With a fluid core, the earth developed a strong magnetic field (much stronger than those of the other inner planets) through dynamo action. This field protected nascent life from space radiation---and here we are! (Hecht, Jeff; "Lunar Link with Life on Planets," New Scientist, p. 40, January 21, 1988.)

Comment. Current scientific opinion declares that the moon was captured by the earth---a rather rare astronomical event. The capture of a very large moon would be even rarer. From this shaky chain of thoughts, we conclude that life in the universe must be exceedingly scarce. However, such long chains of inferences are usually found to be far off the mark.

More recently, calculations suggest still another influence of the moon on the development of terrestrial life. The large size of earth's moon tends to stabilize the planet's spin axis, thus preventing chaotic oscillations of climate which might have been inimical to life. See Science Frontiers #87.

SPACE ORIGINS

TUNNELLING TOWARDS LIFE IN OUTER SPACE

Most books on biology begin the history of life in those apocryphal warm ponds of primordial soup. Leave comfortable earth for a moment and consider the immense, cold clouds of gas and dust swirling between the stars and galaxies. At near-absolute-zero, sunless and waterless, these clouds hardly seem the womb of life. Yet, there may be found the atoms necessary to life---H, C, O, N, etc---and in profusion. Collisions of cosmic rays can promote the synthesis of fairly large molecules. We have already detected molecules as complex as formaldehyde in the interstellar medium. But surely the immensely more complicated molecules of biology cannot be synthesized near absolute zero. This may not be true either because at extremely low temperatures the quantum mechanical phenomenon of "tunnelling" becomes important.

To achieve molecular synthesis, repulsive barriers must be overcome. The warm temperatures in that terrestrial pond can provide the extra kinetic energy to climb over these barriers. In cold molecular clouds we must look elsewhere. The laws of quantum mechanics state that there is always a very low probability that atoms and molecules can tunnel through repulsive barriers---no need to climb over them via thermal effects. "Specifically, entire atoms can tunnel through barriers represented by the repulsive forces of other atoms and form complex molecules even though the atoms do not have the energy required by classical chemistry to overcome the repulsion." Of course, the reaction rates are slow, but the size of the cosmic pond is vastly greater than any terrestrial puddle.

The above quotation is from V.I. Goldanskii who, even before Hoyle, suggested a cold prehistory of life, during which complex organic molecules were synthesized at just a few degrees above absolute zero. His article dwells primarily on the physics of the tunnelling phenomenon, which is well verified in the laboratory, but he does not shrink from the biological implications. (Goldanskii, Vitalii I.; "Quantum Chemical Reactions in the Deep Cold," Scientific American, 254:46, February 1986.)

Comment. In the first few paragraphs you can almost hear the theme music from the movie 2001: A Space Odyssey.

LIFE-CREATION FROM A DIFFERENT PERSPECTIVE

The preceding discussion of life's origin at hydrothermal vents was penned by an oceanographer. Astronomers, it seems, prefer different scenarios. C. Chyba and C. Sagan, in a major review article in Nature, see a two-fold problem: (1) identifying the source of the raw materials; and (2) identifying the source(s) of energy required for the synthesis of complex organic chemicals. First, they point to the steady drizzle of tiny, organic-rich particles drifting down to earth from cometary debris. These particles, which even carry space-synthesized amino acids down to the earth's surface, seem likely chemical precursors of life. However, the atmosphere is also a potential source of prebiotic chemicals---providing energy sources are available. Chyba and Sagan suggest as sources: lightning, ultraviolet radiation, and the shock energy derived from meteorite/asteroid/comet impacts. Together these energy sources, especially ultraviolet light, might synthesize thousands of tons of complex organic compounds each year. (Chyba, Christopher, and Sagan, Carl; "Endogenous Production, Exogenous Delivery and Impact-Shock Synthesis of Organic Molecules: An Inventory for the Origins of Life," Nature, 355:125, 1992. Also: Henbest, Nigel; "Organic Molecules from Space Rained Down on Early Earth," New Scientist, p. 27, January 25, 1992.)

Comment. Little is said in either of the above articles about the nature of and impetus for that final elusive step from organic chemicals to the simplest life forms. Everyone assumes that it happened, but did it? One can always imagine a universe in which matter, energy, and life have always existed.

THE COSMIC CHEMISTRY OF LIFE

C. Ponnamperuma, at the University of Maryland, speculates that the genetic code employed by earth life may, in fact, reflect a universal chemistry. His lab data "suggest that the formation and linking of life's building blocks---amino acids and nucleotides--- may have been all but inevitable, given the starting chemistry of earth's 'primordial soup.'" Going even further Ponnamperuma believes that "if there is life elsewhere in the universe, chemically speaking it would be very similar to what we have on earth." (Raloff, J.; "Is There a Cosmic Chemistry of Life?" Science News, 130:182, 1986.)

ARTIFICIAL MOLECULE SHOWS 'SIGN OF LIFE'

A synthetic molecule has been found that apparently replicates itself. This seems to be a step on the road to artificial life.

> Julius Rebek, Tjama Tjivikua and Pablo Ballester of the Massachusetts Institute of Technology say that their compound, an amino adenosine triacid ester (AATE), acts as a "template" which combines molecular fragments to make a copy of the original compound. This process is very similar to that used by DNA. The difference is that the biological copying usually needs an enzyme to make it work.

(Emsley, John; "Artificial Molecule Shows 'Sign of Life,'" New Scientist, p. 38, April 28, 1990.)

Comment. No one can say that replication is a "spontaneous" property of inorganic matter. It is, though, truly remarkable that base matter is intrinsically self-organizing and replicating. Apparently, this is a consequence of the so-called Anthropic Principle, which states that the laws of nature and the properties of matter are just those that lead to the development of humans. In other words, we are here because that's the way things are! A tautology?

EVOLUTION

THE FOSSIL RECORD

The theory of evolution, as presently formulated, posits small random changes in the genomes of the species. Such small changes, when modulated by natural selection, should have created a fossil record showing gradual, essentially continuous, morphological changes as new species evolved from old. In the actual fossil record, successions of transitional forms are rare. Rather, the fossil record seems better characterized as a collection of sudden appearances of brand new species, classes, even phyla. These discontinuities are usually called "saltations." The prevalence of saltations in the fossil record has led to a proposed modification of the theory of evolution called "punctuated equilibrium." The causes of these saltations remain mysterious. The greatest saltations of all occurred at the beginning of the Cambrian period, about 650 million years ago, when all known phyla (basic body plans) appeared suddenly with few, if any, preceding transitional forms. The discussions of discontinuities in the fossil record, taken from past issues of Science Frontiers, begin with this Cambrian "explosion."

LOP-SIDED EVOLUTION

We risk supersaturating our readers with the anomalies of evolution, but we simply cannot bypass an article that is introduced as follows:

> An analysis of the fossil record reveals some unexpected patterns in the origin of major evolutionary innovations, patterns that presumably reflect the operation of different mechanisms.

The most interesting "unexpected pattern" is the gross asymmetry between the diversification of life in the Cambrian explosion (about 570 million years ago) and that following the great end-Permian extinction (a little over 200 million years ago). Biological innovation was intense in both instances; both biological explosions burst upon a life-impoverished planet. Many niches were unoccupied. Even so, all existing (and many extinct) phyla arose during the Cambrian explosion and none followed the Permian extinction.

> ...why has this burst of evolutionary invention never again been equaled? Why, in subsequent periods of great evolutionary activity when countless species, genera, and families arose, have there been no new animal body plans produced, no new phyla?

Some evolutionists blame the asymmetry on the different "adaptive space" available in the two periods. "Adaptive space" was almost empty at the beginning of the Cambrian because multicellular organisms had only begun to evolve; whereas after the Permian extinction the surviving species still represented a diverse group with many adaptations. (Just how the amount of "adaptive space" available was communicated to the "mechanism" doing the innovation is not addressed.) Scientists contemplating these matters, however, seem to concur that microevolution, which supposedly gives rise to new species, cannot manage the bigger task of macroevolution, in particular the creation of new phyla at the beginning of the Cambrian. (Lewin, Roger; "A Lopsided Look at Evolution," Science, 241:201, 1988.)

Representatives of three basic body plans (phyla): jellyfish (coelenterata); aphid (arthropoda); eohippus (chordata). All phyla appear rather suddenly in the fossil record at the beginning of the Cambrian period.

BIOLOGY'S BIG BANG

The title refers to the so-called "Cambrian explosion," that period that began some 570 million years ago, during which all known animal phyla that readily fossilize seem to have originated. The biological phyla are defined by characteristic body plans. Humans, for example, are among the Chordata. Some other phyla are the Arthropoda (insects, crustaceans), the Mollusca (clams, squids), the Nemotada (roundworms), etc. All of these phyla trace their ancestries back to that biologically innovative period termed the Cambrian explosion. Even at the taxonomic level just below the phylum, the class (i.e., the vertebrates), most biological invention seems to stem from the Cambrian.

J.S. Levinton, in a long article in the November 1992 Scientific American, explores the enigma of the Cambrian explosion. Did some unknown evolutionary stimuli prevail 570 million years ago that made the Cambrian different from all periods that followed? Or, has something damped evolutionary creativity since then? Levinton holds that biological innovation has continued unabated at the species level since the Cambrian explosion, but that new body plans; that is, new phyla; have not evolved for hundreds of millions of years. Therefore, something special and very mysterious ---some highly creative "force"---existed then but is with us no longer. (Levinton, Jeffrey S.; "The Big Bang of Animal Evolution," Scientific American, 267:84, November 1992.)

Comment. If evolution is truly the result of random mutation modulated by natural selection, perhaps mutation was "different-from-random" during the Cambrian! Now that's a heretical thought.

MISSING LINKS: THE BIG ONES STILL ELUDE US

> Under the very best circumstances, however, morphological and stratigraphically graded transitions between classes and subclasses have been found. At the level of phyla and higher categories, any information on transitions as far as the fossil record is concerned is essentially nonexistent.

(Olson, Everett C.; "The Problem of Missing Links; Today and Yesterday," Quarterly Review of Biology, 56:405, 1981.)

HOPEFUL MONSTERS RATHER THAN GRADUAL EVOLUTION?

S. J. Gould, who conducts a monthly column in Natural History reviews the sad history of Goldschmidt and his villification by the scientific establishment. Goldschmidt saw the fossil record as woefully inadequate to justify the assumption of gradual evolution of one form into another. Intermediate forms between separate species do not seem to exist in the fossil record and, if they did, they would probably not have been viable creatures. What good is half a wing?

Gould believes that Goldschmidt's "hopeful monster" concept will ultimately be dusted off. The key to "macromutation," Gould feels, is not to be found in major gene reorganizations that might produce a whole wing, feathers included, all at once, but rather in changes in the genes that control the development of embryos. Embryos in their early stages are pretty much alike regardless of species. Gould hopes further that the ruling neo-Darwinians will not be so hostile to new ideas and eventually acknowledge Goldschmidt's important work. (Gould, Stephen J; "The Return of the Hopeful Monster," Natural History, 86: 22, June-July, 1977.)

FACING UP TO THE GAPS

The textbooks and professors of biology and geology speak confidently of the fossil record. Darwin may have expressed concern about its incompleteness, but, especially in the context of the creation-evolution tempest, evolutionists seem to infer that a lot of missing links have been found. Some scientists, however, are facing up to the fact that many gaps in the fossil record still exist after a century of Darwinism. One has even dispaired that "the stratigraphic record, as a whole, is so incomplete that fossil patterns are meaningless artefacts of episodic sedimentation."

D.E. Schindel, Curator of Invertebrate Fossils in the Peabody Museum, has scrutinized seven recent microstratigraphical studies, evaluating them for temporal scope, microstratigraphical acuity, and stratigraphical completeness. His first and most important conclusion is that a sort of Uncertainty Principle prevails such that "a study can provide fine sampling resolution, encompass long spans of geological time, or contain a complete record of the time span, but not all three." After further analysis he concludes with a warning that the fossil record is full of habitat shifts, local extinctions, and general lack of permanence in physical conditions. (Schindel, David E.; "The Gaps in the Fossil Record," Nature, 297:282, 1982.)

Comment. This candor makes one wonder how much of our scientific philosophy should be based upon such a shaky foundation.

RECENT PULSATIONS OF LIFE

At a recent meeting of scientists at the Lamont-Doherty Geological Observatory, E. Vrba, of the Transvaal Museum, in Pretoria, stated:

> If we eventually are able to establish a good time resolution with the continental record, I expect to be able to discern synchronous pulses of evolution that involve many groups of fauna and flora. Many different lineages in the biota will respond by synchronous waves of speciation and extinction to global temperature extremes and attendant environmental changes. This is my starting hypothesis.

Vrba was speaking mainly about the last 25 million years, a mere flash in geological time. For this brief period, the Deep Sea Drilling Program has provided geologists with a detailed and continuous record of climate changes as they were recorded in deep-sea sediments. By contrast, the faunal history of the continents is rather fragmentary, making it rather difficult to match up pulsations of climate with pulsations of life. Even so, scientists have found rather strong correlations between climatic change and biological speciation and extinction at 15, 5, and 2.4 million years ago. (Lewin, Roger; "The Paleoclimatic Magic Numbers Game," Science, 226: 154, 1984.)

Comment. Note that this is just the period our ancestors seemed to be evolving rapidly. Also interesting is the general agreement between Vrba's statement about the driving forces behind evolution and McClintock's conclusion quoted on p. 170.

CAN YOU GUESS WHERE THIS QUOTATION COMES FROM?

> THE GRADUALIST'S DILEMMA
>
> The basic article of faith of a gradualist approach is that major morphological innovations can be produced without some sort of saltation. But the dilemma of the New Synthesis is that no one has satisfactorily demonstrated a mechanism at the population genetic level by which innumerable very small phenotypic changes could accumulate rapidly to produce large changes: a process for the origin of the magnificently improbable from the ineffably trivial. This leads to skepticism about the microevolutionary approach. Perhaps, as Waddington put it: "the real guts of evolution---which is, how do you come to have horses and tigers, and things---is outside the mathematical theory."
>
> Did you guess a creationist publication? Sorry! (Thomson, Keith Stewart; "Macroevolution: The Morphological Problem," American Zoologist, 32:106, 1992.) (And just the other day, we read that evolution was a proven fact!)

THE FOSSIL RECORD AND THE QUANTIZATION OF LIFE!

A recent article on the possible quantization of galaxies was placed immediately before an interview with S. Stanley, one of the proponents of punctuated evolution. Either fate or the mysterious forces of seriality seemed to be saying that it is now time to broach the subject of the quantization of life itself. The reader can blame Stanley only for the stimulus and his discussions of speciation and the discontinuous (quantized?) fossil record. (Campbell, Neil A.; "Resetting the Evolutionary Timetable," BioScience, 36:722, 1986.)

Editorial. When we suggest quantization in biology, two phenomena come to the fore:

1. The obvious splitting of life into well-defined states---the species---as defined morphologically and/or by the genetic code; and
2. The gaps in the fossil record, which imply a frequent lack of transitional forms from one species to another.

As Stanley asserts repeatedly in his interview, the fossil record is actually quite good in many places, despite the long-voiced claims of the gradualists that transitional forms do not exist merely because of the deplorable state of the fossil record. In physics the analogous phenomena would be: (1) The chemical elements and their isotopes (or an atom's energy levels); and (2) The lack of transitional forms.

Straining the analogy still further, the evolution of one species into another simply means that life-as-a-whole moves from one quantized state to another. There need be no transitional forms, just as there are none when elements are transmuted or galaxies change redshifts (?). Atomic physicists, long since mystical about this whole business, no longer try to explain what happens during a quantum transition. The only observables are the quantum states---or species, if you will. Is life no more than a Table of Isotopes, defined once and forever by eerie quantum selection rules?

MECHANISMS OF EVOLUTION

The apparent quantized or punctuated nature of the fossil record has led to a search for evolutionary mechanisms with the potential for forcing sharp, substantive changes in the inherited characteristics of organisms. Such mechanisms would supplement or even supplant the small, random changes in the genome that supposedly provide the succession of finely graded, transitional forms required by ideal Darwinism. Five such candidates are introduced below:

- **Lamarckism. The inheritance of acquired characters could, in principle, be large and abrupt.**
- **Endosymbiosis. The absorption of one life form by another; viz., the acquisition of mitochondria by animal cells.**
- **Virus and plasmid carriers. The inter- and intraspecies transfer of genetic material by viruses, plasmids, and other biological vectors.**
- **Cell fusion.**
- **Spontaneous self-organization. The consequence of some vaguely defined intrinsic tendency of matter, including genetic materials, to organize into new configurations, leading to new species---a rather mystical concept.**

LAMARCKISM

THE PROPAGATION OF ACQUIRED CHARACTERISTICS

Almost all biologists reject Lamarck's idea that characteristics acquired by a parent can be transmitted to the progeny. In the field of immunology, especially, experimental findings are stimulating a revival of forbidden Lamarckism! Taylor reviews several experiments in which acquired immunity seems to be passed along from generation to generation. This, of course, directly contradicts the Dogma of Evolution and Weissmann's closely related doctrine of the inviolatability of the germ plasm. But Taylor goes on to suggest several ways to circumvent Weissmann's doctrine, the most interesting of which employs viruses to carry acquired genetic information from generation to generation. (Taylor, R.B.; "Lamarckism Revival in Immunology," Nature, 286:837, 1980.)

Comment. The possible role of viruses and other "disease carriers" in the unfolding (rather than "evolution") of life is only now being widely recognized. Could it be that the price of evolution and/or the responsiveness of life to environmental pressures is a certain level of infection?

THE EVOLUTIONARY STRUGGLE WITHIN

Ted Steele, an immunologist, has come up with experimental evidence showing in some cases that acquired immunity may be transmitted to progeny. When Steele's research was announced, many scientists and science writers rushed to the defense of Darwinism. They pointed out with unseeming vigor that a revival of dread Lamarckism or the Inheritance of Acquired Characters was not indicated.

It is true that Steele has proposed a Darwinian interpretation of his findings, but his theory adds a startling new dimension to the development of life. In essence, Steele asserts that an organism's immunological system is really the evolutionary scenario in miniature and compressed in time. The body's immunological system is trying to cope with up to 10 million defensive cells. The only defensive cells that survive and multiply are those that happen to encounter an invader that they can lock onto and destroy. The "fittest" defensive cells are those that have just the right characteristics to knock off invaders, and only they survive permanently in the body's defensive arsenal, giving it acquired immunity. The Lamarckian part of this story occurs when the RNA of the selected defensive cells gets passed on to the organism's progeny. (Tudge, Colin; "Lamarck Lives---In the Immune System," New Scientist, 89:483, 1981.)

Comment. The picture evolving here is one of a hierarchy of evolutionary struggles---say, humans on one level and their contained defensive cells on another level. The levels are not completely independent. The question that arises next is whether there are other evolutionary struggles going on, possibly in the mitochondria and chloroplasts, which possess their own genetic material. Or, waxing speculative, are there hierarchies of evolutionary struggle above humanity of which we know nothing except for perhaps a few anomalies representing cross talk between levels?

ENDOSYMBIOSIS

HERETICAL EVOLUTIONARY THEORY

> Over the past 15 years, away from the limelight of mainstream evolutionary argument, cell biologists have been debating a concept that is fundamental to our understanding of how cells evolved. It is the proposal that some of the structures that are found in the larger cells of animals, plants and fungi (eukaryotic cells) are the descendants of simpler bacteria-like organisms (prokaryotic cells) that had at some stage entered into an intracellular existence, or endosymbiosis. The idea is not a new one, but only in the light of modern experimental evidence has it become acceptable to many biologists. If the hypothesis is correct, then virtually all the major groups of familiar organisms originated 'suddenly' through endosymbiotic associations.

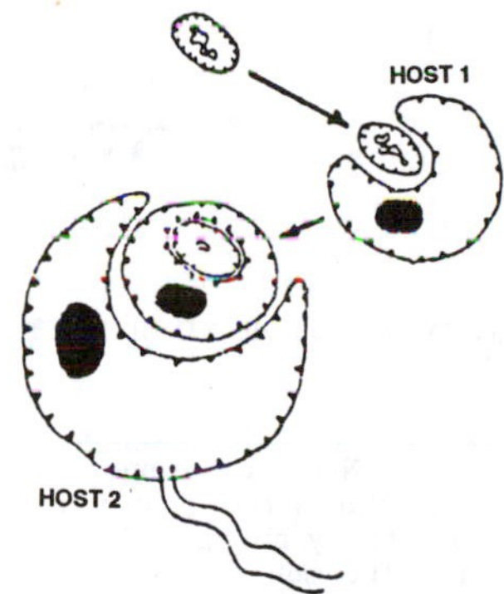

Serial symbiosis: bacterium to alga to alga. This process may have led to the development of the chloroplasts.

Following this lead paragraph, with its paradigm-shaking final sentence, are three pages summarizing the biological evidence favoring evolution by endosymbiosis. (Kite, Geoffrey; "Evolution by Symbiosis; The Inside Story," New Scientist, p. 50, July 3, 1986.)

Comment. We cannot possibly do justice to this exciting idea of evolution forced by the uniting of different organisms in the limited format of Science Frontiers. Instead, we encourage readers to purchase a new book by L. Margulis and D. Sagan (son of Carl Sagan and L. Margulis) entitled Micro Cosmos. In passing, we must also remark on the obvious relationship of endosymbiosis to F. Hoyle's "evolution from outer space." Hoyle believes that microorganisms and other biological information are transported in cosmic debris. Finally, the story of Lynn Margulis's uphill fight against scientific skepticism is related in: (Keller, Evelyn Fox; "One Woman and Her Theory," New Scientist, p. 46, July 3, 1986.)

EVOLUTION THROUGH MERGERS

The overview in Natural History describes how, in theory, the mitochondria in cells were created by bacterial invasion. The presence of chloroplasts in plants, too, may have come about in this way. A case also exists for the alliance of spirochetes with cells to form flagella and cilia. These three "mergers" provided cells with metabolism, photosynthesis, and mobility. Margulis and Sagan obviously do not believe that the "bacterial connection" ended there. They bring their article to a close with an almost poetic manifesto that we now quote in part. The context of the quotation is their assertion that plant and animal evolution would never have taken place unless one life form attacked another and the latter defended itself, all this followed by accomodation and the development of a symbiotic relationship.

> Uneasy alliances are at the core of our very many different beings. Individuality, independence---these are illusions. We live on a flowing pointillist landscape where each dot of paint is also alive. Earth itself is a living habitat, a merger of organisms that have come together, forming new emergent organisms, entirely new kinds of "individuals" such as green hydras and luminous fish. Without a life-support system none of us can survive. It is in this light that we are beginning to see the biosphere not only as a continual struggle favoring the most vicious organism but also as an endless dance of diversifying life forms, where partners triumph.

(Sagan, Dorion, and Margulis, Lynn; "Bacterial Bedfellows," Natural History, 96:26, March 1987.)

Comment. One should observe that there is a strong connection between the Gaia concept of a living planet and the theory of symbiotic evolution. Strong philosophical statements are also inherent in this outlook on life and its development. For example, individuality and free will would seem to be denied. Also, can life forms be "vicious" and yet "cooperative" at the same time?

VIRUS AND PLASMID CARRIERS

VIRUSES AS AGENTS OF EVOLUTION

The role of viruses in transferring genetic material across species barriers is at last getting some serious attention. D. Erwin and J. Valentine of the University of California are now pointing out how a whole colony of "hopeful monsters" might be created en masse by an attack by viruses carrying new blueprints. (Anonymous; "Gene-Swapping Breaks Barriers in Evolutionary Theory," New Scientist, p. 19, February 21, 1985. Also: National Academy of Sciences, Proceedings, 81:5482, 1984.)

Comment. We don't want to get carried away, but it just may be that all life forms are interconnected informationwise by viruses, diseases, symbiotic relationships, and similar "channels." Gaiawise, there could be just a single superorganism in the universe, which is exploring and experimenting!

THE NOMADS WITHIN US

It was originally believed that human chromosomes were fixed at conception and all subsequent organic development proceeded from the instructions encoded on them. Biologists have recently discovered that genes grasshopper about, constantly modifying genetic instructions, at least that's current thinking.

Additional modification of genetic instructions seems to be accomplished by entities called "nomads" or "mobile dispersed genetic elements." One type of nomad is a simple ring of DNA called a plasmid. Plasmids seem to be identical to a kind of virus called a retrovirus, which can penetrate into cells and tamper with gene expression; that is, the way genetic instructions are interpreted. Plasmids have been discovered in maize, fruit flies, bacteria, and now, man---and healthy people, at that. No one is quite sure what these plasmids do. Even though they look like retroviruses, they may not be associated with illness but rather help organisms adapt to changing environments. But no one really knows. (Anonymous; "Human Wandering Genes Can Live on Their Own," New Scientist, 94:18, 1982.)

Comment. So, the human body is not only beset by new genetic instructions and static introduced by invading viruses and other disease agents, but it has an indigenous population of nomads continually fiddling with our cells' genetic instructions. Our bodies seem like Grand Central Station with trains loaded with new biological ideas constantly arriving from far and near---maybe even from outer space a la Fred Hoyle's Diseases from Space!

CELL FUSION

THE UNIVERSAL URGE TO JOIN UP

Take a mouse cell and place it in contact with a human cell. The two separating membranes will dissolve and the cell contents will mix. The once-independent and widely different cell nuclei will fuse, forming a single hybrid cell with a common membrane. Even more astonishing, this totally new biological entity will often divide and produce an endless line of the new hybrid. As might be expected, some hybrids do not remain true and revert to one or the other of the original species.

Although cell fusion has been observed only under laboratory conditions, it seems to represent a near-universal cell phenomenon that might be realized rarely under natural conditions. The implications for the history of life are far-reaching. For example, the mitochondria in human cells that help our bodies use oxygen to obtain energy may well be descendants of bacteria that once fused with primitive cells. The same may be true for the chloroplasts in plant cells. (Thomas, Lewis; "Cell Fusion: Does It Represent a Universal Urge to 'Join Up'?" Science Digest, 86: 52, December 1979.)

Comment. Natural cell fusion might make large evolutionary steps possible and be much faster than endless small genetic changes. Are we all composite creatures?

SPONTANEOUS SELF-ORGANIZATION

SPONTANEOUS ORDER, EVOLUTION, AND LIFE

In our thinking, one of the most remarkable articles ever to appear in Science bears the above title. Most remarkable of all is the use of the word "spontaneous" without philosophical comment.

The stimuli for the research described are such observations as: (1) life exists; (2) life evolves; (3) the fossil record displays stasis, extinctions, and great gaps between phyla and lesser classifications; and (4) disordered molecules move smoothly and surely into the order manifest in the living cell. The question asked in the article is whether science has missed something in its description of the origin and development of life. Just what makes molecules coalesce into cells and humans? The answer given is: spontaneous self-organization! In other words, there is no guiding external force. Molecules do this spontaneously. There are even computer models being developed, based on a branch of mathematics called "dynamical systems," that describe how this all happens---spontaneously, of course. (Waldrop, M. Mitchell; "Spontaneous Order, Evolution, and Life," Science, 247:1543, 1990.)

Comment. When water molecules spontaneously cluster together to form a snowflake, with all its symmetry and order, science explains the process in terms of the properties of water molecules. The same must be true when molecules merge to form life forms. But why do atoms and molecules possess these properties that lead to bacteria, to humans, to who-knows-what's-next? "Spontaneous self-organization" is a cop out!

OTHER CONCEPTS

EVOLUTION'S MOTOR RUNS FAST AND QUIETLY

M. Kimura has been promulgating what he terms The Neutral Theory of Evolution. He concludes that "the most prevalent evolutionary changes that have occurred at the molecular level, that is in the genetic material itself, since the origin of life on Earth are those that have been caused by random genetic drift rather than by positive Darwinian selection." Kimura maintains and can experimentally prove to some extent that the genetic material of all organisms changes rapidly and constantly. The genes change much more rapidly than scientists believed just a few years ago. We observe few if any changes at the phenotype level (organism morphology) because the overwhelming majority of these changes are neutral. They confer no significant advantage or disadvantage on the organism. In fact, Kimura and others have demonstrated that the fastest molecular evolution occurs in the least important genes. It is fastest of all in the pseudogenes or dead genes, which seem to have no discernable functions. In other words, the genetic engine is running, but the gears are in neutral!

In Kimura's thinking, these molecular changes may eventually become important at the phenotype level if the environment changes or there is some other destabilizing influence. Kimura is the author of the 1983 book The Neutral Theory of Molecular Evolution. Although Kimura has some experimental support, his theory is not widely accepted. (Motoo Kimura; "The Neutral Theory of Molecular Evolution," New Scientist, p. 41, July 11, 1985.)

Comment. The intriguing part of Kimura's article concerns that steady hum of gene changes---all to little or no avail when times are stable--but seemingly ready to provide the genetic pressure required to fill new niches. Why do these nonliving molecules have just those properties vital to life? It hardly seems sufficient to say that if nonlife did not have the properties it has we wouldn't be here. It raises again the question of whether the human blueprint is implicit in the electron and other simple particles. If so, what blueprints reside there still unrealized?

GEOPHYSIOLOGY

The concept of the earth-as-an-organism---the so-called Gaia hypothesis---keeps popping up. We now have an excellent progress report on current Gaia research; e.g., the Daisyworld model; by J.E. Lovelock. He defines Gaia in an early paragraph:

> In the early 1970s, Lynn Margulis and I introduced the Gaia hypothesis. It postulated the earth to be a self-regulating system comprising the biota and their environment, with the capacity to maintain the climate and the chemical composition at a steady state favorable for life.

L. Margulis, a coauthor of Micro-Cosmos, is a champion of evolution-via-endosymbiosis, in which diverse organisms uniting to create new species. Going back to Lovelock's review, there is little that is anomalous on a small scale. Of course, on a large-scale, the data supporting the concept of life-as-a-whole manipulating the atmosphere, oceans, etc., to perpetuate and perhaps improve itself are highly anomalous, because the Gaia hypothesis is far out of the scientific mainstream. (Lovelock, James E.; "Geophysiology," American Meteorological Society, Bulletin, 67:392, 1986.)

Comment. Our secret purpose here is to use the Lovelock article as an excuse to out-Gaia Gaia! Lovelock's article plus those preceding on Martian life, cosmic life, "geocorrosion," etc., made us wonder if Gaia as a closed terrestrial system (see diagram), is not too limited. If Hoyle and Wickramasinghe are correct, the diagram should have a box

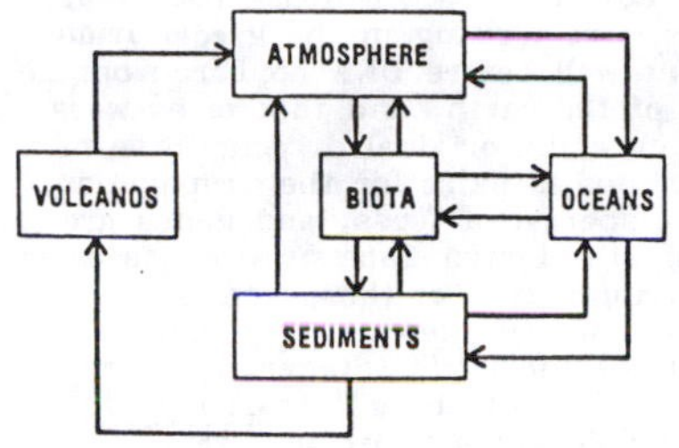

labelled "outer space," with an inwardly directed arrow carrying life-forms (Hoyle's space viruses and bacteria), meteorites, icy comets, etc. Likewise, the earth can contribute life-forms to the cosmos via impact and volcanic ejecta. Where does geocorrosion fit in? Life-as-a-whole could control terrestrial magnetic field reversals geochemically. This sounds more and more like science fiction, but life-as-a-whole must "want" to evolve to make itself more adaptable and capable of controlling and exploring the cosmos. (These are anthropomorphic desires we assign to life-as-a-whole, which may have completely different objectives!) By occasionally reducing the earth's field to zero, bursts of space radiation would be admitted to stir the earth's pot of genes. We could also work in "selfish genes" and God, but it is time to go back to anomalies once more. See p. 235 for "geocorrosion."

EVOLUTION BY NUMBERS

The following paragraph is taken from a letter to Nature by a "practising geneticist."

> In the discussion in your columns about the application of quantitative methodology based on the study of evolutionary processes to the analysis of the development of human culture, there is an unquestioned assumption on both sides of that issue that quantitative theory, as expounded by practitioners such as Fisher, Haldane, Wright, Cavalli-Sforza and Maynard Smith, has been successful in illuminating and explaining the process of biological evolution and the genetic relationships between species. As far as I know, there is no evidence to support this assumption. Indeed, there is a vast number of observations unaccounted for in the extant quantitative evolutionary theories. Many of these observations (inducible mutation systems, rapid genomic changes involving mobile genetic elements, programmed changes in chromosome structure) challenge the most fundamental assumptions which these evolutionary theories make about the mechanisms of hereditary variation and the fixation of genetic differences.

(Shapiro, James A.; "Evolution by Numbers," Nature, 303:196, 1983.)

Comments. The "observations unaccounted for" are buried in such obscure journals as S.B. ges. Morph. Physio. (Munchen). It is pretty obvious that the Sourcebook Project is just scratching the surface.

PHILOSOPHICAL CONFUSION?

In the past, creationists have claimed Sir Karl Popper as their own because he seemed to believe that evolution was not a science because it could not be falsified. More lately, the evolutionists have pronounced that Popper now supports Darwinism. Still more recently, Popper gave the first Medawar Lecture at the Royal Society; and he had something for both sides. How can this be?

First, for those who do not know of Sir Karl Popper, we should state that his philosophical views carry considerable weight in scientific circles. He has described how science should work and how its hypotheses should be tested. In this sense, he has defined what is scientific and what isn't.

Now, back to the Medawar Lecture, as recounted by M. Perutz. Popper declared his detestation of determinism in all its guises. "Popper disputes the existence of historical laws and holds that our future is in our own hands." Zeroing in on evolution, Popper accepts Darwinism in the sense that "organisms better adapted than others are more likely to leave offspring." He then splits Darwinism into passive and active forms. His passive variety of evolution is that which is currently in vogue---the deterministic view that random mutation combined with natural selection invariably leads to higher forms of life. But, as already stated, Popper hates determinism and believes that deterministic mechanisms are noncreative. They lead only to deadends. Instead, he prefers "active" Darwinism in which the "idiocyncracies of the individual have a greater influence on evolution than natural selection" and that "the only creative activity in evolution is the activity of the organism."

There you have it! Whose side is Popper on? Does it really matter? (Perutz, Max; "A New View of Darwinism," New Scientist, p. 36, October 2, 1986.)

Comment. It is unclear how the acts of individuals can modify organisms. Sounds like Lamarckism.

MIND BEFORE LIFE

For some years S.W. Fox, at the Institute for Molecular and Cellular Evolution, University of Miami, has been experimenting with possible precursors of life. So-called "microspheres" are hot items in Coral Gables these days. Fox and his colleagues make microspheres by preparing a heated stew of various amino acids. These amino acids form long polymer chains spontaneously. Then, when water is added and the mixture reheated (or processed in some other way), the polymers organize themselves, again spontaneously, into spheres a few microns in diameter. Each sphere consists of a two-layer membrane with residual material trapped inside. Although thicker, the microsphere membrane is very similar to the lipid bilayer enclosing normal living cells. The relatively stable microspheres could, in theory, have formed sheltered environments for the evolution of the more complicated parts of living cells.

The microspheres absorb sunlight and, with the addition of this energy, display some of the electrical characteristics of biological neurons, like those in the brain. The implication is that some components of "mind" may have existed in the very earliest life forms. (Peterson, Ivars; "Microsphere Excitement," Science News, 125:408, 1984.)

Comment. Two comments here: First, the word "spontaneous" is customarily employed when describing how atoms unite to form molecules and molecules combine into polymers, which then gather into microspheres. The word "spontaneous" seems to imply chance is operating rather than design. Actually, atoms and even subatomic particles must have innate properties which force them to combine into larger structures the way they do. Philosophically, one can ask whether the lowliest subatomic particles are "coded" to combine into molecules, microspheres, and living creatures. In other words, the design of life could be inherent in quarks.

Second, if nonliving microspheres possess some of the properties of neurons, it is possible that natural, nonliving minds can form spontaneously---a sort of "natural" artificial intelligence! Mind, then, could precede life, which is manifestly more complex than computers. The science-fiction possibilities are endless here.

EVOLUTIONARY MISCELLANY

IS A DOG MORE LIKE A LIZARD OR A CHICKEN?

Cladism has been around since at least 1950, when the German entomologist William Hennig began classifying organisms in a new way. "His method was simple: classifications of animals and plants should be based on an assessment of their characters, and only advanced, or derived, characters (synapomorphies) should be used. For example, the character 'possession of feathers' is a synapomorphy shared by all birds that is found in no other organisms; it is used in classification to define Class Aves. All birds have a backbone too, but that is a primitive character for birds because it is seen in all other vertebrates. Primitive characters are of no use in defining a monophyletic group." Early on, cladism ignited many fireworks in the biological community. Today, cladism has become quite respectable.

The cladists, however, are now fighting among themselves. One camp draws their "cladograms" (see illustration) using evolutionary theory as a guide. The other side believes that the cladograms should be drawn up first, based only upon actual characters and ignoring the theory of evolution---let the species fall where they may. When evolutonary theory is omitted in the deliberations, radically different cladograms result. Mammals then seem more closely related to birds than reptiles, for example, as expressed in the illustration. A key connecting character here is mutual warm-bloodedness. The shifting of a few lines in a cladogram may seem trivial to the nonbiologist, but saying that man is more closely related to a chicken than a lizard is pretty controversial stuff to the conservative evolutionist.

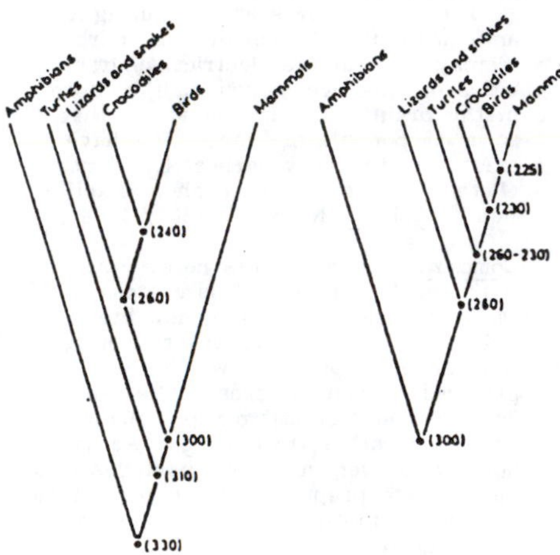

(Left) A standard evolutionary cladogram. (Right) A cladogram based only upon characters. The numbers represent millions of years before the present.

DNA analysis supports the contention that humans are more closely related to birds than reptiles. Paleontological evidence, however, supports the opposite view. The fight rages on. (Benton, Mike; "Is a Dog More Like a Lizard or a Chicken?" New Scientist, 18, August 16, 1984.)

Comment. It all seems to boil down to morphology in the end. Which is a more faithful record of the development of life through the long eons, the phenotype (what the organism looks like) or the genotype (what the DNA looks like)? This may be a dangerous simplification. Why? Because life just may be shaped by more than DNA.

EXTRATERRESTRIAL INFLUENCES ON CHEMICAL AND BIOLOGICAL SYSTEMS

Conventional science shows little interest in the subject indicated by the title, except for some work that is done on circadian rhythms. However, readers of the journals Cycles and the Journal of Interdisciplinary Cycle Research are treated regularly to a wide variety of purported correlations of biological systems with solar and other extraterrestrial influences. The present paper suggests that extraterrestrial forces influence the earth's weather which, in turn controls physiological processes. The physiological processes studied include blood precipitation rate and blood hemoglobin values. Also mentioned are Piccardi's precipitation-rate experiments that seem to show a highly variable behavior of simple chemical systems that bear no obvious relationship to weather conditions. Tromp concludes from these data that unknown forces, probably extraterrestrial in nature, act upon the earth and its inhabitants. (Tromp, Solco W.; "Study of Possibly Extraterrestrial Influences on Colloidal Systems and Living Processes on Earth," Cycles, 28:34, 1977.)

WHICH CAME FIRST?

The advent of complex life keeps getting pushed back farther and farther in time, as evidenced by the following abstract:

Abstract. "Microfossils resembling fecal pellets occur in acid-resistant residues and thin sections of Middle Cambrian in Early Proterozoic shale. The cylindrical microfossils average 50 x 110 µm and are the size and shape of fecal pellets produced by microscopic animals today. Pellets occur in dark gray and black rocks that were deposited in the facies that also preserves sulfide minerals and that represent environments analogous to those that preserve fecal pellets today. Rocks containing pellets and algal microfossils range in age from 0.53 to 1.9 gigayears (Gyr) and include Burgess Shale, Greyson and Newland Formations, Rove Formation, and Gunflint Iron-Formation. Similar rock types of Archean age, ranging from 2.68 to 3.8 Gyr, were barren of pellets. If the Proterozoic microfossils are fossilized fecal pellets, they provide evidence of metazoan life and a complex food chain at 1.0 Gyr ago. This occurrence predates macroscopic metazoan body fossils in the Ediacaran System at 0.67 Gyr, animal trace fossils from 0.9 to 1.3 Gyr, and fossils of unicellular eukaryotic plankton at 1.4 Gyr." (Robbins, Eleanora Iberall, et al; "Pellet Microfossils: Possible Evidence for Metazoan Life in Early Proterozoic Time," National Academy of Sciences, Proceedings, 82:5809, 1985.)

The senior author of the above paper also submitted a unique interpretation of the data by J.C. Stager. Stager begins by noting that the paper of Robbins et al has been criticized because the earliest known fossils of metazoans date back to only about 1 Gyr and, therefore, the supposed fecal pellets were obviously something else. Stager next makes a giant conceptual leap: Quite clearly the data prove that feces evolved before animals did. He goes on, tongue in cheek:

"In standard systematic reasoning, one assumes that the most widespread characteristic represents the primitive state. The fact that feces look so much the same from individual to individual strongly suggests that feces are the primitive condition. The variety of animal bodies, on the other hand, implies that bodies are secondary or derived features of the organisms. The expansion of genetic research in the twentieth century has led to the conclusion among many geneticists that bodies exist sorely for the propagation and dispersal of genes. This perspective has been dubbed the 'selfish gene theory.'

"While the author acknowledges the insight and creativity that went into the development of the selfish gene theory, it must be pointed out that the idea has not been carried far enough by the geneticists. Where did the genes come from in the first place? Who ever heard of a sea bottom made up of DNA ooze? It is obvious from the fossil data that feces were teeming in the Precambrian oceans well before DNA appeared on the face of the earth, and that feces were therefore the original driving force of life. Bodies exist for the propagation and dispersal of feces, and genes are simply the instructions used by feces in the manufacture of those bodies. This concept is best described as 'the selfish feces theory.'" (Stager, J. Curt; "The Origin of Feces," Journal of Irreproducible Results, p. 20, 1986.)

Chapter 4

GEOLOGY

TOPOGRAPHY

GEOLOGICAL ANOMALIES: CHEMICAL, PHYSICAL, BIOLOGICAL

STRATIGRAPHY

INNER EARTH

TOPOGRAPHY

CRATERS AND DEPRESSIONS

Like a snow-capped mountain, a gaping crater in the earth's surface quickly captures the traveler's attention. Meteor Crater, in Arizona, is a prime example, being a favorite tourist attraction. But Meteor Crater, only a mile or so wide, is too small to be called anomalous or even curious anymore. To qualify for inclusion in Science Frontiers, craters must be tens or hundreds of miles in diameter, or present in large numbers, such as the half million Carolina Bays. Why be so concerned about craters? Because they are linked today with massive biological extinctions in the fossil record, as discussed on p. 217. Further, some may not be of meteoric origin after all.

Other "negative" topological features included in this section vary from small "cookie-cutter" holes to strange ring structures hundreds of miles across seen in satellite photos of the American Southwest. This section will adhere to the following course:

- Impact craters. Very large craters, including the effects of the Tunguska event and the nature of the multitudinous Carolina Bays.
- Large ring structures.
- Natural-gas craters. As created by spontaneous explosions and (supposedly) eruptions of methane from the sea floor.
- Cookie-cutter holes. A surprising and mysterious phenomenon.

IMPACT CRATERS

THE RISE OF ASTRONOMICAL CATASTROPHISM

After being ridiculed for well over a century, astronomical catastrophism is now coming into its own. First, there was the admission that a few small craters, like Meteor Crater in Arizona, just might be of meteoric origin; then, more and bigger craters (astroblemes) were recognized; and, recently, the discovery of the iridium-rich layer at the Cretaceous-Tertiary boundary has made the subject very popular, as evidenced by the following three items:

(1) A long, very thorough and scientific review of geological and biological changes caused by meteor strikes throughout the earth's history. (McLaren, Digby J.; "Bolides and Biostratigraphy," Geological Society of America, Bulletin, 94:313, 1983.)

(2) A shorter, popular version of the above. (McLaren, Digby; "Impacts That Changed the Course of Evolution," New Scientist, 100:588, 1983.)

(3) Evidence is growing that the collision of planetary material with the Earth can profoundly affect local geology, and that impacts of very large meteorites may have influenced the evolution of the Earth and the life that exists upon it.

This quotation is from the lead-in to the article referenced below. which also has a nice world map of major impact sites over 1 km in diameter. (Grieve, Richard; "Impact Craters Shape Planet Surfaces," New Scientist, 100:516, 1983.)

EVERGLADES ASTROBLEME?

Astrobleme means "star wound," and the southern tip of Florida seems to have been wounded by an asteroid or some other celestial projectile. At a recent meeting of the Geological Society of America, E.J. Petuch proposed that the Everglades region received a direct hit from an asteroid about 36 million years ago.

The Everglades region is a swampy, forested area surrounded by an oval-shaped system of ridges. Geologists usually maintain that the Everglades represent a collapse feature caused by groundwater dissolving away limestone. (Buildings and cars seem to be swallowed fairly regularly by Florida sinkholes.) Petuch disagrees with the collapse theory and points to the following evidence for an impact origin:

1. The presence of a strong positive magnetic anomaly;
2. Eocene formations, 40 million years old, are missing over the southern Everglades;
3. A network of fractures pervades rock layers older than Eocene;
4. High iridium concentrations, probably of extraterrestrial origin, exist at the Eocene-Oligocene boundary on nearby Barbados; and
5. The oval reef structure that seems to have grown around the impact area as sealevels rose.

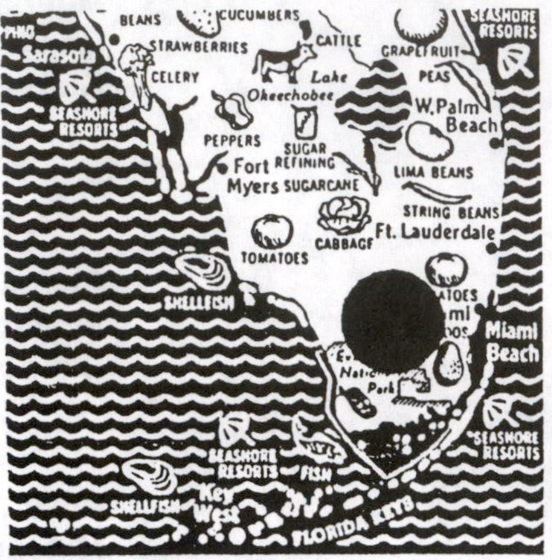

Time, of December 9, 1985, has a nice map of the Everglades asteroid's "footprint," but copyright law prevents us from using it. So, we have made our own! The solid black circle is the collapse basin surrounding the impact point. The elliptical coral reef is tangent to the southern rim of the collapse basin and runs northwest through the tomatoes and loops around Lake Okeechobee between the peas and lima beans.

Some geologists do not concur with the asteroid theory, but they are all reviewing Florida's geological history in a new light. (Weisburd, S.; "Asteroid Origin of the Everglades?" Science News, 128:294, 1985.)

TWO POINTS OF GREAT IMPACT

Geologists have been searching in vain for a large crater that might account for the biological extinctions at the Cretaceous-Tertiary boundary some 65 million years ago. C.J.H. Hartnady believes he had found the culprit. It is somewhat larger than expected (300 kilometers in diameter instead of 100-200), but it is of the right age. Supporting this notion is the observation that the Seychelles Bank and Madagascar suddenly shifted their locations at about this time. (Murray, M.; "Point of Impact: The Indian Ocean," Science News, 129:356, 1986.)

The existence of another terrestrial cataclysm at an earlier date is suggested by a layer of shattered crustal rock fragments stretching over at least 260 kilometers in South Australia. Folded within Precambrian marine shales, these fragments reach 30 centimeters in diam-

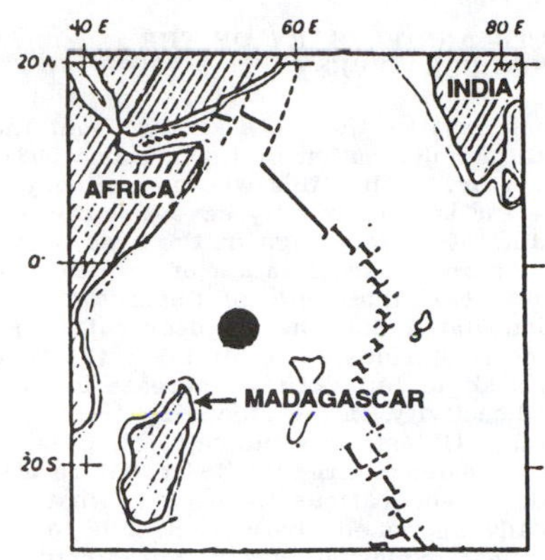

The Amirante Basin (black circle) lies about 500 kilometers northeast of Madagascar.

eter and show evidence of vertical fall. Evidence points to an origin near Lake Acraman, about 300 kilometers west. (Gostin, Victor A., et al; "Impact Ejecta Horizon within Late Precambrian Shales, Adelaide Geosyncline, South Australia," Science, 233:198, 1986.)

TERRESTRIAL MARIA?

Those dark patches that constitute the face of the man-in-the-moon are really huge outpourings of lava. Once considered real seas, they are called "maria." Maria like those on the moon also exist on the earth. This is a surprise because terrestrial maria were never mentioned in my college geology courses. But, in those days, the one-mile-wide Meteor Crater, in Arizona, was the largest accepted consequence of celestial bombardment.

D. Alt and colleagues, at the University of Montana, have identified four large basalt plateaus that might be terrestrial maria:

> Notable examples of large lava plateaus that resemble lunar maria include, among others, the Deccan Plateau of India, the Columbia Plateau of western North America, the Parana Plateau of South America, and the Tunguska Basin of Siberia. All consist mostly of basalt lava flows; those on continents include minor quantities of rhyolite, and variable amounts of sediment. All seem to have appeared suddenly, within plates. No consistent context of plate interactions explains them. We suggest that large lava plateaus are indeed terrestrial maria.

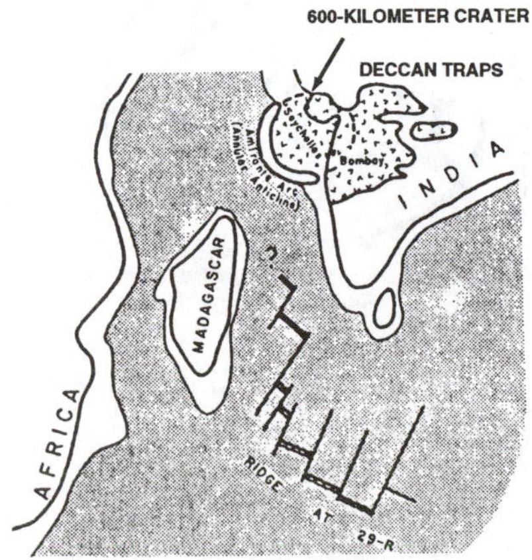

Deccan basalt flows (also called the Deccan Traps) in western India, shown before the formation of the Carlsberg Ridge and the splitting off of the Seychelles.

Alt et al go on to show that these lava plateaus seem to have initiated continental rifts and hotspot tracks where none existed before. A reasonable inference is that these plateaus are the consequence of the impacts of large meteorites. This is particularly the case with the Deccan Plateau, which is agedated as synchronous with the Cretaceous-Tertiary Boundary event, with its legacy of worldwide iridium deposits and the wholesale extinction of life. The paper concludes with:

> It therefore appears that random encounters with vagrant asteroidal objects play an important role in setting the course of plate tectonic events. The earth does not control its own agenda.

(Alt, D., et al; "Terrestrial Maria: The Origins of Large Basalt Plateaus, Hotspot Tracks and Spreading Ridges," Journal of Geology, 96:647, 1988.)

IMPACT CRATER BENEATH LAKE HURON

> With the help of magnetic sensors, scientists have detected a rimmed circular structure, 30 miles in diameter, more than a mile beneath the floor of Lake Huron. They believe the magnetic ring marks a buried crater---blasted by a meteorite at least 500 million years ago.

(Stolzenburg, W.; "Impact Crater May Lie beneath Lake Huron," Science News, 138:133, 1990.)

POSSIBLE CHAIN OF METEORITE SCARS IN ARGENTINA

In the January 16, 1992, issue of Nature, P.H. Schultz and R.E. Lianza describe a curious chain of grooves incised in the Argentine pampas near Rio Cuarto.

> During routine flights two years ago ..., one of us (R.E.L.) noticed an anomalous alignment of oblong rimmed depressions (4 km x 1 km) on the otherwise featureless farmland of the Pampas of Argentina. We argue here, from sample analysis and by analogy with laboratory experiments, that these structures resulted from low-angle impact and ricochet of a chondritic body originally 150-300 m in diameter.

There are ten gouges in all, strung out along 50 kilometers. The scars are young, perhaps only a few thousand years old, well within the time of human habitation. Schultz and Lianza also found pieces of meteoritic rock and glassy fragments of impact melt. (Schultz, Peter H., and Lianza, Ruben E.; "Recent Grazing Impacts on the Earth Recorded in the Rio Cuarto Crater Field, Argentina," Nature, 355:234, 1992. Also: Monastersky, R.; "Meteorite Hopscotched across Argentina," Science News, 141:55, 1992.)

Comment. Note the similarities to the much more numerous Carolina Bays. See ETB1 in the Carolina Bays, Mima Mounds, Submarine Canyons. More recently, doubts have been raised concerning the meteoric origin of these Argentine scars.

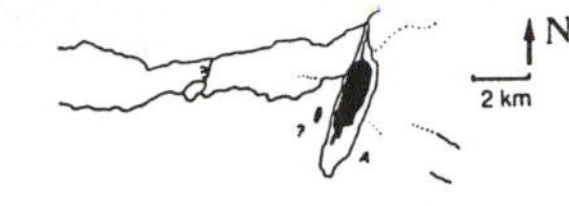

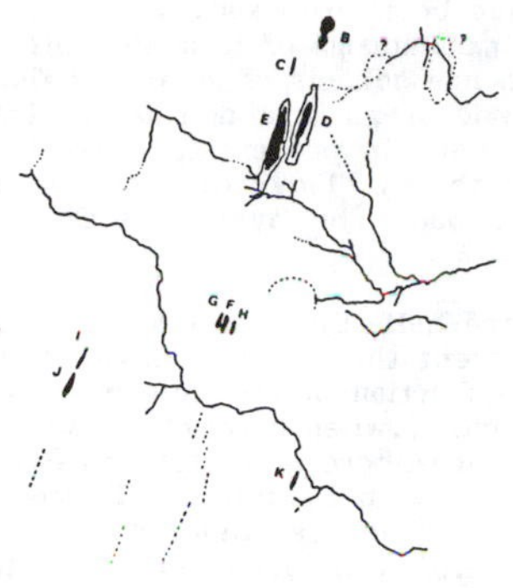

A string of linear depressions characterizes the hypothesized Rio Cuarto crater field in Argentina.

DID A HALF MILLION METEORS FALL ON THE CAROLINAS?

The Carolina Bays get scant notice in the literature these days, but E.R. Randall has rescued them this undeserved obscurity.

> For years, people living along the Carolina coast have marveled at a series of strange, oval-shaped depressions in the ground called "Carolina Bays."
>
> From the air these shallow, marshy bogs created a landscape that resembles the pockmarked surface of the moon. They crisscross each other in a chaotic tapestry, but at ground level are hardly noticeable because of thick forests and semitropical swamplands.
>
> Highways and modern housing developments have all but obliterated thousands of the bays, leaving them visible only to trained eyes.
>
> Still, it is estimated that no fewer than 300,000 such bays, ranging from a few feet across to almost two miles in diameter, dot the East Coast landscape from southern New Jersey to northeastern Florida. One source places the number at more than half a million.

Floyd continues with a brief history

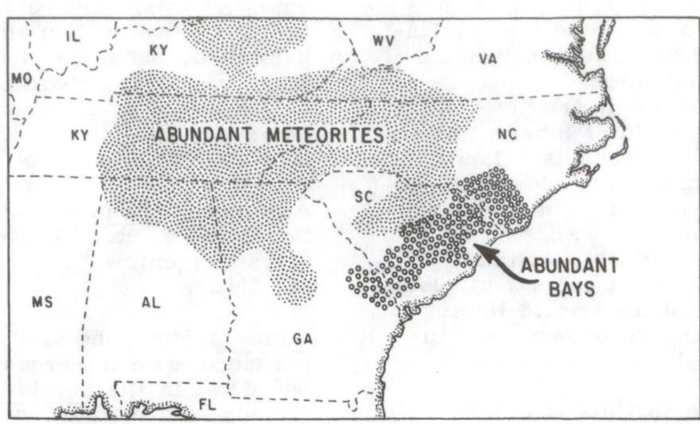

Map showing areas of abundant Carolina Bays and frequent meteorite finds. However, meteorites are rare in the area of the bays themselves.

of the Carolina Bay region and then reviews some of the theories of origin that have been proposed. Two now-discarded mechanisms of formation invoked: (1) immense schools of spawning fish; and (2) icebergs stranded as the Ice Ages waned. In presenting today's favorite theory, Floyd quotes from H. Savage's book The Mysterious Carolina Bays:

> "These half-million shallow craters represent the visible scars of but a small fraction of the meteors that fell to earth...when a comet smashed into the atmosphere and exploded over the American Southeast," Savage wrote. "Countless thousands of its meteorites must have plunged into the sea beyond, leaving no trace; while other thousands fell into the floodplains of rivers and streams that soon erased their scars."

(Floyd, E. Randall; "Comet May Have Created Carolina Bays," Birmingham News, May 16, 1992. Cr. E. Kimbrough.)

Comment. Floyd neglected to mention that D. Johnson, a critic of the comet theory, wrote a whole book (The Origin of the Carolina Bays) based on his own theory of spring-sapping.

TUNGUSKA-LIKE EVENT IN NEW ZEALAND 800 YEARS AGO?

> Very calm and placid have become
> the raging billows,
> That caused the total destruction
> of the Moa,
> When the horns of the Moon fell from
> above down.

Thus have the Maoris sung. Their myths, songs, and poetry clearly link the demise of the moa, not to their own overhunting as others maintain, but rather to a cataclysmic event that occurred some 800 years ago. Maori oral history tells of "the falling of the skies, raging winds, upheaval of the Earth, and mysterious devastating fire from space." Even some of the place names in New Zealand relate to some kind of catastrophe. In the province of Otago, there is Waipahi (place of the exploding fire) and Tapanui (big explosion).

Oral history is entertaining, but scientists want something more palpable before they will entertain Velikovskian ideas about recent history. Well, if you visit Tapanui (big explosion place), you can find Landslip Crater, a 900 x 600-meter depression 130 meters deep. This does not have the appearance of a bona fide meteor crater, but all around it are suspicious signs. For example, tree-fall distribution from 800 years ago was radially away from Tapanui out to 40-80 kilometers. In the same area one finds the trinities, small globules of silicates with tektite overtones. And then there is the extirpation of the moas about this time.

To be sure, there are separate, conventional explanations of all these phenomena. But, if you add the Maori oral traditions to all these suspicious physical signs, a Tunguska-like event does not seem impossible. (Steel, Duncan, and Snow, Peter; "The Tapanui Region of New Zealand: A 'Tunguska' of 800 Years Ago?" paper at the Conference on "Asteroids, Comets, Meteors, '91," Flagstaff, June 1991.)

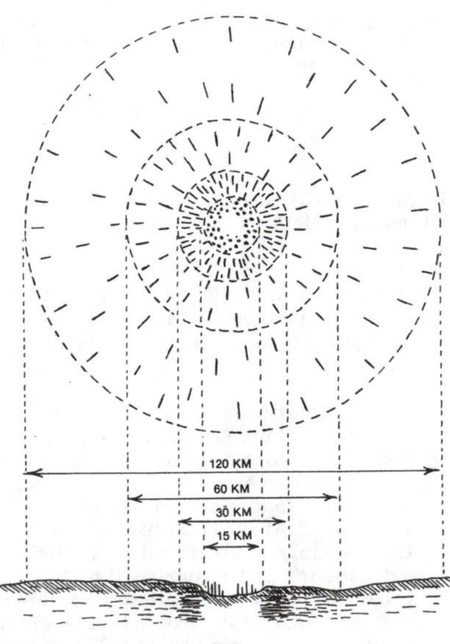

Pattern of fallen trees in the area of the Tunguska Event. Strangely, trees in the center were left standing.

70TH ANNIVERSARY OF THE TUNGUSKA EVENT

Rich reviews the facts known about the fall and detonation of the famous 1908 "meteor." That this was no ordinary meteor is born out by several curious data: (1) tree-rings in the area show an enormous acceleration of growth since 1908; (2) inhabitants of this remote region stated that the reindeer suffered from mysterious scabs in 1908; (3) there is a slight but definite increase in the radioactivity of the surviving trees; and (4) testimony indicates that the meteor changed direction twice before impact. The various theories of what really happened, from black hole to nuclear explosion, are listed without comment. (Rich, Vera; "The 70-Year-Old Mystery of Siberia's Big Bang." Nature, 274:207, 1978.)

GRAVITY ANOMALY RIPPLES CENTERED IN CANADA

When scientists recently examined gravity anomaly data for North America, strange circular ripples appeared to surround a point near Hudson Bay. These ripples seem to have spread out like those from a pebble dropped into a pond, but here the ripples are actually ancient density variations in the earth's crust, now covered over by thick sediments. One hypothesis is that a 60-90 kilometer meteorite smashed into the earth some 4 billion years ago, wrinkling the young surface for several thousand kilometers in all directions around a colossal crater. Magma welling up in the crater solidified creating the nucleus of the North American continent. It is quite possible that the other continents began their existences in this way---meteor impact. The gravity data that led to this hypothesis have been available for some time but apparently no one ever looked at them with continental patterns in mind. (Simon, C.; "Deep Crust Hints at Meteoric Impact," Science News, 121:69, 1982.)

Gravimetrically measured ripples surround a possible impact structure near Hudson Bay.

Comment 1: John Saul has discovered surface indications of immense ring structures in the American Southwest. See p. 195 and fuller treatment in Carolina Bays, Mima Mounds, Submarine Canyons.

Comment 2: If all our continents were initiated by meteor impacts, and if they were once clustered together in a supercontinent, as postulated by Continental Drift, then the incoming meteorites would have to have been focussed on a restricted portion of the earth's surface; that is, where the supercontinent was formed prior to continental drift. Several solar system bodies show just such preferential cratering on one hemisphere.

LARGE RING STRUCTURES

IMMENSE CIRCULAR TERRESTRIAL STRUCTURES OF GREAT AGE

J.M. Saul has analyzed many topographic and geological maps and discovered faint circular terrestrial patterns that have mostly not been described before. These circles measure between 7 and 700 kilometers in diameter and are nearly perfect geometrical figures. The rims of the circles are generally raised and characterized by fracturing and brecciation. These structures can be traced in many geological environments and rocks of all ages. They control to an extraordinary extent regional geology and ore mineralization.

To date, some 1,170 circles have been discovered, of which more than half can be visually traced for 360°. Larger circles may exist; one with a diameter of about 2,200 kilometers seems to encircle the southern end of Africa. In the United States, the centers of the circles fall in a northwesterly trend in Arizona; northeasterly in the Appalachians. These circular structures may have been created about 4 billion years ago by intense meteorite bombardment similar to and perhaps identical with the bombardment that marked the surfaces of the moon and other inner planets. (Saul, John M.; "Circular Structures of Large Scale and Great Age on the Earth's Surface," Nature, 271:345, 1978.)

MORE HUGE TERRESTRIAL RINGS

In 1978, J.M. Saul reported the discovery of hundreds of large terrestrial ring structures, many of them tens of kilometers in diameter. Around the rims were concentrated metal ores. Saul surmised that episodes of meteor bombardment about 4 billion years ago created these nearly perfect circles. Unfortunately for this theory, little of the earth's surface is 4 billion years old!

Now, a new population of circular structures 8-20 km across have been recognized in New Zealand, where the surface is dated at 100-320 million years. How could 4-billion-year-old meteor craters make themselves felt through all the layers of sediment? One possibility is that the rings are not meteoric but diapiric; that is, expressions of upwelling magma from inside the earth itself. (Hawkes, Donald D.; "More Strange Circles on Earth," Open Earth, no. 7, p. 19, February 1980.)

Comment. The "orphans" mentioned on p. 225 may have been forced up from the interior, too. Could the lunar craters also have originated thus?

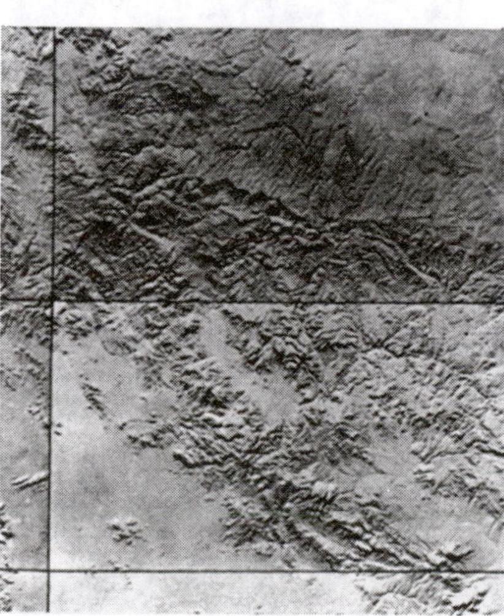

A 70-kilometer ring structure seen on a relief map of Arizona. This ring is the largest seen in the accompanying sketch.

NATURAL-GAS CRATERS

MOON-LIKE CRATERS IN THE NORTH SEA FLOOR

During the exploitation of the North Sea oil fields, geophysicists made detailed surveys of sea-floor topography with seismic instruments called boomers. They were startled to discover thousands of elliptical craters or pockmarks in the sediments. The craters are 30-330 feet across, 6-25 feet deep, and located in water about 500 feet deep. The long axes of the craters point roughly in the same direction; and the craters tend to be arranged in lines. The authors suggest that escaping subsurface gases and fluids may have formed the unusual structures. The possibility was underscored on July 30, 1978, when a very large eruption of sediment was detected by sonar. (McQuillin, Robert, and Fannin, Nigel; "Explaining the North Sea's Lunar Floor," New Scientist, 83:90, 1979.)

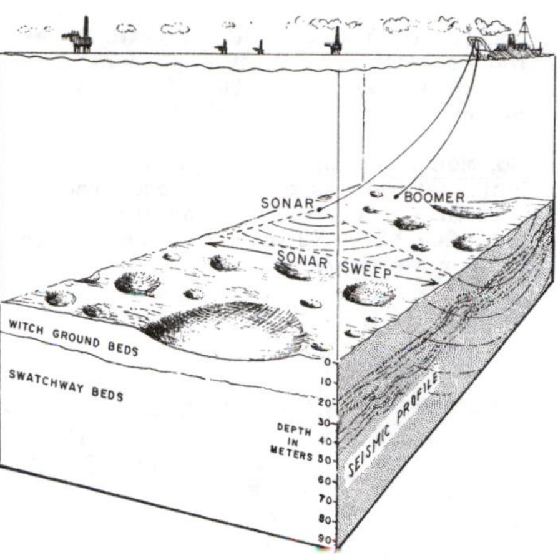

Pockmarks detected by sonar on the floor of the North Sea. These may be due to eruptions of natural gas.

Comment. The North Sea is a prime habitat of mistpouffers (sea-associated booming sounds). There might be a correlation here between natural-gas eruptions and these strange booming sounds. Also, the crude similarity of these sea-floor craters to the Carolina Bays should not be passed over.

NATURAL GAS EXPLOSION?

February 28, 1990. Nowata, Oklahoma. On this date, residents of the Nowata area heard and felt an explosion. Its source was unknown until W. Mitchell checked to find out why Double Creek had backed up. He found a scene of curious devastation, which was then linked to the earlier explosion.

> The explosion blasted up large shale rocks, some estimated to weigh more than a ton, and filled about 150 feet of creekbed with shattered stone. Trees were blown down and the creek was partly blocked.
>
> The water level upstream from the site is 3 to 4 feet higher than below, although water is flowing under the rocks.
>
> .....
>
> "It looks like a giant mole went all under the ground," Mitchell said, ..."There is no telling how big a hole is under there. You can hear water falling. Gas was bubbling up all along the bank. In one place the water was shooting up a couple of feet yesterday, like a fountain, but it has gone down now."
>
> Large pieces of shale landed 20 to 30 feet from the creek bank, mud was blown outward from the explosion and pieces of shale as large as a big tennis shoe were found as much as 200 feet from the creek.

(Smith, Charlotte Anne; "Nowata Creek Blocked after Apparent Explosion," Tulsa World, March 4, 1990. Cr. P.A. Roales.)

Comment. A similar, though more violent, natural gas explosion occurred in 1890 in a river bed near Waldron, Indiana. See ESC4-X3 in the Catalog Anomalies in Geology.

A fissure created by a spontaneous natural gas explosion at Waldron, Ohio, on August 11, 1890.

COOKIE-CUTTER HOLES

THE BIG DIVOT!

Sometime between mid-September and October 18, North-central Washington.

> ...a chunk of earth weighing tons was plucked out of a wheat field, as though someone use a "giant cookie cutter" and put down, right side up, 73 feet away. "All we know for sure is that this puzzle piece of earth is 73 feet away from the hole it came out of," said Greg W. Behrens, a geologist with the Bureau of Reclamation at Grand Coulee Dam. The displaced slab, mostly soil held together by roots, is about 10 feet long and 7 feet wide. Its thickness varies from 2 feet at one end to about 18 inches at the other. The shape and thickness of the piece exactly match the hole that was left behind, just like a piece in a jigsaw puzzle, though it was rotated about 20 degrees.

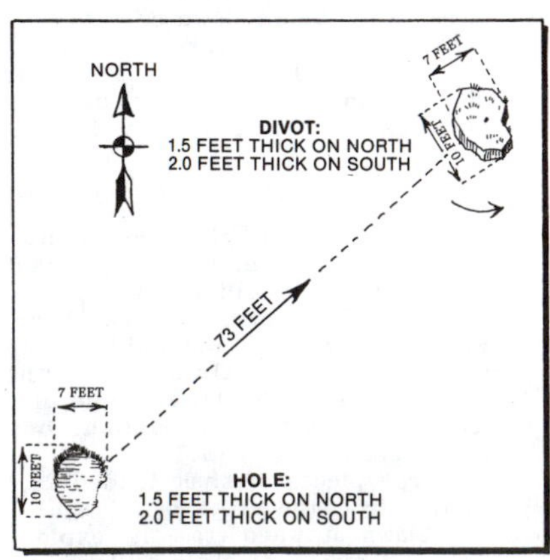

Diagram of the "cookie-cutter" event of October 1984, Grand Coulee, Washington.

There are no marks left by machinery in the area; and the sides of the hole reveal dangling roots, indicating that the slab was torn out rather than cut out. A final item of possible interest: on October 9, there was a small quake, magnitude 3, with the epicenter 20 miles southwest of the hole. (Anonymous; "A Rare Phenomenon Moves Earth," Philadelphia Inquirer, p. 3-A, November 25, 1984. Similar accounts appeared in many other newspapers.

Comment. The quake mentioned was quite small. Large quakes, however, have been known to toss boulders out of the ground, leaving large holes behind. See GQH1 in Earthquakes, Tides, Unidentified Sounds.

ANOTHER "COOKIE CUTTER" HOLE

Back in 1984, the American press had fun with the "cookie cutter" hole found in Washington state. A good-sized chunk of earth or "divot" had been neatly excised intact from the ground and deposited some 73 feet away. (See drawing.) One would think that nature would play only one such bizarre prank, but a remarkably similar occurrence also took place in 1887. A third example of this most curious phenomenon has been resurrected from one of the Middle Ages chronicles:

> 822 A.D.: "In the land of the Thuringians, near a river, a block of earth 50 ft. long, 14 ft. wide, and 1½ ft. thick, was cut out, mysteriously lifted, and shifted 25 ft. from its original location." Royal Frankish Annals.

(Carolingian Chronicles. W. Scholz, translator, Ann Arbor, 1972. Cr. E. Murphy)

THE COOKIE CUTTER STRIKES AGAIN--FOUR TIMES---IN NORWAY

Yes, the cookie-cutter phenomenon has left its mark again: more mysterious divots and holes in the ground. T. Jonassen has sent us a study of the phenomenon published in Ottar, a publication of the Tromse Museum, in Norway. Even better, he has provided a translation, from which we quote a few paragraphs:

> About 1 km SE of Skogvollvatnet (a lake), at Skogvollmyra (a moor), a slab of turf 5.2 m long and 1.8 m wide, has, in an apparently inexplicable manner, torn itself loose from its "mother turf" and placed itself 4-5 m away. The slab of turf is completely undamaged and is placed with the right side up. The piece of turf has rotated 20-30 degrees compared to the original hole. The hole in the moor is absolutely even at the bottom, and the angle between the bottom and its walls is 90 degrees. The hole is 30-35 cm deep, and its edges are nicely cut.
>
> From the hole there is a crack running westwards for about 6 m. Close to the hole this crack is somewhat widened, and one side of the crack twists itself 25-30 cm above the other. This twisting decreases as one gets further from the hole. The crack gradually subsides, and it is hard to tell exactly where it ends.
>
> About 12 m NW of the hole there is an arched crack of about 15 m lying with its concave side towards the hole. It is plainest in the middle. Here the side closest to the hole has been twisted upwards about 15 cm. Here also the crack gradually disappears at both ends. There is an open hollow beneath the part which has been twisted upwards, about 30 cm below the surface.
>
> The slab of turf has an area of about 5 m^2 and this should give a weight of between 1500-1700 kg.

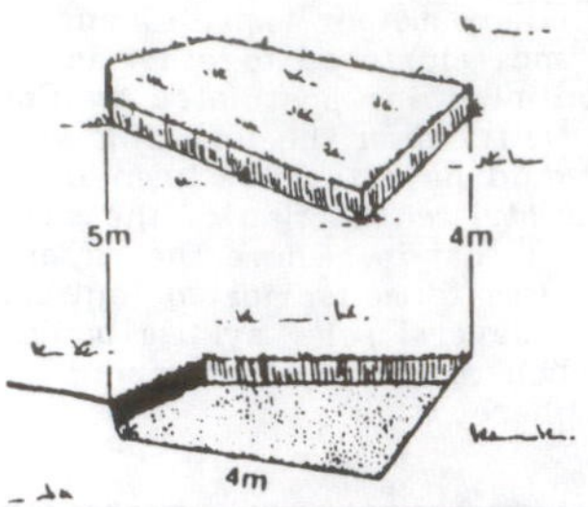

The "divot" excised from Andøya Moor, Norway

The article concludes with a brief description of three similar occurrences of the phenomenon in Norway. (Dybwik, Dagfinn, and Møller, Jakob J.; "Phenomenon in an Andøya Moor---An Insoluble Mystery?" Ottar, no. 5, p. 15, 1988. Cr. T. Jonassen)

Comment. One could easily dismiss (with a knowing smile) a single occurrence of the cookie-cutter phenomenon---but now we have a total of seven! The situation becomes more serious.

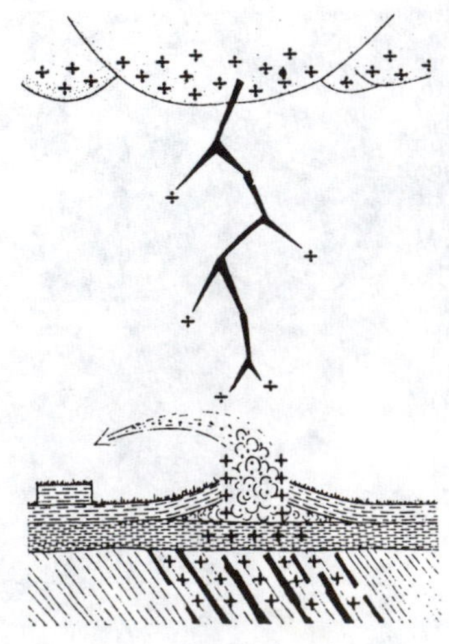

One theory advanced to explain cookie-cutter holes has lightning creating steam explosions from groundwater. However, the divots are intact and show no signs of fusion.

THE GRUYERIZATON OF SWITZERLAND

An impressive and equally inexplicable variety of the cookie-cutter holes have been reported in Switzerland. In these excavation phenomena, all the material removed from the holes has disappeared instead of being translated intact and set down gently nearby.

> The Swiss holes were first observed in 1972. They always occur at night and no one has ever seen them forming. They are circular, drilled deep into the earth, and are created by the total removal of vegetation and soil. A large fleet of lorries would be needed to move the quantities of earth involved, but there are no vehicle marks around the holes. The grass and plants around them exhibit none of the minor damage which would inevitable be caused by any normal process of drilling. The holes have all been excavated directly from above, at an angle of 90 degrees to the surface.

Six Swiss holes have been reported from the environs of Lake Geneva. The two largest are at Begnins (December 17, 1982; 18 feet across, 24.5 feet deep) and Confignon (February 3/4, 1990; 33 feet across, 40 feet deep). Obviously, we are not dealing with minor earth-moving operations here. (Anonymous; "The Gruyerization of Switzerland," The Cerealogist, no. 3, p. 26, Spring 1981.) The Cerealogist is a British publication that focuses on the crop-circle phenomenon.

Comment. Could the "force" flattening the crop circles also gouge out cookie-cutter holes and the Swiss cavernous pits? Additional information on the Swiss excavations and similar events is certainly required. Anomalists know from experience that for every strange phenomenon there exists a hoaxer anxious to reproduce it.

DEGRUYERIZING SWITZERLAND

Some of those astounding "holes" in the integument of Switzerland have been made less mysterious by one of our Swiss readers. Two holes, which we did not mention specifically, have turned out to be a hoax and a mundane sinkhole. The hole at Confignon, which we did pinpoint, was actually 66 feet in diameter and 40 feet deep; but, according to the official geologist of the Geneva Canton, it was simply subsidence due to the drilling of a tunnel. Only the hole at Beginins (actually discovered December 15, 1982) retains an aura of mystery:

> The case was investigated by the official geologist of the Vaud Canton, who found no rational explanation. He put forward the hypothesis of the existence of an old gallery for the harnessing of water. Unfortunately, the verification of his hypothesis would be too expensive, so the hole was filled up. (Mancusi, Bruno; personal communication, September 8, 1991.)

CANYON ENIGMAS

THE GRAND CANYON CONUNDRUM

The Grand Canyon is really "grander" than that which meets the eye when standing on its rim. The immensity of the total excavation job can be appreciated by a trip to Red Butte near the southern rim. This conical hill rises 300 meters and is capped by the remnants of a lava flow, which has a radioisotope date of 9 million years. Thus, in the last 9 million years not only has the entire gorge of the Grand Canyon been cut but also a much greater volume of rock 300 meters thick adjacent to the present north and south rims.

The immensity of the volume of sediment removed is not the real issue. The first problem emerges downstream from Red Butte at Pierce Ferry. There we find 200 meters of Hualapai limestone, which has been dated at 8.7 million years from a thin layer of volcanic ash within the limestone. When this limestone was laid down, the geology is emphatic that no large, sediment-carrying river was in the vicinity. Most geologists agree that the lower end of the Grand Canyon was not active around this time. No one knows where the Colorado River was flowing at this period. Some say southwest from Peach Springs; others point to a northwest route into Utah. All the likely alternate routes face serious geological obstacles such as lava barriers.

What does seem certain is that the stock explanation of the formation of the Grand Canyon is incorrect. It was not steadily ground out, cutting ever-deeper as the whole region was slowly elevated. Actually, geologists believe that the region stopped rising over 50 million years ago. Where, then, did the sediments from a 450-kilometer canyon and the wide areas surrounding its rims go? The present exit of the Colorado near Pierce Ferry was blocked until fairly recently while the Hualapai limestone was being deposited; and other routes don't look too promising. (Rice, R.J.; "The Canyon Conundrum," Geographical Magazine, 55:288, 1983.)

Comment. Moral: beware of facile explanations. The data presented, in fact, make one wonder whether all of the erosion might have occurred quite recently (perhaps when the great submarine canyons were cut?), using the Colorado's present route.

GRAND CANYON SHAMED AGAIN

Exploration and mapping of submarine canyons cut into the continental shelves of Alaska and Siberia emphasize once again the colossal scale of these crustal gashes:

> Erosion of some of the largest known submarine canyons has removed more than 20,000 km^3 of former subduction margin between the Aleutian Islands and Cape Navarin, U.S.S.R. The canyons are incised as deeply as 2,400 m into Tertiary sedimentary and igneous rocks that make up the margin and attendant deep sedimentary basins along the outer Bering shelf. Cutting of the seven major canyons probably occurred during low stands of sea level when the Bering Shelf was exposed to a depth of about -135 m, which allowed the ancestral Anadyr, Yukon, and Kuskokwim Rivers to carry large volumes of sediment to the outer shelf. Although their positions appear to be structurally influenced, the canyons apparently were cut by combinations of massive slumping and sliding of sediment deposited near the shelf edge and of scouring action of the resulting turbidity currents that carried debris to the abyssal sea floor, where deep-sea fans have formed.

(Carlson, Paul R., and Karl, Herman A.; "Ancient and Modern Processes in Gigantic Submarine Canyons, Bering Sea," Eos, 64:1052, 1983.)

Comment. The authors believe that sub-

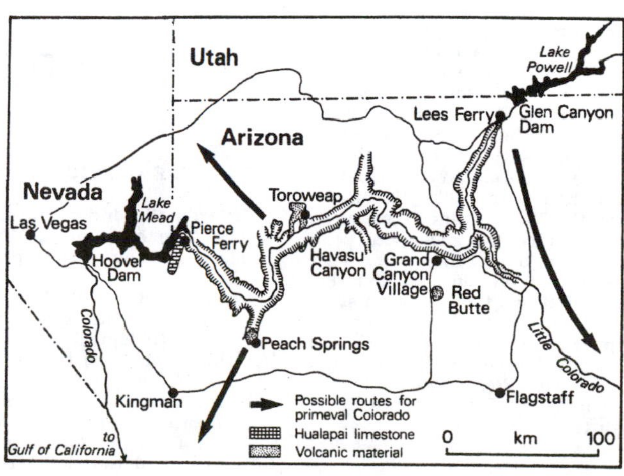

Map of the Grand Canyon of the Colorado. (Adapted from Geographical Magazine, 55:288, 1983.)

marine slumping and turbidity currents were sufficient to have eroded these huge canyons. Other geologists doubt this. The other possibility is that sea level was once a mile or more below present levels and that the canyons were cut by rushing water spilling over the continental shelves.

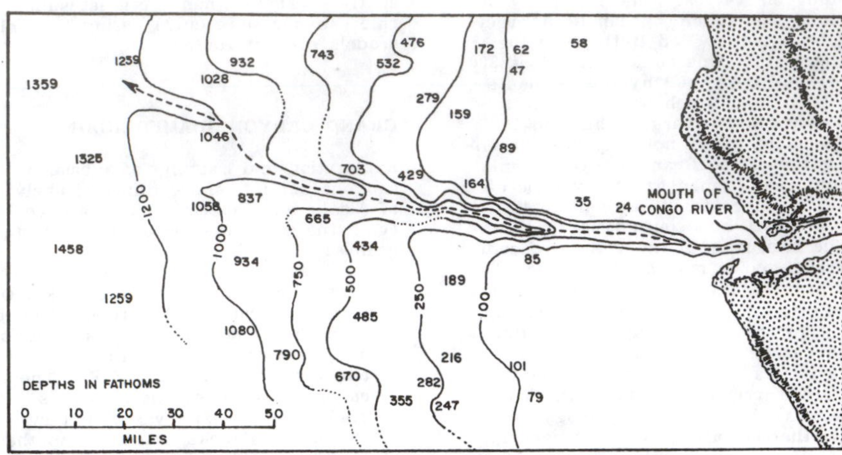

This contour map of the Congo submarine canyon (drawn in 1903) shows how sharply and deeply these canyons are incised. Modern topographic maps are of course more precise.

SUBMARINE CANYONS: A 50-YEAR PERSPECTIVE

F.P. Shepard has devoted most of his professional life to the study of submarine canyons. These great canyons, some cut through thousands of feet of hard crystalline rock, dwarf the largest land-based canyons. No wonder Shepard was fascinated enough to spend his life collecting data and weighing possible explanations. After 50 years, he has concluded that these colossal submarine features have no single cause. Subaerial erosion, turbidity currents, submarine slumping, and faulting have all played roles.

Anomalists will be most interested in Shepard's insistence that the evidence shows that subaerial erosion has played a major part in carving out the submarine canyons. This explanation is definately frowned upon by most geologists because the Pleistocene sea levels dropped only about 100 meters according to current thinking. How could subaerial erosion account for canyons several thousand feet below present sea level? Shepard persists; the evidence is there. The continental margins must have risen and than sank back! He also points out that the Mediterranean seems to have dried up in recent geological times. Could the major oceans have dropped thousands of feet in a similar fashion? Shepard doesn't intimate this, saying only that the submarine canyons still present puzzles. (Shepard, Francis P.; "Submarine Canyons: Multiple Causes and Long-Time Persistence," American Association of Petroleum Geologists, Bulletin, 65:1062, 1981.)

DEEPER MYSTERIES

The first detailed views of vast stretches of the seafloor in U.S. coastal waters have revealed features so immense and unexpected that they defy the imaginations of the scientists who discovered them.

A special sonar device named Gloria is being employed to produce high resolution maps of the seafloor. Apparently previous sonar sounding methods missed startling underwater volcanos, canyons, and immense delta-like deposits. About 170 miles off San Francisco, near a huge volcanic structure, Gloria discovered an underwater canyon comparable in size to the Grand Canyon. No one really knows how it was formed. This great chasm is associated with a delta-like deposit twice the area of Massachusetts. Normally, one expects alluvial fans at the ends of canyons, but in this instance the submarine canyon actually cuts down into the fan. Where such a huge mass of material came from is a mystery rivaling that of the canyon's origin. (Yulsman, Tom; "Mapping the Sea Floor," Science Digest, 93:32, May 1985.)

WHAT'S OK IN THE MEDITERRANEAN IS VERBOTEN IN THE ATLANTIC AND THE PACIFIC

Geologists and geophysicists have now satisfied themselves that a few million years ago the Mediterranean dried up nearly completely. The Deep Sea Drilling Project discovered in 1970 and 1975 that layers of evaporites existed beneath the Mediterranean's floor. In addition, over 80 years ago, the bed of the Rhone River was found to consist of river sands and gravels superimposed upon hundreds of feet of oceanic sediments. Beneath these deposits---some 3000 feet down---was a gorge cut in granitic rock. Other rivers emptying into the Mediterranean had cut similar gorges into solid rock long ago. No one could provide an acceptable explanation for the deep-cut gorges until the evaporites proved that the water level had been low enough for the rivers to cut the gorges subaerially. In other words, the Mediterranean's level fell several thousand feet, allowing the rivers to erode gorges much as the Colorado does today in the Grand Canyon. (Smith, E.G. Walton; "When the Mediterranean West Dry," Sea Frontiers, 28:66, 1982.)

Comment. The Med's buried gorges are obviously close cousins of the many submarine canyons found around the world's continental shelves. Most geologists strongly resist any explanation of the submarine canyons involving subaerial erosion, because no one believes the oceans ever dropped thousands of feet. True, the Med's evaporites confirm a great reduction in water level there, but the flat-topped guyots in the Atlantic and Pacific do the same for the oceans---or at least seem to. The guyot tops are now thousands of feet below the surface, just like the Mediterranean's evaporites. It is thought-provoking to notice that the Mediterranean (and the oceans?) began to dry up about 5½ million years ago, just about when humans are supposed to have split off from the other primates.

CREVICULAR STRUCTURE

CHAIN OF CREVICULAR HABITATS?

Towards the end of an article on the eerie blue holes of the Bahamas appears this intriguing paragraph:

> William Hart, of the Smithsonian Institution, and Tom Iliffe, of the Bermuda Biological Station, believe that blue holes are one link in a chain of crevicular habitats---caves, fissures, rocks of the sea floor---that stretches from one side of the ocean to the other, from the Americas, across the sea floor and the Mid-Atlantic Ridge, to Africa and the Mediterranean. Related Amphipods are not only found in Bahamian caves but in marine caves in Bermuda, the Pacific, and the Yucatan Peninsula.

(Palmer, Robert; "In the Lair of the Lusca," Natural History, 96:42, January 1987.)

Comment. With this, the vision arises of an earth-girdling, biologically and geologically connected stratum of life that we know next to nothing about. How porous is the earth's crust, and how far down in these pores and interstices does life survive?

REALLY-DEEP RIVERS

Ecologists studying rivers have discovered a vast subterranean world filled with dozens of previously un-

> known species of worms, shrimp, insects and microscopic organisms that live in the groundwater below the stream channel and sometimes for miles on each side.

The quotation above once again evokes the concept of "crevicular structure" in the crust. The crevicular world is that immense, unappreciated maze of underground space created by cracks in the rocks, solution channels, permeable gravels, and so on. In the article reviewed here, a crevicular realm has been discovered underneath river beds. But this is just a special case of a subterranean world found many places beneath the surface---even under the continental shelves. The surface waters we see are just (to use an aquatic metaphor) the tip of the iceberg!

Sub-river life lives far under the beds of the great Alaskan rivers and even small desert streams in Arizona. Preliminary exploration has shown that fluid-and life-filled crevicular structure exists at least 30 feet under river beds and may extend several miles to either side. For example, water wells drilled two miles from the Flathead River, in Montana, yield immature stoneflies. J. Stanford, Director of Montana's Flathead Lake Biological Station states, "We have basically enlarged the concept of what a river is." He and his colleagues have found at least a dozen new species in the crevicular world beneath the bed of the Flathead River. (Anonymous; "Life-Filled Subterranean World Found Flowing under Rivers," San Francisco Chronicle, November 24, 1989, Cr. J. Covey)

Comment. There are many instances of fish being found in wells far from surface water.

The Conch Sound blue hole is a shallow example of crevicular structure. Here, extensive passageways reach deep under the Grand Bahama Bank. (Adapted from Sea Frontiers, 32:269, 1986.)

MOUNDS, RIDGES, HILLS

We now turn to "positive" topographical features, as natural opposites of the "negative" craters, canyons, etc., treated above. Most of these "positive" phenomena arise from such natural forces as water currents, ice action, wind, and tectonic forces. Surprisingly, others of no little magnitude, are or seem to be the work of animals! Lastly, some topographical features confound the senses and have no objective reality.

We categorize the relevant items from Science Frontiers as follows:

- **Mima mounds. Are they the work of mammals, water currents, or earthquakes?**
- **Drumlins. Were these mounds really created by the ice sheets?**
- **Termite bands.**
- **Sea-floor structures. Seamounts, sea-floor "dunes."**
- **Topographical illusions. Where objects roll uphill!**

MIMA MOUNDS

A CLASH OF HYPOTHESES

Millions of Mima Mounds dot various terrains west of the Mississippi, from British Columbia south to northern Sonora, Mexico. They are also found in Africa and South America. Mima Mounds are formed of soil and small stones. They are often 2 meters high, with diameters of 20 meters. Mound densities can reach 25-50 per hectare. Mima Mounds, prairie mounds, pimple mounds, or whatever they are called locally, are widely thought to be the work of pocket gophers, although this hypothesis is still contested, as we shall see below.

No one denies that pocket gophers are often associated with the mounds. In fact, a recent scientific paper describes the relationship between the sizes and shapes of the mounds and the numbers of resident gophers. The authors of this paper, G.W. Cox and J. Hunt, state confidently that, "Investigations of Mima mounds in western North America support the hypothesis that mounds are formed by the gradual translocation of soil by pocket gophers ..." (Cox, George W., and Hunt, Jodee; "Form of Mima Mounds in Relation to Occupancy by Pocket Gophers," Journal of Mammalogy, 71:90, 1990.)

In another paper, appearing at almost the same time, A.W. Berg contends that the Mima Mounds could instead be the consequence of seismic activity. In his abstract he says: "Small-scale Mima mounds can be produced experimentally by subjecting a plywood board covered with a thin veneer of loess to impacts that produce vibrations in the board. Experimentally produced mounds have characteristics that are nearly identical to those found in the field. This suggests that most Mima mounds formed as the result of seismic activity in conjunction with unconsolidated fine sediments on a relatively rigid planar substratum." Berg also notes that the regions where the mounds are common have been seismically active. (Berg, Andrew W.; "Formation of Mima Mounds: A Seismic Hypothesis," Geology, 18:281, 1990. Cr. J. Satkoski)

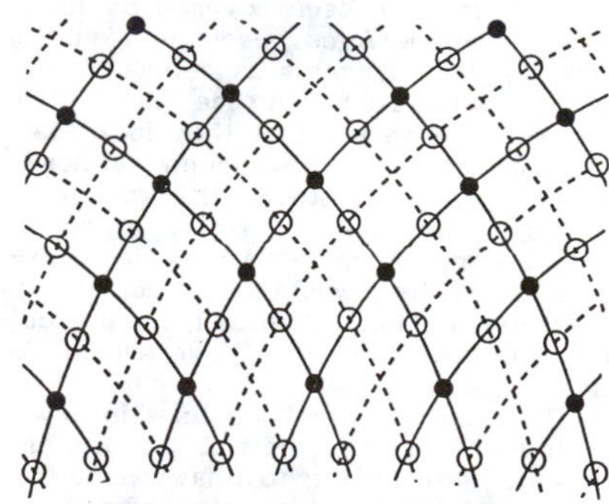

When seismic waves intersect, they create an interference pattern. Loose surface material may occur at those points of minimum surface disturbance (open circles) and form mounds.

MIMA MOUNDS IN THE KENYA HIGHLANDS

Mima Mounds are rarely reported outside the western United States. But Kenya has them, too. At elevations of 1500-3600 meters on Mt. Kenya, fields of mounds up to 6 meters in diameter and 1.5 meters high have been described. Cox and Gakahu have studied some of these African mounds and find them much like the North American Mima Mounds. They are found well within the range of the rhizomyid mole rat, a rodent similar to the North American pocket gopher in size and behavior. Quantitative measurements indicate the mounds to be constructed from dirt immediately surrounding the mounds. In short, the African mounds and probably those of North America seem to be the products of industrious rodents. (Cox, George W., and Gakahu, Christopher G.; "The Formation of Mima Mounds in the Kenya Highlands: A Test of the Dalquest-Scheffer Hypothesis," Journal of Mammalogy, 65:149, 1984.)

Sketch of the San Diego Mima Mounds.

DRUMLINS

DRUMLINS MAY RECORD CATASTROPHIC FLOODS

Drumlins are small, teardrop-shaped hills that occur in large numbers, often aligned in large "fields," in areas thought to have been covered by ice during the Ice Ages. Geologists customarily explain drumlins as debris piled up and sculpted by the ice sheets themselves, despite the fact they look like they might have been shaped by flowing water. As we all know, the word "flood" is anathema in geology, probably because a provable episode of extensive flooding would lend credence to the Biblical Flood! (Actually, many cultures around the world have similar flood legends.)

Canadian geologist J. Shaw is now trying to break out of this philosophical prison. "According to Shaw, heat from the Earth formed huge lakes of meltwater that remained trapped beneath the North American ice sheet. As the sheet began to retreat near the end of the glacial age, the water broke through and flowed in torrents down to the Gulf of Mexico and Atlantic Ocean. While flowing under the ice cap, water would have surged in vast, turbulent sheets that sculpted and scoured drumlins. Each flood lasted until the weight of the ice cap once again shut off the outlet of the covered lake, Shaw says."

Shaw goes on to estimate that one large drumlin field in Saskatchewan was created when 84,000 cubic kilometers of water was discharged. Just this single episode would have raised global sealevels by about 10 inches in a few days or weeks. Imagine what happened as this water flowed across North America.

Many geologists look askance at Shaw's theory of drumlin formation. (Monastersky, R.; "Hills Point to Catastrophic Ice Age Floods," Science News, 136:213, 1989.)

Comment. The famous Channelled Scablands in the Pacific Northwest are thought to have been scoured out when an ice dam broke unleashing the Spokane Flood near the end of the Ice Ages. These land forms are described in Carolina Bays, Mima Mounds, Submarine Canyons.

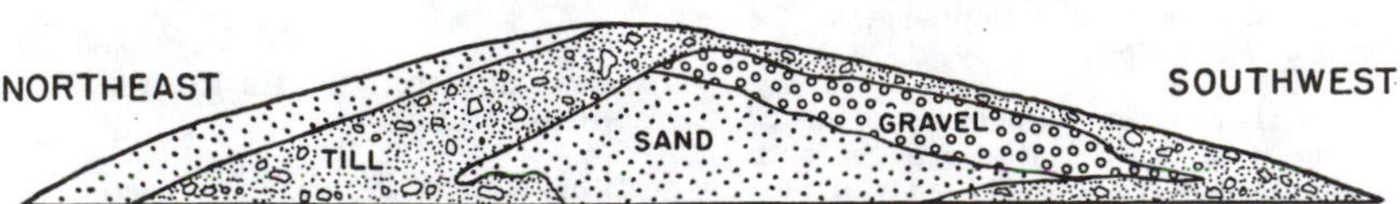

Cross section through a typical drumlin.

TERMITE BANDS

'TERMITE BANDS' IN SOUTH AFRICA

Few animals besides man have significantly affected our planet's geology. To be sure, corals built the Great Barrier Reef and beavers left their mark on the hydrology of parts of Canada and the U.S.; but termites!? It seems that in southern Africa this industrious insect is responsible for enormous striped patterns of ridges and vegetation bands. These amazingly regular patterns are caused by alternating low ridges and gullies. The ridges are about 2 meters high, up to a kilometer long, and separated by about 50 meters.

The ridges themselves are closely spaced termite mounds. Just why the termites choose to build their mounds in long rows is an unanswered question. And how do the termites maintain strict parallelism, especially since they are blind? How could termites in one mound know how their neighbors in the nearest ridge, 50 meters away, are building their mounds? Anyway, the ridges help channel the flow of water and thus the growth of vegetation, giving immense swathes of country a corrugated appearance. (Sattaur, Omar; "Termites Change the Face of Africa," New Scientist, p. 27, January 26, 1991.)

Comment. In Australia, the so-called "magnetic" termites build their slab-like mounds so as to minimize the amount of sun-generated heat.

SEA-FLOOR STRUCTURES

NEW ENGLAND SEAMOUNTS ONCE NEAR SURFACE

Exploration of the New England Seamount chain by the research submarine *Alvin* confirmed that some of these peaks, now all a kilometer or more below the surface, were once at or above the surface of the ocean.

This undersea mountain chain contains more than 30 major peaks and stretches 1,600 miles southeast from the New England coast. Deep-sea dredging has previously brought up Eocene limestone of shallow-water origin from the submerged mountain tops, but the *Alvin* explorations resulted in the first eyewitness accounts of dead coral (which grows only near the surface) and rock samples containing strands of dead algae that grows only within 100 meters of the surface. The New England Seamounts have therefore either subsided on the order of a kilometer since Eocene times or sealevel has altered drastically.

Sketch of the research submersible Alvin inspecting strange undersea buttes.

The *Alvin* dives also discovered a series of very striking and perplexing buttes obviously the results of erosion (see sketch above). The buttes are apparently composed of volcanic rock and are only a few meters high. Some unexplained, extremely vesicular (hole-filled) rocks seen on the sea floor during the dives seem to be identical to samples occasionally dredged up and formerly classified as cinders jettisoned from old steamships. The underwater surveys suggested that these "cinders" have a natural (still mysterious) origin. (Heirtzler, J.R., et al; "A Visit to the New England Seamounts," American Scientist, 65:466, 1977.)

ECHO SOUNDER OUTLINES STRANGE PATCHES OVER UNDERWATER PEAKS

February 11, 1977, on the s.s. Remuera,

in the South Pacific. (00° 24'S, 88° 06' W). The echo sounder clearly showed an undulating sea floor with sharp peaks, some 600 fathoms below the ship. The peaks were recorded distinctly, but above several, faint, elongated, flame-like patches were noted. These patches were interpreted as changes in water density due, perhaps, to volcanic or hydrothermal activity. (Howard, K.E.; "Soundings in a Volcanic Area," Marine Observer, 48:20, 1978.)

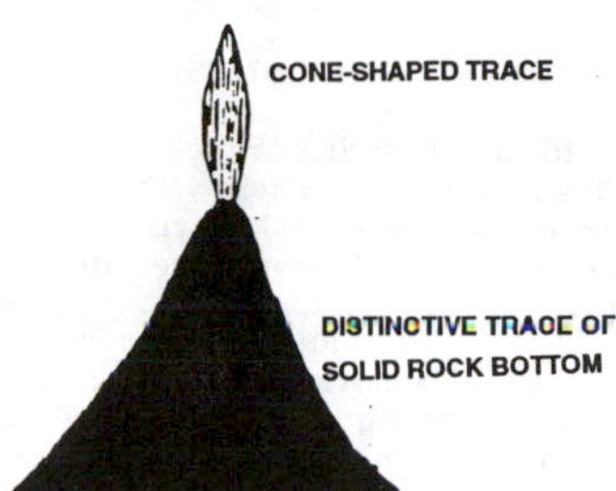

Echo sounders reveal flame-like patches over underwater peaks.

SAND DUNES 3 KILOMETERS DOWN

Side-scan sonar systems are excellent for mapping the major geological features of the sea floor. Some of the scenery revealed by this sonar challenges conventional geological wisdom. To illustrate, the British ship Farnella has been mapping the floor of the Gulf of Mexico with its GLORIA II side-scan sonar. Hundreds of new salt domes, submarine channels, and landslides have been recorded. A big surprise was the discovery of large fields of sand dunes in 3000 meters of water. Similar dune fields were found in the Pacific in 1984. G. Hill, of the U.S. Geological Survey has ventured, "There's something going on in deep water that people just aren't aware of." (Anonymous; "A Systematic Sounding of the Sea Floor," Science News, 128:191, 1985.)

Comment. At issue here is how sand dunes are formed in such deep water. Are they the consequence of colossal bottom currents, perhaps set in motion by huge earthquakes or some other natural catastrophe? Is it possible (gasp!) that the dunes were not formed under water at all? The area involved also boasts salt layers and submarine channels and canyons, both of which are not well-explained.

TOPOGRAPHICAL ILLUSIONS

PINEY PITSTOP OF THE PARANORMAL

Evidently trying to inject some joy into our country's financial capital, the Wall Street Journal recently printed a story on Spook Hill, Lake Wales, Florida. Spook Hill is one of several spots in the U.S. where gravity seems to be defied.

> Sue Robertson motors her maroon Ford Tempo to the white line painted across Fifth Street, shifted into neutral and slowly rolls backward up Spook Hill.
>
> "Eerie, weird, and definitely strange," she says, finally easing to a stop near the top of the rise.
>
> Hers is the same amazed reaction expressed by most tourists who discover this piney pitstop of the paranormal, 50 miles south of Orlando. On a typical Saturday, up to 30 cars an hour line up at the top of the hill for their turn to drive down to the white line and drift back up.

Not only cars roll up the hill. Farmers had to stop planting oranges in the area because visitors pulled them off the trees so they could watch them roll uphill. Skateboarders and cyclists also feel the pull of gravity in the wrong direction.

Scientists who deign to investigate sites like Spook Hill usually end up by claiming them to be merely optical illusions.

> If it's an optical illusion at work here, it's an odd one; a reporter applying a carpenter's level at about the hill's halfway point finds a slope up in the direction the cars are rolling. Joggers report they expend more energy running that way too. "Spook Hill is most definitely a hill," says Paulette Bond, a geologist at the Florida Department of Natural Resources.

(Johnson, Robert; "Just Who, or What, Makes Cars Roll Up a Slope in Florida?" Wall Street Journal, October 25, 1990. Cr. J. Covey)

THE UPS AND DOWNS OF SPOOK HILL

The appearance of an article on Spook Hill in the Wall Street Journal of October 25, 1990, induced G. Wilder to have a look for himself. After all, when the conservative Wall Street Journal admits to being "baffled," there must really be something to the phenomenon!

Spook Hill is located on a section of road in Lake Wales, Florida. Here, at Spook Hill, the road "seems" to slope downhill, but yet cars in neutral apparently roll up the incline. Wilder, a member of the Tampa Bay Skeptics, made several pertinent observations during his investigation:

> (1) Although the phenomenon is striking when one approaches Spook Hill from one direction, if one gets out of the car and looks back, it is quickly apparent that one's senses have been deceived. Because the illusion fails when one looks in the opposite direction, the road has been made one-way, so that tourists will not be disappointed!
>
> (2) A storm drain is positioned at the true low point of the road, and cars seem to roll up to the drain. Water, in its gravitational wisdom, knows where to go! Neither were the city engineers fooled.
>
> Conclusion: Spook Hill is only an amusing illusion; there is no gravitational anomaly. (Wilder, Guss; "Spook Hill: Angular Vision," Skeptical Inquirer, 16:58, 1991.)

ICE PHENOMENA

To geologists, ice is just another kind of rock, although one with a very low melting point. Geologically, ice is most significant in the polar regions where it dominates the scene. In Antarctica, the continent's ice cover has certainly waxed and waned over the millennia; but has Antarctica been nearly ice-free within historical times, and are its ice sheets unstable today and, therefore, threats to the global climate?

The other ice phenomena introduced here do not have such planet-wide ramifications. The Arctic polynas, the Andes ice islands, and those weird ice spikes in bird baths can only be classified as intriguing minor mysteries.

WHO MAPPED ANTARCTICA IN PRE-MEDIEVAL TIMES?

Conventional history has the Antarctic continent being discovered less than 200 years ago. However, the appearance of Terra Australis Re on Orontius Finaeus's map of 1531 and the depiction of a nearly identical continent on Mercator's 1538 map reveal sufficient accurate knowledge of Antarctic features for us to conclude that someone discovered and mapped Antarctica well before 1500. The question is who?

An interesting feature of Finaeus's map is the reduced ice cover compared with what we find today. The Ross Ice Shelf, for example, was almost nonexistent. Such changes in ice cover are consistent with modern theory of Antarctic climate changes. Apparently the seas surrounding Antarctica were a bit warmer and some unidentified early mariners brought knowledge of this continent back to Europe. (Weihaupt, John G.; "Historic Cartographic Evidence for Holocene Changes in the Antarctic Ice Cover," Eos, 65:493, 1984.)

Comment. Obviously missing from Weihaupt's analysis is any consideration of the famous Piri Re'is map and reference to the work of Charles Hapgood; viz., Maps of the Ancient Sea Kings. For a discussion of this map's validity, see p. 41.

ANTARCTICA REVISITED HAPGOOD ACKNOWLEDGED

John G. Weihaupt's paper on possible

recent changes in the Antarctic ice cover (summarized above) evidently stirred up considerable scientific interest. Two long letters and Weihaupt's reply have recently been published in Eos. First and significantly, Weihaupt's omission of any reference to Hapgood's popular work, Maps of the Ancient Sea Kings, was pointed out and belatedly acknowledged by Weihaupt. The second letter was from a French scientist, who concluded that "in spite of some hard facts and in spite of warnings against simplistic theories, the idea of fast change in the Ross Ice Shelf and in its main nourishment area, Marie Byrd Land, is widespread in the United States." Weihaupt responded to this with a massive bibliography supporting the idea of recent, rather extensive changes in the Antarctic ice cover. He stated further that other research suggests that even the East Antarctic Ice Sheet may have undergone deglaciation during the Pleistocene. Those old maps showing Antarctica largely ice-free may not be so crazy after all. (Milton, Daniel J.; "Antarctic Ice Cover," Eos, 65: 1226, 1984.)

Comment. The real mystery is the identity of the ancient map-drawers.

IS THE ARCTIC ICE COVER THINNING?

In May 1987 a British submarine carried out an ice profiling experiment in the Arctic Ocean in which the route closely approximated that of an earlier voyage in October 1976. Over a zone extending more than 400 km to the north of Greenland there is evidence of a significant decrease in mean ice thickness in 1987 relative to that found in 1976. This thinning amounts to a loss of volume of at least 15% over an area of 300,000 km^2.

(Wadhams, Peter; "Evidence for Thinning of the Arctic Ice Cover North of Greenland," Nature, 345:795, 1990.)

In an accompanying discussion of the ice problem, A.S. McLaren et al note that since the late 1800s, Arctic researchers using drills have reported consistently that the Arctic ice thickness averaged 3-4 meters. U.S. submarine surveys concurred with these figures during cruises in 1960 and 1962. Satellite surveys of ice cover from 1978-1987 found no trends. In other words, other sources of data on the Arctic ice reveal little change. The new results, therefore, need further confirmation. (McLaren, A.S., et al; "Could Arctic Ice Be Thinning?" Nature, 345:762, 1990.)

ANTARCTIC ICE SHEETS SLIPPING?

Geologists have generally assumed that the ponderous Antarctic ice sheets do not change their behavior rapidly. But, according to NASA's R. Bindschadler, an ongoing study of the Antarctic coast near the Ross Ice Shelf casts doubt upon this assumption of long-term stability. Measurements of one ice stream flowing down from the mountains to the sea indicate a sudden unexplained, 20% reduction in speed over the past decade. Perhaps even more significant is that, even with this reduction in flow velocity, this particular ice stream carries ice into the sea 40% faster than ice accumulates up in the mountains.

The sudden, rather large velocity change is alarming because it may signify widespread instability in the continent's icy mantle. Researchers state that there is even a chance that much of the Antarctic ice cap could collapse into the sea in the next few centuries----a catastrophic event that would raise global sea levels by 6 meters! (Anonymous; "Antarctic Ice Potentially Unstable," Science News, 137:285, 1980.)

Comment. In addition to looking at future consequences of collapsing Antarctic ice sheets, we should mark that what might happen in the future might also have happened in the past. Obviously, we refer to the often-discussed speculation that the Antarctic was nearly ice-free within historical times. In this connection, we cannot escape mentioning that remarkable ancient map of Piri Re'is that, some say, shows an ice-free Antarctica, mapped presumably by ancient mariners. This was the theme of C.H. Hapgood's book, Maps of the Ancient Sea Kings.

GALLOPING GLACIERS

North America boasts 104 surge glaciers. No one knows why these glaciers behave so differently from normal glaciers; they certainly look the same. But while ordinary glaciers creep along a few inches per day, surging glaciers will sometimes charge ahead at the rate of several yards per hour. The surges may be years apart; and they may occur periodically. The surges start high up on the glacier and propagate down to the foot, which plods along a few inches per day until the surge arrives. Then, it leaps forward, only to return to normality until the next periodic surge. The surges seem to occur when water spreads out under the ice, lubricating its flow. Beyond this we know little.

Why do some glaciers surge while those right alongside behave normally? Are the surges really cyclic? The Variegated Glacier, in Alaska, for example, surged in 1906, probably in 1926, in 1947, in 1964-65, and in 1982---about 20 years between surges. The surges do not seem to be connected to earthquakes, climatic changes, volcanic heat, or anything obvious. (Beard, Jonathan; "Glaciers on the Run," Science 85, 6: 84, February 1985.)

THE POLYNA MYSTERY

"Polyna" is a Russian word meaning "an enclosed area of unfrozen water surrounded by ice." Polynas form for some unknown reason in the Southern Ocean surrounding Antarctica. These transient ice-free "lakes" may cover 300,000 square kilometers, inferring a substantial influx of heat countering the frigid polar temperatures. Although small coastal polynas can be blown free of ice by strong Antarctic winds, the open-ocean polynas are much larger and do not seem to owe their origins to wind.

One suggested explanation is that warm subsurface water rises suddenly to the surface, but it takes a lot of heat to keep hundreds of thousands of square kilometers ice-free. The most recent polyna opened up a region of the Weddell Sea for about three years (1973-1976). (Simon, C.; "Polynas Surrounded by Ice and Mystery," Science News, 122:183, 1982.)

THE ANDES ICE ISLANDS

High in the Bolivian Andes are some shallow, saltwater lakes. From some of these white-shored lakes rise bizarre "islands" of fresh-water ice, neatly layered horizontally, and up to 20 feet higher than the saltwater surface. The ice crystals comprising these islands are vertical, proving that they grew in water and are not pieces of glaciers. Clearly, they did not form from the present saltwater lakes, but they might be relics of the Ice Ages, when the lakes were deeper and fresher. But whence the nicely layered structure? Some scientists have thought that volcanic springs might be the sources of fresh water, but some ice islands occur in lakes where there are no volcanic springs nearby. At the moment, everyone seems stumped by these strange creations of nature. (Anonymous; "Who Made the Andes Islands of Ice?" New Scientist, 96:272, 1982.)

Comment. Could the ice islands be related to the Arctic pingoes---those debris-covered ice hills?

Sketch of a layered ice island, Lake Colorado, in the Bolivian Andes.

SPOOKY SPIKE

The readers of New Scientist continue to supply delightful observations of Nature's quirks. The following is from J. Turner, in Warwickshire:

> I have in my garden a round red plastic container which holds water for the birds. In last winter's first hard frost, I found an odd ice formation in the bowl. Although the weather the previous day had been clement and the water was fluid, we had that night a sharp frost down to about -4°C. The following morning I noticed what appeared to be something sticking up out of the frozen water in the dish. On closer examination, it proved to be a solid "spike"

of ice, ending in an arrowhead. The ice was solid and came out of the side of the frozen water at an angle of about 45°. It was about 9 inches long and solid throughout.

(Turner, Judy; "Spooky Spike," New Scientist, p. 54, November 2, 1991.)

Two weeks later, the same journal published two radically different explanations of the ice spike. G. Lewis called the spike an "ice fountain" and stated that it is due to the well-known expansion of water as it freezes. R. Blumenfeld, on the other hand, attributed the growth of the spike to the fact that water molecules on the surface and in surrounding air are electrical dipoles. In his view, a small defect in the ice's surface attracts polarized water molecules in the air, creating an outwardly growing structure. (Lewis, Geoff, and Blumenfeld, Raphael; "Sprouting Spikes," New Scientist, p. 58, November 16, 1991.)

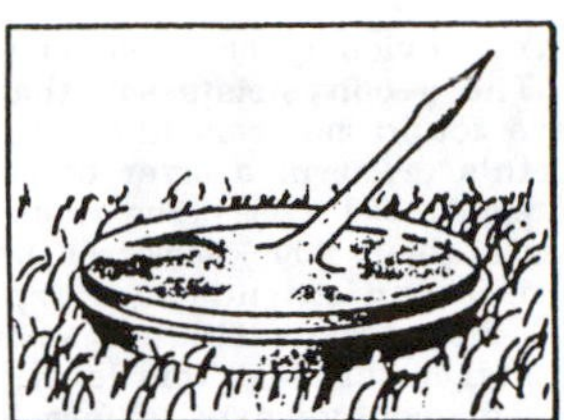

"Spooky spike" found in a Warwickshire garden.

Comment. Seldom does one find such engaging oddities discussed in American scientific publications. American scientists are too stuffy it seems.

GLOBAL PHENOMENA

GEOPHYSICS: THE SICK MAN OF SCIENCE

"In order to be a famed geo-scientist and belong to the inclusive club of fully accepted geophysicists in their unknown thousands, one must kneel on the hassock and swear allegiance to the following tenets regardless of any scientific considerations:

Tenet 1. That the moment-of-ineria of the Earth has never changed.
Tenet 2. That the Earth contains a large central core composed of iron.
Tenet 3. That the continents are drifting as a result of unknown forces.

"These must be held with religious fervour, dissenters are just not to be tolerated, the devotees feeling it their right, and indeed duty, to defend the creed against all criticism by any means of chicanery and of sharp-practice within their power, however crude and improper, so long as they judge they can get away with it, but all the time representing themselves to the world as acting with judicial calm in the best interests of their science. It will be shown that all three of these tenets are wrong, and how their (naive) acceptance has hamstrung the believers from making progress in the deep waters of terrestrial science, though not of course in the worldly world of 'modern science.' Shades of Sir Cyril Burt."

So begins a long technical article by R.A. Lyttleton, author of many scientific books and papers. (He may lose his union card after this paper!) Lyttleton proceeds to demonstrate the incorrectness of the first two tenets above. Lyttleton's reasoning is buttressed by many scientific observations and so much quantitative reasoning that it is impossible to encapsulate it all here. Suffice it to say that it all looks correct, serious, and above-board. (Lyttleton, R.A.; "Geophysics: The Sick Man of Science," ISCDS Newsletter, 5:3, December 1984.)

Comment. Now this is interesting. The ISCDS is the International Stop Continental Drift Society, now defunct. The Society's Newsletter, if you don't already know, is usually a tongue-in-cheek publication. Not so here. Lyttleton is deadly serious. Either that or the joke is lost amid all the equations!

THE EARTH IS EXPANDING AND WE DON'T KNOW WHY

Let us taunt the geologists now with an idea that many of them consider to be nonsense.

The Expanding Earth Hypothesis goes back to at least 1933, a time when the Continental Drift Hypothesis was accorded the same sort of ridicule. Now, Continental Drift is enthroned; and ironically many of its strongest proponents are vehemently opposed to the Expanding Earth, ignoring the lessons of history.

The data that suggest that the earth has expanded significantly over geological time come from the pleasant pastime of continent fitting. If one takes the pieces of continental and oceanic crust and tries to fit them together at various times over the past several hundred million years, taking into account the production of crust at the midocean ridges, the fit gets worse and worse as one works backward in time. Great gaps (or "gores") appear between the pieces of crust which geologists believed existed at these periods. (Of course, one can play this puzzle-piece game only at passive continent-ocean boundaries where the oceanic crust has not slid under the continental crust. The South Atlantic is a good place to work.)

These embarrassing, grotesque gaps can be made to diappear almost as if by magic by assuming that the earth was smaller in the past. This seems, on the surface, to be a crazy idea. Why would an entire planet swell up like a balloon? Hugh Owen answers in this way:

> The geological and geophysical implications of such Earth expansion are so profound that most geologists and geophysicists shy away from them. In order to fit with the reconstruction that seems to be required, the volume of the Earth was only 51 per cent of its present value, and the surface area 64 per cent of that of the present day, 200 million years ago. Established theory says that the Earth's interior is stable, an inner core of nickel iron surrounded by an outer layer that behaves like a fluid. Perhaps we are completely wrong and the inner core is in some state nobody has yet imagined, a state that is undergoing a transition from a high-density state to a lower density state, and pushing out the crust, the skin of the Earth, as it expands.

(Owen, Hugh; "The Earth Is Expanding and We Don't Know Why," New Scientist, p. 27, November 22, 1984.)

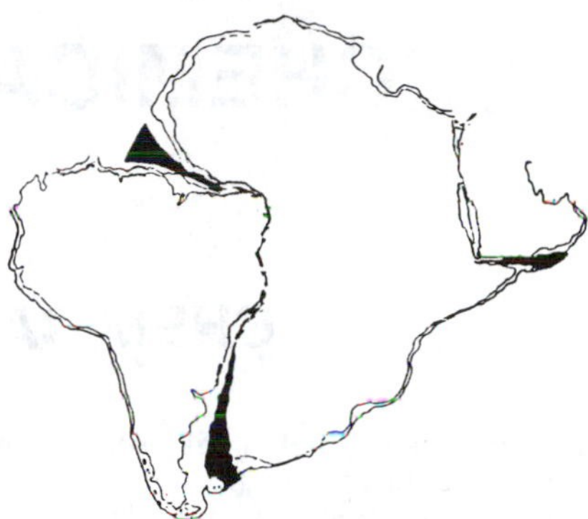

The relatively bad fit of South America and Africa on a globe of modern size is indicated by the black gaps (gores). These black areas disappear on an earth with a diameter of 80% of the modern value.

CHAOS BELOW

In a dive on the submersible Alvin just west of the Mariana trench, scientists discovered a cache of unusual features, including chimneys spewing out mineral-laden cold water on top of submerged mountains that rise 2,500 meters from the seafloor. While volcanic eruptions form most seamounts, these mountains consist of a nonvolcanic rock called serpentinite, and oceanographers are not entirely sure how the serpentinite mountains formed.

The theory of plate tectonics has the Pacific plate diving under the Philippine plate along the Mariana trench. It may be that water trapped in the downgoing crust leaks out, rises, and serpentinizes the crust above. This altered rock, being lighter than that surrounding it, may slowly rise through it, eventually forming undersea mountains. (Monastersky, Richard; "Novel Mountains and Chimneys in the Sea," Science News,

134:333, 1988.

Comment. This all sounds pretty speculative, but those mountains had to come from somewhere.

Perhaps the serpentinite mountains are just one manifestation of a larger phenomenon: the chaotic slithering and popping up and down of crustal material. The following is from New Scientist:

> Geophysicists in California and Illinois say that they have found the Earth's "missing" crust by analyzing shock waves from earthquakes to determine the chemical composition of the Earth's interior. If the researchers are correct, then the view of the interior of the Earth that scientists have previously accepted is wrong.
>
> The geophysicists say that they have found minerals like those in the Earth's crust in a layer of crustal material, 250 kilometres thick, which starts about 400 kilometres below the surface and extends to a depth of 650 kilometres. There is enough crustal material at this level, according to geophysicists to form a crust 200 kilometres thick---the average thickness of the Earth's crust is only 20 kilometres.
>
> .....
>
> The material is not trapped at this depth: the layer acts like a conveyor belt which returns the crustal material to the surface by a process of convection. At the surface, the material cools and sinks along the subduction zones. Below the surface, it reheats and rises to join the crust again, along one of the Earth's mid-ocean ridges.

(Anderson, Ian; "Seismic Waves Reveal Earth's Other Crust,: New Scientist, p. 28, November 26, 1988.)

Comment. An obvious question is: What does this repeated circulation of crustal material do to radiometric and index-fossil dating of the crustal material we can access at the surface? Large sections of the stratigraphic record are missing on our planet; maybe they have now been found.

GEOLOGICAL ANOMALIES: CHEMICAL, PHYSICAL, BIOLOGICAL

CHEMICAL ANOMALIES

The diverse geochemical phenomena recorded in past issues of Science Frontiers have the potential to change the way we think about how such basic substances as petroleum and natural gas were formed. Could these valuable fuels really be abiogenic rather than the emanations from decaying organic materials, as strongly maintained by mainstream scientists? On another tack, chemical "spikes" found at the junctures of major stratigraphic suites suggest that the earth's history and the fate of its cargo of life were modulated powerfully by astronomical catastrophism. Finally, we have the usual residue of curiosa and Forteana, which may not be earth-shaking in their import, but are nevertheless still intriguing.

With these thoughts in mind, we outline this section as follows:

- Oil and gas. Abiogenic or biogenic?
- Chemical spikes. Projectile impacts, global conflagrations.
- Chemical miscellany.

Flames erupting from an English mudflat due to the spontaneous ignition of gases.

OIL AND GAS

IS THE EARTH A GIANT METHANE RESERVOIR?

T. Gold, of Cornell, theorizes that a vast reservoir of methane resides in the earth's crust---a left-over from the formation of the earth. This accumulation of methane, he suggests, has been the major source of carbon at the surface throughout geological time. The existence of subterranean methane is manifested when flames shoot up during earthquakes. Tsunamis or tidal waves are probably caused by the release of immense bubbles of methane during quakes rather than by actual motion of the sea floor. (Lewis, Richard S.; "Is the Earth a Giant Methane Store?" New Scientist, 78:277, 1978.)

IS ALL NATURAL GAS BIOLOGICAL IN ORIGIN?

T. Gold and S. Soter, from Cornell, have championed the theory that earthquake lights, sounds, and precursory animal activities may be due to abiogenic natural gases escaping from deep within the earth. Perhaps some petroleum and natural gas reserves have been created by primordial hydrocarbons working their way outward through the crust rather than by the geochemical alteration of biological materials. Perhaps almost all petroleum is abiogenic---some Russian scientists hold this view!

Western scientists are almost unanimous that natural gas and oil are biogenic with maybe a touch of upwelling abiogenic hydrocarbons. A major reason given for this stance is that the biogenic theory has been so productive in locating hydrocarbon reserves. This, of course, leaves the earthquake lights and sounds still unexplained. (Anonymous; "Abiogenic Methane? Pro and Con," Geotimes, 25:17, November 1980.)

Comment. The moral of this might be that seemingly inconsequential phenomena historically lead to wholesale changes in scientific thinking; viz., the insignificant advance in Mercury's perihelion.

SUBTERRANEAN PETROLEUM FACTORIES?

Sediment samples dredged up from the bottom of the Gulf of California near some hydrothermal vents contain petroleum similar in some ways to commercial petroleum. Apparently organic matter in the vicinity of the vent is thermally converted into oil, or at least something that, like wine, matures into something useful. (Simoneit, Bernd R.T., and Lonsdale, Peter F.; "Hydrothermal Petroleum in Mineralized Mounds at the Seabed of Guayman Basin," Nature, 295:198, 1982.)

Comment. The recently discovered hydrothermal vents are only the external manifestations of what must be extensive chemical factories beneath the crust. The rich assemblages of thermo synthetic life (not photosynthetic life) around the vents makes one speculate about what might be transpiring chemically and biologically in the hot, fluid-saturated crevices and pores of the earth's crust. Carbon dating of petroleum sometimes yields absurdly young ages. Could it be that all the natural gas and petroleum we could ever need is now being manufactured for us subterraneously? The Gaia hypothesis would lead us to expect just such a process. After all, humankind requires abundant fuel if it is to carry earth life out into the reaches of space!

OIL, OIL: EVERYWHERE, EVERY AGE

Geologists have discovered a major deposit of oil in Precambrian rocks in Australia's Northern Territory. Precambrian oil does exist elsewhere---around the Great Lakes, Russia, etc.---but the Australian deposits differ in that they are economically attractive. The oil-bearing strata are dated at between 1.4 and 1.7 billion years; and the oil itself is at least this old. Significantly, the oil contains extremely small amounts of steranes, which are thought to be derived from advanced organisms, but there were plenty of chemicals typical of primitive bacteria. The mere existence of commercially exploitable deposits of Precambrian oil implies that, far from being devoid of life, the ancient earth was host to immense accumulations of bacteria and other simple organisms. (Anonymous; "Ancient Oil in Australia: A New Bonanza?" New Scientist, p. 26, September 11, 1986.)

Comment. As discussed above this Australian oil might have been produced abiogenically.

The surface and near-surface Athabasca oil sands in western Canada constitute a well-known deposit of almost unbelievable size. Geologists have long speculated about where such an immense quantity of biological matter could have originated. (Few dare to suggest non-biological origins!) Now, we learn that below the Cretaceous Athabasca oil sands lies a 70,000 square kilometer "carbonate triangle" estimated to contain about 2 x 10^{11} cubic meters (about 6 cubic kilometers) of bitumen. This bitumen is closely related chemically to the oil sands above it. A common origin seems likely. (Hoffmann, C.F., and Strausz, O.P.; "Bitumen Accumulation in Grosmont Platform Complex, Upper Devonian, Alberta, Canada," American Association of Petroleum Geologists, Bulletin, 70:1113, 1986.)

Comment. Many geologists believe that these incredible accumulations of organic matter migrated from some distant source to their present location. But just where was this prodigious wellspring of biological activity?

OIL & GAS FROM THE EARTH'S CORE

In central Sweden this summer, drillers will be boring into the rocks of the Siljan Ring, Europe's largest known meteor crater. Oil and gas should not be down there in any quantities according to current theory, but that's what they are drilling for. Isn't it futile to fight such a well-established dogma that oil and gas have biological origins and therefore must be looked for only where life once thrived?

Not any longer! Enough anomalies have accumulated to seriously challenge the idea that oil and gas are byproducts of ancient animal life. Here are a few of these anomalies:

1. The geographical distribution of oil seems derived from features much larger in scale than individual sedimentary features.
2. The quantities of oil and gas available are hundreds of times those estimated on the basis of biological origins.
3. The so-called "molecular fossils" found in oil and claimed as proof of a biogenic origin are simply biological contaminants, particularly bacteria that feed upon the petroleum.
4. Petroleum is largely saturated with hydrogen, whereas buried biological matter should exhibit a deficiency of hydrogen.
5. Oil and gas are often rich in helium, an inert gas which biological processes cannot concentrate.
6. The great oil reservoirs of the Middle East are in diverse geolgical provinces. There is no unifying feature for the region as a whole and, especially, no sediments rich in biological debris that could have produced these immense concentrations of oil and gas.

If oil and gas do not come from decaying organic matter, where do they originate? Some scientists, such as T. Gold, say "from the earth's core." As the earth accreted long ago, it collected abundant carbonaceous material from carbonaceous chondrites and comets containing organic sludge. Under the heat and pressure available at great depths, oil and gas were produced abiogenically in immense quantities and driven outwards to where they were trapped in rock reservoirs.

In central Sweden, oil and gas of biogenic origin are highly improbable. If they are found trapped beneath the Siljan Ring, a major tenet of geological thought may have to be revised. (Gold, Tom; "Oil from the Centre of the Earth," New Scientist, p. 42, June 26, 1986.)

GOING FOR GOLD

The name of T. Gold appears often in Science Frontiers. Currently, he is promoting the theory that many of the earth's hydrocarbon deposits (gas, oil, graphite, etc.) are not of biological origin but are formed rather when primordial methane outgases from the planet's interior. A vanishingly small number of geologists buy Gold's theory. Nevertheless, the Swedish State Power Authority and some private investors have been impressed enough to fund a drilling project at the Siljan Ring, a meteorite crater 150 miles north of Stockholm. There are no significant sources of biogenic hydrocarbons nearby, but oil seeps are not uncommon around the Ring. Mainstream theory cannot account for these seeps, but Gold's theory can: primordial methane streaming up through the cracked granite shield is converted, probably with the help of bacteria, into oil and hydrocarbon sludge. "Ridiculous," say the mainstreamers!

Recently, the drilling program, which has reached the 22,000-foot level, brought up 60 kilograms of very smelly black sludge with the consistency of modeling clay. The gunk seems to have a biological origin. In addition to the black sludge, the drillers have been encountering increasing quantities of various hydrocarbon gases as the hole went deeper. All very supportive of Gold's hypothesis.

Establishment geologists are having difficulties explaining these results. They blame contamination by drilling lubricants and/or the surface oil seeps. Gold discounts these explanations. (Anonymous; "Going for Gold," Scientific American, 259:20, August 1988. Also: Begley, Sharon, and Lubenow, Gerald C.; "Gushers at 30,000 Feet," Newsweek, p. 58, June 27, 1988. Cr. C.H. Stiles)

WE LIVE ATOP A CHEMICAL RETORT

Far down there beneath our feet, the earth's chemicals are bubbling away, aided often by bacteria and other life forms. It's a dark world, but it's also hot, permeated by fluids, and possibly the abode of organisms we haven't dreamed of. Three hints of this nether world follow.

Quick oil. Rather than ripening in deep strata for millions of years, as per prevailing theory, some oil is being created in only a few thousand years in the vicinities of ocean-bottom chimneys. Dime-sized oil globules have been sighted floating near these chimneys. Analysis of chunks broken off the chimneys by research submersibles reveal the presence of petroleum-like hydrocarbons that are less than 5000 years old. It is thought that high-temperature fluids percolating up through the sediments convert buried organic matter into oil very rapidly. (Monastersky, R.; "The Quick Recipe for a Soup of Black Gold," Science News, 136:295, 1989.)

Comment. Not mentioned in this article is T. Gold's theory that oil is actually derived from primordial carbon deep in the crust.

Gassy water. Some water wells in Texas also produce much methane. This methane is apparently not related to any oil or gas wells in the region. Rather, surmise has it that bacteria deep in the crust are converting buried organic material into methane and other chemical products. But geologists are confounded by the fact that some water wells are rich in methane while others nearby are devoid of the gas. (Anonymous; "Methane and Ground Water," Geotimes, 34:19, April 1989.)

Comment. As to be expected the possibility of abiogenic methane is ignored.

A really-deep ocean. No, this is not in Tarzan's Pellucidar, but rather an incredible mass of water stored hundreds of kilometers deep in the earth's mantle. Several times the earth's visible surface water may be locked up in water-bearing minerals! Brucite [$Mg(OH)_2$], for example, is 30.86% water. Perhaps such water was released long ago by changes in temperature and pressure to form the present oceans. (Ahrens, Thomas J.; "Water Storage in the Mantle," Nature, 342:122, 1989.)

DEEP-SIXING ANOTHER HYPOTHESIS?

T. Gold once said, "In choosing a hypothesis, there isn't any virtue in being timid." Neither have the Swedes been timid in following Gold's lead by drilling for natural gas in the granite of central Sweden. All the experts predicted this quest would come to naught, because there are no source rocks in the area containing biological materials from which the gas could have been generated. But Gold does not believe that the methane in natural gas comes from buried organic debris. Rather, most methane is primordial and abiogenic, a legacy left deep in the earth's crust when our planet was formed.

The 72-kilometer-diameter Siljan Ring in central Sweden is generally believed to be of meteoric origin. The granite here has been shattered, perhaps to a depth of 40 kilometers. If Gold's hypothesis about the origin of methane is correct, methane might well be found seeping up through this wound in the earth's outer skin. Further, the shattered granite might prove to be a gigantic reservoir of valuable methane. The Swedes decided to drill.

After three years and the expenditure of $40 million, drilling at the Siljan Ring has been terminated. The drill penetrated to 6.8 kilometers before it got stuck. No significant methane had been found. The experts snickered!

But the story is not finished, at least as far as Gold is concerned. He maintains that the drilling stopped just short of an apparent reservoir at 7.2 kilometers (probably located by seismic methods). Another, deeper hole will vindicate him, he believes. After all, there are tantalizing hints:

1. The drillers did find an assortment of hydrocarbons that could have been deposited by upward-seeping methane. Skeptics say they are derived from the drilling fluids.
2. Tons of micrometer-sized grains of magnetite were taken out of the hole. Gold opines that these grains were synthesized by bacteria subsisting upon seeping methane at a depth of 6 kilometers.
3. Russian drillers on the Kola Peninsula report the existence of intriguing circulating fluids as far down as 12 kilometers.

Despite the problems and disappointments at the first hole, some Swedish investors seem ready to finance a second hole at the Siljan Ring. (Kerr, Richard A.; "When a Radical Experiment Goes Bust," Science, 247:1177, 1990.)

BLACK GOLD---AGAIN

The Siljan Ring and T. Gold are back in the news again. A few years ago, at Gold's instigation, private investors and the Swedish govenment put up money to drill for oil and gas at the Siljan Ring, some 200 kilometers northwest of Stockholm. This granitic region is a meteor-created, shattered scar on the earth's crust. It is in just such a spot that Gold expects to find abiogenic petroleum and methane seeping upward from deep inside the earth, where they have resided since the earth was formed. Conventional petroleum geologists have roundly ridiculed the Siljan Ring project; after all, everyone knows that oil and gas derive from buried organic matter.

Three years ago, at a depth of 6.7 kilometers, the "misguided" Swedish drillers pumped up 12 tons of oily sludge from the granite rock. "Just drilling fluids and diesel-oil pumped down from the surface," laughed the experts. This autumn (1991), more oil was struck in a new hole only 2.8 kilometers deep. This time, only water was used to lubricate the drill. How are the skeptics going to explain this? Well, about 20 kilometers away, there are sedimentary rocks; perhaps the oil seeped into the granite from there. Rejecting this interpretation, the drillers are going deeper in hopes of finding primordial methane. (Aldhous, Peter; "Black Gold Causes a Stir," Nature, 353:593, 1991. Anonymous; "Black Gold," The Economist, p. 101, October 19, 1991. Cr. T. Brown)

BABY OIL

Geologists maintain that most of the oil they pump out of the ground was formed tens and hundreds of millions of years ago from biological debris. But in the Guaymas Basin, in the Gulf of California, the oil seeping out of the sediments is only 4240 years old! Actually, it could be even 500-3000 years younger than that for two reasons: (1) The organic debris that was C-14-dated may have taken many years to become incorporated in the sediments; and (2) The dating may be skewed by older material in the sediments. By subtraction, the oil might be as young as 1240 years!

The picture geologists draw of the Guaymas Basin is that of a spreading center covered by perhaps a half kilometer of sediments. Spewing up from the spreading center is hot water at 300-350°C, which "cracks" the organic material in the sediments, converting it into petroleum only 10-30 meters below the sea floor. (Hecht, Jeff; "Youngest Oil Deposit Found below Gulf of California," New Scientist, p. 19, April 6, 1991.)

Comment. Since spreading centers are really cracks in the earth's crust, it is possible that some of the feed materials for this modern "petroleum factory" in the Guaymas Basin could consist of abiogenic, primordial methane and other organics seeping up from deep within the earth.

GAS HYDRATES AND THE BERMUDA TRIANGLE

Gas hydrates are stable, solid lattices of water molecules, in which the water molecules are hydrogen-bonded into hollow spheres or oblate spheroids in which gas molecules are enclosed. They are not true compounds, but rather clathrates or inclusion compounds. They often form in huge masses in ocean-bottom sediments and, if disturbed by slumping or some other stimulus, can release immense quantities of gas bubbles. These gases rise to the surface, subdivide en route, and reach the surface as a huge upwelling of frothy, low-density fluid.

A ship cannot float in such low-density fluids. If a large plume rose above the sea's surface, aircraft passing through it would lose engine power. Quoting the final paragraph:

> Intermittent natural gas blowouts from hydrate-associated gas accumulations, therefore, might explain some of the many mysterious disappearances of ships and planes---particularly in areas where deep-sea sediments contain large amounts of gas in the form of hydrate. This may be the circumstance off the southeast coast of the United States, an area noted for numerous disappearances of ships and aircraft.

(McIver, Richard D.; "Role of Naturally Occurring Gas Hydrates in Sediment Transport," American Association of Petroleum Geologists, Bulletin, 66:789, 1982.)

METHANE HYDRATE: PAST FRIEND OR FUTURE FOE?

What looks like a grayish ice cube, fizzes at its edges, and soon wastes away to a puddle of water? If you wish, you can accelerate the substance's demise by touching a match to it; it is packed with potential energy. The substance is methane hydrate, and it is found in prodigious quantities in oceanic sediments. Each cubic centimeter of methane hydrate contains about 160 cubic centimeters of methane at standard conditions; it is a concentrated source of natural gas. In fact, methane hydrate deposits in the world's oceans hold twice as much carbon as all the coal, oil, and gas reserves on land!

But methane hydrate may be much more than a future fuel source; it may have been humanity's savior in eons gone by; it may be our future nemesis. You see, methane hydrate is very unstable; changes of temperature or pressure on a global basis can trigger the release of immense volumes of this greenhouse gas from oceanic deposits.

For example, when the Ice Ages lowered ocean levels by locking up water in the advancing ice caps, pressures on ocean-bottom methane hydrate lessened and, according to some speculators, released enough gas so that the increased greenhouse heating turned back the Ice Ages. (Was Gaia at work here?)

On the other hand, if present human activities are truly stoking the greenhouse effect, ocean temperatures should rise, possibly destabilizing methane hydrate deposits and thereby aggravating the greenhouse effect. Such positive feedback could cook the biosphere.

(Anonymous; "Did Methane Curb Ice Ages," New Scientist, p. 24, May 25, 1991. Also: Appenzeller, Tim; "Fire and Ice under the Deep-Sea Floor," Science, 252:1790, 1991.)

CHEMICAL SPIKES

IRIDIUM AND MASS EXTINCTIONS

Alvarez and his colleagues at the University of California, while chemically analyzing a series of sedimentary strata from Italy, discovered that one layer had 25 times the concentration of iridium residing in adjacent strata. The iridium-rich layer forms the boundary between the Cretaceous and Tertiary periods, 65 million years ago. During that death-filled interval, 50% of the earth's genera were wiped out. Such are the two correlated facts: iridium increase and mass extinction. But do they have the same cause? Alvarez et al point out that iridium is rare on earth but much more common out in space. The anomalous concentration of iridium could have been injected by a massive solar flare, a big meteor impact, or some other extraterrestrial catastrophe. Thus is catastrophism being resurrected. (Anonymous; "An Iridium Clue to the Dinosaur's Demise," New Scientist, 82:798, 1979.)

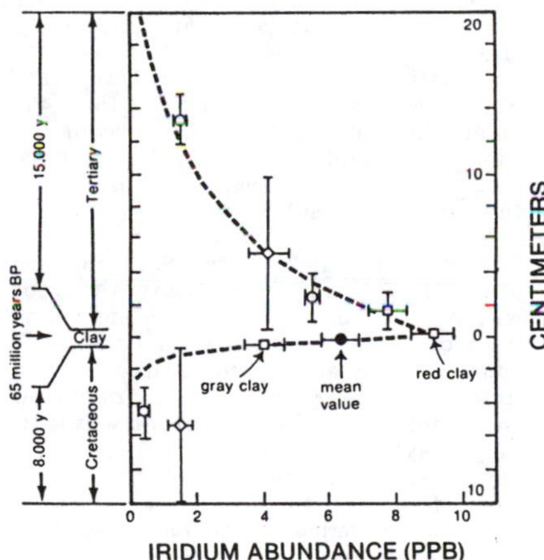

A representative iridium spike. The iridium concentration rises sharply when the clay layer separating the Cretaceous and Tertiary formations is reached. (Adapted from: Mosaic, 12:2 March/April 1981.)

IRIDIUM-RICH LAYERS AND CATASTROPHISM

Kyte et al have discovered a 2.3-million-year-old sedimentary layer under the Antarctic Ocean that contains iridium and gold concentrations comparable to those in the Cretaceous-Tertiary boundary. The noble metals are mostly contained in millimeter-sized grains that resemble ablation debris from a large extraterrestrial object. Unlike the Cretaceous-Tertiary episode, however, the newly found layer is not accompanied by evidence of mass biological extinctions. (Kyte, Frank T., et al; "High Noble Metal Concentrations in a Late Pliocene Sediment," Nature, 292:417, 1981.)

Comment. Perhaps those paleontologists who deny the existence of sudden biological extinctions at the Cretaceous-Tertiary boundary are correct and something else besides catastrophism impacted terrestrial life at that juncture.

ICELAND AND THE IRIDIUM LAYER

The high concentration of iridium between the Cretaceous and Tertiary eras (about 65 million years ago) is widely interpreted as indicating a worldwide catastrophe caused by the impact of a comet or meteor. The increase of iridium concentration over normal levels is much higher in northern latitudes, suggesting that the impact point is in this region. But no impact scar of the proper size and age exists. However, if one looks for scabs rather than scars, one finds that Iceland is formed entirely of volcanic rocks younger than the Cretaceous. To Fred Whipple of the Smithsonian Astrophysical Observatory, these facts dovetail nicely. Iceland was formed by magma welling up from a 100-km hole in the sea floor blasted out by a 10-km meteor. (Anonymous; "The Blow That Gave Birth to Iceland?" New Scientist, 89:740, 1981.)

THE CRETACEOUS-TERTIARY EXTINCTION BOLIDE

The recently discovered worldwide iridium-rich layer is taken by many scientists as evidence of the collision of an asteroid or comet with the earth about 65 million years ago. This cataclysmic event is also blamed (by some, at least) for the apparent sudden biological extinctions recorded on these pages of the fossil record. In this setting, the authors of this paper calculate the effects on the earth of a 10-kilometer-diameter object impacting at about 20/km/sec.

Do the theoretical results jibe with the geological and paleontological data? Very definitely. Crater ejecta rich in extraterrestrial material would be blasted to an altitude of 10 km, where winds would insure global distribution. In terms of biological stress, the 10-km projectile would transfer 40-50% of its kinetic energy to the atmosphere, creating a heat pulse that could raise global temperatures 30°C (50°F) for several days. Many large animals might well succumb to such a temperature transient. In addition, the protective ozone layer might be blown away by shock waves and not reform for a decade. (O'Keefe, John D., and Ahrens, Thomas J.; "Impact Mechanics of the Cretaceous-Tertiary Extinction Bolide," Nature, 298,123, 1982.)

PUNCHING A HOLE IN THE ASTEROID HYPOTHESIS

Scientists have long searched for a cause for the profound geological and biological changes that apparently occurred between the Cretaceous and Tertiary periods. When an iridium-rich layer was found in several areas at this important boundary, many claimed it as proof of an asteroid impact or some other catastrophism that would nicely explain the massive worldwide changes that occurred. With this preamble in mind, consider the following abstract from an article in Science:

> Analyses of the clay mineralogy of samples from the Cretaceous-Tertiary boundary layer at four localities show that the boundary clay is neither mineralogically exotic nor distinct from locally derived clays above and below the boundary. The significant ejecta component in the clay that is predicted by the asteroid-impact scenario was not detected.

(Rampino, Michael R., and Reynolds, Robert C.; "Clay Mineralogy of the Cretaceous-Tertiary Boundary Clay," Science, 219:495, 1983.)

DID AN ASTEROID IMPACT TRIGGER THE ICE AGES?

Asteroids and comets are being blamed these days for more and more of our planet's catastrophism---biological, meteorological, and geological. What a turnabout in scientific thinking in just a decade!

F.T. Kyte et al have now provided additional details on meteoritic debris they first described in 1981. On the floor of the southeast Pacific, about 1400 kilometers west of Cape Horn, about 5 kilometers down, they found high concentrations of iridium in Upper Pliocene sediments about 2.3 million years old. Since the proposed projectile hit in very deep water, no crater was dug out. What did survive is called an "impact melt." This is debris rich in noble metals, such as iridium, and contains particles typical of a low-metal mesosiderite. Some 600 kilometers of the ocean floor received this debris. Kyte and his associates estimate the size of the impacting object at at least 0.5 kilometers in diameter.

No biological extinctions are correlated with the 2.3-million-year date, but there appears to have been a major deterioration of climate at about this time. There was a shift in the marine oxygen isotope records and, more obvious, the creation of the huge loess (sandy) deposits in China. What the impact may have done is to vaporize enough water into the atmosphere to increase the earth's albedo, reflecting sunlight back into space, lowering the average temperature, and thus triggering the Ice Ages. (Kyte, Frank T., et al; "New Evidence on the Size and Possible Effects of a Late Pliocene Oceanic Asteroid Impact," Science, 241:63, 1988.)

Comment. Aficionados of the Ice Age problems will have to add this theory to the already long list of Ice Age hypotheses.

ASTEROID IMPACT OR VOLCANOS?

The debate over the real cause of the terrestrial catastrophism that occurred at the Cretaceous-Tertiary boundary, some 65 million years ago, grinds on. Some physical scientists claim rather imperiously that the dinosaurs and many other species were done in by the impact of a huge asteroid/meteorite. The worldwide iridium spike is conclusive, they say. Many paleontologists and geologists, however, remain unconvinced and prefer widespread volcanism. We have already covered the various arguments in past issues of Science Frontiers; here, we want to advise our readers that a pair of excellent articles by principals in this debate have appeared in Scientific American. Generally speaking, it seems that the proponents of the impact theory are now listening to the other side. For example, multiple impacts are now proposed to account for evidence of the type introduced below. (Alvarez, Walter, and Asaro, Frank; "An Extraterrestrial Impact," Scientific American, 263:78, October 1990. Courtillot, Vincent E.; "A Volcanic Eruption," Scientific American, 263:85, October 1990.)

A spike dulled. The case for a single asteroid/meteorite impact has been weakened by a recent reexamination of the classic exposure of the Cretaceous-Tertiary boundary at Gubbio, Italy. Here, the discovery of an iridium "spike" at the boundary was thought to betoken a sudden, catastrophic, extraterrestrial event. On further study, though, the spike no longer seems so sharp. In fact, it is spread through about 3 meters of sediments at the boundary. Said sediments apparently took 500,000 years to accumulate! This broad smear of iridium no longer seems indicative of a single extraterrestrial impact. (Rocchia, R., et al; "The Cretaceous-Tertiary Boundary at Gubbio Revisited: Vertical Extent of the Ir Anomaly," Earth and Planetary Science Letters, 99:206, 1990.)

Comment. Is it even indicative of multiple impacts? Recall that no one has yet (1990) found a single crater that everyone will accept as evidence.

OSMIUM ISOTOPES SUPPORT METEORIC IMPACT

Comparison of terrestrial and meteoric osmium-isotope abundances tends to confirm the hypothesis of a meteorite strike at the Cretaceous-Tertiary boundary. (Luck, J.M., and Turekian, K.K.; "Osmium-187/Osmium-186 in Manganese Nodules and the Cretaceous-Tertiary Boundary," Science 222:613, 1983.)

THE CRETACEOUS INCINERATION

The worldwide deposit of iridium at the end of the Cretaceous implies, to many at least, that the great biological extinctions of this period were the consequence of a meteorite impact. It has now been discovered that clay samples from the Cretaceous-Tertiary boundary also contain 0.36-0.58% graphitic carbon. It is fluffy stuff and suggests that the planet was once covered by a thick layer of soot. Quantitatively, the soot layer is equivalent to the carbon in 10% of the earth's present biomass. The authors speculate that this soot was created by huge wildfires that consumed much of the earth's vegetation and perhaps fossil fuel as well. Terrestrial life was, of course, devastated---just as it is in the currently popular "nuclear winter" scenarios. The end-of-the-Cretaceous soot is in fact, thicker and more widely spread than nuclear winter theories predict. (Wolbach, Wendy S., et al; "Cretaceous Extinctions: Evidence for Wildfires and Search for Meteoric Material," Science, 230:167, 1985.)

Comment. Questions arise, though: How could a single meteorite impact ignite worldwide wildfires? Why haven't other meteorite impacts, recorded abundantly by large craters and astroblemes, also set fire to the planet and left iridium layers?

GLOBAL FIRE AT THE K-T BOUNDARY

The worldwide deposit of iridium at the K-T (Cretaceous-Tertiary) boundary has been considered very strong evidence that a large astronomical object (asteroid or comet) devastated our planet some 65 million years ago. Some scientists, however, propose that the iridium layer was instead deposited through widespread volcanic activity. The proponents of an astronomical mechanism should be heartened by a recent paper in Nature, by W.S. Wolbach et al. Here is their Abstract:

> Cretaceous-Tertiary (K-T) boundary clays from five sites in Europe and New Zealand are 10^2-10^4-fold enriched in elemental C (mainly soot), which is isotopically uniform and apparently comes from a single global fire. The soot layer coincides with the Ir layer, suggesting that the fire was triggered by meteorite impact and began before the ejecta had settled.

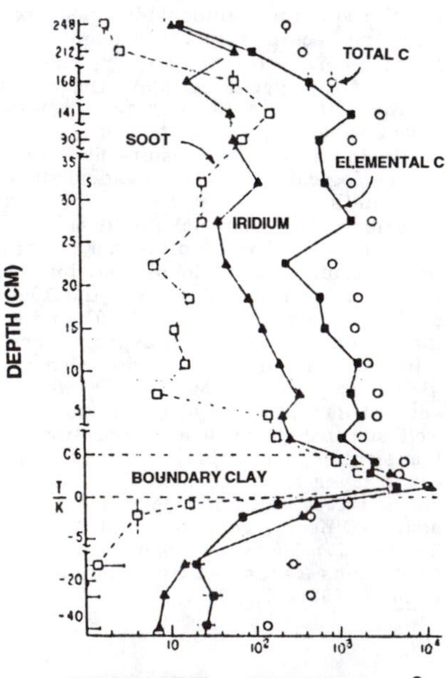

Concentration "spikes" of iridium, elemental carbon, and soot at the Cretaceous-Tertiary (K-T) boundary, from Woodside Creek, New Zealand. (Adapted from: Nature, 334:665, 1988.)

The composition of the hydrocarbons in the sediments points to the earth's biomass (mainly surface vegetation) as the source of the soot. The total quantity of K-T soot is equivalent to that which would be produced by burning 10% of all present terrestrial plant material. (Wolbach, Wendy S., et al; "Global Fire at the Cretaceous-Tertiary Boundary," Nature, 334:665, 1988.)

MORE CONFUSION AT THE K-T BOUNDARY

Just a few years ago, many scientists, especially physicists and astronomers, considered the Book of Science to be closed in the matter of what happened at the Cretaceous-Tertiary (K-T) boundary, 65 million years ago, and why the dinosaurs met their end. It was declared, rather imperiously, that a large asteroid had impacted the earth, causing much physical and biological devastation. Many scientific papers are still being written on this singular period in the earth's history, and the situation is no longer so clear-cut. We select for brief review four papers, each with a different perspective.

Occurrence of stishovite. Stishovite, a dense phase of silica, is widely accepted as an indicator of terrestrial impact events. It is not found at volcanic sites. Now, J.F. McHone et al report its existence at the K-T boundary, at Raton, New Mexico. (McHone, John F., et al; "Stishovite at the Cretaceous-Tertiary Boundary, Raton, New Mexico," Science, 243:1182, 1989.) A plus for the pro-impact side.

Evidence of a global fire. Soot appears at the K-T boundary at many sites, but where did it come from? Chemical analyses of these soots show an enhanced concentration of polycyclic aromatic hydrocarbons over soots above and below the boundary. This is strong evidence of pyrolytic action at the K-T boundary; i.e., widespread fires. (Venkatesan, M.I., and Dahl, J.; "Organic Geochemical Evidence for Global Fires at the Cretaceous/Tertiary Boundary," Nature, 338:57, 1989.) Fires could have been started by either volcanos or impacts.

The evidence of the traps. Traps, like India's famous Deccan Traps, are extensive flood basalts. In this paper, basalt flooding has been correlated with mass extinctions of marine life during the past 250 million years. The Deccan Traps were formed right at the K-T boundary. Traps could, however, be initiated by asteroid impact, which could stimulate eruptions. (Rampino, Michael; "Dinosaurs, Comets and Volcanoes," New Scientist, p. 54, February 18, 1989.)

The dinosaur angle. If dinosaurs were truly susceptible to extinction by either asteroid impact, widespread volcanism, or some combination of both, one would expect to find their numbers and diversity drastically curtailed during Mesozoic impact events. Instead, the dinosaurs not only survived these impacts but prospered. Their demise, which began before the K-T event, was probably not due to either impact or volcanism. (Paul, Gregory S.; "Giant Meteor Impacts and Great Eruptions: Dinosaur Killers?" BioScience, 39:162, 1989.)

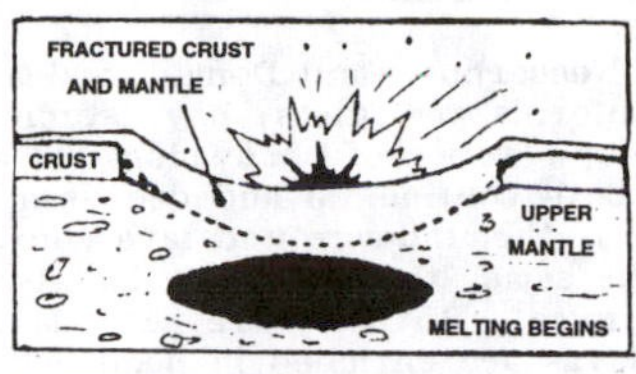

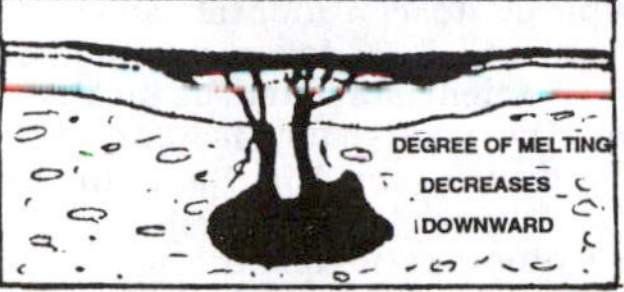

The impact of an asteroid can initiate basalt flooding and trap formation.

CHEMICAL MISCELLANY

CHEMICAL SURPRISES AT THE K-T BOUNDARY

The presence of high iridium concentrations at the Cretaceous-Tertiary (K-T) boundary, some 65 million years ago, has led to the widely accepted notion that an extraterrestrial projectile slammed into the earth at that time, wreaking geological and biological havoc. But the K-T boundary is anything but simple chemically and paleontologically. To illustrate, J.L. Bada and M. Zhao have found unusual amino acids in sediments laid down before and after this geological time marker.

> They find that Danish sediments spanning the narrow boundary layer contain two amino acids, α-aminoisobutyric acid and isovaline, that are relatively uncommon in biological materials but abundant in the organic-rich meteorites. They suggest that the body which collided with Earth 65 million years ago and left the telltale iridium residue may have been organic-rich, perhaps like a C-type asteroid or a comet. Such a possibility has interesting implications for the extinction and related atmospheric effects, and supports the idea that impact events could have supplied the Earth during a much earlier period with the raw materials for organic chemical evolution.

Actually, the above quotation is pretty much in line with present mainstream thinking. Perhaps so, but Bada and Zhao identified two troubling anomalies. First, the amounts of amino acids found were surprisingly high. How could these complex molecules survive the searing temperatures engendered by high-velocity impact? Second, the amino acids may be abundant tens of centimeters above and below the K-T boundary clay containing the iridium, but they are virtually absent in the clay itself! (Cronin, John R.; "Amino Acids and Bolide Impacts," Nature, 339:423, 1989, and Monastersky, R.; "Rare Amino Acids Support Impact Theory," Science News, 135:356, 1989.)

Cross reference. See pp. 119 and 184 for more on "life from space."

FOSSIL UFOs

> Geologists have discovered strange disc-shaped features in slate deposits in California. The features, at Yreka, are between 2 and 7 centimetres across and 2 to 4 millimetres thick; some have centres stained with iron oxides. One geologist, Nancy Lindsley-Griffin of the University of Nebraska, has already dubbed the saucer-shaped features, "unidentified fossil-like objects."
>
> Geologists discoverd the UFOs in bedding planes of the slate, formed from ocean bottom that was deposited between 400 and 600 million years ago. The objects are puzzling because they lack the symmetry that fossils of living organisms usually display. They are also too large to be the droppings of any creature alive at the time, and do not look like concretions, such as agates, formed by natural chemical processes. Lindsley-Griffin says they resemble "very tiny bicycle wheels, with a central core and an outer rim, but with most of the spokes missing."

One thought is that these features may be fossil jellyfish. (Anonymous; "Fossil 'UFOs' Mystify the Geologists," New Scientist, p. 43, July 1, 1989.)

THE EXPLODING LAKE

August 15, 1984. The village of Njindom, Cameroon. About 11:30 PM, the villagers heard a loud explosion coming from Lake Monoun. Early the next morning, people in a van driving past the lake discovered the body of a motorcyclist. The air smelled like battery fluid. One of the van's occupants collapsed. The others ran for their lives toward Njindom. By 10:30 AM authorities had found 37 bodies along a 200-meter stretch of road by the lake. Blood was oozing from the noses and mouths; the bodies were rigid; first-degree chemical burns were present. Also, animals and plants along the shore had been killed. On August 17, the lake turned reddish brown, indicating that it had been stirred up somehow.

Although Lake Monoun is in a volcanic crater, chemical analysis of the water found little of the sulphur and halogens normally associated with volcanic action. However, the analysis did find a tremendously high level of bicarbonate ions, which form from the dissociation of carbon dioxide. One theory is that an earthquake disturbed the carbonate-rich deep water of the lake, which as it rose to the surface and lower pressures, released huge volumes of carbon dioxide---something like opening a soda bottle. The resulting wave of water and cloud of gas caused the deaths and devastation. If there had been some nitric acid in the cloud, the burns could be accounted for. (Weisburd, S.; "The 'Killer Lake' of Cameroon," Science News, 128:356, 1985.)

Comment. The article states that this event is unique, but in our Catalogs similar phenomena are reported. For example, Lake Bosumtwi, Ghana, "explodes" at irregular intervals, changing color, killing fish, and releasing gases (GSD2-X17). We also have the sudden whitening of the Dead Sea (GHC4). Both of these phenomena are to be found in the Catalog Earthquakes, Tides, Unidentified Sounds.

STRANGE BLUE POOL FOUND AT THE BOTTOM OF CRATER LAKE

> A mysterious, small aqua-blue pool of dense fluid has been discovered at the bottom of Crater Lake.
>
> "It is bizarre, it is remarkable," said Jack Dymond, who with Robert Collier heads the three-year Crater Lake exploration project. "I have never seen anything like it," he said.
>
> The Oregon State University oceanographer said the pool, about 6 feet in depth, is approximately 3 feet wide by 8 feet long. It is near the lush white and orange bacteria mats found last summer.

The murky pool of fluid was discovered during a dive in a research submarine. The temperature of the pool was about 4.5°C (40.1°F) which made it 1°C warmer than the surrounding lake water. (Anonymous; "Strange Blue Pool Found in Crater Lake," Sunday Oregonian, August 13, 1989. Cr. R. Byrd.)

Comment. Some lakes in northern climes still retain ancient seawater in their bottoms. Also, we have the well-publicized African lakes that suddenly overturn, producing clouds of poisonous gases. See Anomalies in Geology.

MYSTERIOUS SMOKE IN SRI LANKA

> Mysterious smoke exuding from a dry river bed has produced the highest temperature ever recorded in Sri Lanka, and geologists said Friday they are baffled by the phenomenon.
>
> The 300-degree ground temperature has caused plants to wither in the mountainous region of Diyatalawa, a tourist resort in central Sri Lanka, said D.A. Kathriarachchi, the deputy director of the Geological Survey Department.
>
> He said scientists were puzzled because there is no volcanic activity in Sri Lanka, which lies outside any volcanic zone.
>
> The area, about 75 miles southeast of the capital, Columbo, is 9800 feet above sea level. The villagers have been told to report any other signs of smoke in the area, but no one has been evacuated.

(Anonymous; "Hot Smoke Baffles Geologists," Panama City News Herald, p. 1B, September 5, 1992. Cr. L.B. Peirce)

Comment. Category ESC4, in Anomalies in Geology, describes the "Smoking Hills" of the Canadian Arctic, as well as several other places where the oxidation of iron pyrite and other exothermic chemical reactions create very hot areas in non-volcanic regions.

EARTH'S WATER NOT IMPORTED?

That the earth is continuously bombarded by icy minicomets is unpopular in the Court of Science. Even less acceptable is the notion that over the eons these house-sized chunks of ice contributed substantially to our planet's inventory of water.

In what will surely be hailed as the death knell of the icy comet theory is the discovery by K. Muehlenbachs, of the University of Alberta, and F. Robert and M Javoy, from the University of Paris, that the water contained in the earth's rocks, both ancient and recent, is isotopically different from the water found in meteorites. Meteoric water is assumed to be isotopically the same as cometary water. Conclusion: comets could not have contributed substantially to our planet's water inventory in the geological past. (Anonymous; "Earth's Water Did Not Come from Comets," New Scientist, p. 19, June 20, 1992.)

Comment. Of course, the isotopic measurements have to be weighed against all the data supporting the icy comet theory (See p. 275.) Ice comets could, after all, be a new phenomenon. Also, no one has ever actually analyzed a piece of cometary ice; it is simply assumed that it would be similar to meteoric water.

DINOSAUR FLATULENCE AND CLIMATE CHANGES

Fossilized dinosaur dung contains evidence that flatulence from the giant creatures may have helped warm the Earth's climate millions of years ago, scientists said yesterday.

The researchers detected chemical signs of bacteria and algae in known and suspected dinosaur droppings. That indicates that plant-eating dinosaurs digested their food by fermenting it, a process that gives off methane.

We all know that methane is a "greenhouse gas," so it seems that the dinosaurs may have self-destructed. (Anonymous; "How Dinosaurs May Have Helped Make Earth Warmer," San Francisco Chronicle, October 23, 1991. Cr. D.H. Palmer)

Comment. We are not being facetious here, for it is seriously proposed that much of the greenhouse gas produced today comes from cattle, sheep, and other animals that ferment their food.

PHYSICAL ANOMALIES

Geological "physical" phenomena generally involve light, temperature, sound, radioactivity, gravity, and other facets of that scientific discipline we call physics. Like physics itself, the phenomena corralled here are diverse and often unrelated to one another. Nevertheless, they include some of the most fascinating anomalies and curiosities of nature.

Only a few of the recognized geological physical phenomena are indicated in the outline below. It should be added that the corresponding chapter in the book Anomalies in Geology boasts 21 entries.

- Radiometrical dating problems.
- Anomalous radiohalos. These are the so-called "pleochroic" halos which. according to some investigators, imply that the earth could be very young!
- Physical miscellany. Booming dunes, flattened beach pebbles, the moon's effect on avalanche frequency, etc.

RADIOMETRIC DATING PROBLEMS

THE NUCLEAR THREAT: BAD DATES

Woodmorappe has assembled an impressive and disconcerting collection of anomalous radiometric dates. Over 300 serious discrepancies are tabulated and backed by some 445 references from the scientific literature. To remove trivialities, only dates that were "wrong" by 20% or more were included. This criterion insured that the anomalous dates were off by one or more geological periods. To enhance his case, Woodmorappe excluded data for such troublesome minerals as K-feldspar, which have unreliable records. The surviving discordances will certainly disturb anyone who has long accepted radioactive dating as the near-final word in geochronology.

The lengthy text accompanying the table delves into the geological problems posed by the tabulated anomalies, primarily the severe distortions implied in the supposedly well-established geological time scale. Many attempts have been made to explain away these discrepancies, usually by asserting that the system must have been "open"; that is, contamination and/or removal of materials occurred. But a far more serious situation exists: the reluctance of researchers to publish radiometric dates that fly in the face of expectations. Data selection and rejection are epidemic. Some authors admit tossing out wild points; others say nothing. (Woodmorappe, John; "Radiometric Geochronology Reappraised," Creation Research Society Quarterly, 16:102, 1979.)

AN ANCIENT PLANET BENEATH A YOUTHFUL VENEER

Gerald Wesserburg and Donald de Paolo, two California geologists, have studied the isotopic ratios of neodymium 143 and 144 in both continental and deep-sea lavas. If the underground lava sources were the same, the isotope ratios should be the same. But they are not. Mid-ocean lavas are enriched in neodymium-143 compared to continental lavas. Since neodymium-143 is a decay product of samarium, scientists have been able to establish the neodymium isotope ratio from the time of the Big Bang to the present.

The isotope ratio for the mid-ocean lavas is just what would be expected on a planet where lighter surface materials had come to the surface during a molten state. The continental lavas, though, must tap very ancient reservoirs, possibly those of a true primitive earth. This ancient core is now swathed with younger materials from who knows where! This young envelope wraps around the whole planet, with the present continents being caused by slight protuberances on the ancient core. Whence the young veneer? A rain of material from some recent close encounter? (Anonymous; "Underground Sites of Ancient Earth," New Scientist, 83: 886, 1979.)

CORAL CARBON RATIOS CONFOUND CHRONOMETRY

By measuring the carbon-14/carbon-12 ratios in the annual growth bands of coral, scientists hope to spot natural and man-made changes in global chemistry. For example, the large-scale use of fossil fuels should depress the ratio by adding carbon-12 in undue quantities. The advent of the nuclear age boosts the ratio through the addition of carbon-14 to the environment. Predictably, the carbon ratio rises dramatically after 1950 (the bomb tests, etc.). Before this date, however, anomalies crop up:

1. Coral-ring and tree-ring data differ substantaially when they should not;
2. Coral-ring carbon ratios from relatively close locales, such as Bermuda (solid line) and the Florida Keys (dashed line), also differ significantly.

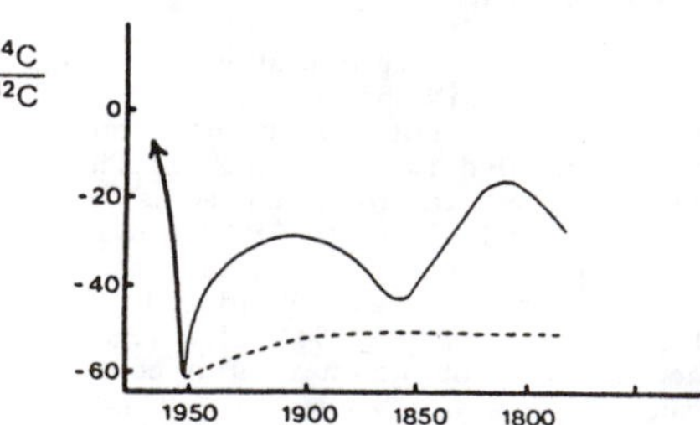

Recent coral-ring carbon ratios for Bermuda (solid line) and the Florida Keys (dashed line).

Item 1 might be due to non-atmospheric carbon upwelling in deep-ocean water; but this would not explain the Bermuda and Florida discrepancies. (Anonymous; "Carbon-14 Variations in Coral," Open Earth, no. 3, p. 30, 1979.)

Comment. These discrepancies are particularly relevant to the carbon-14 dating of seashells, which often produces wildly incorrect ages.

ANOMALOUS RADIOHALOS

WILL RADIOHALOS IN COALIFIED WOOD UPSET GEOLOGICAL CLOCKS?

In some coalified wood, uranium-rich solutions have deposited radioactive particles that subsequently decay and create little rings (halos) that can be seen under high magnification. The radii of the rings depend upon the energies of the particles emitted by the radioactive elements. Each type of radioactive decay has a specific half-life. Thus, the patterns of radiohalos help measure the age of coalified wood. A challenge to geology arises because the radiohalos in coalified wood from Jurassic and Triassic formations, supposedly millions of years old, suggest ages of only a few thousand years. (Connor, Steven J.; "Radiohalos in Coalified Wood: New Evidence for a Young Earth," Creation Research Society Quarterly, 14:101, 1977.)

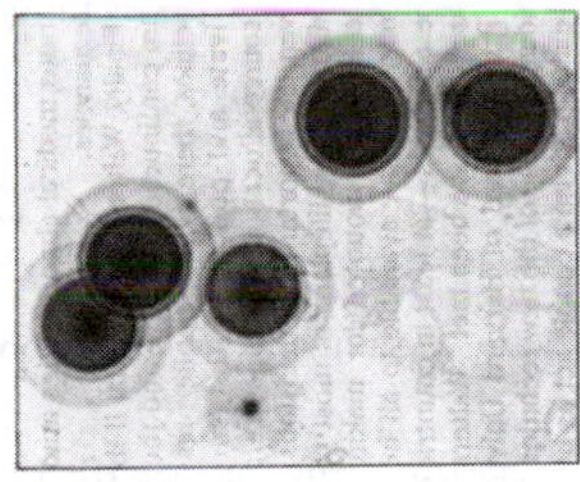

Typical radiohalos as seen through a microscope.

HALOS AND UNKNOWN NATURAL RADIOACTIVITY

R.V. Gentry is well known for his studies of giant halos in minerals, particularly micas from Madagascar. Recently, intense debate seems to have determined that these Madagascar giant halos are not due to naturally occurring superheavy elements. But what did cause them? In this article, Gentry, et al present data for giant halos in Swedish biotite. No conclusion is given as to their possible origin, but it is noted that some of these giant halos have bleached circles around their centers. These circles seem related to the enigmatic dwarf halos known and unexplained for more than 50 years. (Gentry, R.V., et al; "Implications on Unknown Radioactivity of Giant and Dwarf Haloes in Scandanavian Rocks," Nature, 274:457, 1978.)

Comment. The "halo" problem is not as trivial as it may seem because anomalous radioactivity, presumably with a very short half-life, should not be present in billion-year-old rocks such as the Madagascar micas. One implication: geological dating is all wrong!

TARNISHED HALOS?

Pleochroic halos are dark rings of various radii seen in mica and other minerals. There is general agreement that alpha particles emitted by radioactive isotopes create the halos. The radii of the rings are proportional to the alpha particle energy, and can thus identify the isotopes in the mineral. Some halos, however, are apparently formed by very short-lived polonium isotopes without any trace of parent uranium isotopes. How can polonium isotopes with half-lives only seconds long get into geologically old mica sans parents?

York argues the case for selective local chemical concentration of polonium from fluids in the surrounding rocks. The captured polonium atoms decay almost immediately while the fluid containing the parent atoms passes on. R.V. Gentry objects that mica is almost impermeable and that we must consider the possibility that our concepts of geological time are grotesquely wrong. York energetically defends established Geology using radioactive dating and paleontological arguments. His contempt of Gentry's position is scarcely veiled. This paper is an excellent review of the pleochroic halo problem as well as a classic defense of the scientific status quo. (York, Derek; "Polonium Halos and Geochronology," Eos, 60:617, 1979.)

Comment. York does not mention Gentry's years of careful work that led him to his heresy, nor are the many objections to radioactive dating discussed. It reminds one of the confident assertions of the permanency of the ocean basins made only a few years ago. York might be correct, but the "we now know that" approach is disturbing.

THREE "PROOFS" OF A YOUNG EARTH

Responding to letters published in the June 1982 issue of Physics Today that insisted (some, very emotionally) that no evidence exists for a very young earth, Robert V. Gentry summarized three kinds of evidence that certainly seem to undermine current dating schemes.

1. Halos produced by the alpha particles emitted by Po^{218} are found in granite rocks in many areas. Yet, the half-life of Po^{218} is only 3 minutes. Since the Po^{218} has no identifiable precursors in the rock, "...how did the surrounding rocks crystalize rapidly enough so that there were crystals available ready to be imprinted with radiohalos by α-particles from Po^{218}? This would imply almost instantaneous cooling and crystallization of these granitic minerals---and we know of no mechanism that will remove heat so rapidly; the rocks are supposed to have cooled over millennia, if not tens of millennia."

2. In coalified wood dated as older than 200 million years, the ratio between U^{238} and Pb^{206} should be low. It is actually very high. "Thus ages of the entire stratigraphic column may contain epochs less than 0.001% the duration of those now accepted and found in the literature."

3. Diffusion calculations insist that Pb in zircon crystals found in deep granite cores at 313°C should diffuse out of the crystals at the rate of 1% in 300,000 years. No loss of Pb can be detected at all. Therefore, the granite must be younger than 300,000 years. (Gentry, Robert V.; "Creationism Discussion Continued," Physics Today, 35: 13, October 1982.)

Comment. Scientists admit that Gentry's work raises questions but apparently would rather live with the anomalies than with the thought of a young earth!

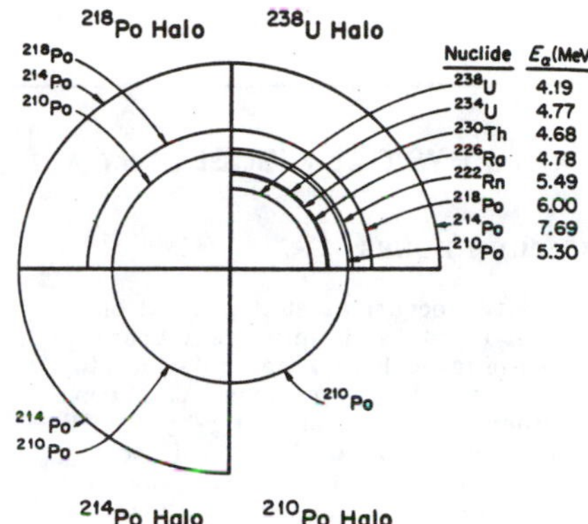

(Right) Normal halo complex associated with the decay of uranium-238. (Left) Polonium halos without precursors.

GENTRY'S TINY MYSTERY---UNSUPPORTED BY GEOLOGY

An article bearing the above title, by J. Richard Wakefield, appeared in the Winter 1987-1988 issue of Creation/Evolution. The title implies that Gentry's "Tiny Mystery" is soon to be demolished. It turns out in the end that a sweeping interpretation of this "Tiny Mystery" is called into question, but the mystery itself, like the smile on the Cheshire cat, remains.

The "Tiny Mystery" is the existence of radiohalos from polonium-210, -214, and -218 in some biotites (micas), sans any detectable precursory uranium or halos thereof. Since the half-lives of the polonium isotopes are 138.4 days, 0.000164 second, and 3.04 minutes, respectively, it is certainly perplexing how the polonium halos got where they are! According to geological thinking, the igneous rocks containing the biotites must have been molten for a good deal longer than 138 days, thus destroying halos of short-lived isotopes.

R.V. Gentry, a creationist, thinks that the polonium isotopes are primordial---created by God some 6,000 years ago, in situ and without precursors. The rocks displaying the halos would, therefore, be among the oldest rocks on earth. What Wakefield does is examine the geology of some of the sites from which Gentry obtained his biotite samples. Basically, he finds that the supposedly primordial rocks are often just dikes and veins that were formed much later than the earth's oldest rocks. Wakefield comments, "This fact alone tells us that the rocks bearing Gentry's halos, even if instantly created, have no bearing on the origin and age of the earth."

But does Gentry's "Tiny Mystery" vanish? Not at all---unlike the Cheshire cat's grin. The polonium halos, seemingly without detectable precursors, are still there. Wakefield states with all honesty, "Still, we must give Gentry his due. Nothing in geology fully explains the apparent occurrence of the polonium halos as described by Gentry. They do remain a minor mystery in the field of physics." Someone may eventually find an explanation in terms of quirkish chemical deposition or misidentification of the halos, but for now the "Tiny Mystery" survives! (Wakefield, J. Richard; "Gentry's Tiny Mystery---Unsupported by Geology," Creation/Evolution, 22:13, Winter 1987-1988.)

PHYSICAL MISCELLANY

BOOMING DUNES

> On two occasions it happened on a still night, suddenly---a vibrant booming so loud I had to shout to be heard by my companion. Soon other sources, set going by the disturbance, joined their music to the first, with so close a note that a slow beat was clearly recognized. This weird chorus went on for more than five minutes continuously before silence returned and the ground ceased to tremble.

P.F. Haff opens his review of booming sands with the above quote from R.A. Bagnold. One would think that since booming sand is not uncommon and scientists can pick it up and take it back to their laboratories, we know all about why it booms so unexpectedly when set in motion down a dune face. Haff relates his own experiences and experiments and ties them into the rather large body of previous work on the subject. The factors of dampness, grain size, cleanliness, grain shape and smoothness, etc., have been examined. But Haff concludes, "In spite of these experiments and the work of other researchers, it is still not known how booming dunes work." (Haff, P.K.; "Booming Dunes," American Scientist, 74:376, 1986.)

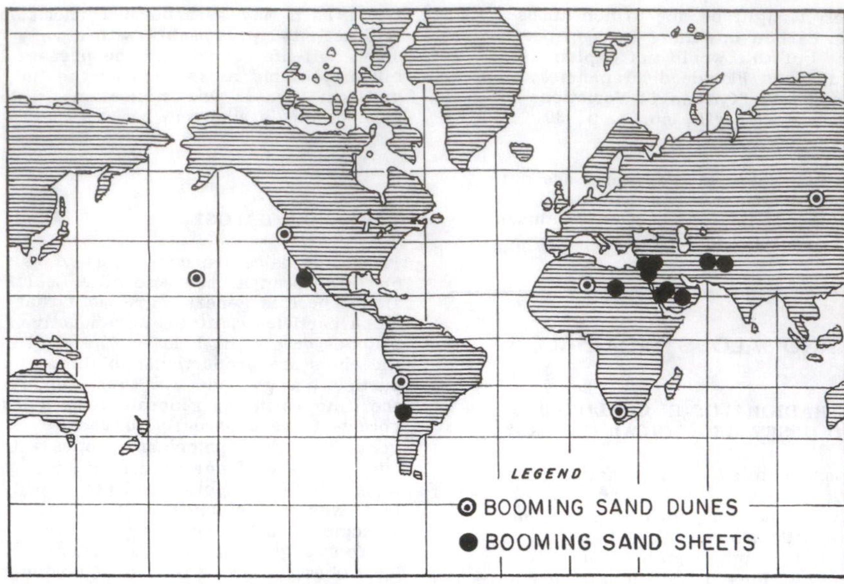

Locations of prominent musical and booming sands. (Adapted from: Geological Society of America, Bulletin, 87: 463, 1976.)

WAGNERIAN SANDS OF THE DESERT

> The sun was sinking in the west African afternoon when we plunged into a wild Land Rover ride across the dunes. Within moments we lost all sense of direction, regaining it only momentarily when we caught a glimpse of the sea, as we sped along the notorious Skeleton Coast of Namibia. When we finally stopped, it was at the top of a crescent-shaped behemoth. On the inside of the curve, the dune was a good 200 feet high and as close to vertical as a sand dune gets.
>
> We gathered timidly at the edge of the precipice. But our guide, seemingly bent on suicide, sat down at the crest and started to slide down the dune face, the seat of his pants sending a small cascade of sand slithering to the bottom. Instantly, the entire dune began to pulsate, groaning and grumbling, as if armies of Frank Herbert's sandworms from Dune were chewing their way to the surface. In a moment, all of us were laughing and scooting down the dune, the unearthly roar echoing in the natural amphitheater.

What a delightful introduction to one of Nature's light-hearted anomalies! Such booming dunes and roaring sands may be found in thirty-or-so localities all over the world, mostly in desert environments. Most of the booming dunes are composed of quartz sands, the main exception being the Barking Sands on Kauai, Hawaii, which are calcium carbonate. Despite over a century of investigation, no one knows exactly why some dunes boom. In fact, the sand grains of booming and silent dunes look pretty much alike. The addition of sand from a booming dune will not make a silent dune roar, but the addition of silent-dune sand to a booming dune will contaminate it and ruin its boomability. Glass beads of the same size as the quartz grains in a booming dune will not boom, despite their smoothness. A lot of experiments have been tried with the booming sands, but though they will boom, they won't talk! (Thompson, Sharon Elaine; "Wagnerian Sands of the Desert," Lapidary Journal, p. 26, July 1990. Cr. R. Calais)

Sand Mountain, Nevada, noted for its booming sand. (J.F. Lindsay)

BLACKENED, BROKEN STONES OF THE MIDDLE EAST

The current issue of Pursuit presents an article by Z. Sitchin in which he expounds his theory about ancient conflicts between gods and men. As evidence of this supposed strife, he reproduced a photograph of a plain in the Sinai Peninsula. Basically white limestone, the plain is thickly strewn with blackish, angular stones. Whence these immense quantities of incongruous stones? Sitchin proposes that they were created when an ancient spaceport was destroyed! (Sitchin, Zecharia; "The Wars of Gods and Men," Pursuit, 18:106, 1985.)

Comment. Whatever one thinks of Sitchin's theory, the stones remain and must be accounted for. They are not the only such deposits in the Middle East. Velikovsky, in his Earth in Upheaval, states that 28 such rock fields are found in Arabia, some with areas of 6 or 7 thousand square miles. These are the "harras." Velikovsky thinks they are meteoric debris! Are there other explanations around?

WHY AREN'T BEACH PEBBLES ROUND?

The next time you walk along an ocean beach, forget the greater anomalies of nature and pick up a few well-worn pebbles. Q.R. Wald collected 200 such at random and measured them with calipers. He found that their three axes tended toward the ratios 7:6:3. No, beach pebbles are not spherical, they are flattened ovoids. We would naturally expect that the ceaseless action of the surf would turn out nearly perfect spheres. So much for intuition! (Wald, Quentin R.; "The Form of Pebbles," Nature, 345:211, 1990.)

IS THERE TRUTH IN THE GRAINS?

Scientific creationists often come up with fascinating minor phenomena in geology, which, as far as most geologists are concerned, have already been satisfactorily explained and are now forgotten. Creationists exhume these tidbits because they can sometimes be interpreted as support for a "young earth" or some other cornerstone in creationist thinking. Of course, the true anomalist has a broader appetite, savoring any unexplained phenomena, regardless of which theories they support or undermine.

But that is anough palaver, creationist D.E. Cox has resurrected the quartz-sand-grain anomaly:

> Uniformitarian geologists, and many creationist geologists as well, interpret sandstones as sedimentary or aeolian accumulations of sand grains. Quartz arenites, for example, are supposed to be detrital accumulations of quartz grains derived from preexisting formations. The source rocks may have either been igneous or sedimentary, but ultimately most quartz grains in sediments must have an igneous parent rock, probably granite, since this is the most abundant igneous rock on the continents. The quartz derived from granite is characterized by the presence of tiny mineral inclusions. Crystals of mica, rutile and tourmaline are common. In sandstones formed from granite-derived quartz grains, mineral inclusions should be evident, but they are rarely present in sandstones.
>
> The accepted explanation of this apparent anomaly is that the inclusions in fresh, granite-derived quartz grains so weakened the grain structure that the flawed grains are quickly broken up during weathering, transportation, and deposition. By the time any sandstone is formed, only flawless bits of sand remain. Case closed!
>
> No anomalist worth a grain of salt would let this delightful phenomenon escape without a bit more study.
>
> 1. Do young sandstones with identifiable granitic sources show more inclusions than older sandstones?
>
> 2. Do desert sands, beach sands, and other unconsolidated quartz grains show any flaws?
>
> 3. Has anyone really examined fresh quartz grains weathered from granite to determine how the number of flaws in a grain varies with the grain size? (Cox, Douglas E.; "Missing Mineral Inclusions in Quartz Sand Grains," Creation Research Society Quartery, 25:54, 1988.)

Comment. Most geologists will complain that we are going out of our way to make trouble. But consider the possibility that some unflawed quartz grains in sandstones may have actually been precipitated from gases and fluids and not be granitic at all. And what about those sandstone dikes and other sandstone intrusive bodies? Where did their quartz grains originate? Not all sandstone is sedimentary!

THERE'S MORE THAN GOLD IN THE KOLAR MINES

When physicists installed nuclear particle detectors deep in a mine in the Kolar Gold Fields in India, they hoped to measure particles created by highly penetrating neutrinos arriving from cosmic sources. They found instead immense showers of nuclear particles coming, not from above as expected, but from the sides and even from below! These huge showers of 1,000 or more different particles are called "anomalous cascades." Neutrinos are the only known particles capable of penetrating the entire earth to create the upwardly directed showers, but ordinary neutrinos do not seem to have enough energy to give birth to the anomalous cascades. (Anonymous; "Particle Shower Sprays Upward," Science News, 118:246, 1980.)

Comment. Are there sources of unrecognized radiation deep within the earth?

THE EARTH AS A COLD FUSION REACTOR

On p. 215, we mentioned the possibility that the helium-3 emanating from the earth might indicate that cold fusion was occurring deep down. In a recent issue of the New Scientist, a short unsigned article reveals that this "excess" helium-3 was an impetus for the cold fusion research at Brigham Young University. In fact, P. Palmer, a geophysicist at Brigham Young, suggested the possibility as long as three years ago! We have not seen Palmer's speculation in print, but the stimulating effect of anomalies on scientific research is reassuring, whatever the final outcome of the cold fusion wars.

The same New Scientist article supports the above speculation as follows:

> Calculations show that more than enough deuterium finds its way into the upper mantle by this route (seawater in subduction zones) to account for the heat emitted by the Earth's core, although the heat obviously comes from other sources as well. The rate of fusion of deuterium nuclei required to produce the observed ratios of helium-3 to helium-4 in rocks, diamonds and metals is similar to that observed by Jones in his experiments with electrolytes. Tritium can also be a product of the fusion of deuterium. Jones and his group say that the tritium detected in the gases from volcanoes is further evidence of cold fusion.

Jones has also wondered whether Jupiter's excess heat could be generated deep within the icy planet via cold fusion. (Anonymous; "Rocks Reveal the Signature of Fusion at the Centre of the Earth," New Scientist, p. 20, May 6, 1989.)

THE MOON AND AVALANCHES

The moon is blamed for many things from earthquake triggering to human crimes of passion. Until now, no one seems to have studied the lunar effect on avalanche frequency; even though avalanches are obvious trigger-type phenomena. We find the following paragraph in an article on snow avalanches in general:

> Another precipitating factor may be the gravitational pull of the moon. In research published last year, Peter Lev of the Utah Highway Department found that based on a statistical study of moon and avalanche cycles in the Wasatch Mountains durthe past 20 years, the chance of an avalanche's occurring on a full and new moon was 100 times greater than it is during other days in the lunar cycle.

(Anonymous; "Full Moon May Contribute to 'Loose' and 'Slab' Avalanches," San Jose Mercury News, p. 3D, December 31, 1985. Cr. P. Bartindale.)

NEWTONIAN GRAVITY MAY HAVE BROKEN DOWN IN GREENLAND

Anomalies in the measurement of gravity in Australian mines have stimulated an experiment in one of the very deep holes drilled in the Greenland ice cap. The hole at location Dye-3 is 2 kilometers deep. Gravity measurements were made at 183-meter intervals between depths of 213 and 1673 meters. Elaborate precautions were taken to assure that proper corrections were made for ice density and the nature of the rock below the ice. Ice-penetrating radar sketched the topography of the ice-rock surface, and surface-gravity measurements assessed density variations in surrounding ice and rock. The results of this finely tuned experiment are found in the final two sentences of the report's Abstract:

> An anomalous variation in gravity totaling 3.87 mGal (3.87×10^{-5} m/s^2) in a depth interval of 1460 m was observed. This may be attributed either to a breakdown of Newtonian gravity or to unexpected density variations in the rock below the ice.

(Zumberge, Mark A., et al; "The Greenland Gravitational Constant Experiment," Journal of Geophysical Research, 95:15483, 1990.)

BIOLOGICAL ANOMALIES

In this chapter on "geology," the adjective "biological" is taken to include fossils (mineralized bones, etc.), unfossilized biological materials (frozen mammoths, undecayed tree stumps, etc.), and even living animals (bacteria laboring below the terrestrial surface. We include here only those phenomena of geological import; that is, those that reveal something anomalous or curious about the earth's history and how its stratigraphic record was created. Those phenomena that have something to say about the development of life throughout our planet's history are assigned to the earlier chapter on Biology. Also note that the formation of oil and natural gas deposits are covered on p. 206.

Not only do life forms (as fossils) help us to understand the earth's history, but they also have and still do contribute to the formation of some geological materials and structures, as suggested in the following outline:

- Biogeology. The role of life forms in creating geologically significant materials and structures.

- Biological extinctions. Anomalies of the fossil record, the possibility of periodical extinctions.

- The cause(s) of biological extinctions. Is astronomical catastrophism (asteroids, comets, etc.) the only cause of extinctions?

- Biological miscellany.

BIOGEOLOGY

BIOGEOLOGY

It is accepted that every cubic centimeter of the topsoil beneath our feet seethes with thousands of microorganisms. It is less well known that life's domain extends down much further. The hard rocks and strata of earth's crust---seemingly sterile and inert---are continuously being transformed by bacteria and other life forms. In fact, it was easy to find three examples of such processes from the literature collected from the past two months.

Although the discoveries reported below may seem dull to anomalists used to more exciting fare, it may well be that life from "inner space" has been and will be more important to humankind than life from "outer space," as implied in third item below:

Bacteria and placer gold.

> Lacelike networks of micrometre-size filiform gold associated with Alaskan placer gold particles are interpreted as low-temperature pseudomorphs of a Pedomicrobium-like budding bacterium. Submicron reproductive structures (hyphae) and other morphological features similar to those of Pedomicrobium occur as three-dimensional facsimiles in high-purity gold in and on placer gold particles from Lillian Creek, Alaska.

In short, bacteria help create placer gold deposits. The author believes that bacterioform gold is widespread. (Watterson, John R.; "Preliminary Evidence for the Involvement of Budding Bacteria in the Origin of Alaskan Placer Gold," Geology, 20:315, 1992.)

Microorganisms and iron deposits. At least 500 million years ago, filamentous bacteria and/or fungi were already playing vital roles in the deposition of iron from hydrothermal fluids. Abundant microbial filaments indicative of biological activity are found in the Cambrian ironstones in Australia's Thalanga deposit. (Duhig, Nathan C., et al; "Microbial Involvement in the Formation of Cambrian Sea-Floor Silica-Iron Oxide Deposits," Geology, 20:511, 1992.)

Deeper implications. The formation of placer gold and ironstone are only part of the repertoire of deep-living microorganisms. A five-year survey of microbial life conducted by the U.S. Department of Energy (DOE) found that bacteria were everywhere---even 3 kilometers deep in a Virginia borehole. F. Wobber, the DOE manager of the project underscored the mystery and probable importance of "biogeology":

> Besides asking how subsurface bacteria affect geology, he wonders how geologic processes could have carried living things so deep into the planet. "When you find these organisms at great depths," he says, "you have to ask, 'Where did they come from?'" Microbes from the soil could easily infiltrate shallow aquifers...but in very deep sediments, like those in the Texaco well, the microbes may have been entombed when the rock was first deposited, tens or hundreds of millions of years ago. If so, the deep Earth might be a den of survivors, toughened by millennia of evolution in their harsh environment. Attacking rock might be just one of their feats.

(Appenzeller, Tim; "Deep-Living Microbes Mount a Relentless Attack on Rock," Science, 258:222. 1992.)

Comment. Is Wobber suggesting that these super-tough, deep-living bacteria might be dangerous to humans, like the microorganism from outer space in the movie The Andromeda Strain?

BIOGENIC MINERALS

We generally think of minerals as having been formed by purely inorganic processes. Only once, on p. 215, where biogenic stalactites were described, have we pursued the idea that minerals, including crystal forms, might be biogenic. We now have at hand a survey of biogenic minerals. It turns out that biogenic minerals are quite common---so common, in fact, that the Gaia concept is recalled, in which biological processes preside over much that happens upon this planet.

Here follows a sampler of some biogenic minerals:

- Much, if not all, travertine (calcite and/or aragonite) and silicious sinter (opal) are deposited through algal action.
- Much pyrite and marcasite in sedimentary rocks comes from bacterial sulfate reduction.
- Bacterial breakdown of oil produces organic complexes that dissolve, transport, and precipitate quartz. The reknowned Herkimer "diamonds" may be of biological origin.
- Living cells synthesize isometric crystals of magnetite.
- Mitochondria manufacture crystals of hydroxylapatite.
- Better known are the apatite in bones and teeth and the aragonite, calcite, or fluorite in the vestibular systems of vertebrates.

(Dietrich, R.V., and Chamberlain, Steven C.; "Are Cultured Pearls Mineral?" Rocks and Minerals, 64:386, September/October 1989. Cr. R. Calais)

AN EXTRAORDINARY PEAT FORMATION

Most of Beauchene Island, in the Falklands, is covered with a tussock-forming grass. During the past 12,500 years, a deep accumulation of exceptionally dense peat has formed. The basal peat is lignitic, but is several hundred times too young to be a true lignite. This peat does not decay as rapidly as it should, given its populations of bacteria, yeasts, and other fungi. The peat accumulates about ten times faster than in other peat-forming regions. The authors conclude that the peat-forming process is poorly understood. (Smith, R.I. Lewis, and Clymo, R.S.; "An Extraordinary Peat-Forming Community on the Falkland Islands," Nature, 309:617, 1984.)

Comment. If we do not understand how present-day peat forms, how can we be so dogmatic about coal-forming processes millions of years ago?

THE CARBON PROBLEM

The "carbon problem" seems to hit the scientific creationists the hardest, but it also has interesting implications for today's earth. Consider first where the carbon in the earth's crust resides:

Petroleum	201×10^{18} grams
Coal	15
Limestone	64200
Biosphere	0.3

In this article, these figures are made more understandable by physical descriptions of some of the truly colossal deposits of oil, coal, and limestone. For example, in the Canadian Rockies, the Livingstone limestone was deposited 2000 feet deep on the margin of the Cordilleran geosyncline but thins eastward to about 1000 feet in the Front ranges. "...it may be calculated to represent at least 10,000 cubic miles of broken crinoid plates."

Two implications are: (1) Even if the earth's biosphere were completely converted into oil, coal, and limestone each year, the earth would have to be far older than the 6000 years desired by the creationists, unless most of the carbon deposits had non-biological origins, which seems unlikely. (2) The immense inventory of carbon tied up in biologically produced deposits was originally abiogenic. Where did it come from? Abiogenic methane and carbon dioxide released from the crust seem the most likely sources. This means that the crust must have once had, and may still have, prodigious supplies of methane. T. Gold and S. Soter have long argued that the earth's crust still retains and sometimes releases methane. (Morton, Glenn R.; "The Carbon Problem," Creation Research Society Quarterly, 20: 212, 1984.)

Comment. Methane gas releases may account for several anomalies, such as earthquake lights and unidentified detonations. See pp. 251 and 283.

NITRATE DEPOSITS DEFY RATIONAL EXPLANATION

"The nitrate deposits in the extremely arid Atacama Desert of northern Chile are among the most unusual of all mineral deposits. In fact, they are so extraordinary that, were it not for their existence, geologists could easily conclude that such deposits could not form in nature. The nitrate deposits consist of water-soluble saline minerals that occur as cement in unconsolidated surficial material---alluvial fill in valleys, loose rocky debris on hillsides, and windblown silt and sand---and as impregnations and veins in porous and fractured bedrock. They are found chiefly along the eastern side of the Coastal Range, but also within the Coastal Range, in the Central Valley to the East, and along the lower Andean front. Features of the deposits that appear to defy rational explanation are their restricted distribution in a desert characterized throughout by saline soil and salt-encrusted playas; the wide variety of topography where they occur; the abundance of nitrate minerals, which are scarce in other saline complexes; and the presence of other, less abundant minerals containing the ions of perchlorate, iodate, chromate, and dichromate which do not exist in any other saline complexes. Iodate, chromate, and dichromate are known to form under such conditions, but no chemical process acting at temperatures and pressures found at the earth's surface is known to produce perchlorate." (Ericksen, George E.; "The Chilean Nitrate Deposits," American Scientist, 71: 366, 1983.)

BIOLOGICAL EXTINCTIONS

THE REHABILITATION OF CUVIER

Cuvier (1769-1832) was a catastrophist. To him, the record of death in the layers of fossiliferous rocks was obviously the consequence of terrestrial convulsions. But Cuvier's ideas were swept aside by the uniformitarians who saw the earth and its cargo of life unfolding with almost agonizing slowness. But Cuvier is making a comeback, as illustrated by the following back-to-back articles in Nature:

Abstract. Closely spaced samples from an uninterupted calcareous pelagic sequence across the Cretaceous-Tertiary boundary reveal that the extinction of planktonic Foraminifera and nannofossils was abrupt without any previous warning in the sedimentary record, and that the moment of extinction was coupled with anomalous trace element enrichments, especially of iridium and osmium. The rarity of these two elements in the crust of the Earth indicates that an extraterrestrial source, such as the impact of a large meteorite may have provided the required amounts of iridium and osmium. (Smit, J., and Hertogen, J.; "An Extraterrestrial Event at the Cretaceous-Tertiary Boundary," Nature, 285:198, 1980.)

Abstract. Evidence is presented indicating that the extinction, at the end of the Cretaceous, of large terrestrial animals was caused by atmospheric heating during a cometary impact and that the extinction of calcareous marine plankton was a consequence of poisoning by cyanide released by the fallen comet and of a catastrophic rise in calcite-compensation depth in the oceans after the detoxification of the cyanide. (Hsu, Kenneth J.; "Terrestrial Catastrophe Caused by Cometary Impact at the End of Cretaceous," Nature, 285:201, 1980.)

HOW REAL ARE BIOLOGICAL EXTINCTIONS IN THE FOSSIL RECORD?

Much has been made by catastrophists of the apparent wholesale extinctions of many forms of life from one geological period to another. Some uniformitarians have been arguing all along that these so-called extinctions are more apparent than real. The present study supports this view with its study of world-wide records of the Triassic-Jurassic transition period. Correlation of U.S. and South African rocks imply that this key geological interface was actually a period of gradual faunal replacement rather than sudden, simultaneous extinction of live forms. (Olsen, Paul E., and Galton, Peter M.; "Triassic-Jurassic Tetrapod Extinctions: Are They Real?" Science, 197:983, 1977.)

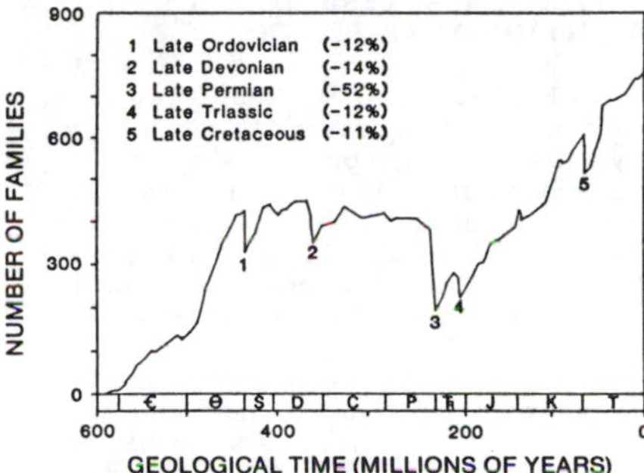

Five important biological extinction events in the fossil record, as measured by the number of families affected.

BONE BED DISCOVERED IN FLORIDA

A new bone bed has been discovered south of Tampa. Paleontologists say it it is one of the richest fossil deposits ever found in the United States. It has yielded the bones of more than 70 species of animals, birds, and aquatic creatures. About 80% of the bones belong to plains animals, such as camels, horses, mammoths, etc. Bears, wolves, large cats, and a bird with an estimated 30-foot wingspan are also represented. Mixed in with all the land animals are sharks' teeth, turtle shells, and the bones of fresh and salt water fish. The bones are all smashed and jumbled together, as if by some catastrophe. The big question is how bones from such different ecological nitches---plains, forests, ocean---came together in the same place. (Armstrong, Carol; "Florida Fossils Puzzle the Experts," Creation Research Society Quarterly, 21: 198, 1985.)

COSMIC DEATH WAVES

In the language of science, W.M. Napier and S.V.M Clube provide a scenario of cyclic terrestrial catastrophism. Their thesis is that the solar system periodically passes through the regularly spaced spiral galaxy arms every few 10^7 years. Planetesimals in these arms crater the solar-system planets at these times and also provide the raw materials for new comets, asteroids, satellites, and even planets.

Supporting their theory is the repeating history of geological revolutions with the accompanying extinctions and reflowerings of life. A remarkable feature of this paper is a table of short-lived solar-system phenomena (comets and rapidly evolving satellite-and-ring systems.) The tenor is one of episodic catastrophism and a rapidly changing solar system; viz., Saturn's rings evolving in only 10^4 years. (Napier, W.M., and Clube, S.V.M.; "A Theory of Terrestrial Catastrophism," Nature, 282:455, 1979.)

Comment. This outlook differs radically from that still disbursed in our schools and colleges.

WANTED: DISASTERS WITH A 26-MILLION-YEAR PERIOD

J. Sepkoskiand and D. Raup, two researchers at the University of Chicago, have drawn up graphs showing the numbers of families of marine organisms that have vanished from the fossil record over the eons. From this overview of manifest mass extinction emerged a puzzling and potentially profound pattern. Roughly every 26 million years over the last 250 million years, the number of extinctions jumped well above the background level. Some cyclic phenomenon seems to have been killing off life forms on a systematic basis. But no natural 26-million-year cycles are known. Although meteors and comets are favored causes of extinctions these days, they display no such cyclic period. (Simon, C.; "Pattern in Mass Extinctions," Science News, 124:212, 1983.)

Comment. Instead of looking outward to astronomical catastrophism, perhaps we should look inward. The earth itself may undergo cyclic paroxysms; or life might undergo intrinsic phases of decline and rejuvenation.

PERIODIC EXTINCTIONS AND EXPLOSIONS IN TERRESTRIAL LIFE

Claims of a 26-million-year periodicity in biological extinctions have been given wide publicity recently. One theory has it that periodic comet showers have created this regular pulse beat in the history of life. A recent issue of Science presents the results of an analysis of the ups and downs of 20,000 marine genera---that is, percent extinctions over a period of 600 million years, as revealed by the fossil record. The graph of ten groups of 1000 genera shows at least two things: (1) strong hints of periodicity; and (2) suggestions that extinctions, whatever they really are, "cut across functional, physiological, and ecological lines." The plotters of these graphs, D. Raup and G. Boyajian, claim that whatever the mechanism, "major pulses of extinction result from geographically pervasive environmental disturbances." What besides powerful, external physical forces (read "comets and asteroids") could affect such wide ranges of marine organisms? (Lewin, Roger; "Pattern and Process in Extinctions," Science, 241:26, 1988.)

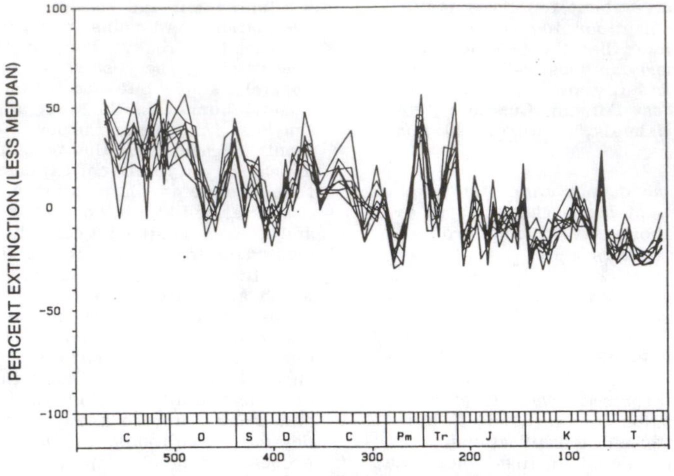

Periodicity in the fossil record, as illustrated by percent extinction for ten random groups of 1000 genera. (Adapted from: Science, 241:26, 1988.)

Comment. This all sounds so reasonable that one must wonder why it is given space in Science Frontiers! The reason is that we have a suspicion that it is all too easy, too simplistic! Could something more subtle be at work? After all, we really know next to nothing about the real workings of life-as-a-whole, its ups and downs. It is so easy to say that a group of organisms was done in by a temperature change or the fall of acid rain brought on by the impact of an asteroid. We always look for external forces, whereas the real cause of "crises" in the history of life may be intrinsic to life itself.

With a tip of the hat to the Gaia hypothesis, let us think of life-as-a-whole as a most complex, interlinked system. What might be the dynamics of such a megasystem? From the mathematical point of view, many of the processes involved, as life copes with the environment, are doubtless nonlinear, which means that chaotic conditions may sometimes prevail. In fact, the graphs presented above could have been taken right out of a book on chaotic systems. Life's extinctions and explosions might have no connection to asteroids, Ice Ages, or global volcanism. If something as simple as a spherical pendulum can lapse into chaotic motion, life-as-a-whole should show occasional wild behavior, too. Take our planet's weather as another example. Idealists once thought that given enough observations and computer power, weather could be predicted. But, alas, our atmosphere has also turned out to be a chaotic system, and long-term predictions are next to impossible. So, too, with life and its development. Those extinction-explosion plots may be just the paleontological expressions of a chaotic system.

COLLISION/ERUPTION/EXTINCTION/ MAGNETIC REVERSAL

An increasingly popular scenario is: (1) Every 34 million years the solar system bobs up and down through the thickly populated disk of our galaxy; (2) The resulting encounters lead to showers of comets and/or asteroids on earth; (3) The mechanical trauma leads to basalt flooding; (4) Great biological extinctions occur in consequence; and (5) The terrestrial magnetic field reverses in step.

Now, if scientists could show that all of these phenomena occur at the same frequency and are roughly in phase, it would constitute one of science's most important syntheses. The stratigraphic record and the estimated ages of meteor craters certainly hint at such synchrony. Recently, two more papers have appeared which support the above scenario. First, M.R. Rampino and R.B. Stothers show that during the past 250 million years, eleven episodes of basalt flooding have occurred with an average cycle time of 32 million years. Second, J. Negi maintains that the earth's magnetic record boasts a similar string of disturbances, with an average period of 33 million years. (Anonymous; "Regular Reversals in Earth's Magnetic Field a Fluke?" New Scientist, p. 32, August 25, 1988.)

THE CAUSE(S) OF EXTINCTIONS

MORE DOUBTS ABOUT ASTEROIDS

In an apparent reaction to the stampede to climb aboard the extinction-by-asteroid bandwagon, dissenting papers have begun to appear in the scientific literature. For example, Van Valen's list of objections to the hypothesis of asteroid impact at the Cretaceous-Tertiary boundary are reproduced in the next item from Science Frontiers. First, though, we record the objections of T. Hallam, as voiced in New Scientist.

1. Tropical plants, mammals, crocodiles, birds, and benthic invertebrates were little affected by whatever happened at the Cretaceous-Tertiary interface. Furthermore, many groups that were extinguished were already well into a decline.
2. Some geologists insist that some of the supposedly synchronous extinctions were probably separated by several hundred thousand years; viz., plankton and dinosaurs.
3. The vaunted iridium anomaly in deep-sea cores is spread through a considerable thickness of sediment. Even after allowing for the mixing of sediments, the iridium-rich layer is thousands of years thick.
4. According to the asteroid scenario, the clay layer separating the Cretaceous from the Tertiary should represent the fallout from impact-raised dust, which would include asteroidal material and a mixed sample of earth rocks. However, in Denmark, the boundary is marked by the so-called Fish Clay, which is almost pure smectite---a single mineral and not a mixture of terrestrial rock flour.

If it wasn't an asteroid impact, why the iridium concentration? At least three hypotheses have been proposed to circumvent the asteroid debacle: (1) volcanic activity; (2) a concentration of micrometeorites, thousands of tons of which fall each day, through extreme reduction of sedimentation; and (3) selective enrichment of iridium by an anoxic environment acting upon kerogen- and pyrite-rich clay.

In short, some geologists at least do not find the asteroid hypothesis compelling at the moment. (Hallam, Tony; "Asteroids and Extinction---No Cause for Concern," New Scientist, p. 30, November 8, 1984.)

THE CASE AGAINST IMPACT EXTINCTIONS

The neocatastrophists seem to be getting overly smug with their iridium-rich deposits at the Cretaceous-Tertiary boundary. Is there really incontrovertible evidence that a large asteroid or comet hit the earth at this point in history, causing widespread biological extinctions? To add some perspective, Leigh M. Van Valen has tossed 15 arguments against impact extinctions on the scales. Ten of these are reproduced below, as taken from Nature. More details may be found in Paleobiology, 10: 121, 1984.

1. Freshwater life was unaffected;
2. In Montana and its vicinity, the last occurrence of dinosaurs was detectably below the crucial boundary;
3. Transitional floras also exist below the boundary;
4. Apparently extraterrestrial material exists below the boundary;
5. The expected effects of eliminating atmospheric ozone are missing;
6. The Cretaceous-Tertiary boundary is coincident with a very large marine regression, suggesting a nonextraterrestrial cause;
7. Marsupials but not placentals were nearly eliminated, while most aboreal multituberculates (a type of vertebrate) and birds survived;
8. The predicted cooling effects on the earth are absent;
9. The predicted effects of acid rain cannot be found; and
10. Assuming a marine impact, no turbidites can be found; assuming a land impact, no large terrestrial crater has been discovered. (Van Valen, Leigh M.; "The Case against Impact Extinctions," Nature, 311:17, 1984.)

NOW, IT'S COMET SHOWERS THAT DID IT

The impact/extinction controversy still rages. A careful evaluation of paleontological evidence has persuaded catastrophists to think in terms of comet showers spread out over a few million years, rather than a single impact per extinction. This short abstract from a Nature article says it all:

> If at least some mass extinctions are caused by impacts, why do they extend over intervals of one to three million years and have a partly stepwise character? The solution may be provided by multiple cometary impacts. Astronomical, geological and palaeontological evidence is consistent with a causal connection between comet showers, clusters of impact events and stepwise mass extinctions, but it is too early to tell how pervasive this relationship may be.

(Hut, Piet, et al; "Comet Showers as a Cause of Mass Extinctions," Nature, 329:118, 1987.)

Comment. In other words, the significance to earth of astronomical catastrophism is still up in the air.

BIOLOGICAL MISCELLANY

VALLEYS OF DEATH AND ELEPHANT GRAVEYARDS

Legend has it that elephants near death separate from their companions and trek alone to ancestral graveyards, dying only when they reach these special places. The truth is that accumulations of elephant bones have indeed been discovered, but no one seems to have followed expiring elephants to these boneyards. We hope someone will tell us otherwise, but the tale seems apocryphal.

The piles of elephant bones could, in fact, be the work of mazukus. (Mazuku means "evil wind" in Swahili.) It seems that there are places on this earth where CO_2 and other deadly gases emitted from volcanic vents accumulate. J. Lockwood and M. Tuttle investigated three mazukus known to natives in East Africa. In these low-lying areas, they came upon the remains of small mammals and birds that had been asphyxiated by concentrations of CO_2 dense enough to snuff out burning kerosene-soaked rags. Unfortunately for the elephant-graveyard legend, they found no elephant bones. (Anonymous; "Elephant Graveyards," Discover, 12:10, May 1991.)

Comment. It would be interesting to know if other species of animals are found in the elephant graveyards. So-called "valleys of death" are found elsewhere in the world, including Yellowstone.

STRATIGRAPHY

POLAR FORESTS

In both the Arctic and Antarctic are found the debris of recent forests ---bits of wood, leaves, stumps, etc. Much of this plant mateial is fresh enough to burn. How old is this debris and what are its implications?

FORESTS FROZEN IN TIME

Axel Heiberg Island in the Canadian Arctic is only 700 miles south of the present North Pole. Little grows there today, but there is on these icy shores the remnant of a forest that flourished

45 million years ago, according to conventional geological dating of the strata. A University of Saskatchewan scientist, J. Bassinger, has been studying the 15-20 layers of stumps, some with diameters of 3 feet, and logs up to 30 feet long. Even rather blackish leaves survive in the soil. This once lush forest boasted trees like dawn redwoods and water firs; being analogous to Florida's Cypress Swamp in the Everglades. So excellent is the preservation of the forest that its wood cuts as if it were recent lumber and burns readily. (Howse, John; "Forestry Frozen in Time," Maclean's Magazine, p. 55, September 8, 1986. Cr. B. Ickes)

Comment. Question 1: Even if the earth was warmer 45 million years ago, could a tropical-type forest survive the nearly six months of total darkness at Axel Heiberg Island? Question 2: Can wood be preserved so well for so long? In the postulated warmer climate, there must have been many chemical and biological agents to promote rotting. Also relevant is the discovery, reported below, that wood that floats and burns with ease has been found in Antarctica. This Antarctic wood has been dated at 3 million years.

WHEN ANTARCTICA WAS GREEN

Something is wrong with our recent history of Antarctica. Conventional wisdom insists that the continent has been ice-covered for over 15 million years. But now Peter Webb and his coworkers have found pollen and the remains of roots and stems of plants in an area stretching some 1300 kilometers along the Transantarctic Mountains. The Antarctic wood is so recent that it floats and burns with ease.

Webb's group postulates that a shrub-like forest grew in Antarctica as recently as 3 million years ago. The dating, of course, is critical, and is certain to be subjected to careful scientific scrutiny. Nevertheless, these deposits of fresh-looking wood do suggest that trees recently grew only 400 miles from the South Pole. Also of interest is the fact that the sedimentary layers containing the wood have been displaced as much as 3000 meters by faults, indicating recent large-scale geological changes. (Weisburd, S.; "A Forest Grows in Antarctica," Science News, 129:148, 1986.)

A PERMIAN POLAR FOREST

An in situ Upper Permian fossil forest in the central Transantarctic Mountains near the Beardmore Glacier includes 15 permineralized trunks in growth position; the paleolatitude of the site was approximately 80° to 85° south. Numerous leaves of the seed fern Glossopteris are present in the shale in which the trunks are rooted. The trunks are permineralized and tree rings reveal that the forest was a rapidly growing and young forest, persisting in an equable, strongly seasonal climate---a scenario that does not fit with some climate reconstructions for this time period.

Some models of the Permian climate, based on astronomical and meteorological parameters, have winter temperatures at the site averaging -30° to -40°C, with the average summer temperature at merely 0°C. This fossil forest is clearly at odds with these models. (Taylor, Edith L., et al; "The Present Is Not the Key to the Past: A Polar Forest from the Permian of Antarctica," Science, 257:1675, 1992.)

METEORITES, GLASSES, TEKTITES

Among the superficial deposits of sand, soil, and stones that cover much of our planet's surface are found objects that scientists think either originated in outer space or were forged when extraterrestrial projectiles impacted the earth.

•Meteorites. Unusally large, dense concentrations. (See p. 74 for some of the astronomical aspects of meteors.)

•Impact glasses. The famous Libyan desert glass (LDG). (The Darwin glass and other impact glasses are discussed in Neglected Geological Anomalies.)

•Tektites. Possible origins, anomalous ages, association with reversals of the earth's magnetic field and biological catastrophes.

METEORITES

THE SOLAR-SYSTEM DUST BIN

For hundreds of thousands of years, miscellaneous rocky debris swirling around the sun has been falling upon the icy wastes of Antarctica. The motion of Antarctica's ice sheet carries these meteorites conveyor-belt fashion out towards the encircling seas. But where Antarctic mountains get in the way, the rocky cargo tends to get concentrated. Several thousand meteorites have already been picked up at these favored spots. In just a few brief summers of searching, these massive finds have posed unexpected questions. Here is a sampling:

1. The terrestrial ages (times since arrival on earth) measure between 1,000 and 700,000 years, implying that the Antarctic ice sheet may be at least 700,000 years old. This is unfortunate for several proposed scenarios of recent catastrophism, which envision an iceless Antarctica.

2. At least 20 amino acids appear in the more than 40 carbonaceous chondrites picked up with sterile equipment. These meteorites are dated as 4.5 billion years old, or 1 billion years older than the earliest terrestrial life found in the rocks. These finds highlight the old question: Did meteorites seed life on earth?

3. The much-publicized "lunar" meteorite, supposedly blasted out of the moon's crust by asteroid impact, thence falling to earth, shows little evidence of mechanical shock. If this meteorite, with a composition so similar to the Apollo samples is not from the moon, where did it come from? (Marvin, Ursula B.; "Extraterrestrials Have Landed on Antarctica," New Scientist, 97:710, 1983.)

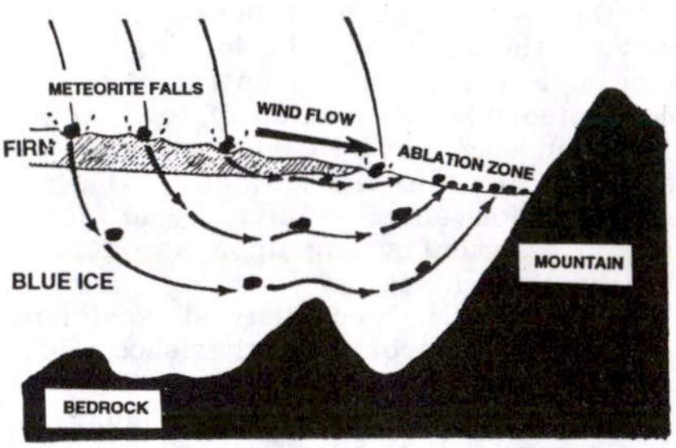

The ice "conveyor belt" that helps concentrate Antarctic meteorites in favored spots.

Cross references. Another impressive concentration of meteorites is to be found on the Nullarbor Plain in South Australia. See p. 76. Below there is additional material on the Antarctic lunar meteorites.

ANTARCTIC METEORITE MAY HAVE BEEN BLASTED OFF THE MOON

Meteorite ALHA 81005, discovered in the snowy wastes of Antarctica about a year ago, clearly resembles some of the rocks brought back from the moon by the Apollo astronauts. First, the meteorite's isotope ratios echo those found in bona fide moon rocks. Second, the meteorite is a breccia, consisting of small chunks cemented together, some of which are pinkish, magnesium-aluminum-rich spinels sometimes seen in lunar rocks but not terrestrial rocks or ordinary meteorites. Anorthosite is also present---a type of rock found on the earth and moon but not ordinary meteorites.

The implication is that ALHA 81005 was blasted off the moon by a comet or big meteorite. It escaped the moon's gravitational field, was captured by the earth, and plunged into the Antarctic snows. (Eberhart, J.; "Early Hints at a Moonish Meteorite," Science News, 123:54, 1983.)

Comment. Geologically speaking, the ice and snow of Antarctica are fairly recent. This meteorite may then be evidence of recent astronomical catastrophism that might also have affected the earth.

IMPACT GLASSES

LIBYAN DESERT GLASS MAY NOT BE THE PRODUCT OF IMPACTS.

The mysterious Libyan Desert Glass (LDG) is almost pure silica. It occurs in pieces weighing up to 16 pounds in the Sand Sea of the Libyan desert, in an area roughly 130 by 53 kilometers. Most scientists have attributed it to meteorite impact.

The results of a thermal, microstructural, and chemical analysis of LDG suggest that it is more likely derived from a low-temperature chemical process rather than meteorite impact on sand. (McPherson, D., et al; "Was Libyan Desert Glass (LDG) Formed by a Low-Temperature Chemical Process?" Eos, 66:296, 1985.)

Comment. This short abstract in Eos is frustrating. What sort of natural chemical process could leave pieces of glass strewn over such a huge area? And what about the Darwin Glass in Australia?

LIBYAN DESERT GLASS

Pieces of Libyan Desert Glass weighing as much as 16 pounds are found in an oval area measuring approximately 130 by 53 kilometers. The clear-to-yellowish-green pieces are concentrated in sand-free corridors between north-south dune ridges. The origin of this immense deposit of glass has been attributed by some to ancient nuclear explosions and alien activities, but investigating scientists have always been satisfied with a meteor-impact hypothesis. A recent study (abstract below) also opts for this explanation, although no one has found a crater of suitable size or other supporting evidence.

Abstract. "Libyan Desert Glass (LDG) represents ~1.4 x 10^9g of natural glass fragments scattered over 6500 km^2 of the western Desert of Egypt. We made a systematic study (employing INAA, microprobe and mass spectrometry techniques) of several varieties of LDG and locally associated sand and sandstone to provide insight into the nature and formation of these enigmatic glass fragments. These studies indicate that:

1. Although the LDG has restricted major element compositions (97.98 wt % SiO_2; 1-2 wt % Al_2O_3) their trace element contents (ppm) (Fe, 490-5200; Co, 0.2-1.2; Cr, 1.2-29 and Sc. 0.46-2.5) vary by as much as a factor of 5 to 30.

The LDG fragments exhibit a factor of three variation in the REE abundances (La, 5.4-15.3 ppm). They all show parallel and steep LREE enriched patterns ($[La/Sm]_N$, 3.8-4.2) and flat HREE ($[Tb/Lu]_N$, 1.1-1.2) and distinct negative europium anomalies (Eu/Eu*, ~0.5).

3. The gases in the vesicles of LDG (N_2, Ar, O_2, CO_2, H_2O and their dissociation products) are present in proportion as consistent with derivation from the terrestrial atmosphere.

4. Dark streaks present in some samples of LDG contain significantly higher siderophile element abundances (Ir, ~0.5 ppb), possibly representing a meteoritic residue.

"Our studies suggest that LDG is the product of meteorite impact into quartz-rich surficial eolion and alluvial sand, and perhaps also into quartz-rich sandstone, of the western Desert of Egypt." (Murall, A.V., et al; Eos, 70: 379, 1989.)

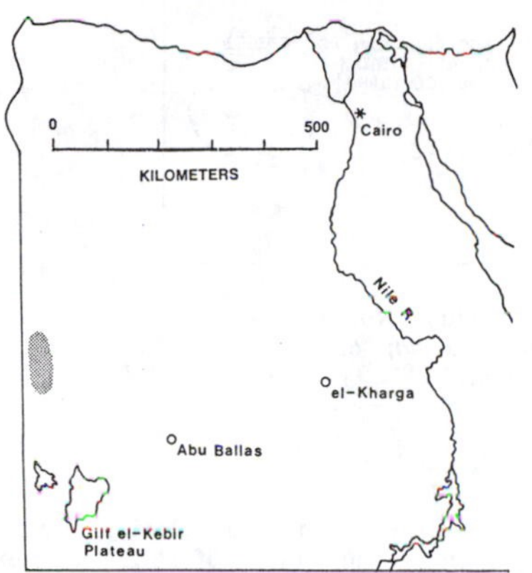

The Libyan desert glass is found in the stippled area of western Egypt.

TEKTITES

REVIEW OF THE TEKTITE PROBLEM

Tektites are small glassy stones with drop-like and button-like shapes. They are found primarily in four strewn fields in Europe, Australia, North America, and Africa. Just about everyone believes that tektites are once-liquid droplets of rock that were solidified in flight. The major question---often intemperately debated---is the location of the tektite source. Are they of terrestrial, lunar, cometary, or some other origin?

A recent study by Shaw and Wasserberg, using element abundances as a guide. strongly favors a terrestrial origin, with meteor impacts serving as the liquifying and splashing agents. Indeed, specific craters have long been associated with the European and African strewn fields. Those who believe that the tektites were splashed all the way from the moon by meteor impacts have not given up yet. One provocative fact stressed in this article is that the ages of the four groups of tektites are 35, 14, 1.3, and 0.7 million years. Tektites are all quite young! (Smith, Peter J.; "The Origin of Tektites---Settled at Last?" Nature, 300:217, 1982.)

Comment. Were there no tektite-forming meteor impacts prior to 35 millions of years ago? Is this observation related to the great rarity of meteorites in sedimentary rocks? Just what is different about the past 35 million years?

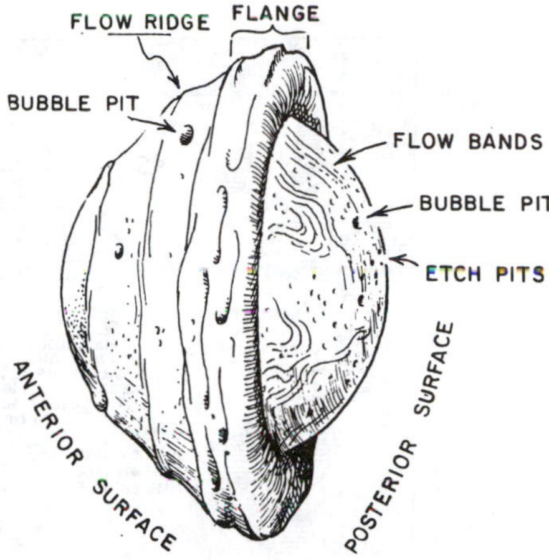

An ablation-sculpted australite; one of several types of tektites found in the Australasian strewn field.

LOOKING FOR THE SMOKING GUN

We already know the victims (the dinosaurs and other fauna and flora), and there is considerable evidence that the bullet was a cosmic projectile of some sort. The absence of a smoking gun (a sufficiently large terrestrial crater with an age of 65 million years) has allowed volcanists to deny the cosmic catastrophists a complete victory. However, the recent identification of tektite-like glasses at the Cretaceous-Tertiary boundary (KTB) on Haiti is leading geological detectives closer and closer to the missing crater.

Elsewhere in the world, the KTB is characterized by an iridium anomaly and a thin layer of "impact clay" consisting of tiny bits of shocked minerals. At Beloc, on Haiti, though, geologists find a 55-centimeter-thick layer of glassy debris. Approximately 25% of this stratum consists of 1-6-millimeter particles of tektite-like glass. Most of the glass particles are spherical, but a few have the splash-forms and dumbbell shapes of bona fide tektites. The thickness of the Haitian deposit and the large sizes of the particles suggest that the smoking gun must be nearby. Ironically, the Haiti stratum was originally classified as of volcanic origin; and we must add that we are presenting here only the conclusions of the asteroid school.

But where oh where is this crater? The Manson crater in Iowa (now buried) is of the right age but too small. The best candidate so far is buried in northern Yucatan. The Chicxulub crater is discernible on gravity- and magnetic-anomaly maps and is probably of the right age. Only drilling will confirm the guilt of the suspect.

Even if Chicxulub is the culprit, much debate prevails over exactly how the dinosaurs were done in. Was it a "cosmic winter" due to dust intercepting sunlight? Or perhaps a "cosmic summer" resulting from a super-greenhouse effect caused by: (1) impact-released methane trapped in sediments, and (2) the CO_2 from zapped carbonate rocks. (Smit, Jan; "Where Did It Happen?" Nature, 349:461, 1991, and Sigurdsson, Haraldur, et al; "Glass from the Cretaceous/ Tertiary Boundary in Haiti," Nature, 349:482, 1991.)

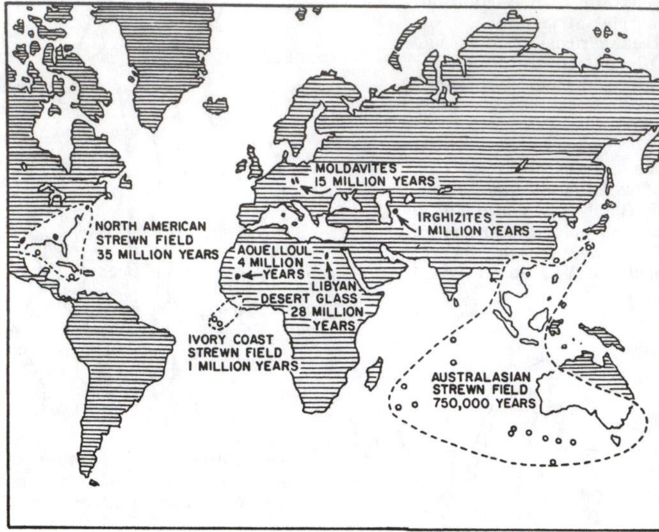

Tektites are concentrated in strewn fields. Not all of the major strewn fields indicated on this map can be associated with specific impact craters.

METEOR-IMPACT WINTERS, MAGNETIC FIELD REVERSALS AND TEKTITES

"Nuclear winter" is a term now in vogue. And, believe it or not, the rains of tektites discussed below may have been the forerunners of climatic catastrophes similar to the postulated nuclear winters. We shall call them "meteor-impact winters."

First, a tad of background: Great meteor impacts and tektite events seem to have occurred nearly simultaneously with deep-cutting biological extinctions and reversals of the earth's magnetic field. Ever since this apparent synchrony was recognized a few decades ago, theorists have been vying in generating scientific scenarios, especially some mechanism that would reverse the earth's magnetic field.

New entrants in the lists are R. Muller and D. Morris, two Berkeley physicists. Here is how they see it:

> A sufficiently large asteroid or cometary nucleus hitting the Earth lofts enough dust to set off something like a "nuclear winter." The cold persists long after the dust settles because of the increased reflectivity of the snow-covered continents. In the course of a few centuries, enough equatorial ocean water is transported to the polar ice caps to drop the sea level about 10 meters and thus reduce the moment of inertia of the solid outer reaches of the Earth (crust and mantle) by a part in a million. "That doesn't sound like much," Morris told us. "But when we realized that this translates into a full radian of slippage between mantle and core in just 500 years, we began to look seriously at the consequences." With the moment of inertia of the crust and mantle "suddenly" decreased, the argument goes, they begin spinning faster than the solid-iron inner core at the center of the Earth. The 2300-km-thick shell of liquid outer core that separates the mantle from the inner core thus acquires a velocity shear, which in the course of about a thousand years destroys the pattern of convective flows that served as the dynamo maintaining the Earth's dipole field.

The field reversal is not immediate according to both calculations and analyses of sediments. After the impact event, the dipole field decreases for a few thousand years, followed by a longer hiatus. Then, there is a sudden reversal. (Schwartzschild, Bertram; "Do Asteroid Impacts Trigger Geomagnetic Reversals?" Physics Today, 40:17, February 1987.)

Comment. In trying to relate all this to our present situation, recall that a major "ice age" is just a few millennia behind us, and our dipole field has been decreasing slowly for as long as we have been able to measure it. Brace yourselves!

NON-LETHAL TEKTITES

It is the fashion these days to blame the many so-called "extinctions" prevalent in the fossil record on extraterrestrial cataclysms. Some deposits of tektites and microtektites have indeed been correlated with the disappearances of some species. Since tektites are supposedly formed during meteor collisions with the earth, many scientists thought the evidence, circumstantial though it may be, very convincing. What has not been publicized as well is the fact that many microtektites, particularly in sediments 30-40 million years old, have no correlations whatsoever with any important biological extinctions. (Anonymous; "Non-Lethal Tektites," New Scientist, 99:345, 1983.)

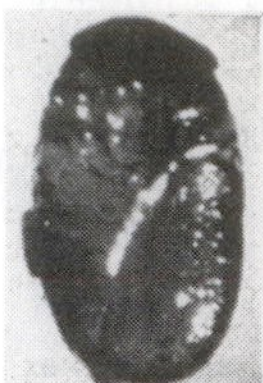
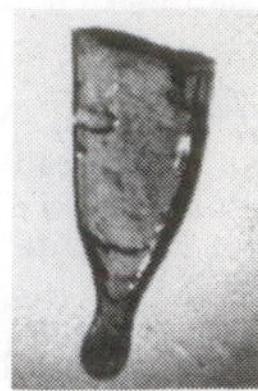

These glassy objects are microtektites recovered in deepsea sediments off the Ivory coast.

TEKTITE-LIKE OBJECTS AT LONAR CRATER, INDIA

Arguments about the origin of tektites persist in the scientific literature. A strong consensus has these small, droplike glassy bodies originating when meteors smash into the earth, liquifying themselves and some of the surface rocks. The resulting liquid droplets solidify in flight and when they descend form "strewn fields" hundreds, even thousands, of miles in extent.

The main argument has been over whether the actual impact craters giving rise to the tektites might actually be on the moon instead of the earth. Looking over the literature, one sees that this debate has been characterized by much invective and scientific infighting. Today, most scientists concur that some tektite strewn fields are definitely associated with specific, although distant, meteor craters on the earth's surface. Unfortunately, the large separations of craters and strewn fields add a circumstantial flavor to the evidence.

However, some tektite-like objects are to be found in the immediate vicinities of terrestrial craters, but not in far-flung strewn fields; viz., the Aouelloul Crater in Mauritania, and the Zhamashin Crater in the USSR. Another example has now come to light: the Lonar Lake Crater, a 50,000-year-old impact crater, in the Deccan flood basalts in India. From the abstract:

> Homogenous, dense glass bodies (both irregular and splash form) with high silica contents (~67% SiO_2) occur in the vicinity of Lonar Crater, India. Their lack of microlites and mineral remnants and their uniform chemical composition virtually preclude a volcanic origin. They are similar to tektites reported in the literature....Our geochemical data are consistent with these high silica glass bodies being impact melt products of two-thirds basalt and one-third local intertrappean sediment (chert). The tektite-like bodies of the impact craters Lonar, Zhamanshin, and Aouelloul are generally similar. Strong terrestrial geochemical signatures reflect the target rock

REE patterns and the abundance ratios and demonstrate their terrestrial origin resulting from meteorite impact, as has been suggested by earlier workers.

(Murali, A.V., et al; "Tektite-Like Bodies at Lonar Crater, India: Implications for the Origin of Tektites," Journal of Geophysical Research, 92B:E729, 1987.)

Comment. Obviously, the glassy droplets at the Lonar Crater strongly support a terrestrial origin for tektites. Proponents of a lunar origin can still point out, however, that some strewn fields cannot be associated with any known terrestrial crater. And why doesn't the very recent Lonar Crater have a strewn field of tektites somewhere? Ditto for Zhamanshin and Aouelloul. There seems to be much more to learn here.

OLD TEKTITES IN YOUNG SEDIMENTS?

A curious little geological debate now going on concerns the Australian tektites. The age of formation for these tektites, as determined by both fission-track and potassium-argon dating lies between 700,000 and 860,000 years. Geological evidence, however, suggests that the tektites fell only 7,000 to 20,000 years ago---a substantial discrepancy. Surely, say some, these old tektites were washed out of some equally old deposits and transported to the young strata where they now reside. Not so, say Australian geologists, because most of these tektites are found in areas devoid of outcroppings 700,000 years old. Furthermore, the rather fragile tektites show little signs of wear, as they should if transported by flood waters for long distances. These and other geological facts militate against the 700,000-year date. Geologists have questioned the two dating techniques, while geophysicists think the geological evidence is shaky. (Chalmers, R.O., et al; "Australian Microtektites and the Stratigraphic Age of the Australites," Geological Society of America, Bulletin, 90:508, 1979.)

Comment. It is important to resolve this issue because the dating methods employed are crucial to the now-dominant theory of plate tectonics. In particular, the 700,000-year figure seems to represent a major crisis in biological and geological history.

DID THE AUSTRALITES FALL RECENTLY?

Many thousands of australites, tektites strewn over Australia and Tasmania, have been found over the years; but, except for five, all were discovered loose on the surface or in unconsolidated sediments. Even the five australites found in rocks were in grit, sandstone, and other rocks that were hard to date and could have been recently lithified. Although the "official" date for the australite fall is 700,000 BP, the authors of this article, and presumably other Australian geologists, find "it difficult to believe that australites fell as long as 700,000 years ago." (Cleverly, W.H., and Kirsch, Steve; Meteoritics, 19:91, 1984.)

Comment. A few geologists even venture that this catastrophic event may have occurred just a few thousand years ago, and may be reflected in myth and legend.

HEAVY BOMBARDMENT OF SOUTHEAST ASIA 700,000 YEARS AGO

Tektites are found all over much of Australasia---an immense area. The tiny, often-illustrated teardrop- and button-shaped tektites clearly seem to have been formed when an extraterrestrial object smashed into the earth, melted terrestrial rock and soil, and splashed the fluid droplets over thousands of kilometers of Australia and Southeast Asia. Solidifying in flight, these particles fell by the millions.

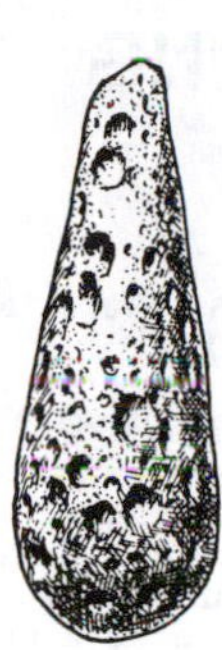

Dumbbell and teardrop tektites found in Indochina.

But another type of tektite is also found in Southeast Asia. These are the layered or Muong-Nong tektites, which are not aerodynamically sculptured. They come instead in large, irregular masses, 3-20 centimeters thick, weighing up to 24 kilograms. Their layered appearance is thought to result from flow and stirrings as they solidified in small pools of melted rock and soil splashed from nearby impact craters. These irregular chunks of solidified melt could not have traveled great distances like their streamlined brothers. They lie at most only a few crater diameters from their parent craters.

Since layered tektites are found over an area 800 x 1140 kilometers in extent, and they are not far-travelers, Southeast Asia must have been peppered with many small cosmic projectiles 700,000 years ago (the disputed age of the event). Whereas geologists have been searching diligently for a single huge crater (perhaps 100 kilometers in diameter) to explain the Australasia strewn field, they should be looking for many 1-kilometer craters.

This scenario is radically different from mainstream thinking about this great event in earth history. (Wasson, John T.; "Layered Tektites: A Multiple Impact Origin for the Australasian Tektites," Journal of Geophysical Research, 102:95, 1991.)

AUSTRALASIAN TEKTITES COUGHED UP BY A MOON OF JUPITER?

The Australasian tektites are glassy blobs found on or near the surface of the ground from the Philippines, through southeast Asia, all the way to Tasmania. Similar but much smaller "microtektites" appear in deep-sea deposits in the adjacent oceans. Radiometric and fission-track dating indicate that the tektites solidified about 700,000 years ago. Yet, their geological age, as measured by the age of the terrestrial sediments in which they are found, is only a few thousand years. This great disparity in age engendered a confrontation between geologists and geophysicists. The latter insisted on their 700,000-year figure; the former said "maybe so" but the tektites are still found only in very young, superficial sediments. The point here is that time-of-solidification may not be the same as time-of-fall.

At stake is the prevailing theory, now dogmatically proclaimed, that tektites are created when a large asteroid impacts the earth, ejecting molten droplets of rock which shower back to earth as solidified tektites. No one has ever found a suitably large crater (~200 miles in diameter) assignable to the Australasian tektite strewn field. Nevertheless, the impact model prevails; and the young geological age of the tektites is dismissed as erroneous.

A Soviet scientist, E.P. Izokh, has recently proposed a radically different scenario that would produce both the young and old dates. If a moon, or Jupiter, or some similar body, explosively ejected the glassy tektites, embedded in an icy cometary body some 700,000 years ago, the tektites could, after cruising through space for millennia, have fallen to earth recently and over a wide area. Thus, both geologists and geophysicists would be satisfied! (Sullivan, Walter; "New Answer Proposed for Tektites: A Comet," New York Times, November 28, 1989. Cr. R. Adams)

Comment. Russian scientists have long suggested that comets may be ejected from solar-system bodies and have been laughed at by American scientists for their trouble!

THE COMING REVOLUTION IN PLANETOLOGY

Current ideas about the moon appear to be mistaken on two fundamental points. First, at least within certain large classes of lunar craters, internal origin (i.e., some form of volcanism) predominates over impact; this result raises questions about the reality of the 'era of violent bombardment.' Second, the origin of tektites by meteoric impact on the earth cannot be reconciled with physical principles and is to be abandoned. The only viable alternative is origin by lunar volcanism, which implies the following: continuance of (rare) explosive lunar volcanism to the present time; existence of silicic lunar volcanism and of small patches of silicic rock at the lunar surface; a body of rock in the lunar interior, probably at great depth, which is closely similar to the earth's mantle and which contains billions of tons of volatiles, probably including hydrogen; and the origin of the moon from the earth after the formation of the earth's core.

The editor of Eos, a publication of the American Geophysical Union, recognized the highly controversial nature of the above quotation and felt impelled to append the following note to the article:

> Editor's Note. This article by John O'Keefe puts forth a viewpoint with which most planetologists disagree strongly. On the ground that a fresh airing of the long-standing discussion on lunar volcanism is appropriate, Eos offers this article, untouched by editors or referees, and awaits reply by readers.

O'Keefe's article reviews considerable evidence supporting his two points. Point One: crater dimensions and frequencies, craters with dark floors, lunar soil constituents; and, for Point Two; tektite analysis. He also remarks that the ages of the terrestrial tektite fields correlate with biological extinctions. This can be explained in terms of lunar volcanism as follows: lunar volcanos expel material violently, some of which escapes the moon's gravitational field and is drawn toward earth. Some falls as tektites; the rest forms a temporary ring arount the earth. The ring shadows parts of the earth, causing radical climate changes and, as a consequence, biological extinctions. (O'Keefe, John A.; "The Coming Revolution in Planetology," Eos, 66:89, 1985.)

Comment. The Editor's Note quoted above does not really convey the depth of the antagonism in the controversy about tektite origin.

The present article deals with the complex stratigraphy in the Western Arbuckle Mountains in southern Oklahoma. Here are located many examples of old-on-young rock as well as completely inverted stratigraphic members. Much attention is paid to the evidence of sliding between beds (breccia, small overfolds, etc.). Some excellent photos of these contact planes are presented. (Phillips, Eric H.; "Gravity Slide Thrusting and Folded Faults in Western Arbuckle Mountains and Vicinity, Southern Oklahoma," American Association of Petroleum Geologists, Bulletin, 67:1363, 1983.)

Comment. Some extensive thrust faults do not show as much evidence of horizontal sliding as those in the Western Arbuckles. Scientific creationists use such examples as evidence that the geological time scale, as determined by the fossil contents of the rocks, is all mixed up.

POLE FLIPPING

EARTH, THE MAGIC TOP

The Journal of Physics is a most respectable British scientific publication, but in a recent issue we find an article that would warm the heart of Ignatius Donnelly, to say nothing of Hapgood, Brown, Velikovsky, and more recent catastrophists. Employing a wide span of data from complex top theory to ancient legend, Warlow suggests that the earth has undergone many violent catastrophes, some of them within the time of man. Flood legends, geomagnetic reversals, tektites, paleoclimatology, salinity crises, and other familiar standbys of the catastrophists force P. Warlow to examine the stability of the earth in the presence of astronomical collisions and near-collisions.

He shows that the earth rotates slowly and that, even with the stabilizing equatorial bulge, our planet is rather sensitive to outside forces. It is, he says, like a tippe top or magic top; a 8,000-mile-diameter top that turns over repeatedly in response to external influences. Did not the ancient Egyptians write that the sun once rose in the west? Are there not massive faunal extinctions? Have not stray solar-system bodies left scars on all the inner planets? (Warlow, P.; "Geomagnetic Reversals," Journal of Physics, 11:2107, 1978.)

FATAL FLAW IN POLE-FLIPPING THEORY

V. Slabinski of the Communications Satellite Corporation claims that there are three separate errors in P. Warlow's theoretical analysis of terrestrial pole-flipping due to the gravitational torques created by a passing celestial body. With these errors corrected, the earth is 200 times less sensitive to pole-flipping. Slabinski does not believe that any known solar system object could turn the earth end-for-end if it passed by. This item proclaims that the discovery of Warlow's errors is a serious blow to Velikovskian catastrophism. (Anonymous; "Fatal Flaw in Pole-Flipping Theory," New Scientist, 92:433, 1981.)

Comment. We shall now wait for a rebuttal by Warlow and/or the Velikovskians. The flipping torques depend, of course, upon the mass and distance of the perturbing body. Whatever the outcome, the reality of astronomical and terrestrial catastrophism depends upon terrestrial geology, the testimony of history and myth, and other sources.

Update. Over a decade has passed and no rebuttal by Warlow has been seen. We must, therefore, consider his hypothesis highly questionable.

Chief Mountain, in Glacier National Park, is a classical example of thrust faulting. The upper part of the mountain (marked by arrows) is solid limestone; the lower part is also limestone, but it is broken by many oblique faults. The famed Lewis Overthrust is under the base of the mountain. It marks the transition from older limestone to much younger shales. (Geological Society of America, Bulletin, 13:307, 1902.)

THRUST FAULTING

THIN-SKINNED TECTONICS

In many regions of the world, older rocks are superimposed on top of younger rocks, just the opposite from what is expected. The usual explanation is that the layers of older rocks were thrust parallel to the bedding planes over the top of the layers of younger rock, sometimes for hundreds of miles. So numerous are these instances of inverted strata that a new branch of geology called Thin-Skinned Tectonics is arising to handle them.

THE MECHANICAL PARADOX IN THRUST FAULTING

In many parts of the world, older rocks are found on top of younger rocks. Obviously the Principle of Superposition is contradicted in such situations. Two possible explanations exist for these inverted strata: (1) The dating of the rocks is incorrect; or (2) Geological forces somehow slid the older rocks over the top of the younger rocks. Few mainstream scientists give any thought to the first possibility because it implies that evolution did not proceed as presently envisioned. (Rocks are often dated by their fossil contents.) So, geologists are left with the problems of sliding great masses of rock over rough surfaces for great distances.

Low-angle thrust faulting is not a trivial geological process. To illustrate, the Lewis Overthrust in Montana and adjacent Canada involves the shoving of a block of old strata hundreds of feet thick, hundreds of miles long, over younger rock for a distance of possibly 50 miles. In contemplating such overthrusts, one immediately comes face to face with the Mechanical Paradox. Briefly, given the coefficient of friction between the layers of rock, the weight of the thrust block, and the mechanical strength of the rock being pushed, it can be shown that pushing the thrust block at the rear edge will crush it long before it begins to slide. For nearly a century, geologists have been trying to resolve this Mechanical Paradox.

Three potential solutions have been proposed: (1) Lubricate the sliding surfaces with water under high pressure (the pore-pressure approach); (2) Allow the thrust block to slide, not as a unit, but in small discrete areas at different times (the dislocation approach); and (3) Push the thrust block not only from the rear edge but along the top surface (the tapered wedge approach). In fact, all three solutions may apply; but there is no consensus so far. Each solution has problems. (Washington, Paul A., and Price, Raymond A.; "The Mechanical Paradox of Large Overthrusts; Alternative Interpretation and Reply," Geological Society of America, Bulletin, 102:529, 1990.)

Comment. Note that scientific creationists consider large thrust faults as arrows in their quiver. If a "geological solution" cannot be found, the order of life's evolution might be at risk, for how else can we explain old rocks superimposed upon younger rocks, especially in those cases where there is little if any evidence of gross sliding? Happily, a "geological solution" does not seem out of reach.

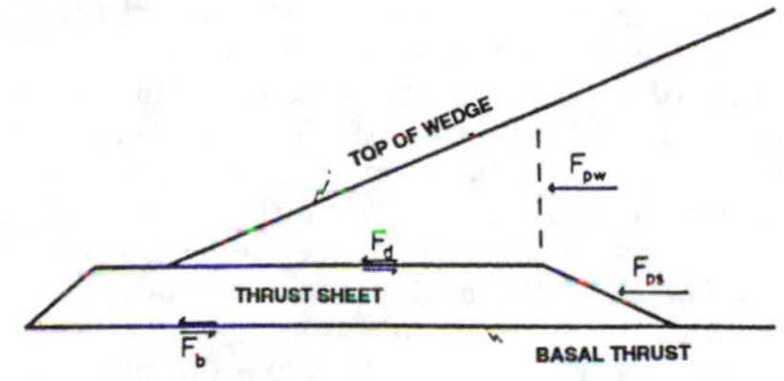

Sketch of the forces acting upon a thrust block being pushed by a wedge-shaped driver. See discussion in text.

STRATUM SHUFFLING AT PLATE BOUNDARIES

In March 1981, the Glomar Challenger was drilling into the oceanic sediments north of Barbados. Here, oceanic sediments are apparently being scraped off the oceanic plate as it thrusts under the North American plate. As a result, older Miocene deposits now overlie younger Pliocene deposits. Direct observations of the pressure in the zone where the shearing occurs showed it to be some 20 bars higher than the equilibrium pressure of 550 bars. This higher pressure may support the theory that low-angle thrust faults many kilometers wide are physically possible because high pressure fluids lubricate the shear zone, allowing massive thicknesses of sediments to slide over one another without resulting in wholesale fracturing and obvious damage. (Anderson, Roger N.; "Surprises from the Glomar Challenger," Nature, 293:261, 1981.)

Comment. Scientific creationists have long thought that the many low-angle thrust faults, where many miles of older rock are superimposed on younger rock, contradict geological dating schemes and therefore the theory of evolution. Establishment geologists, although somewhat amazed at the sizes of some of the overthrusts, especially one in Wyoming, have never despaired of finding a reasonable physical mechanism that would preserve the Law of Superposition and the idea of dating rocks by their included fossils. The Glomar Challenger results should buoy their spirits. Nevertheless, we must wonder how widespread stratum shuffling really is. What stratigraphic sequence is now immune from claims that some of its members were inserted in the wrong order time-wise?

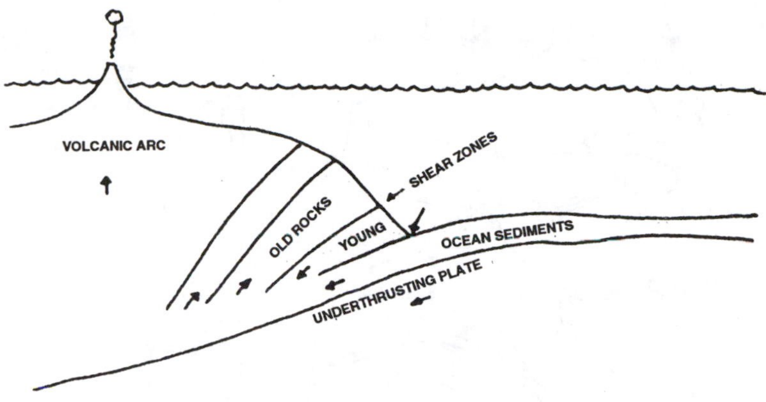

A sketch showing how younger strata may be inserted under older strata at plate boundaries. This process is sometimes termed "thin-skin" tectonics.

EXOTIC TERRANES

Exotic terranes are blocks of continental or oceanic crust that possess characteristics suggesting that they originated at locations far distant from their present locations. Their fossils, remanent magnetic fields, and radiometric ages differ markedly from surrounding geological units to which they are welded. Some terranes seem to have been transported several thousand miles from their places of origin. What sort of transportation system collected and fused together Alaska with its more than 50 exotic terranes? (See map below.)

ORPHANS OF THE WILD WEST

North of San Francisco, all along the Oregon and Washington coasts, the geologically oriented traveller will discover many huge boulders, mostly 10-20 m across, but some 100 m in size. Their constitution varies, but many are coarse-grained basalts that appear to have spent much of their lives at least 30-40 km underground. These boulders are "erratics" in the sense that no one has found surface outcrops that might have given them birth. So, where did they come from? But origin is only part of the problem. The presumably non-glacial erratics occur in a geologically confused area that seems to be upside-down time-wise according to the few fossils that have been found.

One theory is that the erratics were long ago carried to great depths by the conveyor-belt layers that slide eastward and downward under the U.S. Pacific Coast. Later, geological pressures squeezed the rock containing the erratics back to the surface like toothpaste. In the last phase, the matrix rock was eroded away leaving the erratics orphans. (Wood, Robert Muir; "Orphans of the Wild West," New Scientist, 85:466, 1980.)

Comment. Note that this complex scenario is dictated by the dogmas of continental drift and the geological time scale.

THE ALASKAN JIGSAW PUZZLE

Alaska seems to be plastered together from bits and pieces that originated far from the present position of this state. The units that now make up southern Alaska, for example, started north of the equator 250 million years ago, crossed into the southern hemisphere, and started back north about 160 million years ago. However, the itinerary of

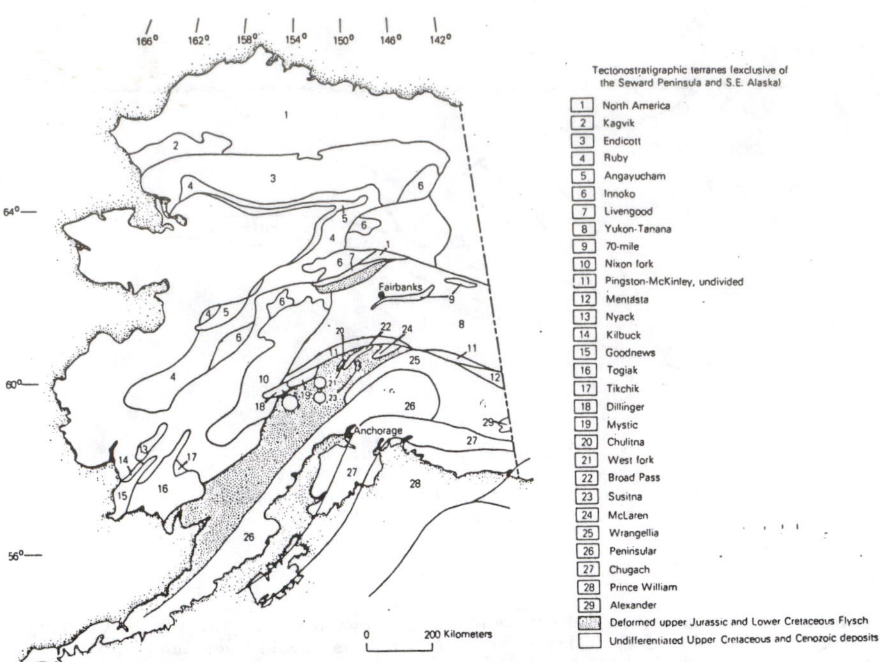

Alaska is a pastiche of some 50 exotic terranes, 29 of which are named on this map. (U.S. Geological Survey)

northern Alaska cannot even be guessed at. Present thinking is that this portion of the state was close to the geographic north pole during the late Cretaceous. But this is the period when the huge coal deposits were formed on the Arctic slope. Much of this coal comes from evergreens, which could not have survived in high latitudes due to the lack of sunlight. So, the pieces of the puzzle are at hand, but their travels are a mystery. (Anonymous; "Fragmented Alaska," Open Earth, no. 17, 1982.)

TERRANES CONTINUE TO PILE UP

More and more the continents seem to be pastiches of rocks from far corners of the earth plastered one atop the other. Specialists in this new field have recognized more than 300 different terranes around the margins of the Pacific. In keeping with this spirit, this article quotes a perceptive "Old Geologist's Saying: I wouldn't have seen it if I hadn't believed it." (Kerr, Richard A.; "Suspect Terranes and Continental Growth," Science, 222:36, 1983.)

FLORIDA MORE EXOTIC THAN THE TRAVEL AGENTS PROMISE

Anyone who has visited Florida knows that it differs in several ways from the rest of North America. Now we find that Florida doesn't even belong to North America; it is an interloper, an "exotic terrane."

How does one know this? Three facts hint that Florida doesn't belong:

1. When pre-Cenozoic land masses are fitted together, assuming the truth of continental drift, an awkward overlap arises that suggests that Florida was not always where it is today;
2. The latest paleomagnetic measurement of Florida's Paleozoic latitude is consistent with it being part of Gondwanaland rather than at its present latitude;
3. Radiometric dating of zircons retrieved from a core extracted from Northern Florida yield an age of 1650-1800 million years. There are no known source rocks in the southeastern U.S. that old; Africa and South America are likely sources of such zircons.

> These (latter) two new lines of geologic data provide strong evidence confirming previous suggestions that Florida was part of Gondwana during the early Paleozoic and that its current configuration is that of an exotic terrane sutured to North America during the fragmentation of Pangea.

(Opdyke, Neil D., et al; "Florida as an Exotic Terrane: Paleomagnetic and Geochronologic Investigation of Lower Paleozoic Rocks from the Subsurface of Florida," Geology, 15:900, 1987.)

A SLICE OF OCEAN CRUST IN WYOMING

Tucked among Wyoming's Wind River Mountains is a region of exotic crustal rocks. The best explanation conventional geology has come up with is that they were formed some 2.5 billion years ago by geological processes not in operation today. G. Harper, however, thinks that these Wyoming rocks look very much like some of the slices of ocean crust (terranes) that continental drift's conveyor belt has plastered against North America's west coast. The conveyor belt is, of course, the ocean floor that dives under the continent. The more he looked, the more Harper was convinced that there, in the middle of the continent, was a substantial chunk of ancient ocean crust. The implications: continental drift and terrane plastering have been in operation for billions of years; "...from their very beginnings continents have been built up from the bits and pieces of plate tectonics." Some other geologists concur and point to similar rocks in northern Canada and around the Great Lakes. (Kerr, Richard A.; "Plate Tectonics Is the Key to the Distant Past," Science, 234:670, 1986.)

Comment. If the continents have been slapped together in such a disorganized manner, have stratigraphy and geological dating been compromised?

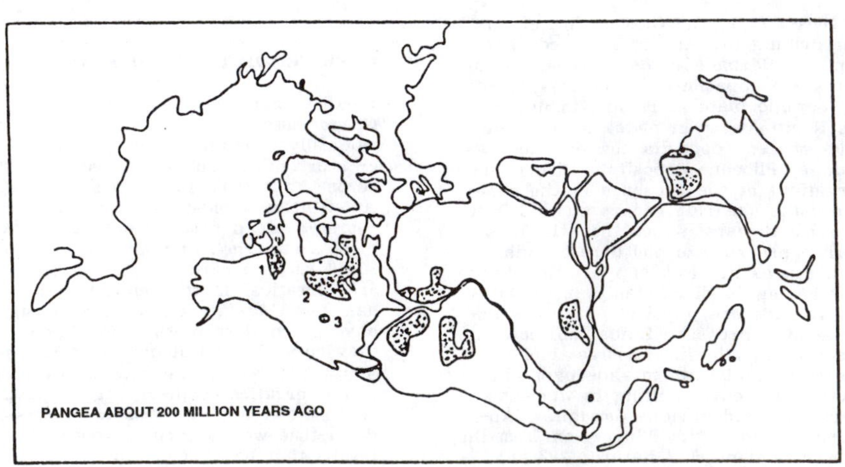

The stippled areas represent ancient terranes in the oldest parts of the continents. Containing greenstone belts, these areas are now thought to mark closures of ocean basins.

MARINE INCURSIONS

EVIDENCE FOR A GIANT PLEISTOCENE SEA WAVE

On the Hawaiian island of Lanai, limestone-bearing gravel blankets the coastal slopes. Called the Hulopoe Gravel, it now reaches a height of 326 meters above sea level. Taking into account the 1000,000-year age of the gravel and the slow subsidence of the Hawaiian Islands, the deposit probably reached 380 meters when it was first formed. The big question is how it got deposited at such great heights. High-sea stands are rejected by the authors in favor of a single episode of catastrophic waves about 100,000 years ago. Earthquake-generated tidal waves are considered unlikely because of the great heights involved. (The highest tsunami ever recorded in historical times reached only 17 meters above sea level.) A great meteor impact or submarine volcanic explosion are good possibilities, but the authors favor a giant submarine landslide on the Hawaiian Ridge, noting that in 1958 a similar event off Alaska produced a wave that reached 524 meters above sea level. (Moore, James G., and Moore, George W.; "Deposit from a Giant Wave on the Island of Lanai, Hawaii, Science, 226:1312, 1984.)

NORTHWEST INDIAN TRADITION OF A LARGE-SCALE SEA INUNDATION

Science, quite wisely, places little value on legend and tradition. The authors of this article stress the pitfalls of using data handed down verbally from generation to generation. With these caveats, they reproduce an Indian tradition originally set down by Judge James Swan back in 1888:

> "A long time ago," said my informant, "but not at a very remote period, the water of the Pacific flowed through what is now the swamp and prairie between Waatch village and Neeah Bay, making an island of Cape Flattery. The water suddenly receded leaving Neeah Bay perfectly dry. It was four days reaching the lowest ebb, and then rose again without any waves or breakers, till it had submerged the Cape, and in fact the whole country, excepting the tops of the mountains at Clyoquot. The water on its rise became very warm, and as it came up to the houses, those who had canoes put their effects into them, and floated off with the current, which set very strongly to the north."

The authors of the present article wonder if the above could be an account of a massive tsunami! They admit that the 4-day recession is inconsistent with tsunami action and that the warm water is hard-to-explain. The height reached by the inundation---some 400 meters---is also incredible. (Heaton, Thomas H., and Snavely, Parke D., Jr.; "Possible Tsunami along the Northwestern Coast of the United States Inferred from Indian Traditions," Seismological Society of America, Bulletin, 75:1455, 1985.)

Comment. See illustration on p. 50.

TWO TSUMANI TALES

Only a few years ago astronomical catastrophism was denied as a major factor in geological change. Now one reads everywhere of huge terrestrial impact craters, iridium layers, and tektite deposits. Even so, one necessary consequence of 70% of the large impacts has been neglected until recently: the giant tsunamis and marine incursions that must have swept over the coasts of the continents following impacts at sea. This oversight is now being rectified.

100,000 BP. The South Pacific. A long-standing geological enigma of the New South Wales coastline is the curious distribution of sand dunes. Those headlands less than 40 meters high have lost most of their dunes, leaving only raw, unweathered rock. On the other hand, the higher headlands have retained these dunes. Australians B. Young and T. Bryant hypothesize that a tsunami 40 meters high swept the lower headlands clean about 100,000 years ago. They can even plot the incoming wave's direction, because a few remnants of the coastal dunes still cling to the southwest corners of the headlands along the NSW coast south of Newcastle. In their scenario, the tsunami came from the northeast, smashed into the Solomons, southeastern Australia, and northeastern New Zealand. The Great Barrier Reef protected northeastern Australia from the full force of the wave. Young and Bryant favor a Hawaiian landslip as the initiator of the tsunamis, but acknowledge that an asteroid impact could also have done the job. If the wave began near Hawaii, it would initially been about 375 meters (about ¼ mile) in height. (Davidson, Garry; "A Tsunamis Tale from Sydney," New Scientist, p. 17, October 17, 1992.)

65,000,000 BP. Northeastern Mexico. The date mentioned is, of course, that of the Cretaceous-Tertiary boundary. This is the time when, many scientists believe, a very large asteroid slammed into northern Yucatan, forming the now-buried Chicxulub crater and wiping out the dinosaurs. Since the impact site was covered with ocean at the time, a powerful tsunami should have surged out from this area. Indeed, debris attributable to a tsunami has been found on the U.S. Gulf Coast and on some Caribbean islands. J. Smit et al now report finding a layer of debris up to 3 meters thick in northeastern Mexico. This layer was apparently deposited in water about 400 meters deep as the giant wave wreaked havoc along Mexico's shore and its backwash piled up debris offshore. This interpretation is supported by the presence of tektites, microtektites, glass spherules, abundant plant material, an iridium anomaly, and near the top ripple beds. (Smit, Jan, et al; "Tektite-Bearing, Deep-Water Clastic Unit at the Cretaceous-Tertiary Boundary in Northeastern Mexico," Geology, 20:99, 1992.)

PARADOX OF THE DROWNED CARBONATE PLATFORMS

The biological processes that build carbonate reefs and platforms are so efficient that platform growth potential is easily several times the rate of average geological subsidence or sea-level rise. Therein lies the paradox: the geological record is full of drowned carbonate platforms, inferring that the sea has frequently engulfed them in episodes that must be termed catastrophic. Since the usual long-term geologic processes are clearly inadequate, Schlager proposes several more violent schemes; including massive submarine volcanism (Middle Cretaceous) and extraterrestrial deterioration of the oceanic biological environment (Lake Devonian). (Schlager, Wolfgang; "The Paradox of Drowned Reefs and Carbonate Platforms," Geological Society of America, Bulletin, 92:197, 1981.)

HOPE FOR ATLANTIS?

That huge vertical movements in the crust occur is not in question. One could cite the deep sea oozes resting on coals of Tertiary Age in Barbados, for example. The coals represent a shallow water, tropical environment which sank to over 4-5 km depth for the deposition of the ooze and was then raised again, all in a very short period.

(James, Peter M.; "A New Model for Crustal Deformation," Open Earth, no. 17, 1982.)

DATING PARADOXES

Dating paradoxes in stratigraphy can arise when radiometric dates conflict with the dates expected from evolutionary considerations and/or from established climate scenarios---as illustrated by several items below. Also subsumed under the present heading are those chronological prob-

lems posed by missing strata (missing pages in the stratagraphic record) and by the discovery of strata that were previously assumed absent (unexpected pages in the record).

GALAPAGOS YOUNGER THAN THOUGHT

Marine stratigraphy, radioactive dating, and paleontology all point to the relatively recent emergence and biological colonization of the Galapagos. These islands are no older than 3-4 million years. The unique terrestrial life forms had to develop in less time than this. (Hickman, Carole S., and Lipps, Jere H.; "Geologic Youth of Galapagos Confirmed by Marine Stratigraphy and Paleontology," Science, 227:1578, 1985.)

Comment. Several remarks seem appropriate here:

1. The varied fauna and flora of the Galapagos did not evolve independently; viz., the bills of the Darwin finches are tailored to specific food sources (plants). Many species changed rapidly and in concert.

2. A recent Science article (228: 1187, 1985) notes that inbred mice often evolve different morphological characteristics very quickly. This observation probably applies to the initial Galapagos populations, which must have been small and inbred.

3. Harking back to the item on the Guadeloupe skeleton (p. 43). The Galapagos display similar strata of limestone, beach rock, etc. Until now, the limestones had been dated from the Miocene to the Pleistocene, but according to Hickman and Lipps they must be much younger than Miocene. The Guadeloupe dates may also be in error. Caveat emptor.

TOO MANY PAGES MISSING

The geological record has often been likened to pages in a book, each rock formation being a page, etc. The problem is that this book is not even close to being complete over most of the earth. Woodmorappe has examined the massive geological literature and drawn an extensive (and most impressive) suite of world maps showing just where the ten major geological periods are represented and where they are absent. The statistics are disturbing. Two thirds of the land surface display five or fewer periods; 15-20% of the earth's surface has three or less periods appearing in the "correct" order. Where are all the missing pages? Why, missing pages mean only that no deposition occurred in an area during the period in question or, if it did, erosion wiped it off the record. The "book" of strata forming the vaunted geologic column is really a composite of a few scraps from here and there. The enormity of what is missing is made all to clear by Woodmorappe's maps and statistics. (Woodmorappe, John; "The Essential Nonexistence of the Evolutionary-Uniformitarian Geologic Column: A Quantitative Assessment," Creation Research Society Quarterly, 18:46, 1981.)

Comment. Do missing geological pages constitute anomalies? Not when taken one by one, for occasional lapses are to be expected. But taken en masse, the record seems so skimpy that one wonders. Woodmorappe turns the knife by emphasizing that most fossils used for dating overlap anywhere from a few to all ten periods, further compounding the uncertainty.

A QUESTIONABLE 200-MILLION-YEAR HIATUS

Hiking down into the Grand Canyon of the Colorado is a geological education. As one descends past the beautifully exposed horizontal strata, one also turns back the geological clock in well-defined ticks. That is what the signs along the way say, and that is what all the textbooks proclaim! But when the juncture between the Redwall Limestone and Muav Limestone is reached, a 200-million-year gap appears. The sign posted here by the National Park Service reads:

> **AN UNCONFORMITY**
>
> Rocks of the Ordovician and Silurian Periods are missing in Grand Canyon. Temple Butte Limestone of Devonian age occurs in scattered pockets. Redwall Limestone rests on these Devonian rocks or on Muav Limestone of much earlier Cambrian Age.

This supposed unconformity is puzzling for several reasons:

1. The two limestone strata "seem" conformable in most places. Both are nicely horizontal, and there is basically no evidence that 200 million years of erosion and tectonic disturbances separate them.

2. In some places, the two limestone strata intertongue or interfinger, such that by moving vertically one flashes back and forth in 200-million-year jumps.

3. In both limestone strata, one finds layers of the same micaceous shale containing the same fossil tubeworms, suggesting near-simultaneous deposition.

4. In one place, the two limestones clearly grade into one another, with no separation at all.

Anyone who walks down the Canyon trails can see that the evidence for a 200-million-year hiatus between the Mississippian and Cambrian limestones is shaky at best. With the accuracy of geological dating through the use of contained fossils at risk, one would expect many professional papers dealing with this situation. Instead, the geological literature says little. One of the few papers mentioning the "unconformity" states that the contact between the two limestones displays ripples 2 feet from crest to trough, as one might expect with a true unconformity. Such ripples do not seem to exist. (Waisgerber, William, et al; "Mississippian and Cambrian Strata Interbedding: 200 Million Years Hiatus in Question," Creation Research Society Quarterly, 23: 160, 1987.)

Comment. Aha, this paper was written by scientific creationists, who have an obvious ax to grind. There's surely nothing to it. However, the senior author is a consulting geologist, and the paper is replete with photographs and diagrams. And you can always go see for yourself! It is the interpretation of the data that is in question. Where is the error?

EARLIER PAGES IN EARTH'S HISTORY REVEALED

Geologists are wont to liken the earth's sedimentary strata to pages in a history book. Well, it seems that seismologists may have discovered a previously unread chapter or two deep beneath the continental United States, in the guise of extensive stratified rocks in the Precambrian basement:

> The extent of the layered rocks became evident last summer as the Consortium for Continental Reflection Profiling (COCORP) completed a major deep seismic reflection traverse across Ohio, Indiana, Illinois, and

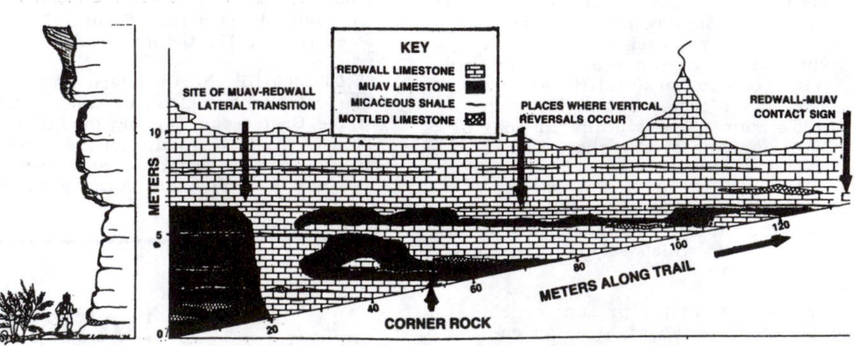

Stratigraphy of the Redwall-Muav contact, North Kaibab Trail, Grand Canyon. Notice the intertonguing. The mottled limestone and the layers of micaceous shale occur in both the Redwall and Muav limestones, even though they are supposedly separated in time by 200 million years. (Adapted from the referenced CRSQ article.)

part of Missouri. The survey was conducted partly because industrial seismic data and studies by the Illinois Geological Survey showed basement layering in southern Illinois, partly because earlier COCORP surveys also showed such layering in Oklahoma and Texas, and partly because COCORP's broad program calls for comprehensive exploration of the entire continental basement of the United States.

Although the composition and precise age of the Precambrian rocks are yet to be determined, their seismic reflection character suggests a sedimentary assemblage, at least in part. These layers occur within the Proterozoic Granite-Rhyolite province, where drilling typically recovers undeformed granite or rhyolite with ages of 1.3 to 1.5 b.y. Such prominent and orderly layering is surprising, given the widespread occurrence of granitic rocks. If the layered rocks are indeed igneous, the volume of silicic volcanic material is spectacular.

(COCORP Research Group; "COCORP Finds Thick Proterozoic (?) Strata under Midcontinent," Eos, 69:209, 1988.)

Some dimensions for these newly discovered pages were given in a report bearing the almost embarrassingly alliterative title indicated in the reference at the end of this item:

> In southern Illinois and Indiana, the layered rocks extend at least 180 kilometers in an east-west direction and average about 6 km in thickness.

(Monastersky, Richard; "Boring Plains Belie Bounty Beneath," Science News, 133:363, 1988.)

OF TIME AND THE CORAL---AND OTHER THINGS, TOO

R. Fairbanks is a paleooceanographer at the Lamont-Doherty Geological Observatory. Recently he has been drilling deeply into the submerged coral reefs off Barbados. During his research, it has been discovered that the radiocarbon (C^{14}) scale is in serious error beyond 10,000 years BP. Radiocarbon dating is widely used in archeology, but it has always been hard to estimate how much radiocarbon was present in the earth's atmosphere thousands of years ago. As a matter of fact, even before Fairbanks' discovery, a major correction to the radiocarbon time scale was made using tree-ring counts as an absolute reference. But tree ring data go back to only about 10,000 years.

The latest correction was made by E. Bard, also at Lamont-Doherty, who took Fairbanks' coral cores and compared the radiocarbon dates with uranium-thorium dates. The result is that at 20,000 BP, the radiocarbon date is 16,500 BP, 3500 years too low. [Of course, all this assumes that the uranium-thorium dates are accurate.] Use of the newly corrected radiocarbon scale has pushed the peak of the Ice Ages back from 18,000 BP to 21,000 BP.

But there is more. The same article in Science, without saying how he came up with the number, has Bard fixing the strength of the earth's magnetic field at only half its present level 20,000 years ago. This is most interesting because over the last 400 years of direct measurements, the geomagnetic field has been steadily decreasing! When and why was there a peak in the intensity of the geomagnetic field?

Back to Fairbanks, who also used his coral data to estimate changes in global sea level versus time. About 12,000 BP, he states, sea level was rising ten times faster than today due to melt water from the polar ice caps. This amounts to 2.5 to 4 meters per century. "...perhaps fast enough to prompt legends of a Great Flood"! (Kerr, Richard A.; "From One Coral Many Findings Blossom," Science, 248:1314, 1990.)

Comment. It is exceedingly rare to find a scientist musing that there really might have been a Flood.

PENNSYLVANIAN TIME-SCALE PROBLEM

The advent of radiometric dating seemed to solve once and for all the problem of assigning dates to the key events in the earth's history. Indeed, all of the reference books confidently label charts with firm dates for the appearance of fishes, the demise of the dinosaurs, and so on. Alas, things are not quite as certain as they appear. Radiometric dating is not all that precise; errors may be large indeed.

Take the Pennsylvanian period for example. It is part of the Carboniferous period, when many of the great coal deposits were laid down. The classical duration of the Pennsylvanian---used in many texts---is 34 million years. A meticulous new study of central European stratigraphy now pegs the Pennsylvanian as spanning only 19 million years. Now that's a 44% change!

This new figure for the duration of the Pennsylvanian has already cast doubt on the origin of the famous Pennsylvanian cyclothems (repetitive strata) in North America. It had been thought that these seemingly cyclic deposits were correlated with sea level changes forced by variations in the earth's orbit (the Milankovitch periods). With this substantial compression of Pennsylvanian time, this correlation falls apart. The cyclothems, which are of impressive area and thickness, now seem to have been created by some other, still unrecognized phenomenon. (Klein, George deV.; "Pennsylvanian Time Scales and Cycle Periods," Geology, 18:455, 1990.)

Comment. Even worse, perhaps, is the fuzziness conferred on the entire geological time scale by this compression of the Pennsylvanian and the possibility of similar revisions for other periods.

CYCLOTHEMS AS SOLAR-SYSTEM PULSE RECORDERS

Geologists can help astronomers look back in time. The sunspot cycle can be seen in variations of varves; i.e., annual layers of sediment; and the growth rings of shells have been used to estimate the number of days in the lunar month when the solar system was younger. Cyclothems may also be useful. Cyclothems are groups or bundles of strata that repeat themselves in stratigraphic columns. A generalized cyclothem from Illinois is shown in the illustration.

In the U.S. western interior, rhythmic sedimentation appears in the Fort Hays Limestone Member of the Niobrara Formation. These cyclothems can be correlated over distances exceeding 800 kilometers and are believed to be the

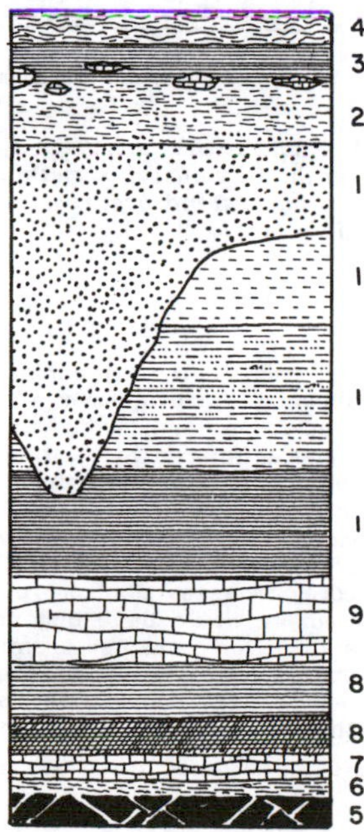

All ten members of this cyclothem are never present at a single location. The most common sequences are: 1, and/or 2, 4, 5, 8, 9, and 10.

consequence of climatic changes associated with the earth's precession and orbital eccentriciy. These rhythms have been captured in bundles of shale-limestone couplets. A bundle of five couplets, for example, is thought to express 21,000- and 100,000-year Milankovitch-type climatic cycles, as impressed by variations in the earth's orbital precession and eccentricity.

Analysis of the Fort Hays Limestone Member, however, reveals that while bundles of five couplets do occur, the number may vary from 1 to 12. Clearly, things are not clear-cut. (Laferriere, Alan P., et al; "Effects of Climate, Tectonics, and Sea-Level Changes on Rhythmic Bedding Patterns in the Niobrara Formation (Upper Cretaceous), U.S. Western Interior," Geology, 15: 233, 1987.)

Comment. As the rather comprehensive title of this article states, purely terrestrial forces may alter the rhythm. It may be, though, that cyclothem changes reflect actual changes in the earth's orbit resulting from astronomical catastrophism. The Cretaceous saw many geological and biological changes. Cyclothems may tell us more about them.

INNER EARTH

EARTHQUAKE PHENOMENA

WHAT IS EXPLODING 400 MILES BENEATH OUR FEET?

The author of the article we review here, C. Frohlich, was also the reviewer of our Catalog volume Earthquakes, Tides, Unidentified Sounds for a scientific journal. He liked the book but pointed out that we had overlooked an important earthquake anomaly: the deep-focus earthquake. He was right; we never realized how anomalous deep quakes are! Frohlich's review and those of other specialists make us realize how many more anomalies there are out there, even though we have produced 25 volumes of descriptions of hard-to-explain phenomena.

Be this as it may, let us see what Frohlich has to say about deep-focus earthquakes. Why are they anomalous? Can't quakes occur at any depth in the earth? No! Because below about 60 kilometers, the rocks should be so hot that they become ductile; instead of breaking catastrophically under stress, they just deform or "flow." It would appear, then, that conditions for earthquakes do not exist below 60 kilometers. Nevertheless, since 1964, more than 60,000 earthquakes have been recorded below 70 kilometers---some as far down as 700 kilometers. Conditions way down there cannot be what we think they are!

Most deep-focus earthquakes occur near subduction zones, where the science of plate tectonics says that the earth's crust is diving below another crustal plate. In addition to this geographical preference, deep-focus quakes are different from shallow quakes in that they produce few if any aftershocks. They are fundamentally different.

We don't really have enough clues as yet to guess just what is going on between 60 and 700 kilometers. If the rocks that far down cannot break to create earthquake shocks, perhaps there are explosions of some sort. There may be something about the relatively cool mass of subducted crust that stimulates explosions when it contacts the hot, deep rocks. Possibly, the descending crust carries water or other chemicals that react explosively. Complicating the problem are those few deep-focus earthquakes that shake the planet's innards in locations where there are no plates being thrust down into the earth's interior. It is becoming more and more apparent that that part of our planet between the crust and core possesses much more structure than we would have believed a decade ago. Even more, some very energetic events transpire "down there." (Frohlich, Cliff; "Deep Earthquakes,," Scientific American, 260:48, January 1989.)

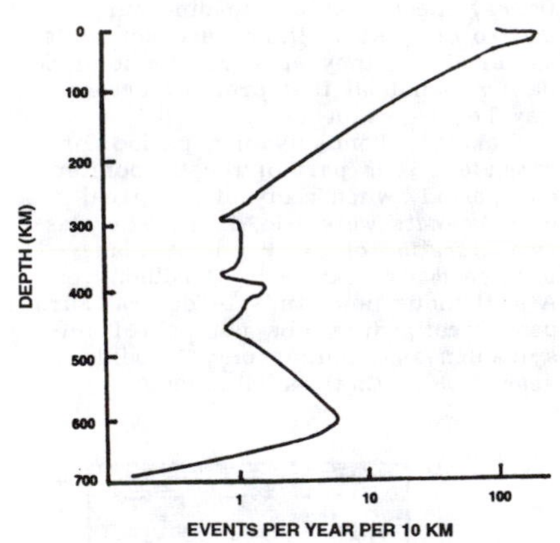

Earthquake frequency as a function of depth. Obviously, something we do not yet understand occurs at about 600 kilometers.

SEISMIC GHOST SLITHERS UNDER CALIFORNIA

From 1973 into 1978, instruments monitoring the San Andreas Fault told geophysicists that north-south distances between observing stations along the fault line were contracting. The Pacific Coast was moving north while the rest of North America was heading south! Then, in 1978, the strain eased in the far south of California only to reassert itself in a few weeks. This strange relaxation of strain seemed to propagate slowly across southern California, as seen by other instruments farther north.

As the wave of relaxation moved ponderously along, like a slow flexure of the earth's strata, microearthquakes almost disappeared while the flow of radon from the ground increased, as one might expect. When the strain reappeared, radon flow diminished and a rash of microearthquakes were detected. It took about a year for the strain wave, called a "seismic ghost" by the geophysicists, to flow north from the Imperial Valley into the Los Angeles area. (Alexander, George; "Quake-watch," Science 82, 3:38, September 1982.)

Comment. Something had to cause this curious disturbance; and where is it located now?

OUTRAGEOUS EARTHQUAKE WAVES

Shock waves produced by earthquakes travel faster through the earth when going in a north-south direction than when they travel east-west.

When these waves, called seismic waves travel east-west they take two seconds longer to reach a point on the other side than if they travel north-south. The measurements of velocity take the slightly wider girth of the Earth from east to west into account.

D. Anderson, director of Cal Tech's seismological laboratory states, "all of the possible explanations of this phenomenon are outrageous."

The "least outrageous" explanation for the anomaly is that the earth's outer core is not a pure liquid, but more like a slurry of aligned solid particles. (Anonymous; "Earthquake Waves Give 'Outrageous' Result," New Scientist, p. 38, February 18, 1988.)

GEYSERS AS DETECTORS OF DISTANT EARTHQUAKES

June 1992. Landers, California. An earthquake of magnitude 7.5 shook this small town. In apparent sympathy with the Landers disturbance, seismic activity appeared from one end of California to the other, as well as in Nevada, Utah, Idaho, and Wyoming.

Yellowstone Park, Wyoming. Here, 1100 kilometers from Landers, the geyser Echinus, which had been erupting on a regular schedule of every 56 minutes, went berserk. It didn't settle down for 34 hours. Geyser eruptions are frequently disturbed by nearby quakes, but Landers was hardly nearby!

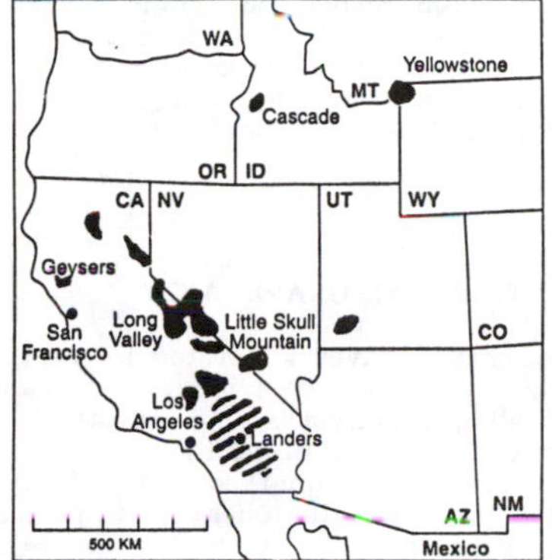

The Landers earthquake stimulated unusual seismicity in the solid black areas.

The seismology community.

Those distant shocks have startled seismologists as well as ordinary residents. Conventional thinking, at least among U.S. researchers, holds that stress generated when a fault slips in an earthquake peters out within a distance equal to a couple of times the length of the ruptured fault. For Landers, where about 70 kilometers of fault ruptured, this would amount to only about one-tenth of the observed reach.

Seismologists are now searching for ways to account for these unexpectedly far-reaching effects. (Monastersky, Richard; "Yellowstone Geyser Shows Quake Effect," Science News, 142:428, 1992. Also: Kerr, Richard A.; "Landers Quake's Long Reach Is Shaking Up Seismologists," Science, 259:29, 1993.)

PRECARIOUSLY BALANCED ROCKS AS EARTHQUAKE DETECTORS

Precariously balanced rocks (PBRs) are rather common where earthquakes have never occurred. In this sense, balanced rocks are measures of seismic stability. For example, says J. Brune, you won't find PBRs within 10 miles of spots where quakes have shaken the ground over the past few thousand years. To illustrate:

> Rocks stacked in piles and balanced on their narrow ends on Yucca Mountain near the Nevada border with California, he said, have not moved in at least 10,000 years and perhaps as many as 100,000 years, judging from the depth of "rock varnish," or weathering, on their exposed surfaces.

Looking for PBRs is not really as useless as it sounds, for they are indicators of stability to construction engineers planning nuclear waste disposal sites and similar projects requiring long-term seismic quiet. (Petit, Charles; "Seismologist Studies Precariously Balanced Rocks," San Francisco Chronicle, December 8, 1992. Cr. J. Covey)

Comment. How do rocks become "precariously balanced" in the first place? Melting glaciers and snow packs are known to ease their cargos of rocky debris gently down into unstable configurations.

PBRs, such as this "rocking stone" near Peekskill, NY, signify a lack of recent, large quakes in the region.

INTERNAL STRUCTURES

One of the most productive ways to plumb the secrets of the earth's interior is simply to listen to the seismic signals generated by earthquakes. The earth is a restless planet; and, if it is not restless enough, additional seismic signals can be created artificially. Listening with geophones, earth scientists have been able to accumulate records of these seismic waves as they are refracted and reflected by the earth's internal structures. Inner earth, it now seems, harbors intriguing velocity discontinuities. inhomogeneities, convection cells, continental roots, and slabs of crust that penetrate deeply into the planet's mantle. Many of these structures are enigmatic or, even when identified, occur where theory excludes them.

COMPLEXITIES OF THE INNER EARTH

So many new and startling facts about the inner earth are emerging from current magnetic, seismic, and gravitic research that a little list is in order. Bear in mind when going down the list that most of these features would have been considered absurd only a decade or two ago.

1. The earth's solid inner core, which "floats" inside the liquid core, is not spherical. Rather, it is anisotropic with its axis of symmetry aligned with the earth's axis of rotation. (Ref. 1 below)
2. "The CMB (core-mantle boundary) is the most dramatic discontinuity in the earth's internal structure in terms of the physical and chemical properties as well as the time scale of the processes that take place on either side of it. Its shape, if different from that predicted by the hydrostatic equilibrium theory, may contain information important to our understanding of geodynamic processes in the mantle or the geomagnetic field generated in the outer core." (Ref. 1, and also item #7 below)
3. The earth's magnetic field possesses four lobes which remain fixed relative to the earth's surface, as demonstrated by 300 years of data. These lobes do not drift westward like the general field. (Ref. 2)
4. "Core-spot pairs" of magnetic intensity seem to move westward and poleward. In the southern hemisphere, they originate under the Indian Ocean and drift under South Africa into the southern Atlantic. This motion reminds one of sunspot motion, except that sunspots move equatorward. There may be a connection here. (Ref.2)
5. The general decrease in the earth's magnetic field over the past few

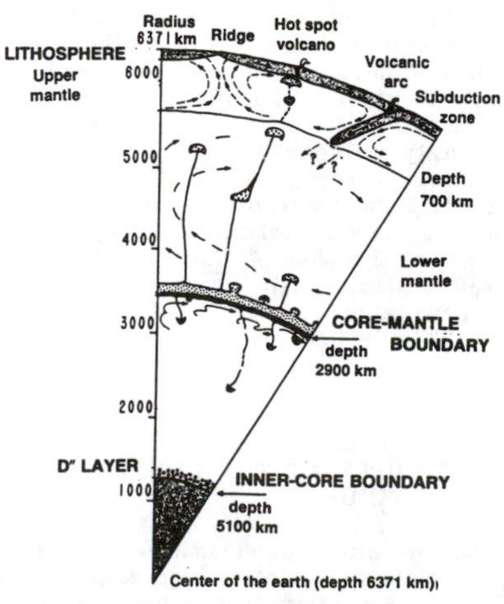

Schematic drawing of the earth's interior based on present thinking.

centuries may be due to intensifying core spots, which are magnetized in a sense opposite that of the main field. (Ref. 2)

6. Large, deep earthquakes in 1983 and 1984 produced slow, wavelike changes in the local gravitational field at the surface, as measured by new superconducting gravity meters. The periods were 13-15 hours. (Ref. 2)

7. Gravity and magnetism measurements from satellites show strong, coincident anomalies in the Indian Ocean (3°N 81°E). In fact the whole ocean surface is depressed in this region. To explain these overlapping anomalies, geophysicists suggest that a "valley" 5-10 kilometers deep exists at the core-mantle boundary. (Ref. 3)

References

Ref. 1. Dziewonski, Adam M., and Woodhouse, John H.; "Global Images of the Earth's Interior," Science, 236:37, 1987.

Ref. 2. Weisburd, Stefi; "The Inner Earth Is Coming Out," Science News, 131:222, 1987.

Ref. 3. Anonymous; "Satellites See Valleys in the Earth's Core," New Scientist, p. 33, May 21, 1987.

Comment. Reviewing Item 5 above, one wonders if the so-called "core spots" might attain such strengths that they locally reverse the terrestrial magnetic field at the surface. If so, geological scenarios relying on paleomagnetism (like continental drifting) would become suspect.

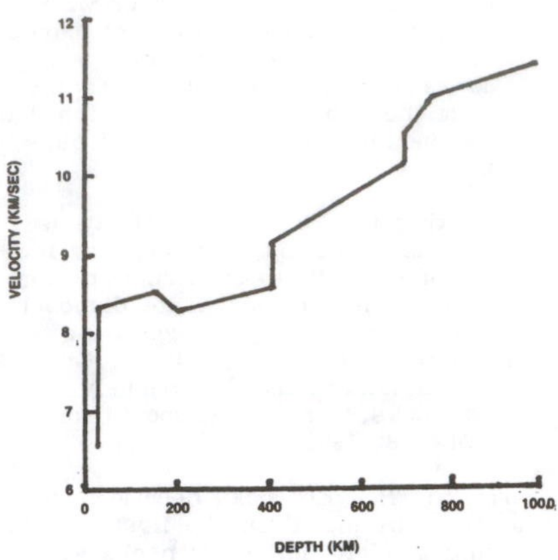

This plot of compressional P-wave velcities versus depth reveals three strong velocity discontinuities.

MOHO VICISSITUDES

For a long time the Moho (Mohorovicic discontinuity) has been considered a stable plane dividing the crust from the mantle. It is at the Moho that seismic wave velocities change abruptly. There is something there, but no one knows just what. At the recent Second International Symposium on Deep Seismic Reflection Profiling of the Continental Lithosphere, a lot of doubts about the stability and character of the Moho surfaced. Under the North American Cordillera, which runs from Alaska to Mexico, the Moho is flat, continuous and oblivious to the faults, terrane plastering, mountain "roots," and the geological phenomena above it. In other areas, though, several Mohos are stacked up. Some Mohos are disconinuous, jumping from one depth to another. Others are strongly influenced by overhead geological structures. Gone is the neat, so simple Moho figured in all the textbooks. (Barton, Penny; "Deep Reflections on the Moho," Nature, 323:392, 1986. Also: Weisburd, S.; "The Moho Is Immutable No More," Science News, 130:326, 1986.)

CONTINUITY AT THE CONRAD DISCONTINUITY

From the study of seismic waves, geophysicists have determined that between 7.5 and 8.6 kilometers below the surface there exists a clear-cut "discontinuity." Practically speaking, this means that above this layer seismic waves travel at a markedly different velocity than they do below it. This discontinuity is so widespread, occurring beneath all of the continents, that it has received a special name: the Conrad Discontinuity.

Ordinarily, a geophysicist would expect to find a significant change in rock type when drilling through such a strong discontinuity. It was widely expected that, at the Conrad Discontinuity, drillers would find the granitic rocks typical of the continents changing suddenly into basalt, which is thought to make up the lower reaches of the earth's crust. However, when Soviet drills pierced the Conrad Discontinuity below the Kola Peninsula, they found no such switchover to basalt at all. In fact, they hadn't even found it when they penetrated to 12 kilometers.

This was a shocker. Now, no one knows what the Conrad Discontinuity represents. It doesn't signal a change in rock type; neither is there a fault or boundary of any kind. It is important to find out what is wrong here, because much of modeling of the unseen structure of the earth's crust depends upon a realistic interpretation of seismic records. (Monastersky, Richard; "Inner Space," Science News, 136:266, 1989.)

CONFUSING SEISMIC DATA FROM THE DEEP CONTINENTAL CRUST

Seismic exploration of the deep continental crust seems to indicate that huge sheets of crystalline rock have been pushed over sedimentary strata. The crystalline sheets, perhaps kilometers in thickness, were forcibly shoved hundreds of kilometers over sedimentary deposits during continental collisions--- so the theory goes. One such crystalline sheet is under the Southern Appalachians. Seismic data say it is about 10 kilometers thick and was pushed westward some 225 kilometers. If it seems intuitively impossible for such a thin sheet to remain intact during 225-kilometers of shoving over other rocks, consider a similar sheet in the Basin and Range province of Utah. This sheet was pulled down an inclined fault without coming apart!

These sliding sheets with remarkable structural integrities are required to explain what geophysicists see in the seismic reflections; namely, transparent zones of crystalline rock sitting on top of rocks that return strong reflections typical of layered sedimentary strata. However, one such situation in Arizona was explored with a drill bit. When the upper crystalline layer was penetrated, the drill found only more crystalline rocks, nothing sedimentary. In fact, the crystalline rock was not layered and was homogeneous. Thus, the source of the misleading seismic reflections is unknown. (Kerr, Richard A.; "Continental Drilling Heading Deeper," Science, 224: 1418, 1984; and "Probing the Deep Continental Crust," Science, 225:492, 1984)

Comment. So-called "low-angle thrust faulting is often invoked to explain situations where large areas of older sedimentary rocks overlie younger rocks. See p. 224.

CONTINENTAL GRAVEYARD?

The seismic waves generated by earthquakes penetrate deeply into our planet and allow geophysicists to, in effect, X-ray the earth. In addition to the hypothesized solid inner core, the fluid outer core, and the encapsulating solid mantle, there seem to be continent-sized inhomogeneities in the vicinity of the mantle and the outer core. The anomalous seismic signals can be interpreted as huge blocks of different composition and/or temperature.

T.H. Jordan, from MIT, ventures "that what they have mapped are 'continents' on the core-mantle boundary. 'What we've seen is something really incredible,' he says. According to Jordan, the anomalies are analogous to continents on the surface of the earth, because they can't be accounted for by temperature variations but must reflect some compositional change as well. These features 'represent the scum or slag that sits up on the outer core boundary, just as continents sit on the outer surface of the earth,' he says."

Some have even speculated that these subterranean chunks of debris are pieces of surface continents that were subducted at the plate boundaries long ago. There are, after all, missing pieces in the continental drift jigsaw puzzle. There might even be substantial chuncks of asteroids and comets down there waiting, like the Titanic, to be explored by scientific instruments. Can once-subducted continents and cosmic debris ever rise again? (Weisburd, Stefi; "Seismic Journey to the Center of the Earth," Science News, 130:10,

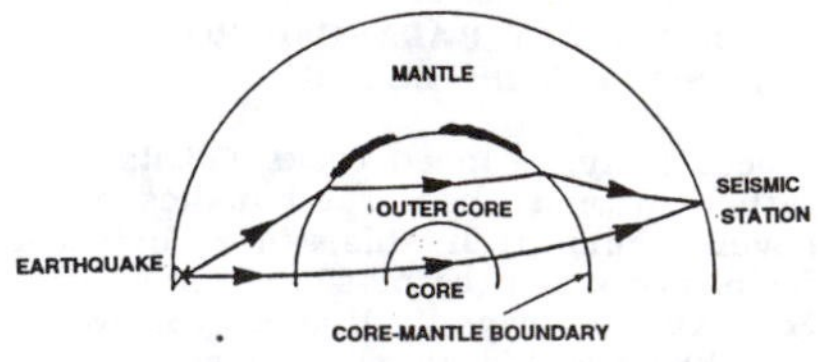

Earthquake waves allow seismologists to construct models of the earth's interior. Continent-sized inhomogeneities have been found near the core-mantle boundary.

1986; and Kerr, Richard A.; "Continents at the Core-Mantle Boundary?" Science, 233:523, 1986.)

WHERE ON EARTH IS THE CRUST?

This long article concludes with an intriguing snippet.

> Plate tectonic processes circulate the entire oceanic crust back into the mantle every 100 million years. Erosion also removes part of the continental crust, and some of the eroded material may eventually find its way to deep oceanic trenches, where it is also returned to the mantle. This implies that at any given time only about 10% of the crust is at the surface. Much of the continental crust, however, is more than half the age of the earth, so one can infer this part has not recirculated recently.

(Anderson, Don L.; "Where on Earth Is the Crust?" Physics Today, 42:38, 1989.)

WATER, WATER: HOW FAR DOWN?

The upper 10-15 kilometers of the earth's continental crust is different in several ways from the lower crust. The top layer is electrically resistive, seismically transparent, the source of almost all earthquakes, and responds to stress elastically. In contrast, the lower crust is electrically conductive, contains many reflectors of seismic energy, provides few quakes, and responds like a ductile material to stress.

The diverse characteristics of both regions can be explained if the entire crust contains saline water. In the upper crust the water is thought to be in separated cavities, while deep down it forms an interconnected film on crystal surfaces. (Gough, D. Ian; "Seismic Reflectors, Conductivity, Water and Stress in the Continental Crust," Nature, 323: 143, 1986.)

In an accompanying commentary, B.W.D. Yardley notes that the Soviet deep borehole on the Kola peninsula has found water down to at least 12 kilometers. (Yardley, Bruce W.D.; "Is There Water in the Deep Continental Crust?" Nature, 323:111, 1986.)

GEOMAGNETIC PHENOMENA

The earth's magnetic field varies from place to place geographically. It also varies in intensity and direction on scales of seconds to years to millennia. Its configuration is always changing, too, in hard-to-fathom ways. Although it predominantly resembles the field of a bar magnet, a substantial portion (about 10%) of the earth's field is nondipolar. This nondipolar portion waxes and wanes here and there over the planet's surface, and in the crust and mantle as well. Obviously, we have here a most complex phenomenon.

Several anomalous aspects of the geomagnetic field have been mentioned in past issues of Science Frontiers:

- Secular variations. Long-term changes, particularly the apparent steady decrease of field strength in historical times.

- Jerks and glitches. Sudden changes in geomagnetic field strength and/or direction.

- Reversals. Polarity changes; their origin, scenarios, and correlations with geological and biological phenomena.

- The geomagnetic dipole's inclination to the axis of rotation.

- Paleomagnetism. Some questions about its utility in deciphering the earth's geological history.

SECULAR VARIATIONS

LARGE CHANGES OF THE EARTH'S MAGNETC FIELD IN HISTORICAL TIMES

By measuring the magnetic properties of bricks and other accurately dated human artifacts, geophysicists can reconstruct the history of the local magnetic field. Near Loyang, China, the field was as much as 54% higher in 300 A.D. than it is now. It was 15% higher in 1500 A.D. In 1000, it was less than today's value. (Wei, Q.Y., et al; "Intensity of the Geomagnetic Field near Loyang, China, between 500 BC and AD 1900," Nature, 296:728, 1982.)

Comment. Direct measurements of the earth's field go back only a few hundred years, but they are consistent with the data reconstructed from artifacts, both showing a steady decrease since 1500. No one has estimated the effects of these substantial changes on radiocarbon dating and, perhaps, human biology.

THE CHANGING MAGNETIC CLIMATE: DOES IT AFFECT CIVILIZATIONS?

> Past values for the geomagnetic intensity may be obtained by laboratory analysis of the thermoremanent magnetization carried by clay baked in ancient times. From global averages of such determinations it is commonly accepted that the intensity in any given region went through a broad maximum about 2000 years ago, reaching a level ~ 50% higher than at present. Here we present results obtained from a wide range of Chinese pottery, spanning the interval from 4000 BC to the present, indicating that the field behaviour was more complex. The intensity was high between 1500 and 1000 BC and again in the first half of the first millennium AD. Comparison with results reported for Western Asia, Egypt and Crete suggests that these high values are due to non-dipole disturbances in the geomagnetic field, consistent with long-term records of the cosmogenic radioisotopes ^{14}C and ^{10}Be.

(Quing-Yun, Wei, et al; "Geomagnetic Intensity as Evaluated from Ancient Chinese Pottery," Nature, 328:330, 1987.)

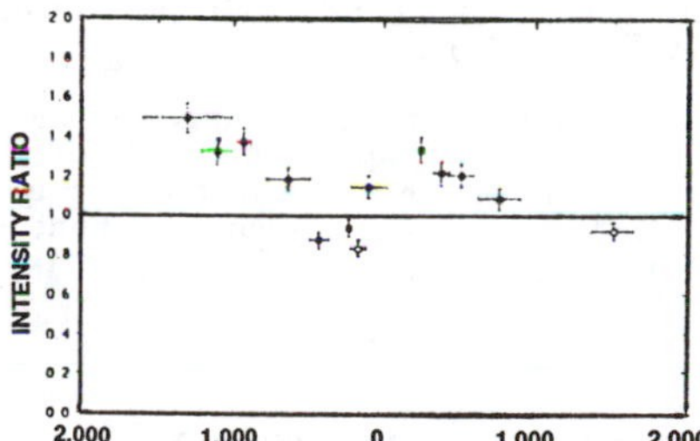

Ratios of ancient geomagnetic field intensity to the present level versus date. Data from China.

Comment. This article stimulates three questions:

1. What caused the geomagnetic changes; could some be of internal origin?
2. Are periods of reduced magnetic fields associated with cultural changes? The graph, for example, reveals a dip during the flowering of Greek civilization.
3. Could such ambient magnetic changes have an effect on human imagination, as reported in laboratory tests? See p. 309.

THE FIELD IS FALLING, THE FIELD IS FALLING

NASA's Magnetic Field Satellite has confirmed a trend that goes as far back as Gauss in 1830; namely, that the terrestrial magnetic field is decreasing in strength. At the rate measured by the satellite, the Earth's field will hit zero in about 1200 years. Of course, NASA's scientists warn that the observed decrease may only be a temporary fluctuation. The geological record seems to register a long history of magnetic field reversals, with great biological changes coinciding with the field flips. (Anonymous; "Magsat Down: Magnetic Field Declining," Science News, 117:407, 1980.)

Comment. Some field reversals have been within the time of man; 12,000 years ago and less. What happens to life forms dependent upon the earth's field for navigation during a reversal? Can they evolve new navigation methods in only a few thousand years?

JERKS AND GLITCHES

EARTH'S MAGNETIC FIELD JERKS

> It now seems almost certain that around 1969 a spectacular change took place in the geomagnetic field. The change was almost synchronous over the whole of the Earth's surface, took place in less than two years, and is now known to have consisted of a "jerk": a step change in secular acceleration of the magnetic field that has its origin inside the Earth.

(Whaler, K.A.; "Geomagnetic Impulses and Deep Mantle Conductivity," Nature, 306:117, 1983.)

Comment. No one really knows just how a "jerk" in the magnetic field is initiated; in fact, the origin of the geomagnetic field as a whole is not well-understood.

THE MAGNETIC JERK PROBLEM

We reported above that the earth's magnetic field "jerked" in 1969; that is, it suddenly accelerated its westward drift. The earth's core, which through dynamo action reputedly generates the magnetic field we detect at the surface, apparently does not keep pace with the outer crust. It is this sluggishness that produces the observed westward drift of the magnetic field of about 1 meter per hour. While most geophysicists acknowledge that something significant happened to the core in 1969, the geographical extent of the "jerk" is unclear. The acceleration of the field was clearcut in Europe but obscure or undetectable over much of North America. If the jerk was geographically limited, the core perturbation probably was, too. The earth's core may, in fact, eddy and swirl like the planet's atmosphere.

Going over past records, geophysicists think they have spotted another jerk in 1912; only that time the field decelerated. (Kerr, Richard A.; "Magnetic 'Jerk' Gaining Wider Acceptance," Science, 225:1135, 1984.)

EPISODE OF STEEP GEOMAGNETIC INCLINATION

K.L. Verosub has reported very steep geomagnetic inclinations in 120,000-year-old sediments in California. The mean inclination in these deposits ranged from 62° to 66°. Because this episode lasted several thousand years, Verosub believes that it opens to question the interpretation of other paleomagnetic data, where it is assumed that samples represent enough time for the geomagnetic field to have averaged out to a geocentric axial dipole. (Verosub, Kenneth L.; "An Episode of Steep Geomagnetic Inclination 120,000 Years Ago," Science, 221:359, 1983.)

Comment. The gist of this rather technical article is that all the scenarios of crustal plate motion may have to be modified substantially.

REVERSALS

ANATOMY OF A MAGNETIC FIELD REVERSAL

> A highly detailed record of both the direction and intensity of the Earth's magnetic field as it reverses has been obtained from a Miocene volcanic sequence. The transitional field is low in intensity and is typically non-axisymmetric. Geomagnetic impulses corresponding to astonishingly high rates of change of the field sometimes occur, suggesting that liquid velocity within the Earth's core increases during geomagnetic reversals.

The time period required for the field to reverse was about 4500 years, as measured at Steens Mountain, Oregon. There were three periods of very rapid change (impulses), which hint at radical changes in the core. The average magnetic field at the earth's surface decreased to 20% of normal during the reversal. (Prevot, Michel, et al; "How the Geomagnetic Field Vector Reverses Polarity," Nature, 316:230, 1985.)

Comment. The illustration reveals that the reversal was far from a clean 180° flip; there was much meandering. Just what was happening in the core during the reversal is a mystery. When the magnetic field dropped to low levels, flux of cosmic rays and other radiation at the earth's surface probably increased drastically. Terrestrial life might have been adversely affected.

THE STEENS MOUNTAIN CONUNDRUM

The layered lava flows of Steens Mountain, in southeastern Oregon, have preserved video-like records of the earth's magnetic field as it switched from one polarity to another about 15.5 million years ago. The scientific "instruments" here are the cooling lava flows. As they solidify from the outside in, a process taking about 2 weeks for a 2-meter-thick flow, the lava is magnetized in the direction of the field prevailing at the moment of solidification. We would thus have a 2-week continuous record of the behavior of the earth's field. Ordinarily, we would not expect to see very much change in 2 weeks; even a reversing field is thought to take thousands of years to complete its flip-flop. However, at Steens Mountain, when the field reversed 15.5 million years ago, the lava flows suggest that the field's axis was rotating 3-8° per day---incredibly fast according to current thinking, in fact a thousand times faster than expected.

The conundrum (one might call it a scientific impasse) arises because the flowing electrically conducting fluids

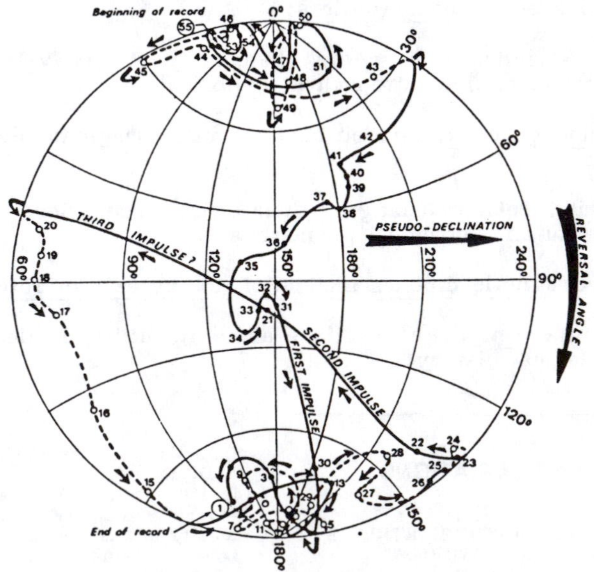

The Steens Mountain geomagnetic field directional record. The numbers refer to the samples used from the volcanic sequence in order of increasing age. The dotted lines represent field directions in the opposite hemisphere.

that supposedly constitute the earth's dynamo would have to flow at speeds of several kilometers/hour. No one has ever contemplated molten rock moving at such speeds in the core! (Appenzeller, Tim; "A Conundrum at Steens Mountain," Science, 255:31, 1992. Also: Lewin, Roger; "Earth's Field Flips Flipping Fast," New Scientist, p. 26, January 25, 1992.)

Comment. Is it possible that the prevailing dynamo theory is incorrect?

To make matters more interesting, it now seems that the paths taken by the reversing poles follow similar routes with each flip-flop. One preferred path is a band about 60° wide running north-south through the Americas; the other path is 180° away cutting through east Asia and just west of Australia. The implication is that some unknown structure in the core somehow guides the reversing poles. (Anonymous; "A New Path to Magnetic Reversals," Eos, 72: 538, 1991.)

GEOMAGNETIC REVERSALS FROM IMPACTS ON THE EARTH

R.A. Muller and D.E. Morris review the evidence tying geomagnetic reversals to the impacts of large bodies with the earth: the tektites and microtektites; the climate changes; the biological extinctions, etc. Then they propose a physical mechanism for geomagnetic reversals:

> The impact of a large extraterrestrial object on the Earth can produce a geomagnetic reversal through the following mechanism: dust from the impact crater and soot from fires trigger a climate change and the beginning of a little ice age. The redistribution of water near the equator to ice at high latitudes alters the rotation rate of the crust and mantle of the Earth. If the sea-level change is sufficiently large (> 10 meters) and rapid (in a few hundred years), then the velocity shear in the liquid core disrupts the convective cells that drive the dynamo. The new convective cells that subsequently form distort and tangle the previous field, reducing the dipole component near to zero while increasing the energy in multipole components. Eventually a dipole is rebuilt by dynamo action, and the event is seen either as a geomagnetic reversal or as an excursion.

(Muller, Richard A., and Morris, Donald E.; "Geomagnetic Reversals from Impacts on the Earth," Geophysical Research Letters, 13:1177, 1986.)

Comment. That the earth's field is generated by internal dynamo action is still a theory, although a widely accepted one.

BEHIND MAGNETIC FLIP-FLOPS

The earth's magnetic field frequently reverses its polarity. Such flips can often be correlated with climate changes, global ice volumes, sea-floor spreading rates, and deposition of black shales, tektite falls, biological extinctions, etc. The frustrating thing is the lack of clear-cut cause and effect; that is, how these phenomena are linked physically to the geomagnetic field. Part of the problem is that we can only guess at how the geomagnetic field is generated.

Let us assume that the earth's magnetic field is created by dynamo action in the planet's fluid core. P. Olson finds analytically that the core dynamo may reverse sign due to fluctuations in core turbulence caused by two competing energy sources: heat loss at the mantle-core boundary and progressive growth of the inner core. In concept, the heat lost at the core-mantle boundary might be linked to climate changes and sea-floor spreading.

Taking a different tack, D. Gubbins has investigated the possibility that field reversals are triggered by ice ages and meteorite impacts (tektite falls). The physical mechanism here would be the increase in pressure upon the core, which affects the rate of freezing in the outer core, and thus the power available to the core dynamo. Gubbins found that these externally caused pressure changes were too small to explain the polarity changes. However, the parameters involved are not well-known, and external triggers cannot yet be written off. Summarizing, very little progress has been made in explaining how the earth's field is generated and how polarity changes are linked to other geophysical parameters. (Jacobs, J.A.; "What Triggers Reversals of the Earth's Magnetic Field?" Nature, 309:115, 1984.)

WHEN THE EARTH SHIFTED GEARS

No one really knows just how the terrestrial magnetic field is generated or why it has reversed its direction so frequently in past geological time. Perhaps there is a clue in the following correlation:

> The Mesozoic-Cenozoic histories of reversals in the earth's magnetic field and of periods of widespread anoxia in the ocean basins show a remarkable correlation; periods of black-shale deposition ('anoxic events') occur during lengthy periods without magnetic reversals ('quiet periods'). My assembly of published work indicates a remote connection between quiet periods and anoxic events and suggests its form: Magnetic quiet periods coincide with fast seafloor spreading. During these periods, buoyant spreading ridges displace seawater into broad shelves, thus decreasing earth's albedo and causing global warming. Temperature gradients, and thus density gradients, from pole to equator decrease in surface waters, and the deep ocean currents of oxygenated polar waters wane. Oxygen minimum zones intensify and widen; anoxic conditions throughout entire basins are indicated by black shales deposited in the deep sea. These relations thus suggest that the earth's interior processes and its climates are related and their status recorded by both magnetic polarity and anoxic event chronologies of the earth.

(Force, Eric R.; "A Relation among Geomagnetic Reversals, Seafloor Spreading Rate, Paleoclimate, and Black Shales," Eos, 65:18, 1984.)

Comment. But what stopped and restarted the magnetic reversals and other concurrent processes? Strangely enough, the quiet, anoxic periods do not seem to coincide with biological extinctions!

GEOCORROSION?

In studying the minute electrochemical cells responsible for metal corrosion, J.G. Bellingham and M.L.A. MacVicar, at MIT, discovered a remarkable effect:

> One interesting and surprising property of electrochemical cells was discovered by accident. Normally, the magnetometer scans each cell as the cell moves horizontally beneath the magnetometer. During one run, the researchers left the cell in a single position for a long time while the magnetometer was still on. After 20 minutes or so, the magnetic field strength began to drop. "It was very dramatic to watch this field collapse," says MacVicar. After about a minute at zero, the magnetic field grew larger again but in the opposite direction.

These reversals occurred over and over again at regular intervals. (Peterson, I.; "Tracing Corrosion's Magnetic Field," Science News, 130:132, 1986.)

Comment. The self-reversal of magnetic specimens has been observed before under some conditions, but here is a periodic reversal of an electrochemical system. Why place it under the heading of Geology? Because the earth's field seems to reverse on a fairly regular basis. Catastrophists have invoked asteroid or cometary collisions to account for these flip-flops, but it might be that the earth contains giant electrochemical cells that spontaneously reverse on a million-year timescale rather than minutes. We know the earth's crust is filled with brines and other conducting fluids. Who knows what electrochemical activity transpires down there? See p. 191.

AXIS INCLINATION

WHAT'S ANOTHER DIPOLE OR TWO?

> Planetary exploration by deep space probes in recent years has shown that the dipole moment of some magnetized planets has a surprisingly large inclination angle with respect to the rotation axis. It is argued that the inclined dipole thus obtained may not be physically realistic. Applying the method we have developed for the source surface magnetic field of the sun (a spherical surface of 2.5 solar radii), it is suggested that the main dipole of the earth and the magnetized planets is actually axial (the magnetic moment is parallel or antiparallel to the rotation axis), and

that two or three smaller dipoles near the core surface are responsible for the apparent inclination of the main dipole.

(Akasofu, S-I., and Saito, T.; "Is the Earth"s Dipole Actually Inclined with Respect to Its Rotation Axis?" Eos, 71: 490, 1990.)

Comment. On p. 70, we see that the magnetic field of Uranus is inclined a whopping 60° to its axis of rotation. Can a few, small additional dipoles distort the main field so much? And just what are these small dipoles anyway---physically and electrically?

PALEOMAGNETISM

PALEOMAGNETIC PITFALLS

> Magnetism in rocks has provided a traditional tool for studies of the Earth's geomagnetic field. These studies have tended to rely on the assumption that the direction of magnetization was "frozen in" during formation of the rock. But many sedimentary rocks formed during the Palaeozoic acquired their remanent magnetization through alteration processes that occurred after deposition of the sediment. The causes and geological significance of this phenomenon have been much debated.

The foregoing paragraph is enough to send shivers throughout the geological world. Does this undermine paleomagnetism and generalizations flowing from it, such as plate tectonics?

The "alteration processes" mentioned in the above quotation include: (1) The chemical conversion of pyrite into magnetite in ancient rocks after they were deposited; and (2) The reorientation of remanent magnetization following exposure to moderately high temperatures. That these processes can be important is evident in a second quotation:

> During the past eight years, however, evidence has accumulated that the remanent magnetization of many carbonate sediments was not acquired at the time of deposition, thereby invalidating some previous interpretations of the palaeomagnetic data. Instead, magnetization seems to have been acquired over a limited time span during the late Palaeozoic, from about 310 to 250 million years ago.

(Reynolds, Richard L.; "A Polished View of Remagnetization," Nature, 345: 579, 1990.)

UNWANTED NOISE ON THE TERRESTRIAL TAPE RECORDER

The hypothesis of continental drift and sea-floor spreading depends heavily upon the strip-like magnetic anomalies that parallel the active ocean ridges. Molten material pushing out along these ridges spreads out, solidifies, and is magnetized by the prevailing terrestrial magnetic field. Thus, the spreading sea floor becomes a "tape recorder" preserving the record of changing terrestrial polarity over the past several hundred million years.

As one drills into this thin conveyor belt/tape recorder, one would expect to encounter only rocks of one polarity. Not so! Some of the holes drilled by the Deep Sea Drilling Project have passed through several polarity zones. To illustrate, core 395A from the mid-Atlantic ridge is magnetized normally for the upper 170 meters, reversely for the next 310 meters, and normally again for 40 meters. Is the tape-recorder idea therefore incorrect? Some scientists argue that it is and that the whole modern edifice of plate tectonics is suspect. (Anonymous; "Testing Vine-Matthews," Open Earth, 28, no. 3, April 1979.)

REVERSED MAGNETIZATION IN ROCKS

A fundamental assumption of paleomagnetism is that the natural remanent magnetism (NRM) of rocks is acquired parallel to the applied magnetic field. There are unsettling exceptions:

> Andesitic pumice, which was hurled several hundred kilometres during the disasterous 1985 eruption of the Nevado del Ruiz volcano (Columbia), carries a stable but reversed NRM with southerly declination and negative inclination. Heating experiments show that this magnetization is due to a self-reversal mechanism which also induces a reversed thermoremanent magnetization (TRM) in the laboratory field.

(Heller, Friedrich, et al; "Reversed Magnetization in Pyroclastics from the 1985 Eruption of Nevado del Ruiz, Columbia," Nature, 324:241, 1986.)

Comment. Much of the evidence for continental drift, especially the paths taken by the continents, is based upon paleomagnetism.

EARTH CURRENTS

SUBTERRANEAN ELECTRIC CURRENTS

We have little appreciation of the immense electrical currents that flow through the rock formations beneath our feet. These "telluric" currents are primarily those induced by the earth's changing magnetic field, as it is affected by the solar wind. Telluric currents do not flow uniformly through the earth's crust. Rather, they seek out low resistance rocks, in accordance with Ohm's Law. Such current concentrations can be detected at the surface with magnetometers.

The present paper announces the discovery of a regional telluric current flowing in the vicinity of the San Francisco Peaks volcanic field in Arizona. The shallow part of the current flows in an unidentifiable "geoelectrical" structure not more than 10 kilometers below the surface. There are no surface hints as to what this geoelectrical structure could be. (Towle, James N.; "The Anomalous Geomagnetic Variation Field and Geoelectric Structure Associated with the Mesa Butte Fault System, Arizona," Geological Society of America, Bulletin, 95:221, 1984.)

Comment. Similar anomalous magnetic fields exist in many areas, indicating a vast subterranean system of poorly understood geoelectrical structures. Some of the channeled earth currents are man-made, being the return paths in electrical power transmission systems. The return paths may be far-removed from the actual power lines because they tend to follow the geoelectrical structures.

POWERFUL EARTH CURRENT ENTERS NORTH AMERICA FROM THE PACIFIC

An immense current of terrestrial electricity originating somewhere in the Pacific enters the North American continent along the Strait of Georgia (between Vancouver Island and the British Columbia mainland) and shoots past Tacoma toward Oregon. The discoverers of the current, John R. Booker and Gerard Hensel, at the University of Washington, traced the flow of electricity through a narrow wedge of porous, water-bearing rock that parallels a fault line. Another branch of this terrestrial circuit enters along the Strait of Juan de Fuca. No estimates are given of the magnitude of the current; and there are no speculations as to the origin of the electromagnetic force driving the current. (Anonymous; "Nature's Hidden Power Line," Science Digest, 90:18, October 1982.)

Comment. In some areas, large artificial earth currents are created by high-power transmission lines.

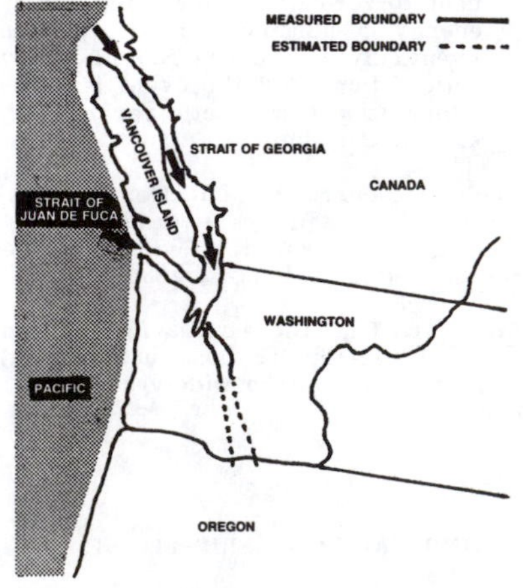

The Strait of Georgia between Vancouver Island and the mainland is the site of a strong current of terrestrial electricity. Arrows mark its flow.

THE NACP ANOMALY

The NACP (North American Central Plains) electrical conductivity anomaly snakes west from Hudson Bay, then south into the States, and wiggles a bit before terminating in Wyoming. As delineated by magnetic surveys, it is over 2000 kilometers long, and may be longer and wider than shown on the map. Since the top of this belt of high electrical conductivity rock is some 10 kilometers below the surface, no one is sure of its constitution---graphite in schistose rocks is one guess. Its meaning for the geology of North America is also a mystery---it could be the edge of a buried tectonic plate. Whatever it is, it is important: "the largest and most enigmatic continental-scale structure discovered to date by electromagnetic induction studies." (Jones, Alan G., and Savage, Peter J.; "North American Central Plains Conductivity Anomaly Goes East," Geophysical Research Letters, 13:685, 1986.)

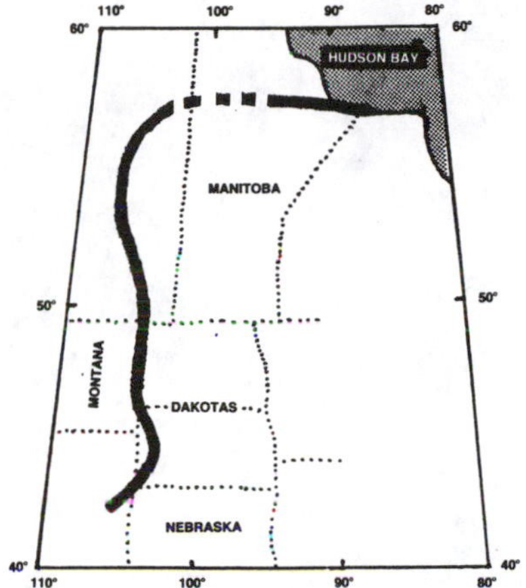

The dark streak looping west and then south from Hudson Bay represents the North American Central Plains (NACP) electrical conductivity anomaly, as mapped by magnetometer surveys.

HUGE UNDERGROUND ELECTRICAL CIRCUIT

> Geophysicists from the Department of Earth Sciences and the Bureau of Mineral Resources have discovered part of a huge underground circuit near Broken Hill (Australia), which contains electric currents of more than a million amps.
>
> The currents are spread too thinly for power production, but their existence helps account for problems experienced generally in interpreting the magnetic data used to produce geological maps.
>
> The circuit was found using a sensor which detects fluctuating electric fields in the earth's crust. These are created in response to electrical events, such as thunderstorms and the movement of dissolved salts in artesian water.

(Anonymous; "Scientists Discover Huge Underground Circuit," Monash Review, p. 10, December 1986. Cr. R.E. Molnar. The Monash Review is an Australian university publication.)

Comment. Could it be that a portion of the earth's "permanent" magnetic field is likewise generated by internal electrical currents? Are the ponderously moving internal convection cells and widely accepted dynamo effect really necessary? In other words, could our planet be a huge natural battery based upon geochemical differences?

UNDERGROUND CURRENT ELECTRIFIES AUSTRALIA

A weak electrical current wends its way for some 6000 kilometers along fracture zones in some of Australia's sedimentary basins. Located some 15-45 kilometers below the surface, the current begins at the continental shelf in Western Australia, runs southward into South Australia, and then loops northward, exiting in the Gulf of Carpentaria near Birdsville, Queensland. The current's path has a width varying between 50 and 200 kilometers. It seems to be flowing in alkaline fluids contained in the broken edges where ancient tectonic plates collided to create the continent. The current is weak and is induced by the earth's changing magnetic field. (Anonymous; "Underground Current Electrifies Australia," New Scientist, p. 10, March 10, 1991.)

Comment. Modern Australia is spotted in the center of a huge tectonic plate.

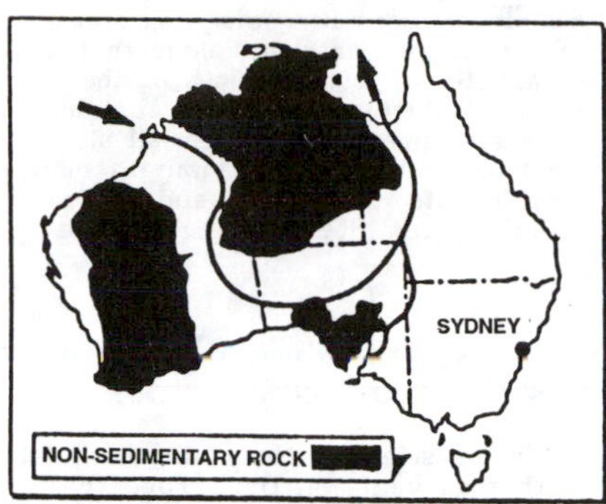

The arrows indicate the path of one of Australia's underground electrical currents.

MISCELLANY

QUIET SUN: VIOLENT EARTH

When R.B. Stothers, at NASA's Goddard Institute for Space Studies, decided to look into the possible correlation of solar activity and terrestrial volcanism, he fully expected to find no connection at all. After all, what force generated by small changes in the sun's output could stir up the earth's magmas from a distance of 93 million miles? Stothers was surprised:

> Stothers analyzed two immense catalogs, published in the early 1980s, that list more than 55,000 known eruptions since the year 1500. Concentrating on several hundred of the moderate-to-large eruptions, he found statistically significant patterns in eruption frequency that match the solar cycle. Eruptions seemed most numerous during the weakest portion of the solar cycle.

Further, there was a 97% confidence that the correlation was not a statistical accident.

The only cause-and-effect explanation offered by Stothers was negative and indirect. During periods of abundant sunspots, increased solar emissions jar the earth's atmosphere slightly. Communicated to the crust, these slight taps trigger tiny earthquakes that relieve stresses beneath volcanos, thus delaying their eruptions until solar activity dies down! (Anonymous; "Volcanoes on Earth May Follow the Sun," Science News, 137:47, 1990.)

Comment. Down the years, many scientists and laymen have tried to correlate sunspots and earthquake frequency. The results have been murky and sometimes contradictory. See GQS1 in Earthquakes, Tides, Unidentified Sounds.

WHAT HEATS THE EARTH

The currently popular model of the earth has its heat generated by the radioactive decay of uranium and other elements. Some of these decay reactions produce helium---so-called radiogenic helium. But, as the following excerpt asserts, the amount of helium actually detected is way out of line with the measured heat flow.

> The present rate of mantle heat loss, however, is out of equilibrium with the rate of helium loss---too large by about a factor of 20. Either radiogenic helium is accumulated in the mantle while heat escapes or current models for the bulk chemistry of Earth are in error and much of the terrestrial heat loss is nonradiogenic.

(Oxburgh, E. Ronald and O'Nions, R. Keith; "Helium Loss, Tectonics and the Terrestrial Heat Budget," Science, 237:1583, 1987.)

Comment. Such data encourage the thought that a portion of the earth's heat may be generated electrochemically. Do we live atop a huge spherical electrochemical hot plate?

A GEOTHERMAL WOMB?

A flurry of papers and at least one TV documentary have widely promulgated the news that many life forms thrive near the thermal vents 2550 m under the sea along the Galapagos Rift. Mollusks, worms, crabs, and other forms of life make up a successful biological community where light never penetrates. Terrestrial heat rather than the sun keeps this life going. The geothermal heat reduces sulfur compounds emitted from the vents and chemosynthesis proceeds up the biological ladder without need for sunlight. (Karl, D.M., and others; "Deep-Sea Primary Production at the Galapagos Hydrothermal Vents," Science, 207:1345, 1980.)

Comment. The implications are far-reaching. Does life exist at great depths in the earth and beneath the apparently lifeless surfaces of the other planets? Photosynthetically sustained life may represent only a small slice of the biological pie. Was sunlight necessary for life to originate and evolve (assuming it did)? See also p. 185.

MYSTERY GLOW ON SEA FLOOR

> Marine scientists have discovered two thermal vents on the seabed that glow in the dark. A group of researchers, led by oceanographers from the University of Washington, detected faint light during an expedition to the Juan de Fuca Ridge. The ridge and the vents, 300 km off the coast of British Columbia, follow an underwater fault formed by the junction of the Juan de Fuca and the Pacific tectonic plates.....John Delaney, leader of the expedition described the glow as a 'flame-like light' that seems to emanate from the superheated water emerging from the thermal vents, 2200 meters below the surface..... 'The source of the light is still unclear,' said Joe Cann, a geologist from the University of Newcastle in Britain. The scientists suspect that the water itself is glowing.

The water temperature is so hot--350°C --that bioluminescance is unlikely. The presence of the glow does, however, imply that photosynthesis is still possible in these sunless depths. Life forms do congregate around these vents. A curious shrimp found in the area where the glow was noted is eyeless but does possess photoreceptors on its back! (Dayton, Sylvia; "The Underwater Light Fantastic," New Scientist, p. 32, August 25, 1988. (Anonymous; "Mystery Glow Emanates from Ocean Bottom," Albuquerque Tribune, p. D3, August 18, 1988. Cr. D. Eccles.)

Comment. The Juan de Fuca Ridge is also the site of a strong flow of subterranean electricity. See p. 236.

THE THROBBING EARTH

The planet earth throbs regularly every 12 sidereal hours according to gravity-wave detectors located in Geneva and Frascati, Italy. The pulsations, presumably expansions and contractions of the earth-as-a-whole, have been recorded at both places for over a year. Pulse amplitudes are about 100 times larger than those that are expected from gravity waves, so planetary pulsations are blamed. Since sidereal time is measured with respect to the fixed stars rather than the sun, an extraterrestrial origin is possible, although no one knows what sort of cosmic force could make our planet throb like this with such precise timing. (Anonymous; "Italians Discover Earth Throb," New Scientist, 98:913, 1983.)

THE ORBITING MOUNTAINS BELOW

Two years ago a Russian scientist suggested that tiny black holes orbiting within the earth might trigger volcanic activity. Now, he has extended the idea to earthquakes.

> A.R. Trofimenko of the Minsk Department of the Astronomical-Geodesical Society of the USSR believes that all cosmic bodies, including the Sun and the Earth, are riddled with "mini" black holes left over from the big bang. Though much smaller than atoms. such black holes would each contain as much mass as a mountain, up to about 2 x 10^{20} grams.
> Trofimenko originally suggested that energy radiated by these mini black holes could make hot spots that produce volcanic outbursts. Now he has investigated the way in which such objects, by orbiting about the Earth's core, would distort the gravitational field at the surface of our planet.

Each time a mini black hole passes beneath a spot on the surface, there would be a "gravitoimpulse" too short to be detected by current instrumentation but sufficient to trigger earthquakes. (Anonymous; "Baby Black Holes Blamed for Earthquakes," New Scientist, p. 18, September 19. 1992.)

BULL'S EYE PATTERN OF MAGNETIC ANOMALIES

On p. 196, concentric rings of gravity anomalies centered on Canada are described. A similar pattern of magnetic rings has shown up in the Yucatan peninsula. The inner ring is 60 kilometers across; the second, 180 kilometers. The rocks causing the magnetic anomalies are about 1100 feet down. Since these rocks are probably Late Cretaceous in age, this potential impact feature may be the eagerly sought scar of the asteroid impact that some think wiped out the dinosaurs and left an iridium-rich layer all over the world. (Anonymous; "Possible Yucatan Impact Basin, " Sky and Telescope, 63:249, 1982.)

Chapter 5

GEOPHYSICS

LUMINOUS PHENOMENA

WEATHER PHENOMENA

HYDROLOGICAL PHENOMENA

EARTHQUAKES

ANOMALOUS SOUNDS

ATMOSPHERIC OPTICS

LUMINOUS PHENOMENA

ORDINARY LIGHTNING

Ordinary lightning is the rapid, concentrated discharge of electricity through the atmosphere. The adjectives "rapid" and "concentrated" must be specified to distinguish ordinary lighting from ball lightning (covered in the following section) and from phenomena involving the slow and diffuse discharge of electricty, such as mountain-top glows.

The overwhelming majority of lightning bolts are not anomalous. They are simply like large sparks---noisy and frequently scary but amenable to explanation. But there are puzzling and bizarre exceptions; and Science Frontiers has recorded several of them, as itemized below:

•Rocket lightning. Discharges from clouds upwards into the ionosphere to some diffuse, unseen terminal.

•Bizarre bolts. Some strange antics and predilections of lighting.

•Triggered lightning. The apparent stimulation of lightning discharges by other discharges, solar activity, and cosmic rays.

•Lightning miscellany. Unusual sounds, superbolts, peculiar forms, and unusual geographical distribution.

It must be emphasized that we have here only a sampling of lightning's anomalies. The Catalog volume Lightning, Auroras, Nocturnal Lights, describes 25 different lightning anomalies and curiosities.

ROCKET LIGHTNING

ROCKET LIGHTNING PHOTOGRAPHED FROM SPACE SHUTTLE

Scientific disbelief in this phenomenon recedes as observations accumulate. We present below a few short excerpts from a scientific report:

> Video images from space showing a single upward luminous discharge into the clear night air above a thunderstorm were recorded for the first time during the space shuttle STS-32 mission, and later during the STS-31 mission and other missions using the shuttle's payload-bay TV cameras.
>
> .....
>
> Figure 1 [impossible to reproduce] shows the upward luminous discharge that was seen to move out of the top of a single thunderstorm during the flight of STS-31. This video image was taken at 0335:59 UTC 28 April 1990 while the shuttle was on its 55th orbit and passing over Mauritania, northwest Africa.
>
> .....
>
> The storm that had the luminous discharge was located at approximately 7.5°N, 4.0°E, and was about 2000 km from the shuttle's position. The lightning discharge was determined to be at least 31 km long.
>
> .....
>
> We are now trying to understand the significance in relationship to the earth's atmosphere and the global electric circuit.

(Vaughan, Otha H., Jr., et al; "A Cloud-to-Space Lightning as Recorded by the Space Shuttle Payload-Bay TV Cameras," Monthly Weather Review, 120: 1459, 1992.)

Comment. Somewhere 31 kilometers above the thundercloud, there must have been a concentration of electrical charge that acted as a "terminal" for the bolt. How did it get there?

An example of rocket lightning observed over the South Atlantic in 1964.

UPWARDLY DIRECTED LIGHTNING FROM CLOUD TOPS

> An image of an unusual luminous electrical discharge over a thunderstorm 250 kilometers from the observing site has been obtained with a low-light-level television camera. The discharge began at the cloud tops at 14 kilometers and extended into the clear air 20 kilometers higher. The image, which had a duration of less than 30 milliseconds, resembled two jets or fountains and was probably caused by two localized electric charge concentrations at the cloud tops.

(Franz, R.C., et al; "Television Image of a Large Upward Electrical Discharge above a Thunderstorm System," Science, 249:48, 1990.)

Comment. Note that the above discharges were diffuse and quite unlike most cloud-to-ground lightning discharges. They were, in fact, much like the mountain-top glows seen along the Andes. Also, one should ask where those "localized electric charge concentrations" came from and why they did not disperse.

BIZARRE BOLTS

LIGHTNING "ATTACKS" VEHICLES

"On Sunday, 17 July, 1988, the usual calm and quiet on the Shelhamer property in rural Hannacroix, New York, was suddenly disrupted by a bright flash and a powerful concussion. The house shook. A picture came off the wall and crashed onto the floor. A contact lens popped out of Mrs Shelhamer's eye. The electric power went out. The kitchen clock froze at 3.30PM. Mr Shelhamer saw three wisps of smoke rise from his parking area. He noticed a strange smell in the air. A pickup truck, parked near the house, was covered with dirt. Two of its tyres were flat, and the hubcaps were lying on the ground. Strange trenches and tracks had appeared in the surface of the parking area. These trenches led Mr Shelhamer to a hickory tree, about 20m from the house. On it a 10cm scar had appeared, spiralling up the trunk toward the sky. A few minutes earlier it had begun to rain, and all this was the result of the lightning strike.

"Perhaps the most extraordinary aspect of this lightning strike was the way the currents flowing from the hickory tree along the surface of the ground apparently leapt out of the ground in places to pass through the automobiles parked in the parking area. Two of the trenches that radiated from the tree ended at an automobile, but reappeared at the opposite side or end of that automobile, where they either terminated in a crater, or continued tracking on. One of these trenches

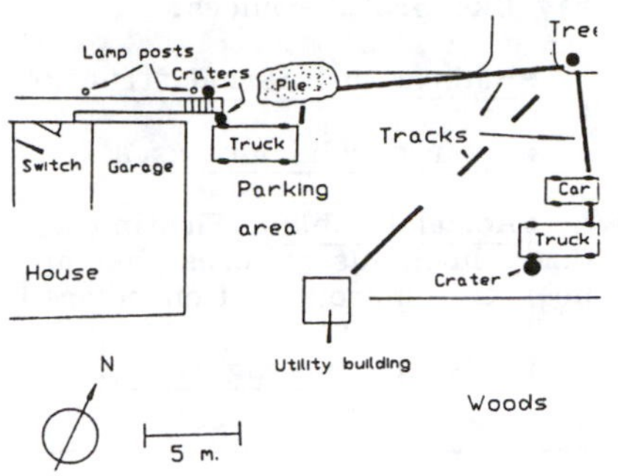

Sketch of the lightning's ground tracks and craters in Shelhamer's parking area.

tracked through two automobiles that were parked side by side."

Electrical currents are supposed to follow the path of least resistance. How did all three vehicles in the parking area end up on paths of least resistance? Some of the phenomena observed can be accounted for by vaporization of moisture in the ground, but what popped those hubcaps off? (Jonsson, H.H., et al; "Unusual Effects of a Lightning Ground Strike," Weather, 44: 366, 1989.)

MAN-KILLER LIGHTNING

From category GLL8 in Lightning, Auroras, Nocturnal Lights, we know of lightning's preference for oaks over other trees in mixed forests; now it seems that a search of death records shows that men are far more often killed by lightning than women. P.E. Brown checked the figures for England and Wales for the years 1974-1989, when 56 people were listed as killed by lightning. Of these 48 were male, 8 female. Brown supposes that the preponderance of male deaths is due to males being more likely to be outside in exposed areas. (Brown, Paul R.; "Lightning Deaths and Sex," Journal of Meteorology, U.K., 16:244, 1991.)

TRIGGERED LIGHTNING

LIGHTNING TRIGGERED FROM THE MAGNETOSPHERE

Whistlers are, as their name implies, curious whistling noises heard on radio receivers. They are caused naturally by lightning, which sends radio noise travelling through natural "ducts" in the earth's magnetosphere. It has recently been discovered that some of the whistlers are synchronized in a way that strongly suggests that some event high up in the magnetosphere triggers some lightning discharges far below near the surface. In other words, lightning is not always a product of activity in the lower atmosphere. (Armstrong, W.C.; "Lightning Triggered from the Earth's Magnetosphere as the Source of Synchronized Whistlers," Nature, 327:405, 1987.)

Comment. Ball lightning has been correlated with solar activity and other extraterrestrial influences. See GLB17 in our Catalog Lightning, Auroras, Nocturnal Lights.)

SYMPATHETIC LIGHTNING

> Video recordings made from the space shuttle at night show large areas of lightning activity in clouds and some flashes that appear to be sympathetic with other flashes. Nearly simultaneous appearances of lightning as far apart as 100 km suggest that widely separated discharges may somehow be related.

(Ahmadjian, Mark, et al; "Video Pictures of Lightning Discharges Taken from the Space Shuttle," Eos, 67:891, 1986.)

THE SUN CONTROLS THE EARTH'S GLOBAL ELECTRICAL CIRCUIT

Data collected from electrosondes (balloons measuring atmospheric electrical currents) over the Antarctic ice caps infer that solar flares stimulate large surges in the flow of electrical charge from the upper atmosphere to the earth's surface. Because this unidirectional flow of fair-weather electricity must ultimately be balanced by thunderstorms somewhere on the planet, it follows that the frequency and severity of terrestrial thunderstorms are dictated, at least on the average, by solar activity. Formerly, global circuit theory had it that the thunderstorms themselves were the driving force behind the fair-weather current flow. Now it seems that the sun calls the tune and that thunderstorms do not arise at random. (Anonymous; "Solar Activity and Terrestrial Thunderstorms," New Scientist, 81:256, 1979.)

COSMIC RAYS MAY TRIGGER LIGHTNING FLASHES

Science has long claimed to have the explanation of lightning discharges well under control. But the discharge paths followed by lightning strokes often seem unnecessarily tortuous when more direct routes are readily available. The mechanism by which large reservoirs of unlike charges are built up is also obscure. Cosmic rays have now been proposed as both a source of charged particles and a provider of low-resistance ionized conduits for lightning to follow. Primary cosmic rays carry considerable energy, most of which appears near the earth's surface in the form of cascades of secondary particles that create complex ionized tracks as they penetrate the dense lower atmosphere. Lightning bolts would tend to follow these precursors along their crooked trails. (Anonymous; "Do Cosmic Rays Trigger Lightning Discharges?" New Scientist, 77: 88, 1978.)

Comment. Thunderstorm frequency has often been linked to solar activity, and cosmic rays could provide the connection. Could meteorites or "thunderbolts" do likewise?

LIGHTNING MISCELLANY

UNUSUAL SOUNDS PRECEDING LIGHTNING

May 13, 1989. Austin, Texas. About 8 PM, during an intense thunderstorm, P. Gunkel heard a whistling sound, like that made by a descending firework rocket. A general pinkish brightening of the surroundings accompanied the sound. Half a second after the sound ceased, there was a tremendous clap of thunder. Discussions with neighbors within 1 hour of the event, elicited additional data: one said that it sounded like rocks falling through the air; another heard a strange humming sound from a windowpane for 2 seconds prior to the lightning; yet another spoke of a sound like that of a whistling teakettle, but with an ascending pitch; and a fourth actually saw the lightning strike the street about 500 feet away. (Gunkel, Patrick; personal communication, May 13, 1989.)

Comment. The most common sound heard prior to nearby lightning strikes is a "vit" sound, or a sound like fabric tearing. Such sounds are thought to be caused by brush electrical discharge from nearby objects as the atmospheric electrical field intensifies. See GLL10 in Lightning, Auroras, Nocturnal Lights.

CONCENTRATED SOURCE OF LIGHTNING IN CLOUD

July 21, 1985. Strait of Malacca. m.v. Staffordshire.

> Between 2000 GMT and 2200 GMT whilst the vessel was transiting the Strait of Malacca in a southeasterly direction, the following phenomenon was observed.
> For several hours lightning had been seen ahead of the vessel. As we approached, it appeared to take on several forms, the most interesting of which is shown in the sketch. It had the appearance of a central point of light with ragged streaks radiating from the centre in a mainly horizontal direction. At no time did this lightning reach the sea surface.

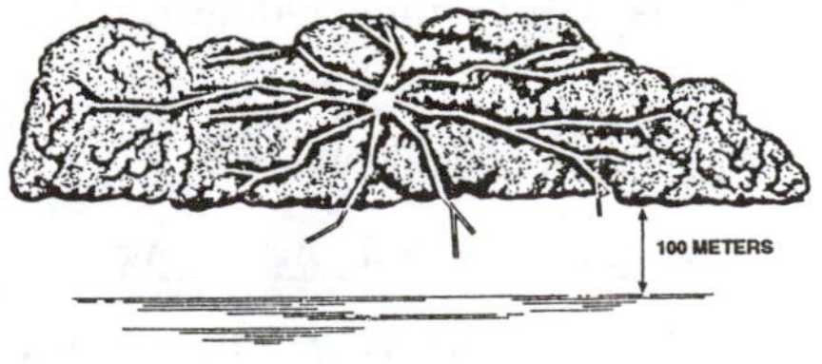

Streaks of lightning emanating from a central region in a cloud deck. No strokes reached the surface.

This type was observed about ten times during the period of observation...

(Thomas, C.O.; "Lightning," Marine Observer, 56:116, 1986.)

WHY SO LITTLE LIGHTNING AT SEA?

Using satellite data, Richard Orville and Bernard Vonnegut have compiled maps showing the global distribution of lightning at night. (At present, satellites can detect only nighttime flashes.) As might be expected, the flashes are strongly concentrated in the earth's tropical regions. The feature of the maps that is most difficult to understand is the very obvious dearth of lightning over the world's oceans. (Anonymous; "Patterns of Thunderbolts," New Scientist, 92:102, 1991.)

Comment. Reinforcing these modern quantitative observations are the centuries-old speculations on why thunder is heard so rarely by mariners.

LIGHTNING SUPERBOLTS DETECTED BY SATELLITES

The Vela satellites carry optical sensors for the detection of terrestrial nuclear explosions. Four Vela satellites keep the entire earth under constant surveillance. In addition to nuclear explosions, these satellites register many intense lightning flashes. Some of the flashes are over 100 times more brilliant than average. Only about five of these "superbolts" occur for every 10 million flashes registered.

Superbolt flashes have relatively long durations (about one thousandth of a second) and do not appear to be confined to the upper levels of the clouds. A large fraction of the superbolts are recorded over Japan and the northeast Pacific during intense winter storms. Ground observations during these storms reveal occasional very powerful discharges of long duration from positively charged regions near the cloud tops to the ground. In contrast, typical lightning arises from negatively charged regions of clouds. (Turman, B.N.; "Detection of Lightning Superbolts," Journal of Geophysical Research, 82: 2566, 1977.)

BALL LIGHTNING

Ball lightning is one of nature's more mysterious manifestations. Basically, ball lightning is a mobile, luminous mass, usually spherical in shape, that accompanies such violent natural phenomena as thunderstorms, tornados, and earthquakes. It is one of those rare natural phenomena that is widely recognized as real but which eludes all attempts at explanation.

One reason ball lightning resists explanation is that it is so variable. It may be as small as a pea or larger than a house. It may be violet, red, or yellow, or even change colors during its brief life. Ball lightning is generally spherical, but rods, dumbbells, spiked balls, and other shapes have been sighted. Sometimes, ball lightning appears to have a snake-like internal structure. Rarely, long crinkled tails may be attached to the balls.

Ball lightning is a dynamic entity. It may glide silently and disinterestedly past an observer, or it may inquisitively explore a room as if directed by intelligence. While a few of these engimatic luminous apparitions dematerialize silently, most explode violently with smoke, the smell of electricity (ozone), and considerable material damage.

No reasonable explanation exists at present for ball lightning. Plasma spheres, antimatter meteorites, intense cosmic radiation, and other theories have been found wanting. The sheer difficulty in accounting for ball lightning has led some scientists to assert that all observations of this phenomenon are illusory. The few photographs that exist are not completely convincing. Nevertheless, to the many thousands who have observed ball lightning, it is very real.

In the first 86 issues of Science Frontiers, one finds some 30 cases of what seems to be ball lightning. These cases actually represent only a small fraction of those reported during this period. I have grouped these 30 cases into nine categories:

- "Ordinary" ball lightning. Examples that are typical in size, color, structure, duration, etc.
- Repeating ball lightning.
- Forced-entry ball lightning. Cases where ball lightning penetrates screens, windows, airplane fuselages, etc.
- Giant ball lightning. Luminous spheres larger than 1 meter.
- Rayed ball lightning. Luminous spheres displaying spikes, knobs, or ray-like protuberances.
- Balls with internal structure.
- Bizarre-behavior cases.
- Aerial bubbles. Floating, dimly luminous spheres that appear transparent. Often colored.
- Artificial ball lightning.

"ORDINARY" BALL LIGHTNING

BALL LIGHTNING IN BAVARIA

August 2, 1921, Hohenschaftlern, Bavaria. 9:00 AM.

"The witness who reported the event was nine years of age at the time of the observation, and was indoors with her uncle on the first floor of a building during a severe morning thunderstorm with heavy rainfall. There was a lull in the storm and the ball lightning appeared on the left side of the window sill about 4-5 m from the observers. The window had been left open because there was a balcony above it which prevented the rain from entering the room.

"The ball fell to the floor where it jumped up and down once or twice. It then started to roll slowly towards the observers across the floor, at about the speed of a dropped ball of wool. Its diameter was about 20 cm, it was translucent, and the rapidly changing colours showed spots of light green, crimson, light blue and pale yellow. It was bright enough to be clearly visible in daylight, and it was uniformly bright over its entire surface. It had protrusions 'like the Andromeda nebula.'

"When it came near the table, where my uncle and I were sitting, I tried to get up to have a closer look. My uncle (fortunately) held me back. It then rolled towards the tiled stove on the right side of the room, crept up the iron parts of the stove leaving (in its path) a deep groove about the width and depth of a thumb, then it exploded in the (airvent) higher up, the sound was like that of a blown up paper bag when (burst) leaving a smell of ozone. The path of the ball was about 5-6 m in length, and it left no marks on the wooden floor." (Stenhoff, Mark; "Torro Ball Lightning Division Report: April 1987," Journal of Meteorology, U.K., 12:200, 1987.)

BALL LIGHTNING IN YORKSHIRE

May 14, 1985. Yorkshire, England.

"At Garton-on-the-Wolds, two miles west-north-west of Driffield and 60 metres AMSL, the electricity went off at 6.15 pm. Half an hour later Mr and Mrs Foster, who were in their paddock tending to the horses during the thunderstorm, heard a 'terrific bang.' On arriving back in their house they found that the television aerial had been blown out of its socket and there were scorch marks on the window sill and curtain lining. The television plug's

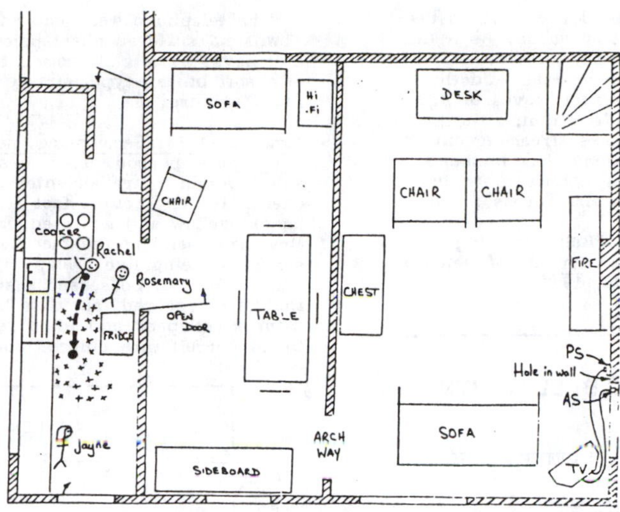

Sketch of the living room in Wakefield in which an orange fireball with orange rays appeared.

negative and positive pins had been blown out of the socket but the earth pin was still intact. A hole some 8 cm by 10 cm across and 4 cm deep was found in the wall by the side of the socket. Several components of the television were damaged and fuses in the main fuse box were blown. Also, at 6.45 pm, Mr and Mrs Foster's daughters, Rachel and Rosemary, were with a friend in the kitchen at the other side of the house. Rachel was standing with her hand on the cooker when, without warning she felt 'a sort of thump' in her back. The other two girls saw an orange, spherical object---about the size of a table tennis ball---moving very quickly. It had no smell, made no noise and seemed to be rotating. The ball of light did not harm Rachel's clothes but made a red, five-pointed star mark on her left shoulder blade which subsequently cleared the following day. The ball then fell onto the wet floor where it exploded 'with the noise of a shotgun' and 'like a firecracker' into many white stars. There were no burn marks on the floor although there was a smell of burning in the air---but this may have been the television." (Sunderland, P.G.; "Ball Lightning in Yorkshire, May 1985," Weather, 43:343, 1988.)

POSSIBLE BALL LIGHTNING IN ANKARA

June 14, 1988. Ankara, Turkey.

"Another phenomenon occurred on 14 June in Ankara: May and June are usually very showery and thundery in central Anatolia and this year is no exception. However, the previous few days had been unusually stormy here in Ankara, and on the 14th the second thunderstorm of the day was in progress with curtains of rain and flickers of lightning, a few kilometres away to the north-west. The storm was moving towards us and the squally wind had already begun. I was again watching the weather from my office, which is on the fifth floor, when I was suddenly distracted by the appearance of a very bright, circular flash of blue-purple light (perhaps one metre or less in diameter), which persisted for about two seconds and then silently 'popped out,' leaving behind a puff of smoke, which then drifted away. The flash of circular light occurred about 500 m away from me: it was about 30 m above the ground, close to, and partly behind, a tall factory chimney. There was definitely no cloud-to-earth lightning over that area at that time, but the edge of the cumulonimbus cloud, giving the storm a few kilometres away, was directly overhead." (Kirvar, Erol; "Thunderstorm and Possible Ball Lightning in Ankara, June 1988," Weather, 44:136, 1989.)

SILVER BALL OF LIGHT

October 21, 1989. Near Glasgow, Scotland. Text of a letter from Mrs. Barr of Dalmuir:

"The day started off as a nice clear autumn day with blue skies. About 11 a.m. the sky became grey and very heavy. It was not cold. By lunchtime there was a steady drizzle of rain which became quite heavy at times.

"I was in a butcher's shop in Dalmuir (west of Glasgow) at 2 p.m. with my husband and about a dozen or so other customers. The rain had stopped. I looked through the windows to see if it had started again when I saw a silver-coloured ball, shoot across from the east, not far overhead but low down. It seemed to be inside a broad ribbon of silver light. The ball looked as if it was still, not rotating on itself. It all happened very quickly. As it shot across the road outside, there was a 'swoosh' sound, something like a firework rocket shooting up, and it landed with a deafening explosion (like a bomb) which shook the ground and the butcher's shop." Torrential rain followed. There was no mark on the road where the ball hit. (Anonymous; "Possible Ball Lightning in 'a Ribbon of Light,'" Journal of Meteorology, 15:176, 1990.)

HOVERING BALL OF FIRE

June 12, 1991. Braintree, Massachusetts.

> One of Earth's rarest and most mysterious weather phenomena occurred in front of Olga Perrow's Braintree home yesterday afternoon.
>
> Ball lightning, an orange-reddish glow of luminosity that Perrow said "looked like a bowling ball," greeted Perrow and her two grandchildren as they drove into the driveway at 665 Commercial St. during the height of yesterday's thunderstorm.
>
> "I was stunned," Perrow said. "It was so smooth-looking. It was like a big ball of fire."
>
> Perrow said the ball moved alongside the car up to the front wheel and "exploded" when the car went into the garage.
>
> "It sounded like a bomb," she said. "We expected to see a hole in the ground, but there was none."
>
> Chase Trowbridge, Perrow's grandson, said the ball was hovering about five or six inches off the ground. "It moved very slowly; we were watching it for about 10 seconds," he said. "It was weird."

(Macrae, Scott; "Powerful Storm Hurls Rare Ball Lightning," Quincy Patriot-Ledger, June 13, 1991. Cr. B. Greenwood)

UNUSUAL ELECTRICAL (?) PHENOMENA

November 24, 1975. Tendele Hutted Camp, Drakensberg Mountains, South Africa. The following observations were made during violent electrical storms.

> Around 10pm, WN observed a luminous vertical column in an easterly direction which appeared suddenly at a location low on the hillside on the far bank of the Tugela River at a distance of about 1km. This stationary light column seemed to have the dimensions of a pencil stub (approx. 50mm x approx. 7mm) held vertically at arm's length. The column, which had a bluish glow like a fluorescent tube, was visible for about 5 to 10 seconds.
>
> .....
>
> At 11.15pm, when the intensity of the storm had abated and the sky was lit intermittently with flashes of sheet lightning, the writer saw a luminous spherical object, seemingly of golf to tennis ball size, moving rapidly with an apparently vertical undulating motion from left (northeast) to right (southwest) on a horizontal course in the general direction of Mont-Aux-Sources (3282m) where the Tugela River has its origin. This sighting lasted 2 to 3 seconds. About 3 minutes later, another similar object crossed the field of view, following the same course as the first object and showing about 2 or 3 undulations in its passage. At mid-

night, a third object was seen having the same characteristics as the first two objects. However it did not arise from the extreme left of the field of view but appeared to originate from a point marked by a small tree close to and in the middle of the window. These objects had a bright yellowish-blue luminescence, and no noise was heard which could be associated with their passage.

(Neish, William J.P.; "Lightning Phenomena in the Drakensberg Mountains of Natal, RSA," Journal of Meteorology, U.K., 15:377, 1990.)

REVOLVING SPHERE OF LIGHT

The following report is from H.D. Mayor, a scientist at Baylor College of Medicine:

> During an incredible electrical storm [in Houston, Texas, on 18 June 1991] in the evening while sitting at a table in the breakfast room, I saw a ball of lightning enter the utility room [an extension of the breakfast room] apparently through the back door. It hovered as a revolving sphere of bright yellow, orange and red light about 10 inches in diameter, in the air about three feet above the floor. It stayed in the same place. After about two or three seconds the globe disappeared with a loud pop rather like a discharge from a champagne bottle. The discharge was followed by a distinct odor of ozone. My Siamese cat also appeared to see the ball; at least he ran toward it.

(Mayor, Heather D.; "Watching the Ball," Nature, 353:496, 1991.)

AN ELECTRICAL VIRTUOSO

August 12, 1992. Conwy, Wales. Here is a carefully observed case of ball lightning with rather spectacular side effects. Mrs. P. Stafford was looking through her front window:

> ...when she saw what she first thought was a 'ball of white fire', larger than a football, about 20 to 30 feet from her, travelling horizontally at a constant height up her drive. There was very heavy rainfall, perhaps with some hail, but no lightning or thunder. The ball was seen against the background of other houses and her view of it was not interrupted. It was round, opaque and predominantly white with some yellow, and surrounded by a blue, irridescent halo. She said it was reminiscent of a meteor or comet and the light from it was like that from a fluorescent tube. It was bright enough to be clearly visible in daylight and appeared to be spinning or rotating. It hit the oak tree, perhaps 12 or 13 feet away, in Mrs. Wignall's front garden, with a terrific crack and explosion.
>
> The ball was in sight for about 10 to 15 seconds, and its appearance did not change until it struck the tree, whereupon it became smaller. It hit the trunk about half way up and split the bark and trunk, showering splinters of wood over a distance of about 50 yards. As it did so, it rolled down the tree and dispersed in flashes---she said that there seemed to be 'waves of lightning' passing from it into the ground and radial sparks streaming out of it in all directions. Her husband, however, thought he saw the ball, now smaller in size, cross the lawn.

(Stenhoff, Mark; "Ball Lightning Reported in Conwy," Journal of Meteorology, U.K., 17:308, 1992.)

REPEATING BALL LIGHTNING

BALL LIGHTNING STRIKES TWICE!

Summer 1977. Haymarket, Virginia. A severe storm was threatening. Mrs. Patricia Townsend was standing in front of her kitchen counter talking on the telephone.

"Several things happened at the same time and the whole incident probably lasted no more than a few seconds at the most. While I was on the phone, I heard a tremendous crack, something like the report of a high-powered rifle or the sound of a bat hitting a baseball. At the same time the outside of my house, meaning the outdoors, lit up brilliantly. A split second later or perhaps at the same time, I heard a loud swooshing or hissing noise and the phone seemed to come alive in my hand. Then my whole kitchen lit up like a floodlight. Lightning or electricity or whatever it was seemed to flow rapidly from the open kitchen door across the expanse of the far end of my kitchen at ceiling level as shown by the jagged line in my drawing. I'm not sure where the red ball came from but I have depicted it as coming from the jagged lightning on my ceiling. Anyhow, almost at the same time as the lightning zoomed across my kitchen and the phone started vibrating in my hand, a large red ball (with yellow and white somewhere) appeared in front of me and hit me on the chest with the force of a large man hitting me with his fist. I fell to the floor and I believe the phone was still in my hand. I'm still not sure if I was knocked unconscious or not. I couldn't swear I was and couldn't swear I wasn't. The ball hit me with the accompanying sounds of smacking and crackling, kind of like a string of firecrackers being set off."

The telephone was dead and Mrs. Townsend suffered chest pains for several days. The ball seemed to be made of a soft burlap-type surface with a fuzzy texture.

June 21, 1978. Same place, same kitchen. Same person. Thunderstorm outside. Again a fireball entered the kitchen. It was about a foot across with jagged yellow and white edges. It hit Mrs. Townsend in the face, with the sensation being like a slap with an open hand. The ball possessed a surface like that of a textured fabric as before. The witness collapsed and when recovering found herself with slurred speech and

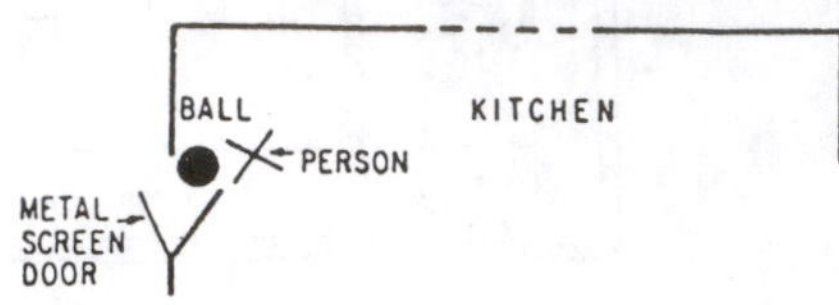

neck pains. She soon regained full faculties and health. (Bailey, B.H.; "Ball Lightning Strikes Twice," Weather, 39:76, 1984.)

FORCED-ENTRY BALL LIGHTNING

BALLS OF FIRE ENTER ROOM THROUGH METAL SCREENS

> Thunderstorms are frequent in the Entebbe Peninsula, Lake Victoria (Uganda). During one of these storms, which usually come at night time, there was a simultaneous flash of lightning and its associated clattering crash of thunder. A second or less later, several balls of brilliant blue light, about 4-6 cm diameter, entered the room through a window on the south side and "floated" across the room to leave by a window on the east side. My wife and I were already awake (it would have been difficult not to be) and independently exclaimed aloud on what we had just seen.

Both windows were open but had metal screens. The same phenomenon occurred again during the same rainy season. (Gillett, J.D.; "Balls of Fire," Nature, 299:294, 1982.)

Comment. Ball lightning is fairly rare; repeat performances unheard of.

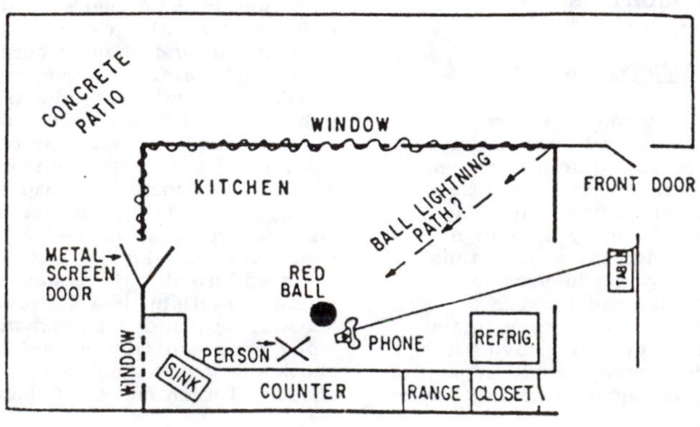

THREE ANOMALIES IN ONE STORM

During the passage of a cold frontal trough between 1030 and 1100 GMT on Monday 21 March 1983, squally thunderstorms affected south Cheshire and north Staffordshire. Two incidents of ball lightning, a fall of seashells and three occurrences of probable tornado damage were reported, mostly within a 10 km radius of Stoke-on-Kent.

At Camillus Road, Knutton. Ball lightning about 40 cm in diameter with a luminous tail 4 m long. One observer saw it descend at an angle of 45° and hit the roadway.

At Kingsley:

A large white luminous ball, probably over a metre in diameter, blasted its way into a factory workshop by shearing an irregular hole through a steel-mesh-reinforced window. There was no evidence of any fusion of the glass. The ball, accompanied by a deafening roar, passed very quickly in a straight line through the processing shop and left by blasting a 2 by 3 metre hole in a wall of 6 mm corrugated asbestos, fragments of which were later found 20 to 30 metres away outside the factory.

(Swinhoe, P.J.; "Unusual Events along the Squally Cold Front of 21 March 1983 in North Staffordshire." Journal of Meteorology, U.K., 8:233, 1983) See p. 274 for the description of the shell fall.

BALL LIGHTNING SPLITS AND RECOMBINES INSIDE SOVIET AIRLINER

An Ilyushin-18 took off from Sochi, on the Black Sea, in fair weather. Soon after takeoff thunderclouds were noted about 60 miles away.

Suddenly, at the height of 1,200 yards, a fireball about four inches in diameter appeared on the fuselage in front of the crew's cockpit. It disappeared with a deafening noise, but reemerged several seconds later in the passenger's lounge, after piercing in an uncanny way through the air-tight metal wall. The fireball slowly flew about the heads of the stunned passengers. In the tail section of the airliner it divided into two glowing crescents which then joined together again and left the plane almost noiselessly.

Upon landing back at Sochi, holes were discovered in the fuselage fore and aft. (Anonymous; "Tass Says Lightning Ball Entered Soviet Airliner," Associated Press dispatch, January 13, 1984, Cr. M.A. Lohr)

Comment. Several examples of splitting ball lightning are to be found in Lightning, Auroras, Nocturnal Lights, but recombination is a much rarer event.

BALL LIGHTNING PUNCHES CIRCULAR HOLE IN WINDOW

In the Autumn 1992 issue of the Journal of Scientific Exploration, A.I. Grigor'ev et al collected 43 eyewitness accounts of ball lightning penetrating into closed rooms. Most of the reports came from the former USSR and are new to Western scientists.

The majority of these balls entered through closed glass windows. Sometimes the balls penetrated the windows without damaging the glass at all, but in a few cases neat circular holes were somehow melted or punched through the glass. The accompanying photograph illustrates an incident in which lightning (supposed to be ball lightning) surgically excised a coin-like piece of glass. (Grigor'ev, A.I. et al; "Ball Lightning Penetration into Closed Rooms: 43 Eyewitness Accounts," Journal of Scientific Exploration, 6:261, 1992.)

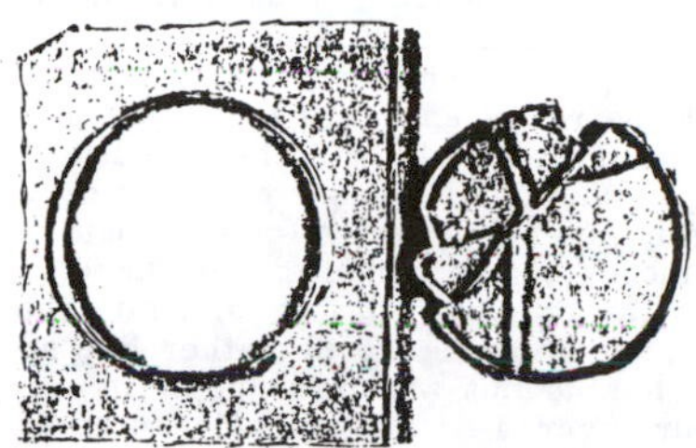

GIANT BALL LIGHTNING

GIANT BALL LIGHTNING

June 8, 1977, Fishguard, Dyfed, West Wales. A brilliant, yellow green, transparent ball, the size of a bus with a fuzzy outline, floated down a hillside. Slowly rotating, it seemed to bounce off projections on the ground. It flickered out after 3 seconds. (Jones, Ian; "Giant Ball Lightning," Journal of Meteorology U.K., 2:271, 1977.)

PRESUMED BALL LIGHTNING

November 24, 1987. Tulsa, Oklahoma.

Circa 3:20 P.M. CST. Location: 2800 Southwest Blvd. in said city. Parents of Keith L. Partain saw a lightning strike near an oil refinery storage tank. Immediately after the strike they saw a bluish sphere with red and yellow highlights, not more than 9 feet in diameter, some 100 yards away, near the tank. The sphere lasted in that form some five seconds before fragmenting in a loud detonation. During the act of detonation the sphere became an irregular spheroid before fragmentation. Mr. Partain reported that he could feel the heat from the detonation. Both individuals, seated in a truck, were quite astounded by the apparition. The weather was quite stormy and violent in its gales, rain and lightning.

(Partain, Keith; personal communication, November 24, 1987.)

Comment. K. Partain checked the Catalog Lightning, Auroras, Nocturnal Lights and classified the phenomenon as GLB1 or Ordinary Ball Lightning.

BALL LIGHTNING STUDIES

The April 1990 issue of the Journal of Meteorology, some 63 pages of it, presents us with a wonderful compendium of ball lightning observations. It is unfortunate that we have room for only a few of the many fascinating descriptions.

Giant ball lightning.

The following display of ball lightning was observed by an officer at the coastguard station at Fishguard, Dyfed, West Wales, on 8 June 1977. The occurrence was at 0227 GMT, grid reference SM(12)895389.

The ball lightning phenomenon was very large and estimated to be about the size of a bus. It was described as a brilliant, yellow green, transparent ball with a fuzzy outline which descended from the base of a towering cumulus over Garn Fawr Mountains and appeared to 'float' down the hillside. Intense light was emitted for about three seconds before flickering out. Severe static was heard on the radio. The object slowly rotated around a horizontal axis, and seemed to 'bounce' off projections on the ground. It was noticed that cattle and seabirds in the immediate vicinity became disturbed.

(Jones, Ian; "Giant Ball Lightning or Plasma Vortex," Journal of Meteorology, 15:178, 1990.)

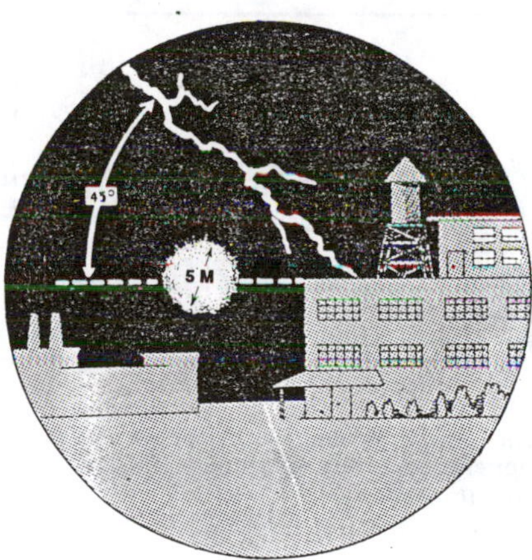

This example of giant ball lightning was observed near a factory in Albany, New York, in 1975.

RAYED BALL LIGHTNING

BALL LIGHTNING AND BLUE FLASHES

May 31, 1982. Wakefield, England.

"We live to the south of Wakefield on the ground floor of a large Victorian house with high ceilings, attics and cellars. Our kitchen and living room face roughly S.S.W. In the late afternoon, a very heavy thunderstorm erupted with torrential rain, and thunderclaps and forked lightning occurring not quite, but almost, simultaneously. Towards the end of the storm, about 5.30 P.M., I was in the kitchen and my mother in the living room, both facing the windows. There was a very loud peal of thunder and a loud crack, rather like the little explosion of a Christmas cracker greatly magnified. I was at the sink, close to the window, on the ledge of which stands an electric Corvette water heater, plugged in but not switched on. Beside me, about four feet from the ground and two feet to the left of me, at the moment of the crack, there appeared for a second or so, electric blue flashes, six to eighteen inches in length coming from a white centre. I felt nothing, but was startled. There was no damage to the water heater or anything else. Just as I exclaimed at the blue flashes, I heard my mother cry out and ran to her in the living room. She was sitting in a chair about ten feet from the window, under which stands the television set, plugged in but not switched on. About eight feet from the floor and four feet in front of the double-glazed window (i.e. between the television and the fireplace, but not quite over or opposite either), appeared an orange fireball, rather smaller than a football, with straight lines of orange light, varying from about one to two feet in length coming from it in all directions. This ball seemed to hover for up to five seconds before disappearing. I did not see it myself. My mother quickly recovered from the shock and there was no damage in the room." (Gilbey, J.C.M.; "Orange-Coloured Ball Lightning.....," Journal of Meteorology, U.K., 9:245, 1984.)

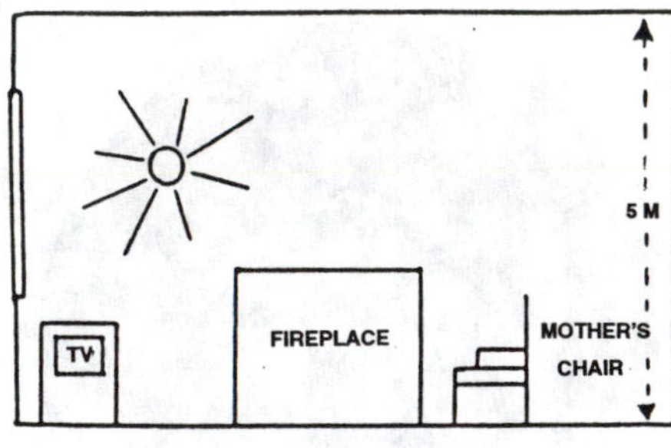

An orange fireball with orange rays appeared in this Wakefield, England, living room.

BALL LIGHTNING BURNS A RAYED CIRCLE ON A SHED WALL

B. Evans sent the following account to the Editor of the Journal of Meteorology:

> Your report of 26th August (1986) about the mysterious five circles which appeared in cornfields near Devel's Punchbowl, near Winchester ---the largest being 42 feet across ---reminded me of an incident during the night shift in 1980 at Shotton steelworks.
>
> A high wind was followed by a bright light which lit up the whole area. When we looked down on the yard from our vantage point we could see that a great ball of lightning had struck. As it bounced from spot to spot, we had to duck to get out of its way, but as soon as it has passed we ran out and saw it strike the side of a scrap shed. When the sun came up, it picked out the shape of a dartboard on the scrap shed. The pattern was clear, with all the segments in place, and it was about 37 feet across.

(Meaden, G.T.; "Rayed Circle Made by Ball Lightning on the Wall of a Shed," Journal of Meteorology, U.K., 11:271, 1986.)

SPIKED BALL LIGHTNING

December 3, 1979. Fleetwood, England.

"On the evening in question there was an intermittent thunderstorm with rain in heavy showers. My son Michael had just come in from the college and had gone into the room and was standing watching the T.V. The time would be a little before 6.00 p.m. I said something to the effect that his meal would be ready and he'd better wash his hands, so he turned the television off, although it remained plugged in... At this point a spherical object about six inches (15 cm) in diameter floated down the (sealed) chimney and into the room. It appeared to be rather like a soap bubble but was dull purple in colour covered or rather made up of a furry/spiky emission all over. The coating seemed to be about one inch (2.5 cm) thick with spikes of two inches here and there but changing all the time. It was quite dim and appeared to be semi-transparent, in so much as I could see through to the inside of the opposite side, which appeared quite smooth---all the spikes pointing outwards from the surface. It appeared to me to be insubstantial and made no sound. It drifted between the two of us towards the television screen at about 30 inches (75 cm) from the floor, covering the six feet (2 m) in about four seconds. When about eight inches from the screen it disappeared (imploded?) with a fairly loud crack/pop sound leaving behind a smell as of an electrical discharge." (Rowe, Michael W.; "Another Unusual Ball Lightning Incident," Journal of Meteorology, U.K., 9:135, 1984.)

RAYED BALL LIGHTNING HITS PLANE

The following material is "reprinted" from CompuServe's Aviation Special Interest Group (AVSIG), with the permission of J. Baum. (Cr. E. Kimbrough) For the uninitiated, we are dealing with a computer bulletin board here!

> Sb: #235852-Ball Lightning
> Fm: Jeff Baum (PHX) 73740, 1302
> To: Emory Kimbrough [TCL] 72777, 1553 (X)

On 8 January 1992 we were in MSP [Minneapolis/St. Paul] ready for pushback at sunrise. Weather was sleet squalls, temperature of +2 degrees C (35 degrees F), ceiling of indefinite 100 obscured, visibility of about 1 and ½ mile variable. We deiced and taxied for the active 11L, airborne in 8 minutes after deicing had ended. The First Officer was flying that leg. Climbing through about 900 feet ABL, this incandescent sphere approximately 10 cm (6 inches) in diameter surrounded by a, what I called, plasma cloud of bluish white approx 1 to 1 and ¼ meter (3 to 4 feet) in diameter with bright white "rays" similar to a fireworks explosion formed just forward and to the left of the radome. We contacted this within ¼ second on our left side, just aft of the attach seam of the radome (namely about in line with my left foot). With this contact there was a sharp bang. The cabin crew reported the loud bang but didn't see any haze or light inside the cabin. One did report seeing a bright light on the left side of the aircraft's exterior.

BALL LIGHTNING WITH INTERNAL STRUCTURE

BALL LIGHTNING WITH INTERNAL STRUCTURE

September 1981. Berkhamsted, England.

> I was resting on my settee listening to music on the Third Programme when there was some interference of a crackling kind. Suddenly, a ball of bright light appeared in front of my radio. It was about the size of a large orange. It was dazzlingly white and gave the appearance of dozens of stick crystals 5.0 mm in length jigging about with a crackling sound. By the time I reached the switch it had disappeared, but a loud burst of thunder broke overhead.

(Cook, M.L.; "Ball Lightning Incident in Berkhamsted, 13 September 1981." Journal of Meteorology, U.K., 7:18, 1982.)

"CRYSTAL" BALL LIGHTNING

June 1, 1984. Nottingham, England. Testimony of Mrs. Elsie Haigh:

> ...at approximately 5.45 p.m., I was in my kitchen. The window and door were both closed. I was standing with my back to the window when I heard a 'boiling noise' and a noise which sounded like glass splintering---a crash sound. I was very scared and turned around to face the window and saw a large glass-looking ball, approximately 10 inches in diameter (25 cm), slightly oblong (oblate), with a white filament in the middle. This was floating on a bowl of water which was in the sink.
> I ran to the bathroom, and seconds later I heard an explosion and splintering glass. When it was quiet---I think a few seconds elapsed---I returned to the kitchen. The ball had gone and there was no damage. I can only describe it as a miracle.

(Meaden, G.T.; "'Crystal' Ball Lightning," Journal of Meteorology, U.K., 9:218, 1984.)

Comment. Other ball lightning observations on file include sounds like "breaking glass." See our Catalog Lightning, Auroras, Nocturnal Lights.

WORMY BALL LIGHTNING

1920. Parkside, Australia. A loud noise was heard several hundred yards away. "It wasn't long till we heard a hissing noise and, looking up to the western sky, saw an object about 12 inches in diameter slowly moving through the air down toward us---about 12 feet away. It was travelling eastward and came down over Mrs. Harris's wooden fence landing on the cement porch floor about 3 feet behind us. It gracefully bounced along the cement floor in a straight path covering the 30 foot length of the verandah at a walking pace. It bounced three or four times rising to a height of 18 inches on each occasion. Each time the spherical ball touched the cement it was flattened at the point of contact, and deformed, but it quickly resumed its globular shape when it left the ground. It was not transparent but, rather, like a ball of smoke with glowing 'comma shaped' electrical 'worms' wriggling about---sizzling, hissing and flickering. It flattened by 1/4 into the egg shape on each bounce. On reaching the far western end of our verandah it accelerated rapidly and rose a steep angle of about 45 degrees clearing the apricot tree, wires, and the house next door. At this stage my mother rushed in the back door of the house where we huddled for about 30 seconds before hearing a resounding crash some 250 yards away off to the east. It had hit Green's house at the far eastern end of Campbell Road. It apparently then bouced all the way to the Salvation Army home and demolished the whole house somewhere near Dawson and Florence Streets at Fullarton." The ball rotated slowly and emitted small sparks. (Illert, Theodore Charles; "The Parkside Lightning Ball," personal communication from C. Illert. To be published in "Speak No Evil: A Case Study of Lives and Times of German Settlers in South Australia," by C. Illert)

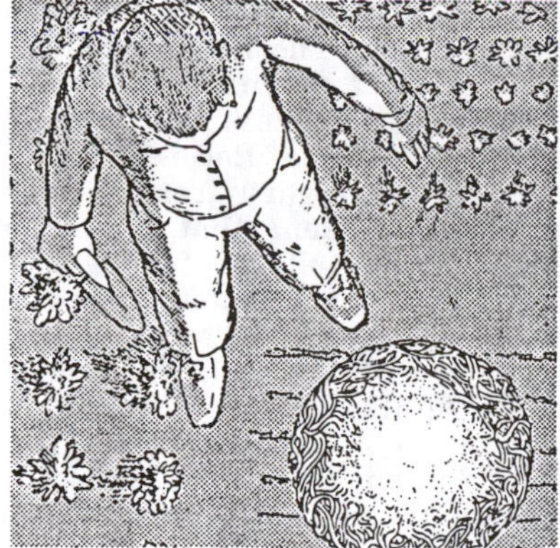

Ball lightning with a wormy or rope-like surface . The example sketched above was observed in an English garden and seems similar to the Parkside case. See Lightning, Auroras, Nocturnal Lights.

BALL LIGHTNING WITH BIZARRE STRUCTURE

An older case of ball lightning with special features has just surfaced. It happened in the afternoon in 1924. There came a large ball of fire---or so it looked---but the thing was that it had chains all the way round. It lasted about five minutes, then all the chains clashed together with a terrific bang; then we had a terrible thunderstorm which lasted quite a long time." In a second letter Mrs. Revell drew the ball lightning as a red globe with 16 rays, composed of links like a chain, issuing from it; the rays were rather longer than the diameter of the ball. She said, "You asked the size. From the ground it looked about four to five yards across, which would be larger than that in the sky. The chains opened from the top to the bottom with a terrific bang.

(Rowe, Michael W.; "Unusual Ball Lightning," Journal of Meteorology, U.K., 8:125, 1983.)

BIZARRE-BEHAVIOR CASES

AGGRESSIVE BALL LIGHTNING

August 17, 1978. Caucasian Mountains, Russia. Victor Kavunenko and four other mountaineers were camped for the night at an altitude of 3900 meters. He reported as follows:

"I woke up with the strange feeling that a stranger had made his way into our tent. Thrusting my head out of the sleeping bag, I froze. A bright-yellow blob was floating about one metre from the floor. It disappeared into Korovin's sleeping bag. The man screamed in pain. The ball jumped out and proceeded to circle over the other bags now hiding in one, now in another. When it burned a hole in mine I felt an unbearable pain, as if I were being burned by a welding machine, and blacked out. Regaining consciousness after a while, I saw the same yellow ball which, methodically observing a pattern that was known to it alone, kept diving into the bags, evoking desperate, heart-rendering (sic) howls from the victims. This indescribable horror repeated itself several times. When I came back to my senses for the fifth or sixth time, the ball was gone. I could not move my arms or legs and my body was burning as if it had turned into a ball of fire itself. In the hospital, where we were flown by helicopter, seven wounds were discovered on my body. They were worse than burns. Pieces of muscle were found to be torn out to the bone. The same happened to Shigin, Kaprov and Bashkirov. Oleg Korovin had been killed by the ball---possibly because his bag had been on a rubber mattress, insulating it from the ground. The ball lightning did not touch a single metal object, injuring only people." (Anonymous; "The Puzzle of Ball Lightning," Journal of Meteorology, U.K., 9:112, 1984. Original source: Soviet Weekly, February 11, 1984.)

Comment. Ball lightning has often been called inquisitive, but this is one of the few reports where it deliberately (?) seemed to attack people. Some Russian English-language publications verge on the sensational, and one must always have some salt on hand.

BALL LIGHTNING OR MIRAGE OF VENUS?

On p. 335 S. Campbell explained a potential UFO sighting in terms of a mirage of a jet landing at Edinburgh. Now he interprets a Russian ball lightning report as a mirage of Venus on the horizon. See what you think:

"Dr. Aleksandr Mitrofanov of the Institute of Physical Problems (sic) of the Academy of Sciences of the U.S.S.R. and two friends were camping on the left bank of the River Oka near Ryazan (at a point where the river makes a sharp bend to the east) on the 23rd July 1974. It had been a clear day, very hot in the afternoon. Together with Muscovites from another encampment they sat up talking and drinking tea (sic) until late in the evening (in fact until early the next morning). At 2:10 a.m. they all saw a light which at first they thought was a torch. It appeared to be 70 metres away in the undergrowth along the bank. As they all stood up the 'ball lightning' (which is what Mitrofanov thought it was) seemed to 'float up' from behind the bushes and move straight towards them, increasing in size. But it did not reach them; it slowly 'swam' horizontally before disappearing after 4 minutes. When it seemed to be at its nearest a ring detached itself, like the ripple of water when a stone is thrown into water. The ring vanished as it expanded, but was followed by a second ring, less bright than the first. Before it vanished the ball took on a pear shape. Just after it vanished the sky in that direction,

Sketch of the luminous phenomenon reported by Mitrofanov. The "ball lightning" seems to be positioned in front of some trees and might be a mirage of a celestial object.

for about 10° of azimuth, became reddish and lighter than the rest of the sky to the north. This illumination lasted no longer than half a minute. The 'ball' had made no sound and there were no traces or smell remaining. Mitrofanov did not mention hearing thunder or seeing lightning." (Campbell, Steuart; "Russian Accounts of Ball Lightning," Journal of Meteorology, U.K., 13:126, 1988.)

SLITHERING PATCH OF LIGHT

September 8, 1981. Te Ngaere, New Zealand. Inside a house during a thunderstorm.

"The next lightning seemed directly overhead and very bright and was accompanied by a simultaneous very loud clap of thunder. I looked up as the whole house shook and then looked down and saw a flow of light come in under the door. It settled in a blob near the edge of the area where the tools were laid out. It was not in any true shape but about 3 or 4 inches long and 2 inches wide, moving along the floor, less than half an inch thick, seemingly fluid in shape and texture. It reminded me of quicksilver, being a bluish-silver colour and it had rounded sides like a blob of mercury. It was brighter at the edges than in the middle, but it did not seem, especially in the light of the room, to glow, nor did it give out sparks. From the central body arms flowed out like runs of oil among the tools. The trails weaved through the tools---not actually over them but round them---moving back into the main body of the blob and then going out doing the same kind of movement over again. There was no sound or smell. The arms finally all went back into the blob which disappeared again suddenly out under the door. There was no bang and when I ventured to touch the tools there was no charge on them."

A subsequent magnetic survey of the area showed a weak correlation between the patch's motion and regions of intense magnetic field. Such a correlation would be expected if the patch contained free magnetic dipoles or current loops. (Burbidge, P.W., and Robertson, D.J.; "A Lightning-Associated Phenomenon and Related Geomagnetic Measurements," Nature, 300:623, 1982.)

AERIAL BUBBLES

ANOTHER LUMINOUS AERIAL BUBBLE

September 1943. On a ship in the South Atlantic enroute from South Africa to Brazil.

During the voyage, a multicolored object about the size of a basketball appeared and the ship changed course to parallel the course of the object. The object was in view for about 20 minutes, moved slowly across the water at a height of 5 feet, and finally disappeared. It looked like a glass ball and appeared to have a membrane enclosing it. Its motion was from the NW to SE and it was seen sometime between 3:00 and 5:00 in the afternoon sometime in September 1943. The color was at times orange and yellow, sometimes green, blue, and red. The sky was overcast and the object was on the starboard side of the ship as it moved towards the NW. The ship's crew, consisting of about 20 men, saw the event and concluded that it might be a 'fireball.'

The original observation was made by Charles L. Reifenhler. (Seal, James; personal communication, June 25, 1986.)

Comment. For more accounts of Luminous Aerial Bubbles, see category GLD7 in LIghtning, Auroras, Nocturnal Lights.

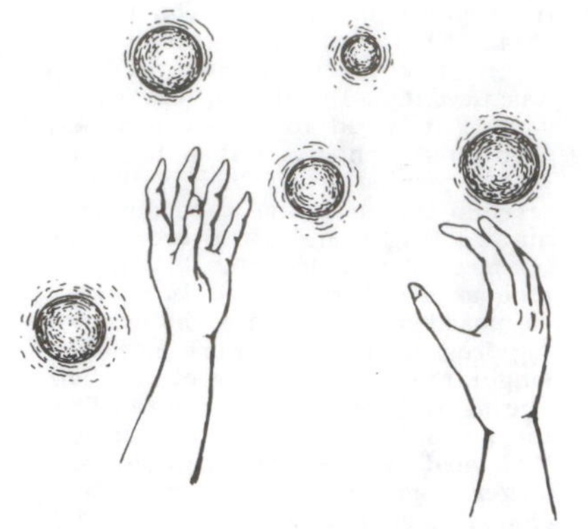

On August 17, 1876. Numerous luminous, multicolored, bubble-like spheres were observed by many at Ringstead Bay, England.

AN ELECTRICALLY CHARGED VORTEX?

Mid-October 1990. Near Bristol, England.

Two ten-year-old girls were playing with a frisbee in a playing field adjacent to Downend sports centre, on the north-east outskirts of Bristol, in the County of Avon. While throwing the plastic disc, one little girl was surprised when it was suddenly flung back at her by 'some invisible force.' Next, the children were enveloped in a 'yellow bubble.' This bubble apparently gave both girls a mild 'shock' and threw them to the ground. While lying on the grass, both children experienced difficulty in breathing. Eventually, the girls managed to 'break out' of the 'bubble' and run home, very frightened indeed.

(Bendall, P.D.; "Did an Electrically-Charged Vortex Strike Two Girls in Bristol, Mid-October 1990?" Journal of Meteorology, U.K.. 15:403, 1990.)

ARTIFICIAL BALL LIGHTNING

PLASMA FIREBALLS MADE IN THE LABORATORY

Many scientists have studied and theorized about ball lightning, but so far no consensus exists as to how it is created and why it behaves as it does. In the lab, fireballs have been produced by electrical discharges in an atmosphere containing dilute propane and other fuel gases. More recently, Japanese scientists have simulated ball lightning in a natural atmosphere by microwave interference. This type of fireball is reported to emulate two of the more mysterious properties of ball lightning: the ability to move against the wind and the capacity to pass through a wall intact. (Ohtsuki, Y.H., and Ofuruton, H.; "Plasma Fireballs Formed by Microwave Interference in Air," Nature, 350:139, 1991.)

NOCTURNAL LIGHTS

Nocturnal lights are not associated in any obvious way with conventional meteors, electrical storms, or other violent weather. The name "nocturnal light" had its origin in the UFO literature; but most nocturnal lights, such as the will-o'-the-wisps, seem to have prosaic origins. The major characteristics of this class of luminous phenomena are a flame-like or globular shape, a wide spectrum of colors, sizes that vary from an inch to many feet, and, most diagnostic of all, erratic motions that are often described as playful, elusive, and inquisitive. All in all, nocturnal lights resemble ball lightning but are less energetic and not associated with electrical storms.

Our treatment of nocturnal lights commences with the well-publicized Marfa Lights seen in southwest Texas. These lights, however, represent only one brand of "ghost lights;" others are seen in specific localities on a fairly regular basis all over the world. Although some nocturnal lights, such as the Foo Fighters of World War II, do not succumb to the usual marsh-gas explanation, it is easy to find examples that do. This section concludes with three such scientifically digestible phenomena.

THE MARFA LIGHTS

Circa 1974. Marfa, Texas.

About 10 years ago, Mr. Whatley was driving home a little before dawn from his night job as a computer operator when he saw what he thought were car lights speeding toward him on a road east of town. The next thing he knew, he says, a cantaloupe-sized globe of orange-red light appeared and hovered a few feet outside the rolled-down window of his pickup.

Understandably, Whatley hit the accelerator, but the light stayed with him for about two miles and then disappeared. Several other incidents are recounted in this article. (Stipp, David; "Marfa, Texas, Finds a Flickering Fame in Mystery Lights," Wall Street Journal, March 21, 1984.)

Comment. What a strange place to find an article on ghost lights!

AIRBORNE OBSERVATIONS OF THE MARFA LIGHTS

The Marfa lights are elusive, and most people lucky enough to see them observe them from the ground. Nevertheless, a few pilots and aircraft passengers have encountered them. In February 1988, R. Weidig was flying at about 8000 feet, some 20 miles from Alpine, Texas, when he noticed white lights in motion around the Alamito Tower's red beacon light.

> We noticed white lights coming up... I don't know how high, but it seemed like several hundred feet. Then the lights would just dissipate ... They moved around that tower for some reason. They'd get on the right hand side of it, the left hand side of it, and go just straight up.

In June 1988, a stranger case was reported by E. Halsell, who was a passenger on a plane flying toward the Chianti Mountains.

> Suddenly a bright light came toward them rapidly, seemingly from a great distance. "It came straight at us til it got to the hood of the plane....It was engulfing us, larger than the plane." It seemed as though they were inside the light. "We couldn't see to fly. It scared us." According to Halsell, as they tried to turn away from it, it moved in front of them. "Always it moved around us, like it was observing us....We made right turns and left turns and it stayed right with us, like it was playing a game." The light was very bright, but "It was kind of fuzzy, like a halo or aura, a ball of light without an obvious center." The light was white in color, was constant rather than pulsating or flickering. There was no unusual sound.

(Brueske, Judith; "Encountering 'The Lights,'" The Desert Candle, 2:1, July-August 1988.)

REMARKABLE PHOTOGRAPH OF THE MARFA LIGHT

The century-old fame of the Marfa, Texas, nocturnal light was greatly enhanced some months ago, when it was written up in the Wall Street Journal, of all places! We now have at hand a time-exposure photograph showing the typical erratic motion and flickering nature of this "spook" light. The photo was taken by James Crocker in September 1986. The location was 10 miles deep in Mitchell Flats, southbound from Highway 90. A single-lens reflex camera mounted on a tripod was used. Exposure was less than 3 minutes, at f/1.8, 50 mm lens, EL 400 color film. Three additional observers were present.

It is interesting that the light's motion resembles that of some observations and photos of ball lightning. The lights in the upper right, just above the right loop of the Marfa Light, are thought to be car lights on Route 67, about 10 miles distant. The photograph and the data printed on it are copyrighted and are used here with permission.

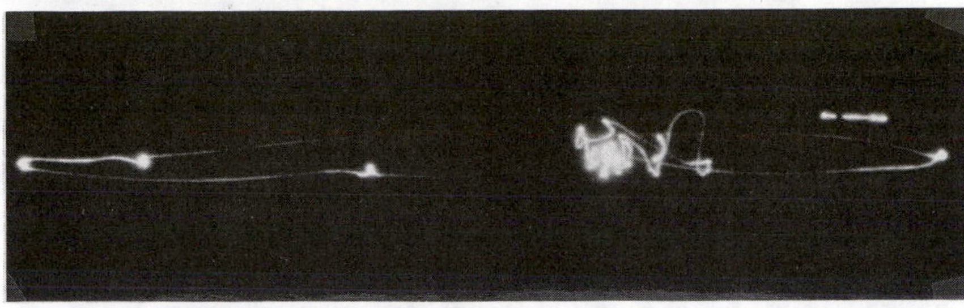

A time-exposure photograph of one of the famous Marfa lights. See text for details. (Copyright James Crocker)

CHECKING OUT SOME TEXAS GHOST LIGHTS

Some members of the Houston Association for Scientific Thinking (HAST) have visited the sites of the famed Marfa Lights (West Texas) and the less-publicized Saratoga Lights (East Texas). With binoculars, telescopes, and road maps, it was fairly easy for them to ascertain that the Saratoga Lights were simply the headlights of automobiles traveling along Route 787. The Saratoga display is a bit eerie but not at all mysterious, according to HAST.

The Marfa Lights turned out to be more impressive and, in consequence, quite a tourist attraction. The favorite viewing site is on Highway 90, 9 miles east of Marfa. HAST logged a total of 9 hours of observation there on three successive nights. All of the lights observed were easily attributed to cars traveling north from Presidio to Marfa. People at the viewing site who knew of the Presidio-Marfa road had no trouble identifying the lights as those of automobiles. But those unaware of the road called the lights mysterious. As for the frequent reports of Marfa lights cavorting and executing strange maneuvers, HAST thought they were probably due to low-flying aircraft in the neighborhood of the Chianti Mountains some 40 miles away. In fact, just such a plane was observed during a daylight trip to Shafter, a town near the mountains.

Admitting that the Marfa Lights are indeed entrancing and even mildly mystical, the report closes (rather incongruously for an admittedly skeptical writer) with:

> A reminder that caution must be taken. Because what we saw four nights in Saratoga and three nights in Marfa did not go out of the bounds of the ordinary does not mean that the extraordinary has never occurred in either place.

(Lindee, Herbert; "Ghost Lights of Texas," Skeptical Inquirer, 16:400, 1992.)

Comment. Previous descriptions of the Marfa lights (See earlier entries.) seem to portray phenomena much more "extraordinary" than automobile headlights! For example, one of the "lights" is said to have approached to within a few feet of a car and then followed it for a couple of miles.

THE MIN MIN LIGHT

The Min Min light occurs in the legends of the aborigines of Australia and in modern traveller's tales. In Queensland:

> ...around Boulia and Winton, there appears from time to time an unmistakable light---a luminous fluorescent shape that fades and brightens, recedes and advances across the flat never-ending plain. It has mystified men for centuries. It fascinates. It begs you to follow. And it can be eerie and frightening on that lonely dark plain at night.

The Min Min light is reputed to be oval in shape and to move in irregular circles and spirals. It has been seen close to the surface and as high as 300 meters. Riders claim that their horses are not disturbed by it. (Shilton, Pam; "The Min Min Light," Journal of Meteorology, U.K., 8:248, 1983.)

Comment. Nocturnal lights rarely make the scientific journals, so it is a pleasure to discover an item on the famous Min Min light.

EXPERIMENTS ON BROWN MOUNTAIN

Brown Mountain, in North Carolina at the end of the Blue Ridge Mountain chain, is famous for its enigmatic nocturnal lights. In this article M.A. Frizzell summarizes the most important attempts to come to grips with this phenomenon during the past 70 years. He concludes by describing recent experiments conducted by The Enigma Project and the Oak Ridge Isochronous Observation Network (ORION). Rather than repeat once again the older published observations, let us concentrate on the Enigma/ORION work.

In May 1977, ORION placed a 500,000 candlepower arc light in Lenoir, 22 miles east of Brown Mountain. Simultaneously, a group of observers gathered on an overlook on Route 181, 3.5 miles west of Brown Mountain, a favorite spot for watching for the Brown Mountain lights. Brown Mountain itself was interposed between the arc light and observers. When the arc light was switched on, the observers saw an orange-red orb hovering several degrees above the crest of Brown Mountain. Conclusion: the majority of the so-called Brown Mountain lights, particularly those seen above the crest, are refractions of artificial lights.

The real Brown Mountain lights, the mysterious ones, are those that flit through the trees well below the crest. These lights are extremely rare. Typically, they commence as a brilliant blue-white or yellow light, which tapers off to dull red before disappearing, all in 2-10 seconds. Horizontal motion is often only a degree or so, although some older reports have the lights wandering greater distances at speeds faster than a human could manage in the difficult terrain.

In an experiment to determine whether the "true" Brown Mountain lights might be seismic in origin, ORION detonated small charges on Brown Mountain in July 1981. No artificially stimulated lights were recorded. (Frizzell, Michael A.; "Investigating the Brown Mountain Lights," INFO Journal, no. 43, p. 22, Jan./Feb. 1984.)

CONCENTRIC, ROTATING LUMINOUS RINGS SEEN IN SWEDEN

January, 1991. Near Borlange, Sweden. One evening near the end of the month, four observers watched what seemed to be a system of luminous rings for about 15-20 minutes. The rings were estimated to be 2-5 kilometers away and 10-30 meters in size. E. Witalis has provided a direct translation of their report:

> The light phenomenon, which was strongly luminous, consisted of scintillating, rotating rings arranged as a system of plates (small rings in the centre and increasingly large ones towards the edges). The rings seemed to consist of sparks running around in a circular or slightly elliptical path with an impressive speed. Spontaneously I associated them with electrical discharges. The plate shape was felt as being in some way static as to location, and the sparks moved in a pattern without spreading out at the edges. It seemed that strong force was controlling the event. The sparks looked like having a core and a "tail" behind. They looked like being able to crackle but no sound was heard, perhaps because of the distance. It was interesting to note that the rings were spinning in opposite directions. The effect clearly illuminated the surrounding air space. The whole light phenomenon took place with knife-sharp edges in clear weather (temperature between -1 and -3 deg. C.

(Witalis. Erik; "An Air Plasma Whirl, Seen in the Atmosphere, Sweden, January 1991," Journal of Meteorology, U.K., 16:273, 1991.)

OFFICIAL FOO-FIGHTER RECORDS REVEALED

The famous foo fighters of World War II were bright balls of light, about a foot in diameter, of different colors, that appeared mostly over Germany to both German and Allied pilots. Although the foo fighters could maneuver around and through bomber formations with apparent ease, they were nuisances rather than physical threats. Most of the foo-fighter reports made by Americans came from the 415th Night Fighter Squadron. Recently a microfilm roll containing the Unit History and War Diary of the 415th was obtained from the U.S. Air Force. We quote below three incidents found on Frames 1613 and 1614. The year is 1944:

> December 18. In Rastatt area sighted five or six red and green lights in a 'T' shape which followed A/C thru turns and closed to 1000 feet. Lights followed for several miles and then went out. Our pilots have named these mysterious phenomena which they encounter over Germany at night 'Foo-Fighters.'
>
> December 23. More Foo-Fighters were in the air last night...In the vicinity of Hagenau saw 2 lights coming toward the A/C from ground. After reaching the altitude of the A/C they leveled off and flew on tail of Beau (Beaufighter---their aircraft, Ed.) for 2 minutes and then peeled up and turned away. 8th mission---sighted 2 orange lights. One light sighted at 10,000 feet the other climbed until it disappeared.
>
> December 28. 1st patrol saw 2 sets of 3 red and white lights. One appeared on the port side, the other on starboard at 1000 to 2000 feet to rear and closing in. Beau peeled off and lights went out. Nothing on GCI scope at the time
>
> Observed lights suspended in air, moving slowly in no general direction and then disappeared. Lights were orange, and appeared singly and in pairs. These lights were observed 4 or 5 times throughout the period.

There is no evidence as yet that any of the World War-II combatants had anything in their arsenals that could have accounted for the foo fighters. (Greenwood, Barry; Just Cause, no. 32, p. 1, June 1992. Address: P.O. Box 218, Coventry, CT 06238)

Comment. If the foo fighters were a natural phenomenon, one would expect at least a few modern reports from the thousands of commercial and military aircraft in the skies.

A CURIOUS SIGHTING

September 29, 1991. Winchester, England. A wet day, but the rain had stopped and the wind velocity dropped. The 'subject' is a B. Brumpton. His actual words are in single quotes, as told to B. Hayes:

> A high easterly breeze was blowing along New Road which runs east-west and is about 400 yards long. The subject was 30 yards from the western end facing east. He first noticed the 'object' at approximately 150 yards, at which his reaction was that he was 'seeing things'. The object 'filled the highway', so this suggests a width of eight metres or so. It was 'on the ground' and 'round-topped', suggesting to me a hemisphere. The 'object' was a 'mass of mist' and 'looked very wet'. It seemed to 'roll' toward the subject at a speed that he estimated at 30 m.p.h. However, his estimate of half-a-minute for travelling 150 yards gives a speed of 20 m.p.h. Let us not forget that time is difficult to estimate after an event.
>
> The object emitted a noise 'like very heavy rain pounding on the road', except that it was not raining at the time, and the subject became concerned about 'getting soaked'. Also, the 'mist' was clearly visible in spite of the darkness. This suggests the possibility that the 'object' was luminous. The observer moved into a driveway on the south side of the road, but when a few yards away 'the object moved over' to the north side and 'just vanished'.

(Hayes, Brian; "A Curious Sighting, 29 September 1991," Journal of Meteorology, U.K., 17:346, 1992.)

IGNIS FATUUS IGNORANCE

A.A. Mills, a British scientist, has had the courage to research will-o'-wisps, those greatly neglected luminous phenomena frequenting marshy places. His literature search confirms the reality of these cold flames, though they seem to be reported only rarely in modern times. Actually, today's science tends to laugh off will-o'-the-wisps as old wive's tales or as misidentifications of St. Elmo's Fire or Ball Lightning. At the best, will-o'-the-wisps are considered simply the spontaneous ignition of marsh gas---a trivial phenomenon not worth wasting time on.

Flames from spontaneously ignited marsh gas erupted from an English mud flat in 1902.

Mills' study, however, shows this condescending attitude to be far off the mark. He has experimented with marsh gases, even constructing his own controlled "swamp," and has been unable to duplicate the established characteristics of will-o'-the-wisps; i.e., spontaneous ignition, cold blue flames, no significant odor, etc. The marsh gas theory does not seem to hold water, despite many chemical variations. (Mills, A.A.; "Will-O'the-Wisp," Chemistry in Britain, 16:69, February 1980.)

EARTHQUAKE AND SUBTERRANEAN FIRE

1738. Month and day are unknown. Pecs, 46° 06' N, 18° 15' E.I = 7.5 - 8°.

> A very mysterious event was observed. Near Pecs, a black cliff named Szennyes (=dirty or filthy) belched forth fire during three days, accompanied by an earthquake that was strongly felt in the town of Pecs. After this shock fires were also seen on many occasions from the cliff. Naturally it was not a volcanic manifestation. The nature of this particular event is perfectly obscure. (Note by the author, P.H. "An earthquake light is improbable because of the long time during which the phenomenon was seen.

(Hedervari, Peter; "Unusual Phenomena Associated with Earthquakes within the Carpathian Basins," Compilation from personal files, 1983.)

Comment. Compare this phenomenon with the detonations and flashes from Old Hannah's Cave, Staffordshire, England, as reported below.

OLD HANNAH'S EXPLOSIONS

December 10, 1899. Staffordshire, England, near Old Hannah's Cave. Two men heard explosions like rifle shots.

> Realizing that no one was shooting, they looked up the cliff and witnessed an explosion which emitted a flash from a hole or fissure in the upper part of the cliff. This had a bluish column 'not of steam or fire or smoke, but apparently of aqueous vapour,' which travelled with immense force across the valley (approximately 12 m wide). Within minutes another discharge from higher up the cliff and then 'several ones with crackling sounds producing semi-transparent wavy streaks in the air, not smokey in appearance.' Next came a very loud explosion which 'we had the good fortune to see plainly.' Wardle describes this as 'like a gun but with crackling, a series of continuous reports, cleaving the air in a zigzag or riverlike course in a narrow band about 15 cm to 20 cm broad, of bluish colour.
>
> Several other reliable descriptions exist of detonations and flame-like discharges around old Hannah's Cave. The supposition is that natural gases liberated by decaying organic material and, perhaps, geochemical reactions are ignited by static electricity. A recent landslip seems to have extinguished this curious phenomenon. (Pounder, Colin; "Speculations on Natural Explosions at Old Hannah's Cave, Staffordshire, England," National Speleological Society, Bulletin, 44:11, 1982.)

Comment. No one should overlook the similarity between Old Hannah's activity and the will-o'-the-wisps, earthquake lights, the Barisal Guns, mistpouffers, the Moodus Sounds, and other sound and light phenomena. See our Catalog Earthquakes, Tides, Unidentified Sounds.

EARTHQUAKE LIGHTS

Earthquake lights are flashes of light in the sky, ordinary lightning, ball lightning, aurora-like streamers, flames issuing from the ground, sky glows, and St. Elmo's fire observed in the general vicinity of an earthquake. Earthquake lights, like ball lightning, have been recorded since ancient times. Until recently, they have not been recognized as legitimate phenomena by scientists. One reason for this neglect is their great variety; another is the fact that earthquakes are frequently accompanied by electrical storms, the arcing of damaged power lines, and the ignition of ruptured gas lines. Real earthquake lights may be explained by the spontaneous ignition of natural gas released by the quake and, perhaps, by luminous phenomena created by piezoelectricity during crustal deformation.

EARTHQUAKE LIGHTS OBSERVED IN CANADA

Numerous earthquake lights (EQLs) were reported between November 1, 1988, and January 21, 1989, in the Saguenay region of Quebec. These luminosities were associated with 54 seismic shocks recorded in this area. Most were small, but a strong foreshock (magnitude 4.8) occurred on November 23; the main quake (magnitude 6.5) hit 60 hours later. Through appeals by radio and newspapers, 52 observers of EQLs were located. They reported a wide spectrum of luminosities, some of which were very strange. In the sky, some observed silent sparkings, diffuse glows, and aurora-like stripes. For an account of the more enigmatic EQLs, we quote M. Ouellet:

> Fireballs a few metres in diameter often popped out of the ground in a repetitive manner at distances of up to only a few metres away from the observers. Others were seen several hundred metres up in the sky, stationary or moving. Some observers described dripping luminescent droplets, rapidly disappearing a few metres under the stationary fireballs. Only two fire-tongues on the ground were reported, one on snow and the other on a paved parking space without any apparent surface fissure. The colours most often identified were orange, yellow, white and green. Some luminosities lasted up to 12 min.

(Ouellet, Marcel; "Earthquake Lights and Seismicity," Nature, 348:492, 1990.)

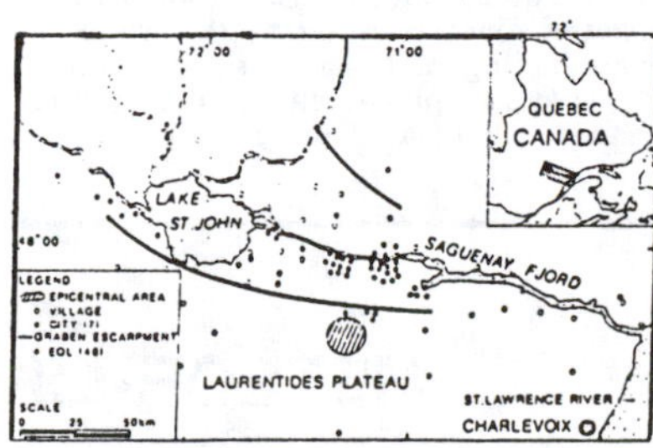

Locations of earthquake lights observed in the Saquenay Lake/Saint John region of Canada near the times of the November, 1988, quakes. The large cross-hatched circle identifies the epicentral area.

ARE NOCTURNAL LIGHTS EARTHQUAKE LIGHTS?

> Nocturnal lights reported from 1972 to 1977 by fire lookouts on the Yakima Indian Reservation, Washington, correlate with earthquake origin time, distance, magnitude, and depth within a 200 km radius of the observations. Photographs and eyewitness accounts show that the luminous phenomena (LP) near ground level appear to be spherical, are colored white to orange, and show little or no internal detail. Locations of the observations preclude explaining them as terrestrial vehicles or other human activity."

This abstract continues and notes that some of the luminous displays seem to be enhanced by geomagnetic activity. (Persinger, M.A., and Derr, J.S.; "Relations among Nocturnal Lights, Geomagnetic Activity and Earthquakes in Southern Washington," Eos, 64:762, 1983.)

THE ZEITOUN LUMINOUS PHENOMENA

> Between April, 1968 and May, 1971, hundreds of thousands of people reported complex luminous events over a church in Cairo (Zeitoun), Egypt. Many of these events were photographed. Most of the luminous phenomena (LP) occurred during 1969 when seismic activity within the radius of less than 500 km was approximately a factor of 10 greater than for any single year before or afterwards. Whereas the distribution of epicenters around Zeitoun was randomly distributed for the years 1966 through 1968 and 1970 through 1972, there was a significant focus of their frequency during the year 1970. Most of them occurred off the coast of Gemsa, approximately 375 km to the southeast of Cairo.

Analysis of the LP and seismic records demonstrated a significant increase in the number of LP during the month of, or the month before, increases in the number of earthquakes per month. The relationship between LP and quakes was not, however, as strong as it had been for episodes of luminous phenomena in Toppenish, Washington; the Uintah Basin, Utah; Carman, Manitoba; and the New Madrid region in the central US. Still, the Zeitoun phenomena must be considered as supportive of the hypothesis that many LPs are associated with tectonic strain in the earth's crust. (Derr, John S., and Persinger, Michael A.; "Temporal Association between the Zeitoun Luminous Phenomena and Regional Seismic Activity," The Explorer, 4:15, October 1987.)

THE ZEITOUN APPARITIONS

The luminous phenomena observed at Zeitoun, Egypt, have attracted the attentions of a wide spectrum of believers and nonbelievers. Each group seems to interpret the phenomena according to its own particular mind-set!

To set the stage for the study reviewed here, we quote first the two lead paragraphs of another paper:

> Between April 1968 and May 1971 hundreds of thousands of people reported seeing apparitions of the Virgin Mary over a Coptic Orthodox Church in Zeitoun, near Cairo, Egypt. When photographed, these phenomena appeared as irregular blobs of light. Primarily there were two types of events: small short-lived, highly kinetic lights ('doves') and more persistent coronal type displays that were situated primarily over the apical structures of the church. More detailed descriptions of the phenomena, such as visions often occurred as 'flashes'; their details usually reflected the religious background of the experient.
>
> The characteristics of these luminous phenomena strongly suggested the existence of tectonic strain within the area. According to the hypothesis of tectonic strain, anomalous luminous phenomena are generated by brief, local changes in strain that precede earthquakes within the region. Psychological factors determine more elaborate details of the experiences because there are both direct stimulations of the observer's brain as well as indirect contributions from reinforcement history.

The authors of the study at hand, J.S. Derr and M.A. Persinger, are well known for their theory associating anomalous, terrain-related, luminous phenomena with tectonic strains. In the Zeitoun case, they have discovered that a year before the phenomena commenced there was an unprecedented increase (by a factor of ten) in seismic activity some 400 kilometers to the southeast. Also, there was a moderate (0.56) correlation between the luminous phenomena and increases in seismicity during the same or preceeding months. Derr and Persinger claim these observations support their hypothesis. (Derr, John S., and Persinger, Michael A.; "Geophysical Variables and Behavior: LIV. Zeitoun (Egypt) Apparitions of the Virgin Mary as Tectonic Strain-Induced Luminosities," Perceptual and Motor Skills, 68:123, 1989.)

Comment. It does seem, however, that seismic activity 400 kilometers away would have little effect at Zeitoun. How can luminosity be produced from an earthquake that far away?

ELECTROMAGNETIC RADIATION FROM STRESSED ROCKS

Anomalous electromagnetic radiation has been detected prior to and during several recent earthquakes. T. Ogawa and his colleagues have surveyed these observations briefly and then repaired to the laboratory where they put rocks through the third degree. They concluded:

> It has been demonstrated by the sample rock experiments that the ordinary crustal rocks produced electricity when they were shocked or fractured and radiated EM waves in the frequency range of 10 Hz to 100 kHz. The electric dipole moment to radiate the EM waves was estimated to be 10^{-14} C m.

(Ogawa, Toshio, et al; "Electromagnetic Radiations from Rocks," Journal of Geophysical Research, 90:6245, 1985.)

Comment. What was not mentioned by Ogawa et al is that it is just this kind of seismoelectricity that may be responsible for many of the sightings of earthquake lights and maybe even some so-called UFOs. See GLD8 in Lightning, Auroras, Nocturnal Lights.

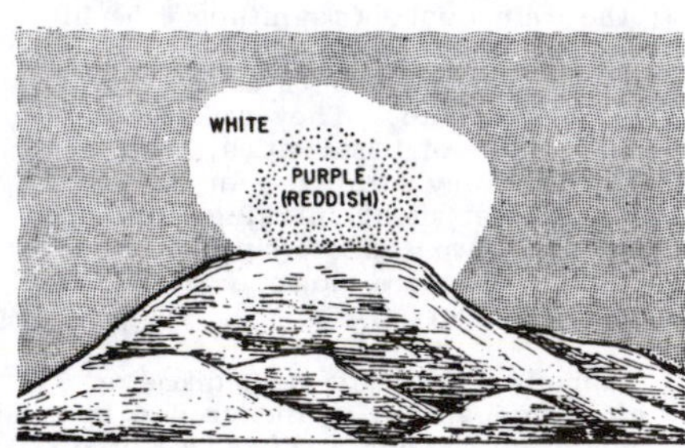

A luminous display atop Mt. Noroshi, Japan, during the 1966 earthquake swarm.

EARTHQUAKE LIGHTS AND CRUSTAL DEFORMATION

Hedervari supports the hypothesis that some earthquake lights, particularly those preceding strong regional quakes, are caused by the release and ignition of gases from the stressed rocks. Several curious features of earthquake lights favor this assertion:

1. Prequake lights are regional in character corresponding to the widespread flexing of the strata. (In the 1933 Japanese quake, earthquake lights were seen along a 1000-km arc);

2. There is no correlation between the earthquake epicenter and the location of earthquake lights. (In the 1977 Romania quake, the epicenter was east of Cluj but the earthquake lights lit up the western horizon.

(Hedervari, Peter; "The Possible Correlations between Crustal Deformations Prior to Earthquakes and Earthquake Lights," Seismological Society of America, Bulletin, 71:371, 1981.)

Comment. In essense, Hedervari is saying that earthquake lights often do not occur where rock stresses are greatest and that the piezoelectric effect may not be the whole story.

FLUID INJECTION CAUSES LUMINOUS PHENOMENA

> The first recorded sighting of earthquake lights (EQL) dates from 373 BC in Greece. The same report mentions extensive underground rivers, but it has taken over 2300 years and the development of statistical methods to suggest a connection among fluid pressure, earthquakes, and geophysical luminosities. Many of the sightings are treated as mystical experiences, depending on local cultural values. In Denver and Rangely, Colorado, and Attica, New York, these sightings correlate with earthquakes and injection of fluid into the earth for waste disposal or secondary oil recovery. In the New Madrid, Missouri, area, luminosities are highly correlated with flooding on the Mississippi River and tend to occur 9 months after high water. Enough luminosities, and radio emissions in the ULF band, are observed weeks to months before earthquakes to suggest that they be tested as a possible forecasting tool for the select places where they occur. The pattern of occurrence may delineate the progress of tectonic strain and so indicate the direction or even location of a future epicenter. Fluid moving through developing cracks may be the source of electrical energy which powers the EQL. A number of potential mechanisms should be considered, involving tectonic strain, exoelectron emission, streaming potential, EM excitation of water droplets, and the fault zone as an EM waveguide.

(Derr, John S., and Persinger, Michael A.; "Fluid Injection Causes Luminous Phenomena," paper presented at the 11th Annual Meeting, Society for Scientific Exploration, Princeton, NJ, June 11, 1992.)

METEOR-LIKE PHENOMENA

Most meteor anomalies occur at very high altitudes and belong, therefore, to the field of astronomy. As such, they are treated in the Astronomy chapter. There remain, however, varieties of meteor-like luminous phenomena that apparently transpire at very low altitudes and can be legitimately be included here. Predominant among these low-level meteor-like phenomena are luminous and nonluminous objects that do not follow the usual meteor protocol by burning up in the atmosphere or impacting the earth's surface. These wayward "meteors" are scientifically perplexing.

MARSH GAS OR THE PLANET VENUS?

October 16, 1976. Aboard the m.t. Farnella, Barents Sea fishing grounds. Observers, Skipper H. Powdrell and Mr. G. Christmas, Radio Officer.

"At 2307 GMT while I was visiting the wheelhouse, the Skipper pointed out to me an object flying across the sky. It had already been in view for some five minutes or more and was first observed on a bearing of 140°T heading due North. I first sighted it on a bearing of 050°T.

"Observation was constantly kept by myself and the Skipper with the aid of binoculars from the time I first sighted the object. It could be described as being a brilliant light travelling at a very high altitude, leaving a bright V-shaped trail of rays which could be likened to the sun's rays as they would appear from behind a cloud. However, they were very much smaller due to the height and were also horizontal. The object followed a course from south to north to be astern of us at 2308. It then commenced to come back along its course while losing altitude. I would point out here that there was no visual evidence of the object actually turning back but rather as though it had been put into reverse.

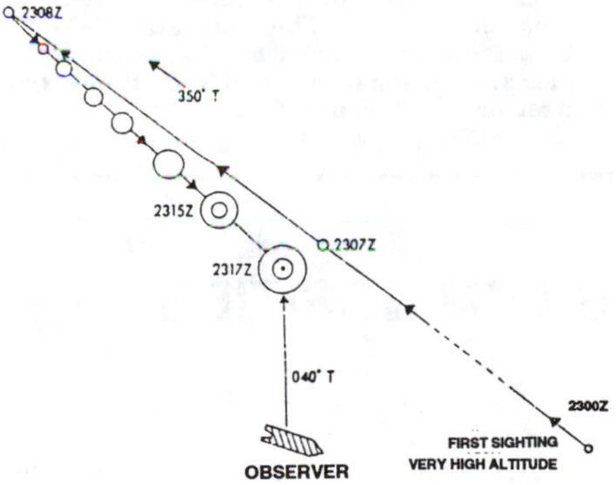

Apparently reversed path taken by an object observed over the Barents Sea.

"The appearance and shape of the object was now changed, becoming totally circular in shape, still losing height and coming closer. The outer edge of the circle I would describe as glowing and within that was another circular object, more intense, and within that was a brilliant pulsating white light as when the object was first sighted. The object reached its closest point to us by 2317 on a bearing of 040°T.

"The object stayed in this position for approx. two minutes and then vanished within the outer glow, this glow finally fading from our sight also. At 2320 nothing was left to be seen of either the object or the glow.

"I have tried to reproduce what the Skipper and I saw in sketch form. The

Phenomenon as seen at 2315Z. Brilliant pulsating light at center.

object was also seen by several other vessels who were fishing in the area with us. The night was fine with a small amount of low cloud, a quarter moon and an average number of stars. Position of ship: 69° 56'N, 33° 46'E."

(Powdrell, H.; "UFO," Marine Observer, 47:177, 1977.)

UNIDENTIFIED OBJECT

April 9, 1983. North Atlantic Ocean, from the n.v. Dorsetshire.

At 2304 GMT, Mr. Haney pointed out a bright white object in the sky. It was bearing approximately 360° (T) at an elevation of about 40°. It was moving rapidly southwards across the sky, leaving a bright trail behind it, like an afterglow. Also trailing astern of the object was a light trail of sparks (possibly large solid particles). The object disappeared behind clouds, bearing about 170° (T) at an elevation of approximately 35°, and lighting the edges of the clouds. The time taken for the passage was around 20 seconds. It was obviously a very large object, judging from its apparent size as seen from sea level. The impression given was that of an object within the atmosphere, easily showing around a one-penny piece held at arm's length.

(Edwards, R.A.F.; "Unidentified Flying Object," Marine Observer, 54:82, 1984.)

Comment. This seems to be a description of a large fireball, but the direction of flight (north-south) is unusual, and the time-of-passage (20 seconds) extremely long.

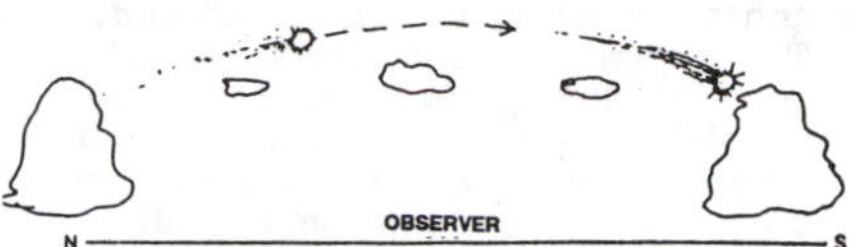

A large, very sluggish fireball trailing sparks over the North Atlantic.

THE SO-CALLED GREEN FIREBALLS OF 1948-1949

Under the recent law making most government records available to the public, B.S. Maccabee obtained the FBI's UFO file. His analyses of this file have been serialized in the APRO Bulletin. One of the most unsettling revelations concerns the FBI data on the notorious "green fireballs" of the 1948-1949 era. According to the verbatim transcript of the FBI record, dated January 31, 1949, File No. 65-58300, the phenomenon was concentrated over top secret facilities in the American southwest, including Los Alamos, New Mexico. Many sightings were made by highly creditable witnesses.

Briefly, the "fireballs" were a brilliant green, sometimes beginning and ending with red or orange flashes. The objects travelled mainly on an east-west line at an average speed of 27,000 miles per hour. They seemed to pass over in level flight at altitudes of six to ten miles. On two occasions vertical changes of course were noted. Size was about one-fourth the diameter of the full moon. Multiple fireballs appeared in two instances. No sound was ever noted. No debris was ever discovered. (Maccabee, Bruce S.; "UFO Related Information from the FBI File," APRO Bulletin, 7, March 1978.)

SKY FLASHES

Two classes sky flashes are covered in this section. The first consists of transient brightenings of all or most of the sky. These all-sky flashes usually last from one to several seconds and are often bluish. The surrounding landscape may be lit briefly as bright as day. It is easy to blame these flashes on bright meteors, perhaps meteors hidden by clouds. In many instances, though, the sky is clear, and this explanation fails. The second type of sky flash is much shorter and point-like. Some of these flashes may have astronomical origins; others have been identified as satellite glints; that is, reflections of the sun from specular surfaces. Even so, the final item in our little collection confirms the existence of still unexplained flashes originating in the upper atmosphere.

ANOMALOUS SKY FLASH

December 28, 1980. In the South Atlantic.

> At approximately 2245 GMT on a moonless night the entire ship and immediate surrounding area were illuminated by what can be best described as a great camera flash. The flash was bluish-white and a small bolt of lightning appeared to be centred just above the vessel's samson posts. No noise was heard and the flash lasted only a second. The sky was clear at the time and stars of all magnitudes were clearly visible. The only clouds that could be seen were two or three small cumulus clouds; one of these was above the vessel and the others were moving towards us from the south, our course being 142°(T) and the wind being S'E, force 3. The cloud above the vessel was at a height of about 600 feet.

(Rutherford, N.W.C.; "Unidentified Phenomena," Marine Observer, 51:186, 1981.)

Comment. This was obviously not ordinary lightning, but the small cloud and small bolt of lightning indicate some sort of anomalous electrical discharge. The literature contains many other reports of bright sky flashes that cannot be attributed to meteors, heat lightning, or other sources.

LIGHT FLASHES OVERHEAD

January 23, 1984. Salonta, Romania.

> During the whole day we experienced snow which finally stopped at 23h 00m. The temperature was 1°C, when I experienced a few pale flashings from the top of the sky. The lights were seen from 18h 15m to 18h 30m, Universal Time. Their number was four, and they appeared with equal intensity with pauses in between, each pause lasting for 3-5 minutes. All were white in color. At 18h 31m, a new flashing was seen; this was the fifth flash in the series, and it had higher intensity than the previous ones. I have not been able to discover the source of these strange lights. Each flashing was very shortlived, lasting less than a second.

The observer in Salonta was Attila Kosa-Kiss. (Hedervari, Peter; "Further Observations of Atmospheric Light Phenomena of Unknown Origin," personal communication, February 15, 1984.)

GREEN SKY FLASHES

March 25, 1984. Indian Ocean.

"Two successive 'green flashes' were observed. The first, at 2100 Ship's Time or 1530 GMT, was a bright green and bore 240° at an altitude of 75°; it moved vertically downwards to an altitude of approximately 20° over a period lasting about 3 seconds. The second flash was observed at 2250 Ship's Time. It was green/white and was first observed at an altitude of 40°, bearing approximately 340°. It moved diagonally across the sky before disappearing behind low cloud at an altitude of 30°, bearing 310°. This time the duration was 1-2 seconds. In both cases the ship's radars were turned on but nothing was observed other than rain showers between 4 and 12 n. mile from the ship, mainly forward of the beam. Both flashes were of about the same brightness as that of lightning, the first being brighter than the second. In both cases it was difficult to judge the distance. The phenomenon was thought to have possibly been some form of lightning as its appearance was unlike that of any flare and in both cases the distance from the ship did not appear great enough to be compatible with a meteor or other object entering the earth's atmosphere." (Aston, A.; "'Green Flashes'"; Marine Observer, 55:30, 1985.)

Comment. Probably low-altitude fireballs.

MYSTERIOUS SPATE OF SKY FLASHES

B. Katz and a small group of Canadian amateur astronomers have accumulated a total of 14 bright flashes in Aries in just a year or so. "Point" meteors (meteors seen head-on) usually appear as flashes like this, but to see 14 in the same region of the sky in such a short span of time is truly remarkable. (Katz, Bill; "Chasing the Ogre," Astronomy, 13:24, April 1985.)

Comment. These flashes were finally determined to be glints (solar reflections) off an artificial satellite. See pp. 59 and 86 for the "Perseus flasher."

ANOMALOUS OPTICAL EVENTS IN THE UPPER ATMOSPHERE

A photodiode was recently lofted to high altitudes (over 400 kilometers) in an experiment to measure the optical power of lightning. During the 10-minute flight, more than 500 lightning-related events were recorded over that part of the globe visible from the rocket's altitude.

> Among these is a class of about 23 events all having an anomalous signature, with obvious clustering of optical impulses or continuous emissions, and resulting durations of several hundred milliseconds. Such durations are much longer than typical for lightning-related events recorded at the rocket, which are more frequent overall. Every anomalous optical event (AOE) was accompanied by broadband VLF signals of a distinctive character...In considering possible sources above 30 km we find that the AOEs do not seem to resemble other natural optical phenomena, such as meteors which burn up well above 30 km in the mid-latitude atmosphere.

(Li, Ya Qi, et al; "Anomalous Optical Events Detected by Rocket-Borne Sensor in the WIPP Campaign," Journal of Geophysical Research, 96:1315, 1991. Cr. C. Rush.)

Comment. Apparently these anomalous "flashes" have not yet been detected from the ground. The implication is that there are many more high-altitude electrical discharges than scientists expect or can account for.

MISCELLANEOUS LUMINOSITIES

LUMINOUS RIPPLES MOVE THROUGH THE NIGHT SKY

A.W. Peterson, during his studies of nighttime airglow in the infrared, has reported three events also invisible to the naked eye. The most spectacular event occurred on the night of April 4-5, 1978, when luminous ripples were observed at about 90 kilometers altitude moving at 91 meters/second, with a crest-to-crest wavelength of 16 kilometers. The precise source of the visible light is still in doubt as is the identity of the stimulus causing the glowing ripples. Peterson has noted some correlation between the ripples, both visible and infrared, and the lunar high tide in the atmosphere. Gravity waves could thus be the stimulus creating the ripples. (Peterson, Alan W.; "Airglow Events Visible to the Naked Eye," Applied Optics, 18:3390, 1979.)

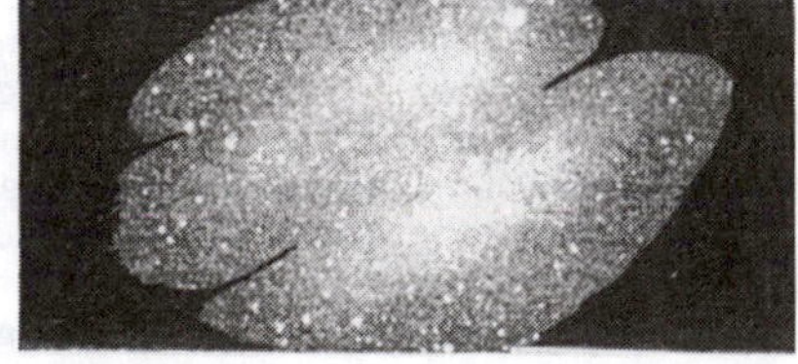

METEORIC NIGHT-GLOW

During some intense meteor showers, such as the Leonids in 1866 and the Bielids in 1872, observers noted a faint diffuse glow of the night sky in the direction of the meteor shower radiant. The glows were aurora-like but no sunspot or magnetic activity was noted. A New Zealand scientist, W.J. Baggaley, has suggested that these strange glows were caused by sunlight scattered from huge clouds of fine meteoric dust accompanying the meteor swarms. (Anonymous; "Meteoric Night-Glow," Sky and Telescope, 55:485, 1978.)

Comment. This item is closely related to the many observations of luminous skies and, in particular, the vivid sky glows following the Tunguska Event of 1908. There may also be a connection with the highly variable behavior of the not-so-well-understood zodiacal light.

LOW-LEVEL AURORA?

June 10, 1982. Near Sturgis, Michigan.

About 3:00 A.M., two young women were driving in a semi-rural area. Fog made visibility poor. It began to rain----a brown jelly-like slime that smeared the windshield. A rotten-egg odor pervaded the area. The car engine stopped, and the two began to walk to find assistance. After 50 yards, they encountered millions of small rays of "lightning" flashing everywhere. They were 2-3 feet long and reached high into the sky. Looking back toward the car, they saw a reddish fluorescent glow with streams of light coming down from the sky to the glowing region. Grass and weeds along the roadside were standing straight up and glowing. Deep-red lines of light were seen dancing on the road. They returned to the car, and it felt hot to the touch! Soon, clouds moved in and the display was over.

The authors of this article personally investigated this event within a few days of its occurrence. They found the two witnesses obviously very shaken, but believe that the accounts are fresh and unadulterated.

Also pertinent is the fact that a large solar flare had just occurred, and intense auroral displays had been predicted. Also, the two women were apparently the only witnesses of this phenomenon. (Swords, Michael D., and Curtis, Edward G.; "Atmospheric Light Show," Pursuit, 16:116, 1983.)

Comment. The article also contains the authors' analysis in some depth. Basically, they thought the phenomenon to be auroral with coincidental rainfall containing organic debris.

GHOSTLY WHITE DISK AND LIGHT BEAM IN SKY

June 22, 1976. 2113 GMT, position 24°N, 09'W. Officers aboard the s.s. Osaka Bay observed a pale orange glow coming from a cumulus bank to the west. At 2115, a ghostly white disk appeared above the cloud bank and began expanding. Ten minutes later, the disk had grown to the point where the lower edge touched the horizon and the upper edge reached an elevation of 24° 30'. As the disk sank below the horizon a searchlight-like beam emerged from the clouds. The disk disappeared completely by 2140, but the light beam remained another five minutes. (Moore, R.; "Unidentified Phenomenon," Marine Observer, 47:66, 1977.)

Comment. Some distant rocket launchings have created similar phenomena.

EXPANDING BALL OF LIGHT (EBL) PHENOMENON

In the latest number of the Journal of Scientific Exploration, R.F. Haines presents a summary table of 15 cases of a luminous phenomenon he has dubbed the Expanding Ball of Light or EBL. EBLs are very large, sometimes occupying much of the sky. They seem to occur everywhere, though rarely. Haines elaborates:

> According to several pilot witnesses, the center of the EBL is at relatively high altitude while it is forming. Its color is evenly whitish or yellowish and becomes increasingly transparent to background stars as it expands. As it enlarges it appears to maintain a sharply defined edge. At some point it fades completely from sight. The rate of boundary interface expansion is impossible to determine without knowing its distance from the observer. It is also of interest to note that most EBL events have taken place after dark. If EBL phenomena are associated with an advanced weapons test, one wonders why it would be conducted (a) after dark, and (b) in so many different geographic areas.

(Haines, Richard F.; "Expanding Ball of Light (EBL) Phenomenon," Journal of Scientific Exploration, 2:83, 1988.)

Comment. In our Catalog volume Lightning, Auroras, Nocturnal Lights, we describe many phenomena of the EBL type under GLA15. The accompanying drawing is taken from this Catalog volume.

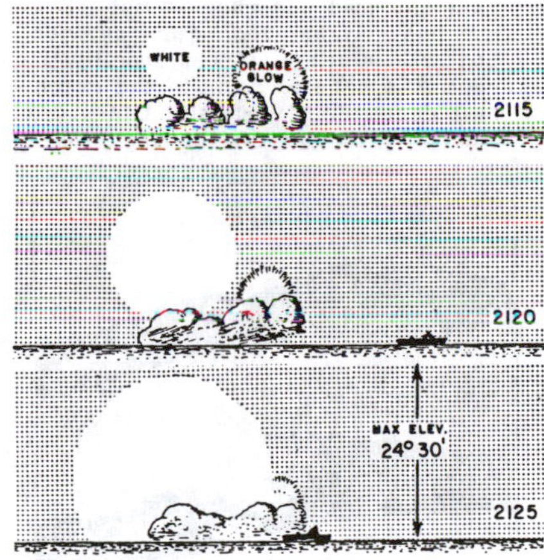

June 22, 1976. North Atlantic. At 2113 GMT, a pale orange glow was seen coming from a bank of cumulus in the west. At 2115, a ghostly white disc appeared at about 10° altitude, bearing 290°. The disc and glow combined as shown. By 2140, the disc had disappeared. Stars could be seen through the disc at all times.

CURIOUS PATCHES OF LIGHT ON THE HORIZON

March 24, 1977. Aboard the m.v. Kinpurnie Castle. Captain M. Brackenbridge. Cape Town to Antwerp. Observers, the Master, Mr. C.A. Neave, 3rd Officer and Mr. T.J. Martel, Radio Officer.

> At 0855 GMT the look-out observed what appeared to be a searchlight shining downwards for about 10 seconds on a bearing of 300°T and 20° above the horizon. This light was extinguished and was replaced by a luminescent patch of approximately one degree in diameter. A semicircular area of overall moderate luminosity formed about the luminescent patch. This took about three minutes to form and the dimensions are shown in the sketch. When this had formed, another luminescent patch was also observed above the semicircular area and after a total period of seven minutes the phenomenon dispersed completely. Weather conditions were as follows: dry bulb 19.0°C, wet bulb 17.0°C, barometer reading 1016.7 mb, good visibility, no cloud. Position of ship: 23° 05'N, 17° 25'W.

(Brackenbridge, M.; "Unidentified Phenomenon," Marine Observer, 48:21, 1978.)

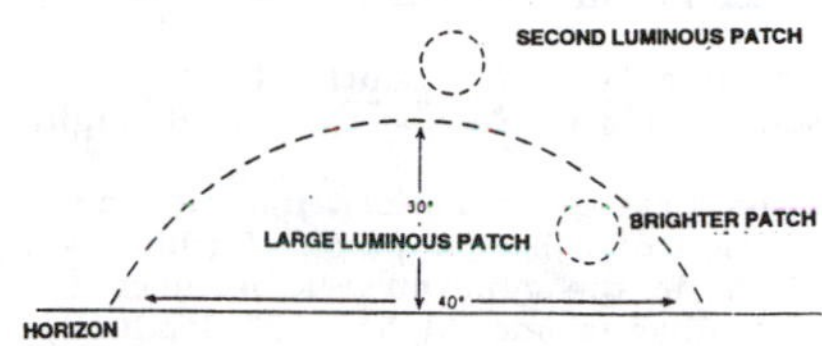

Patches of light seen near the horizon over the Atlantic in 1977.

UNIDENTIFIED LIGHT

January 13, 1991. Caribbean Sea. Aboard the m.v. Trinidad and Tobago.

> At 0210 UTC whilst the ship was proceeding eastwards along the north coast of Trinidad, a relatively bright patch was noticed in the almost cloudless sky and was thought to be a cluster of stars. Its bearing was approximately 300° at an elevation of about 50°, and closer inspection through binoculars revealed a rather strange phenomenon, as shown in the sketch.
>
> The bright patch was a perfect circle of a bright, light-blue colour and was transparent as stars could be seen through it. There was also a trail from the circle which looked like the track a disc would describe if it moved through an arc. This trail was also of light-blue colouration but was not nearly so bright as the circle. The entire phenomenon dissipated after about five minutes.

(Knight, M.; "Unidentified Light," Marine Observer, 62:22, 1992.)

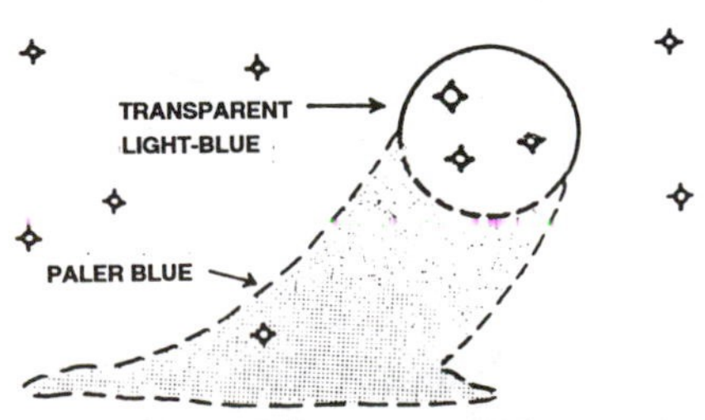

Strange optical phenomenon seen off the coast of Trinidad in 1991.

UNIDENTIFIED LIGHT EXPLAINED?

The strange optical phenomenon reported above may have been the consequence of a barium release from a NASA satellite. At 9:17 PM EST, on January 13, 1991, the Combined Release and Radiation Effects Satellite (CRRES) detonated a small canister of barium over South America. The greenish glow was visible from the U.S. southeast coast in the southwestern sky. (Suplee, Curt; "NASA Light Show 'Paints' Earth's Magnetic Field," Washington Post, January 14, 1991. Cr. D. Kreinbrink)

Comment: The observation in question was logged as occurring at 0210 UTC, January 13, so there is a time discrepancy that needs to be resolved here.

UNIDENTIFIED PHENOMENA

September 17, 1982. South Atlantic Ocean. 2103 GMT on a clear dark night.

> The first thing noticed was the formation of a bright patch of white light in the general area between Rasalhague and Alphecca. Gradually a dark eye formed in the centre of the patch in which shortly afterwards a very bright object appeared like a star of magnitude -2. After one or two seconds this object appeared to undergo a tremendous explosion and became a large bright orange gaseous fireball, which appeared to be hurled earthwards directly down the observer's line of sight, growing constantly larger and larger. One witness described the fireball as resembling rolling orange smoke. The ball then ceased to increase in size, giving the impression that it had stopped. Its orange colour rapidly gave way to rainbow colours which gradually gave way to white and faded in brilliance until all that remained were several patches of luminous white light, although these were impressive in their own right.

A similar phenomenon was noted the following night, although the ship was 7° farther south.

September 18, 1982. South Atlantic Ocean. From a different ship in the same area as the one above.

> The altitude of the first sighting was approximately 24°, level with the planet Jupiter and offset to its right. The six subsequent bursts were above the first, and slightly to the right, leaving a fantail of purple/white lenticular clouds which leaned to the right as shown in the sketch. Although they all kept their lenticular shape, the final burst did break up, giving the appearance of being in a gaseous state. Each burst commenced as a pinprick of bright, white light expanding rapidly to at least 2½ times the diameter of the sun (No. 1)....

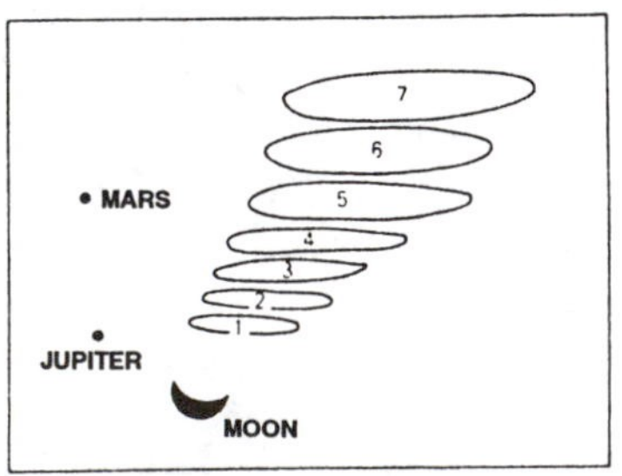

Sequence of optical phenomena seen at sea off Brazil in 1982.

(Anonymous; "Unidentified Phenomena," Marine Observer, 53:132, 1983.)

Comment. The above observations were made aboard ships 300-400 miles off Brazil. The only event correlated with the phenomena was a meteorological rocket said to have been launched on September 18. Rocket launches do provide spectacular luminous phenomena offshore from Cape Kennedy, but the above phenomena do not seem consistent with small meteorological rockets.

A LUMINOUS-TUBE PHENOMENON

Night of July 11-12, 1991, near Alton Barnes, Wiltshire, England. Three individuals were monitoring nearby fields for crop-circle phenomena. Instead, they observed a strange, but possibly related, luminous mass. R.L Goold described it in the following words:

> Suddenly, at 2.55 a.m., birds began singing which heightened our alertness and made us check wrist watches. It was soon quiet again, but at 3.00 a.m., almost exactly, I spotted a tube of light to the north-east descending vertically beneath a cloud in that part of the sky. Most of the remainder of the sky was clear and starry. The tube extended steadily in length as we watched, and its milky-white colour seemed to be due to a self-luminosity like one might expect from the electrical effect known as plasma. As it came down against the black sky and neared the ground, the tube began to broaden, and branched out to give two opposed arms, as indicated in the drawing, forming a design in the air with rounded ends. Then the tube dissipated from the top downwards, and disappeared into the horizontal arms which themselves proceeded towards the ground out of sight beyond the hill peaks. No noise was heard. The whole phenomenon lasted about six seconds.

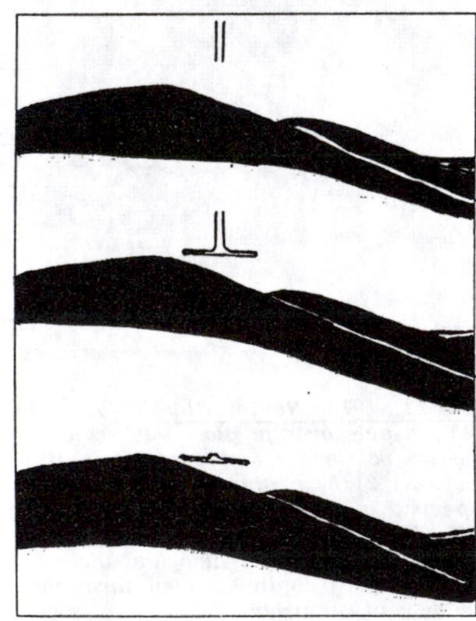

The luminous tube phenomenon. Doesn't the bottom sketch resemble a classical UFO?

The trio of observers used their fingers held at arm's length to estimate angular dimensions of the phenomenon. Using these figures and the known distances of the surrounding hills, G.T. Meaden estimated the distance of the phenomenon at 1,400 meters; the width of the tube at 16 meters; and the width of the entire luminous mass at roughly 100 meters. (Goold, Rita L.; "Observation of a Luminous-Tube Phenomenon at Alton Barnes, 12 July 1991," Journal of Meteorology, U.K., 16:274, 1991.)

INFRARED ATMOSPHERIC WAVES

When the night sky is photographed in the infrared portion of the spectrum, luminous wave-like structures appear in the upper atmosphere. The wave crests are about 50 kilometers apart; lengthwise, they stretch up to 1.000 kilometers; altitude, about 85 kilometers. As many as ten waves may be seen at the same time. Morphologically, these waves resemble noctilucent clouds, which are sun-illumined, high-altitude clouds. The infrared waves, however, appear when the sun is well below the horizon. Since these waves are seen only at low angles over the horizon, some geophysicists propose they are the result of a geometric effect produced by viewing a rippled layer of weakly emitting gases in the upper atmosphere.

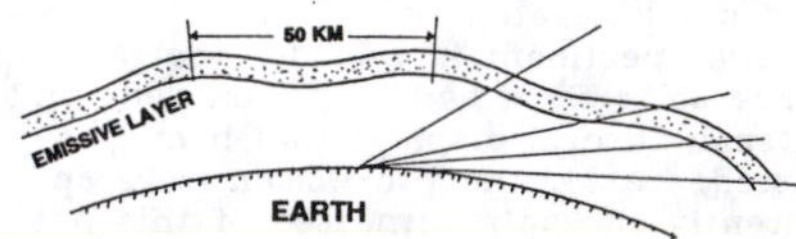

Rippled emissive layer around the earth. The variable optical path at low angles above an observer's horizon could create the effect of luminous ripples to someone on the ground.

When one looks at this rippled layer just above the horizon, one sees alternating thick and thin sections due to the perspective. The thick portions will appear brighter than the thin sections. As for the origin of this postulated rippled layer; no one is sure. Gravity waves may be involved. (Herse, M.; "Waves in the OH Emissive Layer," Science, 225:172, 1984.)

Comment. As described in our Catalog Lightning, Auroras, Nocturnal Lights, luminous atmospheric waves are, on rare occasions, visible to the naked eye. It is possible that the banded-sky phenomenon is related to the infrared waves. See p. 254.

CURIOUS LUMINOUS DISPLAY OVER THE PACIFIC OCEAN

Our general policy admits only those phenomena described in the scientific literature, but here we must make an exception. This account, submitted by a retired Air Force colonel, is unlike anything we have found before---either meteorological or auroral.

"During the last week of August 1970, I was on a trip from Viet Nam to Clark AFB, Philippines. We were cruising at 9000 feet. We had departed Camron Bay between 2 and 3 AM. Upon reaching cruise altitude, I noticed a rather unusual display in the sky. Our magnetic heading from V.N. to Clark was approximately 155°. The display appeared on the horizon in an easterly direction. It consisted of a series of dashes perhaps the color of the moon with a south-to-north directional flow. There was no visible beginning of this display, such as you would see in a comet.....

"The awesome feature of this display was its magnitude. I thought if whatever was creating the display collided with our Earth, then we would no longer exist. I also feel that I was seeing only a small portion of the whole. As hindsight I can only regret that I did not report this sighting through Air Traffic Control channels. The crew discussed and speculated on what we had seen, then promptly forgot about it."

The account continues, relating how on the next night, when the aircraft departed the Philippines for Guam and Wake Island, the same phenomenon appeared in the east, although the plane's heading was then about 60°. The display appeared unchanged on two more nights on the legs from Wake to Hawaii and Hawaii to California. On the fifth night, however, from California to Washington, DC, nothing was seen. (Silva, John J.; personal communication, December 28, 1985.)

GREEN CLOUD WITH LIGHT RAYS

An account of a mysterious "green cloud" sending out powerful shafts of light and flying in tandem with an airliner appeared in a Soviet newspaper today. The strange cloud was seen over Byelorussia by passengers and crew of a flight from Georgia to Tallinn, and by the crew of an airliner from Leningrad, passing 10 miles away according to Trud. Nikolai Zheltukhin, a corresponding member of the Academy of Sciences, said the object was certainly very big. He rejected the idea that the green cloud was an image caused by far-off atmospheric changes because the airman had fixed its location from the shafts of light it sent to the ground.

(Anonymous; "Mysterious 'Green Cloud' Appears near Airliner," Baltimore Sun, p. 4A, January 31, 1988.)

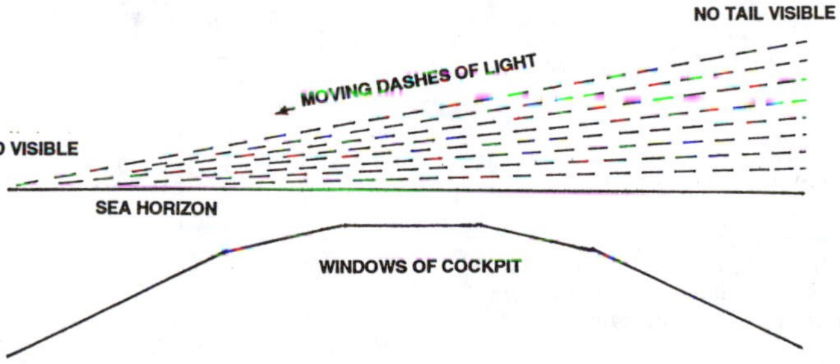

Array of moving dashes of light seen over the Pacific Ocean on each of four successive nights.

MARINE LIGHT DISPLAYS

Luminous ships' wakes, common in tropical seas, are mundane and unimpressive compared to the vast rotating wheels of light and other fantastic luminescent displays encountered from the Persian Gulf, across the Indian Ocean, and into the South China Sea. Ridiculed as wild sailors' tales for centuries, modern ships have reported scores of these weird geometrical light displays. Mariners tell of great spoke-like bars of light seemingly spinning about some distant hub. Occasionally, two or more wheels will overlap while simultaneously turning in different directions, creating a vast tableau of intersecting spokes miles wide. Expanding rings of light and whirling crescents are also seen on occasion. Crews that see these bizarre apparitions do not soon forget them. Scientists, however, have generally ignored them.

One's first reaction is to explain the wheels of light and related geometrical displays in terms of marine bioluminescence that is stimulated by some mechanical forces akin to the gross motion of a ship that causes a luminous wake in the warmer seas. Sound waves emanating from a ship's propeller or perhaps sea-floor earthquakes have been suggested as possible causes, but they cannot account for complex, counter-rotating wheels that spin for a half hour or more. Furthermore, the situation is complicated by well-attested cases of luminous displays occurring well above the sea's surface and by interactions between radars and the displays. This class of phenomena still resists scientific explanation.

Seven of the 14 types of marine light displays described in Lightning, Auroras, Nocturnal Lights have been recorded in Science Frontiers:

- Light bands. Long trains of moving, parallel, luminous bands.
- Light wheels. Spokes of light rotating about a hub.
- Expanding rings.
- Whirling crescents. Arrays of luminous crescents or boomerangs rotating together.
- Milky seas. Also called white water.
- Radar interactions. Detection and/or stimulation of light displays by a ship's radar.
- Aerial manifestations. Displays that seem to occur well above the surface of the water.

LIGHT BANDS

PHOSPHORESCENT BARS AND WHEELS

May 29, 1955. The Java Sea.

"At 0210 LMT I witnessed the start of a bioluminescent display. My first impression was that the ship was being 'attacked' on all sides from different directions by pulsing light-bands. A dull 'strobe-light' effect flashed through a mist, giving the bands a dirty white to gray colouration which was not a 'smooth' colour, but rather grainy in appearance. The bands were about 2 m wide and about 2 m apart and moving at speed. At first it was difficult to discern whether or not the bands were in the water or just above the surface, as no form of reflection or distortion was visible off the hull. In the end, I decided that the effect must be waterbourne if only because nothing was visible in the vessel's wake.

"The most intense activity was observed on the starboard side of the ship where the phenomenon appeared to stretch as far as the horizon. At this stage, it did not appear localized, just a mass of high-speed interacting bands of light. The effect is shown in the first sketch. As is usual on an 'all aft' ship, you become 'deaf' to the constant background noises, but I gradually became aware that the pulses of light seemed to match those of the main engine's throb, that is, about two per second. The radar (3-cm radar, running on the 24 n. mile range), and the echosounder (indicating a water-depth of about 35 fathoms), were switched off in turn to see if any change was discernible, but there was not.

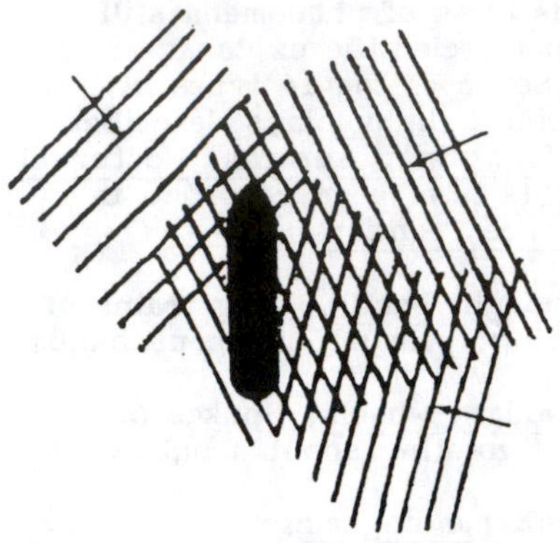

"However, at about this time, the ship passed a localized revolving system, distance off appeared to be about 150 m. My impression was that of a catherine wheel revolving and casting out waves in an angular motion, as shown in the second sketch. How many spokes it had I'm not sure owing to the speed of the pulsations, but I think that there were at least three. If viewed from above, the system rotated in a clockwise direction wheeling itself along the ship's track. No central hub was visible, just a dark area devoid of activity. One or two systems were visible farther out to starboard." (Lakeman, J.D.; "Bioluminescence," Marine Observer, 56:68, 1986.)

LIGHT WHEELS

ANOTHER INDIAN OCEAN LIGHT WHEEL

March 27, 1976. Position 10°N., 101°E.

At 1917 GMT, C.J.A. Cladingbowl, the Second Officer of the s.s. Benattow saw pulsating parallel bands of light rushing toward the ship from 045°T. After two to three minutes, the bands assumed a spoke formation with the center of rotation unseen but in the direction of 315°T. The spokes were about 22 m in width, with 22 m between each spoke. Rotating clockwise, the spokes swept past the vessel at ever increasing speeds, reaching two spokes per second maximum. By 1925, the display had reverted to the parallel band form. Then, the bands changed into a counterclockwise rotating wheel. The performance ended when the display again reverted to parallel bands and faded out altogether. The light from the spokes was white to light green and its intensity increased with the speed of rotation.

(Cowie, R.E.; "Bioluminescence," Marine Observer, 47:17, 1977.)

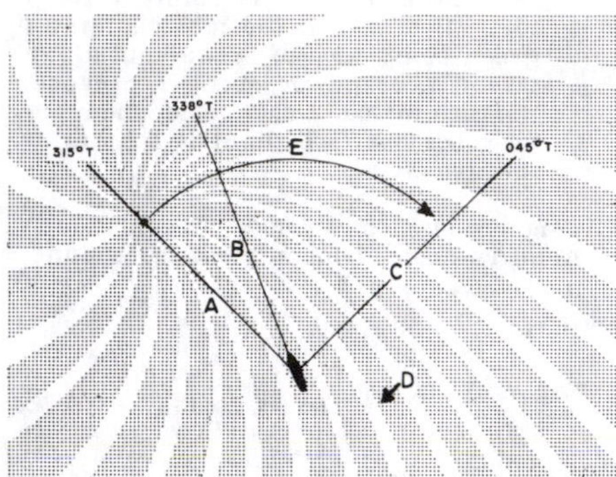

A· LOCUS OF CENTER OF SPOKE FORMATIONS AT BEGINNING
B SHIP'S COURSE
C LOCUS OF CENTER OF SPOKE FORMATIONS AT END
D ADVANCING BANDS OF LIGHT
E DIRECTION OF MOVEMENT OF SPOKE FORMATIONS

The wheel phase of a complex sequence of moving luminous bars and rotating wheels seen in the Gulf of Siam.

BIOLUMINESCENT CARTWHEELS AND WHIRLPOOLS IN THE ARABIAN SEA

6 March 1980. Arabian Sea.

At 1552 GMT bioluminescence in the form of diffused white light in 'whirlpool' and 'cartwheel' formations was observed; within 3 minutes it completely encircled the vessel and extended to the horizon. The 'cartwheel' formations were brightest at the centre with a halo effect surrounding the outer edges. As the vessel passed over 2 such formations the 'spokes' were estimated to be $2-2\frac{1}{2}$ metres in width and the entire concentration, which was more than the width of the vessel (approximately 27 metres), was observed on both sides of the bridge-wing simultaneously. The 'whirlpool' formations, with a distinct central hub, varied from $1\frac{1}{4}$ to 2 metres in width and from 14 to 15 metres in length. The phenomenon was observed for 40 minutes.

(Messinger, P.A.; "Bioluminescence," Marine Observer, 51:13, 1981.)

Comment. This is the first report uncovered that described "whirlpool" formations. The variety of phosphorescent formations and their long durations cast doubt on the usual seismic explanation.

INCREDIBLE PHOSPHORESCENT DISPLAY ON THE CHINA SEA

April 29, 1982. China Sea. The m.v. Siam encountered---or perhaps caused---a most baffling display of marine phosphorescence lasting some 2.5 hours. The complete report is 6 pages long, with 8 diagrams, so only the highlights can be reported here. As is often the case, this display began with parallel phosphorescent bands (2 sets) rushing toward the ship at about 40 mph. They were 50-100 cm above the sea surface. The bands then changed into two rotating wheels; then a third wheel formed. All three rotated counterclockwise, with their hubs 300, 300, and 150 meters from the ship. The spokes stretched to the horizon. The display ceased for about 20 minutes and recommenced with four systems of onrushing parallel bands, which soon metamorphosed into four rotating wheels. Radar, visible

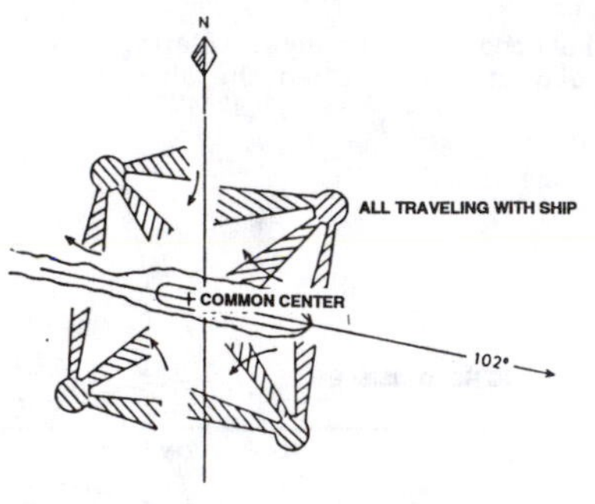

Four rotating light wheels, Actually, spokes surrounded all four hubs and extended to the horizon.

light (from an Aldis lamp), and engine revolution appeared to have no effect on the spectacle. Next, evenly distributed, circular, flashing patches of brilliant blue-white light appeared all around the ship out to a distance of about 150 meters. This system of patches flashed away simultaneously with the wheel display. The patches varied from 15-60 cm in diameter, and flashed 114 times per minute. When an Aldis lamp played steadily on the patches, nothing happened. When the lamp was flashed, the whole array of flashing patches disappeared, only to reappear in about 2 minutes. Each patch seemed to consist of worm-like segments $2\frac{1}{2}$ cm long, $2\frac{1}{2}$ cm apart. The worms were all aligned perpendicular to a vector from the ship. In contrast to the bands and wheels, the worms were located about 5 cm below the surface of the water. Water samples revealed no luminous organisms ---only a few animals a few millimeters long. The sea was calm, visibility excellent, although atmospheric electrical activity could be seen all around. (Kuzmanov, Zoran; "Phosphorescence in the

China Sea," Marine Observer, 53:85, 1983.)

Comment. The luminous "worms" resemble the spinning crescents sometimes associated with radar.

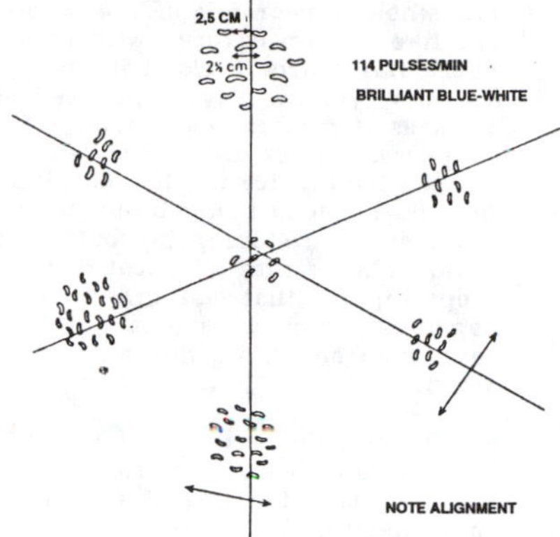

Flashing patches of worm-like shapes. A later phase of the display.

AN UNUSUALLY COMPLEX MARINE LIGHT DISPLAY

May 6, 1991. Straits of Hormuz. Aboard the m.v. Zidona, enroute from Muscat to Ruwais.

> At 1805 UTC a blue-white pattern of fast-moving light was seen around the ship. Initially, it was thought to be a reflection in the bridge windows of the Didimar lighthouse, but on going to the bridge wing, the observers saw an amazing display of flashing lights taking place over 80-90 per cent of the surface of the water. The whole ship was surrounded by a mass of blue and white light forming complex patterns that were visible in all directions as far as the eye could see. Looking almost like an 'electric mist', it moved with such speed and ease, as if it were alive.
>
> At the peak of the activity, there appeared to be two central points of spiralling, each about 150 m off either side of the ship about midships. From these points there seemed to be emerging highly confused patterns of spiralling spokes moving in an anticlockwise direction on the port side and clockwise on the starboard side of the ship. It was difficult to estimate accurately how many spokes were present in each circle, but it was thought that there were three or four at any one time, moving very fast and curving to produce what could only be described as a 'whirlpool' effect.
>
> At the same time, there were pulsating rings expanding from the centres at intervals of about three-quarters of a second. They moved extremely fast, each circle taking about one second to reach a diameter of about 200 m before being lost in the mass of flashing blue and white lights. The thickness of each ring remained constant at about 2 m as the diameter of the circle increased; the formation was always a perfect circle.
>
> About 300 m off the ship's side, large irregular shapes were observed, They were all about 3-m in diameter and changed both size and shape while flashing intensely. By 1817 the effect had completely stopped on the starboard side and only the pulsating rings were left on the port side and, with these, the intensity of the light reduced until 1822 when there was nothing more to observe.

(Bowden, P.; "Bioluminescence," Marine Observer, 62:64, 1992.)

A PARADE OF SPINNING PHOSPHORESCENT WHEELS

October 7, 1991. Gulf of Aden. Aboard the m.v. Wiltshire enroute from Aqaba to Fujayrah.

> At 1745 UTC the glow of bioluminescence was first noted around the hull of the vessel, illuminating the hull above the waterline. The passage of an area of phosphorescent wheels was recorded as follows:
>
> 1750: First large wheel of diameter approximately 15 m passed by vessel. Smell of fish in the air.
> 1806: Continuous wheels passing vessel 6-8 at a time down either side. The larger wheels were of 15 m diameter and the smaller ones were about 6 m in diameter.
> 1811: Wheels stopped but bioluminescence still visible around vessel.
> 1950: Bioluminescence diminished.
>
> The Aldis lamp was shone upon the water but gave no change, then the echo sounder was switched on and off but made no difference either. Several samples of sea water were taken which when shaken contained glowing, luminous, yellow-green specks 1 mm in size.
>
> The wheels were turning in slow clockwise motion and the closest that any came to the ship was about 12 m. There was intense milky-white colouring in the centres which faded to pale white towards the outer limits.

None of the radial spokes so common in phosphorescent-wheel reports were remarked in the Wiltshire report. Wheel rotation was also much slower than normal. One scientist supposed that the wheels were caused by fish swimming in tight circles! (Marsh, C.H.; "Bioluminescence," Marine Observer, 62:177, 1992.)

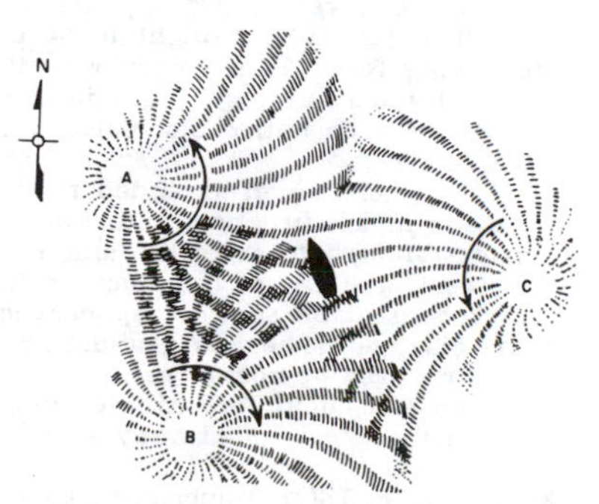

Three marine light wheels turning simultaneously in the Gulf of Thailand.

EXPANDING RINGS

EXPANDING PHOSPHORESCENT RINGS

May 9, 1983. Gulf of Oman. Aboard the m.v. Mahsuri.

> "At 1650 GMT, a pale green glow was seen to emanate from the horizon ahead. This gave the appearance of strong moonlight upon the surface of the water. The moon, however, was not in evidence. At 1700 GMT, rapid flashes of light were observed sweeping across the sea directly ahead of the vessel, giving the initial impression of a sudden increase of wind speed causing excessive spray. By 1715 GMT, the vessel was totally surrounded by completely random movements of light as far as the eye could see. The onset of this phenomenon was so rapid, not to say eerie, that the Master was called to the bridge to witness the event. For the next 15 minutes the sea was at a height of activity, displaying several systems of the most unusual bioluminescence. The most significant of these were what appeared to be Phosphorescent Wheels, which, although they did not seem to rotate, originated from a central hub and spread out rings in rapid succession, forming concentric circles. This was pointed out by many of those who observed them as being similar to the instance of a stone being dropped into a quiet pond and causing waves to spread

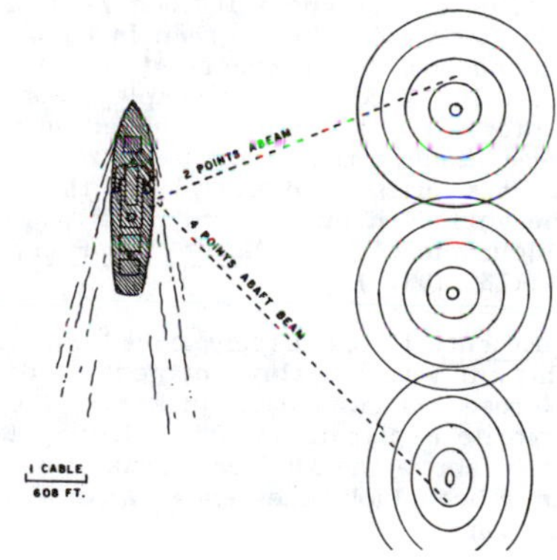

Three sets of expanding luminous rings---two circular, one elliptical.

> out. In this case each wave crest was a band of fantastic light. Each wheel would last for a couple of minutes, continually flashing out bands of light as though a transmitter was located at its centre. Wheels could be observed in all directions. At the same time systems of moving parallel bands could be observed, again travelling in totally random directions which respect to each other and passing off into the distance, only to be followed by another set.
>
> "It must be noted that the sheer complexity of the sea at this stage made observation very confusing to the eye. One particular characteristic of the wheels in question was the fact that their centres appeared to travel along with the ship, that is to say, a wheel on the beam seemed to remain on the

beam, until fading to be replaced by a new pattern. At one stage parallel bands were seen to be emanating from the ship's side, passing away to port and starboard as if the ship had become the centre of a wheel.

"The bands of light themselves were approximately 3-5 metres wide and, as mentioned, showed no preference of direction. The length of a parallel band was at least 160 m, and the diameter of the circles in a wheel ranged from possibly 3 m to 200 m. The period between successive flashes was less than a second, and the flashes were of pale green, yet having an almost golden quality.....At 1740 GMT the vessel passed out of the anomaly and seemed to cross a distinct line, one side of which was active and the other dark. The lights could be seen astern disappearing over the horizon." (Round, G. E.; "Bioluminescence," Marine Observer, 54:78, 1984. Also: Huyghe, Patrick; "Wheels of Light, Sea of Fire," Oceans, 20:20, December 1987.)

WHIRLING CRESCENTS

WHIRLING CRESCENTS MOVE WITH SHIP

July 11, 1980. Malacca Strait. Uniform crescents of luminescence appeared suddenly. Horizontal to the sea surface, they moved around the ship in circles, starting just forward of the bow. The crescents were about 100 meters long, 0.5 meter wide, light green in color, and passed the observers at the rate of three per second. The display was centered on the ship and moved with it. Some thought the crescents were above the sea surface, others placed them on the surface itself. (Lardler, D.A.; "Bioluminescence," Marine Observer, 51:116, 1981.)

Comment. If the display moved with the ship, it was probably not generated by microseisms (tiny earthquakes)---the favorite explanation. If the display was truly above the surface, it may not have been bioluminescence. What was it then?

MORE PHOSPHORESCENT BOOMERANGS

July 10, 1979. The Arabian Sea. Aboard the m.v. Strathelgin.

> At 1200 GMT large patches of milky-grey bioluminescence were observed; the patches appeared to form circular patterns resembling cartwheels, some of the configurations, however, did not have the central hub, see sketch. The patches pulsated at regular intervals (3 or 4 times per second). They moved in an anticlockwise direction until about 3 points abaft the beam where the direction of movement was reversed. On the beam they appeared to be at eye level, at all other times they were just above the surface of the water. The average size of the 'wheels' was 35 metres.

(Penman, B.; "Bioluminescence," Marine Observer, 50:114, 1980.)

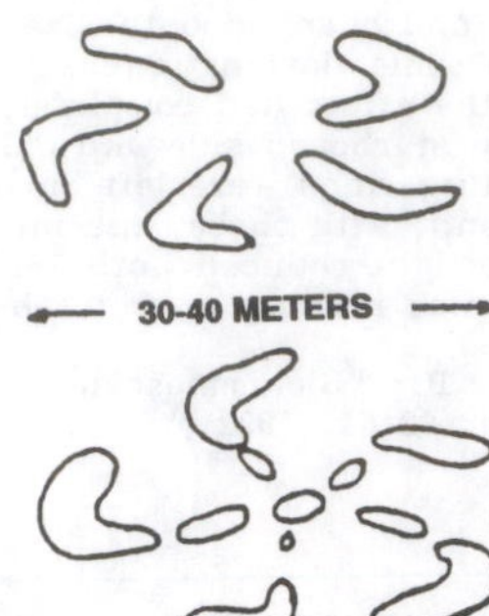

Comment. A similar case of spinning boomerangs was reported in Lightning, Auroras, Nocturnal Lights, where the display was stimulated by switching the ship's radar on and off. Here, one must also ask how a bioluminescent phenomenon can exist at "eye level" many feet above the sea surface.

MILKY SEAS

MORE MILKY SEAS SEEN

August 4, 1977. Indian Ocean. The s.s. British Reknown entered a large area of milky sea. The intensity of the light was so great that the deck appeared to be just a black shadow. During the display, humidity seemed to increase and the radio operator reported a decrease in signal strengths at HF and MF frequencies.

September 6, 1977. Indian Ocean. An area of bioluminescence resembling white sea fog was spotted near the m.v. Wild Curlew. Entering the area, the milky light seemed to hover above the sea's surface. So strong was the light that the clouds above were illuminated. (Anonymous; "Bioluminescence" Marine Observer, 48:118, 1978.)

MILKY SEA

August 18, 1990. Enroute from Singapore to Jeddah. Aboard the m.v. Benalder.

> At 1640 UTC the sea surface was noticed to have a white appearance which at first was thought to be a low-lying fog. This theory was disproved when shining a light in the water gave no noticeable increase in luminance.
>
> The phenomenon extended to the horizon in all directions and was bright enough to make the ship's foredeck and the sky appear much darker than the sea. Its appearance and disappearance was gradual apart from an area of normal sea which was passed about five minutes before the phenomenon faded away at 1725.

(Anderson, F.G.J.; "Bioluminescence," Marine Observer, 61:117, 1991.)

Comment. Luminous bacteria are usually blamed for milky seas, but no one has ever studied the phenomenon.

THE MILKY SEA A.K.A. "WHITE WATER"

June 1854. South of Java. Aboard the American clipper Shooting Star. Captain Kingman reporting:

> The whole appearance of the ocean was like a plain covered with snow. There was scarce a cloud in the heavens, yet the sky...appeared as black as if a storm was raging. The scene was one of awful grandeur; the sea having turned to phosphorus, and the heavens being hung in blackness, and the stars going out, seemed to indicate that all nature was preparing for that last grand conflagration which we are taught to believe is to annihilate this material world.

We selected this account of the milky sea phenomenon because of its vivid verbiage---something absent from the modern reports:

> August 13, 1986. Northwest Indian Ocean. The entire sea surface took on an intense white glow which was not unlike viewing the negative of a photograph.

The milky sea is a rather common phenomenon. In fact, the British Meteorological Office has established a Bioluminescence Database, which presently contains 235 reports of milky seas seen since 1915. P.J. Herring and M. Watson have employed this Database in a review paper on these impressive displays.

Geographical plotting of the reports shows a strong concentration in the northwest Indian Ocean (see figure). Seasonally, there is a strong peaking in August and a secondary blip in January. The phenomenon is independent of water depth and distance from land.

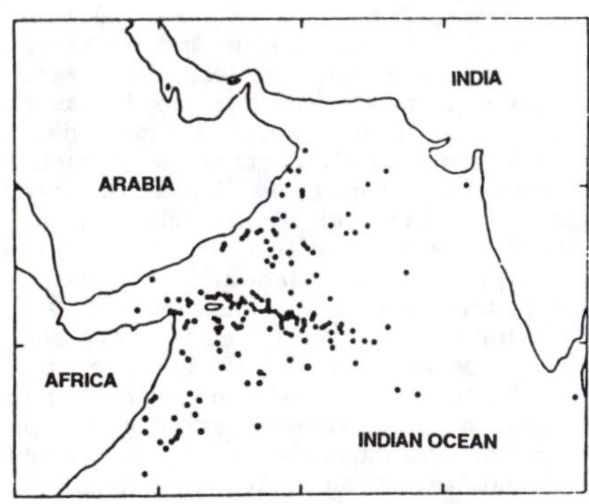

Milky sea reports are concentrated in the northwest Indian Ocean.

Surely bioluminescent organisms must be the explanation for milky seas. But most such organisms simply flash briefly and are incapable of generating the strong, steady glow of the milky sea. Marine bacteria alone glow steadily. However, calculations show that unrealistic concentrations of bacteria would be needed to generate the observed light. Furthermore, samples from the affected waters show no such bacteria. Herring and Watson admit there is no acceptable explanation of the milky sea. What, they ask, is so special about the northwest Indian Ocean? Why do milky seas not occur in the adjacent Red Sea and Persian Gulf? These bodies of water seem equally promising. (Herring, P.J., and Watson, M.; "Milky Seas: A Bioluminescent Puzzle," Marine Observer, 63: 22, 1993.)

RADAR INTERACTIONS

BIOLUMINESCENCE AND SPURIOUS RADAR ECHOES

March 18, 1977. North Atlantic. Aboard the m.v. Ebani. Throughout the day spurious radar echoes had been appearing on the radar screen. Resembling the echoes from small clusters of fishing boats, they would close to within 8 nautical miles, and then disappear. At 2200, echoes appeared, closed to within 5.5 nautical miles, and then spread out around the ship in a circle, all the while maintaining a 5 mile range. At this time, the entire sea took on a milky appearance and a fishy smell was detected. The beam from an Aldis lamp revealed luminescent organisms in the sea. After 45 minutes, both milky sea and spurious radar echoes disappeared together. (Richards, A.W.; "Radar Echoes and Bioluminescence," Marine Observer, 48:20, 1978.)

Comment. Why should radar echoes and bioluminescence be connected?

BIOLUMINESCENT PATCH DETECTED BY RADAR

June 20, 1977. Santos to Buenos Aires. Aboard the m.v. Gambada. Unusual echoes were observed on the radar screen. What was thought to be a patch of rain on the radarscope moved toward the vessel against the wind. The radar echo of the patch had a distinct edge to it unlike that of other rain areas. Yet, when the ship was at the center of the patch there was no precipitation. Throughout the watch, however, the sea was strongly luminescent. (Turney, R.J.; "Bioluminescence," Marine Observer, 48:69, 1978.)

RADAR INTERFERENCE AND LUMINESCENCE

March 8/9, 1989. Arabian Sea. Aboard the m.v. British Esk.

> During the night a particularly strong and distinct patch of radar interference was noted by all observing officers. The sketch shows the phenomenon as seen on the 12-n. mile range of the 3-cm radar. The racon type mark varied in length from 1-3 n.mile at a nearest range of 5-10 n.mile. The effect was minimal on the 10-cm radar.

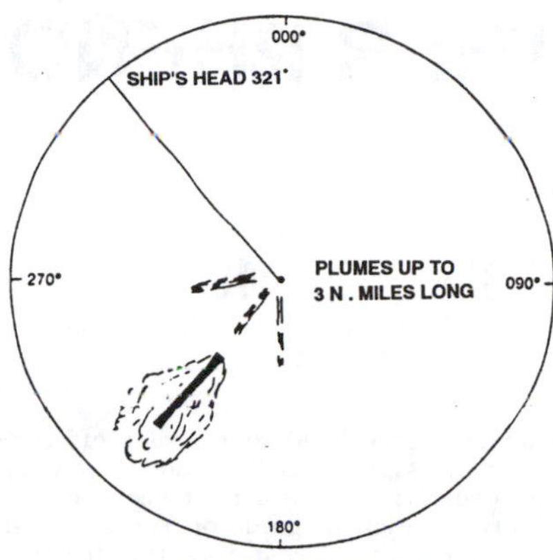

Radar interference in the Arabian Sea associated with bioluminescence.

> The bearing of the mark remained fairly constant at about 20° abaft the port beam or about 230°. Of particular note was that around 1600 GMT to 1700 GMT (about 2 hours after sunset), when the mark on the radar was very distinct, the satellite communication system suffered a loss in signal strength sufficient to prevent transmission or reception, the bearing of the satellite being almost due south of the vessel. It was thought at the time that the signal mast had become aligned between the aerial and the satellite, but alteration of the ship's head to port or starboard did not cure the low signal strength.
>
> .....
>
> Of note, although this may have been a coincidence only, was that the vessel was passing through patches of bioluminescence at the time, mostly only bright enough to show up in the breaking waves of the ship's wake, but during the period of low signal strength, the whole area of white, foamy water along the ship's side frequently shone a bright greenish colour. (St. Lawrence, P.F.; "Radar Interference," Marine Observer, 60: 17, 1990.)

Comment. Other cases of radar interaction with bioluminescence are cataloged under GLW in Lightning, Auroras, Nocturnal Lights. Apparently, some sort of electromagnetic atmospheric disturbance affected not only radar, but also satellite communications and bioluminescent organisms in the water. Could it have been an ionized plasma vortex similar to those supposed responsible for the corn circles?

AERIAL MANIFESTATIONS

AERIAL BIOLUMINESCENCE

January 19, 1991. South China Sea. Aboard the m.v. Benavon. The vessel was heading for Singapore on a body of water noted for bioluminescent displays. Flashes of light were seen in the bow wave and the ship's wake, appearing to be both on the surface and slightly below. This type of display is rather common, but another, much rarer phenomenon was also present:

> At the same time as the above form of bioluminescence, there seemed to be a second type but it was difficult to pinpoint the source. The effect

Off the Malabar coast of India, a vessel was engulfed by great waves of light seemingly floating on the sea surface.

> was that the atmosphere around the ship and extending to the horizon had some form of faint white illumination not provided by the light in the water, which was black apart from the previously described flashes. On the other hand, there was no obvious source in the sky either,

A radar-initiated display of rotating luminous crescents observed in the Gulf of Oman in 1951.

which although virtually cloudless was very dark, and certainly darker than the atmosphere at the level of the ship.

The only conclusion that the observers could come to was that this was a faint example of (to quote The Marine Observer's Handbook), 'luminescence in the air a few feet above the sea surface when there is no light in the water'. This form lasted for about 30 minutes, whereas the bright flashes continued for three or four hours before they too eventually ceased.

(Thompson, P.C.; "Bioluminescence," Marine Observer, 62:14, 1992.)

Comment. Many cases of aerial marine light displays have been cataloged. It is assumed by scientists that bioluminescent particles are somehow carried into the air from the ocean, but there is no evidence at all for this. It is quite possible that some marine phosphorescent displays are electrical rather than biological. See Lightning, Auroras, Nocturnal Lights.

WEATHER PHENOMENA

REMARKABLE WINDSTORMS

SATAN'S STORM

June 1960. Kopperl, Texas. Thunderclouds and lightning gave way to winds in excess of 75 mph, with temperatures of up to 140°F. Surveying the storm damage later:

> Aside from the expected remains of a severe wind storm---uprooted trees, snapped telephone poles, roof damage and banged-up boats docked lakeside ---the area had the ironic appearance of having been stung by a June freeze. Tree leaves, shrubs, hanging plants and crops were curled and wilted, as if frost-bitten. Uncut Johnson grass was dried and ready to bale, although the hay normally required two or three days of drying time after being cut. Perhaps the most startling remains of the storm was in what had been the cotton patch at Pete and Inez Burns' farm. The cotton was about knee high and a 'lucious crop' the day before, according to the couple. The next morning all that was left were carbonized stalks peeping out of the ground. The corn fared little better.

(Glaze, Dean; "Kopperl's Close Encounter with Satan's Storm," Meridian (TX) Tribune, May 12, 1983, p.1. Article appeared originally in the Dallas Times-Herald Westward Magazine. Cr. J. Mohn)

Comment. The consequences of this storm closely resemble the burning and drying effects of some tornados.

SCORCHING AND BURNING IN A TORNADO

August 16, 1985. Annesley Woodhouse, England. A tornado.

"What was interesting about the storm was not only the damage it caused, but also the type of damage. After touching down the tornado uprooted a large oak tree, 15 metres high, in Lawn Road (luckily the residents of the house were away on holiday). The tornado proceeded to rip tiles off several roofs, demolished completely several greenhouses, and next scorched a 4-metre section of gable on the south side of a house in Forest Street (number 9). The gable section was scorched so badly that the gable had already been repainted when I called, although the evidence could still be seen." (Matthews, Peter; "Lightning Inside a Tornado?" Journal of Meteorology, UK., 10:375, 1985.)

Comment. All of the foregoing anomalies are discussed at length in GWT2 of our Catalog, Tornados, Dark Days, Anomalous Precipitation, for tornado burning and dehydration.

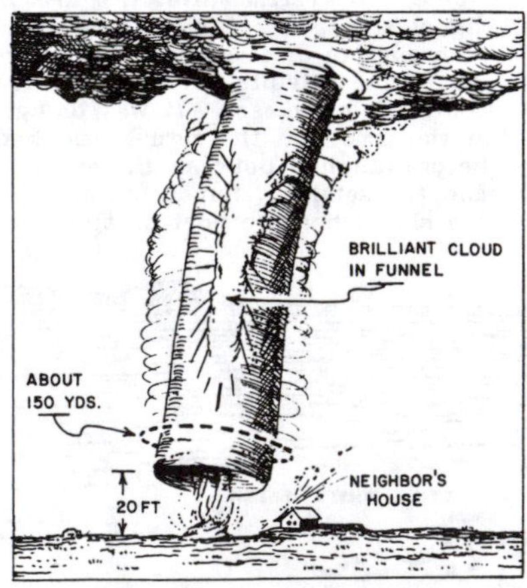

The funnel of the 1955 tornado at Blackwell, Oklahoma, was lit inside like a neon tube, suggesting electrical activity. Cloud-earth electrical currents might cause the scorching and burning effects mentioned above.

INSIDE A TEXAS TORNADO

May 3, 1943. McKinney, Texas. People rarely get the chance to look up into the funnel of a tornado and live to tell about it. R.S. Hall did; and what he saw is very strange:

"The bottom of the rim was about 20 feet off the ground, and had doubtless a few moments before destroyed our house as it passed. The interior of the funnel was hollow; the rim itself appearing to be not over 10 feet in thickness and, owing possibly to the light within the funnel, appeared perfectly opaque. Its inside was so slick and even that it resembled the interior of a glazed standpipe. The rim had another motion which I was, for a moment, too dazzled to grasp. Presently I did. The whole thing was rotating, shooting past from right to left with incredible velocity.

"I lay back on my left elbow, to afford the baby better protection, and looked up. It is possible that in that upward glance my stricken eyes beheld something few have ever seen before and lived to tell about it. I was looking far up the interior of a great tornado funnel! It extended upward for over a thousand feet, and was swaying gently, and bending slowly toward the southeast. Down at the bottom, judging from the circle in front of me, the funnel was about 150 yards across. Higher up it was larger, and seemed to be partly filled with a bright cloud, which shimmered like a fluorescent light. This brilliant cloud was in the middle of the funnel, not touching the sides, as I recall having seen the walls extending on up outside the cloud.

"Up there, too, where I could observe both the front and back of the funnel, the terrific whirling could be plainly seen. As the upper portion of the huge pipe swayed over, another phenomenon took place. It looked as if the whole column were composed of rings or layers, and when a higher ring moved on toward the southeast, the ring immediately below slipped over to get back under it. This rippling motion continued on down toward the lower tip." Hall also reported a peculiar bluish light and blue streamers that appeared to consist of vapor. (Hall, Roy S.; "Inside a Texas Tornado," Weatherwise, 40:73, 1987.)

CHICKEN-PLUCKING BY TORNADOS

In the early 1840s, Professor Loomis of Western Reserve College stuffed dead chickens into an eight pounder and fired the cannon vertically. Object: to see if high velocities would deplume the chickens. It seems that Loomis didn't know whether to believe all those stories about tornados tearing the feathers off chickens. Unfortunately, his experiments were inconclusive because the cannon disgorged clouds of feathers plus well-shredded chickens. He recommended that other scientists try lower muzzle velocities!

Since those more innocent days, science has accumulated many eye-witness accounts of this incredible phenomenon, including actual photographs that do not appear to have been tampered with in the darkroom. There can no longer be any reasonable doubt, tornados do deplume chickens. (Galway, Joseph G., and Schaefer, Joseph T.; "Fowl Play," Weatherwise, 32:116, 1979.)

SAILING THROUGH A WATERSPOUT

Sailing on the Pagan in the mid-Pacific, One July morning J. Caldwell spotted a tropic waterspout. Having heard that spouts had hurricane force winds inside, whirlpools at their bases that could suck a ship under, and a solid wall of water being sucked up into the clouds, Caldwell threw caution to the winds and headed directly for the spout.

> Pagan was swallowed by a cold wet fog and whirring wind. The decks tilted. A volley of spray swept across the decks. The rigging howled. Suddenly it was dark as night. My hair whipped my eyes, I breathed wet air, and the hard cold wind wet me through. Pagan's gunwales were under and she pitched into the choppy seaway. There was no solid trunk of water being sucked from the sea; no hurricane winds to blow down sails and masts; and no whirlpool to gulp me out of sight. Instead, I sailed into a high dark column from 75 to 100 feet wide, inside of which was a damp circular wind of 30 knots, if it was that strong. As suddenly as I had entered the waterspout I rode out into bright free air. The high dark wall of singing wind ran away. For me another mystery of the sea was solved.

(Caldwell, John; "On Sailing through a Waterspout," Journal of Meteorology, U.K., 11:236, 1986.)

RECIPE FOR DUST DEVILS

R.H. Swinn, formerly Chief Instructor for the Egyptian Gliding School, has had much practical experience with those fascinating little (sometimes not so little) swirls of hot air called "dust devils." Under the broiling Egyptian sun, dust devils launched themselves naturally every few minutes from a tented camp near the airfield where Swinn taught. Curiously, the devils often were born in pairs; a big one followed by a modest little chap following behind by 100-150 yards. The devils ranged from just a foot or so in diameter to 500 yards and more. The giants were majestic masses of swirling sand that moved along at leisurely paces. These appeared harmless enough, but stepping through the outer wall into the vortex sucked the air out of the lungs.

> Outside our hangar there is a large stretch of wind-sheltered concrete which becomes intensely hot. In this area, close to the foot of the hangar, one can start up one's own little devils on occasions by a quick sweep of a signalling bat (which is shaped like a large ping-pong bat) from shoulder level in circular and downwards direction to a point almost touching the ground; one must step rapidly back or the vortex that is set up is spoilt. Such a miniature thermal starts about a foot in diameter and quickly assumes a conical shape about two feet high, moving along the ground at a walking pace. Its rotation increases very rapidly, until one has the impression of a whirling snake in front of one. As it reaches the edge of the concrete a little sand is thrown up and the thermal dies away.

(Swinn, R.H., "On Flying Gliders into Wind Devils," Journal of Meteorology, U.K., 10:17, 1985.)

UNUSUAL GUST OF WIND

February 7, 1988. Lancashire, England.

It was a day with modest winds of 5-10 mph, with some gusts to 20 mph. Suddenly at 2100 GMT, the anemometer at Hazelrigg weather station registered a gust at 106 mph. Almost immediately after, the wind dropped to only 5 mph.

A gust of this strength should have caused considerable damage. In a nearby wood a few branches and twigs were down, but the major effect seems to have been the transportation of a 75-kilogram sheep feeding trough across a distance of 5.1 meters! Conclusion: a sudden, small squall had passed through. (Reynolds, David J.; "Unusual Gust of Wind in Lancashire 7th February 1988," Journal of Meteorology, 13:284, 1988.)

Comment. The weather is really playing tricks on the English, with hundreds of mysterious circles cut into crop fields and now this dislocated sheep trough! Or is it just weather?

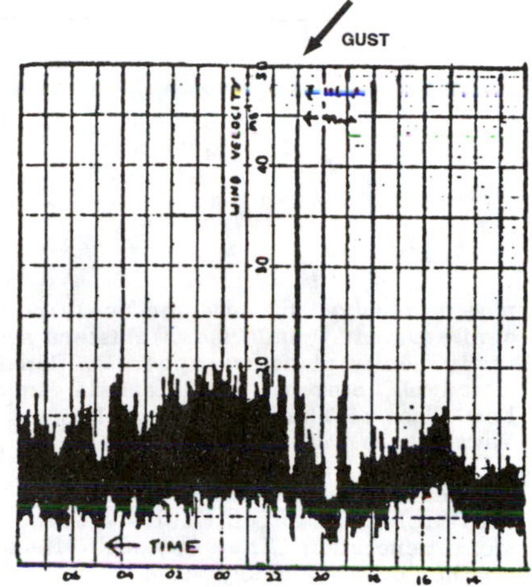

Anemograph trace showing the 106-mph wind gust at Lancashire, 1988.

THUNDERSTORMS

DO LIGHTNING CHANNELS ACCELERATE MATTER?

E.W. Crew suggests in this paper that two rare classes of meteorological observations may be created by the intense electrostatic accelerating forces present in lightning channels. The first class of observations consists of blasts of hot air noted some distance from violent lightning strikes but seemingly associated with the discharges. Second, some superhailstones (hydrometeors) also seem to be correlated with violent lightning. The physical mechanism for the concentration and propulsion of matter is the electrostatic force naturally present in lightning discharge channels; it functions much the same as the particle accelerators in the physics lab.

The observations of hot air blasts and superhailstones collected by Crew to support his theory are indeed suggestive, but more are needed. Crew also feels that some UFO sightings may be produced by the same mechanism. (Crew, E.W.; "Meteorological Flying Objects," Royal Astronomical Society Quarterly Journal, 21:216, 1980.)

Comment. Note also that the fall of thunderstones is usually coincident with lightning discharges; and that some high quality observations of thunderstone falls are on record---despite the tendency of Science to relegate them to myth. One must also consider the possibility that the passage of a meteor or superhailstone through the atmosphere might trigger lightning, thus putting the cart before the horse.

GIANT THUNDERSTORM CLUSTERS

Conventional wisdom has it that thunderstorms are small-scale phenomena 50-100 miles across. However, J.M. Fritsch and R.A. Maddox of NOAA have announced that satellite photos show a radically different situation. The more violent thundersotrms are often organized into roughly circular clusters that may span 1000 miles. Previously, all thunderstorms were considered local convective storms that were regulated by upper air patterns. This view must

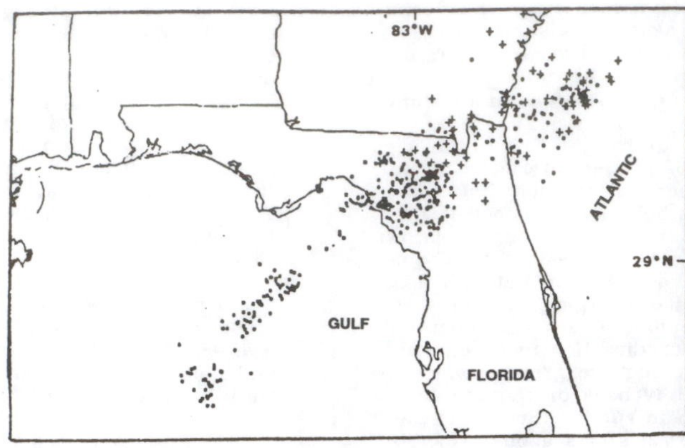

Lightning distribution across Florida on February 22, 1987, 8:00-9:00 PM. The plus signs designate positive lightning strokes.

now be changed because the newly recognized giant thunderstorm clusters actually modify planetary upper air flow. (Bardwell, Steven; "Satellite Data Show New Class of Thunderstorms," Fusion Magazine, p. 50, September 1981.)

Comment. Later on this page, cosmic rays are shown to contribute to thunderstorm generation. They may also affect weather on a planetary scale.

THE LARGE-SCALE STRUCTURE OF ELECTRICAL STORMS

Ground-based observers see only a few of the anvil-shaped clouds comprising a big electrical storm. The entire storm may stretch for hundreds of kilometers ---and it is not a simple structure.

The latest surprise is that all large electrical storms are bipolar; that is, the rare positive lightning strokes are concentrated at the northeast end of the storm complex, while the negative strokes are everywhere else along a northeast-southwest line. This bipolar structure persists for several hours, and it has been found in all North American storms analyzed so far. This insight into the structure of large electrical storms was provided by magnetic lightning detectors that have now been installed over nearly 75% of the United States.

The positive lightning strokes are of longer duration and more liable to start fires than the common negative strokes. But why are they concentrated at one end of the storm complex? R. Orville ventures that in a big mesoscale electrical storm, the prevailing winds blow the positively charged upper portions of the clouds to the northeast, thus establishing bipolarity. (Orville, Richard E., et al; "Bipole Patterns Revealed by Lightning Locations in Mesoscale Storm Systems," Geophysical Research Letters, 15:129, 1988. Also "New Lightning Theory Strikes," Eos, 69:57, 1988.)

SOLAR COSMIC RAYS STIMULATE THUNDERSTORMS

Not so long ago the idea of short-term solar influences on terrestrial weather was treated with contempt. However, meteorologists are now being converted in droves because believable physical links have been found linking sun and earth. A prime example is the bombardment of the terrestrial atmosphere by solar cosmic rays. The cosmic rays and the secondary particles they create ionize enough of the atmosphere to disturb the entire planetary electrical circuit. The details of the circuit changes are still under study, but there seems no question about cosmic rays initiating thunderstorm activity. Plots of global thunderstorm activity peak strongly about three days after any maximum in solar cosmic rays. (Lethbridge, M.D. "Cosmic Rays and Thunderstorm Frequency," Geophysical Research Letters, 8:521, 1981.)

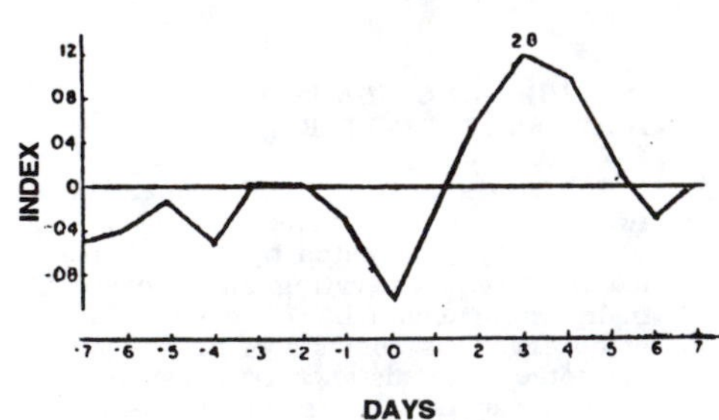

This thunderstorm frequency index reached a maximum 3 days after a cosmic-ray maximum.

Comment. At its present rate of decline the earth's magnetic field will reach zero in 1200 years. With this protective magnetic bottle gone, we see a good future for lightning rod manufacturers.

MAGNETIC PRECURSORS OF LARGE STORMS

On January 22, 1986, a magnetometer at the Fredricksburg Magnetic Observatory, in Virginia, recorded a sudden jump (of 45 gammas) in the earth's horizontal magnetic field component. Alerted to this, G. Wollin, at the Lamont-Doherty Geological Observatory, immediately predicted that a major snowstorm or flooding rains would hit northeastern states within six days. Wollin contacted the weather people in the region, but they discounted the prediction because satellite pictures and conventional weather indicators implied nothing of the sort. A three-day storm began on January 25, depositing 3 feet of snow in northern New England and 4 inches of rain along the coast from Washington to Boston. Wollin has had similar successes, without even looking at a weather map!

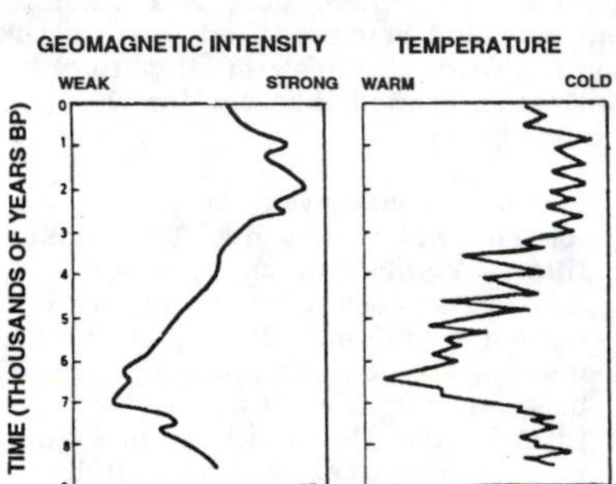

Long-term changes in global temperature tend to follow changes in geomagnetic intensity.

Obviously, Wollin's forecasting techniques are not yet part of the Weather Bureau's arsenal. This is not too surprising because even Wollin does not understand why major storms should be preceded by several days by nervous magnetometers. He talks in a tentative way about solar storms, which do affect terrestrial magnetism, dumping energy into the oceans and thence into the atmosphere. But this is mainly speculation. Historically, we do know that long-term changes in the earth's magnetic field are linked to global temperature levels (see graphs); but here, too, cause and effect are not obvious. (Gribbin, John; "Magnetic Pointers to Stormy Weather," New Scientist, p. 70, December 25, 1986.)

UNUSUAL CLOUDS

THE MORNING GLORY

The Morning Glory is a spectacular roll cloud that frequently sweeps in low over Australia's Gulf of Carpentaria, often around sunrise in clear, calm weather. The cloud is only 100-200 m thick but very long and straight, extending from

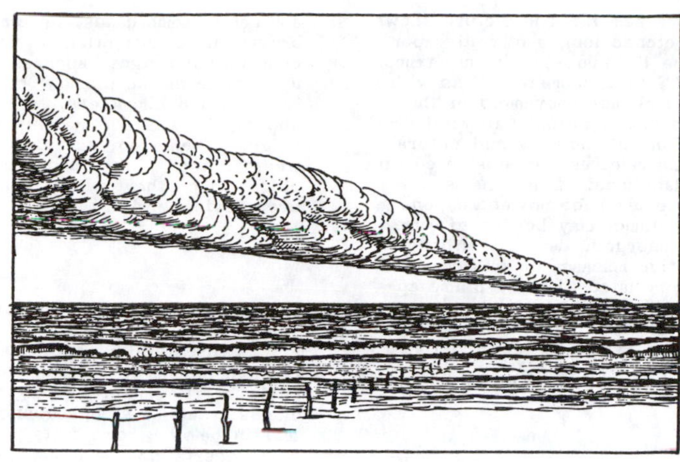

The Morning Glory, a spectacular roll cloud seen occasionally over the Gulf of Carpentaria, Australia.

one horizon to the other. (One pilot followed if for 120 km without finding its end.) Sometimes as low as 50 m, the Morning Glory brings squall-like winds but rarely more than a fine mist. Double Morning Glories are not uncommon. Seven were once reported. Oriented NNW to SSE in the main, they advance east-to-west low and fast (30-50 mph). Convincing explanations are wanting. One meteorologist has proposed that the Morning Glory is a "propagating undular hydraulic jump." (Neal, A.B., et al; "The Morning Glory," Weather, 32:176, 1977.)

A MOST PECULIAR CLOUD ARCH

July 12, 1980. Strait of Gibraltar.

> At 1825 GMT whilst the vessel was transiting the Strait of Gibraltar, a line of low cloud was observed in an otherwise cloudless sky, see sketch. The cloud was in the form of an arc in an east-west line, reaching the surface approximately 2 n. mile ahead and astern of the vessel. Visibility under the cloud was about 10 n. mile in the north-south direction and 2 n. mile to the east and west. Once the vessel reached the point where the cloud touched the surface, the visibility was reduced to approximately 1.5 n. mile. Whilst the vessel was passing the cloud, the barograph trace fell almost vertically and both the air and sea temperatures dropped several degrees.

(Shepherd, F.; "Cloud," Marine Observer, 51:107, 1981.)

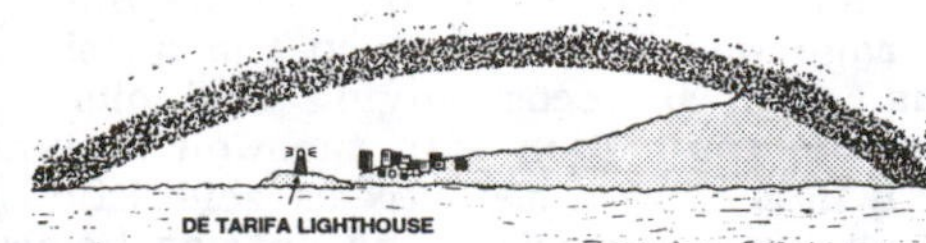

A curious cloud arch in the Strait of Gibraltar. The ends terminated in the ocean at points 4 miles apart.

Comment. This is just one more mysterious cloud arch, but on a very small scale. What bizarre meteorological conditions create such strange structures?

STRANGE HIGH-LEVEL HAZE IN THE ARCTIC

Every March and April, the supposedly pristine air of Alaska is defiled by a peculiar haze concentrated at about 10,000 feet. The sky has a whitish, diffuse look; from an airplane the horizon seems to disappear entirely. Is the haze due to pollutants in this remote region? Recent studies indicate two components in the haze: (1) dust, and (2) sulfuric acid droplets, both of which must be imported because there are no sources of such materials in the arctic. Violent wind storms in the Gobi Desert may carry some dust into the arctic. Strong winds might also transport sulphuric acid from Japanese industries to Alaska. These are speculations, though, and no one is sure where this haze comes from or how far it extends beyond Alaska into the stable, stagnant air over the Arctic Ocean. (Anonymous; "Alaska's Imported Haze," Mosaic, 9:14, September/October 1978.)

MYSTERY CLOUD OF AD 536

"Dry fogs appear in the atmosphere when large volcanic eruptions inject massive quantities of fine silicate ash and aerosol-forming sulphur gases into the troposphere and stratosphere. Although the ash gravitationally settles out within weeks, the aerosols spread around the globe and can remain suspended in the stratosphere for years. Because solar radiation is easily absorbed and backscattered by the volcanic particles, a haziness in the sky and a dimming of the Sun and Moon are produced. Very dense and widespread dry fogs occur, on the average, once every few centuries. The sizes and intensities of some of the largest of them before the modern scientific era have been estimated by several indirect methods. The densest and most persistent dry fog on record was observed in Europe and the Middle East during AD 536 and 537. Despite the earliness of the date, there is sufficient detailed information to estimate the optical depth and mass of this remarkable stratospheric dust cloud. The importance of this cloud resides in the fact that its mass and its climatic consequences appear to exceed those of any other volcanic cloud observed during the past three millenia. Although the volcano responsible remains a mystery, a tropical location (perhaps the volcano Rabaul on the island of New Britain, Papua, New Guinea) can be tentatively inferred." (Stothers, R.B.; "Mystery Cloud of AD 536," Nature, 307:344, 1984.)

Comment. Some of the "dry fogs" were accompanied by luminous nights, as in 1821 and 1831. See GWD4 in Tornados, Dark Days, Anomalous Precipication.

STILL ANOTHER MYSTERY CLOUD

April 9, 1984. Western Pacific. The crews of three airliners en route from Tokyo to Anchorage observed a gigantic mushroom cloud about 180 miles east of Japan. The cloud was moving rapidly up and away from a cloud layer at 14,000 feet. It eventually reached a maximum altitude of about 60,000 feet, at which time its maximum diameter was about 200 miles. No fireball or flash was seen by anyone.

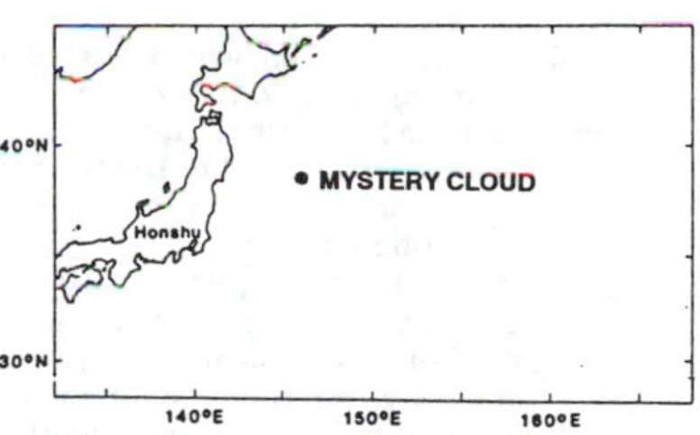

Location of the 1984 "mystery" cloud in the western Pacific.

A nuclear explosion, possibly on a submarine was suspected. One pilot issued a Mayday alert and ordered his crew to don oxygen masks. However, when an F-4 Phantom dispatched from Japan arrived at the scene, it detected no abnormal levels of radioactivity. Wake Island hydrophones, to the southeast, detected some submarine volcanic activity far south of the cloud, but no detonations in the area the cloud were spotted. The distance of the volcanic disturbances and the prevailing winds ruled out volcanic sources of smoke. In the absence of any hydrophonic evidence, the authors concluded that the mysterious cloud came either from a man-made atmospheric explosion (a huge one!) or some as yet unknown natural phenomenon. (Walker, Daniel A., et al; "Kaitoku Seamound and the Mystery Cloud of 9 April 1984," Science, 227: 607, 1985.)

MYSTERY PLUMES AND CLOUDS OVER SOVIET TERRITORY

Following the famous mystery cloud of April 9, 1984, the authors of the original article in Science collected additional observations and now have this to say:

> Our conclusion is that original estimated positions were in error. Additional data, primarily from Van den Berg, place the event between the Kuriles and Sakhalin. The altitude of the center of the halo at the maximum observed size is estimated to have been greater than 200 miles, and the diameter of the halo is estimated to have been at least 380 miles. It seems unlikely that a groundbased explosion could produce this kind of effect. It is surprising to us that no official data have been provided by government agencies and that such a significant observation from a region of demonstrated military sensitivity was, and still remains, a mystery.

(McKenna, Daniel L., and Walker, Daniel A.; "Mystery Cloud: Additional Observations," Science, 234:412, 1986.)

Evidently the mystery cloud mentioned above is only one in a long series:

> Large icy clouds, similar to plumes of gas that rise over volcanoes, have appeared over islands along the coast of the Soviet Union during the past several years, baffling experts, who cannot explain what they are or what causes them.
>
> The clouds dissipate in a few hours vanishing as mysteriously as they appear.
>
> Among the plumes are a series of massive clouds that during the past four years have periodically swelled over Novaya Zemlya, the Arctic island long used by the Soviets for nuclear weapons tests.
>
> However, there appears to be no correlation between the clouds and known Soviet tests, which are usually detected by Western governments. Further, non-governmental scientists said the 200-mile-long plumes appear to be many times larger than the largest conceivable nuclear explosion could produce.

A NOAA satellite detected a large plume coming from the Arctic Ocean near Bennett Island, north of the Soviet Union, in 1983. Three distinct sources were found; one on the island and the other two about 9 miles offshore on the ice-covered ocean. This plume was 6 miles wide, 155 miles long, and 23,000 feet high. (Anonymous; "'Plumes' over Soviet Isles Continue to Baffle Experts," Las Vegas Sun, July 20, 1986. Cr. T. Adams via L. Farish)

MORE ON THE SOVIET PLUME EVENTS

A recent issue of Eos, published by the American Geophysical Union, presents some amazing and at the same time unsettling photographs of immense plumes taken by satellites passing over Soviet Arctic islands. Eleven such events are tabulated from October 12, 1980, to June 12, 1986. Perhaps the most dramatic event occurred on March 12, 1982, over Novaya Zemlya. The picture shows a sharply etched tongue of cold vapor arcing some 175 kilometers at a maximum altitude of 9.5-10 kilometers. As with most of the plumes, movement of the vapor does not correspond to wind direction. Volcanic activity and natural methane gas releases are considered unlikely explanations. Since the islands involved are used for Soviet weapons tests, the plumes may be due to some incredibly energetic devices, although no radioactive releases or seismic activity seem correlated with the plume appearances. Queries to Soviet scientists have gone unanswered. (Anonymous; "Large Plume Events in the Soviet Arctic," Eos, 67:1372, 1986.)

ARE THE SOVIET PLUMES ONLY OROGRAPHIC CLOUDS?

F.C. Parmenter-Holt, a NOAA scientist, has reacted to the recent discussions of Soviet plume events as follows:

> I believe that these clouds are naturally occurring, orographically-induced formations. When winds blow perpendicular to the 2,500-plus foot glacial ridge, along the northern portion of the island, a long gravity-wave pattern is established downwind, on the lee side. The cases collected by Matson show sharp boundaries conforming to the contour of this glacial barrier.

(The Matson reference is Science News, March 28, 1987, p. 204.) (Parmenter-Holt, Frances C.; "Plumes and Peaks," Science News, 131:403, 1987.)

Comment. Parmenter-Holt could well be correct in some cases, for wave-like orographic clouds often form in the lee of mountain ranges, such as the Rockies. Some of the plumes, however, extend for 175 kilometers, as described above. This is pretty long for a glacial ridge. Then, too, one should inquire whether such plumes occur near similar ridges in northern climes and not just over Soviet territory.

MYSTERY AT NOVAYA ZEMLYA

F.C. Parmenter-Holt opined above that the long plume-like clouds detected over Soviet territory were merely orographic clouds; that is, a consequence of the terrain below. Some facts presented by W.O Roberts, in the latest issue of The Explorer, hardly square with that interpretation. For example, the March 12, 1982 plume seen over Novaya Zemlya was 109 miles long and at an altitude of about 6 miles. Its position did not conform to the wind direction at that altitude. Other plumes over Novaya Zemlya have been aligned with the wind, but they too have been at great altitudes. Says Roberts:

> Taken together the data suggest irregular emissions from a single point source near the north end of the Island as the cause of the mysterious episodes.

Just what is being vented, if anything, remains unknown. No active volcanos are in this area, neither are there copious sources of natural gas. There have been no seismic or radioactive signs of nuclear tests. (Roberts, Walter Orr; "Mystery at Novaya Zemlya," The Explorer, 4:6, April 1988.)

CROP CIRCLES

The crop circle saga had two beginnings, both based upon casual observations of large, flattened circles of grass or grain in various parts of the planet but chiefly in Britain. First, meteorologists wondered whether the simple, circular patches might be the consequence of natural atmospheric phenomena, such as whirlwinds and vortexes. Second, UFO enthusiasts were quick to ascribe the circles to UFO landings or "saucer nests." In the mid-1980s, crop circles began to proliferate and, in addition, became more and more complex. Multiple circles appeared; some circles boasted smaller "satellite" circles. Ultimately, fantastic "designs" were impressed upon the fields of British farmers. The crop circle "craze" evolved--almost exponentially as the media became interested. The obvious intelligence involved in some of the crop circle designs led to claims of extraterrestrial involvement. New Agers were ecstatic. However, it soon became obvious that even the most elaborate designs could be constructed simply and easily, and that many such were outright hoaxes. But were (and are) all crop circles hoaxes?

Science Frontiers devoted considerable space to the crop circle phenomenon from 1983 to 1993. Its coverage was slanted toward the possible scientific content of the observations, such as accompanying luminous and acoustic phenomena and the possible connection with the well-known bizarre effects of tornados and whirlwinds. Since mainstream scientific journals generally ignored the whole business, most of the entries below came from the Journal of Meteorology, U.K., which, though edited by a scientist, does not avoid such controversial phenomena as crop circles, ball lightning, and hydrometeors.

The organization of this rather lengthy section is chronological, be-

cause we really have here a tale that has as much social import as scientific value. There are obvious parallels with UFOs, cold fusion, infinite dilution, etc., in that legitimate scientific phenomena may have become overwhelmed by hoaxes, bad science, and scientific opprobrium.

MYSTERY SPIRALS IN CEREAL FIELDS

Late summer 1981. Ross-on-Wye, England. An eye-witness account:

> I live on a ridge 450 feet (135 metres) above sea level, about 100 feet (30 metres) above the adjacent land; it is quite steep in parts on the north side and stretches for about 1½ miles (2.5 kilometres). One day, at about noon, I was inside my cottage when suddenly I heard a very loud roaring sound, not unlike an express train. I ran outside to see what it was, but saw nothing; the noise was something like the sound of a falling bomb. I thought no more of this until the following morning when taking my dog for a walk. Then I saw two large circles, about 25 feet (7.6 metres) in diameter, of flattened barley in a nearby field. A neighbor who lives on the north side of the ridge had also heard the roaring noise but could find no cause for it. I wondered if we had heard some part of an aircraft or satellite, or even a small meteor, coming down and, with the local farmer, we investigated the circles, but found no debris at all---just flattened barley. The farmer said that sometimes growing conditions made barley collapse at its base, though he could not understand the almost perfect circle.

Further investigation turned up people who had seen a whirlwind in the area at the time. (Anonymous; "Mystery Spirals in Cerealfields," Journal of Meteorology, U.K., 8:216, 1983.)

Comment. UFO enthusiasts usually attribute such circles of flattened crops to flying saucers, but apparently whirlwinds are adequate explanations. However, the noise and action of the reputed whirlwind force us to categorize it with the explosive onset of other whirlwinds, as described in GWW1 of Tornados, Dark Days, Anomalous Precipitation.

WHIRLWIND SPIRALS IN CEREAL FIELDS: QUINTUPLET FORMATIONS

"In 1983, as British meteorologists are well aware, Britain had one of its better summers of the century, with July proving to be the hottest in the 300-year-record. At the same time, 1983 proved to be a bumper summer for the production of 'mystery spirals' (and for heat whirlwinds generally). Moreover, and entirely unexpectedly, some of the spiral formations turned out to be symmetrically complex systems in an extraordinary manner: as many as four sets in different parts of southern England were found to consist of a single circle attended by four smaller satellite ones.

"The beauty of these sets of circles caught the attention of the national newspapers, and thence the imagination of the general public. The story about the manner and the sequence of several of the 1983 discoveries has been given by Ian Mrzyglod (Probe Report, vol. 4, 4-11). Here, we shall simply summarize the main facts, many of which have not been detailed before.

"Set 1. Set of five circles at Bratton, Wiltshire (NGR ST 902522, below and northeast of the Westbury White Horse), consisting of one large circle (15 m diameter) and four satellites (each 4 m diameter). The distance between opposite pairs of circles was about 40 m (centre to centre)."

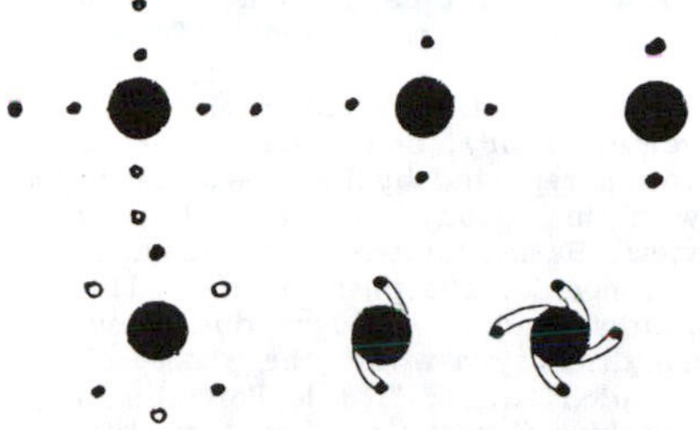

Some dynamically possible patterns of whirlwinds and their satellites. Set 1 conformed to the second pattern.

The other three sets are very similar and are omitted here. The aerial photographs of the quintuplets are remarkable. Meteorologists describe the circles as being the consequence of a large central whirlwind accompanied by four satellites. There seems to be some aerodynamic basis for accepting the reality of large vortexes attended by several smaller ones. (Meaden, G.T.; "Whirlwind Spirals in Cereal-Fields: The Quintuplet Formations of 1983," Journal of Meteorology, U.K., 9:137, 1984.)

Comment. The regularity of whirlwind circles has prompted some to label them "UFO nests." Desert dust devils and steam devils often exhibit coordinated motion and geometrical organization. See GWW0 in Tornados, Dark Days, Anomalous Precipitation.

MULTIPLE WHIRLWIND PATTERNS

English meteorologists are spending some of the lazy summer days out in the countryside tracking down whirlwind patterns engraved on fields of wheat and other crops. One eyewitness account of the formation of a single spiral pattern has been found. However, the multiple spiral patterns excite the most interest because of their geometric regularity. Between 1980 and 1984, eight quintuplet patterns have been found, consisting of a large central circle and four smaller satellite circles. Triplets were also discovered. Although the origin of the multiplet patterns is still unexplained, some interesting generalizations have emerged:

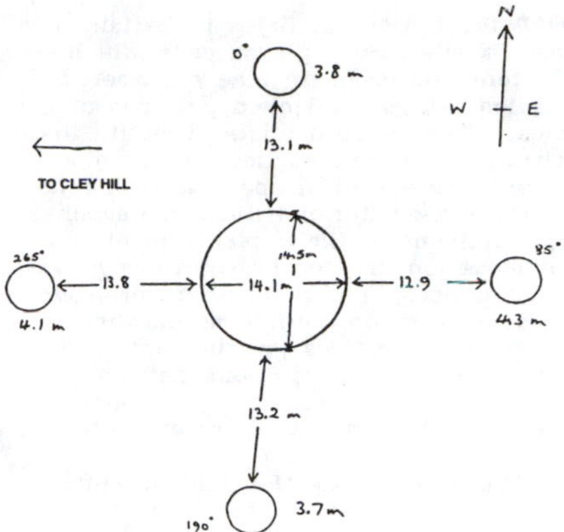

Dimensions of the crop circles found in a grain field near Cley Hill, England.

1. The whirlwinds responsible for the flattened circles of crops have lifetimes of only a few seconds, whereas dust devils may persist for many minutes;
2. These whirlwinds seem to occur around evening time instead of during the heat of the day; and
3. They are all anticyclonic, while tornados are almost all cyclonic and true heat whirlwinds are split about evenly in their spin direction. (Meaden, G.T.; "Advances in Understanding of Whirlwind Spiral Patterns in Cereal Fields," Journal of Meteorology, U.K., 10:73, 1985.)

SPIRAL-CIRCLE GROUND PATTERNS IN FIELD CROPS

G.T. Meaden has recently summarized his research into those marvelously sharp and complex designs cut into English field crops by atmospheric vortices. (Why other countries are not similarly affected is unknown.) The simple circles range from 3 to 30 meters in diameter. The central, flattened vegetation is crushed clockwise about half the time, counterclockwise in the other cases. On two known occasions, the flattening process has actually been observed. The invisible atmospheric vortex does its work in 30 seconds or less and generates a high-pitched humming sound.

The complex nature of these vortices is attested to by the rare ringed circles and multiple patterns. Both single- and double-ringed circles are known. The wind direction always alternates from central circle to first ring to second ring. The most common multiple patterns consist of large central circles flanked by two or four, smaller satellite circles, all nicely spaced. In the quintuplets, all five circles are usually flattened clockwise, but one case has presented theorists with four counterclockwise circles accompanied by a single improbable clockwise circle! (Meaden, G.T.; "The Mystery of Spiral-Circle Ground Patterns in Crops Made by a Natural Atmospheric-Vortex Phenomenon," Journal of Meteorology, U.K., 13:203, 1988.)

VISUAL SIGHTINGS OF VORTICES IN BRITAIN

Of late, the cereal fields of Britain have been visited by a phenomenon which flattens the crops in nicely geometric circles, rings, and even patterns of circles. The meteorologists attribute these circles to unseen vortices in the atmosphere; more radical speculators invoke UFOs and mysterious Russian weapons.

Pertinent to the explanation of this phenomenon are recent sightings of vapor vortices in regions where crop circles are common. While no one has yet seen these vortices gouging out circles, these visual manifestations betoken strong circular winds in the proper locations. Here follows a recent account:

"Looking across the field of winter wheat to the east..., he suddenly noticed at a distance of 80 metres...what he took to be a large puff of white 'bonfire smoke' rising to 15 feet (5m) maximum height. The outer part of this 'smoke' column was scarcely rotating but the middle part, which was too thick to see through, was spinning rapidly. In a couple of seconds the effect had ended; the spinning central column had gone and the residual 'smoke' or cloud of fog drifted gently in the prevailing light north-east wind towards the south-west and dissolved after going several yards. He used the word smoke out of convenience but said that the effect was more likely caused by water vapour, cloud droplets or fog. He further emphasized the swiftness of the appearance and disappearance of the phenomenon. It had arrived suddenly like 'smoke from a distant cannon' or just as if 'a smoke-filled or fog-filled balloon had suddenly burst.' That is to say, it emerged as if from nowhere. He made the further point that the spinning column might have been very much longer than he could judge, for he realized that the only part he could see was the part rendered visible by the smoke or fog. The diameter of the cloud was about the same as its height, viz 4 or 5 metres."

The same phenomenon appeared again a few seconds later, and still again 5 minutes later. Many crop circles have been found in the fields around Yatesbury. (Meaden, G.T.; "The Vortices of Vapour Seen near Avebury, Wiltshire, above a Wheatfield on 16 June 1988," Journal of Meteorology, U.K., 13:305, 1988.)

Comment. There may be a connection between these visible vortices and the curious wind gust reported on p. 263. Even more speculatively, there might be a connection to the strange cloud plumes seen in recent years over the Soviet Union. See p. 265.

EYEWITNESS ACCOUNT OF CROP-CIRCLE FORMATION

UFO enthusiasts have had a field-day speculating that nighttime landings of alien craft have created the famous crop circles. Now, a good daytime description of actual crop-circle formation is at hand. Even so, the whole business still sounds pretty mysterious. The source is a letter to G.T. Meaden, Editor of the Journal of Meteorology, U.K., from R.A. Barnes.

"I have been meaning to write to you for some time on the subject of corn circles. About six or seven years ago I was fortunate to see one of these form in a field at Westbury. It happened on a Saturday in early July just before six in the evening after a thunderstorm earlier that afternoon; in fact it was still raining slightly.

"My attention was first drawn to a 'wave' coming through the heads of the cereal crop in a straight line at steady speed; I have since worked this out to be about fifty miles per hour.

"The agency, though invisible, behaved like a solid object throughout and did not show any fluid tendencies, i.e. no variation in speed, line or strength. There was no visual aberration either in front, above or below the advancing line.

"After crossing the field on a shallow arc the 'line' dropped to a position about 1 o'clock and radially described a circle 75 ft radius in about 4 seconds. The agency then disappeard."

Meaden, a champion of the plasma-vortex theory, believes that the observation reported by Barnes is consistent with this theory. During a later interview, Barnes stated that a hissing noise accompanied the phenomenon. This, thinks Meaden, could be due to electrical discharges within the plasma cloud. (Meaden, G.T.; "Circle Formation in a Wiltshire Cereal-Crop---an Eye-Witness Account and Analysis of a Circles-Effect Event at Westbury," Journal of Meteorology, U.K., 14:265, 1989.)

Comment. Still at issue are the formation of a large, swirling mass of ionized air, its mysterious motion, the precision of the circles, and the diverse, almost too-neat geometrical patterns.

CROP CIRCLE CRAZE CONTINUES

Trying to maintain our sanity during the current crop-circle flap, we have adhered closely to reports in the conservative Journal of Meteorology, U.K. But don't imagine that wilder accounts are not surfacing in the newspapers. Two accounts should suffice.

Australia. "Three frightened farmers believe a gigantic UFO landed last week on their Victorian wheat property.

"A pattern of five perfect circles---in which unbroken wheat stalks swirl anticlockwise---have been found on the 9000-hectare farm, West Park, run by Max and Nancee Jolly and their son Stuart, 29.

"The startling discovery comes only weeks after the family's 700 sheep were panicked by a huge yellow object pulsating above the paddocks in western Victoria's Mallee region." (Pickney, John; "Sheep Panicked by Eerie Light in Fields," Melbourne Truth, December 16, 1989. Cr. P. Norman via L. Farish)

Canada. "Argyle, Man.---There's a mystery on Ray Crawford's land and stumped investigators say anything from bizarre weather phenomena to visitors from outer space could have put it there.

"Sometime in the past year, an almost perfect circle was gouged out of a remote patch of the elderly cattle farmer's property, 30 kilometres north of Winnipeg, on the edge of the rock-strewn scrub and bush that comprise the region between Lake Manitoba and Lake Winnipeg.

"There was no sign anything human had a hand in its creation."

Other crop circles have been found in Manitoba, including one close to a barbed-wire fence that had been partially melted by intense heat. (Blackwell, Tom; "Circle Found Gouged in Field Starts UFO Landing Rumors," London (Ontario) Free Press, December 26, 1989. Cr. B.M Cleveland)

BOOKS ABOUT THE CROP CIRCLES

You know a phenomenon has "made it" when a book is devoted to it. With the English crop circles, we now have two books (bibliographical information below.) We will try to stock at least one of these books, but it will be a few months before can can get them on this side of the Atlantic. Meanwhile, a review of the two books in New Scientist provides some information beyond that already presented in several past issues of SF.

First, the crop circles, spanning 5 to 20 meters, are incredibly precise and sharp. Flattened stems on the periphery of a circle almost touch erect, undamaged stems.

The so-called satellite circles that sometimes array themselves around the main circles may be connected by a narrow ring, thus:

Even more curious, a short radial spur extends outward from some circles, so that from the air the circle resembles a fat tadpole.

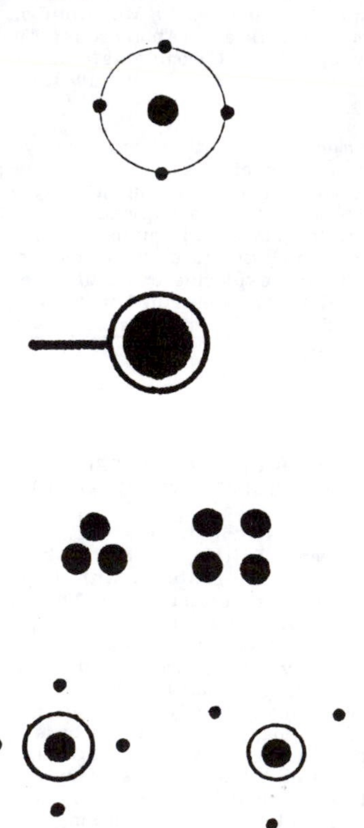

Various "simple" crop circle patterns, which may be aerodynamically possible.

In his book, G.T. Meaden, the Editor of the Journal of Meteorology, U.K., presents his theory of how the circles are incised in field crops:

> He describes the clues that have enabled him to point to the circles being formed by the impact of a body of fast-spinning air that has been partially ionized. He explains how a columnar atmospheric vortex, with a vertical or inclined axis, provides the channel for the formation of a plasma (ionized gas) vortex and for its conduction towards the ground. The ionization of the air ought to be sufficient to make the vortex luminous at night and the fast spin may make the vortex appear ball-shaped. Such a description suggests that Meaden may well have explained some sightings previously reported as UFOs in areas where circles have been found.

(Elsom, Derek; "A Crop of Circles," New Scientist, p. 58, July 29, 1989.) The books are: Circular Evidence, Pat Delgado and Colin Andrews; and The Circles Effect and Its Mysteries, G. Terence Meaden.

Comment. Why are the crop circles so common in England (160 so far this summer alone) so rare elsewhere? Could the luminous phenomenon predicted by Meaden be related to the tornado lights reported under GLD10 in Lightning, Auroras, Nocturnal Lights?

SPINNING BALL OF LIGHT INSCRIBES CROP CIRCLES

In the January 1990 issue of the Journal of Meteorology, U.K., two reports appeared describing eyewitness observations of crop circles-in-the-making. Both involved a self-luminous spinning ball of light. We reproduce here the second of these accounts.

June 28, 1989, north-central Wiltshire, near Silbury Hill.

"Soon after midnight the occupier of the roadside cottage by the path which leads to West Kennett Long Barrow noticed a large ball of light 400 metres distant in a wheatfield to the west. At the time of the observation he was walking from house to garage, and had a clear view to the illuminated part of the field through a gap in a hedge which borders his garden. He described the ball as orange in colour, adding that it was brighter around the periphery, and he guessed the diameter as 30-40 feet (say, 10-13 metres). When first seen, the ball was already low over the field and still descending. The witness watched the base of the ball 'go flat' as it made contact with the crop and/or the ground. The ball then gave 'a little bounce' and after a further 'seven or eight seconds' disappeared in situ.

"Next morning on leaving the house the witness could see via the gap in the hedge a large circle at the place which corresponded to the position of the light source the previous night, and some smaller circles were evident as well."

A flyover the same day revealed a big circle with a ring around it plus smaller circles. G.T. Meaden (the writer) arrived at the site on the morning of the 30th to find that a half dozen additional circles had joined the earlier ones. Five of these formed a quintuplet ---a large central circle with four small evenly spaced outriders. (Meaden, G.T.; "Nocturnal Eye Witness Observation of Circles in the Making. Part 2: North Wiltshire, 29th June 1989," Journal of Meteorology, U.K., 15:3, 1990.)

Comment. Meaden, Editor of the Journal of Meteorology, U.K., is the proponent of the plasma-vortex theory of crop-circle formation. It is remarkable that circles should have been inscribed in the same spot two nights running. It is equally strange that crop circles seem to favor ancient sites, such as Silbury Hill.

CROP CIRCLE CORNER

Every week, it seems, some new facet of the crop circle phenomenon appears. It is reminiscent of the early days of UFOs. Several books have appeared. A periodical, The Cerealogist, now promises to keep everyone up-to-date. People are scouring the old literature and pouring over aerial photos for pre-1980 examples. Theories abound, especially those invoking extraterrestrials!

We have room here for only a few brief items.

Scorched earth in Xenia, Ohio. A hired hand of Gene Eck was harvesting a field of soybeans when he came upon a circle of flattened plants---bent but not broken ---some 80 feet in diameter. Inside this circle was a 40-foot-diameter ring of burnt stubble. Within the ring was a patch of undisturbed foxtail 14 feet in diameter. The soybean circle was a half mile from the nearest road; no tracks led into it. (Williams, Nat; Illinois Agri-News, November 9, 1990. Cr. R. A. Ford)

Column of light in Wiltshire. During the summer of 1990, teams of English observers scanned the cereal fields at night. At 2:30 AM, on July 25, R. Flaherty, an experienced wildlife photographer, saw a single shaft of light descending from high in the sky toward a Wiltshire wheat field. Flaherty's view of the field itself was cut off by a ridge, so he could provide no further data. When morning came, as you probably surmised, the field displayed flattened wheat---not the run-of-the-mill circles, but a scroll of sorts and even a triangle with rounded corners.

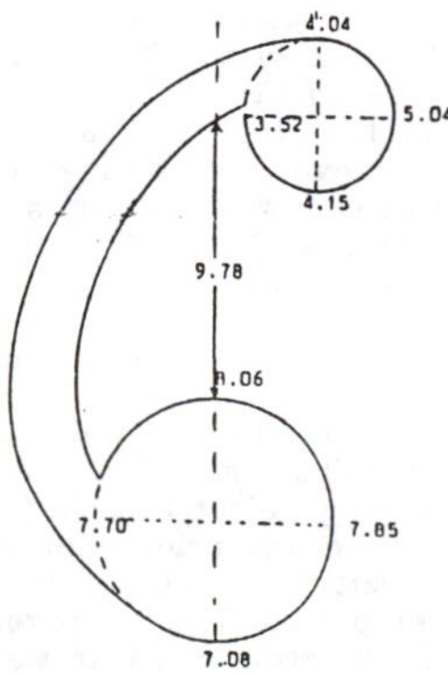

The Wiltshire crop "scroll." Dimensions in meters.

(Meaden, G.T.; "The Beckhampton 'Scroll-Type' Circles, the Beckhampton 'Triangle,' and Strange Attractors," Journal of Meteorology, U.K., 15:317, 1990.)

Comment. Could the English column of light have been created by the same force that made the Ohio burnt circle?

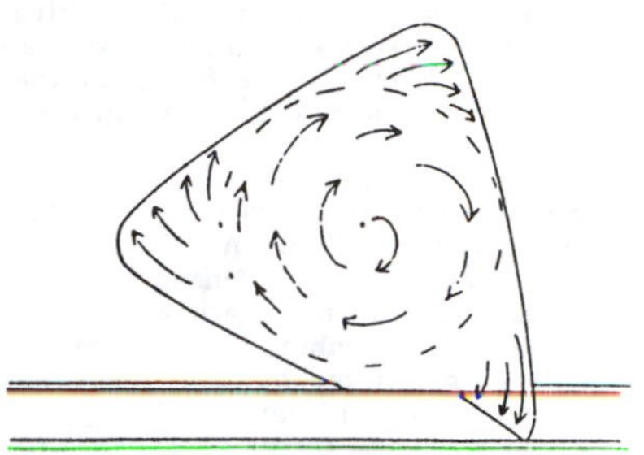

The crop "triangle" located near the "scroll" figured above. The triangle's sides are 10-11 meters long. The more complex the crop figures get, the more difficult they are to explain aerodynamically.

Des Ronds dans le Blé. Yes, the French are chasing crop circles, too. In fact, a team of 8 French observers (designated VECA 90) spent the summer of 1990 in England. After watching by night without success, reviewing the English data, they finally discovered the secret ("ils ont finalement decouvert le pot aux roses"). The crop circles and all the elaborate designs are man-made! In fact the French team demonstrated how one could quickly make circles and more complex designs with a garden roller. Case closed!? (Pinvidic, Thierry; "L'Histoire Folle des Ronds dans le Blé," Science et Vie, no 878, p. 28, November 1990. Cr. C. Maugé.)

Comment. That all crop circles are man-made is debatable, certainly many are, and one can make a case for a meteorological origin for the simpler geometries. One English contact states firmly that a review of old aerial photographs found no complex patterns at all. But who knows what next week's mail will bring?

CROP CIRCLE ROUNDUP

Summary of Britain's 1990 crop circles.

> Over seven hundred circles were found in Britain in 1990, the earliest in April, the latest toward the end of August. They were spread across thirty counties, including Wales and Scotland, besides which there were numerous good reports of circles from Ireland, Holland, Bulgaria, Japan (at least twenty), Canada (over twenty), the U.S.A. and some other countries. The large British total was made possible because of the cooperation of so many enthusiasts via the nationwide CERES organisation.
>
> As usual for Britain most circles were found in wheatfields, but there were some reports from fields of barley, rapeseed, linseed, and long

grass grown for silage. In 1990, as in 1989, Wiltshire dominated the British scene with about 70% of the year's total. This year the leading counties were Wiltshire: over 400; Hampshire: over 50; Norfolk: 18; Devon: 17; Sussex: 16; Oxfordshire: 13; Buckinghamshire: 12; and so on.

In his review of the 1990 phenomena, G.T. Meaden dwelt on the Hampshire dumbbell formations. From these many spectacular "circles" we focus on the one at Seven Barrows, Hampshire.

> The next system [see figure] was at Seven Barrows, north of Litchfield, Hampshire near the A34 highway to Newbury. On the evening of 22 June I pointed out this featureless field to conference members as we drove past following our circles tour, saying that this was a 'repeater' region for circles events (circles are known for these fields for 1976, 1978, 1981, 1983 and 1985). The next morning, the day of the conference, attendees travelling north from Hampshire to Oxford spotted the formation which had appeared overnight. The circles were a hundred metres from a group of Bronze Age barrows which had been there for over three thousand years.

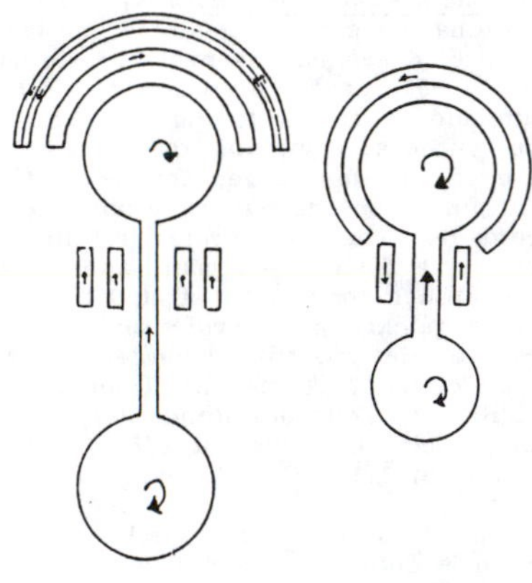

Two of the 1990 Hampshire dumbbell "designs." (Left) Near Seven Barrows. (Right) Near Morestead. Such "designs" are impossible to explain in terms of known, naturally occurring forces.

CERES is the Circles Effect Research Group, operated by G.T. Meaden, who is also the Editor of the Journal of Meteorology, U.K. Meaden, a scientist, strongly contends that all crop circles, despite their complexities and seeming symbology, are natural phenomena; namely the products of atmospheric vortices. Yet, he feels compelled to state that "the details of these vortices, the vortex-crop interaction and the resulting crop-circles display many amazing features which denote an extraordinary phenomenon at work---one which will be shown to have very considerable consequences for physics, meteorology, and other research disciplines in the coming years."

(Meaden, G.T.; "The Major Developments in Crop-Circles Research in 1990, Part 1," Journal of Meteorology, U.K., 16:51, 1991.)

Crop-circle formation observed in 1934. July 1934. Eversden, England. The observer was a Miss K. Skin of Cambridge.

> I witnessed a corn circle being formed in 1934. I was gazing over a field of corn waiting to be harvested when I heard a crackling like fire and saw a whirlwind in the centre of the field, spinning stalks, seeds and dust up into the air for about 100 or more feet. I found a perfect circle of flattened corn, the stalks interlaced and their ears lying on top of each other (some even plaited) on the periphery. The circle was hot to the touch.

An interview with Skin confirmed that the plaiting of the stems was the same as observed today, 56 years later. In addition, the same whirlwind created a second circle about 4 meters in diameter in the corner of the same field. (Meaden, G.T.; "Crop-Circle Formation Witnessed in Cambridgeshire, July 1934." Journal of Meteorology, U.K., 15:389, 1990.)

CROP CIRCLE UPDATE: WHAT ARE "THEY" TRYING TO TELL US?

Several hundred crop circles have afflicted English farmers' fields already this year. The English newspapers haven't neglected them, but where are Nature and Science? Here is a genuine mystery, and most scientists don't dare touch it for fear of being labelled "kooks." Well, we do have one item from New Scientist, but we'll have to rely upon the newspapers for this ever-more-bizarre phenomenon.

Australia now a target.

> Dozens of flattened rings of wheat have been reported recently in Australia over a wide area of arable country in northwestern Victoria. The rings resemble the corn circles found in southern Britain.
>
> Max and Nancee Jolly, farmers at Turiff West, found 12 rings in their wheat crop. Each ring was 12 metres across and rock-hard in the middle. 'The wheat in the rings was bent over at the base but not damaged in any other way.' said Nancee Jolly.
>
> The Australian Sceptics suspect pranksters. (Anonymous; "More Circles," New Scientist, p. 23, August 11, 1990.)

An article in a Perth newspaper puts the number of wheat circles in Victoria at 400, as of July 9. That's a lot of work for hoaxers! It is also said that the soil in the rings is "magnetically altered," whatever that means. (Anonymous; "Outback Martian Rings Riddle," Perth Daily News, July 9, 1990. Cr. P. Norman via L. Farish)

Plasma vortex picked up by radar.

> Japanese and British meteorologists are investigating a link between a fast-moving object crossing the Pacific and the mysterious appearance of crop circles in English fields.
>
> A ship from Tokyo University's Ocean Research Institute was in the Pacific when its radar equipment located a large object travelling more than four times the speed of sound. The radar discounted it as an aircraft because of its size, 400 metres across, and it sped northwards.
>
> The Japanese scientists identified the object as a plasma vortex, caused by freak weather. The phenomenon is similar to ball lightning and believed to be generated by 'mini-tornadoes' of electrically-charged air.
>
> Plasma vortices can be luminous at night. 'They are often mistaken for UFOs,' says Dr Terence Meaden, director of the Oxford-based Tornado and Storm Research Organization.

(Spicer, Andi; "Clue to Mystery of Circles," London Observer, May 20, 1990. Cr. T. Good via L. Farish)

Alien Hieroglyphics?

> The extraordinary variety of circle formations and multiple-ringed circles is quite unlike what one would expect for a natural phenomenon, such as an atmospheric vortex. The complexity has increased through the 1980s, and this year it has developed at a startling pace.
>
> In May there began to appear what researchers call pictograms. Initially these consisted of two circles joined by a straight channel of flattened corn, with extra features such as rectangles or semicircular rings. There have now been about 20 of these. The latest have quadrupled the length to about 150 yards and consist of complex arrangements of up to nine plain or ringed circles with new features, like 'keys,' which can be seen in photographs of the Alton Barnes pictogram. Nothing like this was observed in previous years.

(Wingfield, George; "Ever-Increasing Circles of Bewilderment," London Independent, August 4, 1990. Cr. T. Good via L. Farish)

Comment. Having failed to establish communication with us via UFOs, marine lightwheels, cookie-cutter holes, and other phenomena, "they" are now trying crop circles! But, less flippantly, once hoaxes have been winnowed out, we may have an important phenomenon here.

CROP CIRCLES: DAISY PATTERNS AND A RED BALL OF LIGHT

G.T. Meaden, in the second installment of his review of 1990 crop-circle research, singled out for special attention the so-called "daisy patterns." While these are not as intricate and mysterious as the spectacular nine-circle complex at Alton Barnes, the formation of one of the daisy patterns may have been accompanied by luminous phenomena.

> Circles in a daisy pattern were reported from Devonshire and Somerset County: the first a centre circle with seven regular satellites, evenly spaced, from Bickington in June; the second a circle with six similar satellites from Butleigh Wootton, near Glastonbury in mid-July.
>
> A third daisy-pattern system, one with ten ringed satellites surrounding a central ringed circle, turned up at the end of July in East Anglia.

This last was formed on the night of 30-31 July, possibly in the late evening of 30 July at the time of the observation of a glowing ball of red light. It was seen by the farmer shining above his field at Hopton as viewed from his house on the edge of Gorleston (Norfolk). "He looked at it through his binoculars and described it as a red central glow with a thinner red outer ring...By the time he had passed the binoculars to his son the thing had gone" (Eastern Daily Press).

(Meaden, G.T.; "Major Developments in Crop-Circle Research in 1990: Part 2," Journal of Meteorology, U.K., 16:127, 1991.)

SUBTERRANEAN "CIRCLES"

As if we didn't have enough problems with crop circles on the earth's surface, it now seems that whatever agency (or "entity") that is responsible for them also plies its craft underground!

> Sets of concentric rings, similar to those found last summer in British wheat fields, have been discovered in a Japanese subway tunnel.
>
> .....
>
> Many sets of concentric rings were found drawn in dust that accumulated on the ground and walls inside the tube. The metro versions of the mystery circles are much smaller ---up to 8 centimeters in diameter--- than the British ones, the largest of which measures scores of meters.

Y. Otsuki, a professor of physics at Waseda University, discovered the rings and believes that plasma generated in the air creates them. Subway tunnels, he says, create conditions similar to those in the plasma generators he uses in his fireball research.

A photo of the rings accompanying the article shows six neatly-formed, concentric rings around a central crude circle. (Anonymous; "'Mystery Circle' Found in Tunnel," Asahi Evening News, April 5, 1991. Cr. Y. Matsumura via L. Farish.)

Speculations. Apparently, plasmoids can be of any size: crop circles may be 100 feet in diameter of just a foot or two, and now we may have centimeter-sized expressions of plasmoid activity in the unexpected locale of the subway tunnel!

Building upon these observations, it is not unreasonable to ask whether plasmoids (including plasma vortices) may not exist on larger scales, say, astronomical and geological. We are drawn to those strange swirl markings on the moon. These loop-like patterns are 10-50 kilometers in size and are associated with strong magnetic anomalies. (See ALE5 in The Moon and the Planets.) And right here on earth we have the devastation of the Tunguska Event sans a gaping crater. And how about the shallow Carolina Bays, some of which are associated with magnetic anomalies? Could such phenomena be the handiwork of plasmoids rather than meteorites or comets? Plasmoids could also be involved in such phenomena as ball lightning, the cookie-cutter holes (p. 198), and even spontaneous human combustion!

CROP CIRCLES: HOAXES OR NATURAL PHENOMENA?

The question posed by the above title was answered presumptuously and one-sidedly by Time (September 23) and more objectively by Science (August 30). While Time implies that all crop circles are hoaxes, the Science article states that the "really bizarre" circles are hoaxes and that the simpler circles may have acceptable meteorological explanations. Unfortunately, the ridiculing tone of the Time article will probably set back the budding scientific interest in crop circles reported in Science. The real losers, as we shall see below, are those crop-circle experts who assert that they can always detect hoaxes.

Essence of the Time article. Two crop-circle hoaxers have confessed. D. Chorley and D. Bower have admitted that they have made as many as 25-30 fake crop circles per year, since 1978, including some of the bizarre ones. All they needed was a 4-foot wooden plank, a ball of string, and a baseball cap with a wire mounted on it for sighting purposes. It was all too easy! And, they assure us, other hoaxers were active in fields at night, too. This is indeed damaging evidence to crop-circle enthusiasts. Time concluded that the admissions of Chorley and Bower have "brought to an end one of the most popular mysteries Britain---and the world---has witnessed in years." (Constable, Anne; "It Happens in the Best of Circles," Time, 138:59, September 23, 1991.)

Essence of the Science article. The Science piece was written before the Time exposé, but it presents several points supporting the existence of a genuine natural phenomenon beneath all the obvious hoaxing.

T. Meaden, an English scientist, has personally investigated over 1000 crop circles. He remarks that, before all the media hype, all the corn circles were simple in design---including plain circles, ringed circles, and circles with small satellite circles nearby. In 1991, Meaden organized Operation Blue Hill, during which 40 researchers watched British crop fields. Significantly, one circle formed in a field they had ringed with automatic alarms, making a hoax very unlikely.

All in all, some 1800 crop circles have been recorded, many in other countries, including Japanese rice fields. One doubts that hoaxers could have been this ubiquitous.

Further, Y. Ohtsuki, a Japanese scientist, has been able to reproduce some of the characteristic corn-circle patterns by dropping plasma fireballs into a plate dusted with aluminum powder. Even double rings can be created around central circles in this way. So, the simple crop circles could well have reasonable explanations.

The gist of the Science article is that interesting science might be done with the crop-circle phenomenon. But will it after the Time broadside? C. Church, professor of aeronautics at Miami University in Ohio, says, "Everyone should be open minded, but I wouldn't want to get a reputation among fellow scientists as working on weird and off-beat things." (Anderson, Alun; "Britain's Crop Circles: Reaping by Whirlwind?" Science, 253:961, 1991.)

CROP CIRCLES: IF SOME ARE HOAXES, ARE THEY ALL HOAXES?

One could almost hear a sigh of relief among the skeptics of unusual natural phenomena when two Britons admitted to manufacturing scores of crop circles. After all, the crop circles are about as outrageous as UFOs and toads-entombed-in-stone. However, the crop circles have not gone away. In fact, plant and soil samples from the circles seem to point to bizarre, highly energetic processes at work. This aspect of the phenomenon has been discussed by R. Noyes, Secretary of the Center for Crop Circle Studies (CCCS). First, though, Noyes has asserted that hoaxes cannot explain the large numbers of circles that have been counted---about 1000 between 1980 and 1989. He continued as follows:

> The events of 1990 and 1991 (totalling about a further 1000 over the two years) certainly present a puzzle. Hoax is beyond doubt in some cases, but it seems very unlikely as a general explanation. Many events have been very large and very elaborate; they have occurred widely about the country (sometimes several on the same night in counties far from each other); there have been very few cases of detection of hoax, despite massive surveillance in the Marlborough/Devizes area, where so many of the events took place; circles (including a dumbbell formation) occurred within visual and radar range of a hi-tech watch mounted by [G.T.] Meaden and supported by anti-hoax equipment without a trace of human action. It is clear that hoax cannot account for all we have been seeing.
>
> Very recently, laboratories in the US, acting in collaboration with CCCS's Crop Research Panel, have reported interesting physical changes in crops and soil collected from circle formations as compared with control samples. Other "hard" evidence is accumulating for the action of some short-lived force in the formation of genuine events, and its nature seems to be such that human activity cannot account for it. It may or may not be evidence for the operation of Meaden's "plasma vortex" (in whatever form he decides to develop it).
>
> The only thing to add at this stage is that if the "plasma vortex" is in question, it seems capable of far more elaboration in the creation of crop formations than Meaden has yet allowed. For sheer exuberance and inventiveness, there has been no force in scientific history to match it.

(Noyes, Ralph; Letter to UFO Brigantia, November 23, 1991.)

Comment. We shall be looking forward to documentation of the lab tests mentioned. It is "interesting" that both UFO and crop-circle phenomena have historically begun with relatively simple observations, and now apparently to anomalous physical traces. Will there now be "encounters" and "abductions"?

FOUR LUMINOUS SPINNING VORTICES

July 21, 1991. Wiltshire, England. The following observations were made by D. and E. Haines, about 11 PM, after they had just passed through the village of Hill Deverill:

> We saw what looked like the reflection of the moon from the driver's window (i.e. we were looking in a westerly to south-westerly direction), and, as we travelled on, it then looked like four beams of a high-powered torch, but, as we went still further and were more or less alongside, we could see it was in fact four swirling shapes, shining white (not very bright, but bright against the night sky). We turned off the car engine and could hear a whooshing noise (like a car a distance away, going fast on a motorway---but the sound did not come any closer). The national grid reference was ST 866392 approximately.
>
> These four spinning shapes (like the top of a cotton bud---not dense and solid) went round and round in a clockwise direction. They came together in the middle, out and round and round. They did this several times (once, one went off to the right but came back into 'formation'), and then they came back together and just disappeared.

(Haines, David, and Haines, Elaine; "An Observation of Four Luminous Spinning Vortices, 21 July 1991," Journal of Meteorology, U.K., 17:24, 1992.)

Comment. Could the controversial crop circles, common in Wiltshire, be related to these luminous objects?

CROP-CIRCLE CONTEST

July 11-12, 1992. Buckinghamshire, England. On this dark night, in a barley field, 12 teams assembled in hopes of winning a $5,200 prize provided by the Koestler Foundation and the German magazine PM. This sum was to be awarded to the best crop-circle hoaxers.

First prize went to three engineers from a British helicopter company, who used rope, plastic piping, and a ladder suspended from a trestle. Close behind in the competition was American J. Schnabel, who, working all alone, required only a plank, some rope, and a small garden roller to produce a creditable, rather elaborate design. (Anonymous; "Circle Hoax Contest," Science, 257:481, 1992.)

CROP CIRCLE FOUND INSIDE A FENCED COMPOUND IN JAPAN

Y-H. Ohtsuki, a physicist at the Waseda University, Tokyo, reports (in pretty good English):

> A large circle appeared in the doubly-fenced compound of a Radio Nippon transmitting station near Tokyo. At the same time the survey protection equipment of the transmitting system worked abnormally. This shows that the circles effect is not simply a matter of hoaxing.
>
> It was the night of 31 August 1991 when the survey protection equipment (electromagnetic-noise warning detector) of the transmitting system of the Radio Nippon station in Kisarazu, Chiba prefecture, worked seventeen times during 40 minutes from 2.00 a.m. (that is, on 1 September). This is a wholly abnormal occurrence because the survey protection goes off only once a week on average.
>
> Next morning one of the staff discovered a 10-metre circle in the doubly-fenced ground of the Radio Nippon station. The grass in the ground was pushed down, but without leaving a clear spiral mark. The ground area is approximately 20,000 m^2, and three antenna towers are located in the ground. The fences are formed by 2.5-m-high wire netting and the station was watching for 24 hours. There were only two men in the station, and they were in a watching room for eight hours from 10.00 p.m. to 6.00 a.m.. Moreover, I can add that neither of the men had ever heard of the circles effect at that time, so that after the discovery of the grass circle the next day they did not report it for 40 days. By the way, there are no roads or railway which a hoaxer could have used to approach by car or train.

(Ohtsuki, Yoshi-Hiko; "An Example of the Circles Effect Which Appeared in a Well-Protected, Fenced Compound in Japan," Journal of Meteorology, U.K., 17:115, 1992.)

CROP CIRCLES: A MIDDLE GROUND?

On one hand, mainstream scientists, when they deign to notice them at all, pronounce that all crop circles are the work of hoaxers, as inf the article by J.W. Deardorff references below. On the other hand, several books and a flood of reports in fringe publications claim that the crop circles, particularly the complex ones, are evidence that extraterrestrial intelligences are attempting to communicate with us. There is also a middle ground upon which stands G.T. Meaden, a physicist, and a few other scientists. Meaden has summarized this third position in the following paragraph:

> ...we believe that the formation of real crop circles is a rare phenomenon resulting from the motion of a spinning mass of air which Professor Tokio Kikuchi has modelled by computer simulation and calls a nanoburst. This disturbance could involve the breakdown of an up-spinning vortex of the eddy or whirlwind type. On this theoretical model such a process leads to plain circles and ringed circles---types which are known from pre-hoax times in Britain and other countries, and are the only species which credible eye-witnesses have seen forming. All other so-called crop circles reported in the media news in recent years are likely to be the result of intelligent hoaxing, while the so-called paranormal events to which Deardorff alludes are nothing but the consequence of poor observation and/or exaggeration by susceptible mystics and vulnerable pseudo-scientists. In the absence of hoaxing the subject would still be unknown to the general public because the average number of real-circle reports per annum is small (indeed in some years it may be zero).

(Meaden, G. Terence; "Crop Circles: The Real and the Hoaxed," Weather, 47:368, 1992. Deardorff, J.W.; "Crop Circles: Someone Had to Say It!" Weather, 47:142, 1992.) This item actually appeared in SF#89.

PRECIPITATION AND FALLS

Following Fortean convention, "things" that fall from the sky are divided into two classes: (1) "Normal" precipitation, such as rain and snow, that is colored, of abnormal size, or appears under unusual conditions; and (2) Falls of material not considered indigenous to the atmosphere, such as large ice chunks, frogs, fish, etc. Our outline here follows:

- Large snowflakes.
- Ice chunks and hydrometeors.
- Organic matter. Frogs, fishes, crabs, etc.
- Miscellaneous materials. Coke, unidentifiable, jelly-like blobs, etc.

LARGE SNOWFLAKES

UNUSUALLY LARGE SNOWFLAKES

In a recent review of records of falls of very large snowflakes, W.S. Pike lists eleven instances where flakes more than 5 centimeters (2 inches) in diameter have been observed. Of these, we have already cataloged six in GWP2, in Tornados, Dark Days, Anomalous Precipitation, including the prize of the lot: the 15-inch snowflakes that parachuted down on Fort Keough, Montana, on January 28, 1887. Five of Pike's cases that we did not catalog have diameters of "only" 5 or 6 centimeters. The sixth uncataloged observation would certainly have been worth including if we had known about it:

> March 24, 1888. Shirenewton, Eng-

land. "Snowstorm with extraordinary flakes, some were 3 3/4 in. in diameter, yet only ¼ in. thick, falling like plates. The storm lasted only 2 minutes but in this short period the ground was covered 2 in. deep."

The quotation is from British Rainfall, 1988, as requoted by Pike.

In all cases, huge snowflakes are really aggregations of many thousands of individual flakes. Observers have thought that the big flakes attract individual flakes. (Pike, W.S.; "Unusually-Large Snowflakes," Journal of Meteorology, U.K., 13:3, 1988.)

Snowflakes 38 centimeters in diameter fell at Fort Keogh, Montana, in 1887.

ICE-FLAKE FALL

July 28, 1979, near Norwich, England. On the fringes of a thunderstorm, large ice flakes fell for about 5 minutes along with pea-size hail and a few raindrops. The flakes were about 5 x 10 cm (2 x 4 inches) by 0.3 cm (1/8 inch) thick. They floated down slowly with the flat areas horizontal like falling leaves. (Anonymous; "Remarkable Fall of Large Ice-Flakes," Journal of Meteorology, U.K., 4:280, 1979.)

ICE CHUNKS AND HYDROMETEORS

HAILY ROLLERS

August 1897. Stirling, England. After a heavy thunderstorm with hailstones 'no larger than usual,' a shepherd thought he saw a sheep prostrate in a field. Closer inspection revealed instead a block of ice weighing about 50 kg (110 pounds!). This seems much too heavy for a conventional single hailstone. If it had been an agglomeration of smaller hailstones, it would have been smashed to bits upon impact. One meteorologist has suggested the ice block might have been a hail roller analogous to snow rollers. Snow rollers form when a small bit of snow starts rolling under the influence of the wind and/or gravity, ending up as a substantial natural cylinder of rolled-up snow. However, even the author seemed a bit dubious about hail rollers! (Harrison, S.J.; "A Nineteenth Century Hail Roller?" Journal of Meteorology, U.K., 7:77, 1982.)

MASSIVE ICE LUMP FALLS ON ENGLAND

December 10, 1980. Birmingham. An ice lump weighing 1 pound, 6 ounces (626 grams) fell into a garden at Kings Norton. The lump was almost spherical and had a circumference of 12 inches. (Anonymous; "Ice Ball Falls into a Birmingham Garden," Journal of Meteorology, U.K., 6:46, 1981.)

1500-POUND ICE CHUNK FALLS FROM SKY

June 26, 1985. Hartford, Connecticut.

> Scientists yesterday tried to determine the origin of a 1500-pound sheet of ice that mysteriously dropped from the sky and smashed into a backyard fence. David H. Menke, directory of the Copernican Observatory and Planetarium, said the ice was probably 6 feet long, 8 inches thick and moving at about 200 mph. 'It's unusual in the fact that it fell from the sky,' said Craig Robinson, curator at the planetarium. 'That does not happen often.' A 13-year-old boy was in his backyard Monday with a friend when the ice came 'whirling' from the sky and smashed into the fence about 10 feet away from them.

The remainder of the article gives the opinions of some scientists who were contacted about the fall. The director of the observatory thought the ice probably fell off the wing of an aircraft. The director of the American Meteor Society suggested a cosmic origin, providing the ice were pure. An astronomy professor assured everyone that it couldn't be cometary, because the sun would melt particles of ice in outer space. Instead, he opted for strong thunderstorm winds picking the ice up from "somewhere" and dropping it on Hartford! (Anonymous; "1,500-Pound Ice Chunk Falls from Sky," Manchester (NH) Union Leader, June 27, 1985. Cr. B. Greenwood via L. Farish)

GIANT ICE BLOCK FALLS IN CHINA

April 11, 1983. Wuxi, China. In full view of passersby, a 50-kilogram block of ice hurtled from the sky and splintered on the pavement. The ice was milky white (some say greyish), and apparently of roundish shape before breaking up. Chinese scientists hurried to the scene and were able to preserve specimens. Their study of the meteorological conditions and specimens led them to conclude that the ice was truly meteoric; that is, extraterrestrial. (Wei, Chen; "Giant Ice-Block Falls in East China City," Journal of Meteorology, U.K., 8:188, 1983.)

Comment. In the West, such ice falls are automatically attributed to chunks falling from aircraft overhead.

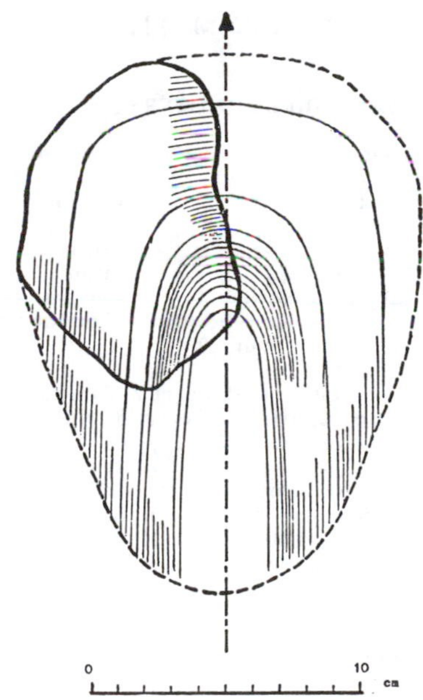

Layered structure of a large hydrometeor that fell in Manchester, England, in 1973 during a thunderstorm. Weight: about 2 kilograms.

50-POUND 'ICE BOMB' FALLS IN WEST VIRGINIA

June 26, 1990. Jerry's Run, West Virginia.

> Heisel and Alice Amos, and their grandson, Aaron Hupp, had just turned on a movie on television when the house was jarred with what Mrs. Amos thought was an explosion.

> Looking out the front door, they saw their son, Donald, 43, looking in the direction of their television satellite dish some 30 yards away where something had hit the ground with a terrific impact.
>
> Inspecting that area, they found a hole some 24 inches long and 18 inches wide, and about four to six inches deep filled with large chunks of broken ice. Amos said pieces of baseball- and marble-size ice were scattered in a 30-foot radius around the hole.

Further facts from this newspaper account:

1. Several other chunks of ice were found in an area about 1 mile long.
2. Some chunks made whistling sounds as they fell.
3. The larger chunks were completely transparent except for a yellowish-brown streak.
4. Many of the chunks had sand in them.
5. Some contained holes.
6. The weather was clear.
7. The Federal Aviation Administration stated that if the ice originated in aircraft toilets it would have been blue from the chemicals used. (Hawk, Harold; "50-Pound 'Ice Bomb' Falls near Jerry's Run," Parkersburg News, June 27, 1990. Cr. M. Frizzell)

ORGANIC MATTER

FISH AND WINKLE SHOWERS

"Loud slapping sounds disturbed Ron Langton as he settled down to watch late-night television at his home in East Ham in London on 26 May 1984. He thought nothing more of the noises until next morning when he went outside and saw half a dozen fish in the backyard and on his roof. They were flounders and whitings, about 10 to 15 centimetres long.

"Two residents of nearby Canning Town also reported between 30 and 40 fish scattered over their gardens. Could a flight of herons returning from the Thames have dropped their catch? The Natural History Museum identified the fish as just what you would expect to find in the lower Thames. Could a waterspout on the Thames have lifted the fish up to cloud level, carried them a few kilometres north, and dropped them on Canning Town and East Ham?"

.....

"On 16 June 1984, a month after the fish falls in London, the owner of a service station near Thirsk in north Yorkshire found winkles and starfish covering the forecourt of his garage and the top of its high canopy. The winkles were salty and many were still alive. Thirsk is 45 kilometres from the sea, and the garage owner thought that this collection of marine life arrived with the torrential thunderstorms during the night. Though proof remains elusive, the winkles and starfish were probably lifted by a waterspout along the east coast and carried aloft for an hour or more within the powerful updrafts of the thunderstorms."

Several more anecdotes of a similar nature can be found in this article. (Elsom, Derek; "Catch a Falling Frog," New Scientist, p. 38, June 2, 1988.)

A TRUE FISH STORY

"It happened during a smelt storm in early spring with the Rhonda K foundering dead in the icy water amid terrifying six- and eight-foot waves off Lake Huron's Thunder Bay light near Alpena."

Wait a minute! A SMELT STORM?

"Spookiest thing I ever saw," Stiles said of the small fish carried from the water by winds. "Something I never heard of before. Smelt started falling from the sky. Thousands of smelt falling like a silver rain, and thousands of gulls going after them. The gulls were all around our heads picking up the smelt." (Barton, John; "A True Fish Story," Ann Arbor News, July 24, 1986, p. B1. Cr. C.R. Engholm)

CRAB FALL AT BRIGHTON

June 5, 1983. Brighton, England. A large spider crab dropped out of a storm cloud in front of Julian Cowan. The crab measured 25 cm across and had a 7-8 cm shell. It was dead and lacked two legs and one claw. The fall was followed almost immediately by wind-driven hailstones. (Meaden, G.T.; "The Remarkable 'Fall' of a Crab at Brighton, 5 June 1983," Journal of Meteorology, U.K., 9:56, 1984.)

FALL OF SEASHELLS IN STORM

At Dilhorne. Sea shells fell with heavy hail:

> They extended for an area of about 50 by 20 metres and occurred in thousands on lawns, flower beds, paths and even the road. Roy was kind enough to give me half-a-dozen specimens for identification. They turned out to be small gastropods, almost certainly of marine origin.

Recourse to field guides did not result in positive identification, so samples were sent to the Bristol Museum. The specimens turned out to be Dove Shells, family Columbellidae, genus Pyrene, which are normally inhabitants of the tropical seas. Their most likely origin was the Philippines! (Swinhoe, P.J.; "Unusual Events along the Squally Cold Front of 21 March 1983 in North Staffordshire," Journal of Meteorology, U.K., 8:233, 1983.)

FALLS OF ALL SORTS OF THINGS

About May 16, 1983. Chippenham, Wiltshire, England. A group of students at Chippenham Technical College reported that a lot of tadpoles had fallen during a thunderstorm. The River Avon is nearby; and a waterspout or small tornado was suspected. (Meaden G.T.; "Shower of Tadpoles....," Journal of Meteorology, U.K.; 9:337, 1984.)

May 26 or 27, 1984. East London. Flounders found on the ground, smelt on the roof.

June 19, 1984. Thirsk, North Yorkshire. After a heavy thunderstorm, a small area was covered with winkles (shellfish), some still alive, and starfish. (Rickard, R.J.M.; "A Remarkable Fall of Fish in East London 26 or 27 May 1984," Journal of Meteorology, U.K., 9:290, 1984.)

July 24, 1984.

> I was in my car waiting at traffic lights in Winton, Bournemouth, when a sheet of off-white or dove-grey liquid fell from the blue sky on to the roof, windscreen and bonnet of my car. A yellow bus next to me and the road around were also affected.

No aircraft could be seen. (Hodge, E.J.; "Fall of a Mysterious Liquid from the Sky," Journal of Meteorology, U.K., 9:340, 1984.)

MISCELLANEOUS MATERIALS

SHOWER OF COKE

"After the severe thunderstorm of 5 June 1983, it was brought to my attention that large amounts of coke had fallen in a gentleman's garden. After being reported in the local press and on local radio that evening, my telephone never stopped ringing with reports of coke haven fallen all over the Bournemouth, Poole and Christchurch area. I investigated several reports and found the pieces to be the same, all having been discovered over lawns, paths, etc. and all found after the storms of the 5th. At one lady's house I picked up 92 pieces of coke and there were still many pieces left. The largest piece of coke measured 6.0 cm by 4.6 cm. At one investigation I was given small roof-like stone chippings which the lady said she saw in melting hailstones." The author is the Bournemouth Meteorological Registrar. (Rogers, P.A.; "Remarkable Shower of Coke from Cumulonimbus," Journal of Meteorology, U.K., 9: 220, 1984.)

BURNING MASS FALLS IN B.C.

March 11, 1984. Duncan, BC. David Thompson was returning home at 8:30 PM, when he spotted a soccer-ball-size burning mass high over the trees. It landed in the road about 200 feet away, sounding like a light bulb popping. For about 3 seconds, it flamed. When approached, it was still sizzling, probably because the road was wet. The fallen substance quickly hardened, but samples were scraped off the asphalt. It turned out to be an odorless, rock-like substance. Left outside overnight, it had become soft by the next day and seemed to be melting. Samples were sent to Victoria for analysis. (Hausch, Karen; Cowichan Leader, March 15, 1984. Cr. L. Farish)

PURPLE BLOBS IN TEXAS

In early September 1979, the Associated Press carried a story about three purple blobs found in a yard in Frisco, Texas. One blob evaporated away, while the remaining two were preserved for analysis by NASA. The blobs were warm when found and had appeared during the height of a meteor shower. At first, NASA scientists did not rule out the possibility that the jelly-like goo might be extraterrestrial, but an AP dispatch the next day (not as widely printed) inferred that the blobs were merely industrial waste! (Anonymous; "NASA Scientists to Probe Mystery of 2 Purple Blobs Found in Texas," Baltimore Sun, September 8, 1979.)

Comment. The blobs closely resemble gelatinous meteors or pwdre ser reported rarely down the centuries. One instance of pwdre ser was reported in 1978 from England in the Journal of Meteorology, U.K., and there are doubtless more that are swept under the rug. We may be sure that NASA will have nothing further to do with something as outrageous as pwdre ser.

VANISHING GOO

> Some time between the 12th and 18th of December (1983), the west end of North Reading in Massachusetts was bombarded with blobs of jelly-like goo, greyish-white and oily-smelling. The first blob---two feet in diameter ---was found by Thomas Grinley in his driveway. He thought something was leaking from his car until he found similar blobs on Main Street and on the gas station pumps. State officials denied that the blobs were dropped by a plane. They were soon absorbed into the pavement, but a little goo was saved and was being studied at the state's Department of Environmental Quality Engineering. Preliminary results showed that they were not toxic.

(Anonymous; "Vanishing Goo," Fortean Times, no. 43, p. 23, Spring 1985. Extracted from USA Today of December 22, 1983.)

Comment. These disappearing blobs represent a typically Fortean phenomenon with a history going back before the first aircraft. The reports are generally ridiculed and quickly written off. Given their historical persistence, perhaps we should pay more attention to them, trivial though they seem.

Speaking of falling goo, a detailed historical study of pwdre ser in folklore and science has just appeared. Pwdre ser, as readers of our Handbooks and Catalogs will know, is the Welsh name for star jelly. That jelly-like lumps of materials have been found in the fields after the fall of a shooting star is an integral part of European folklore. Here is a typical poetic mention by Donne:

> As he that sees a starre fall, runs
> apace,
> And findes a gellie in the place...

(Belcher, Hilary, and Swale, Erica; "Catch a Falling Star," Folklore, 95: 210, 1984.)

ICY MINICOMETS

Icy minicomets are theorized to be house-sized chunks of extraterrestrial ice weighing in the neighborhood of 100 tons on the average. The existence of these unexpected denizens of interplanetary space was first inferred from dark spots seen on the images of the earth's dayglow, as viewed from above by satellite instruments. Some independent measurements of water vapor in the upper atmosphere also tend to support the reality of icy minicomets. The characteristics of icy minicomets that make them so controversial are their large size and the rate at which they are supposed to be impacting the terrestrial atmosphere. If icy minicomets pelt the earth's atmosphere at the suggested rate, and have been doing so down the geological eons, they could have been a major source of the earth's oceans. Such a copious extraterrestrial water source conflicts strongly with current geological theory. In fact, the icy minicomet controversy has been particularly bitter, although unlike the media-dominated crop circle controversy, it has been conducted almost entirely in academic circles.

As with the other scientific controversies followed closely in Science Frontiers, a chronological presentation seems appropriate here.

ATMOSPHERIC FOOTPRINTS OF ICY METEORS

Serendipity triumphs again. From a dayglow experiment aboard NASA's satellite DE-1 (Dynamic Explorer-1) comes an unexpected discovery of considerable potential importance. Looking down on the earth, the DE-1 records the light emitted by atomic oxygen at altitudes of about 200-300 kilometers---the so-called dayglow. The experimenters, L. Frank, J. Sigwarth, and J. Craven, all at the University of Iowa in Iowa City, have found that their dayglow images are speckled with transitory dark spots.

> According to Sigwarth, each hole expands like a drop of dye spreading out in a glass of water; within about 30 seconds the dayglow intensity drops by about 95 percent over an area of about 3000 square kilometers. Then, over the next 3.5 minutes, the dayglow intensity increases toward its normal value as the hole grows to an area of about 25,000 km^2.

The Iowa group thinks that the holes or spots are created by meteors hitting the upper atmosphere because the spots follow the same time distribution as meteors. For example, they are more frequent during the well-known meteor showers. The theory is that the dark spots are formed when ice associated with the meteors is turned into water vapor, which reacts with the atmospheric oxygen producing the dayglow, in effect removing temporarily part of the light source.

So far, everything seems relatively nonanomalous. But when quantities are calculated, though, jaws begin to drop. The sizes of the spots imply that the average meteor weighs 10 kilograms, mostly ice and far larger than has been thought. In fact, they may be characterized as small icy comets; that is, compositionally like the dirty snowballs that comets are now thought to be, but much, much smaller. The implication is that from 1000 to 10,000 times more material is being added to the earth's atmosphere than previously believed---most of it being water. (Weisburd, S.; "Atmospheric Footprints of Icy Meteors," Science News, 128:391, 1985.)

Comment. From this launchpad, one's thoughts can really takeoff. How much water can this bombardment of icy meteors add to the earth, Mars, and other solar system bodies? In the item under "Geology" about the Greenland ice cores, it was indicated that the extraterrestrial dust influx during the Ice Ages might have been as high as 3 x 10^7 tons per year. If 10,000 times this amount of water is added to the atmosphere from icy meteors, we are approaching 10^{12} tons of extraterrestrial water per year ---far from an inconsiderable amount. The effects on the earth's climate could be large. If even greater fluxes of icy meteors were intercepted in the past, one might account for "pluvial episodes" on the planets. And then, comets now seem to transport "primordial organic sludge" around the solar system, as mentioned in the chapter on astronomy, p. 128. We will leave further speculation for the reader.

Note: the 10-kilogram figure for the size of the average icy minicomet mentioned above was greatly increased in later discussions.

OCEANS FROM OUTER SPACE?

In the preceding item, we related how L.A. Frank, at the University of Iowa, had detected dark spots on satellite images of the earth's dayglow. Frank thought that the spots might be due to clouds of water vapor released as small, icy comets hit the atmosphere. Pat Huyghe has recently written more about Frank's discovery, his theory, and its reception by the scientific community.

> "These comets are not occasional vis-

itors, he (Frank) says, like the one that comes by every 76 years and---lucky for us---never actually drops in. No, these are very small, comet-like objects that enter our atmosphere at a rate of 20 per minute, he says. These comets, which he believes must contain about 100 tons of water apiece, vaporize on impact with the atmosphere and fall as rain or snow. Now that may seem like one sizable cold shower, but on a yearly basis he says it's actually only a tiny fraction of the annual precipitation. Then again, over a span of 4.5 billion years, which is about how old the earth is, that's enough water, he says---trumpets blaring---to create the oceans."

Naturally, such a theory is very disturbing because it runs counter to the widely accepted idea that the oceans were created by the outgassing of water vapor from the newly accreted earth. As a consequence, Frank's data are readily accepted, but his explanation of them is not.

"That's as crazy as they come." A noted astronomer.
"...a case of Halley's fever." One geologist.
"...his interpretation is preposterous." Fred Whipple.

Critical as other scientists may be of Frank's theory, they have no other explanation for the dark spots on the earth's dayglow image. Furthermore, scientists are far from united about how the earth's oceans did form.

As serendipity would have it, Frank's theory connects in an interesting fashion with the origin-of-life speculations on p. 184. Frank remarks, in connection with organic sludge in comets, "These objects, because they are like a piston of gas, can bring organic material down without burning it up like a meteor does." (Huyghe, Patrick; "Origin of the Ocean," Oceans, 19:8, August 1986.)

SMALL, ICY COMETS AND COSMIC GAIA

L.A. Frank and his associates at the University of Iowa have speculated that the earth is continuously and copiously bombarded by small, icy comets. Not just a few now and then, but a steady rain so intense that over geological time some major geological consequences must ensue. Some observers commented that surely these scientists had thrown away their careers by suggesting something so ridiculous. But the data are there---in the form of dark spots on satellite images of the earth's dayglow---and late results continue to support this far-out interpretation, ridiculous or not.

"The mass of these objects is estimated at $\sim 10^8$ gm each, and the total flux is $\sim 10^7$ small comets per year. If this flux is representative of the average flux over geologic time, then the water influx is sufficient to fill Earth's oceans. The fluxes of these objects are also large for all the planets outside the orbit of Earth. Considerations of thermal stability imply that the fluxes of comets that impact Venus are considerably less. The outer giant planets may be significantly heated relative to solar insolation by the small-comet impacts. For example the total energy input due to both solar insolation and comet impacts may be similar for Uranus and Neptune. Thus it is possible that the temperatures of these two planets are similar, even though Neptune is farther from the sun." (Frank, L.A., et al; "On the Presence of Small Comets in the Solar System," Eos, 68:343, 1987.)

Comment. What has all this to do with "cosmic Gaia"? By "cosmic Gaia" we mean the cosmic version of the conventional Gaia concept; i.e., earth-as-an-organism. The answer is that small icy comets can in principle transport throughout all of space: (1) Immense quantities of water needed for life-as-we-know-it; (2) The carbonaceous material basic to that same kind of life (See p. 178.); (3) The seeds of life, a la Hoyle and Wickramasinghe; and (4) Energy, as discussed by Frank et al. Observe that Frank et al are saying that the kinetic energy in the flux of small comets is sufficient to raise a planet's temperature as well as supply water. In this light, exobiologists need not confine their search for extraterrestrial life to planets surrounding warm suns. Somewhere, far from stars, there may be places where comets may raise atmospheric temperatures to where life can prosper! Sunlight is, of course, not needed at all, as demonstrated by the profusion of life about deep-sea vents.

OCEANS FROM SPACE

In keeping with the foregoing extraterrestrial flavor, we are happy to report that our oceans may be exogenous; that is, derived from extraterrestrial materials. Once again, comets seem to be the culprits. C.F. Chyba has examined the lunar impact record and derived an estimate of the total mass of objects impacting the moon during the (hypothetical) period of heavy bombardment 3.8 to 4.5 billion years ago. This allowed him to calculate the mass influx for the earth during this period. His conclusion: if only about 10% of the incoming mass consisted of comets (mostly ice), the earth would have acquired all its ocean water. (Chyba, Christopher F.; "The Cometary Contribution to the Oceans of Primitive Earth," Nature, 330:632, 1987.)

Comment. Frank claims that the earth today is continually bombarded by small icy comets, which down the eons may have kept the ocean basins full. So, we have two possible extraterrestrial sources of oceans---both of a cometary nature. It was only yesterday that the idea of ice surviving in outer space was ridiculed; no one even dreamed that our oceans could be composed of space ice!

ICY COMETS EVAPORATING?

To keep the scorecard up-to-date, we here record the probable obliteration of observations of excess hydrogen in the inner solar system. Mentioned below, this excess hydrogen was observed from Voyager 2, as it cruised toward Mars and looked backward towards earth. Although the amount of excess hydrogen detected was only 1/10,000,000-th of that required by the small icy comets postulated by L. Frank et al, the result was surprising and gave a boost to the icy comet theory. Unfortunately, perhaps, the "excess" hydrogen evolved from a clerical error, when a student miscopied a figure during the data analysis. (No PhD for that student!)

Frank's icy comets are in even deeper trouble, since independent analysis hint that his satellite data may be attributable to instrument noise. (Kerr, Richard A.; "Comets Were a Clerical Error," Science, 241:532, 1988. Hall, D.T., and Shemansky, D.E.; "No Cometesimals in the Inner Solar System," Nature, 334:417, 1988.)

ONE OF THE MOST ASTONISHING DISCOVERIES OF MODERN SCIENCE!

We take this title from P. Huyghe's recent overview of the "oceans from space" controversy, printed not surprisingly in Oceans. (See above.) As readers will recall, we have been following this debate for over two years. Rather than retrace all the details, it is sufficient to say that the scientific community has been generally negative and often condemnatory about L. Frank's assertion and evidence that each year some 10 million icy comets, each averaging sixty compact cars in weight, strike the earth's atmosphere and, in the fullness of time, help fill the ocean basins.

In his article Huyghe reviews the considerable evidence that has accumulated supporting Frank's claim:

1. The water in Halley's comet had the same abundances of two key isotopes as the earth's oceans;
2. The rocket detection of unexpected amounts of water vapor in the upper atmosphere;
3. The microwave detection of unusual water-vapor events in the upper atmosphere;
4. The Lyman-alpha detection of hydrogen concentrated near the earth; and
5. The photographic detection of small, incoming objects with the characteristics of the debated icy comets.

(Huyghe, Patrick; "Oceans from Space---New Evidence," Oceans, 21:9, April 1988.)

Item 5. has been reported in other publications:

> Using a telescope with a moving field of view---a difficult technique that required a year of preliminary calculations to plan---physicist Clayne Yeates has found and photographed what seems to be a population of fast-moving objects near earth that range between 8 and 16 feet in size. These previously undetected bodies match Frank's predictions concerning the speed, direction and number of proposed comets flying by earth, says Yeates, a scientist at the Jet Propulsion Laboratory in Pasadena, Calif.

(Monastersky, R.; "Cometary Controversy Caught on Film," Science News, 133:340, 1988. See also: Hecht, Jeff; "Snowballs from Space 'Filled Earth's Oceans'," New Scientist, p. 38, May 12, 1988.)

Comment. Now all this does not mean that Frank's hypothesis is proven in the eyes of all scientists. Far from it, there is too much at stake; namely, our whole view of the small-scale structure of the solar system and, even more important, the heretical notion that the earth's oceans have slowly filled with extraterrestrial water.

It has not been an easy two years for Frank. His reputation has been at risk. Huyghe hinted at this when he recorded Frank's reactions to the new photographic evidence: "'Looking at the data, seeing those streaks, has made a lot of people's hearts stop,' says Frank. He is thrilled at this result, but he dreads what will follow. 'For the past two years I paid the price for being wrong. Now I'll pay an equal price for being right.' After all, you can't just tip the scientific world askew and expect everyone to cheer."

ICY MINICOMETS NOT SO DEAD!

An item in the June 1990 issue of Scientific American is entitled "Death Watch." In it, J. Horgan plays dirges for four phenomena that have received considerable attention in Science Frontiers: (1) minicomets; (2) cold fusion; (3) abiogenic oil; and (4) the fifth force. (Apparently Benveniste's "infinite dilution" work has already been interred.) (Horgan, John; "Death Watch," Scientific American, 262:22, June 1990.)

But wait, there is a microwave flicker of life remaining in the minicomets. J.J. Olivero and his colleagues at Penn State have been monitoring the sky with a microwave radiometer in their search for emissions from high-altitude gases. During more than 500 days of observations, they detected 111 sudden bursts of water vapor. Olivero et al suggest that these bursts occur when small, icy comets vaporize at very high altitudes. These minicomets are of the same size (about 100 tons) and frequency (20 per minute over the whole atmosphere) as those predicted by L.A. Frank. Frank's icy comets have been received with about as much warmth as "cold fusion." One reason for the unpopularity of icy comets is that they would have provided sufficient water to fill the ocean basins, thus undermining the accepted view that our oceans derived from outgassed water vapor from deep within the earth.

Besides this mindset, the minicomets do have some counts registered against them: (1) The effects of all the purported water vapor on the ionosphere should be easily detected but they are not; (2) Seismometers emplaced on the moon have not detected their impacts there; and (3) Military surveillance satellites have not seen these house-sized objects. (Monastersky, Richard; "Small Comet Controversy Flares Again," Science News, 137:365, 1990. Emsley, John; "Are 'Minicomets' Peppering the Earth's Atmosphere?" New Scientist, p. 36, June 9, 1990.)

IMPACT DELIVERY OF EARLY OCEANS

Where did the earth's oceans come from? For decades the stock answer has been: from the condensation of vapors escaping from the planet's cooling crust; that is, "outgassing." The possibility that some terrestrial water might have arrived from extraterrestrial sources after the earth's formation has been discounted. The major reason behind this neglect was the expectation that the erosive effects of large-scale impacts of water-carrying comets and asteroids would preclude any net accumulation of volatiles, and could even reduce any existing inventories of surface water.

C.F. Chyba has recently reexamined this question of cometary water influx vs. impact-caused water losses using the latest estimates of comet/asteroid fluxes during the period between 4.5 and 3.5 billion years ago, when bombardment of the inner solar system was thought to be especially severe. Rather than the expected net loss, Chyba computes that the earth would really have gained more than 0.2 - 0.7 ocean masses in that billion-year period. Venus would have fared equally well, but Mars, more sensitive to impact erosion, would have accreted "only" a layer of water 10-100 meters deep over the whole planet! (This Martian water is now mostly below the surface supposedly.) (Chyba, Christopher F.; "Impact Delivery and Erosion of Planetary Oceans in the Early Inner Solar System," Nature, 343:129, 1990.)

Comment. Not mentioned in this paper is what might have happened after 3.5 billion years ago. The comet/asteroid flux did not drop suddenly to zero. In fact, there may still be some net influx of cometary extraterrestrial water, as suggested by L.A. Frank. Incidently, the work of Frank et al is not mentioned at all in Chyba's article. Too controversial?

WEATHER MISCELLANY

SOIL TEMPERATURES FORECAST RAINFALL PATTERNS

Dig a hole about 40 inches deep, take the soil temperature at that depth, and you can predict future wet and dry periods months ahead of time. To illustrate, warm spring soils are usually followed by rainy summers; cold soils precede dry summers most of the time. At first, American scientists doubted this Chinese discovery, but their research soon proved that the correlation is even stronger in the United States. The best explanation so far is that soil temperatures affect atmospheric convection and modify weather patterns locally. (Anonymous; "Digging for a Forecast," Science Digest, 91;30, September 1983.)

LAGEOS FALLS TOO FAST

In May of 1976, NASA launched a geodetic satellite into an orbit over the poles about 3700 miles out. The satellite, called Lageos, is covered with laser reflectors so that it can be tracked with high precision. At its altitude of 3700 miles, the earth's atmosphere is supposed to be so thin that friction will bring the satellite only 1/250th of an inch closer to the earth each day. The trouble is that Lageos actually falls at ten times this rate. In 1979 it descended 60% faster than it does now. Lageos will stay in orbit several hundred thousand years, but space scientists are understandably concerned about thier theories about the upper atmoshphere. Many suggestions have been made to explain this anomaly. Some say the atmosphere is thicker than expected; others prefer to think there is more helium than predicted; but the "plasma drag" effect seems to fit the situation the best. Lageos may, in fact, be electrically charged and interacting with the surrounding cloud of electrically charged particles and is ever so slightly braked by the electrical forces. (Maran, Stephen P.; "Fall from Space," Natural History, 91:74, December 1982.)

SODIUM SURGES OVER ILLINOIS

An advanced lidar device, located at Urbana, Illinois, has been sending pulses of light up into the atmosphere and measuring the reflections from atmospheric atoms and molecules. To the researchers' surprise, this instrument detected sudden appearances of clouds of sodium atoms at about 85 kilometers altitude. The clouds quickly dissipate, but where do they come from? The best guess is that their source is meteors vaporizing in the upper atmosphere. (Raloff, Janet: "Sudden Sodium Surges Seen over Illinois," Science News, 134: 238, 1988.)

Comment. Could these sodium clouds have any connection with the controversial icy comets mentioned on pp. 79-81.

OZONE HOLE OVER ANTARCTICA

Reviewing ozone-mapping data from the polar-orbiting Nimbus-7 satellite, R. Stolarski and colleagues at NASA's Goddard Space Flight Center have seen the concentration of ozone over Antarctica drop dramatically---some 40%---every October. This disappearing act commences about a month after the sun begins to graze the northern horizon and affects the entire continent. By early November, the sun is high enough to manufacture enough ozone via its ultraviolet radiation to fill the ozone hole up again. An analogous hole does not form over the North Pole in the northern spring. An additional fact of interest: the hole is getting deeper each year; that is, the ozone concentration is less and less each October. Speculations about the deepening seasonal hole involve the widespread use of chlorofluorocarbons and Antarctica's physical isolation from other land masses which would help channel ozone southward from areas where the sun still shines. (Weisburd, S.; "Ozone Hole at Southern Pole," Science News, 129:133, 1986.)

ANTARCTIC OZONE HOLE HAS COMPLEX STRUCTURE

One more mystery has been added to the seasonal loss of ozone in the stratosphere over Antarctica. It now appears that the "hole" is an uneven one, with 2- to 3-kilometer-thick slices of ozone-poor air sandwiched within layers of only minimal depletion.

These new data came from McMurdo Sound, where a series of balloons carrying ozone sensors were released. Ozone depletion seems to be confined to the region 12-20 kilometers altitude and the top of the stratosphere. The overall depletion in this region was 35% at the time the balloons were lofted. However, some zones from 1 to 5 kilometers thick showed depletions as great as 90%. The reason for this stratificaiton is not yet known. (Silberner, J.; "Layers of Complexity in Ozone Hole," Science News, 131:164, 1987.)

A COSMIC CAUSE FOR THE OZONE HOLE?

Wouldn't you know it? Now they are blaming the polar ozone holes on Frank's icy comets---or something very much like them! M. Dubin and I. Eberstein, two NASA scientists think that small icy comets can account for the seasonal ozone hole and the mysterious polar stratospheric clouds that form during the winter. They propose that ozone molecules bond to tiny ice particles in the winter and, when spring arrives, solar ultraviolet radiation converts water (ice) plus ozone into oxygen and hydroxyl ions. (Anonymous; "A Cosmic Cause for the Ozone Hole?" Sky and Telescope, 75: 465, 1988.)

MALODOROUS MYSTERY

A scent squad has been unleashed in Bartlesville, Okla., to trace and identify an elusive odor that has plagued residents for months. The 19-member Bartlesville Odor Mitigation Task Force will distribute about a half dozen devices to trap the smell, which will then be sniffed and characterized by trained noses at a Chicago Research Company. The city has received 60 calls this year about the odor, described as smelling like rotten eggs or butane, but they have been unable to determine the source.

(Newman, Steve; "Malodorous Mystery," Baltimore Sun, p. 5E, July 22, 1990.)

Comment. We rarely come across olefactory anomalies, although they occasionally crop up in the Fortean literature.

HYDROLOGICAL PHENOMENA

WAVES

MASSIVE FREAK WAVE

November 23, 1981. Off Fair Isle, Scotland.

The crew of the stern trawler Clarkwood were hauling the net aboard because of gale-force winds when the wave struck. A fisherman told the inquiry the wave was like 'a big green cliff.' ....Skipper Stewart Thomson told the inquiry: 'The ship was struck by a massive lump of water.'

No estimates were made of the wave height. Three men were swept to their deaths. (Anonymous; "Massive Freak Wave," Mariners Weather Log, 26:79, 1982.)

FREAK WAVE OFF SPAIN

1960s. At sea off Spain.

"...the wind was north-by-west, force 6-7 and the ship was spraying and occasionally shipping water. The weather was not troubling our ship to any extent. The sky was partly cloudy with a full moon in the west. At 0520 hours the moon was blotted out and all turned dark. I looked to port to see what type of cloud could obscure the moon so thoroughly, and was amazed---horrified, rather, to discover it was no cloud, but an immense wave approaching on our port beam. It stretched far north and south, had no crest, nor white streaks, and as it neared at quite a speed, I could see its front was nearly vertical. I yelled to the lookout man to come into the wheelhouse as he was on the starboard side of the bridge and could not see the wave.

"As near as I could judge, about 80 to 100 yards away the wave started to break, and in another few seconds reached our ship and struck us fair abeam with three distinct separate shocks, sweeping our ship for her full length. Fortunately, the vessel rolled away just before the impact and this I am sure saved us from even more serious damage."

.....

"The wave was higher than our foremost track---85 ft above the water. As this wave approached from a direction 90 degrees different from the normal sea and wind, which had been northerly for a few days previously, I put its existence down to a submarine earthquake in the mid-Atlantic ridge. Certainly it appeared so much different from the normal wind-generated sea, of which I have seen thousands. There was no crest, nor white streaks, a nearly vertical front and quite fast approach." (Cameron, T. Wilson; "Treachery of Freak Wave," Marine Observer, 55:202, 1985.)

Comment. Earthquake generated waves or tsunamis are hardly noticeable in deep water. Only when they approach shallow water and the shore do they crest dangerously.

ROGUE WAVES

"Shortly before dawn on Sunday, June 3, 1984, the 117-foot, three-masted Marques sailed into a fierce squall about 75 miles north of Bermuda. Heavy rain began to pelt the ship, and a furious wind sprang up out of nowhere. Squalls were

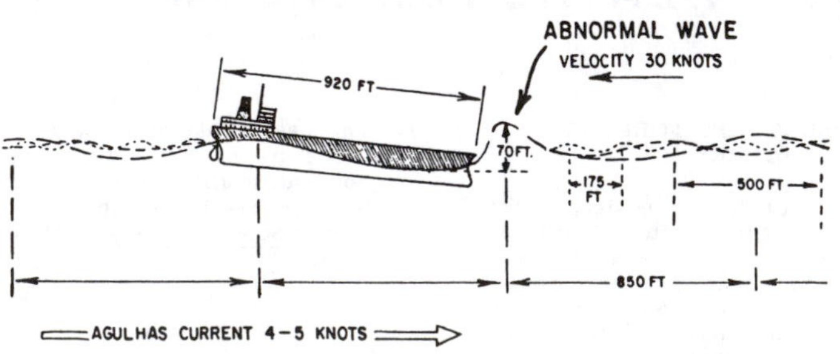

A tanker encounters steep-sided, giant waves in the Aguhlas Current off the South African coast.

nothing new to the Marques, one of 39 tall ships participating in a transatlantic race. But as a precaution, Stuart Finlay, the seasoned 42-year-old American captain of the ship, shortened the sails. The Marques was carrying a crew of 28 ---half of whom were under 25. At the helm, Philip Sefton, 22, fought the angry waves that now confronted them.

"Suddenly a heavy gust of wind pushed the Marques down on its starboard side. At the same instant 'a freakish wave of incredible force and size,' as Sefton later described it, slammed the ship broadside, pushing its masts farther beneath the surging water. A second wave pounded the ship as it went down. The Marques filled with water and sank in less than one minute. Most of the crew were trapped as they slept below deck. Only Sefton and eight shipmates survive."

Accounts such as that above are part of sea lore. Waves 50-100 feet high have been frequently reported over the years. Most often, they are encountered in rough seas, but some walls of water have smashed ships in relatively calm waters.

Until recently, oceanographers were confident that any unusually large wave was just the chance addition of two smaller waves. Now, a consensus is emerging that at least two other factors are important: seabed topography and ocean currents.

To illustrate, perhaps the most dangerous stretch of water in the world lies off southeast Africa, where the fast (8 feet/second) Agulhas Current often runs into storm waves surging up from Antarctica. The African continental shelf is so shaped that it funnels the current directly into the storm waves. Immense, steep-fronted waves have broken many a ship here.

In sum, the old statistical theory about the origin of rogue waves has been jettisoned, but a new approach is still in the formative stages. (Brown, Joseph; "Rogue Waves," Discover, 10:47, April 1989.)

Comment. But can any theory explain giant, solitary waves on calm seas. For more on this subject, see GHW in Earthquakes, Tides, Unidentified Sounds.

THE FLORIDA ROGUE WAVE

Very little has appeared in the scientific literature about the huge wave that crashed ashore at Daytona Beach, Florida, on July 3, 1992. Apparently, the scientific community is happy with the landslide explanation, but there may have been a different sort of disturbance.

First, the basic data:

> A wall of water as much as 18 feet high rose out of a calm sea and crashed ashore, smashing hundreds of vehicles parked on the beach and causing 75 minor injuries, officials and witnesses said.
>
> An undersea landslide apparently caused the 27-mile-long rogue wave late Friday night, a federal seismologist said yesterday.

The seismologist cited, F. Baldwin from the U.S. Geological Survey, estimated that the wave was 18 feet high and 250 feet wide. (Anonymous; "Rogue Wave Smashes into Beach," Hawaii Tribune-Herald, July 5, 1992. Cr. H. DeKalb.)

Rumors of a falling object. The landslide theory sounds good, but there have been rumors that another phenomenon was involved. B. Stein, of Orlando, has reported the testimony of a boater, who was far offshore at the time:

> ...the boater came forward with the information that, shortly before the time of the wave, he was in his boat about eight miles offshore. He watched as a distant object approached across the sky toward the ocean at a high rate of speed, and crossed the bow of his boat at an angle with a "whoosh" (his word). Shortly after, a giant swell made his 41-foot sailboat handle like a large surfboard. Various news sources state that the meteorite, as it is now being called, was anywhere from a meter to 10 feet across. The boater who wished to remain anonymous, gave the professors enough information so that they are hoping that the Navy will retrieve the object, which is presumed to be lying in about 70 feet of water off the Daytona Beach coastline, with plenty of coordinates for locating it.

(Stein, Becky; "Daytona Beach Mini-Tidal Wave," Louisiana Mounds Society Newsletter, no. 52, p. 2, October 1, 1992.

Comment. With all the military and space-vehicle tracking equipment in the area, someone must know more about this event.

THE TSUNAMI TUNE

Tsunamis are giant sea waves set into motion by earthquakes on the sea floor. Some 322 tsunamis have been recorded in the Pacific between A.D. 83 and 1967 ---or about one every six years on the average. The surprising thing is that tsunamis are more common in November, August, and March, but rarer in July and April. Offhand, no good explanation comes to mind why sea floor quakes should favor some months over others. (Anonymous; "The Times for Tsunamis," Science News, 127:88, 1985.)

WAVE-BANDS IN CALM WATERS AND BISCAY BOILS

An excerpt from an article in Nature:

> There are numerous reports of internal waves being "made visible" on the sea surface by their effect on the surface-wave field and the production of bands of steeper, often breaking, waves separated by zones of relatively calm water. The effect is sometimes quite dramatic. There are accounts of a "low roar" as the bands of breaking waves, "walls of white water," pass a vessel. The bands are sometimes visible from aircraft, on ships' radar and are observed from satellites. In the Bay of Biscay "boils" have been reported on the sea surface in the calm zones, and appear to be related to pulses of nutrients from the thermocline.

These surface phenomena are truly delightful and almost always the consequence of internal waves interacting with the surface. The great bulk of the referenced report is concerned with sonar observations of internal waves and their effects along the coast of Scotland. (Thorpe, S.A., et al; "Internal Waves and Whitecaps," Nature, 330:740, 1987.)

Comment. For some remarkable accounts of wave packets, as well as solitary waves, see category GHW in Earthquakes, Tides, Unidentified Sounds.

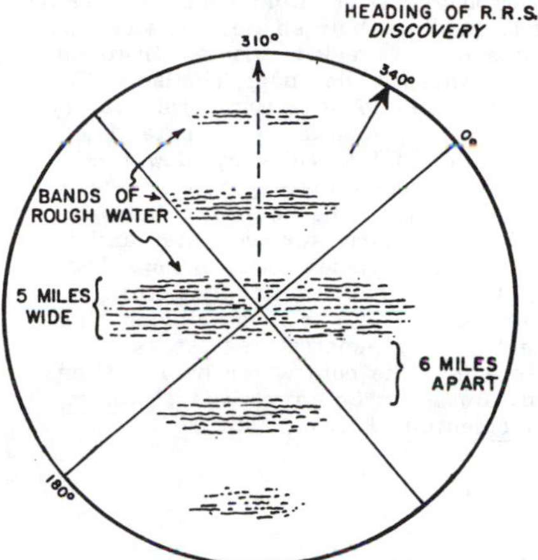

On March 28, 1964, in the Indian Ocean, the R.R.S. Discovery's radar detected five bands of large waves approaching in an otherwise calm sea. When the waves passed the ship, they were about 2 feet high. No wind change was felt.

ATLANTIC'S WAVES GETTING BIGGER

> According to a study by the Institute of Oceanographic Sciences, waves in the northeastern Atlantic are getting bigger. The average wave height in the 1960s was about 7 feet. Now the average is 9-10 feet. (Anonymous; Coming Changes, 13:7, May-June 1981.)

Comment. This is a very large increase (about 30%) for this geophysical variable. Have surface winds increased that much in such a short period?

Above, we inserted a short, rather vague note on this phenomenon. We now have a bit more to report, although no one seems to have any answers.

> The North Atlantic is getting rougher ---much rougher. In the mid-1980s average waves in the ocean were 25 per cent higher than during the 1960s. More recent studies show that by the end of the 1980s the tops of the waves were 50 per cent higher, as measured by both instruments and estimated by sailors.
>
> .....
>
> The cause of the increasing choppiness of the waters of the North Atlantic is unclear. Waves are whipped up by strong winds, yet there has been no corresponding increase

in wind speeds. [S.] Bacon believes that a clue may lie in the persistence of winds from a certain direction.

(Anonymous; "Making Waves in the North Atlantic," New Scientist, p. 10, August 29, 1992.)

SOLITARY WAVES

Unlike the well-known long trains of ocean swells that sweep past ship and swimmer with great regularity, solitary waves move "in splendid isolation, steadfastly holding their shape." Spacecraft photos have revealed curious striations in the Andaman Sea near Thailand. They are presumed to be examples of solitary waves. The Andaman waves extend for many miles and travel very slowly---less than 10 kilometers per hour. They propagate along the boundary between the layer of warm surface water and the great mass of cooler water below. The amplitude of the downwardly pointing wave troughs of warm water along this interface may penetrate as far as 100 meters into the cold water below. (Herman, Russell; "Solitary Waves," American Scientist, 80:350, 1992.)

VIOLENT UNDERSEA WEATHER

Long lines of frothing, turbulent water and transitory packets of large waves occasionally sweep across an otherwise placid sea. Usually dismissed as "rips," satellite photos reveal that these disturbances may be 125 miles long. Often several can be seen criss-crossing an ocean simultaneously from different directions. Some have a 12.5-hour period, linking them to lunar tidal action. The surface manifestations, like the tip of the iceberg, only hint at what transpires beneath the surface. The long corridors of disturbance, moving at about 5 mph, mark where "internal waves" intersect the surface. Down below, submarines and other objects may suddenly rise or fall as much as 600 feet. Internal waves may in fact have caused several submarine disasters.

How are internal waves created? Tidal waters may spill over an undersea sill or ledge, creating a travelling disturbance. Some oceanographers liken the internal waves to the lee waves formed parallel to large mountain ranges. Manifestly, there is much to learn about undersea weather. (Anonymous; "Underwater Waves Held Possible Clue to Disappearances of U.S. Submarines," Baltimore Sun, p. A12, October 5, 1980.)

POWERFUL CONCENTRIC WAVES

We wish we could pass along more data on this phenomenon. The totality of our information resides in two sentences mentioning an observation by the crew of the Soviet spacecraft Mir: "The crew reported seeing an unexplained ocean phenomenon, 'powerful concentric waves going out in the midst of a serene sea.' The cosmonauts did not report where they saw the waves but said the circular features were many miles across." (Anonymous; "Soviets Demonstrate Flight Readiness with Firing of Heavy-Lift Booster," Aviation Week, p. 20, March 16, 1987. Cr. G. Earley)

GEYSERS AND SPRINGS

SOME OLD GEYSERS ARE NOT SO FAITHFUL

Yellowstone's Old Faithful has a namesake in Calistoga, California. This not-so-well-known geyser is usually very dependable, erupting every 90 minutes, shooting 350°F water 60 feet into the air. However, some 60 hours before the October 1989, 7.1-magnitude quake in the San Francisco Bay area, the geyser's period suddenly lengthened to more than 100 minutes. After the quake, it settled back into its usual routine. Prior to two other earthquakes, in 1975 and 1984, the clockwork of Calistoga's Old Faithful also ran slow. (Anonymous; "Unfaithful Geyser," Discover, 12:8, July 1991.)

Comment. Since the quake epicenters were many miles distant from the geyser, how is the geyser's clockwork altered? Somehow, small earth movements must have changed the size of the geyser's water reservoir or, possibly, pressure changes in the surrounding rocks might have reduced the flow of water into the reservoir. Pertinent here are the often-observed changes in well levels and spring flows prior to earthquakes.

WYOMING: IS OLD FAITHFUL A STRANGE ATTRACTOR?

"Eruptions of Old Faithful Geyser are generally perceived as extremely regular events, with variation of eruptive interval being attributed to random noise. The governing equations for such a hydrothermal system are highly non-linear, therefore it is reasonable to assume that such systems are capable of operating in regimes that display chaotic behavior. Three-dimensional state-space reconstruction of eruption time data provides strong evidence of a strange attractor quite similar to the Rossler attractor. Establishing the system as chaotic indicates that while one can predict eruptive intervals in the short term, long term predictions regarding Old Faithful's eruptive behavior are impossible, no matter how carefully and accurately the system is modeled. The mean eruptive interval of Old Faithful has changed over time. This is consistent with the behavior of a chaotic system, which by definition must be nonstationary in the mean. Seismic activity is believed to be a perturbation shifting Old Faithful into a new chaotic state with a different shape to the strange attractor. A simple non-linear dynamic model of geyser behavior is proposed that leads to chaotic behavior and is consistent with the observations of eruption interval data for Old Faithful." (Nicholl, Michael, et al; "Is Old Faithful a Strange Attractor?" Eos, 71:466, 1990.)

Comment. "Strange attractor" is a specialized term employed in chaos analysis. So, Old Faithful is not really faithful; neither are the planets in their orbits (see under Astronomy). Is there nothing left in Nature that is reliable---just about everything is non-linear and therefore a candidate for chaotic behavior.

WYOMING: A PERIODIC SPRING

Near the base of a limestone cliff in Wyoming's Bridger-Teton National Forest, spring water gushes from an opening for several minutes, stops abruptly, then begins a new cycle a short time later. This is Periodic Spring, whose intermittent flow is a rare geologic phenomenon. The water is cold and clear, an indication that this is not a geyser like Old Faithful; such geysers, of volcanic origin, send forth hot water. Through the years various observers have timed the flows at anywhere from four to twenty-five minutes, with similarly varying dry spells. The intermittent flow is especially regular in late summer and autumn. During stormy periods or when there is heavy snow melt-off, the flow fluctuates but does not stop entirely.

At full flow, Periodic Spring discharges about 285 gallons/second into a stream 9½ feet wide and 1¼ feet deep. Periodic Spring, therefore, is fairly impressive, but it is not anomalous. (Mohlenbrock, Robert H.; "Periodic Spring, Wyoming," Natural History, 99:110, April 1990.)

Comment. Siphon action nicely explains periodic springs. Water keeps flowing through the upper loop (see illustration) until the water level in the reservoir drops below the siphon intake. The spring will not flow again until the reservoir fills to the top of the upper loop (level a) again initiating siphon action. Siphon action seems simple and easy to explain; why mention it here? Well, spherical pendulums are ostensibly sim-

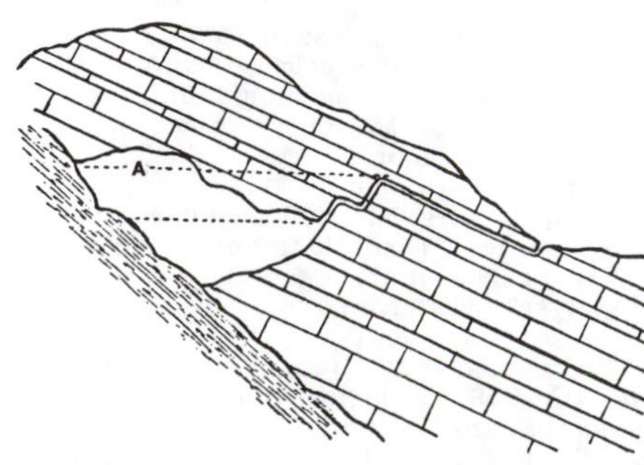

ple, too, but we now know that they sometimes run amok and chaotic motion develops. Like Old Faithful Geyser, there is a possibility that periodic springs also may exhibit chaotic behavior on occasion.

MARINE MISCELLANY

DEAD WATER

Two recent issues of Weather discuss some of the unusual geophysical phenomena encountered by F. Nansen in his epic attempt to reach the North Pole at the end of the last century. Nansen first endeavored to go as far north as he could in his specially constructed vessel, the Fram. It was on the Fram that he struggled with "dead water." J.M. Walker, drawing from Nansen's marvelous written account Farthest North, describes the phenomenon:

> Towards the end of August 1893, when Fram was off the Taymyr Peninsula, near the Nordenskiold Archipelago, 'dead water' was encountered. In the words of Nansen, this is "a peculiar phenomenon", which "occurs where a surface layer of fresh water rests upon the salt water of the sea, and this fresh water is carried along with the ship, gliding on the heavier sea beneath as if on a fixed foundation". It "manifests itself", he observed, "in the form of larger or smaller ripples or waves stretching across the wake, the one behind the other, arising sometimes as far foward as almost midships". When caught in dead water, Nansen reported, Fram appeared to be held back, as if by some mysterious force, and she did not always answer the helm. In calm weather, with a light cargo, Fram was capable of 6 to 7 knots. When in dead water she was unable to make 1.5 knots. "We made loops in our course", Nansen wrote, "turned sometimes right around, tried all sorts of antics to get clear of it, but to very little purpose.

Nansen asked the noted physicist and meteorologist V. Bjerknes to look into the frustrating phenomenon. Bjerknes found that the energy of the ship's propeller was, in essence, being siphoned off to create internal waves along the interface between the light fresh water and dense ocean water underneath rather than propelling the ship forward. (Walker, J.M.; "Farthest North, Dead Water and the Ekman Spiral," Weather, 46:158, 1991.)

WRONG-WAY WATERSPOUT

September 28, 1991. Aboard the m.v. Staffordshire in the western Mediterranean. On this date, between 0555 and 0810 UTC, observers on the bridge counted 15 waterspouts, one of which was anomalous:

> At 0722 the two spouts furthest forward and the one on the beam dissipated leaving one which was of quite a large diameter, about 20 m as seen at a range of about 300 m. The direction of rotation of the water in the spout was clearly seen. Although the observers were aware that the direction of rotation should be anticlockwise in this case, they decided (with great surprise) that the direction of this particular one was clockwise. The only other spout that passed closer, within 15 m, was very weak, but the direction of rotation at the surface was clearly anticlockwise.

(Edwards, R.A.F.; "Waterspouts," Marine Observer, 62:113, 1992.)

Comment. All Northern Hemisphere waterspouts should rotate anticlockwise.

THE CURRENT ANOMALOUS EL NINO

Bad spring weather? It's the El Nino. El Nino is the name given the annual movement of warm water southward along the western coast of South America. Every few years (range 2-10 years, average about 3 years) this current penetrates much farther south, devastating the fishing industry. Usually the catastrophic El Ninos begin in the eastern Pacific and work westward. The current El Nino is out of phase somehow, beginning in the western Pacific and moving east. (The current extreme drought in Australia is part of this phenomenon.) The more powerful El Ninos are usually associated with severe winters in North America; the opposite is true this time. Obviously, something is amiss with the current El Nino. (Philander, S.G.H.; "El Nino Southern Oscillation Phenomena," Nature, 302:295, 1983.)

CURRENT TREADS IN THE NORTH PACIFIC

We are a bit hestitant about including still another eyebrow-raising item, but the source here is Eos, a weekly publication of the American Geophysical Union, and the story is irresistible!

It all began when the container ship Hansa Carrier, enroute from Korea to the U.S., encountered a fierce storm and lost 21 40-foot-long containers to the sea.

> Approximately 80,000 Nike brand shoes were lost overboard on May 27, 1990, in the north Pacific Ocean (~48°N, 161°W). Six months to a year later, thousands of shoes washed ashore in North America from southern Oregon to the Queen Charlotte Islands...We have gathered beachcomber reports and compared the inferred shoe drift with an oceanographic hindcast model and historical drift bottle returns. This spill-of-opportunity provided a calibration point for the model. Computer runs for 1946-1991 suggested that drift of floatable material across the northeast Pacific Ocean for May 1990-January 1991 was farther south than the mean of forty-five simulations.

Well, we can tweak the model a bit; but the authors added a postscript:

> As we were finishing this article, we received reports of shoes arriving at the northern end of the Big Island of Hawaii. These shoes appear to have followed the California current southward, and then traveled westward.

(Ebbesmeyer, Curtis C., and Ingraham, W. James, Jr.; "Shoe Spill in the North Pacific," Eos, 73:361, 1992.)

Comment. In addition to the amusing thought of 80,000 athletic shoes drifting around the north Pacific, the shoes probably took the same course as many pre-Columbian Asian voyagers, some deliberately searching for new worlds and others caught by storms.

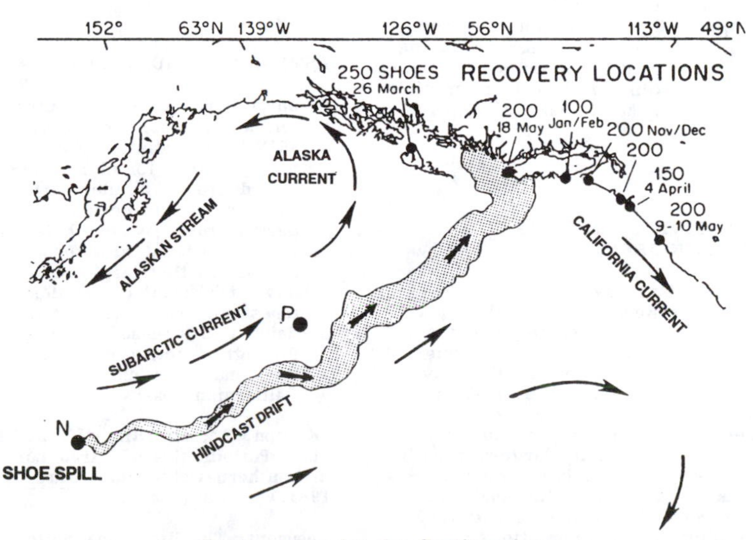

Routes taken by the floating shoes.

EARTHQUAKES

POSITIVE ION EMISSION BEFORE EARTHQUAKES MAY AFFECT ANIMALS

Both folklore and modern observations are emphatic that many animals become agitated prior to earthquakes. Cats carry their kittens outdoors; cattle panic in their barns; dogs bark for no apparent reason; and even some humans become restless. Tributsch notes that similar behaviors also accompany certain weather situations, such as the Alpine foehn and Near East sharav, which are characterized by high concentrations of positive ions. The unusual "fogs" and luminous displays preceding some earthquakes may also have electrical origins. In essence, Tributsch has reviewed many earthquake precursors and suggests that most can be explained in terms of positive ion emission from the earth due to pre-quake strains. (Tributsch, Helmut; "Do Aerosol Anomalies Precede Earthquakes?" Nature, 276;606, 1978.)

ELECTROMAGNETIC NOISE PRIOR TO EARTHQUAKES

March 31, 1980. Tokyo, Japan. Anomalously high electromagnetic noise was recorded during a 30-minute period preceding a magnitude 7 quake, 250 km away and 480 km deep. The emissions were detected between 10 and 1500 Hz and around 81 kHz.

July 24-28, 1976. Tangshan, China. For 3-5 days before the Tangshan earthquake, unusual radio interference was experienced within 250 km of Tangshan.

Several similar cases are also on record, including one in which the radio noise coincided with the appearance of earthquake lights. No generally accepted physical mechanism for producing these electromagnetic emissions has been found. (King, Chi-Yu; "Electromagnetic Emissions before Earthquakes," Nature, 301:377, 1983.)

SOLAR ACTIVITY TRIGGERS MICROEARTHQUAKES

Several scientists have suggested connections between solar activity, geomagnetic activity, and earthquake frequency. Singh has also found strong correlations between geomagnetic activity (definitely sun-triggered) and microearthquakes. He discovered first that the great solar storm of August 1972 was accompanied by large surges of both geomagnetic activity and microearthquakes. Following this lead, he studied records between 1963 and 1969, again finding strong correlations. (Singh, Surendra; "Geomagnetic Activity and Microearthquakes," Seismological Society of America, Bulletin, 68:1533, 1978.)

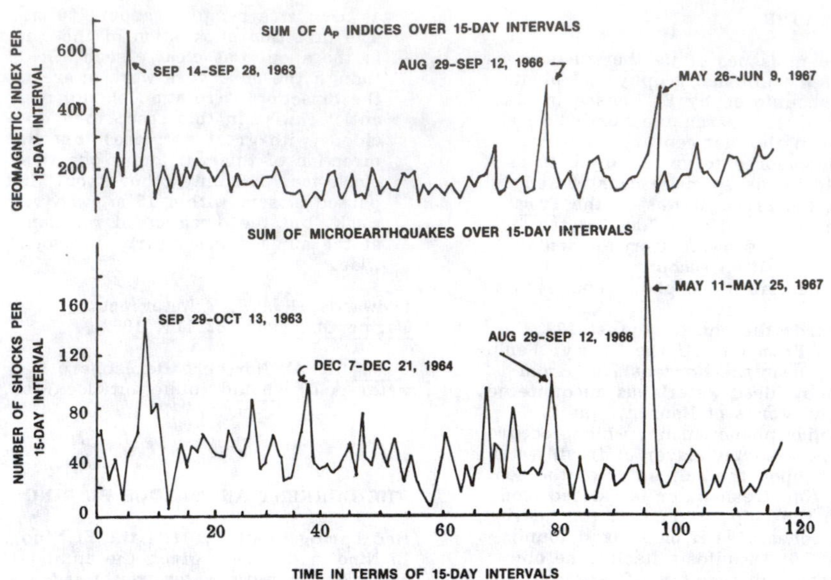

Geomagnetic index correlated with earthquake frequency.

Comment. While one can conceive of ways in which the streams of electrically charged solar plasma can modulate geomagnetic activity, the coupling of solar plasma variations to microearthquakes is more obscure.

ASTRONOMY AND EARTHQUAKES

Large earthquakes in southern California with epicenters between 33 and 36°N have statistically significant 12-hourly, lunar fortnightly and 18.6-yr periodicities. Smaller earthquakes in the same region do not display these periodicities. A search for tidal effects associated with these periodicities shows that large earthquakes have significant correlations with the times and orientations of daily/semidaily tidal stresses while the lunar fortnightly terms are associated with the ocean tides along the Southern Californian coast.

(Kilston, S., and Knopoff, L; "Lunar-Solar Periodicities of Large Earthquakes in Southern California," Nature, 304:21, 1983.)

Comment. The literature contains many similar correlations for quakes of various sizes, for various restricted geographical areas, and for various depths. Many of these studies are summarized in our Catalog volume Earthquakes, Tides, Unidentified Sounds.

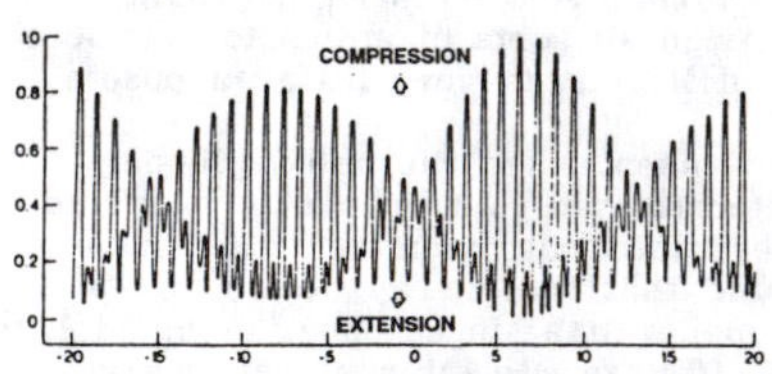

One tidal strain component (normalized) plotted for 20 days before and after the earthquake of June 25, 1925, showing the large fortnightly modulation of the diurnal tide. Curiously, perhaps anomalously, the quakes occurred during a period of minimum strain!

EASTERN QUAKES MAY BE LUBRICATED BY HEAVY RAINFALLS

Some of our continent's most powerful earthquakes have shaken the eastern half rather than the Pacific states, where the edges of tectonic plates grind together. The devastating New Madrid and Charleston quakes did not occur at plate boundaries, and it is hard to find active faults to blame for the crustal commotion. A hint of a possible solution to the dilemma comes with the correlation of earthquakes with heavy rainfalls and high water tables. For example, the Charleston quake of 1886 was preceded by two years of unusually heavy rainfall followed by a short dry spell. Also, seismicity in the New Madrid (MO) area increases 6-9 months after the Mississippi has crested. The theory is that the added water penetrates deep into the earth where it lubricates faults, causing them to become active and jolt the surface above. (Weisburd, S.; "Trickle-Down Theory of Eastern Quakes," Science News, 129:165, 1986.)

Comment. The above correlations and our inability to explain deep-focus earthquakes underscore our ignorance of the mantle. To illustrate, Soviet drillers have found fluids circulating through fractured rocks 11 kilometers down, where one would expect everything to be sealed tight by the weight of the overlying sediments. See p. 230.

QUAKES AND UFOs

"A strong temporal correlation was found between the numbers of reports of UFOs (unidentified flying objects) and nearby seismic activity within the Uinta Basin for the year 1967. The numbers of UFO reports per month during this classic UFO flap were correlated 0.80 with the sum of the earthquake magnitudes per month for events within 150 km of the report area. Numbers of UFO reports were not correlated significantly with earthquake activity at distances greater than 150 km but less than 250 km away. The strongest correlation occurred between UFO reports and nearby seismic activity within the same month but not for previous or consequent months. Close scrutiny of daily shifts of epicenters and reports of UFOs indicated that they occurred when the locus of successive epicenters shifted across the area. These analyses were interpreted as support for the existence of strain fields whose movements generate natural phenomena that are reported as UFOs."

(Persinger, M.S., and Derr, J.S.; "Geophysical Variables and Behavior: XXIII. Relations between UFO Reports within the Uinta Basin and Local Seismicity," Perceptual and Motor Skills, 60:143, 1985.)

Comment. The "natural phenomena" mentioned above are probably close kin or identical to earthquake lights. An earlier paper by Persinger alone in the same journal (60:59, 1985) links transient and very localized geophysical forces to such psychic phenomena as haunts and poltergeist activity. These two papers are the latest in a long series, mostly authored by Persinger, in this psychological journal. See p. 251.

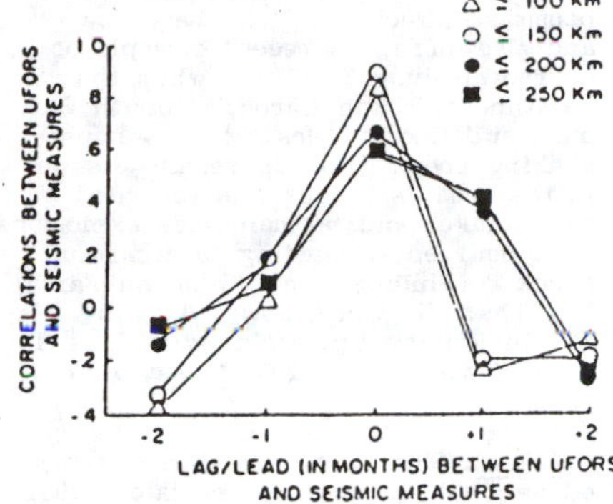

Calculated correlations between seismic activity and UFO observations in the Uinta Basin.

ANOMALOUS SOUNDS

UNEXPLAINED BOOMS

By "booms" we mean natural explosive sounds, distant and muffled, resembling far-off artillery, but coming from directions difficult to specify precisely. Booms are often episodic, sometimes heard several times per day and then not again for months or years. Unexplained booms are heard in all seasons and at all times of day, although warm, calm, hazy summer days seem to be favored. The booms described here prevail along seashores and around large lakes, thus deserving their alternate name: "waterguns." Probably the most famous of all natural booms are the Barisal Guns heard in the Ganges Delta. The so-called "mistpouffers," heard around the shores of the North Sea, are also well-represented in the scientific literature. Waterguns, though, are not confined to salt water. Some of the New York finger lakes have their own waterguns; i.e., the Seneca Guns.

Although the waterguns are similar to sonic booms from aircraft, they predate jet flight. Quite possibly, they arise from seismic activity or, perhaps, the release and natural detonation of natural gases from undersea deposits, including methane hydrate.

MODERN EPISODE OF OFFSHORE BOOMS

Beginning in December 1977, offshore detonations heard along the Atlantic Coast from Canada to South Carolina captured the media's fancy. Newspapers and TV news programs all over the country described these unidentified explosions. However, not a word about the detonations appeared in most of the scientific publications we regularly monitor, with the exception of the British New Scientist and a recent article in Science, 199:1416, March 31, 1978.

Comment. The detonations were rather strong, shaking houses and even causing picture windows to fall out. In some instances, flashes of light and other luminous phenomena were reported. The sounds were characterized as "air quakes" by some scientists because they did not always register on seismographs, although they were usually recorded on air-pressure monitoring equipment.

One's first inclination is to attribute such detonations to supersonic aircraft and missiles, but the U.S. military immediately denied they were to blame. Seismic noises come to mind next, but the frequent failure to register the events on seismographs suggested an atmospheric phenomenon. The National Enquirer (January 24, 1978) rather predictably linked the booms to UFOs. In the federal government, the Naval Research Laboratory (NRL) was assigned the task of tracking down the booms. In March, NRL reported that all of the 183 detonations they investigated were due to supersonic aircraft. That seemed to end the matter---just as the Condon Report signalled the demise of UFOs!!

BRONTIDES BECOME RESPECTABLE

The mystery of natural detonations (Barisal Guns, mistpouffers, etc.) was probed by several scientific groups following the recent episodes of off-shore booms. This paper by Gold and Soter, from Cornell, would have warmed the heart of Charles Fort, for he made much of natural detonations or "brontides," as they are termed in the early literature. Gold and Soter review the long history of brontides, noting that brontide activity is often associated with earthquakes, but not always. Natural booming noises, they contend, may be due to eruptions of natural gas. This would square with the rare observations of earthquake lights. Interestingly enough, the recent off-shore detonations were occasionally accompanied by luminous phenomena.

(Gold, Thomas, and Soter, Steven; (Brontides: Natural Explosive Noises," Science, 204:371, 1979.)

OFFSHORE BOOMS ARE STILL WITH US

Although they don't get the publicity they did a few years ago, powerful booms still rock the U.S. East Coast and elsewhere. A recent example occurred on June 24, 1981, when the coastline of North Carolina, South Carolina and Georgia was hit by a house-shaking boom. No supersonic jets were in the area, seismographs recorded no earthquake, and no man-made explosions had occurred, according to a careful check. (Phillips, Jim; "What on Earth Was That? No One Knows: Theories Fizzle Out on the Upstate's Mysterious 'Big Bang,'" Greenville, SC, News, June 25, 1981.)

Comment. The East Coast booms attracted national attention in the late 1970s. Despite several government-sponsored investigations, there was no consensus of scientific opinion. The booms remain anomalies.

BOOMS STARTLE ARKANSAS

A series of mysterious loud booms reported by residents of Hope, De Queen, Fulton, Mela, Ola, Baresville, Little Rock and other Arkansas cities will remain mysterious, at least for a while. Authorities are baffled about their source.

The noises, which have been described as sounding like an explosion, a sonic boom, a book falling off a shelf and a hand pounding on a wooden door, apparently have been occurring since the beginning of the recent cold weather. Inquiries have produced a number of theories and guesses but no plausible explanations.

No supersonic aircraft could be implicated, so the most popular view was that the extreme cold weather caused house timbers to crack. (Anonymous; "Mysterious Booms Heard around State Baffle Authorities; Some Blame Ice Cold," Arkansas Gazette, p. 11A, December 24, 1983. Cr. L. Farish)

Comment. If popping house timbers were the cause, similar reports would be expected from other states every winter. The Arkansas episode echoes the famous 1977-1978 series of booms heard all along the eastern coast of North America. These detonations also occurred during cold weather and were blamed, by some, on the Concorde SST.

MORE CAROLINA WATERGUNS

Residents of North Carolina's southeastern coast call it the "Seneca Guns" and say it's caused by chunks of the continental shelf dropping off a cliff under the Atlantic Ocean.

"You will feel the house kind of shake and windows rattle," said Walt Workman, assistant chief of police in Long Beach. "It sounds a lot like a sonic boom type of thing."

The rumbling boom with a sound like artillery fire is heard along North Carolina's southernmost beaches, sometimes as often as once or twice a week, and scientists can't explain the phenomenon. The sounds have been heard as far north as Fort Fisher, located just north of Cape Fear.

(Anonymous; "Booms Keep Coastal Areas Guessing," Charlotte Observer, January 26, 1987. Cr. G. Fawcett via L. Farish)

Comment. The real Seneca Guns are, of course, in New York, where they have been heard for years about Lake Seneca. Category GSD, in Earthquakes, Tides, Unidentified Sounds, provides numerous examples of such waterguns, from all around the world.

BOOMS ALONG THE BEACH

Just because we don't report them don't think that unexplained detonations are no longer heard along the world's seacoasts. These "waterguns" are still booming away, as they have for centuries. Take the Carolina beaches for example.

Sunset Beach---Just what is that noise that residents along the coast have been hearing?

Reverberations powerful enough to shake beach cottages are heard several times every autumn along the coast in New Hanover and Brunswick counties.

"It moves the earth, I tell you," Minnie Hunt of Sunset Beach said. "Sometimes you get two or three in a row."

.....

Residents who are now grandparents say their own grandparents remembered the rumbles, so they predate the sonic booms of jets breaking the sound barrier.

.....

The noises clearly emanate from the sea, she said..."It's not a land phenomenon."

.....

The sounds occur most often in the fall and spring, though they occasionally shiver across the beaches in other seasons. Sometimes they shake the coast more than once a day. Sometimes they happen a few days in a row. Sometimes they are weeks apart. They have been reported as far north as Carteret County, but are most frequent near Wilmington and southward.

(Anonymous; "Mysteries, Marvels and Things That Go Boom at the Beach," Asheville Citizen, September 4, 1990. Cr. J. Fisher via L. Farish)

Comment. Waterguns, a.k.a. mistpouffers, are dealt with in depth under GSD in Earthquakes, Tides, Anomalous Sounds.

LIGHTNINGLESS THUNDER?

August 22, 1987. Wotton-under-Edge, England.

"There had been thunder about, but the clouds were high, and there was intermittent sunshine. I was at work with a friend of mine, sawing logs, in one part of the garden. My wife was picking beans, about 80 yards away. Without any kind of warning, there was a violent detonation overhead, at what might have been tree-height. Service in war enables one to describe an explosion better than those whose experience is limited to Guy Fawkes' Day. This seemed to me about the same as an air-burst from a German 88 mm high velocity gun. My friend and I took it to be lightning; but neither of us saw any flash---perhaps because we were both looking downwards at the time.

"Shortly afterwards, my wife appeared, dazed and shaken. The explosion had evidently been closer to her, for she (having served in the WRNS) was reminded of an ammunition ship blowing up 'whoosh,' suggestive of a very high speed aircraft flying very low. That is what she momentarily thought it was, coming from the ridge of the Cotswold escarpment, under which this house lies; and she instinctively ducked. Immediately before the detonation, there seemed to her to be a sound not unlike machine-gun fire; and there was a movement of the air which disturbed the surface of the soil where she was working. She also saw no flash. For some hours afterwards she had a massive headache.

"Two near neighbors of ours observed the explosion, which they too assumed to be lightning. One of them, about a quarter of a mile away, says he saw a flash. The other saw none." (Carter, G.; "Two Unusual Lightning Events on 22 August 1987," Weather, 43:58, 1988.)

UNDERGROUND NOISES

In several localities around the globe, inhabitants are sometimes startled by sudden episodes of sharp, explosive sounds---like rifle-fire, not distant cannon-like booms---which seem to emanate from beneath their feet. Their senses are not deceiving them, because sensitive seismographs correlate these sounds with tiny tremors too slight to be felt kinesthetically. The most famous of these afflicted spots is located near East Haddam, Connecticut, where the so-called Moodus Noises have been heard since before the arrival of white settlers.

In addition to the Moodus Noises, we find the air alive with a wide spectrum of mysterious hums, some of which are accompanied by perceptible vibrations of the ground, some which are not. In some instances, such as the well-publicized English hums and the Taos hum, seismographs detect no vibrations of the earth at all, although microphones monitoring air-borne sounds verify that "something" is humming somewhere.

On the lighter side, nature does provide us with atmospheric pressure changes that cause the earth to breathe through cave mouths and fissures, thus generating a wide panoply of curious sounds. Insects, too, especially the periodical cicadas, and even fish add hums to our audio environment.

FROM FORTEANISM TO SCIENCE

The famous Moodus Noises have long been a Fortean staple---at least since 1923 when good old Charley mentioned them in his New Lands. Recently, perhaps mostly because there is a nuclear power plant right across the Connecticut River, there has been a concerted scientific effort to find out just what is going on in south-central Connecticut.

A brief glimpse of the phenomenon was provided by W. Sullivan in the New York Times:

> From last Sept. 17 to Oct. 22, more than 175 small earthquakes occurred near the town of Moodus, Conn. Many were accompanied by sounds like gunshots; the strongest vibrated a van. The phenomenon was another swarm of Moodus quakes that have puzzled generations of earth scientists. The earliest was recorded in 1568 and Indians knew of them long before then: Moodus is an Indian word meaning "place of noises."

Sullivan's article was derived from a spate of scientific papers delivered at the Spring meeting of the American Geophysical Union. (Sullivan, Walter; "A Connecticut Mystery Still Defying Scientists," New York Times, May 22, 1988. Cr. P. Huyghe, D. Stacy, R.M. Westrum)

Abstracts of all the scientific papers appeared in Eos. Here are excerpts from one of them:

> Since the installation of a six-station microearthquake network in the Moodus, Connecticut, area in 1979, four extensive microearthquake swarms of several months duration each, all accompanied by main shocks of Mc$\sim$2, have been recorded. All of the swarms have occurred at shallow depths ($\sim$ 2.3 km)and have been concentrated primarily in one small source volume....The 1986 swarm was characterized by a number of small bursts of activity culminated by the largest event near the end of the swarm. The 1987 swarm behaved in a very similar temporal manner to that of the 1986 swarm with one strong difference in that the largest event was the first one in 1987...the shallow depths of all the earthquakes there, the small lateral dimensions of the active source volume and the lack of a fault to which the earthquakes can be ascribed make it difficult to argue that this swarm activity indicates the possibility of a large earthquake at the locality.

(Ebel, John E.; "Comparisons of the 1981, 1982, 1986 and 1987 Swarms at Moodus, Connecticut," Eos, 69:495, 1988.)

MORE MOODUS SOUNDS

"Geologists from New Jersey are preparing to bore a 6-inch hole almost a mile into the Earth's crust on farmland off Sillimanville Road near Moodus (Connecticut).

"Once and for all, they hope to determine the exact cause of the 'Moodus Noises'---sounds that have been likened to the crack of a ball on candlepins in a distant bowling alley.

"Indians thought the sounds were the grumblings of an evil spirit, and they named the area 'Machimoodus' or place of noises.

"Geologists today say the sounds stem from earthquakes close to the surface. The quakes are so small that most can be measured only with special seismic instruments. But the reasons for the quakes are still the subject of hypothesis." (Barnes, Patricia G.; "Geologists Will Get to the Bottom of Moodus Noises," New Haven Register, April 30, 1987. Cr. J. Singer)

SHAKE NO QUAKE

January 30, 1987. Much of southern California was beset by a shaking phenomenon that stimulated scores of telephone calls to newspapers, universities, and government facilities. The shaken areas included Long Beach, Pasadena, the San Gabriel Valley, Buena Park, San Pedro, Fullerton, and Newport Beach. Caltech's Seismological Laboratory, at Pasadena, insisted that no seismic activity had been detected. The FAA ruled out sonic booms; the Navy said its ships were not engaged in target practice, and the National Weather Service exonerated weather phenomena. No one seems to know what happened. (Tessel, Harry; "Southland Rattled, But This Mysterious Shake Is No Quake," Long Beach Press-Telegram, January 31, 1987. Cr. L. Farish)

A WEST COAST MOODUS?

Late 1987 to date. Commerce, California.

> Something is going bump in the night under the city of Commerce, rattling nerves and household items alike.
>
> 'When it goes off, it can shake so badly that you feel it up through your feet and it can collapse your spine.' said David Stacy, a resident. 'Sometimes it would rattle stuff in the house like the earthquake did.'
>
> The underground explosions occur every 10 or 20 minutes, say residents in the area between Gage and Zindell avenues. They say the tremors have been forceful enough to wake them from their sleep, shake windows and knock down pieces of china.
>
> The muffled explosions are not accompanied by smoke or luminous phenomena. They may be due only to the subsidence of traffic noise. (Chong, Linda: "Commerce Becomes Reluctant Boom Town," Los Angeles Herald Examiner, January 17, 1988. Cr. K.H. Taylor)

Comment. Will Californians let us know more about these subterranean sounds? They resemble the famous Moodus Sounds in Connecticut, which are thought to be of seismic origin. See Category GSD2 in Earthquakes, Tides, Unidentified Sounds.

MACHINE-LIKE UNDERGROUND NOISES

On p. 286, we report on the "English hums." One theory is that these particular humming sounds emanate from turbines in Britain's network of buried natural gas pipelines. It seems, though, that similar subterranean sounds are heard in many places where buried pipelines are unlikely. Mainstream scientists care little for such phenomena; and the following data were collected by an independent researcher.

Yakima Indian Reservation, Washington. Statement by W.J. Vogel, Chief Fire Control Officer:

> "In September 1978 the sound of underground turbines or engines was heard at Sopelia Tower at the southern boundary of the reservation for seven hours (approximately 9:00 p.m. to 4: a.m., September 3-4). According to Vogel the noise was like a 'turbine' or 'unsynchronized propellers on a multi-engine aircraft.' When let outside the lookout station, the fire lookout's dog displayed anxiety, and the lookout felt barely perceptible vibrations under her feet when standing on a concrete slab. The lookout said she had heard the same sound during the summer of 1978, but always during the daylight."

Colorado. Testimony of T. Adams:

> "Camping on the western slope of the Sangre de Cristos, northwest of Mount Blanca and south of the Sand Dunes, we heard the sounds in 1970. Two or three nights in succession, it 'cranked up' after midnight and subsided shortly before dawn. It wasn't really loud enough to hear over converstion---but definitely sounded like a motor of some sort, with a suggestion of a dynamo-type whine to it. And others had said that it sometimes seems even louder with an ear to the ground. One could easily imagine (Adams' emphasis) the sound was coming from beneath the surface, but whether it did or not remains purely speculative."

The report referenced below contains similar observations from Texas, New Jersey, New York, Pennsylvania, California, Puerto Rico, England, and Italy. Unusual luminous phenomena have also been noted in the areas where the underground sounds have been heard, which explains why these phenomena were reported in a UFO publication.

Some UFO investigators suspect that both the sounds and luminous phenomena are due to low-level seismic activity. (Long, Greg; "Machinelike Underground Sounds and UFO Phenomena," International UFO Reporter, p. 17, November/December 1989.)

HOMING IN ON THE HUM

In 1977 the English Sunday Mirror ran a story about someone who claimed to hear a steady and very annoying humming noise. To everyone's surprise, the article elicited some 800 letters from others who heard hums. Amazed by the magnitude of the problem, doctors began examining some of the afflicted. In a few cases, the hum seemed to be internally generated---something akin to tinnitus, which causes one to hear a high-pitched whine. Many others, however, heard a 40 Hz hum modulated at 1.6 kHz, and apparently of external origin. The hum sufferers were inclined to blame industrial noise, but no obvious sources could be uncovered. The hum investigators have considered sea noise, jet-stream noise, and other natural sources. Whatever the source, most people do not hear it at all. It is possible that a small percentage of the population is abnormally sensitive to sound at 40 Hz. (Wilson, Steve; "Mystery of People Who Hear the Hum," New Scientist, 84:868, 1979.)

Comment. Anomalous natural hums are not unknown; viz., the Yellowstone Lake Whispers and "desert sounds." Also, a select few seem to be able to hear the very low auroras. See p. 287.

MORE MYSTERIOUS HUMS

Summer 1984. Aldershot, England. Several occurrences of unidentified, loud humming.

> The 'indescribable' sounds, which lasted about 20 minutes on each occasion, seemed to cut through the atmosphere, filling the air with a loud humming. Witnesses who wrote to the 'News' telling of their experience, were woken up by the mysterious noise which began suddenly and ended just as abruptly. The latest set of sounds were heard last month in the early hours but the same noises have been noted many times before over the past few years.

(Anonymous; "Mystery Noises Baffle Readers," Aldershot (England) The News, July 27, 1984. Cr. L. Farish and T. Good)

Comment. Southern England, particularly the Bristol region, seems to be frequently afflicted with episodes of steady and transitory hums. Despite considerable searching, no sources have been uncovered.

THINGS THAT GO BUZZ IN THE NIGHT

In a recent issue of the New Scientist, B. Fox mused about the weird world of electromagnetic transmissions and unidentified audio-frequency humming. We and our complex array of high-tech gadgetry are continuously bombarded by all manner of electrical and electromagnetic signals, noise, and transients. A particularly annoying source of unwanted signals impinging upon European radios is the Soviet Woodpecker over-the-horizon radar. In some bands, radio hams are blasted off the air when the Woodpecker is aimed at them.

So much for electromagnetic problems. In the audio range of the sound spectrum, Fox brings up the topic of those still unidentified hums that afflict a small group of people, who are now known as "hummers." Fox himself turns out to be a hummer. "By coincidence, I happen to be blessed, or cursed, with good low-frequency hearing. For several years now, I have intermittently heard a curious low-frequency sound coming from a deep below the high ground around my home in Hampstead Heath in London. Most of the time it is swamped by other noises, because human hearing adjusts sensitivity to compensate for background noise. The noise is an intermittent rumble, like a very distant generator, or the compressor for a pneumatic drill, coming on and off load. Most people cannot hear it at all. I usually hear it only in the still of night." Fox applied considerable effort in trying to find the source of the sound to no avail. The hum has been recorded and analyzed. It peaks at about 48 Hertz, and thus seems unrelated to the British 50 Hertz power mains. (Fox, Barry; "Things That Go Buzz in the Night," New Scientist, p. 72, October 8, 1987.)

THE ENGLISH HUMS: RADAR OR BURIED PIPELINES?

In our catalog volume Earthquakes, Tides, Unidentified Sounds, we recorded many curious natural sounds, including the "desert hum," the "Yellowstone Whispers," and, pertinent to the present discussion, a throbbing, humming sound afflicting some, but not all, residents of the British Isles. Percipients describe the hum as like a "diesel truck with its engine idling." The electronic environment of Britain has been blamed for the hum: transformers, high-voltage transmission lines, and pulsed radars are all candidate hum-makers. For, it has been discovered, some people somehow convert pulses of electromagnetic energy into a perception of sound. This facet of the British "hum problem" was covered above, where the infamous Soviet "Woodpecker Radar" is mentioned specifically. But are electromagnetic pulses really to blame?

The British hum has become a nuisance---to those who can hear it---during the past 20 years. This is just the period during which British Gas has been installing a nationwide gas-distribution system, which employs powerful turbines to pump natural gas through underground pipelines. H. Witherington, an unhappy hum-hearer, has for years driven around Britain at night when things are quieter, plotting places where the hum can be heard. He has found that the sound follows the gas pipelines and extends for several kilometers on each side. Houses, he finds, tend to amplify the sound, because closed rooms sometimes create resonant conditions. (Fox, Barry; "Low-Frequency 'Hum' May Permeate the Environment," New Scientist, p. 27, December 9, 1989.)

THE TAOS HUM

Over the years, we have reported on the British hum and the Sausalito hum The latter has been attributed to mating toadfish in the harbor; the former (tentatively) to an underground network of gas pipelines. We have resisted reporting many other hums that have made newspaper pages. However, a recently reported hum possesses some interesting features. It is called the Taos hum, and it has been bothering some sensitive individuals in the U.S. Southwest:

> More than a dozen people living in an area from Albuquerque to the Colorado border said in July 1992 interviews with the Albuquerque Journal that they had heard the low-level hum.
>
> A Denver audiologist said that she had recorded a steady vibration of 17 cycles per second with a harmonic rising to 70 cycles per second near Taos. The low range of human hearing is 20 to 30 cycles per second. (Anonymous; "Defense Dept. Denies Link to Taos Hum," Albuquerque Journal, April 7, 1993. Cr. L. Farish.)

Some residents of Taos are plagued by this machine-like sound that grinds away 24 hours a day, with only occasional respites. Some cannot sleep; others complain of headaches. Most people, however, cannot hear the hum at all.

Nevertheless, it is there. Instruments pick it up. In fact, they have even recorded a higher-frequency component that pulses between 125 and 300 cycles per second.

The cause of the hum is a mystery. One hint comes from the observation that the hum seems concentrated along the Rio Grande Rift, a fault that also runs into Texas and Colorado. One theory blames the hum on the fault's rock surfaces grinding against each other! (Begley, Sharon, et al; "Do You Hear What I Hear?, Newsweek, May 3, 1993. Cr. J. Covey.) This item actually appeared in SF#88.

UNDERGROUND WEATHER

Most residents of central Oregon can tell you tales about winds blowing out of the ground, about roaring and whistling sounds emanating from their wells, and of mysterious holes in the winter snow on Mount Bachelor. Central Oregon is geologically young and plastered with lava flows and cinder cones. It is like a sponge, with many cubic miles of holes, air channels, and open fissures. The air in this rocky sponge is usually on the move in response to changes in atmospheric pressure. The earth absorbs air during barometric highs and expels it during lows. In some spots the expelled air is captured to cool homes during the summer. In the High Cascades, though, the underground winds pose hazards to skiers by creating blowholes in the snow.

These blowholes are actually mildly

anomalous because they blow out a gentle 40°F breeze regardless of the barometric pressure. Some of Oregon's blowing caves also "breathe" without reregard to barometric pressure. Also, the water wells north of Fort Rock Basin often blow for days during periods of high barometric pressure---times when they should be taking air in. (Chitwood, Larry; "Central Oregon's Underground World Filled with Wind That Roars, Whistles," The Oregonian, October 31, 1985. Cr. R. Byrd)

Comment. See category GHG2, in Earthquakes, Tides, Unidentified Sounds, for material on blowing wells, etc.

THE SAUSALITO HUM

The mysterious underwater hum that has annoyed Sausalito's houseboat community for the past 11 summers is back, and investigators still do not know the cause. "It's a loud and audible mechanical raspy hum," said Waldo Point Harbormaster Ted Rose, who said the vexing noise sounds like an electric razor. "It sounds like this mzmzmzmzmzmzmzm," Rose said. "Sometimes it gets so loud you have to talk above it. It can drown out conversations and wake people from a dead sleep." For reasons no one understands, the noise can be heard only from about 8 p.m. until sunrise, and it goes silent from late September until mid-April, when it begins humming again through the summer.

Acoustical engineers from Berkeley could not pinpoint the source of the sound with the help of instruments and a diver. Biologists believe the noise is made by the singing toadfish, also called the plainfin midshipman. (Leary, Kevin; Sausalito's Weird Hum is Back," San Francisco Chronicle, July 29, 1985. Cr. P. Bartindale. Also: Anderson, Ian; "Humming Fish Disturb the Peace," New Scientist, p. 64, September 12, 1985.)

Comment. See category GSM1 in Earthquakes, Tides, Unidentified Sounds for more instances of underwater sounds, especially those mysterious sounds heard near the mouth of the Pascagoula River, Mississippi.

MYSTERIOUS HUMS: THE SEQUEL

The toadfish mating season is over and quiet has returned to Sausalito, but over at Pacific Heights and the Marina District on San Francisco Bay a new hum is driving people crazy. Speculation is rampant: Is the source a diesel generator, a hospital's portable CAT-scan machine, or underground electrical power lines? We'll just have to await more communications from California. (Rubenstein, Steve; "Detours on the Trail of Mysterious Hum," San Francisco Chronicle, September 25, 1985. Cr. P. Bartindale)

METEORIC AND AURORAL SOUNDS

THE SOUNDS THAT SHOULDN'T HAVE BEEN

November 1984. Several Texas localities.

The spectacular reentry of the Space Shuttle Discovery was observed by many Texans in the pre-dawn skies. Among these were Ben and Jeannette Killingsworth. As they observed the Space Shuttle streak across the sky, "they both heard an unmistakable 'swishing noise' as it passed south of their rural Galveston County home. The sonic boom came several minutes later---but the swishing sound occurred simultaneously with the visual apparition....Ben graphically described the sounds as 'like a skier coming down a slope,' but with a rapid fluctuation in loudness, 'about two or three hertz.' Jeannette compared the faint sound to the noise made by a fast boat as it slaps across waves on a choppy lake. 'But there was no motor noise,' she added, 'just a sound like repeated puffs of air through your mouth.'"

Oberg points out that the mysterious Space Shuttle sounds are basically the same as the anomalous swishes and whizzes attributed by some to meteors. So far, few scientists have accepted meteor sounds as real, preferring to label them "psychological." But now that the Space Shuttles are known to generate similar anomalous sounds, perhaps scientists will install instruments along their well-known reentry paths and find out what is really happening. (Oberg, Jim; "Shuttle 'Sounds' May Provide Answer to Old Puzzle," Houstonian, January 1985, p. 4. A McDonnell Douglas publication. Cr. J. Oberg)

Comment. One possible explanation of anomalous meteor sounds is that a few individuals detect electromagnetic signals as sounds; i.e., they are "electrophonic" sounds. The history of auroral sounds is basically the same as that of anomalous meteor sounds. They may both have the same explanation. See category GSH in Earthquakes, Tides, Unidentified Sounds.

ANOMALOUS SOUNDS FROM AN AUSTRALIAN FIREBALL

On April 7, 1978, a very large fireball passed through the atmosphere above the east coast of New South Wales. Seen by hundreds, it generated many high quality reports. Fifteen of the written reports mentioned anomalous sounds---hisses, hums, swishes, and crackling sounds heard simultaneously with the visual sighting. Such sounds are anomalous because the meteor is tens of kilometers high and real sound would take a minute or more to reach the ground. (The sound from a detonating meteor is often heard several minutes later.)

Keay is convinced of the reality of the anomalous sounds and suggests that the highly turbulent plasma in the meteor wake generates powerful electromagnetic radiation at audio frequencies. This intense radio energy reaches the earth at the same time the visible light does. It may be converted into sound as it interacts with the surface and the observer. (Keay, Colin S.L.; "The 1978 New South Wales Fireball," Nature, 285: 464, 1980.)

AURORAL SOUNDS

March 13, 1989. South Dakota. This night on a cattle ranch, L. Hasselstrom was dazzled by waves of blue auroral light sweeping up from the horizon and meeting at a focal point nearly directly overhead. As the sky blazed, with the blue waves and crimson streamers, she heard:

> ...a distant tinkling, like bells. It came again, louder, just as a curtain of green light swept the entire width of the sky from north to south. Each time green flushed the sky, the bells rang, the sound softening to a gentle tinkle as the light died.

(Hasselstrom, Linda; "Night of the Bells," Readers Digest, p. 185, April 1992. Cr. J.B. Dotson.)

Comment. Note the correlation of the sound with the green portion of the aurora.

July 29, 1990. Coll Island, Centennial Lake. 120 kilometers west of Ottawa. Watching an auroral display, L.R. Morris heard the sound of the aurora:

> It was a faint but distant windlike sound; which, by process of elimination, could not be accounted for by any phenomenon other than the aurora.

(Anonymous; "Auroral Sounds," Sky & Telescope, 83:105, January 1992. Cr. D. Snowhook.)

Comment. Auroras have been heard for centuries, but they "shouldn't be." Current theory restricts auroral activity to altitudes above 50 miles, where a fair vacuum prevails, and sound generation and propagation are impossible. One explanation for auroral sounds is that intense electromagnetic waves created by the auroras sweep through the observer's brain and are rendered as sound (electrophonic sound). But perhaps some auroras reach down lower into the atmosphere than theory allows. See GSH3 in Earthquakes, Tides, Unidentified Sounds.

ATMOSPHERIC OPTICS

REFRACTION PHENOMENA

THE NOVAYA ZEMLYA EFFECT

Rarely, as the long polar night draws to a close, the sun will suddenly burst above the horizon weeks ahead of schedule. This is the Novaya Zemlya Effect, and it is basically a polar mirage. Even when the sun is still 5° below the horizon, its light can become trapped between thermoclines and be transmitted over the usual horizon. The atmospheric ducts act much like flat light pipes. In the Novaya Zemlya Effect, the sun's image is grossly distorted, quite different from the high quality mirages sometimes seen over hundreds of miles in the polar latitudes. (Anonymous; "New Light on Novaya Zemlya Polar Mirage," Physics Today, 34:21, January 1981.)

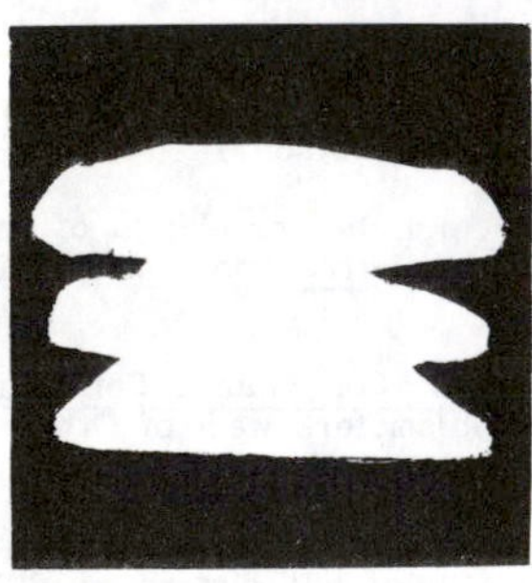

Triple Novaya Zemlya Effect. Three highly distorted images of the sun, which is still well below the horizon.

ZOOM LENS IN THE ATMOSPHERE

June 29, 1981. South Atlantic Ocean. At 1700 GMT, the master and third officer of the Rockhampton Star observed a vessel at a distance of 6 nautical miles. Its appearance was normal for that distance. At 12 n.m., however, it was magnified four times over what it should have been for that distance. At 16 n.m., the apparent magnification was eight. The interesting aspects of this phenomenon are: (1) Increasing magnification with distance; and (2) The magnified image was perfect in two dimensions and not distorted along the vertical axis as is usually the case in abnormal refraction. (White, A.H.; "Abnormal Refraction," Marine Observer, 52:80, 1982.)

Comment. What kind of stable atmospheric lens can do this?

DOUBLE IMAGE OF CRESENT MOON

November 24, 1989. Knoxville, Tennessee.

"Conditions: Clear sky (no clouds, but a slight haze). Waning moon was approx. 60° above the horizon, air temperature 33°F.

"I came out of the house about 6:25 AM to perform a task and, being an amateur astronomer, I looked up at the sky to see what was visible. I noticed the crescent of the waning moon appeared as a double crescent. My eyes kept trying to resolve it into a single image, but it wouldn't resolve. I then looked at several other light sources (radio tower, porch light, & street light) and determined that my vision was probably fine, as these objects appeared as single images. Looking back at the moon, it still appeared as a double image. I covered the right eye and I still saw a double image. I did the same with the left eye and got the same results. I then held out my right arm and extended my thumb to cover one crescent. I saw only one image that way. I moved my thumb and the image was again doubled. I concluded that I was viewing refracted images of the moon.

"Conditions prevented continuous observation, but I was able to return approximately every five minutes. By 6:55 AM the sky was brightening and there was only a single lunar crescent." (Miles, Bob; "Double Image of Crescent Moon," Teknowledgy Press, 1989. Cr. D.K. Hackett)

Comment. Although refraction is usually blamed for such double images, the different layers of air, with different refractive indices, would normally be disposed horizontally, not at the angle shown in the illustration.

Double image of the crescent moon with a faint, thin rim of light around the dark portion.

DOUBLE IMAGE OF LUNAR CRESCENT

A letter from R. Eason states that he, too, has observed this phenomenon, as reported above.

> It was in the early evening a few years back. On seeing the double crescent, I called out my brother who was visiting. It looked like a double image to him, too. Then I got my binoculars for a better look. There was NO double image seen in the binoculars, so the effect was clearly physiological/psychological.

(Eason, R.; personal communication, May 5, 1990.)

Comment. Recently, on a trip to the American Virgin Islands, my wife and I observed a thin crescent moon. I saw a doubling of the ends of the crescent; she did not! Both of us agreed, however, that a thin, bright ring enclosed the dark part of the moon, and that the dark part of the moon was distinctly brighter than the dark sky around the moon.

UNUSUAL DOUBLE SUN

April 30, 1984. Mediterranean Sea.

> Shortly before sunset the phenomenon shown in the sketch was observed.

From the m.v. Stability out of Piombino. (Twiselton, J.; "Mock Sun," Marine Observer, 55:78, 1985.)

Comment. This is actually a case of abnormal refraction. Mock suns are characteristically 22° and/or 46° from the true sun horizontally.

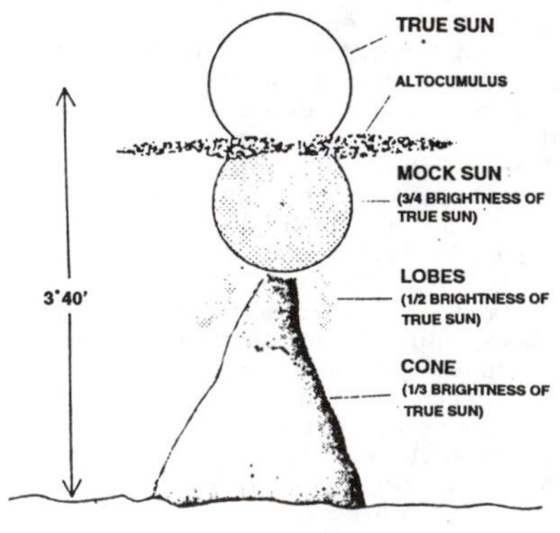

An unusual double sun seen over the Mediterranean.

HALOS AND BOWS

FAST-MOVING DARK BANDS CROSS HALO

December 15, 1976. Chelmsford, Essex, England. At 1505 GMT, two groups of closely spaced grey bands were seen crossing the upper arc of contact of a 22° halo from left to right. The first group lasted 10 seconds, with a 15-second quiescent period before the second group. The second group lasted about 5 seconds. About 30 straight parallel, regularly spaced bands appeared during the first observation. Moving steadily, they took 2 seconds each to cross the arc of contact.

The most likely cause of the phenomenon was thought to be changes in the orientation of the ice crystals that created the upper arc of contact. However, the author could suggest no physical mechanism for producing such unusual motion in the ice crystals. (Burton, B.J.; "Fast-Moving Dark Bands Crossing the Arc of Contact," Journal of Meteorology, (U.K.), 2:233 1977.)

ELLIPTICAL HALOS

In the catalog Rare Halos, Mirages, Anomalous Rainbows, we list nine cases of elliptical halos. Such observations are anomalous because the only well-explained elliptical halo is formed when the lower and upper tangential arcs of a 22° halo join together. Possibly because of the absence of appropriate theory, R. White, in 1981, suggested that the observations recorded in GEH2 were only the consequences of observational error or inaccuracy in representation of the phenomenon. (This assertion is well-known to all anomalists!) Recently, however, several elliptical halos have graced the skies of Finland. We provide below a summary of these observations, as prepared by J. Hakumaki and M. Pekkola. First, though, we express appreciation to Hakumaki and Pekkola for a paragraph headlined SOURCEBOOK PROJECT ANOMALIES, where in effect they vindicate the approach of the Project.

Summary.

"In December 1987 two Finnish amateur astronomers observed and photographed a peculiar vertically elliptical ring surrounding the moon. A literature study carried out soon after this first observation brought to light ten reported historical cases of this type of rare halo phenomenon. It was found out that the existence of these elliptical halos has been uncertain to date due to a lack of photographic evidence. One indication of this is that none of the major modern works on halos mentions such phenomena. During 1988 three more elliptical halos were seen by the Finnish halo observing network. Observations and photographs taken of these phenomena seem to indicate that at least two types of elliptical halo exist. The smaller one was first reported by US astronomer Frank Schlesinger in 1908 and its vertical axis has in all four possible cases been about 7°. No name has been suggested for this halo. The larger one seems to have a vertical axis of about 10° and it has been called the 'halo of Hissink' by Dutch halo observers. Outside the Netherlands this rare halo has received little attention." (Hakumaki, J., and Pekkola, M.; "Rare Vertically Elliptical Halos," Weather, 44: 466, 1989.)

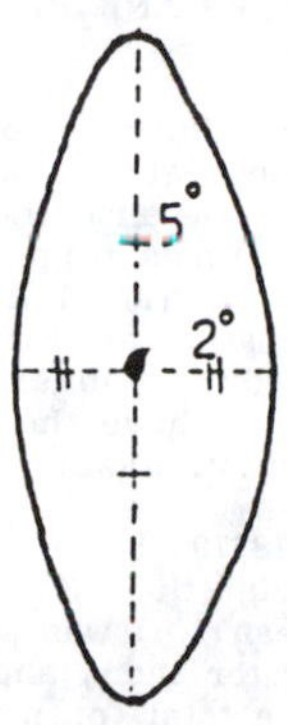

Original drawing of the halo of Hissink, observed at Leiden, January 26, 1977.

ANOTHER ELLIPTICAL HALO

June 6, 1992. Aboard the m.v. British Skill in the Indian Ocean.

Between 1300 and 1345 UTC, a complete halo phenomenon was observed round the moon, as shown in the sketch. The ring was complete although its appearance was elliptical. Its horizontal diameter was 40° with its vertical diameter being 53°.

The illuminated part of the moon was not in the centre of the halo, its altitude at the lower limb (phase, new waxing) being 38° 54'. The altitude of the upper part of the halo was 59° whereas the lower edge was at 6°.

(Anderson, P.R.; "Elliptical Halo," Marine Observer, 63:65, 1993.)

Comment. Once again we have another observation called "impossible" by geophysicists. Halos, they say, must be symmetrical about the sun or moon. Yet, photos and precise measurements, like those above, demonstrate the reality of the phenomenon. This item actually appeared in SF#88.

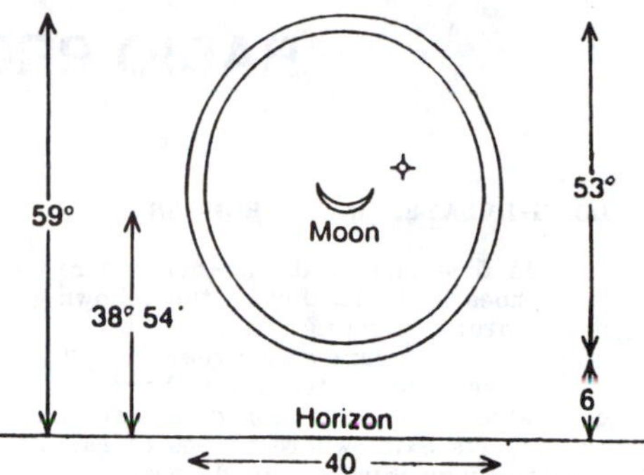

Measurements of an elliptical lunar halo observed in the Indian Ocean.

LUNAR RAINBOW AND UNEXPLAINED WHITE ARC

April 12, 1990. North Atlantic. Aboard the m.v. Canterbury Star.

At 0004 UTC a bright, white arc was seen on the starboard bow and was quickly identified as a lunar rainbow. The moon was one day after its full phase and was just rising; it had little colouration but was unusually bright. A faint, secondary bow became visible outside the main bow at 0010 while the latter, at the same time, began to show colouring; it was possible to see a bluish reddish-orange colour on the outside, merging into yellow, then to a bluish colour on the inside edge. See sketch. Unfortunately, measurement of the width of these bands was not possible as they were not clear enough. During this time, the outer secondary bow together with a third, inner bow remained faint and were white in colour; the inner secondary bow being nearly too faint to see.

Comments from an expert in meteorological optics remarked that the radii of

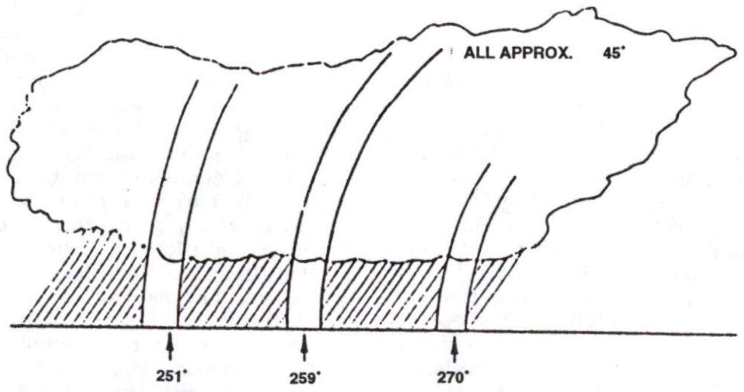

From left to right: A normal secondary lunar rainbow (white); a normal primary bow (colored); and an anomalous secondary bow (white).

the primary and outer secondary bows were less than the theoretical values. He dismissed the inner secondary bow as a misinterpretation, since "theory predicts no such inner secondary bow." (Jackson, C.; "Rainbow," Marine Observer, 61:74, 1991.)

RADIO PROPAGATION

LONG-DELAYED RADIO ECHOES

J. Hals first observed long-delayed radio echoes in 1927. During the following half-century, scientists have been studying this perplexing problem, but it has been the amateurs who have accumulated the bulk of the data. Over 100 reports exist where echoes of radio transmissions were received seconds later at the original transmitting station. Since light travels 186,000 miles per second, any simple radio-wave reflector would have to be well beyond the moon's orbit.

A wide variety of natural phenomena (interplanetary matter) and even artificial devices (alien space probes) have been postulated to explain the long delays. Muldrew's article is first of all an excellent summary of the long and fascinating history of this effect. His bibliography is extensive and apparently nearly complete.

Muldrew next examines the various ionospheric mechanisms that might cause long delays. The ionosphere is a complex structure with ducts in which radio signals can get trapped. Delays of a second or so might be due to such trapping but the longer delays require some other explanation. Muldrew favors a rather complex interaction between signals from separate transmitters that (theoretically at least) can create a long-lived electrostatic wave that travels in the ionosphere---a sort of natural memory device. The coded signals could then be read out much later when the proper natural conditions developed. Delays of up to 40 seconds might be possible with this "ionospheric memory." (Muldrew, D.B.; "Generation of Long-Delay Echoes," Journal of Geophysical Research," 84:5199, 1979.)

LDE PROBLEM STILL UNSOLVED

LDEs (Long-Delayed Radio Echoes) have been known for over 50 years. Radio hams, in particular, will on rare occasions hear their transmissions repeated several seconds later. Various theories have been proposed, including: round-the-world propagation; trapping in ionospheric ducts; reflections from distant plasma clouds; and beam-plasma interactions. R.J. Vidmar and F.W. Crawford, at Stanford, have been studying the LDE problem experimentally and theoretically and conclude that we still don't know which of the proposed mechanisms are valid. (Vidmar, R.J., and Crawford, F.W.; "Long-Delayed Radio Echoes: Mechanisms and Observations," Journal of Geophysical Research, 90: 1523, 1985.)

LUNAR ECLIPSES AND RADIO PROPAGATION

One can understand why long range radio propagation might be affected during a solar eclipse, because the ionizing radiation of the sun is temporarily intercepted by the moon. There is no such obvious explanation for radio propagation problems during lunar eclipses. Nevertheless, we have the following observation by L.M. Nash:

> During 1978/79, I was stationed on Diego Garcia (U.S. Naval base in the Indian Ocean). I was an amateur radio operator then, and one night there was a total (or near total) eclipse of the moon. I was in contact with a station in Utah, on the 15-meter (21.0 to 21.45 MHz) band. When the eclipse started, the Utah station faded out, and all I heard was a sizzling, crackling noise across the entire 15-meter band. This started and ceased within the duration of the eclipse. I then reestablished contact with the Utah station, who was still on the same frequency talking to a friend of his. When I asked him what happened, he stated that my signal had just disappeard.

(Nash, Lemuel M.; personal communication, May 12, 1990.)

SHIP ENVELOPED BY FALSE RADAR ECHO

August 1, 1985. Red Sea. Aboard the m.v. Botany Bay.

"At 1800 a crescent-shaped trace of spurious echoes appeared about 15 n. miles ahead of the vessel. This gradually developed, in an encirculating manner, until, by 1845 the echoes had totally surrounded the vessel. The effect looked like, or could be likened to a plan view of a black island with sandy beaches around its perimeter. The echoes were significant with strong contrast and could even have been mistaken for land on the radar. The effect could not be removed or diminished by changing range scales, motion modes, gain, tuning or perhaps most significantly, altering the pulse lengths. An identical effect was observed on the vessel's independent ARPA radar. The Master commented that although false echoes were invariably encountered in this region, he had not seen one such as this, which actually 'encapsulated' the vessel within the PPI of the radar. By 1935 the false echoes had dissipated into isolated batches splayed randomly across the screen." (Leslie, A.J.; "Radar Echoes," Marine Observer, 56:117, 1986.)

Comment. In the Red Sea and Persian Gulf, similar false echoes are often associated with bioluminescent phenomenon. For details, see p. 261.

MODERN TECHNOLOGY GETS SUNBURNED

During the 400-or-so years we have been counting sunspots and taking other measures of solar activity, the sun has, on the average, been getting more and more rambunctious. The sunspot peaks have been ascending to greater heights every 11-or-so years. Right now, near the peak of the present cycle, the earth is being bombarded by extra-high fluxes of X-rays, ultraviolet light, and other energetic radiation. A century ago, no one would have noticed or cared, but today our technological infrastructure is suffering. K.H. Schatten has listed some of the "symptoms" in a recent article in Nature.

- Fade-outs of over-the-horizon radio communications
- Greater aerodynamic drag on satellites and earlier reentry
- Glitches and outright damage in satellite electrical systems
- Anomalous induced voltages in electrical power systems and long-line communications
- Blackouts of high-frequency polar communications
- Induced errors in VLF (Very Low Frequency) navigation systems
- Occasional radiation levels that are hazardous to humans in high-flying aircraft.

(Schatten, Kenneth H.; "The Sun's Disturbing Behavior," Nature, 345:578, 1990.)

Comment. It would be interesting to learn whether the "computer errors" we encounter so frequently follow the sunspot cycle.

One phenomenon, at least, seems anticorrelated with solar activity: The number of solar neutrinos measured here on earth falls as sunspots multiply. This is particularly puzzling because neutrinos are presumably generated in the solar core, whereas sunspots are supposed to be manifestations of solar surface activity. One phenomenon "should not" affect the other. (Waldrop, M. Mitchell; "Solar Neutrino-Sunspot Connection Found," Science, 248:444, 1990.)

Chapter 6

PSYCHOLOGY

DISSOCIATION PHENOMENA

HALLUCINATIONS

MIND-BODY PHENOMENA

HIDDEN KNOWLEDGE

REINCARNATION

INFORMATION PROCESSING

PSYCHOKINESIS

DISSOCIATION PHENOMENA

OVERVIEW

By far the most important class of "psychic behavior" originates in what is termed "dissociation. Dissociative behavior occurs when normal conscious behavior is modified, sometimes overwhelmed, by subsidiary mental activity. A second self seems to control the hand of the automatic writer and the speaker's tongue in glossolalia. On occasion, the secondary personality will take over all bodily activity, and we have cases of multiple personality and possession. Dreams, sleep walking, fugues, and hypnotism are all manifestations of dissociative behavior in which the waking mind is pushed into the background.

The important anomalous aspects of dissociative behavior are : (1) The curious, innate susceptibility of the human mind to dissociation phenomena; (2) The inescapable suspicion that every human mind harbors many "other selves" residing deep within; and (3) The strong tendency of the human mind to succumb to religious hysteria and exaltation as well as other all-consuming mind-sets and to theories of "what should be" in science as well as everyday life. Why do these propensities exist? What evolutionary value do they offer? Are they symptomatic of imperfections in the human make-up or do they presage an evolutionary leap towards something better---whatever that might be?

IS THE PARANORMAL ONLY A SET OF SUBJECTIVE EXPERIENCES?

Anomalists usually interpret paranormal phenomena as indications that our knowledge of the human mind and how it interacts with other minds and the so-called material world is sadly deficient. But some psychiatrists see paranormal phenomena as merely symptoms of mental disturbance and nothing esoteric at all. Such a view is supported by studies employing interviews with members of society at large.

In a revealing but demystifying study of 502 residents of Winnipeg, C.A. Ross and S. Joshi found: (1) That so-called paranormal experiences are very common indeed, with 65.7% of the interviewees reporting having had them. The most common were deja vu (54.6%), precognitive dreams (17.8%), and mental telepathy (15.6%). Many reported experiencing more than one of the 13 different types of paranormal phenomena included in the survey. But do survey statistics prove that such paranormal phenomena are truly objective?

The real nature of paranormal experiences, according to Ross and Joshi, lies in the close ties these paranormal phenomena have with dissociative phenomena (i.e., automatic writing), hypnotic phenomena, and childhood traumas. They theorize:

> A model is proposed in which paranormal experiences are conceptualized as an aspect of normal dissociation. Like dissociation in general, paranormal experiences can be triggered by trauma, especially childhood physical or sexual abuse. Such experiences discriminate individuals with childhood trauma histories from those without at high levels of significance.

(Ross, Colin A., and Joshi, Shaun; Paranormal Experiences in the General Population," Journal of Nervous and Mental Disease, 180:357, 1992.)

Comment. Perhaps psychiatrists Ross and Joshi hold that those experiencing paranormal phenomena are "treatable." Clearly, they think that the whole of the paranormal is subjective. On a different tack, one must ask why humans are subject to paranormal experiences at all, seeing as they seem to have no survival value and should have been weeded out by natural selection long ago! Wouldn't humans be "fitter" without a proclivity for paranormal experiences? The same question can be asked about motion sickness and other human "weaknesses."

MULTIPLE PERSONALITY

THE ENIGMA OF MULTIPLE PERSONALITY

The multiple personalities of individuals afflicted with this disorder are sharply defined. Each personality has its (his or her) distinctive handwriting, artistic talents, foreign language capabilities, and other aspects of behavior---all in the same body. Such mental and behavioral facets of multiple personality are well-recognized, even if not understood. What is even stranger and more anomalous about multiple personality is the remarkable mind-body relationship manifested. Consider, for example, the case of Timmy:

> When Timmy drinks orange juice he has no problem. But Timmy is just one of close to a dozen personalities who alternate control over a patient with multiple personality disorder. And if those other personalities drink orange juice, the result is a case of hives.
>
> The hives will occur even if Timmy drinks orange juice and another personality appears while the juice is still being digested. What's more, if Timmy comes back while the allergic reaction is present, the itching of the hives will cease immediately, and the water-filled blisters will begin to subside.

How does one explain such phenomena? We cannot obviously, at least not yet. What the phenomenon of multiple personality does do is offer us a "window" for observing that mysterious interface between one's thoughts and bodily functions. Perhaps, if we can heal the mind, we can also heal the body. (Where have we heard this before?) (Goleman, Daniel; "Probing the Enigma of Multiple Personality," New York Times, p. B7, June 28, 1988. Cr. J. Covey)

DIFFERENT PERSONALITIES: DIFFERENT BRAINWAVES

Psychologists have always wondered how a single brain could harbor several personalities, as is the situation in the so-called MPD (Multiple Personality Disorder). Are the different personalities really separate physiologically? Are they faked? Psychological testing of the separate personalities, conducted at different times when each personality was 'in charge,' seemed to indicate discrete personalities, even though one personality might know of the existence of one or more of the others in the same brain. Brain-wave tests were even more conclusive. Each personality had its own distinct set of brain waves in response to identical stimuli. In effect, the brain was operating in a different mode for each personality. (Anonymous; "Multiple Personality Not All in the Mind," New Scientist, 98:290, 1983.)

Comment. Could the ability of the brain to operate in different modes be turned into an asset rather than an affliction? After all, computers can be made more effective when they execute several programs at once.

MPD REVISITED

Above, an item on Multiple Personality Disorder (MPD) appeared along with an editorial comment that perhaps MPD might have some advantages---something like the increases in efficiency in computer multiprogramming. In response, a reader sent in a Washington Post article about a woman with MPD who made just such a claim. People with MPD, she said, "have boundless energy"! Some of the personalities can sleep while others work. (Gregg, Sandra R.; "The Multiperson," Washington Post, June 20, 1983. Cr. W.J. Douglas)

AUTOMATISMS

GLOSSOLALIA: POSSIBLE ORIGINS

N.P. Spanos et al begin their article with a neat encapsulation of the status of psychological research into glossolalia:

"Glossolalia (i.e., speaking in tongues) is vocalization that sounds languagelike but is devoid of semantic meaning or syntax. In the Christian tradition this vocalization pattern is associated with the ideas of possession by the Holy Spirit and communication with God through prayer or prophecy. Some scientific investigators conceptualize glossolalia as the product of an altered or dissociated state of consciousness, whereas others view it as symptomatic of psychopathology.

"The available empirical data fail to support either of these hypotheses. For example, both ethnographic observations and experimental findings indicate that glossolalia can occur in the absence of kinetic activity, disorientation, and other purported indexes of trance, and that experienced glossolalics do not differ from nonglossolalic controls on measures of absorption in subjective experience and hypnotic susceptibility. Relatedly, the available empirical data fail to support the hypothesis that glossolalics suffer higher levels of psychopathology than nonglossolalics."

Spanos et al then go on to detail their own research, in which they tried to teach glossolalia as a learnable skill. First, 60 subjects listened to a 60-second sample of genuine glossolalia. All subjects then tried to speak in tongues for 30 seconds. Some 20% spoke in tongues immediately without further training. The subjects were then divided into a control group and a group that received various kinds of training. Tests then showed that 70% of the trained subjects were now fluent (?) in glossolalia. Glossolalia, therefore, seems likely to be a type of learned behavior rather than a special altered state of mind. (Spanos, Nicholas P., et al; "Glossolalia as Learned Behavior: An Experimental Demonstration," Journal of Abnormal Psychology, 95:21, 1986.)

HYSTERIA

ENVIRONMENTAL STRESS AND ANOMALIES

A succession of outbreaks of so-called mass hysteria in West Malaysia has been investigated by Lee and Ackerman. The victims behaved in bizarre ways, had difficulties in breathing, entered trance-like states, and saw grotesque and mysterious beings. The local populace knew, of course, that the victims (all young students) had offended the spirits. The psychologists knew, of course, that these were typical instances of mass hysteria akin to windshield-pitting episodes, kissing-bug scares, etc. Bomohs (Malay healers) performed adequate sacrifices and passed out suitable talismans, and the episodes were soon under control. The bomohs and the psychologists both explained things successfully in their respective frames of reference. Lee and Ackerman attribute the mass hysteria to social conflicts in a society in transition. (Lee, Raymond L.M., and Ackerman, S.E.; "The Sociocultural Dynamics of Mass Hysteria..," Psychiatry, 43:78, 1980.)

Comment. Could bomohs likewise suppress outbreaks of UFOs, bigfoot sightings, spoonbendings, and sundry anomalies?

FOLIE A FAMILLE

In the psychological literature "folie a deux" is a common topic. It means that psychological illness can be contagious. We now find that the process can extend to whole families.

Reports of folie a famille are rare; this may be the first reported case involving a Vietnam veteran. Expression of his paranoid schizophrenia involved delusions and hallucinations relating to Vietnam, and his wife and children shared his paranoia. Typical of follie a famille were the dominant family member, the threat of violence, and the family's social isolation, frequent crises, and stable membership. (Glassman, Jaga Nath S., et al; "Folie a Famille: Shared Paranoid Disorder in a Vietnam Veteran and His Family," American Journal of Psychiatry, 144: 658, 1987.)

HYPNOTISM

THE MIND'S RHYTHM

Observations of individuals undergoing hypnosis suggest that intervals of particular susceptibility come along about every 90 minutes. Expert hypnotherapists are cognizant of this psychophysical cycle and often wait for it to appear. By closely watching the subject's swallowing, eye-blinking, respiration, etc., the hynotizer can take advantage of these periods of heightened susceptibility. (Rossi, Ernest L.; "Hypnosis and Ultradian Cycles: A New State(s) Theory of Hypnosis?" American Journal of Clinical Hypnosis, 25:21, 1982.)

Comment. The reason for this 90-minute rhythm and the control system behind it are obscure.

IT'S EASIER TO HYPNOTIZE RIGHT-HANDERS

Successful hypnotic induction requires that the subject focus intently upon the hypnotist. Subjects with left-brain dominance (right-handers) are usually able to concentrate their attention better than right-brain people. They therefore enter the trace state more readily. However, once hypnotized, the left-brain-dominated subjects shift into the right-brain mode. It seems that the hypnotic state, with its dream-like quality, altered time sense, etc., is associated with the right brain. Left-handers, who are always in the right-brain mode, possess 'broadened attention' and resist hypnotic induction more than right-handers. (Grist, Liz; "Hypnosis Relies on Left-Brain Dominance," New Scientist, 36, August 2, 1984.)

Comment. As if to balance things out, Nature has apparently made left-handers more talented in the arts and other endeavors. In any case, only trends are involved here; exceptions are everywhere.

Cross reference. The effect of hypnosis upon mind-body phenomena is treated on p. 300.

DREAMS

DREAMS MORE REAL THAN REALITY

Those people who experience so-called lucid dreams say that they are not only vivid in all human senses but completely under the control of the dreamer. Specifically, individuals can be commanded to appear and the action controlled to please the dreamer. As Keith Hearne, a dream researcher, remarks with tongue-in-cheek, the entertainment possibilities are endless if lucid dreaming could be induced in everyone!

Lucid dreams are so real that the dreamer will sometimes believe that he has awakened and answered questions from the researcher, when nothing like that has happened. Lucid dreaming occurs only during periods of REM (Rapid Eye Movements) sleep. The lucid dreamer, however, can signal the dream researcher that lucid dreaming has begun with agreed-upon eye movements and changes in the rate of breathing. (Hearne, Keith; "Control Your Own Dreams," New Scientist, 91:783, 1981.)

Comment. The fact of lucid dreaming encourages many questions. How is it related to out-of-the-body experiences and hallucinations? Pertinent once more is that old philosophical teaser: How do we know that reality is not a dream from which we shall soon awaken? It turns out that lucid dreamers have to devise special tests to ascertain whether they are dreaming or awake!

DREAMS THAT DO WHAT THEY'RE TOLD

A few people can dream and, in their dreams, know that they are dreaming, and then take charge of their dreams, directing them to unfold according to their wishes. This all sounds occultish, to say nothing about far-fetched. It is called "lucid dreaming." F. van Eeden, a Dutch psychiatrist, defined lucid dreaming in this way:

> ...the reintegration of the psychic functions is so complete that the sleeper reaches a state of perfect awareness and is able to direct his/her attention, and to attempt different acts of free volition. Yet the sleep, as I am able confidently to state, is undisturbed, deep and refreshing.

Lucid dreams are real dreams. They occur during REM (Rapid Eye Movements) sleep, usually in the early morning, and they last 2-5 minutes. High levels of physical and emotional activity during the preceding day can encourage lucid dreaming. When lucid dreaming occurs, there are pauses in breathing, brief changes in heart rate, and changes in the skin's electric potential.

There is even a recipe for triggering lucid dreaming. If you awake from a normal dream in the early morning, wake up fully but don't forget the dream. Read a bit or walk about, then lie down to sleep again. Imagine yourself asleep and dreaming, rehearsing the dream from which you awoke, and remind yourself: "Next time I'm dreaming, I want to remember I'm dreaming."

Lucid dreaming, it seems, is not an isolated phenomenon. There are strong similarities between lucid dreaming and out-of-the-body experiences and even the experiences of UFO abductees. S. Blackmore remarks:

> In all these experiences, it seems as though the perceptual world has been replaced by another world, built from the imagination, a hallucinatory replica.

Some people enjoy their lucid dreams; but others fear them and report that objects in this false world are surrounded by a "strong diabolical light." (Blackmore, Susan; "Dreams That Do What They're Told," New Scientist, p. 48, January 6, 1990.)

THE CINEMA OF THE MIND

Many curious mental states occur during sleep, most of which we cannot begin to explain. Horne homes in on REM sleep (Rapid Eye Movements sleep), during which: (1) Both heart and respiration rates become irregular; (2) The flow of blood to the brain increases 50% over relaxed waking levels; (3) Body temperature regulation becomes impaired; and (4) Blood flow to the kidneys and urine production drop markedly. REM sleep is thought to be "primitive" from an evolutionary standpoint because body temperature control is "reptilian." Horne also discusses REM and normal sleep in the context of nightmares, sleeptalking, sleepwalking, night terrors, narcolepsy, and sleep paralysis. (Horne, Jim; "The Cinema of the Mind," New Scientist, 95:627, 1982.)

Comment. All in all, sleep is a complex phenomenon. We know we need it, but why? A really efficient organism would run 24 hours a day---like a computer.

SOLVING PROBLEMS IN YOUR SLEEP

Psychiatrist Morton Schatzman is investigating how some people solve problems in their dreams. Part of his research involves assigning problems (really brain twisters) to students, who are then supposed to solve them in their sleep. One problem posed asked the students to discover how to generate the remaining letters in the following infinite, nonrepeating sequence of letters OTTFF.....Seven students said they discovered the correct answer in their dreams. Another twister concerned the letter sequence HIJKLMNO. What single English word is represented by this string of letters? One student dreamed he was swimming but failed to make the proper connection. (Schatzman, Morton; "Solving Problems in Your Sleep," New Scientist, 98:692, 1983.)

Answers: (1) SSEN..., for Six, Seven, Eight, Nine,... (2) "Water" for H-to-O. Cute!

DO DREAMS REFLECT A BIOLOGICAL STATE?

Scientists have never been able to agree on the meaning of dreams or even if there is one. Mostly dreams were thought to have psychological import, as in the work of Freud and his followers. But there has also been another group of researchers who have considered dreams to be a consequence of one's biological state; that is, one's physical health. The present paper supports this latter belief.

Some 216 patients with heart problems participated in this study. "The patients' dreams were evaluated for the predicted correlations of the number of dream references to death (men) and separation (women) with different levels of severity of heart disease. The severity of heart disease was evaluated with anatomical (coronary angiography) and physiological (ejection fraction) measures obtained at cardiac catheterization, each represented by a 6-point scale of increasing severity. There was no correlation of the number of dream references with the severity of abnormalities on coronary angiography. However, the number of dream references to death and separation correlated with the severity of cardiac dysfunction, as measured by the ejection fraction, which is a more sensitive parameter of disease severity." (Smith, Robert C.; "Do Dreams Reflect a Biological State?" Journal of Nervous and Mental Disease, 175: 201, 1987.)

Comment. One would suppose that the minds (and dreams) of people who knew they had heart problems would normally be filled with dire thoughts.

SCHIZOPHRENIA

SCHIZOPHRENIA AND SEASON OF BIRTH

Several scientists have published data that suggest that schizophrenics are more likely to be born in the winter months. The data have been subjected to much criticism. In this paper, the authors were careful to avoid previous errors in data collecting and analysis. From the Abstract:

> We studied the birth months of 3,556 schizophrenics at a Minnesota Veteran's Administration hospital before and after instituting corrections for year-to-year across-month variations in birthrates in our expected values and the age-prevalence bias toward the January-March seasonality effect described in some earlier studies. Finally, we reanalyzed our data on a subset of patients in whom the age-incidence effect should be minimal. Even after these corrections the results supported the contention that the winter birthrate for schizophrenics is excessive, at least in severe climates.

(Watson, Charles G., et al; "Season of Birth and Schizophrenia: A Response to the Lewis and Griffin Critique," Journal of Abnormal Psychology, 91:120, 1982.)

Comment. What could possibly cause this seasonal effect, assuming it withstands scrutiny?

HALLUCINATIONS

OVERVIEW

Hallucinations or images and sensations that do not exist in the objective sense form the foundation of a large group of strange mental phenomena. Scientific information about hallucinations comes almost entirely from personal testimony. This kind of knowledge is obviously highly subjective and cannot be easily tested for validity like claims of telepathy and dowsing. The idiosyncrasies of human perception, the percipient's imagination, and outright fraud muddy the scientific waters here. Nevertheless, the immense numbers of reports of hallucinations and illusions suggest that many people see, hear, feel, taste, and otherwise sense nonexistent "things."

That some hallucinations are products of the subconscious seems certain. The hypnagogic illusions or "faces in the dark" that appear on the borderland between sleep and wakefulness probably fall into this category. Ghosts, visions, doppleganger, psychic lights, and some UFOs may be joint products of suggestion and the subconscious. So may the images seen in crystal balls, which are in essence optical planchettes. People often see what they want to see and/or what their culture tells them they should see. The human subconscious is a rich lode of strange images which, like the outpourings of the automatic writer, are foreign to the normal conscious person.

Perhaps the biggest mystery in the study of hallucinations is that tendency of people of all cultures and all time periods to say claim they perceive the same sorts of hallucinations; that is, the same kinds of ghosts, religious figures, monsters, UFOs, etc. Why is the human mind made in this way? Did (or does) our susceptibility to hallucinations have survival value---or will it be useful in some way in the future?

OBEs AND NDEs

HOW NDEs DIFFER FROM OBEs

NDEs (Near-Death Experiences) and OBEs (Out-of-the-Body Experiences) are rather common altered states of consciousness that are now the subject of considerable psychological research and public interest. OBEs, where one feels detached from his body and may even view it from afar, occur to many people who are not near death. Yet, the two phenomena have many features in common; so many that some psychologists have claimed that NDEs have no unique features at all.

Gabbard et al have examined hundreds of experiences of both kinds and support the contention that none of the curious features of the NDE are the exclusive province of the NDE. They go a step further, however, by trying to separate NDEs and OBEs statistically. The following experiences occur significantly more often in NDEs: (1) Noises are heard early in the scenario; (2) The sensation of travelling through a tunnel; (3) The physical body is seen from a distance; (4) Other beings in nonphysical form are sensed, especially deceased people emotionally tied to the percipient; and (5) Encounters with communicative entities of a luminous nature. (Gabbard, Glen O., et al; Journal of Nervous and Mental Disease, 169:374, 1981.)

DEATHBED EXPERIENCES LAID TO REST

Over the years, doctors and psychiatrists have accumulated a large lode of deathbed or near-death experiences. Typically, one about to pass through the veil feels exhaltation, meets long-dead relatives, reviews his past life, encounters luminous beings, departs his body to view friends clustering around, and so on. What to make of it all? Is heaven or some sort of afterlife just beyond death's door?

Alcock's analysis is very revealing. He first describes several well-known psychological states: out-of-the-body experiences, hypnagogic sleep, hallucinations, and the so-called mystical experience. He concludes that the full spectrum of deathbed experiences can occur any time, not merely during the final moments of life. Nothing unique happens at death's door, merely the altered states of consciousness expected at such a crucial moment. Thus, deathbed experiences reveal nothing of the territory beyond the grave. Alcock does maintain, however, that these various altered states of consciousness are very curious and well worth further study. (Alcock, James E.; "Psychology and Near-Death Experiences," The Skeptical Inquirer, 3:25, Spring 1979.)

NDEs AND REALITY (WHATEVER THAT IS!)

Several books describing NDEs (Near Death Experiences) have appeared in recent years. The level of objectivity varies widely as do the interpretations of NDEs. Some authors are certain that NDEs confirm life after death; others see NDEs as states of consciousness to be expected from the physiological processes occurring dear death. Now a key medical journal has published a series of opinions by doctors familiar with NDEs and other death-bed events.

Although the articles are quite objective, their thoughts span a wide spectrum. Pure reductionism occupies one end of this spectrum; that is, all NDEs can be explained in terms of known physiological processes. A few doctors, though, point out that NDEs are remarkably consistent regardless of cultural background. One doctor maintains that NDEs are "complex, baffling," and contain "perplexing paranormal features." (Rodin, Ernst, et al; "The Reality of Death Experiences," Journal of Nervous and Mental Disease, 168:259, 1980.)

DISTRESSING NEAR-DEATH EXPERIENCES (NDEs)

Noting that most NDEs are touted as involving "profound feelings of peace, joy, and cosmic unity," B. Greyson and N.E. Bush have collected much contrary testimony, which they organized into three categories:

> (1) Phenomenology similar to peaceful near-death experiences but interpreted as unpleasant, (2) A sense of nonexistence or eternal void, or, (3) Graphic hellish landscapes and entities.

One of these testimonies, from Category 3, is worth reproducing here. The percipient was a woodworker with little interest in religion, although he was married to a "religious fanatic." He had been saving for a vacation for years, but just before they were about to leave, he was arrested for drunk driving and heavily fined, losing his license and vacation savings. Distraught, he tried to hang himself. He testified:

> From the roof of the utility shed in my backyard I jumped to the ground. Luckily for me I had forgot the broken lawn chair that lay near the shed. My feet hit the chair and broke

my fall, or my neck would have been broken. I hung in the rope and strangled. I was outside my physical body. I saw my body hanging in the rope; it looked awful. I was terrified, could see and hear, but it was different---hard to explain. Demons were all around me. I could hear them but could not see them. They chattered like blackbirds. It was as if they knew they had me, and had all eternity to drag me down into hell, to torment me. It would have been the worst kind of hell, trapped hopeless between two worlds, wandering lost and confused for all eternity.

I had to get back into my body. Oh my God, I needed help. I ran to the house, went in through the door without opening it, cried out to my wife but she could not hear me, so I went into her body. I could see and hear with her eyes and ears. Then I made contact, heard her say, "Oh, my God."

His wife then grabbed a knife, ran to the shed, and cut her husband down. An emergency squad revived him. (Grayson, Bruce, and Bush, Nancy Evans; "Distressing Near-Death Experiences," Psychiatry, 55:95, 1992.)

Comments. The above case might well be classified under "telepathy." It is also interesting that UFO contactee tales also have their upsides and downsides, from meeting benevolent "space brothers" to entities that perform vile experiments on the percipient.

OUT-OF-THE-BODY TRAVELLER EXERTS NO INFLUENCE

Many out-of-the-body travellers describe remote scenes observed during their adventures and some are credited with registering their presences on instruments and animals. Tests with a subject using "human detector" instruments a quarter mile away showed no consistent results while the subject was "out-of-the-body." A kitten in the area gave no sign of a presence. Although the subject described some of the remote targets accurately, the results did not differ from chance. (Morris, Robert L., et al; "Studies of Communication During Out-of-Body Experiences," American Society for Psychical Research, Journal, 72:1, 1978.)

OTHER VISIONS

THE MYSTERY OF SPONTANEOUS VISIONS

This fascinating survey of visionary experience during the Middle Ages is more valuable as a thought-provoker than an anomaly collection, although one might claim that any spontaneous vision is anomalous. The authors went back in history and examined a host of documents from the Middle Ages---the lives of saints, histories, biographies, etc. They identified visionary experiences and explored the biological and sociological contexts. Kroll and Bachrach found 134 visions; that is, experiences that had no objective reality. We call such experiences hallucinations today; and their contents were the same then as they are now.

What the authors are after in this study are the perceptions of the visionary experiences by the community. The survey demonstrated immediately that the visions of the Middle Ages appeared to all types of people, not just saints and seers; and, further, that most of the 134 experiences were unrelated to physical and mental health. It was also obvious that the various communities readily accepted these visions as bona fide spiritual and parapsychological experiences. In other words, they were taken as messages from God, predictions of future events, marks of spiritual favor, etc. Kroll and Bachrach concluded that in the Middle Ages visions were culturally supported phenomena and not evidences of psychological illness, as they are today. (Kroll, Jerome, and Bachrach, Bernard; "Visions and Psychopathology in the Middle Ages," Journal of Nervous and Mental Disease, 170:41, 1982.)

Comment. Superficially there is little that is surprising in these results. The people of the Middle Ages had wider spiritual horizons, while we build mental hospitals and consider UFO contactees as nuts. Regardless of the cultural environment, visions keep on occurring. They virtually never have any practical import. Why, then, do we keep on seeing them? Waxing speculative again, the false-head butterflies mentioned on p.157 probably have no inkling about the real value of their markings; is there some yet uncomprehended purpose behind these strange human mental quirks, or are they merely a little snow on our TV screens?

A MENTALLY CREATED REALITY

M. Schatzman, an American psychiatrist, has conducted extensive psychological experiments with a subject named Ruth. Ruth is perfectly sane but is apparently able to create vivid hallucinations at will. Neither the experimenters nor photographic film detect these apparitions, but they are very real to Ruth. Just how real was determined by tests during which Ruth was instructed to create the image of her daughter between herself and a TV screen. The TV screen displayed a reversing checkerboard pattern that normally shows up very distinctly on a subject's electroencephalogram (EEG)---the so-called visual evoked response. Ruth's EEG did not show the visual evoked response when the apparition of her daughter was in the way, although it was normal when she was not hallucinating.

Ruth also had the talent for age regression during memory trances. The Stroop Test, administered when Ruth had regressed to the age of three, confirmed that she had lost the ability to read under these conditions. (Schatzman, Morton; "Evocations of Unreality," New Scientist, 87:935, 1980.)

Comment. This apparent ability of Ruth to distort her own reality has an occult flavor, but perhaps through experiments such as these scientists can get a handle on UFOs, Bigfoot, and similar phenomena. Incidentally, the Stroop Test evokes an involuntary response, making it impossible for the subject being tested to fake a lack of reading ability.

A DELUSION OF DOUBLES

R.J. Berson has reviewed 33 cases of a curious delusion called Capgras' Syndrome. People displaying this syndrome believe that important people in their lives (family members, etc.) have been replaced by exact doubles. No hallucinations or illusions are involved; rather it is a belief. Those afflicted with Capgras' Syndrome may even believe that they themselves are represented somewhere by a double they never see. Not all persons with close emotional ties are believed to be doubled; and these unreplaced persons are always identified accurately. People with these beliefs usually possess normal perceptions and memories but are (obviously) disturbed emotionally with paranoid tendencies. (Berson, Robert J.; "Capgras' Syndrome," American Journal of Psychiatry, 140:969, 1983.)

Comment. This strange mental state is apparently not related to autoscopy, where one hallucinates one's self.

THE KALEIDOSCOPIC BRAIN

It first hit me early in the morning of December 27, 1970. Suspended in the dreamy state just before waking and with my eyes still closed, I experienced an extraordinarily vivid visual image. Extending before me was an infinite array of diamond-shaped amber regions; each was filled with a regular pattern of black spade-like forms, all pointing straight up, or, in alternate diamonds, to the left. A latticework of delicately beveled edges, gleaming like polished gold, framed with diamonds. The whole array shimmered before me in perfect amber and gold splendor for what must have been several seconds.

This kind of geometrical visual experience occurs rarely to some people during "twilight" states just before falling asleep or waking. Direct pressure on the optic nerve produces similar geometers' delights; so can drugs, fever, sleeplessness, and other altered states of consciousness. Migraine headaches, too, are often presaged by floating, semicircular fields of closely spaced parallel lines or bars arranged in zigzag patterns. This geometrical visual phenomenon may, like a berserk TV screen, be diagnostic and betray regularities in the brain's circuitry. The kaleidoscopic patterns seem to occur when imput signals from the eyes are weak or sus-

Visual hallucinations induced during controlled intoxication with cocaine. (From Unfathomed Mind)

pended, leaving the brain to generate its own "favorite" patterns. (Shepard, Roger N.; "The Kaleidoscopic Brain," Psychology Today, 17:62, June 1983.)

Comment. But why the elaborate geometry? Could this apparently "built-in" pattern-generating capacity manifest itself in waking humans as an urge to describe the universe in terms of regular mathematical laws and geometric models?

HOSTAGE HALLUCINATIONS

"The literature on hallucinatory experiences of hostage victims is reviewed. The phenomenology is examined in 30 case studies involving 31 persons, including exprisoners of war and victims of rape, kidnapping, terrorism, robbery, and 'UFO abductions.' The victims were subjected to conditions of isolation, visual deprivation, restraint on physical movement, physical abuse, and the threat of death. For eight victims, these conditions were sufficient to produce a progression of visual hallucinations from simple geometric images to complex memory images coupled with dissociation. The other 23 victims, subjected to similar conditions but without isolation and life-threatening stress, did not experience hallucinations. The hostage hallucinations are compared to those resulting from sensory deprivation, near fatal accidents, and other states of isolation and stress. A common mechanism of action based on entopic phenomena and CNS (central nervous system) excitation and arousal is suggested."

In a typical case, an 18-year-old female college student was kidnapped and held for ransom. She was bound, blindfolded, and denied food, water, and toilet facilities. She was periodically threatened with death. She saw dull flashes of light in front of her eyes and small animals and insects on the periphery of her visual field. Becoming hypervigilant, she heard strange sounds and whispers. Hearing loud noises, she thought her captors were coming to kill her. It was then her whole life ran off like a slide show before her eyes. The noises were the police coming to rescue her. (Siegel, Ronald K.; "Hostage Hallucinations," Journal of Nervous and Mental Disease, 172:264, 1984.)

Comment. Some of the hostages experienced the tunnel hallucination so common in near-death visions. These seemingly 'built-in' or hard-wired images may be related to UFO and sea-serpent phenomena.

THE VOICE OF GOD

Back in 1976, Julian Jaynes promulgated a novel hypothesis about ancient man in his book, The Origin of Consciousness and the Breakdown of the Bicameral Mind.

"According to Jaynes, consciousness, as we know it today, is a relatively new faculty, one that did not exist until as recently as 2000 B.C. He holds that a basic difference between contemporary and ancient man is the process of decision-making. When faced with a novel situation today, man considers alternatives, thinks about future consequences, makes a decision, ruminates over it, and finally acts. He then reconsiders his action, evaluates it, worries about it, feels good or bad about it, makes resolves about future decisions, and so forth. The cerebral activity that precedes and follows an action response is consciousness. Jaynes believes that man of antiquity had no consciousness---that when faced with a novel situation, he simply reacted. He reacted without hesitation by following the directions of a personal voice that told him exactly what to do. Ancient man called this voice God; today it is called an auditory hallucination. To ancient man, God was not a mental image or a deified thought but an actual voice heard when one was presented with a situation requiring decisive action."

You must really read Jaynes' book to appreciate the evidence he has collected in support of his hypothesis. In the present article, J. Hamilton has found additional support for Jaynes' theory. His Abstract follows:

"When a system for communicating with nonverbal, quadriplegic, institutionalized residents was developed, it was discovered that many were experiencing auditory hallucinations. Nine cases are presented in this study. The 'voices' described have many similar characteristics, the primary one being that they give authoritarian commands that tell the residents how to behave and to which the residents feel compelled to respond. Both the relationship of this phenomenon to the theoretical work of Julian Jaynes and its effect on the lives of the residents are discussed." (Hamilton, John; "Auditory Hallucinations in Nonverbal Quadriplegics," Psychiatry, 48:382, 1985.)

MIND-BODY PHENOMENA

OVERVIEW

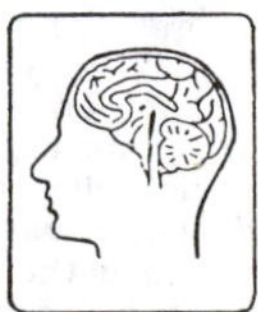

The control of the mind over the body is more amenable to scientific study than most topics in parapsychology. Psychosomatic medicine is, in fact, not usually considered part of parapsychology, although the mystery of placebo action is certainly as strange as the force that drives the hand of the automatic writer or the ouija board user. Perhaps the mechanisms behind all of these phenomena are related.

Everyone knows that mental attitude affects physical well-being; headaches disappear as do other complaints when mental conditions improve. But the mind-body relationship is much deeper. Consider also the phenomena of faith healing, voodoo death, hypnotic deafness, and the raising of stigmata and blisters through suggestion. The mind exerts powerful influences over many bodily functions. Anomalies exist because we know virtually nothing about how the mind accomplishes these things.

BELIEF SYSTEMS AND HEALTH

Can one's mind-set affect one's immunity to disease? Lenard explores (in popular style) the roles of mental attitude, visualization techniques, and placebos in fighting and preventing cancer and other ailments. Placebos are nothing new. Most doctors admit they sometimes work for some people. Why, they don't know. Placebo action seems closely allied to a person's mental attitude. Many doctors will also allow that a positive attitude helps a lot in fighting illness and that depression aggravates it.

Visualization techniques, though, are hotly debated. Will cancer cells be destroyed, or at least stop growing, if the patient visualized them as weak things that are vulnerable to the body's killer cells? Proponents of visualization recommend that a cancer patient visualize his killer cells as protecting knights in armor that swoop down and skewer the enemy cancer cells. In a visualization session, one focuses one's mind on such images and, in essence, wills his body to fight back. There is some evidence that visualization helps. (Lenard, Lane; "Visions That Vanquish Cancer," Science Digest, 89:59, March 1981.)

Comment. The crucial scientific question in all the above methods is: How does a belief system mobilize biological systems?

THE MIND'S CONTROL OF BODILY PROCESSES

T.X. Barber has reviewed the role of the mind in the control of many physiological processes in a chapter appearing in a new book. The chapter is 58 pages long, with 176 references, making it a major contribution to the subject. To give the reader the flavor of this paper, two paragraphs are now reproduced:

"The data presented in this chapter should, once and for all, topple the dualistic dichotomy between mind and body which has strongly dominated Western thought since Descartes. The meanings or ideas imbedded in words which are spoken by one person and deeply accepted by another can be communicated to the cells of the body (and to chemicals within the cells); the cells then can change their activities in order to conform to the meanings or ideas which have been transmitted to them. The believed-in (suggested) idea of being stimulated by a poison ivy-type plant, transmitted to a person who is normally hypersensitive to this type of plant, can affect specific cells (probably in the immunological and vascular systems) so that they produce the same type of dermatitis which results when the person actually is stimulated by a poison ivy-type plant. Similarly, individuals who are viewed as allergic to pollen or house dust may not manifest the allergic reaction when they believe (falsely) that they have not been exposed to the allergic substance.

.....

"Believed-in suggestions can affect specific parts of the body in very specific ways. Suggestions of being burned can give rise to a very specific irregular pattern of inflammation on the hand that closely follows the pattern of a previously experienced actual burn in the same place. Suggestions that a congenital skin disorder will ameliorate step by step first in one area of the body, then in another, can be actualized exactly as suggested. At least one set of investigators found that suggestions that specific warts will regress can be effective in removing just those warts and not others."

In addition to the subjects mentioned in the quotations, the following topics are explored: congenital skin diseases, blisters produced by suggestion, stigmata, mental inhibition of bleeding, fire-walking in safety, control of blood flow and skin temperature, and so on. (Barber, Theodore X.; "Changing 'Unchangeable' Bodily Processes by (Hypnotic) Suggestions: A New Look at Hypnosis, Cognitions, Imagining, and the Mind-Body Problem," in A.A. Sheikh (ed.) Imagination and Healing, Baywood Publishing Co., Farmingdale, NY, 1984, pp. 69.)

THE MIND'S "SCOPE"

Never underestimate the power of the mind. Take "bad breath," for example:

> An example is the case of B.O., white, married, mother of three children (ages 9, 6, and 4), operating room nurse. Her chief complaint was severe bad breath of several years duration. In the past, she had consulted dentists, an E.N.T. surgeon, and a family practice physician who had prescribed two series of antibiotics, then a powerful mouthwash that had denuded the epithelium of her tongue, resulting in severe pain and diet restriction. It took 16 weeks for the tongue to heal. B.O. came to me in January of 1983, when she felt the symptom had worsened.

H.P. Golan, who treated B.O. (sic), employed hypnotic techniques in which the patient was first shown the power of her own mind over her body. B.O. responded well, and was soon able to produce temperature changes in her hand and glove anesthesia.

> It was explained to her that her physical symptom was an expression of emotional problems caused by stress. The feeling of her hand temperature change and the view of her hand anesthetized had made her realize physiological control was possible over one part of her body. It was explained to her that stress often causes excess acid production in the stomach, which can cause bad breath. If she could control her hand physically, she could control the excess physical secretions in her stomach.

B.O. did just that. Case closed! (Golan, Harold P.; "Using Hypnotic Phenomena for Physiological Change," American Journal of Clinical Hypnosis, 28:157, 1986.)

PSYCHOSOMATIC MEDICINE

WARTS ON DEMAND?

Warts can often, but not always, be cured by hypnosis or suggestion. The medical literature is emphatic on this point. Warts can, in fact, be banished from one side of the body, leaving those on the other side intact, if the hypnotist suggests such asymmetry! What has been lacking in the psychosomatic story, according to Gravitz, is evidence that warts can also be produced by suggestion. Evidently modern psychologists have been content with curing warts, for Gravitz had to go all the way back to 1924 to find accounts of warts being induced by suggestion. This scientific neglect of the supply side of wart economics is unfortuanate in the view of Gravitz, because here is a mild affliction involving both a virus and psychology that can be studied easily. The knowledge gained might be applicable to some cancers, which also seem to involve viruses and psychology. (Gravitz, Melvin A.; "The Production of Warts by Suggestion as a Cultural Phenomenon," American Journal of Clinical Hypnosis, 23:281, 1981.)

Comment. The mysterious spontaneous remission of some cancers may have psychosomatic overtones.

HYPNOTIC SUGGESTION SUPERIOR TO SALICYLIC ACID

Forget the stump water, too, for hypnosis is better than both when it comes to chasing away warts. If you don't believe this, read the following Abstract.

Subjects with warts on their hands and/or feet were randomly assigned to a hypnotic suggestion, topical salicylic acid, placebo, or no treatment control condition. Subjects in the three treated groups developed equivalent expectations of treatment success. Nevertheless, at the six-week follow-up interval only the hypnotic subjects had lost significantly more warts than the no treatment controls.

(Spanos, Nicholas P., et al; "Effects of Hypnotic, Placebo, and Salicylic Acid Treatments on Wart Regression," Psychosomatic Medicine, 52:109, 1990.)

Comment. The real import here is not in the vanquishing of these benign tumors caused by papillomaviruses but rather in the potential of altered states of mind for treating more severe afflictions, such as cancer.

MENTAL CONTROL OF ALLERGIES

Most of the effort directed at understanding the problems of allergy has focused on the interacting components of the immune system. The possibility that histamine may be released as a learned response has now been tested. In a classical conditioning procedure in which an immunologic challenge was paired with the presentation of an odor, guinea pigs showed a plasma histamine increase when presented with the odor alone. This suggests that the immune response can be enhanced through activity of the central nervous system.

The article begins by noting that many anecdotal reports suggest that allergic reactions can be induced by suggestion; viz., an allergy to roses induced by an artificial rose. (Russell, Michael, et al; "Learned Histamine Release," Science, 225:733, 1984.)

FALSE PREGNANCIES IN MALES

This is a very rare psychosomatic condition. In the past 45 years, about 100 cases of false pregnancies in females have been reported, but only 3 in males. A fourth has now come to light. It is the story of a 40-year-old, married man, who wished to have another child but his wife didn't. Subsequently, the man's abdomen began to protrude and his weight increased by 20 pounds. Symptoms similar to those of morning sickness also developed. The condition eventually subsided as he and his wife "talked out" their disagreement. The man had a previous history of depression and schizophrenia. (Evans, Dwight Landis; "Pseudocyesis in the Male," Journal of Nervous and Mental Disease, 172:37, 1984.)

HYPNOTICALLY ACCELERATED BURN WOUND HEALING

This study was designed to assess the efficacy of hypnotically induced vasodilation in the healing of burn wounds. Patients were selected on the basis of having symmetrical or bilaterally equivalent burns on some portion of their right and left sides. Since one side only of the body was treated by hypnotically induced vasodilation, the patient served as his own experimental control. In this single blind study, the hypnotist and patient knew the side selected for treatment, the evaluating surgeon and nursing staff did not. Four of the five patients demonstrated clearly accelerated healing on the treated side, the fifth patient had rapid healing on both sides. It is concluded that hypnosis facilitated dramatic enhancement of burn wound healing.

(Moore, Lawrence Earle, and Kaplan, Jerold Zelig; "Hypnotically Accelerated Burn Wound Healing," American Journal of Clinical Hypnosis, 26:16, 1983.)

FIRE-WALKING: ANYONE CAN DO IT

San Pedro Manrique, Spain, 1969, the annual fire-walking ceremony. Arriving too late for the official ceremony, the author and his friends find a bed of coals too hot to stand near. Two members of a French TV crew remaining behind walk across the coals with no adverse effects. A Spanish companion does, too. Thus encouraged, the American takes off his shoes. He is advised: (1) make sure your feet are brushed free of grass and twigs; (2) place your feet firmly and with force; and (3) never hesitate, keep moving. The author walked across the coals without the slightest hint of burning. It was, he states, a "spiritual experience!" (McElroy, John Harmon; "Fire-Walking," Folklore, 89:113, 1978.)

APATHY AND CANCER

Doctors have frequently observed that the "will to survive" is important in controlling the progression of serious diseases. Most of the evidence linking the patient's mood with recovery from illness is anecdotal---little wonder since mood is hard-to-measure. Some statistical evidence has recently been accumulated by S.M. Levy and R. Herberman of the National Cancer Institute; but the situation still seems complex at best.

From a study of 75 women with breast cancer, there appears to be a significant and involved relationship between age, the body's immune function, and a psychological factor called "fatigue." One clear-cut finding was that young patients facing radiation therapy and also reporting high levels of psychological fatigue were the only patients in the surveyed group showing diminished activity by the body's natural killer cells. These killer cells comprise an important part of the defense against cancer. This biological consequence of apathy is confirmed by another study showing that cancer patients with "psychological distress" had better chances of recovery than those who had no "fight." (Herbert, W.; "Giving It Up---At the Cellular Level," Science News, 124:148, 1983.)

Comment. Assuming such mind-body correlations are real, how is mental attitude (supposedly some pattern of nerve signals in the brain) converted into greater or lesser populations of natural killer cells?

PSYCHOTHERAPY MAY DELAY CANCER DEATHS

In previous entries, we have reported that imaging, positive thinking, and other psychological stratagems seemed to have some effect on the progress of cancer in humans. Such positive results have generally been pooh-poohed by the medical establishment. In fact, the results recently reported by Stanford psychiatrist D. Spiegel were obtained during an attempt to show that psychotherapy had no effect whatsoever on cancer.

Thirteen years ago, Spiegel participated in a short-term program in which group therapy was given to 86 patients with advanced breast cancer. The goal was simply to make the patients feel better and "face their mortality." The result was that the patients became less anxious, less fearful, and more positive. They even learned to reduce their pain through self-hypnosis. That was the end of the program.

Recently, Spiegel, fed up with claims that positive thinking could help control cancer, tracked down the patients who had received psychotherapy earlier. He expected to find no difference between their fates and those of a control group that had not received psychotherapy. Not so! Those in the control group had lived an average of 19 more months, compared to an average 37 months for those getting the psychotherapy. Spiegel said, "I just couldn't believe it." "What I am flat out certain of is that something about being in groups helped these women live longer. But what it is, I don't know." (Barinaga, Marcia; "Can Psychotherapy Delay Cancer Deaths?" Science, 246:448, 1989.)

BE HAPPY, BE HEALTHY: THE CASE FOR PSYCHOIMMUNOLOGY

Hints are accumulating from many clinical studies that one's mental state has much to do with the effectiveness of one's immunological system. Happy, unstressed people get fewer colds. Introverts get worse colds than extroverts. Men who have just lost their wives have lowered white-cell responses. Although many physicians and medical researchers think it too early to claim that mental stress significantly suppresses the human immunological system and thus leads to more illness, one can see the pendulum start to swing away from the time-honored belief that mind and body are entirely separate entities.

The foregoing studies and others like them are discussed in a recent survey of psychoimmunology by B. Dixon. Toward the close of the article, Dixon asks why humans (and other animals, too) have evolved an immunological system sensitive to stress. Evolutionists can always find some sort of justification in Darwinian terms, and Dixon's is rather ingenious. Suppose a primitive human was attacked by a saber-toothed tiger (what else?). If the human survived, his immunological system would immediately go into high gear to clean up the wounds and repel invading germs. The trouble is that a revved-up immunological system (especially the white blood cells) can go too far and chomp up healthy tissue, too. However, evolution has constructed animals such that stress (saber-toothed tigers are stressful!) suppresses the immunological system to restrict the consumption of healthy tissues. While this short-term damper on immunological activity may have been useful to primitive humans, the stresses on modern man are less intense and long-term. We are sick a lot, particularly with cancer, because evolution has not yet adjusted or fine-tuned us to the new environment. (Dixon, Bernard; "Dangerous Thoughts," Science 86, 7:63, April 1986.)

PLACEBO EFFECT

CONDITIONED RESPONSES THAT SHORT-CIRCUIT THE CONSCIOUS BRAIN

Don't let the title of this item deter you; this is serious stuff. We all know about the placebo effect. A sick patient improves because he believes he is getting a helpful medicine, even though it is an inert substance. The reverse works, too, at least in experiments with mice. It seems that mice can be conditioned into believing that an ordinarily delectable substance (saccharin and water) gives them stomach pain, by simultaneously injecting them with a pain-producing chemical. Unexpectedly, this chemical also suppressed the immune system of the mice. The mice, of course, knew nothing about the effect on their immune system. Nevertheless, whenever they received saccharin after being conditioned, their immune system was suppressed even though the pain-producing chemical was not administered.

While one can imagine the mice consciously associating saccharin and stomach pain, and their brains somehow sending signals that simulated pain, it seems inconceivable that the the mice knew anything about their immune system. We have always assumed that the placebo effect (and its reverse) worked because of the subjects' logical association of cause and effect, but evidently there is something else going on here! (Wingerson, Lois; "Training the Mind To Heal," Discover, 3:80, May 1982.)

Comment. This all opens a rather large Pandora's Box, because it implies that seemingly innocent signals can trigger unrealized reactions. It's something like a post-hypnotic suggestion. Some cause ---not recognized as a cause---results in an effect---not consciously related to the real cause. We could all be puppets, not even recognizing the strings that control us!

THE SUBTLE PLACEBO

A most interesting series of placebo experiments have been carried out by J.D. Levine and N.C. Gordon, of the University of California at San Francisco. The subjects were all dental patients who were tested when their surgical anesthesia was wearing off. The substances administered were: (1) a placebo; (2) morphine; and (3) naloxone, a substance that blocks the opiates produced in the brain. The doses were administered: (1) openly, when the experimenter knew which substance was being given; (2) by a person hidden from both experimenter and patient; and (3) by a machine.

Two findings are particularly revealing. First, pain always increased after naloxone was administered, implying that the opiates blocked by naloxone are probably the same as those released by placebos. More significant, however, was the fact that both the open and hidden administrations of the placebo reduced pain while the machine-applied placebo resulted in more pain. In other words, when either the experimenter or the hidden administrator knew that the placebo was being given, the placebo worked. Levine and Gordon supposed that there must have been subtle clues, detected subconsciously by the patients, that the hidden person was administering the placebo. (Anonymous; "The Subtle Strength of Placebos," Science News, 127:25, 1985.)

Comment. If no subtle clues existed, wouldn't this be a possible example of telepathy?

Artist's concept of telepathy, a claimed phenomenon discussed on the following pages.

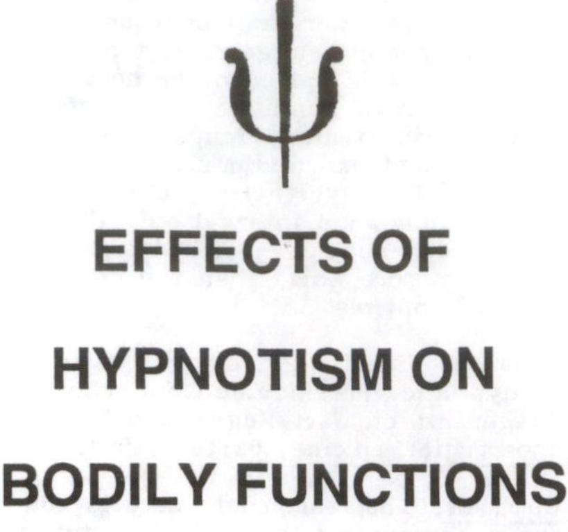

EFFECTS OF HYPNOTISM ON BODILY FUNCTIONS

HYPNOSIS AND SKIN TEMPERATURE

The mind can affect the body in many ways. One of the most bizarre is the use of hypnosis to raise skin temperature on selected parts of the body. A recently published example of this phenomenon was observed by P. Hajek et al, at the Institute of Physiology in Prague:

> Eight patients with atopic eczema and six healthy subjects were given hypnotic suggestion to feel pain in the upper part of the back and in one case on the palm. An average local increase in skin temperature of 0.6°C (detected by thermovision) occurred under this condition. For some patients, cutaneous pain threshold was increased before the experiment by means of repetitive hypnotic suggestion of analgesia. These subjects reported feeling no pain subjectively, but the local change in skin temperature was equal in both cases. The results suggest a central mechanism induced by measuring changes in pain threshold in the skin, which changes are independent of local changes in blood flow.

(Hajek, P., et al; "Increase in Cutaneous Temperature Induced by Hypnotic Suggestion of Pain," Perceptual and Motor Skills, 74:737, 1992.)

HYPNOSIS AND BASKETBALL

Hypnosis has often been applied to improving various aspects of human performance---both intellectual and physical---with debatable success. Basketball may now be added to this file, according to a recent paper by E.H. Schreiber:

> Throughout one basketball season 12 male and 12 female basketball players were interviewed prior to individual and group sessions with hypnosis. The athlete's shooting scores were compared with those of the previous basketball season. The hypnosis groups showed higher cumulative scores for shooting than players in the control group.

(Schreiber, Elliott H.; "Using Hypnosis to Improve Performance of College Basketball Players," Perceptual and Motor Skills, 72:536, 1991.)

HIDDEN KNOWLEDGE

OVERVIEW

Telepathy, precognition, dowsing, divination, etc. are now and always have been cornerstones of parapsychology. Since time immemorial, some individuals have claimed that they could obtain information that was physically hidden, residing in the brains of others, or located in the future. According to the tenets of science, such data acquisition is impossible. Consequently, claims of telepathy, precognition, and the like should always be subjected to the closest scientific scrutiny---even closer scrutiny than that accorded less extraordinary claims, such as the sighting of a new species of mammal. As a matter of fact, reviews of parapsychological experiments have led to many charges of fraud, sloppy design, and bad statistics. Indeed, it seems that almost every phenomenon of parapsychology can be duplicated by a good magician. Even today, in the age of New Agers, we must face squarely the possibility that, marvelous though the human mind is, its capabilities may not embrace telepathy, dowsing, and other purported methods of acquiring hidden knowledge. These phenomena, if they exist at all, certainly do not seem to be pronounced human talents!

OH MAGIC, THY NAME IS PSI

Winkelman has written a remarkable article. Even more remarkable is the fact that it has been published in a mainstream scientific journal. In fact, this contribution is full of statements guaranteed to ruffle the feathers of any rationalistic scientist---and 99.9% of all scientists are rationalistic. Take, for example, the first sentence of Winkelman's summary:

> The correspondences between parapsychological research findings and anthropological reports of magical phenomena reviewed here suggest that magic is associated with an order of the universe which, although investigated empirically within parapsychology, is outside of the understanding of the Western scientific framework.

As a consequence of this inflammatory tone of the article, the comments that followed were rather emotional in many instances. A philosopher could write another paper on the character and prejudices of the "scientific belief system" based on these comments alone!

Enough of the controversial nature of Winkelman's article; what does it say? Basically, it states that many magic systems; such as sorcery, witchcraft, divination, and faith healing; may have originated and still be sustained by the psi abilities of the practitioners of magic. Psi here includes telepathy, precognition, psychokinesis, etc. The body of the paper reviews the characteristics of magic and the areas "correspondence" with parapsychology. Examples of areas of correspondence: altered states of consciousness, visualization, positive expectation, and belief. Indeed, the correspondences are strong; and this fact leads to the sentence from the summary quoted above. (Winkelman, Michael; "Magic A Theoretical Reassessment," Current Anthropology, 23:37, 1982.)

Comment. A conventional rationalistic scientist would, or should, react to Winkelman's paper by saying that: Magic is not based on psi, rather psi is magic and has no scientific basis. Winkelman, on the other hand, tacitly assumes the reality of telepathy, precognition, etc. The real issue, of course, is whether there really is an "order of the universe" beyond the ken of present-day science. The history of science shows that science has eventually accepted one impossible idea after another; viz., meteorites and hypnosis. Psi may not make the grade. Even if psi is real, it cannot become part of the scientific belief system until dogmas about the nature of life and man change.

PARAPSYCHOLOGY: A LACK-OF-PROGRESS REPORT

Speculating, as in the above item, is fun; but we need something to deflate balloons before they drift too high into the wild blue. The Skeptical Inquirer is the perfect "something." In the latest issue, ESP or psi takes it on the chin. J.E. Alcock reviews the last 8 years of parapsychological research. His conclusion:

> The past eight years have been no kinder to those seeking compelling evidence about the reality of paranormal phenomena than were the previous eighty: The long-sought reliably demonstrable psychic phenomenon is just as elusive as it always has been.

Alcock believes that parapsychology is on the ropes and must grasp at straws. One of these straws is the enthusiastic espousal of those quantum mechanical effects which seem to transcend time, space, and even human comprehension. Alcock contends that the admitted enigmas of quantum mechanics are being unfairly twisted by the parapsychologists. [Parapsychologists and their critics will argue interminably about the applicability of quantum mechanics to psi, ceasing only when someone with powerful, undeniable psi powers comes along---the equivalent of a UFO landing on the White House lawn. WRC]

Meanwhile, Alcock identifies an important characteristic of psi, which is truly anomalous, for it is completely foreign to science as we understand it today. This is the generalizability of psi.

> ...psi effects turn up whether one uses cockroaches or college students, whether the effects are to be generated in the present or the future or the past, whether the subjects know that there is a random number generator to be affected, whether a sender and receiver are inches or continents apart...

Alcock believes this generalizability of psi weakens the case for its existence. He attributes any non-chance effects in psi research to bad experiment design and to the vicissitudes of chance itself. (Alcock, James E.; "Parapsychology's Past Eight Years: A Lack-of-Progress Report," Skeptical Inquirer, 8:312, 1984.)

Comment. If psi effects are real and also transcend our usual concept of causality and the space-time framework, the conventional scientific approach really becomes impossible. For example, the results of an experiment could be modified by someone in the future---so-called retroactive psychokinesis! Psi when completely generalized is independent of humans and other life forms. It is then a general property of the cosmos---a certain tendency of lists made by people, random-number generators, and similar sources to match up in non-chance ways. If this tendency is real, the laws of chance do not truly reflect the way the cosmos works. There is, after all, no absolute requirement that mathematics be a faithful mirror of reality. Reality is reality; and theory is, well, something the left side of the brain is good at generating.

ESP

STACKED DECK IN ESP EXPERIMENT

Balanovski and Taylor have assumed that the many purported extrasensory phenomena are very likely effected by electromagnetic forces---the only known action-at-a-distance force they believe could be involved. Therefore, they assembled a wide variety of very sensitive electromagnetic instruments (antennas, EM probes, skin electrodes, magnetometers, etc.) and tried to find electromagnetic fields associated with people claiming paranormal abilities. Despite the high sensitivities of the apparatus, no ESP-connected electromagnetic fields were detected. (Balanovski, E., and Taylor, J.G.; "Can Electromagnetism Account for Extra-Sensory Phenomena?" Nature, 276:64, 1978.)

Comment. J.G. Taylor is the author of Superminds, a rather unabashed pro-paranormal book. He has since recanted. His experiment (described briefly above) certainly has not disproved the existence of ESP, only that there are no accompanying electromagnetic fields. ESP, if it exists, may work through "unknown" fields or, perhaps, no fields at all, as we understand them. Although caveats appear in the article relative to the limited nature of the experiment, such an article in a key scientific journal just makes most doubting scientists say "I told you so."

GEOMAGNETIC ACTIVITY AND PARANORMAL EXPERIENCES

"---25 well-documented (and published by Stevenson in 1970) cases of intense paranormal ('telepathic') experiences concerning death or illness of friends of family were analyzed according to the global geomagnetic activity (the aa index) at the times of their occurrence. The characteristics of these cases were representative of the general literature and occurred between the years 1878 and 1967. All 25 experiences were reported to have occurred on days when the geomagnetic activity was less than the means for those months. Repeated-measures analysis of variance for the daily aa indices for the 7 days before to the 7 days after the experience confirmed the observation that they occurred on days that displayed much less geomagnetic activity than the days before or afterwards. These results are commensurate with the hypothesis that extremely low fields, generated within the earth-ionospheric cavity but disrupted by geomagnetic disturbances, may influence some human behavior." (Persinger, Michael A.; "Geophysical Variables and Behavior: XXX. Intense Paranormal Experiences Occur during Days of Quiet, Global, Geomagnetic Activity," Perceptual and Motor Skills, 61: 320, 1985.)

DREAM ESP AND GEOMAGNETIC ACTIVITY

"The 24-hour periods in which the most accurate telepathic dreams occurred during the Maimonides studies displayed significantly quieter geomagnetic activity than the days before or after. This statistically significant V-shaped temporal sequence in geomagnetic activity was not evident for those periods when less accurate dreams occurred. When geomagnetic activity around the time of the strongest experimental telepathic dreams was compared to the geomagnetic activity around the time of spontaneous telepathic dreams from the Gurney, Myers and Podmore (1886) collection, very similar (statistically undistinguishable) temporal patterns were observed. Analyses of both experimental and spontaneous telepathic experiences indicated that they were more accurate (or more likely to have occurred) during 24-hour intervals when the daily average antipodal (aa) index was approximately 10 ± 3 gammas. When the daily aa index exceeded amplitudes of approximately 20-25 gammas, telepathic experiences became less probable." (Persinger, Michael A., and Krippner, Stanley; "Dream ESP Experiments and Geomagnetic Activity," American Society for Psychical Research, Journal, 83:101 1989.)

Comment. It must be added here that mainstream science does not (yet) admit that telepathy exists as a legitimate scientific phenomenon. Nevertheless, there is an immense literature on telepathy and related parapsychological subjects. Once again we have a "shadow science," with its own journals, conferences, and research institutions---all outside the fold of mainstream science.

GEOMAGNETIC ACTIVITY RELATED TO MENTAL ACTIVITY

Several recent reports have indicated significant relations between extrasensory perception (ESP) experiences and performances and the Earth's geomagnetic field (GMF) activity. ESP experiences are reported more frequently, and accuracy of laboratory ESP is more accurate, on days of relatively quiet GMF activity. On the other hand, there are indications that a complementary paranormal process, psychokinesis, may be enhanced by high GMF activity. We conducted retrospective analyses of possible relations between GMF activity and (a) electrodermal activity (as an index of sympathetic autonomic activity), (b) rate of hemolysis of human red blood cells in vitro, (c) attempted distant mental (i.e., psychokinetic) influence of electrodermal activity, and (d) attempted distant mental (psychokinetic) influence of rate of hemolysis. For each of these four measures, high activity was associated with high GMF values, while low activity was associated with low GMF values. The relations were statistically significant for three of the four analyses and showed a consistent trend in the fourth." (Braud, William G., and Dennis, Stephen P.; "Geophysical Variables and Behavior; LVIII. Autonomic Activity, Hemolysis, and Biological Psychokinesis: Possible Relationships with Geomagnetic Field Activity," Perceptual and Motor Skills, 68:1243, 1989.)

PSI EFFECTS IN THE SACRIFICE OF MARINE ALGAE

It is safe to say that mainstream science will categorically reject the results of the experiments reported below. The reason is simple: no known mechanism exists for ESP---in this instance, the anomalous transfer of information between isolated life forms.

Two scientists at the University of Delaware have designed an experiment that measures the activity of marine algae in a seawater culture. By passing a laser beam through the culture and thence to a photomultiplier tube, they can, utilizing the Doppler shift, measure the collective activity of the cells. (See figure.)

Various experiments were run by the Delaware researchers, but their second series in particular seems worth reporting.

> A second series of experiments used the sacrifice of clones as a distant stimulus. The data appear to show that the marine alga Tetraselmis suecica reacts dramatically to the sacrifice of cells in a physically isolated aliquot of the same culture if the experimenters are aware of the moment of sacrifice, and excited by the novelty of the experiment. In sharp contrast, only marginally significant results were obtained when the same experiment was run entirely automatically, with the time of the sacrifice defined by random number selection, and the experiment activated by computer command in an empty laboratory.

(Pleass, C.M., and Dey, N. Dean; "Conditions That Appear to Favor Extrasensory Interactions between Homo

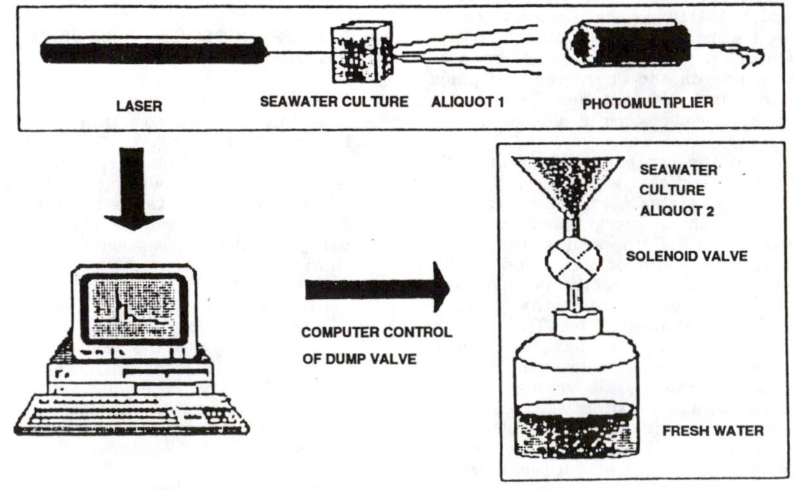

Experimental setup for measuring the ESP-modulated activity of marine algae.

Sapiens and Microbes," Journal of Scientific Exploration, 4:213, 1990.)

MIND-BENDING THE VELOCITY VECTORS OF MARINE ALGAE

"A consciousness experiment in which the Doppler shift of He/Ne laser light was used to describe changes in the velocity and vector of a marine alga, Dunaliella, was reported by Pleass and Dey in 1985. Because the subject of the consciousness experiment is living, we expect strings of baseline velocity and vector data which are, at some level, inextricably time-variant. This complicates the statistical procedures which must be used to analyze the data.

"This paper examines the variation in baseline data strings, and describes two alternative statistical procedures which have been used to determine the probability of consciousness effects. Two levels of control are applied, allowing global comparison of frequency distributions of experimental scores with similar distributions derived artificially from baseline data. In both cases the null hypothesis is that there is no psi effect. The data quite strongly suggest the rejection of the null hypothesis, although the distributions of run scores contain several values beyond 3σ and are nonnormal. This limits the definition of probabilities." (Pleass, C.M., and Dey, N. Dean; "Finding the Rabbit in the Bush: Statistical Analysis of Consciousness Research Data from the Motile Alga Dunaliella," The Explorer, 3:6, no. 2, 1986.)

Comment. Scientifically, these experiments seem to far transcend guessing Zener cards. The Explorer is published by the Society for Scientific Exploration.

TELEPATHIC RABBITS

Well, why not?

> Our experimental study has tried to bring out the existence of a correlation, at a distance, between the physiological reactions of two rabbits from the same litter who had always lived together. We chose photoelectric plethysmography as being the least traumatic method for the rabbits and the one most capable of giving evidence of the physiological reaction specific to stress. Through this method, we studied the coincidences between the outsets of the two rabbits' emotional reactions. Added to the rabbits' isolation through distance, some experiments involved the setting up of sensorial and electromagnetic isolation boxes. We studied the coincidences occurring between the spontaneous emotional reactions of the rabbits as well as the coincidences occurring between the reactions provoked by small stimulae, such as the sound of a bell in one of the boxes. Two series of experiments out of four gave significant results, leading one to think that a conscious or unconscious telepathic link does exist between two rabbits that have close links with each other.

(Thouvenin, Bernard; "A Study of Telepathic Phenomena among Rabbits," Revue Francaise de Psychotronique, 1:15, July-September 1988. As abstracted in Exceptional Human Experience, 9:47, June 1991.)

DOWSING

BPM EQUALS DOWSING

It is remarkable that this article appears in a reputable scientific journal. Williamson was stimulated to write about dowsing by apparent recent Russian successes with BPM (Bio-Physical Method) in locating minerals. BPM has created quite a stir in the USSR, with all the scientific trappings of conferences and journal papers. The Russians evidently use BPM in conjunction with aerial photogeological surveys in pinpointing mineral deposits. BPM anomalies are detected on foot by hand-held BPM detectors (read: divining rods).

Williamson goes on to describe the ridicule heaped on dowsing in the West. The negative experiments of Foulkes with trained dowsers shoved dowsing out to the lunatic fringe. But recently, a little-mentioned American study by Chadwick and Jensen seems to contradict Foulkes. Chadwick and Jensen, highly skeptical at the beginning of their experiments, were surprised to discover that their 150 novice dowsers were actually sensitive to the small magnetic field changes one expects in the neighborhood of mineral concentrations. The dowsing effect is weak but apparently real. (Williamson, Tom: "Dowsing Achieves New Credence," New Scientist, 81:371, 1979.)

MAGNETIC THEORY OF DOWSING

"David Marks, in The Skeptical Inquirer, asserts that there are no theories to account for paranormal effects. This is not true for dowsing. Serious dowsing claims, such as those made by Soviet geologists, which are difficult to account for in terms of the reception of normal sensory cues, may be explained by postulating human sensitivity to small magnetic field gradient changes. The theory is supported by a series of tests involving 150 subjects.

"The magnetic theory predicts that dowsers can achieve above-chance results only if the features they claim to detect are associated with magnetic gradients of at least one nanotesla per metre. This was not the case in Randi's recent experiments, so his chance results are therefore consistent with the magnetic theory, which merits further investigation." (Williamson, Tom; "Dowsing Explained," Nature, 320:569, 1985.)

REMOTE, EXTRASENSORY DESCRIPTION OF MINERAL SAMPLES

"A series of remote viewing experiments were run with 12 participants who communicated through a computer conferencing network. These participants, who were located in various regions of the United States and Canada, used portable terminals in their homes and offices to provide typed descriptions of 10 mineral samples. These samples were divided into an open series and a double-blind series. A panel of five judges was asked to match the remote viewing descriptions against the mineral samples by a percentage scoring system. The correct target sample was identified in 8 out of 33 cases; this represents more than double the pure chance expectation. Two experienced users provided 20 transcripts for which the probability of achieving the observed distribution of the percentage score by chance was 0.04." (Vallee, Jacques,; "Remote Viewing and Computer Communications---An Experiment," Journal of Scientific Exploration, 2:13, 1988.)

DOWSING SKEPTICS CONVERTED

A while back, New Scientist ran an article on the "dowsing sense." Two letters prompted by the article were from scientifically trained people who originally were very skeptical about dowsing.

The first letter from P.L. Younger, a university hydrogeologist, first mentioned that most dowsers are convinced that they are hunting underground streams of water. In actuality, he says, most underground water flow is intergranular and laminar. There are no underground streams to find! Then, he continued:

> Having said all this, while conducting hydrogeological fieldwork in Colorado, I was involved in "dowsing" the exact location of buried metal pipes using two L-shaped metal rods, which were balanced on the fingers (not clutched at all). Surface and subsurface pipes gave clear deflection of the rods. I was led to conclude that the rods operated as a crude magnetometer.

B.W. Skelcher originally did not believe that any variation in the magnetic field or any other natural force would cause a hand-held stick to move. But:

> One day, on the undeveloped plot of land adjacent to my abode, I spied a "nutter" pacing to and fro with hazel in hand. When the fellow assured me that he was seriously checking the site for hidden water mains, power cables, and so on, I expressed my grave doubts. At this he handed me the twigs and after a brief instruction goaded me to try. After a few paces I was astonished to feel the two bent twigs move in my hands. I am not skeptical any more, I know it works.

(Younger, Paul L., and Skelcher, B.W.; "Dowsing-Sense," New Scientist, p. 62, April 9, 1987.)

Comment. Such testimonial evidence, abundant though it is, will not be accepted by the scientific community. Instead they point to their controlled experiments, which are strongly negative. Why do so many individuals experience psi phenomena casually, but when controls are applied, the effects are most elusive? Most people, at one time or another, have had a profoundly shocking psychic experience. Are these events real, and, if so, why don't they manifest themselves in the labs?

PREDICTION

AT LAST: SOMEONE WHO CAN PREDICT THE FUTURE!

Most psychics claiming to know what's ahead down the road of time draw up rather long lists of predicted events. They may score a hit or so, but their records are generally very poor. The present article records the astounding performance of Emory Royce, a New Zealander. "The whole thing is preposterous," says Richard Kammann, the author and noted skeptic. Royce made four predictions, and four only, on a radio talk show. Some of the predictions were a bit vague on details, but the overall outcome was unbelievable: all four events occurred! The predictions were Brezhnev's death (very close timewise); naval disaster in the Falklands (prediction made well before the surprise invasion by Argentina); a New Zealand political scandal; and the completely unexpected cancellation of a New Zealand aluminum factory. (Kammann, Richard; "Uncanny Prophecies in New Zealand: An Unexplained Scientific Anomaly," Zetetic Scholar, no. 11, p. 15, August 1983.)

A HOAX ADMITTED

Above, we summarized a paper entitled "Uncanny Prophecies in New Zealand: An Unexplained Scientific Anomaly." The author of this paper, R. Kammann, has revealed that the whole business was a hoax. He now comes forth with details of the hoax and an explanation of why he chose to withhold data from the readers of the Zetetic Scholar.

> In writing up this episode for readers outside of New Zealand, I chose initially to hide the skeptical origin of these bogus prophecies to allow readers to experience them as they might be presented by paranormal advocates. Although the predictions drew widespread attention in New Zealand media, their impact was undoubtedly dampened by their honest portrayal as an anti-astrology lesson by a skeptical psychologist. For a proper evaluation, it was therefore necessary to hide their true origin for at least one public presentation. Even there (in Zetetic Scholar, #11) I constrained myself to scientifically accurate reporting and quoted my own original wordings of the prophecies and their fulfillments from the tapes of the radio programs. For the miracle mongers of the paranormal press, such loyalty to the facts would be considered a bad precedent.

(Kammann, Richard; "New Zealand Prophecies Exposed as a Hoax," Zetetic Scholar, no. 12, p. 34, August 1987.)

Comment. Kammann's confession is nice to have, but can anyone be certain that he is not now playing a new game with us? Even more so, can we now trust any statements by individuals dedicated to shielding us from the paranormal?

PREDICTIVE PSI

> A case study is reported in which the author, a psychologist, began spontaneously saying an unusual word, coup d'etat, aloud repeatedly, and then received a letter from a Mrs. Coudetat the following day.

(Tart, Charles T.; "A Case of Predictive Psi, with Comments on Analytical, Associative and Theoretical Overlay," Society for Psychical Research, Journal, 55:263, 1989.)

Comment. Verily. psi works in mysterious ways! This amusing abstract came from the periodical Exceptional Human Experience, 8:93, December 1990. Address: Parapsychological Sources of Information Center, 2 Plane Tree Lane, Dix Hills, NY 11746.

REINCARNATION

DOES MAN SURVIVE DEATH?

In this remarkable paper, published in one of the most important medical/psychological journals, the author surveys the history of research into the survival of bodily death. He identifies three historical periods that mirror the scientific thinking of their times. At one point, research waned as many investigators believed that living individuals with paranormal powers were responsible for all the evidence. Now, however, research again proceeds on a broad front; even though hampered by most scientists' outspoken disbelief in the whole business.

The important types of evidence reviewed include the speaking of languages not normally learned, out-of-the-body experiences, and reincarnation memories. [Subjects that 99% of the scientific community would dismiss without examination. Ed.] The author, a professor of psychiatry, feels that this contempt is unwarranted and that most scientists are simply not aware of the vast amount of high quality data available. The long, well-documented paper concludes with the assertion that the data acquired so far do not actually compel the conclusion that life exists after death but that it certainly infers it strongly. (Stevenson, Ian; "Research into the Evidence of Man's Survival after Death," Journal of Nervous and Mental Disease, 185:152, 1977.)

PROOF OF REINCARNATION?

"The authors report a case of the reincarnation type with several unusual features. First, the subject began to have apparent memories of a previous life when she was in her thirties, a much older age than that of the usual subjects of cases of this type; second, the memories occurred only during periods of marked change in the subject's personality; and third, the new personality that emerged spoke a language (Bengali) that the subject could not speak or understand in her normal state. (She spoke Marathi and had some knowledge of Hindi, Sanskrit, and English.) A careful investigation of the subject's background and early life disclosed no opportunities for her to have learned to speak Bengali before the case developed. A final interpretation of this case cannot be made on the basis of present information and knowledge. The authors, however, believe that, as of now, the data of the case are best accounted for by supposing that the subject has had memories of the life of a Bengali woman who died about 1830." (Stevenson, Ian, and Pasricha, Satwant; "A Preliminary Report of an Unusual Case of the Reincarnation Type with Xenoglossy," American Society for Psychical Research, Journal, 74:331, 1980.)

Comment. Most such cases of purported reincarnation quickly collapse under scrutiny, but this one seems a bit more substantial.

RESEARCHES IN REINCARNATION

I. Stevenson, at the University of Virginia, has long studied claims of reincarnation. The method employed (and there are precious few alternatives) focuses on children who claim to have lived before and can provide verifiable details about their past lives. If the details check out, one can at least claim that reincarnation is a possible interpretation of the data. Usually, however, before a researcher can get to the scene of the phenomenon, the parents of the deceased have been found and the way has been left open for much exaggeration.

In his present contribution, Stevenson reports three cases in Sri Lanka where the recollections of the supposedly reincarnated children have been written down in detail and the family of the deceased has not been located. Here is one of his cases:

"The Case of Iranga. The child was born in a village of Sri Lanka near but not on the west coast, in 1981. When she was about 3 years old she spoke about a previous life at a place called Elpitiya. Among other details, Iranga mentioned that her father sold bananas, there had been two wells at her house, one well had been destroyed by rain, her mother came from a place called Matugama, she was a middle sister of her family, and the house where the family lived had red walls and a kitchen with a thatched roof. Her statements led to the identification of a family in Elpitiya, one of whose middle daughters had died, probably of a brain tumor, in 1950. Among 43 statements that Iranga made about the previous life, 38 were correct for this family, the other 5 were wrong, unverifiable, or doubtful. Iranga's village was 15 kilometers from Elpitiya. Each family had visited the other's community, but they had had no acquaintance with each other (or knowledge of each other) before the case developed."

Stevenson's conclusion was that the three children had information about deceased persons that could only have been obtained paranormally. (Stevenson, Ian, and Samararatne, Godwin; "Three New Cases of the Reincarnation Type in Sri Lanka with Written Records Made before Verification." Journal of Nervous and Mental Disease, 176:741, 1988.)

Comment. Our prediction is that science-in-general will remain unimpressed by such data.

THE DEATH OF DEATH

Almost nothing is known about why birthmarks occur in particular locations of the skin. The causes of most birth defects are also unknown. About 35% of children who claim to remember previous lives have birthmarks and/or birth defects that they (or adult informants) attribute to wounds on a person whose life the child remembers. The cases of 309 such children have been investigated. The birthmarks were usually areas of hairless, puckered skin; some were hypopigmented macules; others were hyperpigmented nevi. The birth defects were nearly always of rare types. In cases in which a deceased person was identified, the details of whose life unmistakenly matched the child's statements, a close correspondence was nearly always found between the birthmarks and/or birth defects on the child and the wounds on the deceased person. In 40 of the 46 cases in which a medical document (usually a postmortem report) was obtained, it confirmed the correspondence between wounds and birthmarks (or birth defects).

(Stevenson, Ian; "Birthmarks and Birth Defects Corresponding to Wounds on Deceased Persons," abstract of a paper presented at the Princeton meeting of the Society for Scientific Exploration, June 1992.)

INFORMATION PROCESSING

OVERVIEW

The mind of the normal individual processes information in impressive and not fully understood ways. The abnormal mind, or the normal mind in an altered state, is even more spectacular. Indeed some feats of memory and calculation are incredible. Calculating and memory prodigies, for example, seem to rely upon faculties deep with the brain that may actually belong to everyone if only we knew how to tap them. The apparent ability of the subconscious mind to process information independently of the conscious mind seems to confirm this hidden talent.

Mathematical genius is also a fit subject for this section. Where does it come from? Why do "strokes of genius" seem to explode unexpectedly from the subconscious? Why does genius often seem to fade with maturity like the eidetic-image phenomenon? The fact that genius is often a companion of mental illness is another puzzle.

It is common to think of the human brain as a biological analog of the modern electronic computer. The phenomena described below do not support this view.

CALCULATING PRODIGIES

SOME HIGHLY FOCUSSED MINDS

Here is a modern study of calculating prodigies, idiot savants, or, as Rimland prefers, "autistic savants." Calculating prodigies are rarely idiots; that is, with IQs below 30: rather they are almost always autistic, displaying gross disturbances in communication, and/or motor behavior.

Rimland and his colleagues have studied 5,400 autistic children and found 10% of them to have extraordinary abilities. We hear most often about those prodigies who can multiply large numbers in their heads instantaneously or give us calendar information far in the past or future without the blink of an eye. But autistic savants are also prodigal in the fields of art, music, and mechanics. No one knows how they perform their feats, although psychologists speculate that their minds are intensely focussed on their special skills to the exclusion of most everything else. A few "normal" people, such as Gauss and Ampere, have matched the capabilities of the autistic savants, but the rest of us have our minds spread too thinly. We are in the majority, so the autistic savants usually end up in institutions while we plod along outside. (Rimland, Bernard; "Inside the Mind of the Autistic Savant," Psychology Today, 12:68, August 1978.)

Comment. We may speculate that the capabilities of the autistic savants are inherent in all of us, awaiting only some key.

MYSTERY OF THE IDIOT SAVANT

To describe the enigmas of the idiot savant, we can do no better than quote the first two paragraphs of a review article by D.A. Treffert:

"At the 1964 annual meeting of the American Psychiatric Association, a discussant concluded, 'The importance, then, of the Idiot-Savant lies in our inability to explain him; he stands as a landmark of our own ignorance and the phenomenon of the Idiot-Savant exists as a challenge to our capabilities.' In the years that have followed, the inability to explain the idiot savant has not lessened, and the challenge to our capabilities remains undiminished. However, no model of brain function, particularly memory, will be complete until it can account for this rare but spectacular condition, with its islands of mental ability in a sea of mental handicap and disability.

"Through the past century, since Down's description of this disorder, the several hundred idiot savants reported in the world literature have shown remarkable similarities within an exceedingly narrow range of abilities, given the many possible skills in the human repertoire. Why do so many idiot savants have the obscure skill of calendar calculating? Why does the triad of retardation, blindness, and musical genius appear with such regularity among them? Why is there a 6:1 male-to-female ratio in this disorder? What accounts for the more common occurrence of the idiot savant among patients with infantile autism than among those with other developmental disabilities?"

Other questions that can be framed based on the rest of the paper are: How do some talents arise from injuries? Why do some talents disappear when other, different, skills are learned?

Treffert admits to science's complete bafflement over this phenomenon. No wonder, for how can we, in our present state of knowledge, account of these two cases:

- Twin savants who can instantly name the day of the week over a span of 8000 years, and who may have an unlimited digital span; i.e., an unlimited memory for numbers.

- Blind Tom, possessing a very limited IQ, who played Mozart on the piano at 4, and who could play back flawlessly any piece of music, regardless of complexity. He could also repeat a discourse of any length in any language without the loss of a syllable.

(Treffert, Darold A.; "The Idiot Savant: A Review of the Syndrome," American Journal of Psychiatry, 145: 563, 1988.)

Comment. Note that many child prodigies, who are different from idiot savants, lose their talents as they age.

CALENDAR CALCULATING BY "IDIOT SAVANTS"

M.J.A. Howe and J. Smith have reported on an extensive study of calendar-calculating by individuals with otherwise subnormal intelligence. It is very clear that these so-called "idiot savants" use a variety of mental techniques, all rather different from rote memory, such as employed by memorizers of pi. First, we present Howe-and-Smith's abstract; then, a particularly interesting specific case.

"A number of mentally handicapped individuals are able to solve difficult calendar date problems such as specifying the day of the week for a particular date, sometimes over spans of more than 100 years. These individuals are self-taught and do not follow procedures at all similar to the usual, published, algorithms. An investigation of one individual revealed that he retained considerable information about the structure of days in particular months, probably as visual images. His skill closely depended on the extent and form of his knowledge of calendars, and his errors were often a consequence of lack of knowledge about a particular time period. Mentally retarded individuals who perform calendar date feats are often socially withdrawn and devote considerable periods of time to calendar dates. The most capable calendar-date calculators are usually individuals who have a strong interest in calendars as such."

Although some calendar calculators may use visual imagery---perhaps something like eidetic imagery---at least one calendar calculator was blind from birth.

Example. "One of the few serious attempts that have been made to understand the mental operations underlying calendar skills is described by William Horwitz and others. They examined the abilities of a pair of mentally retarded identical male twins, both of whom performed calendar-calculating feats. During the twins' early childhood, despite severe family difficulties caused partly by the father's alcoholism, the parents, whose efforts to teach numerical and reading abilities to their sons were largely unsuccessful, were impressed by seeing one of the boys looking at a perpetual calendar in an almanac. Subsequently the parents encouraged the boys to acquire calendar skills. The feats of one twin were especially remarkable: he was reported to have had a range of at least 6000 years---beyond the range of any conventional or (so-called) perpetual calendar---although he gave incorrect answers for dates prior to 1582, owing to his ignorance of the 10-day calendar adjustment which was made at that time, when the Gregorian Calendar replaced the Julian one. As well as solving problems requesting the days of the week for specified dates, the more able twin could also answer questions asking, for instance, for those months in a given year in which the first day fell on a Friday, or the dates of specified days, such as 'The fourth Monday in February 1993?' (the 22nd)." (Howe, Michael J.A., and Smith, Julia; "Calendar Calculating in 'Idiot Savants,' How Do They Do It?" British Journal of Psychology, 79:371, 1988.)

MEMORY STRUCTURE OF AUTISTIC IDIOT SAVANTS

The first two paragraphs of this study by N. O'Connor and B. Hermelin provide some interesting background concerning the idiot savant phenomenon.

"There are two basic questions which arise from the phenomenon of the idiot-savant. One is whether the specific memory of the idiot-savant is categorically organized or not and the second is whether this memory is IQ independent or IQ related. The first problem needing consideration is therefore concerned with the nature of the memory system which is involved. In the idiot-savant this has, until recently, been deemed to be a relatively unorganized rote memory. It is supposed to be mechanical, inflexible and extremely concrete. However, in a number of studies we have demonstrated that the outstanding memories of idiot-savant calculators and musicians are founded on strategies which are rule based, and which draw on a knowledge of the structure of the calendar or of music.

"In addition to evidence concerning the organized nature of the specific memories of idiot-savants for dates and music, we have also accumulated evidence that the eliciting of rules and the development of rule-based strategies tended to be specific in the particular areas of ability of these subjects. For example, although able to calculate past and future dates rapidly, calendrical calculators were often quite poor in the addition and subtraction of numbers unrelated to the calendar. Similarly, in one experiment, musical memory in an idiot-savant was found to be confined to compositions based on familiar tonal structure, and did not extend to atonal music."

In their study O'Connor and Hermelin showed that the verbal memory factor in mnemonists was independent of IQ; also, data of special interest to the subjects were stored in their memories in categorized form. (O'Connor, N., and Hermelin, B.; "The Memory Structure of Idiot-Savant Mnemonists," British Journal of Psychology, 80:97, 1989.)

Comment. Psychological studies like the one reported above often do not emphasize the fantastic mental capabilities of idiot-savants. It is apparent that at least some human brains have mental capabilities far beyond and/or radically different from what is needed for survival. What is evolution's purpose here?

MEMORY PHENOMENA

EVERYONE A MEMORY PRODIGY

Our handbook Unfathomed Mind presents many cases of exceptional memory. Without question, some people can reproduce incredible blocks of words and numbers as well as drawings, music, etc. The question is: Is such exceptional memory the consequence of an exceptional brain or just long training? Those who hold the first position believe an anomaly exists because: (1) The difference between normal memory and exceptional memory is so large; and (2) People with exceptional memory seem to employ different mental processes in transferring information into long-term memory, notably visual techniques like 'photographic' memory.

The authors of this article present several cases where subjects with normal memory have been trained to where they perform nearly as well as memory experts. The key seems to be the use of mnemonic devices and other methods of imposing some sort of order or meaning on the information involved. To illustrate, a chess master can usually recall the positions of all the pieces on a chessboard after a quick glance. But if the chessmen are arranged randomly and meaninglessly, his memory is reduced to near-normal. The gist is that long practice and the application of mnemonic devices can vastly improve anyone's memory and, in consequence, memory prodigies are not really so anomalous. (Ericsson, K. Anders, and Chase, William G.; "Exceptional Memory," American Scientist, 70:607, 1982.)

Comment. The real anomaly here may be the fact that the human memory and related memory faculties seem orders of magnitude better than needed for survival. How did such capabilities evolve? Of what use is a prodigious memory to an Ice Age man facing a cave bear? Are we dealing with prescient evolution, like the moth described on p. 162, holding capabilities in reserve until they are really needed.

MNEMONISM NOT SO EASY!

"This paper reports a systematic study of a man (T.E.) with astonishing mnemonic skills. After a brief description of his most favoured mnemonic technique, the 'figure alphabet,' his performance and the mnemonic techniques used on five classical memory tasks are described. These are: one task involving both short- and long-term memory (the Atkinson-Shiffrin 'keeping track' task), two tasks involving just long-term memory (recall of number matrices and the effects of imagery and deep structure complexity upon recall), and two tasks involving just short-term retention of individual verbal items and digit span. Whenever possible, T.E.'s performance was compared with that of normal subjects, and also with other mnemonists who have been studied in the past. There was no evidence to suggest that T.E. has any unusual basic memory abilities; rather he employs mnemonic techniques to aid memory, and the evidence suggests that previous mnemonists who have been studied by psychologists have used very similar techniques."

The "figure alphabet" employed by T.E. was used in Europe as early as the mid-1700s. The Hindus had a Sanskrit version even earlier. Basically, each digit is represented by a consonant sound or sounds:

1	2	3	4	5	6	7	8	9	0
T	N	M	R	L	J	K	F	P	Z
D	Ng				G-soft	G-hard	V	B	S
Th					Ch	Q			C-soft
					Sh	C-hard			

The letters AEIOU and WHY have no numerical value and are used to build up words. Thus, 21 can be NeT, NuT, aNT, auNT, etc. The system is phonetic in that the digits are sounds rather than the letters themselves. Silent letters do not count and double letters count as single.

In one of the tests, T.E. was presented with strings of digits on a computer screen at the rate of one per second. In spite of the rapid rate of presentation, T.E. used the figure alphabet to convert digit strings into several words. Generally, he converted three digits into one two-syllable word. Twelve to 14 digits might be remembered as four or five two-syllable words. In this test, T.E. could remember more than 12 digits in the strings as they flashed by at one string per second. (Gordon, Paul, et al; "One Man's Memory: A Study of a Mnemonist," British Journal of Psychology, 75:1, 1984.)

Comment. Two comments here: (The figure alphabet seems rather cumbersome at first, but its long history suggests that it dovetails nicely with human memory processes; and (2) Several ancient languages were written without the vowels, like the figure alphabet. Could there be a connection?

PI IN THE MIND!

"Srinivasan Mahadevan, a graduate student from India at Kansas State University in Manhattan, is trying to become the fourth person to remember the first 100,000 digits to pi, the theoretically infinite computation that measures the ratio of the circumference of a circle to its diameter. He has already memorized the first 35,000 digits of pi.

"Only three cases of such ability to remember have been documented in 200 years. One of these people became insane when he became unable to forget anything and his reasoning processes drowned in a flood of facts." (Anonymous; "Student's Memory Wins Bets," San Mateo Times, June 2, 1989. Cr. J. Covey)

HYPNOSIS AND MEMORY

Hypnotic hypernesia is the unusually vivid and complete recall of information from memory while under hypnosis. The present article reviews the extensive literature on the subject and the long-standing controversy as to whether hypnosis can enhance memory at all.

One fact does seem clear, hypnosis does not help subjects recall nonsense data or information without meaning, such as random numbers and words. When it comes to meaningful phrases, sentences, paragraphs, etc., hypnosis does aid recall to some extent. If the words evoke considerable imagery, as poetry often does, hypnosis seems to help recall even more. Finally, the recall of meaningful visual images and connected series of images is helped most of all by hypnosis. In fact, there is some evidence that eidetic imagery, that vivid, near-total recall of images, which is almost exclusively a talent of childhood, can be recovered by mature subjects under hypnosis. There do not seem to be any theories that explain all these effects of hypnosis on memory. (Relinger, Helmut; "Hypnotic Hypernesia," American Journal of Clinical Hypnosis, 26:212, 1984.)

Comment. Of course, memory shorn of hypnotic effects cannot really be explained either. The results of Relinger's survey make one wonder whether the human brain is specially "wired" or built to efficiently handle visual imagery that is "meaningful" in the context of human experience and theoretical expectations. This kind of construction is quite different from computer memories which process meaningless data as easily as meaningful data. UFOs, sea monsters, N-rays, etc. might just be eidetic images from human memories evoked by certain stimuli and encouraged by suggestion.

HYPNOTIC MISRECALL

Hypnosis is occasionally employed to help witnesses remember forgotten events in criminal and civil investigations. Scientists have long questioned whether hypnosis really improved recall. After a two-year study of hypnosis experiments, a panel of the American Medical Association has issued recommendations that hypnosis be limited to the investigative stages of the judicial process. This conclusion is based on studies that show that hypnosis may increase the amount of information recalled, but that it also introduces errors. The ability of subjects to recall poems learned years earlier was enhanced by hypnosis. Unfortunately, the subjects also fabricated forgotten sections of the poems in the styles of the poets. Conclusion: Hypnosis should be used only where information recalled can be checked for accuracy. (Anonymous; "Hypnotic Misrecall," Scientific American, 252:73, June 1985.)

RARE BUT THERE: HYPNOTIC ENHANCEMENT OF EIDETIC IMAGERY

Eidetic imaging is a remarkable capability, manifested more often in children, in which complex images can be recalled with great detail and realism in a format similar to a hallucination. This mysterious "talent" can be enhanced by hypnotism, indicating perhaps that it is latent in us all.

"The production of eidetic-like imagery during hypnosis in subjects with high but not low hypnotizability was supported in three separate experiments using nonfakable stereograms. In Experiment 1, 6 (25%) of 24 stringently chosen, high hypnotizables were able to perceive one of the superimposed stereograms (presented monocularly) during conditions of standard hypnosis or age regression, or under both conditions, but not during waking. In Experiments 2 and 3, low and high hypnotizables were presented stereograms in an alternating, monocular fashion (one-half to each eye). In Experiment 2, 10% of the high hypnotizables perceived one or more stereograms in hypnosis or age regression, but not during waking. In Experiment 3, none of the 17 low hypnotizables reported correct stereograms, but 6 of the 23 high hypnotizables (26%) did. Relationships between imagery performance and visuospatial abilities were investigated. Results support the general hypothesis that hypnosis enhances imaginal processing of information to be remembered that is a literal or untransformed representation." (Crawford, Helen J., et al; "Eidetic-Like Imagery in Hypnosis: Rare But There," American Journal of Psychology, 99:527, 1986.)

MATHEMATICAL GENIUS

A "MAGICAL GENIUS"

"If ever there were an exemplar of inborn mathematical ability it would be Srinivasa Ramanujan, a poor, uneducated Indian, born 100 years ago, who was one of the greatest and most unusual mathematical geniuses who ever lived. Although he died young---at age 32---Ramanujan left behind a collection of results that are only now beginning to be appreciated.

"Ramanujan's story is one of the great romantic tales of mathematics, made all the more haunting because of the mystery surrounding the man. No one, no matter how much they try, has ever been able to understand the workings of Ramanujan's mind, how he came to think of his results, or the source of this incredible outpouring of mathematics."

Ramanujan has been termed a "magical genius." In contrast, "ordinary geniuses" are merely an order of magnitude of two smarter than you and me. In Ramanujan's case, no one knows where his voluminous results came from. They appeared as if by magic, in a manner transcending ordinary human mental activity.

Ramanujan did complete high school, but his entire mathematical education seems to have come from the reading of just two books. Nevertheless, he was invited to Cambridge on the basis of a letter he wrote to G.H. Hardy in 1913. The letter contained about 60 theorems and formulas stated without proof. After some study, Hardy concluded that Ramanujan's results must be true because, "if they were not true, no one would have had the imagination to invent them."

Ramanujan lived for mathematics. He would work 24-36 hours and then collapse. He died in 1919, leaving behind three notebooks crammed with some 4000 "results," again stated without proof and again seeming to come from nowhere. Step by step, his results are being proved. Ramanujan evidently saw their truth without going through laborious proofs. (Kolata, Gina; "Remembering a 'Magical Genius,'" Science, 236:1519, 1987.)

REINCARNATION OF RAMANUJAN?

In India, Shakuntala Devi is considered to be the reincarnation of Srinivasa Ramanujan, about whom we heard in the above item. We will not comment on the reincarnation bit, but it does seem that S. Devi's remarkable capabilities are somewhat different from those of Ramanujan. The latter intuitively saw mathematical relationships as expressed in equations and identities; Devi is a mental calculator of no mean talent.

> In 1977, Ms. Devi beat a UNIVAC 1108 computer to the 23rd root of a 201-digit number. The machine, which required two hours to program for the task, took more than a min- to solve the problem. She took 50 seconds.
>
> And, in 1981, she made the Guinness Book of World Records as the "Human Computer" by correctly multiplying two 13-digit numbers---7,686,369,774,870 times 2,465,099,745,779---in 28 seconds. The awesome answer? 18,947,668,177,995,426,462,773,730.

S. Devi is also a calendar calculator, being able to name the day of the week for any date in the past or future, taking into account leap years and calendar changes.

She never attended school or had any formal mathematical training! (Young, Luther; "Numbers Whiz Takes Delight in Beating Computers;" Baltimore Sun, January 21, 1988, p. A1.)

Comment. Such prodigies have appeared regularly down recorded history. What is the meaning of the phenomenon? Why does evolution produce talents that far exceed the "need" of the species? Is there a "need" that we are not aware of? It could be that prodigies are precursors of new evolutionary developments, which will leave poor homo sapiens in the intellectual dust. Surely, science fiction has a story about a secret society of transcendent geniuses living under some mountain or even on some planet! Maybe that's how "the face on Mars" got there!

WHAT MAKES A CALCULATING PRODIGY?

The above title is also that of a new book by Steven Smith. Naturally, the book is full of anecdotes about the phenomenal accomplishments of calculating prodigies, both unlettered children and such famous scientists and mathematicians as Euler, Gauss, and A.C. Aitken. The latter "...had the uncanny power of mentally computing, to a long string of decimals, the values of e and $e^{\pi\sqrt{163}}$. When asked (by his children) to multiply 987...1 by 123...9, he remarked afterwards: 'I saw in a flash that 987...1 multiplied by 81 equals 80 000 000 001, and so I multiplied 123...9 by this, a simple matter, and divided the answer by 81.'" But what, asks Smith, led Aitken to 81? To this question, which is the heart of the mystery, he commendably admits he has no reply. And the same deep mystery confronts us even after all has been said about the surface, as distinct from the underlying, structure of the processing.

At the unlettered end of the spectrum of mental calculators, the "...ignorant vagabond, Henri Mondeux, who at the age of 14 years, before the French Academy of Sciences, was able promptly to state two squares differing by 133."

Of course, some mental feats of calculation can be done consciously employing various shortcuts and mathematical tricks. The really fantastic performances, however, are accomplished unconsciously. No one knows how, even the calculators themselves. (Cohen, John; "What Makes a Calculating Prodigy?" New Scientist, 100:819, 1983.)

INFORMATION PROCESSING MISCELLANY

BRAILLE AND THE BRAIN

The human brain apparently can handle more than one stream of input data simultaneously, as shown by J. Hartley's studies of Braille readers.

"A chance observation of a skilled blind reader led me to think that reading braille might provide a more natural task for studying the way in which people do two things at once. My studies of my colleague Lewis Jones suggest that he appears to use some form of parallel processing. Jones cannot recall whether or not he was taught to read Braille with two hands, but in common with other skilled readers of Braille, he has done so for many years. Jones' method if reading is typically as follows. The left forefinger starts to read the beginning of the line. It then meets the the right forefinger returning from the line above. When the two fingers touch, the right forefinger continues to read the line while the left forefinger returns to the beginning of the next line. The whole operation is quite smooth and cyclical [diagram on next page]. However, I was startled to observe that the left forefinger starts to read the next line before the right forefinger has finished the line above. This overlap, demonstrated by my crude data, suggests that Jones appears to be using some form of parallel processing: it seems as if his brain stores initial information from his left forefinger before using it."

S. Miller, at Oxford, has also investigated Braille readers. She questions whether the left forefinger actually starts processing data before the right forefinger has finished the line above. She thinks the left forefinger may just be homing in on the space preceding the first letter. However, Miller agrees that the two fingers are moving in different places at the same time. (Hartley, James; "Braille and the Brain," New Scientist, p. 34, August 7, 1986.)

TAKING FOOD FROM THOUGHT

M.D. was a stroke victim. The stroke had apparently damaged only a very specific portion of his brain, because his only memory impairment was for the names of fruits and vegetables! When shown the pictures of any fruits or vegetables, M.D. had great difficulty in naming them. However, when given words like pear and broccoli, he easily associated the names with the correct pictures. No other objects were affected. The implication is that the brain stores information like a Thesaurus; that is, like data with like data, all the fruit and vegetable files together. (Bower, B.; "Taking Food from Thought: Fruitful Entry to the Brain's Word Index," Science News, 128:85, 1985.)

THE WOMAN WHO COULDN'T DESCRIBE ANIMALS

The astounding complexity of the human brain was underscored recently by a 70-year-old woman, who could describe anything she saw except animals. The woman had an immune-system disorder that had damaged a small portion of her brain's temporal lobe. Her other mental faculties were intact.

> The woman could name plants, foods, and inanimate objects and describe them without hesitation. The scientists were impressed that when shown a trellis, not exactly your everyday object, she could correctly name it and describe its cross-hatched geometry.
>
> But when shown a squirrel or a dog, she froze. She couldn't find the right name for either, nor could she describe their size or shape or furry coats.
>
> Her deficit, involving only a tiny portion of her language skills, was amazingly narrow.

(Bor, Jonathan; "The Woman Who Couldn't Describe Animals," Baltimore Sun, p. A1, September 7, 1992.)

The implications of this strange case were described in Science News:

> The peculiar inability of a 70-year-old woman to name animals has led scientists to propose that the brain harbors separate knowledge systems, one visual, the other verbal or language-based, for different categories of living and inanimate things, such as animals and household objects.

(Bower, B.; "Clues to the Brain's Knowledge Systems," Science News, 142:148, 1992.)

Comment. Experiments with animals suggest that the brain's memory is "distributed"; that is, like a hologram, both nowhere and everywhere. Also, it may not be pertinent, but R.O. Becker (See below.), hypothesized that the brain utilizes two modes of communication: one digital, the other analog.

HIGHER SIGHT

In the January 4, 1990, issue of Nature, G. Humphreys reviews a book that is just too expensive for us to consider buying. The title is: Synesthesia; A Union of the Senses. It costs $75 and is published by Springer/Verlag. The review, however, is expansive and provides some facts about synesthesia worth passing on to our readers.

Synesthesia is an "oddity" of human perception in which words, musical instruments, objects, concepts, evoke sensations sharply different from what is actually being processed by the brain. For example, specific musical tones elicit specific color sensations; that is, B-flat evokes the color green; A-sharp, yellow, etc. Or the phenomenon may be more complex, with Mozart being green; Wagner, red, etc. Most "synesthetites" seem to experience colors, but geometrical figures sometimes appear in response to particular stimuli. As for the stimuli that call forth these exotic sensations; they are usually music or numbers. To some synesthetites, the cardinal numbers are associated with specific colors.

The book's author is R.E. Cytowic, and he has provided some very interesting observations about synesthetites: There is much consistency among them; that is, if the number 5 evokes a red sensation with one, it does with most others, too. Also, synesthetites seem to run in families. Perhaps most significant is the observation that synesthetic experiences seem to be correlated with changes in cortical blood flow! (Humphreys, Glyn; "Higher Sight," Nature, 343:30, 1990.)

MAGNETIC FIELDS AND THE IMAGINATION

> 12 male and 12 female volunteers were evaluated for their suggestibility before and after an approximately 15-min. exposure to either sham, 1-Hz or 4-Hz magnetic fields that were applied across their mid-superior temporal lobes. During the field application, subjects were instructed to view a green light that was pulsating at the same frequency as the field and to imagine encountering an alien situation. Results were commensurate with the hypothesis that weak brain-frequency fields may influence certain aspects of imaginings and alter suggestibility.

(De Sano, Christine F., and Persinger, M.A.; "Geophysical Variables and Behavior: XXXIX. Alterations in Imaginings and Suggestibility during Brief Magnetic Field Exposures," Perceptual and Motor Skills, 64:968, 1987.)

Comment. The import of this paper is written between the lines of the Summary. One of the authors (M.A.P.) has long been investigating the possibility that UFOs, earthquake lights, and nocturnal lights are basically hallucinatory and are stimulated by natural variations in the magnetic field.

THE TYRANNY OF THE [NORMAL] SENSES

R.O. Becker, author of Cross Currents, has contributed a thought-provoking article on the mechanisms by which humans perceive the cosmos. From the many stimulating ideas he presented, we select his rationale for believing that an electromagnetic basis may exist for the reception and processing of psi signals.

A key concept in Becker's scheme is his belief in the presence, in humans and other organisms, of a dual system for receiving and processing information arriving from the environment. The system we are all aware of and which scientists study in depth is the nerve-impulse system, which transmits digital signals; i.e., 0s and 1s. This system connects to all our everyday senses and controls our motor functions. The second system Becker designates as "primitive." It transmits information in analog (continuously varying) form via electrical currents and magnetic fields, rather than as impulses along neurons. This second system is not recognized by mainstream science.

Becker advances the notions that: (1) Psi-type phenomena are actually handled by the "primitive" analog system; (2) The flood of information normally arriving from our sensory organs via the "modern" digital system masks the psi-type signals; and (3) These assertions are consistent with the elusive nature of psi phenomena in both everyday experience and the parapsychological laboratories. Becker's ideas also jibe with experimental evidence that the psi faculty is suppressed by electromagnetic storms, which (presumably) act only upon the "primitive" analog system.

Becker readily admits that the physical basis for the generation, transmission, and reception of psi signals is unknown. (Becker. Robert O.; "Electromagnetism and Psi Phenomena," American Society for Psychical Research, Journal, 86:1, 1992.)

Comment: We use above the adjective "psi-type" in describing signals traveling along the primitive analog system because we may also be receiving other kinds of signals not yet recognized even by the parapsychologists, and which are also masked by everyday sensory data rushing brainward. After all, we must not let the parapsychologists restrict our vision of the universe!

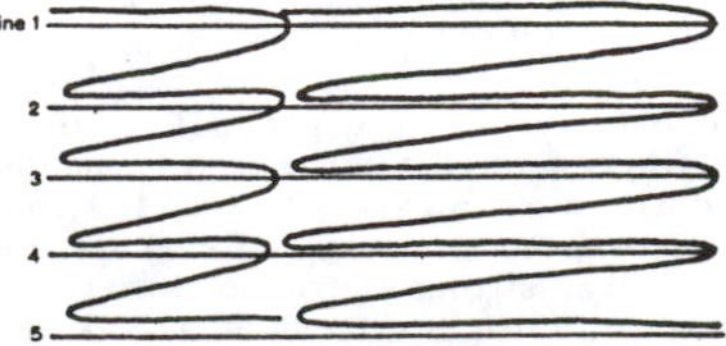

Forefinger traces of a Braille reader.

PSYCHOKINESIS

OVERVIEW

Psychokinesis (PK) occurs when physical objects---popularly spoons or compass needles---are apparently bent, deflected, or otherwise affected by the mental activity of one or more individuals. Many claims for the reality of PK originate under relatively uncontrolled circumstances, such as stage performances and the so-called PK parties. We separate these and other "casual" observations from the more serious laboratory experiments, where efforts are made to: (1) eliminate situations conducive to fraud; (2) avoid observer bias; (3) account for physical effects that might skew results; and (4) apply statistical analysis rigorously.

However, even with the most careful application of scientific controls, experimental observations favoring the existence of psychokinesis are generally rejected by science at large. The reason for this prejudice is not hard to find: The reality of PK undermines the very foundations of scientific observation, because the human mind could then affect all observations of nature from the swinging of a pendulum to the switching of transistors in laboratory instruments.

CASUAL OBSERVATIONS

PK PARTIES: REAL OR SURREAL?

A PK (psychokinesis) party gets into full swing when 15 or more people gather and start bending steel bars, softening metal, and popping soy beans open through PK.

Here follows a description of the prototype PK party:

"Twenty-one people gathered at the author's home on Monday evening, January 19, 1981. All were friends of the author and came from varied backgrounds. After introductions and general discussion, everyone was relaxed and comfortable. The author's grandparents' silverplated silverware was passed out and everyone had either a fork or spoon. Severin stood in the middle of the room with everyone seated in a circle and gave the following instructions:

1. 'Get a point of concentration in your head.'
2. 'Make it very intense and focussed.'
3. 'Grab it and bring it down through your neck, down through your shoulder, down through your arm, through your hand, and put it into the silverware at the point you intend to bend it.'
4. 'Command it to bend.'
5. 'Release the command and let it happen.'

He then instructed the group to use their fingers to test for warmth coming out of the silverware or to feel the metal surface become sticky. Everyone felt pretty silly, sitting there holding the silverware, until the head of a fork being held by a boy (age 14) bent over all by itself! Almost everyone in the room saw this happen and experienced an instantaneous belief system change. Then the silverware in the hands of many people in the room became soft. They easily bent and twisted the silverware into unusual shapes. The period during which the metal remained soft was between five and twenty seconds. Everyone was shouting and extremely excited. During the next hour, nineteen of the party attendees had experienced the metal getting soft and being easily formed into any shape."

Such PK "parties" have been held scores of times since 1981, leaving trails of damaged kitchenware and popped soy beans. It's all a lot of fun. The people attending "feel good" about themselves and their shared experiences. (Houck, Jack; "PK Party History," in Proceedings of a Symposium on Applications of Anomalous Phenomena, C.P Scott Jones, ed., Kaman Tempo, Alexandria, Virginia, 1984.)

Comment. Is mass delusion the foundation of PK parties? Is the above article serious? Houck's paper is in a long collection of rather standard parapsychological fare presented at a conference held under the auspices of Kaman Tempo. The phenomena of PK parties are similar to the audible effects produced by a Toronto group a few years ago. In their case, the participants conjured up "Phillip, the Imaginary Ghost," who communicated via table rapping. In all such group efforts, including the classical seance, there is strong psychological involvement. Skeptics do not do well at PK parties.

TECHNO-JINX

George Gamow once said of his fellow physicist Wolfgang Pauli that "apparatus would fall, break, shatter, or burn when he merely walked into a laboratory." Some people just seem to have adverse effects on machines. When they appear on the scene, computers crash, copying machines jam, and telephones go on the fritz. Robert Morris, an experimental psychologist at Syracuse University, has been collecting such anecdotes and finds them far from rare. On the other side of the coin, other individuals seem to have phenomenal positive rapport with machinery, like those favored few who can fix anything.

Of course, bulging files of anecdotes prove nothing. Many of the stories are likely embellished with each retelling. And some people are singularly clumsy, careless, and ignorant about machines. These types are always pushing the wrong buttons and otherwise mishandling the man-machine interface. Obviously, objective tests are required to determine of there is really anything to this curious business. (Huyghe, Patrick; "Techno-Jinx," Omni, 6:20, May 1984.) Even typewriters can be jinxeeed@+/!

SLI: A SOMEWHAT AMUSING PSI PHENOMENON

Before you snort in derision at SLI, recall that, just as some gardeners have "green thumbs," so do some people appear to induce electronic panic in high-tech equipment just by walking into a laboratory. If the thoughts of a human experimenter can affect marine algae (p. 302), why not electrical equipment? Of course, we know why not! But we have heard some pretty weird stories! For that matter, can you account for everything your computer does?

> A strange new phenomenon has recently been studied in detail for probably the first time. It is street lamp interference, or SLI. Author Hilary Evans, a founder member of ASSAP, has established SLIDE, the Street Lamp Interference Data Exchange, to study the reports.
>
> SLI is the apparent ability that certain people have to switch street lamps on or off merely by being in their vicinity. While it may seem to be a form of psychokinesis, Hilary insists that no conclusion should yet be drawn from the small body of data so far accumulated. The effect is surprisingly common, though most reports so far have come from the USA. ASSAP, in cooperation with SLIDE, now wants to obtain information on the phenomenon in Britain.

(Anonymous; "Standing by the Lamplight...," ASSAP News, no. 39, p. 2, January 1991.)

GEOMAGNETIC STIMULATION OF POLTERGEIST ACTIVITY

"Several researchers have reported that poltergeist episodes frequently began on the day (± 1 day) of a sudden and intense increase in global geomagnetic activity. To test this visual observation, a near-complete account of these episodes for which the inception dates were recorded and verified was examined. Statistical samples clearly indicated that global geomagnetic activity (aa index) on the day or day after the onset of these episodes was significantly higher than the geomagnetic activity on the days before or afterwards. The same temporal pattern was noted for historical cases and for those that have occurred more recently. The pattern was similar for episodes that occurred in North America and Europe. The results were statistically significant and suggest that these unusual episodes may be some form of natural phenomena that are associated with geophysical factors." (Gearhart, Livingston, and Persinger, M.A.; "Geophysical Variables and Behavior: XXXIII. Onsets of Historical and Contemporary Poltergeist Episodes Occurred with Sudden Increases in Geomagnetic Activity," Perceptual and Motor Skills, 62:463, 1986.)

Comment. To put this interesting item in perspective, one must realize that few established scientists recognize the reality of poltergeists.

CONTROLLED EXPERIMENTS

PSYCHOKINETIC CONTROL OF DICE

For 2000 years, people have been saying: "The die is cast," as if the act of casting relinquished the fall of the die completely to fate. Perhaps not!!

"This article presents a meta-analysis of experiments testing the hypothesis that consciousness (in particular, mental intention) can cause tossed dice to land with specified targets face up. Seventy-three English language reports, published from 1935 to 1987, were retrieved. This literature described 148 studies reported by a total of 52 investigators, involving more than 2 million dice throws contributed by 2,569 subjects. The full data base indicates the presence of a physical bias that artifactually inflated hit rates when higher dice faces (e.g., the '6' face) were used as targets. Analysis of a subset of 39 homogeneous studies employing experimental protocols that controlled for these biases suggests that the experimental effect size is independently replicable, significantly positive, and not explainable as an artifact of selective reporting or differences in methodological quality. The estimated effect size for the full data base lies more than 19 standard deviations from chance, while the effect size for the subset of balanced, homogeneous studies lies 2.6 standard deviations from chance. We conclude that this data base provides weak cumulative evidence for a genuine relationship between mental intention and the fall of dice."

(Radin, Dean I., and Ferrari, Diane G.; "Effects of Consciousness on the Fall of Dice: A Meta-Analysis," Journal of Scientific Exploration, 5:61, no. 1, 1991.)

PSYCHING OUT PIEZOELECTRIC TRANSDUCERS

Our title is perhaps too flippant. The experiment described below is serious and was conducted at Stanford Research Institute. Five participants were chosen in an attempt to mentally affect an electronic device.

Each participant was asked to influence one of a pair of piezoelectric transducers, operating in a differential mode, so as to produce an event above a predetermined threshold. During the formal data collection, the transducer enclosure was located in a locked laboratory adjacent to the participants' room. Under these conditions, one of the participants produced a total of 11 events above threshold, distributed in three separate effort periods. Control trials were recorded with no one present in the experimental room but with normal activity in the rest of the building. No equivalent, uncorrelated events above threshold were detected in those control periods."

The author emphasizes the preliminary nature of the results, but believes they warrant further investigaion. (Hubbard, G. Scott, "Possible Remote Action Effects on a Piezoelectric Transducer," The Explorer, 4:10, October 1987.)

WHEN TO BELIEVE AND WHEN NOT TO

Bizarre effects in quantum mechanics. Recently, three "delayed choice" experiments have been consummated in physics labs. In such experiments, the result depends upon what the observer tries to measure; viz., light as particulate-in-nature or light as wavelike-in-nature. The funny thing is that it doesn't matter when the experimenter decides what to measure; he can do this months before the experiment or even afterwards! The effect of the choice is the same---before or after. Now that is weird! But everyone believes it because a theory for it exists. (Thomsen, Dietrick E.; "Changing Your Mind in a Hurry," Science News, 129:137, 1986.)

Bizarre effects in psychokinesis. Recently, several laboratories have been trying to determine if the human mind can affect random physical events, such as radioactive decay.

Surprisingly, the PK effect appears quite independent of physical variables, such as the distance or the complexity of the random generator. The subjects succeed by aiming at the result, regardless of the intermediate steps required to reach this result. Such a goal-oriented, even non-causal feature of PK has been emphasized by PK experiments with pre-recorded, random events. In these experiments, random events were first pre-recorded, and later played back to a PK subject who tried to enforce a certain outcome. These experiments gave positive results, even though the subject's mental effort occurred after the random events had been macroscopically recorded.

Now, that is weird, too! (In fact, one paragraph echos the other.) But of course almost no one believes in PK because there is no theory for it. (Schmidt, Helmut; "The Strange Properties of Psychokinesis," The Explorer, 3:4, January 1986. The Explorer is the newsletter of the Society for Scientific Exploration.)

BEND INTERFEROMETERS NOT SPOONS

Because spoonbending and similar purported mental feats involve so much legerdemain and trickery, scientists generally avoid psychic research. Taking a different tack, R.G. Jahn, at Princeton, has been experimenting with microscopic psychic effects, such as raising the temperature of a thermistor by a few thousandths of a degree or changing the separation of interferometer mirrors by a hundred-thousandth of a centimeter.

Quite unexpectedly (at least to the conventional physicist) the mind seems able to cause such changes at will under controlled conditions. The changes are miniscule to be sure, but cause-and-effect is clear-cut according to Jahn. But don't say that psi power has now been scientifically proven. The effects vary from person to person and, for the same individual, from time to time. The fact that one cannot predict the occurrence of the effects has led Jahn to speculate that the phenomena are inherently statistical. (Anonymous; "Dean Justifies Psychic Research," Science News, 116: 358, 1979.)

Comment. In other words, the effects resemble radioactivity where the behavior of a single atom is unpredictable but en masse the atoms follow the law of radioactive decay.

HIGH TECHNOLOGY EXPERIMENTS IN PARAPSYCHOLOGY

R.G. Jahn, who is Dean of the School of Engineering/Applied Science at Princeton, has written one of the most important papers on parapsychology in recent years. Not the least significant factor is its publication in a top technical journal. This alone will insure wide discussion and debate within the scientific establishment. Probably the key feature of the Princeton work is its "high technology" content. This long, highly technical article is replete with circuit diagrams, photos of shiny equipment, charts, and the complete panoply of modern scientific research. In the section on psychokinesis, we read about Fabry-Perot interferometers, dual thermistors, glow-discharge experiments, Gaussian analog devices, etc. (There is a companion section on remote viewing experimentation!) To round out this overview, the section on historical/philosophical background and the superb bibliography must be mentioned. Although Jahn regards his work as only beginning, he does feel that the early results clearly show the existence of non-chance factors in psychokinesis and remote-viewing experiments. For example, interferometer fringes and strain-gauge readings seem to be changed by the application of "mental forces." But the experiments cannot always be replicated and subjects' abilities are ephemeral. The flavor of the Princeton findings are well put in these sentences from the summary paragraphs:

"...it appears that once the illegitimate research and invalid criticism have been set aside, the remaining accumulated evidence of psychic phenomena comprises an array of experimental observations, obtained under reasonable protocols in a variety of scholarly disciplines, which compound to a philosophical dilemma. On one hand, effects inexplicable in terms of established scientific theory, yet having numerous common characteristics, are frequently and widely observed; on the other hand, these effects have so far proven qualitatively and quantitatively irreplicable, in the strict scientific sense, and appear to be sensitive to a variety of psychological and environmental factors that are difficult to specify, let alone control. Under these circumstances, critical experimentation has been tedious and frustrating at best, and theoretical modeling still searches for vocabulary and concepts, well short of any useful formalisms."

(Jahn, Robert G.; "The Persistent Paradox of Psychic Phenomena: An Engineering Perspective," IEEE, Proceedings, 70:136, 1982.)

Comment. The quotation above could just as well apply to UFO research, some aspects of cryptozoology, and other anomaly research. Clear-cut, reproducible ESP experiments are as rare as captured UFOs and Sasquatches! One entire section of human experience seems to be---well---cagey, sneaky, and beyond logic. How far does this magic land extend, if it exists at all?

MENTAL DEFLECTION OF CASCADING SPHERES

"9,000 polyethylene spheres, 3/4" diameter, cascade downward through 360 nylon pegs to be collected in 19 bins each equipped with real-time counters and LED displays. The 'Baseline' distribution of terminal bin populations is found closely to approximate Gaussian, so that normal statistics can be applied. Operators attempt, on volition or instruction, to shift the distribution mean to the right or left of the baseline value. Results, plotted as cumulative deviations of the mean, display comparable levels of significance and similar individual 'signatures' of achievement to those obtained by the same operators on our microelectronic random event generator." (Jahn, R.G., et al; "A Psychokinesis Experiment with a Random Mechanical Cascade," The Explorer, 1:7, November 1983.)

Comment. The abstract does not come right out and say it, but some subjects do, with high degrees of statistical significance, slightly alter the cascades of falling spheres. These experiments were conducted at Princeton and constitute some of the best modern evidence for the reality of psychokinesis.

NUDGING PROBABILITY

The premiere issue of the Journal of Scientific Exploration, published by the Society for Scientific Exploration, contains an excellent summary of the ESP research conducted at Princeton over the past several years. R.G. Jahn, the leader of the Princeton group, terms the research "Engineering Anomalies Research." This title is apparently more palatable to mainstream science than "Mental Influence on Electronic Devices" or "Affecting Cascading Spheres with Thought Waves." Nevertheless, most of the experimental work is in these two areas. As parapsychological research goes, the Princeton work is of the highest scientific quality.

In the first category, subjects (called operators in the report) were asked to influence the pulses produced by a Random Event Generator (REG). The REG was actually an electronic noise source coupled with circuits that created random positive and negative pulses. The operator mentally tried to increase or decrease the number of counts, or generate baseline data for experiment control. After 33 different operators and over 250,000 trials, there appeared a small but statistically significant indication that the operators were actually able to influence the equipment. Also interesting is the fact that each operator had a private "signature"; that is, individual cumulative deviation graphs (like the one shown) had typical shapes for each operator.

Related experiments were carried out with a Random Mechanical Cascade (RMC). In this device 9,000 3/4-inch polystyrene spheres cascaded down through an array of 330 nylon pegs into 19 bins. According to chance a Gaussian distribution of spheres should be found in the bins. The operators were asked to mentally try and skew the distribution to the right or left, or construct a baseline, as with the REG. Once again, there was statistically significant evidence of a mental influence.

Jahn's group also engaged in remote-viewing experiments, in which subjective factors were suppressed as far as possible. An excerpt from the report's Abstract summarizes this phase of the work nicely:

> Quantitative analysis of a large data base of remote perception experiments reveals similar departures from chance expectation of the degree of target information acquired by anomalous means. Digital scoring techniques based on a spectrum of 30 binary descriptors, applied to all targets and perceptions in the experimental pool, consistently indicate acquisition of substantial topical and impressionistic information about remote geographical locations inaccessible by known sensory channels.

In some trials, the percipient was asked to describe the remote scene even before the location was selected or visited! (Jahn, R.G., et al; "Engineering Anomalies Research," Journal of Scientific Exploration, 1:21, 1987.)

PSICHOTOMY

A great chasm separates the "hard" sciences from parapsychology. In fact, most scientists do not recognize parapsychology as a legitimate science. A recent spate of letters in Physics Today reveals both the depth of the chasm and why it is there. The major letter writers are P.W. Anderson and R.G. Jahn, who work only a few hundred meters apart at Princeton but are light years apart on the matter of parapsychology.

The letter writing commenced after Anderson wrote a column in the December 1990 issue of Physics Today entitled

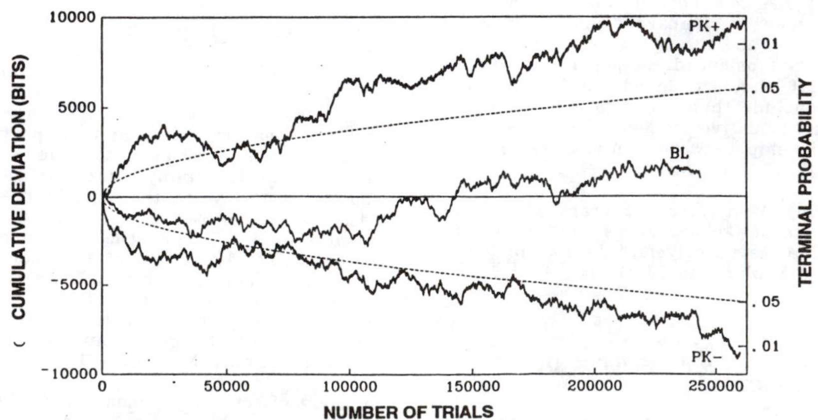

Cumulative deviations from chance for higher numbers of counts (PK^+), lower numbers of counts (PK^-), and baseline (BL). The low probabilities, which were obtained over 250,000 trials, are statistically very significant. Each operator produced curves with distinctive shapes or "signatures."

"On the Nature of Physical Law." Here he recommended the categorical dismissal of all anomalous observations that might tear apart the fabric of science. Although Anderson did not name Jahn specifically, it was obvious to Jahn that his work was the primary target. Jahn's response was a long letter summarizing the stupendous quantity of data he and his colleagues have amassed on psi effects:

> We have in hand several prodigious data bases, acquired over 12 years of continuous, intensive experimentation, that clearly establish the existence, scale and primary correlates of certain anomalous influences of human consciousness on a variety of physical systems and processes. In our Microelectronic Random Binary Generators experiment, 95 selected human operators attempted to shift the output distribution means to either higher or lower values than the chance mean, in accordance with their prerecorded intentions. In 3 850 000 experimental sequences of 200 binary samples, the overall results were that means in high intensity runs exceeded means in low intensity runs by 4.38σ . (The probability of chance occurrence of this outcome is less than 6×10^{-6}.)

Jahn also reviewed the results of other types of psi experiments which also produced positive results.

Replying to Jahn, Anderson admitted that he indeed had Jahn in mind when he wrote his original article.

> What my piece actually said was within my competence as a theorist, which is to make logical connections, and the logical point I made is that physics as it is practiced, and specifically precise mensuration, is not compatible with Jahn's claims; one must choose one or the other, not both, as he also emphasizes. If the "observer effect" as he calls it---or "magic" as one might equally well characterize it---is correct, precise measurement is not possible. His ideas are as incompatible with the intellectual basis of physics as "creation science" is with that of cosmology and biology. It is for this reason that I feel measurements such as Jahn does must be tested with more rigor and more suspicion than their proponents, for some reason, are ever prepared to undergo.

In other words, if psi exists, one cannot measure anything exactly, because the experimenter's mind can skew the results. (Jahn, Robert G., and Anderson, Philip W.; "A Question of Mind over Measurement," Physics Today, 45:13, October 1992.)

Chapter 7

CHEMISTRY, PHYSICS, MATH, ESOTERICA

CHEMISTRY

PHYSICS

MATHEMATICS

ESOTERICA

CHEMISTRY

COLD FUSION

Since the cold fusion debate sank well below the scientific noise level some time ago, we perhaps should refresh our memories a bit. It all began in 1989, when M. Fleischmann and B.S. Pons announced at a press conference that they had achieved cold fusion; cold nuclear fusion, that is; in an electrochemical cell. Nuclear fusion occurs when two nuclei of light elements unite or fuse with the liberation of considerable energy, as in the heart of the sun and thermonuclear bombs. In 1989, the conventional wisdom was that nuclear fusion could take place only at very high temperatures. To achieve fusion at room temperature was incredible. All nuclear fusion projects around the world were based upon "hot" fusion. Needless to say, there was nothing "cold" about the debate that Fleischmann and Pons ignited! In the months that followed the first announcement, Science Frontiers predictably devoted generous space to the subject.

HOW FARES COLD FUSION?

During the past two months, you could have read across a very wide spectrum of conclusions and opinions concerning cold fusion.

Nature. "End of Cold Fusion in Sight." This editorial begins with: "Although the evidence now accumulating does not prove that the original observations of cold fusion were mistaken, there seems no doubt that cold fusion will never be a commercial source of energy." Editor J. Maddox concludes by stating that Pons and Fleischmann should have placed their responsibility to the scientific community above publicity. (Maddox, John; "End of Cold Fusion in Sight," Nature, 340:15, 1989.)

American Scientist. "Cold Fusion Confirmed." Here are reported the observations of neutrons at Los Alamos. (Hively, William; "Cold Fusion Confirmed," American Scientist, 77:327, 1989.)

Science. "Cold Fusion Still in State of Confusion." Some labs have seen neutrons and heat production. (Pool, Robert; "Cold Fusion Still in a State of Confusion," Science, 245:256, 1989.)

Baltimore Sun (Anonymous; "'Cold Fusion' Boils Water, Chemist Says." The chemist is Utah's B.S. Pons.

> In the experiment, a 'boiler' the size of a thermos emitted 15 to 20 times the amount of energy that was being put into it---a reaction that 'cannot be explained by normal chemical reactions we're aware of,' according to Dr. Pons.
>
> .....
>
> He said he was convinced his device could be developed to have practical household applications---providing a home with hot water year-round, for example.

(Uzelac, Ellen; "'Cold Fusion' Boils Water, Chemist Says," Baltimore Sun, July 12, 1989.)

COLD FUSION UPDATE

We gingerly approached the topic of cold fusion above. Sure enough, a lot has happened in the past two months. For one thing, a panel of physicists got together and "voted" down cold fusion almost unanimously. This authoritative declaration seemed to be the beginning of the end for cold fusion---good riddance to those impertinent electrochemists!

At the end of May, scientists assembled at Santa Fe for a Workshop on Cold Fusion Phenomena. Most thought this would be the coup de grace for cold fusion. Not so! More and more researchers reported either anomalous heat production or anomalous emission of neutrons from experiments based on the cold fusion results of Pons and Fleischmann at the University of Utah. Curiously, no one seemed able to get heat and neutrons at the same time and in the amounts Pons and Fleischmann had reported.

We cannot go into all the experiments here. The upshot seems to be that cold fusion is not dead at all. In fact, a lot of people now believe that cold fusion actually does take place in palladium and titanium electrodes. Why, no one is sure. Nor is anyone able to explain the anomalous heat generation. Some think that two separate and distinct phenomena are being observed. One unbelievable anomaly has fissioned into two more-believable anomalies!

Tune in next issue for the latest. Don't believe anything until things cool down a bit. (Pool, Robert; "Cold Fusion: End of Act I," Science, 244: 1039, 1989. Also: Amato, I; "Big Chill for Cold Fusion as Energy Source," Science News, 135:341, 1989.)

COLD FUSION AND ANOMALIES

"I think the physicists have suddenly discovered electrochemistry," said A.J. Bard, from the University of Texas. Ironically, the "cold fusion" experiments making headlines these days are small, simple, and cheap. In contrast, the "hot fusion" physicists have spent some 30 billion dollars since 1951 on huge, complex machines. We don't yet know the specific thoughts and observations that led the Utah electrochemists to cold fusion, but we hope they were anomalous observations of some sort, such as the detection of unexpected neutrons from electrochemical experiments. If such turns out to be the case, the role of anomalies in scientific research will be underscored.

Be that as it may, a genie has been uncorked. Both cold fusion and the recent excitement over high-temperature superconductivity demonstrate that a largely unexplored universe exists in the electrochemistry of the solid state. Favorite theories lie in shambles; faces are very red; the most elite of our scientific institutions were caught with blinders on! Beyond these amusements, the practical import for energy production is enormous, and who know what else will eventuate?

But what about science itself? First, cold fusion will doubtless generate a brand new crop of anomalies which we are only able to guess at now. Pertinent to our effort to catalog anomalies, it is possible that cold fusion may be occurring deep in the earth giving rise not only to heat but the upwelling flux of helium-3, now called "primordial" perhaps in error. The icy planets Jupiter and Saturn generate heat, too, and may also be cold fusion reactors! Even the sun, which is undeniably hot (at least on the surface!), may be fusing light elements in ways we have yet to grasp in our stellar models. After all, no one has yet explained that deficit in solar neutrinos! There will be egg on a lot of faces if our notions of stellar energy generation are wildly in error. We may be overreacting here; but we predict with confidence that the future will bring a good many scientific papers that begin with those all-too-familiar words, "We now know that..."

If physicists must now develop a new appreciation of electrochemistry, should not the biologists, too? At the risk of going too far, we recall that L.C. Kervran talks at length about his findings that the human body converts one element into another. See his book Biological Transmutation. Could he possibly have something? The scientific world utterly rejects Kervran---with some justification.

This is being written in mid-April (1989). By the time it is received, the situation may be radically changed. In any event, it is good to see important work accomplished without huge machines, billions in Federal funds, and a phalanx of white-coated technicians.

COLD FUSION DIED ONLY IN THE MEDIA

When so many scientists ridiculed cold fusion, the media backed off. One reads virtually nothing on the subject today, despite some encouraging developments.

In a panel discussion at the University of Utah, J. Bockris stated that Japan has organized a fusion institute where more than 80 scientists are "rapidly moving forward to develop cold fusion." Bockris himself has achieved positive cold-fusion results in his lab at Texas A&M.

Another panel member, R. Huggins from Stanford, who has also replicated Utah's cold fusion results, remarked that several other U.S. labs have achieved excess power levels of 10-30 watts/cc. (Anonymous; Access to Energy, p. 4, October, 1989. Cr. P.F. Young)

Despite positive results like those mentioned above, a recent report in Nature, by scientists at Cal Tech and the University of California, states emphatically that they can find no excess heat, neutrons, gamma rays, tritium, or helium in their cold-fusion experiments. (Lewis, N.S.; "Searches for Low-Temperature Nuclear Fusion of Deuterium in Palladium," Nature, 340:525, 1989.)

FRACTO-FUSION?

The hot topic in cold fusion research is now fracto-fusion; that is, the inducing of deuterium fusion by means of the electric fields established along microcracks developing in substances charged with deuterium or tritium. Back in 1986, Soviet researchers reported the observation of neutron emission when they violently crushed lithium deuteride in the presence of ice made from heavy water. More recently, they saw the same phenomenon when milling several deuterium-containing metals. Conceivably, deuterium nuclei accelerated by the electric fields along the cracks could be fusing, producing neutrons. (Amato, I.; "If Not Cold Fusion, Try Fracto-Fusion," Science News, 137:87, 1990.)

Pouring cold water on the Soviet results, two American scientists described negative results in the February 15 issue of Nature. They fired small (0.131-gram) steel ball bearings at an ice target made with 99.9% deuterium. Despite the violent shattering of the deuterated ice, no significant numbers of neutrons were measured. (Sobotka, L.G., and Winter, P.; "Fracture without Fusion," Nature, 343:601, 1990.)

Comment. Whatever the fate of fracto-fusion, several labs around the world are still pursuing cold fusion. The scientific mainstream, though, considers cold fusion a dead issue, even though anomalous neutrons and heat emission have been found in several experiments. We are happy to report, however, that cold fusion has definitely generated its first book: Cold Fusion: The Making of a Scientific Controversy, F.D. Peat.

COLD FUSION UPDATE

Over the past six months, we have collected a couple dozen articles on cold fusion. Most authors now dismiss cold fusion as a false trail that leads nowhere interesting, certainly not to small, cheap fusion powerplants. It is time, they say, to stop wasting money and move on. Yet, a small band of researchers insists that "something is going on," something worth persuing just to see what it is. After all, almost 100 laboratories have reported anomalous phenomena; that is, anomalous neutrons, charged particles, heat production, or helium. Can all of these results be in error? Those who would answer "yes" point to more than 100 laboratories with negative results. In the face of all these claims, counterclaims, and contradictions, to say nothing of mean-spirited academic sniping, one must conclude cold fusion is down but not totally out. Good con and pro articles appeared in a recent issue of New Scientist. (Close, Frank; "Cold Fusion I: The Discovery That Never Was," and Bockris, John; "Cold Fusion II: The Story Continues," New Scientist, pp. 46 and 50, January 19, 1991.)

COLD FUSION: NEW EXPERIMENTS AND THEORIES

Cold-fusion research continues in many labs, particularly outside the US, where minds seem more open. At a recent meeting in the Soviet Union, 45 cold-fusion papers revealed intense foreign activity. The Soviets are spending 15 million rubles for further research. In Japan, a Japanese-American team has even set up an experiment a half-mile underground to cut out stray radiation. (R2) However, the US is doing something despite the ridicule from the popular and scientific media.

•B.F. Bush and J.J. Lagowski of the University of Texas in Austin and M.H. Miles and G.S. Ostrom of the Naval Weapons Center in China Lake, Calif., say the helium levels they measured correlate roughly with the amount of heat generated in the fusion reaction.

.....

Those who believe in cold fusion are quite excited. "It's a world-turning experiment, a lollapalooza," says John O'M. Bockris, a physical chemist who has researched cold fusion at Texas A&M University in College Station. (R1)

•According to Dr. Mallove of M.I.T., another provocative set of experiments are those at the Los Alamos National Laboratory in New Mexico where Dr. Howard Menlove has repeatedly detected bursts of neutrons, subatomic particles that are a fusion byproduct. (R2)

•In an interview, [Dr. R.I. Mills] said he had conducted 1,000 experiments with a simple apparatus over the past 18 months and had applied for patents on the process, which differs markedly from the Utah one. He also asserted that he had set a new record for generating heat, saying his apparatus puts out up to 40 times more energy than put in. (R4) (R3)

•Physicists F.J. Mayer and J.R. Reitz theorize that cold fusion as well as several other riddles of physics can be explained with a hypothetical new particle created through the fleeting union of a proton and electron. (R4) (This new "particle" closely resembles the nuclear molecules introduced on p. 324.) In sum, cold-fusion research seems to be rising Phoenix-like after being zapped by most establishment scientists. It will be difficult to overcome the PR campaign orchestrated (by whom?) against cold fusion. However, honest doubts do remain: "It just violates all that we know about nuclear physics," said nuclear chemist J.R. Huizenga. (R1)

References

R1. Pennist, E.; "Helium Find Thaws the Cold Fusion Trail," Science News, 139:180, 1991.

R2. Broad, William J.; "There Still May Be Something Scientific about Cold Fusion," New York Times, April 14, 1991. (Cr. P. Gunkel)

R3. Anonymous, "Pennsylvania Company Claims It Has Solved Cold Fusion Mystery," Butler (PA) Eagle, April 25, 1991. (Cr. E. Fegert)

R4. Broad, William J.; "Two Teams Put New Life in 'Cold' Fusion Theory of Energy," New York Times International, April 26, 1991. (Cr. P. Gunkel)

NEW KIND OF COLD FUSION

Buried among other news items in R&D Magazine for November 1991, we find:

> A research team at Nippon Telegraph and Telephone Corp., Tokyo, claims to have created nuclear fusion at room temperature not by electrolysis, but by placing heavy hydrogen on the surface of a metal in a vacuum and discharging electricity for 14 hours. In five out of 14 tests, the team identified protons apparently emitted as a result of a nuclear fusion reaction.

(Anonymous; "Chrysler, Cold Fusion, Steel," R&D Magazine, p. 5, November 1991. Cr. J.J. Wenskus, Jr.)

JAPANESE CLAIM GENERATES NEW HEAT

While most scientists, especially the hot-fusionists, have been ridiculing cold fusion as "pathological science," more adverturesome researchers have been forging ahead. The most interesting current results, gleaned from many, are those of A. Takahashi, who is a professor of nuclear engineering at Osaka University.

> He says his cold-fusion cell produced excess heat at an average rate of 100 watts for months at a time. That's up to 40 times more power than he was putting into the cell, and more power per unit volume (of palladium) than is generated by a fuel rod in a nuclear reactor.

Takahashi has made several modifications in his cold-fusion cell. Rather than palladium rods, he employs small sheets. In addition, surmising that cold-fusion phenomena might prosper better under transient conditions, he varies cell current. Takahashi, however, measures only a few of the neutrons expected from the usual nuclear fusion reactions. Undaunted, he remarks, "This is a different ballgame, and it could be a different reaction." Indeed, some exotic fusion reactions do not generate neutrons. (Freedman, David H.; "A Japanese Claim Generates New Heat," Science, 256:438, 1992.)

FIRST COLD-FUSION BOMB?

Possibly we are overreacting to the following event:

> Cold fusion researchers are puzzled and worried by an explosion last week that killed one of their colleagues, a British electrochemist. A cold fusion "cell" at SRI International in Menlo Park, California, blew up while Andrew Riley was bending over it, killing him instantly.
>
> Now small explosions in cold-fusion cells are not unknown. At the tops of some cells palladium-wire electrodes are exposed to oxygen and deuterium (heavy hydrogen) gases. If the palladium wires are not protected by films of water, the palladium can catalyze the explosive combination of the oxygen and hydrogen. This sometimes happens if a dry spot develops on a wire. Such detonations, though, cause little damage. The SRI explosion was much more powerful. The detonating cell (only 2 inches in diameter and 8 inches long), not only killed Riley but peppered three other researchers in the lab with debris.

(Charles, Dan; "Fatal Explosion Closes Cold Fusion Laboratory," New Scientist, p. 12, January 11, 1992.)

Comment. One cannot refrain from asking if the explosion involved only chemical energy.

COLD-FUSION UPDATE

SRI explosion due to wayward piece of Teflon? The final report on the fatal explosion of a cold-fusion experiment at SRI International (above) blames a loose piece of Teflon that may have blocked a gas outlet tube. Possible scenario: After many hours, the researchers finally noticed something was awry. When A. Riley lifted the cell from its water bath, it exploded. "The investigators believe that hot palladium ignited the pressurized mixture of oxygen and deuterium. The bottom blew off the cell, turning the rest of it into a rocket which shot upwards at 50 metres per second. It struck Riley in the head." (Charles, Dan; "Piece of Teflon Led to Fatal Explosion," New Scientist, p. 5, June 27, 1992. Also: Holden, Constance; "Fusion Explosion Mystery Solved," Science, 257: 26, 1992.)

Another cold-fusion book: Huizenga, John R.; Cold Fusion: The Scientific Fiasco of the Century, 259 pp., 1992, The title betrays the book's slant. A single sentence from Nature's review will suffice:

> Commenting on the hundreds of millions of dollars of research time and resources that were taken up in showing that there is no convincing evidence for cold fusion as a source of nuclear power, he [Huizenga] notes that "much of this would not have been necessary had normal scientific procedures been followed."

(Close, Frank; "The Cold War Remembered," Nature, 358:291, 1992.)

But what's this from Los Alamos?

> A Los Alamos National Laboratory researcher says he has duplicated the results of a Japanese experiment in which power was generated by cold fusion.
>
> Edmund Storms, a high-temperature chemist at Los Alamos, used palladium metal supplied by Japanese fusion researcher Akito Takahashi of Osaka University. (See p. 316.)

(Anonymous; "Los Alamos Scientist Duplicates Japanese Cold Fusion Experiment," Associated Press, July 28, 1992. Cr. E. Hansen)

Where There's Heat There's Yen. Japan's Ministry of Trade and Industry (MITI) plans to launch a five-year program to study cold fusion. Isn't this folly, since most physicists have declared cold fusion to be impossible?

> Not so, says MITI---it's just Japanese pragmatism. All MITI is interested in is the continuing reports of excess heat generated in the hydrogen-palladium cells studied by Pons and Fleischmann and the possibility of putting any new phenomenon---even if chemical rather than nuclear in origin---to industrial use.

(Myers, Frederick S.; "Where There's Heat There's Yen," Science, 257:474, 1992.)

INFINITE-DILUTION CLAIMS

The rather cryptic title above merits an interpretation. The anomaly claimed here insists that a chemical or biological substance can still affect an organism even after it has been diluted to the point where it is unlikely that even one molecule of the active substance remains in the administered solution. Not only does this claim seem far-fetched itself, but its implication raises the specter of homeopathy and its possible efficacy in medicine. Mainstream science has been engaged in a long battle with the homeopathists, and the research reported below has intensified hostilities.

NOTHING REACTS WITH SOMETHING?

An absolutely delightful event occurred at the end of June, 1988. The authoritative journal Nature published an article that says, in essence, that a solution of antibodies diluted by a factor of 10^{120} can still trigger a strong biological response from basophils (a kind of white blood cell). Now, 10^{120} is such an incredibly large number that it is extremely unlikely that even one antibody molecule could be present in the diluted activating solution. Never-the less 40-60% of the basophil cells reacted.

So unbelievable are the reported experimental results that the editors of Nature felt compelled to add an "Editorial Reservation" stating that, "There is no physical basis for such an activity."

This is all great stuff. The original French work was duplicated by six other laboratories in France, Italy, Israel, and Canada. What makes it even more fun is the homeopathy connection. Homeopathic medicine is based on the theory that substances causing the symptoms of a disease in a healthy person can cure a sick person displaying these symptoms, providing the dose administered is vanishingly small. Science strongly and passionately debunks homeopathic medicine.

The Editor of Nature thinks that there must be a systematic error somewhere. Other scientists suggest that, perhaps, somehow, the antibodies left an "imprint" on the diluting water molecules! So far, we have not read that Sheldrake's "morphic resonance" theory has been invoked.

The first phase of this controversy is about complete, and we now list the references we have used so far: Davenas, E.; "Human Basophil Degranulation Triggered by Very Dilute Antiserum against IgE," Nature, 333:816, 1988. Browne, Malcolm W.; "Impossible Idea Published on Purpose," New York Times, June 30, 1988. Cr. D. Stacy, M Truzzi; Nau, Jean-Yves, and Nouchi, Franck; "La Memoire de la Matiere," Le Monde, p. 1, June 30, 1988. Cr. C. Maugé. Beil, L.; "Dilutions of Delusions," Science News, 134:6, 1988. Vines, Gail; "Ghostly Antibodies Baffle Scientists," New Scientist, p. 39, July 14, 1988; Pool, Robert; "Unbelievable Results Spark a Controversy," Science, 241: 407, 1988.

As a matter of fact, the second phase of the controversy has already begun. The July 28 issue of Nature reports that seven repetitions of the dilution experiment produced four positive and three negative results. The three negative experiments were the only double-blind versions of the basic experiment that have been performed so far. "Double-blind" means that "all test tubes had been randomly coded twice. The person measuring the cells' reaction to the antibodies could not have been influenced by a preconcieved idea of the results."

These seven repetitions were carried out at the University of Paris-Sud laboratory of J. Benveniste. In a reply to the July 28 report in Nature, Benveniste complains that the three double-blind tests, the negative ones, "worked poorly mainly due to erratic controls." (Bell, L.; "Nature Douses Dilution Experiment," Science News, 134:69, 1988.)

Comment. Closely related to the French dilution experiments with antibodies was a clinical trial of a homeopathic treatment for hay fever, carried out by D. Reilly, of the Glasgow Royal Infirmary, and published in The Lancet about two years ago. "In trials with double-blind controls, he and his colleagues found that a solution, so dilute as to contain no molecules of pollen from the original solution, reduced allergic symptoms." (From: Vines, above.)

Naturally, we will be keeping SF readers up-to-date on this matter. Already comparisons are being made with R. Blondlot's famous experiments with N-rays---experiments that were also duplicated in other laboratories. The solution may be psychological rather than physical.

UPDATE ON THE "INFINITE DILUTION" EXPERIMENTS

The above summary of J. Benveniste's "infinite dilution" experiments and Nature's subsequent "investigation" was written in mid-August (1988). When we returned from a long vacation at the end of September, we found eleven new references on this novel development in the progress (?) of science. Basically, there are only two big questions: (1) Is there, despite all the furor and the machinations of Nature's "hit squad," anything of scientific value in the experiments of Benveniste's group? and (2) Should scientific journals police scientific research? The latter question should be answered by Science itself: we shall focus here on the possibility that real scientific anomalies are being concealed by all the media smoke.

What did Nature's hit squad really find? J. Maddox (Nature's editor) et al concluded that Benveniste and his colleagues did not take enough care in their work, that their data did not have errors of the right magnitude (a statistical quibble), that no serious attempt was made to eliminate systematic errors and observer bias, that the climate of the lab was "inimical to an objective evaluation of the exceptional data," and that the phenomenon was not always reproducible. (7) No evidence of fraud was found. The data originally published in Nature were not explained or shown to be invalid. (11) In fact, the Nature investigation actually confirmed some of the original findings. (5) All of the French work and that of the cooperating laboratories were attributed to "autosuggestion"! (4)

Qualifications of the Nature investigators. J. Benveniste pointed out that none of the three members of the Nature team had any experience in immunology. (4, 11) The team consisted of J. Maddox (a physicist), J. Randi (a professional magician), and W. Stewart (an organic chemist).

Curious aspects of Nature's publication and following investigation. Why did Nature accept and publish a paper when fraud and poor science were suspected? (4, 11) Why didn't Nature hold publication of the original Benveniste paper for four weeks until the investigation was completed? (4, 11) Why didn't Nature insist upon prior experiment replication by an independent laboratory? (6) Actually, replications of the experiment were completed before publication, but at labs selected by Benveniste.

Conventional explanations of Benveniste's results. Several letters to Nature have proposed reasonable explanations for the supposedly impossible results of the "infinite dilution" experiments. (8, 9) It is therefore possible that Benveniste's data are valid and not due to "autosuggestion."

Has the "infinite dilution" anomaly been exorcised? Not in our opinion. Too many unexplained data survive. We doubt, however, that many scientists will rush to their labs to explore this subject. It would be too risky in the present scientific environment. Nature has, in effect, relegated "infinite dilution" research to pseudoscience, whether deserved or not.

References.

1. Anonymous; "Now You See It..., Scientific American, 259:19, September 1988.
2. Vines, Gail; "The Ghostbusters Report from Paris," New Scientist, p. 30, August 4, 1988.
3. Anonymous; "Inhuman Nature," New Scientist, p. 19, August 18, 1988.
4. Pool, Robert; "More Squabbling over Unbelievable Result," Science, 241: 658, 1988.
5. Benveniste, Jacques; "Benveniste on Nature Investigation," Science, 241: 1028, 1988.
6. Plasterk, Ronald H.A., et al; "Explanation of Benveniste," Nature, 334:285, 1988.
7. Maddox, John; "'High-Dilution' Experiments a Delusion," Nature, 334: 287, 1988. Written in conjunction with J. Randi and W.W. Stewart, with a reply by J. Benveniste.
8. Metzger, Henry, et al; "Only the Smile is Left," Nature, 334:375, 1988.
9. Seagrave, JeanClare, et al; "Evidence of Non-Reproducibility," Nature, 334:559, 1988.
10. Stanworth, D.R., and Johnson, Mark; "Outlandish Claims," Nature, 335: 392, 1988.
11. Anonymous, "Quand l'Eau Fait Fremir les Scientifiques," La Recherche, no. 202, p. 1005, September 1988.

ARE YOU SATURATED WITH DISCUSSIONS ABOUT THE "INFINITE-DILUTION" CAPER?

Another pile of references has accumulated concerning J. Benveniste's experiments with (almost) infinitely diluted solutions. This time we shall be brief. After all, the journal Nature has now cut off debate on the subject; shouldn't everyone else? The Book of Science is closed on this one.

Despite replications of Benveniste's experimental results at other laboratories and the existence of similar results from several meticulously conducted experiments over the past few decades, Nature's official investigative team (formerly called the "hit team" in these pages) has labelled Benveniste's results a "delusion." Now begins the dirty work of completely destroying the reputation of Benveniste and the believability of any work done in this field. First, in the New York Times, J. Maddox, editor of Nature, stated that Benveniste's positive results were "nonexistent." Then J. Randi, the magician member of the investigative team, called the positive results "fraudulent" in the Lisbon Expressor. This means that five independent laboratories all produced fraudulent results! (Benveniste, Jacques; "Benveniste on the Benveniste Affair," Nature, 335:759, 1988. Also: Maddox, John; "Waves Caused by Extreme Dilution," Nature, 335:760, 1988.)

Comment. Regardless of the merits of the scientific work done by Benveniste and his coworkers, it now appears, to some outsiders at least, that Benveniste was set-up, entrapped, and sandbagged. A similar campaign is being waged to discredit M. Gauquelin's Mars Effect. (See item on p. 113.) So, you heretics beware, the Inquisition lives!)

HOW FARES BENVENISTE?

J. Benveniste is in the news again!

The July 20 issue of Nature has a news item entitled emphatically "INSERM Closes the File." (INSERM is the French institute of health and medical research.) Two INSERM committees recommended that Benveniste cease his work on high dilutions; work that seems to support the principles of homeopathy. The directory-general of INSERM, P. Lazar, however, did not endorse these recommendations. Benveniste was asked to look for errors in his experiments that might account for his "unusual results." Without question Benveniste is under the gun and future funding in jeopady. (Coles, Peter; "INSERM Closes the File," Nature, 340:178, 1989.)

The headline in Science's news item was less emphatic: "Benveniste Criticism Is Diluted." Here, Lazar is reported as saying that he did not want to stifle research on new ideas and that Benveniste had been treated badly by Nature.

Benveniste has not remained silent. In Le Monde, he stated that the results he had published in Nature have now been confirmed by two French teams, two American teams, and one in the USSR. (Dickson, David; "Benveniste Criticism Is Diluted," Science, 245:248, 1989.)

WATER'S MEMORY OR BENVENISTE STRIKES BACK

J. Benveniste has broken a two-year drought in the "water-memory" or "infinite-dilution" saga.

> Working with colleagues at INSERM, the French medical research council, in Paris, Benveniste has completed fresh experiments to test his assertion that solutions of antibody diluted to the point where they no longer contain any antibody molecules continue to evoke a response from whole white blood cells, as if they possess "ghosts" of the original molecules. If proven, this would shatter the laws of chemistry and vindicate homeopaths, who say that extremely dilute drugs can have a physical effect.

Benveniste's latest scientific paper was published in Comptes Rendus after being rejected by both Nature and Science. Benveniste states that he has corrected the flaws in his original research that evoked passionate responses from the scientific world. However, Benveniste's latest paper prompted one of Nature's reviewers to charge that Benveniste was "throwing out data because they don't fit the conclusion."

This story is not yet finished, because Benveniste promises to reveal new research that demonstrates that a solution of histamine, from which all traces of histamine were subsequently diluted out, can still affect blood flow in the hearts of quinea pigs! Furthermore, this phenomenon can be inhibited by the application of weak magnetic fields!! (Concar, David; "Ghost Molecules' Theory Back from the Dead," New Scientist, p. 10, March 16, 1991.)

Comment. One would expect that the effects of clustering would be important in biology, too. We obviously have a lot to learn about this new realm between single atoms/molecules and bulk materials.

STRUCTURE OF MATTER

MAGIC NUMBERS IN HELIUM ATOM CLUSTERS

Because spheres can be packed snugly together in certain configurations, one would expect the atoms of inert gases, such as helium, to cluster together in similar tightly packed groups. The numbers of spheres or atoms in these geometrically and energetically favored clusters are termed "magic." One would expect, for example, the number 13 to be magic because this would be the number of spheres fitting tightly into a 20-sided solid. Sure enough, when clusters of helium atoms are "weighed" with a mass spectrometer, the number 13 is strongly favored; so are 19, 25, 55, 71, 87, and 147.

Some of these experimentally derived magic numbers can be predicted theoretically, but others were surprises. Some theoretical magic numbers did not turn out to be magic in reality; notably 7, 33, and 43. Clearly nature has other criteria for "magicness" than the physicists. (Anonymous; "Magic Numbers Do Hold Atoms Together," New Scientist, 92:598, 1981.)

KRYPTON-CLUSTER MAGIC NUMBERS

Ordinarily, krypton, being a noble gas, does not enter into any combinations with other atoms---even other krypton atoms. However, P. Lethbridge and T. Stace, at the University of Sussex, have coaxed krypton atoms to cluster together in large, crystal-like clumps with icosohedral symmetry; that is, each clump possesses 20 regular faces. The coaxing occurs when gaseous krypton trickles into a vacuum chamber through a hole only 200 micrometers in diameter. The expansion of the gas cools it so that when krypton atmos collide, relative velocities are low, and the weak Van de Waals forces between the atoms are sufficient to hold the clumps together.

So far, clumps of 147 and 309 atoms have been detected with a mass spectrograph. One theory of atomic "packing" predicts clumps should have "magic numbers" of 13, 55, 147, 309, 561, 923 .... So far, the "magic" has been working! (Baggott, Jim; "Krypton Atoms Cling Together in 'Shells,' New Scientist, p. 31, March 3, 1990.)

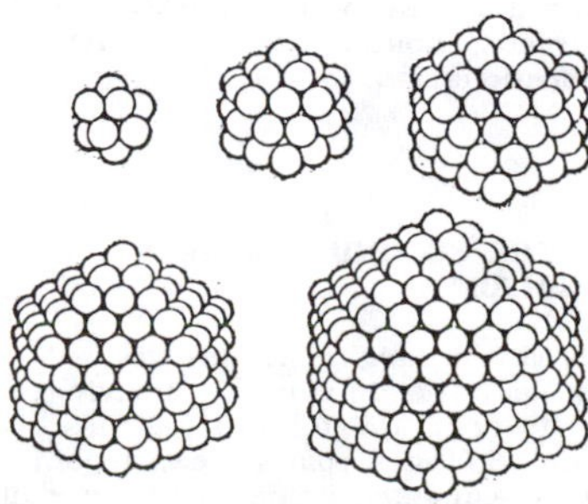

Krypton atoms cluster together in stable clumps of 13, 55, 147, 309, 561,...atoms.

Comment. One would anticipate that the smaller clumps of 13 and 55 atoms would be easier to assemble.

NOVEL FORMS OF MATTER

Clusters. What is the difference between "vanilla" and "chocolate" niobium? To begin with, these two forms of niobium are neither single atoms nor crystalline arrays of atoms. Both "flavors" of niobium consist of 19 niobium atoms (Nb_{19}^{+}), but the atoms are clustered differently, as illustrated. The different clusters react quite differently chemically. The "chocolate" niobium, on the right, is a capped icosahedron and reacts readily with hydrogen. The "vanilla" double pyramid (left) has flatter surfaces and does not readily combine with hydrogen.

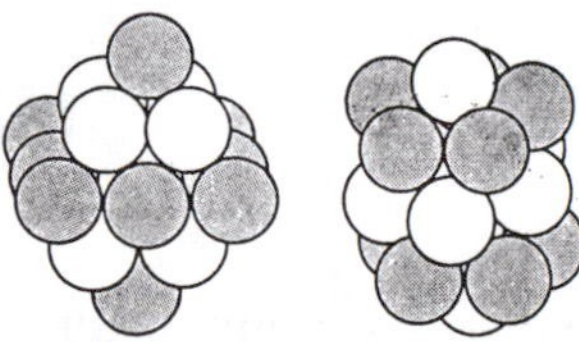

Cluster research is embryonic, with new surprises popping up almost every day. Once a cluster size exceeds a few hundred atoms, its properties begin to resemble those of the bulk material. However, "cluster-assembled materials" have been made by attaching cluster-to-cluster in a sort of patchwork quilt. Such materials have unique properties that are quite different from those of the normal crystalline and amorphous materials. (Pool, Robert; "Clusters: Strange Morsels of Matter," Science, 248:1186, 1990.)

TWO SNOWFLAKE ANOMALIES

Rarely is there anything in the scientific literature suggesting that anything about snowflakes could possibly be mysterious. Surprisingly, two articles on snowflake anomalies have appeared recently.

To form at all above -40°F, snowflakes supposedly require a solid seed or nucleus around which ice can crystallize---or so scientists have assumed for many years. It was long believed that airborne dust, perhaps augmented by extraterrestrial micrometeoroids, served as the necessary nuclei. But cloud studies prove that there are about a thousand times more ice crystals than dust nuclei. Now, some are convinced that bacteria blown off plants and flung into the air by ocean waves are the true nuclei of atmospheric ice crystals. Remember this the next time you taste a handful of snow! (Carey, John; "Crystallizing the Truth," National Wildlife, 23:43, December/January 1985.)

Comment. The possibility that the fall of snow and all other forms of precipitation is largely dependent upon bacteria brings to mind the Gaia Hypothesis; that is, all life forms work in unison to further the goals of life.

The second item is from Nature and is naturally more technical. After reviewing the great difficulties scientists are having in mathematically describing the growth of even the simplest crystal, the author homes in on one of the fascinating puzzles of snowflake growth:

> The aggregation of particles into a growing surface will be determined exclusively by local properties, among which surface tension and the opportunities for energetically advantageous migration will be important. But the symmetry of a whole crystal, represented by the exquisite six-fold symmetry of the standard snowflake, must be the consequence of some cooperative phenomenon involving the growing crystal as a whole. What can that be? What can tell one growing face of a crystal (in three dimensions this time) what the shape of the opposite face is like? Only the lattice vibrations which are exquisitely sensitive to the shape of the structure in which they occur (but which are almost incalculable if the shapes are not simply regular)!!

(Maddox, John; "No Pattern Yet for Snowflakes," Nature, 313:93, 1985.)

Comment. It is amusing that this usually fairly open-minded journal Nature once blasted Sheldrake's A New Science of Life as a good candidate for burning. It is in this book that Sheldrake proposed morphogenetic fields as the explanation of crystal growth. Morphogenetic fields seem at least as reasonable as "vibrations." (See p. 132.)

FORBIDDEN MATTER

When a hot mixture of aluminum and manganese, iron or chromium is squirted onto a spinning water-cooled copper wheel, the molten metal freezes into a thin, metallic ribbon. If it is cooled too fast, a metallic glass results; cooled too slowly, it forms normal metal crystals. But when conditions are just right, icosahedral crystals cluster together in nodules a few microns in size. These icosahedral crystals are not normal in the sense that they have five-fold symmetry. In fact, to a crystallographer, these crystals are the equivalent to ESP in psychology.

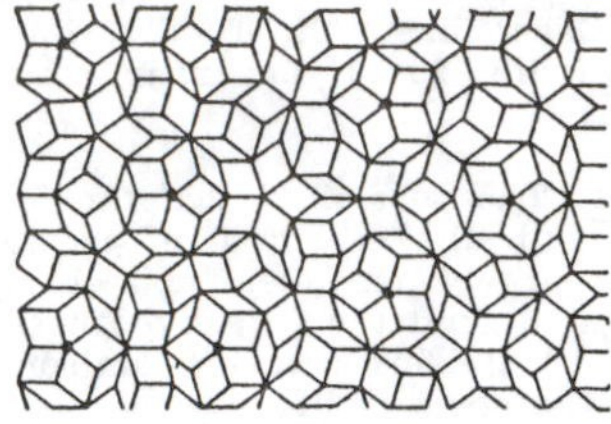

Two-dimensional quasiperiodic geometry (Penrose tiling) with five-fold symmetry formerly thought to be impossible in nature.

All the rules of crystallography insist that icosahedral crystals should not exist. One scientist reacted in this way, "All my training has been with the assumption that crystals are periodic. Now, almost everything has to be reexamined." Actually, the icosahedral crystals are "quasi-periodic"; that is, they are completely regular only over small distances. Nevertheless, there are hints that these materials that should not exist have remarkable structural and electronic properties. (Peterson, Ivars; "The Fivefold Way for Crystals," Science News, 127:188, 1985.)

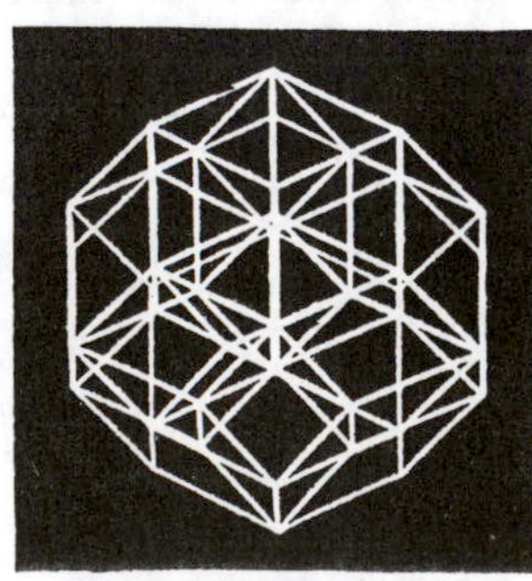

The tricontahedron with 30 faces is the basis of three-dimensional quasiperiodic structures with five-fold symmetry.

Quasicrystals. Quasicrystals, once considered physically impossible, have been found easy to grow---once one's mindset is corrected. At Bell Labs, for example, quasicrystals of an aluminum-cobalt-copper alloy reveal atoms packed together in pentagonal arrays---the very geometry that crystallographers assured us could not exist. As one of the Bell researchers remarked, "Most of the applications are unimagined." This goes for the basic properties of quasicrystals, too. (Keller, John J.; "Bell Labs Confirms That New Form of Matter Exists," Wall Street Journal, February 8, 1990. Cr. J. Covey.)

Nitrogen molecules that shouldn't exist.

> Chemists in West Germany have discovered a compound of nitrogen which breaks one of the fundamental rules of chemistry. The molecule has five bonds and is "an extremely stable species." According to the textbooks, a nitrogen atom cannot form more than four bonds.

The unruly nitrogen atom lurks at the center of a tribonal bipyramid of 5 gold atoms. Should you ask to synthesize a few of these molecular anomalies, take some (phosphine-gold) ammonium tetrafluoroborate from your shelf and add (triphenylphosphine) gold tetrafluoroborate! (Emsley, John; "The Nitrogen Molecule that Shouldn't Exist," New Scientist, p. 30, May 26, 1990.)

RESTLESS GOLD

Thanks to the development of high-resolution electron microscopes and video recorders, we can now watch the bizarre behavior of tiny solid particles, which, it turns out, are not so solid after all. Ultrafine particles of gold about 18 Angstrom units across, containing only about 500 atoms, are not static aggregations. The shapes of the particles are always changing. The gold atoms move cooperatively to shift kaleidoscope-like into various crystal structures. They have, in fact, been dubbed 'quasi-solids.' A large gold particle may even ingest smaller gold particles. The phenomena have no explanations as yet. (Anonymous; "Japanese Gold in Atomic Motion," Nature, 315:628, 1985.)

SPECULATIONS FROM GOLD

The item "Restless Gold" has now been amplified by J.-O. Bovin et al. Using a high-resolution electron microscope, magnification 30,000,000, with a real-time video recorder, this group has obtained startling pictures of gold crystals and their environs.

> At this magnification, individual columns of atoms in the gold crystals are clearly revealed; it appears that not only are atoms of the surfaces of small crystals in constant motion, hopping from site to site, but also that the crystals are surrounded by clouds of atoms in constant interchange with atoms on the crystal surface. The clouds of gold atoms extended up to 9A from the crystal surface, continually changing their shape and density.

The remarkably dynamic nature of solid surfaces, as now revealed, has many implications. (Bovin, J.-O. et al; "Imaging of Atomic Clouds Outside the Surfaces of Gold Crystals by Electron Microscopy," Nature, 317:47, 1985.)

Comment. The problem of snowflake growth (See p. 319.) is probably solvable in terms of clouds of water molecules surrounding crystal nuclei, with electrostatic fields guiding the symmetric deposition of molecules. Biological structures, too, are probably encompassed by clouds of atoms and molecules; viz., the crystal-like, polyhedral viruses. Does the highly ordered DNA structure also possess an aura of molecules constantly swapping places? Such would not be inconsistent with "jumping genes" and M. Kimura's Neutral Theory of Evolution. (See pp. 183 and 189.)

MISCELLANY

HOW CAN THE SUN INFLUENCE CHEMICAL REACTION RATES?

When water is added to bismuth trichloride, a precipitate, bismuth oxychloride, forms. The precipitation rate seems to vary with time, being different from one month to the next. This time variation has been confirmed at many laboratories around the world and is not dependent on any obvious meteorological condition. Instead, some investigators claim to have found a rather good correlation with the sunspot cycle! (Majorino, Gianfranco, and Zecca, Luigi; "Period Analysis of the Picardi P-Test," Cycles, 33:78, 1982.)

Comment. This precipitation test, called the Piccardi P-Test, has been offered for years by cycle students as proof of extraterrestrial influences on earthly chemistry, including biochemistry. No explanation has been suggested; and one reads about it only in "fringe" publications.

HAS THE SECOND LAW BEEN REPEALED?

> From the largest to the smallest scales, the universe is evolving. Matter, in the form of galaxies, is undergoing a colossal expansion. Gas, condensed into stars, is radiating thermonuclear energy out across an infall of matter, drawn by gravity. The simplest of chemical reactions and the most complex of

biological activities are occurring on the surface of the earth in a state far from equilibrium; they are heated by the sun and cooled by the vacuum of space. This pervasive cosmic imbalance is the driving force in producing an environment conducive to the formation of structure and complexity.

This sweeping statement seems to apply to the entire universe. The Second Law of Thermodynamics, however, insists that, on the average, for the entire universe, the above paragraph cannot be true.

The article introduced by this unqualified assertion about the evolution of the universe is really about self-organizing chemical reactions. We might well classify it under Biology because the authors imply that some biological phenomena are self-organizing.

The famous Belousov-Zhabotinskii reaction is used as the prime example of chemical self-organization. First, one takes a shallow dish filled with a solution of bromate ions in a highly acidic medium. Here's what happens:

> A dish, thinly spread with a lightly colored liquid, sits quietly for a moment after its preparation. The liquid is then suddenly swept by a spontaneous burst of colored centers of chemical activity. Each newly formed region creates expanding patterns of concentric, circular rings. These collide with neighboring waves but never penetrate. In some rare cases rotating one-, two-, or three-armed spirals may emerge. Each pattern grows, impinging on its neighboring patterns, winning on some fronts and losing on others, organizing the entire surface into a unique pattern. Finally, the patterns decay and the system dies, as secondary reactions drain the flow of the primary reaction.

Spectacular, evolving forms erupt in the Belousov-Zhabotinskii reaction. Waves of chemical activity propagate through a receptive liquid medium.

From this starting point, the implication is made that all manner of biological "reactions" are analogous and therefore reducible to nought but physics and chemistry. Some examples given of self-organizing biological phenomena are: (1) the sequencing of amino acids into self-replicating structures; (2) slime-mold organization; and (3) the origin of the lens structure of the firefly. All of these claims are accompanied by computer simulations of self-organizing reactions. (Madore, Barry F., and Freedman, Wendy L.; "Self-Organizing Structures," American Scientist, 75:252, 1987.)

Comment. While we believe that science is the best way yet discovered to search for truth, we have to admit that scientists sometimes get carried away in their zeal to explain things, especially with computer graphics. The Belousov-Zhabotinskii reaction is certainly impressive. So is crystal growth. But are the atoms falling together to form a crystal analogous to soldiers falling into ranks, or the assembly of genetic information into the genotypes for our planet's multitudinous species? How far can we apply reductionism?

MUTANT MOLECULES FIGHT FOR FOR SURVIVAL

In our Astronomy chapter, p. 106, we saw how universes might mutate and evolve. Shifting dimensions by just a few-score orders of magnitude, from the cosmological to the molecular, we find that molecules, too, may mutate and evolve. A group of scientists at MIT, led by J, Rebek, believes that it has discovered the chemical equivalent of biological evolution. This is the same team that claimed the synthesis of the first self-replicating molecule in 1990.

> Now the same chemists have carried out experiments with two more self-replicating molecules, and discovered that they can cooperate, catalysing each other's formation. Furthermore, when one of the molecules is exposed to ultraviolet light and is "mutated", it becomes "aggressive" and takes over the system. According to Rebek and his colleagues, this is evidence that evolution can be modelled at the molecular level.

(Emsley, John; "How 'Mutant' Molecules Fight for Survival," New Scientist, p. 22, February 29, 1992.)

Comment. We wonder if the "passive" molecules are "uphappy" about being bullied in this way!

PHYSICS

PARTICLE PHYSICS

A FUNNY THING HAPPENED ALONG THE MEAN FREE PATH

A little anomaly may go a long way. Accelerator experiments at Berkeley have again focussed attention on those few fragments from nuclear reactions that have unexpectedly short trajectories. About 6% of these fragments travel only about one tenth as far as prevailing physical laws say they should. These anomalously short mean free paths are not new, having first cropped up in 1954, but they have gone unexplained for 26 years. Current speculation is that the anomalous fragments somehow change their identities, making them more susceptible to collision (i.e., their collision cross sections spontaneously increase by ten times). But no known transformations of matter can do this! Consequently, we are left with the possiblity that some entirely new form of matter exists. (Robinson, Arthur L.; "A Nuclear Puzzle Emerges at Berkeley," Science, 210:174, 1980.)

Comment. Just a few weeks ago, some nuclear physicists were saying that the advent of quark theory explained everything in their field. (See next page for more on "anomalons."

MAGICAL COMMUNICATION IN THE SUBATOMIC WORLD

How do physically separated subatomic particles (and devices based on them) communicate with one another? Somehow one subatomic particle knows what a distant compatriot is doing and reacts accordingly. Physical experiments have confirmed this seemingly impossible situation. No fields, wires, or any other sort of communication line connect the two particles; yet they behave as though there were. Quantum mechanics has an explanation of sorts but it still leaves the situation with an aura of mystery. (Mermin, N.D.; "Bringing Home the Atomic World: Quantum Mysteries for Anybody," American Journal of Physics, 49:940, 1981.)

Comment. This digest is greatly oversimplified, and readers are encouraged to read the whole article. It is an important type of physical experiment because some have suggested it may help explain ESP, assuming ESP exists.

SCHIZOPHRENIC NEUTRINOS

As the concept of the neutrino has developed since the early 1930s, it has developed a split personality and put on weight. The neutrino is now thought to come in three varieties---electron neutrino, muon neutrino and tau neutrino. And a number of experiments are showing hints that a neutrino has a small mass and that it can oscillate from one variety to another.

These experiments are not yet conclusive; and if the neutrino mass is not zero, it hardly weighs more than the grin of a Cheshire cat. But taken together, the laboratory results confirm that neutrinos are perplexing particles. Are they different entities or a single species wearing different costumes? The implications of the recent measurements are far-reaching:

1. Physicists believe that there are a billion neutrinos around for each nucleon (proton, neutron, etc.) so that if neutrinos possess just a hint of mass, they will dominate the mass of the universe; and
2. Measurements of solar neutrinos fall short by a factor of three of what theory says the sun should spew out. This discrepancy could be explained if the solar neutrinos change from electron neutrinos to another form during their flight from the sun to earth, for the terrestrial neutrino detectors measure only electron neutrinos.

(Anonymous; "Do Neutrinos Oscillate from One Variety to Another?" Physics Today, 33:17, July 1980.)

THOSE DARN QUARKS

There's no escaping it, those fractionally charged niobium balls just can't be swept under the rug. In fact, more recent experiments have served only to accentuate the anomaly. Researchers at Stanford University have been magnetically suspending superconducting niobium spheres in a modern version of Millikan's oil-drop experiment. With the niobium spheres thus suspended, their net electrical charges can be measured. The trouble is that several of the spheres have fractional electrical charges---+1/3 or -1/3 electronic charges.

For decades the charge on the electron was supposed to be the basic, indivisible natural unit of electrical charge. In 1964, however, theorists began muddying the waters with talk of new fundamental constituents of matter called quarks, which could possess 1/3 or 2/3 electron charges. No one really expected that quarks, if they existed at all, would be floating around free. But the niobium balls tell us that not only are quarks free but that we could have detected them with relatively simple experiments decades ago if we had not been so blinded by the idea of integral electronic charges. (Robinson, Arthur L.; "Evidence for Free Quarks Won't Go Away," Science, 211:1028, 1981.)

ANOMALONS ARE LAZY OR FAT

No, an "anomalon" is not an animal unknown to science.

"Anomalon" is the name that has been given to unusual fragments that are created in high-energy collisions of atomic nuclei. The fragments are peculiar because they appear not to travel as far as expected in the special "nuclear" emulsion used to study the interactions of high-energy nuclei from heavy-ion accelerators or in cosmic rays. This suggests that the anomalons are either much larger than conventional nuclei, and are more likely to interact in the emulsion and therefore do not travel so far, or are some unusually long-lived form of matter, lasting for around 10^{11} seconds or more.

One thought is that anomalons may be constructed of two triplets of quarks. These sextets are called "demon deuterons." Another hypothesis has small nuclei bound loosely together---they don't say by what. The whole thing is up in the air, or should we say in the emulsion? (Sutton, Christine; "Anomalon Data Continue to Baffle Physicists," New Scientist, 96:160, 1982.)

Comment. One thing is sure, nuclear physicists have a lot of fun naming their newly found particles.

MAYBE THERE'S ONE STABLE PARTICLE!

The new Grand Unified Theories (GUTs) of physics predict that the proton decays radioactively---contrary to what you may have been taught in physics class. Several experiments in deep mines and tunnels seem to have registered proton decays. Is nothing stable anymore? There is hope. A huge cubical detector, 21 meters on a side, is now operating 2000 feet under Lake Erie in a salt mine. The water-filled cube is monitored by 2048 photomultipler tubes, and is serviced by divers! After 80 days of operation, no events resembling proton decays have been detected, whereas many would have been expected if the other reports were true. The GUTs may be in trouble; and there may be something stable in the universe that we can count on! (Thomsen, D.E.; "Decay-Resistant Protons in Ohio," Science News, 123:85, 1983.)

Comment. See p. 323 for some of the strange particles detected in the Kolar Gold Fields, in India, where some of the supposed proton decays were reported.

ZETA NOT A HIGGS: TOO BAD!

It should have been a Higgs particle but it wasn't---at least not quite. So they called it a "zeta." About eight times the mass of a proton, the zeta particle was created when electrons and positrons collided at about 10 Gev (gigavolts of energy), where it appeared among the decay products of the upsilon particle. Physicists needed a Higgs particle to bolster the latest theory of particles. Unfortunately, the zeta's properties don't quite match those predicted for the Higgs particle. There are similarities, but at the moment the zeta is definitely anomalous. It turns out that there is a similar anomalous particle produced by the decay of psi particles, so the zeta is not alone. (Thomsen, Dietrick E.; "Zeta Particle: Physicists' New Mystery," Science News, 126:84, 1984.)

Comment. It is easy to become jaded by all the confusing particles flying around physics labs these days. But we must appreciate that physicists absolutely must find that Higgs particle. Theory says that if the Higgs doesn't exist, all other particles will have either zero or infinite masses, neither of which makes much sense. Such is the power of theoretical expectations.

BLOOMS IN THE DESERT?

Particle physicsts have recently observed anomalous events in data from the proton-antiproton collider at the European Laboratory for Particle Physics (CERN). These unexpected events occur when a proton hits an antiproton head-on at very high energies. Highly collimated jets of charged particles shoot out in one direction, while an unrecognized "something" takes off in the opposite direction. The reactions take place at energies just beyond the masses of the newly discovered W and Z particles.

Till now, this energy region has been dubbed a "desert" because, according to the so-called Grand Unification Theory of particle interactions, nothing is supposed to happen there. But there is something there after all; and whatever it is, it does not seem to be remotely like any known or predicted particle. (Waldrop, M. Mitchell; "Blooms in the Desert?" Science, 224:589, 1984.)

SQUARKS AND PHOTINOS AT CERN?

At the CERN lab, in Geneva, physicists shoot protons and antiprotons at each other so that they collide head-on. The colliding particles usually annihilate one another and in the process release a variety of subatomic debris and energy. Large arrays of detectors surrounding the collision site record the particles as they streak away. Usually the escaping particles can be easily identified; but in 1983 nine strange events were recorded, and more have occurred in 1984. Something both invisible and inexplicable carried off large amounts of energy during these "strange" events. Physicist Carlo Rubbia, of CERN and Harvard, said, "There is no sensible way to explain the missing energy by known

particles."
Some theorists believe that these anomalous events will be explained only by invoking what is termed "supersymmetry" theory. Supersymmetry predicts that twice as many particles as those known today must exist. Already physicists are rushing to name the new, though unverified, particles. The symmetric partner of the quark will be the squark; the photon will be paired with the photino; there will be a selectron for the electron; and so on. (Thomsen, D.E.; "Strange Happenings at CERN," Science News, 126:292, 1984.)

"AND SO ON INFINITUM"

Connoisseurs of facetious scientific poetry will recognize the above title as coming from a poem about vortices which have littler vortices preying upon them, etc. Well, it seems that matter may not have a basement of truly fundamental, indivisible particles either. If one does not count the rather primitive notion of Air, Fire, Earth, and Water, there are five basic levels of compositeness: (1) molecules; (2) atoms; (3) nuclei; (4) nucleons; and (5) quarks and leptons. But now physicists are beginning to see regularities in the lowest accepted layer, quarks and leptons, that betoken a sixth layer of compositeness or subdivisibility. In other words, quarks and leptons are not really fundamental and instead are composed of something else, which will undoubtedly eventually receive fanciful names.
In this article, O.W. Greenberg delves into this sixth stratum and the "regularities" it engenders. The article is really too technical for Science Frontiers, but we thought our readers might like to be warned that our concepts of matter are based on infinite quicksand. (Greenberg, O.W.; "A New Level of Structure," Physics Today, 38:22, September 1985.)

Comment. With ever-more-gigantic galactic superclusters being charted and the possibility of Big Bangs occurring "somewhere else," matter may also be infinitely aggregative, too.

DOUBLE NUCLEI AT DARMSTADT

At the GSI Darmstadt (West Germany) heavy ion facility, particle physicists have been firing heavy nuclei at each other. Some very suggestive evidence has turned up that giant nuclei are being formed with atomic weights and numbers approximately double those of the colliding nuclei. The words "giant nuclear molecule" and "giant di-nuclear system" are being bandied about. Two nuclei of uranium-238 might, for example, unite for 10^{-19} second in a super-nucleus. As the author of this article says, "What is going on?" (Silver, Joshua; "Giant Nucleus at Darmstadt?" Nature, 315:276, 1985.)

ANOMALOUS ANOMALONS

Anomalons are fragments of atomic nuclei that interact with other nuclei more readily than expected. They seem to represent a previously unknown and highly reactive state of nuclear matter. Not all physicists can find them experimentally; and far from all believe they exist. Until now, only large nuclear fragments have been found to be anomalons. But some Indian physicists working in the USSR have bombarded carbon-12 nuclei with carbon-12 nuclei and found anomalously active alpha particles in the debris from the collisions.
Not all of the alphas were anomalous, which makes the situation all the more mysterious. Just what makes a law-abiding alpha particle (a combination of two protons and two neutrons) into a highly reactive anomalon? (Anonymous; "More Anomalous Nuclear Fragments," Science News, 127:105, 1985.)

Comment. This is the first case of very small anomalons. There does not seem to be much one could do to something as simple as an alpha particle to make it more reactive. See also p. 322.

ESP OF ATOMS?

Preamble. Theosophy is an occult doctrine with three professed goals:

(1) To form a nucleus of the universal brotherhood of humanity, without distinction of race, creed, sex, caste, or color.
(2) To encourage the study of comparative religion, philosophy, and science.
(3) To investigate the unexplained laws of nature and the powers latent in man. (From: Encyclopedia Americana)

Just before the turn of the century, two leaders of the Theosophical movement, Annie Besant and C.W. Leadbeater, decided to collaborate on Goal 3 and investigate the micro-structure of matter. They eschewed the physics laboratory, preferring instead ESP. S. Phillips has now summarized their discoveries in a compact little paper. He concludes as follows:

"This article has presented a few examples of the many correlations between modern physics and psychic descriptions of sub-atomic particles published over seventy years ago. Scientists and laypersons alike may find it difficult to believe that Besant and Leadbeater could in some way unknown to science describe the structure of objects at least as small as atomic nuclei, which are about one ten-thousand-billionth of an inch in size. But they cannot in all sincerity dismiss the Theosophists' claims as fraudulent for the obvious reason that they finished their investigations many years before pertinent scientific knowledge and ideas about the structure of sub-atomic particles and the composition of atomic nuclei became available to make fraud possible in principle.
"Nor can critics plausibly reject their claim to possess micro-psi powers at its face value and interpret, alternatively, their observations as precognitive visions of future ideas and discoveries of physics. If they had been merely looking into the future, they might reasonably, have been expected to describe atoms or atomic nuclei or both, not more exotic objects formed from two nuclei. The Rutherford-Bohr model of the atom was formulated after they had finished most of their work on MPAs. Yet none of its features can be found in their publications (this, indeed, is the main problem that Occult Chemistry presented to scientists for seventy two years). How can one, therefore, account for the remarkable research of these two people except by admitting that they did, truly, observe the microscopic world by means of ESP?" (Phillips, Stephen; "ESP of Atoms?" Theosophical Research Journal, 3:93, December 1986. Cr. G. Oakley)

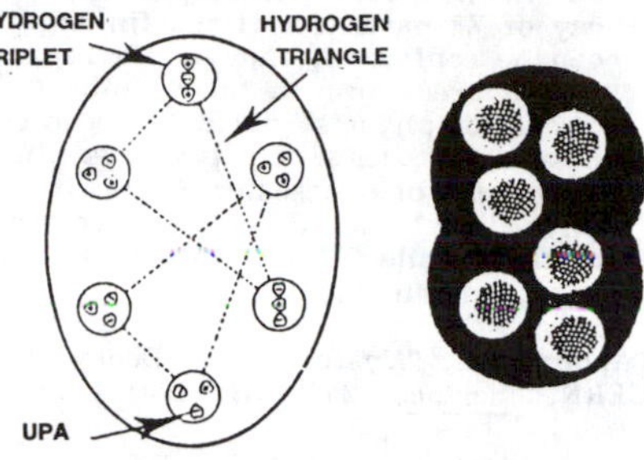

Note the similarities between the old Besant-Leadbeater, ESP-derived model of hydrogen (left) and the modern dibaryon "bag model" (right.)

Comment. Can any of the above be true? The old published works of Besant and Leadbeater are there for anyone to leaf through. Theosophists certainly see connections between their visions, acquired through ESP, and modern models of the microscopic world. Unfortunately, we will probably never get any modern physicist to even look at this occult material, much less venture an opinion. As in the case of "evolution" (see earlier discussion under Biology), one's philosophical predilections have a lot to do with what one sees and believes in a collection of data.

THROUGH A PEEPHOLE TANTALIZINGLY

> The flood of data that comes out of the type of physics experiment in which two subatomic particles collide at high energy is often so copious that physicists need some time to notice and interpret some of the strange new things that appear. This is especially true if the strange new things are of a sort that nobody was looking for.
> Thus, some anomalous events that occurred at the PETRA colliding beam apparatus of the German Electron Synchrotron Laboratory (DESY) in Hamburg back in 1984 are now being interpreted as what Harald Fritzsch of DESY calls "a peephole" into a possible new domain of physics...

What happened in 1984 was that one detector saw unexplainable particles---that is, unexplainable in the context of current theories. But since so other detectors in operation saw the event, the data were forgotten. But later, five more such events were seen on a different detector. (Anonymous; "Through a Peephole Tantalizingly," Science News, 132:219, 1987.)

Comment. Just when we were getting used to fractionally charged quarks and particles of different "colors," this has to happen!

PERHAPS Z° PARTICLES DECAY CHAOTICALLY

In an extremely controversial result, physicists at CERN's Large Electron Positron (LEP) collider have discovered an unexpected anomaly in the decay of Z° particles. If the findings can be confirmed, they would represent a breakdown in the "standard model" of physics, a widely accepted bestiary of elementary particles. In the words of one senior Stanford physicist: It would be the discovery of the decade." Most physicists, however, remain dubious.

(Anonymous; "Mysterious Z° Decay at CERN," Science, 252:1241, 1991.)

NEW INSIGHTS AS TO THE STRUCTURE OF MATTER

Inside the atom. Physicists have long visualized the atomic nucleus as being a shell-like arrangement of its constituent protons and neutrons. Tantalizing experiments suggest otherwise. Magnesium-24, for example, may under some circumstances exist as two carbon-12 nuclei in tight orbit, as in the illustration. Even more startling is the "sausage" form of magnesium-24, in which six helium-4 nuclei (alpha particles) are lined up in a row. This "hyperdeformed" state has not yet been detected in the lab, but it demonstrates new thinking among the physicists. (Kenward, Michael; "Are Atoms Composed of Molecules?" New Scientist, p. 21, April 6, 1991.)

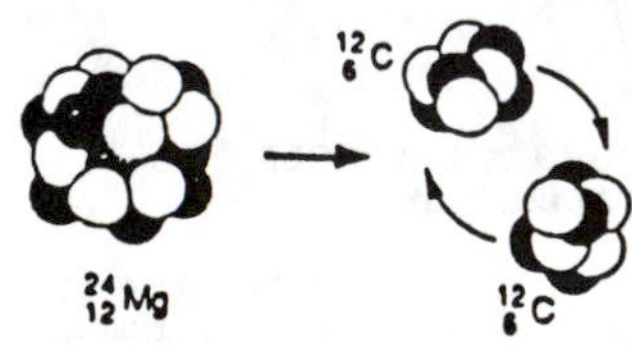

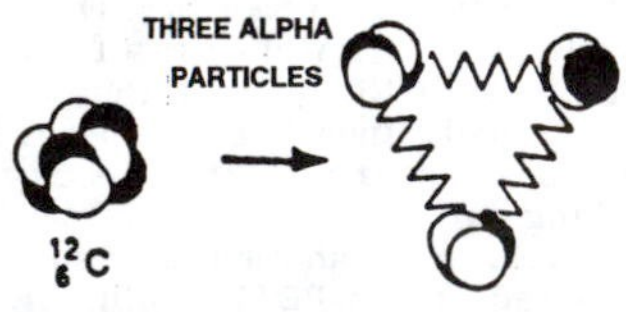

Possible "nuclear-molecular" forms of magnesium-24 and carbon-12.

Comment. Evidently we do not know everything about nuclear physics, as implied by J.R. Huizenga as the end of the cold-fusion item on p. 317.

Beyond the molecule. We are used to seeing atoms and molecules arranging themselves into mathematically regular crystals. Now it appears that particles consisting of thousands of atoms also spontaneously organize themselves.

A.S. Edelstein et al find that molybdenum particles assemble themselves in cubes with two prominent edge lengths: 4.8 and 17.5 nanometers. The larger cubes show up in micrographs as 3x3x3 groupings of the smaller cubes. The smaller cubes each contain about 7000 atoms. (Edelstein, A.S., et al; "Self-Arrangement of Molybdenum Particles into Cubes," Science, 251:1590, 1991.)

Comment. What are the "organizing forces" here? Why cubes? Why the heirarchy of cubes? Why 3x3x3 supercubes?

THE "RESIDUE FALLACY" DOES NOT APPLY TO ALL RESIDUES!

While preparing our latest Catalog volume in the field of geology, we have been struck by how slavishly some mainstream scientists worship the Residue Fallacy. Briefly, this "fallacy" states that a single type of discrepant observation should not be considered viable if it contradicts a large body of well-established, internally consistent observations. In geology, the Residue Fallacy is employed to dismiss the precursorless polonium halos found by R.V. Gentry, as well as some other radiometric discordances. These scientists seem to have forgotten about the anomalous advance of Mercury's perihelion and a few other obvious residues that ultimately stirred up revolutions in our thinking. Anyway, it is now satisfying to find the Editor of Nature, mainstream science's preeminent journal, acknowledging the value of anomalies.

The stimulation in this case is the more-than-decade-old inability of astronomers and physicists to explain the missing solar neutrinos. Two new, more sophisticated neutrino detectors have come on line, in Japan and the U.S.; and they have confirmed the results obtained in a huge vat of cleaning fluid in the Homestead Mine, in South Dakota. For some reason, everyone measures only about one-third the number of solar neutrinos expected. Either something is wrong with our model of the sun's (and other star's) energy-producing mechanism or our knowledge of nuclear physics is faulty. Recently, the solar neutrino anomaly has been complicated by the fact that the Homestead Mine detector seems to "see" more nutrinos during violent solar flares, although the two newer detectors find no connection.

J. Maddox, Nature's Editor, closes his discussion of these problems with this sentence: "However this tale comes out, it will remain a marvel that so much work, experimental as well as theoretical, has been stimulated by a single discrepant observation." (Maddox, John; "More Sideshows for Solar Neutrinos," Nature, 336:615, 1988.)

Cross reference. The problem of the missing solar neutrinos is detailed on p. 84.

QUANTUM MECHANICS

THE MOST PROFOUND DISCOVERY OF SCIENCE

This is what one scientist calls Bell's Theorem. Certainly not all scientists would agree with such an absolute declaration. Since Bell's Theorem lurks in the fog-shrouded country of quantum mechanics, most biologists probably haven't even heard of it. In any event, they would probably think the discovery of the genetic code more profound.

Why all the fuss over Bell's Theorem? In the laboratory, Bell's Theorem is associated with an admittedly spooky effect: the measurements made on one particle affect the measurements made on a second, far-removed particle. In theory, the second particle could be on the other side of the galaxy, with absolutely no physical connection between the two---unless you admit to spooky action-at-a-distance forces. (Some overly zealous think-tankers have even contemplated applying this effect to long distance, untappable, unjammable communications with submarines!)

The article (referenced below) in which this apparent magic is discussed also dwells on another profundity associated with quantum mechanics: does that which is not observed exist? Einstein felt intuitively that it did; and one of his remarks on the subject led to this article's title. Unfortunately for Einstein, all recent laboratory experiments demonstrate that spooky action-at-a-distance forces do exist and that Einstein's intuition was incorrect. (Mermin, N. David; "Is the Moon There When Nobody Looks? Reality and the Quantum Theory," Physics Today, 38: 38, April 1985.)

Comment. The laboratory experiments discussed in the article prove only that quantum mechanics is correct, not that it is spooky. After all, radioactivity was pretty mysterious not too many years ago. It still is, but we are accustomed to it now.

A WATCHED ATOM IS AN INHIBITED ATOM

Strange as it may sound, the act of observing atoms to determine their energy states interferes with their quantum jumps between atomic energy levels. This is another "spooky" prediction of quantum mechanics theory.

This prediction was recently verified by W.M. Itano et al, at the National Institute of Standards and Technology, in Boulder. They employed radio waves to drive beryllium ions from one energy level to another. While the beryllium ions were jumping from one level to another, the researchers sent in short pulses of light to determine the ion's state. The more frequently they interrogated the ions, the less apt they were to jump to new energy states, despite the stimulating radio waves. (Peterson, I.; "Keeping a Quantum Kettle from Boiling," Science News, 136:292, 1989. Also: Pool, Robert; "Quantum Pot Watching," Science, 246:888, 1989.)

Comment. It is logical, but perhaps not practical, to contemplate delaying or stopping radioactive decay by interrogating poised radioactive atoms.

OPTICS

GOETHE'S OPTICS REEVALUATED

It is beyond dispute that the main objective of the polemical part of Goethe's Farbenlehre, namely, the refutation of Newton's Opticks, was a misguided one. Many consider it to be inexplicable that a man of Goethe's intellectual standing should have behaved in such an apparently irrational manner. It so happens, however, that the characteristics of the subjective spectrum are more akin to Goethe's model than to Newton's. It is true that Goethe put an incorrect interpretation upon what he saw---and was the first to see---but a careful scrutiny of his scientific method reveals that his reasoning was far from irrational.

(Duck, Michael; "The Bezold-Bruecke Phenomenon and Goethe's Rejection of Newton's Opticks," American Journal of Physics, 55:793, 1987.)

Comment. Goethe just did not see what Newton saw, and their feud was rather bitter. To illustrate, Goethe considered the subjective aspects of his optical experiments, while Newton neglected them. For example, in the Bezold-Bruecke phenomenon, reds became yellower with increasing brightness---or seem to with human observers. Goethe's theory of color took such effects into account. Once again, one person's reality can be different from another's.

HAS THE SPEED OF LIGHT DECAYED?

In a recent technical report, The Atomic Constants, Light, and Time, T. Norman and B. Setterfield answer this question affirmatively. Scientific creationists have in the main welcomed this report, because its findings are consistent with their desire to prove the earth very young. However, G.E. Aardsma, at the Institute for Creation Research, in California, urges caution:

> Measurements of the speed of light have been made for the past three hundred years which could potentially provide the required empirical basis. Norman and Setterfield tabulate the results of 163 speed of light determinations in The Atomic Constants, Light, and Time, and claim clear support for the decay-of-c hypothesis from this data set. [c= velocity of light] My inability to verify this claim when this data set was subjected to appropriate, objective analyses is the motivation for this article which is intended to caution creationists against a wholesale, uncritical acceptance of the Norman and Setterfield hypothesis. At the present time, it appears that general decay of the speed of light hypothesis is not warranted by the data upon which the hypothesis rests.

(Aardsma, Gerald E.; "Has the Speed of Light Decayed?" ICR Impact Series no. 179, May 1988.

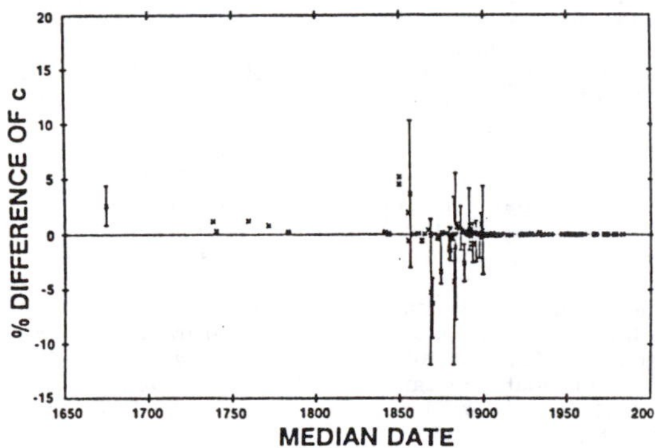

Measurements of the velocity of light, c, versus year, as plotted by Aardsma. Vertical lines are error bars.

GRAVITATION

FALLING MASSES SWERVE SOUTH

In 1901, Florian Cajori had a paper published in Science with the title: "The Unexplained Southerly Deviation of Falling Bodies." Cajori reviewed the pertinent measurements that had been made prior to 1901 on falling bodies, emphasizing that the anomaly described in the title of his paper truly existed.

In a recent letter to the American Journal of Physics, A.P. French brings the record up to date. (It should be pointed out here that a slight easterly deflection of falling bodies is predicted, but that a southerly deflection should be negligible, although not zero.) In the post-1901 experiments, small southerly and northerly deflections have been detected. These should not occur for an ideal rotating sphere---which the earth isn't. French ends his brief review by stating that the earth's gravitational field is now known well enough so that further experiments with falling objects might once-and-for-all determine the nature (and reality) of the delightful anomaly. (French, A.P., "The Deflection of Falling Objects," American Journal of Physics, 52:199, 1984.)

THE GRAVITATIONAL "CONSTANT" IS NOT!

For too many years, physicists have been content with laboratory determinations of G (the gravitational constant) using the old Cavendish Balance. In this paper, Stacey and Tuck offer a disturbing collection of values of G determined from geophysical measurements; i.e., measurements in mines, boreholes, and under the sea. These measurements are unanimous in producing G's that are larger than the usually accepted value by about 1%. Furthermore, the deeper the experiment, the greater the departure from the standard value. (Stacey, F.D., and Tuck, G.J.; "Geophysical Evidence for Non-Newtonian Gravity," Nature, 292:230, 1981.)

Comment. These geophysical measurements must be added to recent laboratory experiments indicating that gravity may not be best described by an inverse square law. See Mysterious Universe.

GRAVITY DOWN, MASS UP

The variation of the gravitational constant, G, with time would not be considered seriously were it not for the surprising coincidence of two enormous dimensionless numbers: (1) the ratio of the electrical to gravitational force between the electron and the proton in a hydrogen atom; and (2) the ratio of the age of the universe and the atomic unit of time. If these two ratios are truly equal, then G must decrease with time.

Beyond the unstable feeling one gets, there is nothing in physics or cosmology to discourage a belief in time-varying gravity. Indeed, some astronomical data weakly support the idea. It is geophysics, though, where one finds strong evidence. Measurements of the decreasing length of the day and the expansion of the earth give about the same value for a decreasing G---after other contributing factors have been eliminated.

An interesting consequence of all this is that astrophysical theory seems to require that a decreasing G be balanced by increasing mass. Experiments are now underway to detect the continual creation of mass in terrestrial objects. (Wesson, Paul S.; "Does Gravity Change with Time?" Physics Today, 33: 32, July 1980.)

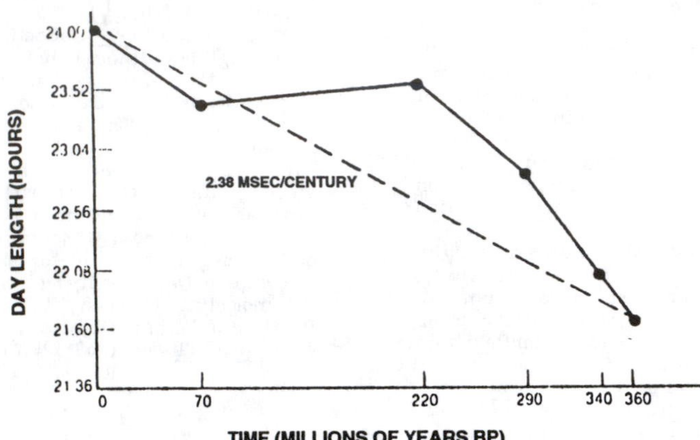

The length of the terrestrial day has apparently increased down the eons.

HIGH G VALUES IN MINES

Measurements of G, the constant in Newton's Law of Gravitation, made in mines are always significantly higher than those made in surface laboratories. No one is quite sure why. It has been suggested that a lack of knowledge of the densities of surrounding rocks might account for this discrepancy. But Holding and Tuck report new measurements in Australian mines, where the densities are very well known. The G values are still high. Mine measurements of G differ in the fact that the masses employed can be much farther apart. (Holding, S.C., and Tuck, G.J.; "A New Mine Determination of the Newtonian Gravitational Constant," Nature, 307:714, 1984.)

ON THE TRAIL OF THE FIFTH FORCE

Well publicized lately has been the modern reanalysis of the old Eotvos balance experiments. Some think they show the presence of that famous "fifth force" which is supposed, according to some theories, to modify Newton's Law of Gravitation, and become measurable at distances of about 100 to 1000 meters. The Eotvos results may also be explicable in terms of laboratory air currents. But the fifth force may be showing up in geophysical experiments.

One of the most comprehensive geophysical experiments so far has been conducted by Frank Stacey at University of Queensland in Brisbane, Australia, and his colleagues. Working in two metal mines, the researchers have measured a gravitational constant that is 0.7 percent greater than that measured in the laboratory---suggesting the presence of a fifth force.

(Weisburd, Stefi; "Geophysics on the Fifth Force's Trail," Science News, 131: 6, 1987.)

MORE CHALLENGES TO NEWTON'S LAW OF GRAVITATION

Two experiments reported at the 1988 meeting of the American Geophysical Union in San Francisco can be added to the others that question Newton's venerable Law of Gravitation. The abstracts of these papers are short and to-the-point, so we quote them:

> We have performed an experimental test of Newton's inverse-square law of gravitation. The test compared accurately measured gravity values along the 600 m WTVD tower near Raleigh, North Carolina, with upward, continued gravity estimates calculated from ground measurements. We found a significant departure from the inverse-square law, asymptotically approaching -547 ± 36 μGal at the top of the tower. If this departure is derived from a scalar Yukawa potential, the coupling parameter is α = 0.023, the range is λ = 280 m, and the Newtonian Gravitational Constant is G_{∞} = (6.52 ± 0.01) x 10^{11} m^3 kg^{-1} s^{-2}. We do not yet have adequate resolution to discriminate this scalar model from a scalar-vector model.

(Eckhardt, D.H., et al; "Experimental Evidence for a Violation of Newton's Inverse-Square Law of Gravitation," Eos, 69:1046, 1988.)

> In the late summer of 1987, an experiment was performed to determine the value of the Newtonian gravitational constant, G, by measuring the variation of the earth's gravity, g, with depth in the Greenland ice-sheet. The site for the experiment---the radar station at Dye-3, Greenland---was selected because of the existing 2000-m-deep ice borehole there. Previous analysis of ice-cores from the borehole indicate that the ice density can be accurately modeled. Gravity measurements were made to a depth of 1673 meters in the ice, the sub-ice topography was mapped with high-precision radar echo sounding over a 10-km-diameter region, and a series of 24 locations in a 32-km-diameter network around the hole were surveyed with gravity, leveling, and GPS positioning.
>
> When corrected for the sub-ice topography, a gravity anomaly that accumulated to nearly 4 mGal in 1.4 km was observed. We find measured anomalies can be taken as evidence for non-Newtonian gravity, but can also be accounted for in terms of Newtonian physics if a suitable distribution of high densiity masses exist beneath the borehole.

(Zumberge, Mark A., et al; "Results from the 1987 Greenland G Experiment," Eos, 69:1046, 1988.)

The paper on the Greenland experiment led to a short article in Science in which differing opinions among the research team members about the experiment's significance were aired. Some opted for an unusual density distribution of the rock beneath the experiment to explain the results; others thought that the required density distribution was too unlikely and contrived and consequently favored a modification of Newton's inverse-square law. (Poole, Robert; "'Fifth Force' Update: More Tests Needed," Science, 242:1499, 1988.)

Comment. Should we permit this tiny residue of anomalous observations to cast doubt upon a law verified in countless experiments and astronomical observations? Those who believe in the "residue fallacy" will say, "No! Discard the wild points!"

SCIENCE WAITS FOR---ALMOST BEGS FOR---REFUTATION

Two Japanese scientists, H. Hayasaka and S. Takeuchi, have spun up some gyroscopes, weighed them and---Horrors!---found that they weighed less when spinning in one direction than the other. They admit the heresy of their results: "The experimental result cannot be explained by the usual theories."

The gyroscopes employed are small, weighing about 175 grams when not spinning. When spun clockwise, as viewed from above, no weight changes were observed. But rotating at 13,000 rpm counterclockwise, the 175-gram gyroscope lost about 10 milligrams. The balance's sensitivity was 0.3 milligram. This is a very large effect; and the weight loss increased linearly with increased speed of rotation. Obviously, the physicists are most perplexed by this "antigravity" effect.

Perplexity has been accompanied by outright disbelief. R.L. Park, a physics professor at Maryland, remarked: "It would be revolutionary if true. But it is almost certainly wrong. Almost all extraordinary claims are wrong." R.L. Forward, an Air Force consultant, concurs: "It's a careful experiment. But I doubt it's real, primarily because I've seen so many of these things fall apart." (Anonymous; "Anti-Gravity Effect Claim by Japanese," San Francisco Chronicle, December 28, 1989. Cr. J. Covey. Anonymous; "A Gyroscope's Gravity-Defying Feat," Science News, 137:15, 1990.)

Comment. The amazing thing---the anomaly---is that such "misguided" research got funded at all and the results published. But then, maybe Japanese research proposals do not have to get by 7 (that's seven) reviewers, as required by the U.S. National Science Foundation!

GRAVITY-DEFYING GYROS COME DOWN TO EARTH

It didn't take long for physicsts to rush into their labs to repeat the Japanese gyroscope experiments. The thought that a spinning mass might lose weight was just too horrible to contemplate. Two replications of the Japanese experiment have been reported so far.

James E. Faller and his colleagues at the Joint Institute for Laboratory Astrophysics in Boulder, Colo., repeated the Japanese experiment by looking for signs of weight loss in a spinning gyroscope consisting of a brass top about 2 inches in diameter sealed in a small plastic chamber. "We conclude that within our experimental sensitivity, which is approximately 35 times larger than needed to see the effect reported...there is no weight change of the type...described." (Anonymous; "An Absence of Antigravity," Science News, 137: 127, 1990. Cr. F. Hanisch)

Now T.J. Quinn and A. Picard of the International Bureau of Weights and Measures in Sevres Cedex, France, have repeated the experiment. They find changes in the apparent mass of their gyroscope that depend on the speed and sense of rotation, but they amount to only about 5 per cent of the effect reported by Hayasaka and Takeuchi. (Anonymous; "Experiments Weaken Japanese Gyro Claim," New Scientist, p. 32, March 3, 1990.)

The French scientists think that the Japanese results can be explained as functions of friction and temperature on the gyro. On the other hand, S.H. Salter makes a case for gyro vibrations compounded by nonlinearity in the weighing mechanisms being the culprits. (Salter, S.H.; "Good Vibrations for Physics," Nature, 343:509, 1990.)

RELATIVITY

HERBERT IVES AND THE ETHER

Herbert Ives was a top scientist at Bell Laboratories who performed some largely forgotten experiments on relativity and space-time a few decades ago. His experimental prowess and reputation were so good that his work on relativity was published in great detail in the Journal of the Optical Society of America. Ives would have had a more difficult time getting his results published today, for he showed quite clearly that Einstein's Special Theory of Relativity did not correspond to lab results. At the time, such results were not so shocking. Indeed, some philosophers had shown that Special Relativity led to undesirable paradoxes, and experiments by Sagnac and Michelson/Gale had cast additional doubt on this aspect of Relativity.

Such experiments by Ives and other key scientists suggested that an ether actually did exist and that it could serve as an absolute reference frame. Another implication was that time was an independent entity unaffected by motion and that the infamous Twin Paradox was a fiction.

Ives himself believed his work proved that so-called relativistic effects could be easily explained by phenomena appealing more to the common sense, such as the change of a light source's frequency with motion (over and above the Doppler Effect), rather than revamping space-time concepts. In short, Ives thought he had proved Special Relativity untenable experimentally and an unnecessary distortion of science's worldview. (Barnes, Thomas G., and Ramirez, Francisco S.; "Velocity Effects on Atomic Clocks and the Time Question," Creation Research Society Quarterly, 18:198, 1982.)

Comment. Why do the textbooks neglect to mention the Ives experiments and why should a review of Ives' work appear in a creationist publication? The answers are easy: Special Relativity now has the status of scientific dogma, which one questions at his own peril. The creationists, on the other hand, vehemently reject relativitism in favor of absolute standards in space-time as well as other features of human existence. It would be amusing if the real world conformed to neither model, both of which are defended so passionately.

THE THORNY WAY OF TRUTH: PART II

This item was discovered by accident among the paid ads in Nature---a place we do not routinely examine for scientific anomalies. Stefan Marinov or someone acting for him evidently inserted the ad. We will meet Marinov again below, where the Editor of Nature admits that some of Marinov's ideas might have some scientific support. We missed Part I of the Marinov advertising campaign, in which he presented his fight with the scientific establishment to restore absolute space-time concepts. In other words, Marinov wants to dump Relativity, and believes he has experimentally disproved it.

Part II, the present ad, is entitled "The Perpetuum Mobile Is Discovered." It is replete with equations, diagrams, and references, like a scientific paper, but still a paid ad! Apparently, no scientific or engineering journal will publish Marinov's work. (Marinov, Stefan; "The Thorny Way of Truth: Part II," Nature, 317:unpaged, September 26, 1985.)

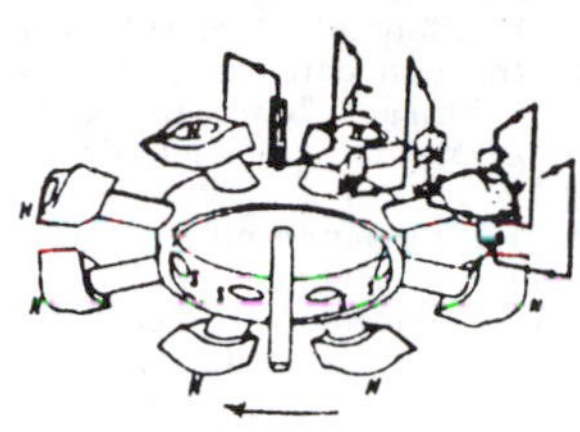

Marinov's perpetuum mobile, as illustrated in his Nature advertisement.

A POSSIBLE CRACK IN THE WALL OF THE TEMPLE OF RELATIVITY

Stefan Marinov is a remarkable iconoclast who is convinced that Einstein's special theory of relativity is mistaken.

Marinov apparently has been expelled from Russia because of his scientific and political opinions. So infuriated is he by the reluctance of mainstream scientific journals, such as Nature, to print his anti-relativity papers that he has threatened to immolate himself outside the British embassy in Vienna. Happily, he he didn't strike the match, because it may be that he has something.

Marinov claims that he has demonstrated experimentally that the velocity of light is not the same in all directions in all reference frames, as Einstein insisted. He says he can even detect the motion of the earth through absolute space and time, contrary to most Michelson-Morley-type experiments. Based upon some recent theoretical analysis, the journal Nature has bent a bit and now calls for repetitions of Marinov's experiments. (Maddox, John; "Stefan Marinov Wins Some Friends," Nature, 316:209, 1985.)

Comment. Recently, three books highly critical of relativity have been published: (1) Turner, Dean, and Hazelett, Richard, eds.; The Einstein Myth and the Ives Papers; (2) Santilla, Ruggero Maria; Il Grande Grido: Ethical Probe on Einstein's Followers in the U.S.A.; (3) Dingle, Herbert; Science at the Crossroads.

WIN $2000: CHALLENGE EINSTEIN

H. Hayden and P. Beckmann are offering $2000 to anyone who can cite, not necessarily perform, an experiment proving that light travels westward at the same velocity that it travels eastward on the earth's surface (to an accuracy of 50 meters/second). If the speeds are indeed the same, then Einstein's assumption that the speed of light is the same in all directions regardless of the motion of the observer will be proven. Then skeptical scientists like Hayden and Beckmann, will rest easier.

But suppose the east and west velocities of light are different? Then Special Relativity would collapse. Hayden and Beckmann do not dread this at all. In fact, they (and others) point out that some of the vaunted experimental "proofs" of Special Relativity can be explained in other ways. For example: (1) The bending of starlight passing close to the sun can easily be accounted for using Fermat's Law; and (2) The advance of Mercury's perihelion was explained by P. Gerber, 17 years before Einstein's 1915 paper on the subject, using classical physics and the now-accepted assumption that gravity propagates at the speed of light.

As for the famous Michelson-Morley experiment, Michelson (an unbeliever in Relativity) believed that he and Morley failed to detect ether drift because the ether was entrained with the earth as it orbited the sun. It is rarely mentioned that Michelson and H.G. Gale repeated the experiment in 1925 to see if ether drift could be detected as the earth rotated on its axis. They did! And Einstein was sorely tried explaining the result. A 1979 repeat of the experiment at the University of Colorado, using lasers found "unexpected perturbations," which were blamed on "other causes." (After all, Relativity and Einstein are sacrosanct!)

The gist of all this is that Hayden and Beckmann suspect that Special Relativity is founded upon quicksand. The reader should not be surprised to learn that Beckmann himself has a theory to supplant Special Relativity once it is discredited. (Bethell, Tom; "A Challenge to Einstein," National Review, p. 69, November 5, 1990. Cr. P. Gunkel.)

CHAOS

UNPREDICTABLE THINGS

Simple spherical pendulums are fixtures of physics labs, to say nothing of grandfather clocks. It is now widely recognized that pendulums can behave chaotically; that is, unpredictably. As the pendulum bob swings farther away from its rest position, the restoring force becomes nonlinear; i.e., not proportional to the displacement. At some combination of displacement and driving frequency, a region of chaos may develop, in which theory is powerless to tell what is going to happen next.

> It is not just the behavior of pendulums that has sprung this surprise. Systems as diverse as simple electrical circuits, dynamos, lasers, chemical reactions and heart cells behave in an analogous way and the implications extend far beyond these examples---to matters such as weather forecasting, populations of biological species, physiological and psychiatric medicine, economic forecasting and perhaps the evolution of society.

(Tritton, David; "Chaos in the Swing of a Pendulum," New Scientist, p. 37, July 24, 1986.)

Comment. Some of the anomalies we record may be the consequence of simple systems gone wild. Chaotic motions of some asteroids and at least one solar system moon are already suspected. Imagine what might happen in much more complex systems, such as biological evolution (hopeful monsters?), brain development (idiot savants?), etc.

DRIP, DROP, DRUP, DR**

A dripping faucet is usually conceived as a well-ordered dependable phenomenon. You simple turn the faucet a bit counterclockwise and the drip rate increases. It's so simple.

Surprise! Dripping faucets are chaotic systems, as described in the following Abstract:

> The dripping water faucet is a simple system which is shown in this article to be rich in examples of chaotic behavior. Data were taken for a wide range of drip rates for two different faucet nozzles and plotted as discrete time maps. Different routes to chaos, bifurcation and intermittency, are demonstrated for the different nozzles. Examples of period-1, -2, -3, and -4 attractors, as well as strange attractors, are presented and correlated to the formation of drops leaving the faucet.

(Dreyer, K., and Hickey, F.R.; "The Route to Chaos in a Dripping Water Faucet," American Journal of Physics, 59:619, 1991.)

Comment. O.K., so faucets dribble a bit.

THE RECIPROCAL SYSTEM

Dewey B. Larson did not live to see mainstream science decisively evaluate his Reciprocal System. Even if he had lived to 500, he probably would not have seen physicists follow his lead and dismantle the present framework of theory based upon the familiar dimensions of mass, time, distance, etc. The Reciprocal System, you see, requires a complete change in thinking. Nevertheless, this revolutionary scheme possesses many appealing features. Even better for anomalistics, Larson's books brim with unexplained phenomena in physics and astronomy, which, of course, is why he felt impelled to revise physical theory. Here is an introduction to Larson's approach in his own words.

RECIPROCAL SYSTEM AVOIDS TAINT OF REDUCTIONISM

"Some of the readers of my latest book, The Neglected Facts of Science, are apparently interpreting the conclusions of this work as indicating that the Reciprocal System of theory leads to a strict mechanistic view of the universe, in which there is no room for religious or other non-material elements. This is not correct. On the contrary, the clarification of the nature of space and time in this theoretical development removes the obstacles that have hitherto prevented science from conceding the existence of anything outside the boundaries of the physical realm.

"In conventional science, space and time constitute a framework, or setting, within which the entire universe is contained. On the basis of this viewpoint, everything that exists, in a real sense, exists in space and in time. Scientists believe that the whole of this real universe is now within their field of observation, and they see no indication of anything non-physical. It follows that anyone who accepts the findings of conventional science at their face value cannot accept the claims of religion, or any other non-material system of thought. This is the origin of the long-standing antagonism between science and religion, a conflict which most scientists find it necessary to evade by keeping their religious beliefs separate from their scientific beliefs.

"In the Reciprocal System, on the other hand, space and time are contents of the universe, rather than a container in which the universe exists. On this basis, the 'universe' of space and time, the physical universe, to which conventional science is restricted, is only one portion of existence as a whole, the real 'universe' (a word which means the total of all that exists). This leaves the door wide open for the existence of entities and phenomena outside (that is, independent of) the physical universe, as contended by the various religions and many systems of philosophy." (Larson, Dewey B.; "A Note on Metaphysics," Reciprocity, 12:11, Summer 1983.)

Comment. To fully appreciate Larson's distinctions between his Reciprocal System and the conventional views, you must read his Nothing But Motion.

MORPHIC RESONANCE

THE HYPOTHESIS OF FORMATIVE CAUSATION LIVES!

Rupert Sheldrake's hypothesis has been roundly condemned by many scientists, presumably because it departs so radically from current thinking. Basically, the hypothesis maintains that the forms of things (from crystals to life forms) and the behavior of organisms is influenced by "morphic resonance emanating from past events." Convergent evolution, wherein human eyes closely resemble squid eyes, might well be explained by the hypothesis.

Sheldrake has been testing his idea in various ways. One experiment involves the accompanying illustration

Find the hidden image in this drawing. Solution on p. 330.

containing a hidden image. Once the solution of this illustration is learned, it is hard to forget, but few people see the answer right off. The hypothesis of formative causation insists that once one or more persons learn the drawing's secret, the easier it will be for others to see the solution. Actual tests consisted of broadcasting the illustration and its solution (that is, the hidden image) on English television combined with before-and-after checks elsewhere in the world outside TV range. The results strongly supported the hypothesis, for it was far easier for people outside England to identify the hidden images after the broadcast. (Sheldrake, Rupert; "Formative Causation: The Hypothesis Supported," New Scientist, 100:279, 1983.)

Comment. With all this prior publicity, the hidden image should pop immediately into the reader's mind. But if it doesn't, look under "The Big Picture," on the next page---an appropriate spot since telepathy might be at work instead of "formative causation."

MORPHIC RESONANCE IN SILICON CHIPS

Silicon chips can be made to assume various electronic configurations. In this, they are morphic, like the cattle guards on p. 133; and Sheldrake's hypothesis of morphic resonance should also apply. Two French researchers, F.J. Varela and J.C. Letelier, have applied a microcomputer in testing Sheldrake's theory. Briefly, they cleared the computer's memory and had it "grow a crystal" in the form of a unique, or at least very rare electronic state in the computer memory. Once this has been done, the time taken for the same pattern to be grown in subsequent attempts should become less and less.

One of Sheldrake's major claims is that once a new crystal is synthesized it thereafter becomes easier and easier to resynthesize it---due to the presence of morphogenic fields. But Varela and Letelier found that, even after 100 million crystallizations, no acceleration of the growing process was detectable. The authors conclude that either Sheldrake's hypothesis is falsified or that it does not apply to silicon chips. (Varela, Francisco, and Letelier, Juan C.; "Morphic Resonance in Silicon Chips," Skeptical Inquirer, 12:298, 1988.)

Comment. At least one other interpretation is possible: the particular "crystal" grown in the computer had actually been synthesized many times before by other computers within the range of morphogenic fields.

THE BIG PICTURE

DUE TO A FORTUNATE COINCIDENCE YOU CAN READ ABOUT A FORTUNATE COINCIDENCE

There are embedded in the fabric of our universe a number of curious coincidences among the so-called physical constants. Two amusing examples are: (1) the size of a planet is roughly the geometric mean of the size of the universe and the size of the atom; and (2) the mass of man is the geometric mean of the mass of a planet and the mass of the proton. Less hilarious is the observation that the age of the universe is of the order of the quotient of the electron time scale and the gravitational fine structure constant; and that only at the present time are physical conditions in the universe favorable to the existence of life-as-we-know-it!

The surprising number of coincidences that have been identified suggests that we exist and are aware of the universe around us only when certain coincidences prevail among physical constants. Is "now" a magic moment in the history of the universe during which we have "happened" as a natural coincidence of blindly drifting physical constants, or did some metaphysical force tune the universe specially for us? This long, rather mathematical article is redolent with metaphysics and mystery. (Carr, B.J., and Rees, M.J.; "The Anthropic Principle and the Structure of the Physical World," Nature, 278:605, 1979.)

Comment. One might speculate that at other times different coincidences prevail that would permit life-as-we-do-not-know-it or something equally unimaginable!

REPENT! THE PHASE CHANGE IS COMING!

The world as we know it may not end in a nuclear holocaust or even in the greenhouse effect. Rather, suggest M. M. Grone and M. Sher, the universe-as-a-whole may undergo a phase change. Such an event has already happened once and it may again. Approximately 10^{-10} seconds after the Big Bang, the force laws changed discontinuously as the universe cooled. Some models of the cosmos predict that another such phase change may occur when photons suddenly acquire mass. Grone and Sher have sketched the effects on terrestrial civilization:

> The most dramatic effect would be the elimination of all static electric and magnetic fields over a range greater than 1 cm, and the elimination of all electromagnetic radiation with frequencies smaller than a few hundred gigahertz. We have shown that there would be relatively little impact on atomic structure and on solar radiation. The absence of electrostatic fields would force a redesign of current power plants (to use smaller solenoids); the absence of radio and television waves would force a much greater use of cables. The elimination of solar and geomagnetic fields would have a significant meteorological impact. The potentially most devastating effect could be on the propagation of neural impulse along motor neurons; it appears that the effects might be small, but they do depend on the precise value of the photon mass.

Crone and Sher conclude that the effect would be devastating to humanity but probably not fatal. (Crone, Mary M., and Sher, Marc; "The Environmental Impact of Vacuum Decay," American Journal of Physics, 59:25, 1991.)

Comment. No TV; now that is fatal!

ISLANDS OF HOPE FOR LIFE ETERNAL

Frautschi examines expanding "causal" regions in the universe, where entropy (disorder) does not increase as fast as the maximum predicted by Thermodynamics. The conclusion of this highly theoretical paper is that, even in these "islands of hope," life and the order it requires cannot survive indefinitely if it is restricted to solid substances. But, "it stands as a challenge for the future to find dematerialized modes of organization (based on dust clouds or an e^-e^+ plasma?) capable of self-replication. If radiant energy production continues without limit, there remains hope that life capable of using it forever can be created." (Frautschi, Steven; "Entropy in an Expanding Universe," Science, 217:593, 1982.)

Comment. Who said Science was a conservative journal? This smacks of sci-fi tales of electrolife and Hoyle's Black Cloud.

Hidden face in the drawing on p. 329.

WHAT DOES IT ALL MEAN?

W.G. Pollard, a distinguished physicist, has written a very philosophical, almost mystical article on the nature of the cosmos. Let us begin with his abstract:

> There are several hints in physics of a domain of external reality transcendent to three-dimensional space and time. This paper calls attention to several of these intimations of a real world beyond the natural order. Examples are the complex state functions in configuration space of quantum mechanics, the singularity at the birth of the universe, the anthropic principle, the role of chance in evolution, and the unaccountable fruitfulness of mathematics for physics. None of these examples touch on the existence or activity of God, but they do suggest that external reality may be much richer than the natural world which it is the task of physics to describe.

Pollard then elaborates:

Example 1. Quantum mechanics, a mathematical formulation of reality, has been extraordinarily successful in describing and predicting many things in the microscopic world. Pollard notes that quantum mechanics contains no hint of God per se and possesses no numinous quality, but its great complexity and multidimensionality provide evidence for "the reality of the transcendent order in which the natural universe is embedded."

Example 2. The singularity at the beginning of the universe. Science is at a loss to explain creation and what happened before. (Pollard assumes that creation occurred like most scientists.)

Example 3. The anthropic principle can be interpreted as a restatement of the religious contention that the universe was made-for-humankind. In Pollard's words: "If one imagines an ensemble of universes of different size and duration and equipped with different values of the fundamental constants G, h, c, e and others, this principle selects only that member of the ensemble for which life and its evolution to man is a possibility. But merely stating the problem in this way suggests a creator with a mysterious plan or purpose of his own. And certainly by any standard for judging a creative artist, to carry life from bacteria to man in three billion years is a startlingly immense creative achievement. Even the creation of a planet like the Earth is also a remarkable creative achievement. All such considerations are clearly beyond the competence of science to either affirm or deny."

Example 4. The role of chance in evolution is assumed by Pollard, as it is by most scientists. The atoms and molecules had to have just the right properties as well as enough time and room to fall together into humankind. "Could it be that whoever or whatever started this universe, some 18 billion years ago in the big bang, designed it to last that long, and therefore to be as big as it is, in order to have an opportunity to create man?"

Example 5. The fruitfulness of mathematics. "Since the seventeenth century, we have had at least four major and numerous minor examples of mathematical systems which were produced initially as pure products of the human mind simply for our delight in their inner beauty, but which later turned out to mirror the workings of the natural world accurately and precisely in every detail in ways completely unforeseen and unexpected by their originators." In other words, God is a geometer.

(Pollard, William G.; "Rumors of Transcendence in Physics," American Journal of Physics, 52:877, 1984.)

Comment. Pollard's article is laced with reductionism; that is, he feels that everything can be reduced to physics, and that whatever physicists have found out about their corner of reality applies everywhere!

THE DEATH OF MEMORY

With increasing entropy and decaying protons on their minds, it comes as no surprise that physicists likewise believe that when one dies, that's it. An afterlife is impossible. How do physicists conclude this? In a letter to the American Journal of Physics, J. Orear proffered an interesting sort of "proof":

> One such proof: human memory is stored in the circuitry of the brain and after death this circuitry completely decomposes.

But not all physicists were satisfied with this simplistic view. In a follow-on letter, J.B.T. McCaughan asked how Orear knew that memory is limited to the brain's neuron circuitry. Perhaps there is something that the reductionists are missing. McCaughan then states that Orear's assertion would be negated if people really did return from the dead. He refers to the numerous accounts in the Scriptures in which witnesses attested that some individuals did indeed come back to life. (Yes, this is all printed in the American Journal of Physics!!) After all, concludes McCaughan, with respect to witnesses, "...so much in life depends on such evidence, even the credibility of physicists themselves."

(Orear, Jay; "Religion vs. Science," American Journal of Physics, 60:394, 1992. McCaughan, J.B.T.; "Scientific Faith," American Journal of Physics, 60:969, 1992.)

THE DEATH OF MATTER

Physicists have maintained for over a century that the Second Law of Thermodynamics guarantees that our universe will run down one day and that life must cease. This cold reductionist view is seconded by recent evidence that protons, long considered immortal, may after all decay. The consequences of proton decay are even more dismal than the dire predictions of thermodynamics:

> Perhaps the most disturbing piece of speculation to come out of theoretical physics recently is the prediction that the whole universe is in decay. Not only do living things die, species go extinct, and stars burn out, but the apparently immutable protons in the nucleus of every atom are slowly dissolving. Eventually---in more than a quadrillion years---nothing will be left of the universe but a dead mist of electrons, photons, and neutrinos.

(Flam, Faye; "Could Protons Be Mortal after All?" Science, 257:1862, 1992.)

MATHEMATICS

INNATE KNOWLEDGE

In the early 1800s, the mathematician Gauss dispatched observers to the tops of three mountains to determine whether the sum of the angles in a real triangle was truly 180°. Gauss was not certain that mathematics really matched reality perfectly. (His experiment was inconclusive.) Today, in most scientific education and practice, it is customary to assume that mathematics is not only a faithful mirror of the real world but that it can actually lead us to new insights into reality.

Unfortunately, two facts mar this idealistic picture. (1) Mathematics itself contains contradictions and does not have a solid foundation; that is, it is "impure." (2) Some portions of reality seem to confound mathematics; for example, Einstein found Riemannian geometry and tensor analysis imperfect for formulating the Theory of Relativity. Despite this disappointment, Einstein maintained his belief that God does not play dice with the universe. Some more recent scientists suggest that God not only plays dice but throws them where they cannot be seen!

Despite the acknowledged deficiencies, it is clearly more than a stroke of luck that mathematics describes so much of reality so accurately. And here is the spooky part of the whole business. In formulating their web of logic, mathematicians make many more or less "artistic" decisions that are colored by reality and their expectations of reality. To illustrate, "symmetry" is a human passion that reality may disdain. In other words, because mathematicians are prejudiced by their experience in the real world and are an integral part of that world, it is not surprising that their artistic renditions mirror reality to some extent. (Little, John; "The Uncertain Craft of Mathematics," New Scientist, 88:626, 1980.)

Comment. It might also be that mathematics predicts unknown realities because of the human mind's subconscious knowledge or appreciation of them---a sort of innate appreciation of that portion of the universe still unknown to the conscious mind.

TRANSCENDENTAL TRIVIA?

Both e (2.7182...) and pi (3.1416...) are transcendental numbers of great significance in mathematics and the scientific description of nature. Instead of being neat and orderly (as we devoutly hope nature will be), the decimal expansions of these two numbers are patternless, some say ugly. Faint hope arises at the 710,150th digit of pi where a satisfying string of seven consecutive 3s appears (....353733333338...). More reassuring is the observation that $(pi)^4 + (pi)^5$ almost exactly equals e^6.

We are sure that great truths lie hidden in these two numbers (despite their unattractive decimals) when we find that a 5x5 magic square (first row: 17, 24, 1, 8, 15) can be transformed by the alchemy of pi into an unmagic but very strange square. To do this, replace the 17 by the 17th digit of pi (this is 2); 24 by the 24th digit (this is 4); and so on. The rows and columns of the new square add up to the same numbers: columns; 17, 29, 25, 24, 23; rows; 24, 23, 25, 29, 17. (Yes, the order given is correct.)

Gardner maintains that this astounding transformation is merely a coincidence, like all of the other peculiar relationships between e and pi. Millions and millions of relationships are possible and a few will certainly be remarkable, just as only a few of the many possible mathematical equations describe natural phenomena. (Gardner, Martin, "Mathematical Games," Scientific American, 241:22, September 1979.)

PI SURPRISE

Consider the positive integer, 8. It can be written as $m^2 + n^2$, a sum of two squares of integers, in just 4 ways, namely when the pair (m, n) is (2, 2), (2, -2), -2, 2), and (-2, -2). The integer 7, on the other hand, cannot be written as the sum of any squared integers. On the average, over a very large collection of integers from 1 to n, in how many ways can an integer be written as the sum of such squares? The answer is little short of astounding: closer and closer to pi! (Anonymous; "Closing Pi Surprise," Algorithm, p. 7, n.d. Cr. C.H. Stiles)

TEMPTATIONS OF NUMEROLOGY

> Too much innocent energy is being spent on the search for numerical coincidences with physical quantities. Would that this Pythagorean energy were spent more profitably.

Following this admonition, John Maddox conceded that numerology, on rare occasions, has provided useful insights. Musings about Bode's Law are not complete wastes of time; and Prout's hypothesis that the masses of the elements would be found to be integral multiples of the mass of the hydrogen atom was not far off the mark.

Certainly an entertainment factor exists, too, for Maddox cannot resist printing a curious little contribution by Peter Stanbury, entitled "The Alleged Ubiquity of π." Stanbury has discovered a large number of relations between the masses of the fundamental particles that are closely related to π. Four representative examples follow: (1) The proton-to-electron mass ratio is almost exactly $6\pi^5$; (2) The sum of the masses of the basic octet $\pi^0, \pi^+, \pi^-, K^+, K^-, \bar{K}^0, K^0, \eta$ is 3.14006 times the proton mass; (3) The sum of the masses of the baryon octet is very close to π^2 times the proton mass; and (4) The reciprocal of the fine structure constant, 137.03604 is close to $4\pi^3 + \pi^2 + \pi$, or 137.03630.

There are many more such relationships. Further, the ratios 1.0345 and 1.1115 keep popping up more frequently than coincidence would seem to allow. What could these ratios be? At least π has geometrical significance. (Maddox, John; "The Temptations of Numerology," Nature, 304:11, 1983.)

THE SECRET OF IT ALL IS IN THE PI

On p. 330, Pollard discoursed on the meaning of it all and how mathematics seemed to mirror reality so marvelously. Now, one fixture of mathematics is the transcendental number. The adjective "transcendental" is most appropriate here given the title of Pollard's article. Of the transcendental numbers, pi is a great favorite. Mathematicians like pi so much that they have computed it out to well beyond 10 million decimals. Are there any inklings to the meaning of it all in these 10 million-plus decimals? Well, at decimal 710,100 there are seven 3s in a row. At decimal 1,526,800, we find the digits 2718281, the first seven digits of e, the base of natural logarithms. Then at decimal 52,638 there is 14142135, the first eight digits of the square root of 2. But all these discoveries are hardly profound, for they could occur by chance---nothing really "transcendental" so far.

A more astounding discovery is that:

$$22\pi^4 = 2143$$

A few multiplications, and the 10 million-plus decimals of pi have vanished. (Can this remarkable relationship mirror some as yet undiscovered facet of physical reality?)

While it is difficult to squeeze the meaning of the universe out of pi's 10 million-plus decimals, one has to admit that pi is everywhere. To grasp the insidiousness of this number, write the alphabet out beginning with J, as follows:

J K L M N O P Q R S T U V W X Y

Z A B C D E F G H I

Now cross out all those letters with right-left symmetry, such as M, O, etc. The remaining letters are in five groups, with populations of (you guessed it) 3, 1, 4, 1, 6. Surely this must mean something! (Gardner, Martin; "Slicing Pi into Millions," Discover, 6:50, January 1985.)

FRUITFULNESS OF MATH NOT AN INTIMATION OF A TRANSCENDENT MIND!

W.G. Pollard's article "Rumors of Transcendence in Physics," (p. 330) produced an interesting response in the

American Journal of Physics by Jan Portnow. Portnow worries about Pollard's assertion that the "remarkable" tendency of mathematics to mirror physical reality betokens the existence of something beyond our ken: i.e., transcendence. He puts it this way:

> All we can ever know about the world is what our mind lets us know ---our perception and awareness of external reality are limited by our mind. We can never know more than the mind can assimilate and process, nor can we discuss any aspect of the world for which there is no language. The fact that our mathematical laws of nature explain the world is no miracle---it can't be otherwise. The laws of nature describe the world we know and that world is a reflection of our thinking and our language---of our mind."

He goes on to maintain that the fruitfulness of math for physics hints more at the limitations of the human mind than the existence of a transcendent mind! (Portnow, Jay; "Letter to the Editor," American Journal of Physics, 53:299, 1985.)

Comment. Apparently we'll be forever boxed in by the limitations of mathematics---to say nothing of computer programs.

THE FABRIC OF PRIME NUMBER DISTRIBUTION

Mathematicians and many nonmathematicians have soft spots in their hearts for numbers that cannot be subdivided; that is, the prime numbers. No one has ever been able to figure out any foolproof system to their occurrence or how to generate them all by formula. Some primes, such as 11 and 13, 17 and 19, etc., come in pairs; but most don't. And some formulas are particularly good at generating primes, but they all fail somewheres. One such formula was discovered by Leonhard Euler, the great mathematician:

$$n^2 + n + 41$$

This formula works for n = 0, 1, 2,.... 39; but fails at n = 40.

An interesting consequence of Euler's formula can be made apparent when all numbers from 41 to 440 are written in a square spiral, like so:

53 52 51 50
54 43 42 49
55 44 41 48
56 45 46 47
57 →

All of the numbers on the diagonal indicated are Euler formula primes, even when the spiral is expanded to 20 x 20. However, when the 20 x 20 spiral is examined closely, many of the other primes---those not generated by Euler's formula---also tend to line up on diagonals. This is the most intriguing characteristic, one which goes far beyond the 20 x 20 array mentioned above.

The computer-generated display shown below lays out a huge square spiral, with each prime a bright dot. The picture has a pronounced diagonal texture. Why this is so is a mathematical mystery. (Crypton, Dr.; "Prime Numbers and National Security," Science Digest, 93:86, October 1985.)

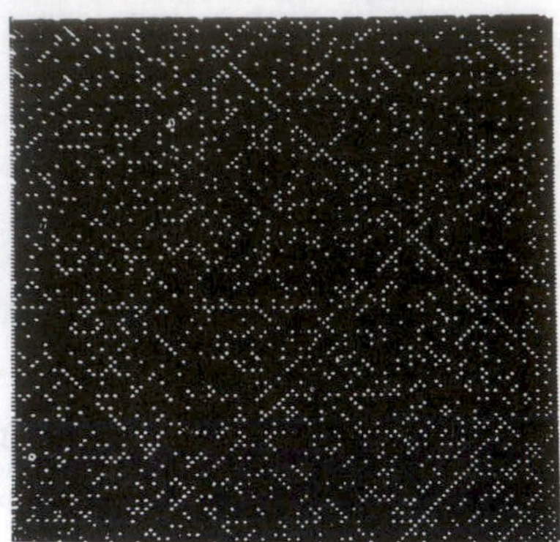

Computer-generated display of numbers. Each white dot represents a prime. Note the pronounced diagonal trends.

Comment. Does the diagonal fabric of primes have any practical significance? At the moment no one knows. Again and again, abstruse mathematical structures have turned out to mirror phenomena in the world we call "real."

FRACTALS, FRACTALS EVERYWHERE

Anyone who follows the popular scientific literature knows that fractals are now "in." Commonly employed to "explain" patterns in nature, fractals are, from a simplistic viewpoint, mathematical ways to predict the development of a growing structure, be it a crystalline mass, a plant, or the universe-as-a-whole.

Yes, the universe-as-a-whole, the clouds of stars and clusters of galaxies, may be mimicked by cellular automata (i.e., fractals). Imagine the universe as a cubical lattice, and start in one corner, adding one layer of cubes after another. Galaxy distribution could be simulated by using a rule telling us which of the added cubical cells had galaxies in them and which did not.

> The rule actually used supposes that the question whether each point in a newly added layer will (or will not) be occupied by a galaxy is mostly determined by the occupancy of the five nearest neighbors in the previous layer, but for good measure, there is a random variable to introduce an element of white noise to the system. To make the process a little more interesting, the determination whether a new site is occupied depends on whether a number characteristic of that site, and calculated by simple arithmetic from the corresponding number for the five nearest neighbors in the preceeding layer, exceeds an arbitrarily chosen number.

Comparing this fractal simulation with the observed universe is startling. The agreement is "spectacularly successful." (Maddox, John; "The Universe as a Fractal Structure," Nature, 329:195, 1987.)

Comment. In biology, too, fractal modelling can be very impressive. But, doesn't it all verge on numerology? The existence of a galaxy at a point in space is simply dependent on its neighbors; and the law of gravitation is not even mentioned. Are the physical laws that we usually assume as controlling the dynamics of matter now "hidden" beneath a more general property of the universe? One is reminded of R. Sheldrake's theory of morphic resonance, in which the mere existence of a certain structure makes it more easy for the same structure to be duplicated elsewhere in the universe!

PI AND RAMANUJAN

Someone has finally complained about the equality sign on p. 331; namely,

$$22\pi^4 = 2143$$

D. Thomas has correctly pointed out that we have here only a very good approximation. Of course, one need not do the actual calculation to prove that it is an approximation, because 2143/22 is a rational fraction which can be expressed as a repeating decimal; whereas pi is irrational.

The number $(2143/22)^{\frac{1}{4}}$ is a discovery of Ramanujan, about whom we heard on p. 308. How did he ever stumble upon this extremely accurate approximation of pi---one that is accurate to 300 parts in a trillion? N.D. Mermin suggests that Ramanujan may have taken it from the expansion:

$$\pi^4 = 97 + 1/(2 + 1/(2 + 1/(3 + 1/(1 + 1/(16539 + ...$$

If 16,539 is replaced by infinity, Ramanujan's result follows. (Mermin, N. David; "Pi in the Sky," American Journal of Physics, 55:584, 1987.)

DOES NATURE COMPUTE?

Back in the 1960s, kids used to watch the TV series Lost in Space. Starring on this show was a robot which, when asked a stupid or answerless question replied, "It does not compute!" More seriously, we now ask, "Does Nature compute?"

Science believes very deeply that mathematics reflects the real world, that we live in an ordered universe where everything can be reduced to mathematical expressions. The progress of science, particularly physics, seems to bear out this symbiotic relationship between mathematics and the physical world.

However, P. Davies points out that that there are uncomputable numbers and operations. In fact, there are infinitudes of them. All the world's computers could chug away forever and not come up with answers in these cases. So far, Nature has been kind, or we have been lucky, because we have been able to nicely mirror Nature with "do-able" math. Davies wonders if it has been entirely a matter of luck:

> Einstein said that God is subtle but not malicious, and we must hope that the laws of physics will turn out to be computable after all. If so, that fact alone would provoke all sorts of interesting scientific and philosophical questions. Just why is the world structured in such a way that we can describe its basic principles using "do-able" mathematics? How was this mathematical ability evolved in humans?

Are our minds and, therefore, our computers so structured that we can understand (compute) only a limited portion of Nature? Have other entities evolved in ways such that what we know of Nature is uncomputable to them? (Davies, Paul; "Is Nature Mathematical?" New Scientist, p. 25, March 21, 1992.)

MATH'S MYSTERY

In extolling J. Barrow's new book, Pi in the Sky, T. Siegfried first reiterates a point made in past issues of SF, namely, that mathematics is a logical system and that we have no right to expect it is correspond structurely with a physical system. In other words, math and nature are fundamentally different entities. Nevertheless, as Barrow stated in a recent interview:

> If we were just inventing mathematics from our everyday experience, we would find that it would work really rather well in those areas from which that intuition was gained. But we find almost the opposite...It works most powerfully and persuasively in areas that are farthest removed from the everyday experience that has led to it.

Mathematics, for example, leads to verities in quantum mechanics far outside the realm of daily experience. Why is this so?

The puzzle deepens when one discovers that there are different kinds of math based upon different forms of logic (as in Euclidian and non-Euclidian geometries). Some brands of mathematics mirror reality better than others. Why?

In trying to dispose of these "whys," both mathematicians and scientists fall back on the Anthropic Principle with all its unsatisfying tautological overtones:

> ...the universe is the way it is because that's the way it has to be for anybody to be around to study it. And perhaps math works so well in studying the universe because math, too, must be the way it is in order for anybody to be around to do the calculations. So maybe the existence of communicating creatures requires a correspondence between the physical universe and mathematics. [??]

(Siegfried, Tom; Dallas Morning News, p. 7D, January 4, 1993. Cr. L. Anderson)

ALL ROADS LEAD TO 123

> Start with any number that is a string of digits---, say, 9 288 759 ---and count the number of even digits, the number of odd digits, and the total number of digits it contains. These are 3 (three evens), 4 (four odds), and 7 (seven is the total number of digits), respectively. Use these digits to form the next string or number, 347. If you repeat the process with 347, you get 1, 2, 3. If you repeat with 123, you get 123 again. The number 123, with respect to this process and universe of numbers, is a mathematical black hole.

We have a black hole because we cannot escape, just as spaceships are doomed when captured by a physical black hole! You end up with 123 regardless of the number you start with. Other sorts of mathematical black holes exist, such as the Collatz Conjecture, but we must not fall into them because our printer awaits. (Ecker, Michael; "Caution: Black Holes at Work," New Scientist, p. 38, December 19/26, 1992.)

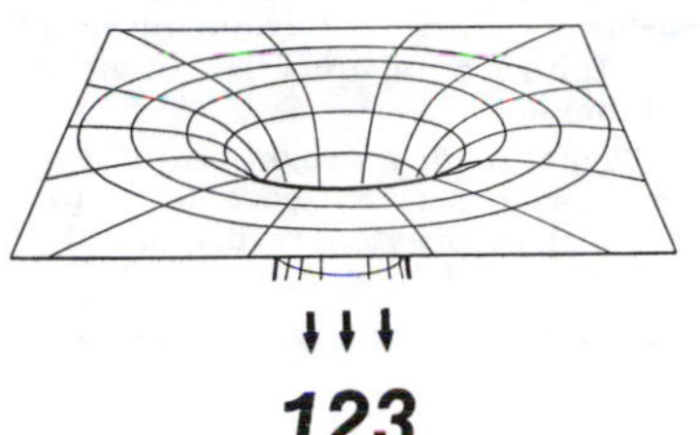

ESOTERICA

EXTRATERRESTRIAL CIVILIZATIONS

DON'T BUILD VON NEUMANN MACHINES

Back in April 1981, Frank J. Tipler published a comment in Physics Today entitled, "Extraterrestrial Beings Do Not Exist." A key element of his argument was that the first intelligent civilization would inevitably colonize the entire universe with themselves or their self-reproducing (von Neumann) machines. Since we can detect neither kind of colonist, we must be the first and only intelligent civilization. Essentially Tipler was criticizing the prevailing notion that, from a statistical point of view, there must be many other civilizations among the numberless stars and galaxies. Tipler's thoughts evoked a richly provocative group of letters, some of which were published in the March 1982 issue of Physics Today along with a closely related article by Ornstein.

Ornstein tackled the problem of estimating the probability that intelligent life would evolve on other planets from a biologist's perspective. Whereas some physical scientists have set this probability at about 1.0, Ornstein inclines toward 10^{-9}, believing that intelligent life is probably unique to the earth. But concludes Ornstein, the 15 separate developments of eyes among disparate terrestrial organisms may infer some unrecognized directing factor in evolution that would force him to revise his estimate upwards drastically.

On the other hand, those 15 eyes might indicate a common, but still undiscovered, eye-possessing ancestor far back along evolution's track. The fossil record might be mute on this matter because eyes are soft tissues that are rarely preserved or perhaps because many eyes were jettisoned because the organisms didn't need them, as cave dwellers are wont to do with surprising rapidity.

Some of the letters responding to Tipler questioned whether an intelligent civilization would be stupid enough to build self-reproducing von Neumann machines for galactic exploration. Wouldn't it be far more fun to go in person rather than by proxy? And, some pointed out, von Neumann machines would be ravenous consumers of energy and materials and might turn on man as an unnecessary competitor. Machines are not immutable. Space radiation and other environmental factors might alter computer programs and memories to drastically affect the behavior and objectives of such machines. Acutally, as one letter writer observed, the earth has already been invaded by a self-reproducing, energy-hungry machine with exploratory tendencies---man! (Anonymous; "Extraterrestrial Intelligence: The Debate Continues," Physics Today, 35: 26, March 1982, Ornstein, Leonard; "A Biologist Looks at the Numbers," Physics Today, 35:27, March 1982.

THE 'GREAT SILENCE'; OR WHY AREN'T ALIENS LANDING ON THE WHITE HOUSE LAWN?

It is anomalous that despite the widespread belief that other civilizations must abound "out there," not one has yet contacted us. G.D. Brin has conducted an analysis of this puzzle and has come to these conclusions:

"The quandry of the Great Silence gives the infant study of xenology its first traumatic struggle, between those who seek optimistic excuses for the apparent absence of sentient neighbors and those who enthusiastically accept the Silence as evidence for humanity's isolation in an open frontier.

"Both approaches suffer greatly from personal bias, and from lack of detailed comparative study. In this article we have attempted to deal with a subject that, for all of its great importance, is almost ghostly in its intangibility. We have broken the subject into its logical elements and attempted a morphological discussion of the possibilities. Table I (not reproduced because of its size) presents an overview of many of the ideas discussed here and their respective effects on the equations....

"Some of the branch lines discussed here serve the optimists, while others seem pessimistic to an unprecedented degree. We have laid out only the outline of a full analysis of the problem. Further work should consider every experimental test that could be applied to this fundamental question of humanity's uniqueness.

"This survey demonstrates that the Universe has many more ways to be nasty than previously discussed. Indeed, the only hypotheses proposed which appear to be wholly consistent with observation and with non-exclusivity---'Deadly Probes' and 'Ecological Holocaust'---are depressing to consider.

"Still, while the author does not accept that elder species will necessarily be wiser than contemporary humanity, such noble races might have appeared. If such a culture lived long, and retained much of its vigor of youth, it might have instilled a tradition of respect for the hidden potential of life in subsequent space-faring species.

"It might turn out that the Great Silence is like that of a child's nursery, wherein adults speak softly, lest they disturb the infant's extravagant and colourful time of dreaming." (Brin, Glen David; "The 'Great Silence,' The Controversy Concerning Extraterrestrial Intelligent Life," Royal Astronomical Society, Quarterly Journal, 24:283, 1983.)

Comment. It would be unrealistic not to expect an editorial comment after this article, perhaps to the point that any really intelligent entities would consider rocketry and physical space travel as crude and demeaning. Fred Hoyle may have been closer to the mark in seeing in life, its forward development and unplumbed potentials (mental calculation) proof positive of intelligent entites "out there."

ANOMALOUS SIGNALS FROM SPACE

RADIO SIGNALS FROM THE STARS

Curious signals have been picked up from 12 stars by the 300-foot radio telescope at Green Bank, WV. The signals took the form of strong bursts at a wavelength of 21 cm, one of the wavelengths characteristic of the hydrogen molecule. Unfortunately, the signals were so short that their information content, if any, could not be recorded. Since the bursts were not repeated (except for a second burst from Barnard's Star), some natural phenomenon may be at work rather than intelligent communicators, who would presumably be more persistent. The peculiar signals, which have never been recorded before, were discovered as part of Project Ozma II, in which radio astronomers have been listening to 21-cm radio waves from hundreds of nearby stars. (Anonymous; "Possible Messages from Space Reported," Baltimore Sun, p. A3, January 29, 1978.)

THE 1977 "WOW" SIGNAL

Over the years, several large radio astronomy antennas have listened for "intelligent" radio signals from outer space. The acronym SETI is customarily applied to such searches; SETI standing for Search for Extraterrestrial Intelligence.

There have been a few exciting false alarms during these listening periods, but most could be attributed to known natural radio sources or manmade interference. All in all, it has been rather disappointing to those who are sure someone else is out there.

The major exception in the SETI record was the so-called "WOW" (like Egad!) signal picked up in 1977 by a radio telescope at Ohio State University, in Columbus. The bandwidth of the signal was narrower than those of most natural sources; there was also some evidence of periodic and drifting features. The signal never recurred, nor could it be correlated with any manmade or natural radio sources. (Eberhart, Jonathan; "Listening for ET," Science News, 135:296, 1989.)

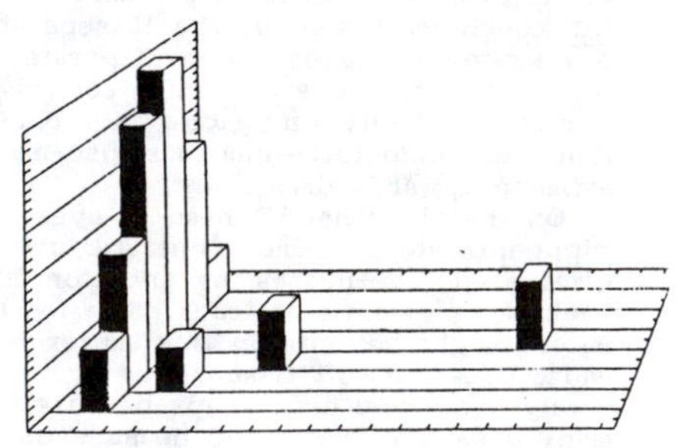

Anatomy of the "WOW" signal. Vertical ordinate represents intensity; horizontal axis is frequency in 10 kilohertz intervals. Time axis runs into the paper with 12-second intervals.

Comment. We can only speculate as to what alien intelligence might mean. Then, too, aliens have probably progressed far beyond primitive radio communication!

UFO-TYPE OBSERVATIONS

STRANGE OBJECT IN THE SKY

January 20, 1983, 0515 GMT:

The m.v. Baron Pentland was drifting off Christmas Island in the Indian Ocean. A report was heard on the Christmas Island Radio, that an object had been spotted high in the sky to the west of the Island. The officers of the Baron Pentland picked up the object with binoculars and sextant.

"First thoughts were that this was the satellite Cosmos 1402, but this was dismissed as it was a day too early and was not moving fast enough. In fact it appeared stationary to the naked eye. Another school of thought was that this was a weather balloon. As can be seen from the simple sketch, it was unlike

any weather balloon previously seen by the observers. It was of a squat cylindrical shape, wider than it was tall. The circle at the bottom appeared to be dimly lit with a pale blue colour. The 'torso' was almost invisible, even with binoculars, giving the impression that there were two distinct and separate lights. The top 'light' appeared to be a dome atop the main body and it was extremely bright. By wedging the binoculars in the bridge doorway, it was possible to gain a very steady view as the vessel's main propulsion system was shut down and the seas were slight. Thus a clearer picture was obtained and in this way the object was identified as being cylindrical. The bottom circle also appeared to be 'webbed' with dark radial lines."

At 0643 GMT, the object had disappeared in the west. (Strachan, C.; "Unidentified Flying Object," Marine Observer, 54:27, 1984.)

UNIDENTIFIED FLASHING OBJECT

May 6, 1984. Equatorial Eastern Atlantic.

"Whilst the vessel was approaching

the equator, on a course of 023° and at a speed of 10.3 knots, flashing white lights were observed. The sea was rippled, with a low NW'ly swell and the wind light airs. The visibility was good, with the moon in its first quarter. At first it was thought that there were three lights, one being bright and the other two relatively dim, but as the vessel approached it it was decided that there were only two, one bright and one dim.

"The radar, a 10-cm S-band Decca, was switched on and a single echo was detected initially at a range of 5 n. mile. By this time the lights had already been observed for half-an-hour, so it was estimated that they had first been observed when they were 10 n. mile distant. The time of the first sighting was 2220 GMT. The target, once detected, gave a very strong echo and gave the impression of being a large target. It was plotted and found to be stationary. The initial course of 023° was altered to 028° in order to enable the vessel to close the passing distance.

"Throughout the observations neither light followed any set characteristic. Instead they just flashed at random, but never together. The intensity of the bright light would have put many a lighthouse to shame. As the light came on the beam they both seemed to be on one and the same object---perhaps at each end of a light float was one possibility suggested. As the lights came on the beam at a distance of 1.3 n. mile the Aldis lamp was shone in their direction but with no success in identifying them.....

"The two previous voyages had been to Brazil and France and the lights were observed on every occasion near the equator; they were identical in every way, but in the previous encounters the lights, both dim and bright, were more numerous."

(Guy, M.E.; "Unidentified Flashing Object," Marine Observer, 55:79, 1985.)

Comment. Similar lights have been reported in this area and elsewhere at sea. See GLN1 in Lightning, Auroras, Nocturnal Lights.

EDINBURGH UFO A MIRAGE?

September 30, 1986. Edinburgh, Scotland.

> Yvonne Westgarth looked out of a north-facing window of her house in south Edinburgh. She was amazed to see a white cylindrical object like a missile travelling westwards just above the roofs of houses opposite (as she thought). She called her husband who also saw the object. Their sketches of what they saw are shown in Fig. 1. Although their descriptions differ slightly, they agree that the "missile" had a black band around its centre. They watched the object for a period of between 0.5 min and 1.5 min (period uncertain). It was first seen almost due north and it disappeared in the west-north-west. No noise was heard.

No one else reported seeing the object. A real missile was considered very unlikely. However, the object appeared in the direction of the glide path of the Edinburgh airport, where two aircraft

The "missile" seen over Edinburgh on September 30, 1986, by Y. Westgarth (top sketch) and her husband (bottom sketch).

had landed at about the time of the sighting. The witnesses were adamant that the UFO did not look at all like a plane; and that it was much higher in the sky than planes on normal glide paths, which were to be seen just above the horizon between the houses.

S. Campbell, who investigated this event, suggests that the Westgarths saw an enlarged distorted mirage of a Boeing 757 landing at Edinburgh. The timing and direction were right. Mirage action would elevate the image. By assuming that both an erect and inverted image of the aircraft were projected one above the other, something looking like a missile could be imagined. The black band would have been the doubled image of the wing. (Campbell, Steuart; "Mirage over Edinburgh," Journal of Meteorology, U.K., 12:308, 1987.)

Comment. As in many explanations of UFOs, one must decide whether a string of somewhat strained scientific assumptions is preferable to believing that a "real" UFO was sighted. In this instance, however, probability seems to be on the side of the scientific explanation.

A HUNGARIAN UFO

Somehow an interesting UFO report snuck into the Baltimore Sun---a newspaper normally very conservative about such things. The report was embedded in a syndicated review of the week's "natural" phenomena from around the world.

> Meteorologists and military pilots in the western Hungarian town of Papa reported seeing four large, and bright orange unidentified flying objects after midnight on November 25. Government meteorologist Gyula Bazso said the objects were spherical and about 50-100 yards wide. He said one flew at the speed of 2,626 miles per hour. Bazso contacted authorities at the local military airbase who sent up an experienced pilot to investigate. He located the four objects at a height of around four miles. All the UFOs were said to have disappeared suddenly after 2 a.m.

(Anonymous; no title, Baltimore Sun, December 3, 1989.)

THE BELGIAN FLYING TRIANGLE

"The Belgian air force has been on alert for three nights running, writes Lucy Kellaway. Two Hawker Siddeley aircraft equipped with infrared cameras and sophisticated electronic sensors have been patrolling the skies. Down below, the Belgian police force has kept a constant watch, helped by more than 1,000 concerned civilians. Along the border with Germany, 20 lookout posts have been set up. Their target: an Unidentified Flying Object.

"Since December, there have been 800 reported sightings, and even though some resemble a lamp-post more closely than a UFO, many of the others are being earnestly examined by SOBEPS, the Belgian Society for Studying Special Phenomena.

"More surprising is how seriously the army is taking the whole thing. For the time being it says it is viewing the matter as a 'technical curiosity' as the intruder has shown no aggressive signs. Should it turn nasty, it will be a different matter altogether."

.....

"Scientists on the ground appear in the past few days to have produced a clear image of the object, which is said to correspond to the reports of eye-witnesses. It is a triangle 30m-50m in diameter, with red, green and white lights at the corners, 10 times brighter than any star. It has a convex underbelly and makes a sharp whistling noise." (Anonymous; "Flying Triangle Has Belgians Going around in Circles," London Financial Times, April 18, 1990. Cr. T. Good via L. Farish)

Comment. This is a strange place to find an even stranger report. Could there be a hoax here? Perhaps some of our readers in Europe will enlighten us.

THE BELGIAN FLYING TRIANGLE UPDATED

Above we printed a brief item from an English newspaper about a rash of Belgian sightings of "flying triangles." Our invitation to our European readers to expand on this "flap" brought numerous articles from European papers and magazines. Normally, we bypass UFO-type observations, but the Belgian flying triangles are so remarkable that they deserve a little space.

Very briefly, we have roughly 1000 observations by several thousand people, beginning in October 1989 and still continuing. Most witnesses report a dark, triangular object with three bright lights plus a flashing red one in the middle. Size estimates vary from the size of a football field to that of conventional aircraft. The object sometimes hovers for minutes at a time. It also can move very slowly and then suddenly accelerate to high speeds. Some observers report a faint humming sound; others say that it is noiseless.

The American Stealth Fighter (F-117) is roughly triangular, and there has been much speculation that people have been seeing this craft on night missions. The characteristics reported for the flying triangle, however, are hardly those of a jet aircraft. But one must always remember that human observers are imperfect.

The July 5, 1990, issue of Paris

Match presented a remarkable account of an encounter between two Belgian F-16s and one of the flying triangles. We use here those portions of a translation provided by R.J. Durant to the International UFO Reporter (15:23, July/August 1990).

"On the night of March 30th, one of the callers reporting a UFO was a Captain of the national police at Pinson, and [Belgian Air Force] Headquarters decided to make a serious effort to verify the reports. In addition to the visual sightings, two radar installations also saw the UFO. One radar is at Glons, southeast of Brussels, which is part of the NATO defense group, and one at Semmerzake, west of the Capitol, which controls the military and civilian traffic of the entire Belgian territory. The range of the two radars is 300 kilometers, which is more than enough to cover the area where the reports took place...Headquarters determined to do some very precise studies during the

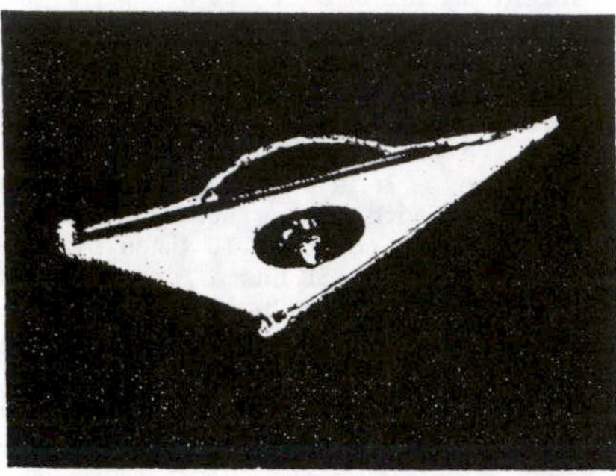

Artist's concept of the Belgian Flying Triangle, with three white lights on the corners and one red light in the center. (From: Télérama, February 21, 1990.)

next 55 minutes to eliminate the possibility of prosaic explanations for the radar images. Excellent atmospheric conditions prevailed, and there was no possibility of false echoes due to temperature inversions.

"...at 0005 hours the order was given to the F-16s to take off and find the intruder. The lead pilot concentrated on his radar screen, which at night is his best organ of vision. The F-16 is equipped with very sophisticated equipment, including chase radar, which is not fixed directly ahead of the airplane, but makes a wide search in an arc of 90 degrees left and right of the nose...

"Suddenly the two fighters spotted the intruder on their radar screens, appearing like a little bee dancing on the scope. Using their joy sticks like a video game, the pilots ordered the onboard computers to pursue the target. As soon as lock-on was achieved, the target appeared on the screen as a diamond shape, telling the pilots that from that moment on, the F-16s would remain tracking the object automatically.

"[Before the radar had locked on for six seconds] the object had sped up from an initial velocity of 280 kph to 1,800 kph, while descending from 3,000 meters to 1,700 meters...in one second! This fantastic acceleraton corresponds to 40 Gs. It would cause immediate death to a human on board. The limit of what a pilot can take is about 8 Gs. The trajectory of the object was extremely disconcerting. It arrived at 1,700 meters altitude, then it dove rapidly toward the ground at an altitude under 200 meters, and in doing so escaped from the radars of the fighters and the ground units at Glons and Semmerzake. This maneuver took place over the suburbs of Brussels, which are so full of man-made lights that the pilots lost sight of the object beneath them...

"Everything indicates that this object was intelligently directed to escape from the pursuing planes. During the next hours the scenario repeated twice...

"This fantastic game of hide and seek was observed from the ground by a great number of witnesses, among them 20 national policemen who saw both the object and the F-16s. The encounter lasted 75 minutes, but nobody heard the supersonic boom which should have been present when the object flew through the sonic barrier. No physical damage was reported. Given the low altitude and speed of the object, many windows should have been broken."

FIVE REASONS WHY UFOs ARE NOT EXTRATERRESTRIAL MACHINES

Regardless of what mainstream science thinks of them, UFO observations continue to pile up---by the tens of thousands! In fact, like the crop-circle events, UFO reports are increasing in number and strangeness. It doesn't matter that the UFOs and their alleged occupants may not be physically real. There are tens of thousands of people who think that they have observed something strange---even after all hoaxes and misinterpretations of natural phenomena have been culled out. Most of those who are willing to accept UFOs as valid phenomenon think they are real hardware piloted by extraterrestrials.

J. Vallee, a computer scientist and prolific writer on the subject, demurs, and he gives five reasons why:

> "(1) Unexplained close encounters are far more numerous than required for any physical survey of the earth; (2) The humanoid body structure of the alleged 'aliens' is not likely to have originated on another planet and is not biologically adapted to space travel; (3) The reported behavior in thousands of abduction reports contradicts the hypothesis of genetic or scientific experimentation on humans by an advanced race; (4) The extension of the phenomenon throughout recorded history demonstrates that UFOs are not a contemporary phenomenon; and (5) The apparent ability of UFOs to manipulate space and time suggests radically different and richer alternatives."

If not extraterrestrial hardware, what are the UFOs? Vallee has three suggestions: (1) they are "earth lights" a la P. Devereux; that is, an unappreciated terrestrial phenomenon that impresses mental images on the minds of observers; (2) They are artifacts of a "control system" operated by a nonhuman intelligence or, perhaps, a Gaia-like manifestation of supernature (of which we are a tiny part) that is trying to modify our behavior; and (3) They are apparitions caused by entities manipulating space and time; viz,. time travellers from our own past and/or future. (Vallee, Jacques F.; "Five Arguments against the Extraterrestrial Origin of Unidentified Flying Objects," Journal of Scientific Exploration, 4:105, 1990.)

MEN IN BLACK (MIBs)

A.K. Bender seems to be one of the first humans contacted by MIBs. In 1953, just after he had written a letter to a friend stating that he had learned the origin and ultimate goal of extraterrestrial visitations to earth, he was approached by three men dressed in black suits. They had his letter! After this contact, Bender ceased all his UFO-related activities. So goes this classic MIB tale.

> MIB activity flourished with the increased sightings of UFOs during the "flap" of 1966-67, and numerous UFO researchers claimed MIB experiences. MIB have been reported to arrive unannounced, sometimes alone or in twos, traditionally in threes, at the homes or places of employment of selected UFO witnesses and investigators or their research assistants, usually before the witness or researcher has reported the UFO experience to anyone; or in the case of some investigators, before they have even undergone a UFO experience of any kind. People have reported that MIB know more about them than the average stranger could possibly know, and thus MIB can possess an omniscient air."

The central thesis of this lengthy article is the close relationship of MIBs and the ancient figure of the Devil. Also treated are the similarities between older folklore traditions and the UFO/MIB phenomena. The author also notes that UFO percipients also often see Bigfoot-like creatures and other "monsters." (Rojcewicz, Peter M.; "The 'Men in Black' Experience and Tradition: Analogues with the Traditional Devil Hypothesis," Journal of American Folklore, 100:148, 1987.)

Comment. Little is said in this item about the objective reality of MIBs, UFOs, and the Devil, but the reader is left with the impression that all these phenomena are aspects of a continuum of experience as old as the human race.

NEW SPECIES EMERGING?

> A San Diego science writer named Anne Cardoza is sending out flyers asking anyone who has given birth to a child by an extraterrestrial to submit a 3,000-word, first-person account.
>
> She's compiling a book on breeding between humans and inhabitants of UFOs. She won't pay for the stories but she promises confidentiality.

(Anonymous; "Talk about Mixed Marriages!" San Diego Tribune, January ?, 1990. Cr. D. Clements via L. Farish)

MISCELLANEOUS FORTEANA

Forteana are observations of nature that are eyebrow-raising and often seem bizarre. Named after the American iconoclast, Charles Fort, modern Forteana are gleaned from scientific journals, magazines, and newspapers alike, just as Fort did in the 1920s and 1930s. Forteana are not necessarily anomalous, though many are. Their chief attribute is their supposed value in taunting and embarrassing science. Falling fish are Fortean, but gaps in the fossil record are not, although both are anomalous. Two-headed snakes are Fortean but hardly anomalous. The semantics are really not important. Basically, Fort considered most scientists pompous and chose his data with their deflation in mind. Despite this predilection, his methodology and four thick volumes of observations and wry theories have greatly influenced off-mainstream thinking.

CATTLE MUTILATIONS CALLED EPISODE OF COLLECTIVE DELUSION

During the past several years, farmers in the western states have been reporting dead cattle that seemed strangely mutilated. Soft, exposed parts, such as the ears and genitals, were apparently removed with surgical precision. Some corpses seemed bloodless. Local papers blamed satan worshippers and UFO occupants.

This paper analyzes the 1974 mutilation "flaps" in South Dakota and Nebraska, with special attention to the rapid rise and equally rapid decline of public interest as measured by newspaper coverage. In the opinion of the author, these two episodes are classic cases of mild mass hysteria, similar to the occasional crazes of automobile window-pitting. In all cases where university veterinarians examined the corpses, the mutilations were ascribed to small predatory animals. The veterinarians also pointed out that blood coagulates in a couple days after death, accounting for the frequent "bloodless" condition. With such expert reassurances, the "mass delusions" subsided quickly. Cattle mutilation flaps are thus seen by the author as episodes when people interpret the mundane in bizarre new ways, due perhaps to cultural tensions.

It is noted, however, that expert veterinarians examined only a few of the dozens of mutilations, and that some people rejected the above commonplace explanations. (Stewart, James H.; "Cattle Mutilations: An Episode of Collective Delusion," The Zetetic, 1:55, Spring/Summer 1977.)

BERMUDA TRIANGLE IN ORBIT

Every time the British research satellite Ariel 6 passes over British Columbia and the Caspian Sea, something turns off the high voltage power to two of its experiments, leaving a third power supply unaffected. Even more eerie is the discovery that the sun must be shining on the ground for the phenomenon to occur. The radio commands controlling the switching are coded on a 5 kHz subcarrier superimposed on a 148.25 MHz carrier. The frequencies and coding are so highly specific that it is hard to imagine how the spurious commands arise. Also peculiar is the finding that the undesired switching can be prevented by simply beaming the pure carrier at the satellite just before it enters the two mystery zones. (Schwartz, Joe; "Mystery Beams Affect UK Satellite," Nature, 280:25, 1979.)

Comment. This is only the latest in a long series of mysterious spacecraft electronic problems.

SOUTH OF THE BERMUDA TRIANGLE

On August 4, 1944, the 85-foot staysail schooner Island Queen departed Grenada bound for St. Vincent, carrying 75 passengers to a wedding. The black-hulled Providence Mark accompanied the Island Queen, and a friendly rivalry kept them neck and neck, often only a half mile apart.

Off the western tip of Carriacou, a light, windless rain belt passed overhead. Observers aboard the Providence Mark saw the Island Queen enter a rain shower ahead of them. When the Providence Mark emerged from the shower, the Island Queen was nowhere to be seen. Thinking the Island Queen's more powerful motor had enabled it to pull far ahead, the Providence Mark captain figured he had lost the race. When he arrived at St. Vincent at 2350, the Island Queen was not there. In fact, a thorough search by the U.S. and British navies never found any trace of the ship or its 75 passengers. (Anonymous; "The Riddle of the 'Island Queen!'" Nautical Magazine, 219:26, 1978.)

LIGHTNING IN THE FAMILY

What follows is hardly a scientific report, but we have no reason to doubt its accuracy.

On last Saint Patrick's Day, G. Patterson of Phoenix, Maryland, north of Baltimore, was in bed sick during a hard rainstorm, when the bulb in her bedside lamp exploded. Lightning had struck her house.

She got out of bed and rushed over to her daughter's house nearby to find a red ball of fire on her baseboard outlet. The daughter's house had also been struck by lightning! Worse yet, the TV and VCR had been destroyed.

Later on the same day, Patterson's daughter in Bel Air, northeast of Baltimore, called to say that the chimney of her house had been struck by lightning, scattering fireplace bricks all over the floor! (Simon, Roger; "After Lightning Strikes a Family Thrice, Call Priest," Baltimore Sun, April 9, 1990.)

SPOOKY STATS IN MARYLAND

Easton, Maryland.

Justin Kyle Russum was born at 6:53 PM on October 31, 1988---Halloween. Nothing particularly spooky about that.

> What's so amazing about a Halloween baby? Consider that both the Russums [parents] were born on Halloween (1954 and 1963, respectively). Then consider that the physician who delivered the child---a doctor who had just come on duty a few hours earlier---was also born on October 31.

(Jensen, Peter; "Birth of Halloween Baby Offers Some Spooky Stats," Baltimore Sun, November 4, 1988.)

A TRULY FORTEAN HOUSE

A host of Fortean forces descended upon a house in Orland Hills, Illinois, in 1988.

> Once, a blue flame an inch in diameter shot out of a wall socket for more than 30 seconds, and the outlet still worked. That incident was witnessed by two police officers. Another time, a similar flame set a mattress afire while investigators were prowling outside.
>
> In all, there were 26 separate incidents, all of them witnessed by either police or fire investigators...
>
> "I was there one night when the room was filled with a white haze. I couldn't see my hand in front of my face," Smith said. "There was a strong sulfur smell and my eyes were burning. I took a sample (of the vapor) in a vacuum canister. We came up with nothing."
>
> Neither did engineers, chemists and geologists.

Understandably, the occupants of this hexed house had moved out long ago with such goings-on. Teams of experts ruled out arson, natural gas, methane leaks, sewer gas, and electrical malfunctions. The house was finally bulldozed in October 1988. (Quote from: Elsner, David; "Bulldozers Lay House to Rest," Chicago Tribune, October 16, 1988. Cr. K. Fabian. Also: Anonymous; "Strange Phenomena Force Bulldozing of House," Lorain (Ohio) Journal, October 16, 1988. Cr. J.K. Wagner. Also: Anonymous; "Mists, Fires in Dwelling Defy Logic," Oregonian, October 26, 1988. Cr. R. Byrd.)

DEEP-SIXING 666s

On May 1, 1991, the British Driver and Vehicle Licensing Agency announced that it would no longer issue automobile license plates bearing the number 666. Cars bearing such plates, it seems, have been involved in too many accidents! (Grossman, Wendy M.; "No More 666s," Skeptical Inquirer, 16:128, 1992.)

Comment. Recall that 666 is the "Number of the Beast" in the Bible.

ELECTRONIC CHANNELING

Those ubiquitous solar-powered calculators may be the latest mechanism by which other "intelligences" are trying to communicate with us. Here is how this novel information channel works. You put your solar-powered calculator (a cheap one will do) in your desk drawer and close it. When you again open the drawer and light hits the calculator, a number, perhaps a letter, or even an unrecognized symbol may appear. What's going on here? What do these "messages" mean?

C. Bentley, in a letter to the New Scientist, related how his calculator most frequently flashes the number 5, but many other numbers may also appear. His calculator works perfectly after he has disposed of the gratuitous information. Something must be generating these strange data. In his final paragraph, Bentley muses:

> It has occurred to me that perhaps someone or something is trying to communicate but I fear that if this is the case the message has so far eluded me. The numbers don't work on pools either.

(Bentley, Chris; "Dark Secret," New Scientist, p. 52, October 10, 1992.)

TRANSCENDENTAL MESSAGES IN TRANSCENDENTAL NUMBERS

Those solar-powered calculators and complex crop circles that are supposedly conveying messages from extraterrestrial entities have not been too helpful. We must, therefore, search out other sources of transcendental signals. Fortunately, a brand-new, unhoaxable communication channel has opened up.

Forget standard numerology, the Number of the Beast (666), and all that. Instead, give the letter A the value 1, B = 2, C = 3, etc. Next, add to your scheme a breakthrough discovery of L. Sallows, let 0 = _, and interpret _ to be a space, so that we can make sentences out of words. Finally, discard our usual base of 10 and adopt as a base 27---the number of letters in our alphabet plus _, the space. In this system, B_C decodes as $2 \times 27^2 + 0 \times 27^1 + 3 \times 27^0$, which equals 1461 in decimal. Now we have a way to convert numbers into words in a novel, though tedious, way, and vice versa.

For example, CHAT + TALK = WIND, which is not an unlikely word equation. Really fantastic word-number equalities can be found with the help of a computer. Who would have ever guessed that the following magic square of meaningful words could be constructed?

DIM	OWE	TUG	RAP
RIG	TAP	DOT	RAY
THE	TIP	NAP	DID
PAP	DUD	SPY	TOW

The magic constant is BEAN, and all horizontal, vertical, and diagonal rows add up to this constant.

The real value of this system of numerology is apparent when we turn to eternal verities: the transcendental numbers such as pi, e, and the Golden Mean (1.618034....). The latter converts to: A.PRNTPFCUCRKDYGRYLLC-QNBIG... Ah, BIG, part of a message, no doubt! Sallows remarks, "Perhaps the first message to appear in pi is ...GOD_EXISTS..., While that in e might be ...PROVE_IT..." Repair to your computers to find the meaning of it all. (Stewart, Ian; "Number Mysticism for the Modern Age," New Scientist, p. 16, July 10, 1993.) This item actually was printed in SF#89.

WHEN THE CHIPS ARE DOWN

M.A. Persinger, an indefatigable investigator of terrestrial correlations, has identified another:

> The hypothesis that sudden commencements of global geomagnetic activity ("sudden impulses") could induce anomalous changes in onboard computers and facilitate commercial aircrashes was investigated. During the years 1988 and 1989 the mean daily occurrence of a commercial disaster somewhere in the world increased from 0.06 to 0.12 within 24 hr. of a sudden commencement. When numbers of sudden commencements per month were correlated with eight major categories of catastrophes (including air disasters) only aircrashes, primarily occurring during maximum computer-dependent flight conditions, were significantly correlated (.54) with numbers of sudden commencements but not with the average monthly geomagnetic (aa) activity.

(Persinger, M.A.; "Geophysical Variables and Behavior: LXVI. Geomagnetic Storm Sudden Commencements and Commercial Aircrashes" Perceptual and Motor Skills, 72:476, 1991.)

DID CHARLES DARWIN BECOME A CHRISTIAN?

It has long been claimed by some Christians that Charles Darwin, who helped lay the intellectual foundations of secular humanism, reembraced Christianity as he neared death. A central figure in this tale is a Lady Hope, who supposedly visited Darwin in the months before he died. What is the basis for the Lady Hope story; and what do Darwin's own writings reveal about his religious beliefs?

Alas, Darwin's return to the fold seems an apocryphal tale. W.H. Rusch, Sr., and J.W. Klotz, well-known scientific creationists, have prepared a 38-page historical study of the question---quoting at length from Darwin himself. They conclude about Darwin:

> He had made the human mind his authority, and it led him from orthodoxy to theism to agnosticism. Indeed it appears he might well be characterized as an atheist, a doubter of the very existence of God. His caution, however, and his recognition of the impossibility from a scientific standpoint of proving a negative led him to characterize himself as an agnostic which he says he is content to remain.

(Rusch, Wilbert H., Sr., and Klotz, John W.; "Did Charles Darwin Become a Christian?" Emmett L. Williams, ed., Norcross, 1988.)

CLOUD PLUMES NATURAL BUT STILL A BIT ANOMALOUS

During the mid-1980s, satellites photographed strange cloud plumes that stretched hundreds of kilometers downwind of some nothern islands, especially Bennett Island, in the Soviet Arctic. Some wondered if perhaps the Soviets were conducting tests of some new type of weapon in these remote locations. (See p. 265.) With the end of the Cold War, flights of instrumented aircraft over the islands were permitted. Data from these flights support the idea that the mystery cloud plumes are formed by air currents passing over the islands. In other words, they are only orographic or mountain-caused clouds, like those sometimes seen over the Rockies.

But puzzles persist:

(1) Why are the plumes so long?
(2) Why do they form at such high altitudes---more than 3 kilometers above the tops of the relatively small mountains on the islands?

(Monastersky, R.; "Mountains Give Rise to Perplexing Plumes," Science News, 141:422, 1992. Also: Fett, Robert W.; "Major Cloud Plumes in the Arctic and Their Relation to Fronts and Ice Movements," Monthly Weather Review, 120: 925, 1992.)

Cross reference. This item slipped between the cracks. It actually belongs on p. 266. What better place than among Forteana?

GENERAL ANOMALISTICS

BLINDED BY THE NIGHT

Ron Westrum is a sociologist who specializes in cases where scientific data are rejected out-of-hand because they challenge prevailing paradigms too strongly. In this article, Westrum describes several classical cases where science has ultimately admitted its errors and embraced the formerly rejected data:

1. The fall of stones from the sky;
2. The existence of thousands of parent-battered children; and
3. The reality of the coelacanth.

In connection with meteorite falls, he provides a wonderful quote from James Pringle, of the Royal Society:

> I venture to affirm that, after perusing all the accounts I could find of these phenomena, I have met with no well-vouched instance of such an event; nor is it to be imagined, but that, if these meteors had really fallen, there must have been long ago so strong evidence of the fact as to leave no room to doubt of it at present.

Next, Westrum tackles spontaneous human combustion and ball lightning, neither of which have been assimilated by science. He closes with a very complimentary paragraph on the Sourcebook Project and our Catalog of Anomalies, for which we thank him. (Westrum, Ron; "Blinded by the Night," The Sciences, 25:48, May-June 1985.)

THE UNCERTAINTY OF KNOWLEDGE

> Human beings of all societies in all periods of history believe that their ideas on the nature of the real world are the most secure, and that their ideas on religion, ethics and justice are the most enlightened. Like us, they think that final knowledge is at last within reach. Like us, they pity the people in earlier ages for not knowing the true facts. Unfailingly, human beings pity their ancestors for being so ignorant and forget that their descendants will pity them for the same reason.

E. Harrison, who penned the above, sees knowledge as perpetually uncertain and always changing. Scientists will always be surprised, he says, and scientific laws are never final. He concludes:

> I feel liberated by this philosophy. I find comfort in the thought that the creative mind fashions the world in which we live. For it means that the mind and reality are more profound than we normally suppose.

(Harrison, Edward; "The Uncertainty of Knowledge," New Scientist, p. 78, September 24, 1987.)

THE NEW HOLISM---BUT IS IT WHOLE ENOUGH?

In the short space of two weeks, the New Scientist printed two articles that confront the obvious complexity of nature. Not only is this complexity persistent under the attack of the Second Law of Thermodynamics, but it seems to actually increase with time. Do formative or guiding principles exist that science does not take into account? The two articles have very different answers.

The creative cosmos.

> Most people accept without question that the physical world is coherent and harmonious. Yet according to the traditional scientific picture, the Universe is just a random collection of particles with blind forces acting upon them. There is, then, a deep mystery as to how a seemingly directionless assembly of passive entities conspires to produce the elaborate structure and complex organisation found in nature.

The author of this introductory paragraph, P. Davies, asks, as we all do, "What is the origin of this creative power?" In groping for an answer, he presents first a common example of "blind" organization: the hexagonal convection cells in a pan of heated water. Using for a stepping stone the cooperative action of atoms in a laser, he leaps to the development of an embryo from a single strand of DNA! All such systems are "open"; that is, energy can flow in and out. They are also nonlinear, which means that chaotic, unpredictable action may occur. Davies implies that such action can be "creative," almost as if they possessed free will!

His final example is that of the network with large numbers of interacting sites or nodes. With random inputs, large networks do exhibit self-organization. Network theory is now very popular in the field of artificial intelligence. (Remember the computer Hal in 2001?) Davies's conclusion: "...Neo-Darwinism, combined with the mathematical principles emerging from network theory and related topics, will, I am convinced, explain the 'miracle' of life satisfactorily." (Davies, Paul; "The Creative Cosmos," New Scientist, p. 41, December 17, 1987.)

The superorganism. One week later, O. Sattaur expanded on the Gaia concept. He quotes J. Lovelock's definition:

> ...the physical and chemical condition of the surface of the Earth, of the atmosphere and of the oceans has been, and is, actively made fit and comfortable by the presence of life itself...in contrast to the conventional wisdom which held that life adapted to planetary conditions as it, and they, evolved their separate ways.

Mainstream science has shown scant love for the Gaia concept, probably because of its holistic nature. The idea of the earth being greater than the sum of its organic and inorganic parts---a superorganism---is foreign to reductionistic science. In Gaia, our planet is a giant, self-regulating entity, something larger than and independent of humanity. Is this scientific?

D. Abram deplores modern, mechanistic, reductionistic science as "immature." He thinks that the Gaia hypothesis may well signal the growing up of science. Sattaur concludes the article with Lovelock's assertion that the fate of humanity is interlocked with that of the earth, and that we are not the masters. If we reject Gaia's imperative, she may retaliate! (Sattaur, Omar; "Cuckoo in the Nest," New Scientist, p. 16, December 24/31, 1987.)

Comment. God is not mentioned in either article. Extrapolating the Gaia hypothesis to cosmic dimensions, we get closer to God. At the reductionist end of the spectrum, we could assume that everything the universe (life and all) is and will be is encoded into the smallest particles known---the quarks. The properties of the quarks, after all, must be consistent with the development of the cosmos. Here, God would be only a quarksmith, and everything would evolve from them!

ANOMALISTICS AT THE AAAS MEETING

At the recent Boston meeting of the AAAS (American Association for the Advancement of Science), at a session entitled "The Edges of Science," P. Sturrock dealt with the "gray areas" of science---what we call "anomalistics." In particular, Sturrock explained why mainstream science does not clasp anomalistics to its breast.

> "They are uncomfortable," he said. "Your friends may doubt your judgment. You may lose the respect of some of your colleagues. You will get no funding. You will have difficulty publishing your work. Your boss may think you are wasting your time."
>
> And, he added, "If you don't have tenure, don't even consider it."
>
> But the reasons why there should be serious consideration of at least some anomalous phenomena, Dr. Sturrock said, include the fact that "the gray area of science is the crucial area... You may---perhaps without knowing it---start a scientific revolution."
>
> Also, he added, "You may be honored---posthumously."

(Orndorff, Beverly; "Scientist Stresses 'Gray Areas,'" Richmond Times-Dispatch, February 16, 1988. Cr. L. Farish)

MEMOIRS OF A DISSIDENT SCIENTIST

> The Editors have set aside their ordinary scruples to publish the following recollections, in which Hannes Alfven defends a theory that is now rejected by virtually all working astrophysicists.

This first sentence from a (sort of) disclaimer by the Editors of the American Scientist really seems unscientific. After all, Alfven shared a Nobel Prize in 1970; is that not recommendation enough? Apparently not---at least not when Alfven believes that cosmic rays have a local rather than galactic origin.

The reflections of Alfven on the development of cosmic-ray theory are rather amusing; they reveal how much people and ideas change. Alfven originally maintained that cosmic rays were of galactic origin. But when he met E. Teller, who favored a local origin (within the sun's domain), Alfven was swayed. He is now among that tiny minority that defends the local origin view. Ironically, Teller switched sides, too, and now espouses a galactic origin.

Alfven concludes his reminiscing with a paragraph that says much about today's scientific environment:

> The mentioned conditions and quite a few other factors have led to a disagreement between a very strong establishment (E) and a small group of dissidents (D) to which the present author belongs. This is nothing remarkable. What is more remarkable

and regrettable is that it seems to be almost impossible to start a serious discussion between E and D. As a dissident is in a very unpleasant situation, I am sure that D would be very glad to change their views as soon as E gives convincing arguments. But the argument "all knowledgeable people agree that..." (with the tacit addition that by not agreeing you demonstrate that you are a crank) is not a valid argument in science. If scientific issues were decided by Gallup polls and not by scientific arguments science will soon be petrified forever.

(Alfven, Hannes; "Memoirs of a Dissident Scientist," American Scientist, 76: 249, 1988.)

Comment. If you do not climb on the scientific bandwagon, you won't get funding, papers published, or even a handshake from your colleagues. Look what has happened to Arp, Gold, Hoyle, and other "dissidents" in the past few years.

CONFORMITY STRIKES AGAIN

It is difficult to believe how hard it is to get new ideas funded in science. T. Gold has contributed a rather plaintive article to the Journal of Scientific Exploration detailing some of his experiences down the years.

His first "bad" experience occurred just after the end of World War II when, fresh from intensive work in radar signal processing, he proposed that the human ear is an active rather than passive receiver; that is, it actually emits sound itself. This self-generated tone aids the ear in signal processing. The thought that the ear could be a sound source was patently ridiculous, and Gold's idea got nowhere. However, recent experiments confirm that the human ear does indeed emit a tone at about 15,000 Hz.

Another, more recent, proposal for research on the behavior of hydrocarbons under high temperatures and pressures got very high marks from reviewers on all points but one: Should the proposal be funded? Several reviewers thought not; one saying that the whole idea was "misguided." In what way was Gold misguided? Well, it seems that his proposed work on hydrocarbons related to his idea that primordial hydrocarbons deep in the earth's crust contribute heavily to the reservoirs of oil and methane we tap on the planet's surface. And everyone knows that all oil and gas is biogenic; that is, derived from buried organic matter!

Gold has concluded that "not all is well" with American science. (Gold, Thomas; "New Ideas in Science," Journal of Scientific Exploration, 3:103, 1989.)

Comment. The idea that the human ear actually emits sound is treated further on p. 125.

SUCCESSFUL PREDICTIONS MEAN LITTLE IN SCIENCE

Many maverick scientists make the pages of Science Frontiers because they are the ones who deviate from the mainstream. They pursue those anomalies that the Sourcebook Project filters from the great river of scientific literature. But enough fluvial allusions! The maverick here is, again, H. Alfven. We first met Alfven on p. 339, where we commented on his paper "Memoirs of a Dissident Scientist." Alfven is still a dissident, a scientist who has the temerity to claim that cosmic rays have a local rather than galactic origin. Even more heretical is his assertion that electromagnetic forces have shaped the universe rather than the Big Bang!

The subject of this entry is not so much Alfven's conflicts with accepted scientific views, but rather whether correct scientific predictions really influence the scientific community's acceptance of theories. This, after all, is what science is all about. It turns out that Alfven has made many correct scientific predictions. (He even shared a Nobel Prize in 1970.) But, as S.G. Brush has related in a detailed article in Eos, being correct is not the same as being accepted.

> According to some scientists and philosophers of science, a theory is or should be judged by its ability to make successful predictions. This paper examines a case from the history of recent science---the research of Hannes Alfven and his colleagues on space plasma phenomena---in order to see whether scientists actually follow this policy. Tests of five predictions are considered: magnetohydrodynamic waves, field-aligned ("Birkeland") currents, critical ionization velocity and the existence of planetary rings, electrostatic double layers, and partial corotation. It is found that the success or failure of these predictions had essentially no effect on the acceptance of Alfven's theories, even though concepts such as "Alfven waves" have become firmly entrenched in space physics. Perhaps the importance of predictions in science has been exaggerated; if a theory is not acceptable to the scientific community, it may not gain any credit from successful predictions.

Brush concludes that the continuing resistance to Alfven's work is due to the widely held opinion that his theory is not plausible; that is, it does not conform to the dominant paradigm. (Brush, Stephen G.; "Prediction and Theory Evaluation," Eos, 71:19, 1990. Cr. L. Ellenberger)

Comment. In other words science does not work as it is supposed to.

WAS BURT STITCHED UP?

Sir Cyril was, until a few years ago, the Grand Old Man of British Psychology. But then a scandal erupted. Burt was charged with manufacturing data and even research assistants. A devastating biography by L. Hearnshaw seemed to ruin Burt's reputation for good. With all the attendant publicity, the label "scientific fraud" stuck in everyone's mind.

But, hold on, another book, one with a completely different conclusion, has hit the bookstores. The book, The Burt Affair, is by R.B. Joynson. Joynson demonstrates in great detail that Hearnshaw's biography is seriously deficient in places. If Joynson is correct, Hearnshaw and Burt's other enemies are guilty of selective reporting and misreporting. The case for fraud was a fraud itself. (Blinkhorn, Steve; "Was Burt Stitched Up?" Nature, 340:439, 1989.)

HOW GENIUS GETS NIPPED IN THE BUD

At once amusing and tragic, the article bearing the above title treats the reader to the musings of a university researcher who has a brilliant new idea---or so it seems to him. At first he is ready to rush to the literature and check out the idea's originality and feasibility. But wait, this groundwork will take considerable time, and the idea may turn out to be old hat. Is it all worth the risk? Will the university support research on this new untried idea? This would be unlikely since funds are thin, the idea is not part of the overall research plan, and worse yet, an enemy sits on the research fund committee. Getting external funds is impossible unless one can show that your university has enough confidence in the idea to support it financially. A way out is to publish the idea at a conference in hopes of raising interest and money. Hold it, the referees will ask why a few obvious points have not been checked out in the lab.

> Suddenly you realize that there is a far more serious problem. The current scientific system is based on the assumption that there is no such thing as a radically new idea. Each new paper is required to climb on the back of a plethora of earlier published papers, and does little more than add another layer of gloss to the cited references. Genuinely new ideas no not have a heap of existing papers to support them. This means that the standard of proof expected will be very much higher than for a typical "me too" paper. You glance across the old medieval library and note with alarm that the bust of Aristotle seems to be laughing at you.

And what does our hero do after confronting these realities of university life? He plays it safe. He will publish a few third-rate "me too" papers every year to fill his quota and try to get his name on a friend's research grant application to show his concern for his institution's cash flow. (Reynolds, Chris; "How Genius Gets Nipped in the Bud," New Scientist, p. 68, July 14, 1988.)

IS THERE A SCIENCE OF ANOMALIES?

Westrum and Truzzi term their paper a "bibliographical introduction" to anomalies. Indeed, the article is laced with references to the great classics on scientific anomalies and worth reading for this aspect alone. But the authors go farther. First, they define three different kinds of scientific anomalies: accepted, validated, and alleged. Next, they discuss the criteria used in judging anomalies by the scientific establishment. In dealing with anomalies, the focus inevitably narrows down to the reliability of the data and, quite reasonably, the honesty of those collecting, reproducing, and otherwise manipulating these data. Westrum and Truzzi then turn to the great anomaly collectors (Fort, Gould, Sanderson, Heuvelmans). What did these men produce and is it valuable to science? (Westrum, Ron, and Truzzi, Marcello; "Anomalies: A Bibliographical Introduction with Some Cautionary Remarks," Zetetic Scholar, 1:69, 1978.)

TWO SPACE FILLERS

This final page of our compilation has so much extra space that we felt impelled to append two items from more recent issues of Science Frontiers. We trust that readers will find them "esoteric" enough to fit in here!

"ALREADY, NOW, WE ARE FORGOTTEN ON THOSE STELLAR SHORES"*

Humans have many ways to predict the future: animal entrails, Tarot cards, and the Copernican Principle. The Copernican Principle, in particular, leads to all sorts of profound prophecies.

The Copernican Principle states that the earth does not occupy a special place in the cosmos. To this we add Darwinism, which asserts that, in the realm of biology, human origin is not special either; i.e., we enjoy no special place among life forms. Building upon these two general "beliefs," J.R. Gott, III, proceeds to estimate the longevities of various observables, such as the lifetime of a particular species. What follows is a long, highly technical computation of various probabilities, such as the evolution of intelligent life in the universe. All this (and there is a lot of it) leads to the following:

> Making only the assumption that you are a random intelligent observer, limits for the total longevity of our species of 0.2 million to 8 million years can be derived at the 95% confidence level. Further consideration indicates that we are unlikely to colonize the Galaxy, and that we are likely to have a higher population than the median for intelligent species.

Why won't we colonize the Galaxy? Not because we are not able to, Gott says, but because "living things do not usually live up to their maximum potential."

Also of interest here is Gott's assessment of SETI (our Search for Extraterrestrial Intelligence). Will our big radio telescopes pick up intelligent murmurings arriving from outer space? Gott's calculations are very pessimistic here:

> Thus, we do not expect to see a Dyson sphere civilization within our Galaxy, or a Karadashev type III civilization within the current observable horizon.

(Gott, J. Richard, III; "Implications of the Copernican Principle for Our Future Prospects." Nature, 363:315, 1993.)

*A line from Stephen Spender's poem From All These Events, from the Slump, from the War, from the Boom.

Comment. But perhaps the Copernican Principle, Darwinism, or some other of Gott's assumptions are in error. This item originally appeared in Science Frontiers #88.

IS NOTHING CERTAIN ANYMORE?

It was discouraging enough to learn that many natural systems, from simple pendulums to our weather, are basically chaotic; that is, tiny changes in the initial conditions upon which predictions are based can lead to highly unpredictable outcomes. Chaotic systems are usually qualitatively predictable but not quantitatively predictable. We have no choice but to live with this chaos; it seems that that's the way the cosmos is constructed! However, it now seems that the situation is even worse than chaotic! Some systems, perhaps most systems, are also indeterminate, meaning that we cannot predict their qualitative behavior either. A simple example is the water swirling down the bathtub drain. This is not only chaotic but it has two qualitative final states: clockwise and counterclockwise. Regardless of which hemisphere you are in, you can change the direction of swirl with negligible effort. Each of the two final states of motion is still quantitatively unpredictable. Systems that are more complex will possess many different final states, all chaotic. Can nature really be fundamentally chaotic as well as qualitatively uncertain?

J.C. Sommerer and E. Ott have mathematically examined a relatively simple system consisting of a single particle moving in a force field, experiencing friction, and being periodically jolted. Besides settling into chaotic motion, this particle may also be forced away to infinity---two radically different final states. The analysis revealed that for any set of initial conditions leading to the first type of behavior, there was an infinite number of slightly different initial conditions that would lead to the second type of behavior. In other words, systems that we have long thought to be deterministic, like the motions of the planets, may be not only chaotic but indeterminate.

Since Sommerer and Ott found their indeterminate system easily, we must face the possibility that the future behavior of just about everything is beyond our capability to predict, even with our best instruments and computers. Apparently the universe is built in such a way that exact science is impossible. (Sommerer, John C., and Ott, Edward; "A Physical System with Qualitatively Uncertain Dynamics," Nature, 365:138, 1993. Also: Peterson, I.; "Finding Riddles of Physical Uncertainty," Science News, 144:180, 1993.)

Comment. This discovery is even more profound than Heisenberg's Uncertainty Principle, which is merely quantitative in character and for practical purposes rules only the atomic world. From the above we see that the entire cosmos is uncertain quantitatively and qualitatively. A heck of a way to construct a universe! This item originally appeared in Science Frontiers #90.

SUBJECT INDEX

THE UNCLASSIFIED RESIDUUM

THE UNCLASSIFIED RESIDUUM WAS DEFINED BY WILLIAM JAMES, THE GREAT AMERICAN PHILOSOPHER, IN THESE WORDS:

> ***"ROUND ABOUT THE ACCREDITED AND ORDERLY FACTS OF EVERY SCIENCE THERE EVER FLOATS A SORT OF DUST-CLOUD OF EXCEPTIONAL OBSERVATIONS, OF OCCURRENCES MINUTE AND IRREGULAR AND SELDOM MET WITH, WHICH IT ALWAYS PROVES MORE EASY TO IGNORE THAN TO ATTEND TO .... ANYONE WILL RENOVATE HIS SCIENCE WHO WILL STEADILY LOOK AFTER THE IRREGULAR PHENOMENA. AND WHEN THE SCIENCE IS RENEWED, ITS NEW FORMULAS OFTEN HAVE MORE OF THE VOICE OF THE EXCEPTIONS IN THEM THAN OF WHAT WERE SUPPOSED TO BE THE RULES."***

TO CLASSIFY THE UNCLASSIFIED RESIDUUM, THE SOURCEBOOK PROJECT IS COMPILING AN OBJECTIVE, UNSENSATIONALIZED CATALOG OF ANOMALOUS PHENOMENA.

The Catalog of Anomalies is in effect an encyclopedia of the unknown and puzzling that is based primarily upon recognized scientific research. It is the only organized, indexed, unsensationalized collection of difficult-to-explain phenomena. The Catalog is supplemented by several "Handbooks" containing more voluminous descriptions of some of the phenomena.

The first thirteen volumes of the *Catalog of Anomalies.* An incomparable collection of difficult-to-explain observations and curiosities of nature.

REVIEWS IN SCIENTIFIC AND LIBRARY PUBLICATIONS

The Catalogs and Handbooks have been favorably reviewed in many scientific journals, such as Nature, American Scientist, and New Scientist. In addition, library publications such as Choice, Booklist, and Science Books have recommended them. Four have been book club selections.

DATA BASE

40,000 articles from the scientific literature, the results of a 25-year search through more than 12,000 volumes of scientific journals, including the complete files of Nature, Science, Icarus, Weather, etc.

USES FOR THE CATALOGS AND HANDBOOKS

(1) Librarians will find these books to be unique collections of source materials and bibliographies; (2) Scientists will find research ideas as well as unexpected observations and many references; (3) Students can use these books to select and develop research papers and theses; (4) The science-oriented layman will find thousands of those mysteries of nature that make science exciting.

COMPILER

All Catalogs and Handbooks have been compiled by William R. Corliss

ORDERING INFORMATION

Prices are in U.S. dollars. Canadian dollars and pounds sterling are accepted at prevailing exchange rates. U.S. customers should add $1 for each order under $30. Foreign customers add $1.50 per book for surface mail.

ORDER FROM:

The Sourcebook Project
P.O. Box 107
Glen Arm, MD 21057

BIOLOGY CATALOGS

BIOLOGICAL ANOMALIES: HUMANS I; A Catalog of Biological Anomalies

This volume, the first of three on human biological anomalies, looks at the "external" attributes of humans: (1) Their physical appearance; (2) Their anomalous behavior; and (3) Their unusual talents and faculties.

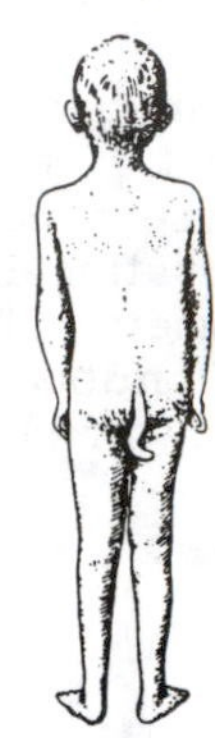

A Moi boy with a nine-inch tail

TYPICAL SUBJECTS COVERED

- Mirror-image twins
- The sacral spot
- The supposed human aura
- Baldness among musicians
- Human tails and horns
- Human behavior and solar activity
- Cycles of religiousness
- Cyclicity of violent collective human behavior
- Handedness and longevity
- Wolf-children
- The "Mars Effect"
- Telescopic vision
- Dermo-optical perception
- Hearing under anesthesia
- Human navigation sense
- Asymmetry in locomotion
- Sex-ratio variations

COMMENTS FROM REVIEWS

All I can say to Corliss is carry on cataloging.
NEW SCIENTIST

304 pages, hardcover, $19.95
52 illus., 3 indexes, 1992
548 references, LC 91-68541
ISBN 0-915554-26-7, 7x10

BIOLOGICAL ANOMALIES: HUMANS II: A Catalog of Biological Anomalies

The second Catalog volume on human biological anomalies focuses upon the "internal" machinery of the body: (1) Its major organs; (2) Its support structure (the skeleton); and (3) Its vital subsystems (the central nervous system and the immune system).

TYPICAL SUBJECTS COVERED

- Enigma of the fetal graft
- Phantom limbs
- Blood chimeras
- Anomalous human combustion
- Bone shedders
- Skin shedders
- "Perfection" of the eye
- Dearth of memory traces
- Sudden increase of hominid brain size
- Health and the weather
- Periodicity of epidemics
- Extreme longevity
- AIDS anomalies
- Cancer anomalies
- Human limb regeneration
- Nostril cycling
- Voluntary suspended animation
- Male menstruation

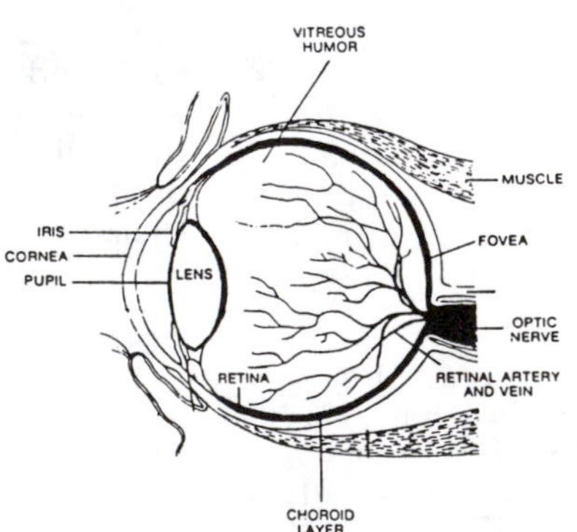

Is the complexity of the human eye anomalous?

297 pages, hardcover, $19.95
40 illus., 3 indexes, 1993
494 references, LC 91-68541
ISBN 0-915554-27-5, 7x10

BIOLOGY HANDBOOK

INCREDIBLE LIFE; A Handbook of Biological Mysteries

Even with its 1000-plus pages, this Handbook barely does justice to the immense number of biological anomalies in the scientific literature.

Crow "anting" with a lighted match

TYPICAL SUBJECTS COVERED

- Human health and astronomy
- Yeti and Sasquatch
- DNA: the ultimate parasite
- Luminous plants
- Diseases from outer space
- The strange synchronous flowering of bamboos
- The problem of excess DNA
- Sea and lake serpents
- Unexplained senses of ants
- Water-breathing in mammals
- Life and thermodynamics
- Is evolution a tautology?
- Unusual behavior of animals
- Cryptobiosis or latent life

COMMENTS FROM REVIEWS

...the collection is endlessly fascinating. NATURE
...it certainly does pique the interest of the reader.
LIBRARY J.

1024 pages, hardcover, $24.50
100 illustrations, index, 1981
References, LC 80-53971
ISBN 915554-07-0, 6x9

GEOPHYSICS CATALOGS

LIGHTNING, AURORAS, NOCTURNAL LIGHTS; A Catalog of Geophysical Anomalies

Nothing catches the human eye and imagination as quickly as a mysterious light. All down recorded history, scientists and laymen alike have been seeing strange lightning, sky flashes, and unaccountable luminous objects.

TYPICAL SUBJECTS COVERED

- Horizon-to-horizon sky flashes
- Episodes of luminous mists
- Mountain-top glows (Andes glow)
- Earthquake lights
- Ball lightning with tails
- Rocket lighting
- Lightning from a clear sky
- Ghost lights; ignis fatuus
- Darting streaks of light (sleeks)
- The milky sea and light wheels
- Radar-stimulated phosphorescence of the sea
- Double ball lightning
- Luminous phenomena in tornados
- Black auroras

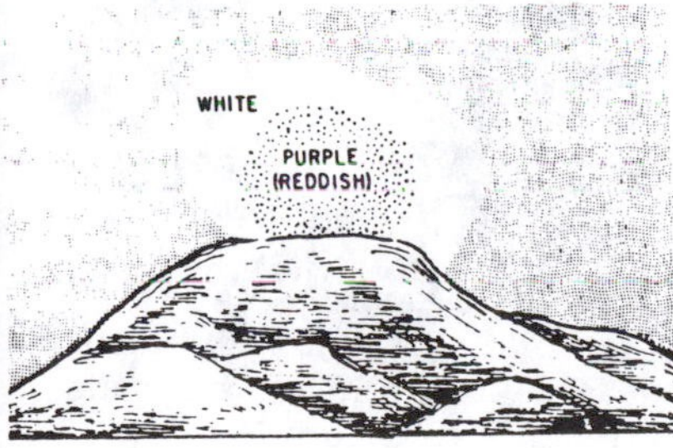

Luminous display over Mt. Noroshi during earthquake swarm.

COMMENTS FROM REVIEWS

...the book is well-written and in places quite fascinating. SCIENCE BOOKS

248 pages, hardcover, $14.95
74 illustrations, 5 indexes, 1982
1070 references, LC 82-99902
ISBN 915554-09-7, 7 x 10 format

TORNADOS, DARK DAYS, ANOMALOUS PRECIPITATION; A Catalog of Geophysical Anomalies

Here is our "weather" Catalog. As everyone knows, our atmosphere is full of tricks, chunks of ice fall from the sky, tornado funnels glow at night. The TV weathermen rarely mention these "idiosyncracies".

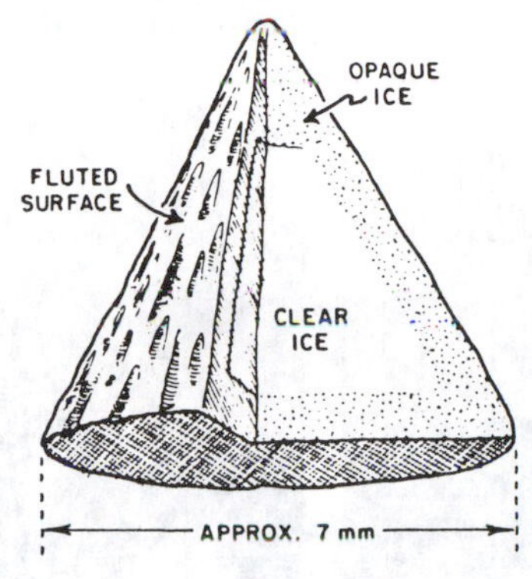

Conical hailstones with fluted sides.

TYPICAL SUBJECTS COVERED

- Polar-aligned cloud rows
- Ice fogs (the Pogonip)
- Conical hail
- Gelatinous meteors
- Point rainfall
- Unusual incendiary phenomena
- Solar activity and thunderstorms
- Tornados and their association with electricity
- Multiwalled waterspouts
- Explosive onset of whirlwinds
- Dry fogs and dust fogs
- Effect of the moon on rainfall
- Ozone in hurricanes
- Ice falls (hydrometeors)

COMMENTS FROM REVIEWS

...can be recommended to everyone who realizes that not everything in science has been properly explained. WEATHER

202 pages, hardcover, $14.95
40 illustrations, 5 indexes, 1983
745 references, LC 82-63156
ISBN 915554-10-0, 7 x 10 format

EARTHQUAKES, TIDES, UNIDENTIFIED SOUNDS; A Catalog of Geophysical Anomalies

Quakes and monster, solitary waves and natural detonations; these are the consequences of solids, liquids, and gases in motion. In our modern technological cocoon, we are hardly aware of this rich spectrum of natural phenomena.

Sand craters created by earthquakes.

TYPICAL SUBJECTS COVERED

- Periodic wells and blowing caves
- Sun-dominated tides
- Immense, solitary waves
- Animal activity prior to earthquakes
- Earthquake geographic anomalies
- Earthquake electricity
- The sound of the aurora
- Musical sounds in nature
- Mysterious detonations
- Anomalous echos
- Slicks and calms on water surfaces
- Periodicities of earthquakes
- The vibrations of waterfalls
- Unusual barometric disturbances

COMMENTS FROM REVIEWS

...surprisingly interesting reading. NATURE

220 pages, hardcover, $14.95
32 illustrations, 5 indexes, 1983
790 references, LC 83-50781
ISBN 915554-11-9, 7 x 10 format

RARE HALOS, MIRAGES, ANOMALOUS RAINBOWS; A Catalog of Geophysical Anomalies

Most of us have seen rings around the moon, but what does it mean when such rings are not circular or are off-center? Neither are rainbows and mirages devoid of mysteries. And the Brocken Specter still startles Alpine climbers!

TYPICAL SUBJECTS COVERED

- Rainbows with offset white arcs
- Sandbows
- Offset and skewed halos
- The Brocken Specter
- The Alpine Glow
- Unexplained features of the green flash at sunset
- Fata Morgana
- Telescopic mirages
- Long-delayed radio echos
- Eclipse shadow bands
- Geomagnetic effects of meteors
- Intersecting rainbows
- The Krakatoa sunsets
- Kaleidoscopic suns

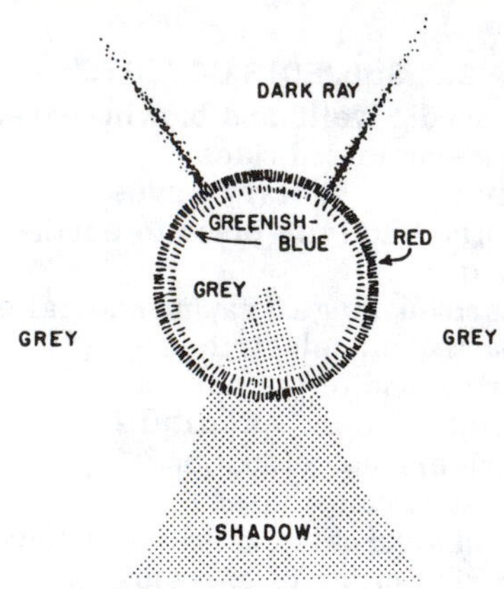

Shadow of Adam's Peak with glory and radial rays.

COMMENTS FROM REVIEWS

...all in all it's a fascinating book. SKY AND TELESCOPE

...any student of the physical sciences will find it fascinating. SCIENCE BOOKS

244 pages, hardcover, $14.95
111 illustrations, 5 indexes, 1984
569 references, LC 84-50491
ISBN 915554-12-7, 7 x 10 format

GEOPHYSICS HANDBOOK

HANDBOOK OF UNUSUAL NATURAL PHENOMENA

This is our first Handbook, as rewritten in a more popular style for publication by Doubleday in paperback form. It deals with most of the subjects mentioned in the preceding four Catalog volumes.

A low-level aurora descends below mountain peaks.

TYPICAL SUBJECTS COVERED

- Nocturnal lights and will o' the wisps
- Oceanic light wheels
- Non lunar tides
- Falls of ice, fish, grains, etc.
- Strange hums and hisses
- Unexplained mirages
- Low-level auroras
- Ball lightning
- Cloudless rain and snow
- The Barisal Guns and other "water guns"
- Freak whirlwinds
- Dark days, yellow days, etc.
- Anomalous solar and lunar halos

COMMENTS FROM REVIEWS

...fascinating reading may be found at almost any point in the book. BOOKLIST

...full of fascinating morsels. NATURE

431 pages, hardcover, $9.95
133 illustrations, index, 1983
References, LC 78-22625
ISBN 517-60523-6, 6x9 format

ARCHEOLOGY HANDBOOK

ANCIENT MAN; A Handbook of Puzzling Artifacts

Now in its third printing, our archeology Handbook reproduces hundreds of items from the difficult-to-obtain archeological literature.

TYPICAL SUBJECTS COVERED

- Ancient Florida canals
- The Maltese "cart tracks"
- New England earthworks
- Ancient coins in America
- Ancient Greek analog computer
- Inscriptions and tablets in unexpected places
- The great ruins at Tiahuanaco
- Zimbabwe and Dhlo-dhlo
- Huge spheres in Costa Rica
- The Great Wall of Peru
- Ancient batteries and lenses
- Mysterious walls everywhere
- Pacific megalithic sites
- European stone circles and forts

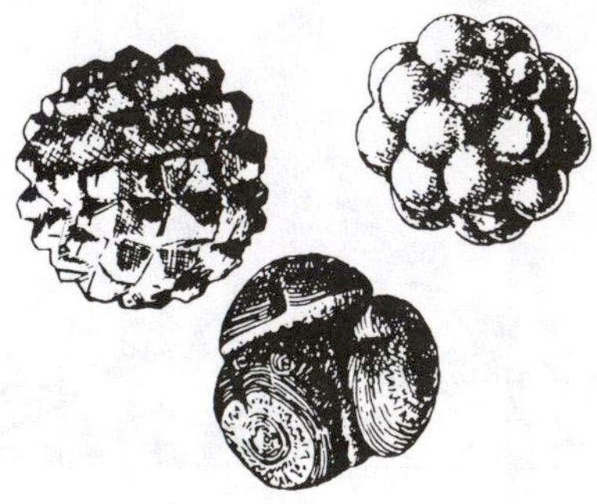

Scottish carved stones from circa 1000 B. C.

COMMENTS FROM REVIEWS

...a useful reference in undergraduate, public, and high school libraries. BOOKLIST

792 pages, hardcover, $21.95
240 illustrations, index, 1978
References, LC 77-99243
ISBN 915554-03-8, 6 x 9 format

GEOLOGY CATALOGS

CAROLINA BAYS, MIMA MOUNDS, SUBMARINE CANYONS; A Catalog of Geological Anomalies

Topographical phenomena are the subject of this Catalog. The ups and downs of the earth's surface betray many anomalies. Could continental drift be inferior to the expanding earth hypothesis? Have ocean levels fluctuated wildly down the eons?

TYPICAL SUBJECTS COVERED

- Carolina Bays and oriented lakes
- Large circular structures
- Immense craters
- Raised beaches
- Guyots (flat-topped seamounts)
- Island arcs
- Doubts about plate tectonics (continental drift)
- Mima mounds
- Drumlin anomalies
- Patterned ground
- Esker problems
- Lake walls and ramparts
- Crevicular structure
- Submarine canyons

Pyramid of frozen foam on the Bozenkill, New York State

COMMENTS FROM REVIEWS

. . . enough terrestrial intrigue to keep us thinking for years.
PURSUIT

245 pages, hardcover, $17.95
84 illustrations, 5 indexes, 1988
682 references, LC 87-63408
ISBN 915554-22-4, 7 x 10 format

ANOMALIES IN GEOLOGY: PHYSICAL, CHEMICAL, BIOLOGICAL; A Catalog of Geological Anomalies

Journey here into ice caves, exhume Siberian mammoths, see animals perish in gas-filled valleys---a little media hype is justified here. But more serious questions involve the origins of oil, coal, and natural gas.

TYPICAL SUBJECTS COVERED

- Biological extinction events
- Musical sands, ringing rocks
- Anomalies of oil's origin
- Ice caves, frozen wells
- Natural fission reactors
- Marine organisms and fossils found far inland
- Siberia's frozen mammoths
- Radiometric dating problems
- Anchor ice, frazil ice
- Violent lake turnovers
- Flexible rocks
- Origin of ocean water
- Skipping in fossil record
- Valleys of death

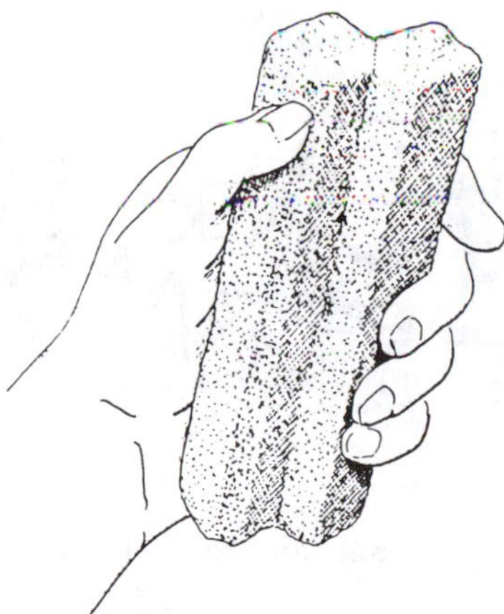

Prismatic sandstone from Missouri

335 pages, hardcover, $18.95
55 illustrations, 5 indexes, 1989
1260 references, LC 89-90680
ISBN 915554-23-2, 7 x 10 format

NEGLECTED GEOLOGICAL ANOMALIES; A Catalog of Geological Anomalies

Neglected but far from insignificant are the anomalies cataloged here. Do we really know how concretions and geodes form, where tektites come from, whence the immense deposits of superficial debris all over our globe?

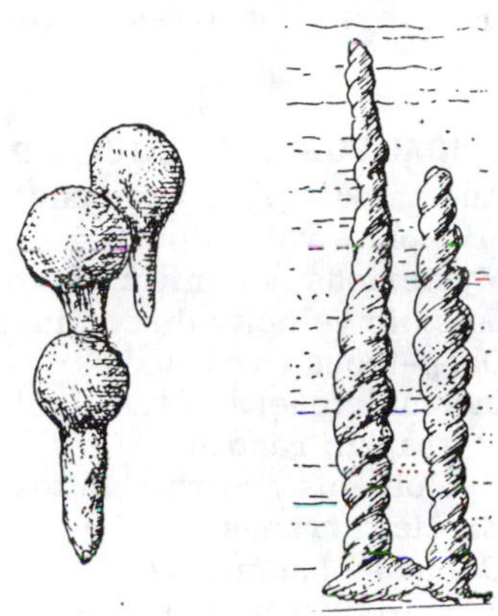

Mace-shaped and sand-spike concretions from the Colorado delta.

TYPICAL SUBJECTS COVERED

- Concretions and geodes
- Tektites and microtektites
- Erratic boulders and gravels
- Polystrate fossils
- Bone caves and bone beds
- Giant basalt flows
- Natural glasses
- Surging glaciers
- Driftless regions
- Stretched pebbles
- Crystal inclusions
- Rarity of fossil meteorites and tektites
- Elevated erratics
- Stone rivers and rock glaciers

333 pages, hardcover, $18.95
80 illustrations, 5 indexes, 1990
1030 references, LC 90-60568
ISBN 915554-24-0, 7 x 10 format

INNER EARTH: A SEARCH FOR ANOMALIES;
A Catalog of Geological Anomalies

The focus of this, the eleventh volume in the Catalog of Anomalies, is the earth's interior, which is revealed to us mainly through seismic signals, magnetic variations, and the flow of heat from great depths. Hundreds of kilometers below the surface lurk huge pieces of foundered continental crust and bizarre structures of unknown origin.

TYPICAL SUBJECTS COVERED

- Anomalous gravity signals
- Mid-plate volcanism
- Mysterious seismic reflectors
- Seismic velocity discontinuities
- Deep-focus earthquakes
- Incompleteness of the stratgraphic record
- Cyclothems and rhythmites
- Exotic terranes
- Compass anomalies
- Earth-current anomalies
- Problems of paleomagnetism
- Polarity reversals

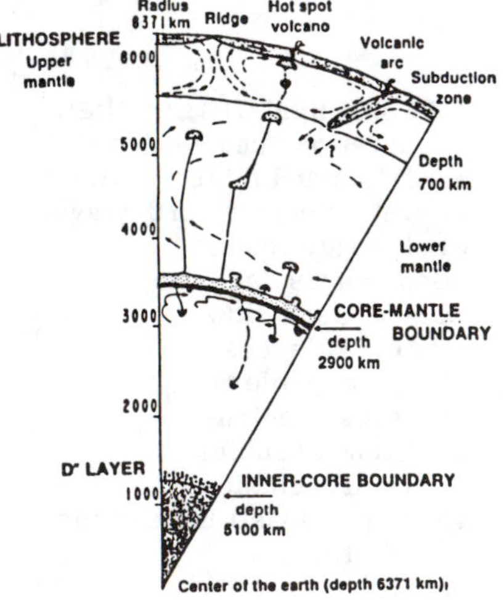

Model of the earth's interior

230 pages, hardcover, $18.95
52 illustrations, 5 indexes, 1991
619 references, LC 90-92347
ISBN 915554-25-9, 7 x 10 format

GEOLOGY HANDBOOK

UNKNOWN EARTH; A Handbook of Geological Enigmas

Two additional geological Catalogs are still in preparation. Until they are published, anomalists must rely on this Handbook for data on geodes, bone beds, thrust faulting, rhythmic sedimentation, geomagnetism, and other subjects noted below.

TYPICAL SUBJECTS COVERED

- Catastrophic flood events
- Old tektites in young sediments
- Bone caves and bone beds
- Immense craters and other circular structures
- Gaps in the fossil record
- Rock rivers and boulder fields
- Frozen wells and ice caves
- Mounds and pimpled plains
- The Carolina Bays and Alaskan oriented lakes
- Deep ocean-floor channels
- Boulder trains
- Cylinders in strata
- Problems of the glacial drift

A Kansas "rock city".

COMMENTS FROM REVIEWS

...belongs in every college geology library. SCIENCE BOOKS (Natural Science Book Club selection)

839 pages, hardcover, $21.95
125 illustrations, index, 1980
References, LC 80-50159
ISBN 915554-06-2, 6 x 9 format

ASTRONOMY HANDBOOK

MYSTERIOUS UNIVERSE; A Handbook of Astronomical Anomalies

Our Astronomy Handbook covers much the same ground as the three preceding Astronomy Catalogs, but in more detail. For example, the quotations are much more extensive.

Unexplained rift in the zodiacal light.

TYPICAL SUBJECTS COVERED

- The lost satellite of Venus
- Transient lunar phenomena
- Ephemeral earth satellites
- Venus' radial spoke system
- Relativity contradicted
- Cosmological paradoxes
- Changes in light's velocity
- Vulcan; the intramercurial planet
- Knots on Saturn's rings
- Bright objects near the sun
- The sun's problematical "companion star"
- "Sedimentary" meteorites
- Life chemistry in outer space
- Planet positions and sunspots

COMMENTS FROM REVIEWS

...highly recommended...excellent value for money. NATURE (Astronomy Book Club selection)

716 pages, hardcover, $19.95
103 illustrations, index, 1979
References, LC 78-65616
ISBN 915554-05-4, 6 x 9 format

ASTRONOMY CATALOGS

THE MOON AND THE PLANETS; A Catalog of Astronomical Anomalies

From our own moon's cratered surface to the red, rock-strewn plains of Mars, the Solar System is a fertile field for scientific research. Despite centuries of observation, each new spacecraft and telescope provides us with new crops of anomalies.

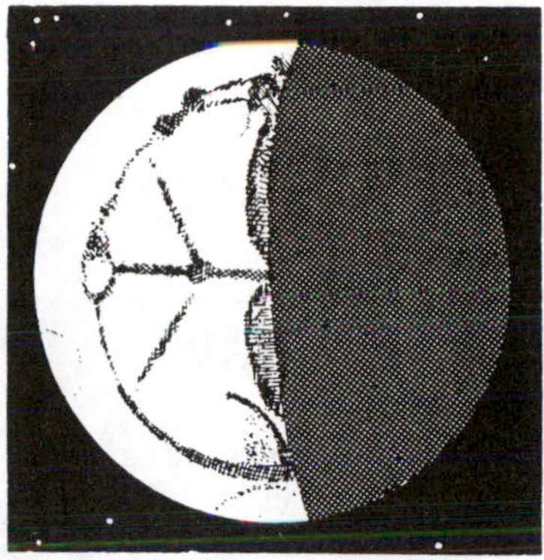

One drawing of the Venusian radial spoke system.

TYPICAL SUBJECTS COVERED

- The ashen light of Venus
- The Martian 'pyramids'
- Kinks in Saturn's rings
- Continuing debate about the Voyager life-detection experiments
- Neptune's mysterious ring
- Evidence of water on Mars
- The strange grooves on Phobos
- The two faces of Mars
- Lunar clouds, mists, "weather"
- Ring of light around the new moon
- Dark transits of Jovian satellites
- Io's energetic volcanos
- Jupiter as a "failed star"
- Venus-earth resonance

COMMENTS FROM REVIEWS

The author is to be commended for his brilliantly conceived and researched volume. SCIENCE BOOKS

383 pages, hardcover, $18.95
80 illustrations, 4 indexes, 1985
988 references, LC 85-61380
ISBN 915554-19-4, 7 x 10 format

THE SUN AND SOLAR SYSTEM DEBRIS; A Catalog of Astronomical Anomalies

Our sun, powerhouse of the Solar System and an enigma itself, is orbited by clouds of asteroids, comets, meteors and space dust. These "minor objects" cause "major headaches to astronomers searching for explanations.

TYPICAL SUBJECTS COVERED

- Solar system resonances
- Bode's Law and other regularities
- Blackness of comet nuclei
- Cometary activity far from solar influences
- Unidentified objects crossing sun
- The 'missing' solar neutrinos
- Pendulum phenomena during solar eclipses
- Observations of Planet X
- Meteorite geographical anomalies
- Meteorites from the moon
- Long fireball processions
- Very long duration meteorites
- Zodiacal light brightness changes

One of the many possible modes of solar surface oscillation.

COMMENTS FROM REVIEWS

It is an unusual book, nicely executed, and I recommend it highly...ICARUS

288 pages, hardcover, $17.95
66 illustrations, 4 indexes, 1986
874 references, LC 86-60231
ISBN 915554-20-8, 7 x 10 format

STARS, GALAXIES, COSMOS; A Catalog of Astronomical Anomalies

Did the Big Bang really begin the existence of all we know? Do we honestly know how the stars (and our sun) work? Can we rely on Newton's Law of Gravitation? According to this volume the answer seems to be: "Probably not!"

TYPICAL SUBJECTS COVERED

- Optical bursters and flare stars
- Historical color change of Sirius
- Infrared cirrus clouds
- Quasar-galaxy associations
- The red-shift controversy
- Quantization of red shifts
- The quasar energy paradox
- Apparent faster-than-light velocities in quasars and galaxies
- Evidence for universal rotation
- Swiss cheese structure of universe
- Is the "missing mass" really missing?
- Superluminous infrared galaxies
- Shells around elliptical galaxies

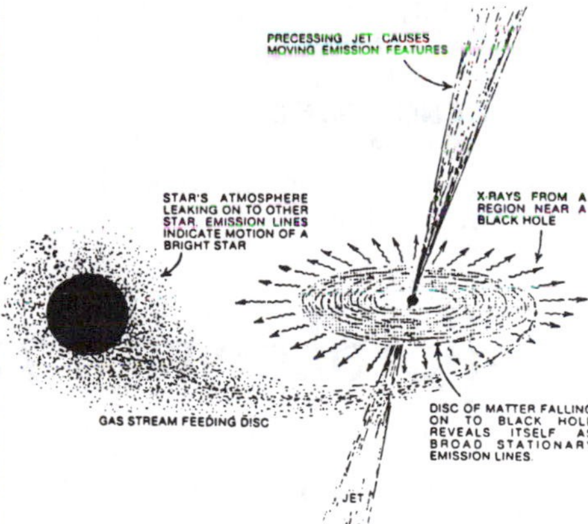

Model of the mysterious star SS 433.

COMMENTS FROM REVIEWS

...it never fails to be interesting, challenging and stimulating. NEW SCIENTIST

246 pages, hardcover, $17.95
50 illustrations, 4 indexes, 1987
817 references, LC 87-60007
ISBN 915554-21-6, 7 x 10 format

SOURCEBOOKS

The first publications of the Sourcebook Project appeared in the 1970s. These were loose-leaf notebooks called "Sourcebooks." In these notebooks were reproduced articles and excerpts of articles dealing with anomalous phenomena. Although the Sourcebooks were superceded by the Handbooks and Catalogs, the continuing demand for them has encouraged us to keep most of them in print, as detailed below.

STRANGE UNIVERSE: vol. A2
W.R. Corliss, 286 pp., 1977, $16.95
Astronomical anomalies. Xeroxed text, original printed binder.

STRANGE PLANET
W.R. Corliss, Geological anomalies
vol. E1, 289 pp., 1975, $9.95
Printed text, printed binder
vol. E2, 275 pp., 1978, $16.95
Xeroxed text, plain binder

STRANGE PHENOMENA
W.R. Corliss, Geophysical anomalies.
vol. G1, 277 pp., 1974, $16.95
Xeroxed text, plain binder.
vol. G2, 270 pp., 1974, $9.95
Printed text, plain binder.

STRANGE ARTIFACTS
W.R. Corliss, Archeological anomalies
vol. M1, 268 pp., 1974, $16.95
Xeroxed text, printed binder.
vol. M2, 293 pp., 1976, $16.95
Xeroxed text, printed binder.

STRANGE MINDS: vol. P1
W.R. Corliss, 291 pp., 1976, $9.95
Psychological anomalies. Printed text, plain binder.

PHOTOCOPIED CLASSICS

LEGENDARY ISLANDS OF THE ATLANTIC
W.H. Babcock, 196 pp., 1922, $12.95p

The title of this book immediately conjures up thoughts of Atlantis; but many other Atlantic islands were once thought to exist, were placed on maps, and then disappeared. The island of Brazil (or Hy Brazil) is one of these phantom islands. Babcock has written an engrossing, scholarly treatise, with many old maps, and hints of pre-Columbian contacts with the New World. Here follow some chapter titles: •Atlantis; •The Island of the Seven Cities; •The Problem of Mayda; •Estotiland and the Other Islands of Zeno; •The Sunken Land of Buss and Other Phantom Islands. This is a reprint of our xeroxed classic.

THE MAMMOTH AND THE FLOOD: An Attempt to Confront the Theory of Uniformity with the Facts of Recent Geology
H.H. Howorth, 1887, 498 pp., $19.95p

Sir Henry Howorth was one of the great synthesizers of science in the late 1800s. In this book, he brought together all of the available evidence on recent catastrophic flooding on the earth: the bone caves, the Siberian mammoth carcasses, the masses of fresh moa bones in Australia, and a host of other geological and biological puzzles. Most of Howorth's attention, however, is focussed on the mammoths and their recent demise. This book is one of the classics of catastrophe literature. Our high-quality xerox edition is bound with heavy covers.

ANCIENT MONUMENTS OF THE MISSISSIPPI VALLEY
E.G. Squier and E.H. Davis, 376 pp., 1848, xeroxed classic, $29.95p

One of the most remarkable archeological books ever published in America! This book was Volume 1 in the Smithsonian Contributions to Knowledge series. Its appearance in 1848 created a sensation. For, as America moved west, the remnants of the great civilization of the Moundbuilders raised much speculation. Even today we marvel at their immense, flat-topped temple mounds, the huge earthen enclosures, and the meticulously wrought artifacts of copper, mica, and clay. Squier and Davis objectively described the features of this New World civilization in words and drawings. It is the drawings, though, that really capture the reader. They are superb, almost overwhelming. (Hardcover rereprints of this book run over $80.)

DOUBT/FORTEAN SOCIETY MAGAZINE

During the 30s, 40s, and 50s, the work of Charles Fort was promoted by the Fortean Society. The Society initially published the Fortean Society Magazine, later changing its name to Doubt. These publications are delightful collections of Forteana of the period and also include reproductions of some of Fort's original notes. Curious and fun to read. All numbers are available in photocopied format bound as listed below.

Nos. 1-10 (152 pp.) $16.95
Nos. 11-20 (160 pp.) $16.95
Nos. 21-30 (160 pp.) $18.95
Nos. 31-40 (160 pp.) $16.95
Nos. 41-50 (160 pp.) $16.95
Nos. 51-61 (184 pp.) $18.95

RUDE STONE MONUMENTS IN ALL COUNTRIES: Their Age and Uses
J. Fergusson, 1872, 578 pp., $19.95p

Fergusson's famous compilation of worldwide megalithic monuments is a fit complement to our photocopied edition of Ancient Monuments of the Mississippi Valley, from 1848. Fergusson has filled his book with 233 line drawings of artifacts from the megalithic period. The emphasis is on the massive monuments, but you'll also see some sketches of pottery and inscribed stones. Naturally, there are long chapters on the British Isles, Ireland, and Europe; but the author also demonstrates how the megalithic culture extended into North Africa, the Middle East, and India. It is a pleasure to page through this old classic and read how our parents' parents interpreted these edifices.

Our photocopied edition is of high quality---no mean feat considering the yellowing, crumbling pages we had to work with. 8½ x 11 format, with heavier printed covers.

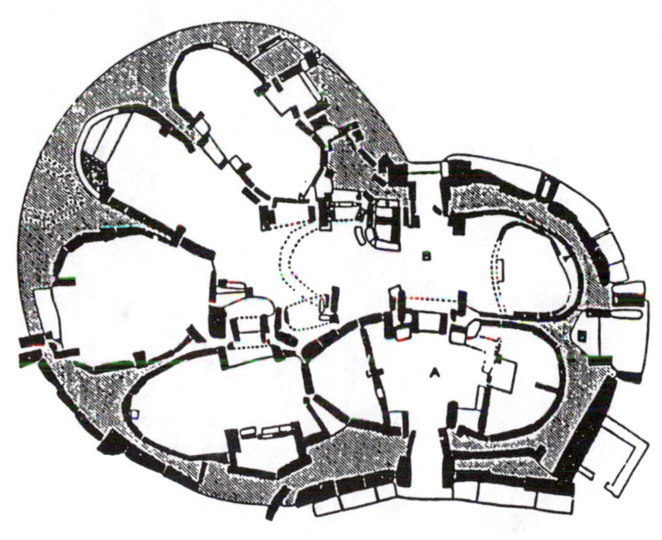

Plan of Haglar Khem, Malta, as drawn in Rude Stone Monuments.

JOURNAL REPORTS (order by number)

#1. Fuller, Myron L.; "The New Madrid Earthquake," U.S. Geological Survey Bulletin 494, 1912. A meticulous account of the great cataclysms of 1811-1812, including many eye-witness accounts of remarkable phenomena. Scientific analysis of the shocks. 119 pp., $9.00.

#2. Olmsted, Denison; "Observations on the Meteors of November 13th, 1833," American Journal of Science, 1:25:363, 1834. One of the greatest meteor showers of historical times lit up the skies of North America in 1833. This report records many eye-witness accounts of unusual and spectacular phenomena. 90 pp., $7.00.

#3. Fryer, A.T.; "Psychological Aspects of the Welsh Revival," Society for Psychical Research, Proceedings, 19:80, 1905. A thorough study of the great religious revival of 1905 and the abundant luminous phenomena seen by the participants. To be compared to modern UFO flaps. 82 pp., $7.00.

#4. McAtee, Waldo L.; "Showers of Organic Matter," Monthly Weather Review, 45:217, 1917. PLUS: Gudger, E.W.; "Rains of Fishes," Natural History, 21:607, 1921. These two classical papers review historical falls of animals and organic matter. 21 pp., $4.00.

#5. Mansfield, George Rogers; "Origin of the Brown Mountain Lights in North Carolina," U.S. Geological Survey Circular 646, 1971. Maps, history of the phenomena, and discussion of similar occurrences elsewhere. 18 pp., $4.00.

#6. Middlehurst, Barbara M., et al; "Chronological Catalog of Reported Lunar Events," NASA Technical Report R-277, 1968. An exhaustive descriptive catalog of 579 transient lunar phenomena, with complete references. 59 pp., $8.00.

SCIENCE FRONTIERS

A bimonthly newsletter reporting on scientific anomalies discussed in the current literature. (For more details, see the Preface of this book.) It is free to regular customers; otherwise, $7.00 for six issues. For a sample copy, send your request to: Sourcebook Project, P.O. Box 107, Glen Arm, MD 21057, USA.